S0-ATI-886

Trace Elements in Man and Animals – 9

Proceedings of the
Ninth International Symposium on
Trace Elements in Man and Animals

Inquiries: Monograph Publishing Program, NRC Research Press, National Research Council of Canada, Ottawa, Ontario K1A 0R6, Canada

Correct citation: Fischer, P.W.F., L'Abbé, M.R., Cockell, K.A., and Gibson, R.S. (editors). 1997. *Trace Elements in Man and Animals – 9: Proceedings of the Ninth International Symposium on Trace Elements in Man and Animals*. NRC Research Press, Ottawa, Canada. 677 pages.

Trace Elements in Man and Animals – 9

Proceedings of the Ninth International Symposium on Trace Elements in Man and Animals

Editors

P.W.F. Fischer
M.R. L'Abbé
K.A. Cockell
R.S. Gibson

NRC Research Press
Ottawa 1997

Printed in Canada on acid-free paper ♾

This publication was produced from electronic copy provided by the editors.

ISBN 0-660-16404-3
NRC No. 40702

Canadian Cataloguing in Publication Data

International Symposium on Trace Elements in Man and Animals (9th: 1996: Banff, Alta.)
Proceedings of the Ninth International Symposium on Trace Elements in Man and Animals

"Held in the Canadian Rockies at Banff, Alberta, from May 19–24, 1996." — Preface.

Includes an index.
Issued by the National Research Council of Canada.
ISBN 0-660-16404-3

1. Trace elements in the body — Congresses. 2. Trace elements in nutrition — Congresses. 3. Trace elements in animal nutrition—Congresses. I. Fischer, Peter Wilhelm Fritz, 1948- . II. National Research Council of Canada. III. Title.

QP534.I56 1997 572.'5151 C97-980062-5

International Parent Committee

J. McC. Howell, Chairman, Murdoch University, Perth, Australia
I. Bremner, Treasurer, Rowett Research Institute, Aberdeen, UK
C.F. Mills, General Secretary, Rowett Research Institute, Aberdeen, UK
M. Anke, Friedrich Schiller University, Jena, Germany
N.D. Costa, Murdoch University, Perth, Australia
P.W.F. Fischer, Health Canada, Ottawa, Canada
M. Kirchgessner, Technical University of Munich, Freising, Germany
B. Momčilović, USDA, Grand Forks, USA
J.R. Prohaska, University of Minnesota, Duluth, USA
R.A. Sunde, University of Missouri, Columbia, USA

Canadian Local Organizing Committee

P.W.F. Fischer, Chairman, Health Canada, Ottawa, Ontario
M.R. L'Abbé, Secretary/Treasurer, Health Canada, Ottawa, Ontario
E.R. Chavez, McGill University, Ste. Anne de Bellevue, Quebec
R.W. Dabeka, Health Canada, Ottawa, Ontario
J.K. Friel, Memorial University of Newfoundland, St. John's, Newfoundland
R.S. Gibson, University of Otago, Dunedin, New Zealand
L.M. Klevay, USDA, Grand Forks, North Dakota, USA

Acknowledgements

The financial support of the following sponsors is gratefully acknowledged. Without their generous support the meeting would not have been possible.

International Lead Zinc Research Organization, Inc.
International Copper Association, Ltd.
Kellogg Canada Inc.
Medical Research Council of Canada
Nestec S.A.
The Procter and Gamble Company
Ross Laboratories, Division of Abbott Laboratories, Ltd.
United States Pharmacopeial Convention, Inc.
Wyeth-Ayerst Research
Zinpro Corporation

Contents

Trace Elements and Genetic Regulation: Fe, Zn (Part 1)

Trace Elements in Pregnancy and Lactation

Assessment of Trace Elements Status: Opinions and Prejudices

Kinetic Modelling

Trace Elements in the Environment and Food Supply

Trace Elements, Brain Function, and Behaviour

Trace Elements, Membrane Function and Cell Signalling

Analytical, Experimental and Isotopic Trace Element Techniques

Ethics of Trace Element Research in Human and Animal Nutrition

Defining Trace Element Requirements of Preterm Infants

Trace Element Intervention Studies: Design and Interpretation

Trace Elements and Animal Production

Trace Elements and Free Radical Mediated Disease

Trace Elements and Genetic Regulation (Part II)

Trace Elements, Food and Nutrition Policy

Analytical Quality Control

Trace Elements, Infection and Immune Function: Research Developments at Diverse Levels of Biological Organization

Trace Element Intervention Strategies in Animal and Human Nutrition

Trace Element Binding Proteins: Their Regulation and What They Regulate

Role of Trace Elements in Growth and Development

Trace Elements in Bone Metabolism

Mechanisms of Trace Element Toxicity

Newer Trace Element Interactions in Animal and Human Nutrition

Metabolic and Physiological Consequences of Marginal Trace Element Deficiencies

Preface

The Ninth International Symposium on Trace Elements in Man and Animals was held in the Canadian Rockies at Banff, Alberta, from May 19–24, 1996. This meeting was the latest in the series of symposia which began in Aberdeen, Scotland, in July, 1969. The TEMA meetings are held every three years in a relaxed, informal atmosphere and are a unique gathering of specialists from a variety of disciplines sharing an interest in trace elements. TEMA–9 attracted 309 scientists from 38 countries, who, along with another 46 guests of delegates, enjoyed the pristine air of the snow covered mountains in addition to the science.

The scientific program was designed to highlight work at the "cutting-edge" of trace element research. The meeting was opened by Dr. John Arthur from the Rowett Research Institute in Aberdeen, who presented the Underwood Memorial Lecture entitled, "Selenium Biochemistry and Function". This was followed by 21 oral sessions, poster presentations, and five workshops, with a total of 322 abstracts submitted. Each oral session consisted of two invited papers followed by four papers selected from the submitted abstracts. The posters were grouped into sessions with the same subject titles as the oral presentations. The workshops were generally designed to be interactive. All sessions and workshops, even those held in the evenings, were extremely well attended.

Of the 355 attendees, the majority came from the United States, followed by Canada, the United Kingdom, and Japan. Despite extreme financial and/or political difficulties, there were participants from Croatia, Slovenia, Yugoslavia, India, Hungary, Lithuania, Poland, China and Russia. Others travelled long distances to attend from as far away as South Africa, Australia, New Zealand, and Malaysia. The symposium truly was international. The setting of the meeting at the Banff Centre for Conferences, with most delegates living and eating in the excellent on-site facilities, allowed for much interaction outside of the scientific sessions.

The social program was one that was enjoyed by all. The opening Western BarBQ in the Donut Tent, with the hips of beef carried in from the charcoal cooking pits skewered on pitchforks, will long be remembered. The tour of the Banff National Park was, unfortunately, curtailed by the snow remaining in the mountains and by the only rainy day of the week. The fabulous English high tea served at the Chateau Lake Louise made up for the disappointing weather. The Closing Banquet at the Banff Springs Hotel ended the week in grand style. In addition to the planned activities, many will remember the impromptu gatherings around the piano in Bourgeau Lounge.

Many people contributed to making the meeting a success. The Local Organizing Committee who had overall responsibility for the planning of the meeting consisted of Peter Fischer, Mary L'Abbé, Eduardo Chavez, Bob Dabeka, James Friel, Rosalind Gibson, and Les Klevay. They were assisted by the session chairs who chose the invited speakers and selected the oral speakers from the submitted abstracts. The chairs were mainly responsible for making the scientific program so stimulating. The continuous encouragement and helpful suggestions, as well as the initial financial support, which we received from the Parent Committee were very much appreciated. The projectionists, Janet-Marie Huddle and Jeannette Molnár contributed greatly to the smooth running of the oral sessions. Thanks are due to the rapporteurs, who spent many hours listening to tape recordings trying to capture the discussions for inclusion in the Proceedings. These individuals are Kevin Cockell, Christine Hotz, Mark Levy, Janis Randall-Simpson, and Carla Taylor. Finally, Moira Walman, Lyn More, and Keith Trick, all from Health Canada, are thanked for their assistance in dealing with the large amount of clerical work.

The meeting would not have been possible without the generous financial support of the sponsors. The companies and organizations that contributed financially to the symposium are: the International Lead Zinc Research Organization, Inc., the International Copper Association, Ltd., Kellogg Canada, the Medical Research Council of Canada, Nestec S.A., The Procter and Gamble Company, Ross

Laboratories, Division of Abbott Laboratories, Ltd., the United States Pharmacopeial Convention, Inc., Wyeth-Ayerst Research, and Zinpro Corporation.

We certainly hope that everyone who attended this meeting took back many pleasant memories of TEMA–9, Banff and Canada. We wish the organizers of TEMA–10 much success and look forward to the next meeting in the French Alps in 1999.

Peter W.F. Fischer
Mary R. L'Abbé
Kevin A. Cockell
Rosalind S. Gibson

December 1996

Underwood Memorial Lecture

Selenium Biochemistry and Function

J.R. ARTHUR

Division of Biochemical Sciences, Rowett Research Institute, Bucksburn, Aberdeen AB21 9SB

Keywords selenium, selenoproteins, review, function

Introduction

Since the first TEMA meeting in 1969 there have been several major advances in the understanding of the biochemistry and function of selenium (Se). Glutathione peroxidase was the first selenoprotein to be identified, in 1973, and contains stoichiometric amounts of selenium (Rotruck et al. 1973). Subsequently, *in vivo* labelling experiments with ^{75}Se have indicated there may be upwards of 30 selenoproteins in mammalian systems (Wu et al. 1995, Behne et al. 1988, 1995, Evenson and Sunde 1988). On the basis of the ratio of abundant and non-abundant proteins in the genome it has been suggested that there may actually be up to 100 selenoproteins (Burk and Hill 1993). Thus, the identification and characterisation of many of these selenoproteins has provided a framework for major advances in our understanding of selenium biochemistry and function. This review will consider some of these advances in the context of my own research.

Selenium and Vitamin E Deficiencies in Myopathies

The identification of selenium as an essential component of cytosolic glutathione peroxidase provided a possible explanation for the interaction between selenium/vitamin E deficiencies in the pathogenesis of various degenerative disorders in animals. Vitamin E was thought to protect the cell membrane against reactive oxygen metabolites whereas glutathione peroxidase prevented similar oxidative damage in the soluble fraction by metabolising hydrogen peroxide (Hoekstra 1975). Although this mechanism provided a logical explanation for the interaction between selenium and vitamin E it seems more likely that the membrane-associated phospholipid hydroperoxide glutathione peroxidase is the basis of the selenium/vitamin E interaction (Maiorino et al. 1991). Since cytosolic glutathione peroxidase activity in selenium and vitamin E-deficient animals could decrease by at least 90–99% without any overt symptoms other additional functions for selenium seemed likely (Burk 1989, Arthur et al. 1987, Reiter and Wendel 1983). Furthermore, selenium deficiency did not invariably cause disease in experimental animals and pathological changes were, almost invariably, associated with additional forms of stress. For example, a combination of unaccustomed exercise, increased dietary intake of polyunsaturated fatty acid and unsaturated fatty acid composition of membranes all underlie the pathogenesis of myopathy in selenium and vitamin E-deficient cattle (Arthur 1988). Keshan disease, a cardiomyopathy in humans, provides another example of the requirement for additional stresses to induce adverse effects of selenium deficiency. Although Keshan is associated with selenium deficiency in the Chinese population, additional stresses such as viral infection may be required to induce clinical symptoms of disease (Levander et al. 1995, Xia et al. 1989). However, a selenium deficiency of similar severity to that in Keshan disease, does not cause cardiomyopathy in the population of Zaire. In this case the deficiency interacts with iodine deficiency and may be involved in the development of myxoedematus cretinism (Dumont et al. 1994, Goyens et al. 1987).

Correct citation: Arthur, J.R. 1997. Selenium biochemistry and function. In *Trace Elements in Man and Animals – 9: Proceedings of the Ninth International Symposium on Trace Elements in Man and Animals. Edited by* P.W.F. Fischer, M.R. L'Abbé, K.A. Cockell, and R.S. Gibson. NRC Research Press, Ottawa, Canada. pp. 1–5.

Evidence for Multiple Biochemical Functions for Selenium

The variable effects associated with selenium deficiency and cytosolic glutathione peroxidase activity allied with findings from *in vivo* labelling studies with ^{75}Se all stimulated the search for other biological functions of selenium. Thus we demonstrated that three completely different biochemical systems had similar sensitivity to selenium repletion in selenium-deficient rats and mice. These systems were the induction of glutathione S transferases, the ability of neutrophils to ingest and kill yeast cells, and decreases in pyruvate kinase due to muscle damage in selenium and vitamin E-deficient rats (Arthur et al. 1987). Since all these changes could have been caused by a metabolic control dependent on hormone levels, and because hypothyroidism could increase glutathione S transferase activity and have adverse effects on neutrophil function, we investigated the role of selenium in thyroid hormone metabolism (Beckett et al. 1987, 1989, Arthur et al. 1987).

Selenium and Thyroid Hormones

The characteristic response of thyroid hormone metabolism to acute selenium deficiency is an increase in plasma thyroxine (T4) with a smaller decrease in plasma 3,3′,5-triiodothyronine (T3) consistent with lower type I iodothyronine deiodinase (IDI) activity in liver and kidney (Beckett et al. 1987, 1989). We confirmed that there was indeed decreased deiodinase activity in liver, kidney and brain of selenium-deficient rats (Beckett et al. 1989). Labelling of IDI with ^{75}Se *in vivo* and ^{125}I containing substrate analogues *in vitro* indicated that the enzyme was a selenoprotein (Arthur et al. 1990, 1991). This was consistent with cloning of a cDNA for IDI which had an in frame stop codon specifying a protein with selenocysteine at its active site (Berry et al. 1991).

The identification of a role of selenium in type I deiodinase and thyroid hormone metabolism has many implications. In particular, the interactions of selenium and iodine deficiencies have received much attention. Selenium deficiency can either exacerbate or ameliorate some of the effects of a concurrent iodine deficiency. As indicated above, selenium deficiency was proposed to have a causal role in the development of myxoedematus cretinism in iodine-deficient humans in Zaire (Dumont et al. 1994). However, selenium supplementation of selenium- and iodine-deficient subjects had adverse effects on thyroid hormone metabolism (Contempre et al. 1991). Thus when correcting a selenium and iodine deficiency it is essential to provide both micronutrients simultaneously or precede selenium treatment with iodine or thyroid hormone treatment.

As well as implications for human health, interactions between selenium and iodine deficiency may also have significance for animal health and livestock production. In combined selenium and iodine deficiency there are decreased levels of uncoupling protein in brown adipose tissue of rats (Geloen et al. 1990). A similar effect on the brown adipose tissue of farm animals would decrease its potential for thermogenic activity and thus increase susceptibility to cold stress (Nicol et al. 1994, Trayhurn et al. 1993). The potential interaction between selenium and iodine deficiencies in control of survival of neonatal animals therefore warrants further investigation.

Further emphasising the importance of selenium for thyroid hormone metabolism, type II and type III deiodinases have also been cloned and shown to have in frame TGA codons specifying selenocysteine (Croteau et al. 1996, Davey et al. 1995, Salvatore et al. 1995). Although these deiodinases are selenoproteins much remains to be learned about their responses to changes in selenium status and consequent effects on thyroid hormone metabolism. Initial evidence suggests that both type II and type III deiodinase are less likely than type I deiodinase to decrease in activity in selenium deficiency. Unusually, type II deiodinase activity can increase in selenium deficiency, perhaps indicating its critical importance for thyroid hormone metabolism and maintenance of normal neurological development in young animals (Arthur et al. 1996).

"New" Selenoproteins

The discovery of three selenoproteins involved in thyroid hormone interconversions emphasise the advances made in establishing the critical role of selenium in cell metabolism. Furthermore, several other selenoproteins have been characterised and cloned. These include extracellular, phospholipid hydroperoxide and gastrointestinal glutathione peroxidases. Selenoprotein P, the major form of selenium in plasma, selenoprotein W and thioredoxin reductase have been reviewed (Arthur et al. 1996). A specific selenoprotein has also been identified in prostate tissue and there may be a selenoprotein in sperm capsule from mice although this has not been confirmed in other species (see Arthur et al. 1996, Kalcklosch et al. 1995).

The expanding number of known selenoproteins gives rise to several important questions about their biochemistry and function. It is essential to understand which are the most biologically-relevant of the selenoproteins and how this will be elucidated. Only when the functions of each selenoprotein have been characterised will it be possible to predict the adverse or beneficial effects of changes in selenium status. Although diseases caused by severe selenium deficiency have almost been eliminated from man and animals, more subtle subclinical changes associated with variations of selenium status in the normal nutritional range may prove to be of public health importance. For example, recent research showing increased toxicity of viruses after passage through selenium-deficient hosts and decreased incidence of cancers caused by selenium supplementation require further investigation (Clark and Alberts 1995, Levander et al. 1995). Such observations emphasise the need to monitor selenium status in populations and to be able to predict how this relates to possible incidence or risk of disease.

Regulation of Selenoprotein Synthesis

Approximately 30% to 40% of selenium within rats is in cytosolic glutathione peroxidase (Bielstein and Whanger 1988), the remaining 70% of the selenium is distributed in at least 30 other selenoproteins. Understanding the mechanisms whereby selenium is distributed between these selenoproteins, particularly when supplies of the micronutrient are limiting is an important area of research. In selenium deficiency different organs retain selenium with greatly different avidities. Brain and thyroidal selenium concentration decrease by less than 50% whereas liver selenium concentration may decrease by over 95% in selenium deficiency (Chanoine et al. 1993). As well as these gross changes in organ selenium retention, more subtle mechanisms exist to regulate the levels of selenoproteins within the organs. Of the selenoproteins so far characterised phospholipid hydroperoxide glutathione peroxidase is most efficiently retained in the liver in selenium deficiency (Bermano et al. 1995, Lei et al. 1995). In contrast, type I iodothyronine deiodinase activity is more efficiently retained than phospholipid peroxidase in the thyroid (Bermano et al. 1996b). Regulation of these processes probably occurs through tissue-specific changes in mRNA levels for the different selenoproteins. In liver, selenium deficiency does not affect transcription of the different selenoproteins (Bermano et al. 1995). Therefore the decreases in selenoprotein mRNA levels may be a reflection of their inherent stabilities, with cytosolic glutathione peroxidase mRNA being much less stable than the mRNA for phospholipid hydroperoxide glutathione peroxidase. These different stabilities may result from the protection of the mRNAs from ribonucleases by binding to polysomes with different efficiencies as well as structures in the 3′ untranslated regions of the molecules (Bermano et al. 1996a). In the thyroid gland however, such differences in mRNA stabilities cannot explain all changes in selenoprotein levels in selenium deficiency. Transcriptional regulation of selenoprotein mRNA levels under the influence of hormones such as TSH may therefore cause maintenance of deiodinase activity and thus compensatory regulation of thyroid hormone metabolism in selenium deficiency (Bermano et al. 1996b).

Is Selenium Deficiency Always Harmful?

It is generally considered that selenium deficiency has an adverse effect on health in man and animals. However, on occasion selenium deficiency may be beneficial to the animal. Selenium-deficient rats are more resistant to *Salmonella typhimurium* infection and to aflatoxin B_1 hepatotoxicity (Boyne et al. 1984). This latter effect of selenium deficiency can be related to induction of a glutathione S transferase (Yc_2) which specifically conjugates aflatoxin metabolites (Hayes et al. 1996). This selenium deficiency-dependent increase is protective against toxicity from foreign compounds and can be considered to be either beneficial or a reflection of how the body is trying to compensate for adverse decreases in important selenoproteins.

Conclusions

The last 25 years have seen a major increase in our understanding of the function of selenium in animals. Many of these advances have derived from understanding the biochemical function of selenium in an ever increasing variety of metabolic processes. This approach to trace element research, mechanistic rather than observational, is one that was pioneered and encouraged by Underwood. In conclusion, many questions still remain to be answered as to the biochemical functions of selenium. Further characterisation of the mechanisms which control selenoprotein expression in selenium deficiency will help to identify the most important selenoproteins and our recognition of subtle responses to selenium deficiency. More selenoproteins will be identified and characterised, perhaps revealing other cell functions and metabolic pathways which may be influenced by selenium. In the immediate future much remains to be understood about the role of selenium in influencing immune function, prevention of cancer and interactions with thyroid hormone metabolism. The biochemical approach to trace element research advocated by Underwood will provide important information for the clarification of these issues.

Acknowledgements

I am grateful to my many colleagues at the Rowett Institute and elsewhere with whom I have collaborated in the research covered in this review. Their contributions are clear from the reference list. I am also grateful to Colin Mills, Ian Bremner and John Chesters for encouraging research into thyroid hormone metabolism and the search for "non-glutathione peroxidase" functions of selenium. The financial support of the Scottish Office Agriculture, Environment and Fisheries Department (SOAEFD) is gratefully acknowledged.

Literature Cited

Arthur, J. R., Nicol, F., Boyne, R., Allen, K. G. D., Hayes, J. D. & Beckett, G. J. (1987) Old and new roles for selenium. In: Trace Substances in Environmental Health (Hemphill, D.D., ed.), vol. 21, pp. 487–498. Univ. Missouri, Columbia, MO.

Arthur, J. R. (1988) Effects of selenium and vitamin E status on plasma creatine kinase activity in calves. J. Nutr. 118: 747–755.

Arthur, J. R., Nicol, F. & Beckett, G. J. (1990) Hepatic iodothyronine deiodinase: The role of selenium. Biochem. J. 272: 537–540.

Arthur, J. R., Nicol, F., Grant, E. & Beckett, G. J. (1991) The effects of selenium deficiency on hepatic type-I iodothyronine deiodinase and protein disulphide-isomerase assessed by activity measurements and affinity labelling. Biochem. J. 274: 297–300.

Arthur, J. R., Bermano, G., Mitchell, J. H. & Hesketh, J. E. (1996) Regulation of selenoprotein gene expression and thyroid hormone metabolism. Biochem. Soc. Trans. 24: 384–388.

Beckett, G. J., Beddows, S. E., Morrice, P. C., Nicol, F. & Arthur, J. R. (1987) Inhibition of hepatic deiodination of thyroxine caused by selenium deficiency in rats. Biochem. J. 248: 443–447.

Beckett, G. J., MacDougall, D. A., Nicol, F. & Arthur, J. R. (1989) Inhibition of type I and type II iodothyronine deiodinase activity in rat liver, kidney and brain produced by selenium deficiency. Biochem. J. 259: 887–892.

Behne, D., Hilmert, H., Scheid, S., Gessner, H. & Elger, W. (1988) Evidence for specific selenium target tissues and new biologically important selenoproteins. Biochim. Biophys. Acta. 966: 12–21.

Behne, D., Weissnowak, C., Kalcklosch, M., Westphal, C., Gessner, H. & Kyriakopoulos, A. (1995) Studies on the distribution and characteristics of new mammalian selenium-containing proteins. Analyst 120: 823–825.

Bielstein, M. A. and Whanger, P. D. (1988) Glutathione peroxidase activity and chemical forms of selenium in tissues of rats given selenite or selenomethionine. J. Inorg. Biochem. 33: 31–46.

Bermano, G., Nicol, F., Dyer, J. A., Sunde, R. A., Beckett, G. J., Arthur, J. R. & Hesketh, J. E. (1995) Tissue-specific regulation of selenoenzyme gene expression during selenium deficiency in rats. Biochem. J. 311: 425–430.

Bermano, G., Arthur, J. R. & Hesketh, J. E. (1996a) Selective control of cytosolic glutathione peroxidase and phospholipid hydroperoxide glutathione peroxidase mRNA stability by selenium supply. FEBS Lett. 387: 157–160.

Bermano, G., Nicol, F., Dyer, J. A., Sunde, R. A., Beckett, G. J., Arthur, J. R. & Hesketh, J. E. (1996b) Selenoprotein gene expression during selenium-repletion of selenium-deficient rats. Biol. Tr. Elem. Res. 51: 211–223.

Berry, M. J., Banu, L. & Larsen, P. R. (1991) Type-I iodothyronine deiodinase is a selenocysteine-containing enzyme. Nature 349: 438–440.

Boyne, R., Mann, S. O. & Arthur, J. R. (1984) Effect of Salmonella typhimurium infection on selenium-deficient rats. Microbios-Letters. 27: 83–87.

Burk, R. F. (1989) Recent developments in trace element metabolism and function: Newer roles of selenium in nutrition. J. Nutr. 119: 1051–1054.

Burk, R. F. & Hill, K. E. (1993) Regulation of selenoproteins. Ann. Rev. Nutr. 13: 65–81.

Chanoine, J. P., Braverman, L. E., Farwell, A. P., Safran, M., Alex, S., Dubord, S. & Leonard, J. L. (1993) The thyroid gland is a major source of circulating-T(3) in the rat. J. Clin. Invest. 91: 2709–2713.

Clark, L. C. & Alberts, D. S. (1995) Selenium and cancer: Risk or protection? J. Natnl. Cancer Inst. 87: 473–475.

Contempre, B., Dumont, J. E., Ngo, B., Thilly, C. H., Diplock, A. T. & Vanderpas, J. B. (1991) Effect of selenium supplementation in hypothyroid subjects of an iodine and selenium deficient area — The possible danger of indiscriminate supplementation of iodine-deficient subjects with selenium. J. Clin. Endoc. Metab. 73: 213–215.

Croteau, W., Davey, J. C., Galton, V. A. & St Germain, D. L. (1996) Cloning of the mammalian type II iodothyronine deiodinase — A selenoprotein differentially expressed and regulated in human and rat brain and other tissues. J. Clin. Invest. 98: 405–417.

Davey, J. C., Becker, K. B., Schneider, M. J., St Germain, D. L. & Galton, V. A. (1995) Cloning of a cDNA for the type II iodothyronine deiodinase. J. Biol. Chem. 270: 26786–26789.

Dumont, J. E., Corvilain, B. & Contempre, B. (1994) The biochemistry of endemic cretinism — Roles of iodine and selenium deficiency and goitrogens. Mol. Cell. Endoc. 100: 163–166.

Evenson, J. K. & Sunde, R. A. (1988) Selenium incorporation into selenoproteins in the Se-adequate and Se-deficient rat. Proc. Soc. Exp. Biol. Med. 187: 169–180.

Geloen, A., Arthur, J. R., Beckett, G. J. & Trayhurn, P. (1990) Effect of selenium and iodine deficiency on the level of uncoupling protein in brown adipose tissue of rats. Biochem. Soc. Trans. 18: 1269–1270.

Goyens, P., Golstein, J., Nsombola, B., Vis, H. & Dumont, J. E. (1987) Selenium deficiency as a possible factor in the pathogenesis of myxoedematous cretinism. Acta. Endoc. (Copenh). 114: 497–502.

Hayes, J.D., McLeod, R., Pulfird, D.J., Judah, D.J., Nguyen, T., Arthur, J.R. and Neal, G.E. (1996) Regulation of rat Alpha glutathione S-transferases and the contribution of GST Yc2 to resistance to aflatoxin B1. In: Glutathione S-Transferases: Structure Function and Clinical Implications. (Verneulen, N.P.E., et al. eds.), pp. 73–78. Taylor and Francis, London.

Hoekstra, W. G. (1975) Biochemical function of selenium and its relation to vitamin E. Fed. Proc. 34: 2083–2089.

Kalcklosch, M., Kyriakopoulos, A., Hammel, C. & Behne, D. (1995) A new selenoprotein found in the glandular epithelial cells of the rat prostate. Biochem. Biophys. Res. Commun. 217: 162–170.

Lei, X. G., Evenson, J. K., Thompson, K. M. & Sunde, R. A. (1995) Glutathione peroxidase and phospholipid hydroperoxide glutathione peroxidase are differentially regulated in rats by dietary selenium. J. Nutr. 125: 1438–1446.

Levander, O. A., Ager, A. L. & Beck, M. A. (1995) Vitamin E and selenium: Contrasting and interacting nutritional determinants of host resistance to parasitic and viral infections. Proc. Nutr. Soc. 54: 475–487.

Maiorino, M., Thomas, J. P., Girotti, A. W. & Ursini, F. (1991) Reactivity of phospholipid hydroperoxide glutathione peroxidase with membrane and lipoprotein lipid hydroperoxides. Free Rad. Res. Commun. 12–3: 131–135.

Nicol, F., Lefranc, H., Arthur, J. R. & Trayhurn, P. (1994) Characterization and postnatal development of 5′-deiodinase activity in goat perirenal fat. Am. J. Physiol. 267: R144-R149.

Reiter, R. & Wendel, A. (1983) Selenium and drug metabolism-I, Multiple modulations of mouse liver enzymes. Biochem. Pharmacol. 32: 3063–3067.

Rotruck, J. T., Pope, A. L., Ganther, H. E., Swanson, A. B., Hafeman, D. G. & Hoekstra, W. G. (1973) Selenium: biochemical role as a component of glutathione peroxidase. Science 179: 588–590.

Salvatore, D., Low, S. C., Berry, M., Maia, A. L., Harney, J. W., Croteau, W., St Germain, D. L. & Larsen, P. R. (1995) Type 3 iodothyronine deiodinase: Cloning, *in vitro* expression, and functional analysis of the placental selenoenzyme. J. Clin. Invest. 96: 2421–2430.

Trayhurn, P., Thomas, M. E. A., Duncan, J. S., Nicol, F. & Arthur, J. R. (1993) Presence of the brown fat-specific mitochondrial uncoupling protein and iodothyronine 5′-deiodinase activity in subcutaneous adipose tissue of neonatal lambs. FEBS-Letters. 322: 76–78.

Wu, L., Mcgarry, L., Lanfear, J. & Harrison, P. R. (1995) Altered selenium-binding protein levels associated with selenium resistance. Carcinogenesis 16: 2819–2824.

Xia, Y. M., Hill, K. E. & Burk, R. F. (1989) Biochemical studies of a selenium-deficient population in China — Measurement of selenium, glutathione peroxidase and other oxidant defense indices in blood. J. Nutr. 119: 1318–1326.

Trace Element Interactions with Nutrients and Non-Nutrients in Foods and the Food Supply

Chair: P.W.F. Fischer

Trace Element Interactions in Food Crops

R.M. WELCH

USDA-ARS, U.S. Plant, Soil and Nutrition Laboratory, Ithaca, NY 14853-2901, USA

Keywords bioavailability, plant foods, antinutrients, promoters

Nearly half the world's 5.4 billion people are malnourished with more than two billion people consuming diets that are inadequate in several micronutrients including Fe, vitamin A, I, Zn, Se, and possibly other trace elements and vitamins. At greatest risk are the underprivileged (Mason and Garcia 1993). Until recently, eliminating global micronutrient malnutrition (i.e., "hidden hunger") was of relatively low priority internationally. Within the last decade, studies have shown that "hidden hunger" affects a far greater number of people than those suffering from protein/energy malnutrition (ACCSN 1992). Additionally, "hidden hunger" has been shown to have even more serious consequences on human health and sustainable development efforts than previously believed (Anonymous 1992). Now, eliminating micronutrient malnutrition is a high priority globally (Buyckx 1993).

Primary interventions used to eliminate micronutrient malnutrition have included various diet supplementation and food fortification programs. While many of these programs have been successful initially, they have not proven to be sustainable in many developing countries for various economic, political, social and logistic reasons (McGuire 1993). Food systems need to be studied from holistic viewpoints to ascertain what sustainable interventions are possible (Combs et al. 1996). Ideally, agricultural systems should be linked to human nutrition and health goals.

Because staple plant foods (rice, wheat, maize, beans, and cassava) are thought to be relatively poor dietary sources of micronutrients, increasing the density of bioavailable Fe, vitamin A, and Zn in these foods would contribute greatly to finding sustainable solutions to "hidden hunger". A recent review suggests that significant improvements in staple food crop Fe, vitamin A, and Zn bioavailability and density can be made using traditional plant breeding and/or modern gene modification approaches (Graham and Welch 1996).

How can we increase the density of bioavailable Fe, vitamin A and Zn in staple plant food products in sustainable ways? To answer this question requires knowledge of the interacting factors that affect the absorption and utilization of these micronutrients in major plant foods in meals as eaten (Graham and Welch 1996).

Trace Element – Antinutrient Interactions. Staple plant foods can contain substances that inhibit the absorption and/or utilization (i.e., bioavailability) of trace elements. Table 1 gives examples of some antinutrients and their major dietary sources.

Phytic acid forms insoluble precipitates with many polyvalent trace metal ions (e.g., Zn^{2+} and Fe^{3+}) at the basic pH that occurs within the small intestine, thus, lowering their availability for absorption by mucosal cells. While phytic acid inhibits Fe and Zn bioavailability to humans, its naturally occurring form,

Correct citation: Welch, R.M. 1997. Trace element interactions in food crops. In *Trace Elements in Man and Animals – 9: Proceedings of the Ninth International Symposium on Trace Elements in Man and Animals. Edited by* P.W.F. Fischer, M.R. L'Abbé, K.A. Cockell, and R.S. Gibson. NRC Research Press, Ottawa, Canada. pp. 6–9.

Table 1. Major antinutrient substances in plant foods reported to reduce the bioavailability of Fe and/or Zn to humans under most, but not necessarily all, circumstances (modified from Graham and Welch 1996).

Antinutrient	Examples of major dietary source
Phytic acid or phytin	Whole legume seeds and cereal grains
Fiber (e.g., cellulose, hemicellulose, lignin, cutin, suberin, etc.)	Whole cereal grain products (e.g., wheat, rice, maize, oat, barley)
Tannins and polyphenolics	Tea, coffee, beans, sorghum
Oxalic acid	Spinach leaves, rhubarb
Hemaglutinins (e.g., lectins)	Most legumes and wheat
Heavy metals (e.g., Cd, Hg, Pb, Ag)	Plant foods obtained from crops grown on metal polluted soils (e.g., Cd in rice)

phytin, may actually increase Fe bioavailability to humans under some circumstances (e.g., monoferric-phytate in whole wheat products when fed to humans; see Morris 1983, cited in Graham and Welch 1996).

Not all plant-fibers inhibit Fe and Zn bioavailability. Some fibers (e.g., sugar beet fiber) can actually enhance Zn bioavailability (see Fairweather-Tait and Wright 1990, cited in Graham and Welch 1996). Fiber usually acts through its ability to bind tightly metal ions and by its effects on decreasing the sojourn time of digesta through the alimentary tract allowing less time for metal ion absorption by intestinal mucosal epithelial cells (Graham and Welch 1996).

Tannins, polyphenols, and hemaglutinins can also tightly bind certain trace metal ions making them unavailable for absorption. While oxalic acid has been shown to reduce calcium bioavailability to humans, it has not been proven to inhibit Fe and Zn bioavailability (Graham and Welch 1996).

Trace Element – Promoter Interactions. Certain foods are rich in substances that promote the absorption and/or utilization of trace elements by humans even when plant antinutrient substances are present in the meal (see Table 2). For example, ascorbic acid (vitamin C) is a promoter of non-heme Fe(III) bioavailability in meals containing antinutrients (e.g., phytin) from plant foods. Animal meats are important sources of substances (i.e., "meat factors") that promote the bioavailability of both non-heme Fe(III) and Zn when included in meals containing plant food antinutrient substances (Welch and House 1995).

Beard et al. (1996) studied the bioavailability of Fe(III) in phytoferritin. They reported that Fe(III) in phytoferritin was bioavailable to monogastric animals even when fed to rats provided diets containing plant foods high in phytin (soy bean meal). Additionally, a recent study suggests that the sulfur amino acids, methionine and cysteine, are promoters of Zn bioavailability from plant foods high in phytin, i.e., maize kernels (House et al. 1996). Many of these factors (e.g., phytoferritin and the S-amino acids) could be greatly increased in staple plant foods through plant breeding and genetic engineering techniques.

Both Se and Zn promote the bioavailability (i.e., utilization) of either I or vitamin A, respectively in certain population groups. Individuals deficient in Se cannot utilize I, while Zn-deficient individuals cannot

Table 2. Promoter substances in foods reported to enhance the bioavailability of iron, zinc, and vitamin A to humans eating meals containing plant foods in mixed diets under some, but not necessarily all circumstances (modified from Graham and Welch 1996).

Substance	Micronutrient	Major dietary sources
Certain organic acids (e.g., ascorbic acid or vitamin C, fumarate, malate, citrate)	Fe and/or Zn	Fresh fruits and vegetables
Phytoferritin (plant ferritin)	Fe	Legume seeds and leafy vegetables
Certain amino acids (e.g., methionine, cysteine, histidine, and lysine)	Fe and/or Zn	Animal meats (e.g., beef, pork and fish)
Long-chain fatty acids (e.g., palmitic acid)	Zn	Human breast milk
Fats and lipids	Vitamin A	Animal fats, vegetable oils
Se	I	Sea foods, tropical nuts
Zn	Vitamin A	Animal meats
Vitamin E	Vitamin A	Vegetable oils, green leafy vegetables

utilize vitamin A efficiently. Therefore, interventions that only target I and vitamin A deficiencies may not be effective strategies for people deficient in Se or Zn.

Trace Element – Trace Element – Heavy Metal Interactions. Heavy metals (e.g., Cd) can interfere with the absorption and/or utilization of Fe, Zn and other trace element metals in people eating highly heavy metal-contaminated foods, especially if these individuals are already deficient in other mineral nutrients (i.e., Ca). Additionally, major imbalances in dietary trace element intakes can also lead to undesirable interactions between trace elements that reduce their bioavailability. Therefore, food-based strategies (e.g., dietary diversification) to increase micronutrient density in diets run less risk of inducing micronutrient imbalances in people than do other interventions such as supplementation and fortification programs.

Food-based Agricultural Strategies to Improve Micronutrient Bioavailability. It is highly likely that staple food crops could be developed, via plant breeding and/or modern genetic engineering techniques, that accumulate more micronutrients in their edible parts having higher bioavailability then modern high yielding cultivars. Strategies could include designing genotypes that i) accumulate more micronutrients, ii) have lower concentrations of antinutrient factors, and/or iii) contain increased levels of promoter substances in their edible portions.

A recent review suggests that this could be achieved using the genetic diversity that already exists within the gene banks of the world's collections of staple food crops (Graham and Welch 1996). Breeding for increased micronutrient density and levels of promoter substances was viewed as a more desirable strategy then reducing the levels of antinutrient substances. Antinutrients carry out important functions as plant metabolites. Thus, reducing their levels may decrease crop yields. Furthermore, antinutrients may provide health benefits to humans (e.g., possibly acting as anti-carcinogens or by reducing the risk of heart disease). Once the improved genotypes were identified, and their genetic characteristics incorporated into high yielding varieties, the new "super micronutrient" varieties could be used to "field fortify" the diets of poor people in developing countries in sustainable ways, i.e., via their food systems.

Literature Cited

ACCSN (1992) Second Report on the World Nutrition Situation, Vol. I. Global and Regional Results, Micronutrient Deficiency — the Global Situation. United Nations, New York, NY.

Anonymous (1992) Enriching Lives, Overcoming Vitamin and Mineral Malnutrition in Developing Countries. World Bank, Washington, D.C.

Beard, J. L., Burton, J. W. & Theil, E. C. (1996) Purified ferritin and soybean meal can be sources of iron for treating iron deficiency in rats. J. Nutr. 126: 154–160.

Buyckx, M. (1993) The international community's commitment to combating micronutrient deficiencies. Food Nutr. Agr. 7: 2–7.

Combs, G. F., Welch, R. M., Duxbury, J. M., Uphoff, N. T. & Nesheim, M. C. (1996) Food-Based Approaches to Preventing Micronutrient Malnutrition: an International Research Agenda, pp. 1–68. Cornell Internat. Inst. for Food, Agr., and Development, Cornell University, Ithaca, NY.

Graham, R. D. & Welch, R. M. (1996) Breeding for staple-food crops with high micronutrient density. Agricultural Strategies for Micronutrients Working Paper 3. pp. 1–72. Internat. Food Policy Res. Inst., Washington, D.C.

House, W. A., Van Campen, D. R. & Welch, R. M. (1996) Influence of dietary sulfur-containing amino acids on the bioavailability to rats of zinc in corn kernels. Nutr. Res. 16: 225–235.

Mason, J. B. & Garcia, M. (1993) Micronutrient deficiency — the global situation. SCN News 9: 11–16.

McGuire, J. (1993) Addressing micronutrient malnutrition. SCN News 9: 1–10.

Welch, R. M. & House, W. A. (1995) Meat factors in animal products that enhance iron and zinc bioavailability: implications for improving the nutritional quality of seeds and grains. pp. 58–66. 1995 Cornell Nutr. Conf., Depart. Animal Sci, and Div. of Nutr., Cornell University, Ithaca, NY.

Discussion

Q1. Peter Fischer, Health Canada, Ottawa, ON, Canada: What is the role of bioengineering in increasing the bioavailability of minerals in plants?

A. There are many people who are thinking about it, with two general approaches — decreasing anti-nutrients or increasing promoters of bioavailability. For example, Teale and colleagues at North Carolina are working on a gene for phytoferritin.

Q2. Lindsay Allen, University of California at Davis, CA, USA: What is the validity of using animal models where intestinal phytases are high?

A. We looked at rat phytase 20 years ago and didn't find much loss of fecal phytate. Even if the rat is a less sensitive model, if there is any effect in rats, there is likely to be one in humans. The model may not be ideal and it would be better to use humans or pigs, but we don't have access to them, so we have to make do with what is available.

Q3. Neville Suttle, Moredun Research Institute, Edinburgh, Scotland: I am surprised that you have dismissed the approach of reducing anti-nutrients. For example, consider goitrogens, as studies where there have been glucosinolates and selection against them has not had any negative impact on yield.

A. Well yes, I agree in this circumstance.

Food Processing Influencing Iron and Zinc Bioavailability

A.-S. SANDBERG

Department of Food Science, Chalmers University of Technology, S-402 29 Göteborg, Sweden

Keywords iron, zinc, bioavailability, food processing

Food processing can have both positive and negative effects on iron and zinc bioavailability, because enhancers/inhibitors can be formed or degraded during the process. Food additives can also be inhibitors or enhancers of absorption. In addition, the chemical form of iron can be altered during food processing and storage, e.g., the relative amounts of ferrous and ferric iron and the relative amounts on non-heme and heme iron can be altered, thereby affecting bioavailability. The bioavailability of iron is related to its solubility, with soluble forms having a low bioavailability. Changes in solubility of iron during processing therefore can affect the bioavailability (Clydesdale et al. 1982, Johnson 1991, Lee 1982, Sandberg 1991).

Food Additives

A variety of substances are added to foods during processing for several purposes: to maintain freshness and safety; to help in processing or preparation, and; to make food more appealing.

Some of these additives have an effect on iron bioavailability because they are inhibitors or enhancers of iron absorption. Preservatives and antioxidants include substances such as calcium lactate, calcium sorbate, citric acid, EDTA, ascorbic acid and phytic acid. Calcium and phytic acid are well known inhibitors of iron absorption (Hallberg et al. 1991, Rossander-Hultén et al. 1992), while ascorbic acid (Hallberg et al. 1986, Sayers et al. 1973), and under certain conditions, citric acid enhances iron absorption (Gillooly et al. 1983). Iron is easily complexed by EDTA and Fe(III) EDTA is well absorbed (MacPhail et al. 1981).

Other food additives are used as emulsifiers, stabilizers, leavening agents, pH control agents, humeactants, dough conditioners and anti-caking agents. These include a number of calcium compounds, polysaccharides and organic acids. Food additives which affect appeal characteristics of foods include flavour enhancers, flavours, colours and sweeteners. Food dyes are complex organic molecules which have the potential of forming complexes with trace minerals (Johnson 1991).

Heat Processing

There are a number of processes which may affect iron and zinc bioavailability, e.g., heat processes, such as flaking, puffing, extrusion cooking, cooking and canning. Heat processing can inactivate the enzyme phytase, which hydrolyses phytate, while vitamin C can be destroyed during heat treatment and the structure of polyphenols can be altered. Losses of trace elements can occur by leaching into the cooking medium. These effects are dependent on processing conditions, such as time, temperature, pH and moisture. Both baking and microwave cooking increased non-heme iron in ground beef and a linear relationship was observed between non-heme iron in meat and the time of exposure to heat treatment (Schricker et al. 1983). The change was large enough to cause changes in absorbable iron in the meat. Added nitrite appeared to protect against heat induced changes in non-heme iron. Carpenter and Clark (1995), found a significant difference in the relative amounts of heme iron to non-heme iron in raw and cooked meat of different sources, which was cooked to an internal temperature of 71°C.

Changes in relative amounts of ferric and ferrous iron as a consequence of a wet-heat process such as thermal processing in glass of spinach puree, were reported (Lee and Clydesdale 1981). Cooking in iron utensils can also considerably increase the iron content in the food (Johnson 1991).

Correct citation: Sandberg, A.-S. 1997. Food processing influencing iron and zinc bioavailability. In *Trace Elements in Man and Animals – 9: Proceedings of the Ninth International Symposium on Trace Elements in Man and Animals. Edited by* P.W.F. Fischer, M.R. L'Abbé, K.A. Cockell, and R.S. Gibson. NRC Research Press, Ottawa, Canada. pp. 10–14.

Heat treatment during extrusion cooking of wheat bran caused inactivation of phytase and was found to decrease apparent absorption of zinc, magnesium and phosphorus in ileostomy subjects (Kivistö et al. 1986) and also zinc absorption measured by isotope technique (Kivistö et al. 1989) when compared with raw bran. This effect was ascribed to the lack of phytate hydrolysis of the extruded bran in the stomach and small intestine of humans (Sandberg et al. 1987). By reducing the phytate content of bran before extrusion cooking, the zinc absorption was increased (Kivistö et al. 1989). Recent results demonstrate that extrusion cooking has such an effect on the microstructure of the food that it affects digestibility in the gut (Autio and Sandberg, unpublished results).

Physical Processing

Physical processing, e.g., milling of the grain by removing the outer layers of the grain, reduces the polyphenol and phytate content. Trace elements are also removed by this process. Thus, the bioavailability may be improved while the amount of trace elements is significantly reduced. Removing the outer layers of sorghum grains reduced the polyphenol and phytate contents by 96% and 2%, respectively, and resulted in an increase of iron absorption from 1.7 to 3.5% (Gillooly et al. 1984).

Bioprocessing

Bioprocessing of cereals and vegetables is a means to improve the bioavailability of minerals by degrading specific inhibitors that prevent the absorption from the food and by formation of promoting factors for the absorption. Phytate hydrolysis can occur in food production, by phytase from either plants, yeasts or microorganisms. Examples of bioprocesses activating the intrinsic enzymes of cereals and legumes are soaking, malting, lactic fermentation and hydrothermal processing. During the germination step of the malting process, enzymes are synthesised or activated. Lactic fermentation leads to lowering of pH as a consequence of bacterial production of organic acids, mainly lactic acid, which is favourable for phytase activity. Some organic acids formed during fermentation are enhancers of iron absorption. The microorganisms of the starter culture used in fermentation can, in some cases, exert phytase activity. Phytase can also be added in the food process.

Optimal Conditions for Phytase Activity

To optimize food process to increase iron bioavailability by phytate degradation, it is essential to know optimal conditions for phytase activity in cereals and legumes. Phytase activity is strongly dependent on pH and temperature and most phytases are active only in a narrow pH and temperature range. There are differences in optimal conditions for phytase activity between different plant species. Peers (1953) showed that optimal conditions for wheat phytase were pH 5.1 and 55°C and activity declines rapidly when changing pH. We recently found that the oat phytase has its optimal conditions at pH 4.3–4.5, 38°C which thus differs from that of wheat (Larsson and Sandberg 1992). These findings partly explain why oats were suggested to have a low phytase activity.

Scott (1991) demonstrated an alkaline phytase activity of phytase extracted from different varieties of *Phaseolus vulgaris*. We found that brown beans have phytase activity at pH 4.5 and at pH 8 and 37°C, but the optimal activity was at pH 7.0 and 55°C. Microbial phytase activity works in a broader pH range. A pH optimum of yeast phytase was found at pH 3.5, but high activity occurs between 3.5 and 4.5 (Türk et al. 1996). Phytase produced by *Aspergillus niger* has one optimum at pH 5 and one at pH 2.5–3.0, probably due to additional acidic phosphatase activity (Simell et al. 1990), but activity occurs at all pH between 1.0 and 7.5.

Soaking and Malting

Soaking of wheat bran, whole wheat flour and rye flour at optimal conditions for wheat phytase activity (pH 4.5–5, 55°C) resulted in complete phytate hydrolysis (Mellanby 1950, Sandberg and Svanberg 1991) and to a marked increase in iron availability estimated *in vitro* (Sandberg and Svanberg 1991), if phytate is reduced to levels beyond 0.5 μmol/g.

Malting is a process during which the whole grain is soaked and then germinated. The amount of phytate in malted grains of wheat, rye and oats intended for the production of flour was only reduced slightly or not at all. However, when the malted cereals were ground and soaked at optimal conditions for wheat phytase, there was a complete degradation of phytate (Larsson and Sandberg 1992), except for oats, which under the conditions studied, had a low phytase activity. By germinating oats for five days at 11°C followed by incubation at 37–40°C, it was possible to reduce phytate content of oats by 98% (to 0.5 µmol/g) (Larsson and Sandberg 1992). The iron and zinc absorption from oat porridge made of untreated flour was compared to that from oat porridge made of malted flour, with a phytate reduction of 77%. Both iron and particularly zinc absorption, was significantly improved from the porridge made of malted flour (the fractional iron absorption increased from 4.4 to 6.0 and zinc absorption from 11.8 to 18.3) (Larsson et al. 1996).

Fermentation

Lactic fermentation of maize, soya beans and sorghum reduces the phytate content (Lopez et al. 1983, Sudarmadji et al. 1977). We have shown that lactic fermentation of white sorghum and maize gruels can yield an almost complete degradation of phytate. At optimal conditions, the effect on *in vitro* estimation of iron availability was an almost tenfold increase (Svanberg et al. 1993). The presence of tannins in sorghum not only decreased the availability of iron *per se*, but also seemed to inactivate the enzyme phytase. Further studies are therefore needed in order to find methods to increase mineral availability in tannin rich foods.

Sauerkraut, a lactic fermented vegetable, was found to markedly increase iron absorption from a meal (Hallberg et al. 1982). Lactic acid-fermented vegetables added to a meal were found to increase the iron absorption in humans (Rossander-Hultén et al. 1992). The fresh vegetables (carrots, turnips, onions) contained small amounts of phytate, which was hydrolysed during the fermentation. The amount of iron absorbed was increased when the fermented vegetables were added to a white wheat roll (the fractional absorption increased from 13.6 to 23.6) and also when added to phytate-rich meals such as wholemeal rye and wheat rolls (the fractional absorption increased from 5.2 to 10.4). This indicates the formation of iron-promoting factors in lactic acid-fermented vegetables. No differences were found in zinc absorption between a meal containing raw or fermented vegetables.

We investigated the phytate reduction in scalded bread and bread with varying amounts of sourdough baked with rye bran or oat bran addition. The most marked phytate reduction of 96–97% occurred in bread made with 10% sourdough (pH 4.6) or in the breads in which the pH had been adjusted in the mild scalding with lactic acid, resulting in a pH between 4.4 and 5.1 (Larsson and Sandberg 1991). We found that the percentage iron absorption from bread meals with such sour dough fermented bread was similar to that of white wheat bread, not containing inhibitory factors (Brune et al. 1992). Consequently, the amounts of iron absorbed from whole meal bread with its high content of iron, would be greater than from white bread, provided the fermentation is optimized.

Addition of milk to the dough during bread-making, inhibited enzymatic phytate hydrolysis (Türk et al. 1992) resulting in depressed human iron absorption from the bread (Hallberg et al. 1991). The results were considered an effect of formation of insoluble calcium phytate complexes. Fermented milk did not significantly affect enzymatic hydrolysis during fermentation, probably depending on the presence of lactic acid lowering the pH (Türk et al. 1992). Addition of acetic acid or lingonberries to the dough increased the phytate reduction to 96% dough (pH 4.5–5) and 83%, respectively compared to 55% in control bread without additives (Türk et al. 1996).

Addition of Enzymes

Microbial phytase enzyme preparations are now available commercially, making their use in food processing technically feasible. A very effective phytate degradation was achieved by adding *A. niger* phytase to an oat-based nutrient solution fermented by *Lactobacillus plantarum*, but the added enzyme had a negative influence on the viable counts of *Lactobacilli* as well as the aroma (Marklinder et al. 1995), while addition of *A. niger* during bread-making was less effective (maximum 88% hydrolysis) unless the pH was lowered to 3.5. The effect of reducing the phytate in soy-protein isolates on iron absorption was investigated in humans (Hurrell et al. 1992). Addition of *A. niger* phytase, in amounts which almost completely removed phytate in soy isolates, increased iron absorption four- to five-fold. We also found that this enzyme added to a meal increased iron absorption presumably by degrading phytate in the stomach (Sandberg et al. 1996).

Conclusions

Increased bioavailability of iron and zinc by food processing can be achieved by degradation of phytate to low levels and for iron by addition/formation of ascorbic and certain other organic acids. The bioavailability can also be improved by addition of microbial phytase, which can exert an effect in the stomach and small intestine. Methods for degradation of polyphenols during food processing should be developed.

References

Brune, M., Rossander-Hulthén, L., Hallberg, L., Gleerup, A. & Sandberg, A.-S. (1992) Human iron absorption from bread: inhibiting effects of cereal fiber, phytate and inositol phosphates with different numbers of phosphate groups. J. Nutr. 122: 442–449.

Carpenter, C.E. & Clark, E. (1995) Evaluation of methods used in meat iron analysis and iron content of raw and cooked meats. J. Agric. Food Chem. 43;1824–1827.

Clydesdale, F.M. (1982) The effects of physicochemical properties of food on the chemical status of iron. In: Nutritional bioavailability of iron, ACS Symposium Series 203. (Kies, C., ed.), pp. 55–84.

Gillooly, M., Bothwell, T.H., Charlton, R.W. et al. (1984) Factors affecting the absorption of iron from cereals. Br. J. Nutr. 51: 37–46.

Gillooly, M., Bothwell, T.H. & Torrance, J.D. (1983) The effects of organic acids, phytates and polyphenols on the absorption of iron from vegetables. Br. J. Nutr. 49: 331–342.

Hallberg, L., Brune, M. & Rossander, L. (1986) Effect of ascorbic acid on iron absorption from different types of meals. Studies with ascorbic acid-rich foods and synthetic ascorbic acid given in different amounts with different meals. Hum. Nutr. Applied Nutr. 40A: 97–113.

Hallberg, L., Brune, M., Erlandsson, M., Sandberg, A.-S. & Rossander-Hultén, L. (1991) Calcium effect of different amounts on nonheme and heme iron absorption in humans. Am. J. Clin. Nutr. 53: 112–119.

Hallberg, L. & Rossander, L. (1982) Absorption of iron from Western-type lunch and dinner meals. Am. J. Clin. Nutr. 35: 502–509.

Hurrell, R.F., Juillerat, M.-A., Reddy, M.B., Lynch, S.R., Dassenko, S.A. & Cook, J.D. (1992) Soy protein, phytate, and iron absorption in humans. Am. J. Clin. Nutr. 56: 573–578.

Johnson, P.E. (1991) Effect of food processing and preparation on mineral utilization. In: Nutritional and toxicological consequences of food processing. Advances in experimental medicine and biology (Friedman, M., ed.), pp. 483–498.Plenum Press, New York, NY.

Kivistö, B., Cederblad, A., Davidsson, L., Sandberg A.-S. & Sandström, B. (1989) Effect of meal composition and phytate content on zinc absorption in humans from an extruded bran product. J. Cereal Sci. 10: 189–197.

Kivistö, B., Andersson, H., Cederblad, G., Sandberg A.-S. & Sandström, B. (1986) Extrusion cooking of a high-fibre cereal product. II. Effects on apparent absorption of zinc, iron, calcium, magnesium and phosphorus in humans. Br. J. Nutr. 55: 255–260.

Larsson, M., Rossander-Hulthén, L., Sandström, B. & Sandberg, A.-S. (1996) Improved zinc and iron absorption from malted oats, with reduced phytate content. Br. J. Nutr. (accepted).

Larsson, M. & Sandberg, A.-S. (1992) Phytate reduction in oats during malting. J. Food Sci. 57: 994–997.

Larsson, M. & Sandberg, A.-S. (1991) Phytate reduction in bread containing oat flour, oat bran or rye bran. J. Cereal Sci. 14: 141–149.

Lee, K. & Clydesdale, F.M. (1981) Effect of thermal processing on endogenous and added iron in canned spinach. J. Food Sci. 46: 1064–1068.

Lee, K.M. (1982) Iron chemistry and bioavailability in food processing. In: Nutritional bioavailability of iron (Kies, C., ed.) ACS Symposium Series 203, pp. 27–54.

Lopez, Y., Gordon, D.T. & Fields, L. (1983) Release of phosphorus from phytate by natural lactic acid fermentation. J. Food Sci. 48: 953–954.

MacPhail, A.P., Bothwell, T.H., Torrance, J.D., Derman, D.P., Bezwoda, W.R., Charlton, R.W. & Mayet, F. (1981) Factors affecting the absorption of iron from Fe(III) EDTA. Br. J. Nutr. 45: 215–227.

Marklinder, I.M., Larsson, M., Fredlund, K. & Sandberg, A.-S. (1995) Degradation of phytate by using varied sources of phytases in an oat-based nutrient solution fermented by *Lactobacillus plantarum* 299 V. Food Microbiology 12: 487–495.

Mellanby, E. (1950) Some points in the chemistry and biochemistry of phytic acid and phytase. In: A Story of Nutritional Research, pp. 248–282. Williams and Wilkins, Baltimore, MD.

Peers, F.G. (1953) The phytase of wheat. Biochem J. 53: 102–110.

Rossander-Hultén, L., Sandberg, A.-S. & Sandström, B. (1992) The effect of dietary fibre on mineral absorption and utilization. In: ILSI Human Nutrition Reviews. Dietary fibre — A component of food-nutritional function in health and disease. (Schweizer, E., ed.) pp. 197–216. ILSI, Washington, DC.

Sandberg, A.-S. & Svanberg, U. (1991) Phytate hydrolysis by phytase in cereals. Effects on *in vitro* estimation of iron availability. J. Food Sci. 56: 1330–1333.

Sandberg, A.-S., Rossander-Hultén, L. & Türk, M. (1996) Dietary Aspergillus niger phytase increases iron absorption in humans. J. Nutr. 126: 476–480.

Sandberg, A.-S. (1991) The effect of food processing on phytate hydrolysis and availability of iron and zinc. In: Nutritional and Toxicological Consequences of Food Processing (Friedman, ed.), pp. 499–508. Plenum Press, New York, NY.

Sandberg, A.-S., Andersson, H., Carlsson, N.-G. & Sandström, B. (1987) Degradation products of bran phytate formed during digestion in the human small intestine. Effect of extrusion cooking on digestibility. J. Nutr. 117: 2061–2065.

Sayers, M.H., Lynch, S.R., Jacobs, P. et al. (1973) The effect of ascorbic acid supplementation on the absorption of iron in maize, wheat and soya. Br. J. Nutr. 24: 209–217.

Schricker, B.R. & Miller, D.D. (1983) Effects of cooking and chemical treatment on heme and nonheme iron in meat. J. Food Sci. 48: 1340–1349.

Scott, J.J. (1991) Alkaline phytase activity in non-ionic detergent extracts of legume seeds. Plant Physiol 95: 1298.

Simell, M., Turunen, M., Piironen, J. & Vaara, T. (1990) In: Proceedings of VIIth Symposium 122 on Bioconversion of Plant Raw Materials — Biotechnology Advancement, pp. 145–61. Finland.

Sudarmadji, S. & Markakis, P. (1977) The phytate and phytase of soybean tempeh. J. Sci. Fd. Agric. 28: 381–383.

Svanberg, U., Lorri, W. & Sandberg, A.-S. (1993) Lactic fermentation of non-tannin and high-tannin cereals: effect on *in vitro* estimation of iron availability and phytate hydrolysis. J. Food Sci. 58: 408–412.

Türk, M., Carlsson, N.-G. & Sandberg, A.-S. (1996) Reduction in the levels of phytate during wholemeal breadmaking; effect of yeast and wheat phytases. J. Cereal.Sci. (in press).

Türk, M. & Sandberg, A.-S. (1992) Phytate hydrolysis during bread-making: effect of addition of phytase from *Aspergillus Niger*. J. Cereal Sci. 15: 281–294.

Discussion

Q1. Rosalind Gibson, University of Otago, Dunedin, New Zealand: I am wondering about the effect of ascorbic acid on phytate as I am not aware of any interaction.

A. Ascorbic acid does not break down phytic acid but it has been shown in several studies that it may increase the solubility of phytase and indirectly increase the bioavailability of minerals.

Q2. Les Klevay, USDA-ARS, Grand Forks, ND, USA: I wonder what you can tell us about effects of processing on copper bioavailability.

A. We haven't studied copper, so I cannot comment.

Calcium Does Not Inhibit Iron Absorption after Adaptation to a High Calcium Diet in the Infant-Piglet Model

INE P.M. WAUBEN, FILOMINA INCITTI, AND STEPHANIE A. ATKINSON
Department of Pediatrics, McMaster University, Hamilton, Ontario, Canada L8N 3Z5

Keywords calcium–iron interaction, calcium supplementation, adaptation, infant-piglet

Based on findings in animals and adult humans studies, which demonstrated that a diet high in calcium (Ca) inhibited iron (Fe) absorption, it has been suggested that infants fed a diet with high Ca content may be at risk for developing Fe deficiencies (Barton et al. 1983, Hallberg et al. 1992).

Supplementation of Ca to premature infants fed their mother's milk is common practice in order to prevent osteopenia. Supplementation with Ca and phosphorus of formulas intended for feeding of premature infants after discharge from hospital is gaining interest in order to optimize growth and development. In premature infants, Fe stores are low at birth. Therefore, the issue of Ca:Fe interaction may be important for this pediatric population.

However, results from human or animal studies cannot be readily extrapolated to premature infants for several reasons. 1) Other factors in human adult or animal diets such as phytates and fibre, which are not present in infant formula or breast milk, might partially account for inhibiting Fe absorption. 2) Premature infants have higher Fe needs and might therefore have a greater efficiency of absorption. 3) Most of the reported studies were performed with single test meals, ignoring possible adaptation mechanisms.

Since Ca supplementation in premature infants is maintained over a prolonged period of time, we investigated whether chronic exposure to a high Ca diet inhibits Fe absorption in the infant-piglet model.

Materials and Methods

Eighteen piglets were randomized to a regular liquid piglet formula (Norm-Ca, Ca:Fe molar ratio = 230, N = 9) or to a high Ca piglet formula (Hi-Ca, Ca:Fe molar ratio = 540, N = 9) at three to four days of age. The increase in Ca:Fe (mmol:mmol) ratio from Norm-Ca to Hi-Ca was designed to be similar to the increase in Ca:Fe ratio in premature infants fed mother's milk supplemented with Ca. At three to four days of age, an iron dextran injection providing 100 mg elemental Fe was administered intramuscularly, in combination with modified Fe content in the experimental diets, to achieve a marginal state of Fe repletion to simulate the Fe status of the premature infant. Diets were fed for two and a half to three weeks. During the feeding protocol, weight was determined daily and length weekly and Fe status was determined by hemoglobin and hematocrit (determined by Coulter Counter). Whole body Fe retention from the diet was determined with tracers of ^{55}Fe and ^{59}Fe, as described before (Davidson et al. 1990). Ca:Fe interactions at absorption sites were investigated by measuring time dependent apparent ^{59}Fe absorption in response to molar Ca:Fe ratios of 230 and 540 injected into ligated loops of the jejunum of anaesthetized piglets. At necropsy, Fe status was assessed by Fe content in organs and in serum determined by flame atomic absorption spectrometry (Perkin Elmer, Norwalk, CT).

Results

No differences were found in body size, growth, whole body Fe retention and Fe status between piglets fed Norm-Ca and Hi-Ca. *In situ* apparent Fe absorption in the jejunum of the piglets fed Norm-Ca and Hi-Ca is shown in Figure 1.

Correct citation: Wauben, I.P.M., Incitti, F., and Atkinson, S.A. 1997. Calcium does not inhibit iron absorption after adaptation to a high calcium diet in the infant-piglet model. In *Trace Elements in Man and Animals – 9: Proceedings of the Ninth International Symposium on Trace Elements in Man and Animals. Edited by* P.W.F. Fischer, M.R. L'Abbé, K.A. Cockell, and R.S. Gibson. NRC Research Press, Ottawa, Canada. pp. 15–16.

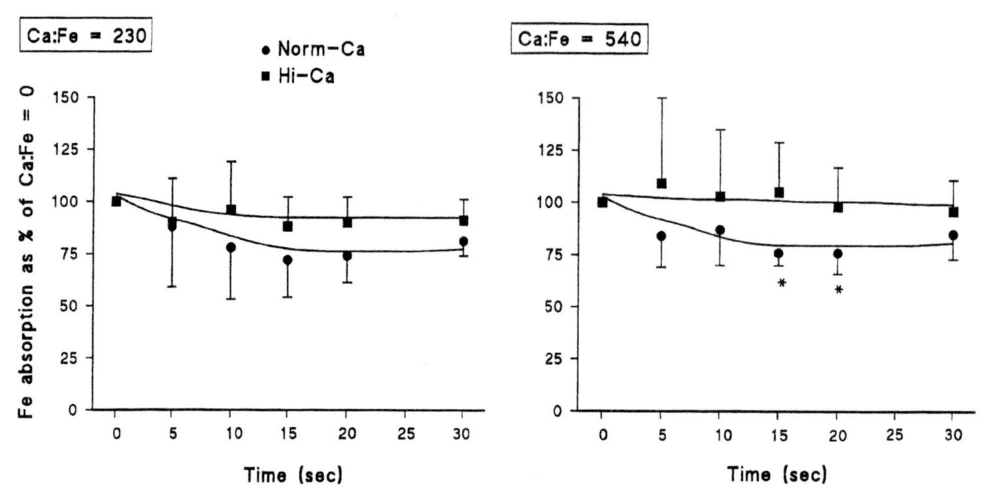

*Figure 1. Time dependent apparent ^{59}Fe absorption (as percent of control, Ca:Fe = 0) in response to Ca:Fe ratios of 230 and 540 injected into ligated loops of the jejunum of anaesthetized piglets adapted to Norm-Ca or Hi-Ca. *P <0.05.*

Summary

Theoretically, the sites at which Ca:Fe interactions may occur in order to inhibit Fe absorption are: 1) the luminal site; 2) the brush border membrane; 3) intracellular pathways in the enterocyte or 4) the basolateral membrane. Adaptive mechanisms to alleviate Ca:Fe interactions may be present at any site where Ca:Fe interactions can occur. Our observations suggest that the piglets fed Hi-Ca may adjust Fe absorption to counteract the inhibitory effect of Ca on Fe absorption. This adaptation may be, either in the lumen or at the brush border membrane. The latter site is supported by previous findings in our laboratory that Ca and Fe compete for uptake in brush border membrane vesicles of piglets (Bertolo et al. 1994). Future investigations in in-vitro and/or in-situ models will be necessary to determine the specific site(s) of Ca:Fe interactions and possible adaptation mechanisms in a developing animal-model.

In conclusion, Ca does not inhibit Fe absorption in the infant-piglet after adaptation to a high Ca diet. Since the piglet is a suitable model for the human neonate, it can be speculated that at concentrations currently used in premature infant diets, Ca supplementation will likely not compromise Fe status.

Literature Cited

Barton, J.C., Conrad, M. & Parmley, R.T. (1983) Calcium inhibition of inorganic iron absorption in rats. Gastroenterology 84: 90–101.

Bertolo, R.F.P., Bettger, W.J. & Atkinson, S.A. (1994) Calcium, magnesium, iron, copper and manganese compete with zinc for a multidication channel in brush border membrane vesicles from the piglet. FASEB J. 8: A697.

Davidson, L.A., Litov, T.E. & Lonnerdal, B. (1990) Iron retention from lactoferrin-supplemented formulas in infant rhesus monkeys. Ped. Res. 27: 176–180.

Hallberg, L., Rossander-Hulten, L., Brune, M. & Gleerup, A. (1992) Bioavailability in man of iron in human milk and cow's milk in relation to their calcium content. Ped. Res. 31: 524–527.

Discussion

Q1. Peter Fischer, Health Canada, Ottawa, ON, Canada: I would like to know the details of the brush border membrane vesicles.

A. This work was done by a former student in our lab and basically involves mucosal scrapings, isolation with $MgCl_2$ to isolate the vesicles which are amenable to *in vitro* studies for time-dependent uptake at the brush border membrane. My study only looked at ligated loops.

Q2. Wilhelm Windisch, Technical University of Munich, Germany: What are the Ca effects if the Fe status is marginal?

A. The pigs had marginal Fe status although they did have iron dextran injections and therefore the absorbability of Fe was likely optimized. However, the model reflects the real life situation of marginal status.

Calcium and Zinc Exchange in ^{65}Zn Labelled Rats at Different Levels of Calcium Supply

WILHELM WINDISCH AND MANFRED KIRCHGESSNER
Institute of Nutrition Physiology, TU Munich-Weihenstephan, 85350 Freising, Germany

Keywords calcium–zinc interaction, metabolism, bone, rat

High dietary Ca contents may principally affect the bioavailability of Zn, as has been shown widely by the use of performance data and parameters of Zn status. However, investigations on quantitative effects are rare, especially for lower ranges of Ca supply, where Zn homeostasis may cover the Ca effects. Therefore, the present experiment was designed to quantitate Zn metabolism at levels of Ca supply ranging from deficiency to a moderate excess.

Material and Methods

Thirty-two female Sprague Dawley rats were reared for 4 weeks from 75 to 186 g of body weight with a semisynthetic diet containing 0.41% Ca, 18 µg/g Zn and 16 Bq ^{65}Zn/µg Zn. Subsequently 5 of the young adult animals were killed for reference. Each 9 of the remaining animals were fed restrictively (7.7 g/d) a diet with modified Ca contents (deficiency: 0.01%; normal: 0.41%; moderate excess: 0.91%). The Zn content was uniformly 19 µg/g without radiozinc. Fecal and renal excretions were collected daily. After 16, 23, and 30 d, 3 animals of each treatment group were killed and completely dissected into organs and tissue fractions. The excrements and tissues were analyzed for Ca, Zn, and ^{65}Zn.

Results and Discussion

The body weight of the animals remained unchanged (187 g). Ca deficiency stopped the excretion and retention of Ca. The excessive Ca intake was completely compensated for, by an increased fecal Ca excretion which resulted in a Ca retention of 16 mg/d for both normal and excessive Ca supply (Windisch and Kirchgessner 1994a). The Ca amounts in the soft tissues were of quantitative minor importance. The quantity of Ca in the skeleton of Ca deficient animals was equal to the reference level (1927 and 1919 mg), while it increased in normal and excessive Ca supply (2179 and 2180 mg). Therefore, Ca deficiency inhibited the Ca deposition into the skeleton (Windisch and Kirchgessner 1994b).

The Zn intake was constant among treatment groups (Table 1). Ca deficiency increased fecal Zn excretion by a rise in endogenous fecal Zn, while true absorption of dietary Zn remained unchanged. Also urinary Zn increased slightly. According to the pronounced excretions, the Zn retention fell from 17 µg/d at normal Ca supply, to –2 µg/d. In Ca excess, the fecal Zn excretion was also increased. However, this effect resulted from a depressed true absorption of dietary Zn. The drop in true absorption was not fully compensated for by a respective reduction of endogenous fecal Zn, as would have been expected according to Zn homeostasis at normal feeding conditions. Obviously, Ca excess depressed the absorption of dietary Zn and also that of endogenous Zn, which is assumed to be secreted into the digestive tract and then partially reabsorbed together with dietary Zn (Windisch and Kirchgessner 1995). The negative effect of Ca on Zn absorption seemed to exceed the compensatory capacity of the Zn homeostasis and to induce a Zn deficit in the intermediate metabolism, as can be seen by the reduction in urinary Zn to the level of the obligatory losses and the depressed Zn retention.

The Zn in soft tissues was not affected by different Ca supply (Table 1). In the skeleton of Ca deficient animals, the Zn quantity remained at the reference level, while it increased in normal Ca supply. Obviously,

Correct citation: Windisch, W., and Kirchgessner, M. 1997. Calcium and zinc exchange in ^{65}Zn labelled rats at different levels of calcium supply. In *Trace Elements in Man and Animals – 9: Proceedings of the Ninth International Symposium on Trace Elements in Man and Animals. Edited by* P.W.F. Fischer, M.R. L'Abbé, K.A. Cockell, and R.S. Gibson. NRC Research Press, Ottawa, Canada. pp. 17–18.

Table 1.

	Reference level	Ca deficiency (0.01% Ca)	Normal Ca (0.41% Ca)	Ca excess (0.91% Ca)
Zn balance (μg/d)				
Intake		133	134	134
Feces, total		107[a]	92[b]	107[a]
Feces, endogenous		86[a]	70[b]	66[b]
True absorption		111[a]	113[a]	94[b]
Urine		19[a]	17[a]	10[b]
Retention		–2[b]	17[a]	9[ab]
Zn quantity in tissues (μg)				
Organs and blood	335	328	330	312
Coat (skin + hair)	1239	1219	1208	1213
Muscle and fat tissue	1952	1949	1996	1924
Skeleton	1942	1976	2157*	2145*

[a,b,c]) significant differences among Ca levels; *) significant difference compared to the reference level.

the stop of Ca deposition into the skeleton also blocked the incorporation of Zn into the bone. This mechanism seemed to depress the intermediate utilization of Zn and may explain the pronounced excretion of absorbed Zn. With Ca excess, the amount of skeleton Zn did not differ compared to the level at normal Ca supply. However, the Zn concentration of the femora was significantly reduced (Windisch and Kirchgessner 1994b), which is another indication of a deficient Zn supply to the intermediate metabolism due to an excessive Ca intake.

Whole body Zn was exchangeable at 40% regardless of the level of Ca supply. The half-life of exchangeable Zn was 26 d for normal and excessive Ca supply (Windisch and Kirchgessner 1994b). Ca deficiency reduced the half-life to 20 d which reflects the pronounced Zn excretion in these animals.

In total, Ca deficiency seems to primarily reduce intermediate Zn utilization by inhibiting the Zn deposition into the skeleton. As a secondary effect, it may provoke symptoms of an excessive Zn supply such as an increased Zn excretion via feces and urine. On the other hand, Ca excess seems to depress the Zn supply to the intermediate metabolism by reducing the intestinal availability of the Zn. This mechanism might be independent of the level of dietary Ca. Therefore any increase in dietary Ca content may enhance the risk of Zn deficiency if dietary Zn supply is close to the minimum requirement.

Literature Cited

Windisch, W. & Kirchgessner, M. (1994) Calcium- und Zinkbilanz ^{65}Zn-markierter Ratten bei defizitärer und moderat hoher Ca-Versorgung. J. Anim. Physiol. a. Anim. Nutr. 72: 184–194.

Windisch, W. & Kirchgessner, M. (1994) Verteilung von Calcium und Zink sowie Zinkaustausch im Körpergewebe bei defizitärer und hoher Calciumversorgung. J. Anim. Physiol. a. Anim. Nutr. 72: 195–206.

Windisch, W. & Kirchgessner, M. (1995) Anpassung des Zinkstoffwechsels und des Zn-Austauschs im Ganzkörper ^{65}Zn-markierter Ratten an eine variierende Zinkaufnahme. J. Anim. Physiol. a. Anim. Nutr. 74: 101–112.

Discussion

Q1. Donald Oberleas, Texas Technical University, Lubbock, TX, USA: What was the protein source in the diet?

A. We used 20% casein in a semi-synthetic diet.

Q2. Manfred Anke, Friedrich Schiller University, Jena, Germany: What is the biochemical basis for the interaction between Ca and Zn?

A. In summary, Ca deficiency had a primary effect of decreasing Ca incorporation into the skeleton and a secondary effect of decreasing Zn retention by increasing the excretion of Zn into the intestine. Ca excess had a primary effect of decreasing Zn absorption in the intestine. The high Ca content of the diet was borderline for Zn homeostasis.

Effect of Phytate on Endogenous Zinc and Fractionation of Pancreatic/Biliary (P/B) Fluid in Rats*

DONALD OBERLEAS AND IN-SOOK KWUN
Food and Nutrition, Texas Tech University, Lubbock, TX 79409-1162, USA

Keywords zinc homeostasis, pancreas, phytate, rat

Practical zinc homeostasis is intimately related to the phytate content of a diet. This was demonstrated and a theoretical model presented to describe zinc homeostasis in TEMA–8 (Oberleas, 1993). The purpose of this report is to provide experimental data to support that theoretical model.

Materials and Methods

Mature (>300 g) male Sprague-Dawley rats (48) were fed 20% casein-based semipurified diets with added sodium phytate. The diets contained either 0.8 or 1.6% calcium. The rats were depleted for four weeks to reduce the body zinc pool. After depletion, the animals were reallotted by weight to phytate and non-phytate diets within their respective Ca level. Radioactive ^{65}Zn (370 kBq) was injected IP to label the endogenous pool. The animals were immediately placed in metabolic cages and feces were collected daily for 3 weeks (2 weeks of the initial collection and 1 week after dietary crossover). Ratios of fecal radioactivity >1 phytate/non-phytate represent the phytate effect on endogenous zinc.

At termination, the animals were anaesthetized and the common bile duct cannulated with small bore tubing. The P/B fluid was collected with protein stimulation (Finley and Johnson 1992).

P/B fluid (1.5 mL) was fractionated on Sephadex G–75 with 0.01 mol/L Tris buffer, pH 8.1. The effluent was monitored at 280 nm. Fractions (2.5 mL) were collected for analysis of protein, total Zn, and ^{65}Zn radioactivity.

Results and Discussion

Diets were designed to contain Na phytate at a final phytate:zinc molar ratio of 30. This was confirmed by analysis: 33 for the low calcium diet and 27 for the high calcium diet.

Mean fecal ^{65}Zn label was higher for phytate than non-phytate fed rats during the 14-d collection periods ($p < 0.0001$). Though the pattern of excretion varied somewhat from previous reports, the total excretion and the ratios agree with excretion patterns reported earlier (Oberleas 1993, Oberleas 1996). The ratios were higher for phytate diets during the 14-d collection period and were confirmed by atomic absorption analysis.

Flow rates of P/B fluid ranged from 0.5 to 0.7 mL/h. Total protein of the P/B fluids were not different. It was necessary to pool some of the P/B fluid samples for animals from the same dietary group to provide sufficient fluid for protein fractionation. Protein fractions usually showed a high molecular weight (MW) >66 kDa, a carboxypeptidase (35 kDa) peak, and a large peak at <6.5 kDa. Though the zinc was somewhat dispersed, there were frequently zinc peaks associated with each fraction. The presence of small MW compounds in P/B fluid associated with zinc may serve as loosely associated Zn-binding ligands for secretion of endogenous Zn into the duodenum which then may dissociate and become vulnerable to phytate complexation (Figure 1).

*Supported by the TTU Nutrition Institute.

Correct citation: Oberleas, D., and Kwun, I.-S. 1997. Effect of phytate on endogenous zinc and fractionation of pancreatic/biliary (P/B) fluid in rats. In *Trace Elements in Man and Animals – 9: Proceedings of the Ninth International Symposium on Trace Elements in Man and Animals. Edited by* P.W.F. Fischer, M.R. L'Abbé, K.A. Cockell, and R.S. Gibson. NRC Research Press, Ottawa, Canada. pp. 19–21.

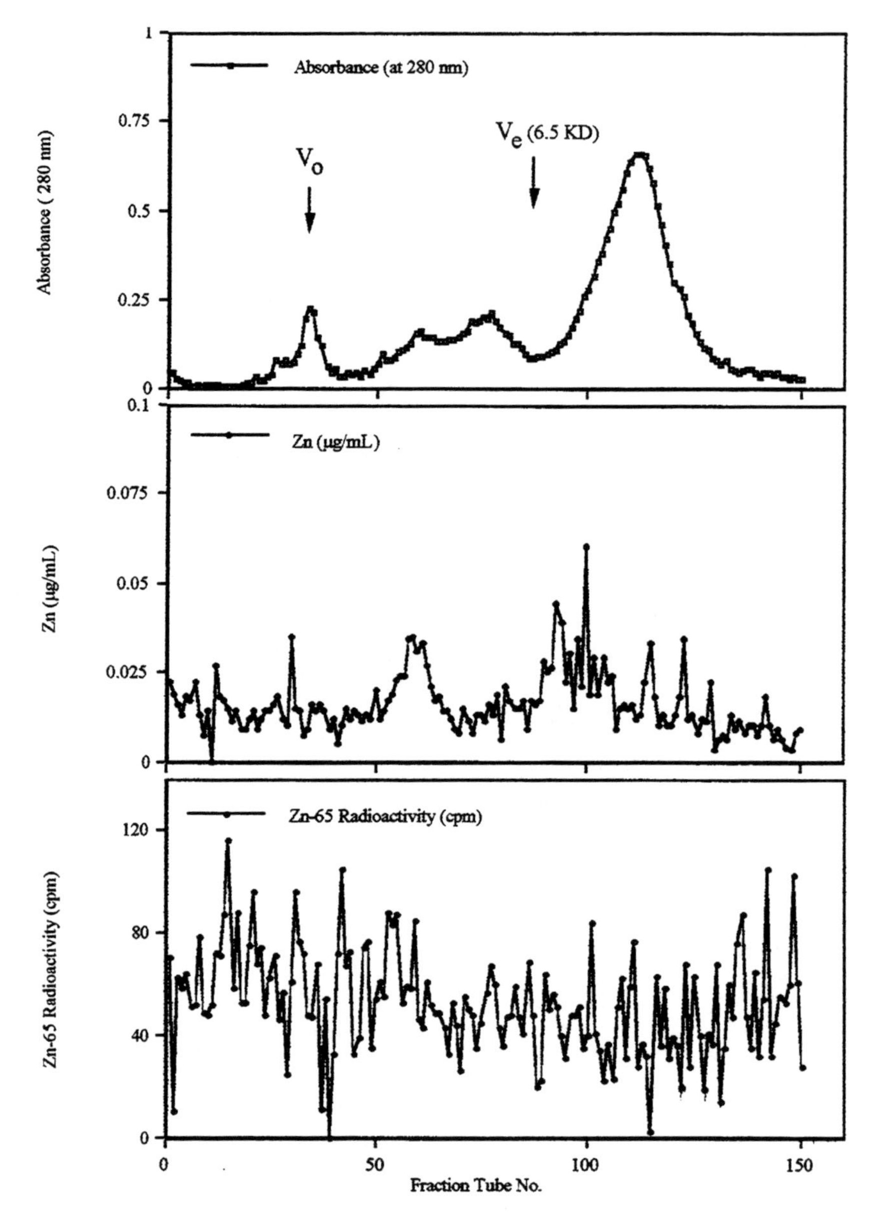

Figure 1. Separation of rat P/B fluid on Sephadex G-75. From low-Ca-phytate/low Ca-nonphytate dietary group. Sample size: 1.5 mL, fraction volume: 2.5 mL, buffer: 0.01 mol/L Tris/HCl, pH 8.1, flow rate 60 mL/h.

Summary

The mean dietary Zn intake was 108 µg/d. The pancreatic pool was estimated by a mean flow rate of 0.63 mL/h, a tubing correction of 3.74, and Zn concentration of 4.13 µg/mL in P/B fluid. This provided for 76 to 82% absorption/reabsorption uncorrected for urinary and insensible losses. Homeostasis required 68% absorption from the duodenal pool. These data, Figure 2, reflect a fair estimate of the balance provided by this model.

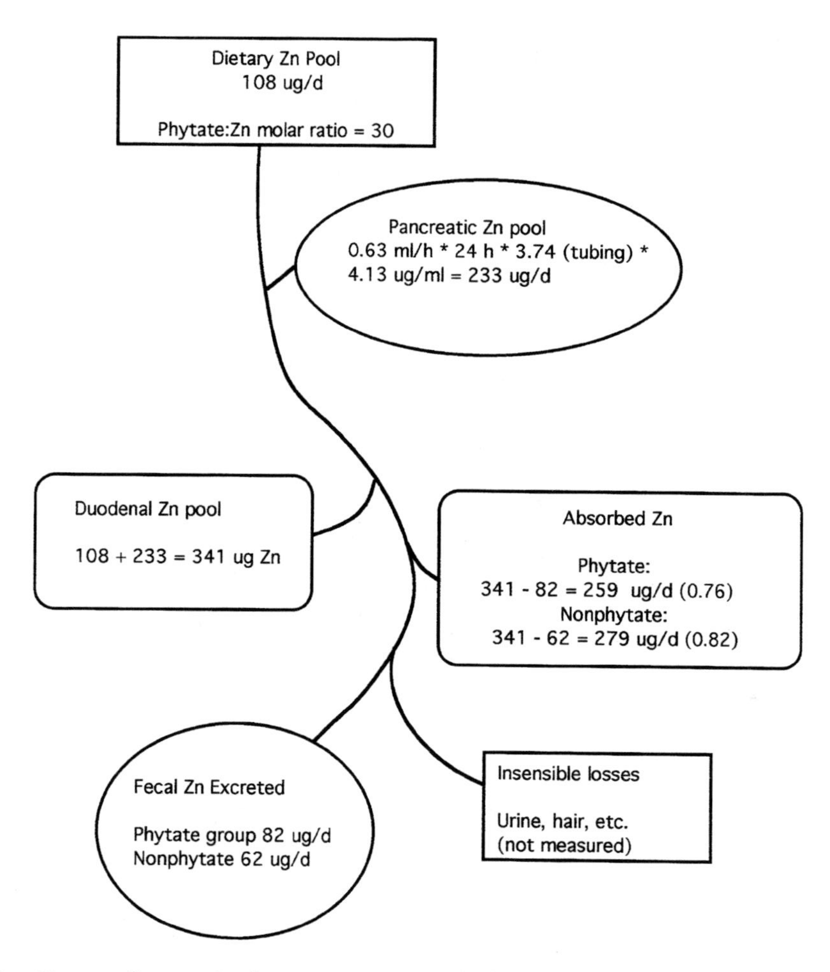

Figure 2. Phytate effect on zinc homeostasis using calculated means.

References

Oberleas, D. (1993) Phytate:zinc interaction in the maintenance of zinc homeostasis. In: Proceedings of the 8th International Symposium on Trace Elements in Man and Animals (TEMA–8), (Anke, M., Meissner, D. & Mills, C.F., eds.), pp. 630–633, Verlag Media Touristik, Badstrasse, FDR.

Finley, J. & Johnson, P.E. (1992) Relative influence of amount of dietary zinc and infusion of protein into the duodenum on the amount of zinc in biliary/pancreatic secretions. Nutr. Res. 12: 1217–1228.

Oberleas, D. (1996) Mechanism of zinc homeostasis. J. Inorg. Biochem. 62: 231–241.

Acute and Subchronic Effects of Dietary Cadmium on Iron Absorption in Rats

K.J.H. WIENK,[†] M. VAN ZONNEVELD,[†] A.G. LEMMENS,[†] J.J.M. MARX,[‡] AND A.C. BEYNEN[†]

[†]Department of Laboratory Animal Science, Utrecht University, P.O. Box 80.166, 3508 TD Utrecht, The Netherlands, and [‡]Eijkman-Winkler Institute and Department of Internal Medicine, University Hospital Utrecht, The Netherlands

Keywords: mucosal iron uptake, mucosal iron transfer, cadmium, rats

Dietary cadmium can reduce the bioavailability of iron as indicated by its lowering effect on iron absorption and organ iron levels (Huebers et al. 1987, Kozłowska et al. 1993). It has been suggested that cadmium competes with iron for the same binding sites at the luminal side of the mucosal cell, thus inhibiting mucosal iron uptake (Hamilton and Valberg 1974). However, it cannot be excluded that cadmium diminishes iron transfer from the mucosal cells into the body. In order to discriminate between mucosal iron uptake and transfer we applied a technique by which both variables can be measured separately. In the first experiment with rats, we monitored the effects of subchronic dietary administration of cadmium. In the second experiment, the effects of a single dietary exposure to cadmium were studied. It was hypothesized that dietary cadmium inhibits both mucosal uptake and transfer of iron, and that subchronic cadmium feeding, as compared to a single cadmium dose, triggers compensatory mechanisms in order to meet the body's need for iron.

Materials and Methods

***Experiment 1: Subchronic Cadmium Exposure*.** Male Wistar (U:WU) rats, aged 5 weeks, were fed for 15 d on iron-adequate purified diets containing either 0 or 25 mg added cadmium/kg (n = 10 per group). Cadmium was added in the form of $CdSO_4$. The molar ratio of Fe:Cd in the cadmium-containing diet was approximately 3.5 : 1. All rats were fed 80% of their normal daily intake, which was determined during the pre-experimental period. On day 8 of the experiment, mucosal uptake and transfer of $^{59}Fe(II)$ were measured according to Marx (1979). All rats received a suspension of their respective diets after an overnight fast. The suspensions were spiked with $^{59}Fe(II)$ and $^{51}Cr(III)$, the latter radioisotope serving as a non-absorbable marker. Radioactivity was measured with the use of a whole-body counter with separate detection windows for ^{59}Fe and ^{51}Cr. Final ^{59}Fe retention was determined at day 15. Mucosal uptake and transfer of ^{59}Fe were determined at 48 h after dosing. Apparent iron absorption was measured for a 3-d balance period (day 12–14) during which faeces were collected quantitatively. At the end of the experiment the animals were killed and liver iron was determined. Iron in all samples was determined using flame atomic absorption spectrometry.

***Experiment 2: Single Cadmium Exposure*.** Male Wistar (U:WU) rats, aged 4 weeks, were fed for 15 d on a purified iron-adequate diet without added cadmium. On day 8, mucosal uptake and transfer of ^{59}Fe were measured after supplying a diet suspension to which ^{59}Fe was added, without or with 70 μg cadmium per rat (n = 8 per group). In the cadmium-containing diet suspension, the molar ratio of Fe:Cd was 1:1. Mucosal $^{59}Fe(II)$ uptake and transfer, ^{59}Fe retention, apparent iron absorption and liver iron were measured as in Experiment 1.

Correct citation: Wienk, K.J.H., Van Zonneveld, M., Lemmens, A.G., Marx, J.J.M., and Beynen, A.C. 1997. Acute and subchronic effects of dietary cadmium on iron absorption in rats. In *Trace Elements in Man and Animals – 9: Proceedings of the Ninth International Symposium on Trace Elements in Man and Animals. Edited by* P.W.F. Fischer, M.R. L'Abbé, K.A. Cockell, and R.S. Gibson. NRC Research Press, Ottawa, Canada. pp. 22–23.

Results

Within both experiments, there were no significant differences in feed intake, growth, and final body weights. Mucosal iron uptake, but not transfer, and iron retention were significantly decreased when cadmium was fed for 8 d. When a single dose of cadmium was given, both mucosal iron uptake and transfer were significantly decreased. The single administration of cadmium also reduced the retention of ^{59}Fe. Apparent iron absorption was decreased when cadmium was fed subchronically, but was unaffected at 3 to 5 d after a single cadmium dose. Liver iron had decreased significantly, but only when cadmium was fed for 15 d.

	Experiment 1 Subchronic Cd exposure		Experiment 2 Single Cd dose	
Treatment	Control n = 10	Cadmium n = 10	Control n = 8	Cadmium n = 8
Mucosal Fe uptake (% of intake)	35.9 ± 13.1	20.6 ± 6.9*	74.8 ± 10.4	34.6 ± 7.0*
Mucosal Fe transfer (fraction of mucosal uptake)	0.97 ± 0.05	0.98 ± 0.08	0.95 ± 0.01	0.90 ± 0.03*
Fe retention (% of intake)	34.2 ± 11.9	19.9 ± 6.2*	70.8 ± 9.6	31.2 ± 6.3*
Apparent Fe retention (% of intake)	36.9 ± 11.7	19.7 ± 10.7*	62.0 ± 3.9	62.0 ± 1.4
Liver Fe (μg/g)	316 ± 50	202 ± 69*	273 ± 31	304 ± 62

Results are expressed as means with their standard deviation. Treatment values were compared with their respective controls: *$P < 0.05$ (Student's t test).

Discussion

Both single and subchronic exposure to dietary cadmium reduced mucosal ^{59}Fe uptake as well as ^{59}Fe retention. Mucosal ^{59}Fe transfer was significantly decreased when rats were given cadmium only once, but not if they were fed cadmium subchronically. Subchronic cadmium administration seems to elicit some kind of intramucosal adaptation resulting in a restoration of maximal mucosal iron transfer capacity. Single cadmium exposure does not appear to affect iron absorption or at least not irreversibly.

Literature Cited

Hamilton, D.L. & Valberg, L.S. (1974) Relationship between cadmium and iron absorption. Am. J. Physiol. 227: 1033–1037.

Huebers, H.A., Huebers, E., Csiba, E., Rummel, W. & Finch, C.A. (1987) The cadmium effect on iron absorption. Am. J. Clin. Nutr. 45: 1007–1012.

Kozłowska, K., Brzozowska, A., Sulkowska, J. & Roszkowksi, W. (1993) The effect of cadmium on iron metabolism in rats. Nutr. Res. 13: 1163–1172.

Marx, J.J.M. (1979) Mucosal uptake, mucosal transfer and retention of iron, measured by whole-body counting. Scand. J. Haematol. 23: 293–302.

Cadmium Accumulation in Rats and Pigs Influenced by Dietary Phytate and Microbial Phytase

G. RIMBACH AND J. PALLAUF

Institute of Animal Nutrition and Nutritional Physiology, Justus Liebig University, D-35390 Giessen, Germany

Keywords: cadmium, phytic acid, microbial phytase, rats and pigs

Introduction

In previous experiments it was demonstrated that a diet rich in phytogenic phytase or supplemented with phytase from *Aspergillus niger* considerably improves the utilization of phytate-phosphorus, calcium and zinc in pigs (Pallauf et al. 1994a, 1994b) and rats (Rimbach and Pallauf 1993). *In vitro* studies indicate that there are also interactions between phytic acid (PA) and cadmium (Wise and Gilburt 1981). Cadmium is a potentially dangerous industrial and environmental pollutant and its concentration in food and feed-stuffs can still be high. Therefore it is important to establish nutritional factors that reduce cadmium bioavailability and protect against cadmium intoxication. However the effect of PA and microbial phytase on the carry over of cadmium *in vivo* is still not clear.

Materials and Methods

Three experiments with rats (trials A and B) and pigs (trial C) were conducted. In trials A (Rimbach et al. 1995a) and B (Rimbach et al. 1995b) 3 × 6 male albino rats (initial average weight 50 g) were fed albumin-cornstarch diets over a 4 week period, containing 5 mg Cd (from $CdCl_2$) and 15 mg Zn (trial A) or 100 mg Zn (trial B) per kg diet. Controls were fed the basal diets free of PA and microbial phytase. In groups II and III, cornstarch was replaced by 0.5% NaPA. In group III, 2000 U of *Aspergillus*-phytase per kg diet was added. In trial C (Rimbach et al. 1996) 2 × 6 pigs, continuously housed in metabolic cages from 25–100 kg weight, were fed N-reduced diets based on barley, soybean meal and maize. Diet I contained 0.56%, 0.48% and 0.46% P (feeding phases A, B, C) and 0.76%, 0.71% and 0.68% Ca. Diet II was low in phosphorus (0.46%, 0.40%, 0.32%) and calcium (0.69%, 0.62%, 0.52%) and 800 U *Aspergillus*-phytase per kg was added. Analyzed native cadmium concentrations in diet I were 23, 20 and 14 μg/kg and 21, 15 and 13 μg/kg in diet II respectively.

Results

In rats fed the diet containing PA, liver and kidney Cd accumulation was higher than in the controls (Table 1). Phytase supplementation lowered liver and kidney cadmium accumulation. The increase from 15 mg (Exp. A) to 100 mg Zn per kg diet (Exp. B) led to a reduction of 10–50% in Cd-accumulation in kidney and liver but the phytate/phytase effects could still be recognized. In Exp. A, live weight gain and zinc bioavailability (Zn in plasma and femur, plasma alkaline phosphatase), was reduced by PA and enhanced due to the phytase supplementation. In group I (control), the highest apparent zinc absorption (58.2%) was measured. The addition of 0.5% PA (group II) significantly decreased apparent zinc absorption to 23.4%. In rats receiving the diet containing 2000 U phytase per kg (group III), 46.5% of ingested zinc was apparently absorbed. In Exp. B, apparent zinc absorption and zinc status remained unchanged by the different dietary treatments.

Correct citation: Rimbach, G., and Pallauf, J. 1997. Cadmium accumulation in rats and pigs influenced by dietary phytate and microbial phytase. In *Trace Elements in Man and Animals – 9: Proceedings of the Ninth International Symposium on Trace Elements in Man and Animals. Edited by* P.W.F. Fischer, M.R. L'Abbé, K.A. Cockell, and R.S. Gibson. NRC Research Press, Ottawa, Canada. pp. 24–25.

Table 1. Effect of phytic acid and microbial phytase on zinc status and cadmium accumulation in growing rats.

	Dietary concentration				Parameters investigated				
	Cd (mg/kg)	Zn (mg/kg)	PA (g/kg)	Phytase (U/kg)	Plasma-Zn (μg/mL)	Femur-Zn (μg/g FM)	ALP (mU/mL)	Liver-Cd (μg/g FM)	Kidney-Cd (μg/g FM)
Exp. A									
I	5	15	–	–	0.98[a]	70.6[a]	319[a]	241[a]	398
II	5	15	5	–	0.60[b]	45.3[b]	230[b]	515[b]	629
III	5	15	5	2000	0.93[a]	68.9[a]	297[ab]	301[ab]	487
Exp. B									
I	5	100	–	–	1.25	98.1	350	164[a]	306
II	5	100	5	–	1.22	95.7	368	463[b]	544
III	5	100	5	2000	1.25	96.8	355	141[a]	317

Within a column, values with different superscripts are significantly different ($p < 0.05$).

In pigs, at 100 kg weight in both treatment groups, low cadmium concentrations in liver (11.8 vs. 17.3 μg Cd/kg FM) and kidneys (59.6 vs. 102 μg/kg FM) were found. Contrary to findings for rats fed semisynthetic diets enriched with high $CdCl_2$ levels, phytase supplementation to the P- and Ca-reduced pig diet, with a low Cd concentration, significantly enhanced liver and kidney Cd accumulation. Irrespective of the dietary treatment, liver and kidney cadmium concentrations in both groups were considerably lower than maximal permitted values.

Discussion

In Exp. A, liver and kidney Cd accumulation in rats fed the diets containing PA were significantly higher than those in the controls. However in Exp. B with high dietary zinc unlikely to adversely affect zinc bioavailability, a moderate increase in cadmium accumulation by PA was also evident. These data indicate, that beside Zn–Cd interactions induced due to PA and microbial phytase, other minerals, especially calcium, might have additionally influenced the carry over of Cd in the rat experiments. In pigs, Ca concentration of diet II was marginal. In accordance with previous results (Pallauf et al. 1994a, 1994b) microbial phytase enhanced calcium absorption (in % of intake), but during the third fattening period, a higher absolute Ca retention (5.9 g vs. 4.5 g Ca/d) in the controls in comparison to the phytase group was evident. Thus, a possible explanation for the enhanced Cd accumulation in the phytase group with a lower absolute calcium absorption and retention compared with the controls might be due to an increase in the calcium binding protein activity accompanied by an increase in the mucosal Cd uptake. Furthermore, differences in dietary Cd, the binding form of Cd in the diets and the duration of the experimental trials may partially explain the differences found between rats and pigs.

Literature Cited

Pallauf, J., Rimbach, G., Pippig, S., Schindler, B., Höhler, D. & Most, E. (1994a) Dietary effect of phytogenic phytase and addition of microbial phytase to a diet based on field beans, wheat, peas and barley on the utilization of phosphorus, calcium, magnesium, zinc and protein in piglets. Z. Ernährungswiss. 33: 128–135.

Pallauf, J., Rimbach, G., Pippig, S., Schindler, B., Most, E. (1994b) Effect of phytase supplementation to a phytate-rich diet based on wheat, barley and soya on the bioavailability of dietary phosphorus, calcium, magnesium, zinc and protein in piglets. Agribiol. Res. 47: 39–48.

Rimbach, G. & Pallauf, J. (1993) Enhancement of zinc utilization from phytate-rich soy protein isolate by microbial phytase. Z. Ernährungswiss. 32: 308–315.

Rimbach, G., Brandt, K., Most, E. & Pallauf, J. (1995a) Supplemental phytic acid and microbial phytase change zinc bioavailability and cadmium accumulation in growing rats. J. Trace Elem. Med. Biol. 9: 117–122.

Rimbach, G., Pallauf, J., Brandt, K. & Most, E. (1995b) Effect of phytic acid and microbial phytase on cadmium accumulation, zinc status and apparent absorption of Ca, P, Mg, Fe, Zn, Cu and Mn in growing rats. Ann. Nutr. Metabol. 39: 361–370.

Rimbach, G., Pallauf, J. &; Walz, O. P. (1996) Effect of microbial phytase on cadmium accumulation in pigs. Arch. Anim. Nutr. (in press).

Wise, A. & Gilburt, D. J. (1981) Binding of cadmium and lead on the calcium-phytate complex *in vitro*. Toxicol. Lett. 9: 45–50.

Effect of Phytic Acid and Phytase on the Bioavailability of Iron in Rats and Piglets

J. PALLAUF AND S. PIPPIG

Institute of Animal Nutrition and Nutritional Physiology, Justus Liebig University, D-35390 Giessen, Germany

Keywords iron bioavailability, phytic acid, microbial phytase, rats, piglets

Introduction

Phytic acid (PA; myo-inositol hexakisphosphate), the major plant phosphorus store, is known to impair the intestinal absorption of divalent cations such as Ca^{2+} and Zn^{2+} in monogastric species by formation of insoluble metal-phytate complexes. There is controversy regarding the effects of dietary phytic acid on iron absorption. Experimental results on the influence of phytase supplementation on iron bioavailability are scarce. The purpose of this study was 1) to investigate whether PA diminishes the bioavailability of iron in rats and piglets and 2) to elucidate whether microbial phytase reduces the antinutritive effects of PA on the absorption of iron and other essential minerals and trace elements.

Materials and Methods

In trial 1, a total of 5 × 7 male weaned rats were fed a cornstarch-casein diet with 35 mg Fe/kg. Group I received the basal diet. In groups II and IV, a dietary molar PA:Fe ratio of 18:1 and in groups III and V of 36:1 was generated by supplying the diets with NaPA. In groups IV and V, 1000 U Aspergillus phytase per kg diet were added. In trial 2, 5 × 6 male castrated weaned piglets received a corn-soybean diet with 60 mg Fe/kg and 0.84% PA. By adding $FeSO_4 \times 7\ H_2O$, dietary Fe in groups III and IV was increased to 80 mg, in group V to 100 mg/kg diet. Groups II and IV were supplied with 1000 U phytase/kg. Balance studies were carried out and Fe concentration in spleen, liver and femur (trial 1 only), plasma Fe, plasma total (TIBC) and latent (LIBC) iron-binding capacity, transferrin saturation, hemoglobin, hematocrit, erythrocyte count, mean corpuscular hemoglobin (MCH), mean corpuscular hemoglobin concentration (MCHC), mean corpuscular volume (MCV), concentration of free protoporphyrin (FPP) and zinc protoporphyrin (ZPP) in erythrocytes were analyzed to estimate the absorption of iron and the iron status of rats and piglets as influenced by dietary PA and phytase.

Results

In both trials, feed intake, feed conversion and growth rate were not influenced by the different dietary treatments. In trial 1, dietary PA reduced the apparent iron absorption of rats fed diets with a molar PA:Fe ratio 18:1 (Table 1). Liver and femur iron concentrations, hemoglobin, hematocrit and MCH values were lowered in rats fed diets with PA, while plasma TIBC and LIBC increased. In trial 2, the daily iron absorption (10.4 mg/d in the live weight range of 10.7–12.8 kg; 12.8 mg/d from 15.6–19.4 kg live weight) of piglets was neither significantly influenced by varied Fe intake nor by supplying corn-soybean diets containing 60 mg or 80 mg Fe/kg with dietary phytase. The balance studies did not show significant effects of iron and/or phytase supply on iron absorption, and there were no statistically significant differences between the five groups for blood and plasma parameters of iron status. As shown in Table 2, the iron status of the 30 piglets was influenced by the point in time at which the blood and plasma samples were taken. From d_0 to d_{14}, plasma iron concentration, transferrin saturation, hemoglobin, MCH, MCV and erythrocyte free protoporphyrin values decreased considerably.

Correct citation: Pallauf, J., and Pippig, S. 1997. Effect of phytic acid and phytase on the bioavailability of iron in rats and piglets. In *Trace Elements in Man and Animals – 9: Proceedings of the Ninth International Symposium on Trace Elements in Man and Animals. Edited by* P.W.F. Fischer, M.R. L'Abbé, K.A. Cockell, and R.S. Gibson. NRC Research Press, Ottawa, Canada. pp. 26–28.

Table 1. Influence of dietary phytic acid and Aspergillus phytase on iron absorption and parameters of iron status in growing rats (trial 1).

Group	I	II	III	IV	V
Phytic acid (%)	0.00	0.75	1.50	0.75	1.50
molar PA:Fe ratio	0	18	36	18	36
Phytase (U/kg)	0	0	0	1000	1000
Iron absorption (% of intake)	58.2[b]	46.2[a]	49.8[ab]	52.9[ab]	49.0[a]
Spleen Fe (μg/g DM)	670	603	679	605	643
Liver Fe (μg/g DM)	218[b]	167[a]	149[a]	161[a]	143[a]
Femur Fe (μg/g DM)	77.8[b]	59.0[a]	64.7[ab]	67.9[ab]	61.4[ab]
Plasma Fe (μmol/L)	18.6	13.5	13.2	13.5	17.0
Plasma TIBC (μmol/L)	76.1[a]	85.1[ab]	95.3[c]	81.9[ab]	89.2[bc]
Plasma LIBC (μmol/L)	57.5[a]	71.7[ab]	82.1[b]	68.4[ab]	72.2[ab]
Transferrin saturation (%)	24.8	16.2	14.0	16.7	19.4
Hemoglobin (mmol/L)	7.92[ab]	7.75[a]	7.06[a]	9.19[b]	8.15[ab]
Hematocrit (L/L)	0.428[b]	0.414[ab]	0.398[a]	0.428[b]	0.414[ab]
Erythrocyte count (10^{12}/L)	6.60	6.78	6.73	6.78	6.68
MCH (fmol)	12.1[ab]	11.5[ab]	10.5[a]	13.6[b]	12.3[ab]
MCHC (mmol/L erythrocytes)	18.5[a]	18.7[ab]	17.7[a]	21.5[b]	19.7[ab]
MCV (fL)	65.1	61.2	59.3	63.4	62.2
FPP (nmol/L erythrocytes)	127[a]	159[ab]	151[ab]	162[ab]	177[b]
ZPP (nmol/L erythrocytes)	701	741	767	717	707

Table 2. Kinetics of blood and plasma parameters of iron status of piglets in the live weight range from 10–20 kg (trial 2).

Days on experiment	d_0	d_7	d_{14}	d_{21}	d_{28}
Average live weight (kg)	9.5	10.7	12.8	15.6	19.4
Plasma Fe (μmol/L)	21.0[c]	16.9[b]	12.5[a]	14.0[ab]	11.7[a]
Plasma TIBC (μmol/L)	74.1	76.6	78.3	80.5	80.7
Plasma LIBC (μmol/L)	53.1[a]	59.7[ab]	65.8[bc]	66.5[bc]	69.0[c]
Transferrin saturation (%)	28.7[c]	22.0[b]	16.2[a]	17.3[a]	14.7[a]
Hemoglobin (mmol/L)	7.24[b]	7.71[c]	7.09[ab]	6.72[a]	6.91[ab]
Hematocrit (L/L)	0.390	0.389	0.378	0.387	0.377
Erythrocyte count (10^{12}/L)	6.63	6.86	6.74	6.64	6.69
MCH (fmol)	10.9[bc]	11.3[c]	10.5[ab]	10.2[a]	10.4[a]
MCHC (mmol/L erythrocytes)	18.6[b]	19.9[c]	18.8[b]	17.4[a]	18.4[b]
MCV (fL)	58.9[b]	56.7[ab]	56.1[a]	58.4[ab]	56.4[ab]
FPP (nmol/L erythrocytes)	317[b]	–	154[a]	–	144[a]
ZPP (nmol/L erythrocytes)	1355[b]	–	1583[c]	–	1157[a]

Discussion

The results of the iron balance study indicate that dietary phytic acid impairs iron absorption in growing rats fed a semi-synthetic diet containing 35 mg Fe/kg (according to the recommendations of the NRC (1995)) at molar PA:Fe ratios of >18:1. Contrary to experimental results of Fairweather-Tait and Wright (1990), House and Welch (1988) and Hunter (1981) but in agreement with data presented by Akhtar et al. (1987), Ali and Harland (1991) and Fairweather-Tait (1982), from this observation it can be concluded that, under normal gastrointestinal pH conditions, phytic acid and iron form insoluble complexes which are not available for uptake into the cells of the brush border membrane. Microbial phytase partly counteracted the antinutritive effects of dietary phytic acid on iron bioavailability. Recently, Sandberg et al. (1996) showed a similar positive effect of Aspergillus phytase on iron absorption in humans. From the results of the balance studies it can be concluded that under the conditions investigated (high parenteral Fe pretreatment) even the lowest dietary iron concentration of 60 mg/kg, as recommended by the ARC (1981), meets the iron requirement of piglets in the live weight range of 10–20 kg. Higher dietary iron supplementation (80 mg/kg according to NRC (1988) or 100 mg/kg as postulated by GfE (1987)) did not significantly increase total iron absorption. The tendency of piglets to develop a temporary hypochromic-microcytic anemia in the live weight range of 10–20 kg was also observed by Furugouri and Tohara (1971) and attributed to a disproportional development of live weight and of blood and plasma volume.

References

Akhtar, D., Begum, N. & Sattar, A. (1987) Effect of dietary phytate on bioavailability of iron. Nutr. Res. 7: 833–842.

Ali, H.I. & Harland, B.F. (1991) Effects of fiber and phytate in sorghum flour on iron and zinc in weanling rats: A pilot study. Cereal Chem. 68: 234–238.

ARC (1981) The Nutrient Requirements of Pigs. Commonwealth Agricultural Bureaux, Farnham.

Fairweather-Tait, S.J. (1982) The effect of different levels of wheat bran on iron absorption in rats from bread containing similar amounts of phytate. Br. J. Nutr. 47: 243–249.

Fairweather-Tait, S.J. & Wright, A.J.A. (1990) The effects of sugar-beet fibre and wheat bran on iron and zinc absorption in rats. Br. J. Nutr. 64: 547–552.

Furugouri, K. & Tohara, S. (1971) Studies on iron metabolism and anemia in piglets. II. Blood volume and mean corpuscular constants. Bull. Nat. Inst. Anim. Nutr. Ind. 24: 75–82.

GfE (Gesellschaft für Ernährungsphysiologie, 1987) Energie- und Nährstoffbedarf landwirtschaftlicher Nutztiere, Nr. 4: Schweine, DLG-Verlag, Frankfurt.

House, W.A. & Welch, R.M. (1988) Bioavailability to rats of iron in six varieties of wheat grain intrinsically labelled with radioiron. J. Nutr. 117: 476–480.

Hunter, J.E. (1981) Iron availability and absorption in rats fed sodium phytate. J. Nutr. 108: 497–505.

NRC (1988) Nutrient Requirements of Swine, 9th rev. ed.; Washington D.C.

NRC (1995) Nutrient Requirements of Laboratory Animals, 4th rev. ed.; Washington D.C.

Sandberg, A.-S., Rossander Hulten, L. & Türk, M. (1996) Dietary Aspergillus niger phytase increases iron absorption in humans. J. Nutr. 126: 476–480.

Influence of Citric Acid in Milk on Intestinal Zinc Absorption Inhibited by Phytic Acid

M. DE VRESE AND S. DRUSCH

Institut für Physiologie und Biochemie der Ernährung, Bundesanstalt für Milchforschung, 24103 Kiel, Germany

Keywords zinc, citric acid, phytic acid, milk

Introduction

Phytic acid is able to bind divalent cations, especially zinc. Milk improves intestinal zinc absorption previously inhibited by phytic acid. It was shown by de Vrese et al. (1993) that inclusion of milk powder in a diet containing 0.5% Na-phytate cause a significant improvement of zinc absorption (11.5 ± 1.0 vs. 19.5 ± 1.0%) and femur zinc concentration (1.37 ± 0.02 vs. 1.64 ± 0.05 µmol/g DM; mean ± SEM; $p < 0.05$), whereas casein does not. There are data suggesting that citric acid, with a level of 2 g/L, one of the dominant organic acids in milk, has the potential to improve absorption of zinc. The present study examined whether citric acid is the factor responsible for the improved absorption of zinc when milk is included in phytic acid-containing diets. In addition, the mechanism whereby citric acid exerts its effect on the absorption of zinc was tested.

Materials and Methods

Six hundred thirty-three male Fischer-344 rats received one of four dietary regimens shown in Table 1. Basis for each dietary regimen was a ultra-filtrated, spray-dried milk protein concentrate with a low citric acid content (New Zealand Milk Products, Rellingen, Germany). A recombined experimental milk powder was created by adding all components except for citric acid, which had been lost in the permeate during ultra-filtration. In order to avoid diarrhea, which might occur in rats if lactose concentration in the diet is above 13%, no lactose was included. Experimental diets were supplemented with citric acid, phytate or both. As phytic acid supplementation increased Na concentrations in diet III and IV, this was compensated for by adding appropriate amounts of $NaHCO_3$ to the other diets.

Equal amounts of mineral mix (118 g/kg), vitamin mix (10 g/kg), Soya oil (30 g/kg) and Cellulose (50 g/kg) were added to each diet, The average zinc level was 48.75 mg/kg. Animals were fed 8.5 g of diet per day in a restricted feeding regimen. Doubly distilled water was available ad libitum.

Apparent zinc absorption and retention was determined in a 9-d balance period, beginning after 40 d on the experimental diets. At 49 d, animals were anaesthetized and blood was drawn from the abdominal

Table 1. Composition of experimental diets (g/kg) and number of animals per group.

Component:	I Control (n = 15)	II Citric acid (n = 16)	III Phytic acid (n = 16)	IV Phytic & citric acid (n = 16)
Milk powder	487.0	487.0	487.0	487.0
Corn starch	286.7	278.9	288.2	280.4
$NaHCO_3$	18.3	18.3	–	–
Na-phytic acid	–	–	16.8	16.8
Citric acid added	–	7.79	–	7.79

Correct citation: de Vrese, M., and Drusch, S. 1997. Influence of citric acid in milk on intestinal zinc absorption inhibited by phytic acid. In *Trace Elements in Man and Animals – 9: Proceedings of the Ninth International Symposium on Trace Elements in Man and Animals. Edited by* P.W.F. Fischer, M.R. L'Abbé, K.A. Cockell, and R.S. Gibson. NRC Research Press, Ottawa, Canada. pp. 29–30.

aorta. Liver, left and right femur and 1st to 4th lumbar vertebrae were removed. Activity of alkaline phosphatase was determined in plasma. Zinc concentrations in plasma, liver, bones, urine and feces were determined by AAS. Statistical comparisons of groups were done by one way analysis of variance (ANOVA) followed by Newman-Keuls test. Differences were considered significant at $p < 0.05$. Data given are means ± SEM.

Results and Discussion

Both food intake and growth of rats were not different between groups.

The high phytate content of 1% in diets III and IV distinctly inhibited zinc absorption, both in absence (group I vs. III) and in presence of citrate (group II vs. IV). As a consequence, there was a decreased zinc concentration and alkaline phosphate activity in plasma and significantly decreased zinc concentrations in femora and lumbar vertebrae due to phytate. There was also an increased fecal and a decreased renal excretion due to phytate, and thus a negative zinc balance.

Impact of citrate on the parameters is elicited by comparison of group I to II and group III to IV. In the absence of phytate, there were significantly higher zinc concentrations in plasma and liver, whereas there was only a (non significant) trend towards lower zinc concentrations in femur and lumbar vertebrae. One explanation for this phenomenon would be that citrate also affects absorption of other minerals (e.g., Ca) and that these minerals compete with zinc for transport for bone mineralization.

Under the inhibitory action of phytic acid, citrate increased zinc concentrations of femora, lumbar vertebrae, plasma and liver, although this effect was significant only in femora and liver. Possibly this citrate effect was not more pronounced because citrate was not entirely removed during ultrafiltration and thus citrate levels in diet I + III were already comparatively high. There may have been other yet unknown synergistic factors.

Apparent zinc absorption and retention were calculated from the difference between intake and fecal excretion. That zinc in the intestinal lumen is bound to phytic acid is proven both by the increased fecal zinc loss and the improved absorption in the presence of citrate. These effects were less pronounced than in previous experiments when whole milk was compared with casein. This indicates that the positive effect of milk on zinc absorption in the presence of phytate is not only due to citrate. An unexpected finding was the fact that all balances were negative.

Table 2. Experimental results.

		Control	+ Citrate	+ Phytate	Citrate+Phytate
Plasma zinc	µg/dL	165.5±3.9[a]	179.7±3.9[b]	152.8±5.9a	169.6±6.8[a]
Alkaline phosphatase	IU/L	394.4±11.4[a]	391.6±7.7[a]	347.2±6.9[b]	338.6±6.2
Femur [Zn]	µg/g	307.2±3.2[a]	286.0±3.3[b]	206.6±2.4[c]	223.7±3.0[d]
Lumbar vertebra [Zn]	µg/g	360.5±5.6[a]	349.6±2.8[a]	253.6±5.1[b]	263.0±5.6[b]
Fecal Zn-excretion	µg/9d	48.5±0.5[a]	46.6±0.3[b]	50.5±0.4[c]	48.4±0.6[a]
Renal Zn-excretion	µg/9d	46.9±1.2[a]	45.2±1.1[a]	24.9±1.4[b]	27.9±0.9[b]
% Zn-absorption	%	–30.5±1.4[a]	–25.3±0.8[b]	–35.7±1.0[c]	–29.3±1.0[a]
% Zn-retention	%	–31.7±1.4[a]	–26.5±0.8[b]	–36.4±1.0[c]	–30.4±1.7[a]

Conclusions

Citrate in milk improves absorption of zinc and is able to release zinc from zinc-phytate complexes. This means that citrate naturally present in foods and citrate added during food processing have a positive effect with respect to zinc metabolism.

References

de Vrese M, et al. (1993) Antagonistic effects of milk and milk-components and phytate concerning the bioavailability of zinc, iron and copper. In: Trace Elements in Man and Animals – 8 (Anke, M., Meissner, D., Mills, C.F., eds.), pp. 656–657. Verlag Media Touristik, Dresden.

Efficacy of a Michaelis-Menton Model for the Availability of Zinc, Iron and Cadmium from an Infant Formula Diet Containing Phytate

K.R. WING,[†‡] A.M. WING,[‡] R. SJÖSTRÖM,[¶] AND B. LÖNNERDAL[§]

Departments of [†]Oral Cell Biology and [‡]Environmental Health and [¶]Biophysics Laboratory, Umeå University, S-901 87 Umeå, Sweden, and [§]Department of Nutrition, University of California, Davis, CA 95616, USA

Keywords zinc, iron, cadmium, phytate

Background

At TEMA-7 in Dubrovnik we presented a poster in which we critically examined the use of the Ca · phytate / Zn molar ratio as an expression for the availability of Zn in diets fed to rats. As an alternative we suggested a Michaelis-Menton model, which we tested together with the Ca · phytate / Zn molar ratio on data from the literature. We found the Michaelis-Menton model to be more accurate than the molar ratio in describing the effect of phytate on the availability of Zn from the diet and its subsequent effect on growth in rats (Wing et al. 1990).

Purpose

The purpose of the present study was to further develop the Michaelis-Menton model and to test its efficacy in describing the effect of phytate on the availability of Zn, Fe and Cd from diets based on infant formula. In the Michaelis-Menton model, the mineral available for absorption from the diet, {M}, is related to the mineral and phytate concentrations and the apparent dissociation constant for the mineral-phytate complex, K_d, as follows

$$\{M\} = 0.5 \cdot ([M] - [\text{phytate}] - K_d + \sqrt{([M] - [\text{phytate}] - K_d)^2 + 4 \cdot K_d \cdot [M]}) \quad (1)$$

As the absorption of these minerals is regulated by binding sites/receptors in the intestinal mucosa, this should also be taken into account in the model. In an approximation to the Michaelis-Menton model, the amount of absorbed mineral/kg diet, |M|, is related to the maximum absorption, $|M|_{max}$, the available mineral, {M}, and the apparent dissociation constant for the mineral-receptor complex, K_d , as follows

$$|M| = |M|_{\max} \cdot \{M\}/(\{M\} + K_d) \quad \text{(Humle \& Birdsall 1992, Werner et al. 1982) (2)}$$

The model tested in this study was constructed by substituting Equation 1 into Equation 2.

Materials and Methods

Three-week-old, weanling, male, Sprague-Dawley rats were fed diets based on infant milk formula supplemented with vitamins and minerals. Phytate and Fe were added at different concentrations. The rats were housed individually in metabolic cages to allow measurement of food intake.

After four weeks, the rats were killed and samples of the diets, blood, femur, liver, kidney and the remaining carcass, after the removal of the contents of the gastro-intestinal tract, were taken for the determination of [Zn], [Fe] and [Cd] using atomic absorption spectrometry.

Correct citation: Wing, K.R., Wing, A.M., Sjöström, R., and Lönnerdal, B. 1997. Efficacy of a Michaelis-Menton model for the availability of zinc, iron and cadmium from an infant formula diet containing phytate. In *Trace Elements in Man and Animals – 9: Proceedings of the Ninth International Symposium on Trace Elements in Man and Animals. Edited by* P.W.F. Fischer, M.R. L'Abbé, K.A. Cockell, and R.S. Gibson. NRC Research Press, Ottawa, Canada. pp. 31–32.

The relationships between the accumulation of Zn, Fe and Cd in the rats and the dietary [Zn], [Fe], [Cd] and [phytate] and the amount of diet consumed were studied using non-linear regression techniques with the model described above. The non-linear regression calculates best fits for the constants which determine the outcome of the competition for the mineral between dietary phytate and binding sites/receptors in the intestine. The model used for this competition can be depicted as

Intestinal contents: Phytate–Mineral $\underset{k_{12}}{\overset{k_{21}}{\rightleftharpoons}}$ Phytate, Mineral $\underset{k_{31}}{\overset{k_{13}}{\rightleftharpoons}}$ Phytate, Mineral–Receptor

Intestinal mucosa: — Receptor ········ Receptor ········ Receptor —

$\sum = [\text{phytate}]_{diet}$

$\sum = [M]_{diet}$

$\sum = |M|_{max}$

where $k_d - \text{phytate} = k_{21}/k_{12}$, $|M|_{max}$ = apparent number of binding sites (maximum absorption) and $k_d - \text{receptor} = k_{31}/k_{13}$. The dietary phytate concentration was varied between 2400 and 20000 $\mu mol \cdot kg^{-1}$.

Results and Discussion

The apparent dissociation constants for the phytate and binding sites/receptors obtained in the non-linear regressions are presented below together with the concentrations of the minerals in the diet.

Constant*	Zinc	Iron	Cadmium
$[M]_{diet}$	871	461, 1446, 3349	0.145
K_d-phytate	1320	65000	29900
$\lvert M\rvert_{max}$	186	465	0.030
K_d-receptor	11.6	346	3.03

*$\mu mol \cdot kg^{-1}$; the apparent association constant for binding site/receptor-mediated absorption $K = 1/K_d \cdot 10^6$ $kg \cdot mol^{-1}$.

The Michaelis-Menton model adequately describes the decreases in Zn, Fe and Cd availability caused by increasing the phytate concentration in the infant formula diet. In addition, the approximation to the Michaelis-Menton model provides a relatively accurate description of the absorption and retention of the available Zn, Fe and Cd. The combination of the two models is linearly related to the absorption and retention of the minerals from the formula diet over a wide range of phytate concentrations (corrected r^2 = 0.80, 0.93 and 0.62 respectively).

Conclusions

The Michaelis-Menton model accurately described the availability of Zn, Fe and Cd from infant formula diets in which the phytate concentration was varied by a factor of more than 8. In addition, the model separated the effects of the binding of minerals to phytate and the regulation of intestinal mineral absorption in measurements of mineral bioavailability. As the diet used in this study is similar to breakfast cereal, porridge or a sandwich with milk, this model could also be useful in predicting mineral availability from such diets.

References

Humle, E., C. & Birdsall, J.M. (1992) Strategy and tactics in receptor-binding studies. In: Receptor-Ligand Interactions: A Practical Approach (Humle, E. C., ed.) Oxford University Press, Oxford, UK, pp. 65–70.

Werner, E., Roth, P., & Kaltwasser, J.P. (1982) Relationship between the dose administered and the intestinal absorption of iron from ferrous sulfate in humans. In: The Biochemistry and Physiology of Iron. Proceedings of the Fifth International Conference on Proteins and Iron Storage and Transport (University of California, San Diego, August 24–26, 1981) (Saltman, P. & Hegenauer, J., eds.) Elsevier Biomedical, Amsterdam, pp. 821–823.

Wing, K. R., Sjöström, R. & Moberg Wing A. (1991) The calcium:phytate:zinc molar relation as a measure of zinc bioavailability — A critique. In: Proceedings of the Seventh International Symposium on Trace Elements in Man and Animals TEMA 7 (Dubrovnik, Yugoslavia, 1990), 25-23 – 25-24.

High Iron Formulas May Compromise Zinc but not Copper Status in Term Infants in Early Life

JANIS A. RANDALL SIMPSON, BOSCO PAES, AND STEPHANIE A. ATKINSON
Department of Pediatrics, McMaster University, Hamilton, Ontario, Canada L8N 3Z5

Keywords zinc, infants, copper, hair, serum

Iron (Fe) deficiency anemia may impact on growth and impairments in cognitive function in infants may occur (Walter et al. 1989). Although the prevalence of Fe-deficiency anemia in Canadian infants is unknown, the Canadian Pediatric Society (CPS) recommends that formula fortified with 7 or 12 mg Fe/L be given from birth to 9–12 months of age (CPS, 1991).

Whether high Fe formulas are necessary to maintain normal hematological status of Canadian infants was investigated by us in a double-blind controlled trial in infants randomized to one of four concentrations of Fe in formula which was fed for the first 12 months of life. There were no clinically significant effects of Fe concentration on Fe status (Atkinson et al. 1995). We hypothesized, however, that trace element status would be compromised by the consumption of high Fe formulas.

The objective of this study was to assess the status of copper (Cu) and zinc (Zn) by measuring concentrations in hair and serum of infants who were randomized to formula with varying Fe concentrations.

Methods

Infants were randomized to study formula containing 1.5, 3, 7 or 12 mg Fe/L within 7 d of birth. The study protocol was approved by ethics committees and informed parental consent was obtained.

Two hundred and sixty-seven (122 male, birth wt = 3.5 ± 0.4 kg; 145 female, 3.5 ± 0.4 kg) term-born, appropriate for gestational age, mostly Caucasian infants were recruited; 222 completed the study. Trace mineral analyses of hair and serum were conducted on subsets of the population.

Hair, cut from the sub-occipital area of the scalp, was stored in plastic bags. Samples > 5 mg (6 months, n = 77; 12 months, n = 169) were washed, dry ashed and analyzed by flame atomic absorption spectrophotometry (AAS) (Friel et al. 1986). Accuracy and precision (CV = 10%) were assured by use of a reference hair (Community Bureau of Reference, Brussels) and a pooled hair sample, respectively.

Capillary heel-stick blood was collected into trace-element free containers; serum was separated and stored at –20°. Serum samples (not hemolyzed) > 50 µL (6 months, n = 85; 12 months, n = 81) were lyophilized, dry ashed, reconstituted in acid and analyzed by flame AAS (Veillon et al. 1985). Accuracy and precision (CV = 11%) were assured with a bovine reference material (National Institute of Standards and Technology).

Quattro Pro (Borland) and Sigma Stat (Jandel Corporation) were used for data manipulation and analysis.

Results

Median hair Zn at 6 months was lower, but not significantly, in infants consuming high Fe formulas (7 and 12 mg/L) compared to those consuming low iron formulas (1.5 and 3 mg/L). Further, median hair Zn in the 12 mg/L group was <1.07 µmol/g, considered to indicate Zn deficiency in older children (Figure 1).

No statistically significant differences among treatment groups for serum or hair Cu or Zn at either age were noted. There were no gender differences. Overall median (25–75th %ile) hair Cu (µmol/g) decreased (0.35(0.24–0.56) vs. 0.16(0.11–0.31), $p < 0.05$), whereas median hair Zn (µmol/g) increased

Correct citation: Randall Simpson, J.A., Paes, B., and Atkinson, S.A. 1997. High iron formulas may compromise zinc but not copper status in term infants in early life. In *Trace Elements in Man and Animals – 9: Proceedings of the Ninth International Symposium on Trace Elements in Man and Animals. Edited by* P.W.F. Fischer, M.R. L'Abbé, K.A. Cockell, and R.S. Gibson. NRC Research Press, Ottawa, Canada. pp. 33–34.

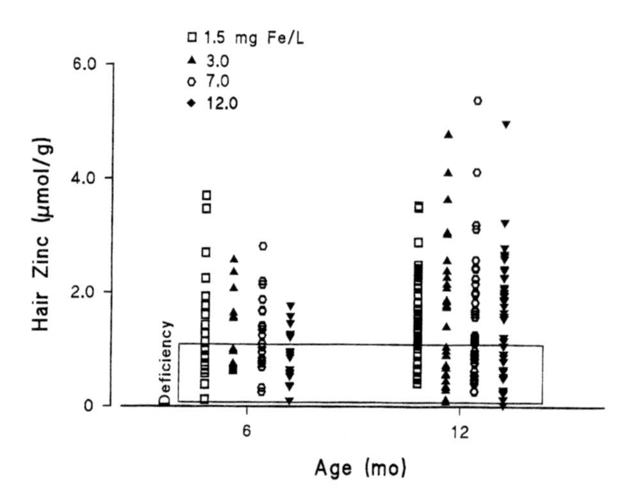

Figure 1. Hair zinc values for Canadian infants consuming infant formulas fortified with 1.5, 3, 7, or 12 mg iron/L. Values within the box are <1.07 μmol/g which is indicative of zinc deficiency in older children.

(1.07(0.69–1.67) vs. 1.43(0.79–2.110), p = 0.05) from 6 to 12 months. Overall median serum Cu (μmol/L) increased (16.0(14.0–18.8) vs. 20.4(17.2–23.6), p < 0.05) whereas median serum Zn (μmol/L) remained stable (18.1(13.7–24.4) vs. 16.8(13.7–20.5)) from 6 to 12 months.

Discussion

Consumption of high Fe infant formulas within the first 6 months of life may have a slight, albeit statistically and clinically insignificant, effect on Zn status but not on Cu status. During the second 6 months of life, when solid foods are introduced into the diet, there appears to be no effect of Fe content of infant formula on trace element status. To our knowledge, there are no other published data on hair analyses to evaluate the effect of dietary iron on Zn status in infants.

Our results for serum Cu and Zn are similar to those of other investigators (Bradley et al. 1993). Lonnerdal and Hernell (1994) reported similar results for serum Zn, although serum Cu was lowest in infants fed the highest concentration of Fe (7 mg/L).

As the formulas used in this study provided generous amounts of Cu and Zn, the likelihood of interactions with dietary Fe were possibly minimized. Nevertheless, given that Zn status, as indicated by hair Zn, may be compromised by high formula Fe and that hematological status is not compromised by low formula Fe (1.5 and 3 mg/L), we suggest that formula Fe at 3 mg/L may be more appropriate for infants up to 6 months of age.

Literature Cited

Atkinson, S.A., Randall Simpson, J.A., Thompson, P., & Paes, B. (1995) Normal hematological status is maintained in term infants randomized to formula containing 1.5, 3, 7 or 12 mg iron /L. J. Am. Coll. Nutr. 14: Abst 37.

Bradley, C.K., Hillman, L., Sherman, A.R., Leedy, D., & Cordano, A. (1993) Evaluation of two iron-fortified milk-based formulas during infancy. Pediatrics. 91: 908–914.

Canadian Pediatric Society, Nutrition Committee. (1991) Meeting the iron needs of infants and young children: an update. Can. Med. Assoc. J. 144: 1451–1454.

Friel, J.K. & Ngyuen, C.D. (1986) Dry- and wet-ashing techniques compared in analyses for zinc, copper, manganese, and iron in hair. Clin. Chem. 32: 739–742.

Lonnerdal, B. & Hernell, O. (1994) Iron, zinc, copper and selenium status of breast-fed infants and infants fed trace element fortified milk-based infant formula. Acta Paed. 83: 367–373.

Veillon, C., Patterson, K.Y., Reamer, D.C. (1985) Preparation of a bovine serum pool to be used for trace element analysis. In: Biological Reference Materials: Availability, Uses, and Need for Validation of Nutrient Measurement. (Wolf, W., ed.), pp. 167–177. John Wiley & Sons, Inc, New York, NY.

Walter, T., De Andraca, I., Chaduad P., & Perales, C.G. (1989) Iron deficiency anemia: adverse effects on infant psychomotor development. Pediatrics 84: 7–17.

Evaluation of Iron Bioavailability in Infant Weaning Foods Fortified with Heme Concentrate*

C. MARTINEZ,† T.E. FOX,‡ J. EAGLES,‡ AND S.J. FAIRWEATHER-TAIT‡

‡Institute of Food Research, Norwich NR4 7UA, UK, and †Universidad de Murcia, 30071 Murcia, Spain

Keywords infant weaning food, iron fortification, heme iron

Iron deficiency in infants is considered to be one of the most common nutritional disorders worldwide. Most cases are directly related to low levels of bioavailable iron. Dietary iron enters the gastrointestinal mucosal cell by two independent pathways, as non-heme and heme iron. Non-heme iron absorption is influenced by a large number of promotors and inhibitors, but heme iron is absorbed into the mucosal cell as the intact porphyrin complex and is not affected by the non-heme iron promotors and inhibitors (Lynch et al. 1985). The prevention of iron deficiency by fortification of infant foods is less costly and more effective than the detection and treatment of anemia in individual infants. The most common method of iron fortification in industrialized countries is the addition of non-heme iron to formula and weaning foods, however heme iron deserves more consideration as a compound for use in iron fortification. The aim of this study was to compare the absorption of iron as heme concentrate with the well-absorbed salt ferrous sulphate when added to infant weaning foods, and to determine whether or not the heme concentrate has an enhancing effect on iron absorption from the non-heme iron pool.

Materials and Methods

Six month old non-anemic infants (n = 16) were recruited from the Child Health Register in Norwich and allocated to one of two groups. Each infant was given two 100 g meals per day (on 7 consecutive days) consisting of a meat and vegetable weaning food to which was added 2.5 mg iron/100 g plus 40 mg ascorbic acid as heme or ferrous sulphate. The heme concentrate was obtained from Aprocat S.A., Granollers, Spain, and contained 1.89 mg Fe/g dried powder. A mixture of three different commercial weaning foods were selected by the parents from "chicken with rice" (0.42 mg Fe/100 g), "lamb with vegetables" (0.72 mg Fe/100 g) and "veal with carrots" (0.91 mg Fe/100 g), all suppled by Hero S.A., Alcantarilla, Spain. On two consecutive days the meals were labelled with ^{57}Fe (1.2 mg/d). The bioavailability of iron in the heme concentrate was assessed by chemical balance and compared with the well-absorbed iron salt. The effect of heme concentrate on non-heme iron absorption was determined from the luminal disappearance of ^{57}Fe. Duplicate diets, feces and breast milk were analyzed for total iron by thermal ionization quadrupole mass spectrometry (Fairweather-Tait et al. 1995). Apparent iron retention was calculated from the difference between total iron intake and total iron excretion, and ^{57}Fe absorption calculated from the difference between isotope dose and fecal excretion.

Results and Discussion

Breast milk contributed less than 3% to the total iron intake of breast-fed infants, as shown in Table 1.

Mean daily iron intake was 8.6 mg (SD ± 1.2), which was significantly lower than that of infants fed formula (14.7 ± 2.8 mg); infant formulas in the UK are fortified with ferrous sulphate. The amount of iron consumed was positively correlated with iron retention ($p < 0.05$), and also with iron excretion ($p < 0.05$), therefore iron retention can be improved by increasing the total iron intake from the diet.

*Funding: this work was supported by EC and BBSRC.

Correct citation: Martinez, C., Fox T.E., Eagles, J., and Fairweather-Tait, S.J. 1997. Evaluation of iron bioavailability in infant weaning foods fortified with heme concentrate. In *Trace Elements in Man and Animals – 9: Proceedings of the Ninth International Symposium on Trace Elements in Man and Animals. Edited by* P.W.F. Fischer, M.R. L'Abbé, K.A. Cockell, and R.S. Gibson. NRC Research Press, Ottawa, Canada. pp. 35–37.

Body weight and hematological indices were similar in the two groups. The 7-d balance data for the heme concentrate and ferrous sulphate groups are shown in Table 2.

Iron retention was similar in the two groups. Mean retention over the 7 d was 28.8% (3.5 mg/d) in the group fed ferrous sulphate and 24.4% (3.0 mg/d) in the group fed heme concentrate.The type of milk (breast or formula) had no effect on iron retention, nor was there any relationship between hemoglobin concentration (which ranged from 102 to 130 g/L) and the amount of iron retained.The latter observation has been made in other studies (Fairweather-Tait et al. 1995, Kastenmeyer et al. 1994) and indicates that the efficiency of iron absorption and body iron levels are not directly related in non-anemic infants. Heme concentrate did not enhance the absorption of ^{57}Fe administered as ferrous sulphate solution (Table 2) which suggests that the enhancing effect of animal protein, sometimes referred to as the "meat factor", is not related to blood proteins and must therefore be linked to myosin, actin, collagen or other constituents of animal tissue.

Table 1. Iron intake from breast milk during the 7-d balance period.

Subject	Breast milk intake (mL)	Iron concentration (mg/mL)	Total breast milk (mg)	% of total iron intake
2	3890	0.4	1.56	2.7
3	3356	0.3	1.01	2.3
9	2230	0.4	0.89	1.3
11	4110	0.4	1.64	2.8
12	3170	0.6	1.91	2.7
15	880	0.4	0.35	0.6
16	2740	0.3	0.82	1.4

Table 2. Iron retention and ^{57}Fe absorption in infants fed with heme concentrate (1–8) or ferrous sulphate (9–16) throughout a 7-d balance period.[1]

Subject	FeI (mg/d)	FeE (mg/d)	FeR (mg/d)	FeR (%)	^{57}FeA (mg)	^{57}FeA (%)
1	14.5	9.1	5.4	37.2		
2	8.3	5.3	3.0	36.1		
3	6.4	3.0	3.4	53.1	0.7	34.3
4	11.0	9.4	1.6	14.5	0.3	15.4
5	15.3	11.6	3.7	24.2	0.3	13.4
6	11.1	10.2	0.9	8.1	0.0	0.3
7	13.6	11.5	2.1	15.4	0.3	17.4
8	17.9	10.4	7.5	41.9	0.4	21.5
Mean±SD	12.2±3.8	8.8±3.0	3.5±2.1	28.8±15.7	0.3±0.2	17.0±11.1
9	9.6	8.0	1.6	16.7		
10	15.9	9.6	6.3	39.6		
11	8.3	4.7	3.6	43.4	0.9	45.0
12	10.0	10.6	–0.6	–6.0	0.1	8.0
13	13.9	10.3	3.6	25.9	0.2	12.9
14	19.4	13.5	5.9	30.4	1.0	53.2
15	8.9	8.5	0.4	4.5	0.5	27.2
16	8.4	5.0	3.4	40.5	0.4	24.2
Mean±SD	11.8±4.1	8.8±2.9	3.0±2.4	24.4±18.0	0.5±0.4	28.4±17.7

[1]Abbreviations used: FeI, iron intake; FeE, iron excretion; FeR, iron retention; ^{57}FeA, ^{57}Fe absorption.

Conclusion

The results demonstrate a promising future for the heme concentrate as a fortificant in infant foods since it is as well absorbed as ferrous sulphate when added to commercial mixed weaning foods. However, because the present product is coloured, further research and development is needed to clarify the technical feasibility of using heme concentrate to fortify foods.

References

Fairweather-Tait, S., Fox, T., Wharf, S.G., & Eagles, J. (1995) The bioavailability of iron in different weaning foods and the enhancing effect of a fruit drink containing ascorbic acid. Paed. Res. 37:1–6.

Kastenmayer, P., Davidsson, L., Galan, P., Cherouvrier, F., Hercberg, S., & Hurrell, R.F. (1994) A double stable isotope technique for measuring iron absorption in infants. Br. J. Nutr. 71:411–424.

Lynch, S.R., Dassenko, S.A., Morck, R.A., Beard, J.L., & Cook, J.D. (1985) Soy protein products and heme iron absorption in humans. Am. J. Clin. Nutr. 41:13–20.

The Short and Long Term Effects of Calcium Supplements on Iron Nutrition — Is there an Adaptive Response?*

A.M. MINIHANE, S.J. FAIRWEATHER-TAIT, AND J. EAGLES
Institute of Food Research, Norwich NR4 7UA, UK

Keywords iron absorption, iron status, calcium, interaction

Recent concern regarding the high prevalence and economic consequences of osteoporosis in an ever-aging population has resulted in the promotion of calcium products as a potential means of preventing or delaying the onset of the condition. Acute studies in experimental animals and humans have shown a significant negative effect of calcium on non-heme iron absorption, which has implications for iron nutrition. However before any dietary recommendations can be made the long term effects of calcium on body iron levels must be evaluated. In this study the effects of daily supplementation for 6 months with 1200 mg calcium (as carbonate), given as 400 mg with each meal, on non-heme iron absorption, and on body iron stores was determined.

Materials and Methods

Iron-replete volunteers (hemoglobin (Hb) > 120 g/L men, > 110 g/L women), aged 20–70 years were recruited for the studies. Iron status was assessed from measurements of Hb, hematocrit (Hct) (MD8 Coulter Counter), zinc protoporphyrin (ZPP) (hematofluorimeter), and plasma ferritin (in-house ELISA). No volunteer was taking mineral or vitamin supplements, or antacids, or had donated blood for at least 1 month prior to commencing the study.

(a) The Effect of Taking 400 mg Calcium with each Main Meal on Non-heme Iron Absorption

The daily absorption of dietary non-heme iron (13 mg/d) was determined from a low calcium diet (250 mg/d), with or without 1200 mg of supplemental calcium, in 14 volunteers. On day 1, after a 12 h overnight fast, each volunteer was given breakfast (wholemeal bread, jam, and margarine) labelled extrinsically with 1 mg ^{57}Fe (as ferrous sulphate) at 8:30 h, followed by a similarly labelled lunch (chicken, chips, and salad) and dinner (chili, rice, and flapjack) at 13:30 h and 18:00 h respectively. The iron isotope was administered in a drink of orange juice (breakfast) or cola (lunch and dinner). On day 2 identical meals were consumed, but each meal was labelled with 1 mg ^{58}Fe-enriched iron. Two indigestion tablets (Setlers Tums, SmithKline Beecham), containing 400 mg Ca, were taken with each of the three daily meals. An oral reference dose of iron (3 mg ^{54}Fe and 30 mg ascorbic acid) in 200 mL cola was consumed by the fasting subjects at 9:00 h on day 3, and then each resumed his/her normal diet 4 h later. Radio-opaque markers were taken with breakfasts and the cola drink containing the reference dose in order to check the completeness of the fecal collections. Capsules containing carmine (500 mg) were consumed with the evening meal on day 3, to mark the end of the collection period, and all fecal samples were collected from 8:30 h on day 1 until all the carmine had been excreted. The total iron content of the foods and fecal samples was determined by atomic absorption spectrometry (AAS), and the isotope ratios of the feces and administered isotopes were measured by thermal ionization quadrupole mass spectrometry

*Funding: this work was supported by the EC and the BBSRC.

Correct citation: Minihane, A.M., Fairweather-Tait, S.J., and Eagles, J. 1997. The short and long term effects of calcium supplements on iron nutrition — Is there an adaptive response? In *Trace Elements in Man and Animals – 9: Proceedings of the Ninth International Symposium on Trace Elements in Man and Animals. Edited by* P.W.F. Fischer, M.R. L'Abbé, K.A. Cockell, and R.S. Gibson. NRC Research Press, Ottawa, Canada. pp. 38–40.

(TIQMS). The absorption of iron was calculated as the difference between the dose of isotope and the amount excreted in the feces. In addition, the observed absorption data were corrected using the methods of Magnusson et al. (1981) and Cook et al. (1991), to allow for inter-individual variability in efficiency of iron absorption relating to iron stores. Iron absorption (%) in the absence (day 1) or presence of 1.2 g of supplemental calcium (day 2) was compared.

(b) The Effect of Taking Calcium Supplements for 6 Months on Iron Stores

Eleven test subjects consumed 1200 mg calcium (as carbonate), as described above, for 6 months. Iron stores (ferritin) and functional iron (Hb, Hct, ZPP) were monitored. Thirteen control subjects were also recruited to monitor any seasonal changes in blood status parameters. A blood sample was taken every 2 weeks and the plasma was removed and stored at –18°C until the end of the study period and ferritin concentrations were determined in each sample. Measurements of Hb, Hct, and ZPP were taken at 0, 3, and 6 months and transferrin receptor concentration was determined by ELISA (R & D Systems, Abingdon, UK) in the test subjects at 0 and 6 months.

All statistical analyses were carried out with the SPSS statistical package. As Hb, Hct, ZPP, and ferritin data are not normally distributed, statistical tests (t-tests, ANOVA, and correlations) were performed on logarithmically transformed data.

Results

The consumption of 400 mg calcium with each meal significantly reduced daily iron absorption approximately 4-fold. Mean % absorption (± SEM) decreased from 15.8(± 2.1) to 4.7(± 1.4)% ($p < 0.001$) (Figure 1). When the observed absorption data were corrected to allow for differences in iron stores, highly significant differences between day 1 (no calcium) and 2 (calcium supplemented) were still observed.

The consumption of 1200 mg supplemental calcium/day (with meals) for 6 months did not reduce iron status. No significant changes in measures of functional iron such as Hb, Hct, and ZPP or storage iron (plasma ferritin) or plasma transferrin receptor (TfR) levels were observed (Table 1). At 0 and 6 months the mean plasma ferritin and Tfr concentrations of the calcium-supplemented group were 39.7 and 38.0 µg/L and 1.67 and 1.70 mg/L, respectively. Similarly, no significant changes in indices of iron status were observed in the control group.

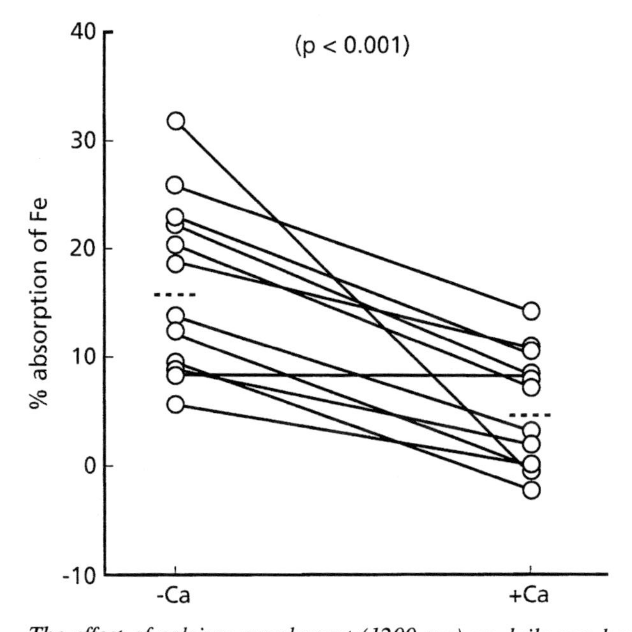

Figure 1. The effect of calcium supplement (1200 mg) on daily non-heme iron absorption.

Table 1. Indices of iron status (mean ± SEM).

Group	Month	Hb (g/L)	Haematocrit (%)	ZPP (μg/L)	Ferritin (μg/L)	Tfr (mg/L)
Calcium-supplemented	0	139±4	41.2±1.3	222±18	46.9±7.2	1.67±0.16
	3	138±4	41.3±1.3	232±14	45.8±6.2	
	6	136±4	41.6±1.3	225±19	49.5±6.7	1.70±0.12
Control	0	143±3	42.3±1.0	226±18	39.7±7.4	
	3	143±4	42.7±1.0	286±17	42.3±7.2	
	6	139±4	42.4±0.9	277±19	38.0±7.0	

Discussion

The results from this study demonstrate a highly significant short-term effect of calcium supplementation on daily non-heme iron absorption. Based on similar findings, Hallberg et al. (1992) suggested that high calcium foods should not be consumed with the main iron meals, because of an adverse effect on iron nutrition. This recommendation may lead to a decrease in total calcium intake and have adverse effects on the bone health of individuals with high physiological requirements for calcium. However the fact that the consumption of relatively high quantities of calcium with meals does not have any long term adverse effect on body iron levels demonstrates that such modifications in dietary patterns are unnecessary. We suggest that there is an adaptive response to the reduction in absorbable iron in the presence of calcium, possibly involving an up-regulation in the efficiency of iron absorption.

References

Cook, J.D., Dassenko, S.A. & Lynch, S.R. (1988) Assessment of the role of nonheme-iron bioavailability in iron balance. Am. J. Clin. Nutr. 54: 717–722.

Hallberg, L., Rossander-Hulten, L., Brune, M. & Gleerup, A. (1992) Calcium and iron absorption: mechanism of action and nutritional importance. Eur. J. Clin. Nutr. 46: 317–327.

Magnusson, B., Bjorn-Rasmussen, E., Hallberg, L. & Rossander, L. (1981) Iron absorption in relation to iron status. Model proposed to express results to food iron absorption measurements. Scand. J. Haematol. 27: 201–208.

Trace Elements and Genetic Regulation: Fe, Zn (Part 1)

Chair: W.J. Bettger

The Copper-Transporting ATPases Defective in Menkes Disease and Wilson Disease

DIANE W. COX

Research Institute, The Hospital for Sick Children, Toronto, Canada M5G 1X8. Present address: Department of Medical Genetics, University of Alberta, Edmonton, Canada T6G 2H7

Keywords Wilson disease, Menkes disease, copper transport, mutations

Introduction

Menkes disease is a rare X-linked disease with features of copper deficiency. Wilson disease (hepatolenticular degeneration) is an autosomal recessive inherited copper overload, with an incidence of about 1/30,000 in most populations. These two genetic diseases are of considerably greater importance than their frequency would indicate. The cloning of the two genes defective in these diseases has provided a major advance in our understanding of copper transport and the mechanisms involved in providing this essential element,while preventing toxicity. Furthermore, transport via a membrane ATPase could be a mechanism, as yet unidentified, for other essential trace metals.

The transport, utilization and elimination of copper involves absorption through the intestinal epithelial cells, transfer to proteins for blood transport, and uptake by tissues, particularly the liver. Excess copper can be stored in the low molecular weight, metal-inducible protein, metallothionein. In the hepatocyte, copper is incorporated into ceruloplasmin and the excess of ingested copper is excreted into the bile. The position of the defects in copper transport are indicated in the simplified pathway in Figure 1. Ceruloplasmin functions as a ferroxidase in liver and other tissues, and an inherited deficiency leads to iron storage (Miyajima et al. 1987).

Menkes Disease

Menkes disease is due to a defect of transport of copper across the intestinal membrane, which prevents exit of copper from intestinal cells. Excess copper can be stored in metallothionein induced in cells of the intestinal mucosa. Lack of intestinal transport leads to widespread copper deficiency, and insufficient levels of all of the copper containing enzymes: the connective tissue and elastin cross linking enzyme lysyl oxidase, the free radical scavenging superoxide dismutase, the electron transfer protein cytochrome oxidase, the pigment pathway enzyme tyrosinase, and the neurotransmitter enzime dopamine β-monooxygenase. Characteristic features of Menkes disease: twisted hair and arteries, abnormal connective tissue, neurological abnormalities, developmental delay and hypopigmentation (Danks 1995) can be explained by the absence of these enzymes. Copper histidine shows some success in treatment of this condition (Sarkar et al. 1993), which, without treatment, results in death in early childhood. The gene for Menkes disease (designated *ATP7A*, called here *MNK*) was cloned from a female affected due to

Correct citation: Cox, D.W. 1997. The copper-transporting ATPases defective in Menkes disease and Wilson disease. In *Trace Elements in Man and Animals – 9: Proceedings of the Ninth International Symposium on Trace Elements in Man and Animals. Edited by* P.W.F. Fischer, M.R. L'Abbé, K.A. Cockell, and R.S. Gibson. NRC Research Press, Ottawa, Canada. pp. 41–46.

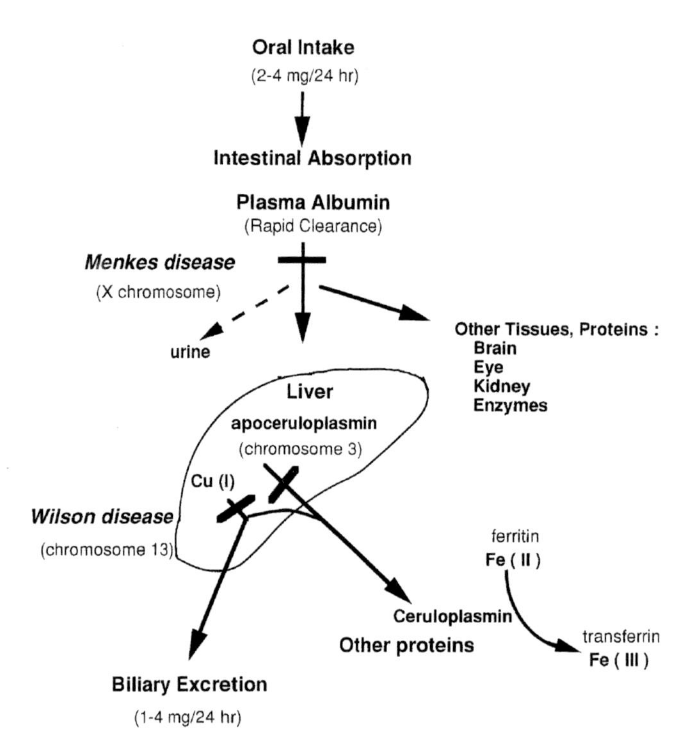

Figure 1. Simplified overview of the pathways for copper ion transport. Modified from Cox 1995, with permission from the American Journal of Human Genetics.

interruption of the gene by a chromosome translocation breakpoint (Vulpe et al. 1993, Chelly et al. 1993, Mercer et al. 1993, references in Bull and Cox 1994). The recognition of gene function was assisted by information on the copper transporting ATPase in bacteria, initially thought to be involved in potassium transport (Odermatt et al. 1993a).

Approximately 20% of the mutations in the Menkes disease gene are deletions several kb in length (Vulpe et al. 1993, Chelly et al. 1993), usually associated with lethality within the first few years of life. Milder forms of the disease, including cutis laxa, are in some cases due to splice site defects (Das et al. 1994). The mottled mouse mutants (blotchy, dappled, brindled) are due to defects in the orthologous gene in mice.

Wilson Disease

Wilson disease results in copper accumulation, particularly in brain, liver, kidney, and cornea (Danks 1995). When the capacity of hepatic metallothionein binding for copper is exceeded, excess copper accumulates in and damages hepatocytes, is effluxed into the urine, and accumulates in a variety of body tissues. Copper is not incorporated into the serum protein ceruloplasmin, and is not excreted into bile. Copper accumulation in the hepatocytes leads to cell damage. The varied signs and symptoms of Wilson disease, occurring between the ages of 3 and 50 years, include cirrhosis of the liver, tremor, loss of speech, erythrocyte hemolysis, kidney abnormalities and occasionally psychiatric disorders. Treatment involves the use of chelating agents such as penicillamine or triethylene tetramine, which remove the excess copper from plasma and in part from the liver, promoting its excretion via the urine. Administration of zinc salts is also used to block copper absorption and prevent copper accumulation. Antioxidants may be helpful.

Cloning the Gene

The *WND* gene was localized to human chromosome 13 by co-segregation with the red cell enzyme marker, esterase D, in Middle Eastern kindreds (Frydman et al. 1985). Further linkage analysis using more polymorphic DNA markers subsequently defined a region close to the retinoblastoma (RB) locus as the candidate region for the Wilson disease gene (Farrer et al. 1991). We used a positional cloning strategy to identify the gene, beginning with genetic linkage studies in our series of 25 Canadian families with Wilson disease, to narrow the candidate region (Thomas et al. 1993). We developed new markers in the region, derived a long-range restriction map using pulsed field gel electrophoresis, then used these markers to identify 19 yeast artificial chromosomes (YACs). At this time, the Menkes disease gene was cloned and was shown to be expressed in many tissues but not the liver. Reasoning that *WND* might be related, we used *MNK* gene probes, and identified a region of homology, first on YACs in the region of the gene, then on cosmids from a chromosome 13 library. Through cDNA selection, we assembled the cDNA for Wilson disease. The resulting gene (designated *ATP7B,* called here *WND*) was 57% identical in amino acid sequence to the Menkes disease gene (Bull et al. 1993). Another group obtained a cDNA from a brain library, which represented a shorter alternative transcript expressed in low amounts in the brain (Tanzi et al. 1993). Major functional regions of the gene are conserved between the *MNK* and *WND* genes, and genes from bacteria and yeast. The transduction domain transfers energy of ATP hydrolysis to cation transport. The aspartate residue within a conserved motif forms a phosphorylated intermediate during the cation transport cycle. A conserved hinge region lies at the C-terminal end of a large cytoplasmic ATP-binding domain. The Cys-Pro-Cys motif, characteristic of the heavy metal transporting ATPases, is predicted to lie within the cation channel. There are six copper binding and eight transmembrane domains predicted. Several alternatively spliced shorter products can be produced.

Haplotype Studies with Microsatellite Markers

Highly polymorphic markers, stretches of di- or trinucleotide repeats, were useful during cloning of the gene in initially identifying the region of the *WND* gene (Thomas et al. 1993). The same markers are useful in helping to identify which mutation might be present in any given patient. Haplotypes are combinations of marker alleles located together on the same chromosome. A specific haplotype tends to be associated with a specific mutation. Because no simple test is yet available to identify which mutation might be present, we use haplotypes (markers D13S316 - D13S301 - WND) - D13S114) to guide mutation testing (Thomas et al. 1995). Some haplotypes of repeats are unique to Wilson disease chromosomes and can support the diagnosis of Wilson disease.

Testing of close markers has a major practical application. A small percentage of heterozygotes have biochemical features indistinguishable from presymptomatic homozygotes. Such heterozygotes apparently never develop symptoms of the disease and would therefore not need to be exposed to the risks of lifelong treatment with chelating agents. Completely reliable diagnosis is possible in most families and is currently the most reliable way to identify the genotype of sibs of an affected patient.

Mutations of the Wilson Disease Gene

We identify mutations using single strand confirmation polymorphism (SSCP) analysis, in which changes from normal in the mobility of 200 to 300 base pair fragments of each exon, amplified by PCR, indicate the presence of a mutation. The appropriate fragment is then sequenced to identify the specific mutation present. No large gene deletions, as found in Menkes disease, have been identified. We have identified 26 mutations in our patient series (Thomas et al. 1994), and additional mutations have been identified (Petrukhin et al. 1993, Figus et al. 1996). All types of mutations are present throughout the gene: 22 small insertions or deletions, 3 nonsense, 4 splice site, and 20 amino acid substitutions. Each ethnic group has its own spectrum of mutations. Mutations predicted to completely destroy gene function (deletions, duplications, nonsense) are associated generally with an early age of onset, as early as three years of age. The effects of splice site mutations are variable, depending upon the location, as alternative splicing occurs and products may be functional missense mutations, which can alter function to a variable degree, depending upon the specific alteration, are found mainly in the ATP and transmembrane regions (Figure 2). Knowledge of the mutations present in various ethnic groups will dramatically improve diagnosis of Wilson disease, particularly in the early stages when biochemical assays may not be definitive.

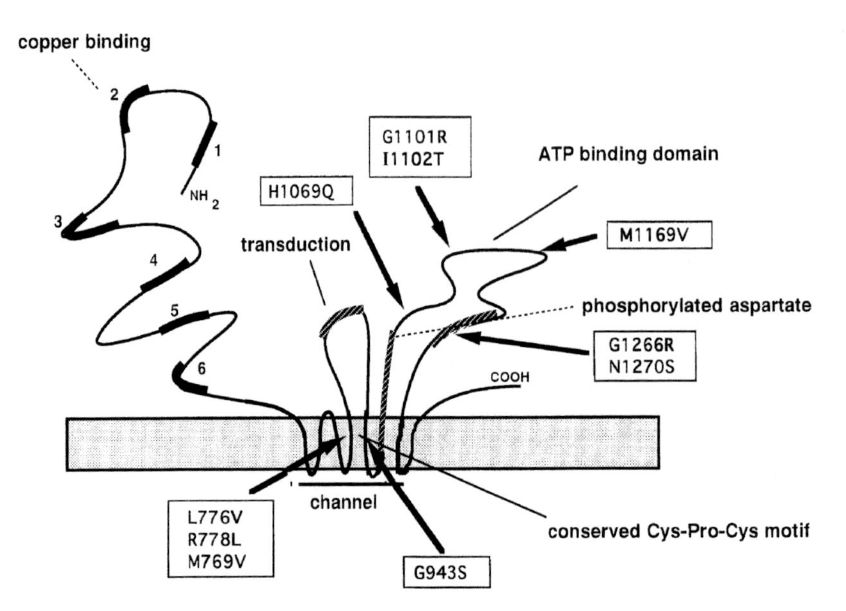

Figure 2. A model of the predicted product of the Wilson disease gene (modified from Bull and Cox 1994, with permission from Trends in Genetics) showing missense mutations (Thomas et al. 1994). The functional regions conserved in the Menkes disease gene and in bacterial genes are indicated. Shaded transmembrane domains are alternatively spliced.

Animal Models

The Long-Evans Cinnamon (LEC) rat, a mutant identified in Japan from the Long-Evans (LE) strain, shows biochemical alterations in copper as in patients with Wilson disease, and develops liver disease. We have shown the these rats have a deletion of the last 25% of the gene, eliminating key functional regions, including the ATP binding region (Wu et al. 1994). This rat is therefore an excellent model for the study of both pathology and treatment of Wilson disease.

The toxic milk mouse (tx) has a missense mutation in the Wilson disease gene (Theophilus, Cox and Mercer, unpublished). The infant mouse is born copper-deficient and dies if suckled on an affected mother. Death is due to copper depletion, so apparently transport of copper to the fetus during pregnancy and from the mother's milk is impaired.

Conclusion

Discovery of the basic genetic defect in Wilson and Menkes diseases has increased our understanding of copper transport. Many questions still remain, e.g., the role of the ATPase in incorporating copper into ceruloplasmin, the mechanism for cellular updake.

In addition to the high degree of identity between *MNK* and *WND* genes, a high degree of homology is shown with other metal transporting ATPases, in bacteria (Odermatt et al. 1993b, Vulpe et al. 1993) and more recently in yeast (Yuan et al. 1995). The bacterial genes are generally present on plasmids, and replicate in the presence of high metal concentration in the surrounding environment, as provided by pollution, fungicides, or other situations in which the metal content of the environment is increased. A particularly high degree of homology is seen with those bacteria which are resistant to either copper or cadmium. Bacteria resistant to chromium, cobalt, nickel, zinc, arsenic, silver have been identified, and equivalent genes in higher organisms await discovery.

References

Bull, P.C., Thomas, G.R., Rommens, J.M., Forbes, J.R. and Cox, D.W. (1993) The Wilson disease gene is a putative copper transporting P-type ATPase similar to the Menkes gene. Nature Genet. 5:327–337.

Bull, P.C. and Cox, D.W. (1994) Wilson disease and Menkes disease: new handles on heavy-metal transport. Trends Genet 10:246–252.

Chelly, J., Tumer, Z., Tonnesen, T., Petterson, A., Ishikawa-Brush, Y., Tommerup, N., Horn, N., et al. (1993) Isolation of a candidate gene for Menkes disease that encodes a potential heavy metal binding protein. Nature Gene. 3:14–19.

Danks, D.M. (1995) In: The Metabolic Basis of Inherited Disease: Disorders of copper transport. (Scriver, C.R., Beaudet, A.L., Sly, W.S., Valle, D., eds.) Seventh Edition. McGraw-Hill, New York.

Das, S., Levinson, B., Whitney, S., Vulpe, C., Packman, S. and Gitschier, J. (1994) Diverse mutations in patients with Menkes disease often lead to exon skipping. Am. J. Hum. Genet. 55:883–889.

Farrer, L.A., Bowcock, A.M., Hebert, J.M., Bonné-Tamir, B., Sternlieb, I., Giagheddum, M., St. George-Hyslop, P., et al. (1991) Predictive testing for Wilson's disease using tightly linked and flanking DNA markers. Neurology 41: 992–999.

Figus, A., Angius, A., Loudianos, G., Bertini, C., Dessi, V., Loi, A. and Deiana, A. (1996) Molecular pathology and haplotype analysis of Wilson disease in Mediterranean populations. Am. J. Hum. Genet. 57:1318–1324.

Frydman, M., Bonné-Tamir, B., Farrer, L.A., Conneally, P.M., Magazanik, A., Ashbel, A. and Goldwitch, Z. (1985) Assignment of a gene for Wilson disease to chromosome 13. Proc. Natl., Acad. Sci. USA 82:1819–1821.

Mercer, J.F.B., Livingstone, J., Hall, B., Paynter, J.A., Begy, C., Chandrasekharappa, S., Lockhart, P., et al. (1993) Isolation of a partial candidate gene for Menkes disease by positional cloning. Nature Genet. 3:20–25.

Miyajima, H., Nishimura, Y., Mizoguchi, K., Sakamoto, M., Shimizu, T. and Honda, N. (1987) Familial apoceruloplasmin deficiency associated with blepharospasm and retinal degeneration. Neurology 37:761–767.

Odermatt, A., Suter, H., Krapf, R. and Solioz, M. (1993a) An ATPase operon involoved in copper resistance by *Enterococcus hirae*. Ann N.Y. Acad. Sci. 484–486.

Odermatt, A., Suter, H., Krapf, R. and Solioz, M. (1993b) Primary structure of two P-type ATPases involved in copper homeostasis in *Enterococcus hirae*. J. Biol. Chem. 268:12775–12779.

Petrukhin, K.E., Fischer, S.G., Pirastu, M., Tanzi, R.E., Chernov, I., Devoto, M., Brzustowicz, L.M., et al. (1993) Mapping, cloning and genetic characterization of the region containing the Wilson disease gene. Nature Genet. 5:338–343.

Sarkar, B., Lingertat-Walsh, K. and Clarke, J.T.R. (1993) Copper-histidine therapy for Menkes disease. J. Pediat. 123:828–830.

Tanzi, R.E., Petrukhin, K.E., Chernov, I., Pellequer, J.L., Wasco, W., Ross, B., et al. (1993) The Wilson disease gene is a copper transporting ATPase with homology to the Menkes disease gene. Nature Genet. 5:344–350 .

Thomas, G.R., Bull, P.C., Roberts, E.A., Walshe, J.M. and Cox, D.W. (1993) Haplotype studies in Wilson disease (1993) Am. J. Hum. Genet. 54:71–78.

Thomas, G.R., Forbes, J.R., Roberts, E.A., Walshe, J.M. and Cox, D.W (1994) The Wilson disease gene: the spectrum of mutations and their consequences. Nature Genet. 9:210–217.

Thomas, G.R., Roberts, E.A., Walshe, J.M. and Cox, D.W. (1995) Haplotypes and mutations in Wilson disease. Am. J. Hum. Genet. 56:1315–1319.

Vulpe, C., Levinson, B., Whitney, S., Packman, S. and Gitschier, J. (1993) Isolation of a candidate gene for Menkes disease and evidence that it encodes a copper-transporting ATPase. Nature Genet. 3:7–13.

Wu, J., Forbes, J.R., Shiene Chen, H. and Cox, D.W. (1994) The LEC rat has a deletion in the copper transporting ATPase gene homologous to the Wilson disease gene. Nature Genet. 7:541–545.

Yuan, D.S., Stearman, R., Dancis, A., Dunn, T., Beeler, T. and Klausner, R.D. (1995) The Menkes/Wilson disease gene homologue in yeast provides copper to a ceruloplasmin-like oxidase required for iron uptake. Proc. Natl. Acad. Sci. USA .92:2632–2636.

Discussion

Q1. Maria Linder, California State University, Fullerton, CA, USA: I was wondering whether you could update us on the intracellular localization of proteins in Menkes and Wilson diseases, and abnormal cell systems?

A. We're working on this right now, and I don't have a final answer to give you, but there have been presentations from others to suggest that it is in the transgolgi. That, however, in both cases, for Menkes and Wilson's diseases, has been presented from abnormal cells, where there is an overproduction in a particular cell type. In the natural system, I think what we are expecting, partly from the work on the ATPase, not shown to be this particular Cu-transporting ATPase, though I believe this will be found, that's in the canalicular membrane. I believe we also have to expect it in the endoplasmic reticulum, because copper is being incorporated into ceruloplasmin someplace early in that pathway. So I think we can expect that it could be in more than one location. It's going to be membrane-bound. However, there are no good data that I know of from normal physiological cells.

Q2. Ed Harris, Texas A&M University, College Station, TX, USA: Regarding that last comment you made about the protein found in the transgolgi being involved in Cu incorporation into ceruloplasmin, and then also in the biliary canaliculi where it's involved in excretion through the bile, did you say you think these may be two different proteins? My question is do you have a Wilson's patient who would then show ceruloplasmin synthesis but no biliary excretion?

A. No, I think that the same protein is involved in two different places, a CuATPase in the endoplasmic reticulum towards the golgi as well as in bile canaliculi.

Q3. Ed Harris, Texas A&M University, College Station, TX, USA: Have we reached the stage where you think we can correlate the appearance of this protein with the synthesis of ceruloplasmin?

A. If ceruloplasmin is measured immunologically and not by oxidase activity, then it can be overestimated. Depending on the mutation, some patients may have normal amounts of ceruloplasmin, that is they may have the part of the molecule which can incorporate but not transport Cu. Or it may be that there's another mechanism when there's a lot of copper available, then it may be able to get in through glutathione transport, or some other system. I believe that's true, particularly in infants, that there's some other way to get it into ceruloplasmin, so I don't think it's all dependent on the ATPase, but that's really speculation.

Q3. Ed Harris, Texas A&M University, College Station, TX, USA: Just one other comment. Why would there be an energy driven mechanism for Cu when Ca transport or Na/K ATPase so clearly involves a gradient across the membrane?

A. You'd need to be careful if Cu gets out because of its involvement in free radical generation, which can then kill the hepatocytes. Perhaps this is a reason why some people get early liver disease, due to impairment of their free radical destruction system, and there is some evidence, for example that it is beneficial to give exogenous Vitamin E in severe liver disease.

Q4. John Sorenson, University of Arkansas, Little Rock, AR, USA: As I understood what you had to say to us with regard to interpreting high liver copper levels, it is an impairment of this ATPase which is required for transport. My question is, in patients with high liver Cu, did they have liver inflammatory disease?

A. Every patient with Wilson's disease has high liver Cu. For reasons we don't yet understand, some succumb to liver damage much more quickly than others. Some can go on with a high amount of liver copper, with very little damage and eventually get neurological disease. Some of this may be due to the specific mutation, which is why we're looking at all the mutations and correlating them with the disease. Some of it may perhaps be due to the fact that some people can produce metallothionein, which helps to prevent toxicity. So there may be genetic reasons in addition to dietary.

Q5. John Sorenson, University of Arkansas, Little Rock, AR, USA: Do you interpret the presence of Cu in the liver or other tissues as being there because of ionic binding?

A. I interpret it as being there because it's not transported out by the Cu transporter, and therefore binds to metallothionein, and of course, if it exceeds the metallothionein binding capacity, then it can start binding to other things as well, but in general it's bound to metallothionein before it reaches a toxic stage.

Differential Regulation of Selenoprotein Gene Expression by Selenium Supply

G. BERMANO, J.R. ARTHUR, AND J.E. HESKETH
Division of Biochemical Sciences, Rowett Research Institute, Bucksburn, Aberdeen AB2 9SB, UK

Keywords selenium, selenoproteins, glutathione peroxidases, mRNA stability

Background

Most of the biological functions of selenium in mammals are probably mediated by selenoproteins. The functions of the selenoproteins are essential for cell antioxidant systems, thyroid hormone metabolism, maintenance of fertility and possibly some anti-cancer activity. In particular, cytosolic glutathione peroxidase (cGSH-Px) and phospholipid hydroperoxide glutathione peroxidase (PHGSH-Px) are involved in the cell antioxidant system (Weitzel and Wendel 1993) whereas type 1 iodothyronine deiodinase (IDI) is involved in thyroid hormone metabolism (Larsen and Berry 1995). These three selenoproteins contain the selenium (Se) as Se-cysteine (Se-cys), which is incorporated into proteins by recognition of the stop-codon UGA as a codon for Se-Cys and this requires specific stem-loop structures. In addition, reduced availability of Se leads to termination of translation due to the recognition of the UGA codon as a normal stop codon (Burk and Hill 1993).

When dietary Se supply is limiting, the activity and concentration of some selenoproteins are decreased and there is a differential effect on cGSH-Px, PHGSH-Px and IDI activities in liver (Arthur et al. 1990, Sunde et al. 1993). With a variety of biological roles for the micronutrient, it is important to define the mechanisms which regulate selenoprotein gene expression particularly in Se deficiency. The aim of the present work was to investigate how the differential effects of Se deficiency on expression of the various selenoenzymes arise and whether there is a transcriptional and/or translational control of their synthesis.

Differential Expression of Selenoproteins in Se-deficient Rats

The regulation of the synthesis of hepatic and thyroidal cGSH-Px, PHGSH-Px and IDI was studied using animals which were maintained on a severely Se-deficient (0.003 mg/kg Se) and Se-adequate (0.104 mg/kg Se) diet. Severe Se deficiency caused almost total loss of cGSH-Px activity and mRNA in liver. Decreases of 95% in hepatic IDI activity and 50% in mRNA also occurred in Se deficiency. In contrast, PHGSH-Px activity decreased approximately 75% in liver but its mRNA content was unchanged. However in thyroid, PHGSH-Px activity was not affected by Se depletion and its mRNA was increased by 52%, whereas thyroidal IDI activity was increased by 15% and mRNA by 95% by Se deficiency (Figure 1) (Bermano et al. 1995).

These results demonstrate a differential regulation of mRNAs and subsequent protein synthesis for cGSH-Px, PHGSH-Px and IDI within and between organs. Thus mechanisms exist to channel Se for synthesis of specific enzymes and there is regulation of selenoenzyme mRNAs which differs between tissues.

In order to elucidate the mechanisms of regulation of the synthesis of cGSH-Px, PHGSH-Px and IDI, and in particular, the controls which allow preferential synthesis of one protein rather than another, and how this is regulated between tissues, we investigated if selenoenzyme gene expression is influenced by gene transcription or mRNA stability.

Correct citation: Bermano, G., Arthur, J.R., and Hesketh, J.E. 1997. Differential regulation of selenoprotein gene expression by selenium supply. In *Trace Elements in Man and Animals – 9: Proceedings of the Ninth International Symposium on Trace Elements in Man and Animals. Edited by* P.W.F. Fischer, M.R. L'Abbé, K.A. Cockell, and R.S. Gibson. NRC Research Press, Ottawa, Canada. pp. 47–49.

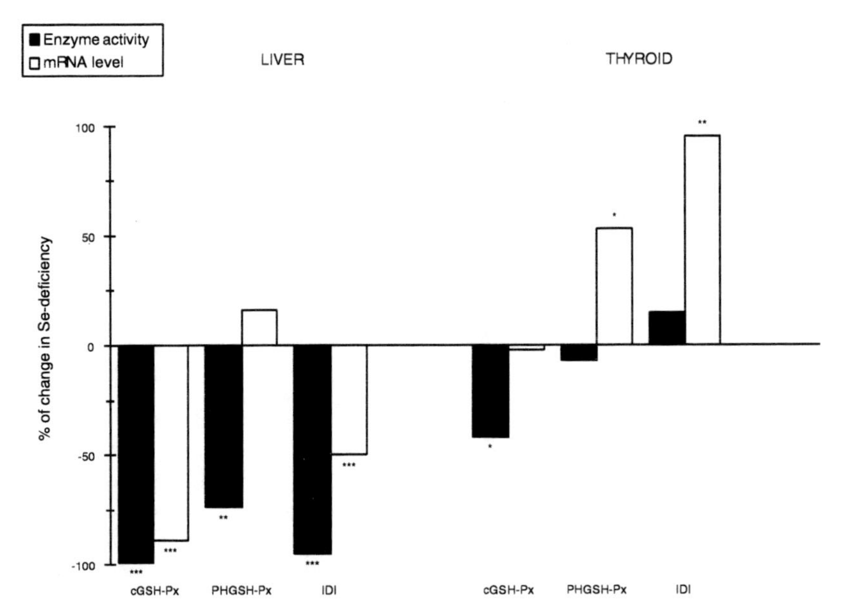

Figure 1.

Run-off assays with hepatic rat nuclei showed that severe selenium deficiency did not have any effect on the transcription of cGSH-Px, PHGSH-Px and IDI genes: this suggests that there is a differential regulation of their stability, at least in the liver (Bermano et al. 1995).

A Cell Culture Model of Se Deficiency

In order to investigate the effect of Se deficiency on selenoenzyme mRNA stability, a cell culture model was developed which allowed the study of the effect of Se supply on the regulation of cGSH-Px and PHGSH-Px gene expression.

Rat hepatoma H4 cells (H4 II EC3) were maintained for three days in serum free medium to which was added either insulin (5 µg/mL) and transferrin (5 µg/mL) (Se-deficient cells) or insulin, transferrin and Se as Na_2SeO_3 (7 ng Se/mL) (Se-replete cells). Measurements of cGSH-Px and PHGSH-Px activity and of their mRNA abundances in cells grown in Se-deficient medium for 3 days showed that cells were Se-depleted. Furthermore, in Se-deficient cells, as in rat liver during Se deficiency, there was a differential effect of Se-depletion on the activity and mRNA abundance of cGSH-Px and PHGSH-Px mRNAs. The regulation of these enzymes during Se-depletion of H4 cells appeared to be similar to that in liver of Se-deficient animals. Thus H4 cells are a suitable model to study the control of these hepatic selenoprotein genes by Se supply (Bermano et al. 1996a).

Differential Stability of Glutathione Peroxidase mRNAs in Se-deficient Cells

To investigate whether the effect of Se deficiency on cGSH-Px and PHGSH-Px expression is due to differences in stability of the two selenoprotein mRNAs under conditions of low Se supply, the rat hepatoma H4 cell culture model has been used. The stability of the cGSH-Px and PHGSH-Px mRNAs was assessed by measuring their abundances over a 12 hour period following inhibition of transcription by addition of Actinomycin D (5 µg/mL). Se depletion had no effect on the stability of PHGSH-Px mRNA but decreased cGSH-Px mRNA stability (Table 1) (Bermano et al. 1996b).

The most likely explanation for the decreased stability of cGSH-Px mRNA is that the effect is secondary to decreased translation. The selective effect of Se deficiency on cGSH-Px and PHGSH-Px mRNA stability is probably due to a differential effect of Se deficiency on translation of the two mRNAs: when Se is limiting in liver or hepatoma cells, translation of PHGSH-Px mRNA is maintained more than that of cGSH-Px.

Table 1.

	mRNA half life (h)	
	Se-replete	Se-deficient
cGSH-Px	13 ± 2	8 ± 1*
PHGSH-Px	10 ± 2	10 ± 3

References

Arthur, J.R., Nicol, F., Hutchinson, A.R. and Beckett, G.J. (1990) The effect of selenium depletion and repletion on the metabolism of thyroid hormones in the rat. J. Inorg. Biochem. 39: 101–108.

Bermano, G., Nicol, F., Dyer, J.A., Sunde, R.A., Beckett, G.J., Arthur, J.R. & Hesketh, J.E. (1995) Tissue-specific regulation of selenoenzyme gene expression during selenium deficiency in rats. Biochem. J. 311: 425–430.

Bermano, G., Arthur, J.R. & Hesketh, J.E. (1996a) A cell culture model to study regulation of selenoprotein gene expression by selenium. Biochem. Soc. Trans. 24: 224S.

Bermano, G., Arthur, J.R. & Hesketh, J.E. (1996b) Selective control of cytosolic glutathione peroxidase and phospholipid hydroperoxide glutathione peroxidase mRNA stability by selenium supply. FEBS Lett. In press.

Burk, R.F. & Hill, K.E. (1993) Regulation of selenoproteins. Annu. Rev. Nutr. 13: 65–81.

Larsen, P.R. & Berry, M.J. (1995) Nutritional and hormonal regulation of thyroid hormone deiodinases. Ann. Rev. Nutr. 15: 323–352.

Sunde, R.A., Dyer, J.A., Moran, T.V., Evenson, J.K. & Sugimoto, M. (1993) Phospholipid hydroperoxide glutathione peroxidase: full length pig blastocyst cDNA sequence and regulation by selenium status. Biochem. Biophys. Res. Commun. 193: 905–911.

Weitzel, F. & Wendel, A. (1993) Selenoenzymes regulate the activity of leukocyte 5-lipoxygenase via the peroxide tone. J. Biol. Chem. 268: 6288–6292.

Discussion

Q1. Ed Harris, Texas A&M University, College Station, TX, USA: Has anyone looked for the 3′end regulator protein?

A. No RNA binding protein has been found or the effect of selenium deficiency determined. In a different system, the 5′end and 3′end of the phospholipid glutathione peroxidase was shown to be sensitive to selenium deficiency. Again, the 3′ untranslated region was important in regulation.

Q2. Roger Sunde, University of Missouri, Columbia, MO, USA: This is excellent, and very exciting work. Have you done additional experiments on the thyroid? In particular when the message levels are going up, does this mean that there's more transcription, or an increase in the half-life of the message, or can you speculate?

A. We need a suitable cell culture model to have sufficient quantity of thyroid cells. However, consideration must be given to TSH hormone and whether the rate of transcription may be influenced by TSH hormone.

The IRT/ZRT Family of Eukaryotic Metal Transporters

D.J. EIDE, M. BRODERIUS, T. SPIZZO, AND H. ZHAO

Department of Biochemistry and Molecular Biology, University of Minnesota, Duluth, MN 55812, USA

Keywords iron, zinc, transporter, regulation

Our understanding of the molecular biology of iron uptake in eukaryotes has benefited greatly from recent studies of *S. cerevisiae*. In yeast, extracellular Fe(III) is reduced to Fe(II) by plasma membrane ferrireductases (Dancis et al. 1992). The Fe(II) product is then taken up by either of two transport systems. One system has a high affinity for iron (apparent K_m of 0.15 µM), is necessary for iron-limited growth, and requires the *FET3* gene for activity (Askwith et al. 1994). *FET3* encodes a multicopper oxidase that may drive iron uptake by oxidizing Fe(II) back to Fe(III) at some point in the uptake process. Iron-replete yeast cells obtain iron through a second, low affinity uptake system with an apparent K_m of 30 µM. This system requires the *FET4* gene for activity and our results suggest that Fet4p is the low affinity Fe(II) transporter (Dix et al. 1994).

A *fet3fet4* mutant strain lacking both high and low affinity systems is extremely sensitive to iron-limiting growth conditions because of its reliance on additional and apparently less efficient iron uptake pathways. This strain was the basis of a functional expression cloning method we have used to identify Fe(II) transporters from the plant *Arabidopsis thaliana*. An *Arabidopsis* cDNA library constructed in a yeast expression vector was screened for clones that, when expressed in a *fet3fet4* strain, could restore iron-limited growth. A single gene was obtained in this screen that we have designated *IRT1* for iron-regulated transporter (Eide et al. 1996). Yeast expressing *IRT1* possess a novel iron uptake activity that is specific for Fe(II) over Fe(III) as well as other transition metals. *IRT1* is predicted to encode an integral membrane protein with eight transmembrane domains and a possible metal-binding domain. In *Arabidopsis*, *IRT1* is expressed in roots, is induced by iron deficiency, and has altered regulation in plant lines bearing mutations that affect the iron uptake system. From these data, we propose that the physiological role of *IRT1* is the uptake of iron from the soil across the plasma membrane of the root epidermal cell layer.

IRT1 is a member of a new gene family of proteins found in a diverse variety of eukaryotes (Table 1). Database comparisons and Southern blot analysis indicated that additional related genes are present in the *Arabidopsis* genome. Moreover, similar genes were also identified in rice, yeast, nematodes, and humans. We have designated the two *IRT1*-related yeast genes *ZRT1* and *ZRT2* and our results indicate that these genes encode zinc transporters. Given that three members of the IRT/ZRT gene family have been implicated in metal transport, it seems likely that related genes play similar roles in the organisms in which they are found.

Biochemical assays of zinc uptake in yeast indicated the presence of two separate uptake systems (Zhao and Eide 1996). One system has high affinity for zinc with an apparent K_m of 0.5–1 µM and is required for zinc-limited growth. Our results suggested that *ZRT1* encodes the transporter protein of this system. The level of *ZRT1* expression correlated with activity of the high affinity system; overexpression of *ZRT1* increased high affinity uptake whereas disruption of the *ZRT1* gene eliminated high affinity activity. More recently, we have demonstrated the Zrt1p is a glycosylated integral membrane protein localized to the plasma membrane. The second system for zinc uptake in yeast has a lower affinity for substrate (apparent K_m of 10 µM) and it is active in zinc-replete cells. Low affinity uptake was unaffected by the *zrt1* mutation, indicating that this system is a separate uptake pathway for zinc. Data similar to that described above for *ZRT1* suggest that *ZRT2* encodes the low affinity zinc transporter.

Correct citation: Eide, D.J., Broderius, M., Spizzo, T., and Zhao, H. 1997. The IRT/ZRT family of eukaryotic metal transporters. In *Trace Elements in Man and Animals – 9: Proceedings of the Ninth International Symposium on Trace Elements in Man and Animals.* *Edited by* P.W.F. Fischer, M.R. L'Abbé, K.A. Cockell, and R.S. Gibson. NRC Research Press, Ottawa, Canada. pp. 50–52.

Table 1. The IRT/ZRT gene family.

Gene	Organism	GenBank Accession No.	% Similarity to *IRT1*	Substrate
IRT1	*A. thaliana*	U27590	100	Fe
IRT2	*A. thaliana*	T04324	85	?
IRT3	*A. thaliana*	M35868	68	?
R-IRT	Rice	D49213	82	?
ZRT1	*S. cerevisiae*	P32804	58	Zn
ZRT2	*S. cerevisiae*	X91258	65	Zn
CE–IRT	*C. elegans*	U28944	47	?
Eti–1	Humans	998569*	53	?
H–IRT	Humans	H20615	43	?

*Entrez Accession Number, not yet submitted to GenBank.

The activity of high affinity zinc uptake in yeast is regulated by two different mechanisms. First, this system is induced in activity greater than 100-fold in response to zinc-limiting growth conditions (Zhao and Eide 1996). When cells were grown in media containing different zinc concentrations, high affinity uptake and *ZRT1* mRNA levels were closely correlated, as was the β-galactosidase activity generated by a reporter gene in which the *ZRT1* promoter was fused to the *E. coli lacZ* gene. Detailed analysis of the *ZRT1* promoter indicates that two zinc-responsive promoter elements (ZREs) are responsible for induction in zinc-limited cells. The *ZRT1-lacZ* fusion gene showed a similar pattern of regulation in response to cell-associated zinc levels in both wild type and *zrt1* mutant cells despite the 75-fold higher extracellular zinc level required to down-regulate the promoter in the mutant. These results indicated that the activity of the high affinity system is controlled by transcriptional regulation of the *ZRT1* gene in response to a regulatory pool of intracellular zinc.

The high affinity system is also regulated by a second mechanism to prevent overaccumulation of zinc. When zinc-limited cells are transferred to a medium containing high levels of zinc, there is a rapid decrease in high affinity uptake activity. This inactivation occurs at a post-translational step because similar effects of excess zinc were observed when the *ZRT1* gene was fused to a heterologous promoter that is not responsive to zinc or in cells treated with cycloheximide, an inhibitor of protein synthesis. Immunoblot analysis indicated that inactivation of the high affinity system is mediated by degradation of the Zrt1p protein in the plasma membrane. Thus, the high affinity system is tightly regulated to ensure an adequate supply of zinc for growth but guard against the toxic effects of zinc overload.

References

Askwith, C., Eide, D., Van Ho, A., Bernard, P. S., Li, L., Davis-Kaplan, S., Sipe, D. M. & Kaplan, J. (1994) The *FET3* gene of *S. cerevisiae* encodes a multicopper oxidase required for ferrous iron uptake. Cell 76: 403–410.

Dancis, A., Roman, D. G., Anderson, G. J., Hinnebusch, A. G. & Klausner, R. D. (1992) Ferric reductase of *Saccharomyces cerevisiae*: Molecular characterization, role in iron uptake, and transcriptional control by iron. Proc. Natl. Acad. Sci. USA 89: 3869–3873.

Dix, D. R., Bridgham, J. T., Broderius, M., Byersdorfer, C. A. & Eide, D. J. (1994) The *FET4* gene encodes the low affinity Fe(II) transport protein of *Saccharomyces cerevisiae*. J. Biol. Chem. 269: 26092–26099.

Eide, D., Broderius, M., Fett, J. & Guerinot, M. L. (1996) A novel, iron-regulated metal transporter from plants identified by functional expression in yeast. Proc. Natl. Acad. Sci. USA in press.

Zhao, H. & Eide, D. (1996) The yeast *ZRT1* gene encodes the zinc transporter of a high affinity uptake system induced by zinc limitation. Proc. Natl. Acad. Sci. USA 93: 2454–2458.

Discussion

Q1. John Chesters, Rowett Research Institute, Aberdeen, Scotland: A fascinating story. It's fairly easy to think of how an excess of a metal or something could shut down a promoter. Do you have any ideas how the deficiency of metal could act in shutting down the promoter?

A. Well, I could draw analogies to other metal-regulated transcription factors. The model we have for this is a DNA binding protein that has an activation function, and binding of metal to this DNA-binding protein prevents it from interacting with DNA.

Q2. Harry McArdle, Rowett Research Institute, Aberdeen, Scotland: Why are there so many similarities in the transporters, given the different chemistries of Fe and Zn? How does the structure of the IRT/ZRT relate to data by Findlay?

A. The second question I find much easier to answer. There is no clear relationship between IRT/ZRT other than histidines. As for the differences in metals, I think that's a very good point, but I just can't give you a good answer to that.

Q3. Roger Sunde, University of Missouri, Columbia, MO, USA: With metallothionein, Palmiter has demonstrated metal regulation with MTF. This mammalian regulator has two Zn fingers and putative information suggests that an inhibitor interacts with MTF. Is that kind of activator and inhibitor story applicable here?

A. As it turns out, that inhibitor that he was reporting on turns out to be an efflux transporter, and not actually involved in transcriptional regulation. But yes, I think that it's a very good possibility that the metal could be controlling the interaction between two proteins.

Poly(ADP-Ribose) Polymerase: A Model Zn Finger Protein which is Sensitive to Zn Status

J.B. KIRKLAND, C.M. GRAUER, M.M. APSIMON, AND W.J. BETTGER
Department of Human Biology & Nutritional Sciences, University of Guelph, Guelph, Ontario, Canada N1G 2W1

Keywords poly(ADP-ribose) polymerase, zinc fingers, zinc deficiency, strand breaks

Poly(ADP-ribose) polymerase (PARP) is a nuclear enzyme which uses NAD^+ to synthesize poly(ADP-ribose) (pADPr) on nuclear proteins. PARP is catalytically activated through the binding of two Zn finger motifs to DNA strand breaks. Addition of pADPr to histones causes a local relaxation of chromatin which is thought to encourage DNA repair or help to prevent recombination events. PARP itself is a major pADPr acceptor, and this automodification causes PARP to leave the strand break due to electrostatic repulsion, halting activity and allowing the completion of the repair process. The two Zn fingers have different functions; FII is sufficient for DNA binding, while FI is required to initiate catalytic activity (Ikejima et al. 1990). The Zn ion in FI is labile under conditions of oxidant stress, and if lost, will create a protein which binds strand breaks, but is catalytically inactive (Buki et al. 1991).

This led to our hypotheses that 1) *in vivo* oxidant stress will also encourage the loss of zinc from finger I of PARP and create an inactive protein which hinders repair of strand breaks and 2) that concurrent zinc deficiency will discourage the correct replacement of the metal due to a lack of storage reserves in metallothionein (MT). In this study, weanling male Wistar rats were fed a Zn deficient diet for 18 d followed by 3 d of Zn repletion at 0.5 (zn–), 100 (zn+) or 3500 (zn++) mg Zn/kg diet, using a meal feeding protocol which provides equal growth in all groups. While this model produces very little change in total liver Zn, MT levels were increased by 5 and 25 fold in the zn+ and zn++ groups relative to zn–.

pADPr is measured by HPLC following partial purification and digestion of the polymer to the unique nucleoside, ribosyladenosine. At 10 h after treatment with diethylnitrosamine (DEN, 200 mg/kg, i.p.), hepatic pADPr levels increased about 10 fold above those generally seen in untreated liver tissue. In the zn+ group pADPr increased by 33% more than that seen in the zn– or zn++ groups (Figure 1).

Because pADPr synthesis is dependent on DNA strand breaks, it was important to determine the relative strand break frequency between the groups, and examine the possibility that the differences in pADPr levels were due to changes in DEN bioactivation. Using alkaline unwinding, the response to increasing doses of DEN was determined (Figure 2). DEN causes a linear increase in single stranded DNA, but a significant basal level of unwinding is present, mainly due mainly to the induction of DNA damage during the isolation of nuclei.

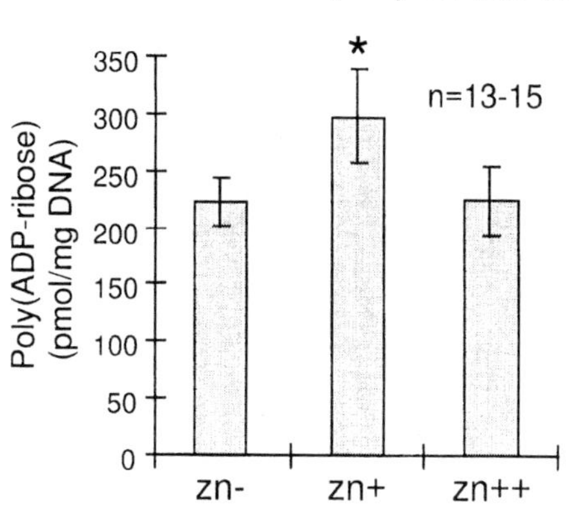

Figure 1

In rats of varying Zn status treated with DEN (200 mg/kg bw), the incidence of DNA strand breaks was found to be highest in the zn– rats, followed by zn+ and then zn++ (Figure 3). Although we only have 3 replicates in this experiment at this point, similar results of Zn deficiency on strand break frequency have been reported by other authors (Castro et al. 1992). We used the standard curve in Figure 2 to estimate an arbitrary frequency of DEN-induced strand breaks in the three dietary groups.

Correct Citation: Kirkland, J.B., Grauer, C.M., ApSimon, M.M., and Bettger, W.J. 1997. Poly(ADP-Ribose) polymerase: A model Zn finger protein which is sensitive to Zn status. In *Trace Elements in Man and Animals – 9: Proceedings of the Ninth International Symposium on Trace Elements in Man and Animals. Edited by* P.W.F. Fischer, M.R. L'Abbé, K.A. Cockell, and R.S. Gibson. NRC Research Press, Ottawa, Canada. pp. 53–54.

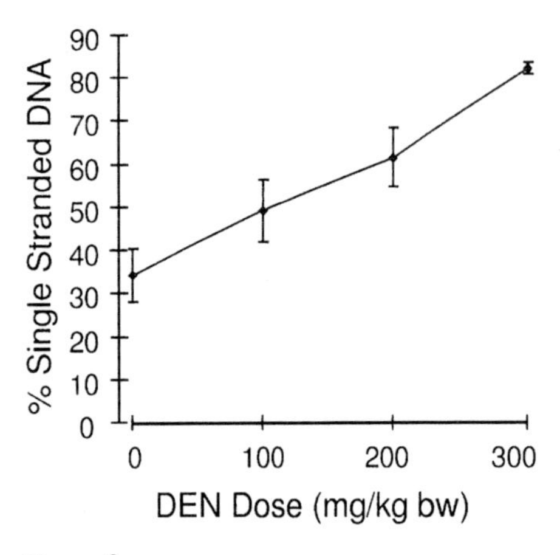

Figure 2

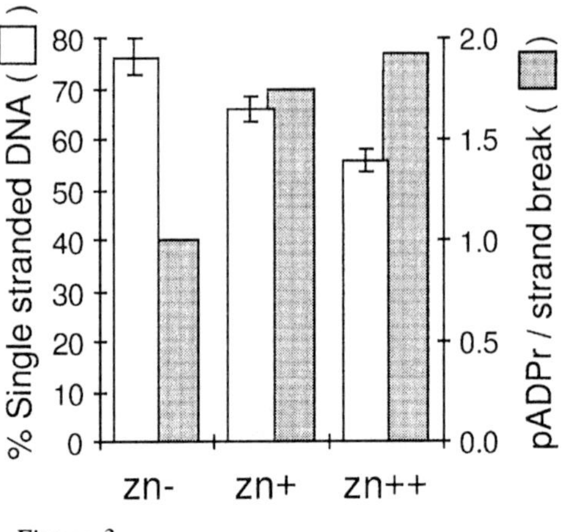

Figure 3

When corrected for the relative level of PARP-activating DNA strand breaks, DEN-induced pADPr levels appear to be 1.75 and 1.93-fold higher in the zn+ and zn++ than zn– livers.

One explanation for these results would involve a loss of PARP polypeptide in the deficient livers, perhaps resulting from a loss of zinc and increased proteolysis. Immunoblot analysis of PARP, at 10 h following treatment with DEN showed that PARP protein actually decreased as Zn status increased. Interestingly, there was a clear negative correlation between MT and PARP levels spanning the range of Zn status.

In conclusion, DEN-induced polymer synthesis is increased when Zn deficient rats are meal-fed a Zn adequate diet for 3 d. This is exaggerated when polymer levels are expressed in terms of the DNA strand breaks which activate the enzyme. While PARP function does appear to be compromised, these experiments do not demonstrate whether PARP is directly responsible for the increase in strand break frequency in the Zn deficient liver.

References

Ikejima, M., Noguchi, A., Yamachita, R., Ogura, T., Sugimura, T., Gill, D., & Miwa, M. (1990) The zinc fingers of human poly(ADP-ribose)polymerase are differentially required for the recognition of DNA breaks and the consequent enzyme activation. Other structures recognize intact DNA. J. Biol. Chem. 265: 21907–21913.

Buki, K.G., Bauer, P.I., Mendeleyev, J., Hakam, A. & Kun. E. (1991) Destabilization of Zn^{2+} coordination in ADP-ribose transferase (polymerizing) by 6-nitroso-1,2-benzopyrone coincidental with inactivation of the polymerase but not the DNA binding function. FEBS. Lett. 290: 181–185.

Castro, E.C., Kaspin, L.C., Chen, S. & Nolker, S.G. (1992) Zinc deficiency increases the frequency of single-strand DNA breaks in rat liver. Nutr. Res. 12: 721–736.

Discussion

Q1. Tammy Bray, The Ohio State University, Columbus, OH, USA: Does DEM metabolism cause oxidative, free radical or oxygen radical production?

A. DEM is activated to a carbon-centered radical. This is oxidative in nature, but I don't think that there's a lot of oxygen radicals produced per se. It is a nitroso compound and Coon's work with destabilization of zinc and copper has also been with nitroso compounds.

Q2. John Sorenson, University of Arkansas, Little Rock, AR, USA: With regard to that last question, nitroso compounds inhibit Cu,Zn-superoxide dismutase and so you can imagine an increase in superoxide accumulation in association with nitroso compounds.

A. This would improve the model because there would be increased oxygen radical damage and DNA damage in an oxidizing environment.

Essentiality of Cellular Zinc Supply for Steroid-Induced Gene Expression*

G.E. BUNCE, K.D. SIMON, B. STORRIE, AND P.K. BENDER
Department of Biochemistry, Engel Hall, Virginia Tech, Blacksburg, VA 24061-0308, USA

Keywords zinc, induction, steroids, protein synthesis

The discovery of the zinc finger as one of the three major motifs utilized by eukaryotic regulatory proteins to bind unique DNA sequences and thus bring about induced gene expression, has raised the possibility that zinc deficiency can severely compromise induction. It has been difficult, however, to discriminate between inductive and constitutive protein synthesis, since zinc metalloenzymes are also widely represented in the latter process. Moreover, zinc appears to be tightly bound to most cellular proteins and it has been argued that it is not easily lost from its protein linkages. The present study was performed to address these questions. Specifically, we have compared steroid-induced and constitutive protein synthesis in cultured HeLa S3 cells transfected with either the steroid-responsive or -unresponsive gene for chloramphenicol acetyl transferase (CAT). We have then created a zinc-deficient state by the addition to the incubation medium of the cell impermeable chelator diethylenetriaminepentaacetic acid (DTPA) and repeated the measurements of inductive and constitutive CAT activity.

Materials and Methods

HeLa S3 cells were grown as monolayer cultures in a 5% CO_2 humidified atmosphere in Joklik's minimal essential medium (JMEM) plus 2 mM glutamine and 10% supplemented bovine serum. Steroid sensitivity was achieved by transfection with plasmid pGMCS in which the CAT reporter gene is fused downstream of the mouse mammary tumor virus promoter and glucorticoid regulatory element (GRE) and upstream of the murine sarcoma virus GRE. Steroid-insensitive constitutive protein synthesis was evaluated using cells transfected with plasmid pBLCAT2 containing a thymidine kinase promoter from herpes simplex virus fused to a CAT reporter gene. Both plasmids were the generous gifts of Dr. John Cidlowski of the NIEHS at Research Triangle Park, NC. CAT activity was measured with a Promega CAT Enzymes Assay System. Cells were depleted of exchangeable zinc by addition to the medium of 10 μM DTPA for 24 h. Total zinc was analyzed by ICP following wet ashing. Inductive protein synthesis was tested by addition of 1 μM dexamethasone to the incubation media. Steroid receptor concentrations were determined by the technique described by Cidlowski et al. (1990).

Results

DTPA caused an outflow of zinc from HeLa cells without affecting viability. Cellular zinc content was reduced from 32 pmol/10^6 cells to 14 pmol/10^6 cells within 4 h and to a plateau of 4 pmol/10^6 cells after 24 h. The cells showed normal morphology and dye exclusion properties as compared to controls. This level of depletion must reflect not only loss of free zinc in solution but also reservoirs of ionically bound zinc in the nucleus and plasma membrane. The results of the CAT assays are shown in Table 1.

The results are expressed as percentage conversion of chloramphenicol to its butylated forms by chloramphenicol acetyltransferase. RU 486 is a glucocorticoid antagonist. All data represents the average of 5 independent experiments along with the standard deviations.

*Supported by USDA Grant No. 93-37200-9168.

Correct citation: Bunce, G.E., Simon, K.D., Storrie, B., and Bender, P.K. 1997. Essentiality of cellular zinc supply for steroid-induced gene expression. In *Trace Elements in Man and Animals – 9: Proceedings of the Ninth International Symposium on Trace Elements in Man and Animals. Edited by* P.W.F. Fischer, M.R. L'Abbé, K.A. Cockell, and R.S. Gibson. NRC Research Press, Ottawa, Canada. pp. 55–56.

Table 1. Plasmid responses to zinc concentration and dexamethasone.

	% Conversion of CAM to acetylated form	
	Plasmid	
Treatment	pGMCS	pBLCAT2
Control (no addition)	0.97 + 0.12	41.13 + 3.03
+ 1 μM dexamethasone	14.11 + 2.16	43.0 + 3.14
+ 10 μM DTPA, 1 μM dexamethasone	1.7 + 0.012	35.56 + 2.65
+ 10 μM DTPA, 1 μM dexamethasone, 50 μM Zn^{2+}	14.02 + 1.06	42.89 + 3.12
+10 μM DTPA	0.68 + 0.045	35.3 + 2.84
+ 1 μM RU 486, 1 μM dexamethasone	1.86 + 0.097	42.01 + 2.95

In essence, 85% depletion of cellular zinc lowered steroid-induced protein synthesis by 88% and constitutive synthesis by only 15%. The steroid receptor pool was reduced by 50% in the presence of 10 μM DTPA. A dose response study showed that inducible-CAT activity was reduced 25% by a depletion of only 20% of the total zinc (0.1 μM DTPA).

Conclusions

We conclude that the results are consistent with our hypothesis that zinc is critically necessary for the induction of proteins under regulation by steroid hormones. Loss of zinc from steroid receptor protein is probably a consequence of receptor turnover rather than dissociation. Constitutive protein synthesis is much less affected by even severe zinc depletion.

References

Cidlowski, J.A., et al. (1990) Novel antipeptide antibodies to the human glucocorticoid receptor. Molec. Endocrinol. 4: 1427–1437.

Discussion

Q1. Xavier Alvarez-Hernandez, Louisiana State University, Shreveport, LA, USA: Are you depleting the cells of Zn alone? Have you done enough controls to make sure that all other trace elements are added back?

A. With TSQ and adding other metals (Fe, Cu, Mn, etc.), the only response was seen with Zn added back.

Q2. John Sorenson, University of Arkansas, Little Rock, AR, USA: What functional groups on dexamethasone are binding metals?

A. Others have studied this, but I'm not that familiar with the literature to be able to answer this for you.

Q3. John Sorenson, University of Arkansas, Little Rock, AR, USA: Is corticoid activity increased?

A. The corticoid binds, enters, and transcript is increased.

Q4. John Sorenson, University of Arkansas, Little Rock, AR, USA: But there is a flock of corticoids. Are specific functional groups important?

A. I haven't really been studying this, but my supposition is that there are a number or variables in terms of the binding affinities. I don't know how many specific receptors there are. Surely there are specific ones for eg. androgens, estrogens and others, but I couldn't tell you what the specificities are, because that's quite a different direction from what we've been doing.

Selenium Regulation of Glutathione Peroxidase mRNA Levels Requires Sequences within the mRNA 3′ Untranslated Region in Transfected Chinese Hamster Ovary Cells*

S.L. WEISS AND R.A. SUNDE

Nutritional Sciences Group, University of Missouri, Columbia, Missouri 65211, USA

Keywords selenium, mRNA, glutathione peroxidase, gene expression

Steady state levels of glutathione peroxidase (GSH:H_2O_2 oxidoreductase, EC 1.11.1.9) (GPX) mRNA can decrease to less than 10% in Se-deficient rat liver (Lei et al. 1995), but the mechanism is unclear. We have previously demonstrated that the GPX 3′ untranslated region (3′UTR) is required for Se regulation of GPX mRNA levels in Chinese hamster ovary (CHO) cells (Weiss and Sunde 1994). It is not known if this regulatory function for the GPX 3′UTR is related to the selenocysteine insertion sequence (SECIS) motif which has also been localized to the 3′UTR of GPX and other eukaryotic selenoproteins (Berry et al. 1991 and 1993). Selenocysteine-specific elongation factors are hypothesized to form a complex with the SECIS stem-loop secondary structure and for the present study, we have hypothesized that this complex could also function to stabilize the GPX mRNA under Se-adequate conditions.

The purpose of the present experiment was to test a series of GPX mutations, designed to disrupt or retain the predicted secondary structure of the GPX SECIS region, to investigate the requirement of the GPX SECIS region for Se regulation of transfected GPX mRNA levels.

Materials and Methods

The rat GPX cDNA clone corresponding to bp 318-1161 of the rat GPX cDNA sequence (Ho et al. 1988) was subcloned into the XbaI/ApaI restriction sites of the pRc/CMV vector (Invitrogen) using the pBluescript II SK(-) multiple cloning site (Stratagene) as a linker. Oligonucleotide-directed mutagenesis was conducted and all mutations were confirmed by sequencing. The secondary structures of the mutant 3′UTRs were predicted using the PC/Gene RNA FOLD program (Table 1). Primer 44 (gacggtgtttcctctgga-gaaacacctg) deleted 10 nucleotides (Δ10) beginning at position 1034, which are predicted to form the GPX SECIS loop. While confirming the Δ10 mutation, a GPX clone with a 55 bp insertion (I55) beginning at position 1048 was identified. This insertion, composed of three partially overlapping #44 primers, is predicted to form an additional stem-loop structure which extends from the GPX SECIS loop. Primer 49 (ccagaaaaatgacgtcagatggg) introduced an AatII restriction site at position 1075 (3′AatII) and should disrupt base pairing within the GPX SECIS stem. Primer 48 (ggctgccctccgacgtcaggtttttcc) introduced a second AatII restriction site at position 1005 (3′5′AatII) and should restore base-pairing within the GPX SECIS stem, but with a total of seven mutant nucleotides in the primary sequence. CHO cells were transfected with calcium phosphate-precipitated DNA and G418-resistant colonies on each 10 cm plate were pooled 14 d after transfection and plated 1–2 d later (6×10^5 cells per plate) in media containing 1% FBS (2 nM Se) with 0 or 20 nM Se as Na_2SeO_3. After 4 d of growth, total RNA was isolated and subjected to Northern blot analysis for GPX mRNA and 18S rRNA as described previously (Lei et al. 1995).

*Supported by the University of Missouri Agricultural Experiment Station and by NIH grants no. DK 43491 and CA 45164.

Correct citation: Weiss, S.L. and Sunde, R.A. 1997. Selenium regulation of glutathione peroxidase mRNA levels requires sequences within the mRNA 3′ untranslated region in transfected Chinese Hamster Ovary cells. In *Trace Elements in Man and Animals – 9: Proceedings of the Ninth International Symposium on Trace Elements in Man and Animals. Edited by* P.W.F. Fischer, M.R. L'Abbé, K.A. Cockell, and R.S. Gibson. NRC Research Press, Ottawa, Canada. pp. 57–58.

To determine if GPX mRNA levels were significantly lower in Se-deficient cells, paired t-test analysis was first used to show that Se status had no effect on 18S rRNA levels. Se-deficient GPX mRNA levels for each transfection pool were then subjected to one-way ANOVA, and Duncan's multiple range analysis was used to identify significantly different means for 18S rRNA, endogenous GPX mRNA and transfected GPX mRNA for different constructs.

Results

Table 1. 18S rRNA and normalized GPX mRNA levels in Se-deficient transfected CHO cells.

Construct	3'UTR	18S rRNA	GPX mRNA[2] End.	GPX mRNA[2] Trans.
GPX		104±13 A	35± 6 C	68±10 B
Δ10		88± 9 A	59±14 B	117±10 A
I55		110±10 A	42± 6 B	87± 9 A
3'AatII		97±21 A	36± 4 B	114±22 A
5'3'AatII		119±15 A	40±10 C	75± 4 B

[1]Values (mean ± SEM) are expressed relative to Se-adequate RNA levels. In a row, values with a common letter are not significantly different ($P < 0.05$).

[2]Se-deficient GPX mRNA levels were normalized (GPX mRNA/18S rRNA) within each sample. Endogenous (End.) GPX mRNA levels and transfected (Trans.) wild-type GPX and 5'3'AatII GPX mRNA levels were significantly reduced in Se-deficient cells. Se-deficient media had no effect on 18S rRNA levels.

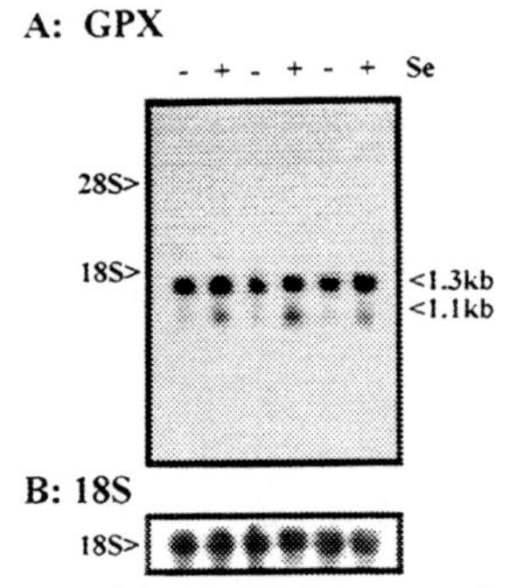

Figure 1. Northern blot autoradiograms of RNA isolated from CHO cells stably-transfected with wild-type GPX and grown in Se-deficient (–Se) or Se-adequate (+Se) media. (A) The membrane was hybridized with the GPX probe. Endogenous GPX mRNA in CHO cells migrates as a 1.1 kb mRNA species and transfected GPX mRNA migrates as a 1.3 kb mRNA species due to additional vector termination sequences. (B) The membrane was hybridized with the 18S rRNA probe.

Discussion

The current experiments demonstrate that mutations within the GPX 3′UTR can eliminate Se regulation of transfected GPX mRNA levels. The results for I55 show that an alteration in GPX 3′UTR secondary structure, without removing any GPX 3′UTR sequences, can be sufficient to block the Se regulatory mechanism. Mutations similar to the Δ10 and 3′AatII have been shown to abolish the selenocysteine insertion ability of the GPX and 5′ deiodinase 3′UTRs (Berry et al. 1991 and 1993) and these mutations also abolished Se regulation of GPX mRNA levels in the current experiment. The results for the 5′3′AatII construct shows that the sequences within this region of the GPX SECIS stem can be changed without affecting Se regulation, provided that the wild-type secondary structure is retained. We conclude that the GPX SECIS motif might also function as a Se responsive element for GPX mRNA; perhaps a SECIS-binding factor masks a specific endonucleolytic cleavage site within the GPX mRNA when Se is present.

References

Berry, M.J., Banu, L., Chen, Y., Mandel, S.J., Kieffer, J.D., Harney, J.W. & Larson, P.R. (1991) Recognition of a UGA as a selenocysteine codon in type I deiodinase requires sequences in the 3′ untranslated region. Nature 353: 273–276.

Berry, M.J., Banu, L., Harney, J.W., & Larsen, P.R. (1993) Functional characterization of the eukaryotic SECIS elements which direct selenocysteine insertion at UGA codons. EMBO J. 12: 3315–3322.

Ho, Y-S., Howard, A.J. & Crapo, J.D. (1988) Nucleotide sequence of a rat glutathione peroxidase cDNA. Nucl. Acids Res. 16: 5207.

Lei, X.G., Evenson, J.K., Thompson, K.M. & Sunde, R.A. (1995) Glutathione peroxidase and phospholipid hydroperoxide glutathione peroxidase are differentially regulated in rats by dietary selenium. J. Nutr. 125: 1438–1454.

Weiss, S.L. & Sunde, R.A. (1994) Selenium regulation of glutathione peroxidase expression in transfected Chinese hamster ovary (CHO) cells. FASEB J. A541.

Determination of Dietary Selenium Requirement in Female Turkey Poults using Glutathione Peroxidase

KEVIN B. HADLEY[†] AND ROGER A. SUNDE[‡]

[†]Department of Biochemistry and [‡]Nutritional Sciences Program, University of Missouri, Columbia, MO 65211, USA

Keywords glutathione peroxidase, requirement, selenium, turkey

Se-deficiency in turkeys is characterized by impaired growth, pathological signs of gizzard and heart myopathy, and reduced hatchability (Scott et al. 1967, NRC 1994). The nutrient requirement of turkeys (NRC 1994) for Se is 0.20–0.28 µg Se/g diet for prevention of gizzard myopathy and 0.23 µg Se/g diet for optimum hatchability and mortality. In contrast to turkeys, the Se requirement for most other species is 0.1 µg Se/g diet when margins of safety or other increments are not included in the requirement (Sunde 1990). In rats, we have shown consistently that 0.1 µg Se/g diet is required for plateau levels of liver glutathione peroxidase (E.C. 1.11.1.9, GPX1) enzyme activity (Knight and Sunde 1987, Lei et al. 1995). The purpose of this experiment was to use several Se-dependant parameters to evaluate the dietary Se requirement for growing female turkey poults.

Methods

Day-old female turkey poults were divided randomly into seven groups and fed a Se-deficient (–Se) semi-purified diet (Knight and Sunde 1987) that was supplemented with 64 mg/g crystalline amino acids to balance the amino acid profile and raise the protein content to more closely match the NRC (1994) requirements. The basal diet contained 0.007 µg Se/g and 11.4 mg/g methionine + cystine, and was supplemented with 100 µg D,L-α-tocopheryl acetate/g diet. Poults were fed the basal diet supplemented with 0, 0.05, 0.1, 0.2, 0.3, 0.4, or 0.5 µg Se/g diet as Na_2SeO_3 for 28 d. As described previously (Lei et al. 1995), diet and liver Se concentration were analyzed by neutron activation analysis, and plasma and 10,000xgx15-min liver supernatant were assayed for GPX1 using H_2O_2 as substrate.

Results and discussion

In female poults, there was no effect of Se supplementation on growth, indicating that under these conditions <0.007 µg Se/g is necessary for growth (Figure 1, left). Liver Se concentration (Figure 1, right) in poults fed the basal diet was 11% of levels in poults fed 0.2 µg Se/g diet. Se supplementation increased liver Se linearly from 0 to 0.2 µg Se/g diet, followed by plateau levels from 0.2 to 0.4 µg/g. This suggests that 0.2 µg Se/g is the minimum dietary level necessary for optimum liver Se concentration.

Plasma GPX1 enzyme activity was not detected in –Se birds (Figure 2, left). Plasma GPX1 enzyme activity increased linearly in response to Se supplementation and reached a maximum at 0.3 µg Se/g diet. Statistical analysis suggests that the plateau is reached between 0.2 and 0.3 µg Se/g diet. Liver GPX1 (H_2O_2) activity in –Se birds was 18% of levels in birds fed 0.2 µg Se/g diet (Figure 2, right). The liver GPX1 activity response curve was sigmoidal, and reached a plateau at 0.2 µg Se/g diet, demonstrating that above 0.2 µg Se/g diet, Se was no longer rate-limiting for GPX1 (H_2O_2) activity. Preliminary experiments suggest that some of this H_2O_2 activity may be due to a second intracellular selenoperoxidase, phospholipid hydroperoxide glutathione peroxidase (E.C. 1.11.1.12, data not shown).

These results clearly demonstrate that under these conditions female poults from commercial suppliers have sufficient body stores such that the apparent dietary Se requirement for growth is <0.007 µg/g diet.

Correct citation: Hadley, K.B., and Sunde, R.A. 1997. Determination of dietary selenium requirement in female turkey poults using glutathione peroxidase. In *Trace Elements in Man and Animals – 9: Proceedings of the Ninth International Symposium on Trace Elements in Man and Animals. Edited by* P.W.F. Fischer, M.R. L'Abbé, K.A. Cockell, and R.S. Gibson. NRC Research Press, Ottawa, Canada. pp. 59–60.

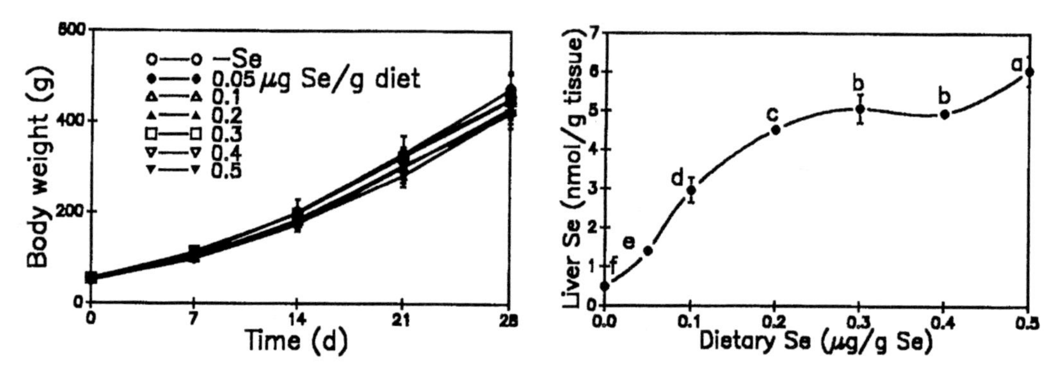

Figure 1. The effect of dietary Se on growth (left) and liver Se concentration (right).

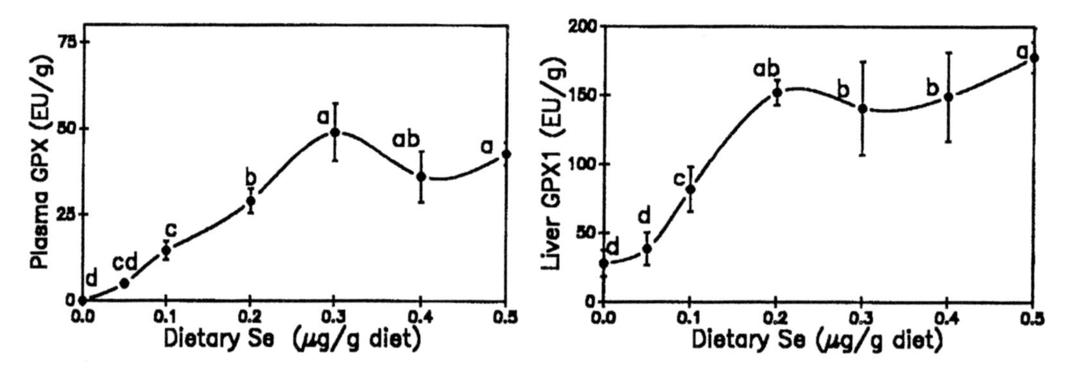

Figure 2. The effect of dietary Se on plasma (left) and liver GPX1 (right) enzyme activity.

Liver Se, plasma GPX1 activity and liver GPX1 activity all decrease to ≤18% of Se-adequate levels without apparent impact in birds supplemented with adequate vitamin E. Importantly, these studies show that the minimum dietary Se necessary to reach plateau levels of GPX1 expression in growing female turkey poults is 0.2 µg Se/g diet. This is twice the level required for rats and many other species. Thus liver GPX1 activity can be a reliable indicator of Se status and useful for evaluating Se requirements in turkeys.

References

Knight, S.A.B. & Sunde, R.A. (1987) The effect of progressive selenium deficiency on anti-glutathione peroxidase antibody reactive protein in rat liver. J. Nutr. 117, 732–738.

Lei, X.G., Evenson, J.K., Thompson, K.M. & Sunde, R.A. (1995) Glutathione peroxidase and phospholipid hydroperoxide glutathione peroxidase are differentially regulated by dietary selenium. J. Nutr. 125: 1438–1446.

National Research Council. (1994) Nutrient Requirements of Poultry, 9th ed. National Academy Press, Washington, D.C.

Scott, M. L., Olson, G., Krook, L. & Brown, W.R. (1967) Selenium-responsive myopathies of myocardium and smooth muscle in the young poult. J. Nutr. 91: 573–583.

Sunde, R.A. (1990) Molecular biology of selenoproteins. Ann. Rev. Nutr. 10: 451–474.

Interactions of Mimosine and Zinc Deficiency on the Transit of BHK Cells through the Cell Cycle*

J.K. CHESTERS AND R. BOYNE

Rowett Research Institute, Bucksburn, Aberdeen AB2 9SB, UK

Keywords zinc, cell cycle, mimosine, S phase

One of the earliest effects of Zn deficiency in animals is a decrease in DNA synthesis (Williams and Chesters 1970) and the activities of several enzymes involved in the synthesis of DNA including thymidine kinase and DNA polymerase α are reduced by Zn deficiency (Chesters et al. 1990, Watanabe et al. 1993). This led to the concept that DNA synthesis is a Zn-dependent process. However, the reduction in DNA synthesis observed with 3T3 cells was associated with a requirement for Zn during the mid-G1 phase of the cell cycle (Chesters et al. 1989). The present investigations were designed to determine whether the mid-G1 requirement was sufficient to explain the decrease in DNA synthesis in Zn-deficient cells or whether Zn was also required during S phase *per se*.

BHK cells were rendered quiescent by culturing in serum-free medium for 72 h and then stimulated to enter G1 synchronously by addition of serum at time zero. The onset of thymidine incorporation after approximately 10 h was inhibited by addition of the chelator DTPA (600 μM) to the medium at T0 but this inhibition was preventable by simultaneous addition of Zn^{2+} (400 μM) along with the DTPA. In each case, a secondary requirement of the cells for Fe was satisfied by including 200 μM Fe^{2+} along with the

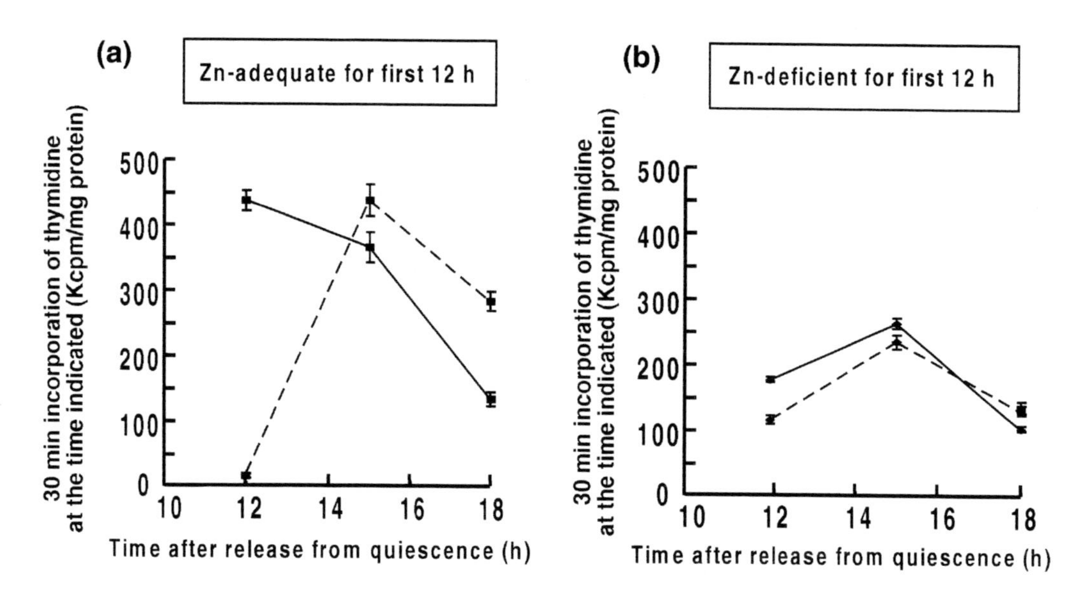

Figure 1. Effects of mimosine and Zn during G1. Cells were cultured in the presence (– – –), or absence (——) of 0.2 mM mimosine and with or without DTPA for the initial 12 h and then transferred to control medium. ^{3}H-thymidine incorporation was measured over 30 min from the times indicated.

*The support of the author's work by the Scottish Office Agriculture, Environment and Fisheries Department is gratefully acknowledged.

Correct citation: Chesters, J.K., and Boyne, R. 1997. Interactions of mimosine and zinc deficiency on the transit of BHK cells through the cell cycle. In *Trace Elements in Man and Animals – 9: Proceedings of the Ninth International Symposium on Trace Elements in Man and Animals. Edited by* P.W.F. Fischer, M.R. L'Abbé, K.A. Cockell, and R.S. Gibson. NRC Research Press, Ottawa, Canada. pp. 61–62.

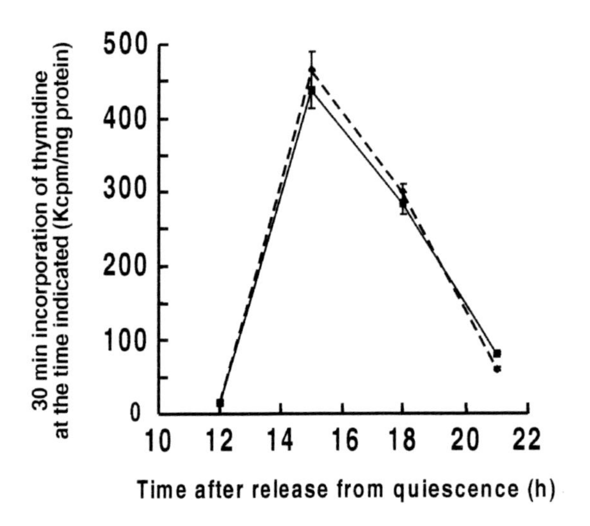

Figure 2. Effects of Zn availability on S phase. Cells were cultured in the presence of mimosine for 12 h and then in control medium with (——) or without (– – –) Zn.

DTPA. Manipulation of the availability of Zn to BHK cells by variously timed additions of DTPA and Zn indicated that, as with 3T3 cells, Zn was required during mid-G1 to permit normal entry into S phase some 5–6 h later.

Addition of mimosine, a toxic amino acid, to BHK cells for the initial 12 h after they were stimulated from quiescence reversibly inhibited the onset of S phase and delayed subsequent cycles of DNA synthesis by 2–3 h (Figure 1a). This suggested that mimosine blocked the progress of the cells at around 9–10 h post stimulation from quiescence, which agrees with the concept that the mimosine block occurs at the G1/S boundary (Mosca et al. 1995).

Addition of DTPA to the culture medium for the first 12 h decreased and delayed thymidine incorporation on release into control unsupplemented medium but supplementation of the Zn-deficient medium with mimosine during the initial 12 h had no additional inhibitory effect on subsequent thymidine incorporation in control medium (Figure 1b).

However, the entry of cells into and passage through S phase after release from a initial 12 h exposure to mimosine was not influenced by Zn availability during S phase provided the cells had adequate Zn during the initial 12 h (Figure 2).

These observations confirm a requirement for Zn which occurs during mid-G1 and thus precedes the mimosine block, suggest that this critical process can be accomplished even in the presence of mimosine and indicate that Zn is not required for passage through S phase after release from a mimosine block.

References

Chesters, J.K., Petrie, L. & Vint, H. (1989) Specificity and timing of the Zn^{2+} requirement for DNA synthesis by 3T3 cells. Exp. Cell Res. 184: 499–508.

Chesters, J.K., Petrie, L. & Travis, A.J. (1990) A requirement for Zn^{2+} for the induction of thymidine kinase but not ornithine decarboxylase in 3T3 cells stimulated from quiescence. Biochem. J. 272: 525–527.

Mosca, P.J., Lin, H-L & Hamlin, J.L. (1995) Mimosine, a novel inhibitor of DNA replication, binds to a 50 kDa protein in Chinese hamster cells. Nuc. Acid Res. 23: 261–268.

Watanabe, K., Hasegawa, K., Ohtake, H., Tohyama, C. & Koga, M. (1993) Inhibition of DNA synthesis by EDTA and its cancellation by Zn in primary cultures of adult rat hepatocytes. Biomed. Res. 14: 99–110.

Williams, R.B. & Chesters, J.K. (1970) The effects of early Zn deficiency on DNA synthesis and protein synthesis in the rat. Br. J. Nutr. 24: 1053–1059.

Trace Elements in Pregnancy and Lactation

Chair: G. Lockitch

Interaction of Selenium and Iodine Deficiency in the Development of Human Iodine Deficiency Diseases

ANTHONY T. DIPLOCK,[†] BERNARD CONTEMPRÉ,[‡] JACQUES DUMONT,[‡] NGO BEBE,[¶] AND JEAN VANDERPAS[§]

[†]Free Radical Research Group, Division of Biochemistry and Molecular Biology, United Medical and Dental School, Guy's Hospital, London SE1 9RT, UK; [‡] IRIBHN University of Brussels, Brussels, Belgium; [¶]UNIKIN Kinshasa, Zaire; [§] Hopital Ambroise Paré, University of Mons, Mons, Belgium

Keywords selenium, iodine, deficiency, Zaire

Introduction and Background

The work described in this paper was carried out with the objective of investigating the possible existence, in Zaire, of combined iodine and selenium deficiency in human populations and to endeavour to assess the likely public health implications of these findings. There is extensive knowledge of the deficiency of iodine which is found in Zaire, and in Malawi, and this has been studied over a long period of time (see Thilly et al. 1993 for review). Goitre, at various degrees of severity, is endemic in these areas of Central Africa and much has been done to correct the incidence of this disabling condition by the administration of iodine. However cretinism and other iodine-deficiency disorders have been described in the areas of severe goitre incidence and their appearance and severity has generally been linked to the severity of the iodine deficiency. The different relative frequencies of the so-called neurological and myxedematous forms of cretinism that have been described in the different endemic areas cannot readily be attributed to a relationship with the iodine deficiency alone, and it is clear that other complementary factors must be invoked to explain the different etiology of cretinism. The pathological development of cretinism, and its main complications which are neurological handicap and mental deficiency, have been thought to be linked to hypothyroidism in the mother or in the fetus, or to a combination of both pathogenicities (Thilly et al. 1978, DeEscobar et al. 1987). It is now thought however that the perinatal period is of critical importance for thyroid adaptation to hypothyroidism, and that the development at a relatively late stage in the perinatal period, or after weaning, of thyroid function degradation may be critical factors in the etiogenesis of cretinism (Delange 1985, Vanderpas et al. 1984).

One factor which was thought to contribute to the complete picture of human iodine deficiency disease in Africa was suggested to be selenium deficiency (Goyens et al. 1987, Corvilain et al. 1993). The hypothesis at that time was that iodine deficiency would be expected to involve an increase in H_2O_2 levels in the thyroid gland, which was undergoing continual stimulation caused by the concommitant iodine deficiency which in turn would cause elevated levels of thyroid stimulating hormone (TSH). It was thought

Correct citation: Diplock, A.T., Contempré, B., Dumont, J., Bebe, N., and Vanderpas, J. 1997. Interaction of selenium and iodine deficiency in the development of human iodine deficiency diseases. In *Trace Elements in Man and Animals – 9: Proceedings of the Ninth International Symposium on Trace Elements in Man and Animals. Edited by* P.W.F. Fischer, M.R. L'Abbé, K.A. Cockell, and R.S. Gibson. NRC Research Press, Ottawa, Canada. pp. 63–68.

that the absence of protection, by selenium-containing glutathione peroxidase, of thyroid cells from oxidative damage might involve the progressive involution of thyroid function which is a characteristic of the development of myxedematous cretinism, and that selenium deficiency might be a more widespread phenomenon in Zaire than that discovered in Kivu (Goyens et al. 1987). This hypothesis was given great stimulus by the finding that the key enzyme in thyroid hormone metabolism, Type I iodothyronine-5′-deiodinase (IDD), is a selenium-containing and selenium-dependent enzyme (Beckett et al. 1987, Arthur et al. 1988, Beckett et al. 1989, Arthur et al. 1990, Berry et al. 1991, Arthur et al. 1992). The function of Type I IDD in the liver and thyroid gland is to catalyse the conversion of thyroxine (T_4) which is the storage form of the thyroid hormone, to the metabolically active hormone known as T_3. Thus, in the absence of this key selenium-dependent enzymic function the metabolically active hormone cannot be formed and the result is thyroid inadequacy. The original hypothesis (Goyens et al. 1987) was modified to suggest that the influence of selenium deficiency, through its role in glutathione peroxidase, might also embrace a more complex role of the trace mineral mediated through Type I IDD. More recently, a further enzyme, Type II IDD has become established as a selenoenzyme (Davey et al. 1995) despite initial controversy as to the role of selenium in this enzyme.This finding may be of crucial significance in the possible role of selenium in the control of T_3 release by the pituitary; thyroid-hypothalamo-pituitary adjustment undergoes major changes during childhood and adolescence.

Earlier Studies in Zaire

Our early work was devoted to study of the selenium deficiency that we found in North Western Zaire and its relationship to the endemia there of goiter and cretinism, and to study of the effects of selenium supplemenation in normal schoolchildren and in cretins in this area (Vanderpas et al. 1990, Contempre et al. 1991, Contempre et al. 1992, Vanderpas et al. 1992, Contempre et al. 1995). The results of these studies are summarised in (Vanderpas et al. 1993) so that it is only necessary to give an outline here. Normal schoolchildren, who were nevertheless severely iodine-deficient, and cretins, from villages in the environs of Karawa were given a physiological supplement of selenium for two months (50 mg/d Se as selenomethionine orally). Table 1 gives a summary of the results obtained when various parameters of iodine and selenium biochemistry were measured; the normal range usually encountered in Belgium is also given for comparison.

Severe selenium deficiency in this region of Nothwestern Zaire is clearly revealed by this study; this adds to the knowledge that this is also a region of severe iodine deficiency, so that the effects of the combined deficiency of these two essential minerals in a human population can be studied. A selenium supplementation programme, with a dose level that approximated to the US RDA for this element, was carried out for a short period of time to evaluate the effect of supplementation with this mineral on thyroid disease. It is clear from the data presented here, and from other work that we have published elsewhere, that selenium plays a definitive role in thyroid metabolism in human subjects. Even during this short period of supplementation the functional indicator of selenium metabolism, erythrocyte GSH-peroxidase, was normalized. The hypothesis that selenium deficiency is associated, together with iodine deficiency, with the etiology of myxedematous cretinism is supported by these data, because there is a high level of incidence of myxedematous cretinism in this part of Zaire. However, other factors may also play a part in the development of this disease. Serum thyroid hormone parameters in clinically euthyroid subjects were modified by this treatment, and there was a dramatic fall in the already impaired thyroid function of clinically hypothyroid patients. It is thought that selenium deficiency in this iodine-deficient area could lead to conflicting clinical consequences; it could protect the general population and the fetus against iodine deficiency and brain damage, but favour the degenerative processes of the thyroid gland that lead to myxedematous cretinism. A further finding from the work described here was that, since selenium deficiency appears to protect against some of the consequences of iodine deficiency, restoring the selenium level by supplementation before restoring the iodine levels could be expected to be unwise or even dangerous.

Table 1. Comparison of selenium and iodine status in normal schoolchildren and cretins given a placebo or selenium supplement (50 mg/d) for 2 months.

Parameter	Before placebo or Se supplement	After placebo or Se supplement	Paired t–test	Normal range in Belgium
Normal with placebo (n = 22)				
Serum Se ng/mL	22.96±10.99	34.23±16.80	$p < 0.001$	46–175
Rbc-Gpx U/gHb	3.6±3.0	3.8±3.0	NS	9–20
SerumT_4 nmol/L	54.8±35.1	55.1±38.0	NS	58–160
Serum T_3 nmol/L	1.73±0.57	1.49±0.51	NS	1.4–3.0
Serum rT_3 pmol/L	95±66	78±55	NS	140–540
Serum TSH mU/L	16.7 (1.8–151)	19.0 (1.6–221)	NS	0.2–10
Normal with selenium supplement (n = 23)				
Serum Se ng/mL	23.87±13.33	66.02±19.95	$p < 0.001$	46–175
Rbc-Gpx U/gHb	3.0±1.9	5.8±2.2	$p < 0.001$	9–20
Serum T_4 nmol/L	73.1±45.4	48.3±23.7	$p < 0.001$	58–160
Serum T_3 nmol/L	2.05±0.48	2.24±0.45	NS	1.4–3.0
Serum rT_3 pmol/L	124±115	90±72	$p < 0.05$	140–540
Serum TSH mU/L	9.6 (7.0–13.0)	7.2 (5.6–9.3)	NS	0.2–10
Cretins with selenium supplement (n = 9)				
Serum Se ng/mL	20.37±9.16	120.54±38.08	$p < 0.001$	46–175
Rbc-Gpx U/gHb	1.4±0.7	10.33±5.31	$p < 0.001$	9–20
Serum T_4 nmol/L	12.8±5.1	23.5±2.5	$p < 0.001$	58–160
Serum T_3 nmol/L	0.98±0.72	0.72±0.29	NS	1.4–3.0
Serum rT_3 pmol/L	3.0±0.0	3.0±0.0	NS	140–540
Serum TSH mU/L	262 (218–316)	363 (304–432)	$p < 0.001$	0.2–10
Serum TBG mg/L	30.4±2.7	31.8±2.5	NS	16–32

RbcGpx is erythrocyte glutathione peroxidase: values given are Mean ± Standard Deviation, except for TSH which is range and geometric mean.

More Recent Studies in Zaire

In order to investigate further the possibility that selenium deficiency is found in Zaire alongside iodine deficiency in other parts of the country we embarked in 1991 on a geographical survey of other regions representative of most of rural Zaire. The results of this work have now been submitted for publication in detail (Ngo et al. 1996)

Pregnant women were recruited in rural prenatal clinics, or in delivery rooms; it was found that this enabled easy recruitment of the women who were subject to routine blood sampling. They gave their informed consent and the collaboration of the supervising doctor, and the agreement of the local medical authorities, were also obtained. The areas chosen were as representative as possible of the entire country, but a major constraint in choice of site was that there was a need to be sure that the samples could be kept cool and transferred to cold storage as quickly as possible after they were obtained. Twenty-nine health areas in seven Regions were studied; the samples were routinely frozen within 3 h of blood collection and kept frozen during transport to Belgium where most of the analyses were performed. The serum selenium measurements were done in London. Table 2 gives the results of the selenium and iodine parameter measurements that were made on blood derived from the the various villages.

Mean urinary iodide was above the lower limit used to define deficiency (5.0 μg/dL) only in Bas-Zaire, indicating that iodine deficiency is widespread throughout Zaire. Mean serum TSH was in the normal

Table 2. Serum and urinary selenium and iodine parameters in pregnant women in rural Zaire.

Region and number of subjects	Urinary I mg/dL Mean and Range	Serum TSH mU/L Mean and Range	Serum T_4 mg/dL Mean±SD	Serum T_3 ng/dL Mean±SD	Serum TBG mg/L Mean±SD	Serum Se ng/mL Mean±SD
Bas-Zaire (60)	6.90 (3.5–13.6)	1.49 (0.87–2.54)	11.7±2.8	132±22	51.1±8.3	110.8 ±20.9
Shaba (61)	2.81 (1.39–5.69)	2.04 (0.95–4.38)	11.1±2.5	147±45	51.7±11.0	71.7±19.9
Kasai (97)	4.70 (2.34–9.45)	1.51 (0.47–4.80)	10.3±3.6	142±32	49.6±10.0	63.6±12.3
Kivu (72)	3.93 (1.93–8.02)	1.48 (0.72–3.00)	11.6±2.9	152±28	53.7±7.0	62.5±18.1
Bandunu (52)	4.35 (2.31–8.21)	1.85 (0.76–4.47)	10.0±3.3	161±37	51.8±8.3	57.3±10.7
Haut-Zaire (45)	3.36 (1.64–6.69)	1.79 (0.86–3.37)	10.2±3.5	143±27	47.7±6.2	46.4±15.2
Equateur (109)	3.21 (1.52–6.79)	1.84 (0.71–4.76)	10.0±3.7	130±39	53.9±10.1	38.8±12.4

range in all regions although the level found was significantly lower in Bas-Zaire than in other regions ($p < 0.001$). Serum T_4, T_3, and TBG did not differ between the regions. The serum selenium values in the various villages showed that all of Equateur and Haut-Zaire is profoundly selenium deficient; some parts of Shaba were also found to have a low selenium level. Further statistical analysis of the results reported here has been undertaken and the following conclusions were made (Ngo et al. 1996). (i) The low urinary iodide excretion found in most of rural Zaire shows little correlation with serum TSH and T_4 levels, suggesting a great capacity for hormonal adaptation, probably mainly by increasing the thyroid uptake and preferential synthesis of T_3 versus T_4. (ii) The frequency of low urinary iodide excretion found in Kasai et al. reflect some degree of iodine deficiency; endemic goitre has not been reported in Kasai and Bandundu, although some limited data are avilable for Kasai. Bas-Zaire was the only region with urinary iodide excretion in the normal range, and no thyroid dysfunction is reported in this region. (iii) Selenium deficiency has been shown to be associated with iodine deficiency in Haut-Zaire and Equateur where myxedematous cretinism is endemic. The combined deficiency was also seen in some parts of Shaba, where no cretinism, either myxedematous or neurological, has been documented, although data are at present scanty.The severe selenium deficiency seen in China associated with Keshan Disease and Kachin-Beck Disease (Diplock 1987) has not been seen in Zaire and there are no reports of these diseases from Zaire. However the involvement of a concommitant Coxsackie viral infection in China may be the important determining factor in the etiology of these diseases (Beck et al. 1994).

The work in Zaire has for the time being been discontinued because political instability renders further field work impossible.

References

Arthur, J. R., Morrice, P. C., and Beckett, G. J. (1988) Thyroid hormone concentrations in selenium-deficient and selenium-sufficient cattle. Res. Vet. Sci., 45: 122–123.

Arthur, J. R., Nichol, F., and Beckett, G. J. (1990) Hepatic iodothyronine 5′-deiodinase. The role of selenium. Biochem. J., 272: 537–540.

Arthur, J. R., Nicol, F., and Beckett, G. J. (1992) The role of selenium in thyroid hormone metabolism and effects of selenium deficiency on thyroid hormone and iodine metabolism. Biol. Trace. Elem. Res., 34: 321–325.

Beck, M. A., Kolbeck, P. C., Rohr, L. H., Shi, Q., Morris, V. C., and Levander, O. A. (1994) Benign human enterovirus becomes virulent in selenium-deficient mice. J. Med. Virol., 43: 166–170.

Beckett, G. J., Beddows, S. E., Morrice, P. C., Nichol, F., and Arthur, J. R. (1987) Inhibition of hepatic iodination of thyroxine is caused by selenium deficiency in rats. Biochem. J., 248: 443–447.

Beckett, G. J., MacDougall, D. A., Nichol, F., and Arthur, J. R. (1989) Inhibition of type I and type II iodothyronine deiodinase activity in rat lever, kidney and brain produced by selenium deficiency. Biochem. J., 259: 887–892.

Berry, M. J., Banu, L., and Larsen, P. R. (1991) Type I iodothyronine deiodinase is a selenocysteine-containing enzyme. Nature, 349: 438–440.

Contempre, B., Duale, N. L., Dumont, J. E., Bebe, N., Thilly, C., Diplock, A. T., and Vanderpas, J. (1992) Effect of selenium supplementation on thyroid hormone metabolism in an iodine- and selenium-deficient population. Clin. Endocrinol., 36: 579–583.

Contempre, B., Dumont, J. E.,Bebe, N., Thilly, C. H., Diplock, A. T., and Vanderpas, J. (1991) Effect of selenium supplementation in hypothyroid subjects of an iodine- and selenium-deficient area: the possible danger of indiscriminate supplementation of iodine-deficient-subjects with selenium. J. Clin. Endocrinol. Metab., 73: 213–216.

Contempre, B., Dumont, J. E., Denef, J. F., and Many, M. C. (1995) Effects of selenium deficiency on thyroid necrosis, fibrosis and proliferation: a possible role in myxoedematous cretinism. Eur. J. Endoc., 133: 99–109.

Corvilain, B., Contempré, B., Longombé, A. O., Goyens, P., Gervy-Decoster, C., Lamy, F., Vanderpas, J. B., and Dumont, J. E. (1993) Selenium and the thyroid: how the relationship was established. Amer. J. Clin. Nutr., 57: 244S–248S.

Davey, J. C., Becker, K. B., Schneider, M. J., Germain, D. L. S., and Galton, V. A. (1995) Cloning of a cDNA for the type II iodothyronine deiodinase. J. Biol. Chem., 270: 26786–26789.

DeEscobar, G. M., Obregon, M. J., and DelRey, F. E. (1987) Foetal and maternal thyroid hormones. Horm. Res., 26: 12–27

Delange, F. (1985) Adaptation to iodine deficiency during growth: etiopathogenesis of endemic goiter and crtinism. Pediatr. Adolesc. Endocrinol., 14: 295–326.

Diplock, A. T. (1987) WHO Environmental Health Criteria No 58:Selenium. Geneva, Switzerland: World Health Organisation.

Goyens, J., Golstein, J., Nsombola, B., Vis, H., and Dumont, J. E. (1987) Selenium deficiency as a possible factor in the pathogenesis of myxoedematous endemic cretinism. Acta Endocrinol., 114: 497–502.

Ngo, D. B., Dikassa, W., Okitolonda, W., Kashala, T. D., Gervy, C., Dumont, J., Vanovervelt, N., Contempré, B., Diplock, A. T., Peach, S., and Vanderpas, J. (1996) Iodine and selenium status in a rural population in Zaire. J. Internat. Health Trop. Med. Submitted.

Thilly, C.-H., Swennen, B., Bourdoux, P., Ntambue, K., Moreno-Reyes, R., Gillies, J., and Vanderpas, J. B. (1993) The epidemiology of iodine-deficiency disorders in relation to goitrogenic factors and thyroid-stimulating hormone regulation. Amer. J. Clin. Nutr., 57: 267S–270S.

Thilly, C. H., Delange, F., and Lagasse, R. (1978) Foetal hypothyroidism and maternal thyroid status in severe endemic goitre. J. Clin. Endocrinol. Metab., 47: 354–360.

Vanderpas, J., Bourdoux, P., and Lagasse, L. (1984) Endemic infantile hypothyroidism in a severe endemic goiter area of Central Africa. Clin. Endocrinol., 20 : 327–340.

Vanderpas, J. B., Contempre, B., Duale, N. L., Deckx, H., Bebe, N., Longombe, A. O., Thilly, C. H., Diplock, A. T., and Dumont, J. E. (1993) Selenium deficiency mitigates hypothyroxinaemia in iodine-deficient subjects. Amer. J. Clin. Nutr., I57: 271–275.

Vanderpas, J. B., Contempre, B., Duale, N. L., Goossens, W., Bebe, N., Thorpe, R., Ntambue, K., Dumont, J., Thilly, C., and Diplock, A. T. (1990) Combined iodine and selenium deficiency in Northern Zaire in relation to endemic myxoedematous cretinism. Amer. J. Clin. Nutr., 52: 1087–1093.

Vanderpas, J. B., Dumont, J. E., Contempré, B., and Diplock, A. T. (1992) Iodine deficiency in Northern Zaire. Amer. J. Clin. Nutr., 56: 957–958.

Discussion

Q1. Pierre Bourdoux, University of Belgium, Brussels, Belgium: How do you reconcile the observation that selenium supplementation worsens the condition of some patients.

A. Quite rightly the situation was made worse in those cretins who had almost no thyroid glands, and that their T4 levels are quite low to start with. What has happened is that as selenium is given, type I deiodinase levels are presumably increased in the thyroid and the small amounts of T4 are then converted to T3 almost immediately. That will presumably be reflected in other parts of the body as well as the thyroid. Exactly how that can be interpreted in the whole body context remains to be determined. Indeed these patients did in fact become very ill and were supplemented with iodine immediately and the situation was quickly rectified. There is large inter-individual variation.

Q2. Noel Solomons, Guatemala: Many possibilities have been developed to measure iodine analytes that can now be done without requiring any cooling or freezing of the samples.

A. Yes, a colleague of mine in China has been developing and using methods of which you speak.

Q3. Barbara Golden, University of Aberdeen, Scotland: Could cassava toxicity have anything to do with the results you have been getting?

A. We are absolutely sure that there is an overlay of cassava toxicity over the whole picture in Zaire, but it is difficult to predict the effects cassava and thiocyanate would have had on the results we have obtained. It would be nice to know in more detail what the people are eating in the different areas. But, unfortunately, that isn't known in very great detail.

Q4. Rosalind Gibson, University of Otago, Dunedin, New Zealand: From a public health perspective, how does the government of Zaire plan to combat the problem of iodine deficiency.

A. I wish I could answer that question. The problem is that the government of Zaire is almost non-existent. There are very profound difficulties indeed. The School of Public Health is existing under very difficult circumstances, so I guess that the answer is that nothing can be done.

Q5. Rosalind Gibson, University of Otago, Dunedin, New Zealand: Would iodization of salt be beneficial, or do rural people of Zaire not eat enough salt?

A. The technique that they have developed in Zaire is to use the Lipiodol, a single injection, which is very much better. So that when people come in to the hospital from the bush they can be given the injection which may last several months.

Ceruloplasmin Expression by Mammary Gland and its Concentration in Milk

R.A. SHULZE, L. WOOTEN, P. CERVEZA, S. COTTON, AND M.C. LINDER
Department of Chemistry and Biochemistry, California State University, Fullerton CA 92634, USA

Keywords ceruloplasmin, copper, mammary gland, milk

Ceruloplasmin is well recognized as the principal copper-containing protein in blood plasma. An α_2-glycoprotein, it accounts for about 65% of the copper in normal serum (Linder and Hazegh-Azam 1996), or about 750 ng of Cu per mL. Like most plasma proteins, it is thought to be produced mainly by the liver (Linder 1991; Linder and Hazegh-Azam 1996). Though the copper in ceruloplasmin is non-dialysable and does not exchange with the rest of the copper in the plasma, it has been shown that it is a source of copper for cells in various tissues, including the placenta and fetus during pregnancy (Lee et al. 1993).

Since clones for ceruloplasmin mRNA have become available, it has also become apparent that the gene for ceruloplasmin is expressed not only in the liver but also in other tissues, particularly those engaged in the production of other body fluids. This includes the choroid plexus that produces cerebrospinal fluid (Schreiber 1987) and the mammary gland that produces milk (Jaeger et al. 1991). We had independently postulated that ceruloplasmin would be in milk because it is present in amniotic fluid (which is ingested by the fetus), and because when we injected ^{67}Cu-labeled ceruloplasmin into the amniotic sac of rats, the radioisotope appeared fairly rapidly in fetal liver (Wooten et al. 1996).

The goals of the studies summarized here were to determine whether ceruloplasmin is indeed present in milk, and to determine whether there is a relationship between expression of ceruloplasmin mRNA by mammary gland and concentrations of the protein in the milk at different stages of lactation. We began by testing milk from different species for ceruloplasmin oxidase activity, with the commonly used substrates, p-phenylene diamine and o-dianisidine (Wooten et al. 1996). Oxidase activity was detected in all milk samples examined from pigs, cows, humans and mice; and more than 90% of the activity was inhibitable by azide (another characteristic of ceruloplasmin). When compared with data for serum, there was no correlation between milk and serum levels across species, levels in milk being fairly similar, while those in serum varied more than 10-fold. Immunoassays of human and pig milk, using antibody against confirmed its presence in significant amounts and were used to quantitate the protein in samples taken at different stages of lactation.

Serial breast milk samples were obtained from 9 women admitted to St. Jude's Hospital, Fullerton, CA, with informed consent (Wooten et al. 1996), on days 1, 3, 5 and 30 d post partum. Fat was removed by centrifugation and aliquots were stored frozen at –20° until use. Total copper in the samples was determined by furnace atomic absorption spectrometry, and ceruloplasmin was assayed immunologically as well as by oxidase activity against p-phenylene diamine. Mean concentrations of ceruloplasmin as well as total copper were steady during the first 5 days, although values varied for and between individuals, and to about the same extent. Mean values for ceruloplasmin during the first week were about 5 mg ceruloplasmin per L by immunoassay, and total copper concentrations about 600 ng/mL (Table 1). (Coefficients of variation were about 25%.) Values for these parameters fell to about 2 mg/L and 300 ng/mL, respectively, by day 30; and two samples of milk obtained from women breast feeding for more than a year showed about the same levels of ceruloplasmin but less total copper (about 180 μg/dL). Ceruloplasmin measured as oxidase activity moved in parallel, except that it had not declined to the same degree by day 30 (Wooten et al. 1996).

Correct citation: Shulze, R.A., Wooten, L., Cerveza, P., Cotton, S., and Linder, M.C. 1997. Ceruloplasmin expression by mammary gland and its concentration in milk. In *Trace Elements in Man and Animals – 9: Proceedings of the Ninth International Symposium on Trace Elements in Man and Animals. Edited by* P.W.F. Fischer, M.R. L'Abbé, K.A. Cockell, and R.S. Gibson. NRC Research Press, Ottawa, Canada. pp. 69–70.

Parallel studies were conducted in pigs with similar findings. Serial milk samples were taken from 9 sows at 3 d and 30 d of lactation. In this case, biopsies of mammary tissue were obtained simultaneously for analysis of ceruloplasmin mRNA expression. By immunoassay, pig milk ceruloplasmin levels were about 26 mg/L and total copper concentrations 1670 ng/mL at day 3 (Table 1). The ratio of oxidase activity to antigen was almost identical for the human and porcine samples, supporting a high structural homology. Values for all three measured parameters fell about 5-fold by day 33 (P. Cerveza, S. Cotton, and M.C. Linder, unpublished). On the assumption that, as for plasma ceruloplasmin, there are 6 Cu atoms per molecule, it was calculated that ceruloplasmin accounted for 3–5% of the copper in the human and porcine milks (Table 1) during the first month of lactation.

Slot blot hybridization of 30 μg portions of total mammary gland RNA with cDNA for human ceruloplasmin indicated that mRNA concentrations were higher on day 3 than 1 month after birth, and that the decrease was of the same order of magnitude as for the concentrations of milk ceruloplasmin (P. Cerveza and M.C. Linder, unpublished). Nevertheless, ceruloplasmin mRNA was already expressed by the mammary gland at the end of gestation, in the absence of lactation, and lower levels of expression were observed in virgin sows. This suggests that not only transcriptional but also translational controls may be regulating production of ceruloplasmin by the mammary gland.

Table 1. Ceruloplasmin in the milk of humans and pigs during the first week.[1]

	Ceruloplasmin			
	Oxidase activity (nmol/min/mL)	Antigen (mg/L)	Total copper (ng/mL)	Ceruloplasmin copper (% of total)
Human milk	1.1 (0.16)[2]	4.7	602	2.5
Sow milk	7.8 (0.0)	26.5	1670	4.8

[1]Mean values for 7–10 determinations, from Wooten et al. (1996) and P.Cerveza, S. Cotton, and M.C. Linder, unpublished.
[2]in the presence of N_3^-

Literature Cited

Jaeger, J.L., Shimizu, N. & Gitlin, J.D. (1991) Tissue-specific ceruloplasmin gene expression in the mammary gland. Biochem. J. 280: 671–677.

Lee, S.H., Lancey, R.W., Montaser, A., Madani, N. & Linder, M.C. (1993) Transfer of copper from mother to fetus during the latter part of gestation in the rat. Proc. Soc. Exp. Biol. Med. 203: 428–439.

Linder, M.C. (1991) The Biochemistry of Copper. Plenum Press, New York, NY.

Linder, M.C. & Hazegh-Azam, M. (1996) Copper biochemistry and molecular biology. Am. J. Clin. Nutr. 63: 797S–811S.

Schreiber, G. (1987). Synthesis, processing, and secretion of plasma proteins by the liver and other organs and their regulation. In: The Plasma Proteins V (F.W. Putnam, ed.), pp. 293–363. Academic Press, New York, NY.

Wooten, L. Shulze, R.A., Lancey, R.W., Lietzow, M. & Linder, M.C. (1996) Ceruloplasmin is found in milk and amniotic fluid and may have a nutritional role. Submitted.

Discussion

Q1. Harry McArdle, Rowett Research Institute, Aberdeen, Scotland: We saw that in many instances when ceruloplasmin decreased, so too did copper. Does ceruloplasmin regulate copper, or vice versa?

A. The amount of copper in ceruloplasmin is about 25% in lactation and the ratio remains constant. I can't speculate but there is likely some relationship to hormone regulation. (Comment from Maria Linder, California State University, Fullerton, CA, USA that prolactin has effects on ceruloplasmin.)

Q2. Bo Lönnerdal, University of California at Davis, CA, USA: One must be careful about developmental patterns as many nutrients behave in the same way.

A. There is a consistent drop in protein synthesis and less need for copper. The regulation may even be set at the level of the brush border membrane of the nursing infant.

Serum Elemental Concentrations in the Pregnancy Disease Pre-Eclampsia

MARGARET P. RAYMAN,[†*] FADI R. ABOU-SHAKRA,[†]
CHRISTOPHER W.G. REDMAN,[‡] AND NEIL I. WARD[†]

[†]Department of Chemistry, University of Surrey, Guildford, GU4 8DJ, UK, and [‡] John Radcliffe Hospital, University of Oxford, Oxford, OX3 9DU, UK

Keywords pre-eclampsia, trace elements, calcium, magnesium

Pre-eclampsia is a disorder of pregnancy which is estimated to cause five million maternal and fetal deaths every year (Lindheimer and Katz 1981).

In this condition, deficient placental implantation gives rise to an ischemic placenta which may release free-radicals thereby causing the observed high levels of lipid peroxidation (Hubel et al. 1989). Circulating lipid peroxides, in women whose antioxidant status may be compromised (Davidge et al. 1992), cause damage to the vascular endothelium including that of the glomerular capillaries (Roberts et al. 1989), resulting in the symptoms of hypertension, proteinuria, and often sudden edema. Slow fetal growth occurs in almost 50% of cases and is an outcome of the poorly-functioning placenta.

Certain elements have been proposed as being relevant to this condition: calcium (Ca) and magnesium (Mg), because of their known relationship to blood pressure and because some studies have demonstrated an effect of supplementation on the incidence of pre-eclampsia (Conradt 1984, Lopez-Jaramillo et al. 1990); copper (Cu) and zinc (Zn), because both have anti-oxidant functions (though Cu^{2+} can also behave as a pro-oxidant) and because supplementation with Zn has been shown to reduce the incidence of PIH (Hunt 1984); selenium (Se), because of its role at the active centre of the peroxide-scavenging enzyme, glutathione peroxidase.

We have investigated the levels of these elements in the serum of pre-eclamptic women and matched controls to ascertain if there are any significant differences.

Subjects and Methods

Serum was prepared from 15 (19 in the case of selenium) severely pre-eclamptic patients, of average gestation 33 weeks, and controls, matched for age, parity and gestation. Analysis of serum for Ca, Mg, Cu and Zn, following 25-fold dilution with 0.0016 M HNO_3, was carried out by inductively coupled plasma mass spectrometry (ICP-MS). Se was determined by hydride-generation ICP-MS in diluted acid-digested serum as previously described (Rayman et al. 1996). Results were validated by analysis of both Nycomed Serum and Second-generation Human Serum. Gestation at delivery, birthweight and sex of the baby were recorded.

Results

Results of serum analyses expressed as means and standard deviations are shown in Table 1. There was no significant difference between cases and controls for Ca, Cu and Se, although there was a trend towards higher levels in the pre-eclamptics. Mg ($P < 0.05$) and Zn ($P < 0.002$) concentrations were significantly higher in the pre-eclamptics.

---

*Daphne Jackson Memorial Fellow, funded jointly by the Leverhulme Trust, Scotia Pharmaceuticals and the University of Surrey.

Correct citation: Rayman, M.P., Abou-Shakra, F.R., Redman, C.W.G., and Ward, N.I. 1997. Serum elemental concentrations in the pregnancy disease pre-eclampsia. In *Trace Elements in Man and Animals – 9: Proceedings of the Ninth International Symposium on Trace Elements in Man and Animals. Edited by* P.W.F. Fischer, M.R. L'Abbé, K.A. Cockell, and R.S. Gibson. NRC Research Press, Ottawa, Canada. pp. 71–73.

Table 1. Elemental concentrations in blood serum of pre-eclamptic women and matched controls.

Element	Number of pairs	Pre–eclamptics*	Controls*	Significance
Mg (mmol/L)	15	0.810 ± 0.078	0.720 ± 0.058	P < 0.05
Ca (mmol/L)	15	2.46 ± 0.19	2.38 ± 0.19	NS
Cu (µmol/L)	15	36.8 ± 7.2	32.3 ± 6.0	NS
Zn (µmol/L)	15	16.7 ± 2.1	13.0 ± 1.5	P < 0.002
Se (µmol/L)	19	0.683 ± 0.170	0.643 ± 0.219	NS

*mean ± SD.

Birth weights, corrected for gestation at delivery and sex of the baby were significantly lower (P < 0.0001) in the pre-eclamptic group, 68% being less than the 10th centile as opposed to 5% in the control group.

Discussion

In our study, all the elements investigated appeared to be more concentrated in the pre-eclamptic than in the control sera, although the difference was significant only for Zn and Mg. There are a number of possible explanations for these findings.

Firstly, plasma volume expansion, which may be as much as 50%, occurs in the course of a normal pregnancy but is smaller or non-existent in pre-eclampsia. This would tend to give higher serum concentrations in the pre-eclamptics in the case of all elements.

Secondly, the extent of fetal growth may have a marked influence on elemental levels at this point in pregnancy. The fetus accumulates its major stores of elements in the last trimester, as much as 80% in the case of some elements. Fetal growth is severely compromised in our pre-eclamptic patients whose placental function is poor, as demonstrated by the significantly lower birthweights. Inefficient placental transfer of nutrients would result in the observed higher maternal concentrations in the pre-eclamptics. This effect would be especially noticeable in the case of Zn where the normal fetal to maternal serum concentration ratio is 1.41 and least noticeable in the case of copper where the ratio is 0.27 (average of six studies in both cases). Glomerular malfunction in the pre-eclamptics will affect the efficiency of elemental excretion. In the case of magnesium, where homeostatic control is exercised by the kidney, serious impairment of glomerular function has been shown to cause a rise in serum Mg despite compensating factors in the tubules (Randall et al. 1964) and this may be an explanation for the higher values of Mg observed in the pre-eclamptics. This mechanism is unlikely to explain the higher Zn values, since Zn is likely to be excreted with its carrier proteins in proteinuria.

Decreased estrogen production has been demonstrated in pre-eclampsia (Klopper 1968). Estrogens are known to depress circulating concentrations of Zn (Halstead et al. 1968); therefore, Zn levels would be expected to be lower in the controls than in the pre-eclamptics, in line with what we have found.

The sum of these factors, most of which will peak in late pregnancy, may explain the observed higher concentrations of Zn and Mg and the tendency towards higher levels of the other elements in the pre-eclamptics. The results of this study highlight the fact that at this point in pregnancy, with this level of disease and fetal-growth retardation, serum concentrations are likely to be an outcome of disease pathology, and cannot be relied upon to give meaningful data about pre-existing elemental deficiencies which may contribute to the etiology of this condition.

Investigation of elemental levels in early pregnancy, in a group of women identified as being at high risk of developing pre-eclampsia, would perhaps be more informative.

Literature Cited

Davidge, S.T., Hubel, C.A., Brayden, R.D., Capeless, E.C. & McLaughlin, M.K. (1992) Sera antioxidant activity in uncomplicated and preeclamptic pregnancies. Obstet. Gynecol. 79: 897–901.

Conradt, A., Weidinger, H. & Algayer, H. (1984) On the role of magnesium in fetal hypertrophy, pregnancy induced hypertension and pre-eclampsia. Magnesium Bull. 6: 68–76.

Halstead, J. A., Hackley, B.M. & Smith J.C. (1968) Plasma zinc and copper in pregnancy and after oral contraceptives. Lancet 2: 278.

Hubel, C.A., Roberts, J.M., Taylor, R.N., Musci, T.J., Rodgers, G.M. & McLaughlin, M.K. (1989) Lipid peroxidation in pregnancy: new perspectives on pre-eclampsia, Am. J. Obstet. Gynecol. 161: 1025–1034.

Hunt, I.F., Murphy, M.S., Cleaver, A.E., Faraji, B., Swendseid, M.E., Coulson, A.H., Clark, V.A., Browdy, B.L., Cabalum, M.T. & Smith, J.C. (1984) Zinc supplementation during pregnancy: effects on selected blood constituents and on progress and outcome of pregnancy in low-income women of Mexican descent. Am. J. Clin. Nutr. 40: 508–521.

Klopper, A. (1968) The assessment of feto-placental function by oestriol assay. Obstet. Gynecol. Surv. 23: 813.

Lindheimer, M.D. & Katz, A.I. (1981) Pathophysiology of pre-eclampsia. Ann. Rev. Med. 32: 273–289.

López-Jaramillo, P., Narváez, M., Felix, C. & Lopez, A. (1990) Dietary calcium supplementation and prevention of pregnancy hypertension. Lancet 335: 293.

Randall, R.E., Cohen, M.D., Spray, C.C. Jr. et al. (1964) Hypermagnesemia in renal failure. Etiology and toxic manifestations. Ann. Intern. Med. 61: 73–88.

Rayman, M.P., Abou-Shakra, F.R. & Ward, N.I. (1996) Determination of selenium in blood serum by hydride generation inductively coupled mass spectrometry. J. Anal. Atomic Spectrom. 11: 61–68.

Roberts, J.M., Taylor, R.N., Musci, T.J., Rodgers, G.M., Hubel, C.A. & McLaughlin, M.K. (1989) Pre-eclampsia: an endothelial cell disorder. Am. J. Obstet. Gynecol. 161: 1200–1204.

Discusion

Q1. Paul Saltman, University of California at San Diego, CA, USA: How do you define preeclampsia?

A. It is defined as high blood pressure, a positive rollover test, or first pregnancy.

Q2. Ine Wauben, McMaster University, Hamilton, ON, Canada: Did you correct for plasma volume and what would you have expected to find if you had? Are your results not just an effect of lower plasma volume?

A. No we did not correct for plasma volume, but it wasn't likely a factor as all elements did not react the same way. If you make the correction, you are looking at other factors.

Q3. John Sorenson, University of Arkansas, Little Rock, AR, USA: The cupric ion was present in such low concentrations that it would not cause oxidation of ascorbic acid and doesn't represent a risk.

A. If you have tissue damage, you would have more. Other people in audience are likely more competent to answer this than I.

Q4. Harvey Gonick, Los Angeles, CA, USA: The problem was that only one tissue (plasma) was investigated and one can get erroneous ideas from that — it seems that you may have fallen into the trap and have not looked at whole body. Zinc is in platelets, etc. There can be redistributions that have nothing to do with trace element status. Part of the phenomena that you have found is a result of this.

A. I accept the comment but am happy with the Se interpretation and the samples that we had were the only ones available. It would be nice to repeat the study with more tissues, but we realize the limitations of the study as presented.

Q5. Harvey Gonick, Los Angeles, CA, USA: The fact that GFR changes cause an increase in Mg is wrong. GFR has to be reduced by 25% before Mg goes up.

A. It's in the literature that if GFR drops, Mg goes up.

Q6. Ed Harris, Texas A&M University, College Station, TX, USA: You mentioned accumulation of Cu in placentas. Did you measure amniotic fluid?

A. No, though it would have nice to have those data.

Q7. Harry McArdle, Rowett Research Institute, Aberdeen, Scotland: With regard to the level of Cu and transport across the placenta, it appeared that the rates of accumulation for Cu and Zn were the same. Would you expect to see the same pattern in fact, when you have high levels of Cu as you did for the Zn?

A. My point is that the amount of Cu in fetus would be so tiny that whatever the rate of transport, the effect of the fetus would not be noticeable.

Measurements of Blood Lead and Stable Lead Isotopes During Pregnancy in a Non-Human Primate (*Macaca fascicularis*)*

M.J. INSKIP,[†] C.A. FRANKLIN,[‡] C.L. BACCANALE,[†] C.M.H. EDWARDS,[†] W.I. MANTON,[¶] E.J. O'FLAHERTY,[§] AND M. TOCCHI[†]

[†]Health Protection Branch, Health Canada, Ottawa, Ontario, Canada K1A 0K9, [‡]Pest Management Regulatory Agency, Health Canada, Ottawa, Ontario, Canada K1A 0L2, [¶]Mass Spectrometry Laboratory, University of Texas at Dallas, Richardson, TX 75080, USA, [§]Department of Environmental Health, University of Cincinnati College of Medicine, Cincinnati, OH 45220, USA

Keywords lead, bone, blood-lead, pregnancy

Because lead can be transferred to the fetus, there is concern about the particular susceptibility of the fetal brain as well as about accumulation of lead in developing fetal bone. The long residence time of lead in bone means that many adult women exposed to lead in childhood and adolescence still have elevated skeletal lead concentrations. During normal processes of bone turnover, lead is remobilized and returns to blood, but a question remains as to whether this remobilization is enhanced during pregnancy (Silbergeld et al. 1988) and during other altered physiological states. This study examined transplacental transfer of lead in the cynomolgus monkey.

Materials and Methods

Healthy, mature female monkeys were maintained on a certified diet (Purina Enriched Monkey Chow 4047 (1% calcium, <0.2 μg/g lead and 2 I.U. $VitD_3$/g), with free access to water. All procedures were carried out in accordance with national recommendations issued by the Canadian Council for Animal Care (1984). The animals were grouped according to previous dosing history: (i) "high lead" (5 animals) which had received lead acetate via the oral route over a lifetime (approximate blood lead of 30–60 mg/100 g) (To convert to μg/100mL, or μmol/L, multiply by 1.031 (density of monkey blood) or 0.049, respectively.); and (ii) a "low lead" group (4 animals) (mean blood lead < 5 mg/100 g). The isotopic "fingerprint" or "signature" of the historic lead present in the animals' skeleton was predominantly "common lead". Measurements of lead concentration and stable isotope ratios were by thermal ionization mass spectrometry (T.I.M.S.) carried out on (a) serially collected blood samples (bi-monthly, prior to, during, and after pregnancy), (b) bone biopsy samples (distal tibia, taken before and after pregnancy) and (c) animal feed consumed over the study period. The special focus on contamination containment in sampling and analysis has been described earlier (Inskip et al. 1996). The isotope signature of the lead dose administered to the "high lead" animals was switched (e.g., for animal CF158, ^{204}Pb-enriched for 200 d, then ^{206}Pb-enriched at confirmation of pregnancy). Using a data transformation procedure, 'end-member un-mixing' (Inskip et al. 1996), we were able to quantify the contribution of 'historic' lead which had originated from maternal bone, to maternal blood lead.

*Supported by NIEHS Contract NO1-ES-05285 to Dr. C.A. Franklin.

Correct citation: Inskip, M.J., Franklin, C.A., Baccanale, C.L., Edwards, C.M.H., Manton, W.I., O'Flaherty, E.J. and Tocchi, M. 1997. Measurements of blood lead and stable lead isotopes during pregnancy in a non-human primate (*Macaca fascicularis*). In *Trace Elements in Man and Animals – 9: Proceedings of the Ninth International Symposium on Trace Elements in Man and Animals. Edited by* P.W.F. Fischer, M.R. L'Abbé, K.A. Cockell, and R.S. Gibson. NRC Research Press, Ottawa, Canada. pp. 74–76.

Results

Figure 1 shows the $^{206}Pb/^{207}Pb$ ratio for serial samples of maternal blood during pregnancy in one of the "low lead" animals. As the $^{206}Pb/^{207}Pb$ ratio for feed (1.23) was quite different to that for bone (cortical (C) = 1.16 and trabecular (T) = 1.17), bone lead remained the predominant influence on the maternal blood isotope fingerprint as seen prior to pregnancy. A change in the contribution coming from lead in the feed would have been reflected in a change in the blood $^{206}Pb/^{207}Pb$ ratio 'towards' the feed.

In Figure 2, we present data on the relative contributions from exogenous (^{206}Pb dose) and endogenous (common lead) sources in one of the "high lead" animals. Compared with the 'steady state' blood lead level observed 50 d prior to pregnancy, there was a decline in the flux of bone lead to blood lead during early pregnancy and a rise during mid- to late-pregnancy until just before C-section. The net effect observed in all but one of the dosed animals was an approximate increase of 73 to 179% of the bone lead contribution to blood lead from early to late pregnancy (calculated from averages of total bone lead

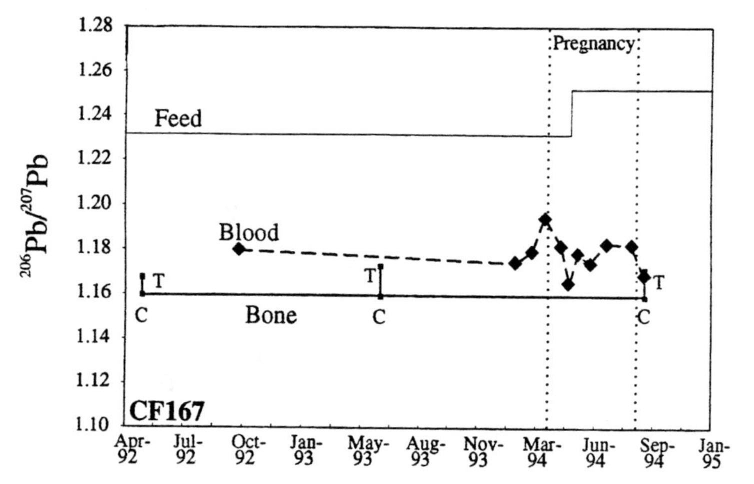

Figure 1. Ratio of lead isotopes $^{206}Pb/^{207}Pb$ in samples of blood (diamonds), bone (T = trabecular, C = cortical) and animal feed, over time in a "low lead" animal.

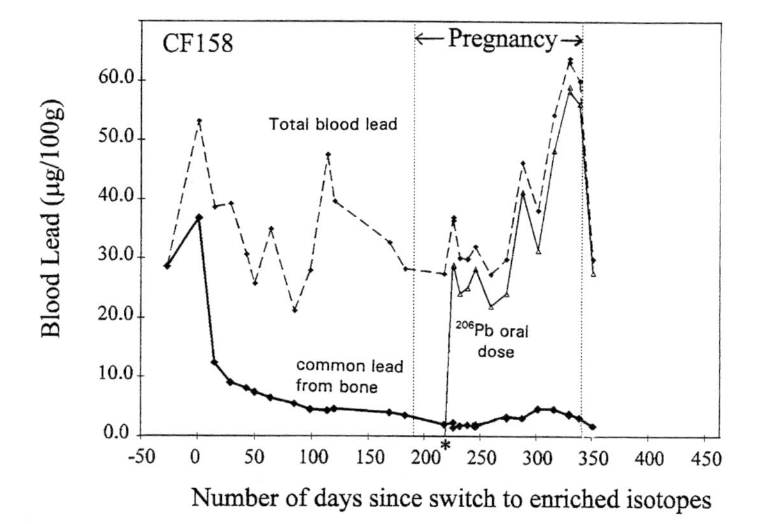

Figure 2. Blood lead during pregnancy in a "high lead" animal. Contribution is from historic lead in bone (common lead = diamonds) and fron the oral dose (^{206}Pb = open triangles). Switch to enriched doses was at day 0 (^{204}Pb) and at day 220 (^{206}Pb) [Data for ^{204}Pb dose not shown].*

contribution to blood lead). The contribution to blood lead from the oral dose during pregnancy (^{206}Pb) increased during late gestation in 3 of the 5 animals, as shown for animal CF158 in Figure 2.

Discussion

The use of sequential stable isotope tracers and TIMS analysis is an effective approach for investigating endogenous (and exogenous) sources of lead to blood in cynomolgus monkeys. The rate of bone-lead mobilization in late pregnancy is approximately double that of early pregnancy. The method is suitable for examining bone:blood fluxes in other physiological states (lactation, bone diseases such as osteoporosis) and also for other elements with suitable stable isotopes.

Acknowledgements

Thanks are extended to Drs. R. Goyer and W. Jameson (U.S. NIEHS) and to C. Ferrarotto, Dr. E. Edwards, D. Schanzer and D. Johnston (Health Canada).

References

Inskip, M.J., Franklin, C.A., Baccanale, C.L., Manton, W.I., O'Flaherty, E.J., Edwards, C.M.H. , Blenkinsop, J.B., & Edwards, E.B. (1996) Measurement of the flux of lead from bone to blood in a non-human primate (*Macaca fascicularis*) by sequential administration of stable lead isotopes. Fund. Appl. Toxicol. (in press).

Rabinowitz, M.B., Wetherill, G.W. & Kopple, J.D. (1976) Kinetic analysis of lead metabolism in healthy humans. J. Clin. Invest. 58: 260–270.

Silbergeld, E., Schwartz, J. & Mahaffey, K.R. (1988) Lead and osteoporosis: Mobilization of lead from bone in post-menopausal women. Environ. Res. 47: 79–94.

Discussion

Q1. John Bogden, New Jersey Medical School, Newark, NJ, USA: All of the Pb in the fetus was from current ingestion and not from bone mobilization. If you hadn't maintained the high lead ingestion, more would have come from skeleton.

A. Yes, we continued dosing with Pb during pregnancy to maintain a steady state and to avoid the unknown. We did not want to change lead exposure as they already had 10 years of Pb exposure.

Q2. John Bogden, New Jersey Medical School, Newark, NJ, USA: But the real experience is that people have more lead exposure as children from paint chips, etc. and less in adolescence and in pregnancy.

A. There was a control animal where 80–90% of the lead came from the bone.

Receptors for Milk Ceruloplasmin in the Brush Border of Piglet Intestinal Mucosa

M.C. LINDER AND D. PAREKH

Department of Chemistry and Biochemistry, California State University, Fullerton, CA 92634, USA

Keywords ceruloplasmin, milk, receptors, brush border

In studies on the transport of copper during gestation in rats (Lee et al. 1993), we observed that copper entering the placenta and fetus is preferentially absorbed from the ceruloplasmin fraction of maternal blood plasma (Lee et al. 1993). Ceruloplasmin is an α_2-glycoprotein, with 6 Cu atoms per molecule. It accounts for 65–70% of the copper in the plasma of normal humans and rats (Wirth and Linder 1985, Linder 1991), or about 750 Cu ng/mL. A versatile protein in terms of function, ceruloplasmin appears to be involved in copper transport, iron fluxing, and in the scavenging of oxygen radicals (Linder and Hazegh-Azam 1996). The copper in ceruloplasmin is incorporated during its synthesis and does not exchange with other copper in the plasma. Plasma ceruloplasmin is synthesized by the liver, but it has become increasingly evident that it is also synthesized by other tissues and appears in other body fluids.

During our gestational studies, we rediscovered that ceruloplasmin is also present in amniotic fluid. This made us consider the possibility that copper in ceruloplasmin might be an additional nutritional source of this metal ion for the fetus, which drinks this fluid. By analogy, we also wondered whether ceruloplasmin might be present in milk, and whether this might represent another way of delivering copper to the newborn.

To begin to test these ideas, ^{67}Cu-labeled ceruloplasmin was injected into the amniotic sac of fetuses a few days before birth, or was fed in cow's milk to newborn and weanling rats (Wooten et al. 1996). Uptake of radioactivity into tissues (and particularly the liver) was followed for one or more hours and compared with uptake of similar doses of ^{67}Cu-labeled Cu(II)-nitrilotriacetate that had alternatively been administered. Copper was absorbed from either source, but in the fetal and newborn rats, copper from ceruloplasmin was accumulated preferentially. In contrast, by 4 weeks of age, there was no difference in uptake of copper from the two sources.

We also tested milk for the presence of ceruloplasmin with positive results (Shulze et al. 1996, Wooten et al. 1996). These findings lead us to hypothesize that ceruloplasmin in milk might be for copper what lactoferrin in milk is for iron, which may be to sequester the metal from bacteria and deliver it to specific receptors in the intestinal brush border (Lönnerdal 1997). One aspect of this hypothesis has now been tested with positive results, and the results are summarized on these pages. Further details will be found in Parekh et al. (1996).

Pigs were used as a source of milk for isolation of ceruloplasmin, and brush border membranes were prepared from the intestinal mucosa of piglets less than one month of age. For isolation of ceruloplasmin, casein was first precipitated from the fat-free milk by titrating to pH 4.6 in the presence of added Ca(II), as described by the Lonnerdal group. The casein-free supernatant was then subjected to two rounds of ion-exchange chromatography, using DEAE-Sepharose CL-6B and 50–200 mM K phosphate, pH 6.8. The resulting material gave a single band in non-denaturing polyacrylamide gel electrophoresis and was labeled with ^{125}I by the Bolton-Hunter technique. Sucrase-rich brush border membranes were prepared as described by Kawakami et al. (1990). To test for the presence of receptors, portions of membrane (600 μg membrane protein) were incubated with various concentrations of ^{125}I-ceruloplasmin. Binding of iodinated casein was also evaluated and compared. Samples were in 1 mM Hepes buffer, pH 7.5, and allowed to incubate overnight at 4°, with gentle agitation. Bound and free radioactivity were separated by centrifugation at high speed, in a microcentrifuge.

Correct citation: Linder, M.C., and Parekh, D. 1997. Receptors for milk ceruloplasmin in the brush border of piglet intestinal mucosa. In *Trace Elements in Man and Animals – 9: Proceedings of the Ninth International Symposium on Trace Elements in Man and Animals.* *Edited by* P.W.F. Fischer, M.R. L'Abbé, K.A. Cockell, and R.S. Gibson. NRC Research Press, Ottawa, Canada. pp. 77–78.

Binding of milk ceruloplasmin to the brush border membranes increased over the range of concentrations tested, which was from 0.05 to 50.0 nM. At the lower concentrations, the percent of the ceruloplasmin bound increased rapidly before reaching a plateau in the range of 0.3 nM and beyond. This was evidence of saturable binding and binding of high affinity. The percent bound then remained constant to the highest concentrations tested, actual binding increasing in proportion to the increase in ceruloplasmin concentrations. This was a sign of non-specific binding at the higher concentrations. Casein only showed non-saturable non-specific binding, namely a contant proportion at all concentrations tested.

Further analysis of the binding data for ceruloplasmin showed that there was an inflection point for the saturable binding at about 0.15 nM ceruloplasmin by Klotz plot; and a similar value for binding affinity (Ks) was obtained from double reciprocal plots. Non-radioactive ceruloplasmin, added in a large excess (50–500) markedly reduced binding of the radioactive ceruloplasmin, indicating same-site competition. In contrast, there was very little competition with ^{125}I-ceruloplasmin binding from excess casein.

Receptors for serum ceruloplasmin are present on the surface of cells on internal organs (Linder 1991), and binding of ceruloplasmin to these receptors can be prevented by an excess of Cu(II), but not Zn(II) (Orena et al. 1985). We therefore tested whether this was also the case for the brush border receptors. Neither of the ions, in concentrations up to 1500 μM, had a substantial effect.

We conclude that there are specific, high affinity receptors for milk ceruloplasmin in the brush border membranes of piglet small intestine, and that these may have a specific function in the efficient absorption of copper from the milk by the neonate.

Literature Cited

Kawakami, H., Shunichi, D. & Lonnerdal, B. (1990) Iron uptake from transferrin and lactoferrin by rat intestinal brush border membrane vesicles. Am. J. Physiol. 258: G535-G541.

Lee, S.H., Lancey, R.W., Montaser, A., Madani, N. & Linder, M.C. (1993) Transfer of copper from mother to fetus during the latter part of gestation in the rat. Proc. Soc. Exp. Biol. Med. 203: 428–439.

Linder, M.C. (1991) The Biochemistry of Copper. Plenum Press, New York, NY.

Linder, M.C. & Hazegh-Azam, M. (1996) Copper biochemistry and molecular biology. Am. J. Clin. Nutr. 63: 797S–811S.

Lönnerdal, B. (1997) Involvement of lactoferrin and lactoferrin receptors in iron absorption in infants. In: Trace Elements in Man and Animals – 9: Proceedings of the Ninth International Symposium on Trace Elements in Man and Animals. (Fischer, P.W.F., L Abbé, M.R., Cockell, K.A. & Gibson, R.S., eds.). NRC Research Press, Ottawa, Canada. pp. 476–480.

Orena, S., Goode, C.A. & Linder, M.C. (1986) Binding and uptake of copper from ceruloplasmin. Bioch. Biophys. Res. Comm. 139: 822–829.

Parekh, D., Shulze, R.A., Cotton, S., Fonda, E., Wickler, S. & Linder, M.C. (1996) Receptors for milk ceruloplasmin in brush border membranes of piglet intestine. Submitted.

Wirth, P.L. & Linder, M.C. (1985) Distribution of copper among components of human serum. J. Nat. Cancer Inst. 75: 277–284.

Wooten, L. Shulze, R.A., Lancey, R.W., Lietzow, M. & Linder, M.C. (1996) Ceruloplasmin is found in milk and amniotic fluid and may have a nutritional role. Submitted.

Selenite Supplementation to Lactating Mothers Enhances the Selenium Concentration in Milk and the Selenium Intake by Breast-Fed Infants*

BRONISŁAW A. ZACHARA,[†] URSULA TRAFIKOWSKA,[†] EWA SOBKOWIAK,[‡] MACIEJ WIACEK,[†] AND MIECZYSŁAWA CZERWIONKA-SZAFLARSKA[‡]

[†]Department of Biochemistry, and [‡]Department of Paediatrics and Gastroenterology, University School of Medical Sciences, 85-092 Bydgoszcz, Poland

Keywords glutathione peroxidase, lactation, milk, selenium supplementation

Human milk is the only source of food for infants during the first months of life. It is therefore important that the milk should contain all trace elements in adequate amounts. Selenium (Se) is of particular interest because the requirement for this element in infants is higher due to their rapid growth (Kumpulainen 1989). As a component of glutathione peroxidase (GSH-Px), Se is involved in protecting cell membranes against peroxidative damage (Zachara 1992).

Se concentration and GSH-Px activity in human milk appear to be influenced directly by the Se intake of the mother (Mannan and Picciano 1987). Since the daily dietary Se intake of the adult population in Poland is below 40 µg, the Se concentration in milk of lactating women is below 10 ng/mL and the calculated Se intake by breast-fed infants is far below the U.S. recommended intakes for full-term infants (Zachara et al. 1994). The aim of this study was to investigate the effect of selenite supplementation to lactating mothers on blood and milk Se concentrations, and to calculate the Se intake by exclusively breast-fed infants.

Subjects and Methods

Ten lactating women were supplemented for 3 months with 200 µg selenite-Se/d mixed with brewer's yeast. A control group (10 lactating mothers) was administered plain brewer's yeast with no added Se. Blood samples were taken from mothers and their infants prior to the start of the study and after 3 months of Se supplementation. Milk samples were collected on the blood sampling day.

Se concentration in whole blood, plasma and milk was assayed by the fluorometric method of Watkinson (1966) with 2,3-diaminonaphthalene as the complexing reagent. The accuracy of the method was tested using whole blood and serum (Seronorm Nycomed, Norway) as reference materials. The Se intake by breast-fed infants was calculated from the amount of milk consumed and the analyzed Se concentration in milk samples collected from mothers on the same day. Statistical significance was assayed using the Student t-test.

Results and Discussion

Se concentrations in the studied material are presented in Table 1.

Our regular studies on Se concentration in blood carried out in Poland in the last few years show that Se levels of healthy subjects are low. This is also true for lactating women (Zachara et al. 1994). Such low levels are comparable only to subjects from Slovenia (Madaric et al. 1994) and previous Yugoslavia (Maksimovic et al. 1992). After 3 months of selenite supplementation to lactating mothers, the Se levels

*Supported by the State Committee for Scientific Research (KBN) No. 4 S405 017 05.

Correct citation: Zachara, B.A., Trafikowska, U., Sobkowiak, E., Wiacek, M., and Czerwionka-Szaflarska, M. 1997. Selenite supplementation to lactating mothers enhances the selenium concentration in milk and the selenium intake by breast-fed infants. In *Trace Elements in Man and Animals – 9: Proceedings of the Ninth International Symposium on Trace Elements in Man and Animals.* *Edited by* P.W.F. Fischer, M.R. L'Abbé, K.A. Cockell, and R.S. Gibson. NRC Research Press, Ottawa, Canada. pp. 79–80.

Table 1. Selenium concentration in blood components of lactating mothers and breast-fed infants, and selenium concentration in milk, and selenium intake by breast-fed infants.

	Lactating mothers		Breast-fed infants	
Months of study	0	3	0	3
Whole blood Se (ng/mL)	76.7±18.1	128 ±21.9[c]	67.6±20.3	90.6±22.5[a]
Plasma Se (ng/mL)	55.1±17.8	97.8±14.8[c]	40.2±19.4	65.9±18.9[b]
Milk Se (ng/mL)	9.8± 2.7	15.1± 3.6[b]	–	–
Se intake (μg/d)	–	–	7.0± 2.8	13.6± 2.7[c]

Statistical significance: the data are compared with the initial values; a, $p < 0.05$, b, $p < 0.01$, c, $p < 0.0001$.

increased significantly ($p < 0.001$) in the whole blood (by 67%) and plasma (by 77%) (Table 1). Our data are in accord with the observation of Kumpulainen et al. (1985) who have shown that after 3 months of selenite supplementation, the Se levels in plasma of lactating women increased significantly by about 80%. Organic Se, however, exerts a more pronounced effect in increasing plasma Se levels in lactating women (Kumpulainen et al. 1985, McGuire et al. 1993).

Our results showing that selenite supplementation produces a significant increase in milk Se concentration are in contrast to the observation of Kumpulainen et al. (1985). They have shown that in lactating mothers supplemented with 100 μg selenite-Se/day, the Se levels in milk decreased linearly during the 6-month period of their study. The Se intake of breast-fed infants is directly related to the Se concentration in milk consumed. Due to the increment of Se level in milk, the daily Se intake by infants increased significantly, reaching the U.S. recommended intakes for full-term infants during the first year (1989). The Se levels in whole blood and plasma of infants also significantly increased as a consequence of milk consumption with higher Se concentration. These results are also in full agreement with other authors (Kumpulainen et al. 1985).

In conclusion, our study shows that the Se concentration in human milk and the Se intake by breast-fed infants can be effectively increased by supplementation of lactating women with inorganic selenium.

References

Food and Nutrition Board. Recommended Dietary Allowances (1989) 10th ed., pp. 217–224. National Academy of Sciences, Washington, DC.

Kumpulainen, J. (1989) Selenium: Requirement and supplementation. Acta Paediatr. Scand. 351: 114–117.

Kumpulainen, J., Salmenpera, L., Siimes, M.A., Koivistoinen, P. & Perheentupa, J. (1985) Selenium status of exclusively breast-fed infants as influenced by maternal organic or inorganic selenium supplementation. Am. J. Clin. Nutr. 42: 829–835.

Madaric, A., Kadrabova, J. & Ginter, E. (1994) Selenium concentration in plasma and red cells in a healthy Slovak population. J. Trace Elem. Electrolytes Health Dis. 8: 43–47.

Maksimovic Z.J., Djujic I., Jovic V. & Rsumovic M. (1992) Selenium deficiency in Yugoslavia. Biol. Trace Elem. Res. 33: 187–196.

Mannan S. & Picciano M.F. (1987) Influence of maternal selenium status on human milk selenium concentration and glutathione peroxidase activity. Am. J. Clin. Nutr. 46: 95–100.

McGuire M.K., Burgert S.L., Milner J.A., Glass L., Kummer R., Deering R., Boucek R. & Picciano M.F. (1993) Selenium status of lactating women is affected by the form of selenium consumed. Am. J. Clin. Nutr. 58: 649–652.

Watkinson J.H. (1966) Fluorometric determination of selenium in biological material with 2,3-diaminonaphthalene. Anal. Chem. 38: 92–97.

Zachara B.A. (1992) Mammalian selenoproteins. J. Trace Elem. Electrolytes Health Dis. 6: 137-151.

Zachara B.A., Trafikowska U., Czerwionka-Szaflarska M. & Sobkowiak E. (1994) Selenium and selenoenzyme — glutathione peroxidase — in human milk during various stages of lactation. A preliminary study. Ped. Pol. 69: 839–844.

Relative Concentrations of Essential and Toxic Metals in Human Milk

S. SIVARAM,[†] V.M. SADAGOPA RAMANUJAM,[‡] AND N.W. ALCOCK[‡]

[†]Clear Lake High School, Houston, TX 77058, [‡] Division of Human Nutrition, Department of Preventive Medicine and Community Health, University of Texas Medical Branch, Galveston, TX 77555-1109, USA

Keywords breast-milk, metals, essential, toxic

Introduction

Human breast-milk is considered to provide optimum nutrition for development of the normal infant during early months of life. Advantages of this source, delineated by the American Academy of Pediatrics (1980) include its availability and safety, its contribution to intestinal development, to provide resistance to infection and the development of a bond between mother and infant. The relative merits and disadvantages of breast-feeding are discussed by Heird (1994). The possibility of transferring agents detrimental to the infant must be considered. The present study analyzed milk obtained from 25 lactating mothers who lived within a 25 mile radius of an area where there was a high density of petroleum refineries and other chemical plants. The specimens were analyzed for the toxic metals Pb and Cd, and in addition the concentration of Na, K, Mg, Ca, Cu, Fe and Zn are also reported and compared with those in cows' milk and a commercial infant formula. An assessment of adequacy of the various minerals for infants is made.

Materials and Methods

Subjects

Twenty-five Caucasian or Hispanic lactating mothers who breast fed an infant for a period of 8 weeks to 7 months were studied. All lived within 25 miles of an area in Texas, USA, where there is a high density of petroleum refineries and other chemical industry plants. Approval for the study was obtained from the Science and Engineering Fair of Houston, Scientific Review Committee and the Institutional Review Board, School or District Level for the first author's Science Fair Project. The participants signed a consent form. In addition a questionnaire was completed to provide relevant information. Each subject hand or pump expressed approximately 30 mL of milk into a sterile 50 mL polypropylene tube, 4 h after the previous feeding. The samples were stored at –20°C until ready for analysis.

Sample Preparation

Five or ten mL aliquots of the milk samples were lyophilyzed, digested with concentrated nitric acid on a water bath to dryness. The digestion was completed with the addition of a minimum volume of hydrogen peroxide (30%). The residue was reconstituted to the original volume in 3 N-nitric acid. After thoroughly mixing on a Vortex, the specimen was centrifuged and the clear supernatant transferred to a polypropylene tube. Further dilution, appropriate for the analyses of the various elements is described below. Duplicate samples of an infant formula (Similac) and of cow's milk were treated identically to the breast milk.

Correct citation: Sivaram, S., Sadagopa Ramanujam, V.M., and Alcock, N.W. 1997. Relative concentrations of essential and toxic metals in human milk. In *Trace Elements in Man and Animals – 9: Proceedings of the Ninth International Symposium on Trace Elements in Man and Animals. Edited by* P.W.F. Fischer, M.R. L'Abbé, K.A. Cockell, and R.S. Gibson. NRC Research Press, Ottawa, Canada. pp. 81–83.

Elemental Analyses

Dilution of the digestate was made with Millipore purified deionized water to give a final concentration of 0.5 N-nitric acid. Analyses were determined by either flame or flameless atomic absorption spectrophotometry using a Perkin Elmer model 5100 Zeeman atomic absorption spectrophotometer. Graphite furnace atomic absorption spectrophotometry was performed for Pb and Cd analyses using electrodeless discharge Pb and Cd lamps respectively. Ammonium dihydrogen phosphate-magnesium nitrate was used as matrix modifier. Zn, Fe and Cu were determined by flame atomic absorption spectrophotometry using hollow cathode lamps at appropriate wavelengths. Calibration standards were prepared in 0.5 N-nitric acid. Ca and Mg were measured by flame atomic absorption in the presence of 1% lanthanum, 4% TCA, and Na and K in the presence of 0.1% CsCl.

Results

The concentration of Pb and Cd in all specimens were below the detection limit (0.1 µg/L for Cd and 1 µg/L for Pb). The median (range of concentrations) of Na, K, Mg, Ca, Cu, Fe Zn in individual specimens at various post-partum intervals and for cow's milk and Similac are shown in Table 1. A histogram of the distribution of Fe, Cu and Zn is shown in Figure 1. Distribution of the other elements measured showed a similar degree of variation. The % increase in weight of the infants from birth to time of collection of specimen, expressed as a % of the birth weight is shown in Figure 2. The correlation coefficient was 0.71.

*Table 1. Concentrations of elements in breast milk — median (range) for various post-partum periods.**

Element	<3 months	3–5.9 months	6–7 months	Cow's milk	Formula
Na (mg/L)	110.35 (55–222)	119 (62–149)	119 (73–166)	386	173
K (mg/L)	607 (125–681)	529 (444–726)	491 (388–669)	1494	792
Mg (mg/L)	46 (26–67)	51 (41–77)	39 (19–80)	107	48.70
Ca (mg/L)	302 (237–385)	329 (291–426)	233 (107–523)	912	448
Cu (µg/L)	232 (94–344)	277 (171–292)	207 (177–353)	64	758
Fe (µg/L)	165 (110–395)	318 (165–495)	218 (65–360)	235	1868
Zn (µg/L)	1153 (285–2075)	1056 (382–2125)	1000 (217–2168)	4245	6555

*Multiplication factor for conversion to SI units Na = 0.043; K = 0.026; Mg = 0.042; Ca = 0.025; Cu = 0.016; Fe = 0.019; Zn = 0.015.

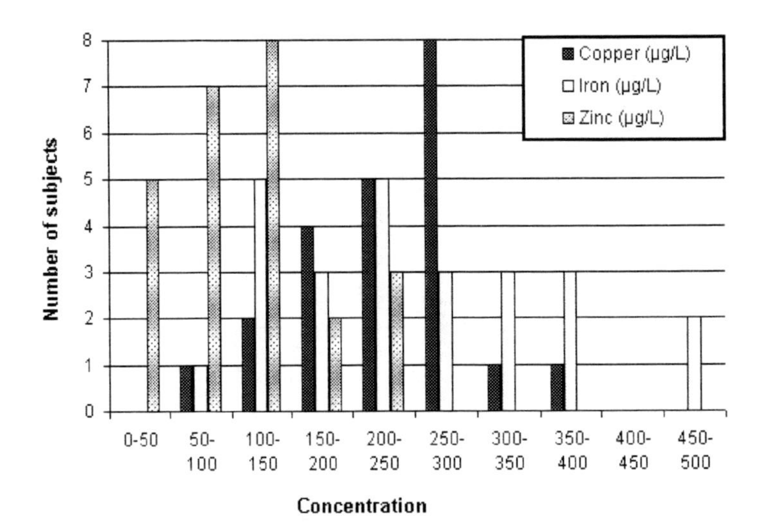

Figure 1. Distribution of iron, copper and zinc.

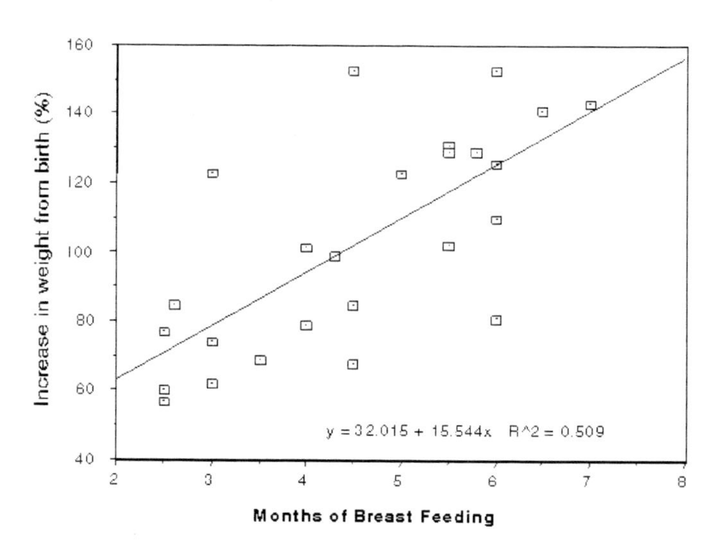

Figure 2. Change of wt (%) vs. period of feeding.

Discussion

There was no evidence of contamination of breast milk with the toxic elements Pb and Cd. The results show a wide range of concentrations of the various elements measured in breast milk. Similar results have been reported in a longitudinal study by Anderson (1993), Butte et al. (1987). The marked differences in concentrations between breast milk, cow's milk and infant formula accentuate the diversity of these as nutrient sources for the new-born infant and for the first year of life. Differences in bioavailability due to interactions between components of the nutrients or specific factors which affect absorption from the gut must be considered. The increase in birth weight with age is consistent with progressive growth (Dewey et al. 1983). However, if it is assumed that the volume of breast milk consumed in a 24 h period is approximately 594–880 mL (Casey et al. 1985) the intake of Cu, Fe and Zn indicates a deficiency from the RDA. This further supports the concept of differences in bioavailability between breast milk and other food sources.

References

Anderson, R.A. (1993) Longitudinal changes in trace elements in human milk during the first 5 months of lactation. Nutr Res. 13: 495–510.

Butte, N.F., Garza, C., O'Brian Smith, E., Wills, C. & Nichols, B.L. (1987) Macro and trace-mineral intakes of exclusively breast-fed infants. Am J Clin Nutr 45: 42–48

Casey, C.E., Hambidge, K.M. & Neville, M.C. (1985) Studies in human lactation: zinc, manganese and chromium in human milk in the first month of lactation. Am J Clin Nutr 41: 1193–1200.

Committee on Nutrition, AAP (1980) Pediatrics 65: 657–6586.

Dewey, K.G. & Lonnerdal, B. (1983) Milk and Nutrient intake in breast-fed infants from 1 to 6 months: relation to growth and fatness. J Ped Gastroent Nutr. 2: 497–506.

Heird, W.C. (1994) Nutritional requirements during infancy and childhood. In: Mod Nutr in Health and Disease (Shils ME, Olsen JA, Shike M, ed.) 8th Ed. pp. 740–758, Lea & Febiger, Philadelphia.

Impact of Social Factors and Heavy Pollution on the Incidence of Anencephaly at the Eastern Coast of Maracaibo Lake, Venezuela*

L.CH. BARRIOS,[†] O. SALGADO,[‡] J.E. TAHAN,[†] L. MARCANO,[¶] R.A. ROMERO,[†] H.S. CUBILLAN,[†] J.M. SANCHEZ,[†] M.C. RODRIGUEZ,[†] M. HERNANDEZ,[†] B.I. SEMPRUN,[†] AND V.A. GRANADILLO[†]

[†]Analytical Instrumentation Laboratory, University of Zulia, Maracaibo 4011, [‡]Renal Service, University Hospital, Maracaibo, [¶]Department of Pathological Anatomy "Pedro García Clara" Hospital, Ciudad Ojeda, Venezuela

Keywords anencephaly, lead, mercury, vanadium

Introduction

Anencephaly is a congenital life-incompatible neural tube defect that originates in early pregnancy (Barrios et al. 1995). The incidence of anencephaly has been found to be lower than 1:1000 births in most epidemiological studies done worldwide (Teixeira et al. 1994). In Venezuela, the incidence of anencephaly among mothers living in Caracas has been reported to range between 0.52 and 0.63:1000 (Uzcategui et al. 1987). The incidence of anencephaly at the Eastern Coast of Maracaibo Lake (ECML), a highly polluted region of Venezuela, has shown a worrisome upward trend in the last decades. It was reported to be 0.9:1000 from 1969 to 1982 (Avila et al. 1984), increasing to 1.5 between 1982 and 1992 (Pineda et al. 1993). The incidence was 5:1000 in 1994 (Barrios et al. 1995). Most of the Venezuelan oil industry is concentrated at the ECML, resulting in an overwhelming environmental pollution of this geographic region. We investigated social factors and heavy metal exposure of mothers living at the ECML giving birth to anencephalic offspring. We also determined the metal content of different organs obtained by autopsy from these fetuses, with the aim of evaluating the impact of social factors vs. environmental metal pollution, on the increasing incidence of anencephaly.

Patients and Methods

Complete organs were obtained by autopsy from both anencephalic and control fetuses within 36 h of delivery. Autopsies were performed at the Department of Pathological Anatomy of Pedro García Clara Hospital, Ciudad Ojeda, Venezuela. Hg, Pb and V were determined using the cold vapor atomic absorption spectrometry, differential pulse anodic stripping voltametry and graphite furnace atomic absorption spectrometry, respectively. In addition, social information such as maternal socioeconomic status, degree of literacy, monthly income, site of residence, etc, was collected by social workers by means of a special questionnaire.

Results and Discussion

Common socioeconomical and cultural features of most of these mothers is shown in Table 1, and included deficient housing conditions, crowding and lacking sanitary facilities, inadequate dietary habits

---

Partially supported by CONDES-LUZ, CONICIT and Consejo Central de Postgrado-LUZ.

Correct citation: Barrios, L.Ch., Salgado, O., Tahan, J.E., Marcano, L., Romero, R.A., Cubillan, H.S., Sanchez, J.M., Rodriguez, M.C., Hernandez, M., Semprun, B.I., and Granadillo, V.A. 1997. Impact of social factors and heavy pollution on the incidence of anencephaly at the eastern coast of Maracaibo Lake, Venezuela. In *Trace Elements in Man and Animals – 9: Proceedings of the Ninth International Symposium on Trace Elements in Man and Animals. Edited by* P.W.F. Fischer, M.R. L'Abbé, K.A. Cockell, and R.S. Gibson. NRC Research Press, Ottawa, Canada. pp. 84–86.

and a history of consumption of contaminated fish or Crustacea from the Maracaibo Lake. Pb and Hg content of the liver and kidneys were higher in anencephalic than in non-anencephalic fetuses ($p < 0.001$) (Table 2). Chemical analysis of organs revealed deposits of mercury and lead in liver, lungs, kidneys and brain of both anencephalic and non-anencephalic fetuses. V was found only in cerebral parenchyma of both groups. Hg might affect embryo development and cause fetal malformations by interfering with the transport of nutrients from mother to the fetus. V, as stated above, was found only in brain and, unlike Hg and Pb, its concentration was higher in non-anencephalic than in anencephalic fetuses, indicating that a V induced teratogenic effect in anencephaly is unlikely.

Table 1. Socioeconomic and cultural data of mothers in 26 cases of anencephalic.

Site of Residency	No.	%
Ojeda City	12	46.1
Bachaquero	5	19.2
Others	9	34.7
Housing conditions		
Crowding	7	27
Drinking water supply	25	96
Electricity	26	100
Sanitary facilities		
Sewerage	13	50
Septic well	9	34.6
Latrine	3	11.5
None	1	3.8
Dietary habits		
Adequate	7	27
Inadequate	19	73
Consumption of contaminated fish	26	100
Degree of literacy		
Illiteracy	7	27
Primary School	18	69.2
High school	1	3.8

Table 2. Dry-weight metal concentrations (mean ± 1 SD, μg/g) found in brain, kidney, liver and lung of anencephalic fetuses and controls evaluated spectrometrically (Hg and V) and voltammetrically (Pb).

Metal	Groups	Brain	Kidney	Liver	Lung
Hg	Anencephaly	0.07 ± 0.01	0.29 ± 0.05*	0.33 ± 0.03*	0.45 ± 0.40
	Controls	UD	0.07 ± 0.03	0.17 ± 0.08	0.12 ± 0.01
			1.3 ± 0.15		
		0.011	0.23		
		(range 0.002–0.029)	(range 0.021–0.81)		
Pb	Anencephaly	UD	1.9 ± 0.3*	2.1 ± 0.3*	3.3 ± 0.4*
	Controls	UD	0.7 ± 0.3	1.1 ± 0.5	0.5 ± 0.1
		0.06 ± 0.08	0.23 ± 0.16	0.54 ± 0.47	0.09 ± 0.06
			0.97 ± 0.81	0.90 ± 0.49	0.44 ± 0.31
		0.17 ± 0.26	0.77 ± 0.41	0.95 ± 0.48	0.27 ± 0.20
V	Anencephaly	0.25 ± 0.18**	UD	UD	UD
	Controls	0.64 ± 0.28	UD	UD	UD
		12.12 ± 3.61			

UD = Undetectable by CVAAS (for Hg), DPASV (for Pb) and ETA-AAS (for V); Statistical differences with respect to controls as follows: *$p < 0.001$; ** $p < 0.05$.

Our results suggest that the social factors play an indirect role in the incidence of anencephaly. The socioeconomic status of the studied mothers led them to live in polluted areas and consume contaminated food. Maternal exposure to potentially teratogenic metals such as Hg, Pb or V was not assessed in the latter study. Whether anencephaly is the result of the toxicity of one metal alone or of the combined effect of all, and to what extent other socials factors, such as maternal malnutrition has contributed to enhance the embryo teratogenecity or toxicity of Pb, Hg and V needs to be further investigated.

Literature Cited

Avila, A., Ferrer, E., & Sevilla, R. (1984) Anencephaly in the hospital Adolfo D'Empaire of Cabimas. Epidemiology and anatomic pathology in 63 cases with special reference to associated defects. Act. Med. Venez. 31: 13–25.

Barrios, L. Ch., Tahán, J.E., Granadillo, V.A., Marcano, L., Cubillán, H.S., Sánchez, J.M., Rodríguez, M.C., Gil de Salazar, F., Salgado, O., & Romero, R.A. (1995) Socio-sanitary factors of anencephaly at the eastern of lake Maracaibo. Ciencia 3: 49–58

Pineda, L., Navarro, G., & Del Villar, A. (1993) Neural tube defects in the Pedro García Clara hospital, Zulia state, Venezuela. Invest. Clin. 34: 41–52.

Teixeira, A.L., Loreto, H., Machado, M.C., & de Almeida, J.M. (1994) Neural tube defects. The 10 year experience of a central maternity hospital. Acta. Med. Port. 7: 419–425.

Uzcategui, O., Mastrolonardo, M., & Villalobos, A. (1987) Anencephaly in the Hospital General "José Gregorio Hernández", Caracas. Rev. Obstet. Ginecol. Venezuela. 47: 104–106.

Effects of Prior Lead Exposure and Diet Calcium on Fetal Development and Blood Pressure During Pregnancy

S. HAN, M.L. EGUEZ, M. LING, X. QIAO, F. KEMP, AND J.D. BOGDEN
Department of Preventive Medicine and Community Health, UMDNJ–New Jersey Medical School, Newark, NJ 07103-2714, USA

Keywords pregnancy, calcium, lead, blood pressure

In a previous study, we found that an increase in dietary calcium could protect the dam and fetus from the toxic effects of exogenous lead exposure via the maternal diet during pregnancy (Bogden et al. 1995). The present investigation was conducted to determine if an increase in dietary calcium could protect the dam and fetus from endogenous lead stored in the skeleton as a result of lead exposure prior to mating and pregnancy.

Materials and Methods

Weanling female Sprague-Dawley rats (Taconic Farms, Germantown, NY) (n = 36, 4 weeks old) were housed individually in plastic cages. Initially, rats were fed a modified AIN–76 diet containing 0.5% calcium (Bogden et al. 1995). Half of the rats (n = 18) were given drinking water containing 250 mg/L of lead as the acetate (Fisher Scientific, Fair Lawn, NJ). Glacial acetic acid was added to the drinking water at a concentration of 12.5 μL/L to prevent the precipitation of lead carbonate. Drinking water consumption was monitored daily during the period of lead exposure. Lead-containing drinking water was administered for 5 weeks, followed by a 4 week period without additional lead exposure. Three female rats were then caged for 3 d with a 14-week-old male of the same strain. Of the 36 females, 16 were successfully impregnated. After being separated from the males, the pregnant females were randomly assigned to receive a diet with a low (0.1%) or normal (0.5%) calcium content during gestation. Calcium carbonate was used to prepare the two dietary calcium concentrations. Systolic blood pressures were measured in triplicate prior to pregnancy and during pregnancy by tail cuff plethysmography.

The dams delivered litters of 3–14 pups after a 21–22 d gestation period. Pup gender was identified at 1 d of age from the anal-genital distance. Pup body weights and lengths were also determined. Blood was obtained from anesthetized dams by cardiac puncture. Blood and organs (liver, femur, kidneys, brain) were obtained from four pups (two males and two females when possible) from each litter within 1 d of birth. Blood and liver lead concentrations were determined by electrothermal atomic absorption spectrophotometry (Bogden et al. 1995). Two-way ANOVA was used to evaluate the effects of lead exposure and dietary calcium on rats in the four treatment groups.

Results

Lead exposure prior to mating influenced fertility. Eleven of 18 (61.1%) of the non-exposed rats but only 5 of 18 (27.8%) lead-exposed animals became pregnant; these percentages are significantly different (Fisher's exact test, $p < 0.05$).

Table 1 contains data on dam blood pressure, pup body weights and lengths within one day of birth, dam blood lead concentrations within 24 h of delivery, and pup blood and liver lead concentrations. Dam blood pressures did not differ significantly among treatment groups when measured on day 5 or 6 of

Correct citation: Han, S., Eguez, M.L., Ling, M., Qiao, X., Kemp, F., and Bogden, J.D. 1997. Effects of prior lead exposure and diet calcium on fetal development and blood pressure during pregnancy. In *Trace Elements in Man and Animals – 9: Proceedings of the Ninth International Symposium on Trace Elements in Man and Animals. Edited by* P.W.F. Fischer, M.R. L'Abbé, K.A. Cockell, and R.S. Gibson. NRC Research Press, Ottawa, Canada. pp. 87–88.

Table 1. Effect of lead exposure prior to pregnancy and diet calcium during pregnancy on dam blood pressure, pup body weight and length, dam blood lead, and pup blood and liver lead concentrations.[1]

	Treatment group			
	Low calcium diet		Normal calcium diet	
	No lead	Lead exposure	No lead	Lead exposure
Dam Blood Pressure — First Week of Gestation[2] (mm Hg)	142±9[A]	157±26[A]	144±5[A]	149±5[A]
Dam Blood Pressure — Third Week of Gestation[3] (mm Hg)	121±6[A]	169±14[B]	143±6[A]	140±7[A]
Pup Body Weight[4] (g)	7.00±0.18[A]	7.09±0.16[A]	6.81±0.09[A]	6.86±0.20[A]
Pup Body Length[4] (mm)	53.6±0.68[A,B]	54.7±0.12[A]	53.7±0.27[A,B]	52.3±0.59[B]
Dam Blood Lead Concentration[5] (μmol/L)	0.12±0.08[A]	1.41±0.14[B]	0.02±0.005[A]	0.92±0.02[C]
Pup Blood Lead Concentration[5] (μmol/L)	0.06±0.02[A]	1.24±0.16[B]	0.02±0.005[A]	0.82±0.13[C]
Pup Liver Lead Concentration[5] (μmol/L)	0.85±0.45[A,B]	2.58±0.35[C]	0.35±0.07[A]	1.51±0.24[B]

1. All data are mean ± SEM. Values in the same row with different superscripts are significantly different (Duncan's test, $p < 0.05$).
2. Dam blood pressure on day 5–6 of gestation. n = 2–7.
3. Dam blood pressure on day 19–20 of gestation. n = 2–7.
4. Within 24 h of delivery. n = 20–62.
5. Sample obtained within 24 h of delivery. n = 2–7 for dam whole blood, and 8–27 for pup whole blood and liver lead concentrations.

gestation. However, the dams exposed to lead and fed the low calcium diet had significantly higher blood pressures on day 19 to 20 of gestation than dams in the other treatment groups. Pup body weights were not significantly influenced by lead exposure or dietary calcium, and there were only small effects of treatment status on pup body lengths. Dam and pup blood lead concentrations and pup liver lead concentrations were highest in the group exposed to lead prior to pregnancy and fed the low calcium diet during pregnancy.

Discussion

The results of this preliminary study demonstrate that a diet low in calcium during pregnancy, when compared to a diet adequate in calcium, will increase both blood lead concentrations and blood pressure late in pregnancy in dams with prior lead exposure. The data also demonstrate that pup blood and liver lead concentrations will also be higher when dams with previous lead exposure are fed diets low in calcium during pregnancy. The source of this lead is likely to be bone lead mobilization from the maternal skeleton, the site of >90% of the body lead burden of rats and humans. The maternal skeleton serves as a source of calcium to meet the demands for fetal growth and development (Pitkin 1992). When the maternal diet is low in calcium, then skeletal stores of calcium will be mobilized. The present study suggests that the mobilization of skeletal calcium stores will be accompanied by the release of the lead that also resides there. The results further suggest that this release will increase maternal and fetal blood and organ lead concentrations and maternal blood pressure late in pregnancy. The timing of these changes is consistent with the greater fetal demand for calcium late in pregnancy than earlier in the course of gestation (Pitkin 1992). The results of this study must be considered as preliminary because of the relatively small number of non-pregnant (n = 20) and pregnant (n = 16) rats studied. Additional studies are in progress to assess the reproducibility of these initial findings.

Literature Cited

Bogden, J.D., Kemp. F.W., Han, S., Murphy, M., Fraiman, M., Czerniach, D., Flynn, C.J., Banua, M.L., Scimone, A., Castrovilly, L., & Gertner, S.B. (1995) Dietary calcium and lead interact to modify maternal blood pressure, erythropoiesis, and fetal and neonatal growth during pregnancy and lactation. J. Nutr. 125: 990–1002.

Pitkin, R.M. (1992) Calcium metabolism in the pregnant and lactating female. In: Calcium Nutriture for Mothers and Children (Tsang, R.C. & Mimouni, F., ed.), pp. 27–37. Raven Press, New York, NY.

Effect of Zinc and Taurine Status during Prenatal and Postnatal Periods on Oscillatory Potentials in the Mature Rat Retina

P.G. PATERSON,[†] K.T. GOTTSCHALL-PASS,[†] B.H. GRAHN,[‡] AND D.K.J. GORECKI[†]
[†]College of Pharmacy and Nutrition and [‡]Western College of Veterinary Medicine, University of Saskatchewan, Saskatoon, Saskatchewan, Canada S7N 5C9

Keywords zinc, taurine, oscillatory potentials, retina

It has been proposed that zinc interacts with taurine in biological systems. Zinc and taurine synergistically protect isolated frog rod outer segments exposed to ferrous sulfate from disc membrane disorganization (Pasantes-Morales and Cruz 1984). Taurine concentration is reported to increase in plasma (Griffith and Alexander 1972) and urine (Hsu and Anthony 1970) of severely zinc-deficient rats. We proposed that the developing retina would be a sensitive tissue in which to explore the physiological significance of this interaction. Kittens and rat pups demonstrate photoreceptor degeneration as a result of taurine deficiency imposed during gestation and postnatal life (Bonhaus et al. 1985, Imaki et al. 1986); the electroretinogram is depressed in rats exposed to this treatment during retinal development (Shimada et al. 1992). The role of zinc in the retina is less clear. However, severely zinc-deficient rat fetuses show retinal folds at term (Rogers and Hurley 1987), and ultrastructural alterations have been described in the retinas of weanling rats fed zinc-deficient diets for seven weeks (Leure-duPree and McClain 1982). Our objective was to investigate whether zinc interacts with taurine to alter retinal development in rats as measured by oscillatory potentials (OP).

Materials and Methods

Virgin female Sprague-Dawley rats (220–250 g) were bred overnight and assigned to 1 of 4 groups. Treatments consisted of four diets containing 2 levels of dietary zinc (50 or 15 µg/g) and taurine (2 or 0 µmol/g) fed *ad libitum* in a 2 × 2 factorial design. Zinc content of the 15 µg/g diets was reduced to 7.5 µg/g at parturition. Guanidinoethyl sulfonate (1% w/v), a taurine transport antagonist, was added to the drinking water of the groups receiving no dietary taurine. At postnatal day 23, male pups (n = 10/group) were weaned onto their respective diets.

Oscillatory potentials were measured on anaesthetized pups at 7.5–8.5 weeks of age after overnight dark adaptation. Oscillatory potentials 1–5 were recorded from a corneal electrode in response to ten 1.2 ms flashes of white light (intensity = 45 J/cm^2) provided by a Ganzfeld stimulator. Responses were recorded using low- and high-cut filter settings of 30 Hz and 200 Hz, respectively.

Plasma and tibia were analyzed for zinc (Clegg et al. 1981) and liver was analyzed for taurine (Gottschall-Pass et al. 1995) as indicators of zinc and taurine status, respectively.

Results

Two-way ANOVA demonstrated a significant interaction between zinc and taurine for plasma (P = 0.0029) and tibia zinc (P = 0.0001) concentrations. Feeding animals marginally zinc-deficient diets significantly depressed plasma and tibia zinc concentrations, but the effect was less pronounced when the animals were also deficient in taurine. An interaction was also apparent for liver taurine (P = 0.0546).

Correct citation: Paterson, P.G., Gottschall-Pass, K.T., Grahn, B.H., and Gorecki, D.K.J. 1997. Effect of zinc and taurine status during prenatal and postnatal periods on oscillatory potentials in the mature rat retina. In *Proceedings of the Ninth International Symposium on Trace Elements in Man and Animals. Edited by* P.W.F. Fischer, M.R. L'Abbé, K.A. Cockell, and R.S. Gibson. NRC Research Press, Ottawa, Canada. pp. 89–90.

Animals fed no dietary taurine and GES had depressed liver taurine independent of zinc status, but zinc deficiency elevated liver taurine in taurine-adequate animals.

Interactive effects were observed in the amplitudes of OP_2 (P = 0.0346) and OP_3 (P = 0.0357). Marginal zinc deficiency decreased the amplitude of these oscillatory potentials only when the animals were also deficient in taurine. Independent effects of zinc and taurine were also observed. Marginal zinc deficiency depressed the amplitude of OP_1 (P = 0.0145) and increased the latency of OP_1 (P = 0.0236) and OP_3 (P = 0.0367). Taurine deficiency depressed the amplitudes of OP_1 (P = 0.0252), OP_4 (P = 0.0006), and OP_5 (P = 0.0639).

Discussion

The results of this study provide electrophysiological evidence that zinc interacts with taurine in the developing retina. While the site for generation of the oscillatory potentials is not certain, these waveforms are generated independently from the a- and b-waves of the electroretinogram in the inner retina. A different origin is suggested for individual oscillatory peaks (el Azazi and Wachtmeister 1990). As interaction was present in only certain oscillatory peaks, zinc and taurine may function synergistically in specific regions of the inner retina. Tissue zinc and taurine concentrations suggest that these nutrients also interact in other tissues, and that taurine will influence the zinc status of the animal. The biochemical mechanisms responsible for this interaction are unknown.

Independent effects of taurine on the development of specific oscillatory potentials were also demonstrated. While previous research has emphasized the dramatic influence of taurine on photoreceptor morphology and function (Bonhaus et al. 1985, Imaki et al. 1986, Shimada et al. 1992), our results suggest that taurine also functions in the inner retina.

The independent influence of marginal zinc deficiency on certain oscillatory potential waves also suggests biochemical functions for zinc in the retina which deserve further investigation.

Acknowledgments

This research was funded by the Hospital for Sick Children Foundation Grant #XG 93-002 and the Natural Sciences and Engineering Research Council of Canada.

Literature Cited

Bonhaus, D.W., Pasantes-Morales, H. & Huxtable, R.J. (1985) Actions of guanidinoethane sulfonate on taurine concentration, retinal morphology and seizure threshold in the neonatal rat. Neurochem. Int. 7: 263–270.

Clegg, M.S., Keen, C.L., Lonnerdal, B. & Hurley, L.S. (1981) Influence of ashing technique on the analysis of trace elements in animal tissue. I. Wet ashing. Biol. Trace Elem. Res. 3: 107–115.

el Azazi, M. & Wachtmeister, L. (1990) The postnatal development of the oscillatory potentials of the electroretinogram. I.Basic characteristics. Acta Ophthal. 68: 401–409.

Gottschall-Pass, K.T., Gorecki, D.K.J. & Paterson, P.G. (1995) Effect of taurine deficiency on tissue taurine concentrations and pregnancy outcome in the rat. Can. J. Physiol. Pharmacol. 73: 1130–1135.

Griffith, P.R. & Alexander, J.C. (1972) Effect of zinc deficiency on amino acid metabolism of the rat. Nutr. Rep. Int. 6: 9–20.

Hsu, J.M. & Anthony, W.L. (1970) Zinc deficiency and urinary excretion of taurine-^{35}S and inorganic sulfate-^{35}S following cystine-^{35}S injection in rats. J. Nutr. 100: 1189–1196.

Imaki, H., Moretz, R.C., Wisniewski, H.M. & Sturman, J.A. (1986) Feline maternal taurine deficiency: Effects on retina and tapetum of the offspring. Dev. Neurosci. 8: 160–181.

Leure-duPree, A.E. & McClain, C.J. (1982) The effect of severe zinc deficiency on the morphology of the rat retinal pigment epithelium. Invest. Ophthalmol. Vis. Sci. 23: 425–434.

Pasantes-Morales, H. & Cruz, C. (1984) Protective effect of taurine and zinc on peroxidation-induced damage in photoreceptor outer segments. J. Neurosci. Res. 11: 303–311.

Rogers, J.M. & Hurley, L.S. (1987) Effects of zinc deficiency on morphogenesis of the fetal rat eye. Development 99: 231–238.

Shimada, C., Tanaka, S., Hasegawa, M, Kuroda, S., Isaka, K., Sano, M. & Araki, H. (1992) Beneficial effect of intravenous taurine infusion on electroretinographic disorder in taurine deficient rats. Japan. J. Pharmacol. 59: 43–50.

Intensity-Response Functions of the Electroretinogram in Rats Deficient in Zinc and Taurine throughout Gestation and into Postnatal Life

K.T. GOTTSCHALL-PASS,[†] B.H. GRAHN,[†] D.K.J. GORECKI,[‡] H.A. SEMPLE,[‡] AND P.G. PATERSON[‡]

[‡]College of Pharmacy and Nutrition and [†]Western College of Veterinary Medicine, University of Saskatchewan, Saskatoon, Saskatchewan, Canada S7N 5C9

Keywords zinc, taurine, electroretinogram, retina

Rats made taurine-deficient using guanidinoethyl sulfonate (GES), a structural analogue of taurine, are known to have decreased a- and b-wave electroretinogram amplitudes (Cocker and Lake 1987, Rapp et al. 1987). The influence of zinc deficiency on retinal function is less clear. Karcioglu et al. (1984) described depressed plasma zinc concentrations in patients with retinitis pigmentosa, a disease characterized by a reduced electroretinogram. When patients with alcohol-induced cirrhosis are given zinc supplements, their elevated dark adaptation thresholds are lowered (Morrison et al. 1978, Russell et al. 1978). Whether a zinc-deficient rat model would produce similar functional changes is unknown. We hypothesized that marginal zinc deficiency imposed throughout gestation and into postnatal life would act synergistically to worsen the electrophysiological response that occurs in taurine deficiency.

Materials and Methods

Virgin female Sprague-Dawley rats were bred overnight and assigned to 1 of 4 treatments in a 2 × 2 factorial design with two levels of dietary zinc (15 or 50 μg/g) and taurine (0 or 2 μmol/g). Animals were provided with free access to modified AIN-93 diets and distilled deionized water; 1% (w/v) GES was added to the drinking water of the animals receiving 0 μmol/g taurine. Zinc content of the 15 μg/g diets was reduced to 7.5 μg/g at parturition. At postnatal day 23, male pups (n = 10) were weaned onto their respective diets. Dark-adapted electroretinograms were recorded as a function of stimulus intensity on 7½–8½-week-old anaesthetized pups. Flashes were provided by a Ganzfeld stimulator and the light attenuated with neutral density filters over 8 log units. The maximum intensity corresponding to 0 log relative intensity was 45 J/cm^2. Trough to peak b-wave amplitude was plotted as a function of log stimulus intensity and an iterative curve fitting procedure used to determine the maximum response (Vmax), slope (n) and half-saturation constant (σ) for each animal. Liver and eye were analyzed for taurine (Gottschall-Pass et al. 1995) and tibia and eye were analyzed for zinc (Clegg et al. 1981) as indicators of taurine and zinc status, respectively.

Results

Marginal zinc deficiency significantly depressed tibial (P = 0.0001) and eye (P = 0.0006) zinc concentrations. When animals were fed taurine-deficient diets and GES, liver (P = 0.0001) and eye (P = 0.0001) taurine concentrations were also significantly depressed. Two-way ANOVA revealed treatment effects on Vmax for taurine (P = 0.0006) and zinc (P = 0.0533). No treatment effects were found on σ or n for zinc or taurine. No interaction between these nutrients for Vmax, σ, or n was noted.

Correct citation: Gottschall-Pass, K.T., Grahn, B.H., Gorecki, D.K.J., Semple, H.A., and Paterson, P.G. 1997. Intensity-response functions of the electroretinogram in rats deficient in zinc and taurine throughout gestation and into postnatal life. In *Trace Elements in Man and Animals – 9: Proceedings of the Ninth International Symposium on Trace Elements in Man and Animals. Edited by* P.W.F. Fischer, M.R. L'Abbé, K.A. Cockell, and R.S. Gibson. NRC Research Press, Ottawa, Canada. pp. 91–92.

Discussion

Electroretinographic responses were influenced by both zinc and taurine status, but the effects were not synergistic. The results of this study provide evidence of a role for zinc in retinal function. These results are of particular interest for the human population given the mild degree of zinc deficiency. Tibial and whole eye zinc concentrations were depressed in the animals fed the marginal zinc-deficient diets. The depression in Vmax observed in zinc-deficient rats suggests a decrease in functioning retina (Wu et al. 1983). Hirayama (1990), using histochemical localization techniques, previously reported a decreased zinc content in the photoreceptors of severely zinc-deficient rats. The influence of zinc status on retinal function deserves further research.

Our results support earlier findings of a decrease in liver and whole eye (Bonhaus et al. 1985) taurine concentration in rats treated with 1% GES. In addition, our data confirm the previous reports of depressed a- and b-wave amplitudes of the electroretinogram (Cocker and Lake 1987, Rapp et al. 1987) and Vmax (Cocker and Lake 1987, Quesada et al. 1988) in taurine-deficient rats.

Acknowledgements

This research was funded by the Hospital for Sick Children Foundation Grant #XG 93-002 and the Natural Sciences and Engineering Research Council of Canada.

Literature Cited

Bonhaus, D.W., Pasantes-Morales, H., & Huxtable, R.J. (1985) Actions of guanidinoethane sulfonate on taurine concentration, retinal morphology and seizure threshold in the neonatal rat. Neurochem. Int. 7: 263–270.

Clegg, M.S., Keen, C.L., Lönnerdal, B. & Hurley, L.S. (1981) Influence of ashing techniques on the analysis of trace elements in animal tissue. I. Wet ashing. Biol. Trace Element Res. 3: 107–115.

Cocker, S.E., & Lake, N. (1987) Electroretinographic alterations and their reversal in rats treated with guanidinoethyl sulfonate, a taurine depletor. Exp. Eye Res. 45: 987–997.

Gottschall-Pass, K.T., Gorecki, D.K.J. & Paterson, P.G. (1995) Effect of taurine deficiency on tissue taurine concentrations and pregnancy outcome in the rat. Can. J. Physiol. Pharmacol. 73: 1130–1135.

Hirayama, Y. (1990) Histochemical localization of zinc and copper in rat ocular tissues. Acta Histochem. 89: 107–111.

Karcioglu, Z.A., Stout, R. & Hahn, H.J. (1984) Serum zinc levels in retinitis pigmentosa. Curr. Eye Res. 8: 1043–1048.

Morrison, S.A., Russell, R.M., Carney, E.A. & Oaks, E.V. (1978) Zinc deficiency: A cause of abnormal dark adaptation in cirrhotics. Am. J. Clin. Nutr. 31: 276–281.

Rapp, L.M., Thum, L.A., Tarver, A.P. & Wiegand, R.D. (1987) Retinal changes associated with taurine depletion in pigmented rats. Prog. Clin. Biol. Res. 247: 485–495.

Russell, R.M., Morrison, S.A., Smith, F.R., Oaks, E.V. & Carney, E.A. (1978) Vitamin A reversal of abnormal dark adaptation in cirrhosis: Study of effects on the plasma retinol transport system. Ann. Intern. Med. 88: 622–626.

Quesada, O., Picones, A. & Pasantes-Morales, H. (1988) Effect of light deprivation on the ERG response of taurine-deficient rats. Exp. Eye Res. 46: 13–20.

Wu, L., Massof, R.W. & Starr, S.J. (1983) Computer-assisted analysis of clinical electroretinographic intensity-response functions. Doc. Ophthal. Proc. 37: 231–239.

The Effects of Lead Levels during Gestation on the Somatometric Parameters of the Neonate

M.J. PUENTE,[†] M.J. TAPIA,[†] I. NAVARRO,[†] A. MARTIN,[†] AND I. VILLA[‡]

[†]Pediatric Research Unit, University of Navarra, 31.080 Pamplona, and [‡]Department of Pediatrics, Gregorio Marañón Hospital, 28.009 Madrid, Spain

Keywords lead, gestation, neonate

Introduction

Lead is a toxic trace element to which various adverse effects have been attributed, especially in the development of the fetus and neonate. One of these effects is a delay in intrauterine growth. One postulated hypothesis by which lead exerts this effect is by producing anemia in neonates (Al-Saleh et al. 1995). Bellinger et al. (1991) suggested a lead serum concentration limit in umbilical cord of 7.2 nmol/L. Some authors, however, did not find a relation between lead levels and neonate weight (West et al. 1994; Loiacono et al. 1992).

The present study was designed to determine if different serum (maternal and umbilical), as well as placental, lead levels affect the weight, size, cranial and thoracic circumferences of the newborn.

Material and Methods

Two hundred and sixty-six pregnant subjects, from Navarra (Spain), were included in this study. This region is characterized by its low environmental contamination, due to the limited number of industries and its abundance of green space. Information about the subjects' life style, supplemental vitamin use during pregnancy and previous illnesses was obtained. Maternal and umbilical blood was collected, as well as a placental fragment, from each subject. Lead was analysed using electrothermal atomization atomic absorption spectrophotometry (Graphite furnace-AAS). The weight, size, Apgar 1′ and 5′, hematological parameters, cranial and thoracic circumferences were recorded at the time of birth. The relationship between variables was determined from Spearman's coefficient.

Results and Discussion

The results demonstrate negative correlations between lead levels in umbilical cord serum and the weight of the newborn ($R = -0.15$, $p < 0.05$), the size of the neonate ($R = -0.17$, $p < 0.05$) and thoracic circumerence ($R = -0.18$, $p < 0.05$). These results prove the hypothesis that elevated levels of lead in the fetus result in a reduction of weight and size of the neonate. There are no reports in the literature on the relationship between fetal lead load and thoracic and cranial circumferences. The decrease in thoracic circumference may be due to the reducted weight of the neonate. The data do not indicate that the lead in maternal serum and placenta correlates with the parameters studied (Table 1). The hematological parameters measured were not significantly correlated to lead levels, suggesting that the mechanism causing reduced growth is not anemia.

Correct citation: Puente, M.J., Tapia, M.J., Navarro, I., Martin, A., and Villa, I. 1997. The effects of lead levels during gestation on the somatometric parameters of the neonate. In *Trace Elements in Man and Animals – 9: Proceedings of the Ninth International Symposium on Trace Elements in Man and Animals. Edited by* P.W.F. Fischer, M.R. L'Abbé, K.A. Cockell, and R.S. Gibson. NRC Research Press, Ottawa, Canada. pp. 93–94.

Table 1. Correlation between neonate cuantitative parameters and lead levels.

Neonate parameters	Number	Correl. coef.	Probability
Pb maternal serum			
Weight	192	–0.01	0.93
Size	153	–0.09	0.29
Apgar 1′	192	0.04	0.61
Apgar 5′	192	0.09	0.19
Red blood	96	0.04	0.68
Hemoglobin	96	0.11	0.27
Leucocite	96	0.09	0.39
Med. Corp.Volume	85	0.09	0.41
Cranial Circumference	149	0.15	0.07
Thoracic Circumference	136	–0.13	0.13
Pb umbilical cord			
Weight	190	–0.15	0.04*
Size	154	–0.18	0.03*
Apgar 1′	190	0.01	0.87
Apgar 5′	190	0.02	0.79
Red blood	94	0.06	0.56
Hemoglobin	94	0.11	0.32
Leucocite	94	–0.05	0.63
Med. Corp. Volume	83	0.12	0.28
Cranial Circumference	150	–0.15	0.07
Thoracic Circumference	138	–0.18	0.03*
Pb placenta			
Weight	200	0.08	0.24
Size	154	0.04	0.61
Apgar 1′	200	–0.02	0.81
Apgar 5′	200	–0.08	0.23
Red blood	101	0.06	0.58
Hemoglobin	101	0.09	0.35
Leucocite	101	0.05	0.62
Med. Corp. Volume	90	0.03	0.77
Cranial Circumference	151	0.09	0.23
Thoracic Circumference	139	0.08	0.36

* $p < 0.05$.

References

Al-Saleh, I., Abdulkarim, M. & Taylor, A. (1995) Lead, erythrocyte protoporphyrin, and hematological parameters in normal maternal and umbilical cord blood from subjects of the Riyadh Region, Saudi Arabia. Arch. Environ. Health. 50 (1): 66–72.

Bellinger, D., Leviton, A., Rabinowitz, M. et al. (1991) Weight gain and maturity in fetuses exposed to low levels of lead. Environ. Res. 54: 151–158.

Loiacono, N.J., Graziano, J.H., Kline, J.K. et al. (1992) Placental cadmium and birthweight in women living near a lead smelter. Arch. Environ. Health. 47 (4): 250–255.

West, W., Knight, E., Edwards, C. et al. (1994) Maternal low level lead and pregnancy outcomes. J. Nutr. 124: 981S–986S.

Influence of the Intake of Calcium and Iron Supplements during Pregnancy and its Effects on Cadmium and Lead Levels

M.J. PUENTE,[†] M.J. TAPIA,[†] I. NAVARRO,[†] A. MARTIN,[†] AND I. VILLA[‡]

[†]Pediatric Research Unit, University of Navarra, 31.080 Pamplona, and [‡]Department of Pediatrics, Gregorio Marañón Hospital, 28.009 Madrid, Spain

Keywords calcium, iron, cadmium, lead, pregnancy

Introduction

Lead and cadmium are toxic elements which have been shown to increase probability of miscarriage, premature births, low birth weight and malformations (Kristensen et al. 1993, Kuhnert et al. 1987, Needleman et al. 1984). These elements are divalent metals which can compete with other elements such as calcium and iron. Iron deficiency, for instance, results in a higher absorption of cadmium and lead from the gastrointestinal tract (Singh et al. 1993). During pregnancy, reduced iron stores are common, a fact that could be related to increased cadmium and lead absorption. Consequently, iron supplementation during pregnancy could protect the mother from increased exposure to the toxic metals. During conditions of bone demineralization, such as pregnancy, lactation and menopause, lead stores, which normally have a half life of approximately 20 years (Rabinowitz et al. 1976), are mobilized to increase the blood levels (Bogden et al. 1995). The aim of this study was to determine whether supplementation with calcium and iron during pregnancy results in changes in lead and cadmium levels in maternal serum, umbilical cord serum and in placenta.

Material and Methods

Two hundred and sixty-six pregnant women were given a questionnaire about calcium and/or iron supplement use. At the time of birth, maternal blood, umbilical cord blood and a piece of placenta was collected to determine the lead and cadmium concentrations. The blood samples were centrifuged and kept frozen (–20°C). Digestion of the placenta samples was done in hot nitric acid in a closed acid-decomposition system using a high-pressure teflon digestion bomb and microwave energy (Milestone mls 120). Lead and cadmium concentrations were determined by electrothermal atomization atomic absorption spectrophotometry (graphite furnace-AAS). Statistical analysis was carried out using the U Mann Whitney non parametric test.

Results and Discussion

The data shows a statistically lower concentration of lead in the umbilical cord serum of neonates whose mothers had taken calcium ($p < 0.05$) and iron ($p < 0.01$) supplements compared to the group which did not take the supplements. There was also a higher placental lead level in those who took iron ($p < 0.05$). No significant differences were detected in cadmium concentrations (Tables 1 and 2). These results indicate that intakes of calcium and iron by the mother can protect the fetus from lead exposure. No significant changes were detected in the lead levels in the maternal serum, suggesting that lead competes with iron and calcium at the level of the placenta (Goyer et al. 1990). Goyer et al. (1990) has identified a calcium transport protein in placenta, similar to the one found in the intestinal tract which could also move lead across the placental barrier.

Correct citation: Puente, M.J., Tapia, M.J., Navarro, I., Martin, A., and Villa, I. 1997. Influence of the intake of calcium and iron supplements during pregnancy and its effects on cadmium and lead levels. In *Trace Elements in Man and Animals – 9: Proceedings of the Ninth International Symposium on Trace Elements in Man and Animals. Edited by* P.W.F. Fischer, M.R. L'Abbé, K.A. Cockell, and R.S. Gibson. NRC Research Press, Ottawa, Canada. pp. 95–96.

Table 1. The effect of iron supplementation during pregnancy on lead and cadmium levels in different compartments.

Concentration	Iron Yes	Iron No	Probability
Pb serum (nmol/L)	24.6±1.9	29.4±2.1	0.09
Pb umb. cord (nmol/L)	11.1±0.7	15.0±1.1	0.007**
Pb placenta (nmol/kg)	149.1±17.6	115.8±16.3	0.04*
Cd serum (nmol/L)	7.1±1.0	0.9±0.17	0.36
Cd umb. cord (nmol/L)	3.6±0.3	3.6±0.3	0.14
Cd placenta (nmol/kg)	56.0±5.3	43.6±3.9	0.24

* $p < 0.05$.
** $p < 0.01$.

Table 2. The effect of calcium supplementation during pregnancy on lead and cadmium levels in different compartments.

Concentration	Calcium Yes	Calcium No	Probability
Pb serum (nmol/L)	28.0±4.2	26.5±1.5	0.84
Pb umb. cord (nmol/L)	11.1±2.12	13.5±0.7	0.048*
Pb placenta (nmol/kg)	114.8±30.8	131.8±13.1	0.38
Cd serum (nmol/L)	8.0±1.7	7.1±0.1	0.23
Cd umb. cord (nmol/L)	3.6±0.6	3.6±0.2	0.4
Cd placenta (nmol/kg)	49.8±7.7	49.8±3.7	0.6

* $p < 0.05$.

References

Kuhnert, B.R., Kuhnert, P.M., Debanne, S. & Williams, T.G. (1987) The relationship between cadmium, zinc and birth weight in pregnant women who smoke. Am. J. Obstet. Gynecol. 157: 1247–1251.

Bogden, J., Kemp, F., Han, S. et al. (1995). Dietary calcium and lead interact to modify maternal blood presure, erythropoiesis, and fetal and neonatal growth in rats during pregnancy and lactation. J. Nutr. 125: 990–1002.

Kristensen, P., Irgens, L., Daltveit, A. & Andersen, A. (1993) Perinatal outcome among children of men exposed to lead and organic solvents in the printing industry. Am. J. Epidemiol. 137:134–144.

Rabinowitz, M.B., Wetherill, G.W. & Kopple, J.D. (1976) Kinetic analysis of lead metabolism in healthy humans. J. Clin. Invest. 58: 260–270.

Needleman, H.L., Rabinowitz, M., Leviton, A. et al. (1984) The relationship between prenatal exposure to lead and congenital anomalies. J.A.M.A. 251: 2956–2959.

Goyer, R.A. (1990) Transplacental transport of lead. Environ. Health. Perspect. 89: 101–105.

Singh, C., Saxena, D.K., Murthy, R.C. & Chandra, S.V. (1993) Embryo-fetal development influenced by lead exposure in iron-deficient rats. Hum. Exp. Toxicol.12(l): 25–28.

Lead and Cadmium Transference through the Placenta in Gestations

M.J. PUENTE,[†] M.J. TAPIA,[†] I. NAVARRO,[†] A. MARTIN,[†] AND I. VILLA[‡]

[†]Pediatric Research Unit, University of Navarra, 31.080 Pamplona, and [‡]Department of Pediatrics, Gregorio Marañón Hospital, 28.009 Madrid, Spain

Keywords lead, cadmium, placenta, transfer

Introduction

The determination of trace and ultratrace toxic elements in the placenta, mother's serum and newborns is of great interest due to the fact that the placenta is considered to be the first environment of a living organism. The toxic element levels in the placenta seem to reflect the contamination during the gestation (Schramel et al. 1988). As some of the trace elements can be transferred from the mother to the fetus through the placenta, it is important to clarify the possible function of the placenta as a fetal protector against toxic elements such as lead and cadmium. The aim of this study was to determine the average concentration of cadmium and lead in the mothers' and fetuses' sera, and placentas in subjects from Navarra (Spain); and try to establish the role of the placenta in protecting the fetus from a possible accumulation of lead and cadmium. The influence of both internal and external factors on the metal levels has been considered.

Material and Methods

For this study 266 pregnant subjects from Navarra were used; 167 from Clínica Universitaria and 99 from Hospital Virgen del Camino. The births took place between October 1992 and October 1993. All the patients included in this study filled out a questionnaire about alcohol consumption, tobacco use and the ingestion of calcium and iron during pregnancy. Information on place of residence, socioeconomic status, medical history and previous pregnancies was also obtained from the questionnaire. Samples of mother's blood, umbilical cord blood and the external part of the placenta were taken from each subject. The samples were prepared for analysis in the Clean Laboratory of the Trace Elements Unit of the Chemistry Department. The placenta samples were digested by nitric acid. Cadmium and lead concentrations were determined by graphite furnace atomic absorption spectrophotometry. A modifier matrix solution of 0.5 mL Triton X-100, 1 mL of hot nitric acid and 0.5 mg of phosphate di-ammonium hydrogen phosphate in 100 mL was used to dilute the samples. The standard solutions of 8.9 and 17.8 nmol/L of cadmium and 4.8 and 9.6 nmol/L of lead were prepared by dilution of 0.89 and 0.48 μmol/L commercial standards of these elements, respectively. The relationship between different variables was established by the Spearman Coefficient.

Results and Discussion

The average, minimum and maximum levels of cadmium and lead in mother's serum, umbilical cord serum and placenta, as well as the standard errors are shown in Tables 1 and 2.

There is a significant decrease ($p < 0.001$) in cadmium and lead levels in umbilical cord serum compared to mother's serum. The decrease is 4.4 ± 0.9 nmol/L for cadmium and 9.7 ± 1.4 nmol/L for lead. Similar results are reported by Soong et al. (1991). These results suggest the possible existence of a protective mechanism to protect the fetus against toxic elements. This protection could be found in the

Correct citation: Puente, M.J., Tapia, M.J., Navarro, I., Martin, A., and Villa, I. 1997. Lead and cadmium transference through the placenta in gestations. In *Trace Elements in Man and Animals – 9: Proceedings of the Ninth International Symposium on Trace Elements in Man and Animals. Edited by* P.W.F. Fischer, M.R. L'Abbé, K.A. Cockell, and R.S. Gibson. NRC Research Press, Ottawa, Canada. pp. 97–98.

placenta, the amniotic fluid or the amniotic membrane. It was not possible to establish a connection between the cadmium and lead levels in the placenta and the mother's and umbilical cord sera. Nevertheless, it appeared that an increase in cadmium level in the placenta occurred when the cadmium concentration in the umbilical cord decreased ($p < 0.10$), as shown in Tables 3 and 4. Further experimentation is necessary to clarify this point. An accumulation of cadmium in membranes and amniotic fluid was reported by Korpela et al. (1986). As the levels of these elements were not determined in these compartments, these results could not be corroborated.

Table 1. Mean concentration, standard error, minimum and maximum of cadmium.

Concentration	Mean value	Standard error	Minimum	Maximum
Cd in mother serum (nmol/L)	8.0	0.9	0.2	117.4
Cd in placenta (nmol/kg)	49.8	3.4	2.7	266.9
Cd umb. cord (nmol/L)	3.6	0.2	0.2	14.2

Table 2. Mean concentration, standard error, minimum and maximum of lead.

Concentration	Mean value	Standard error	Minimum	Maximum
Pb in mother serum (nmol/L)	26.5	1.4	0.2	96.5
Pb in placenta (nmol/kg)	133.7	12.0	2.9	1322.4
Pb umb. cord (nmol/L)	13.0	0.7	0.7	58.4

Table 3. Correlations between cadmium levels in the placenta and serum (maternal and umbilical cord).

	Cadmium in placenta	
Compartments	Correlation coefficient	Probability
Cd in maternal serum	– 0.07	0.4
Cd in umbilical cord	– 0.12	0.1

Table 4. Correlations between lead levels in the placenta and serum (maternal and umbilical cord).

	Lead in placenta	
Compartments	Correlation coefficient	Probability
Pb in maternal serum	0.03	0.75
Pb in umbilical cord	– 0.05	0.52

References

Korpela, H., Loueniva, R., Yrjdnheikki, E. & Kauppila, A. (1986) Lead and cadmium concentrations in maternal and umbilical cord blood, amniotic fluid, placenta, and amniotic membranes. Am. J. Obstet. Gynecol. 155: 1086–1088.

Schramel, P., Hasse, S & Ovcar-Pavlu, S. (1988) Selenium, cadmium, lead and mercury concentrations in human breast milk, in placenta, maternal blood, and the blood of the newborn. Biol. Trace Elem. Res. 15: 111–124.

Soong, Y.K., Tseng, R., Liu, C.H. & Lin, P.W. (1991) Lead, cadmium, arsenic, and mercury levels in maternal and fetal cord blood. J. Formosan Med. Assoc. 90: 59–65.

The Effect Maternal of Nutritional Copper Deficiency on the Skeletons of Newborn Rats. Radiography Study

J. CHAVARRI,[†] I. NAVARRO,[†] AND I. VILLA[‡]

[†]Pediatric Research Unit, University of Navarra, 31.080 Pamplona, and [‡]Department of Pediatrics, Gregorio Marañón Hospital, 28.009 Madrid, Spain

Keywords copper deficiency, skeleton, newborn, rat

Introduction

Many studies have shown that a copper deficient diet fed to pregnant rats affects the growth and development of their pups. It has been speculated that copper affects the regulatory mechanisms of ossification, specifically that it is required for completion of calcification (Dimauro 1990, Allen and Klevay 1988, Dolwett and Sorenson 1988). Others, however, suggest that copper does not have an effect on bone calcium and phosphorus levels (Chapman 1987, Mason 1979). The objective of this study was to determine the role of maternal copper deficiency on calcification of the bones of the offspring.

Material and Methods

The neonates (20 per group) of three groups of female Sprague-Dawley rats were examined radiographically. The copper deficient group was fed with a copper free diet ad libitum (AIN-76). The control group, received a diet supplemented with copper at 6 ppm ad libitum. The pair-fed group (n = 7), received a similar diet but with calories equivalent to the copper-deficient group. The radiological study was carried out using a Mamomat 2 X-ray machine (Siemens) with an 0. 15 mm extrafine lens and a 2 to 1 amplification with 25 kW and 20 mAs. The criterion of calcification was the number of ossification nuclei. Analysis of the variance (ANOVA) and Scheffe (parametric) and Kruskal Wallis (not parametric) tests were used to determine the significance of the data.

Results and Discussion

Following radiographic analysis, no malformations were evident, however there was a notable increase in the transparency of the bones in pups from the copper deficient rats. This suggests a decrease in the calcification of these bones. Furthermore, the copper deficient fetal rat tibias are significantly shorter in length in comparison to the control group and the pair-fed group (Table 1). The number of ossification nuclei was also reduced in the pups from the copper deficient rats (Table 2). These results suggest that maternal copper deficiency affects the development of the bones of the neonates.

Table 1. Radiological measurement of neonatal tibia lengths of the control (C), deficient (D) and pair-fed (P) groups.

	Deficient	Control	Pair-fed	ANOVA	Scheffe		
Parameter	(n = 16)	(n = 19)	(n = 20)	(p)	D/C	D/P	C/P
Length*(mm)	15.25±1.24	17.11±0.94	16.20±0.83	0.0001	**	**	ns

*Length is amplified 3 times.

Correct citation: Chavarri, J., Navarro, I., and Villa, I. 1997. The effect maternal of nutritional copper deficiency on the skeletons of newborn rats. Radiography study. In *Trace Elements in Man and Animals – 9: Proceedings of the Ninth International Symposium on Trace Elements in Man and Animals. Edited by* P.W.F. Fischer, M.R. L'Abbé, K.A. Cockell, and R.S. Gibson. NRC Research Press, Ottawa, Canada. pp. 99–100.

Table 2. Ossification nuclei of the control (C), deficient (D) and pair-fed (P) groups.

Parameter	Deficient (n = 15)	Control (n = 18)	Pair-fed (n = 20)	ANOVA (p)	Scheffe D/C	Scheffe D/P	Scheffe C/P
Ossif. nuclei	5.80±1.52	7.17±0.92	6.45±0.95	0.0045	**	ns	ns

References

Allen, K. & Klevay, L. (1988) Copper deficiency and cholesterol metabolism in rat. Atherosclerosis 31: 259–271.
Dimauro, S. (1990) Cytochrome c oxidase deficiency. Ped. Res. 28: 536–541.
Dolwett, H.H.A. & Sorenson, J.R.J. (1988) Roles of copper in bone maintenance and healing. Biol. Trace Element Res. 18: 39–47.
Chapman, S. (1987) Child abuse or copper deficiency. A radiological view. British Medical Journal 294: 1370.
Mason, K.E. (1979) Conspectus of research on copper metabolism and requirements of man. J. Nutr. 109: 1981–2037.

Copper Levels in Post-Term Newborns

J. QUINTERO,[†] I. NAVARRO,[†] A. MARTIN,[†] AND I. VILLA[‡]

[†]Pediatric Research Unit, University of Navarra, 31.080 Pamplona, and [‡]Department of Pediatrics, Gregorio Marañón Hospital, 28.009 Madrid, Spain

Keywords copper, status, post-term infants

Introduction

Copper is present in the human serum in both protein bound and free forms. It is involved in many metabolic processes, being required for the activation of enzymes. As an example, it is involved in collagen synthesis and defence against superoxide radicals. Copper levels in the umbilical cord of newborns are much lower than those in the adult. This may be due to the low levels of ceruloplasmin present in the newborn serum. After delivery, copper levels raise slowly in term infants as well as in premature ones.

Material and Methods

In a study of 266 pregnant women who gave birth in hospitals in Pamplona (Spain), eleven were post-term, with a gestational age of more than 41 weeks. The levels of copper in the maternal and cord arterial and venous sera, collected at birth, were analysed, as was a sample of placental tissue. The control group consisted of the 255 remaining mothers who gave birth to babies whose gestational age was under 41 weeks. All of the women had a normal nutritional status. Immediately before delivery, a sample of maternal blood was taken by puncture from the antecubital vein with a teflon intravenous cannula. The samples were collected in low density polyethylene tubes previously washed in a solution of ultra pure nitric acid and rinsed with ultra pure water. Umbilical cord blood was collected in a similar fashion. The blood was allowed to clot before being centrifuged. The placental tissue was obtained by cutting a fragment of approximately 5 g from the cotyledon side, and placing it in a small acid washed vessel. The tissue was then stored at –20°C until analysis.

Sample Preparation

All procedures were carried out in a clean laboratory with laminar flow. Serum was separated by centrifuge and was decanted into small polyethylene tubes with screw tops. It was then frozen at –20°C until analysis. A 1 g sample of placental tissue was cut from the centre of the collected tissue, using titanium scissors; it was weighed and then dried in an oven at 75°C for 48 h. The samples were then weighed again to determine the humidity level, and digested in hot ultrapure nitric acid in a high pressure digestion bomb (Milestone MLS 1200).

Determination of copper

The concentration of copper in the sera and placenta samples were determined using beam atomic absorption spectrophotometry (GBC 902 Double Beam Atomic Absorption spectrophotometer). All necessary precautions against environmental contamination were taken. The determination of trace elements was carried out in duplicate, using standard commercial serum (SRM 1598 NIST) as a quality control. Analysis of the data obtained was carried out using a Student's t-test.

Correct citation: Quintero, J., Navarro, I., Martin, A., and Villa, I. 1997. Copper levels in post-term newborns. In *Trace Elements in Man and Animals – 9: Proceedings of the Ninth International Symposium on Trace Elements in Man and Animals. Edited by* P.W.F. Fischer, M.R. L'Abbé, K.A. Cockell, and R.S. Gibson. NRC Research Press, Ottawa, Canada. pp. 101–102.

Results

As shown in Table 1, the copper levels in the umbilical cord sera collected from the post-term newborns is significantly higher than that collected from the control group. The maternal serum and placental levels were not different. These results are in agreement with the findings of Kiilholma et al. (1984). This elevation in cord blood copper levels may reflect the physiological phenomenon of increasing concentrations normally observed in the two months after delivery.

Table 1. Copper levels in placental tissue, maternal and umbilical sera of post-term newborns.

Sample	Controls <41 weeks	Post-term >41 weeks	P
Maternal serum (µmol/L)	35.3 ± 5.8	34.8 ± 6.9	0.751
Umbilical arterial (µmol/L)	8.3 ± 1.9	9.9 ± 2.7	0.009
Umbilical venous (µmol/L)	8.3 ± 1.9	10.0 ± 2.7	0.005
Placenta (µmol/kg)	14.0 ± 3.0	14.4 ± 3.3	0.688

References

Kiilholma, P., Erkkola, R., Pakarinen, P., Gronroos, M. (1984) Trace metals in postdate pregnancy. Gynecol. Obstet. Invest. 18: 45–48.

Assessment of Trace Elements Status: Opinions and Prejudices

Chair: N.W. Solomons

The Effect of Zinc Deficiency on Erythrocyte Deformability as Measured as a Function of Increasing Shear Stress in the Ektacytometer

L.J. ROBINSON,[†] R.T. CARD,[‡] AND P.G. PATERSON[†]

[†]College of Pharmacy and Nutrition and [‡]Department of Medicine, University of Saskatchewan, Saskatoon, Saskatchewan, Canada S7N5C9

Keywords zinc, erythrocyte deformability, plasma membrane, ektacytometer

The detection of marginal zinc deficiency has been difficult due to a lack of a specific sensitive biochemical indicator of zinc status (Lee and Nieman 1993). We proposed that erythrocyte deformability would be a sensitive functional indicator of zinc status. This is based on the hypothesis that zinc has a physiological role in the plasma membrane of mammalian cells (Bettger and O'Dell 1993). Although the biochemical functions of zinc in the membrane have not been fully elucidated, abnormalities in the erythrocyte membrane in zinc deficiency have been reported (Bettger and O'Dell 1993). These changes could alter erythrocyte deformability by influencing membrane flexibility. While previous studies have demonstrated minimal effects of zinc status on red blood cell deformability (Paterson and Card 1993), the techniques used have been limited. Our objective was to investigate the influence of zinc deficiency on erythrocyte deformability measured as a function of increasing shear stress in the ektacytometer. It was anticipated that this technique would be more sensitive to changes in membrane flexibility.

Materials and Methods

Weanling, male Sprague-Dawley rats (48–54 g) were fed *ad libitum* modified AIN-93 (Reeves et al. 1993) diets containing 3 mg zinc/kg diet (–Zn; n = 10) for six weeks. Control animals were either pair-fed (+ZnPF; n = 10) or fed *ad libitum* (+ZnAL; n = 9) the same diet supplemented with 50 mg zinc/kg diet. Feed intake was measured daily and weight gain weekly. At 6 weeks, tibia and plasma were collected for zinc analysis. Erythrocyte deformability was measured on whole blood samples as a function of increasing shear stress in the ektacytometer. Erythrocyte suspensions were also incubated at 48°C for 6 min; this treatment decreases membrane elasticity (Rakow and Hochmuth 1975). Elongation index, the ratio of length to width of the diffraction pattern of deforming cells, was measured.

Results

Weight gain, feed intake and feed efficiency were significantly depressed in the –Zn group as compared to the +ZnAL group (Table 1). Dermatological lesions were present in the zinc-deficient group.

Correct citation: Robinson, L.J., Card, R.T., and Paterson, P.G. 1997. The effect of zinc deficiency on erythrocyte deformability as measured as a function of increasing shear stress in the ektacytometer. In *Trace Elements in Man and Animals – 9: Proceedings of the Ninth International Symposium on Trace Elements in Man and Animals. Edited by* P.W.F. Fischer, M.R. L'Abbé, K.A. Cockell, and R.S. Gibson. NRC Research Press, Ottawa, Canada. pp. 103–104.

*Table 1. The effect of dietary zinc deficiency on growth.**

	–Zn group	+ZnPF group	+ZnAL group
Weight gain (g)	173 ± 4[a]	183 ± 6[a]	329 ± 10[b]
Feed intake (g)	518 ± 10[a]	511 ± 10[a]	808 ± 20[b]
Feed efficiency	0.333 ± 0.004[a]	0.357 ± 0.007[b]	0.407 ± 0.007[c]

*Results expressed as mean ± SEM. Statistical analysis was a one-factor ANOVA followed by Fisher's Protected LSD. Means with different superscripts (a, b and c) are significantly different ($P < 0.0001$). Zinc analysis is ongoing, but mean (± SEM) plasma zinc values (μg/mL) from a pilot study (n = 3–4) were: –Zn = 0.38 ± 0.05; +ZnPF = 1.50 ± 0.09 ; +ZnAL = 1.83 ± 0.11.

As analyzed by one-factor ANOVA, maximum elongation index of the erythrocytes was not significantly altered by zinc deficiency. The mean (± SEM) values of EImax for the three groups were: –Zn = 0.51 ± 0.01, +ZnPF = 0.51 ± 0.01 and +ZnAL = 0.52 ± 0.01. No effect of zinc deficiency on maximum elongation index was found for the heat treated cells.

Discussion

The results of this study suggest no effect of moderate zinc deficiency on erythrocyte deformability measured as a function of increasing shear stress in the ektacytometer. Maximum elongation index was unaltered, suggesting that zinc status did not influence average deformability of the cell population (Johnson 1989). There was also no effect of zinc when the erythrocytes were heated to increase the sensitivity of the assay by decreasing membrane elasticity (Rakow and Hochmuth 1975). Further analysis of the slope of elongation index plotted as a function of log shear stress will provide more specific information on the contribution of the membrane to deformability (Johnson 1989). Future assays based on membrane deformability of resealed ghosts may be sensitive to zinc status.

Acknowledgments

This work was funded in part by the Natural Sciences and Engineering Research Council of Canada.

Literature Cited

Bettger, W. J. & O'Dell, B. L. (1993) Physiological roles of zinc in the plasma membrane of mammalian cells. J. Nutr. Biochem. 4: 194–207.

Lee, R. D. & Nieman, D. C. (1993) Nutritional Assessment. Wm. C. Brown Communications, Madison, WI.

Heath, B. P., Mohandas, N., Wyatt, J. L. & Shohet, S. B. (1982) Deformability of isolated red blood cell membranes. Biochim. Biophys. Acta 6: 439–453.

Johnson, R. M. (1989) Ektacytometry of red blood cells. Methods Enzymol. 173: 35–54.

Paterson, P. G. & Card, R. T. (1993) The effect of zinc deficiency on erythrocyte deformability in the rat. J. Nutr. Biochem. 4: 250–255.

Rakow, A. L. & Hochmuth, R. M. (1975) Effect of heat treatment on the elasticity of human erythrocyte membrane. Biophys. J. 15:1095–1100.

Reeves, P. G., Nielsen, H. & Fahey, G. C. (1993) AIN-93 purified diets for laboratory rodents final report of the American Institute of Nutrition ad hoc writing committee on the reformulation of the AIN-76A rodent diet. J. Nutr. 123: 1939–1951.

Erythrocyte Membrane Enzymes as Indicators of Zinc Status

T. LARSEN

Danish Institute of Animal Science, Department of Animal Health and Welfare, P.O. Box 39, 8830 Tjele, Denmark

Keywords phosphatases, 5′-nucleotidase, pigs

Introduction

The essentiality of zinc in nutrition has been recognized for decades. Diagnosis of mild deficiency of zinc in the organism, however, still is an unsolved problem in man and animals. Assessment of zinc status suffers from lack of an accessible tissue which reflects the zinc content of the organism. No single, specific, and sensitive index of zinc status exists. Therefore, a combination of biochemical and physiological functional indices is frequently used.

However, Johanning and O'Dell (1988) reported that the activity of 5′-nucleotidase in the erythrocyte membrane reflected the zinc intake in rats and pigs. Similarly, Ruz et al. (1992) tested several parameters during a zinc depletion–repletion study with humans. They reported that alkaline phosphatase activity in the erythrocyte membranes might be a potential index of zinc status in humans.

The present study, a controlled animal experiment, was carried out in order to investigate whether different erythrocyte membrane enzymes (alkaline and neutral phosphatases and 5′-nucleotidase), could be useful in assessing zinc index in swine.

Materials and Methods

Eight crossbred castrated male pigs were fed equal rations just beneath the expected *ad libitum* intake. The animals were allocated to two dietary groups. The basal diet was similar. One group was fed the basal diet not further supplemented with dietary zinc (the Zn0 group, 28 mg Zn per kg). The other group was fed the basal diet supplemented with 100 mg zinc oxide per kg (the Zn100 group, 113 mg Zn per kg) to reach the recommended level of feed zinc (Danish recommendations 1991) for the animals concerned.

Blood was drawn from the pigs once a week in heparinized glass tubes for 8 weeks. Plasma was harvested after centrifugation and blood cells were repeatedly washed in an equal volume of isotonic phosphate buffer. Erythrocyte membranes were prepared principally according to the method outlined by Steck et al. (1970). Blood cell suspensions were lysed in cold phosphate buffer, pH 8.0. The lysate was centrifuged, the supernatant was discarded and the process repeated four times. The resulting EM suspension was disintegrated by sonication for 15 s immersed in an ice cold water bath, before it was used in assays.

Protein in the erythrocyte membrane (EM) suspension was determined according to standard laboratory methods. EM 5′-nucleotidase activity (EM-AMP) was measured as AMP hydrolysing activity at pH 7.4. Liberated ortho-phosphate was analyzed according to the vanado-molybdate method. EM alkaline phosphatase (EM-AP, pH 10.3) and EM neutral phosphatase (EM-NP, pH 7.4) were measured in buffers supplied with 0.5 mmol/L co-factor (Mg^{++}) and 10 mmol/L substrate (pNPP). Incubations were accomplished immersed in 37°C agitated water. Enzyme activities were assessed as nmol product (pNP or orthophosphate)/mg protein · min.

Correct citation: Larsen, T. 1997. Erythrocyte membrane enzymes as indicators of zinc status. In *Trace Elements in Man and Animals – 9: Proceedings of the Ninth International Symposium on Trace Elements in Man and Animals. Edited by* P.W.F. Fischer, M.R. L'Abbé, K.A. Cockell, and R.S. Gibson. NRC Research Press, Ottawa, Canada. pp. 105–106.

Results

The final weight of the animals was, after 60 d on the experimental diets, on average 45.2 kg (SD 3.8). This represents a gross daily weight gain of 490 g/d and 520 g/d for the unsupplemented diet and the Zn supplemented diet, respectively. The difference in weight gain is not statistically significant.

EM-AP activity was not affected by time in experiment, neither considered all together nor individually for the dietary groups. The Zn0 group and the Zn100 group revealed an activity of 14.0 and 11.0 nmol P liberation/mg protein · min, respectively. Statistical tests did not expose differences between groups. A similar conclusion holds for the neutral phosphatase activity in the erythrocyte membrane. The Zn0 group on average liberated 14.7 nmol P/mg protein · min, whereas the Zn100 group liberated 12.3 nmol P/mg protein · min under the experimental conditions.

5′-nucleotidase activity in the erythrocyte membrane corresponded on average to 0.55 nmol and 0.62 nmol P liberation/mg protein · min for the Zn0 and Zn100 groups, respectively. The observations from the Zn100 group revealed a significantly declining activity with time in experiment ($p < 0.05$), whereas the decline in the Zn0 group was less pronounced ($P < 0.10$). The statistical test between dietary groups did not support the view that differences were significant.

The EM-AP, EM-NP and EM-AMP activities, however, were highly correlated mutually ($p < 0.001$).

Discussion

None of the parameters tested in this study revealed significant differences between the two dietary groups. At least two interpretations of this fact are possible i) none of the parameters tested are useful as sensitive indices of zinc status, or ii) both groups of animals were well supplied with dietary zinc, so that body zinc stores were never depleted and a depression of zinc dependent enzymes never became actualized.

It is striking, however, how highly the three EM parameters in this study are mutually correlated, leading to considerations that the activity measured is from one enzyme only. Against this hypothesis stands the fact that the gap in pH levels, i.e., from 7.4 to 10.3 is enormous, wider than commonly seen for one enzyme.

Animal trials of the present category suffer from methological difficulties that may end in vicious circles: If the feed is too low in (especially) zinc, the animals lose appetite and refuse to eat — subsequently they will not grow at a satisfactory rate (compared to controls) and their zinc requirement will accordingly decrease. Artificial diets, formulated to be low in zinc, may trigger this pattern. On the other hand, natural occurring feeds, like the present, consisting of adequate protein of a high biological value will inevitably contain a relatively high level of intrinsic zinc. Although the present intrinsic zinc level is below 30% of the recommended level for the animals under consideration, it is obviously not sufficiently low to give any effects on weight gain of the animals or parameters believed to be zinc indices.

Literature Cited

Johanning, G.L. & O'Dell, B.L. (1988) Effect of zinc deficiency on erythrocyte (RBC) plasma membrane enzyme activities. FASEB J. 2: A 636.

Ruz, M., Cavan, K.R., Bettger, W.J. & Gibson, R.S. (1992) Erythrocytes, erythrocyte membranes, neutrophils and platelets as biopsy materials for the assessment of zinc status in humans. Br. J. Nutr. 68: 515–527.

Steck, T.L., Weinstein, R.S., Straus, J.H. & Wallach, D.F.H. (1970) Inside-out red cell membrane vesicles: preparation and purification. Science 168: 255–257.

Zinc Metabolism in Insulin Dependent Diabetic Women

E.B. FUNG,† L.D. RITCHIE,† L.R. WOODHOUSE,† J.C. KING,† AND R. ROEHL‡

†Department of Nutritional Sciences, University of CA, Berkeley, CA 94720, and the ‡CA Public Health Foundation, Berkeley, CA 94704, USA

Keywords zinc, diabetes, zinc absorption, women

Introduction

Studies have shown that diabetics excrete more protein, glucose and minerals in their urine compared to non-diabetics. Hyperzincuria is a common finding in both well and poorly controlled diabetics. Factors governing renal zinc (Zn) losses in diabetic patients have not been fully elucidated, nor is it clear that this excessive Zn excretion affects overall Zn status. Inconsistent reports of plasma Zn values do not add clarification (Kumar et al. 1974, Heise et al. 1988). To maintain Zn status in the presence of chronic hyperzincuria, dietary Zn intake and/or Zn absorption would need to be increased and/or endogenous losses would need to be decreased. Zn intake does not differ between diabetic and nondiabetic individuals (Heise et al. 1988). Apparent Zn absorption in diabetic rats was 3-fold higher compared to non-diabetic rats (Craft and Failla 1983), but the absorption of an oral dose of ^{65}Zn in ten insulin dependent diabetic (IDDM) subjects tended to be lower ($p < 0.01$) compared to controls (Kiilerich et al. 1990). The purpose of this study was to measure Zn absorption, using stable isotopes of Zn, in IDDM and non-diabetic women. Other indices of Zn status and glycemic control were also measured and related to Zn absorption.

Methods

Six insulin dependent diabetic (IDDM) and ten non-diabetic women (Control) were recruited from the San Francisco Bay Area and invited to participate in the study. Each subject gave informed written consent. Average age and body mass index for the IDDM and Control groups were: 28 ± 2 vs. 31 ± 1 years (Mean ± SEM) and 24.7 ± 1.1 vs. 21.9 ± 0.8 kg/m^2, respectively. The average age of diabetes diagnosis was 16 ± 4 years and average insulin dose, 0.54 ± 0.12 U/kg. Glycosylated hemoglobin averaged 10.2 ± 1.7% for the IDDM group. Two weeks prior to and two weeks following the test day, subjects were advised to take a 4 mg Zn tablet (as $ZnSO_4$) in the evening. All subjects consumed an identical standard meal the night prior to each test day and fasted from 10 pm. On the test day, height and weight were taken, a fasting blood sample drawn and a 24-h urine sample collected. The oral isotope (^{68}Zn, 3.0 mg) was given with breakfast, and the intravenous (IV) isotope (^{70}Zn, 0.8 mg) infused 25 minutes after the meal. Spot morning urine samples were collected from days 1–10 and blood samples drawn on day 3, 5 and 7 post infusion.

Food and beverage intakes were weighed and recorded for 3 nonconsecutive days. Nutrient intake was estimated using a computerized nutrient database (Nutritionist III). The concentration of Zn in plasma (PZn), erythrocyte (RBCZn) and urine samples (UZn) was determined using AAS after dilution with HNO_3. The fraction of plasma Zn bound to albumin (PZn_{alb}) was determined by a modified ultra-filtration technique previously described by Lowe et al. (1996).

Fractional absorption of Zn (FZA) was determined using a dual stable isotope technique (Friel et al. 1992). Zn was purified from urine and plasma samples using ion exchange chromatography (Friel et al. 1992). Isotopic ratios were determined using Inductively Coupled Plasma Mass Spectrometry. FZA was calculated from both urine and plasma enrichments and data averaged. The size of the metabolic pools of

Correct citation: Fung, E.B., Ritchie, L.D., Woodhouse, L.R., King, J.C., and Roehl, R. 1997. Zinc metabolism in insulin dependent diabetic women. In *Trace Elements in Man and Animals – 9: Proceedings of the Ninth International Symposium on Trace Elements in Man and Animals. Edited by* P.W.F. Fischer, M.R. L'Abbé, K.A. Cockell, and R.S. Gibson. NRC Research Press, Ottawa, Canada. pp. 107–109.

Zn that exchange with the plasma in 2 d (EZP) was determined as previously described (Miller et al. 1994). The mass of the IV isotope administered was divided by the y-intercept of the linear regression of a semi-log plot of the plasma enrichment data from day 3 to 7.

Data from the two groups were compared using a two-tailed, unpaired Student's T-test. Correlations were determined using Pearson's correlation coefficient. All data were considered significant at $p < 0.05$, and presented as Mean ± SEM.

Results

Energy intake was not different between the two groups. Dietary Zn intake was similar (9.0 ± 1.4 vs. 9.5 ± 0.6 mg/d) for IDDM and Control respectively, though calcium intake was 37% lower ($p = 0.006$) in the IDDM group. There was a strong positive correlation observed between FZA and UZn ($r = 0.904$, $p = 0.01$) and FZA and PZn_{alb} ($r = 0.829$, $p = 0.04$) as well as a negative correlation between FZA and DZn ($p = -0.831$, $p = 0.039$) for the IDDM group.

Discussion

This is the first study to measure Zn absorption using dual stable isotopes in IDDM women along with other measures of Zn status. Our results support previous studies which found urinary Zn excretion to be elevated in IDDM compared to Control subjects (Table 1). The elevation was not associated with an increase in urine volume or creatinine clearance but was positively associated with fractional zinc absorption. It has been hypothesized that glycosylated amino acids or proteins in the urine chelate Zn and decrease its reabsorption within the kidney tubule. The form of Zn which exists in abundance in the urine of diabetics has not been identified.

The elevation of RBCZn in the IDDM group may be secondary to either an acute phase response of metallothionein, which is known to bind Zn, and/or an elevated level of the Zn dependent enzyme, carbonic anhydrase. If Zn is sequestered in tissues of diabetics, we might expect an increase in the proportion of plasma Zn bound to albumin. However, an increase in PZn_{alb} was not observed, nor was there an increase in the size of the exchangeable pools of Zn.

There was considerable variation in FZA among the IDDM subjects, though 70% may be explained by the variability in dietary Zn intake. We did observe a trend towards a reduction in absorbed Zn (FZA*dietary Zn intake) in the IDDM group compared to the Control. Crude Zn balance was calculated from these data combined with literature values for endogenous Zn losses (Baer and King 1984) modified for our subjects body weight. We found IDDM subjects to be in negative Zn balance (–11.8 μmol/d) compared to Control subjects (3.4 μmol/d).

From these data, it does not appear that IDDM reduced the Zn status of these 6 women. However, we did not measure functional indices of Zn status, ie. immune function, nor did we measure endogenous fecal Zn, the largest route of Zn excretion. Endogenous GI Zn excretion should be measured with Zn absorption in future studies to explore how diabetics conserve zinc and maintain homeostasis with chronic hyperzincuria.

Table 1. Circulation, excretion, and intestinal absorption of zinc.

	IDDM (n = 6)	Control (n = 10)	P-value
PZn, μmol/L	11.8 ± 0.4	12.0 ± 0.6	NS
PZn_{alb}	73 ± 3	76 ± 5	NS
RBCZn, μg Zn/g protein	34.5 ± 2.7[b]	20.8 ± 1.1[a]	0.0001
UZn, μmol/d	12.1 ± 1.6[b]	5.7 ± 0.9[a]	0.005
Urine Volume, mL/d	2394 ± 292	2021 ± 231	NS
EZP, μmol/d	3.14 ± 0.11	3.17 ± 0.23	NS
FZA, %	13.3 ± 3.0	16.1 ± 1.9	NS
Absorbed Zn, μmol/d	15.3 ± 1.8	23.7 ± 3.3	0.07

Literature Cited

Baer, M.T., King, J.C. (1984) Tissue Zn levels and Zn excretion during experimental Zn depletion in young men. Am. J. Clin. Nutr. 39:556–570.

Craft, N.E., Failla, M.L. (1983) Zn, iron and copper absorption in the streptozotocin-diabetic rat. Am. J. Physiol. 244:E122–128.

Friel, J.K., Naake, V.L., Miller, L.V., Fennessey, P.V., Hambidge, M.K. (1992) The analysis of stable isotopes in urine to determine fractional absorption. Am.J. Clin. Nutr. 55:473–477.

Heise, C.C., King, J.C., Costa, F.M., Kitzmiller, J.L. (1988) Hyperzincuria in IDDM women. Diabetes Care 11:780–786.

Kiilerich, S., Hvid-Jacobsen, K., Vaag, A., Sorensen, S.S. (1990) 65Zn absorption in patients with IDDM assessed by whole body counting technique. Clinica Chimica Acta 189:13–18.

Kumar, S., Rao, K.S.J. (1974) Blood and UZn levels in diabetics. Nutr. Metab. 17:231–235.

Lowe, N.M., Lam, M.V., King, J.C. (1996) Effect of Zn depletion on the distribution of Zn between plasma-binding proteins by an ultrafiltration method. Am. J. Clin. Nutr.(submitted).

Miller, L.V., Hambidge, K.M., Naake, V.L., Hong, Z., Westcott, J.L., Fennessey, P.V. (1994) Size of the Zn pools that exchange rapidly with plasma Zn in humans. J. Nutr. 124:268–276.

The Copper and Zinc Status of Sheep under Different Management Systems in Malaysia

MUSTAPHA M. NOORDIN AND ABU B. ZUKI

Faculty of Veterinary Medicine and Animal Sciences Universiti Pertanian Malaysia, 43400 UPM, Selangor Darul Ehsan, Malaysia

Keywords copper, zinc, grazing sheep, management system

Introduction

Copper and zinc are two important trace elements required for the well being of animals with excessive and inadequate intake leading to toxicities and deficiencies respectively. In Malaysia, little attention has been given to the mineral status of sheep, particularly Cu, Fe and Zn (Noordin 1995). Thus, the present study was undertaken in an attempt to assess the Cu and Zn status of small ruminants under different management systems.

Materials and Methods

The study was carried out on five different farms adopting three different management systems. Only adult sheep were selected. Basically the management systems were identified as intensive when there is cut and carry, herd separation, controlled mating and concentrate supplementation, semi-intensive where there is free grazing for 6 h without herd separation, uncontrolled mating system and with or without concentrate supplementation; and extensive where there is confined grazing for at least 8 h without herd separation, uncontrolled mating and concentrate supplementation.

Glassware and other containers were soaked in detergent,washed three times with double distilled water (DDW), soaked in 0.002 M EDTA, rinsed in DDW, dried, and then kept in sealed plastic bags until further used.

Plasma, serum, wool (dorsum) and feed (grass and concentrate) were taken from each farm, digested using the double acid method (Ishmael et al. 1972) and the concentration of Cu and Zn were analysed using an Atomic Absorption Spectrophotometer (Varian 400, Varian Australia Pty Ltd). Results obtained were analysed for significance using a one-way ANOVA.

Results

The concentration of Cu and Zn in the wool, serum and plasma of sheep are given in Tables 1 and 2 respectively. The concentration of Cu in the wool of sheep from Ulu Langat was markedly higher ($p < 0.05$) than in any other farm. The concentration of Cu in the wool was approximately 3 to 5 times

Table 1. The concentration of Cu in the wool, plasma and serum of sheep in different farms (μg/g).

Sample	Ladang 2 (n = 5)	Ulu Langat (n = 5)	Puchong (n = 4)	Dengkil (n =7)
Wool	4.6 ± 1.3^a	11.8 ± 6.8^a	4.7 ± 0.8^a	9.9 ± 6.5^b
Serum	1.2 ± 0.3^a	2.1 ± 0.8^b	1.3 ± 0.2^a	4.2 ± 6.4^a
Plasma	1.1 ± 0.2^a	2.2 ± 1.2^b	0.9 ± 0.3^a	2.2 ± 1.5^a

*Values between columns having similar superscripts do not differ at $p < 0.05$, Mean ± SD.

Correct citation: Noordin, M.M., and Zuki, A.B. 1997. The copper and zinc status of sheep under different management systems in Malaysia. In *Trace Elements in Man and Animals – 9: Proceedings of the Ninth International Symposium on Trace Elements in Man and Animals. Edited by* P.W.F. Fischer, M.R. L'Abbé, K.A. Cockell, and R.S. Gibson. NRC Research Press, Ottawa, Canada. pp. 110–112.

Table 2. The concentration of Zn in the wool, serum and plasma of sheep in different farms (µg/g).

Sample	Ladang 2 (n = 5)	Ulu Langat (n = 5)	Puchong (n = 4)	Dengkil (n = 7)
Wool	267.3 ± 134.9[a]	103.9 ± 24.9[b]	73.4 ± 47.6[b]	115.9 ± 34.9[b]
Serum	3.4 ± 1.0[a]	1.4 ± 0.5[a]	2.4 ± 0.8[a]	2.6 ± 3.5[a]
Plasma	3.4 ± 1.4[a]	2.5 ± 3.9[b]	6.7 ± 5.8[b]	1.8 ± 2.6[a]

*Values between columns having similar superscripts do not differ at $p < 0.05$, Mean ± SD.

higher than those found in the serum and plasma. The wool Zn concentration of sheep from Ladang 2 was significantly different ($p < 0.05$) than in any other farm. Sheep from Puchong had higher ($p < 0.05$) serum Zn concentration than in any other farm. Plasma Zn concentration between each farm was similar. The concentration of the Zn in the wool of sheep was about 35 to 100 times higher than those found in serum and plasma.

The Cu and Zn concentration of feed are given in Table 3. The Zn concentration of grass from Ladang 14 was higher and the grass Cu concentration in Dengkil was much higher than in any other farm. The concentration of Zn in feed was about 3 to 9 times higher than Cu.

The concentration of Zn in feed did not correlate ($p > 0.05$) with that in wool ($r = -0.301$), serum ($r = -0.169$ and -0.004) and plasma (-0.348 and 0.297). However, there were correlations between the concentration of Cu in feed ($p < 0.05$) and that in wool ($r = 1.000$), serum ($r = 0.999$) and plasma ($r = 0.987$) of sheep. The concentration of Cu in serum and plasma ($r = 0.870$), and between wool and serum ($r = 0.974$), were positively correlated ($p < 0.05$). The concentration of Zn in serum and plasma also showed a positive correlation ($r = 0.868$) ($p < 0.05$).

Table 3. The type of feed and concentrations of Cu and Zn in different farms (µg/g, dry weight).

Location	Type of feed	Cu	Zn
Ladang 2	Grass + concentrate	9.3	30.1
Ladang 14	Grass	9.3	88.7
Puchong	Grass	9.7	27.2
Dengkil	Grass	29.7	76.2

Discussion

This is the first study on the status of Cu and Zn in small ruminants under different management systems in Malaysia. Biopsy samples of the liver and kidney are better indicators of the trace element status (McDowell 1987), but could not be carried out due to lack of facilities.

Beck (1963) reported that the concentration of blood Cu in sheep is between 0.5 to 1.5 µg/mL, similar to this study, but, slightly higher than the value reported by Grace (1983) (0.8 µg/mL). This could also indicate that serum and plasma are stable pools of Cu.

Grace (1983) reported Cu concentration in wool of sheep as 7.0 µg/g and 9.0 µg/g, lower than those reported by Stevenson and Wickham (1976) (33 µg/g). Thus, it appeared that there are marked differences in the concentration of Cu in wool of sheep in different parts of the world. Although proven to be true for Zn (Underwood 1977), the same reason i.e great proportion of wool or hair covering does indicate more Cu is required and accumulated.

Perdomo et al. (1977) reported that the concentration of Cu and Zn in four grasses of different ages were between 11–16 mg/kg and 26–37 respectively. However, the values obtained in this study had a slightly higher Zn concentration, but were within safety limits of consumption, 5–30 µg/g (Howell 1977, Humpreys 1996).

Kirchgesner (1993) claimed that the concentration of trace elements in the feed should not be used as indicator for the status of animals. This is proven to be true in our study, as tissues taken for the assessment of Cu and Zn did not give good correlation with those in the feed.

The concentration of Zn in wool found in this study was much higher than the Cu concentration, and the value slightly lower than that reported by Grace (1983), but was similar with that found by Stevenson and Wickham (1976). Thus, the Zn concentration of wool differs according to the breed, age, sex and locality. Underwood (1977) reported that in animals covered with hair or wool, a substantial proportion of total body Zn is present in these tissues which is in accord with the finding of this study.

Since this study did not reveal any deficiency in the concentration of Cu and Zn in any farm, no suggestion for improving the status of these elements was made. It could also indicate that the level of these elements in feed in each farm was adequate. This study also suggested that the serum and plasma did not give a good reflection of the concentration of Cu and Zn in the body of the animal. Further studies should be carried out in order to assess the status of other trace elements of small ruminants in other locations and in other tissues particularly liver and kidneys.

References

Beck, A.B. (1963) The copper metabolism of warm-blooded animals which special reference to the rabbit and the sheep. Aust. J. Agric. Res. 14: 129–141.

Grace, N.D. (1983) Amounts and distribution of mineral elements associated with fleece-free empty body weight gains in the grazing sheep. N.Z. J. Agric. Res. 26: 59–70.

Howell, J. McC. (1977) Chronic copper toxicity in sheep. Vet. Ann. 17: 70–73.

Ishmael, J., Gopinath, C. & Howell, J. McC. (1972) Experimental chronic copper toxicity in sheep; Histological and histochemical changes during the development of lesions in the liver. Res. Vet. Sci. 12: 358–366.

McDowell, L.R. (1987) Assessment of mineral status of grazing ruminants. World Rev. Ani. Prod. 4:19–32.

Noordin, M.M. (1995) The status of trace mineral research in Malaysia: Problems and Solutions. ACIAR Proc. (in press).

Perdomo, J.T., Shirly, R.L. & Chicco, C.F. (1977) Availability of nutrient minerals in four tropical forages fed chopped to sheep. J. Ani. Sci. 45: 114–1119.

Underwood, E.J. (1977) Copper. In: Trace Element in Human and Animal Nutrition. 4th Ed., pp. 56–233. Academic Press, New York, NY.

Polymorphonuclear Leukocyte Copper Declines with Low Dietary Copper

J.R. TURNLUND,[†] T.M. SAKANASHI,[‡] AND C.L. KEEN[‡]

[†]USDA/ARS, Western Human Nutrition Research Center, San Francisco, CA 94129, and
[‡]Department of Nutrition, University of California, Davis, CA 95616, USA

Keywords copper, status, polymorphonuclear leukocyte

The search for a reliable, sensitive index of marginal copper status has continued for many years. Studies conducted with multiple levels of dietary copper demonstrated that copper intake must be very low to observe changes in the traditional indices of copper status, such as plasma levels of copper, ceruloplasmin, and superoxidase dismutase. No differences in these indices were observed when dietary intake ranged from 0.8 to 7.5 mg Cu/d (Turnlund et al. 1990). However, these indices declined when diets of young men contained only 0.4 mg Cu/d (Turnlund et al. 1994). Therefore, the sensitivity of polymorphonuclear leukocyte (PMN) copper concentration as an index of copper status was evaluated in blood samples collected at the same times as samples were collected for determination of traditional indices of copper status.

Methods

Eleven young men between the ages of 21 and 32 years were confined to a metabolic research unit for 90 d. They were supervised by nursing staff at all times. Their diet, a 3 d rotating menu, contained 0.66 mg Cu/d (marginal copper) for 24 d, 0.38 mg/d for 42 d (depletion) and 2.48 mg/d (repletion) for 24 d. Copper status, as assessed by plasma copper and ceruloplasmin concentrations, declined during the 42 d of copper depletion and increased after 21 d of repletion (Turnlund et al. 1994). Samples were collected for PMN copper concentration determinations at the beginning of the study, the midpoint and end of the copper depletion period, and the end of the study. PMN were isolated (Vruwink et al. 1991), washed and analyzed for copper and total protein. Several samples were contaminated with copper, with values much higher than the others and literature values. Those values are not included. Statistical analysis was performed with SAS (Anonymous 1989) using an ANOVA model and the LSD test.

Results

PMN copper concentrations are shown in Table 1. PMN copper was 125 ± 61 nmol/g protein (mean ± SD) at the beginning of the study, 91 ± 42 nmol/g at the midpoint of depletion, 54 ± 27 nmol/g at the end of depletion, and 79 ± 35 nmol/g at the end of repletion. PMN copper declined significantly during depletion and was significantly lower ($p < 0.05$) at end of depletion than the beginning of the study. The mean value was higher at the end of repletion than at the end of depletion, but was not significantly different ($p = 0.055$).

Discussion

The response of PMN copper appears to be slower than that of plasma copper and ceruloplasmin. Sample processing for the assay is considerably more difficult and is susceptible to contamination. However, PMN copper concentration may be the assay of choice when assessing copper status during

Correct citation: Turnlund, J.R., Sakanashi, T.M., and Keen, C.L. 1997. Polymorphonuclear leukocyte copper declines with low dietary copper. In *Trace Elements in Man and Animals – 9: Proceedings of the Ninth International Symposium on Trace Elements in Man and Animals. Edited by* P.W.F. Fischer, M.R. L'Abbé, K.A. Cockell, and R.S. Gibson. NRC Research Press, Ottawa, Canada. pp. 113–114.

Table 1. PMN copper concentration (nmol Cu/g protein) during copper depletion and repletion.

Subject	Study day 1	Mid-depletion	End depletion	End repletion
1	224	88	67	53
2		93	31	58
3		151	58	64
4		158	113	154
5		40	26	82
6	164	90	66	116
7	94	43	46	56
8		110	26	84
9	127		33	83
10	85	101	82	93
11	57	38	39	126
Mean ± SD	125 ± 61	92 ± 42	54 ± 27	79 ± 35

acute disease or infection. It would be less susceptible to the rapid increases observed in plasma copper and ceruloplasmin concentrations during acute disease.

References

Anonymous (1989) SAS/STAT Users Guide, Version 6. Fourth: SAS Institute, Inc. Cary NC.

Turnlund, J.R., Keen, C.L., Sakanashi, T.M., Jang, A.M., Keyes, W.R. & Peiffer, G.L. (1994) Low dietary copper and the copper status of young men. FASEB J. 8: A816 (abs.).

Turnlund, J.R., Keen, C.L. & Smith, R.G. (1990) Copper status and urinary and salivary copper in young men at three levels of dietary copper. Am. J. Clin. Nutr. 51: 658–664.

Vruwink, K.G., Fletcher, M.P., Keen, C.L., Golub, M.S., Hendrickx, A.G. & Gershwin, M.E. (1991) Moderate zinc deficiency in rhesus monkeys: an intrinsic defect in neutrophil chemotaxis corrected by zinc repletion. J. Immunol. 146: 244–249.

Urinary Selenium Excretion as Indicator of Selenium Status

I. LOMBECK,[†] H. MENZEL,[‡] F. MANZ,[¶] AND V. MOSTERT[†]

[†]University Children's Hospital and [‡]Institute of Toxicology, Düsseldorf, and [¶]Institute of Child Nutrition, Dortmund, Germany

Keywords selenium, urine, human milk, formula

The Se state is mainly dependent on variations in the urinary Se excretion and not on the intestinal Se absorption. The regulating mechanism is not known. Urinary Se excretion is influenced by the amount and chemical form of the ingested Se, the contribution of other constituents in food or body fluids as well as body stores and growth. An increase in the Se supply often leads to an increased urinary Se excretion.

After acute Se intoxication, urinary Se parallels plasma Se. Peak excretion occurs within the first 6 h (Lombeck et al. 1987). Only scarce information about the chemical compound excreted in the urine is available. Chronic Se poisoning may be monitored also by measuring the renal Se output.

Due to the difficulties in blood sampling from infants, we investigated if, and under which conditions, urinary Se is an indicator of the actual Se state.

Materials and Methods

Se was measured by hydride generation atomic absorption spectrometry after stepwise wet digestion of the samples in duplicates. Accuracy and precision was checked by using reference material (e.g., urine Nycomed 621 ± 24 nmol/L, n = 6) or urine pool (28 ± 2 nmol/L, n = 6).

In 8 male preterm infants (mean age 4 weeks, weight 2400 g), urine was collected in portions between the 6 meals.

During routine examination, spot urine samples were obtained from 116 healthy infants (age: 125 ± 9 d) fed either human milk (M), commercially available formula (F) or partially hydrolysed formula (PF) (Table 1).

Table 1. Data of the investigated term infants.

Feeding regimen	M	F	PF
Number of infants	43	36	35
Weight (kg)	7.02 ± 0.86	7.05 ± 0.73	7.02 ± 0.72
nmol Se/L milk	125 ± 6	77 ± 24	75 ± 8
Number of milk samples	30	81	23

Results

Formula-fed preterm infants showed a small variation (4%) in the mean urinary Se excretion (0.67 ± 0.03 μmol/g creat) during day and night (Figure 1).

The daily urinary excretion was significantly higher ($p < 0.0001$) in healthy 4 months old breast-fed infants than in bottle-fed infants. Despite of the difference of the plasma Se (F vs. PF, $p < 0.001$) the urinary Se excretion did not differ between the infants fed F or PF.

Assuming a mean daily milk consumption of 800 mL and a mean daily creatinine excretion of 12 mg/kg body weight, 49–56% of the ingested dose was excreted (Figure 2). There was no difference of the Se clearance rate between the three groups (0.050–0.056 mL/min).

Correct citation: Lombeck, I., Menzel, H., Manz, F., and Mostert V. 1997. Urinary selenium excretion as indicator of selenium status. In *Trace Elements in Man and Animals – 9: Proceedings of the Ninth International Symposium on Trace Elements in Man and Animals. Edited by* P.W.F. Fischer, M.R. L'Abbé, K.A. Cockell, and R.S. Gibson. NRC Research Press, Ottawa, Canada. pp. 115–117.

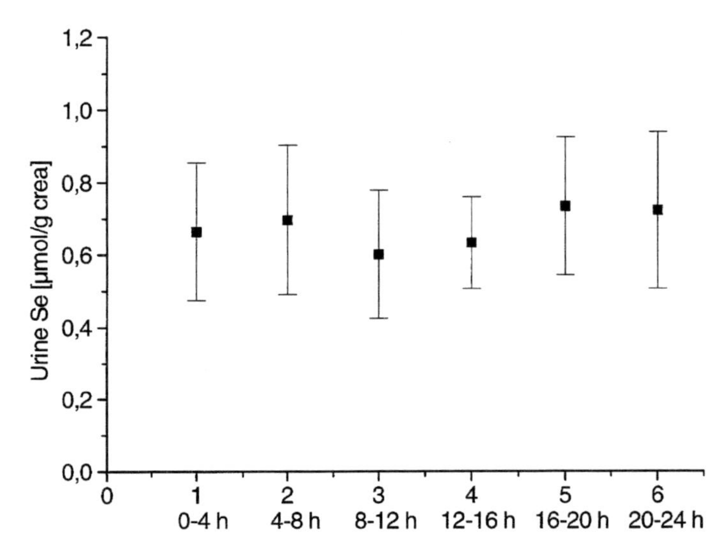

Figure 1. Time-course of urinary Se excretion during day and night.

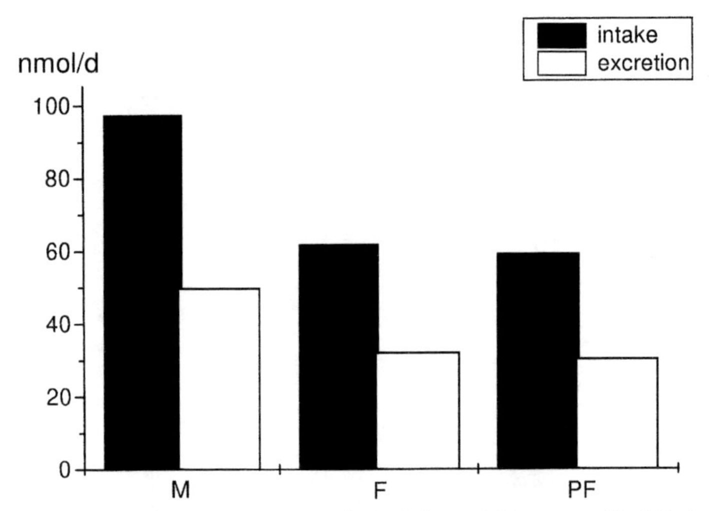

Figure 2. Mean urinary Se excretion vs intake in infants fed human milk (M), formula (F) or partially hydrolysed formula (PF).

Urinary Se excretion was significantly correlated with plasma Se (not whole blood Se or erythrocyte glutathione peroxidase) in all 3 feeding groups (Figure 3, $p < 0.05$). Only in PF-fed infants, was urinary Se and plasma glutathione peroxidase also correlated ($p < 0.01$, data not shown).

Discussion

In healthy people not only blood Se values but also urinary Se excretion show regional variation and age dependency. The latter is mainly due to the low Se intake, rapid growth and increasing glomerular and tubular function during infancy. Because of low creatinine excretion, infants have higher urinary Se excretion/g creatinine than adults (Lombeck et al. 1991, Oster et al. 1990). Our investigation of prematures show that this ratio (urinary Se per g creat) — by compensating the influence of different water supply — gives valuable results. Infants fed human milk, different formula or a hydrolysate had low Se values

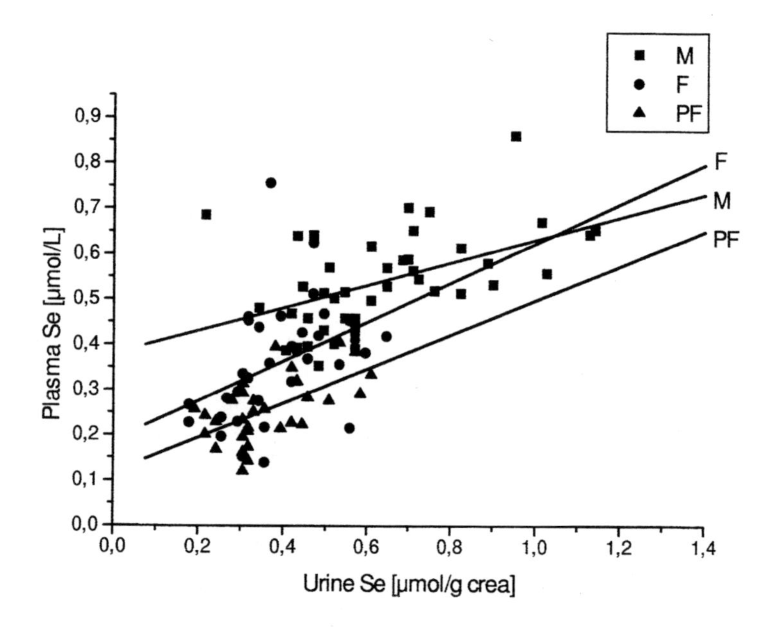

Figure 3. Correlation between urine Se and plasma Se in infants fed in infants fed human milk (M), formula (F) or partially hydrolysed formula (PF).

including urinary Se excretion. All children thrived and had a similar weight gain. Despite of the low Se intake, especially of the bottle-fed infants, about 50% of the ingested dose was excreted in the urine. This finding is in agreement with reports about adults. The clearance rate was also similar, although plasma Se was lower in PF vs. F-fed infants ($p < 0.01$). Maybe different Se fractions contribute to this. It is not known so far, which Se compound in plasma is correlated with the urinary Se excretion. Although the formulae (PF vs. F) differ in various compounds (e.g., peptide, mineral and trace element content including zinc, Jochum et al. 1995), urinary Se excretion did not differ. In former balances of preterm infants (Lombeck et al. 1991), we found an influence of the zinc intake as well as of growth on urinary Se loss.

In contrast to findings in literature, we found a significant correlation between plasma Se and urinary Se excretion (µmol/g creat). The present investigation was performed in infants, who were fed one kind of milk within the first 4 months providing a constant Se intake per joule.

We conclude that under these conditions the analysis of spot urine samples may be an indicator of the present plasma Se values.

Literature Cited

Jochum, F., Fuchs, A., Cser, A., Menzel, H. & Lombeck, I. (1995) Trace mineral status of full-term infants fed human milk, milk based formula or partially hydrolysed whey protein formula. Analyst 120: 905–909.

Lombeck, I., Manz, F. D., Diekmann, L., & Stock, G. (1991) Selenium balance in preterm infants on cow's milk formula. In: TEMA-7 (B. Momcilovic, ed.), pp. 13–15.

Lombeck, I., Menzel, H. & Frosch, D. (1987) Acute selenium poisoning of a 2-year-old child. Eur. J. Pediatr. 146: 308–312.

Oster, O. & Prellwitz, W. (1990) The renal excretion of selenium. Biol. Trace Elem. Res. 24: 119–146.

Urinary Iodine Excretion in New Zealand Residents

C.D. THOMSON,[†] A.J. COLLS,[†] M. STYLES,[‡] AND J. CONAGLEN[‡]

[†]Department of Human Nutrition, University of Otago, Dunedin, and [‡]Department of Medicine, Waikato Hospital, Hamilton, New Zealand

Keywords iodine, urinary excretion, thyroid hormones

Before the introduction of iodised salt in the early 1940s, the iodine deficiency disorder goitre was endemic in many parts of New Zealand (Purves 1974). More recent surveys of the iodine status of New Zealanders indicated more than adequate iodine intakes as a result of iodisation of salt and a substantial contribution to the iodine content of dairy products from the use of iodophors in the dairy industry (Cooper et al. 1984, North and Fraser 1965, Simpson et al. 1984). However, with the recommendations to reduce salt intake and the reduction in the use of iodophors in New Zealand, there is a concern that iodine intakes may be decreasing. This paper reports preliminary results of a study designed to evaluate urinary measures of iodine status and to assess urinary iodine status and thyroid hormone status of residents of two areas of New Zealand where, prior to the iodisation of salt, goitre was endemic due to low soil iodine levels.

Methods

Subjects were recruited from the Blood Transfusion Centres in Otago (195) and Waikato (143) between November 1993 and June 1994. Blood was taken for assays of Free T_3, Free T_4 and TSH. Subjects collected: (i) a fasting urine specimen; (ii) a double-voided fasting urine (i.e., 30 min after fasting urine); and (iii) a complete 24-h collection. Subjects completed a questionnaire on use of vitamin and mineral supplements and iodine-containing medications. Iodide was measured in urine using the method of Dunn et al. (1993) adapted for use on a Cobas Fara autoanalyser. Thyroid hormones were assayed in the Endocrine Laboratory, Health Waikato. Urinary creatinine was measured on the Cobas Fara using a creatinine Uni-Kit II, creatinine calibrator and control sera (Roche Diagnostic Systems Inc). The association between urinary measures was examined using the Spearman's rank correlation coefficient. Differences between means were determined using student's T-test.

Results

Table 1 summarises mean and median values for total daily output of iodine, iodine concentrations in 24-h urines and in fasting and double-voided urine specimens, and the iodine/creatinine ratios for these measures in Otago and Waikato residents. Results for all subjects and for nonsupplementers are presented. Four Otago and 16 Waikato residents reported using supplements containing iodine resulting in daily urinary excretions greater than 200 μg/d. Waikato residents excreted more iodine in urine and all measures were significantly greater than for Otago residents. Males excreted more iodine than females. Differences were significant ($P < 0.05$) for urinary iodine concentrations in all samples and for iodine/creatinine ratio in fasting urines. Thyroid hormones were within the normal range and there were no correlations with 24-h urinary iodide.

Discussion

Urinary iodide excretions of Otago and Waikato residents were considerably lower than those reported previously for New Zealanders (Cooper et al. 1984, North and Fraser 1965, Simpson et al. 1984), and also considerably lower than the recommended dietary intake of 150 μg/d (Truswell et al. 1990). However they do confirm low values found in more recent studies carried out in our laboratory (Thomson et al. 1995, Thomson et al. 1996). Median iodine concentrations fell within the range of 2.0–4.9 μg/dL of moderate stage of IDD severity (Dunn et al. 1993) and 23 of those subjects had levels <2.0 μg/dL within the range for severe IDD. The median total 24 h excretion was 60 μg/d, with the majority (86%) less than 100 μg/d

Correct citation: Thomson, C.D., Colls, A.J., Styles, M., and Conaglen, J. 1997. Urinary iodine excretion in New Zealand residents. In *Trace Elements in Man and Animals – 9: Proceedings of the Ninth International Symposium on Trace Elements in Man and Animals. Edited by* P.W.F. Fischer, M.R. L'Abbé, K.A. Cockell, and R.S. Gibson. NRC Research Press, Ottawa, Canada. pp. 118–119.

Table 1. Urinary iodide excretion.

	Otago			Waikato		
	All subjects	Non-Supplementers	Median	All subjects	Non-Supplementers	Median
n	194	190		140	124	
24 hour iodide (μg/day)	70±52	65±34	60	90±54	76±32	77
24 hour iodide (μg/dL)	4.9±3.5	4.6±2.4	4.0	6.1±3.8	5.3±2.6	5.5
Fasting iodide	5.1±4.0	4.9±3.0	4.0	6.3±4.1	5.7±3.3	5.5
DV fasting iodide	5.0±4.1	4.7±2.8	4.3	5.5±4.2	4.8±2.7	4.5
Iodide/Creatinine ratio						
24 hour urine (μg/g Cr)	44±37	40±22	36	57±32	50±22	49
Fasting urine	36±27	34±22	30	48±32	43±28	39
DV fasting urine	32±24	30±21	26	43±38	37±22	35

and 37% less than 50 μg/d, considered to be a marginal iodine intake. Iodine values for Waikato residents were higher with 71% below 100 μg/d and 23% below 50 μg/d.

These low iodide excretions in New Zealand residents were surprising in view of earlier reported values. They could result from improved sensitivity of the analytical method used. Measures of thyroid hormone status were within the normal range and therefore the physiological significance of the low iodine excretions is unclear. These assays are the most appropriate diagnostic test for hypothyroidism, but may not be true indicators of iodine status, as in mild iodine deficiency when urinary iodide is low, circulating levels of T_3, T_4 and TSH are still normal (Bourdoux 1993).

One aim of this study was to evaluate the fasting and double-voided fasting urine samples for assessing urinary iodine status, since collection of complete 24-h urine specimens is difficult and inconvenient for subjects. Correlations between 24-h urinary iodide and iodide concentrations in fasting urines confirms previous observations by us (Thomson et al. 1996). The latter study indicated that fasting urines were adequate for assessing iodine status of population groups when assessing the risk of iodine deficiency disorders, but not for predicting individual daily iodine excretion. Thus 24-h urines remain the most suitable and recommended measure.

These preliminary observations suggest that iodine status of some New Zealand residents may no longer be considered adequate, and that further surveillance is necessary. Investigations of the reasons for and the clinical significance of the decreasing iodine status is also of interest.

Acknowledgments

We wish to thank Mrs. Margaret Waldron, Dr. Jim Faed, Mr. Les Milligan and the Dunedin and Waikato Blood Transfusion Centres for their assistance, Mr. John Speed for thyroid hormone assays, and the subjects for their participation. The study was funded by an Otago Medical Research Foundation Laurenson Award and the Ministry of Health.

Literature Cited

Bourdoux, P.P. (1993) Biochemical evaluation of iodine status. In: Iodine Deficiency in Europe. (Delange, F. et al. eds.), pp. 119–125. Plenum Press, New York.

Cooper, G.J.S., Croxson, M.S., & Ibbertson, H.K. (1984) Iodine intake in an urban environment: a study of urine iodide excretion in Auckland. NZ Med. J. 97: 142–145.

Dunn, J.T., Crutchfield, H.E., Gutekunst, R., & Dunn, A.D. (1993) Methods for measuring iodine in urine. International Council for the Control of Iodine Deficiency Disorders, The Netherlands

North, K.A.K., & Fraser, S. (1965) Iodine intake as revealed by urinary iodide excretion. NZ Med. J. 65: 512–513.

Nutrition Task Force (1991) Food For Health. Department of Health, Wellington.

Purves, H.D. (1974) The aetiology and prophylaxis of endemic goitre and cretinism. NZ Med. J. 80: 477–479.

Simpson, F.O., Thaler, B I., Paulin, J.M., Phelan, E.L., & Cooper, G.J.S. (1984) Iodide excretion in a salt-restriction trial. NZ Med. J. 97: 890–893.

Thomson, C.D., Packer, M.A., Duffield, A.J., O'Donaghue, K.L., Butler, J.A., & Whanger, P.D. (1995) Urinary iodine during pregnancy and lactation. Proc. Nutr. Society NZ 20.

Thomson, C.D., Smith, T.E., Butler, K.A., & Packer, M.A. (1996) An evaluation of urinary measures of iodine and selenium status. In press.

Truswell, A.S., Dreosti, I.E., English, R.M., Rutishauser, I.H.E., & Palmer, N. (1990) In: Recommended Nutrient Intakes. Australian Papers. Australian Professional Publications, Sydney.

Nutritional Status of Trace Elements in Traditional Populations Inhabiting Tropical Lowland, Papua New Guinea

T. HONGO,[†] R. OHTSUKA,[†] M. NAKAZAWA,[†] T. INAOKA,[‡] AND T. SUZUKI[¶]

[†]Department of Human Ecology, University of Tokyo, Tokyo 113, Japan, [‡]Department of Public Health, Kumamoto University School of Medicine, Kumamoto 860, Japan, and [¶]National Institute for Environmental Studies, Ibaraki 305, Japan

Keywords: zinc, serum, malaria, Papua New Guinea

This study analyzed serum concentrations of essential trace elements (Fe, Cu, Zn, and Se) of the Gidra-speaking people in the lowland Papua New Guinea in connection with food consumption and malaria prevalence, paying special attention to their bioavailability.

Subjects and Methods

In four ecologically diversified villages (Rual, Wonie, Ume, and Dorogori), blood samples were collected from 210 adults (94 men and 116 women) in 1989. Serum Fe was measured by the bathophenanthroline method, Cu and Zn by inductively coupled plasma atomic emission spectrometry, and Se by the fluorometric method. Long-lasting serum antibody titer levels against *Plasmodium falciparum* and *P. vivax* were examined using the indirect fluorescent antibody test (Nakazawa et al. 1994), and those whose antibody titers against either of two *Plasmodium* species were equal to or higher than 1:64 were judged malaria positives. The proportion of malaria positives in Rual, Wonie, Ume, and Dorogori villagers were 72, 33, 88, and 100%. Their daily intake of energy and nutrients per adult male was estimated by using food composition tables for Gidra food (Hongo et al. 1989a, 1989b, Ohtsuka et al. 1984, Yoshinaga et al. 1991).

Results

Table 1 shows the daily intake of Fe, Cu, Zn, and Se. Their mean serum concentrations by village are shown in Table 2 (Hongo et al. 1993, Nakazawa et al. 1996), and these levels were compared between malaria negatives and positives (Table 3). To elucidate variation of serum element levels, multiple regression analysis was conducted using gender (male: 0, female: 1), malaria infection (negative: 0, positive: 1), and four dummy variables for each village as independent variables; for example, Rual villagers were applied "1" for "Rual" variable and "0" for three other village variables (Table 4).

Discussion

Cu and Se status of the Gidra was adequate. Serum ferritin and transferrin saturation levels also indicated an adequate Fe status, though slightly low in serum Fe level (Nakazawa et al. 1996). Thus, low serum Zn and its inter-village variation aroused our interest.

The inter-village difference of Zn intake was reflected in serum Zn concentration. Zinc intake of relatively modernized Ume and Dorogori villagers was about half of the recommended dietary intake in the USA (15 mg). This low level is mostly attributed to a decreased consumption of local animal foods

Correct citation: Hongo, T., Ohtsuka, R., Nakazawa, M., Inaoka, T., and Suzuki, T. 1997. Nutritional status of trace elements in traditional populations inhabiting tropical lowland, Papua New Guinea. In *Trace Elements in Man and Animals – 9: Proceedings of the Ninth International Symposium on Trace Elements in Man and Animals. Edited by* P.W.F. Fischer, M.R. L'Abbé, K.A. Cockell, and R.S. Gibson. NRC Research Press, Ottawa, Canada. pp. 120–122.

Table 1. Daily intake of energy, protein, crude fiber, and trace elements.

	Rual	Wonie	Ume	Dorogori
Energy (kcal)	3024	3769	2620	2656
Protein (g)	105.9	92.3	47.8	60.0
Crude fiber (g)	6.8	25.0	26.3	15.1
Fe (mg)	70.2	62.8	37.9	19.6
Cu (mg)	1.99	2.94	1.69	1.56
Zn (mg)	15.3	16.8	7.2	7.0
Se (μg)	103	95	78	94

Table 2. Serum element concentration (μmol/L) by village.

		Rual		Wonie		Ume		Dorogori	
		Mean	n	Mean	n	Mean	n	Mean	n
Fe	Male	15.4	15	19.3	17	14.7	41	15.8	21
	Female	13.6	26	14.3	20	13.4	35	12.9	32
Cu	Male	21.9	15	18.7	17	21.4	41	23.3	21
	Female	25.0	28	22.7	20	26.1	36	26.3	32
Zn	Male	11.6	15	11.2	17	9.5	41	9.2	21
	Female	10.1	28	10.1	20	9.2	36	9.2	32
Se	Male	1.66	15	1.82	17	1.66	41	1.67	21
	Female	1.74	28	1.81	20	1.60	36	1.70	32

Table 3. Comparison of serum element concentrations (μmol/L) between malaria negatives and positives.

		Negatives		Positives		
		Mean	n	Mean	n	t-test
Fe	Male	17.9	17	15.2	68	ns
	Female	14.3	27	14.0	63	ns
Cu	Male	20.0	17	21.4	68	ns
	Female	24.1	28	25.8	65	ns
Zn	Male	11.9	17	9.8	68	$p < 0.05$
	Female	10.2	28	9.3	65	$p < 0.05$
Se	Male	1.77	17	1.67	68	$p < 0.05$
	Female	1.75	28	1.67	65	ns

Table 4. Result of multiple regression analysis.

	Independent variables	β		F-value	R^2
Fe	Gender	–0.147	*		
	Wonie	0.177	*	4.7 †	0.04
Cu	Gender	0.391	‡		
	Wonie	–0.225	†		
	Dorogori	0.109		16.3 ‡	0.21
Zn	Gender	–0.184	*		
	Malaria	–0.204	*		
	Rual	0.234	†		
	Wonie	0.136		7.0 ‡	0.12
Se	Wonie	0.199	*		
	Ume	–0.217	†	12.3 ‡	0.11

‡, †, *; $p < 0.001$, $p < 0.01$, and $p < 0.05$.

through modernization, characterized by increased dependency on imported foods such as rice and wheat flour and decreased time spent in subsistence activities including hunting.

Compared to data from developed countries, serum Zn concentration was low even in Rual and Wonie villagers. Hair Zn concentrations of the Gidra measured in 1981 were also low compared to those in developed countries, despite the fact that Zn intakes were almost identical (Hongo et al. 1989b, 1990). This discrepancy was explained by inhibition of Zn absorption due to their high intake of crude fiber and nonheme Fe.

On the other hand, the malaria parasite exerts an oxidant stress on infected erythrocytes and Zn has antioxidant functions to stabilize cell membranes. In vitro analysis disclosed that membranes of malaria-infected erythrocytes contained higher Zn compared to uninfected cells. Thus, there is a possibility that the Zn requirement is increased by malaria infection and, as the result, serum Zn level is depressed.

References

Hongo, T., Suzuki, T., Ohtsuka, R., Kawabe, T., Inaoka, T. & Akimichi, T. (1989a) Compositional character of Papuan foods. Ecol. Food Nutr. 23: 39–56.

Hongo, T., Suzuki, T., Ohtsuka, R., Kawabe, T., Inaoka, T. & Akimichi, T. (1989b) Element intake of the Gidra in lowland Papua: inter-village variation and the comparison with contemporary levels in developed countries. Ecol. Food Nutr. 23: 293–309.

Hongo, T., Suzuki, T., Ohtsuka, R., Kawabe, T., Inaoka, T. & Akimichi, T. (1990) Hair element concentrations of the Gidra in lowland Papua: the comparison with dietary element intakes and water element concentrations. Ecol. Food Nutr. 24: 167–179.

Hongo, T., Ohtsuka, R., Nakazawa, M., Kawabe, T., Inaoka, T., Akimichi, T. & Suzuki, T. (1993) Serum mineral and trace element concentrations in the Gidra of lowland Papua New Guinea: inter-village variation and comparison with the levels in developed countries. Ecol. Food Nutr. 29: 307–318.

Nakazawa, M., Ohtsuka, R., Kawabe, T., Hongo, T., Suzuki, T., Inaoka, T., Akimichi, T., Kano, S. & Suzuki, M. (1994) Differential malaria prevalence among villages of the Gidra in lowland Papua New Guinea. Trop. Geogr. Med. 46: 350–354.

Nakazawa, M., Ohtsuka, R., Kawabe, T., Hongo, T., Inaoka, T., Akimichi, T. & Suzuki, T. (1996) Iron nutrition and anaemia in a malaria-endemic environment: haematological investigation of the Gidra-speaking population in lowland Papua New Guinea. Br. J. Nutr. (in press)

Ohtsuka, R., Kawabe, T., Inaoka, T., Suzuki, T., Hongo, T., Akimichi, T. & Sugahara, T. (1984) Composition of local and purchased foods consumed by the Gidra in lowland Papua. Ecol. Food Nutr. 15: 159–169.

Yoshinaga, J., Suzuki, T., Ohtsuka, R., Kawabe, T., Hongo, T., Imai, H., Inaoka, T. & Akimichi, T. (1991) Dietary selenium intake of the Gidra, Papua New Guinea. Ecol. Food Nutr. 26: 27–36.

Lead Concentrations in the Sweat of Lead Workers

K. SUZUKI,* T. ESASHI, M. NISHIMUTA, H. TOJO, AND Y. KANDA
The National Institute of Nutrition, Shinjuku-ku, Tokyo 162, Japan

Keywords lead, trace elements, sweat, mineral metabolism

Lead concentrations were analysed in the sweat of lead workers to determine whether these levels differed from routine values. We previously reported that some 20–30 mg of lead per litre of cell-free sweat was normally found in metabolic studies on young female volunteers living in the Tokyo area (Suzuki et al. 1977–1983).

Experimental Procedure

Seven workers from two lead factories, who served as the subjects of this study, volunteered to participate in the sweating experiment. Diaphoresis was induced in a heated artificial environment, established in a clean room. The day before the experiment the subjects bathed, thoroughly washing the entire body surface. Just before beginning sweat collection, the entire surface of each subject's body was carefully cleansed with detergent (sugar ester) and rinsed thoroughly with purified water (Milli-Q system). The subjects remained in the experimental room for about 30 min until they perspired sufficiently, and the sweat was collected by rinsing the entire body surface, except the scalp, with purified water. All of the water containing sweat was used to re-rinse the body surface, then collected and filtered through a 0.1-μ pore membrane filter. The amount of sweat was calculated as the difference in body weight before and after perspiring. Lead and other trace elements were analyzed in sweat, urine and blood samples by atomic absorption spectrometry. Solvent extractions were performed before these determinations to exclude interference from coexisting macroelements in the samples.

Results and Discussion

The results obtained are shown in Table 1. Experiments were performed in March, April and August of 1981. A, B, C, E, F, H and I were the seven volunteer subjects. D and G were staff members who participated in the experiment as controls. All seven volunteers had high blood lead levels and low blood δ-ALAD activity. The lead concentrations in sweat varied, ranging from 130 to 300 μg/L sweat, but the sweat lead levels of all seven lead workers were significantly higher than our previously reported values (Suzuki et al. 1977–1983) and the routine values reported by Omokohdion et al. (1991), Stauber et al. (1988), Lilley et al. (1988), and by other investigators. On April 9 and 10, perspiration experiments were performed twice daily, once in the morning and once in the afternoon, such that the subjects perspired four times in 2 d. It is noteworthy that the sweat lead levels of subjects C, E and F diminished as the number of perspiration experiments increased on these 2 d. Generally speaking being absorbed orally, by the lungs from airborne dust and in soluble form as organolead compounds and inorganic compounds via the skin (the stratum corneum, sweat glands, etc.), lead is also absorbed through the skin as finely powdered metallic lead (Lilley et al. 1988, Stauber et al. 1994). Whatever the route of absorption, the lead absorbed ultimately reaches the blood stream, and is rapidly taken up by erythrocytes. Lead in plasma then passes into the sweat glands and salivary glands and is excreted externally. Higher sweat lead levels were found in the subjects with high blood lead levels, but no direct correlation between sweat and blood lead levels

*Present address: 1-8 Tokiwadai-3 chome, Itabashi-ku, Tokyo 174, Japan.

Correct citation: Suzuki, K., Esashi, T., Nishimuta, M., Tojo, H., and Kanda, Y. 1997. Lead concentrations in the sweat of lead workers. In *Trace Elements in Man and Animals – 9: Proceedings of the Ninth International Symposium on Trace Elements in Man and Animals. Edited by* P.W.F. Fischer, M.R. L'Abbé, K.A. Cockell, and R.S. Gibson. NRC Research Press, Ottawa, Canada. pp. 123–125.

Table 1. Lead concentration in sweat, urine and blood samples from lead workers (µg/l).

Exp. No.	Subjects	Age (yr)	Work place	Exp. day		Cell-free sweat	Whole sweat	Urine	Whole blood	serum	δ-ALAD*
1	A	38	a	Mar. 23		58	88	6.9	140	–	785
2				Mar. 26		63	88				
3	B	36	a	Mar. 23		43	51	6.9	130	–	965
4				Mar. 26		72	81				
5	C	38	a	Mar. 23		240	310	23	210	–	380
6				Mar. 26		320	350				
7	D	42	Control	Mar. 26		20	26	7.5	–	–	
8	C	38	a	Apr. 9	am	170	190	28	270	27	378
9					pm	150	150				
10				Apr. 10	am	55	56				
11					pm	42	42				
12	E	50	a	Apr. 9	am	29	170	97	300	3.6	123
13					pm	19	33				
14				Apr. 10	am	9.8	20				
15					pm	4	16				
16	F	54	a	Apr. 9	am	87	180	26	220	0.8	213
17					pm	29	70				
18				Apr. 10	am	20	32				
19					pm	19	26				
20	D	42	Control	Apr. 9	am	16		4.2	41	0.0	1020
21					pm	10	13				
22				Apr. 10	am	12	15				
23					pm	11	14				
24	G	37	Control	Apr. 9	am	10	22	4.9	38	1.4	1088
25					pm	10	17				
26	H	35	b	Aug. 12		140	160	21	240	40	–
27				Aug. 13		120	130	26			
28	I	42	b	Aug. 12		60	120	24	230	37	–
29				Aug. 13		77	120	17			

Control, staff member; *, mmolePBG/h/L mL RBCx100.

was observed in this experiment. It is assumed that there are certain rates at which lead is carried from the blood to sweat glands (v1) and other rates at which lead-containing sweat is excreted by sweat glands (v2). Lead levels in sweat are mainly affected by the latter rates of excretion (v2). The diminishing sweat lead levels in subjects C, E and F in the experiments conducted on April 9 and 10, shown in the Table 1, appeared to depend mainly on the balance between the rate at which lead is transported to sweat glands within a given period and the rate at which the lead is excreted by the sweat glands within a comparable period. The concentrations of other trace elements, zinc, iron, copper, manganese, nickel, and cadmium, in sweat are shown in the Table 2. These concentrations did not exhibit the clearly diminishing pattern observed in the lead measurements. Further studies are needed to determine whether this is a characteristic of lead alone.

Acknowledgments

The authors are indebted to Prof. H. Anzai, Kitazato University, Dr. T. Utsunomiya and Dr. H. Kudo, Japan Industrial Safety and Health Association.

Table 2. Trace element concentrations in cell-free sweat (µg/dL).

Exp. No.		Zn	Fe	Cu	Mn	Ni	Cd	Pb
1		33	6.6	9.9	0.34	1.6	0.39	5.8
2		41	4.6	12	0.41	2.9	0.5	6.3
3		64	6.5	6.0	0.31	6.3	0.64	4.3
4		47	3.3	6.1	0.22	0.92	0.71	7.2
5		37	9.8	4.7	0.47	1.6	3.2	24
6		47	4.9	6.8	0.49	2.0	2.0	32
7		18	3.0	2.2	0.19	0.6	0.06	2.0
8	am	37	4.1	6.0	0.36	2.6	0.96	17
9	pm	59	2.6	6.8	0.31	3.8	0.77	15
10	am	28	2.5	2.4	0.18	0.55	0.27	5.5
11	pm	52	2.5	3.5	0.31	0.67	0.51	4.2
12	am	5.0	2.4	3.6	0.13	1.3	0.04	2.9
13	pm	8.1	3.5	2.3	0.11	1.9	0.04	1.9
14	am	4.4	2.7	2.6	0.15	0.61	0.02	1.0
15	pm	9.2	5.7	4.5	0.14	0.90	0.04	0.4
16	am	18	5.6	6.0	0.18	2.5	0.08	8.7
17	pm	12	1.1	2.7	0.11	1.5	0.05	2.9
18	am	13	2.1	2.7	0.15	0.68	0.03	2.0
19	pm	12	1.6	3.6	0.16	0.87	0.05	1.9
20	am	22	3.0	3.2	0.19	0.66	0.07	1.6
21	pm	19	5.3	2.9	0.17	0.49	0.07	1.0
22	am	35	3.0	2.9	0.23	0.52	0.08	1.2
23	pm	32	2.5	3.0	0.24	0.54	0.20	1.1
24	am	15	4.3	6.6	0.39	1.7	0.05	1.0
25	pm	27	3.7	7.3	0.33	1.7	0.09	1.0
26		51	9.1	4.7	0.68	2.1	0.14	14
27		110	8.0	7.9	0.99	1.5	0.17	12
28		72	11.0	6.6	1.90	2.7	0.12	6.0
29		59	6.6	6.8	0.81	1.7	0.15	7.7

Literature Cited

Lilley, S.G., Florence, T.M., and Stauer, J.L. (1988) The use of sweat to monitor lead absorption. Sci. Total Envir. 76: 267–278.

Omokohdion, F.O. & Howard, J.M. (1991) Sweat lead levels in persons with high blood lead levels: lead in sweat of lead workers in the tropics. Sci. Total Envir. 103: 123–128.

Rabinowitz, M.B. & Wetherill, G.W. (1973) Lead metabolism in the normal human: Stable isotope studies. Science 182: 725–727.

Stauber, J.L. & Florence, T.M. (1988) A comparative study of copper, lead, cadmium and zinc in human sweat and blood. Sci. Total Envir. 74: 235–247.

Stauber, J.L., Florence, T.M., Gulson, B.L. & Dale, L.S. (1994) Percutaneous absorption of inorganic lead compounds. Sci. Total Envir. 145: 55–70.

Suzuki, K. (1977–83) Dermal excretion of metallic elements. Annual Report of the National Institute of Nutrition (Japanese ed.) 1978: 23.

Age and Exposure Variations to Arsenic Excretion

JANE L. VALENTINE

Department of Environmental Health Sciences, University of California, Los Angeles, CA 90095, USA

Keywords arsenic, excretion, aging

Human biological monitoring of arsenic status is of interest. Concentrations of arsenic in human tissues and fluids are known to be affected by occupation and environmental exposure (Valentine et al. 1979, Cebrian et al. 1983, Harrington et al. 1978, Landau et al. 1977). Diet and age effects may be related also.

Studies on the relationship of age to trace element status have been extensively studied for the element selenium (Valentine et al. 1989, Westermarck et al. 1977, Verlinden et al. 1983a,b). Selenium concentrations excreted in urine were higher in persons aged less than 20 years and in the 20 to 29 year groups as compared to those aged 30 and greater (Valentine et al. 1988, Robberecht and Deelstra 1984, Verlinden 1981). Fewer reports exist for age effects and arsenic excretion via urine or storage in other tissues and body fluid. This report evaluates age-related excretion of arsenic in human urine.

Method

A data file of arsenic concentrations for populations exposed to the element via drinking water has been maintained. This data set contains also concentrations of arsenic in urine for persons not exposed (low arsenic) to the element in water. Previous publications on this data set have been done (Valentine et al. 1979, Valentine et al. 1992, Valentine 1995). Body burden concentrations were generated using the hydride generation technique for arsine gas development with atomic absorption spectrophotometry (Kang and Valentine 1977).

An evaluation of arsenic concentrations in urine versus age was made on four exposed populations — Virginia Foothills, Hidden Valley, Edison, and Fallon — and two non-exposed controls — Sun Valley and Casper.

Results

The mean urinary excretion values by age are shown in Table 1. It appears that the arsenic concentration in children less than 10 years of age was in all exposure situations concentrations of arsenic less than those of other higher age groups. The arsenic concentrations in age groups greater than 10 years showed no consistent trend across age strata. Any differences currently observed for the groups may have been related to dietary intake more than to metabolic function.

Correct citation: Valentine, J.L. 1997. Age and exposure variations to arsenic excretion. In *Trace Elements in Man and Animals – 9: Proceedings of the Ninth International Symposium on Trace Elements in Man and Animals. Edited by* P.W.F. Fischer, M.R. L'Abbé, K.A. Cockell, and R.S. Gibson. NRC Research Press, Ottawa, Canada. pp. 126–127.

Table 1. Arsenic concentration (μg/L) in urine by age groups.

Communities (As concentration in water, μg/L)	1–10 n mean±SD	11–20 n mean±SD	21–30 n mean±SD	31–40 n mean±SD	41–50 n mean±SD	51–60 n mean±SD	Over 60 n mean±SD
Low-exposed / control							
Sun Valley (2.0)	3 2.39±0.58	18 4.32±6.77	8 6.3±7.58	18 5.64±7.84	11 4.78±6.00	9 6.01±8.10	6 6.78±7.41
Casper (0.6)	3 2.78±2.26	13 3.29 ±1.81	9 2.86±1.05	14 2.48±1.31	9 3.16±2.27	13 3.79±5.87	6 2.00±1.52
Exposed							
Edison (393)	9 147.88±60.05	5 209.85±121.57	7 82.64±49.89	9 125.72±84.84	3 158.33±18.93	5 187.20±168.75	1 46.00
Hidden Valley (100)	1 22.80	4 40.80±13.57	1 5.10	5 42.22±17.77	4 56.32±8.87	6 62.50±27.87	2 35.25±39.10
Virginia Foothills (100)	0 —	3 29.57±12.63	5 40.80±7.36	7 59.81±78.22	7 39.30±10.98	9 37.66±34.71	3 20.00±11.99
Fallon (100)	0 —	9 53.79±29.85	3 73.47±7.64	12 50.68±19.49	8 50.68±16.85	5 39.34±7.64	9 44.20±7.78

References

Cebrian, M.E., Albores, A., Aguilar, M., and Blakely, E. (1983) Chronic arsenic poisoning in the north of Mexico. Hum. Toxicol. 2:121–133.

Harrington, J.M., Middaugh, J.P., Morse, D.L., and Housworth, J. (1978) A survey of a population exposed to high concentrations of arsenic in well water in Fairbanks, Alaska. Amer. J. Epidemiol. 108:377–385.

Kang, H.K., and Valentine, J.L. (1977) Acid interference in the determination of arsenic by atomic absorption spectrometry. Anal. Chem. 49:1829–1831.

Landau, E., Thompson, D.J., Feldman, R.G., Goble, G.J., and Dixon, W.J. (1977) Selected non-carcinogenic effects of industrial exposure to inorganic arsenic. Final Report. EPA 560/6–77–018.

Robberecht, H.J., and Deelstra, H.A. (1984) Selenium in human urine: concentration levels and medical implications. Clinica Chimica Acta 136:107–120.

Valentine, J.L. (1995) Body burden concentrations in humans in response to low environmental exposure to trace elements. To be published in ACS volume based on Environmental Biomonitoring and Specimen Banking. Presented at International Chemical Congress of Pacific Basin Societies, December 17–22 (1995) Honolulu, Hawaii, USA.

Valentine, J.L., Kang, H.K., and Spivey, G. (1979) Arsenic levels in human blood, urine, and hair in response to exposures via drinking water. Environ. Res. 20:24–32.

Valentine, J.L., Kang, H.K., Faraji, B. and Lachenbruch, P.A. (1989) Selenium Status and Age Effects. In: Selenium in Biology and Medicine A. Wendel (Ed). Springer-Verlag, Germany.

Valentine, J.L., He, S.-Y., Reisbord, L.S., and Lachenbruch, P.A. (1992) Health response by questionnaire in arsenic-exposed populations. J. Clin. Epidemiol. 45:487–494.

Verlinden, M. (1981) The determination of selenium by atomic absorption spectroscopy. Methodology and biomedical implications. Ph.D. thesis. University of Antwerp, Wihyk, Belgium.

Verlinden, M., Van Sprundel, M., Van der Anwera, J.C., and Eylenbosch, W.J. (1983a) The selenium status of Belgian population groups. I. Healthy adults. Biol. Trace Elem. Res. 5:91–102.

Verlinden, M., Van Sprundel, M., Van der Anwera, J.C., and Eylenbosch, W.J. (1983b) The selenium status of Belgian population groups. II. Newborns, children, and the aged. Biol. Trace Elem. Res. 5:103–113.

Westermarck, T., Raunu, P., Kijarinta, M., and Lappalainen, L. (1977) Selenium content of whole blood and serum in adults and children of different ages from different parts of Finland. Acta pharmacol. et toxicol. 40:465–475.

Lead and Cadmium Contamination of Calcium Supplements Does Not Raise Blood Lead Levels in Dialysis Patients

E. BURGESS,[†] R. AUDETTE,[‡] R. HONS,[†] K. TAUB,[†] H. MANDIN,[†] AND S. SCHORR[†]

[†]Division of Renal Medicine, Foothills Hospital, Calgary, Alberta, Canada T2N 2T9.
[‡]Environmental Toxicology Laboratory, University of Alberta Hospital, Edmonton, Alberta, Canada

Keywords lead, dialysis, calcium, renal disease

Introduction

Patients with end-stage renal disease have disturbed calcium (Ca) and phosphate (PO_4) homeostasis. As an intervention to increase serum Ca levels and as an intestinal binder of dietary PO_4, patients are given oral Ca supplements in large doses. Bone biopsies from dialysis patients have been reported to have as much stainable lead as bone samples from occupationally-exposed workers.

A recent report raised concerns of lead (Pb) and cadmium (Cd) contamination in Ca supplements. The contamination varied according to the source of the Ca used in the supplements, with oyster shell-derived Ca supplements having more than chelated Ca sources (Bourgoin et al. 1993).

Method

In an effort to screen hemodialysis patients (HD) for accumulation of Pb and Cd, blood samples for trace metals were collected prior to a hemodialysis procedure. For peritonal dialysis patients (PD), blood was drawn at the time of clinic visits. Patients were asked to clarify whether they were taking Ca supplements, and the type of supplement they were taking. Blood was collected into a 7 mL royal blue stopper trace element blood collection tube (#307022-Sherwood Medical, Canada). Phlebotomists wore powder-free gloves. Samples were assayed using a Perkin Elmer Elan 5000A ICP-MS.

ANOVA was used to assess differences across the 4 groups, and a T-test when comparison of only 2 groups was being done.

Results

Hemodialysis patients were separated into 4 groups according to their use of Ca supplements. Of the 160 hemodialysis patients (HD), accurate information could be ascertained from only 123 patients with regard to their Ca supplements.

	# of patients	Blood Pb μmol/L	Blood Cd nmol/L
No Ca-supp	14	0.41 ± 0.14	17 ± 12
Ca lactate	14	0.27 ± 0.18	13 ± 18
TUMS	15	0.30 ± 0.15	10 ± 8
Oyster shell	80	0.30 ± 0.13	12 ± 11
All Ca-supp	109	0.29 ± 0.14	12 ± 11

Correct citation: Burgess, E., Audette, R., Hons, R., Taub, K., Mandin, H., and Schorr, S. 1997. Lead and cadmium contamination of calcium supplements does not raise blood lead levels in dialysis patients. In *Trace Elements in Man and Animals – 9: Proceedings of the Ninth International Symposium on Trace Elements in Man and Animals. Edited by* P.W.F. Fischer, M.R. L'Abbé, K.A. Cockell, and R.S. Gibson. NRC Research Press, Ottawa, Canada. pp. 128–129.

In the HD patients, comparison across the 4 groups demonstrated a difference for blood Pb levels ($p = 0.035$ ANOVA). Paradoxically, the blood Pb level in non-Ca supplement patients were higher compared to all Ca-supplement using patients ($p = 0.005$).

Patients on peritoneal dialysis (PD) ($n = 44$) were also assessed. In this group, all of whom were taking Ca supplements, the mean blood Pb level was 0.22 ± 0.15 µmol/L. This was significantly different from the mean of all the hemodialysis patients ($p < 0.001$)

Blood levels of Cd were not abnormal in any of the patient groups, and were not different in Ca-supplement users. Also, there was no difference in mean blood Cd levels between HD and PD (12 ± 12 vs. 14 ± 17 nmol/L).

Discussion

HD patients using Ca supplements did not have elevated levels of blood Pb compared to the group of patients not taking Ca supplements. In fact the non-users had significantly higher levels. Normal Pb is absorbed via the gastro-intestinal tract, and excreted through the kidneys. The gastro-intestinal absorption of Pb is blocked by calcium. Therefore, even if the Ca supplements have traces of Pb, the competition for GI absorption appears to protect the patients.

The signs and symptoms of Pb toxicity are similar to those of uremia, and include anemia, fatigue, neuropathy, and bone abnormalities. Screening of patients using clinical symptomatology would not be useful. However, this observational data would suggest that hemodialysis and peritoneal dialysis patients do not have elevated blood Pb levels. Closer examination for Pb accumulation would require bone biopsies.

Cadmium is also absorbed via the gastro-intestinal tract and excreted through the kidneys. The GI absorption of Cd may be inhibited by iron, rather than Ca. In this assessment, there is no indication that the blood Cd levels were high in any group of patients, regardless of the use of Ca supplements.

The difference in mean blood Pb levels in HD versus PD patients can not be easily explained. On average, the HD patients had been on dialysis longer (3.8 ± 3.9 vs. 2.0 ± 3.0 years), and duration of dialysis has been noted to be related to serum aluminium levels.

In summary, blood Pb and Cd levels were not elevated above normal in hemodialysis patients taking any form of Ca-supplements. Paradoxically, blood Pb levels were higher in patients no taking Ca-supplements than those taking them.

References

Bourgoin, B.P., Evans, D.R., Cornett, J.R., Lingard, S.M., Quattrone, A.J. (1993) Lead content in 70 brands of dietary calcium supplements. Am J Public Health 83: 1155–1160.

Trace Metals in Hemodialysis Patients

E. BURGESS,[†] R. AUDETTE,[‡] R. HONS,[†] K. TAUB,[†] S. SCHORR,[†] AND H. MANDIN[†]
†Division of Renal Medicine, Foothills Hospital, Calgary, Alberta, Canada T2N 2T9.
‡Environmental Toxicology Laboratory, University of Alberta Hospital, Edmonton, Alberta, Canada

Keywords aluminum, dialysis, renal disease, calcium

Introduction

Aluminum toxicity was first reported in dialysis patients almost 20 years ago. At that time the major source of Al was water used in dialysis. Now the major source is from medication, specifically Al-containing antacids. Due to the accumulation of phosphate (PO_4) in patients with minimal renal function, antacids have been used as intestinal binders of phosphate. Although Ca-containing binders have become widely used, some patients may still require Al-containing antacids.

Methods

Hemodialysis patients had blood samples for trace metals collected prior to a hemodialysis procedure. Patients were asked to clarify whether they were taking antacids, and the type of antacid they were taking. Blood was collected into 7 mL royal blue stopper trace element blood and serum collection tubes (Sherwood Medical, Canada). Phlebotomists wore powder-free gloves. Samples were assayed using a Perkin Elmer Elan 5000A ICP-MS. ANOVA and T-tests were used to assess differences across groups.

Results

Reliable history about current antacid use could only be obtained from 101 patients; 81 non-users and 20 users. All but 14 patients were taking calcium supplements as well. Non-users of Al-containing antacids had lower levels of serum Al (S Al) (0.07 ± 0.15 vs. 0.33 ± 0.49 μmol/L , $p < 0.001$), serum Nickel (S Ni) (89 ± 16 vs. 98 ± 19, = 0.032) and blood Molybdenum (B Mo) (72 ± 29 vs. 92 ± 36 nmol/L, $p = 0.01$). Patients currently taking Al-containing antacids had been on dialysis significantly longer than those patients not taking them.

	Not taking	Taking	p value
S Al (μmol/L)	0.07 ± 0.15	0.33 ± 0.49	<0.001
S Ni (nmol/L)	89 ± 16	98 ± 19	0.032
B Mo(nmol/L)	72 ± 29	92 ± 36	0.01
Dialysis(years)	3.2 ± 3.6	6.4 ± 4.2	<0.001

Serum levels of Al were related to the number of years on dialysis ($r = 0.282$, $p = 0.011$), as was Ni ($r = 0.286$, $p = 0.01$). Serum levels of Al were related to Ni ($r = 0.391$, $p < 0.001$), and B Mo ($r = 0.283$, $p = 0.005$).

Discussion

Al-containing antacids and Al-containing medications such as sucralfate are the most important current sources of Al for dialysis patients. The antacids are used as intestinal PO_4 binders. Although most of the patients surveyed in this dialysis unit were taking calcium-based PO_4 binders, some patients were

Correct citation: Burgess, E., Audette, R., Hons, R., Taub, K., Schorr, S., and Mandin, H. 1997. Trace metals in hemodialysis patients. In *Trace Elements in Man and Animals – 9: Proceedings of the Ninth International Symposium on Trace Elements in Man and Animals. Edited by* P.W.F. Fischer, M.R. L'Abbé, K.A. Cockell, and R.S. Gibson. NRC Research Press, Ottawa, Canada. pp. 130–131.

also using the older Al-containing antacids. In the hemodialysis patients taking antacids in this study, serum Al levels were significantly increased, as were levels of Ni and Mo. It is unclear what the source of these latter trace elements was.

Aluminum toxicity includes a painful fracturing bone disease and dementia. Anemia may also be associated with Al toxicity. Anemia is a hallmark of chronic renal failure, and has been noted as a cause for resistance to treatment with human recombinant erythropoietin. No patient in this survey was currently being treated for aluminum toxicity.

Molybdenum is an essential element for humans and animals, and is required for the function of xanthine oxidase. The daily requirement is about 25 µg. It is found mostly in legumes and other vegetables, as well as dairy products. Generally, only a limited intake of these foods is recommended for dialysis patients because of their high content of phosphate. Toxicity of Mo is associated with increased serum urate levels and gout. However, virtually all patients on dialysis have increased serum urate levels, presumably on the basis of reduced renal excretion.

Similar to molybdenum, the foods high in nickel are those which are limited for dialysis patients, eg. legumes and nuts. Smoking can be a significant source of nickel, as could be the use of metal cooking utensils. Serum Ni levels were increased in these hemodialysis patients taking Al-containing antacids, and were related to the years on dialysis. Previous studies have noted elevated levels of Ni in dialysis patients, and estimated that there was a net uptake of 100 µg of Ni with each hemodialysis treatment (Scaller et al. 1994). Acute toxicity has occured secondary to nickel contamination of the water used to mix dialysate. Chronic toxicity can result in chronic sinusitis, rhinitis, and asthma, and an increase in nasal and lung cancers has been reported in occupationally-exposed workers.

In conclusion, intake of Al-containing antacids is associated with increased levels of Al, Ni, and Mo. The levels of Al and Ni were close to the upper limit of normal, and the levels of Mo were above normal. When patients were categorized according to serum levels of Al, there was a pattern of increased levels of Ni, Mo, V, Zn, and Cu. The source of the trace elements other than Al is not clear.

References

Scaller K-H, Raithel H-J, Angerer J. (1994) Nickel, In: Handbook on Metals in Clinical and Analytical Chemistry (Seiler HG, Sigel A, Sigel H, eds.), pp 651–666. Marcel-Dekker, Inc. New York. 1994.

Trace Metals in Peritoneal Dialysis Patients and Hemodialysis Patients

E. BURGESS,† R. AUDETTE,‡ R. HONS,† K. TAUB,† S. SCHORR,† AND H. MANDIN†

†Division of Renal Medicine, Foothills Hospital, Calgary, Alberta, Canada T2N 2T9

‡Environmental Toxicology Laboratory, University of Alberta Hospital, Edmonton, Alberta, Canada

Keywords dialysis, aluminum, renal disease, zinc

Introduction

Most trace elements are eliminated from the body via renal excretion. When their renal function falls to approximately 5–10% of normal, patients are placed onto dialysis. Initially, the combination of the patient's own renal function and dialysis maintains a total creatinine clearance of approximately 7–12%, but over time the patient's own renal function deteriorates, and only the modest clearance provided by dialysis is present (2–5% of normal clearance). At some point during this time period, there may not be adequate clearance to maintain normal levels of trace elements.

Hemodialysis offers better clearance to small molecules which are not bound to protein, compared to peritoneal dialysis which allows better clearances for larger molecules. Zinc (Zn) deficiency has been noted in dialysis patients, and postulated as a cause for low gonadotropin levels and related clinical symptomatology. Serum aluminum (Al) levels have been noted to be increased in dialysis patients who are exposed to aluminum in medication and in water used for dialysis.

Methods

One hundred thirty-eight hemodialysis and 44 peritoneal dialysis patients had trace metal screening done as a quality assurrance measure due to concerns over potentially contaminated calcium supplements. Blood was collected into 7 mL royal blue stopper trace element blood and serum collection tubes (Sherwood Medical, Canada). Phlebotomists wore powder-free gloves. Samples were assayed using a Perkin Elmer Elan 5000A ICP-MS. ANOVA and T-tests were used to assess differences across groups.

Results

The mean age of the two groups was similar, 54.7 ± 16 years for peritoneal dialysis patients vs. 57.1 ± 15.9 years for the hemosialysis patients.

Metal	Normal Values	PD (n=44)	HD (n=138)	p value
Aluminum	0.05–0.37 μmol/L	0.38 ± 0.72	0.12 ± 0.27	0.022
Vanadium	0–200 nmol/L	288 ± 57	310 ± 36	0.018
Molybdenum	5–50 nmol/L	62 ± 28	78 ± 35	0.003
Nickel	0–100 nmol/L	150 ± 41	91 ± 17	<0.001
Zinc	8–20 μmol/L	7.9 ± 2.1	6.5 ± 1.2	<0.001

Although only 2 PD patients were on Al-containing antacids, the mean level of serum Al for PD was higher than in HD where 20 patients were confirmed to be taking Al-containing antacids. However, no patients in either group were currently being treated for Al toxicity.

Correct citation: Burgess, E., Audette, R., Hons, R., Taub, K., Schorr, S., and Mandin, H. 1997. Trace metals in peritoneal dialysis patients and hemodialysis patients. In *Trace Elements in Man and Animals – 9: Proceedings of the Ninth International Symposium on Trace Elements in Man and Animals. Edited by* P.W.F. Fischer, M.R. L'Abbé, K.A. Cockell, and R.S. Gibson. NRC Research Press, Ottawa, Canada. pp. 132–133.

The mean levels of vanadium and molybdenum were elevated in both groups of patients, but significantly greater in the HD. On the other hand, mean levels of nickel were significantly higher in PD vs. HD, and were significantly above the normal range for only PD. Serum Zn levels were significantly higher in PD; mean Zn levels were below normal range for the HD.

Discussion

There is not a simple pattern of elevation of all trace elements in this patient population. The primary barrier to over-exposure and toxicity of trace elements is the gastro-intestinal tract. In general, only a very small proportion of ingested trace elements are absorbed. The small amount which has been absorbed is then primarily excreted through the kidneys.

Human studies have demonstrated an increased gastro-intestinal absorption of Al in renal failure patients. Absorption of Al may be augmented by co-administration of citrate, or citrate-containing juices, foods, or medications. It may be that the primary barrier to trace metal poisoning, the gastro-intestinal tract, is compromised in renal failure, and allows an increased absorption. The reduced total body excretion due to markedly reduced renal excretory capacity, although augmented with dialytic removal, can not restore a normal level of the elements in the body.

The subnormal serum Zn levels have been previously described, and reported to be associated with clinical symptomatology. Zinc deficiency may be partially the result of protein restricted diets, used to limit deterioration of renal function and symptomatology of uremia. Acute Ni intoxication has been described in hemodialysis patients, but chronic Ni toxicity has not been noted in peritoneal dialysis. Attempts are being made to locate the source of the nickel in this patient population. Aluminum toxicity in dialysis patients was first described almost 20 years ago. The major source of Al now is Al-containing antacids and medications. These antacids have been used for intestinal binding of phosphate, and have been largely replaced by calcium-based binders.

Elevations of vanadium and molybdenum have not been previously reported. Vanadium toxicity is thought to cause changes in the conjuntivae and respiratory tract, and in severe cases may cause hemolytic anemia, and renal, pulmonary, and nervous system changes. Molybdenum is necessary for the enzyme xanthine oxidase, and chronic exposure to high levels of Mo may cause increased serum urate levels, with resultant gout. However, increased urate levels are commonplace in dialysis patients, and are thought to be a result of reduced renal urate excretion.

In summary, an inconsistent pattern of trace element status is present in dialysis patients. Vanadium and molybdenum levels are significantly elevated, more so in hemodialysis patients.

Do Horses and other Non-Ruminants Require Different Standards of Normality from Ruminants when Assessing Copper Status?

N.F. SUTTLE,[†] J.N.W. SMALL,[†] D.G. JONES,[†] AND K.L. WATKINS[‡]

[†]Moredun Research Institute, Edinburgh, Scotland EH17 7JH. [‡]Royal Hong Kong Jockey Club, Veterinary Department, Sha Tin Racecourse, New Territories, Hong Kong

Keywords copper status, equine, ruminant

Species differences in reference and normal ranges for copper (Cu) concentrations in blood have not been generally recognised and the use of high ranges (18–24 µmol/L) for horses in the United Kingdom may overestimate the incidence of Cu deficiency (Suttle et al. 1995). A worldwide survey showed that for horses ranges varied from 8–14 to 22–28 µmol Cu/L serum (Mee and McLaughlin 1995). The lower ranges equates the horse with ruminants and avoid classifying the majority of healthy horses as 'Cu-deficient'. In ruminants, the relationship between liver and serum Cu provides the basis for assessing Cu status (Vermunt and West 1994) and Cu can be used to differentiate 'deficient', 'marginal' and 'sufficient' bands for serum Cu, providing a more realistic assessment than a single threshold value (Suttle 1994) or reference range. However, ruminants differ from non-ruminants in storing Cu in their livers at low Cu intakes. A relationship between serum and liver Cu was therefore obtained for a population of Thoroughbreds in Hong Kong and used to classify the Cu status of horses in E. Scotland.

Materials and Methods

Hong Kong Survey

Liver and serum samples were obtained from 48 Thoroughbreds, humanely destroyed while out of training or retired after incurring a variety of acute or chronic (mainly musculo-skeletal) injuries during their racing careers in Hong Kong. They were mostly aged between 4 and 11 years old. Immediately prior to euthanasia and exsanguination, blood was obtained by venepuncture using disposable polypropylene syringes, allowed to clot and the serum removed and stored at –20°C until analysis. Liver was sampled by taking approximately 50 g from the surface of one lobe. To characterise the dietary Cu supply, 30 random feed samples (mostly cereals and alfalfa hays) were obtained from local stables. Duplicate fresh liver and dry feed (1.0 g) and serum samples (2.5 mL) were digested with concentrated nitric acid and brought to volume (10 mL) with dilute HNO_3 (1% w/v) (Marrella and Milanino 1986). Copper concentrations were determined by atomic absorption spectrophotometry (AAS) with deuterium background correction. Duplicate Cu determinations were within 4 per cent of the mean and matrix effects were negligible.

Scottish Survey

Serum samples submitted for metabolic profiles from 56 active 'performance' equines in East Scotland in 1995 were also analysed for Cu by AAS, using samples diluted 1:5 with 6% v/v n butanol.

Correct citation: Suttle, N.F., Small, J.N.W., Jones, D.G., and Watkins, K.L. 1997. Do horses and other non-ruminants require different standards of normality from ruminants when assessing copper status? In *Trace Elements in Man and Animals – 9: Proceedings of the Ninth International Symposium on Trace Elements in Man and Animals. Edited by* P.W.F. Fischer, M.R. L'Abbé, K.A. Cockell, and R.S. Gibson. NRC Research Press, Ottawa, Canada. pp. 134–136.

Results

The Cu concentrations in serum, liver and feed varied widely (Table 1). Values for Hong Kong foodstuffs were abnormally distributed but mostly fell between 61–234 µmol/kg DM and the median was only 62.5% of the NRC Cu requirement of horses. Unfortunately, values for feeds could not be matched with serum and liver values. The distribution of serum Cu values in East Scotland was similar to that in Hong Kong, values ranging from 11–28.8 µmol/L and the median values for both populations were 'deficient' by the current UK standard (i.e., <18 µmol/L). The overall relationship between serum (S µmol/L) and liver (L, µmol/kg fresh weight (fw)) Cu in Hong Kong Thoroughbreds was almost significant ($P = 0.054$) but had a high and physiologically unlikely intercept of 12 ± 2.5 µmol/L. Since extremely high values for either serum or liver Cu have a major influence on such relationships while the lowest values have the most diagnostic relevance, the data set was restricted to liver Cu < 190 µmol/kg fw and/or serum Cu < 29 µmol/L which removed four pairs of values. The linear relationship between the remaining values had a much lower intercept (Equation 2) and was highly significant ($P = 0.006$; 42 d.f)

$$S\ (\mu mol/L) = 3.0 \pm 2.49 + 0.125 \pm 0.0244\ L$$

Table 1. Copper concentrations in equine serum, liver and feed samples from retired Hong Kong Thoroughbreds (HK) and in serum from 'performance' horses in E. Scotland (ES): standard deviations given in parentheses and Q1 and Q3 are the lower and upper quartile values.

Source		n	Mean	Median	Q1	Q3
ES	Serum (µmol/L)	56	17.5 (3.67)	16.4	15.1	19.5
HK	Serum (µmol/L)	48	16.7 (4.64)	15.7	13.5	19.3
HK	Liver (µmol/kg FW)	48	113.7 (31.38)	105.7	78.9	239.8
HK	Feed (µmol/kg DM)	30	172.5*	98.6	61.1	233.8

*Feed Cu concentrations were not normally distributed.

Discussion

The mean serum and liver Cu for the sampled Thoroughbred population are similar to those in the literature (Stubley et al. 1993, Cymbaluk and Christensen 1986) and may therefore be considered to be representative of a wider population. The linear relationship described by Equation 2 differs from the curvilinear relationships described in cattle (Vermunt and West 1994), deer (Paynter 1987) and sheep (Suttle 1994) in which plasma or serum values plateau and rarely exceed 16 µmol/L even with liver at >800 µmol/kg fw. Other non-ruminant species such as the mouse (D.G. Jones, unpublished data) maintain serum Cu at or above 16 µmol/L at relatively low liver Cu concentrations, like the horse.

If one assumes that all species have a similar basic need for Cu in their livers which is safely met at a concentration of 105 µmol/kg fw (20 mg/kg DM), that would correspond to an equine serum Cu of 16.0 µmol/L while the threshold of 52.5 µmol/kg fw, proposed to distinguish deficient from marginal liver Cu status in ruminants (Suttle 1994), is equivalent to 11.5 µmol Cu/L in equation. The result of applying this interpretation to the equine samples from Scotland is shown in Table 2: the incidence of 'deficiency' is reduced from 63% to 5%. The interpretation of equine serum Cu values may vary according to the class of horse examined. In the absence of a large liver Cu reserve, horses in training and regularly raced may perform better if maintained with serum and liver Cu concentrations at the upper end of the normal ranges.

Table 2. Alternative diagnostic interpretation of sera from 56 equines sampled in East Scotland for metabolic profiles in 1995.

Cu status	Old UK system	New proposals
Sufficient	37% (>18 µmol/L)	68% (>16µmol/L)
Marginal	–	27% (11.5–16 µmol/L)
Deficient	63% (<18 µmol/L)	5% (<11.5 µmol/L)

However, such refinements in diagnostic interpretation need to be supported by carefully recorded clinical observation and dose/response trials.

It is therefore concluded that

1. Non-ruminant species such as the horse which do not readily store Cu in the liver have higher normal ranges for serum Cu than ruminants.
2. Use of a marginal band to separate the deficient from the sufficient will limit the overdiagnosis of deficiency.
3. For equines, an interim marginal band of 11.5–16.0 μmol Cu/L serum is proposed, pending more detailed investigations.

Literature Cited

Cymbaluk, N.F. and Christensen, D.A. (1986) Copper, zinc and manganese concentrations in equine liver, kidney and plasma. Can. Vet. J. 27: 206–210.

Marella, M. and Milanino, R. (1986) Simple and reproducible method for acid extraction of copper and zinc from rat tissue for determination by flame atomic absorption spectroscopy. At. Spectrosc. 7: 40–42.

Mee, J.F. and McLaughlin, J.G. (1995) 'Normal' blood copper levels in horses. Vet. Rec. 275.

Paynter, D.I. (1987) The diagnosis of copper insufficiency. In: Copper in Animals and Man (Howell, J. McC. and Gawthorne, J.M., eds.), vol. 1, pp. 101–119. CRC Press Ltd. Baca Raton, FL.

Stubley, D., Campbell, C., Dant, C. Blackmore, D.J. and Pierce, A. (1983) Copper and zinc levels in the blood of Thoroughbreds in training in the United Kingdom. Eq. Vet. J. 15: 253–256.

Suttle, N.F. (1994) Meeting the copper requirements of ruminants. In: Recent Advances in Animal Nutrition (Garnsworthy, P.C. and Cole, D.J.A., eds.) Pp. 173-187. Nottingham University Press, Nottingham.

Suttle, N.F., Small, J.N.B. and Jones, D.G. (1995) Overestimation of copper deficiency in horses? Vet. Rec. Feb 4, p.131.

Vermunt, J.H.J. and West, D.M. (1994) Predicting copper status in beef cattle using serum copper concentrations, N.Z. Vet. J. 42: 194–195.

The Use of Caeruloplasmin Activities and Plasma Copper Concentrations as an Indicator of Copper Status in Ruminants

ALEXANDER M. MACKENZIE, DOREEN V. ILLINGWORTH, DAVID W. JACKSON AND STEWART B. TELFER

Department of Animal Physiology and Nutrition, University of Leeds, Leeds, UK, LS2 9JT

Keywords copper, deficiency, ceruloplasmin, ruminants

Introduction

Copper deficiency is an important nutritional disease in ruminants throughout the World. Clinical deficiency signs include alteration in pigmentation, wool keratinization, myelination, infertility, scouring, bone disorders and impaired immune function. However, recent research on the aetiology of ruminant copper deficiency has shown that clinical copper deficiency is not primarily due to lack of copper in the diet. Work by Humphries and Phillippo (unpublished data) has shown that plasma copper concentrations can be lowered down to about 2 μmolar on a purified copper deficient diet without clinical signs of copper deficiency appearing. Hypocupraemia was also produced in cattle fed normal levels of copper but supplemented with iron (250 mg/kg DM), and, again there was no clinical signs of copper deficiency. When molybdenum was introduced into the diets, plasma copper levels did not decrease as drastically, but clinical deficiency signs appeared (Phillippo et al. 1985). It has therefore been shown that clinical copper deficiency involves a complicated interaction between copper, molybdenum, iron and sulphur with molybdenum-sulphur (thiomolybdates) and iron-sulphur compounds being formed in the rumen. When the copper binds to the thiomolybdate or to the iron sulphur complex in the rumen, copper is not absorbed by the animals. However, if the thiomolybdate compound does not bind with copper in the rumen it is absorbed through the rumen wall, into the blood and subsequently the liver where it can then react with copper thus reducing the activities of the copper containing enzymes. It is only when this happens that clinical signs of copper deficiency occur. The diagnosis of copper deficiency has been based on blood or liver copper concentrations but these can give a false indication of adequacy when thiomolybdates are absorbed by the animal. In such cases a significant proportion of the metabolic copper pool is biologically unavailable as it is in the copper-thiomolybdate complex. This study investigated a novel method to assess copper status in both cattle and sheep. It involves the measurement of the activity of a plasma copper enzyme, caeruloplasmin and the plasma copper concentration. Bovine and ovine caeruloplasmin has a molecular weight of about 132,000 dalton, contains 6 copper atoms per molecule and has a turnover of two to three days (Linder 1991). Copper present in caeruloplasmin accounts for about 88% of plasma copper (range 86 to 90%). The model predicts a CP/Pl-Cu ratio of 2.0:1 (CP in mg/dL and Pl-Cu in μmol/L) for the normal animal. If the ratio is less than 2.0 then there is free thiomolybdate being absorbed into the blood which then renders copper unavailable and reduces the activities of the copper enzymes. We propose that a method for determining whether cattle will respond to copper therapy can be based on the relationship between caeruloplasmin (CP) activity and plasma copper (Pl-Cu) concentration.

Correct citation: Mackenzie, A.M., Illingworth, D.V., Jackson, D.W., and Telfer, S.B. 1997. The use of caeruloplasmin activities and plasma copper concentrations as an indicator of copper status in ruminants. In *Trace Elements in Man and Animals – 9: Proceedings of the Ninth International Symposium on Trace Elements in Man and Animals. Edited by* P.W.F. Fischer, M.R. L'Abbé, K.A. Cockell, and R.S. Gibson. NRC Research Press, Ottawa, Canada. pp. 137–138.

Materials and Methods

This study involved over 1500 cattle from dairy or suckler beef herds and over 300 sheep from around the United Kingdom. Heparinised blood and serum samples were obtained from veterinary surgeons for routine trace element evaluation. Plasma copper concentrations were measured by atomic absorption spectrophotometry (1:10 dilution SP9 AA spectrophotometer at 324.8 nm) and caeruloplasmin activities were been determined by the method described by Henry et al. (1974) adapted for the COBAS MIRA (Roche).

Results

Regression analysis shows that caeruloplasmin activity, as expressed as milli-grams of caeruloplasmin per decilitre (mg/dL), is correlated with plasma copper concentration (μmol/L plasma) ($r = 0.55$; $p < 0.01$: $r = 0.67$, $p < 0.01$ for cattle and sheep respectively).

Discussion

As predicted, these results show that plasma caeruloplasmin activity is highly correlated with the plasma copper concentration. The theory predicts that there are 2 Units of CP activity for every 1 μmol Cu per litre plasma but the results of this study show that there is a wide variation in this ratio. This was assumed to be due to less than the 88% of copper present in plasma in the form of caeruloplasmin or by the inhibition of caeruloplasmin activity. Caeruloplasmin is a constitutive protein, being produced and released at a constant rate by the liver under normal circumstances. Caeruloplasmin is recognised as an acute phase protein and elevated activities can be found at time of infection but this response cannot relate to levels of activities lower than expected. Therefore, in cases of a lower than predicted CP/Pl-Cu ratio, the decreased activity of caeruloplasmin is likely to be due to an increased secretion of apo-caeruloplasmin (CP that does not have copper at the active site) or an inhibition of the activity by thiomolybdate. The recognition that clinical copper deficiency signs are due to a thiomolybdate toxicity and not a pure dietary deficiency of copper, mean that this CP/Pl-Cu ratio can be used as a method of identifying cattle which, despite an apparently "normal" copper status, will respond to copper therapy, the clinical problem being one of thiomolybdate toxicity rather than a lack of copper for metabolic function. Mackenzie et al. (1997) have shown that 75% of cattle in this study would be classified as requiring copper supplementation based on the CP/Pl-Cu ratio whereas with the current diagnostic method of measuring blood copper concentrations only 20 to 30% were diagnosed as deficient. Clinical evidence suggests that many of the cattle which were diagnosed as normal by conventional techniques still had clinical signs normally associated with copper deficiency whereas when the decision to supplement with copper was based on the CP/Pl-Cu ratio, an improvement in performance and fertility was observed.

References

Henry, R.J., Cannon, D.C. and Winkelman, J.W. (1974) Clinical Chemistry, Principles and Technics. pp. 860. Harper & Row, Publishers, London.

Humphries, W.R., Bremner, I. and Phillippo, M. (1985) In: TEMA–5. (Mills, C.F., Bremner, I. and Chesters, J.K., eds.). pp. 371–374. CAB, Slough.

Linder, M.C. (1991) Biochemistry of copper. Plenum Press, New York and London.

Mackenzie, A.M., Illingworth, D.V., Jackson, D.W. and Telfer, S.B. (1997) A comparison of methods of assessing copper status in cattle. In: Trace Elements in Man and Animals – 9: Proceedings of the Ninth International Symposium on Trace Elements in Man and Animals. (Fischer, P.W.F., L'Abbé, M.R., Cockell, K.A. and Gibson, R.S., eds.). NRC Research Press, Ottawa, Canada. pp. 301–302.

Phillippo, M., Humphries, W.R., Bremner, I., Atkinson, T. and Henderson, G. (1985). In: TEMA–5. (Mills, C.F., Bremner, I. and Chesters, J.K., eds.). pp. 176–180. CAB, Slough.

Kinetic Modelling

Chair: L. Miller

Zinc Exchange in ^{65}Zn Labeled Rats at Physiologically Adequate Zinc Supply

WILHELM WINDISCH AND MANFRED KIRCHGESSNER
Institute of Nutrition Physiology, TU Munich-Weihenstephan, 85350 Freising, Germany

Keywords zinc, absorption, tissue Zn exchange, rat

Quantitative studies on Zn metabolism are focused predominantly on the effect of extreme nutritional conditions, while the normal situation of Zn metabolism is used only as a control. The present experiment, however, was designed to study the quantitative Zn metabolism under physiologically adequate conditions with special emphasis on the homeostatic interplay of Zn absorption and excretion and the mode of exchange of Zn within the whole body and individual tissues.

Material and Methods

In the present experiment, 32 female Sprague Dawley rats were homogeneously labeled with ^{65}Zn by the alimentary labeling procedure (Windisch and Kirchgessner 1994a, 1995a). For this purpose, the animals were reared for 9 weeks from 50 to 229 g of body weight with a semisynthetic diet containing 23 µg/g Zn and 8.5 Bq ^{65}Zn/µg Zn. Subsequently 5 of the adult animals were killed for reference. The remaining animals were divided into 9 groups (n = 3) and received restrictively (8.4 g/d) diets with Zn contents of 19, 23, 29, 37, 45, 58, 73, 92 or 114 µg/g without radiozinc. Fecal and renal excretions were collected daily. After 23 d, the animals were killed and completely dissected into organs and tissue fractions. The excrements and tissues were analyzed for Zn and ^{65}Zn.

Results and Discussion

The body weight of the animals remained unchanged (229 g). The activity of alkaline phosphatase in blood plasma was not affected by the treatment and ranged at levels typically low for adult animals (72 U/L) (Windisch and Kirchgessner 1995a).

The increasing Zn intake (160 to 958 µg/d) was completely compensated for by a respective rise in fecal Zn excretion (143 to 947 µg/d). The apparent Zn absorption, urinary Zn excretion and Zn retention were not affected by treatment (15 µg/d; 5 µg/d and 0 µg/d) (Windisch and Kirchgessner 1995a). Therefore, the Zn metabolism of the animals was well balanced at all levels of Zn supply. True Zn absorption and endogenous fecal Zn excretion, as calculated by the isotope dilution technique, increased asymptotically toward threshold values of 163 and 151 µg/d. The mirror imaged course and the asymptotic shape of both Zn fluxes indicate that true absorption of dietary Zn and fecal excretion of endogenous Zn are the result of a reciprocal displacement within the digestive tract. Obviously, considerable amounts of endogenous Zn are secreted into the digestive tract and mixed with dietary Zn. This mixture of Zn is submitted to absorption, irrespective of its origin (dietary/endogenous). With increasing Zn intake, the proportion of

Correct citation: Windisch, W., and Kirchgessner, M. 1997. Zinc exchange in ^{65}Zn labeled rats at physiologically adequate zinc supply. In *Trace Elements in Man and Animals – 9: Proceedings of the Ninth International Symposium on Trace Elements in Man and Animals.* *Edited by* P.W.F. Fischer, M.R. L'Abbé, K.A. Cockell, and R.S. Gibson. NRC Research Press, Ottawa, Canada. pp. 139–140.

dietary Zn in the mixture rises at the expense of endogenous Zn. Thus, more dietary Zn is absorbed and more Zn of endogenous origin is excreted with the feces. According to this model, the maximum true absorption of dietary Zn is equivalent to the capacity of total absorption (dietary and endogenous) and the maximum endogenous fecal excretion is equivalent to the amount of endogenous secretion into the digestive tract. Estimates of these maximum values are given by the threshold values of true absorption and endogenous fecal excretion derived in the present study.

According to the time course of ^{65}Zn elimination via feces and urine, whole body Zn was exchangeable at 40%, irrespective of the level of Zn supply (Windisch and Kirchgessner 1995a) This observation confirms former experimental results (Windisch and Kirchgessner 1994b, d). The half-life of mobile Zn fell from 28 to 16 d as dietary Zn content rose from 19 to 114 μg/g. The decrease of half-life was asymptotic and tended towards a threshold of 13.5 d. This value may denote the maximum speed of Zn exchange within the mobile Zn pool and can be interpreted as an estimate of Zn turnover under the condition of a physiologically adequate Zn supply.

The amount and the concentration of Zn in the tissues remained unchanged among the treatment groups and also in comparison to the reference level (Windisch and Kirchgessner 1995b). In all tissues, the content of ^{65}Zn decreased significantly from the reference level to any of the treatment groups. Thus, Zn exchange was present in all tissues. Within treatment groups, the extent of Zn exchange was significantly pronounced by increasing levels of Zn supply. Since this reaction was asymptotic, the mobile Zn pool within individual tissues could be calculated by extrapolation (Windisch and Kirchgessner 1995b). For liver, spleen, lung, reproductive organs, pancreas, digestive tract and blood the estimate of mobile Zn pool was about 100% of total tissue Zn, for brain and kidney 76% and 77%, for the complete muscle and fat tissue 71%, and for the complete coat (skin + hair) and skeleton 27% and 23%. The decreasing order in mobility of tissue Zn from organs and blood across muscle and fat tissue to coat and skeleton confirms former findings (Windisch and Kirchgessner 1994c). However, the skeleton and coat contained considerable quantities of mobile Zn (15% of whole body Zn), because large amounts of whole body Zn were localized in these tissue fractions (58% of whole body Zn). Since in Zn deficiency, quantitative relevant amounts of Zn are mobilized only from the skeleton and coat, the mobile Zn from this tissue may be an estimate for the size of the mobilizable Zn reserve of the organism.

Literature Cited

Windisch, W. & Kirchgessner, M. (1994a) Zur Messung der homöostatischen Anpassung des Zinkstoffwechsels an eine defizitäre und hohe Zinkversorgung nach alimentärer ^{65}Zn-Markierung. J. Anim. Physiol. a. Anim. Nutr. 71: 98–107.

Windisch, W. & Kirchgessner, M. (1994b) Zinkexkretion und Kinetik des Zinkaustauschs im Ganzkörper bei defizitärer und hoher Zinkversorgung. J. Anim. Physiol. a. Anim. Nutr. 71: 123–130.

Windisch, W. & Kirchgessner, M. (1994c) Verteilung und Austausch von Zink in verschiedenen Gewebefraktionen bei defizitärer und hoher Zinkversorgung. J. Anim. Physiol. a. Anim. Nutr. 71: 131–139.

Windisch, W. & Kirchgessner, M. (1994d) Verteilung von Calcium und Zink sowie Zinkaustausch im Körpergewebe bei defizitärer und hoher Calciumversorgung. J. Anim. Physiol. a. Anim. Nutr. 72: 195–206.

Windisch, W. & Kirchgessner, M. (1995a) Anpassung des Zinkstoffwechsels und des Zn-Austauschs im Ganzkörper ^{65}Zn-markierter Ratten an eine variierende Zinkaufnahme. J. Anim. Physiol. a. Anim. Nutr. 74: 101–112.

Windisch, W. & Kirchgessner, M. (1995b) Zinkverteilung und Zinkaustausch im Gewebe ^{65}Zn-markierter Ratten. J. Anim. Physiol. a. Anim. Nutr. 74: 113–122.

Compartmental Model of Zinc Metabolism in Healthy Women

N.M. LOWE,[†] D.M. SHAMES,[‡] L.R. WOODHOUSE,[†] J.S. MATEL,[†] AND J.C. KING[¶]
[†]Department of Nutritional Sciences, University of California at Berkeley, CA 94720. [‡]Department of Radiology, University of California at San Francisco, CA 94143, and [¶]USDA Western Human Nutrition Research Center, San Francisco, CA 94129, USA

Keywords zinc, stable isotopes, tracer kinetics, compartmental model

Compartmental analysis of tracer kinetics has provided valuable insights into the metabolism of many nutrients, including trace minerals. Using radioactive isotopes a detailed compartmental model of zinc metabolism was developed in humans (Wastney et al. 1986). Stable isotopes provide more limited information, but are useful for populations where radioactive tracers are discouraged. Our research group is interested in defining the mechanisms regulating zinc homeostasis in pregnant and lactating women. Thus, we developed a simple model of zinc metabolism in young women, using stable isotopes of zinc, that can be used in the future to define the relationship between zinc status and homeostasis in pregnant women.

Materials and Methods

Subjects: Six women, aged 30 ± 11 years (mean ± SD), were recruited for the study. All of the women were Caucasian and none had acute or chronic health problems. The weight of the women averaged 54.2 ± 8.9 kg, and their body mass index was 20.7 ± 2.6 kg/m^2. The usual dietary zinc intakes of this group, assessed prior to the study with 3-d weighed food intake records, averaged 8.3 ± 3.4 mg/d. The experimental design was approved by the University of California at Berkeley Committee for the Protection of Human Subjects. All participants gave written, informed consent.

Experimental Design: Subjects were maintained on a constant diet containing 7.0 ± 0.1 mg Zn/d for a 7-d equilibration period prior to tracer administration. On the morning of day 8 an indwelling catheter was placed in an arm vein and a fasting blood sample (8 mL) taken in a heparinized zinc-free polypropylene syringe ("Monovet", Sarstedt, Hayward, CA). Fifteen minutes after consuming a breakfast meal containing 1 mg zinc, each subject drank 213 g orange juice containing 1.3 mg of the oral tracer ^{67}Zn (enriched to 91.8% abundance; Cambridge Isotope Laboratories, Woburn, MA). Immediately thereafter, 0.4 mg of a second tracer, ^{70}Zn (enriched to 85.03% abundance; Oakridge National Laboratory, Oak Ridge, TN), was administered intravenously in the arm opposite to that used for sampling. Blood samples (8 mL) were taken via the catheter at timed intervals during the 7 h immediately post ^{70}Zn administration, and daily for the next 7 d. The constant diet was continued for this 7-d period. All urinary and fecal output was collected for 7 and 11 d following tracer administration, respectively. All plasma, urine and fecal samples were stored at –20° until analysis.

Sample Preparation and Analysis: Total zinc concentrations of plasma, urine and fecal samples were measured using atomic absorption spectroscopy (AAS) (Smith-Hieftje-22, Thermo Jarrell Ash, Franklin, MA). Zinc isotope ratios were determined using inductively coupled plasma mass spectrometry (ICP-MS). For mass spectrometer analysis, plasma samples (3–4 mL) and fecal samples (0.3–0.5 g) were wet ashed in 5 mL concentrated HNO_3 by microwave digestion, and inorganic salts were removed from urine samples using a chelating resin (Chelex 100 resin, 100–200 mesh, sodium form, Bio-Rad Laboratories, Hercules, CA). Zinc was purified from the mineral solutions using ion exchange chromatography (Type AG1X-8,

Correct citation: Lowe, D.M., Shames, D.M., Woodhouse, L.R., Matel, J.S., and King, J.C. 1997. Compartmental model of zinc metabolism in healthy women. In *Trace Elements in Man and Animals – 9: Proceedings of the Ninth International Symposium on Trace Elements in Man and Animals. Edited by* P.W.F. Fischer, M.R. L'Abbé, K.A. Cockell, and R.S. Gibson. NRC Research Press, Ottawa, Canada. pp. 141–143.

BioRad Laboratories). All acids used for the preparation of samples for ICP-MS were ultrapure (HCl: Optima brand, Fisher Scientific, Pittsburg, PA; HNO_3: Seastar brand, Seastar Chemicals Inc., Seattle, WA). Zinc isotope ratios were measured using a Perkin-Elmer Sciex ELAN 500 inductively coupled plasma mass spectrometer (Perkin-Elmer, Norwalk, CT). The data acquisition parameters were described previously (Roehl et al. 1995).

Kinetic Analysis: Stable isotope enrichment and total zinc mass in the plasma, urine and feces were analysed concurrently using SAAM/CONSAM (Berman and Weiss 1978). The model was numerically identified in each subject by using least squares parameter estimation.

Results and Discussion

The model developed to describe the isotope kinetics and steady state data is shown in Figure 1, along with the average values for the rate constants and compartment masses calculated for this subject group. The model fits the data from each subject well and has the least number of compartments required to account for the dynamic properties of our data and to remain consistent with known physiology. A linear array of three compartments was used to describe the gastrointestinal (GI) tract, compartment 4 corresponding to the stomach, compartment 5 the intestine and compartment 6 the colon. Compartment 1 is the plasma. It exchanges bi-directionally with the intestine (absorption and endogenous secretion). The fractional absorption of zinc (FZA), given by the ratio of $k_{1,5}$ to the sum of $k_{1,5}$ and $k_{6,5}$, was 0.28 ± 0.04 (MEAN ± SEM). The rates of endogenous secretion (Mass of compartment 1 × $k_{5,1}$) and excretion [(Mass of compartment 1 × $k_{5,1}$) × (1 – FA)] were 2.8 ± 0.5 mg/d and 2.0 ± 0.3 mg/d, respectively. The sum of the masses of the compartments that exchange rapidly with the plasma (compartments 1–3 and 5) are referred to as the exchangeable zinc pool (EZP). Those compartments may be a measure of zinc status

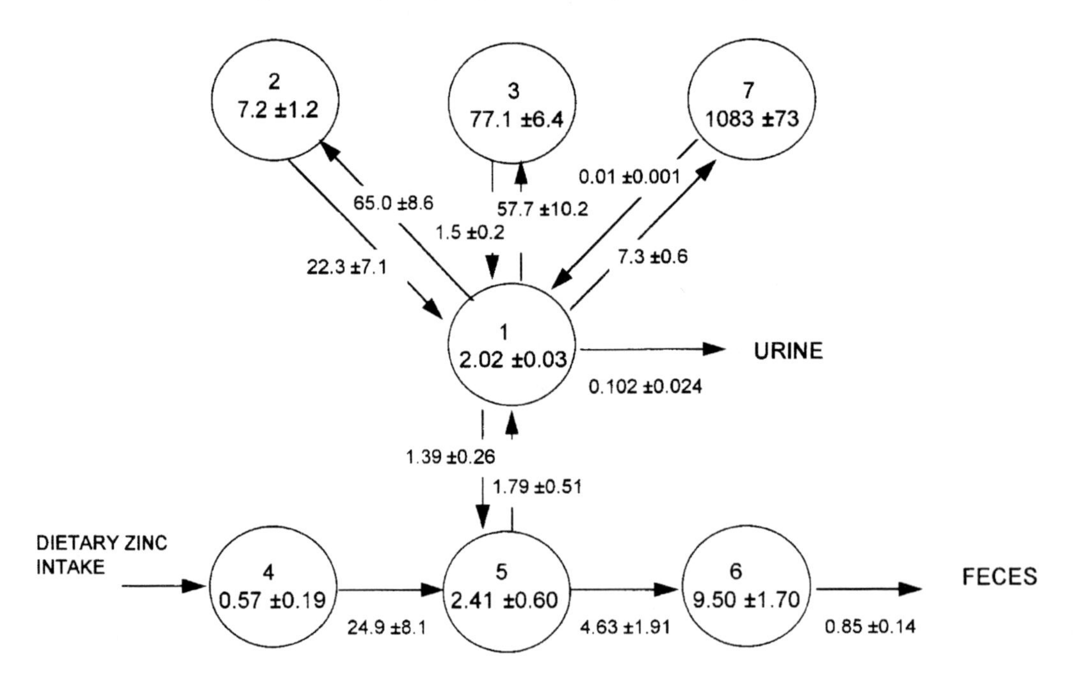

Figure 1. The circles indicate compartments representing kinetically distinct pools of zinc. The compartments are identified by numbers given at the top of each circle; the lower number is the compartment mass (mg). The arrows between the compartments represent the parameters, $k_{i,j}$, (/day), i.e., the transfer rate constants of zinc form compartment j to compartment i. The rate constants and compartment masses represent the mean ± SEM for the six subjects.

(Miller et al. 1994). The EZP averaged 89 mg in these six women, which is estimated to be about 8% of total body zinc. There was a significant negative correlation (r = 0.69) between FZA and EZP when EZP is expressed on a per kg body weight basis. This inverse relationship suggests that the EZP may play a role in the homeostatic mechanisms that determine the proportion of dietary zinc absorbed. We are currently applying this compartmental modeling technique to data from a human zinc depletion-repletion study to more fully describe the relationship between Zn status, absorption and gastro-intestinal secretion.

Literature Cited

Berman, M. & Weiss, M.F. (1978) SAAM Manual. Washington, DC: US Printing Office, [DHEW Publication (NIH) 78–180].

Miller L.V., Hambidge, K.M., Naake, V.L., Hong, Z., Westcott, J.L., & Fennessey P.V. (1994) Size of the zinc pools that exchange rapidly with plasma zinc in humans: alternative techniques for measuring and relation to dietary zinc intake. J. Nutr. 124: 268–276.

Roehl, R., Gomez, J., & Woodhouse, L.R.. (1995) Correction of mass bias drift in inductively coupled plasma mass spectrometry measurements of zinc isotope ratios using gallium as an isotope ratio internal standard. J. Anal. At. Spectrom. 10: 15–23.

Wastney, M.E., Aamondt, R.L., Rumble, W.F., & Henkin, R.I. (1986) Kinetic analysis of zinc metabolism and its regulation in normal humans. Am. J. Physiol. 251: R398–408.

Compartmental Modeling of Human Zinc Metabolism: Evaluation of Method for Estimating the Size of the Rapidly Exchanging Pool of Zinc*

L.V. MILLER,[†] N.F. KREBS,[†] M. JEFFERSON,[†] D. EASLEY,[‡] AND K.M. HAMBIDGE[†]

[†]Center for Human Nutrition, University of Colorado Health Sciences Center, Denver, CO 80262, and [‡]Scott & White Hospital, Temple, TX 76508, USA

Keywords zinc metabolism, compartmental modeling, tracer kinetics, exchanging pools

Zinc performs a variety of physiological functions as a structural ion, catalytic ion, and control ion. Since zinc must be readily available for many of the roles it plays, it is likely that the <10% of total body zinc that exchanges relatively rapidly with plasma zinc is responsible for many of the known physiological functions of this micronutrient. Furthermore, this mobilizable pool of zinc is likely a central participant in zinc homeostatic mechanisms. The ability to measure the rapidly exchanging pool (EZP) may, therefore, provide valuable insight into zinc metabolism and prove to be a useful index of zinc status. We have previously reported our initial investigation of a relatively simple method for estimating the size of the EZP using isotope tracer kinetics (Miller et al. 1994). The EZP is defined as the combined pools of zinc that exchange (intermix) with the plasma zinc within 48 h. The method uses an extrapolation (to the time of tracer administration) of tracer enrichment data from between 2 and 10 d to estimate the enrichment resulting from the dilution of tracer into only the EZP. While our preliminary data demonstrated great promise, we also discussed in detail the inherent limitations of the method. Because the method effectively treats a system of pools as a single pool, it may produce inaccurate results. Two sources of inaccuracy are: (1) the method does not account for initial rapid losses of tracer from the EZP and (2) the individual pools comprising the EZP may not be homogeneously enriched. Here we report on the evaluation of this method using compartmental modeling of data from multiple isotope tracer studies and modifications of the method to improve its accuracy.

Methods

Doses of enriched ^{67}Zn, ^{68}Zn, and ^{70}Zn were administered orally (with meals and in the fasting state) and intravenously to 4 healthy adults. Isotope enrichment was measured in plasma, red cells, urine, and feces to at least 9 d after administration. Data for all isotope tracers from all sampling sites were modeled simultaneously using the SAAM/CONSAM computer programs. Steady state data derived by other tracer kinetic analysis methods were also used to fit the model. The resulting models have 14 or 15 compartments and are similar in structure to a previously reported model of human zinc metabolism based on extensive radioisotope data (Wastney et al. 1986).

Results and Discussion

EZP data from the models were compared to EZP estimates obtained with the method being evaluated, using both plasma and urine data (Table 1). In all but one case the method overestimates EZP size, as previously predicted (Miller et al. 1994). The overestimation is typically around 30%. Several modifications

*Supported in part by National Institutes of Health grants DK12432, DK07658, DK02240, DK48520, and RR00051.

Correct citation: Miller, L.V., Krebs, N.F., Jefferson, M., Easley, D., and Hambidge, K.M. 1997. Compartmental modeling of human zinc metabolism: Evaluation of method for estimating the size of the rapidly exchanging pool of zinc. In *Trace Elements in Man and Animals – 9: Proceedings of the Ninth International Symposium on Trace Elements in Man and Animals. Edited by* P.W.F. Fischer, M.R. L'Abbé, K.A. Cockell, and R.S. Gibson. NRC Research Press, Ottawa, Canada. pp. 144–145.

Table 1. Comparison of EZP size data.

Subject	Body Wt. (kg)	Diet. Zn (mg/d)	$EZP_{estimated}$ (mg)	EZP_{model} (mg)	% Dev. from model
A (plasma)	47	7	181	138	31
(urine)			184	138	33
B (plasma)	59	8	257	197	30
(urine)			188	197	–5
C (plasma)	74	20	223	187	19
(urine)			239	187	28
D (plasma)	73	13	270	201	34
(urine)			212	201	5
					Mean = 22±15

Table 2. Corrections to EZP estimation data.

Subject	EZP (corrected for losses	% Dev. from model	EZP (correct w/exp param)	% Dev. from model
A (plasma)	174	26	134	–3
(urine)	177	28		
B (plasma)	246	25	200	2
(urine)	180	–9		
C (plasma)	211	13	187	0
(urine)	226	21		
D (plasma)	257	28	137	–32
(urine)	202	00		
		Mean = 17±14		

to correct inaccuracies and inconsistencies were explored. First, all urine and fecal data were used to calculate total tracer losses via those routes and correct for the initial rapid losses of tracer from the EZP. This did improve the accuracy slightly, decreasing the average deviation from 22% to 17%, as shown in columns 2 and 3 of Table 2. The second effort to improve the method's accuracy involved the development of a correction factor using additional enrichment data and based on the manipulation of kinetic parameters from a previously published model (Wastney et al. 1986). This model was used because it incorporates data from a much larger study population. The results demonstrate a strong correlation ($r = 0.95$) between a particular algebraic combination of parameters from a sum of exponentials regression of the plasma enrichment data from 6 h to 10 d and the deviation of estimated EZP values. When this relationship is applied to our plasma data, the EZP estimations are adjusted to within 3% of the models' EZP values in 3 out of 4 cases (columns 4 and 5 of Table 2).

Conclusions

Although the correction techniques investigated require additional data, they do improve the accuracy of the EZP estimation method relative to the model data and, furthermore, are consistent with our understanding of the method and its limitations. Additional subjects are being studied to further our understanding of these EZP estimation techniques, as well as to confirm the validity of the compartmental modeling.

Literature Cited

Wastney M.E., Aamodt R.L., Rumble W.F., & Henkin, R.I. (1986) Kinetic analysis of zinc metabolism and its regulation in normal humans. Am. J. Physiol. 251:R398–R408.

Miller L.V., Hambidge K.M., Naake V.L., Hong Z., Westcott J.L., & Fennessey P.V. (1994) Size of the Zinc Pools that Exchange Rapidly with Plasma Zinc in Humans: Alternative Techniques for Measuring and Relation to Dietary Zinc Intake. J. Nutr. 124:268–276.

On-Line Access to Published Models of Biological Systems*

M.E. WASTNEY,[†] N. BROERING,[‡] J. BLUMENTHAL,[‡] AND R. BOSTON[¶]

[†]Department of Pediatrics and [‡]Dahlgren Memorial Library, Georgetown University Medical Center, Washington, DC 20007, and [¶]University of Pennsylvania School of Veterinary Medicine, New Bolton Center, Kennett Square, PA 19348, USA

Keywords kinetics, mathematical modeling

Biological systems are being studied increasingly using mathematical modeling. Models are being used to describe systems and to represent theories on how systems function. The reason that they are being applied is first, because biological systems are inherently complex (Garfinkel 1984) and second, because the tools for modeling, computers and modeling software (Van Milgen et al. 1996) have been developed and are readily available to investigators. In terms of trace elements and minerals, the number of kinetic studies has increased with the recent development of techniques for stable isotope analyses in biological samples (Barnes 1993).

Models are developed to predict and explore systems, and models for many trace elements and minerals including calcium, copper, iodide, iron, chromium, selenium, mercury, and zinc have been published (see Siva Subramanian and Wastney 1995). Once a model has been developed to fit a particular set of data, it forms a 'knowledge base' about the system and it can be used to analyze data obtained under different conditions, to make predictions, and to design further studies on the system. By using an existing model to design new studies, experimental resources (such as supplies and subjects) can be used more efficiently (Southgate 1995). The model can be used to identify areas where data are lacking, determine the length of study and the number of samples and optimal sampling times.

In spite of their utility, it is difficult to locate and obtain working versions of published models; models on a similar topic are often published in diverse journals and although models are developed on computers they are published in a written format. Access to a working, or electronic version of a published model currently relies on co-operation of the developer to supply the model or for the user to reconstruct the model from the paper. This can be time-consuming and error-prone. Therefore, we are developing a facility of published models for on-line access via the Internet, called a *model library* (Wastney et al. 1995). The address of the library is

http://gopher.dml.georgetown.edu/model/model.html

The library contains models of metabolism, endocrinology, physiology, biochemistry, and toxicology. Models of interest can be located by subject, author, model type, or software. Once a model is located, it can be viewed graphically, in terms of its equations, or the plots of the model simulation can be viewed. Limited simulations can be performed in the library but if a user wishes to use a model more extensively, it can be downloaded to their system and run with their own modeling software. Users may upload published models to the library electronically and send comments or enquiries on particular models. The purpose of the library is to facilitate location and access to models so they can be used for their intended purpose, to explore and test systems through simulation.

*Supported by NSF BIR-9503872.

Correct citation: Wastney, M.E., Broering, N., Blumenthal, J., and Boston, R. 1997. On-line access to published models of biological systems. In *Trace Elements in Man and Animals – 9: Proceedings of the Ninth International Symposium on Trace Elements in Man and Animals. Edited by* P.W.F. Fischer, M.R. L'Abbé, K.A. Cockell, and R.S. Gibson. NRC Research Press, Ottawa, Canada. pp. 146–147.

Literature Cited

Barnes, R.M. (1993) Advances in inductively coupled plasma mass spectrometry: human nutrition and toxicology. Analyt. Chim. Acta 283: 115–130.

Garfinkel, D. (1984) Modeling of inherently complex biological systems: Problems, strategies, and methods. Math. Biosci. 72: 131–139.

Siva Subramanian, K.N., & Wastney, M.E. (1995) Kinetic Models of Trace Element and Mineral Metabolism During Development. P 416. CRC Press, Boca Raton.

Southgate, D.A.T. (1995) Design Models. Br. J. Nutr. 73: 1–2.

Van Milgen, J., Boston, R., Kohen, R., & Ferguson, J. (1996) Comparison of available software for dynamic modeling. Annales de Zootechnie (In Press).

Wastney, M.E., Broering, N., Ramberg, C.F.R., Jr., Zech, L.A., Canolty, N., & Boston, R.C. (1995) World-wide access to computer models of biological systems. Info Serv Use 15: 185–191.

Trace Elements in the Environment and Food Supply

Chairs: R.S. Gibson and R.W. Dabeka

Impact of Prenatal and Early Infant Feeding Practices of Native Indians in the Moose Factory Zone on Lead, Cadmium and Mercury Status

R.M HANNING,[†] R. SANDHU,[‡] A. MACMILLAN,[†] L. MOSS,[†] AND E. NIEBOER[‡]
[†]Departments of Pediatrics and [‡]Biochemistry, McMaster University, Hamilton, Ontario, Canada L8N 3Z5

Keywords indigenous peoples, infant nutrition, lead, cadmium

Among the First Nations communities there has been growing concern about the risks associated with environmental contamination of the food supply, including human milk. Chemicals which bind to milk proteins, such as the heavy metals, lead (Pb) , cadmium (Cd) and mercury (Hg), may also be present in human milk. Since these metals are persistent and accumulate in the food chain, they may be found in some traditional foods consumed by aboriginal women and transferred to the fetus or breastfed infant. Although the long term effects of infants' exposure to contaminants via breastmilk are largely unknown, the presence of potentially toxic chemicals in milk must be of concern.

Infant formulas and foods may also be a source of Pb and Cd exposure (Dabeka 1989, Dabeka and McKenzie 1988). However, with the virtual elimination of Pb-soldered cans over the past two decades, intakes of formula or evaporated-milk fed infants appear to be below accepted limits.

Exposure to Pb in utero and in early life presents a risk of nephropathy, hematological, and neurological impairment (Lockitch 1993). Although these disabilities have most often been associated with high dose exposures, e.g., blood Pb > 1.93 μmol/L, there is no defined "safe" level. Blood Pb levels as low as 0.48–0.96 μmol/L have been associated in some studies with cognitive impairment in infancy and childhood and reduced childhood stature (e.g., Baghurst et al. 1992, Needleman and Bellinger 1991). The longterm developmental effects have been debated (Pocock et al. 1994, Wolf et al. 1994). Nevertheless, there was sufficient basis for concern that the U.S. Centers for Disease Control have, over the years, lowered the blood Pb level considered to represent a risk to health to the current level of 0.48 μmol/L (100 μg/L) (1991).

Prior to initiating the current study, the Northern Ontario blood Pb study conducted in 1987 by the Ontario Ministries of Health (M.O.H.) and of the Environment (M.O.E.), had reported blood Pb levels of 0.44 ± 0.02 μmol/L in 76, 3–6-year-old children in nearby Moosonee (Goss et al. 1989). Surprisingly, in this remote setting, values were as high as in urban areas of the province. The explanation for the elevation wasn't apparent, though lead in air, soil and drinking water and socioeconomic status were the major predictors for the study at large (n = 838).

Cadmium is also persistent and high levels have been found to accumulate in Arctic mammals and fishes (Muir et al. 1987). Levels in traditional foods eaten in the Moose Factory Zone are unknown.

Correct citation: Hanning, R.M., Sandhu, R., MacMillan, A., Moss, L., and Nieboer, E. 1997. Impact of prenatal and early infant feeding practices of native Indians in the Moose Factory zone on lead, cadmium and mercury status. In *Trace Elements in Man and Animals – 9: Proceedings of the Ninth International Symposium on Trace Elements in Man and Animals. Edited by* P.W.F. Fischer, M.R. L'Abbé, K.A. Cockell, and R.S. Gibson. NRC Research Press, Ottawa, Canada. pp. 148–151.

Cadmium is higher in soy than milk-based formulas, but infant intakes have been below the FAO/WHO Provisional Tolerate Weekly Intake (P.W.T.I.) (Dabeka and McKenzie 1989). Cigarette smoking is another major source of Cd (Frierg et al. 1985). It has been estimated that as many as 50% of pregnant women living on reserves in the Moose Factory area smoke (Health and Welfare Canada 1990). Cd is known to affect placental structure and influence fetal development in animals, and acts as a potent renal toxin in humans (Frierg et al. 1985), therefore, Cd exposure was of interest.

Mercury bioaccumulates as methylHg in arctic and subarctic fish and marine mammals. Since conditions in the Moose River favour the accumulation of methylHg in fish flesh, excessive placental transfer could occur in women with high fish intakes (P.T.W.I. of Hg = 3.3 μg/kg body weight). The fetal nervous system is especially sensitive to Hg toxicity (Clarkson et al. 1985).

As part of a larger study of the nutritional, immunological and environmental impact of infant feeding practices of Native Indians living in the Moose Factory Zone, the objectives of this study were: 1) to describe Pb, Hg and cadmium Cd in maternal and cord blood at birth and infant blood at four months of age; and 2) to determine the influence of diet on Pb, Hg and Cd status.

Methods

The study was approved by the McMaster University Research Advisory Group and the (then) Moose Factory Zone Board of Health. Healthy term aboriginal infants (n = 91) living in the northern Ontario communities of Peawanuk, Kashechewan, Fort Albany, Attawapiskat, Moose Factory and Moosonee were recruited at birth and studied over their first four months postnatally according to their feeding cohort: breastmilk, commercial infant formula or evaporated milk-based feedings.

Whole blood samples were collected at the Moose Factory Hospital from mothers and the umbilical cord. Infant blood samples were obtained by heel stab capillary sample at the time of the infant's 4 month health check up . Extreme caution was taken to minimize contamination.

Samples of breastmilk, expressed by electric or manual pump from the contralateral breast during the first morning feedings, were obtained at two and four months. Infant formula or evaporated milk-based feedings and representative water samples were also collected. Blood and milk samples were analyzed for metals using cold vapour (Hg) and electrothermal (Pb, Cd) atomic absorption spectrometry. The detection limit for each metal was 2.0 μg/L or less.

Maternal recall of traditional animal foods consumed over the previous year, by season, was recorded by interview using a 188-item scale. Energy intake from these foods was estimated using Canadian Nutrient File Data (1991) to give a gross index of consumption.

Results (Hanning et al. 1996)

Maternal and cord blood Pb levels were similar (22 ± 13 and 22 ± 17 μg/L, respectively, X ± SD) and strongly correlated ($r = 0.77$, $p < 0.001$). Values above 0.48 μmol/L (100 μg/L) were observed in 3% of cord blood samples. Breastmilk Pb ranged from 1.0 to 7.7 μg/L (n = 22) and correlated with maternal blood lead ($r = 0.66$, $p < 0.003$). Infant blood lead (17 ± 10 μg/L, n = 25) correlated with matched cord blood Pb ($r = 0.73$, $p < 0.0001$). Not surprisingly, the Pb levels in formula or evaporated milk-based feedings, as prepared, at 5 ± 2 μg/L and 6 ± 3 μg/L, were slightly higher than has been reported for formulas analyzed as manufactured (Dabeka 1989, Thompson 1993). Nevertheless, no infant exceeded the P.T.W.I..

Maternal blood Cd was higher than cord blood Cd (1.7 ± 1.1 vs. 0.8 ± 0.9 μg/L), but was correlated ($r = 0.48$, $p < 0.0001$). In the 62% of women who smoked during pregnancy, maternal, but not cord blood Cd, correlated with the number of cigarettes smoked ($r = 0.34$, $p < 0.05$). Maternal and cord blood Hg were, respectively, 5 ± 4 and 3 ± 3 μg/L.

Maternal and cord blood Pb were significantly correlated with maternal total traditional animal food intake ($r = 0.46$, $p < 0.0001$) as were the individual contributions of wild fowl, mammals and, to a lesser extent, fish. In contrast there were no significant associations for blood Cd and traditional food consumption. For Hg, a weak association with mammal consumption was observed ($r = 0.30$, $p < 0.03$), but there was no relation with fowl or fish consumption (the latter of which was generally low).

Discussion

The levels of Pb, Cd and Hg observed in maternal, cord and infant blood and in breastmilk samples were generally within an acceptable range. Although in the past it was thought that the fetus was protected from Cd exposure, high Cd has been observed in prenatally exposed versus non-exposed babies (Lagerkvist et al. 1992). The association between maternal blood Cd and smoking and between maternal and cord blood Cd provides yet another compelling reason to discourage smoking during pregnancy. The low Hg levels observed in the study may serve as a baseline for monitoring change in the region, for example, if increased damming of the Moose River, occurs.

The levels of Pb observed in a small number of neonates (3%) were above the level identified as a health concern, though below levels requiring treatment. Interestingly, of 95 cord blood samples collected in Toronto (Koren et al. 1990) none exceeded 0.34 μmol Pb/L. A sequel to the 1987 study of Pb in Moosonee and Moose Factory school children in the Fall of 1992, found that Pb levels had declined over the five years, likely due to the elimination of leaded gasoline (Ontario Ministry of Health 1993). However, similar to our findings, 4.6% of the 396 Moosonee school children sampled had blood Pb > 0.48 μmol/L. While this is below the percentages of children at risk in urban Canadian centres (Jin et al. 1995, O.M.H. 1993), it remains unexpectedly high for a remote area.

The strong association between traditional fowl, mammal and fish consumption and blood Pb may offer some explanation for Pb elevation in this remote northern region. We have speculated that Pb absorption from gunshot may be a potential source of Pb exposure (Hanning, et al 1994). Hunting is part of the culture of the First Nations Cree of the Zone. Anecdotally, the presence of lead pellets in the digestive system of adults when examined by X-rays is not uncommon at the Moose Factory Hospital. Skeletal muscle tissue samples from fowl, though not mammals, harvested in the region show Pb concentrations above the level fit for human consumption (L. Tsuji D.D.S., Ph.D., personal communication). The use of steel versus lead shot is an option for consideration by the regional First Nations Assembly.

References

Baghurst, P., McMichael, A., Wigg, N., et al. (1992) Environmental exposure to lead and children's intelligence at the age of seven years. New Engl. J. Med. 327: 1279–1284.

Centers for Disease Control (1991) Preventing lead poisoning in young children. C.D.C., U.S. Department of Health and Human Services, Atlanta.

Clarkson, T.W., Nordberg, G.F., & Sager, P.R. (1985) Reproductive and developmental toxicity of metals. Scand. J. Work Environ. Health 11: 145–154.

Dabeka, R.W. (1989) Survey of lead, cadmium, cobalt and nickel in infant formulas and evaporated milks and estimation of dietary intakes of the elements by infants 0–12 months old. Sci. Total Envir. 89: 279–289.

Dabeka, R.W. & McKenzie, A.D. (1988) Lead and cadmium levels in commercial infant foods and dietary intakes by infants 0–1 year old. Food Addit. Contam. 5: 333–342.

Frierg, L., Elinder, C-G., Kjellerstrom, T., & Nordberg, G.F. (eds.). (1985) Cadmium and Health: A Toxicological and Epidemiological Appraisal. Vol. 1, CRC Press, Boca Raton, Fl.

Goss, Gilroy and Assoc., Ltd. (1989) Blood lead concentrations and associated risk factors in a sample of Northern Ontario Children 1987. Report to the Ontario Ministries of Health and the Environment, Toronto, ON.

Hanning, R.M., Sandhu, R.S., Moss, L.A., MacMillan, A.B., McComb, I., & Nieboer, E. (1994) Cord blood mercury (Hg), lead (Pb), and cadmium (Cd) in relation to maternal game and fish consumption in the Moose Factory (MF) Zone. Proc. Can. Fed. Biol. Soc. 39: 394.

Hanning, R.M., Nieboer, E., Moss, L., McComb, K., & MacMillan, A. (1996) Impact on lead (Pb) cadmium (Cd) and mercury (Hg) of prenatal and early infant feeding practices of native Indians in the Moose Factory Zone (MFZ), Program and Abstracts, TEMA–9, pp. 39, NRC Research Press, Ottawa.

Health and Welfare Canada (1990) National database on breastfeeding among Indian and Inuit women: survey of infant feeding practices from birth to six months. Canada 1988 (Langer, N., consultant), Health and Welfare Canada, Ottawa.

Jin, A., Hertzman, C., Peck, S.H.S., & Lockitch, G. (1995) Blood lead levels in children aged 24 to 36 months in Vancouver. Can. Med. Assoc. J. 152: 1077–1086.

Koren, G. Chang, N., Gonene, R., et al. (1990) Lead exposure among mothers and their newborns in Toronto. Can. Med. Assoc. J. 142: 1241–1244.

Lagerkvist, B.J., Nordberg, G.F., Soderberg, H.A., et al. (1992) Placental transfer of cadmium. In: Cadmium in the Human Environment: Toxicity and Carcinogenicity (Nordberg, G.F., Herber, R.F.M., Alessio, L., eds.), pp. 287–291. International Agency for Research on Cancer, Lyon.

Muir, D.C.G., Wagemann,R., Lockhart, W.L., et al. (1987) Heavy metal and organic contaminants in arctic marine fishes. Environmental Studies No. 42, Minister of Indian Affairs and Northern Development, Ottawa.

Needleman, H. & Bellinger, D. (1991) The health effects of low level exposure to lead. Ann. Rev. Public Health 12: 111–140.

Ontario Ministry of Health (1993) Blood lead in Moosonee and Moose Factory Children 1992. Public Health Branch, Toronto.
Pocock, S.J., Smith, M., & Baghurst, P. (1994) Environmental lead and children's intelligence: a systematic review of the epidemiological evidence. B.M.J. 309: 1189–1197.
Wolf, A.W., Jiminez, E., & Lozoff, B. (1994) No evidence of developmental ill effects of low-level lead exposure in a developing country. J. Dev. Behav. Pediatr. 15: 224–231.

Discussion

Q1. John Bogden, New Jersey Medical School, Newark, NJ, USA: Your data suggest that meat consumption correlates with higher blood lead, and you suggested that this is due to the use of lead ammunition. Have you considered the possibility that dietary calcium can reduce lead absorption, and that it would be possible that mothers with high meat consumption would have lower calcium intakes?

A. No, we didn't specifically address that possibility. Calcium consumption among these women is quite possibly low. They tend to only use evaporated milk in their tea. Unlike the elders, the younger women in this study tend to no longer use bones in soups, which would have traditionally provided a source of calcium. However, we don't have any hard data on this issue.

Q2. Noel Solomons, Guatemala: Have you considered the possibility of a curvilinear regression of consumption levels, instead of the straight line regression that you employed?

A. We did consider the possibility of skewness in the data. It could be interesting to examine the data using a curvilinear regression, but our use of a straight linear regression met with the approval of our statistician.

Q3. James Friel, Memorial University of Newfoundland, St. John's, NF, Canada: I was a bit surprised that you found no socioeconomic differences between your breast-feeding group and evaporated milk users. This doesn't seem to be consistent with findings in most other Canadian populations that have been examined.

A. Well, there was a tendency towards that. Our socioeconomic categorization was applied only to the fathers in this study. We found significant results only at two months, but little else.

Q4. Rosalind Gibson, University of Otago, Dunedin, New Zealand: Did you include offal on your food frequency questionnaire?

A. We consulted with elders in developing our list of traditional foods for inclusion on our food frequency questionnaires. Although the elders would have consumed offal, the 25-year olds in our study group would turn up their noses at this traditional foodstuff.

Q5. Rosalind Gibson, University of Otago, Dunedin, New Zealand: How do your prevalence data for high blood lead levels compare to somewhere like Toronto?

A. The prevalence of high blood lead levels in our study group is actually comparable to that in Toronto, or even slightly lower.

Q6. Bob Dabeka, Health Canada, Ottawa, ON, Canada: Your values for high lead in cord blood were rather higher than the other blood lead levels. Did these represent mothers with a lead pellet burden in the appendix?

A. It's certainly a possibility. These tended to be the mothers with high lead, and high traditional meat intakes. However, we do not have X-ray data available to be able to answer your question about lead pellet burden.

Role of Goitrogens in the Etiology of Iodine Deficiency Disorders

P.P. BOURDOUX

Université Libre de Bruxelles, Laboratoire de Pédiatrie, Hôpital Universitaire des Enfants, B-1020 Brussels, Belgium

Keywords goitrogens, iodine deficiency, thiocyanate, thyroid

Iodine deficiency is the main but not the sole cause of iodine deficiency disorders (IDD). However, goiter, the most visible manifestation of iodine deficiency remains a problem of public health in areas without apparent deficiency or even with an excess of iodine. In fact, other environmental factors may cause goiter. Because most of them are present in food and/or water they are usually referred to as dietary goitrogens. These notably include sulfurated compounds, flavonoids, phenolics, pyridines, phthalic acid derivatives, polycyclic aromatic hydrocarbons and inorganic compounds (Gaitan 1989). However, their relative role in the development of goiter is extremely different.

1. Goitrogens Acting on the Active Transport of Iodide

These include ions with physico-chemical properties similar to halides. Indeed the ionic size and the electric charge of thiocyanate, nitrate and perchlorate ions are similar to iodide. For this reason, thiocyanate, cyanide, cyanate and selenocyanate are commonly termed pseudo-halides. Dietary goitrogens producing thiocyanate have been extensively studied and their role in the etiology of endemic goiter is certainly the most prevalent and the most convincing (Ermans and Bourdoux 1989).

1.1. Thioglucosides or Glucosinolates

Thiocyanate does not occur in plants as a free ion. It results from the catabolism of precursors named glucosinolates which are present in variable amounts in a series of cruciferous plants (cabbage, turnips, swedes, rapeseeds and mustard seeds). Highest concentrations of glucosinolates are usually observed in roots or tubers and seeds. Any process that ruptures the cell walls (wounding, grinding, chemical agents) causes contact between the substrate and an enzyme resulting in hydrolysis of the glucosinolate. Almost invariably such hydrolysis produces thiocyanate (SCN).

Among glucosinolates, progoitrin (a non-goitrogenic substance), under the action of its enzyme is converted into a cyclic compound L-5-vinyl-2-thiooxazolidone which has been called "goitrin" (Langer and Greer 1977). Contrary to most glucosinolates, goitrin is a potent antithyroid compound (see 2.2). Progoitrin is found in cabbage, turnip and other Brassicae. Cooking which is supposed to eliminate or reduce goitrogenic properties of the food actually results in the destruction of the enzyme causing cyclization into goitrin. It has been shown that even in the absence of the enzyme progoitrin exhibits a goitrogenic action probably through hydrolysis by enterobacteriaceae (*E. coli, Proteus vulgaris*).

1.2. Cyanogenic Glycosides

This class of compounds represents another important group of dietary goitrogens (Ermans et al. 1980). As for SCN, the free cyanide ion is not present in plants but results from the catabolism of cyanogenic glycosides. These have been identified in many plants (cassava, lima beans, flax, sorghum, and bitter almonds). Beans, and above all, cassava are the main source of food for hundreds of millions of human beings living in the tropics. Again, contact between the cyanogenic glycoside and its specific enzyme is made by disruption of cell walls and the resulting hydrolysis produces hydrocyanic acid (HCN).

Correct citation: Bourdoux, P.P. 1997. Role of goitrogens in the etiology of iodine deficiency disorders. In *Trace Elements in Man and Animals – 9: Proceedings of the Ninth International Symposium on Trace Elements in Man and Animals. Edited by* P.W.F. Fischer, M.R. L'Abbé, K.A. Cockell, and R.S. Gibson. NRC Research Press, Ottawa, Canada. pp. 152–155.

In the presence of sulfur-containing amino acids HCN is converted into less toxic substances, principally thiocyanate. The main detoxification pathway is controlled by rhodanese.

The methods of preparing foods are supposed to reduce the toxicity of cassava. In fact they result in a destruction of the specific enzyme (linamarase) whereas the cyanogenic substrate (linamarin) is stable up to 150°C (Bourdoux et al. 1982). In the absence of linamarase, linamarin can also be hydrolyzed by enterobacteriaceae (Klebsiella).

1.3. Smoking

Large amounts of cyanide-rich derivatives are produced by combustion of tobacco. Cyanogenic glycoside resorption of these derivatives produces thiocyanate. Serum thiocyanate is indeed regarded as a good marker of tobacco consumption. Daily consumption of a dozen of cigarettes produces serum thiocyanate levels as high as those observed in subjects eating large quantities of cabbage or cassava.

1.4. Isothiocyanates (R–N=C=S)

These compounds are present in several plants (Ermans and Bourdoux 1989). They produce a strong and characteristic odour (onions, garlic, mustard). Either they are converted into thiocyanate or they spontaneously react with some amino groups to provide thiourea-like compounds (see 2.1). The Bhopal disaster released large amounts of isocyanate derivatives. The release of isocyanate was associated with very high levels of thiocyanate in the population.

1.5. Other Sources

High concentrations of thiocyanate (17 mmol/L) have also been found in water effluents of coal-conversion processes.

2. Goitrogens Acting on the Intrathyroidal Mechanisms of Iodine

This category includes several compounds with different chemical structures.

2.1. Thiourea and Thionamide-like Compounds

These goitrogens do not act on the active transport of iodide but inhibit the processes of organification of iodide and the coupling of iodotyrosines (MIT and DIT). Typical representatives are antithyroid compounds (methylmercaptoimidazole or MMI, propylthiouracil or PTU) used in the treatment of hyperthyroid patients.

2.2. Goitrin (see 1.1)

Goitrin also belongs to this group. *In vitro* it inhibits the action of thyroid peroxidase (TPO) resulting in a very low incorporation of iodine into thyroglobulin. *In vivo* it has a potent antithyroid activity similar to the drugs of the thionamide group,

2.3. Phenolics

Resorcinol is a typical representative of phenolic compounds. It is a potent inhibitor of thyroid peroxidase (Gaitan 1989). Flavonoids are polyhydroxyphenols with a structure consisting of a C15 skeleton called flavone. Their occurrence is limited in animals but extensive in plants where they exist in living tissues usually in combination with sugar molecules forming glycosides. The latter inhibit thyroid peroxidase and exert an antithyroid activity which is enhanced by conversion to aglycones. The role of flavonoids is mainly studied in millet, the staple food for millions of people in Asia and Africa.

Millet-based diets have a definite goitrogenic effect even in rats with sufficient iodine intake. The fermentation process, the traditional way of preparing foods, increases the antithyroid activity of millet. The mechanism of action of phenolic compounds is not yet fully elucidated.

2.4. *Disulfides*

Small aliphatic disulfides are the major components of onion and garlic. They have also been identified as water contaminants in the United States (Kentucky) and in water supplying a Colombian district. They exert a marked antithyroid activity in the rat.

3. Goitrogens Acting on the Proteolysis and Release of Thyroid Hormones

3.1. *Excess of iodine*

Excessive intake of iodine significantly inhibits the synthesis and release of thyroid hormones. This is known as the Wolff-Chaikoff effect. It causes transient hypothyroidism spontaneously resolving when the excess is removed or "iodide goiter" in case of prolonged excess (Suzuki et al. 1965).

3.2. *Lithium*

Lithium is mainly used in the treatment of maniac or depressive patients. Its action is likely due to an iodide-like effect which hampers thyroid secretion. Other mechanisms have also been suggested.

Biochemical Effects

The goitrogenic action of thiocyanate is mainly due to its inhibitory action on the uptake of iodide by the thyroid. In iodine deficiency, the direct competition between thiocyanate and iodine for the uptake further reduces the amount of iodine available for hormonal synthesis.

Within the thyroid, thiocyanate is oxidized into sulfate and this is accelerated by thyrotropin (TSH). Very likely this oxidation also competes with the intrathyroidal mechanism of iodine by using part of the thyroid peroxidase reducing the incorporation of iodine into thyroglobulin.

In contrast, thionamide, goitrin, polyphenols and disulfides have no inhibitory action on the uptake of iodine by the thyroid. Their antithyroid or goitrogenic action is similar to synthetic drugs used in the treatment of hyperthyroidism (MMI, PTU). Their action is mainly on the organification of iodine and on the coupling process of iodotyrosines (MIT and DIT).

If for any reason (food shortage, malnutrition) populations consuming. large amounts of cassava (or any foodstuff containing cyanogenic glycosides), do not have a sufficient intake of sulfur-containing amino acids then epidemics occur as has been described in Nigeria for tropical ataxic neuropathy or in Central Africa for Konzo (Tylleskar et al. 1992). Whatever the mechanisms, this is a direct consequence of a reduced capacity to convert cyanide into thiocyanate.

In most cases, long-term administration of these substances to animals has allowed the production of goitrogens and elucidation of their mechanism(s) of action. It is important to keep in mind that the goitrogenic effect of goitrogens acting on the trapping of iodide by the thyroid can be antagonized by adequate iodine administration. For those acting on the organification of iodine and/or coupling of iodotyrosines, an increase in the amount of iodine ingested will not cause a significant reduction of the goitrogenic effect.

Metabolism of Thiocyanate

In the absence of goitrogens, serum and urinary levels of thiocyanate are below 86 and between 103 and 215 μmol/L, respectively. Measurement of thiocyanate is commonly used in searching for the presence of goitrogens. For this purpose, we have repeatedly advocated the measurement of urinary thiocyanate. Indeed, we have shown that for concentrations up to about 172 μmol/L there is a proportionality between serum and urinary thiocyanate levels. Above this threshold value, urinary thiocyanate concentration further increases whereas serum concentration plateaus. Urinary thiocyanate concentration is therefore the best index of the presence of goitrogens producing thiocyanate. Low or normal urinary thiocyanate levels do not preclude the presence of other goitrogens (flavonoids).

In normal conditions, the biological half-life of thiocyanate ranges between 5 and 7 d. It can be significantly shortened in situations of thiocyanate overload.

It should be kept in mind that thiocyanate freely crosses the placenta and that in case of thiocyanate excess deleterious effects of thiocyanate also affect the fetus. Noteworthy is the observation that in humans, contrary to iodine, thiocyanate is not concentrated by the mammary gland resulting in low levels of thiocyanate in breast milk. Breast feeding is therefore a protection for the newborn since it is preserved from the effect of goitrogens producing thiocyanate.

Conclusions

When iodine deficiency alone does not account for the severity of endemic goiter it is mandatory to look for the presence of goitrogenic factors.

Because thiocyanate derivatives or compounds producing thiocyanate are the most relevant factors in the etiology of endemic goiter, the measurement of urinary thiocyanate concentrations should be part of any evaluation of iodine deficiency. If these concentrations do not indicate that large amounts of thiocyanate producing foodstuffs are consumed, it might be useful to took for other goitrogenic substances.

Literature Cited

Bourdoux, P., Seghers, P., Mafuta, M., Vanderpas, J., Vanderpas-Rivera, M., Delange, F. & Ermans, A.M. (1982) Cassava products: HCN content and detoxification processes. In: Nutritional Factors Involved in the Goitrogenic Action of Cassava (Delange, F., Iteke, F.B. & Ermans, A.M., eds.), pp. 51–58. IDRC, Ottawa.

Ermans, A.M., Mbulamoko, N.M., Delange, F. & Ahluwalia R. (1980) Role of Cassava in the Etiology of Endemic Goitre and Cretinism. IDRC, Ottawa.

Gaitan, E. (1989) Environmental Goitrogenesis. CRC press, Inc., Boca Raton, FL.

Langer, P. & Greer, M.A. (1977) Antithyroid Substances and Naturally Occurring Goitrogens. S. Karger, Basel.

Suzuki, H., Higuchi, T., Sawa, K., Ohtaki, S. & Horiuchi, Y. (1965) Endemic coast goitre in Hokkaido, Japan. Acta Endocrinol. 50: 161–176.

Tylleskar, T., Banea, M., Bikangi, N., Cooke, R.D., Poulter, N.H. & Rosling, H. (1992) Cassava cyanogens and konzo, an upper motoneuron disease found in Africa. Lancet 339: 208–211.

Discussion

Q1. Les Klevay, USDA-ARS, Grand Forks, ND, USA: Are there any public health implications of the eating of plants of the genus Brassica in North America, like Brussels sprouts and cauliflower? Do you have public health reports in relation to this in your populations?

A. Sicily was an important area of goitre even relatively recently. Attempts have been made to change eating habits, or to alter the plant strains of Brassicas, to attempt to reduce the problem.

Q2. Ross Welch, USDA-ARS, Ithaca, NY, USA: What are the relative prevalences of iodine deficiency diseases caused by iodine deficiency versus those caused by goitrogens? This would seem to be quite relevant with regard to the attempts to iodize salt, and so on, to try to reduce the prevalence of iodine deficiency diseases.

A. There are considerable levels of goitrogens in places like Finland, Central Africa, Malaysia and so on, and iodine excess causes thyroid problems in places like Japan and China. This brings us back to the point I made earlier in the presentation, that iodine deficiencies brought about by thiocyanates, which arise mechanistically from interference with iodine uptake by the thyroid, can be overcome by iodine supplementation, while diseases caused by goitrogens, flavonoids, phenolics, and disulfides cannot be overcome by iodine (or any other kind of) supplementation. This does not seem to have gotten proper consideration by the appropriate authorities.

Q3. Neville Suttle, Moredun Research Institute, Edinburgh, Scotland: In lactating farm animals, goitrogens reduce iodine levels in the milk, by inhibiting iodine uptake by the mammary gland. Does this happen in humans?

A. Apparently not. We see normal iodine levels in human milk, despite high loads of thiocyanates. As I mentioned in the talk, thiocyanates may freely cross the placenta, but they are not concentrated by the mammary gland of humans. Rats are not a good model, as they do concentrate thiocyanates through the mammary gland.

Q4. Danny Goodwin-Jones, Carmarthen Dyfed, Wales: Rapeseed is used at a high level in the UK, both the oil and the meal. We also see selenium deficiency in UK cattle and sheep. Do you have any ideas on the possibilities for interactions between these?

A. I don't know enough about the mechanisms of interaction between iodine and selenium in animals to comment on that.

Trace Elements in Wheat, Flour and Bread, and their Relations*

N.K. ARAS,[†] R. ERCAN,[‡] W. ZHANG,[¶] AND A. CHATT[¶]

[†]Department of Chemistry, Middle East Technical University, 06531 Ankara, Turkey. [‡]Department of Food Engineering, Ankara University, Ankara, Turkey. [¶]Department of Chemistry, Dalhousie University, Halifax, Nova Scotia, Canada B3H 4J3

Keywords trace elements, wheat, flour, bread, dietary intake

At present 26 naturally occurring elements appear to be essential to life. These consist of 11 major and micro minerals which are present at g/kg or mg/kg levels in tissues. The remaining 15 elements, namely As, Cr, Co, Cu, F, I, Fe, Mn, Mo, Ni, Si, Sn, V, and Zn, which are called essential trace elements, are present at mg/kg or µg/kg levels. Since the main source of these elements in humans is their total diet, studies on these elements in staple foods and their relation with the total diet and daily intakes are very important (Aras and Olmez 1995). Wheat is by far the most used staple food in Turkish diets. Estimates show that about 400 g wheat or wheat products are used daily by the average person (Koksal 1990). In this work we will present data for 1990, 1993 and 1994 the trace element content of wheat, as well as 1995 data on bread and flour samples, measured by Instrumental Neutron Activation Analysis, INAA.

Collection of Wheat Samples

Wheat samples were collected with the help of the Turkish Grain Board, TGB. They collected a 5 kg sample from each of the over 800 collection centers and silos. We selected four regions for the purpose of this work and collected 5 kg samples corresponding to five silos each, from Istanbul, Ankara, Konya and Yskenderun regions. For each region we combined wheat samples and made one representative sample for that region.

Collection of Bread and Flour Samples

Bread and flour samples were collected with the help of the Ankara Bakeries Association. We selected six major bakeries in Ankara and collected the most widely consumed white bread samples during February 1, March 15, April 15, and May 25, 1995. We also collected flour samples from the same bakeries during the February 1, 1995, bread collection time. In each case, we collected three fresh breads, stored in clean polyethylene bags. We dried them for 8 h at 50°C, then powdered them with a Titanium blade homogenizer.

Determination of Trace elements by Instrumental Neutron Activation Analysis, INAA

We used the Dalhousie University SLOWPOKE-2 Reactor for INAA. About 800 mg wheat, 700 mg flour and 250 mg bread samples, as well as several certified reference materials were used in these experiments for quality control purposes. The neutron flux was 5×10^{11} n/s.cm^2. The Se was determined by using ^{77m}Se (half life 17.4 s) isotope. The same samples were subjected to a 3 min irradiation followed by 3 min cooling and 10 min counting periods for the determination of the concentrations of Ca, V, Mn, and Cu. An anticoincidence spectrometer was used for improving the detection limits of the elements.

---

*This work is partially supported by a NATO Research Grant No: CRG.940429.

Correct citation: Aras, N.K., Ercan, R., Zhang, W., and Chatt, A. 1997. Trace elements in wheat, flour and bread, and their relations. In *Trace Elements in Man and Animals – 9: Proceedings of the Ninth International Symposium on Trace Elements in Man and Animals. Edited by* P.W.F. Fischer, M.R. L'Abbé, K.A. Cockell, and R.S. Gibson. NRC Research Press, Ottawa, Canada. pp. 156–157.

Results and Discussion

The average concentrations of trace elements in wheat, flour and bread are given in Tables 1 and 2, respectively.

As seen in Table 2, the ratios of trace element concentrations of Flour/Bread are between 0.8–1.1 for Ca, Mn, Mg and Cu and 1.5 for Se. This means that there is not much change in concentrations between flour and bread except Se, which is reduced by almost 50% in bread. This could be due to the loss of Se during the high temperature baking of the bread. Estimates show that (Koksal 1990) average daily bread consumption by the Turkish population aged between 20–60 years is about 400 g/person/day. This shows that the daily intake from bread alone will be Ca: 126 mg, Mn: 3.44 mg, Mg: 111 mg, Cu: 0.52 mg and Se: 20 μg.

Table 1. Concentration of trace elements in white bread.

Bakery	Ca (mg/kg)	Mn (mg/kg)	Mg (mg/kg)	Cu (mg/kg)	Se (μg/kg)
Zaman	303±30	7.20+1.03	265±33	1.57±0.30	55±27
Merkez	271±59	8.82±1.59	240±51	1.41±0.44	48±17
Kardes	333±87	10.3±1.45	300±121	1.22±0.34	46±23
Ye-Ye	484±169	9.01±1.67	351±172	1.19±0.39	68±35
Halk	252±27	8.42±1.54	259±45	1.01±0.42	46±19
Altineller	251±25	7.60±2.08	252±58	1.43±0.73	32±17

Each result is the average of samples collected in February, March, April and May, 1995.

Table 2. Trace element content of wheat, flour and bread.

	Ca (mg/kg)	Mn (mg/kg)	Mg (mg/kg)	Cu (mg/kg)	Se (μg/kg)
Wheat	336±95	36.1 ±4.4	916±62	3.45±0.76	45±20
Flour	230±33	9.60±3.20	331±62	1.74±0.26	77±40
Bread	292±91	8.66±1.93	282±106	1.56±0.47	52±16
F/B	0.79	1.11	1.17	1.11	1.48
W/F/B	1/0.68/0.86	1/0.27/0.24	1/0.36/0.31	1/0.50/0.45	1/1.71/1.16

Row 1 shows the average concentration of trace elements in wheat. Rows 2 and 3 show flour and bread results taken from six different bakeries during the February 1, 1995, collection. Rows 4 and 5 show the Flour/Bread and Wheat/Flour/Bread ratios respectively.

References

Aras, N.K. & Olmez, I. (1995) Human Exposure to Trace Elements Through Diet. Supp. to Nutrition, 11: 506–511.
Koksal, O. (1990) National Nutrient-Health and Food Consumption Survey of Turkey. Hacettepe University, Ankara, Turkey.

Discussion

Q1. Roy Scott, Royal Infirmary, Glasgow, Scotland: I'm a urologist. About 30 years ago there were observations regarding bladder stones in Turkish children, which I believe was related to high intakes of bulgur wheat. Have you by any chance tried to correlate the trace metal contents you have measured with some health effects like that?

A. We haven't looked for such correlations, but we will take the idea into account in the future.

Q2. Manfred Anke, Friedrich Schiller University, Jena, Germany: You noted high nickel concentrations in wheat. Do these arise from the geological origin of Turkish soils? For example, serpentine is a nickel-rich rock.

A. The nickel concentration is high, both in wheat and in the total diet, in Turkey compared to the rest of Europe. We have some observations that the levels in soils are also high, and could probably relate these observations.

Q3. Rosalind Gibson, University of Otago, Dunedin, New Zealand: You report regional variations in zinc. Does this relate to lower soil zinc in some areas? And are there any plans for zinc fertilization in such areas?

A. Yes, the variations appear to relate to variations in soil zinc. Fertilization as a means of raising zinc concentrations has been tried, but not on a large scale as yet.

Selenium Intake of Adults in Germany Depending on Sex, Time, Living Area and Type of Diet

C. DROBNER, B. RÖHRIG, M. ANKE, AND G. THOMAS
Institute of Nutrition and Environment, Friedrich Schiller University, D-07743 Jena, Germany

Keywords selenium, intake, duplicate method, vegetarian

The Federal Republic of Germany belongs to the Se deficiency countries. Low Se levels in soils lead to a low Se content in forage crops and locally produced foodstuffs. In order to discover the risk of a possible Se deficiency in the human population, the Se intake of adults in Germany was investigated.

Material and Methods

Duplicate experiments were carried out in 13 test populations each consisting of seven women and seven men in 1988 (4 test populations), 1991 (6 test populations) and 1995 (3 test populations). The test persons collected duplicates of their freely choosen mixed diet over seven subsequent days. In 1995 the Se intake of 10 female and 10 male vegetarians was investigated also. All samples were dried at 60°C, homogenized and digested with concentrated nitric acid. After addition of $Mg(NO_3)_2 \cdot 6\,H_2O$ (1.56 mol/L) and heating to dryness the samples were dry-ashed at 500°C. The ashes were dissolved in hydrochloric acid (6 mol/L) and heated at 80°C. The Se content was determined with hydride generation atomic absorption spectrometry (AAS 3, HS 3, Carl Zeiss Jena). ARC/CL Total diet reference material and ARC/CL Wheat flour reference material were used.

Results

Selenium Intake of Adults Depending on Time of Testing

At the moment German women consume 30 µg Se/day and German men 42 µg Se/day (Table 1). This means an increase of 60% in the Se intake during the last seven years. Men consumed significantly more Se than women. This significant difference resulted from the higher dry matter intake of men (nearly 30%). Sex-dependent differences between the Se concentrations of the consumed food could not be observed.

Table 1. Selenium intake of adults depending on time of testing and sex (µg/d).

		Women		Men			
year	(n, n)	s	$\bar{x}$	$\bar{x}$	s	p	%
1988	(196, 196)	12.3	19.3	25.0	20.4		130
1991	(294, 294)	18.0	24.8	31.0	24.7	<0.001	125
1995	(168, 168)	15.8	30.3	42.0	25.5		139
p				<0.001		–	
%			157		168	–	

Selenium Intake of Adults Depending on Living Area

In order to exclude the influence of time on the Se supply of adults, the Se intake in 1988 and 1991 will be discussed separately. In 1988 a highly significant influence of the living area on Se intake, as well

Correct citation: Drobner, C., Röhrig, B., Anke, M., and Thomas, G. 1997. Selenium intake of adults in Germany depending on sex, time, living area and type of diet. In *Trace Elements in Man and Animals – 9: Proceedings of the Ninth International Symposium on Trace Elements in Man and Animals. Edited by* P.W.F. Fischer, M.R. L'Abbé, K.A. Cockell, and R.S. Gibson. NRC Research Press, Ottawa, Canada. pp. 158–159.

as on the Se content of the duplicates, could be proved. The populations in the north of the former GDR, where the soils are more acid, showed lower Se intakes than populations living in the south. The difference in Se intakes is caused by the consumption of more regionally produced foodstuffs in the former GDR. However, only a slight influence of living area on the Se concentration in the consumed dry matter was found in 1991. This influence on Se intake was no longer significant (Table 2). This result is due to the consumption of more imported foods since the reunification of Germany.

Table 2. Selenium intake of adults in 1991 depending on living area (μg/d).

		Women		Men			
living area	(n, n)	s	$\bar{x}$	$\bar{x}$	s	p	%
Brandenburg/ Mecklenburg-Western Pomerania	(98,98)	14.9	26.5	29.7	33.8		112
Saxony	(98,98)	20.9	24.4	35.9	21.6	<0.01	147
Thuringia	(98,98)	17.7	23.4	27.2	13.7		116
p			>0.05				–
%			88	92			–

Influence of Omnivorous and Vegeterian Diets on the Se Intake

The vegetarian food contained significantly lower Se levels with 76 ng Se/g dry matter compared to that consumed by the omnivorous test persons with 104 ng Se/g dry matter. Due to the higher dry matter intake of the vegetarians the difference of the Se intake remained only slightly significant for men and insignificant for women (Table 3).

Table 3. Selenium intake of adults in 1995 depending on type of diet and sex (μg/d).

		Women		Men			
Type of diet	(n, n)	s	$\bar{x}$	$\bar{x}$	s	p	%
Omnivorous diets	(168,168)	15.8	30.3	42.0	25.5	<0.001	139
Vegetarian diets	(70,70)	20.5	29.6	33.8	28.0	>0.05	114
p		>0.05		<0.05			–
%		98		80			–

Conclusions

After the reunification of Germany, women consumed 25 μg Se/day in 1991 and 30 μg Se/day in 1995 and men 31 and 42 μg Se/day, respectively. The German Society for Nutrition recommends a daily Se intake of 20–100 μg for juveniles and adults (Deutsche Gesellschaft für Ernährung 1991). Nearly 86% of the daily intake (on the weekly average) of women and 94% of that of men are in the desirable daily Se supply but only in the lower range. The Recommended Dietary Allowances of 55 μg Se/day for women and 70 μg/day for men (Food and Nutrition Board 1989) were only met by two male test persons.

Literature Cited

Food and Nutrition Board, Commission on Life Sciences, National Research Council (1989). Recommended Dietary Allowances, 10 th Ed., pp. 217–224. National Academy Press, Washington.

Deutsche Gesellschaft für Ernährung (1991) Empfehlungen für die Nährstoffzufuhr, 5. überarbeitete Auflage, pp. 75–81.Umschau Verlag, Frankfurt/Main.

Zinc Intake and Zinc Balance of Adults in Central Europe — Depending on Time, Sex, Living Area, Eating Habits, Age and Live Weight

M. ANKE, M. MÜLLER, M. GLEI, H. ILLING-GÜNTHER, A. TRÜPSCHUCH, M. SEIFERT, O. SEEBER, AND B. RÖHRIG
Institute of Nutrition and Environment, Friedrich Schiller University, D-07743 Jena, Germany

Keywords zinc, balance, intake, Central Europe

The essentiality of zinc for the fauna was detected in rats in 1934 (Todd et al. 1934) and in humans in 1963 (Prasad et al. 1963). In 1988, the zinc intake of adults was considerably lower in Germany (Anke et al. 1995) than that recommended by the WHO (anonymous 1973). However, zinc deficiency symptoms did not occur in the test populations. Therefore, the zinc intake of adults in Central Europe was systematically investigated between 1988 and 1995.

Material and Methods

The zinc supply of adults was systematically investigated in duplicate in 14 test populations on seven subsequent days, with time, sex, living area, age, weight and type of diet being taken into consideration. Subjects consisted of 7 women and 7 men consuming mixed diets, and 10 women and 10 men eating vegetarian diets. Zinc was determined with flame AAS or ISP OES. Both procedures delivered corresponding values (analytical error ± 5) (Anke et al. 1995).

Dry Matter Intake of Adults in Germany

The zinc consumption of adults depends on the individual dry matter content. The investigations showed that adults with mixed diets took in significantly less dry matter than the different types of vegetarians. On average, male as well as female vegetarians consumed 25% more dry matter than adults with ordinary diets (Table 1). Sex also has a significant effect on the dry matter intake. Men, in both diet groups, consumed 25% more dry matter than women. Therefore, the kind of diet and sex vary the zinc intake considerably.

The duplicate studies were carried out in the Eastern States of Germany in 1988, 1991 and 1995 (n = 4; n = 6; n = 4). They allowed the comparison of the dry matter and zinc intake of women and men before the reunification of Germany with a local food and beverage consumption with that of the

Table 1. Dry matter intake of adults with mixed diets and vegetarians (g/d).

		Women		Men			
Type of diet	(n)	s	$\bar{x}$	$\bar{x}$	s	p	%[1)]
Adults with mixed diets	(658; 658)	89	308	386	110	<0.001	125
Vegetarians	(70; 70)	93	384	479	172	<0.001	125
p		<0.001		<0.001		–	
%[2)]		125		124		–	

[1)]women ≙ 100%; men ≙ x%; [2)]adults with mixed diets ≙ 100%; vegetarians ≙ x%.

Correct citation: Anke, M., Müller, M., Glei, M., Illing-Günther, H., Trüpschuch, A., Seifert, M., Seeber, O., and Röhrig, B. 1997. Zinc intake and zinc balance of adults in central Europe — depending on time, sex, living area, eating habits, age and live weight. In *Trace Elements in Man and Animals – 9: Proceedings of the Ninth International Symposium on Trace Elements in Man and Animals. Edited by* P.W.F. Fischer, M.R. L'Abbé, K.A. Cockell, and R.S. Gibson. NRC Research Press, Ottawa, Canada. pp. 160–165.

world-wide food available after the reunification. It was demonstrated that time did not have a significant effect on the dry matter intake of adults with mixed diets.

When the consumed dry matter is related to body weight, it can be demonstrated that women and men had a mean dry matter intake of 5 g per kg body weight whereas vegetarians consumed about 7 g/kg body weight.

Zinc Content in the Consumed Food and Beverage Dry Matter

The zinc content in the consumed food and beverage dry matter is not varied by the individual dry matter intake. Therefore, it is an important indicator of the zinc available (Table 2). It was astonishing that female vegetarians consumed a diet significantly richer in zinc than did males. The reason for this sex-specific difference might be the difference in the consumption of chocolate and cocoa products. Cocoa delivers much zinc.

The zinc content in the consumed dry matter decreased significantly by about 20% between 1988 and 1995 (Table 3). This surprising finding in both sexes can be explained by the lower consumption of beef and the higher consumption of poultry (the latter is zinc-poor).

The zinc content in the consumed dry matter is of considerable individual importance. It is subjected to extreme variations in the daily analyses and only allows limited conclusions on the average of 7 d. Figure 1 presents the average zinc concentration in the dry matter which was individually consumed by women over a 7 d period. The new guidelines of the WHO/FAO/IAEA recommend a daily consumption of 7 mg zinc for both sexes. The recommendation, which does not take sex differences into consideration, means that, in the case of a dry matter consumption of 308 g/d by women and of 386 g/d by men, a zinc concentration of 20 mg/kg dry matter delivers 6.2 mg zinc/day to women and 7.7 mg/d to men. Only 25 of 94 women or 27% of the female subjects consumed dry matter with <20 mg zinc/kg, while 15% of the men, 14 of 94, consumed dry matter with too little zinc.

Zinc Intake of Adults in Germany

The WHO/FAO/IAEA recommend an intake of 7 mg zinc/day (Sandström 1989). On average, this zinc consumption is reached by persons with mixed diets and vegetarians (Table 4). Due to their higher dry matter intake, vegetarians consume more zinc than people with mixed diets. The 25% higher dry matter consumption and the 2% higher zinc content in the dry matter consumed by male adults with mixed diets

Table 2. Zinc content in the consumed food and beverage dry matter of adults with mixed diets and of vegetarians (mg/kg dry matter).

		Women		Men			
Type of diet	(n)	s	$\bar{x}$	$\bar{x}$	s	p	%
Adults with mixed diets	(658, 658)	7.9	23.3	23.8	7.2	>0.05	102
Vegetarians	(70, 70)	5.0	22.3	19.8	5.6	<0.01	89
p		>0.05		<0.001			–
%		96		83			–

Table 3. Mean zinc content in the consumed food and beverage dry matter depending on time and sex (mg/kg dry matter).

		Women		Men			
Year	(n)	s	$\bar{x}$	$\bar{x}$	s	p	%
1988	(196, 196)	6.9	25.8	26.8	8.3	>0.05	104
1991	(294, 294)	7.6	23.3	23.8	6.9	>0.05	102
1995	(168, 168)	8.5	20.6	20.5	4.5	>0.05	100
Fp			<0.001				–
%[1)]		80		76			–

[1)]1988 $\underline{\Delta}$ 100%; 1995 $\underline{\Delta}$ x%.

leads to a zinc intake which is 28% higher than that of women. The sex difference remains insignificant in vegetarians. The abundant chocolate and cocoa consumption of female vegetarians might be the reason for this surprising result.

Time had a significant effect on zinc intake (Table 5). Zinc consumption decreased significantly by 17–19% from 1988 to 1995. One reason for this negative trend could be the lower beef consumption. In 1995, the zinc intake of women reached the marginal range of the recommendation. Zinc intake was reduced by >1.3 mg/d in both sexes between 1988 and 1995.

The individual zinc consumption/day is of particular importance for meeting the requirements.

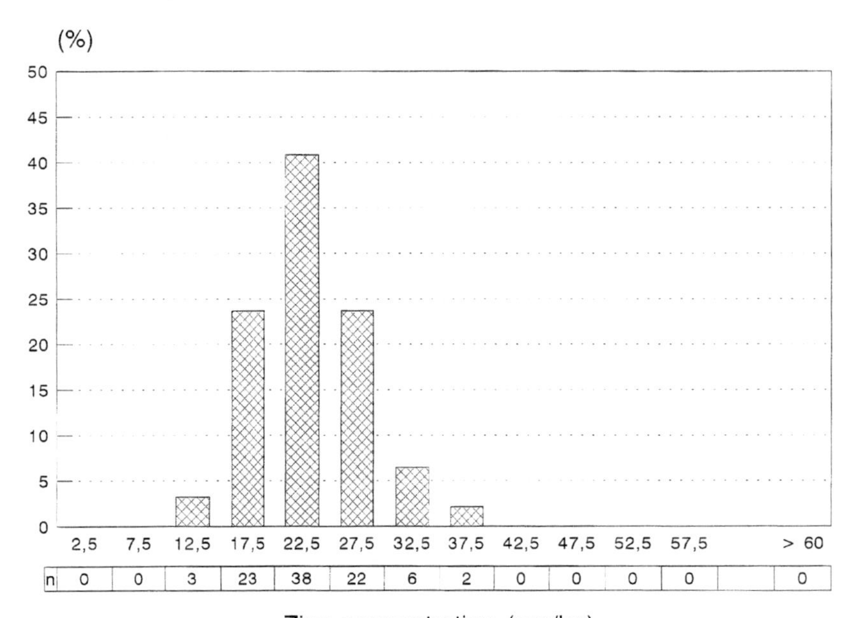

Figure 1. The zinc concentration of the consumed food dry matter (women).

Table 4. Zinc intake of adults in Germany depending on type of diet and sex (mg/d).

		Women		Men			
Type of diet	(n)	s	$\bar{x}$	$\bar{x}$	s	p	%
Adults with mixed diets	(658, 658)	2.9	7.1	9.1	3.6	<0.001	128
Vegetarians	(70, 70)	2.6	8.4	9.5	3.9	>0.05	113
p		<0.001		>0.05			–
%		118		104			–

Table 5. Zinc intake of adults in Germany depending on time and sex (mg/d).

		Women		Men			
Year	(n)	s	$\bar{x}$	$\bar{x}$	s	p	%
1988	(196, 196)	2.6	7.6	10.1	3.4	<0.001	133
1991	(294, 294)	3.1	7.1	9.0	3.4	<0.001	127
1995	(168, 168)	2.9	6.3	8.2	3.7	<0.001	130
Fp			<0.001				–
%		83		81			–

The mean weekly individual zinc consumption of men showed that only 3 of 94 men or 3.2% consumed <5 mg/d (Figure 2). The corresponding proportion amounted to 10% in women. Zinc deficiency symptoms are a possibnility in these individuals.

Only 22 men (35%) and 3 women (3%) took in as much zinc as it was recommended by the WHO (anonymous 1973). Only one of the 20 tested male vegetarians took in <5 mg zinc/day on the average of the week. All female vegetarians had a mean weekly zinc intake of >5 mg/d.

Zinc Intake of Adults per kg Body Weight

The zinc consumption of adults with mixed diets was slightly more than 100 μg/kg body weight in both sexes. Vegetarians consumed about 150 μg/kg body weight (Table 6). Both populations did not show any zinc deficiency symptoms. This could be the recommended amount for the zinc intake of adults with mixed diets and vegetarians.

The zinc intake of adults with mixed diets per kg body weight decreased continuously from 1988 to 1995 and amounted to 100 μg zinc/kg body weight in 1995 (Table 7).

Zinc Excretion and Zinc Balance of Adults

The zinc excretion via urine and faeces offers valuable hints as to the extent of zinc absorption in humans and animals. Renally excreted zinc is absorbed by the body and is available for the enzyme and

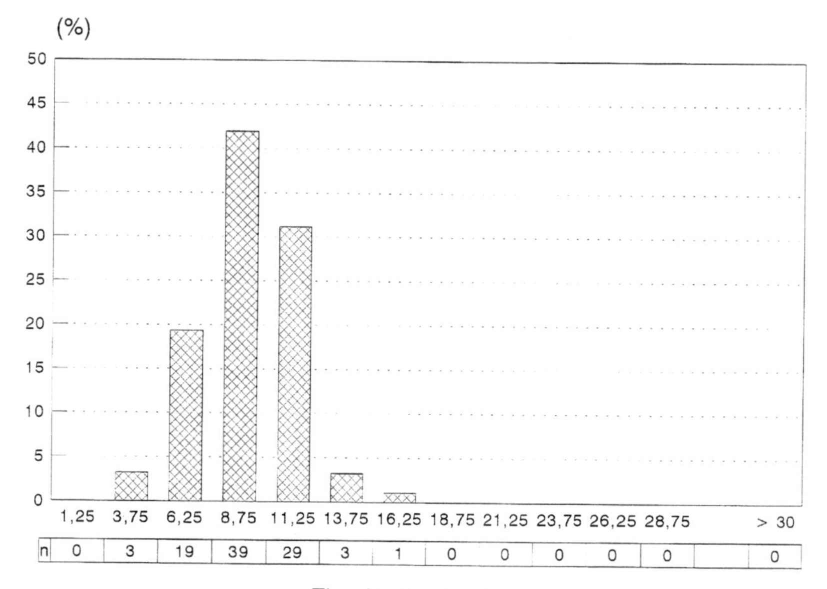

Figure 2. Mean daily zinc intake of men (mg/d).

Table 6. Zinc consumption per kg body weight depending on type of diet (μg/kg body weight).

Type of diet	(n)	Women s	Women $\bar{x}$	Men $\bar{x}$	Men s	p	%
Adults with mixed diets	(658, 658)	46	109	121	52	>0.05	111
Vegetarians	(70, 70)	49	151	138	59	>0.05	91
p			<0.001				–
%			139	114			–

Table 7. Zinc consumption per kg body weight depending on time (µg/kg body weight).

		Women		Men			
Year	(n)	s	$\bar{x}$	$\bar{x}$	s	Fp	%
1988	(196, 196)	46	109	141	54	<0.001	129
1991	(294, 294)	44	114	116	47	<0.001	102
1995	(168, 168)	47	95	107	53	<0.001	113
Fp			<0.001			–	
%		87		73		–	

hormone production or the activation of essential compounds. On the average of 4 test populations, both sexes excreted 11% of the consumed zinc renally. The highest proportion of zinc, however, left the body faecally, either unabsorbed or excreted into the intestine via bile or other secretions. In the case of zinc deficiency, this low absorption rate might allow the body to make zinc absorption more efficient. Marginal zinc deficiency could be compensated for.

On average, the zinc balance of the 4 investigated test populations was balanced. Women excreted 0.3 mg zinc more than they took in.

To sum up, there is a positive correlation between dry matter consumption and zinc intake ($r = 0.66$). Zinc intake increased with rising dry matter consumption.

Comparison of Zinc Consumption with the Duplicate and the Basket Method

There is considerable difference between the zinc intake of adults analytically determined with the duplicate method and that calculated using the basket method. The basket method overestimates the zinc intake by about 40%. Thus, it assumes a higher zinc intake than is actually consumed.

Summary

The zinc content in the consumed dry matter decreased by about 20% between 1988 and 1995. A zinc content of 20 mg/kg in the consumed dry matter is regarded as sufficient to meet the requirements. About 20% of women and 15% of men did not reach these levels in their weekly mean intakes. The zinc intake of women and men decreased significantly between 1988 and 1995 and reached 6.3 mg/d in women and 8.2 mg/d in men without the occurrence of any deficiency symptoms. The zinc consumption per kg body weight was 100 µg in persons with mixed diets and about 150 µg in vegetarians. This amount approximately corresponds to the recommended daily zinc intake of 7 mg.

Literature

Anke, M., Müller, M., Illing-Günther, H., Glei, M., Anke, S., Röhrig, B. & Kräuter U. (1995) Zinkversorgung Erwachsener Deutschlands. Mengen-und Spurenelemente, 15: 741–756.

Anonymous (1973). WHO Technical Report, Series 532.

Prasad, A.S., Miale, A., Fariad, Z., Sandstead, H.H., Shuler, A.R. & Darby, W.J. (1963) Biochemical studies on dwarfism, hypogonadism and anaemia. Arch. Int. Med. 111: 407–428.

Sandström, B. (1989) Zinc requirement. Solicited paper for FAO/WHO/IAEA. Expert Consultation on "Trace Elements in Human Nutrition", Geneva, WHO.

Todd, W.R., Elvehjem, C.A. & Hart, E.B. (1934) Zinc in the nutrition of rat. Am. J. Physiol. 107: 146–150.

Discussion

Q1. Kevin Cockell, Health Canada, Ottawa, ON, Canada: You have told us that since reunification, the selenium intakes of the populations you have studied have increased, while zinc intakes have declined over the same period. On balance, would you say that reunification was beneficial or detrimental to this aspect of the population health of your study populations?

A. With the freeing of markets, and increased import of foods like meat, milk and beer produced elsewhere, the selenium intakes improved. The study region is typified by soils with low pH, which is associated with low selenium intakes. The decline in zinc intakes relates to the improvements in food processing techniques, and the same influences of importing foreign foods. The health status of our populations is quite similar before and after reunification.

Q2. Donald Oberleas, Texas Technical University, Lubbock, TX, USA: Have you looked at phytate levels?

A. Phytate levels were much higher in the vegetarian diet, as would be expected. So we see effects of phytic acid on zinc, magnesium, calcium, phosphorus and iron intake.

Q3. Gerry Combs, Cornell University, Ithaca, NY, USA: Have you looked at blood selenium levels? The selenium intakes you reported seem low compared to previously reported blood selenium levels.

A. We have blood samples from some, but not all of our study populations. These analyses are not finished yet, but no differences have been seen yet in zinc. I am not certain about selenium or glutathione peroxidase levels. Anyone who likes, I can send them a reprint of the data next year.

Trace Element Levels in Australian Cows' Milk

C.M. PATTERSON, U. TINGGI, AND C. REILLY

Centre for Public Health Research, Queensland University of Technology, Brisbane, Queensland, Australia, 4059

Keywords trace elements, cows' milk, Australia, seasons

There is minimal data in Australia on the variation in trace element content in cows' milk both with respect to regional differences which may reflect differences in soil types and in forage or with respect to seasonal change. In order to provide details of such variation samples of pasteurised standardised milk were collected from milk factories across the milk producing areas in Australia over two years.

The Australian Dairy Industry Corporation provided details of the major milk factories across Australia. Nineteen such factories which could be grouped into four major regional areas as determined by climate and geographical location provided milk samples every three months for two years using a randomised sampling system for the homogenised product. These regions were designated as western (WPA), north eastern (EPA), south eastern (SEPA) and southern (SPA).

The samples were first digested by taking a five mL sample and mixing this with 15 mL of nutric acid and 2 mL of sulphuric acid in a 250 mL flask. These mixtures were then digested and made up to 20 mL with deionised water. Zinc levels were then determined using flame atomic absorption spectrophotometry with instrument optimisation according to the manufacturers instructions. Blanks and standards were aspirated first followed by the samples. Copper, iron, manganese, chromium and aluminium were determined using a Zeeman Gas Flame Atomic Absorption Spectrophotometer. Aqueous standards with similar concentrations of acid were used to optimise conditions for ashing and atomisation temperatures for greater sensitivity. The linear ranges for standard curves for Cu, Fe and Al were 10 to 100 μg/L and 1–20 μg/L for Cr and Mn. Results were automatically calculated by a microprocessor coupled to the instrument. In the first instance simple statistical analysis using Students's t-test was performed. For any elements for which significant variation was noted, principal component analysis was undertaken to illustrate graphically the contribution of different components to the variation.

Numerous statistically significant differences were seen between individual regions for individual seasons. For example, for aluminium there were significant differences between the seasons for regions SPA, EPA and SEPA in year 1 and WPA and SPA in year 2. In year 1 Cu showed variation between the seasons for regions WPA and EPA and only for EPA in year 2. However when the results were aggregated for all seasons over the two years no regional significant differences were noted for any of the above trace elements.

The ranges are summarised in the following table.

Trace element (concentration)	Range of individual samples
Aluminium (μg/L)	32 – 154
Chromium (μg/L)	1.4 – 11.4
Copper (μg/L)	30.9 – 108.3
Iron (mg/L)	0.13 – 0.92
Manganese (μg/L)	17.3 – 47.9

Correct citation: Patterson, C.M., Tinggi, U., and Reilly, C. 1997. Trace element levels in Australian cows' milk. In *Trace Elements in Man and Animals – 9: Proceedings of the Ninth International Symposium on Trace Elements in Man and Animals. Edited by* P.W.F. Fischer, M.R. L'Abbé, K.A. Cockell, and R.S. Gibson. NRC Research Press, Ottawa, Canada. pp. 166–167.

However there were seasonal variations for zinc where the spring and winter milk contained higher Zn levels (range 2.1 – 4.9 mg/L) than other seasons in one year.

Reported levels of trace elements in cows' milk in worldwide published studies vary widely which is in part due to the variation in sampling and analytical techniques utilised as well as in some instances, contamination.

The wide range of values for the trace elements Al, Cr, Cu, Fe and Mn found in this study did not translate into significant overall regional or seasonal variation. Zinc showed minimal seasonal variation. Excluded from the results reported here are the results determined for selenium which showed extensive seasonal and regional variation (Tinggi et al. 1996) which may be particularly significant for selenium intake in infants. Milk could generally be considered a poor source of trace element intake for the remaining trace elements considered in this study.

References

U.Tinggi, C.M. Patterson and C.Reilly (1996) In press.

Placental and Mammary Transfer of Lead and Cadmium by Ewes Exposed to Pb and Cd-Enriched, Sewage Sludge-Treated Soils*

N.F. SUTTLE,[†] J. BREBNER,[†] B. STARK,[‡] N. SWEET,[‡] AND J. HALL[‡]

[†]Moredun Research Institute, Edinburgh, Scotland EH17 7JH. [‡]Water Research Centre, Marlow, England SL7 2HD

Keywords lead, cadmium, sewage sludge, pasture, food chain

The use of sewage sludge on land over many years can result in the accumulation of potentially toxic elements (PTE) in soil. Regulations limit their maximum concentrations but there is concern that the limit values for Pb and Cd in particular, may not provide sufficient protection of the human food chain from the consumption of meat from animals which graze sludge-treated pasture. Pb in particular is poorly taken up by plants and the principal route of entry is likely to be the ingestion of soil-contaminated grass after application to pasture (Suttle and Hall 1993). However, PTE uptake from pasture by the grazing animal may also be low (Suttle et al. 1991). If accumulation does occur, its extent in the fattened lamb might be determined by PTE transfer from mother to offspring as well as by direct soil ingestion. The soil ingestion route of entry was therefore simulated in long-term feeding trials with sheep taken through two reproductive cycles.

Materials and Methods

Four groups of six, weaned female lambs were given a basal diet of 1 kg pelleted grass meal per day containing either no soil or 50 g/kg of one of three top soils (H, N and W) from sites where sewage sludge had been applied over many years at high rates. The PTE composition of the soil-contaminated diets is given in Table 1. Soil N contained Cd and Pb at levels similar to the current EC statutory limits and UK guidelines (3 and 300 mg (14.5 and 1450 µmol)/kg DM, respectively), soil H contained excessive Cd while soil W contained excessive concentrations of both elements. The experiment lasted 2.8 years and the females were mated twice after about 450 and 780 d treatment. The additional nutrient requirements for late pregnancy and lactation were met by feeding supplements of a proprietary compound, low in Pb and Cd, to all groups. After the first lambing, one lamb was reared per ewe and the remainder of the litter (usually 2 or 3) was killed at birth. After the second lambing, all lambs born were reared. PTE accumulation was monitored in the live animal (mother and offspring) by taking liver biopsy, heparinised blood and milk samples and at slaughter (at birth or as fat lambs) in liver, kidney and muscle.

Pb and Cd in blood were determined by atomic absorption spectrophotometry (AAS) in supernatants obtained by centrifuging samples diluted 1:5 with 0.6% w/v nitric acid, containing 0.2% w/v Triton × 100. Tissue samples were digested with concentrated HNO_3 prior to AAS and assessed against standard reference materials (horse or porcine kidney and bovine liver) with an acceptability of ±10% of the reference value. Milk samples were analysed by graphite furnace AAS in samples diluted 10-fold in 0.2% v/v ammonia with 10% v/v HNO_3 as matrix modifier and a non-fat milk powder for reference purposes. Data with heterogenous variances were subject to log transformation.

*The work was funded by the UK Government's Department of the Environment.

Correct citation: Suttle, N.F., Brebner, J., Stark, B., Sweet, N., and Hall, J. 1997. Placental and mammary transfer of lead and cadmium by ewes exposed to Pb and Cd-enriched, sewage sludge-treated soils. In *Trace Elements in Man and Animals – 9: Proceedings of the Ninth International Symposium on Trace Elements in Man and Animals. Edited by* P.W.F. Fischer, M.R. L'Abbé, K.A. Cockell, and R.S. Gibson. NRC Research Press, Ottawa, Canada. pp. 168–170.

Table 1. Mean (with standard deviation) Pb and Cd concentrations (µmol kg-1 DM) control (O) and three soil-supplemented diets fed during the first pregnancy and lactation. Composition of the three soils (H, N and W) and of the kidneys of reared lambs at commercial slaughter are also given, the latter as medians with ranges beneath in parentheses.

			Soil Supplement		
		Control	H	N	W
Soil	Cd	–	15.2	29.5	53.6
	Pb	–	2860	1473	4063
Diet	Cd	0.8(0.14)	1.4(0.11)	1.79(0.27)	3.6(0.28)
	Pb	4.8(4.69)	153.6(21.26)	83.1(9.21)	214.0(11.59)
Kidney	Cd	0.39^{ab}	0.26^{b}	0 43^{ab}	3.79^{b}
		(0.29–0.64)	(0.21–0.30)	(0.32-0.48)	(3.53–4.15)
	Pb	0.48^{c}	10.34^{a}	4.22^{ab}	3.79^{b}
		(0.34–1.64)	(3.62–15.36)	(1.84–11.69)	(3.52–4.15)

Superscript differences within rows (or columns in Table 2) denotes treatment differences with $P < 0.05$ after log transformation.

Results

Placental Transfer: Maximum tissue concentrations of Pb were found in the livers of new born lambs from the first lamb crop from soil-supplemented ewes (Table 2) but Cd concentrations were below detection limits, indicating that placental transfer was only significant for Pb. There were no significant differences between the three soils in liver Pb at birth. Kidney Pb was marginally raised in groups H and W but not N (median values 0.56, 0.56 and 0.35 µmol/kg FW) respectively.

Mammary Transfer: Cd concentrations were below detection limits (0.045µmol/L) in all milk samples indicating negligible transfer. Milk Pb concentrations were increased by soil ingestion, particularly by soils H and W at the mid-point of the first lactation (Table 3). Confirmation of mammary transfer of Pb is found in the marked rise in blood Pb concentrations in lambs by mid lactation (Table 3). By late lactation, significant differences were apparent between soils with soil N giving smaller increases in blood Pb than

Table 2. Median (ranges in parentheses) liver Pb concentrations (µmol kg-1 fw) in first crop lambs.

	Group	Birth	10 weeks (biopsy)	34 weeks
Liver	O	0.94^{b}	0.96^{c}	0.48^{c}
		(0.063–1.40)	(0.53–1.07)	(0.43–0.63)
	H	13.50^{a}	6.81^{b}	3.86^{ab}
		(2.66–25.56)	(6.67–8.79)	(2.75–4.93)
	N	6.04^{a}	3.71^{b}	2.49^{b}
		(2.03–7.73)	(2.57–4.79)	(1.98–3.96)
	W	6.76^{a}	8.60^{a}	4.13^{a}
		(2.90–13.62)	(7.25–11.35)	(3.57–6.09)

Table 3. Mean Pb concentrations (pmol L-1) in milk sampled at three stages of the first lactation and in the whole blood of suckling lambs at those times.

	Stage	Group O	Group H	Group N	Group W	s.e.m.
Ewes milk	Early	130^{a}	179^{e}	68^{e}	333^{e}	97.1
	Mid	77^{c}	285^{b}	121^{c}	488^{a}	59.0
	Late	<39^{b}	266^{ab}	97^{b}	478^{a}	77.6
Lambs blood	Early	193^{b}	326^{ab}	290^{ab}	489^{a}	74.7
	Mid	232^{c}	855^{b}	749^{b}	1126^{a}	98.5
	Late	203^{c}	874^{a}	449^{b}	899^{a}	61.5

Superscript differences within rows = $P < 0.05$.

soils H and W. Results from the second crop generally did not differ qualitatively or quantitatively from those of the first crop and are not given in detail.

Discussion

Pb ingested in sludge-treated topsoil by ewes can be transferred to their offspring via the placenta and milk even from a soil (N) that meets the current statutory limits and guidelines. Maximum blood Pb concentrations in the lamb coincided with maximum values in the milk and probably reflect the high availability of Pb in milk. Development of a functional rumen probably lowers Pb absorption and hence blood Pb. The attainment of plateau concentrations for blood Pb in ewe and lamb indicate homeostatic control of lead accumulation which probably prevented excessive accumulation in the tissues. Liver Pb values did not exceed the maximum acceptable concentration (MAC) for human consumption (<2 mg Pb (9.7 µmol)/kg FW) in any individual but values in kidney occasionally exceeded the MAC (<1 mg (4.8 µmol)Pb/kg FW) even in Group N. Soil ingestion did not raise Cd or Pb concentrations in muscle. We conclude that current EC statutory limits and UK guidelines for soil Cd and Pb provide sufficient protection of the food chain.

Literature Cited

Suttle, N.F., Brebner, J., Hall, J. (1991) Faecal excretion and retention of heavy metals in sheep ingesting topsoil from fields treated with metal-rich sewage sludge. In: Proceedings of the 7th International Symposium on Trace Elements in Man and Animals (Momcilovic, B., ed.) pp. 32–7–32–8. IMI, Zagreb.

Suttle, N.F. and Hall, J. (1993) Soil ingestion as a route of entry for potentially toxic elements into the food chain following the application of sewage sludge to pasture In: Proceedings of the Fifteenth International Congress of Nutrition, Adelaide (Smith, R.M., ed.).

Mercury Levels in Muscle of Some Fish Species from the Dique Channel, Colombia*

J. OLIVERO, V. NAVAS, A. PEREZ, B. SOLANO, R. SALAS, R. VIVAS, AND E. ARGUELLO

Applied Chemistry Group, Facultad de Ciencias Químicas y Farmacéuticas, Universidad de Cartagena, A.A. 6541, Cartagena, Colombia (Sur América)

Keywords mercury, fish, muscle, atomic absorption

Introduction

In a recent study we have found that there is a high incidence of mercury contamination among inhabitants of the South of Bolivar, which is the main gold mining zone in Colombia. It was found that hair mercury concentration in these people was related to fish intake (Olivero et al. 1995). The objective of this research is to investigate the mercury levels of several fish species that are consumed in some populations along the Dique Channel, a waterbody which receives contaminated waters coming from the South of Bolívar and the central part of the country, where these species constitute the most important food for almost two hundred thousand people who are living in the Channel zone. The purpose of this paper is to show the preliminary results of the determination of mercury levels in four river fish species commonly consumed by the population living along the Dique Channel to establish if these levels are below the FAO/WHO guidelines.

Materials and Methods

Four species of fish were purchased from local fishermen in three fishing locations along the Dique Channel, a waterbody which connects the contaminated Magdalena river to Cartagena Bay. The sampling was performed in the towns of Soplaviento, Maria La Baja and Gambote, during June-December 1995 (first sampling period). The fish species included in this study were *Prochilodus reticulatus Magdalenae* (Bocachico), *Rhamdia sebae* (Barbudo), *Pseudoplatystoma fasciatum* (Bagre pintado), *Triportheus magdalenae* (Arenca). A sample of fish dorsal muscle (>10 g) was cut with Teflon knives and the samples were packed into plastic flasks and frozen for later analysis. Total fish length was measured for all fish analyzed.

Total mercury concentrations were determined by cold vapor atomic absorption spectroscopy after acid digestion (Sadiq et al. 1991); three replicate determinations were made for each sample. The calculated detection limit was 0.006 μg/g. The analytical procedure was evaluated using samples spiked with mercury chloride. The recovery of mercury in these samples oscillated between 91 and 106%.

Mercury analysis for each fish was reported as a mean for three subsamples, with its respective standard deviation. Fish length and mercury concentrations were correlated by means of simple lineal regression (Walpole et al. 1992). The evaluations of the different mean mercury concentrations among species were performed by simple variance analysis and the Newman Keuls Test (Tallarida et al. 1986), a p-value of ≤ 0.05 was considered significant.

---

*The authors wish to thank Universidad de Cartagena for financial assistance.

Correct citation: Olivero, J., Navas, V., Perez, A., Solano, B., Salas, R., Vivas, R. and Arguello, E. 1997. Mercury levels in muscle of some fish species from the Dique Channel, Colombia. In *Trace Elements in Man and Animals – 9: Proceedings of the Ninth International Symposium on Trace Elements in Man and Animals. Edited by* P.W.F. Fischer, M.R. L'Abbé, K.A. Cockell, and R.S. Gibson. NRC Research Press, Ottawa, Canada. pp. 171–172.

Results and Discussion

The results of mercury analysis in the tested species are shown in Table 1. It can be seen that mean mercury concentrations increase in the order: *Prochilodus reticulatus Magdalenae* (Bocachico) < *Rhamdia sebae* (Barbudo) < *Pseudoplatystoma fasciatum* (Bagre Pintado) < *Triportheus magdalenae* (Arenca).

Total mercury concentrations in the fishes oscillated between non-detectable (<0.006 μg/g) and 0.22 μg/g, for *Prochilodus reticulatus Magdalenae* and *Triportheus magdalenae,* respectively. According to Hakanson (1984) mercury concentrations in aquatic species >0.075 μg/g should be attributed to water pollution due to human activities. In our particular case the observed mercury levels in *Pseudoplatystoma fasciatum* and *Triportheus magdalenae* may indicate the presence of man-made contamination in the Dique Channel.

According to the variance analysis, at a p-value of 0.05, there are significant differences among mercury muscle means in different species that were checked. The Newman Keuls test was performed, and when the groups were compared ($p < 0.01$), there were significant differences among their mean concentrations.

There was no statistical correlation between fish length and total mercury concentrations ($R < 0.01$). Similar results were also reported by Jackson (1991) who detected poor correlations between fish fork length and mercury content in some species from lakes in northern Manitoba, Canada.

The mercury fish concentrations did not reach levels higher than International guidelines established to determine whether fish is safe to eat. Bocachico, the most popular species among river communities and on all the North Coast of Colombia, contains low levels which make this species the best for consumption. Nevertheless, permanent mercury monitoring of the species studied here is fundamental to guarantee nutritional quality and safety.

Table 1. Mercury levels in representative fish of the Dique Channel.

Specie	Mean μg/g	Standard Deviation	n
Prochilodus reticulatus Magdalenae	0.019	0.01	15
Rhamdia sebae	0.040	0.03	28
Pseudoplatystoma fasciatum	0.080	0.05	23
Triportheus magdalenae	0.104	0.04	4

References

Hakanson, L. (1984) Metals in fish and sediments from the river Kolbäcksan water system, Sweden. Arch. Hidrobiol. 101: 373.

Jackson, L. (1991) Biological and environmental control of mercury accumulation by fish in lakes and reservoirs of Northen Manitoba, Canada. Can. J. Fish. Aquat. Sci. 48: 2449–2470.

Olivero, J., Mendoza, C., & Mestre, J. (1995) Hair level of mercury in people from the southern Bolivar. Revista de Saude Publica. 45: 45–48.

Sadiq, M., Zaidi, T., & Al-Mohana, H. (1991) Sample weight and digestion temperature as critical factors in mercury determination in fish. Bull. Environ. Contam. Toxicol. 47: 335–341.

Tallarida, R., & Murray, R. (1987). Manual of Pharmacological Calculations with Computer Programs. 2nd. Edition. Springer-Verlag. New York.

Walpole, R., & Myers, R. (1992). Probabilidad y Estadística. Cuarta Edición. México. McGraw Hill.

Mercury in Fish in the Oldman River Reservoir and its Tributaries

S. WU, L.Z. FLORENCE, H.V. NGUYEN, K.L. SMILEY, AND K. SCHWALME
Alberta Environmental Centre, Bag 4000, Vegreville, Alberta, Canada T9C 1T4

Keywords mercury, fish, reservoir, methylation

The Oldman Reservoir, created in 1991 by construction of a dam near the confluence of the Oldman, Crowsnest, and Castle rivers in southern Alberta, began full operations in 1992. Mercury levels in fish from the reservoir have been monitored from 1991 to 1995 to assess the changing mercury levels in fish as part of a much larger environmental monitoring program and fisheries mitigation strategy for the Oldman Dam project being implemented by the Alberta Government.

In Canada and elsewhere, river impoundment has been known to increase mercury concentrations in fish as a result of enhanced activities of methylating bacteria in the newly inundated soil. Methyl mercury, a potent neurotoxin, is the dominant form of mercury found in fish muscle tissues. To protect public health, a guideline of 0.5 mg kg^{-1} for total mercury (THg) concentration in fish muscle tissue has been developed by Health Canada.

The rate at which mercury is methylated and subsequently taken up by fish and other aquatic organisms depends on numerous factors, including the pH and redox potential of the sediment and overlying water, binding of Hg(II) ion to sulfide, binding of organic mercury (OHg) to organic matter, FeOOH and MnOOH, microbiological activity, mercury concentration in water, temperature, and trophic conditions (Jackson 1988, 1993). Because of the complex chemistry of Hg, its concentration can vary greatly between reservoirs. In the only other Alberta reservoir (Gleniffer Lake) studied to date, significant increases in fish mercury content were not observed (Moore 1989). Nevertheless, it was considered prudent to monitor mercury in fish in the aquatic system of the Oldman Reservoir.

Methods

Fish were sampled once each year during late September or early October in 1991–1995 from two sites in the reservoir, two sites below the dam and two control sites upstream of the reservoir. The most-frequently caught species were bull trout, rainbow trout, mountain whitefish, longnose sucker and white sucker.

Total mercury (THg) was determined by gradually digesting fish tissue at 250°C in a mixture of concentrated sulphuric and nitric acids. The THg concentration in the digested sample, after reduction by stannous chloride, was analyzed using a cold-vapour mercury analyzer.

The statistical model used for each population was one-way analysis of covariance with fish length being the covariant to adjust THg for variability in fish size. To overcome year-to-year random variation in the size of fish collected, the adjusted (i.e., length-specific) least squares means of THg were used for detecting trends in mercury levels from 1991 to 1995. The significance of linear and curvelinear (quadratic and cubic) trends was tested at the 5% significance level.

Results

The temporal changes in mercury levels of bull trout and mountain whitefish are shown in Table 1, where the most two significant p-values for the three trends (linear, quadratic and cubic) tested and the 1986 pre-impoundment data are also listed.

Correct citation: Wu, S., Florence, L.Z., Nguyen, H.V., Smiley, K.L., and Schwalme, K. 1997. Mercury in fish in the Oldman River Reservoir and its tributaries. In *Trace Elements in Man and Animals – 9: Proceedings of the Ninth International Symposium on Trace Elements in Man and Animals. Edited by* P.W.F. Fischer, M.R. L'Abbé, K.A. Cockell, and R.S. Gibson. NRC Research Press, Ottawa, Canada. pp. 173–174.

Table 1. Temporal changes in length-specific THg level in bull trout and mountain whitefish.

Group	Mean Length (cm)	Adjusted Parameter	THg (mg kg^{-1}) 1986	1991	1992	1993	1994	1995	Time Trend (1991–1995)	p-value
Bull Trout										
Upstream on Oldman River	25.7	MEAN	–	0.228	0.297	0.352	0.339	–	linear	0.3927
		SE	–	0.063	0.056	0.077	0.116	–	quadratic	0.6395
Reservoir	33.8	MEAN	0.189	0.192	0.324	0.293	0.395	0.319	linear	0.0001
		SE	0.036	0.011	0.030	0.020	0.026	0.026	quadratic	0.0006
Mountain Whitefish										
Upstream on Crowsnest River	31.5	MEAN	–	0.052	0.041	0.042	0.047	0.039	cubic	0.0206
		SE	–	0.003	0.004	0.004	0.004	0.003	linear	0.0509
Reservoir	23.8	MEAN	0.151	0.120	0.160	0.138	0.228	0.164	linear	0.0001
		SE	0.018	0.012	0.012	0.014	0.014	0.012	cubic	0.0239
Below the Dam	28.7	MEAN	0.143	0.143	0.138	0.127	0.104	0.129	linear	0.0410
		SE	0.014	0.009	0.015	0.016	0.010	0.008	cubic	0.1563

Discussion

Irrespective of species, site or year, mean THg levels were always less than 0.5 mg kg^{-1} (Wu et al. 1996), the Canadian Guideline for consumable fish.

Below the dam, the increases of mean THg levels in the above mentioned five species are not observed. Within the reservoir, increases in mean THg levels are not seen in longnose sucker and white sucker. However, there has been an increase in mean THg levels in bull trout, which also contained higher mercury levels than any other fish species in the Oldman River system. This probably reflects the bull trout's position near the top of the aquatic food chain. Modest increases in mean THg levels in mountain whitefish and rainbow trout from the reservoir, compared to the pre-impoundment level, have also been observed. However, the mean THg levels in these species collected from the reservoir were lower in 1995 than in 1994.

There may be several reasons why mercury has not increased appreciably in fish after the impoundment of the Oldman River. The area has a drier climate and the aquatic system is more alkaline compared to reservoirs in eastern Canada where elevated mercury levels in fish have been observed. Also, about 440 hectares of riparian forest and 150,000 cubic metres of top soil were removed from the impounded area prior to impoundment, reducing the supply of organic matter which is closely related to mercury methylation rates. Finally, being at the confluence of three tributaries, the Oldman River Reservoir has had high yearly flushing rates, which would be expected to lower methyl mercury concentrations in the water column.

This study suggested that, by removing the organic matter from the area prior to the impoundment, increases in Hg in fish can be minimized in certain aquatic systems.

Literatures Cited

Wu, S., Florence, L.Z., Nguyen, H.V., Smiley, K.L. & Schwalme, K. (1996) Oldman River Dam: Mercury in Fish — 1994 Interim Report. AECV96-R3. pp. 23. Alberta Environmental Centre, Vegreville, AB, Canada.

Moore, J.W. (1989) A five-year study of mercury in fish from a newly formed reservoir. AECV89-R4. pp. 24. Alberta Environmental Centre, Vegreville, AB.

Effect of a Zinc Bolus on Absorption of ^{70}Zn by Infants*

E.E. ZIEGLER,[†] R.E. SERFASS,[‡] S.E. NELSON,[†] AND J.A. FRANTZ[†]

[†]Department of Pediatrics, University of Iowa, Iowa City, Iowa 52242, and [‡]Department of Food Science and Human Nutrition, Iowa State University, Ames, Iowa 50011, USA

Keywords zinc, absorption, infants, bolus

Fractional zinc absorption is inversely related to zinc intake. We wished to determine whether this is due to involvement of different mechanisms, or whether it is an adaptive response that requires time to occur. We determined absorption of ^{70}Zn given without a zinc bolus or together with two different zinc boluses providing 0.5 mg of zinc or 5 mg of zinc.

Methods and Procedures

Six normal infants (4 females, 2 males) who ranged in age from 52 to 295 d were fed a milk-based formula with zinc concentration of approximately 6 mg/L. In a randomized crossover design, zinc absorption was determined in each infant on three separate occasions two weeks apart. Each time, a test dose of ^{70}Zn was given orally 3 h after a feeding and 1 h before the next feeding. The ^{70}Zn test dose of about 18 µg/kg was given with no additional zinc (0 Zn bolus), or with 0.5 mg of zinc (0.5 Zn bolus) or with 5.0 mg of zinc (5 Zn bolus). The order of the different boluses was random.

Collection of feces was accomplished by in-house 72-h metabolic balance studies (2 infants) or by 96-h home stool collections, for which infants were also admitted during the daytime, where stool was collected, with some admixture of urine, with the use of acid-washed cloth diapers. All stools were collected between carmine markers. On the first day of an absorption study, a precisely weighed amount (about 0.5 mL) of a $^{70}ZnCl_2$ solution was given directly from a syringe into the mouth of the infant, followed by 0.5 mL or 5.0 mL of a solution containing zinc sulfate heptahydrate (1 mg/ml), followed by 2 rinses of 5 mL of 5% glucose solution.

$^{70}Zn/^{67}Zn$ isotope ratios were determined in fecal ashes and diaper extracts by inductively coupled plasma mass spectrometry (ICP/MS) using a Finnigan Sola (Finnigan MAT, San Jose, CA) instrument. The amount of ^{70}Zn label in feces was calculated as described previously (Ziegler et al. 1989). Absorption of ^{70}Zn was calculated as ^{70}Zn ingested minus ^{70}Zn label excreted in feces and expressed as percent of the dose. Absorption of bolus zinc was calculated as the product of ^{70}Zn% absorption and bolus zinc.

Results

Mean weight and age at the time of study are indicated in Table 1. Mean net total zinc absorption was similar. Percent absorption of ^{70}Zn was high (mean 64.4%) when no bolus was given (0 Zn bolus), and low (mean 32.8%) when a large bolus (5 Zn bolus) was given (Figure 1). The difference was significant at $p < 0.001$. Results with a small bolus (0.5 Zn bolus) showed a dichotomous distribution, with half the studies showing high absorption and the other half low absorption. Absorption was significantly ($p < 0.01$) greater with the 0.5 Zn bolus than with the 5 Zn bolus. Bolus zinc absorption, on the other hand, increased with increasing bolus size. There was no difference in post-collection enrichment in relation to bolus size.

*Supported by U.S. Public Health Service grant HD 26872.

Correct citation: Ziegler, E.E., Serfass, R.E., Nelson, S.E., and Frantz, J.A. 1997. Effect of a zinc bolus on absorption of ^{70}Zn by infants. In *Trace Elements in Man and Animals – 9: Proceedings of the Ninth International Symposium on Trace Elements in Man and Animals. Edited by* P.W.F. Fischer, M.R. L'Abbé, K.A. Cockell, and R.S. Gibson. NRC Research Press, Ottawa, Canada. pp. 175–176.

Table 1.

	0 Zn	0.5 Zn	5 Zn
Weight (kg)	6.85 ± 1.77	6.75 ± 1.42	6.79 ± 1.89
Age (d)	182 ± 87	169 ± 92	177 ± 108
Total Zinc intake (μg/kg/d)	1017 ± 312	1055 ± 297	1287 ± 274
Total Zinc excr. (μg/kg/d)	818 ± 252	863 ± 154	1085 ± 351
Total Zinc net abs. (%)	14.2 ± 32.9	15.5 ± 14.2	12.3 ± 35.6
^{70}Zn absorption (%)	64.4 ± 6.6	56.0 ± 13.2	32.8 ± 7.1
Bolus zinc abs. (μg)	110 ± 18	373 ± 102	1693 ± 307

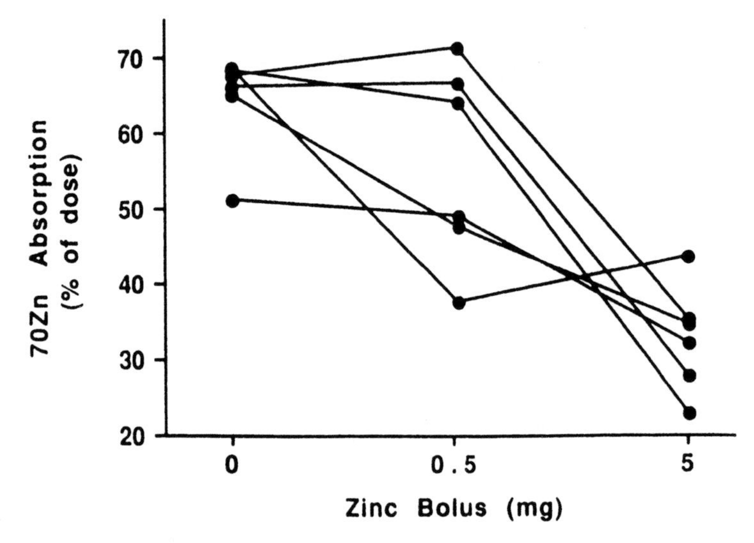

Figure 1.

Discussion

Our results clearly demonstrate that a prefeeding period is not required for differences in percent zinc absorption in response to different zinc loads to become effective. Our data are consistent with active transport at low zinc intakes (0 Zn bolus) and predominantly passive diffusion at high intakes (5 Zn bolus). The dichotomous distribution of absorption values with the small bolus (0.5 Zn bolus) suggests that, at this dose, absorption was active in some infants and passive in others. This dichotomy was not related to age.

Literature

Ziegler, E.E., Serfass, R.E., Nelson, S.E., Figueroa-Colón, R., Edwards, B.B., Houk, R.S., and Thompson, J.J. (1989) Effect of low zinc intake on absorption and excretion of zinc by infants studied with ^{70}Zn as extrinsic tag. J. Nutr. 119: 1647–1653.

Oral Aluminum Exposure of Adults in Germany — A Long-Term Survey

M. MÜLLER, M. ANKE, AND H. ILLING-GÜNTHER
Friedrich Schiller University, Institute of Nutrition and Environment, D-07743 Jena, Germany

Keywords aluminum, double portion technique, oral exposure, toxicological risks

Introduction

The neurotoxicity of aluminum has been known since the end of the 19th century. But until the middle of the seventies of the present century there was no interest in aluminum in the fields of medicine, biology and nutrition. Proving an etiological function for aluminum in diseases related to chronic dialysis (microcytic anaemia, osteomalcia, encephalopathy) and discussing the possible role of higher aluminum bioavailability due to acid rain resulted in a greater research focus on the biological importance of this metal. One aspect in this field is the oral aluminum exposure of adults and its toxicological risks.

Material and Methods

The aluminum intake of adults in Germany was investigated by means of the so-called double portion technique. The experiments were carried out in four towns of East Germany in 1988, in six towns between 1990 and 1992 and in three towns in 1995. At least seven men and seven women took part in each trial. They collected a visually estimated duplicate of their daily meals including all beverages and snacks on seven subsequent days. A total of 1312 duplicates were available for analysis. Food intake and choice of foods were not restricted. Medications were recorded but not added to the duplicate.

The duplicates as well as the beverages were collected in plastic vessels and stored at –18°C until use. The samples were dried at 105°C, ashed in a muffle furnace at 450°C, dissolved in 25% HCl and diluted to make up a 2.5% solution. Aluminum determination was carried out by graphite furnace atomic absorption spectrometry using a combination of the AAS 3030, HGA 400 and AS 40 (Perkin Elmer).

Results

The results of all duplicate experiments showed that the aluminum intake of adults in Germany amounted to 4.9 mg/d in men and 4.4 mg/d in women, respectively (Table 1). Sex had a significant effect on the aluminum intake as well as on the other parameters.

While the higher dry matter intake of men is well-known, the significant difference in the aluminum content of the dry matter consumed by men and women was surprising. Nevertheless, due to the higher dry matter intake of men their aluminum intake exceeded that of women.

Table 1. Results of the duplicate studies.

	Men		Women			
Parameter	s	$\bar{x}$	$\bar{x}$	s	p	%[1]
DM[2] intake (g/d)	110	386	308	89	<0.001	125
Al concentration (µg/g DM)	8.8	13	15	10	<0.01	87
Al intake (mg/d)	3.5	4.9	4.4	3.2	<0.01	11

[1]....women = 100%, men = x%; [2] ....dry matter.

Correct citation: Müller, M., Anke, M., and Illing-Günther, H. 1997. Oral aluminum exposure of adults in Germany — A long-term survey. In *Trace Elements in Man and Animals – 9: Proceedings of the Ninth International Symposium on Trace Elements in Man and Animals. Edited by* P.W.F. Fischer, M.R. L'Abbé, K.A. Cockell, and R.S. Gibson. NRC Research Press, Ottawa, Canada. pp. 177–178.

A strong, time-dependent decline in aluminum consumption was observed (Table 2). compared to 1988, when the highest levels were observed, the aluminum intake was reduced to 49% in men and 57% in women in 1995.

Since the dry matter intake of men and women remained nearly constant from 1988 to 1995, the observed aluminum intakes reflect a decrease in the aluminum concentration. In 1988 men and women took in dry matter with 17 and 18 µg Al/g. Between 1990 and 1992 the aluminum concentrations amounted to 12 and 15 µg/g dry matter in men and women respectively. In 1995 the dry matter contained 8.3 (men) and 10 µg Al/g (women). Related to the figures obtained in 1988, the aluminum content of the dry matter decreased to 49% in men and 56% in women. The reasons for this decline remains hypothetical.

The living area of the subjects was the third criterion which influences aluminum intake. Due to variations in the aluminum content in the dry matter, the effect of the living area on aluminum intake was particularly striking between 1990 and 1992. Persons living in Thuringia consumed dry matter with the highest aluminum amounts. These differences were more obvious in men than in women. Reports indicate that a safe upper intake of Al for adults is 1 mg/kg body weight/d (anonymous 1989). The toxicological risk was evaluated by relating the aluminum intake of our test persons to their body weight. The subjects consumed a mean of 0.06 (men) and 0.07 mg Al/kg/d (women) in the duplicate studies, only 6 and 7% of the upper safe level respectively. These data show that toxicological risks does not arise from the oral aluminum intake via foods and beverages by these healthy subjects.

Table 2. Influence of the time of investigation and sex on the aluminum consumption of adults in Germany (mg/d).

Time of investigation	Men		Women		
	s	$\bar{x}$	$\bar{x}$	s	Fp_{SEX}
1988	3.6	6.5	5.4	3.5	<0.01
1990 – 1992	3.5	4.9	4.6	3.1	
1995	2.2	3.2	3.1	1.9	
Fp_{TIME}		<0.001			–

Summary and Conclusions

The aluminum intake of adults in Germany was investigated by means of the double portion technique between 1988 and 1995. It amounted to 4.9 mg/d in men and to 4.4 mg/d in women. A strong and significant time-dependent decrease of the aluminum intake occurred which resulted from the decline of the aluminum concentration in the dry matter. Related to body weight the test persons consumed only 0.06 (men) and 0.07 mg Al/kg/d (women). Thus, Al poses no toxicological risk for healthy subjects.

References

Anonymous (1989) WHO Food Additives Series 24.

Biological Importance, Analysis and Supply of Vanadium

H. ILLING-GÜNTHER, M. ANKE, M. MÜLLER, A. TRÜPSCHUCH, AND S. ANKE
Friedrich Schiller University, Institute of Nutrition and Environment, D-07743 Jena, Germany

Keywords vanadium, ICP AES, foodstuffs, vanadium deficiency

Introduction

Vanadium (V) is a heavy metal which is often used as an alloying component in steels as well as a catalyst in chemical syntheses. Therefore it has the potential to get into the food chain. In biological material it belongs to the group of ultra trace elements. Its toxic and/or essential properties are discussed in the literature (Jakubke and Jeschkeit 1987).

Material and Methods

Foodstuffs, faeces and urine were collected in duplicate in 1988 and 1991 in the former GDR and the New Lands of Germany. They were dried at 60°C and 105°C, crushed, homogenised and dry ashed at 450°C (muffle furnace). The samples were dissolved in 25% HCl and diluted to form a 2.5% solution. Analyses were carried out by means of sequential ICP AES (SPECTROFLAME-D, SPECTRO Analytical Instruments) at the line position of 292.402 nm. Fox Pro (Version 2.6, Microsoft) and SPSS/PC+ for Windows (Version 6.0, SPSS Inc.) were used for data handling and statistics (Illing-Günther 1995).

The analyzed V concentrations of single foods were compared to nutritional intake data (MONICA survey 1988 and the nutritional survey 1991/92, resp.) (Müller 1993) and the V intake calculated using the market basket.

Daily duplicates of all consumed foods were collected from seven men and seven women on seven subsequent days and their daily urine and faeces were analyzed. The V intake was determined and V balance was calculated.

In addition, the effect of V deficiency was studied in 14 goats which were fed a V-poor semisynthetic ration (Table 1) (Anke et al. 1989).

Table 1. Vanadium content of control and vanadium deficiency rations.

	Control ration		V poor ration			
n;n	s	[µg/kg DM]	[µg/kg DM]	s	p	%[1)]
41;9	1434	2357	20.38	11.40	<0.001	0.9

1)...control ration = 100%.

Results

Flour and farinaceous products, pulse, noodles, bread and pastries, milk and dairy products are V-poor (2...40 µg/kg dry matter). Medium V concentrations are found in vegetables, fruit, meat and sausages as well as in fish (10...90 µg/kg dry matter). Cocoa and cocoa products (50...300 µg/kg dry matter) as well as spices and herbs (20...2400 µg/kg dry matter) are V-rich.

A comparison of the V intakes calculated using the market basket and the duplicated analysed results showed remarkable differences. Men consumed 27 µg/d (market basket) and 19 µg/d (duplicates) in 1988;

Correct citation: Illing-Günther, H., Anke, M., Müller, M., Trüpschuch, A., and Anke, S. 1997. Biological importance, analysis and supply of vanadium. In *Trace Elements in Man and Animals – 9: Proceedings of the Ninth International Symposium on Trace Elements in Man and Animals. Edited by* P.W.F. Fischer, M.R. L'Abbé, K.A. Cockell, and R.S. Gibson. NRC Research Press, Ottawa, Canada. pp. 179–180.

women 12 µg/d (market basket) and 9 µg/d (duplicates). In 1988 the market basket method indicated a 41% higher intake for men and a 24% higher one for women. In 1992 both methods showed the reverse phenomenon. Men consumed 30 µg/d (market basket) and 36 µg/d (duplicates); women 11 µg/d (market basket) and 25 µg/d (duplicates). The V intake was underestimated by the market basket method (–16% for men and –58% for women). Reasons for this are difficult to find. One possibility could be the higher portion of tinned products consumed after the reunification of Germany.

The V concentration of foodstuff dry matter consumed by both sexes was not significantly different in 1988 and 1991. A daily V intake from foods by men of 19 µg/d (1988) and 36 µg/d (1992) and by women of 9.4 µg/d (1988) and 25 µg/d (1992) was found.

The total V excretion amounted to 48 µg/d (men) and 20 µg/d (women). 94% of the excreted V left the body faecally and 6% renally. The V intake and excretion were balanced with 5 µg (men) and 3 µg (women). Our investigations showed a mean apparent absorption of 17% in average.

The experimental female goats showed a decreased feed intake, a decreased growth after intrauterine V depletion, a slowed rate of first insemination and conception, an increased abortion rate, a shifted sex ratio (female:male = 1:0.86) and an increased mortality in the first year of life after a V poor semisynthetic ration was fed. The 16 organs and body fluids analysed after death had a lower V content than in the controls. Particularly kidney, uterus, lung, spleen, rib and carpal bone had reduced V concentrations of up to 15%.

Summary and Conclusion

The ICP AES is suited to analyze the V content among other elements in biological material after dry ashing at 450°C. Foods and beverages had a great variation in their V content. The calculation (market basket) and the analytical (duplicate) methods of determining V intake gave different results. The V intake and excretion were balanced. The experiments with goats fed a V-poor semisynthetic ration showed several deficiency symptoms. All analysed organs and tissues as well as body fluids showed a lower V concentration than those of controls. V seems to be essential. The V requirement for animals could be <35 µg/kg feed dry matter and that of adult humans about 10 µg/d. A V deficiency is improbable under natural conditions.

References

Anke, M., Groppel, B., Gruhn, K., Langer, M., Arnold, W. (1989) The essentiality of vanadium for animals, In: M. Anke et al.: 6th International Trace Element Symposium, Vol. 1, Mo, V, pp. 17–27. University of Leipzig and University of Jena, Germany

Illing-Günther, H. (1995) Bestimmung, biologische Bedeutung und Versorgung des Menschen mit Vanadium. (Determination, biological importance and human supply of vanadium) Dissertation, Friedrich Schiller University Jena, Germany

Jakubke, H.D. and Jeschkeit, H.(1987) Brockhaus ABC Chemie (Brockhaus ABC Chemistry), Vol. 2, VEB Brockhaus Verlag, Leipzig, Germany

Müller, M. (1993) Cadmiumaufnahme und Cadmiumausscheidung Erwachsener nach der Marktkorb- und Duplikatmethode (Cadmium intake and cadmium excretion by means of market basket and duplicate method, Dissertation, Friedrich Schiller University Jena, Germany.

Magnesium Intake and Magnesium Balance of Adults Eating Mixed or Vegetarian Diets

M. GLEI, M. ANKE, AND B. RÖHRIG

Institute of Nutrition and Environment, Friedrich Schiller University, D-07743 Jena, Germany

Keywords magnesium, intake, balance, adults

Many people in industrialized countries are reported to get an only marginal Mg supply, which may be related to the onset of numerous disease symptoms. World-wide Mg intake of adults range from 160 to 680 mg/d (Parr et al. 1992). The Recommended Dietary Allowances for adults is 300–350 mg/d and intakes are often below this level. Therefore, the Mg supply of adults in Germany was investigated by analysing a duplicated dietary portion.

Material and Methods

The level and changes over time of Mg intake by adults in Germany were determined by analysis of one portion of duplicated meals. The subjects consisted of four (1988), six (1991) and three test groups (1995) consuming freely chosen mixed diets. In 1995, there was one additional test group consisting of vegetarians. Each test group was made up of 7 healthy men and 7 healthy women who were between 20 and 60 years old. At the time of the trials, all subjects lived in the New States of Germany. They collected a complete daily duplicate of their food for 7 subsequent days. The food intake was not subjected to any restrictions. In addition to the complete food duplicates, the daily urine and faeces of 2 test populations were also collected and analysed. All samples were first dried air dried at 60°C, followed by grinding and homogenization. They were further dried at 105°C to constant weight and dry ashed at 450°C. The ash was dissolved in 2.5% hydrochloric acid and the Mg content was determined by flame atomic absorption spectrometry. ARC/CL total diet and wheat flour reference material was used.

Results

Magnesium Intake of Adults with Mixed Diets

During the three test periods, men consumed 247–296 mg/d, which was about 30% more Mg than women, who consumed 193–214 mg/d (Table 1). This difference mainly resulted from the higher dry matter intake of the male subjects (Glei 1995). The Mg intake has steadily increased by 120 and 111% for men and women respectively during the last 7 years.

Table 1. Magnesium intake of adults depending on time of testing and sex (mg/d).

		Women		Men			
Year	n	s	$\bar{x}$	$\bar{x}$	s	p	%
1988	196, 196	57	193	247	70	<0.001	128
1991	294, 294	74	211	259	91	<0.001	123
1995	168, 168	68	214	296	143	<0.001	138
p	–	<0.01		<0.001			–
%	–	111		120			–

Correct citation: Glei, M., Anke, M., and Röhrig, B. 1997. Magnesium intake and magnesium balance of adults eating mixed or vegetarian diets. In *Trace Elements in Man and Animals – 9: Proceedings of the Ninth International Symposium on Trace Elements in Man and Animals. Edited by* P.W.F. Fischer, M.R. L'Abbé, K.A. Cockell, and R.S. Gibson. NRC Research Press, Ottawa, Canada. pp. 181–182.

Magnesium Excretion and Balance of Adults

On average, women excreted less Mg/day (88 mg via urine and 145 mg via faeces) than men, who excreted 110 and 163 mg Mg. This, however, did not affect the excretion pattern. On average 39% of the Mg was excreted via urine and 61% via faeces.

Comparison of the intake and excretion data (Table 2) shows that the subjects in the two test groups consumed an average of 215 and 283 mg Mg/d. Female subjects excreted 18 mg more than they ingested, thus indicating a negative balance of 8%. Men had a mean positive balance of 10 mg or 3%. The data indicate an apparent Mg absorption of 33% for women and of 42% for men.

Table 2. Magnesium balance of women and men.

	Women		Men			
Parameter	s	$\bar{x}$	$\bar{x}$	s	p	%
Mg intake (mg/d)	39	215	283	54	<0.001	132
Mg excretion (mg/d)	54	233	273	59	<0.05	117
Mg balance (mg/d)		–18	+10		–	
Mg balance (%)		–8	+3		–	

Magnesium Supply of Vegetarians

A comparison of Mg concentration in the consumed foods and beverages of adults with omnivorous or vegetarian diets shows that vegetarian diets have a significantly higher Mg content (45%). An influence of sex was not observed.

Due to the higher Mg content of the vegetarian diet and the higher dry matter intake of vegetarians, this group consumed highly significantly more Mg than adults with mixed diets. Independent of sex the difference is almost 80% (Table 3).

Table 3. Magnesium intake of adults depending on type of diet and sex (mg/d).

		Women		Men			
Type of diet	n	s	$\bar{x}$	$\bar{x}$	s	p	%
Mixed diet	658, 658	71	207	266	109	<0.001	128
Vegetarian diet	70, 70	101	376	474	199	<0.001	126
p	–		<0.001	<0.001		–	
%	–		182	178		–	

Conclusions

The changed eating habits after the reunification of Germany led to an increased Mg intake of adults. Adults with omnivorous diets consumed less Mg than recommended. Vegetarians consumed significantly more Mg than adults eating mixed diets.

Literature Cited

Glei, M. (1995) Magnesium in der Nahrungskette unter besonderer Berücksichtigung der Magnesiumversorgung des Menschen. Habilitationsschrift Friedrich-Schiller-Universität Jena.

Holtmeier, H.-J. (1995) Das Magnesiummangelsyndrom beim Menschen. In:Magnesium und Calcium, (Holtmeier, H.-J., ed.), pp. 3–32. Wissenschaftliche Verlagsgesellschaft mbH, Stuttgart.

Parr, R. M., Crawley, H., Abdulla, M., Iyengar, G.V. & Kumpulainen, J. (1992) Human dietary intakes of trace elements: a global literature survey mainly for the period 1970–1991, International Atomic Energy Agency, Vienna.

The Effect of Dietary Magnesium on the Transformation Process in Mice

N.A. LITTLEFIELD, W.G. SHELDON, B. LYN-COOK, AND B.S. HASS

Division of Nutritional Toxicology, National Center for Toxicological Research, FDA, Jefferson, AR 72079, USA

Keywords magnesium, transformation, mice, tumorigenesis

While there is little doubt that Mg plays a central role in the influence of specific oncogenic processes, the mechanisms have largely been undefined. This study will attempt to delineate the influence of Mg in its relation to modulation of tumor expression. It has recently been determined that Mg deficiency in humans is not uncommon in the general population and especially in persons undergoing hypotensive therapy.

Anti-tumorigenic potential of Mg was demonstrated by Mills et al. (1984), when a Mg-deficient diet was imposed on Fischer 344 rats with established tumors of the kidneys and the tumor weight decreased. Inhibition of the tumorigenic effect of heavy metals by Mg was demonstrated by Poirier et al. (1983), when Mg acetate was injected into Wistar rats along with $CdCl_2$. Injection site tumors from Cd were prevented. Mg was also shown to have an effect on the tumorigenesis of Ni and lead (Pb) acetate (Poirier et al. 1984). The presence of Mg reduced the average number of lung tumors from the Ni and Pb.

Mg may influence the transformation process. In studies conducted by Rubin et al. (1981), Rubin (1981), transformed 3T3 cells that were maintained in media with low Mg concentrations lost their transformed characteristics. The cells became serum dependent, resembled non-transformed cells in appearance, and the cell density changed. Chick embryo fibroblast cells kept at a reduced rate of proliferation for three days by Mg deprivation were restored to rapid proliferation upon addition of Mg to the growth serum (Rubin and Koide 1977, Rubin and Chu 1978). These combined influences of Mg are indicative of probable Mg-DNA interactions. Increased cellular proliferation is a characteristic of the transformed cell.

There is some evidence to support the concept of an association between Mg availability to living cells and the development of leukemia (Kasprzak et al. 1986). Studies conducted by Battifora et al. (1968) showed that Mg in the presence of N-2-fluorenyacetamide reduced the incidence of leukemia and that these male rats on a Mg-deficient diet, even in the absence of a carcinogen, tended to have a higher incidence of leukemia than animals on a diet with adequate levels of Mg. In a literature review, Kasprzak and Waalkes (1986) showed that Mg-deficient diets lead to immunologic defects and to the occurrence of thymomas, lymphomas, and to leukemias even without exposure to other carcinogenic agents. Alexsandrowvicz (1982) demonstrated an association between the lack of Mg and the appearance of lymphomas, and Bois (1964) suggested an association of thymus tumors with Mg deficiency.

Materials and Methods

Lymphocytes were isolated from spleens of BALB/c male mice and grown in RPMI 1640 medium. These lymphocytes were co-transfected with the *ras* and *fos* oncogene and a pRSVneo selection plasmid. After 48 h the cells were placed in the selection media containing G418 and cultured for growth. The choice of *ras* and *fos* oncogenes was based on their association with leukemia (Padua 1988) and T lymphoma development (Guerrero and Pellicer 1987), on its varied cellular location (Lyn-Cook et al. 1990). These transfected cells were inoculated into both Mg-deficient and regular RPMI 1640 media.

Correct citation: Littlefield, N.A., Sheldon, W.G., Lyn-Cook, B., and Hass, B.S. 1997. The effect of dietary magnesium on the transformation process in mice. In *Trace Elements in Man and Animals – 9: Proceedings of the Ninth International Symposium on Trace Elements in Man and Animals. Edited by* P.W.F. Fischer, M.R. L'Abbé, K.A. Cockell, and R.S. Gibson. NRC Research Press, Ottawa, Canada. pp. 183–185.

At the appropriate time, 4 groups each with 20 male weanling BALB/c mice were started on respective diets in which 2 groups were fed regular diets and 2 fed Mg-deficient diets. After 4 weeks on these respective diets, 2 of the groups (Mg– and Mg+) were injected through the tail vein with the transfected lymphocytes and the other 2 groups (Mg– and Mg+) injected with the regular lymphocyte cell line.

Necropsies and histopathological examinations were made on all of the animals as they died or were terminated at the end of the study. The study was terminated when one group of animals demonstrated an increased death rate. Examinations were made for the appearance of leukemia, lymphoid tumors, thymomas, and lung lesions.

Results

Body weights taken weekly showed no differences between the Mg+/transfected and the Mg+ groups while body weights of the Mg– group was slightly lower and the Mg–/transfected group had the lowest body weights.

Deaths from the respective treatment groups are shown in Table 1. Mortality was effected by both the Mg and by the oncogenes. Mortality increased in both Mg-deficient groups and the presence of Mg effectively prevented death from the oncogenes. There was only one death in the initial 7 months of treatment. After 9 months on treatment, there were 3 random deaths. Onset of mortality of the Mg–/transfected group was rapid, with only one death at the start of the ninth month, followed by 65% mortality after the 11th month. None of the other treatments groups, including the Mg+/transfected group exhibited this mortality rate. Almost all deaths occurred between 9 and 11 months on study. Since 25% of the Mg–/transfected group died within month 11, the study was terminated so that reasonable pathology assessments could be made. There were no neoplasms attributable to the treatment; in particular there was no leukemia associated with the transfected genes, possibly because of the short study duration. Microscopic examination failed to reveal the cause of death in most animals. Hyperplasia and/or degeneration of costosternal, costovertebral or fibulotibial articular cartilage was the most prominent lesions in both groups of Mg– mice, affecting 6/19 and 8/19, respectively. Focal mineralization occurred in the myocardium in 36% of Mg– mice, and in the testicular capsule in 56% (5–10% in controls). Although the histopathology was similar in the two groups of Mg– mice, the mortality was not, suggesting that these lesions were not the proximal cause of death.

Table 1. Mortality at 11 months.

Treatment		Mortality (%)
Mg+		5
Mg+	/transfected	0
Mg–		20
Mg–	/transfected	65

Bibliography

Aleksandrowicz, G. & Skotnicki, A.B. (1982) "Leukemia Ecology," National Library of Medicine, Washington, D.C.

Battifora, H.E., McCreary, PEA., Hahneman, B.M., Laing, G.H., & Hass, G.M. (1968) Chronic magnesium deficiency in the rat. Arch. Path. 86: 610–620.

Bois, P. (1964) Tumour of the thymus in magnesium-deficient rats. Nature 204;1316.

Guerrero, I. & Pellicer, A. (1987) Mutational activation of oncogenes in animal model systems of carcinogenesis. Mutat. Res. 185: 293–308.

Kasprzak, K.S. & Waalkes, M.P. (1986) The role of calcium, magnesium and zinc in carcinogenesis. In: Essential Nutrients in Carcinogenesis (Poirier, L.A., Newberne, P.M. & Periza, M.W., eds.). Plenum Pub. Corp.

Kasprzak K.S., Waalkes, M.P., & Poirier, L.A. (1986) Antagonism by essential metals and amino acids of nickel(II)-DNA binding *in vitro*. Toxicol. Appl. Pharmacol. 82: 336–343.

Lyn-Cook, B.D., Siegal, G.P. & Kaufman, D. (1990) Malignant transformation of human endometrial stromal cells by transfection of c-*myc*: Effects of pRSVneo co-transfection and treatment with MNNG. Pathobiology 58: 146–152.

Mills, B.J., Broghamer, W.L., Higgins, P.J. & Lindeman, A.D. (1984) Inhibition of tumor growth by magnesium depletion of rats. J. Nutr. 114: 739–745.

Padua, R.A. (1988) Oncogene activation and function. Leukemia 2: 821–823.

Poirier, L.A., Kasprzak, K.S., Hoover K.L., & Wenk, M.L. (1983) Effects of calcium and magnesium acetates on the carcinogenicity of cadmium chloride in Wistar rats. Cancer Res. 43: 4575–4581.

Poirier, L.A., Theiss, J.C., Arnold, L.J., & Shimkin M.B. (1984) Inhibition by magnesium and calcium acetates of lead subacetate- and nickel acetate-induced lung tumors in strain A mice. Cancer Res. 44: 1520–1522.

Rubin, H., & Koide, T. (1977) Mutual potentiation by magnesium and calcium of growth in animal cells. Proc. Nat. Acad. Sci. USA 73: 168.

Rubin, H. & Chu, B. (1978) Reversible regulation by magnesium of chick embryo fibroblast proliferation. J. Cell Physiol. 94: 13–20.

Rubin, H., Vidair, C. & Sanui, H. (1981) Restoration of normal appearance, growth behavior, and calcium content to transformed 3T3 cells by magnesium deprivation. Proc. Natl. Acad. Sci. USA. 78: 2350–2354.

Rubin, H. (1981) Growth regulation, reverse transformation, and adaptability of 3T3 cells in decreased Mg^{2+} concentration. Proc. Natl. Acad. Sci. USA. 78: 328–332.

Rubidium in the Food Chain of Humans: Origins and Intakes

M. ANKE, L. ANGELOW, M. GLEI, M. MÜLLER, U. GUNSTHEIMER, B. RÖHRIG, C. ROTHER, AND P. SCHMIDT

Institute of Nutrition and Environment, Friedrich Schiller University, D-07743 Jena, Germany

Keywords rubidium, origins, intake, requirements

In spite of its abundant occurrence in the earth's crust (17th place in the frequency list) and its existence as a stable (72.2%) and a radioactive (27.8%) isotope, rubidium (Rb) belongs to the forgotten ultra trace elements.

Rubidium Transfer

The geological location has a significant effect on the rubidium content of the flora (Table 1). Granite and gneiss weathered soils produce by far the rubidium-richest plant populations.

Due to their high water solubility, sediment weathered soils (new red sandstone, shell, keuper) and diluvial formations (boulder clay, diluvial sand, loess) deliver little rubidium into the food chain (Table 2). The rubidium content of the drinking water in these living areas follows the same rules. A mean and median rubidium concentration of 11 μg/L and 8.1 μg/L respectively were determined. Rubidium is transported from one level of the food chain to another without difficulties. Herbivores store most rubidium whereas carnivores and omnivores accumulate significantly less (Table 3).

Rubidium Content of Foodstuffs and Beverages

The analysis of 137 foods and beverages, repeated 15 times, showed that the starch- and sugar-rich cereals, pasta, bread and confectionary are rubidium-poor (1 mg Rb/kg dry matter (DM)) whereas several spices contained between 1 and 33 mg Rb/kg DM. Coffee (40 mg/kg DM) and black tea (100 mg Rb/kg DM) store much rubidium, 85% of which get into the beverage. Fruit and vegetables accumulate between

Table 1. Influence of the geological origin of the site on the rubidium content of the flora in Central Europe (n = 873).

Geological origin of the site	Relative number
Gneiss weathered soils	100
Granite weathered soils	78
Weathered soils of the lower strata of new red sandstone	65
Phyllite weathered soils	60
Alluvial riverside soils	58
Slate weathered soils (Devonian, Silurian, Culm)	48
Moor, peat soils	37
Loess	33
Boulder clay	32
Weathered soils of new red sandstone	30
Shell weathered soils	27
Diluvial sands	26
Keuper weathered soils	21

Correct citation: Anke, M., Angelow, L., Glei, M., Müller, M., Gunstheimer, U., Röhrig, B., Rother, C., and Schmidt, P. 1997. Rubidium in the food chain of humans: Origins and intakes. In *Trace Elements in Man and Animals – 9: Proceedings of the Ninth International Symposium on Trace Elements in Man and Animals. Edited by* P.W.F. Fischer, M.R. L'Abbé, K.A. Cockell, and R.S. Gibson. NRC Research Press, Ottawa, Canada. pp. 186–188.

5 and >60 mg Rb/kg (asparagus). Boiling reduces the rubidium content of vegetables dramatically. Animal foodstuffs are relatively rubidium-poor. During cheese production, more than two thirds is lost in the whey. Poultry meat, as well as freshwater fish, are relatively rubidium rich (Table 4).

The Rubidium Intake of Adults

The rubidium consumption by adults was analyzed in 10 test groups (7 women, 7 men) during 7 consecutive days. It amounted to 2 mg Rb/day on gneiss and granite weathered soils and to 1.4 and 1.5 mg

Table 2. *Rubidium content of the drinking water in the living areas of different geological origin.*

Geological origin (44 samples)	$\bar{x}$	s
Gneiss	18	11
Granite	14	13
Phyllite	14	3.2
Lower strata of new red sandstone	12	6.2
Moor	10	6.7
Syenite	8.7	5.0
Loess	7.7	4.0
Shell	3.5	2.5
Diluvial sand	3.1	0.5

Table 3. *Rubidium content of mouse and shrew species (mg/kg dry matter).*

Level of food chain	Kind (n)	$\bar{x}$		s
Herbivores	Wood mouse (10)	8.80		2.27
	Field mouse (10)	9.29		2.74
Carnivores	Dwarf shrew (7)	4.55		1.02
	Common shrew (9)	5.26		1.40
p			<0.001	
%			55	

Table 4. *Rubidium content of meat, innards, sausage and hen's eggs in mg/100 g fresh matter (FM) or mg/kg dry matter (DM).*

			mg/kg DM	
Kind	DM%	mg/100 g FM	$\bar{x}$	s
Blood sausage	51.8	0.084	1.63	0.98
Liver sausage	51.8	0.154	2.98	1.24
Salami	64.1	0.204	3.18	0.88
Bockwurst	44.9	0.158	3.53	1.63
Meat loaf**	45.8	0.179	3.91	1.24
Mortadella**	41.2	0.173	4.20	0.86
Kidneys, 1988	26.6	0.125	4.70	2.38
Liver, 1988	31.3	0.151	4.82	1.74
Hen's egg	25.7	0.154	5.99	1.81
Beef	27.2	0.186	6.84	3.58
Mutton	33.1	0.229	6.92	3.27
Kidneys, 1991	20.7	0.196	9.48	3.01
Pork	27.9	0.301	10.80	4.64
Liver, 1991	28.9	0.357	12.40	3.21
Poultry meat, 1988	31.2	0.475	15.20	4.08
Poultry meat, 1991	31.2	0.705	22.60	5.55

in areas having rubidium-poor loess, boulder clay, diluvial sands, new red sandstone, Muschelkalk and keuper (Table 5).

The calculated rubidium intake of adults, using a market basket, overestimates the analysed rubidium consumption by 58 to 105%.

Table 5. Rubidium intake of adults in Germany (µg/d).

	Women		Men		Sex influence	
Living area (n)	s	$\bar{x}$	$\bar{x}$	s	P	%
Rb-poor areas (392;392)	313	1356	1493	323	<0.01	110
Granite, gneiss (98;98)	783	1967	2002	820	>0.05	102
p	<0.001		<0.001			–
%	145		143			–

Conclusion

Due to its potential essentiality, its toxicity, its radioactivity, its widespread occurrence in the food chain and its peculiarities, the forgotten ultra trace element rubidium will need more scientific attention and reference material for its correct determination in the future. Rubidium deficiency experiments with goats showed that their growth was depressed and that >80% of them aborted their kids. The rubidium requirement of humans could be <1 mg/d or <10 µg/kg body weight per day. It is recommended that adults ingest 20 µg Rb/kg body weight daily.

Rubidium — An Essential Element for Animals and Humans

M. ANKE, H. GÜRTLER,[†] L. ANGELOW, J. GOTTSCHALK,[†] C. DROBNER, S. ANKE, H. ILLING-GÜNTHER, M. MÜLLER, W. ARNHOLD, AND U. SCHÄFER
Institute of Nutrition and Environment, Friedrich Schiller University, D-07743 Jena, [†]Institute of Veterinary Physiological Chemistry, University of Leipzig, D-04103 Leipzig, Germany

Keywords rubidium, essentiality, goats, pregnancy

The essentiality of rubidium was tested in goats. The experiments were repeated six times. The rubidium-deficient rations contained <280 ng Rb/g and the control rations 10 μg Rb/g dry matter (DM).

Feed Intake, Rubidium Content of the Milk, Growth

On average, adult goats with rubidium-poor rations, studied over six generations, consumed 16% less feed (Table 1).

Mature milk of rubidium-deficient goats contained 38% less rubidium than that of controls (Table 2).

The weight of the rubidium-deficient kids was 14% lower at birth and 22% lower on the 92nd day of life than that of controls (Table 3).

Table 1. The influence of rubidium deficiency on the feed intake of barren, gravid and lactating adult goats (g/d).

Stage of development	Control goats		Rb-deficient goats			
(experimental days)	s	$\bar{x}$	$\bar{x}$	s	p	%
Barren (84)	296	778	565	159	<0.001	73
Gravid (140)	198	613	567	137	<0.001	92
Lactating (112)	255	696	590	223	<0.001	85
July, 1st – June, 1st	228	682	574	176	<0.001	84

Table 2. Influence of rubidium deficiency on the rubidium content of colostrum and mature milk (mg/kg dry matter).

	Control goats		Rb-deficient goats			
Kind of milk (n)	s	$\bar{x}$	$\bar{x}$	s	p	%
Colostrum (19;9)	0.46	1.14	1.14	0.66	>0.05	100
Mature milk (91;49)	1.25	2.84	1.77	1.02	<0.001	62
p	<0.001		<0.05		–	
%	249		115		–	

Table 3. Weight of control and rubidium-deficient kids during the milking period (kg).

	Control kids		Rb-deficient kids			
Day of life (n;n)	s	$\bar{x}$	$\bar{x}$	s	p	%
1st (223;52)	0.8	2.9	2.5	0.7	<0.05	86
49th (128;12)	2.5	10.5	8.0	1.3	<0.001	76
91st (122;11)	4.4	17.5	13.6	3.1	<0.01	78

Correct citation: Anke, M., Gürtler, H., Angelow, L., Gottschalk, J., Drobner, C., Anke, S., Illing-Günther, H., Müller, M., Arnhold, W., and Schäfer, U. 1997. Rubidium — An essential element for animals and humans. In *Trace Elements in Man and Animals – 9: Proceedings of the Ninth International Symposium on Trace Elements in Man and Animals. Edited by* P.W.F. Fischer, M.R. L'Abbé, K.A. Cockell, and R.S. Gibson. NRC Research Press, Ottawa, Canada. pp. 189–191.

After weaning, intrauterine Rb-depleted kids only had 43% of the body weight of control kids without intrauterine rubidium depletion (Table 4).

Table 4. Weight gain of control and rubidium-deficient kids without and with intrauterine rubidium depletion from the 101st to the 268th day of life (g/d).

	Control goats		Rb-deficient goats			
Rb-depletion	s	$\bar{x}$	$\bar{x}$	s	p	%
Minus	25	95	99	23	>0.05	104
Plus			41	18	<0.01	43
p		–	<0.01		–	
%		–	41		–	

Reproduction Performance, Milk Production

Aside from effects on abortion and conception rates, disturbed reproduction performances were not found (Table 5). The mean abortion rate was 80%. The fetuses were regularly aborted from the 3rd to the 5th month of pregnancy, most frequently in the 5th month.

The rubidium-poor nutrition, however, also had an on milk production via a significantly reduced feed consumption (Table 6). Rubidium-deficient goats produced 31% less milk than control animals.

Table 5. The influence of rubidium-deficiency on the reproduction performance of female goats.

Parameter	Control goats	Rb-deficient goats	p
Without heat, %	2	11	>0.05
Success of first insemination, %	61	46	>0.05
Conception rate, %	88	68	<0.01
Services per pregnancy, %	1.8	2.2	>0.05
Abortion rate, %	2	80	<0.001
Kids carrying to terms	1.3	1.8	<0.001
Sex ratio, ♀ = 1	1.4	1.4	>0.05
Dead kids, 7th – 91st day	2	33	<0.001

Table 6. Influence of rubidium-deficiency on the production and composition of milk (n = 473;163).

	Control goats		Rb-deficient goats			
Parameter	s	$\bar{x}$	$\bar{x}$	s	p	%
Milk production, mL/d	420	927	636	384	<0.001	69
Milk protein, %	0.58	2.95	3.33	1.07	<0.001	113
Milk protein, g/d	12.7	27.6	20.5	14.0	<0.001	74
Milk fat, %	1.68	3.52	4.85	2.42	<0.001	138
Milk fat, g/d	21.3	32.3	28.6	19.9	>0.05	89

Blood Parameters

The analysis of 35 components of the blood and blood plasma only showed significant changes in the creatinine, phosphorus and hormone metabolism. Rubidium is very likely to play a role in the preservation of pregnancy. The control and rubidium-deficient goats pregnant in the 3rd and 4th months had a normal progesterone and estradiol content in blood plasma (Table 7). The rubidium-deficient goats who aborted their fetuses only had a progesterone level of 7% of the normal value. The estradiol content in the blood plasma of goats aborting varied between 37 and 280 nmol/L.

Table 7. Progesterone and estradiol content in the blood plasma of control and rubidium-deficient goats (nmol/L).

	Progesterone		Estradiol	
Group of goats (n)	s	$\bar{x}$	$\bar{x}$	s
Control goats, gravid (7)	3.5	17	221	40
Rb-deficient goats, gravid (3)	4.0	15	209	17
Rb-deficient goats with abortions (3)	0.7	1.2	119	140
Fp	<0.001		>0.05	
%	7		54	

Conclusion

Rubidium is probably an essential element which is available in sufficient amounts in the food chain to prevent deficiency from occurring in animals and humans.

Effects of Fluorine-Poor Diets in 10 Generations of Goats

M. ANKE, H. GÜRTLER,† E. NEUBERT,† M. GLEI, S. ANKE, M. JARITZ, H. FREYTAG, AND U. SCHÄFER

Institute of Nutrition and Environment, Friedrich Schiller University, D-07743 Jena, † Institute of Veterinary Physiological Chemistry, University Leipzig, D-04103 Leipzig, Germany

Keywords fluorine, pregnancy, development, skeletal deformities

Fluorine deficiency, studied in 10 generations of growing, pregnant and lactating goats since 1985, showed skeletal deformities in old goats, increased mortality of kids and smaller thymus glands. Skeletal deformities occurred both in the fore and hind legs. All other fluorine deficiency symptoms were less striking.

The control and fluorine deficient goats were given semi-synthetic diets (2.0 and <0.3 mg F/kg). The feed was available ad libitum. The animals received about 200 g cellulose every day as litter and roughage, about 100 g of which was consumed. The drinking water was distilled. The experiments were repeated 11 times.

Feed Intake

The adult goats given fluorine-poor diets consumed significantly more feedstuff than the control animals (Table 1). This surprising finding cannot be explained.

Table 1. Influence of fluorine deficiency on feed intake of adult goats (g/d).

Stage of development	Control goats		F-deficient goats			
(experimental days)	s	$\bar{x}$	$\bar{x}$	s	p	%
Barren (84)	246	642	795	273	<0.001	124
Gravid (140)	148	618	702	141	<0.001	114
Lactating (112)	186	627	817	298	<0.001	130

Growth

Fluorine deficient goats regularly gave birth to a significantly higher proportion of kids with a birth weight of <1.6 kg which did not survive. The percentage of kids with a birth weight of >3.5 kg was significantly lower than in control animals. Control kids had a higher weight at birth and continued to weigh more than fluorine-deficient animals (Table 2), consequently slower development continued in the postnatal phase.

Table 2. Weight of control and fluorine deficiency kids during the milking period (kg).

	Control kids		F-deficient kids			
Day of life (n; n)	s	$\bar{x}$	$\bar{x}$	s	p	%
1. (223;110)	0.8	2.9	2.5	0.6	<0.001	86
49. (128; 51)	2.5	10.5	8.7	2.3	<0.001	83
91. (122; 46)	4.4	17.5	14.8	3.6	<0.001	85

Correct citation: Anke, M., Gürtler, H., Neubert, E., Glei, M., Anke, S., Jaritz, M., Freytag, H., and Schäfer, U. 1997. Effects of fluorine-poor diets in 10 generations of goats. In *Trace Elements in Man and Animals – 9: Proceedings of the Ninth International Symposium on Trace Elements in Man and Animals.* Edited by P.W.F. Fischer, M.R. L'Abbé, K.A. Cockell, and R.S. Gibson. NRC Research Press, Ottawa, Canada. pp. 192–194.

Reproduction Performance, Milk Production

The fluorine-poor diet had a significant effect on the number of kids carried to term, the sex ratio of the offspring and the mortality of kids (Table 3). It was astonishing that fluorine deficient goats gave birth to an average of 0.4 more kids than control goats and that they had fewer billy-goats than the mothers of control kids. The mortality of fluorine-deficient kids amounted to 37% and differed significantly from that of control kids.

The fluorine-poor diet, however, also had a significant effect on milk production (Table 4). The fluorine-deficient goats produced 9% less milk than control goats. Milk protein and milk fat content only differed insignificantly between control and fluorine-deficient goats.

Table 3. Influence of fluorine deficiency on the reproduction performance of female goats.

Parameter	Control goats	F-deficient goats	p
Success of first insemination, %	71	58	>0.05
Conception rate, %	90	84	>0.05
Services per gravidity	1.5	1.8	>0.05
Abortion rate, %	1	4	>0.05
Kids carried to terms	1.4	1.8	<0.001
Sex ratio, ♀ = 1	1.7	1.2	<0.001
Dead kids, 7th – 91st day, %	13	37	<0.001

Table 4. Influence of fluorine deficiency on production and composition of milk (n = 473;403).

	Control goats		F-deficient goats			
Parameter	s	$\bar{x}$	$\bar{x}$	s	p	%
Milk production, mL/day	502	1031	940	643	<0.05	91
Milk protein, %	0.56	2.94	3.02	1.02	>0.05	103
Milk protein, g/day	15.0	30.4	29.6	19.3	>0.05	97
Milk fat, %	1.56	3.64	3.73	2.05	>0.05	102
Milk fat, g/day	27.0	37.9	36.2	31.3	>0.05	96

Blood Parameters

Thirty-five consituents of blood and plasma were determined at four different time points (July and December 1993, June and November 1995). At the last two times disturbances in Ca and P metabolism were observed. The calcium concentration (not measured at time points 1 and 2) was below the value observed in control animals. The difference was only significant at the 4th time ($p < 0.01$). The blood plasma of fluorine-deficient goats contained more inorganic phosphate than that of control animals (Table 5). The activity of the alkaline phosphatase of fluorine-deficient goats was distinctly lower than that of control animals at all time points (Table 6). The differences, however, only became significant at the third time point ($p < 0.01$). The lack of significance at the other times was due to the high scattering of the alkaline phosphatase in the animals of the control group.

Except for unesterified acids, which were significantly above the values of control animals in fluorine-depleted goats, significant changes in the other parameters did not occur. The fluorine-poor diet of goats did not have a significant effect on their mortality.

Table 5. Influence of fluorine deficiency on the inorganic phosphate concentration of blood plasma (mmol/L).

	Control goats		F-deficient goats			
Time (n)	s	$\bar{x}$	$\bar{x}$	s	p	%
June 1995 (19; 10)	1.00	2.28	3.40	1.03	<0.01	146
November 1995 (14; 8)	0.57	1.27	2.67	0.30	<0.001	210

Table 6. Influence of fluorine deficiency on the activity of alkaline phosphatase (U/L).

	Control goats		F-deficient goats			
Time (n)	s	$\bar{x}$	$\bar{x}$	s	p	%
July 1993 (7; 6)	257	387	218	87	>0.05	56
December 1993 (8; 5)	269	300	114	22	>0.05	38
June 1995 (12; 4)	664	832	203	151	<0.01	24
November 1995 (9; 4)	609	526	192	86	>0.05	36

Conclusion

Goats fed a fluorine-deficient diet consumed significantly more feed, suffered from significant pre- and post-natal growth retardation, had a significantly higher mortality rate of the offspring, had skeletal and joint deformities in old animals, had a higher phosphorus content in the blood plasma, and had a reduced calcium content and alkaline phosphatase activity.

Distribution of Arsenic in the Tissues of Laying Hens Fed with a Diet Containing As_2O_3

VEKOSLAVA STIBILJ,[†] ANTONIJA HOLCMAN,[‡] AND MARJAN DERMELJ[†]
[†]J. Stefan Institute, University of Ljubljana, [‡]Zootehnical Department, Biotechnical Faculty, University of Ljubljana, Slovenia

Keywords arsenic, hens, tissues, distribution

Introduction

The main source of As for animals is their feed which in normal conditions contains only small amounts. In different industrial areas and in agricultural ones where protective agents or herbicides based on As are used, its concentration levels in animal feed can be increased enormously. Since high levels of residual As in food articles of animal origin can be a threat to human health, reliable knowledge of its quantities in animal tissues and their products and of its mode of accumulation are of great importance for different disciplines such as medicine, nutrition, veterinary science etc. In poultry farming, As compounds are applied for growth promotion and increasing feed efficiency, for better pigmentation, increasing laying, therapeutic purposes and to reduce mortality (Daghir and Hariri 1977, Proudfoot et al. 1991, Donoghue et al. 1994). Residues of these compounds pose a human food safety concern, and a tolerance level of 500 ppb elemental arsenic has been established in uncooked muscle tissue and eggs (Donoghue et al. 1994).

From our previous work (Holcman et al. 1995) it was evident that the As concentration in eggs is clearly related to the As content of feed. Accordingly, the aim of the present work was to study As levels and their distribution in the tissues of laying hens fed with a diet containing As_2O_3.

Material and Methods

Eight Rhode Island red hens took part in the experiment. They were 49 weeks old at the time of caging and were randomly divided into two groups of six (test group) and two (control group), and caged individually. The control group was fed with standard feed for laying hens, while the treated group was fed with the same feed in which of 30 mg As/kg had been added in the form of As_2O_3. The As content in drinking water was 155 ng/g. During the experiment the hens were fed with a quantity of feed which was determined on the basis of consumption estimated in a pretrial study. The duration of the trial was nineteen days. During the experiment no animals died or were poisoned by As_2O_3 and no change in the behaviour of the treated group in comparison with the control group was noticed. The addition of arsenic to feed did not affect the body weight of hens nor egg production during the trial period.

At the end of the trial, blood was collected by anterior heart puncture and then freeze dried. Hens were killed by cervical dislocation and liver, kidneys, lungs and breast muscle were excised, frozen and lyophilized. Plumage samples for As determination were prepared according to an IAEA protocol (1978).The arsenic content in feathers and homogenised freeze dried samples was determined in duplicate by radiochemical neutron activation analysis as described by Byrne and Vakselj (1974) and Byrne (1987).

Results and Discussion

The reliability of the method was tested by analysis of certified reference materials (CRM). In NIST Bovine liver 1577a and BCR Mussel Tissue we obtained 49 ± 3 ng/g and 6.09 ± 0.02 μg/g for four aliquots,

Correct citation: Stibilj, V., Holcman, A., and Dermelj, M. 1997. Distribution of arsenic in the tissues of laying hens fed with a diet containing As_2O_3. In *Trace Elements in Man and Animals – 9: Proceedings of the Ninth International Symposium on Trace Elements in Man and Animals. Edited by* P.W.F. Fischer, M.R. L'Abbé, K.A. Cockell, and R.S. Gibson. NRC Research Press, Ottawa, Canada. pp. 195–196.

Table 1. Arsenic content in feathers and other tissues of laying hens (ng/g fresh weight).

	Control group			Treated group				
Tissue	mean	SD	CV(%)	mean	max	min	SD	CV(%)
Liver	1.9 (n=4)	0.3	15.8	83.1 (n=12)	121.5	43.0	28.0	33.7
Kidney	2.7 (n=4)	0.1	3.7	144.4 (n=14)	207.8	78.7	45.3	31.4
Lungs	2.3 (n=6)	0.4	17.4	150.1 (n=11)	271.8	96.3	62.0	41.3
Muscle	0.9 (n=4)	0.1	11.1	82.7 (n=13)	129.7	53.1	26.2	31.7
Blood	2.3 (n=4)	0.7	30.4	46.3 (n=12)	57.5	23.3	11.4	24.6
Feathers	31.8 (n=4)	5.3	16.7	1032.2 (n=14)	1572.2	594.6	288.2	27.9

n = no. of analyses; at least duplicate samples from each animal (1 exception).

which are in good agreement with the certified values (47 ± 6 ng/g and 5.9 ± 0.2 µg/g). Also, we determined the As content in the trial poultry feedstuff with and without added As_2O_3, and the results were 32.43 µg/g and 0.054 µg/g, respectively. Table 1 shows the average arsenic content, CV and maximum and minimum values in tissues of laying hens on a fresh weight basis. The CV for all determinations is about 30%. The highest As concentration was in lungs, then in kidneys and liver. The concentration of As in muscle of exposed hens is 92 times higher than in the control group, in lungs 65 times, in kidney 53 times, in liver 44 times, in feathers 32 times and in blood only 20 times. It can be concluded that muscle, due to its mass, contains the greatest amount of As.

In Slovenia, the maximum allowable amount of As in poultry meat is 0.1 mg kg^{-1} and for liver and kidney 0.5 mg kg^{-1}. From our trial it is evident that the content of As in the treated group was in some cases higher than that allowed.

Differences in arsenic concentrations between tissues in the control group were not statistically significant, while differences in the experimental group were significant, with three exceptions (liver:muscle, liver:blood, muscle:blood). There were significant differences in concentrations of arsenic between the tissues of the control and the treated group.

Literature

Byrne, A.R. &Vakselj, A. (1974) Rapid neutron activation analysis of arsenic in a wide range of samples by solvent extraction of the iodide. Croat Chem Acta 46: 225–235.

Byrne, A.R. (1987) Low-level simultaneous determination of As and Sb in standard reference materials using radiochemical neutron activation analysis with isotopic ^{77}As and ^{125}Sb tracers. Fresenius Z Anal Chem. 326: 733–735.

Daghir, N.J. & Hariri, N.N. (1977) Determination of total arsenic residues in chicken eggs. J Agric Food Chem. 25: 1009–1010.

Donoghue, D.J., Hairston, H., Cope, C.V., Bartholomew, M.J. & Wagner, D.D. (1994) Incurred arsenic residues in chicken eggs. J Food Prot. 57: 218–223.

Holcman, A., Stibilj, V., Knez, V. & Smodi , B. (1995) The relationship between arsenic content in feed and in hens eggs. In: Proceedings of the 6th European Symposium on the Quality of Eggs and Egg Products, pp. 419–424. Facultad de Veterinaria, Zaragoza.

IAEA Report (1978) RL - 50, Vienna.

Proudfoot, F.G., Jackson, E.D., Hulan, H.W. & Salisbury, C.D.C. (1991) Arsanilic acid as a growth promoter for chicken broilers when adminstered via either feed or drinking water. Can J Anim Sci. 71: 221–226.

Distribution of Toxic and Essential Trace Elements in Soils, Surface Waters, Crops, and Agrochemicals in the Palliser Triangle, Southwestern Saskatchewan

L. SONG AND R. KERRICH

Department of Geological Sciences, University of Saskatchewan, Saskatoon, Saskatchewan, Canada S7N 5E2

Keywords trace elements, ICP-MS, soils, agrochemicals

Recent studies have identified that repeated intensive application of agrochemicals, such as phosphorus fertilizers and pesticides is a significant non-point source of inorganic contamination in rural areas (Kabata-Pendias and Pendias 1992, Alloway 1995). Heavy metal concentrations in the soil–water–plant system is receiving wide attention due to the potential threat of adverse effects on ecosystems and human health.

The Canadian prairie is one of the major agricultural regions in the world; many areas have been subjected to intensive application of agrochemicals for over 50 years. However, limited data is available on background levels of trace elements in unbroken land, farmed soils, surface waters, and crops due to long term agrochemical usage. Over sixty elements in soils, surface waters, crops, agrochemicals, and precipitation have been determined in samples from the Palliser Triangle, Southwestern Saskatchewan.

Materials and Methods

The study area encompasses farmed land, and natural unbroken land, which has been designated as a natural prairie ecosystem for several decades. Surface horizon soils (10 to 15 cm) were taken from 32 sites from the three soil associations; Foxvalley, Haverhill, and Willows. To prevent possible contamination during sampling, clean tools were employed. Samples with a particle size <2 mm were analyzed. An aliquot of 0.1 g finely ground powder was placed in screw top teflon beakers, and digested with ultrapure $HClO_4$ – HF – HNO_3 . Approximately, 0.2 g aliquots of fertilizers and granular pesticides were digested using HF – HNO_3 . Surface waters were collected from dugouts and sloughs, filtered on site using 0.45 μm membrane filters, stored in precleaned polythlene bottles, and then acidified immediately with ultrapure HNO_3 to pH < 2. Crops (grain) were collected during harvest, dried at 40°C, and ground and digested with HNO_3 and H_2O_2. Seven wild plants were collected from natural unbroken land, and treated in the same way as crop samples.

An inductively coupled plasma mass spectrometer (ICP MS), model 5000 by Perkin Elmer Ltd., was employed to analyse all samples. Stock elemental solutions were used to prepare calibration ration standards. Be, In and Tl were employed as internal standards to correct for drift and matrix effects. To ensure accuracy, reference materials, SO-4 (a chernozemic A horizon soil from Saskatoon), SLRS-2 (a river water from the Ottawa River), and RM-8436 (a Durum Wheat flour), were analysed along with unknowns in each batch of analysis. Results for reference materials are in agreement with the recommended values by ±2%, < ±15%, and ±5% respectively. Duplicate analyses of the three reference materials show precision of ±5%, ±10%, ±5%, respectively.

Correct citation: Song, L., and Kerrich, R. 1997. Distribution of toxic and essential trace elements in soils, surface waters, crops, and agrochemicals in the Palliser Triangle, southwestern Saskatchewan. In *Trace Elements in Man and Animals – 9: Proceedings of the Ninth International Symposium on Trace Elements in Man and Animals. Edited by* P.W.F. Fischer, M.R. L'Abbé, K.A. Cockell, and R.S. Gibson. NRC Research Press, Ottawa, Canada. pp. 197–198.

Results and Discussion

Soils are from the Brown soil zone, with silt loam and clay silt loam texture. Willow association soils contain more clay contents than soils from other associations. The results indicate that soils in the study area are depleted in Na, Mg, Ca, Mn, and Fe, but enriched in trace elements such as Rb, Se, Zr, Sb, Cs, Ba, La, Ce, Pb, Th, and U, as compared with the mean of upper crustal abundances (Taylor and McLennan 1985). Comparison of selected trace elements in soils of the same soil association collected from farmed land and natural unbroken land suggests that Cu, Co, Zn, Cr, Y, Cd, Mo, Pb, and U are enriched in farmed land soils by 9, 5.6, 25, 2.4, 1.7, 66, 5, 6, and 17% respectively.

Analysis of fertilizers commonly used in the study area indicate that Cd, U, Zn, and Mo concentrations are significantly higher than in regional soils from unbroken land. For example, an ammonium phosphate (27–27–0) contains cadmium at 95.4 mg/kg, zinc at 1163 mg/kg, chromium at 257 mg/kg, and uranium at 65.9 mg/kg, which are 10 to 500 times the concentrations of respective elements in soils from unbroken land. In general, 25 kg of P_2O_5 in fertilizer is added per ha in Saskatchewan, which would introduce 8.8 g Cd, 107 g Zn, 23.8 g Cr, and 61 g U to soils annually, a significant contribution to the metal budget in the ecosystem, considering > 50% bioavailablility of Cd in P fertilizers (Mermurt et al. 1996). Concentrations of Cu, As, and Pb in some herbicides used in the area are up to 92, 15, and 50 mg/kg respectively.

Surface waters from dugouts and sloughs are mainly snow melted water, and surface runoff. Trace element concentrations are highly variable between individual samples. However, comparison between means of trace element concentrations in surface waters and the average of World River Water (Taylor and McLennan 1985) signifies that overall surface waters in the study area are enriched in Mg, Sr, Fe, P, Ba, Mn, Co, Ni, As, Y, Sb, and U, but depleted in Ca, Al, Zn, Cu, and Pb. Surface waters in farmed land have higher concentrations of Mn (1.2 times), Fe (2), Co (7.3), Ni (2.2), Cu (1.7), Zn (2), Mo (7.8), Cd (2), Sb (2.9), Ba (1.8), Pb (4), and U (2) than waters in natural unbroken land.

Concentrations of 30 trace elements have been determined in crop grains (Durum Wheat, Flax, Oilseed, Barley, etc.). Crops appear to possess higher levels of Cd, Zn, and P relative to wild plants from unbroken land, by 50%, 74%, and 35%, respectively, whereas there is a relative deficiency of some elements such as Mg, Ca, Na, Cu, Cr, Ni, and Ba.

These results suggest that fertilizers commonly used in the study area apparently have high concentrations of some trace elements, such as Zn, Mo, Cd, and U, enriched relative to average crustal rock by 6 times (Mo) to 970 times (Cd). Soils from farmed land are enriched in Zn, Cd, and U compared with soils from natural unbroken land. Such enrichment may be related to long term application of agrochemicals. Enrichment of Cd, Zn, Mo, U in crops and surface waters may provide information on the source of trace elements in region.

References

Alloway, B. J. (1995) Heavy metals in soils. 2nd edition. Chapman and Hall, UK.

Kabata-Pendias, A. & Pendias, H. (1992) Trace Elements in Soils and Plants. 2nd edition. Boca Raton, FL, USA.

Mermut, A. R., Jain, J. C., Song, L., Kerrich, R., Kozak, L., Jana, S. (1996) Trace element concentrations of selected soils and fertilizers in Saskatchewan, Canada. J. Environ. Qua. (In press).

Taylor, S. R. & McLennan, S. M. (1985) The continental crust: Its composition and evolution. Blackwell, USA.

Effect of Dietary Zinc on Absorption and Tissue Accumulation of Cadmium in Romney Sheep

J. LEE,[†] J.R. ROUNCE,[†] N.D. GRACE,[†] AND D.C. GREGOIRE[‡]

[†]New Zealand Pastoral Agriculture Research Institute, Private Bag 11008, Palmerston North, New Zealand, and [‡]Geological Survey Canada, Ottawa, Canada K1A 0E8

Keywords cadmium, absorption, sheep, stable isotope

Cadmium (Cd), present in pasture swards from the use of phosphatic fertilisers, accumulates primarily in liver and kidney tissue of grazing ruminants. In controlled grazing experiments we have previously shown that the rate of accumulation of Cd by these tissues is greatest in young animals (Lee et al. 1996). Although net tissue retention of dietary Cd ingested by sheep is low (approximately 1%), many animals accumulate in excess of the maximum permissible concentration for kidney tissue (1 mg Cd/kg fresh tissue). The aim of this study was to measure the effect of supplementary zinc (Zn), administered using a slow release intraruminal bolus of ZnO (1 g Zn/day), as given for pharmacological control of facial eczema, on the absorption of Cd, removal from plasma by the liver and accumulation in the kidney. Dietary Zn may inhibit Cd absorption, although previously we have shown that high intakes of zinc did not reduce copper in liver tissue of grazing Romney sheep (Lee et al. 1993).

Method

Data was obtained from two trials. In the first a stable isotope, ^{110}Cd, was administered intravenously and continuously at 16.1 µg Cd/day over a period of 6 d, to two groups of 6 month old Romney lambs fed lucerne pellets. One group was given an intraruminal ZnO bolus. Intake of Cd from the diet for both groups was 160.2 µg Cd/day. The ratios ^{110}Cd/^{111}Cd in plasma, liver and kidney were obtained using plasma emission mass spectrometry and total concentration of Cd by Zeeman atomic absorption spectrometry (Gregoire and Lee 1994). Isotope enrichment was calculated as described by Pierce et al. (1987). Individual sheep were slaughtered at intervals over the infusion period and a single exponential fitted to describe the increase in enrichment of ^{110}Cd in liver and kidney tissue with time, and to calculate plateau enrichment and the half-life of Cd in the liver. Entry rate of Cd in liver was calculated by dividing the rate of infused ^{110}Cd (µg/d) by the calculated plateau enrichment of the tracer. This estimate was assumed equivalent to total absorption of Cd with the assumption that the liver behaved as a single primary pool and that the majority of Cd from the plasma was removed on first pass through the liver. The second experiment consisted of a larger, replicated grazing trial on ryegrass/clover pasture containing two concentrations of Cd, with or without supplementary dietary Zn given as in trial one. Two intraruminal ZnO boluses were administered successively, each effective for a period of about 45 d. Cadmium accumulation in various tissues was measured after 42 and 84 d. Details of the experimental replicates and Cd concentrations in the pasture are given by Lee et al. (1996).

Results and Discussion

Enrichment of ^{110}Cd in liver increased rapidly with a half-life of approximately 2.5 d. The isotope reached a 'pseudo'-plateau (65% ^{110}Cd enrichment) within 8 d. Kidney tissue accumulated Cd at a slower rate (Table 1). Over 50% of the tracer appeared in the liver, a much smaller proportion in the kidney and the remainder, apart from small amounts in other tissues, secreted back into the gut. Unfortunately enrichment in the faeces (calculated to be about 3.5%) was not determined in this study to confirm this.

Correct citation: Lee, J. Rounce, J.R., Grace, N.D., and Gregoire, D.C. 1997. Effect of dietary zinc on absorption and tissue accumulation of cadmium in Romney sheep. In *Trace Elements in Man and Animals – 9: Proceedings of the Ninth International Symposium on Trace Elements in Man and Animals. Edited by* P.W.F. Fischer, M.R. L'Abbé, K.A. Cockell, and R.S. Gibson. NRC Research Press, Ottawa, Canada. pp. 199–200.

Although total absorption of Cd was shown to be relatively large (Table 1), endogenous losses must also be large as net retention of Cd by the organs is small (Lee et al. 1996). Zinc in the diet significantly increased absorption in the liver (decreased enrichment of ^{110}Cd). This is contrary to some other studies and requires further work to fully explain these results. No treatment effects were observed for entry of Cd into the kidney. The amount of tracer infused was deliberately high so that enrichment in the plasma pool could be measured (pool size 0.1 ng Cd/ml plasma). However, ratios of ^{110}Cd/^{111}Cd in plasma, were lower at plateau than those for liver and similar to those of kidney. This indicates the presence of a much smaller and faster turnover pool in plasma, which was not able to be isolated in this work. There was no measurable treatment effect on ^{110}Cd enrichment in plasma.

However data from trial 2 (Table 2) showed that Zn dosages were observed to reduce net accumulation of Cd in liver, particularly after 84 d, for animals grazing both the low and high Cd pastures. The effect on Cd accumulation in the kidney was the reverse. This higher retention of Cd in kidney with the high Zn doses is interpreted as an increase in metallothionein synthesis and hence improved efficiency of Cd sequestered in the kidney. Thus, although the data from the isotope study indicates that total absorption of Cd may be increased by Zn, net processes, including differences in Cd endogenous loss and induced metallothionein synthesis, results in decreased concentrations in liver tissue.

Table 1. Effect of Zn on enrichment of Cd-110 in liver and kidney and absorption of cadmium in sheep.

	Total Cd pool (μg)	Enrichment (144 h)	% of label infused	Plateau enrichment[1]	Absorption (μg/day)	% of intake	Half-life (d)
Liver							
+Zn	122±181[1]	0.41±0.015[a]	55.5±4.8	–	12.7±0.95[a]	7.9±0.80[a]	
Control	122± 13	0.52±0.02[b]	65.3±4.6	0.64±0.11	9.1±0.48[b]	5.7±0.30[b]	2.5±1
Kidney							
+Zn	19± 0.6	0.13±0.01	4.2±0.26	–	–	–	
Control	19± 2.4	0.16±0.01	3.2±0.5	0.25±0.10	–	–	4.3±3

[1]Values are means ± sem and different superscript letters are significantly different ($P < 0.05$).

Table 2. Effect of zinc on the cadmium content in liver and kidney of Romney sheep grazing ryegrass pasture.

	Days	Liver	Kidney
Low Cd pasture (0.12–0.3 μg Cd/g DM)			
Control	42	20.9±1.3[1]	4.7±0.28
	84	30.9±5.7	8.5±3.12
+Zn	42	20.6±2.8	7.0±1.74
	84	23.8±2.7	6.6±0.38
High Cd pasture (0.5–0.6 μg Cd/g DM)			
Control	42	27.1±4.1	7.1±0.55
	84	49.5±6.0	12.0±0.88
+Zn	42	24.2±2.45	10.3±2.02
	84	34.4±2.5	15.6±2.6

[1]Values are ANOVA means ± sem ($n = 5$). Treatment effects were: Liver; Cd, $P < 0.001$; Zn, $P < 0.02$: Kidney; Cd, $P < 0.001$; Zn, $P < 0.049$.

Literature Cited

Gregoire, D.C. and Lee, J. (1994) Determination of cadmium and zinc isotope ratios in sheep's blood and organ tissue by electrothermal vaporisation inductively coupled plasma mass spectrometry. J. Analyt. Atomic Spec. 9: 393–397.

Lee, J.,. Treloar, B.P. and Grace, N.D. (1993) Metallothionein and trace element metabolism in sheep tissues in response to high and sustained zinc dosages II. Expression of metallothionein m-RNA. J. Aust. Agric. Sci. 45: 321–332.

Lee, J., Rounce, J.R., Mackay, A.D. and Grace, N.D. (1996) Accumulation of cadmium with time in Romney sheep grazing ryegrass-white clover pasture: effect of cadmium from pasture and soil intake. Aust. J. Agric. Res. 47(6): in press.

Pierce, P.L., Hambidge, K.M., Goss, C.H., Millar, L.V. and Fennessey, P.V. (1987) Fast atom bombardment mass spectrometry for the determination of zinc stable isotopes in biological samples. Anal. Chem. 59: 2034–2037.

Enhancing the Consumers' Perception of Wheat and Wheat Products, Especially with Regard to Essential Trace Minerals

C.M. PATTERSON AND E. MORRISON

Centre for Public Health Research, Queensland University of Technology, Brisbane, Queensland, Australia 4059

Keywords trace minerals, wheat products, wheat, nutrition

Australian nutrition authorities' current recommendations include an increased consumption of whole-grain cereal products, including wheat products primarily in order to increase consumption levels of dietary fibre. However, the essential trace mineral contribution of such products often tends to be underestimated. The trace mineral data for Australian wheat and wheat products, moreover, tend to be lacking or out of date. This survey was undertaken to obtain up to date information on the nutritional value of wheat and wheat products sampled across Australia over two years. Additionally a new modified microwave digestion procedure was tested and validated.

The Australia Wheat Board (AWB) co-ordinates the collection of wheat crops around Australia from the five major wheat producing states. Receival depots are located throughout the wheat belts and samples were collected from a random sample of depots in each state with the number of samples requested being directly related to the proportion of the total wheat yield received from that state in the previous year. The AWB classifies crops into grades based on a number of quality criteria but primarily on the protein content of the grain. The number of samples of each grade collected was related to the proportions produced in a state.

Samples were digested in PFA-Teflon pressure relief digestion vessels in a programmable microwave oven with subsequent reduction of the acid volume in a Speedvac apparatus. Quantitative measurements of the elements were performed by inductively-coupled plasma emission spectrometry ICP-AES. Complete details of the analytical procedure are available from the authors. Table 1 summarises the gross data collected in the study.

Prime hard grade wheat was generally the highest in trace mineral and magnesium content, followed by Australian hard grade wheat. Australian soft grade was consistently lowest in magnesium and trace mineral levels followed by Australian standard wheat. These trends indicate a positive relationship between mineral and protein content of the wheat since the gradings are primarily determined by protein.

Table 1. Summary of descriptive statistics for the entire wheat sample set (n = 233).

Variable	Mean (mg/kg)	Standard deviation	Range (min–max)
Magnesium	1160	104	778–1622
Zinc	17.7	6.47	8.01–35.5
Iron	30.2	6.80	16.7–52.2
Manganese	33.4	6.14	16.7–53.7
Copper	3.23	0.814	1.14–5.81
Aluminium	3.67	3.55	0–26.0
Nickel	0.299	0.141	0–1.13
Chromium	0.041	0.066	0–0.537
Cobalt	0.029	0.040	0–0.173

Correct citation: Patterson, C.M. and Morrison, E. 1997. Enhancing the consumers' perception of wheat and wheat products, especially with regard to essential trace minerals. In *Trace Elements in Man and Animals – 9: Proceedings of the Ninth International Symposium on Trace Elements in Man and Animals. Edited by* P.W.F. Fischer, M.R. L'Abbé, K.A. Cockell, and R.S. Gibson. NRC Research Press, Ottawa, Canada. pp. 201–202.

A number of milled products were also analysed. The mineral content of stoneground flour followed by wholemeal flour or the grain concentrate was consistenly highest as expected from products with higher proportions of the mineral-rich grain. The highly refined products baker's and 'superlite' flours had consistly lower mineral content. Similar trends in mineral content were apparent with the wheat bread samples reflecting the cereal ingredients from which the breads were manufactured. Wholemeal bread was generally followed in trace mineral content by multigrain and then wholegrain with white bread containing the lowest levels. The ultra-trace mineral content did not appear to follow any particular trends.

A formula was developed to translate the levels of trace minerals found in wheat and flour samples in this study to theoretical bread products. These levels were then used to calculate the contribution of usual consumption of these products by men and women in Australia as indicated by the Australian Dietary Survey (Cashel et al. 1986) and the recommendations from the core food groups model (Cashel and Jeffreson 1995).

The results of this analysis indicated that if bread were consumed as recommended in the core food groups model, both in terms of type of product and the amounts recommended, it would contribute over 50% of the recommended daily intake for magnesium and 20% of zinc for both men and women. For iron, if consumed according to these recommendations, bread could contribute 80% of the recommended level for men and 30% for women.

Such comprehensive trace mineral analyses provide authorities with data which can then be used as a marketing tool for encouraging the consumption of these products.

References

Cashel, K., English, R., Bennett, S.L., Berzins. J., Brown, G. And Magnus, P.M. (1986) National Dietary Survey of Adults:1983. Report Nr 1: Foods Consumed, AGPS, Canberra.

Cashel, K. And Jeffreson, S. (1995) The Core Food Groups, AGPS, Canberra.

Loss of Canadian Wheat Imports Lowers Selenium Intake and Status of the Scottish Population

A. MACPHERSON,[†] M.N.I. BARCLAY,[†] R. SCOTT,[‡] AND R.W.S. YATES[‡]

[†]SAC, Auchincruive, Ayr, Scotland KA6 5HW and [‡]Glasgow Royal Infirmary, Glasgow, Scotland G31 2ER

Keywords selenium, dietary intake, plasma concentration, Scotland

Up until the mid 1970's some 50% of the UK's breadmaking flour was imported from North America and particularly Canada. This paper studies the effect of the progressive decline in these imports since then on the dietary provision of selenium and consequently on the selenium status of the human population.

Table 1 shows the change in level of N American wheat imports into the UK since 1970. They are now down to <10% of their 1970 level. Table 2 shows the origin of wheat used in Scotland. Table 3 shows the relationship between selenium in flour and bread and place of origin.

Surveys were conducted into the dietary provision of selenium by Thorn et al. (1978) and by the present authors in 1985, 1990 and 1994 (Barclay et al. 1986, 1992 and 1995 respectively). The results reflected a 50% fall in dietary selenium as can be seen in Table 4.

Table 1. Wheat imports (tonnes) to the UK.

	1970	1993	1994	1995
Total	4 921 000	1 629 000	1 146 000	795 000
EC	883 000	1 384 000	899 000	575 000
US	684 000	0	0	0
Canada	1 533 000	227 000	225 000	205 000
Others	1 821 000	17 000	22 000	15 000

Table 2. Origin of wheat purchased by Scottish flour millers.

	1986/87	1987/88	1988/89	1994/95
UK	288 000	277 000	260 000	250 000
EC	120 000	127 000	128 000	130 000
Canada	88 000	68 000	62 000	57 000

Table 3. Selenium content of flour and bread.

Flour	(µg/100 g)	Bread	(µg/100 g)	Origin
Low protein	4.8	White	3.6	EC
Medium protein	16.3	White	11.2	EC
Medium protein	28.0	White	55.0	Canadian
Medium protein	39.0	White	30.0	S Dakota, USA
Wholewheat	35.6	Wholemeal	18.8	Canadian blend
Wholewheat	61.0	Wholemeal	68.0	Canadian
Wholewheat	87.0	Wholemeal	41.0	S Dakota, USA

Correct citation: MacPherson, A., Barclay, M.N.I., Scott, R., and Yates, R.W.S. 1997. Loss of Canadian wheat imports lowers selenium intake and status of the Scottish population. In *Trace Elements in Man and Animals – 9: Proceedings of the Ninth International Symposium on Trace Elements in Man and Animals. Edited by* P.W.F. Fischer, M.R. L'Abbé, K.A. Cockell, and R.S. Gibson. NRC Research Press, Ottawa, Canada. pp. 203–205.

Such a dramatic decline in selenium intakes would be expected to produce a parallel decline in plasma selenium status of the population. Figure 1 shows that this has in fact occurred from 1985 to the present. Figure 2 shows the distribution of the individual plasma selenium concentrations in 1985 and 1994 and reveals almost all above the minimum threshold (90 μg/L) in 1985 and almost all below it in 1994. There has thus been a remarkable shift in status in the space of a single decade.

Table 4. Change in daily selenium intake with time.

	1974	1985	1990	1994
Se intake (μg/d)	60	43	30	32

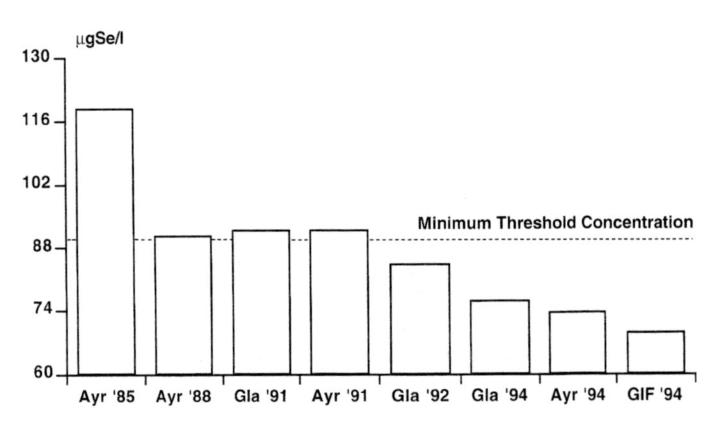

Figure 1. Change in plasma Se with time.

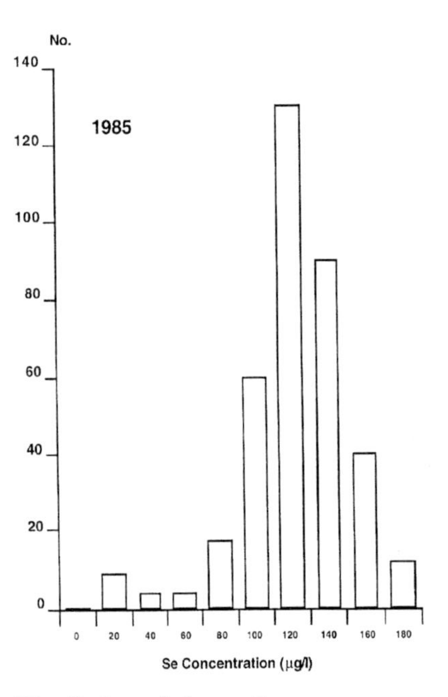

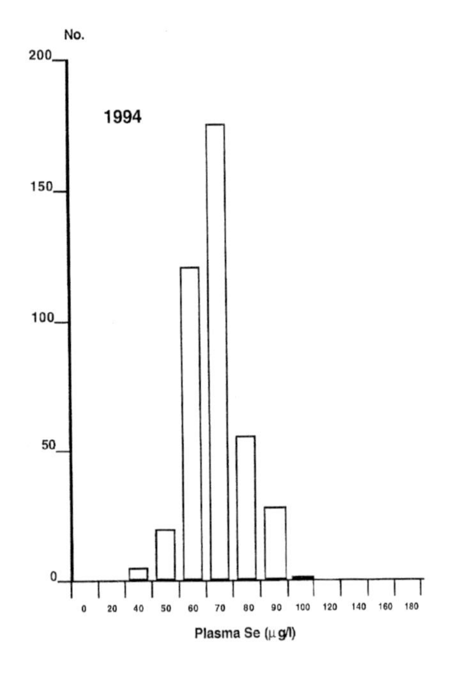

Figure 2. Distribution of plasma Se.

Conclusions

Imports of Canadian wheat are now less than 10% of their 1970 level. The markedly lower selenium concentration of European wheat (due to a lower availability in more acid soils) has led to a 50% decline in daily selenium intakes which has been reflected in a consequent 42% decline in plasma selenium concentrations since 1985. Present dietary and plasma levels are well below recommended values and may well be having adverse effects on susceptibility to cancer and cardiovascular disease and on human fertility.

References

Arthur, D. (1972) Selenium contents of Canadian foods. Can Inst. Food Sci. Technol. J. 5: 165–169.
Barclay, M.N.I. & MacPherson, A. (1986) Selenium content of wheat flour used in UK. J. Sci. Fd. Agric. 37: 1133–1138.
Barclay, M.N.I. & MacPherson, A. (1992) Selenium content of wheat for breadmaking in Scotland. Brit. J. Nutr. 68: 261–270.
Barclay, M.N.I., MacPherson, A. & Dixon, J. (1995) Selenium content of a range of UK foods. J. Food Comp. & Anal. 8: 307–318.
Morris, V.C. & Levander, O.A. (1970). Selenium content of foods. J. Nutr. 100: 1383–1388.
Thorn, J., Robertson, J., Buss, D.H. & Bunton, N.G. (1978) Selenium in British food. Brit. J. Nutr. 39: 391–96.

Absolute Bioavailability of Lead from Soil

K. GROEN, R. NIJDAM, AND A.J.A.M. SIPS

National Institute of Public Health and Environmental Protection, Bilthoven, P.O. Box 1, 3720 BA, Bilthoven, The Netherlands

Keywords lead, bioavailability, soil, *in vitro*

Introduction

Elevated blood lead levels in children may induce neurological development disorders, such as a lower intelligence. In the past risk assessment has been carried out in order to estimate public health risks in case of ingestion of lead contaminated soil. In this risk assessment, bioavailability is assumed to be equal to 100% or, in the most sophisticated approach, is assumed to be equal to bioavailability from a solution. Recent studies on absolute bioavailability of dioxins from fly-ash in cows and arsenic from bog-ore in dogs have demonstrated that these assumptions may result in an overestimation of risk by about a factor 10 to 30 (Olling et al. 1992, Groen et al. 1992). As a consequence remedial activities on soil will be performed unnecessarily. Research on the oral bioavailability of other environmental contaminants, such as heavy metals, from soil is therefore essential.

In assessing oral bioavailability three processes can be distinguished: (a) extraction from the soil matrix by the digestion process, (b) absorption from the intestine into the blood compartment and (c) first-pass effect in the liver. Especially the first process is dependent on the composition of the soil matrix. Since it is impossible to perform animal studies for each type of soil, a simple experimental tool is required for determining the extent of extraction from a certain matrix. Such a tool might be found in an '*in vitro*' digestion simulation model as recently proposed by Rotard et al. (1995).

By combining the results of the '*in vitro*' digestion simulation model with experiment-derived or literature-derived data on absorption from a solution, absolute bioavailability from soil may be estimated.

The aims of this reported study were to determine the effect of a soil matrix on the absolute bioavailability of lead and to obtain information on the applicability of the '*in vitro*' digestion simulation model as a tool for estimating absolute bioavailability of a contaminant from soil.

Materials and Methods

Six dogs (Beagles) received, according to a three-way cross-over design, a single dose of 2 mL lead nitrate solution intravenously (2 mg Pb), 5 mL lead nitrate solution orally (5 mg Pb), and 4.6 g lead contaminated soil (5 mg Pb).

Absolute bioavailability of lead from a solution or from soil was calculated on the basis of the following formula

$$F = \frac{\text{AUC}_{\infty}\,(\text{oral})}{\text{AUC}_{\infty}\,(\text{i.v.})} \cdot \frac{\text{Dose (i.v.)}}{\text{Dose (oral)}}$$

in which AUC_{∞} represents the area under the plasma concentration-time curve extrapolated to infinity following oral or intravenous dosing.

In an additional experiment the applicability of an '*in vitro*' digestion simulation model (Rotard et al. 1995) of the gastro-intestinal tract was tested. In this model, 1 g soil was exposed to semisynthetic saliva (10 mL, 1/2 h), stomach juice (20 mL, 3 h), bile and duodenal juice (10 + 20 mL, 3 h). After centrifugation the lead content was determined by ICP-MS in the supernatant representing the amount of lead extracted from the soil matrix.

Correct citation: Groen, K., Nijdam, R., and Sips, A.J.A.M. 1997. Absolute bioavailability of lead from soil. In *Trace Elements in Man and Animals – 9: Proceedings of the Ninth International Symposium on Trace Elements in Man and Animals. Edited by* P.W.F. Fischer, M.R. L'Abbé, K.A. Cockell, and R.S. Gibson. NRC Research Press, Ottawa, Canada. pp. 206–207.

Results

Absolute bioavailability after oral administration of lead nitrate solution was 12 ± 12%, whereas this was only 1.5 ± 1.7% following ingestion of the lead contaminated soil.

In the '*in vitro*' digestion simulation model 5.6 ± 2.2% of the lead content in soil was present in the supernatant. Based on the bioavailability data from a lead nitrate solution, approximately 12% of lead extracted from soil will be absorbed. Combination of these data sets results in an absolute bioavailability from soil of $0.12 \times 5.6 = 0.7\%$.

Conclusions

Absolute bioavailability of lead from soil is approximately 8 times lower than absolute bioavailability of lead from a lead nitrate solution. The data from the '*in vitro*' digestion simulation model combined with the data on absolute bioavailability from a solution are comparable to absolute bioavailability data derived in dogs fed lead contaminated soil.

The '*in vitro*' digestion simulation model seems to be applicable, although further optimalization and validation of the method is required.

References

Groen, K.G., Vaessen, H.A.M.G., Kliest, J.J.G., de Boer, J.L.M., van Ooik, T., Timmerman, A., Vlug, R.F. (1994) Bioavailability of inorganic arsenic from bog ore-containing soil in the dog. Environ. Health Persp. 102: 182–184.

Olling, M., Berende, P.L.M., Derks, H.J.G.M., Everts, H., Jong de, A.P.J.M. and Liem, A.K.D. (1992) Bioavailability of dioxines and furans from fly-ash in the cow. RIVM-report 328904.004, Bilthoven.

Rotard, W., Christmann, W., Knoth, W., and Mailahn, W. (1995) Bestimmung der resorptionsverfügbioavailabilityren PCDD/PCDF aus Kieselrot; Umweltwissenschaften und Schadstoff-Forschung 7: 3–9.

Trace Elements, Brain Function, and Behaviour

Chair: S.C. Cunnane

Persistent Neurochemical and Behavioral Abnormalities in Adult Rats Following Recovery from Perinatal Copper Deficiency

J.R. PROHASKA

Department of Biochemistry & Molecular Biology, University of Minnesota, Duluth, MN 55812, USA

Keywords copper-deficient, rat, brain, behavior

Copper (Cu) plays a major role in all biological systems including the central nervous system (CNS). Observations in domestic animals (see Smith 1983) and in humans with Menkes' Disease (Danks et al. 1972) underscore the importance of Cu for normal development of the brain. Nutritional Cu deficiency in humans, as described by Cordano et al. (1964), reported hypotonia along with characteristic hematological changes. Shaw (1980) indicated another feature of nutritional Cu deficiency was psychomotor retardation. Neurological consequences of Cu deprivation in humans have not been routinely evaluated because treatment of Cu-deficient infants rapidly corrects their hematological profile (hypocupremia, anemia, and neutropenia).

Studies in mice and rats have been extensively employed to investigate the neurochemical functions of Cu. Many model studies, particularly in rodents, demonstrated that severe Cu deficiency during early development leads to alterations in the biochemistry of the CNS. Recovery from such nutritional insults have been studied less frequently. Following 1 month of repletion, reversal of altered catecholamines was noted but brain Cu levels were not restored to control values (Feller and O'Dell 1980, Prohaska and Bailey 1993, Prohaska and Wells 1975). Experiments investigating perinatal Cu deficiency in rats have shown that there are regional changes in Cu levels (Prohaska and Bailey 1994) and cuproenzyme activities (Prohaska and Bailey 1995a) in month-old offspring. More recent experiments have followed the long-term nutritional repletion of these offspring with dietary Cu. Biochemical studies indicate that the restoration of brain Cu and cytochrome c oxidase activity are not complete after 4 months (Prohaska and Bailey 1995b). Recently, an extended repletion study demonstrated altered sensory-motor behavior of rats supplemented with dietary Cu for as long as 5 months (Prohaska and Hoffman 1996). Behavioral testing included assessment of the mammalian startle response and several other neurotoxicological screening protocols (Beattie et al. 1996). These results imply collectively that there are long term consequences to perinatal Cu deficiency. Women with poor Cu status, especially those who deliver prematurely, could have infants at risk for abnormal CNS development.

Correct citation: Prohaska, J.R. 1997. Persistent neurochemical and behavioral abnormalities in adult rats following recovery from perinatal copper deficiency. In *Trace Elements in Man and Animals – 9: Proceedings of the Ninth International Symposium on Trace Elements in Man and Animals. Edited by* P.W.F. Fischer, M.R. L'Abbé, K.A. Cockell, and R.S. Gibson. NRC Research Press, Ottawa, Canada. pp. 208–212.

Materials and Methods

Sperm-positive Sprague Dawley rats were purchased commercially (Harlan Sprague Dawley, Indianapolis, IN). Rats were offered a Cu-deficient purified diet (Teklad Laboratories, Madison, WI) and either low Cu drinking water or Cu-supplemented drinking water, respectively. The purified diet was similar to the AIN-76A diet used in previous studies (Prohaska and Bailey 1994, 1995a, 1995b) and contained 0.43 ± 0.06 mg Cu/kg and 48 ± 4 mg Fe/kg by chemical analysis. Cu-deficient treatment groups drank deionized water, whereas Cu adequate treatment groups drank water supplemented with $CuSO_4$ (20 mg Cu/L).

Pregnant dams began the Cu-deficient treatment 7 d into gestation. Two days following parturition, litter size was adjusted to eight pups. Offspring were weaned onto the same treatment as their respective dams until 30 d of age. This paradigm is similar to that recently used to study neurochemical changes in young rats. To study Cu repletion, 30-d-old offspring (both Cu-adequate and Cu-deficient) were offered a commercial diet, Purina Laboratory Rodent Chow 5001 (Ralston Purina Co., St. Louis, MO), and tap water. The nonpurified diet contained 13 ± 1 mg Cu/kg and 180 ± 5 mg Fe/kg. Rats from four treatment groups (Cu-adequate and Cu-repleted, female and male) were sampled to evaluate biochemical recovery (Prohaska and Bailey 1995b) or were tested behaviorally (Prohaska and Hoffman 1996).

Rats were killed by decapitation. Livers and brains were removed and processed for biochemical analyses. Brains were dissected on a chilled glass plate following the guidelines of Glowinski and Iversen (1966). Cu status of the rats was evaluated by several criteria including hemoglobin, plasma ceruloplasmin activity, liver Cu and Fe, and Cu levels in six brain regions.

Behavioral testing followed standard laboratory rat neurotoxicity protocols (Beattie et al. 1996). Tests included a modified Functional Observational Battery similar to Moser et al. (1988) using figure 8 mazes for motor activity, grip strength, accelerating rotarod performance, landing foot splay, and startle responses. Automated startle testing utilized a computer controlled sound generation system with on-line data recording (San Diego Instruments, San Diego, CA). Startle responses to three types of stimuli (auditory, tactile, and inhibitory auditory prepulse of tactile startle) were recorded. Startle stimuli were administered to each animal in random order six times. Average response amplitude (voltage) was selected for data analysis.

Data were analyzed by factorial ANOVA for main effects of diet, gender, and age. For behavioral testing diet and gender main effects were examined at each age. Significant ($P < 0.05$) interaction terms were evaluated by Fisher's PLSD test (Steel and Torrie 1980).

Results and Discussion

Injury to the CNS from nutritional Cu deficiency is dependent upon the time during brain development that the deficiency occurs and the severity of the deficiency. In rodents, the largest accumulation of Cu by brain occurs during the suckling period during the last two weeks of lactation (Prohaska and Wells 1974). When a Cu-deficient diet is fed to pregnant rats starting the last two weeks of gestation, the reduction in brain Cu of 1-month-old pups is approximately 80% (Prohaska and Bailey 1994). In a similar paradigm with the same strain of rats and the same diet, a post-weanling Cu deficiency decreased brain Cu pools only approximately 30% (Prohaska et al. 1995). If a Cu-deficient diet is fed to the rats prior to pregnancy there are no births from the Cu-deficient dams (Dutt and Mills 1960). If the Cu-deficient diet is fed throughout gestation and lactation, there is severe growth impairment of the offspring (Prohaska and Wells 1975). Therefore, it is clear that timing of the Cu-deficient diet is critical to the survival of the offspring and to the severity of the Cu deficiency.

The second factor in the outcome of perinatal Cu deficiency is the dietary Cu level. Subtle differences in the level of dietary Cu during perinatal development can have dramatic effects on the outcome of the offspring. Dutt and Mills (1960) fed a diet containing 0.4 µg/g Cu to rats 1 month before mating. This resulted in the pregnancy being aborted with no live litters born to dams. In contrast, when a Cu-deficient diet was fed approximately 3 weeks before mating and the level of dietary Cu was 1 µg/g, normal delivery and size of pups was observed (Jankowski et al. 1993). This subtle nature of the level of dietary Cu can be observed in our perinatal model. The average liver Cu of the 1-month-old male and female offspring in a recent behavior study was 18 nmol/g (Prohaska and Hoffman 1996). In two previous studies using a similar paradigm the 1-month-old male and female offspring had lower liver Cu concentrations. In one study, the average value was 9 nmol/g (Prohaska and Bailey 1994). In a recent follow-up study, the average value for males and females was 10 nmol/g (Prohaska et al. 1995). The difference between these three studies is that the level of dietary Cu in the behavior experiments was approximately 0.1 µg/g higher. Brain Cu concentrations, nevertheless, indicated severe Cu depletion in all these perinatal depletion studies. With the exception of the hypothalamus, mean brain regional Cu concentrations in Cu-deficient offspring were 80% lower than values in Cu-adequate offspring (Prohaska and Bailey 1994). The deficit in hypothalamic Cu was less than other regions.

In the same study, norepinephrine concentrations were lower in all brain regions of Cu-deficient rats except the hypothalamus. Further analysis of regional dopamine indicated that dopamine-β-monooxygenase (DBM) was limiting in brain of young Cu-deficient rats (Prohaska and Bailey 1994). Extended studies found significant reductions in other brain cuproenzymes in all six regions including cytochrome c oxidase (CCO), superoxide dismutase (SOD), and peptidylglycine-α-amidating monooxygenase (PAM) (Prohaska et al. 1995a).

Dietary Cu deficiency in the perinatal rat model has produced other reproducible characteristics in the Cu-deficient month-old offspring including significantly higher brain to body weight and heart to body weight ratios, nondetectable ceruloplasmin diamine oxidase activity, little change in body weight, hemoglobin concentration, or liver Fe concentration.

Brain activity of the cuproenzymes SOD, DBM, and PAM were, for the most part, restored to control levels in Cu-deficient rats that were supplemented with the nonpurified Cu-adequate diet for 1 month (Prohaska and Bailey 1995a 1995b, Prohaska et al. 1995). A study in female rats followed recovery from Cu deficiency for a 4-month period (Prohaska and Bailey 1995b). The 5-month-old female rats still had low Cu in all brain regions and CCO in all regions, except the hypothalamus, even after 4 months of repletion. This suggests that certain neurochemical changes are difficult to overcome following perinatal Cu deficiency.

Following Cu repletion with the nonpurified diet the concentration of Cu in liver and the low ceruloplasmin activity was rapidly restored to control levels following a single month of Cu repletion (Prohaska and Bailey 1995b, Prohaska and Hoffman 1996). In contrast, repletion of brain Cu did not follow a similar rapid recovery profile. In the repleted offspring the concentration of Cu in the cerebrum, midbrain, corpus striatum, and cerebellum (until 6 months of age) and medulla-pons (until 4 months of age) remained significantly lower than in controls (Prohaska and Hoffman 1996). This indicated that the biochemical recovery, as assessed by brain Cu concentration, was not complete in the repleted rats studied. Male rats repleted for 1 year still had lower cerebral cortex Cu levels, 2.16 ± 0.16 (6) µg/g, compared to control levels, 2.71 ± 0.11 (4), $P < 0.01$. The failure to restore brain Cu concentrations may be related to the very slow turnover of Cu in the CNS (Levenson and Janghorbani 1994).

At age 2, 4, and 6 months (that is, 1, 3, and 5 months after Cu repletion was initiated) 32 rats underwent behavioral neurotoxicology assessments ($n = 8$ per group). No reproducible differences were observed in grip strength, figure 8 maze activity, or accelerating rotarod performance. Two particular procedures indicated significant treatment effects, acoustic startle response and foot-splay (Prohaska and Hoffman 1996). Rats that were Cu-deficient during early development had a wider foot-splay upon being dropped than the Cu-adequate control counterparts. Mean foot-splay of 4-month-old Cu-repleted males was 9.3 cm compared to 6.3 cm for the control males. For females, the mean for the Cu-repleted rats was 6.8 cm compared to 5.3 for the control females, $P < 0.01$.

However, the more striking observation was the altered auditory startle responses of rats that were previously Cu-deficient. At 2, 4, and 6 months of age there was a significant effect of diet on mean acoustic startle response. The auditory startle response of the Cu-repleted rats was markedly attenuated (45 ± 8%). Habituation occurred during the six trials but all six times the response of the Cu-repleted rats was lower than the Cu-adequate rats. The same population of animals with blunted acoustic startle exhibited normal responses to tactile startle and the prepulse inhibition of tactile startle (Prohaska and Hoffman 1996). Rats who were of normal body weight and normal hematological status, but who were Cu-deficient during perinatal development, had altered acoustic startle responses even after 5 months of Cu repletion! The mammalian auditory startle response is a short-latency reflex, reliably elicited by abrupt and intense auditory stimuli (Davis 1984).

Cu-deficient post-weaning male Holtzman rats with a higher heart-to-body weight ratio, modestly lower hemoglobin concentration, markedly lower ceruloplasmin diamine oxidase activity, a 75% lower liver Cu concentration, and a 36% lower concentration of midbrain Cu, exhibited severe Cu deficiency compared to Cu-adequate rats. Despite this, there were no significant differences in the neurotoxic screening protocol that was employed, including footsplay and acoustic startle (Prohaska and Hoffman 1996). Thorne et al. (1983) also failed to detect behavioral differences following postweanling Cu deficiency in rats.

Perinatal Cu deficiency has at least one long-term specific behavioral consequence in rats. The mechanism for the auditory startle impairment in the repleted animals is not known. It may be specific to the sensory perception modality as normal motor responses were observed for both tactile and inhibitory pre-pulse reflexes. Dopaminergic pathways might play a role in the altered acoustic startle response of the Cu-repleted rats in this study. Altered noradrenergic pathways may also be involved. Electrolytic and chemical lesioning studies indicate that damage to the locus ceruleus, an area enriched in Cu and cell bodies of noradrenergic neurons, attenuates startle responses (Davis 1984). Additional research is needed to elucidate the mechanisms and long-term consequences of altered acoustic startle following perinatal Cu deficiency.

Acknowledgments

Skillful technical assistance of Dr. Richard Hoffman, William Bailey, M. Kate Beattie, Patricia M. Lear, and Cathy Stoll is appreciated. Research was supported by grant 93-372008756 from NRI Competitive Grants Program/USDA.

LIterature Cited

Beattie, K., Gerstenberger, S., Hoffman, R. & Dellinger, J.A. (1996) Rodent neurotoxicity bioassays for assessing contaminated Great Lakes fish. Environ. Toxicol. Chem. 15: 313–318.

Cordano, A., Baertl, J.M. & Graham, G.G. (1964) Copper deficiency in infancy. Pediatrics 34: 214–226.

Danks, D.M., Campbell, P.E., Walker-Smith, J., Stevens, B.J., Gillespie, J.M., Bloomfield, J. & Turner, B. (1972) Menkes' kinky-hair syndrome. Lancet 1: 1100–1103.

Davis, M. (1984) The mammalian startle response. In: Neural Mechanisms of Startle Behavior (Eaton, R.C., ed.), pp. 287–351. Plenum Press, New York, NY.

Dutt, B. & Mills, C.F. (1960) Reproductive failure in rats due to copper deficiency. J. Comp. Pathol. 70: 120–125.

Feller, D.J. & O'Dell, B.L. (1980) Dopamine and norepinephrine in discrete areas of the copper-deficient rat brain. J. Neurochem. 34: 1259–1263.

Glowinski, J. & Iversen, L.L. (1966) Regional studies of catecholamines in the rat brain. J. Neurochem. 13: 655–669.

Jankowski, M.A., Uriu-Hare, J.Y., Rucker, R.B. & Keen, C.L. (1993) Effect of maternal diabetes and dietary copper on fetal development. Reprod. Toxicol. 7: 589–598.

Levenson, C.W. & Janghorbani, M. (1994) Long-term measurement of organ copper turnover in rats by continuous feeding of a stable isotope. Anal. Biochem. 221: 243–249.

Moser, V.C., McCormick, J.P., Creason, J.P. & MacPhail, R.C. (1988) Comparison of chlordimeform and carbaryl using a functional observational battery. Fund. Appl. Toxicol. 11: 189–206.

Prohaska, J.R. & Bailey, W.R. (1993) Persistent regional changes in brain copper, cuproenzymes, and catecholamines following perinatal copper deficiency in mice. J. Nutr. 123: 1226–1234.

Prohaska, J.R. & Bailey, W.R. (1994) Regional specificity in alterations of rat brain copper and catecholamines following perinatal copper deficiency. J. Neurochem. 63: 1551–1557.

Prohaska, J.R. & Bailey, W.R. (1995a) Alterations of rat brain rat peptidylglycine α-amidating monooxygenase and other cuproenzyme activities following perinatal copper deficiency. Proc. Soc. Exp. Biol. Med. 210: 107–118.

Prohaska, J.R. & Bailey, W.R. (1995b) Persistent neurochemical changes following perinatal copper deficiency in rats. J. Nutr. Biochem. 6: 275–280.

Prohaska J.R., Bailey, W.R. & Lear, P.M. (1995) Copper deficiency alters rat peptidylglycine α-amidating monooxygenase activity. J. Nutr. 125: 1447–1454.

Prohaska J.R. & Hoffman, R.G. (1996) Auditory startle response is diminished in rats after recovery from perinatal copper deficiency. J. Nutr. 126: 618–627.

Prohaska, J.R. & Wells, W.W. (1975) Copper deficiency in the developing rat brain: Evidence for abnormal mitochondria. J. Neurochem. 25: 221–228.

Shaw, J.C.L. (1980) Trace elements in the fetus and young infant. Am. J. Dis. Child. 134: 7481.

Smith, R.M. (1983) Copper in the developing brain. In: Neurobiology of the Trace Elements (Dreosti, I.E. & Smith, R.M., eds.), pp. 1–40. Humana Press, Clifton, NJ.

Steel, R.G.D. & Torrie, J.H. (1980) Principles and Procedures of Statistics: a Biometrical Approach. McGraw-Hill, New York, NY.

Thorne, B.M., Lin, K., Weaver, M.L., Wu, B.N. & Medeiros, D.M. (1993) Postweaning copper restriction and behavior in the Long-Evans rats. Pharmacol. Biochem. Behav. 19: 1041–1044.

Discussion

Q1. John Howell, Murdoch University, Murdoch, Western Australia: Joe, this is very elegant work. You've shown that there's a delay in myelination. You've shown that there's necrosis in the early stages, and you've shown that there are these long term alterations in behaviour. Have you done any cell counts comparing, for example, neural populations in the cortex, between Cu deficient and control groups?

A. That's a superb question, and I wish I could give you an answer.

Q2. Carl Keen, University of California, Davis, CA, USA: You've kind of discounted peripheral neuropathy, have you done any histopathology to assure yourself of this?

A. No.

Q3. Carl Keen, University of California, Davis, CA, USA: If you do NVT neurobehavioural testing the zero to 30 d time period is critical. Have you collected data within the first 30 d, to see if these were worse before you started rehabilitating?

A. No, we didn't. I think it's certainly a valid question. I was more interested in just seeing if there were permanent physiological changes.

Q4. Diane Cox, The Hospital for Sick Children, Toronto, ON, Canada: I thought your observations were very interesting, and just wanted to comment that the relevance this would have for the early treatment of Menkes disease, which is now being carried out. No consideration of this has been given to this in people with Wilson's disease who can be made fairly Cu deficient. I think that this should be considered rather seriously. Women should be assessed for Cu status prior to pregnancy.

A. Yes, especially in teenage pregnancies where women are taking Fe and Zn supplements, Cu should be included in that repertoire to prevent the possibility of this occurring.

Q5. Tomas Walter, University of Chile, Santiago, Chile: The effects seen here are similar to what is seen in Fe deficient rats. Can you comment on the similarity?

A. I think this is not unique to the nervous system. The same is true for the immune system and many other things. The only other comment I would have is that I do not believe this is due to secondary Fe deficiency. Brain Fe was perfectly normal, and there was no evidence for abnormal Hb.

Trace Elements, Brain Function and Behavior: Effects of Zinc and Boron

J.G. PENLAND

USDA, ARS Grand Forks Human Nutrition Research Center, Grand Forks, ND 58202, USA*

Keywords zinc, boron, cognition, behavior

The studies reported represent collaborative research efforts of [zinc studies] the author and H.H. Sandstead (Univ TX Med Br, Galveston), Chen X.C. (Chinese Acad Prev Med, Beijing), Li J.S. (Qingdao Med Col, Qingdao), Yang J.J. (3rd Mil Med Univ, Chongqing), Zhao F. (2nd Mil Med Univ, Shanghai), and many unnamed colleagues, with financial support from the International Lead-Zinc Research Organization and General Nutrition Products; and, [boron studies] the author and F.H. Nielsen (USDA ARS Grand Forks Hum Nutr Res Ctr, Grand Forks ND).

The relationship between trace element nutrition and behavioral function in humans is illustrated by several controlled studies showing that intakes of zinc (Zn) and boron (B) affect performance of cognitive and psychomotor tasks. Zn is known to be essential for normal brain development and behavior in animals (Sandstead 1985), and many studies have shown that deprivation during periods of rapid growth results in behavioral deficits (Halas et al. 1987). In humans, Zn supplementation (Darnell et al. 1991) and deprivation (Penland 1996, Tucker et al. 1984) were shown to affect cognition of adult men and women, but supplementation failed to benefit cognition of adolescent boys and girls in two other studies (Gibson et al. 1989, Cavan et al. 1993). A follow-up study with Chinese children is reported here.

The effects of Zn and micronutrient supplementation on cognition and psychomotor function of Chinese children were determined in 372 children (aged 6–9 years) living in poor urban areas of Chongqing, Qingdao and Shanghai, China. At each location, intact classrooms received a treatment of 20 mg Zn/d (Zn), Zn plus micronutrients (Zn + M; 50% Recommended Dietary Allowances or mean Estimated Safe and Adequate Dietary Intakes, excluding iron, calcium, magnesium (Mg) and phosphorus, with folate at 25% RDA), or micronutrients alone (M) in a double-blind manner for 10 weeks. Treatments were administered by teachers to promote compliance. Cognition and psychomotor function were assessed at baseline and after 10 weeks of treatment by measuring performance (typically response times and accuracy) on a battery of computer-administered tasks specifically designed to emphasize attention, perception, memory and reasoning, and the motor and spatial skills necessary for successful performance.

Performance on Design Matching (perception), Delayed Design Matching (memory) and Oddity (reasoning) tasks showed treatment effects at all three study locations, while performance on Object Pair Search (attention) showed no effects at any location (Table 1). Table 2 shows selected treatment effects on performance pooled across study locations. Typical of findings from individual locations, treatment with Zn and/or Zn + M resulted in the most improved performance, when compared to treatment with M alone. However, treatment effects were not always consistent from location to location and the treatment frequently showed a statistically significant effect on performance at one or two locations, but not the other(s). Future analyses will help determine whether location differences in effects resulted from pre-existing differences among locations in Zn and lead status or from location differences in growth and nutritional status responses to the treatment.

These findings complement those from recent studies of adult humans and adolescent monkeys (Darnell et al. 1991, Golub et al. 1994, Penland 1996) and have important implications for the world's

*US Department of Agriculture, Agricultural Research Service, Northern Plains Area, is an equal opportunity/affirmative action employer and all agency services are available without discrimination.

Correct citation: Penland, J.G. 1997. Trace elements, brain function and behavior: effects of zinc and boron. In *Trace Elements in Man and Animals – 9: Proceedings of the Ninth International Symposium on Trace Elements in Man and Animals. Edited by* P.W.F. Fischer, M.R. L'Abbé, K.A. Cockell, and R.S. Gibson. NRC Research Press, Ottawa, Canada. pp. 213–216.

Table 1. Task battery and results summary from zinc (Zn) supplementation studies.

Function	Task	Chongqing	Qingdao	Shanghai
Attention	Continuous Performance (12)[1]	X[2]	–	X
Attention	Object Pair Search (2)	–	–	–
Perception	Design Matching (12)	X	X	X
Perception	Count Matching (6)	–	X	X
Memory	Delayed Object Search (4)	X	X	–
Memory	Object Recognition (4)	X	–	X
Memory	Delayed Design Matching(14)	X	X	X
Reasoning	Oddity (8)	X	X	X
Psychomotor	Tapping (3)	–	–	X
Psychomotor	Tracking (4)	–	X	X

[1]Number of outcome measures generated by task; [2]One or more measures showing significant ($P < 0.05$) effect of Zn and micronutrient supplementation.

Table 2. Effects of Zn and micronutrient supplementation on performance[1] of selected cognitive and psychomotor tasks (pooled).

Task[2]	Zn	Zn+M	M
Continuous Performance	5.30 ± 2.19[3,a]	2.56 ± 2.29[a,b]	–3.32 ± 2.17[b]
Design Matching	3.72 ± 0.98[a]	1.89 ± 0.99[a,b]	0.23 ± 0.97[b]
Delayed Design Matching	–4.17 ± 2.43[a,b]	–8.33 ± 2.33[a]	0.05 ± 2.23[b]
Oddity	–6.10 ± 2.44[a]	–6.76 ± 2.31[a]	1.97 ± 2.21[b]
Tapping	11.6 ± 1.6[a]	12.3 ± 1.6[a]	6.0 ± 1.6[b]
Tracking	26.1 ± 2.6[a]	34.0 ± 2.5[b]	21.0 ± 2.5[a]

[1]Standardized difference scores ((week 10 – baseline) / (week 10 + baseline)); [2]Continuous Performance & Design Matching (% correct), Delayed Design Matching (reaction time), Oddity (trials to learning criterion), Tapping (# taps), Tracking (% time on target); [3]Mean ± SEM; [a-b]Means with different superscripts differ significantly ($P < 0.05$).

population with suboptimal Zn intakes and status. The failure of previous studies (Cavan et al. 1993, Gibson et al. 1989) to find improved cognition in children supplemented with Zn may be because those studies measured a limited number of cognitive processes and response characteristics. Golub et al. (1995) recently summarized possible mechanisms of action for effects of Zn nutrition on behavior.

Although the trace element boron (B) has yet to be recognized as an essential nutrient for humans, data from numerous animal studies and several human studies suggest that B may play a role in mineral metabolism and membrane function (Nielsen 1991). To further investigate possible functional roles of B, performance on a battery of computer-administered tasks (Table 3) was assessed in response to dietary manipulation of B in three studies with healthy post-menopausal women and men aged >45 years (Penland 1994).

In all three studies, data were collected while subjects were fed a conventional diet containing ≤ 0.25 mg B/d (low B) and contrasted with data collected from those same subjects fed the low B diet

Table 3. Task battery and results summary from dietary boron (B) studies.

Study[1]	Function/Task	Study	Function/Task
I II[2] III[2]	Psychomotor/Tapping	I[2] II[2] III[2]	Memory/Symbol-Digit
I II[2] III	Tracking	I	Shape Recognition
I	Trails	I	Cube Recognition
I[2] II[2] III[2]	Attention/Search-Count	I	Letter Recognition
I	Continuous Vigilance	I[2]	Word Recognition
I II III[2]	Perception/Stroop Color-Word	I	Spatial/Maze
I II III	Time Estimation		

[1]Study in which named task was administered; [2]Studies showing significant ($P < 0.05$) effect of dietary B on performance.

supplemented with 3 mg B/d (high B), as sodium borate. In Study I, data were collected from 13 women participating in a 6-month metabolic study of B and Mg nutrition. High and low B were each fed for two 42-d dietary periods, once each under conditions of low and adequate dietary Mg, 115 and 315 mg/d, respectively. Diets were fed in a double-blind, crossover fashion. Studies II and III were similar to each other except that Mg intakes of 9 women and 5 men were restricted to 115 mg/d in the former, whereas Mg intakes of 10 women and 4 men were adequate (315 mg/d) in the latter. In both studies, data were collected during a 4-month community study which included 63 d of low B followed by 49 d of high B.

As shown in Table 3, all three studies were consistent in showing an effect of dietary B on performance of two tasks, Search-Count and Symbol-Digit. Performance on three other tasks, Tapping, Tracking and Stroop, was affected by dietary B in one or two, but not all three studies. There were no significant effects of B intake on the performance of seven other tasks; however six of these tasks were administered only in Study I (Table 3). One additional task, Word Recognition, showed an effect of dietary B, but was administered during only one study (Study I).

Table 4 shows effects of dietary B on performance of selected tasks administered in all three studies. In all studies when contrasted with the high B intake, low dietary B resulted in increased response times during search and increased times to encode and recall symbol-digit pairs. Both studies II and III found that low B slowed tapping. Study II found that low B resulted in decreased % time on target in visual tracking, while Study III found that low B resulted in increased response times to identify color names regardless of presentation color (Stroop). There were no significant effects of dietary B on error rates.

These performance effects complement the effects of dietary B on brain electrophysiology observed in these same studies (Penland 1994); low B intakes resulted in increased low-frequency EEG activity consistent with decreased alertness. Together, these findings indicate a role for B in brain function and behavior, and provide additional evidence that B should be considered an essential nutrient for humans (Nielsen 1991).

In conclusion, further experimental study of the consequences of Zn and B nutriture for cognition and psychomotor function is necessary to determine the precise roles of these trace elements for optimal health and function in humans, particularly adolescents and older adults. The findings also underscore the need to examine a variety of cognitive processes and response characteristics when assessing the impact of nutritional intervention or suboptimal nutrition on cognitive function. Acceptance of cognition and other aspects of psychological function and behavior as endpoints for determining optimal dietary intakes for humans requires the ongoing development of sensitive and reliable measures.

Table 4. Effects of B intake on performance of selected cognitive and psychomotor tasks.

	Study I Boron		Study II Boron		Study III Boron	
Task[1]	0.25[2]	3.25	0.25	3.25	0.25	3.25
Tapping	23.0[3]	23.0	22.1[5]	24.2	23.2[4]	24.6
	(0.3)	(0.3)	(1.0)	(1.0)	(1.0)	(0.8)
Tracking	19.2	18.5	22.8[5]	26.4	16.0	17.5
	(0.6)	(0.6)	(2.5)	(2.3)	(1.7)	(1.6)
Search	3.34[5]	3.15	3.43[5]	2.63	3.52[5]	3.02
	(0.08)	(0.07)	(0.32)	(0.21)	(0.33)	(0.29)
Stroop	772	806	729	713	731[5]	680
	(18.2)	(18.1)	(27.2)	(24.5)	(25.7)	(27.5)
Symbol-Digit	2.30[4]	2.23	2.14[5]	1.88	2.27[5]	1.98
	(0.02)	(0.02)	(0.06)	(0.05)	(0.11)	(0.08)

[1]Tapping (# taps/30 s), Tracking (% time on target/60 s), Search & Symbol-Digit (response time (s)), Stroop (reaction time (ms)); [2]B (mg/d); [3]Mean (SEM); [4]$P < 0.05$; [5]$P < 0.01$.

Literature Cited

Cavan, K.R., Gibson, R.S., Grazioso, C.F., Isalgue, A.M., Ruz, M., & Solomons, N.W. (1993) Growth and body composition of periurban Guatemalan children in relation to zinc status: a cross-sectional study. Am. J. Clin. Nutr. 57: 334–343.

Darnell, L.S. & Sandstead, H.H. (1991) Iron, zinc and cognition. Am. J. Clin. Nutr. 53: S16.

Gibson, R.S., Vanderkooy, P.D.S., MacDonald, A.C., Goldman, A., Ryan, B.A., & Berry, M. (1989) A growth-limiting, mild zinc-deficiency syndrome in some South Ontario boys with low height percentiles. Am. J. Clin. Nutr. 49: 1266–1273.

Golub, M.S., Takeuchi, P.T., Keen, C.L., Gershwin, M.E., Hendrickx, A.G., & Lonnerdal, B. (1994) Modulation of behavioral performance of prepubertal monkeys by moderate dietary zinc deprivation. Am. J. Clin. Nutr. 60: 238–243.

Golub, M.S., Keen, C.L., Gershwin, M.E., & Hendrickx, A.G. (1995) Developmental zinc deficiency and behavior. J. Nutr. 125: 2263S–2271S.

Halas, E.S. & Eberhardt, M.J. (1987) A behavioral review of trace element deficiencies in animals and humans. Nutr. Behav. 3: 257–271.

Nielsen, F.H. (1991) The saga of boron in food: from a banished food preservative to a beneficial nutrient for humans. Cur. Topics Plant. Biochem. Physiol. 10: 274–286.

Penland, J.G. (1994) Dietary boron, brain function and cognitive performance. Environ. Hlth. Perspect. 102 (Suppl. 7): 65–72.

Penland, J.G. (1996) Short-term, moderate dietary zinc deprivation affects cognitive performance in healthy adult men. Am. J. Clin. Nutr., in press.

Sandstead, H.H. (1985) Zinc: essentiality for brain development and function. Nutr. Rev. 43: 129–137.

Tucker, D.M., Sandstead, H.H. (1984) Neuropsychological function in experimental zinc deficiency in humans. In: The Neurobiology of Zinc, Part B (Frederickson, C.J., Howell, G.A., & Kasarskis, E.J., eds.), pp. 139–152. Alan R. Liss, New York.

Discussion

Q1. Margaret Elmes, University Hospital Wales, Cardiff, Wales: How did you screen for pre-existing anxiety states or depressive states in your adults that would cause impaired cognition?

A. We routinely have people take a variety of specific measures/tests to ensure people coming into the study are not suffering from any gross psycho-pathology.

Q2. Stan Zlotkin, University of Toronto, Toronto, ON, Canada: In listening to your description of the groups in China, it sounded like there are three intervention groups. I would like to suggest that without having had a control group, including a placebo, it's impossible to interpret the results of your study, because it's certainly possible that what you're seeing would have been observed if there was a placebo group, and there may have been a learning effect with time over the ten week period. Why was a control group not included? I would like to suggest that without having included a control group, that this may not have been an ethical study to do, because it's impossible to determine the results.

A. Cost, and the possibility that we're failing to provide beneficial treatment to the children. However, a new study is in the works, in which a placebo group is being included, and I acknowledge that this is important. But, you can also look at the relative rates of improvement between week 10 and week zero.

Q3. Harold Sandstead, University of Texas Medical School, Galveston, TX, USA: I'd like to comment on not having a placebo group. Our IRB at our institution looks askance at studies in humans where there's a possibility of benefit or you don't give a benefit of any kind. It's very difficult to get approval to do a placebo-controlled study of this type. We're working with them because of the findings that we have. The Chinese didn't mind one way or the other. They felt the placebo would be alright. They have, on their own, looked at the growth of children not getting any treatment, and they found that the growth in children receiving no treatment at all was the same as in those receiving zinc by itself, which is not too surprising, as there's a lot of literature saying that if you give zinc by itself in the face of other missing nutrients, it's very unusual to find a beneficial effect. So, I understand that there are differences in the way that ethical groups look at things, but at the time we set these studies up, our IRB felt that to have a placebo group would be unethical, so there you go.

Q4. James Friel, Memorial University of Newfoundland, St. John's, NF, Canada: I appreciate how you feel about psychological testing being a functional test, as opposed to a biochemical test. But you start with boron in the diet, and then you end up with improved cognition. Somewhere in between there, there's a reason for that. Do you have a biochemical hypothesis for the role of boron in the improvement of cognition.

A. No, there are indications boron is involved in cell function, whether it's permeability or something else I don't know, but no, we don't know the biochemical basis. Clearly, when you can hypothesize certain mechanisms of action, those can help guide you to know the kinds of functions to look at. On the other hand, we can't always anticipate mechanisms of action, so we may be limiting ourselves if we only look first to the biochemical function and then to match a behavioural outcome.

Effect of Iron Deficiency Anemia on Cognitive Skills and Neuromaturation in Infancy and Childhood

TOMAS WALTER, PATRICIO PEIRANO, AND MANUEL RONCAGLIOLO
Institute Of Nutrition and Food Technology, University of Chile, Santiago 138-11, Chile

Keywords iron, deficiency, cognitive development, infants

Introduction

When iron deficiency anaemia ensues during the first 2 years of life it has been associated with delayed psychomotor development and changes in behaviour. These effects have been shown to persist after several months of iron therapy, despite complete correction of iron nutritional measures. Moreover, it is still uncertain whether or not and to what extent they are reversible after a extended period of observation, since the long term prospective follow-up studies reported to date, to be discussed below, show the persistence of cognitive deficits at 5–6 and at 10 years of age in those who during infancy had anaemia.

The inherent difficulties of identifying intervening variables in the complex field of mental development, coupled in some cases with suboptimal design have prevented significant progress in the investigation of iron deficiency. However, two studies, one conducted in Costa Rica (Lozoff et al. 1987) and the other in Santiago, Chile (Walter et al. 1989), taking into careful consideration the potential pitfalls, confirm conclusions arising from previous work.

The Santiago study was performed in association with a field trial of fortified infant foods. One hundred ninety-six healthy, full-term infants were assessed with the Bayley scales of infant development (Bayley 1969) at 12, 12½ and 15 months of age. This well-known and accepted tool is used to determine psychomotor development from ages 3 to 30 months. It consists of a mental scale to evaluate cognitive skills, such as language acquisition and abstract thinking, and a motor or psychomotor scale to evaluate gross motor abilities, such as co-ordination, body balance, and walking.

The Costa Rica study enrolled 191 12- to 23-month-old otherwise healthy infants with heterogeneous iron status. The infants were divided into groups ranging from most to least iron deficient. The Bayley scales of infant development were administered before, and after one week and after three months iron treatment with appropriate placebo controls.

At what stage of iron deficiency is infant behaviour adversely affected? It was clear in the Santiago study that a decrease in haemoglobin leading to overt anaemia was necessary to significantly affect mental and psychomotor development scores. The performance of the nonanemic iron-deficient infants as a whole was indistinguishable from that of the controls.

Among anaemic infants, haemoglobin concentration was correlated with performance. The lower the Hgb, the lower the developmental scores. Similarly in the Costa Rica study infants with moderate iron deficiency anaemia (Hgb < 100 g/L) had lower mental and motor test scores than appropriate controls.

The Santiago study also evaluated the effect of chronic anaemia. Infants whose anaemia had a duration of three or more months had significantly lower mental and motor development indices that those with anaemia of shorter duration.

The results of other research published to date support the conclusion of the Santiago and Costa Rica studies — iron deficiency severe enough to cause anaemia is associated with impaired achievement in

Correct citation: Walter, T., Peirano, P., and Roncagliolo, M. 1997. Effect of iron deficiency anemia on cognitive skills and neuromaturation in infancy and childhood. In *Trace Elements in Man and Animals – 9: Proceedings of the Ninth International Symposium on Trace Elements in Man and Animals. Edited by* P.W.F. Fischer, M.R. L'Abbé, K.A. Cockell, and R.S. Gibson. NRC Research Press, Ottawa, Canada. pp. 217–219.

developmental tests in infancy and as anaemia becomes more severe or prolonged deficits are more profound.

What is the effect of iron treatment? Consistent results have been obtained in studies that have included a placebo treatment group. Together, these studies indicate that short-term increases in test scores observed among iron-treated anaemic infants are not significantly greater than those among placebo-treated anaemic infants, thus related likely to a practice effect.

Although separating the effects of iron deficiency from those of anaemia is important a more pertinent question from a clinical perspective is whether iron therapy completely corrects behavioural abnormalities regardless of how soon the changes are detectable.

Studies in Costa Rica (Lozoff et al. 1987), Chile (Walter et al. 1989), the United Kingdom (Aukett et al. 1986) and Indonesia (Idjradinata and Pollitt 1993) included an iron treatment period of 2 to 4 months after which psychomotor development tests were repeated. Despite the improved iron status most of the formerly anaemic infants were unable to improve their psychomotor performance. The only study to date that showed a reversal of lower Bayley Scale of Infant Development is the Indonesian study.

Notwithstanding, in most of the studies iron therapy even complete iron repletion, was ineffective in improving the psychomotor scores of anaemic infants to the level of nonanemic controls. The protocol in Indonesia (Idjradinata and Pollitt 1993) goes to prove that studies in this field may give conflicting results and that newer and more imaginative techniques must be used to elucidate current controversies.

What are the specific patterns of failure? The Chilean study found that when examining the mental scale, items that required comprehension of language but did not involve a visual demonstration were passed by fewer anaemic infants than controls. In the psychomotor scale balance in the standing position and walking (sits from standing stands alone and stands up) were accomplished by significantly fewer anaemic infants than controls. Similar findings were reported in the Costa Rica study.

Long Term Effects of Iron Deficiency Anemia

Cognitive Performance of Children who were Anaemic during Infancy

The long term persistence issue has been addressed by two follow-up studies recently described in five-year-old Costa Rican (Lozoff et al. 1991) and Chilean children (Walter et al. 1989) who had been well characterised as infants in iron status environmental variables and their psychomotor development. These children were the subjects of respective reports during their infancy described above (Lozoff et al. 1987, Walter et al. 1989). At five years of age an evaluation with a comprehensive set of psychometric tests showed that those who as infants had presented with iron deficiency anaemia had lower scores on many of these tests compared to children with higher haemoglobin in infancy. These disadvantages persisted after statistical control of many potentially confounding variables (De Andraca et al. 1991).

Neuromaturational Studies

Auditory Brain Stem Evoked Potentials

The auditory Brainstem responses (ABR) represent the progressive activation of different levels of the auditory pathway from acoustic nerve to inferior colliculus. This interval corresponds to the Central Conduction Time (CCT). As a function of maturation of nerve fibres and synaptic relays, the CCT shows an exponential reduction until stabilisation at 18–24 months of age. In 17 anaemic infants and 18 iron sufficient controls ABR's were measured at 6, 12 and 18 months of age. The ABR's were longer in anaemic infants at all time points despite iron therapy to whole haematological reconstitution initiated at 6 months (Peirano et al. 1996).

Sleep Studies. Autonomic Nervous System Development

Maturational patterns of heart rate variability (HRV) provide non invasive tools for the investigation of central nervous integrity during early human development and are likely to reflect brain function alterations earlier and more closely than tests of behaviour and psychomotor development. Patterns of heart rate and HRV were measured in 18 anaemic 6-month-old infants and corresponding controls from polygraphic recordings during quiet and active sleep and wakefulness. Iron deficient anaemic infants presented lower amplitude in all sleep wake states. It is proposed that delayed myelination of the vagal nerve results in decreased parasympathetic influences that may underlie behavioural effects of iron deficiency in infancy (Roncagliolo et al. 1996).

References

Aukett, M.A., Parks, Y.A., Scott, P.H., & Wharton, B.A. (1986) Treatment with iron increases weight gain and psychomotor development. Arch Dis Child 61: 849–57.

Bayley, N. (1969) Bayley scales of infant development. New York: Psychological Corporation.

De Andraca, I., Walter, T., Castillo, M., Pino, P., Rivera, P., & Cobo, C. (1991) Iron deficiency anemia and its effects upon psychological development at preschool age: a longitudinal study. In: Nestle Nutrition Annals, pp 53–62.

Idjradinata, P. & Pollitt, E. (1993) Reversal of developmental delays in iron-deficient anemic infants treated with iron. Lancet 341: 1–4.

Lozoff, B., Brittenham. G.M., & Wolf, A.W. (1987) Iron deficiency anemia and iron therapy: Effect on infant developmental test performance. Pediatrics 79: 981–995.

Lozoff, B., Jimenez, E., & Wolf A. (1991) Long-term developmental outcome of infants with iron deficiency. N Engl J Med 325: 687–94.

Peirano, P., Williamson, A., Garrido, M., Rojas, P., Rodriguez, E., Ehijo, A., De Andraca, I., & Lozoff, B. (1996) Evidence of delayed autonomic system development in iron deficient anaemic infants: a potential mechanism for altered behaviour and development. Submitted.

Roncagliolo, M., Garrido, M., Williamson, A., Rojas, P., Contreras, J., Lozoff, B., Peirano, P., & Walter, T. (1996) Evidence of altered central nervous system development in iron deficient anaemic infants: Delayed maturation of the Auditory Brainstem responses. Submitted.

Walter, T., De Andraca, I., Chadud, P., & Perales, C.G. (1989) Iron deficiency anaemia: Adverse effects on infant psychomotor development. Pediatrics 84: 7–17.

Discussion

Q1. John Howell, Murdoch University, Murdoch, Western Australia: The basic pathogenesis of iron deficiency anemia is hypoxia at a vulnerable period in development with loss of synapses and possible loss of neurons. Is the prevention of iron deficiency anemia more important than treatment, which will never effect a complete cure?

A. I don't know the mechanism of the damage. Anemia at this period (from six to nine months) is very mild and is very well compensated by other mechanisms. I don't think it has to do with oxygen, but it may have to do with function of brain iron.

Q2. James Friel, Memorial University of Newfoundland, St. John's, NF, Canada: I was looking at your graph showing the difference in anemic vs non-anemic but iron-deficient infants, and I'm trying to get a picture in my head of what's happening. If iron in the body is going preferentially into hemoglobin, there must be some point where you're iron deficient enough that you don't have iron going into all the other processes in the brain for which iron is required, apart from hemoglobin. So I'm really surprised you don't see at some level of iron deficiency, some level of neurological developmental effects.

A. I think that the one method we used here is a crude measure of function. Actually, this has been replicated by every investigator with a control group. We need anemia to show a change.

Q3. Stan Zlotkin, University of Toronto, Toronto, ON, Canada: Tomas, I was going to ask you the same question with respect to ferritin, but I think that it's been discussed. I've always been somewhat curious that the one study that was inconsistent with your current finding (the one published in Lancet in, I believe it was 1993), where they also had very similar groups of infants, and those that had moderate anemia, when given iron as opposed to a placebo, they saw a significant increase in Bayley scores. Why are these results inconsistent with yours and with other similar studies?

A. That's the only study that shows that after 4 months of Fe or placebo changed not only hemoglobin, but also the Bayley scores. I don't have a way to explain that change, but there are maybe half a dozen studies that showed no change.

Q4. Barbara Golden, University of Aberdeen, Scotland: I liked your study, but I'm not sure about your interpretation. Have you taken other aspects of poverty into account sufficiently to be able to say that this is due to Fe deficiency?

A. As far as nutrition is concerned, these infants were followed by us, and were all healthy and well nourished. The study was randomized, most of the anemics came from the non-fortified group, and most of the controls came from the fortified group. We looked at a very large variety of confounding variables.

Q5. Gordon Klein, University of Texas, Galveston, TX, USA: Have you looked at anemia from other causes?

A. We don't have other causes of anemia. For example, we don't have that severe a copper deficiency, and we don't have sickle cell disease.

Impaired Startle Response in Growing Rats Marginally Iron-Depleted without Overt Anemia

J.R. HUNT AND J.G. PENLAND

USDA, ARS, Grand Forks Human Nutrition Research Center, Grand Forks, ND 58202, USA*

Keywords iron, startle response, neurological function

Iron deficiency anemia adversely affects mental and psychomotor development in human infants (Walter 1994) and has been associated with altered central nervous system neurotransmission in animals (Youdim et al. 1989). There has not been a clear demonstration of adverse functional effects of low iron (Fe) stores without anemia. Startle response provides a direct evaluation of central nervous system responsiveness that is sensitive to experimental treatments, including dietary treatments, that modify neurotransmitter metabolism (Davis 1984).

Methods

Startle response was measured in growing animals after marginal iron depletion and subsequent, partial repletion. Weanling, female Sprague-Dawley rats were fed AIN-93G diets modified to contain 15 mg or 90 mg Fe/kg of diet (by analysis) for 53 d (depletion or control animals, respectively). Both groups of rats were fed 90 mg Fe/kg for another 28 d (repletion or control animals, respectively). Six of 12 rats per group were killed for tissue analysis at the end of the depletion period. The other six were tested for startle response during the 12-h dark cycle at the end of depletion and repletion, then killed.

Startle response was measured with a strain gauge transducer/coupler and integrator (Coulbourn Instruments, Lehigh Valley, PA†) with custom-developed software. A total of 120 acoustic startle trials, employing 90 dBC, 5 kHz tones with a duration of 40 ms, were administered in a counterbalanced manner, with an intertrial interval of 30 s (range 21–40 s). The amplitude, measured as the downward force generated by the initial response, and the latency or delay of the response were recorded. Responses were excluded if movement during the 50 ms pre-stimulus period $\geq$ 10 g, if the amplitude $\leq$ 60 g, or if the response latency $\geq$ 20 ms.

Results and Discussion

Fe-depleted rats had marginal Fe status without overt anemia, as indicated by reductions of 7% in hemoglobin, 38% in serum Fe, and 87% in liver nonheme Fe, without significant differences in hematocrit, Fe binding capacity, or brain Fe in comparison to control animals (Table 1). Transferrin saturation after depletion was significantly reduced, but remained above 16% (Table 1), a commonly used threshold for iron deficiency. These marginally Fe-depleted rats displayed increased startle response latency without a significant difference in response amplitude (Table 1). Iron depletion did not affect brain or body weights (data not shown). Fe-repletion was not fully achieved despite 4 weeks of dietary iron that exceeded requirements by 3–4 fold (Siimes et al. 1980); compared with controls, repleted rats had reductions of 5% in hemoglobin and 42% in liver nonheme Fe, without differences in hematocrit, serum Fe, Fe binding

*The U.S. Department of Agriculture, Agricultural Research Service, Northern Plains Area, is an equal opprotunity/affirmative action employer and all agency services are available without discrimination.

†Mention of a trademark or proprietary product does not constitute a guarantee or warranty of the product by the U. S. Department of Agriculture and does not imply its approval to the exclusion of other products that may also be suitable.

Correct citation: Hunt, J.R., and Penland, J.G. 1997. Impaired startle response in growing rats marginally iron-depleted without overt anemia. In *Trace Elements in Man and Animals – 9: Proceedings of the Ninth International Symposium on Trace Elements in Man and Animals. Edited by* P.W.F. Fischer, M.R. L'Abbé, K.A. Cockell, and R.S. Gibson. NRC Research Press, Ottawa, Canada. pp. 220–221.

Table 1. Iron status indices and startle response in growing rats marginally depleted and partially repleted with iron.

	End of Depletion		End of Repletion	
Variable [a]	Fe-depleted	Control	Fe-repleted	Control
Iron Status Indices				
Hemoglobin, g/L	142±7 **	153±4	149±5 *	157±5
Hematocrit, %	40.2±1.7	41.8±1.7	41.4±1.9	43.7±0.8
Serum Iron, μmol/L	25±5 **	40±5	43±6	42±6
Iron Binding Capacity, μmol/L	121±11	111±8	111±18	114±8
Transferrin Saturation, %	21±4 **	36±5	40±7	37±6
Liver Nonheme Iron, μmol/g dw	0.7±0.2 **	5.5±1.4	4.0±0.6 *	6.9±2.3
Brain Iron, μmol/g dw	1.2±0.2	1.2±0.1	1.1±0.2	1.1±0.2
Startle Response				
Amplitude, g	112±42	108±17	99±18	96±31
Latency, ms	14.4±0.5 *	13.5±0.6	14.2±1.0	13.5±1.0

[a] Mean ± SD, with 4–6 rats per group
* $p < 0.05$, ** $p < 0.01$, by student's t-test for differences between diet groups measured at the same time (end of depletion or repletion).

capacity, or brain Fe (Table 1). Unexpectedly, serum Fe concentrations returned to normal before hemoglobin concentrations (Table 1). Response latency, although not completely corrected, was no longer significantly elevated in the repleted rats (Table 1).

This initial test of startle response and iron deficiency suggests that startle response may be a sensitive (though not necessarily specific) functional measure of marginal Fe depletion. Further animal studies will be conducted to validate this initial observation. Startle response testing can be reliably applied in human subjects. If verified to be sensitive to marginal iron status, startle response results could serve as a useful functional indicator to help define optimal iron status in humans.

References

Davis, M. (1984) The mammalian startle response. In: Neuronal mechanisms of startle behavior (R. C. Eaton, ed), 287–343. Plenum Press, New York.

Siimes, M.A., Refino, C., & Dallman, P.R. (1980) Manifestation of iron deficiency at various levels of iron intake. Am J Clin Nutr 33: 570–574.

Walter, T. (1994) Effect of iron-deficiency anaemia on cognitive skills in infancy and childhood. Baillieres Clin Haematol 7: 815–827.

Youdim, M.B.H., Ben-Shachar, D., & Yehuda, S. (1989) Putative biological mechanisms of the effect of iron deficiency on brain biochemistry and behavior. Am J Clin Nutr 50: 607–617.

Discussion

Q1. James Friel, Memorial University of Newfoundland, St. John's NF, Canada: You didn't find any difference in the brain Fe content. I would have thought, if you're using the startle response as an indicator of neurological function, that you might see some. Is it possible there might be the same Fe content, but just be distributed differently throughout the brain?

A. Well, I think that we would need to look at specific sites in the brain. These measurements were done on whole brain, so I think that these were just not sensitive enough to show differences.

Q2. Tomas Walter, University of Chile, Santiago, Chile: I'll just mention that our infants are put in an acoustic room for measuring the evoked potentials. You could measure evoked potentials in rats as well using an acoustic chamber.

A. (no comment)

Metabolic Analysis of Vanadate and Effect on Neurochemical Behavior as a Result of Chronic Oral Administration of Vanadate

MIEKO KAWAMURA,† TATSUO IDO,‡ REN IWATA,‡ YUKIKO NAKANISHI,† AND SHUICHI KIMURA¶

†Faculty of Agriculture, ‡Cyclotron and Radioisotope Center, Tohoku University, Sendai, Japan; ¶Showa Women's University, Tokyo, Japan

Keywords vanadium, neurotransmission, blood-brain barrier, ascorbic acid

In recent years, our interest has been focussed on the role of vanadium as an essential biological element. Vanadium has been shown to be an essential trace element in experimental animals and has been suggested to be a possible etiolological factor in manic depressive psychosis. In this study, the biodistribution of ^{48}V, the behavior of vanadium in the brain, the effect of vanadate on neurotransmission, and the effect of the nutritional environment on the blood-brain barrier permeability of the vanadium were investigated.

Effect on Neurotransmission and Metabolic Analysis of Vanadate

Weanling rats were divided into two groups. One group was given drinking water containing 100 ppm Na_3VO_4 and the control group received distilled water alone for 3~7 months. Compared to the control, a 58.4% reduction in the release of ACh in the striatum was observed, while no change was detected in the hippocampus. There was a reduction in HVA release but no change in DOPAC release was observed. The concentration of both mACh-R and D-2 receptors did not change but the binding affinity of mACh-R decreased. Biodistribution of ^{48}V revealed that the majority of vanadate in the blood was reduced to vanadyl (4+) and bound to transferrin. A fraction of unchanged vanadate was transferred into the brain through the anion channels. Intracellular analysis of brain cells indicated that the cytosol contained transferrin-vanadyl, free vanadate and no free vanadyl (summarized in Figure 1).

Level of Ascorbic Acid (AsA) in the Brain after Long Term Administration of Vanadate

A solution of sodium orthovanadate (100 ppm) was administered in drinking water to Wistar rats for 7 months. AsA concentration was measured by HPLC-ECD. In the blood serum, liver, spleen, kidney, and adrenal glands, the concentration of ascorbic acid remained unchanged. The level of ascorbate decreased in the brain only after long term oral administration of vanadate. The reduction in the concentration of ascorbic acid was different in each part of the brain and the extracellular areas. This phenomenon indicates that AsA is consumed at high quantities by redox systems involving vanadium in the brain.

Effect of AsA or Zinc Deficiency on the Blood-Brain Barrier Permeability to Vanadate

In addition to the chronic oral administration of excess vanadate, the effect of the deficiency of two nutrients, AsA and zinc, on the blood-brain barrier permeability to vanadium was also investigated *in vivo* using radioluminography. Observations indicated a high uptake of ^{48}V in the hypothalamus and ventricular regions in AsA or zinc deficient rats. The uptake of ^{48}V in the hypothalamus of the AsA deficient rats was 2.4 times the control and the uptake of ^{48}V in the hypothalamus of the zinc deficient rats was 3.4

Correct citation: Kawamura, M., Ido, T., Iwata, R., Nakanishi, Y., and Kimura, S. 1997. Metabolic analysis of vanadate and effect on neurochemical behavior as a result of chronic oral administration of vanadate. In *Trace Elements in Man and Animals – 9: Proceedings of the Ninth International Symposium on Trace Elements in Man and Animals. Edited by* P.W.F. Fischer, M.R. L'Abbé, K.A. Cockell, and R.S. Gibson. NRC Research Press, Ottawa, Canada. pp. 222–223.

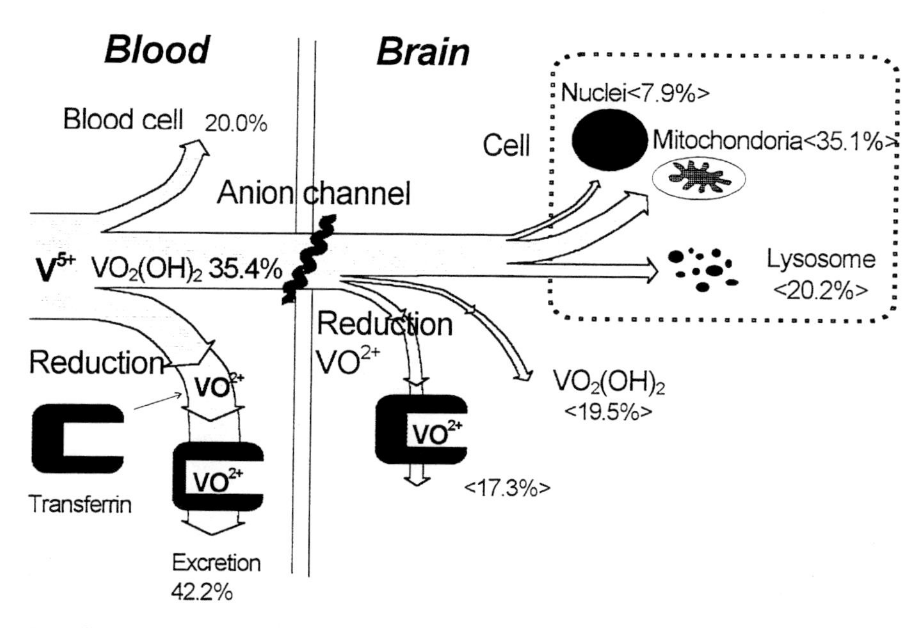

Figure 1. Vanadium transport in the brain.

times the control. As for the venticular region of the AsA deficient rats, the uptake of ^{48}V was 27.6 times the control. These results suggest that the deficiency significantly altered the blood-brain barrier permeability of vanadate.

Conclusion

1. Chronic oral administration of vanadium increases vanadium concentration in the rat brain. This increase also affected the concentration of neurotransmitters, binding properties of receptors, and the metabolism of Ach and dopamine in the brain of Wistar rats.
2. The levels of AsA decreased in the brain of Wistar rats (able to synthesize AsA) after long-term oral administration (7 months) of vanadate.
3. The blood-brain barrier permeability of vanadate changed as a result of AsA or zinc deficiency. This observation indicates that there was an alteration in the function of the blood-brain barrier as a result of changes in the nutritional environment.

Trace Elements, Membrane Function and Cell Signalling

Chair: P. Patterson

Role of Zinc in Maintaining Endothelial Integrity*

BERNHARD HENNIG,[†‡] PATRICE CONNELL,[‡] MICHAL TOBOREK,[¶]
SHIRISH BARVE,[§] AND CRAIG J. MCCLAIN[§]

[‡]Departments of Nutrition and Food Science, [¶]Surgery, and [§]Medicine, University of Kentucky, Lexington, KY 40506-0054, USA

Keywords zinc, endothelial cells, lipids, cytokines, atherosclerosis

Introduction

The vascular endothelium, which is situated at the vital interface between the circulating blood and the body's tissues, plays an important role in cardiovascular functions in health and disease (Ross 1986). Disruption of endothelial cell integrity and barrier function, as well as activation of the vascular endothelium, have been suggested to be critical events leading both to chronic diseases, such as atherosclerosis, and to acute vascular events, such as lung edema in adult respiratory distress syndrome (ARDS) (reviewed in Hennig et al. 1994 and McClain et al. 1995). There is evidence that selected nutrients may be involved in metabolic events that protect the vascular endothelium or maintain endothelial integrity (reviewed in Hennig et al. 1994). Of particular interest is zinc because it is a critical component of biomembranes and is essential for proper membrane structure and function and the activity of numerous enzymes (Bettger and O'Dell 1981). Little is known about the requirements and functions of zinc in maintaining the integrity of the vasculature and particularly the vascular endothelium. A low zinc concentration in the plasma or vascular tissues may be involved in initiation of cell injury by possible potentiation of oxidative stress and related proinflammatory events. This may have important implications during the inflammatory response in the pathogenesis of atherosclerosis (Hennig et al. 1994), and during times of infections and other stressors when plasma zinc is depressed because of possible redistributions of body zinc pools. With a loss of zinc from the plasma during an inflammatory response, it is likely that the vascular endothelium may be deprived of this element as well. This may be sufficient to induce endothelial cell dysfunction.

Properties of Zinc as an Antioxidant and a Membrane Stabilizer

In addition to being an essential component of biomembranes (Bettger and O'Dell 1981), zinc also can participate in protein, nucleic acid, carbohydrate and lipid metabolism, as well as in the control of gene transcription and other fundamental biological processes (Vallee and Falchuk 1993). For example, isolated lysosome membranes were protected from oxidative injury in the presence of zinc (Ludwig and

*Supported in part by grants 1P01 HL36552 from the National Institutes of Health, the Veterans' Administration, the General Clinical Research Center (MO1 RR02602-08), and the Kentucky Agricultural Experiment Station.

[†]To whom correspondence and reprint requests should be addressed.

Correct citation: Hennig, B., Connell, P., Toborek, M., Barve, S., and McClain, C.J. 1997. Role of zinc in maintaining endothelial integrity. In *Trace Elements in Man and Animals – 9: Proceedings of the Ninth International Symposium on Trace Elements in Man and Animals. Edited by* P.W.F. Fischer, M.R. L'Abbé, K.A. Cockell, and R.S. Gibson. NRC Research Press, Ottawa, Canada. pp. 224–228.

Chvapil 1980, Pfeiffer and Cho 1980), and zinc deficiency resulted in oxidative damage to proteins, lipids and DNA in rat testes (Oteiza et al. 1995). Zinc is an essential component of copper zinc superoxide dismutase and is associated with metallothionin, a protein rich in thiolate groups (Sato and Bremner 1993, Vallee and Falchuk 1993). Zinc also can compete with copper and iron for membrane binding sites, thus reducing the potential for hydroxyl radical formation via redox cycling. These mechanisms may be involved in zinc-mediated inhibition of the oxidation of LDL by cells or iron (Wilkins and Leake 1994). In addition, zinc decreased reperfusion injury in isolated rat hearts, presumably by its ability to suppress significantly hydroxyl radical formation after regional ischemia. (Aiuto and Powell 1995). Therefore, in addition to its function as a membrane stabilizer, zinc may have a physiological role as an antioxidant by protecting sulfhydryl groups against oxidation and inhibiting the production of reactive oxygen by transition metals (Bray and Bettger 1990, Oteiza et al. 1995). Furthermore, dietary zinc deficiency was reported to decrease plasma concentrations of vitamin E (Bunk et al. 1989), suggesting that dietary zinc deficiency increases the nutritional requirement for vitamin E necessary to maintain adequate plasma concentrations.

Zinc and Endothelial Cell Integrity

Due to its unique properties, zinc may play a critical role in maintaining endothelial cell integrity. In our laboratory, we studied methods to determine zinc deficiency in endothelial cells. These methods included the culture of cells in low-serum medium (where serum was the only source of zinc), as well as in media previously exposed to different types of chelating agents (1,10-orthophenanthroline or chelex). All these techniques resulted in depletion of intracellular zinc levels and similar metabolic changes (Clair et al. 1995, Hennig et al. 1993, Hennig et al. 1992).

The fact that the vast majority of intra- and extracellular zinc is bound to proteins with varying degrees of affinity and that free zinc ion concentrations are usually very low can complicate zinc depletion and supplementation studies in cultured endothelial cells. Zinc levels used for supplementation studies (as well as deficiency studies), i.e., zinc transport or uptake and saturation into endothelial cells, will thus depend on the amount of serum in the culture media. We found that when zinc is depleted by culture in 1% serum-containing media, cellular zinc levels can be replenished significantly by supplementation with zinc acetate. Zinc acetate appears to be the best tolerated zinc preparation with the highest levels of absorption after oral intake in human studies (Prasad et al. 1993).

Using several methodologies of zinc depletion and supplementation, we have shown that zinc is vital to endothelial integrity and that zinc deficiency causes a severe impairment of endothelial barrier function (Clair et al. 1995, Hennig et al. 1992). In zinc deficient endothelial cells, barrier function was significantly decreased compared with controls. Media supplemented with physiological concentrations of zinc completely restored the cell integrity. Supplementation with calcium or magnesium, however did not restore this function. This suggests that zinc plays a unique role in maintaining normal endothelial integrity. Our data also indicate that in zinc deficiency, disruption of endothelial barrier function is related to a change in cell membrane characteristics secondary to altered cytosolic compositional changes. A redistribution of intracellular zinc may be sufficient to alter activity of membrane-bound enzymes. Indeed, we showed that the activity of the membrane-bound zinc-dependent angiotensin-converting enzyme (ACE) decreased in zinc deficient endothelial cell cultures (Hennig et al. 1992). Furthermore, supplementation with zinc completely restored ACE activity. The observed decrease in ACE activity during zinc deficiency may be due to a total endothelial cell zinc loss or to a significant imbalance in the intracellular exchange among subcellular zinc pools (Bobilya et al. 1991).

Zinc and Lipid or Cytokine-mediated Endothelial Cell Dysfunction

Among different factors, the vascular endothelium is exposed to high concentrations of triglyceride-rich lipoproteins. Hydrolysis of these lipoproteins, mediated by lipoprotein lipase, provides excessive local concentration of fatty acid anions which may cause endothelial injury (Zilversmit 1973). We have extensively studied the effects of fatty acids on endothelial barrier function (reviewed in Hennig et al. 1994) and found that among all fatty acids tested, linoleic acid (18:2n – 6) caused the most marked endothelial cell dysfunction.

To test the hypothesis that zinc may protect endothelial cells against fatty acid-induced injury, we exposed endothelial cells to selected fatty acids using media with and without supplemental zinc. Injury to endothelial cells induced by linoleic acid was prevented when culture media were supplemented with zinc but not with calcium or magnesium (Hennig et al. 1990). The mechanism of zinc protection is not clear and may be accounted for in part by its antioxidant property.

Plasma levels of cytokines are elevated during inflammation, infection or cell injury. In addition, inflammation is an integral part of the development of atherosclerosis, suggesting that inflammatory cytokines, in particular tumor necrosis factor-α (TNF), play a critical role in atherogenesis. TNF may act as a potent endothelial-activating factor, and the action of TNF, like those of other inflammatory cytokines (e.g., IL-1, IL-6, IL-8), may serve to promote coagulation and inflammation (Pober and Cotran 1990). In addition, endothelial cell integrity may be directly compromised by TNF. TNF can directly injure endothelial cells and initiate events which result in increased endothelial permeability (Goldblum et al. 1989).

Pathological conditions related to increased activity of TNF, such as inflammation or infection, may significantly influence zinc metabolism. Our data indicate that there is a depletion of cellular zinc in association with TNF-mediated endothelial cell injury which may lead to disruption of normal membrane integrity (Hennig et al. 1993). The nutritional status of the endothelium is likely to influence its response to TNF (Hennig et al. 1994, Klasing 1988), and a marginal status of protective nutrients (e.g., zinc) may increase the susceptibility of endothelial cells towards TNF-induced injury. In fact, we have evidence that a disruption of endothelial cell monolayer integrity by TNF can be prevented by preenriching cells with zinc (Hennig et al. 1993). However, the specific target cells for zinc and mechanisms of action are still uncertain. For example, in different cell systems zinc has been reported to both decrease or increase the expression of the adhesion molecule ICAM-1 (Gueniche et al. 1995, Martinotti et al. 1995). Zinc also has been shown to diminish the ability of human monocytes to be activated by lipopolysaccharide (Leibbrandt and Koropatnick 1994). Our data support the hypothesis that zinc may prevent TNF-induced endothelial cell dysfunction, at least in part, due to its antioxidant properties. Although this hypothesis requires further clarification, it is possible that supplementation with zinc can provide an adequate antioxidant protection and prevent oxidant (such as TNF)-mediated depletion of cellular antioxidants.

Even though detailed processes of lipid- and cytokine-mediated endothelial cell activation or dysfunction are only speculative at the present time, one of the mechanisms may involve the generation of reactive oxygen intermediates with subsequent activation of oxidative stress responsive transcription factors and genes. A transcription factor, which is critical in regulating the cytokine network and which has been implicated in many endothelial cell activation responses to injury and stress, is nuclear factor κβ (NF-κβ) (Gerritsen and Bloor 1993). Zinc may play a role in NF-κβ binding to DNA (Yang et al. 1995). Recently, we showed that certain lipids, such as linoleic acid, can activate Nf-κβ (Hennig et al. 1996). A significant activation of NF-κβ by linoleic acid was correlated with depletion of cellular glutathione. Modulation of NF-κβ by antioxidants may have a significant impact on the overall inflammatory cytokine response and endothelial cell dysfunction. We now have evidence that zinc deficiency can activate NF-κβ in endothelial cells, and that zinc supplementation attenuates TNF-mediated activation of NF-κβ (Connell et al. 1996). These data suggest that zinc-mediated inhibition of specific transcription factors could attenuate an inflammatory reaction associated with cell injury and thus preserve endothelial cell integrity.

Summary

The vascular endothelium plays an active role in the physiologic processes of vessel tone regulation, inflammatory responses, and vascular permeability. Endothelial cell dysfunction may disturb the normal communication of these cells with plasma components, blood-borne cells, and abluminal tissues. There is evidence that zinc is vital to endothelial integrity and that zinc can protect endothelial cells against lipid- or inflammatory cytokine-mediated insults. These observed protective properties may be due in part to the ability of zinc to act as an antioxidant and a membrane stabilizer. Most of all, zinc may have specific antiatherogenic properties by inhibiting oxidative stress-responsive transcription factors that are activated during an inflammatory response in atherosclerosis.

References

Aiuto, L.T. & Powell, S.R. (1995) Characterization of the antiarrhythmic effect of the trace element zinc and its potential relationship to inhibition of oxidative stress. J. Trace Elem. Exp. Med. 8: 173–182.

Bettger,W.J. & O'Dell, B.L. (1981) A critical physiological role of zinc in the structure and function of biomembranes. Life Sci. 28: 1425–1438.

Bobilya, D.J., Briske-Anderson, M., Johnson, L.K., Reeves, P.G. (1991) Zinc exchange by endothelial cells in culture. J. Nutr. Biochem. 2: 565–569.

Bray, T.M. & Bettger, W.J. (1990) The physiological role of zinc as an antioxidant. Free Rad.Biol. Med. 8: 281–291.

Bunk, M.J., Dnistrian, A.M., Schwartz, M.K. & Rivlin, R.S. (1989) Dietary zinc deficiency decreases plasma concentrations of vitamin E. Proc. Soc. Exp. Biol. Med. 190: 379–384.

Clair, J., Talwalkar, R., McClain, C.J. & Hennig, B. (1995) Selective removal of zinc from cell culture media. J. Trace Elem. Exp. Med. 7: 143–151.

Connell, P., Barve, S., Joshi-Barve, S., McClain, C.J. & Hennig, B. (1996) Zinc attenuates TNF-mediated activation of NF-κβ in endothelial cells. FASEB J. 10: A623.

Gerritsen, M.E. & Bloor, C.M. (1993) Endothelial cell gene expression in response to injury. FASEB J. 7: 523–532.

Goldblum, S.E., Hennig, B., Jay, M., Yoneda, K. & McClain, C.J. (1989) Tumor necrosis factor α-induced pulmonary vascular endothelial injury. Infect. Immun. 57: 1218–1226.

Gueniche, A., Viac, J., Lizard, G., Charveron, M. & Schmitt, D. (1995) Protective effect of zinc on keratinocyte activation markers induced by interferon or nickel. Acta Derm. Venereol. 75: 19–23.

Hennig, B., Diana, J.N., Toborek, M. & McClain, C.J. (1994) Influence of nutrients and cytokines on endothelial cell metabolism. J. Am. Coll. Nutr. 13: 224–231.

Hennig, B., McClain, C.J., Wang, Y., Vasudeva, N., Ramasamy, S. (1990) Zinc protects against linoleic acid-induced disruption of endothelial barrier function in culture. J. Am. Coll. Nutr. 9: 535.

Hennig, B., Toborek, M. & Cader, A.A. (1994) Nutrition, endothelial cell metabolism and atherosclerosis. Crit. Rev. Food Sci. Nutr. 34: 253–282.

Hennig, B., Toborek, M., Joshi-Barve, S., Barger, S.W., Barve, S., Mattson, M.P. & McClain, C.J. (1996) Linoleic acid activates nuclear transcription factor-κβ (NF-κβ) and induces NF-κβ-dependent transcription in cultured endothelial cells. Am. J. Clin. Nutr. 63: 322–328.

Hennig, B., Wang, Y., Ramasamy, S. & McClain, C.J. (1992) Zinc deficiency alters barrier function of cultured porcine endothelial cells. J. Nutr. 122: 1242–1247.

Hennig, B., Wang, Y., Ramasamy, S. & McClain, C.J. (1993) Zinc protects against tumor necrosis factor-induced disruption of porcine endothelial cell monolayer integrity. J. Nutr. 123: 1003–1009.

Hunte, G.C., Dubick, M.A., Keen, C.L. & Eskelson, C.D. (1991) Effects of hypertension on aortic antioxidant status in human abdominal aneurysmal and occlusive disease. Proc. Soc. Exp. Biol. Med. 196: 273–279.

Klasing, K.C. (1988) Nutritional aspects of leukocyte cytokines. J. Nutr. 118: 1436–1446.

Leibbrandt, M.E. & Koropatnick, J. (1994) Activation of human monocytes with lipopolysaccharide induces metallothionein expression and is diminished by zinc. Toxicol. Appl. Pharmacol. 124: 7281.

Ludwig, J.C. & Chvapil, M. (1980) Reversible stabilization of liver lysosomes by zinc ions. J. Nutr.110: 945–953.

Martinotti, S., Toniato, E., Colagrand, A., Alless, E., Alleva, C., Screpanti, 1. Morrone, S., Scarp, S., Frati, L. & Hayday, A.C. (1995) Heavy-metal modulation of human intercellular adhesion molecule (ICAM–1) gene expression. Biochim. Biophys. Acta 1261: 107–114.

McClain, C.J., Morris, P. & Hennig, B. (1995) Zinc and endothelial function. Nutrition 11: 117–120.

Oteiza, P.I., Olin, K.L., Fraga, C.G. & Keen, C.L. (1995) Zinc deficiency causes oxidative damage to proteins, lipids and DNA in rat testes. J. Nutr. 125: 823–829.

Pfeiffer, C.J. & Cho, C.H. (1980) Modulating effect of zinc on hepatic lysosomal fragility induced by surface-active agents. Res. Comm. Chem. Path. Pharmacol. 27: 587–598.

Pober, J.S. & Cotran, R.S. (1990) Cytokines and endothelial cell biology. Physiol. Rev. 70: 427–452.

Prasad, A.S., Beck, F.W.J. & Nowak, J. (1993) Comparison of absorption of five zinc preparations in humans using oral zinc tolerance test. J. Trace Elem. Exp. Med. 6: 109–115.

Ross, R. (1986) The pathogenesis of atherosclerosis — an update. N. Engl. J. Med. 314: 488–500.

Sato, M. & Bremner, I. (1993) Oxygen free radicals and metallothionein. Free Rad. Biol. Med. 14: 325–337.

Vallee, B.L. & Falchuk, K.H. (1993) The biochemical basis of zinc physiology. Physiol. Rev. 73: 79–118.

Wilkins, G.M. & Leake, D.S. (1994) The oxidation of low density lipoprotein by cells or iron is inhibited by zinc. FEBS Letters 341: 259–262.

Yang, J.P., Merin, J.P., Nakano, T., Kato, T., Kitade, Y., Okamoto, T. (1995) Inhibition of the DNA-binding activity of NF-κβ by gold compounds *in vitro*. FEBS Lett. 361: 89–96.

Zilversmit, D.B. (1973) A proposal linking atherogenesis to the interaction of endothelial lipoprotein lipase with triglyceride-rich lipoproteins. Circ. Res. 33: 633–638.

Discussion

Q1. John Chesters, Rowett Research Institute, Aberdeen, Scotland: What concentration of Zn was used in the cultures?

A. We usually use between 5-15 μM Zn.

Q2. John Chesters, Rowett Research Institute, Aberdeen, Scotland: That slightly concerns me, I must admit, because the best estimates of ionic Zn in cells and plasma are about 10^{-9} M, and the so-called physiological level of Zn of 15 μM is in the presence of complete serum. This makes an enormous difference to the amount of free Zn. Does this work represent physiological or pharmacological effects of Zn? Because you're using free Zn concentrations that are way above pharmacological levels and you have taken away 99% of the proteins that bind Zn. Its the free Zn that is going to have a lot of these effects.

A. We have observed the same results when both Zn and albumin were supplemented. This is a difficult issue to deal with experimentally. Carl Keen has evidence that inflammatory cytokines produced during the acute phase response can result in decreased plasma Zn. However, it is difficult to study endothelial cells in an intact system. Dr. Bunce has proposed that under housekeeping conditions Zn levels may be okay, however, when cells are under stress, it is possible to see major changes, for example, during sepsis when cytokines are produced.

Q3. George Bunce, Virginia Polytechnic Institute, Blacksburg, VA, USA: You're proposing that Zn might be protecting by antioxidant capabilities. Have you tested an antioxidant cocktail such as Vitamin E and C, or any other source, say of pharmacologic agents, that might mimic that effect?

A. We have done extensive work with Vitamin E and N-acetyl cysteine, a GSH precursor, and have observed protective effects greater than observed with Zn alone, especially with Vitamin E.

Q4. Maria Linder, California State University, Fullerton, CA, USA: Have you considered that some sort of a prostaglandin-mediated process may be involved to explain the effects with linoleic and linolenic acids? Did linolenic acid have the same kind of oxidative or proliferative effect?

A. We have focused on fatty acids and barrier dysfunction. Prostaglandins could be involved. Linoleic may be cytotoxic to these cells, as they have low levels of desaturases and elongases and so some of these fatty acids accumulate, and there is decreased arachidonic acid metabolism.

Q5. Xavier Alvarez-Hernandez, LSUMC, Shreveport, LA, USA: What does DCF measure and have you correlated it with other methods?

A. This method has been reported to measure primarily hydrogen peroxide and is able to quantitate lipoperoxides. We have confirmed with other measurements such as TBARS and lipid hydroperoxides. Of course you have to use more than one index for oxidative stress to confirm your data.

Q6. Ed Harris, Texas A&M University, College Station, TX, USA: Bernie, I want to go the other direction from Ed Bunce's question. Have you tried inducing free radical production, and see if you can use zinc to reduce or prevent the effects of that?

A. We have tried to use lipid hydroperoxides, but this is a very difficult system. We prefer to induce by adding lipids.

Q7. John Sorenson, University of Arkansas, Little Rock, AR, USA: A number of the agents that you used to prevent or restrict the translocation of albumin form complexes with Zn. Is it possible that protective effect of is due to such a complex that doesn't allow, say linolenic acid to cause its effect?

A. We used free fatty acid-albumin molar ratios of 0.15 to 4; these are physiological concentrations that have been reported in Type 1 diabetics, for example. The very exciting part is how Zn protects and the induction of transcription factors.

Q8. Paul Saltman, University of California at San Diego, CA, USA: There's a mechanism which we and others have shown, which you've not spoken about today. Zn can displace Cu from sites where Fenton chemistry takes place. Thus, in several systems at least, Zn is an antioxidant by moving out redox metals from sites where site-specific Fenton chemistry takes place, and damage is done. What does Cu do in your system? Can you provoke oxidative damage with Cu or Fe?

A. We have used Cu to induce oxidative damage to lipoproteins but not in this system.

Vanadium as a Modulator of Cellular Regulation — The Role of Redox Reactivity

ALLAN DAVISON, LINDA KOWALSKI, XUEFENG YIN, AND SIU-SING TSANG

Department of Chemistry, University of Northern BC, 3333 University Way, Prince George, British Columbia, Canada V2N 4Z9, and School of Kinesiology, Simon Fraser University, Burnaby, British Columbia, Canada V5A 1S6, BC Cancer Centre, Epidemiology and Environmental Carcinogenesis, Vancouver, British Columbia, Canada V5Z 1L3

Keywords vanadium, cell regulation, tyrosine kinases, oncogenes

Diversity of Biological Effects of Vanadium

Vanadium has a remarkable range of effects in biological systems (reviewed in Nechay 1988, and updated by us (Stern et al. 1993)). Let us briefly review these effects. In the mammalian organism it is a growth factor (Lau et al. 1988), a mitogen (Ramanadham 1983), and an anti-diabetic agent. Mitigating against possible clinical use as an antidiabetic agent, vanadium is genotoxic and a suspected co-carcinogen. It is a mitogen that stimulates DNA synthesis (Smith 1983), cell proliferation and differentiation in human fibroblasts and embryonic chicken bone cells (Wice 1987), Swiss mouse 3T3 cells, human breast cancer cells, lung cancer cells, and murine lymphocytes. At the cellular level, vanadium stimulates prostaglandin E2 release from rat Kupffer cells, and modulates ion transport by cell membranes. Many of these effects are observed at vanishingly small intracellular concentrations. It seems important to ask if the cellular actions of vanadium have biological relevance in free-living organisms, or if they represent merely pharmacological curiosities? From the concentration dependence of the actions (Table 1), cellular concentrations are below those required to produce responses *in vitro*. Since, however, cells do respond, one might infer that the cellular effects are induced by local high concentrations at or close to docking sites, or selective concentration in specific cell compartments. Even making allowances for this, effects such as NAD(P)H oxidation (Liochev and Fridovich 1990) are unlikely to have biological significance except under artificial conditions (Stankiewicz 1991).

What amplification mechanisms allow such dramatic actions at micromolar concentrations? The search for amplification mechanisms led directly to the cellular regulatory cascades (Ramasarma 1981), to actions on gene expression, and free radical generation through redox cycling. Cell regulatory effects include tyrosine esterification (Tracey and Gresser 1986), inhibition of tyrosine phosphatases (Feldman et al. 1990), and stimulation of phosphatidyl inositol turnover. It increases expression of proto-oncogenes c-myc, c-fps/fes, c-fos, and c-jun (Feldman et al. 1990, Yin et al. 1992). Consistent with these actions, vanadium acts synergistically with papillomaviruses in neoplastic transformation of cultured C3H 10T½ cells (Kowalski et al. 1992). How can a single metal have such widespread cellular actions? The answers must be sought in the range of chemical reactivities of vanadium.

Chemical Versatility of Vanadium Leads to a Variety of Biological Interactions

The variety of cellular actions of vanadium is matched by its enormous capacity for chemical speciation. In nature, vanadium exists in (III), (IV), and (V) valence states. These include cationic forms such as V^{3+}, oxycations like VO^{2+} or VO_2^+ , or oxyanions like VO_4^{3-}, each of which exhibits a striking capacity for redox reactivity, selective binding of biological ligands, and cellular docking sites extending to protein tyrosine residues, and DNA (Nechay 1986). The redox reactivity is likely integral to its cellular

Correct citation: Davison, A., Kowalski, L., Yin, X., and Tsang, S.-S. 1997. Vanadium as a modulator of cellular regulation — The role of redox reactivity. In *Trace Elements in Man and Animals – 9: Proceedings of the Ninth International Symposium on Trace Elements in Man and Animals. Edited by* P.W.F. Fischer, M.R. L'Abbé, K.A. Cockell, and R.S. Gibson. NRC Research Press, Ottawa, Canada. pp. 229–233.

Table 1. Dose/response profile for biological actions of vanadium compared with biological concentrations.

[V] M availability / effect	10^{-10}	10^{-9}	10^{-8}	10^{-7}	10^{-6}	10^{-5}	10^{-4}	10^{-3}
[V(IV)](free) intracellular	–	–						
[V(V)](free) in plasma		–						
[V](total) in serum			–					
[V](total) in muscle			–					
↓ Na^+/K^+ ATPase			–	–				
↑ [V(V)]chromatid exchange in human diet				–				
↑ adenylate cyclase					–			
↑ tyrosine phosphorylation					–			
↑ Na^+/H^+ exchange					–			
↑ cell proliferation					–			
↑ expression of ODC					–			
↑ neoplastic transformation					–			
↑ glucose uptake					–			
↑ promotes growth of rats					–	–		
↓ fibroblast proliferation					–	–		
↑ liver alkaline phosphatase					–	–		
↑ actin *c*-Ha-*ras* genes					–	–		
↓ Na^+/K^+ ATPase						–	–	
↓ actomyosin interaction							–	
↑ NAD(P)H oxidation								–
↓ STZ diabetes in rats (water)								–
↑ chromosomal aberrations								–
↓ oxidative phosphorylation								–
[V(III)] found in tunicates								→

actions, many of which are amplified dramatically by the presence of oxygen-derived active species. Vanadate resembles phosphate in electronic structure and properties, but unlike phosphate it readily forms esters non-enzymically (Tracey and Gresser 1986). Some of its cellular properties are likely mediated by esterification of tyrosine residues on tyrosine kinase and its cellular targets.

The oxidation state of vanadium is crucial to its ability to form organic esters. Most researchers compare the actions of vanadium(IV) and vanadium(V) and for many actions find vanadium(V) or vanadate slightly more effective. We would contend that such a view underestimates the capacity of vanadium for facile changes in redox state. It is likely that vanadate enters cells more readily than vanadyl, because the normal phosphate channels are accessible to it. At the same time, the redox state of vanadium in cultured cells or culture media depends more directly on availability of reductants or oxidants than on the chemical character — V(IV) or V(V) of the salt added to the medium. V(IV) is readily oxidized to V(V) by ambient oxygen, and intracellularly V(V) is readily reduced even by weak reductants like cytochrome c (Stern 1992). The presence of oxygen in the extracellular and intracellular media is likely to be a steady state reflecting local ligands, concentrations of oxygen, ascorbate, and other oxidants and reductants. Undoubtedly, the biological actions of vanadium depend directly on its state of oxidation. Nevertheless the distinctions often seen in the literature between V(IV) and V(V) are likely to be spurious if they are based on the oxidation state of the compound added, rather than that existing 15 or 20 min later.

In the presence of the ligands available in a living cell, vanadium is capable of existing in so broad a range of complexes that it is probably meaningless to suggest its structure, and the ambiguities of V(V) or V(IV) are preferable to more precise specifications like vanadate or V^{4+}. All of these complications are part of the fascinating diversity of actions and mechanisms available to vanadium in biological systems.

Vanadium and Neoplastic Transformation

Vanadium is a potent genotoxin, but it cannot be stated with any confidence either that vanadium causes human cancers, or that it does not. Certainly its chemical siblings arsenic and chromium are established carcinogens and vanadium is suspect. In bacterial assays, ammonium metavanadate (NH_4VO_3), vanadyl oxychloride ($VOCl_2$) and vanadium pentoxide (V_2O_5) are mutagenic. Moreover, vanadyl sulfate ($VOSO_4$), vanadium oxide (V_2O_3) and ammonium metavanadate induce sister chromatid exchange in mammalian cells (Shi et al. 1985). Vanadium is a tumor promoter (Jamieson 1988), and this is supported by its impact on ornithine decarboxylase gene expression (Davison et al. 1991). All this evidence of genotoxicity and mitogenicity does not comprise strong evidence for carcinogenicity. The genotoxicity evidence is, however, complemented by several studies of cellular neoplastic transformation, as follows.

Vanadate increases multilayered foci in NIH 3T3 cells transfected with DNA coding for tyrosine kinases (Shi et al. 1985). In our own studies, vanadium cooperates with DNA tumor viruses in inducing neoplastic transformation of several cell lines (Kowalski et al. 1992). Both V(IV) and V(V) increase the incidence of transformed foci produced by bovine papillomavirus DNA-transfected C3H 10T½ cells. In this respect it is an order of magnitude more effective than either arsenic or chromium. In cultured mouse embryo fibroblasts, BALB/3T3 cells, vanadium induces neoplastic transformation in a dose dependent manner. Diminishing cellular glutathione levels increased the induction of transformed foci by vanadium (Sabbioni et al. 1993). In the ensuing sections of this review, we turn our attention to the mechanisms of these potentially carcinogenic actions of vanadium, looking in turn at effects on oncogene expression, cell regulatory cascades, and redox cycling with concomitant free radical generation.

Vanadium and Oncogene Expression

Vanadium stimulates oncogene expression (Feldman 1990, Hanauske 1990, Itkes 1990). In our own studies, among 15 cellular genes studied in cultured mouse C127 cells, vanadium increased levels of mRNA for the actin, c-Ha *ras*, and c-*jun* genes. Vanadate at 10 µM increased the mRNA for actin and c-Ha *ras* genes to four times the control values. These increases represented de novo synthesis of mRNA, since actinomycin D inhibited. Vanadate did not increase mRNA corresponding to c-src, c-fos, c-myc, p53, HSP70, pODC or RP genes, and expression of c-erb A, c-erb B, c-sis and c-fes genes was undetectable whether vanadium was present or not (Yin 1991). Our failure to observe stimulation of c-fes or c-myc likely reflects variation among cell types.

Expression of c jun, was augmented by addition of a reductant or oxidant. Addition of NADH or H_2O_2 dramatically enhanced the effect of vanadate on c-jun gene expression. Catalase inhibited most of the effect of NADH or H_2O_2. At least one other example of synergism between H_2O_2 and vanadium in growth factor action, has been reported. In contrast, the effects of vanadate on levels of actin and c Ha *ras* mRNA were unaffected by oxidants, reductants, metal chelators, or anti-oxidant enzymes. The wish for a simple mechanism for the biological effects of vanadium is frustrated by evidence that vanadate acts by more than one mechanism of different categories of genes. Some actions of vanadium are not duplicated by insulin, suggesting that not all its actions are mediated by tyrosine phosphorylation at the receptor for insulin-like growth factors. Moreover, the hypothesis that the actions of vanadate on actin and c Ha ras were mediated by a protein kinase cascade was inconsistent with the following observations. Neither insulin nor epidermal growth factor increased mRNA levels of c Ha ras or actin gene. Neither genistein (a tyrosine kinase inhibitor) nor pretreatment with 12 O tetradecanoylphorbol-13 acetate blocked these actions of vanadate (Yin 1992).

Another class of explanations is rooted in the potential vanadium has for redox cycling with the concommitant generation of reactive oxygen species (Stankiewicz et al. 1991). Intracellular or extracellular pro-oxidants reportedly mediate many types of cancer through oncogene amplification, particularly at the jun-fos locus (Cerutti et al. 1990). The participation of oxidant species in c-jun gene activation suggests a mechanism for the actions of vanadate and other pro-oxidants in neoplastic transformation, most likely in the promotion phase. Yet a different set of actions may underlie other mitogenic actions. For example, vanadium activates and increases esterification of mitogen activated protein kinases (MAP kinases) and ribosomal protein kinases.

Biochemical Mechanisms Underlying the Actions of Vanadium

In the foregoing analysis, we have argued that vanadium has available to it, several plausible mechanisms of biochemical amplification involving both protein phosphorylation and free radical generation. There is evidence that some of its actions on glucose metabolism lie in its ability to modulate cellular regulatory cascades through the ability to esterify alcoholic (Ser, Thr) hydroxyl groups non-enzymically. Some of its growth factor actions, and perhaps its carcinogenic potential lie in its ability to esterify phenolic hydroxyls of tyrosine residues in regulatory proteins. The unfavorable thermodynamics of these mechanisms are overcome by the amplification mechanisms of autophosphorylation, and the flip-flop mechanisms that evolved to prevent futile cycles. These mechanisms may well be synergistic with oxidants like hydrogen peroxide that can oxidize intracellular vanadyl to vanadate. They are, however, presumably unaffected by free radical scavengers. Some of the actions of vanadium involve redox cycling where amplification lies in the fact that one vanadium ion can redox cycle many times through V(IV)/V(V) redox states and produce many free radicals. This would account for numerous reports of inhibition of the actions of vanadium by scavengers of reactive oxygen species, like catalase.

Speculating regarding unifying generalizations, one can say that many of the actions of vanadium on glucose metabolism can be accounted for on the basis of esterification of protein serine or threonine residues. Many of the growth factor-like actions resemble the consequences of esterification of protein tyrosines. Ester formation by vanadium requires that it be in the V(V) or vanadate form, and this explains why oxidants stimulate these reactions and reductants oppose them. Certain actions, including expression of some oncogenes are mediated by free radicals. These actions involve redox cycling of vanadium through V(IV) and V(V) states and should be stimulated by both reductants and oxidants. A full understanding of the actions of vanadium on cell regulation will require careful distinction of these three modes of action.

References

Cerutti, P., Peskin, A., Shah, G., & Amstad, P. (1990) Cancer and oxidative stress. Free Radic. Biol. Med. 9(S1):167.

Davison A., Stern A., Fatur D., & Tsang S-S (1991) Vanadate stimulates ornithine decarboxylase activity in C3H/10T½ cells. Biochem. Int. 24:461–466.

Feldman, R.A., Lowy, D.R., & Vass, W.C. (1990) Selective potentiation of c-fps/fes transforming activity by a phosphatase inhibitor. Oncogene Res. 5:187–197.

Glover, D.M., & Hames, B.D. (1989) Oncogenes. Oxford University Press, New York.

Hanauske, U., Hanauske, A.R., Marshall, M.H., Muggia, V.A., & Von Hoff, D.D. (1987) Biphasic effect of vanadium salts on *in vitro* tumor colony growth. Int. J. Cell. Cloning 5:170–178.

Itkes, A.V., Imamova, L.R., Alexandrova, N.M., Favorova, O.O., & Kisselev, L.L. (1990) Expression of c-myc gene in human ovary carcinoma cells treated with vanadate. Exp. Cell. Res. 188:169–171.

Jamieson, G.A., Jr., Etscheid, B.G., Muldoon, L.L., & Villereal, M.L. (1988) Effects of phorbol ester on mitogen and orthovanadate stimulated responses of cultured human fibroblasts. J. Cell. Physiol. 134:220–228.

Kowalski L., Tsang, S-S., Davison, A. (1992) Vanadate enhances transformation of bovine papillomavirus DNA-transfected C3H 10T½ cells. Cancer Lett. 64:83–90.

Lau, K.H.W., Tanimoto, H., & Baylink, D.J. (1988) Vanadate stimulates bone cell proliferation and bone collagen synthesis *in vitro*. Endocrinology 123:2858–2867.

Leavitt, J. and Bushar, G. (1982) Variations in expression of mutant β actin accompanying incremental increases in human fibroblast tumorigenicity. Cell 28:259-268.

Liochev, S.I., & Fridovich, I. (1990) Vanadate-stimulated oxidation of NAD(P)H in the presence of biological membranes and other sources of O2-. Arch. Biochem. Biophys. 279:1–7.

Lu, K., Levine, R.A., & Camposi, J. (1989) c-ras-Ha Gene expression is regulated by insulin or insulinlike growth factor and by epidermal growth factor in murine fibroblasts. Mol. Cell. Biol. 9:3411–3417.

Moon J, Davison AJ, Smith TJ and Fadl S. (1988) Correlation clusters in the accumulation of metals in human scalp hair: Effects of age, community of residence, and abundances of metals in air and water supplies Sci. Total Env. 72:87-112.

Nechay, B.R., Nannicga, L.B., Nechay, P.S.E., Post, R., Grantham, J.J., Macara, I.G., Kubena, L.F., Phillips, T.D., & Nielsen, F.H. (1986) Role of vanadium in biology. Fed. Proc. 45:123–132.

Ramanadham, M., & Kern, M. (1983) Differential effect of vanadate on DNA synthesis induced by mitogens in T and B lymphocytes. Mol. Cell. Biochem. 51:67–71.

Ramasarma, T., & Crane, F.L. (1981) Does vanadium play a role in cellular regulation? Curr. Top. Cell. Regul. 20:247–301.

Smith, J.B. (1983) Vanadium ions stimulate DNA synthesis in Swiss 3T3 and 3T6 cells. Proc. Natl. Acad. Sci. (USA) 80:6162–6166.

Stankiewicz P., Stern A., & Davison A. (1991) Oxidation of NADH by vanadium: Kinetics, effects of ligands, and roles of H_2O_2 and O_2 Archiv. Biochem. Biophys. 287:8–17.
Stern, A., Davison, A., Wu, Q., & Moon, J. (1992) Effects of ligands on reduction of oxygen by vanadium(III) and vanadium(IV) Arch. Biochem. Biophys. 299:154–163.
Stern, A., Yin, X., Tsang, S-S., Moon, J., & Davison A. (1993) Vanadium as modulator of cellular regulatory cascades and oncogene expression. Molec. Cell. Biochem. 71:103–112.
Tracey, A. S., & Gresser, M. J. (1986) Interaction of vanadate with phenol and tyrosine: Implications for the effects of vanadate on systems regulated by tyrosine phosphorylation. Proc. Nat. Acad. Sci. (USA) 83:609–613.
Viola, M.V.; Fromowitz, F.; Oravez, S.; Deb, S.; Finkel, G.; Lundy, J.; Hand, P.; Thor, A. and Schlom, J. (1986) Expression of ras oncogene p21 in prostate cancer. N. Engl. J. Med. 314:133-137.
Way, M. and Weeds, A. (1990) Actin-binding proteins. Cytoskeletal ups and downs. Nature 344:292-294.
Weinberg, R.A. (1984) ras Oncogenes and the molecular mechanisms of carcinogenesis. Blood 64:1143-1149.
Weitzman, S.A. and Gordon, L.I. (1990) Inflammation and cancer: Role of phagocyte-generated oxidants in carcinogenesis. Blood 76:655-663.
Wice, B., Milbrand, J., & Glaser, L. (1987) Control of muscle differentiation in BC3H1 cells by fibroblast growth factor and vanadate. J. Biol. Chem. 262:1810–1817.
Yin, X., Tsang, S-S., & Davison, A. (1992) Vanadate-induced gene expression in mouse C127 cells: Roles of oxygen derived active species. Molec. Cell. Biochem. 115:85–96.
Yin, X. (1991) Modulation of cellular gene expression by vanadium in cultured mouse fibroblasts: Possible role of active oxygen species. M.Sc. Thesis, Simon Fraser University. Burnaby, B.C., Canada.

Discussion

Q1. Katherine Thompson, University of British Columbia, Vancouver, BC, Canada: I think it's important to point out that inhalation of vanadium leads to vastly different effects than oral intake. In the studies with diabetic rats, the drinking water contained mM concentrations of vanadium, however, tissue concentrations were in the µM range or lower.

A. Agreed. Absorption is very low, approximately 1%.

Q2. Harry McArdle, Rowett Research Institute, Aberdeen, Scotland: If I recall from when I was a student, vanadium was first identified as being involved in physiology as it co-isolated with ATP from equine muscle. What was its role in equine muscle?

A. I don't know, but speculate that in the 1971 work, the procedure of isolation was probably a factor.

Q3. Roger Sunde, University of Missouri, Columbia, MO, USA: Your presentation is very interesting, and it shows how vanadium is involved in a series of nonenzymatic reactions that can interfere with phosphorylation and phosphatases etc. What do molluscs and plants that make vanadium-containing proteins do differently so that vanadium does not interfere with those processes?

A. One of the nice things about being at a conference like this is that there are often people who know more about these kinds of things than I do. Is there anyone here who knows more about tunicates? (comment by Paul Saltman, San Diego: I probably ought to be the last person to be telling you, even though I live near a marine biology institute. Tunicates represent the transition between invertebrates and vertebrates. Perhaps they have developed a nifty transport mechanism for vanadium.) (comment by Katherine Thompson, Vancouver: I can address that question a little further. Vanadium in these organisms is in the +3 oxidation state, and somewhat mimics the function of hemoglobin, at an earlier evolutionary stage. It seems that vanadium has intracellular calcium effects as well as effects on cytosolic protein kinases. There may be a number of other actions, because most intracellular vanadium is in the +4 state and is bound to glutathione.

Manganese Stimulates the Oxidative Burst of HL-60 Cells Through Protein Phosphorylation

K.R. SMITH AND S.S. PERCIVAL

Food Science and Human Nutrition Department, University of Florida, Gainesville, Florida 32611, USA

Keywords manganese, oxidative burst, HL-60 cells, phosphorylation

Although manganese (Mn) is classified as an essential trace element, its role in the immune system is unknown. Studies have shown that Mn is involved in macrophage function (Rabinovitch and Destefano 1973, Smialowicz et al. 1985). In 1994, we showed that Mn may also be involved in neutrophil function (Percival and Smith 1994). Results from this study demonstrated that incubating HL-60 cells with Mn during differentiation enhanced their ability to generate superoxide anions by two fold.

The generation of the superoxide anion, also known as the oxidative burst (OB), is an important microbicidal function of phagocytes. The enzyme responsible for the OB is the NADPH oxidase. In dormant cells, the NADPH oxidase is disassembled. Upon activation, the two cytosolic components, p47phox and p67phox, translocate to the plasma membrane to combine with the two membrane components (Clark 1990). In addition to translocation, p47phox must be phosphorylated for the NADPH oxidase to be activated (Nunoi et al. 1988). Our objective was to determine if the mechanism by which Mn was enhancing the OB in HL-60 cells involved phosphorylation of a component required for NADPH oxidase activation.

Materials and Methods

Cell Culture

HL-60 cells were induced to differentiate by continuous incubation with 1.0 μmol/L of retinoic acid (RA) for 96 h to obtain neutrophil-like cells. Cells received 2.0 μmol/L Mn at the same time as RA.

Cellular Concentration and Subcellular Distribution of Mn

RA cells were equilibrated at 37°C and then the cells were incubated with 37 kBq of ^{54}Mn for 6 h. The cells were washed and ruptured by nitrogen cavitation. Differential centrifugation was used to obtain the subcellular organelles. All pellets and cytosol were counted in a gamma counter with the results expressed as percent of cell associated ^{54}Mn. The Mn concentration in RA cells and their cytosol was determined using atomic absorption spectrophotometry.

Whole Cell Phosphorylation

Manganese treated and untreated RA cells were loaded with 37 MBq/mL of o-^{32}P and stimulated with 90 nmol/L of phorbol 12-myristate, 13-acetate (PMA), which activates the OB. The control cells were treated as above except that either 90 nmol/L of phorbol 12,13 didecanoate (PDD) or 0.1% of DMSO was added. The cells were ruptured by sonication and the cytosol was obtained. All samples were run on SDS-PAGE and the phosphorylated proteins were visualized by the Phosphor Imager system.

Correct citation: Smith, K.R., and Percival, S.S. 1997. Manganese stimulates the oxidative burst of HL-60 cells through protein phosphorylation. In *Trace Elements in Man and Animals – 9: Proceedings of the Ninth International Symposium on Trace Elements in Man and Animals. Edited by* P.W.F. Fischer, M.R. L'Abbé, K.A. Cockell, and R.S. Gibson. NRC Research Press, Ottawa, Canada. pp. 234–236.

Results

The subcellular distribution of ^{54}Mn in RA organelles was evenly distributed with 23% in the cytosol, 26% in the mitochondria, 26% in the nuclear/whole cells, and 11% in the microsomes (Figure 1). RA cytosol contained 39.3 ± 9.3 nmol/L. Phosphorylation studies revealed a 200% increase in phosphorylation of a 47 kD protein in the Mn treated cells as compared to the untreated cells when both treatments were stimulated with PMA (Figure 2).

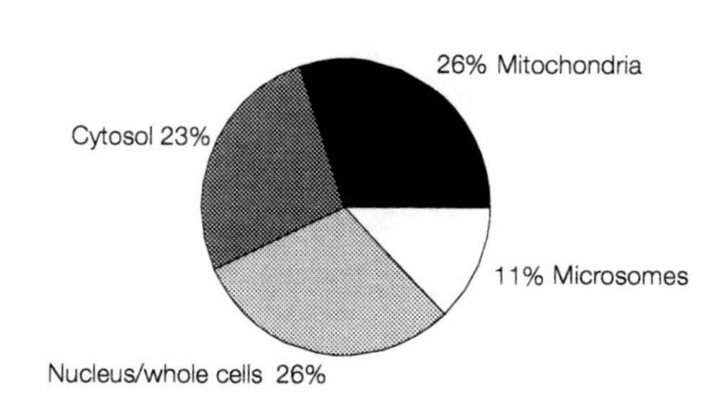

Figure 1.

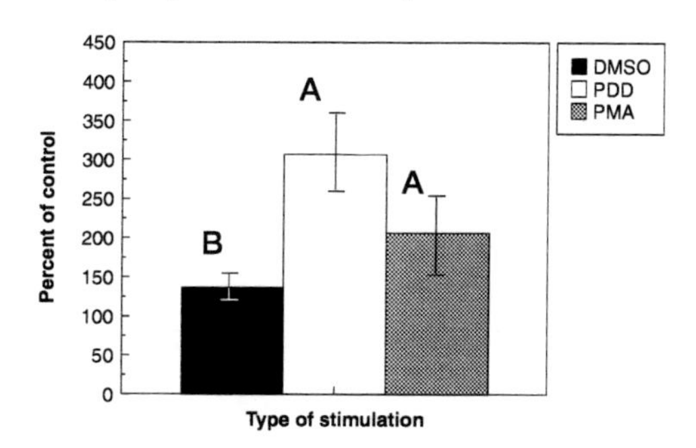

Figure 2.

Discussion

As we have previously published, morphological studies revealed that the two fold increase in the OB seen in Mn treated RA cells was not due to an increase in the proportion of differentiated cells (Figure 3). In addition, the results showed that Mn needed to be present during the entire differentiation process for the enhanced OB to occur. From these studies, we concluded that Mn may be required as a cofactor for a protein or enzyme synthesized during differentiation and required for OB activation. As a result, we hypothesized that Mn enhances phosphorylation of a component required for OB activation through a cofactor role. This hypothesis was based on studies that showed phosphorylation of p47phox was required for activation (Nunoi et al. 1988) and studies that identified a unique Mn-dependent kinase in HL-60 cells (Elias and Davis 1985). The results from the atomic absorption spectrophotometry studies showed that the cytosolic Mn concentration was two times serum physiological levels. Therefore, the cell has the ability to accumulate Mn in sufficient quantities. Also, Mn treatment resulted in an increase in phosphorylation of a 47 kD protein. From this, we concluded that the mechanism by which Mn enhances the OB is through phosphorylation of a protein that resembles p47phox.

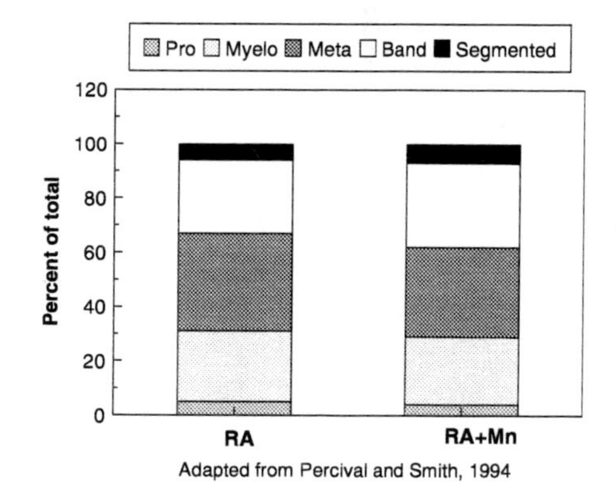

Figure 3.

References

Clark, R.A. (1990) The human neutrophil respiratory burst oxidase. J. Infect. Dis. 161: 1140–1147.

Elias, L. and Davis, A. (1985) Manganese-phospholipid-stimulated protein kinase activity of human leukemic cells. J. Biol. Chem. 260: 7023–7028.

Nunoi, H., Rotrosen, D., Gallin, J.I. and Malech, H.L. (1988) Two forms of autosomal Chronic Granulomatous Disease lack distinct neutrophil cytosol factors. Science 242: 1298–1301.

Percival, S.S. and Smith, K.R. (1994) Manganese stimulates the oxidative burst of differentiated HL-60 cells. Biol. Trace Elem. Res. 42: 243–252.

Rabinovitch, M. and Destefano, M.J. (1973) Macrophage spreading *in vitro*: II. Manganese and other metals as inducers or as co-factors for induced spreading. Exp. Cell Res. 79: 423–430.

Smialowicz, R.J., Luebke, R.W., Rogers, R.R., Riddle, R.M. and Rowe, D.G. (1985) Manganese chloride enhances natural cell-mediated immune effector cell function: Effects on macrophages. Immunopharm. 9:1–11.

Discussion

Q1. Roger Sunde, University of Missouri, Columbia, MO, USA: Is this the only protein phosphorylated, or is it just a unique one? Can you tell us just a little more about the specificity of what you've found?

A. This is the only one we found that changed. We have looked at other proteins that had been shown in the literature to have had increased phosphorylation with manganese treatment in other systems, and we didn't find any differences. In our system, only the p47 is changed.

Q2. John Sorenson, University of Arkansas, Little Rock, AR, USA: Is Mn involved in kinase activity that leads to the NADPH oxidase?

A. I'm not sure that we can definitely say that it's a kinase activity. We do know that there's increased phosphorylation. It may be involved in kinases or some phosphatases may be shut off.

Q3. Jim Kirkland, University of Guelph, ON, Canada: Did you find the same interaction of magnesium with the respiratory burst with both DMSO and retinoic acid treated HL-60 cells?

A. Magnesium did enhance the oxidative burst in RA-treated HL-60 cells. What we found with the initial kinetic studies was that manganese and magnesium seemed to be going through different mechanisms to get to this enhanced oxidative burst.

Q4. Jim Kirkland, University of Guelph, ON, Canada: Actually, I meant to ask you about Mn in my previous question.

A. I only looked at one differentiating agent, which was retinoic acid. DMSO was used in the phosphorylation studies as a control since the PMA is dissolved in DMSO. No other inducers of differentiation in HL-60 cells were used.

Q5. Jim Kirkland, University of Guelph, ON, Canada: Why does Mn have to be present so early for the oxidative burst which is a late event?

A. I speculate that Mn is incorporated into a protein synthesized early on in the differentiation process and that it may not be possible for Mn to bind later on.

Q6. John Sorenson, University of Arkansas, Little Rock, AR, USA: What is the HL-60 cell line?

A. It is a human leukocyte cell line.

Iron Inhibits Copper Uptake by Rat Hepatocytes by Down-Regulating the Plasma Membrane NADH Oxidase*

P. WHITAKER† AND H.J. MCARDLE†‡
†Department of Child Health, University of Dundee, DD1 9SY, and ‡Rowett Research Institute, Aberdeen AB21 9SB, UK

Keywords iron, copper, interactions, hepatocytes

It has been known for many years that the metabolism of copper and iron are interrelated. Bremner and Young (1981) suggested that the effect of increased iron may be due to increase excretion of copper from the liver. However, more recent work showed that animals on high iron diets had decreased copper absorption, associated with a decrease rather than an increase in biliary copper excretion (Yu et al. 1994a, b). In this paper we have tested the hypothesis that iron exerts its effect by inhibiting copper uptake into the hepatocyte.

Materials and Methods

Hepatocytes were isolated from male Hooded rats (250–400 g) by the method previously described (McArdle et al. 1988). Cells were loaded with iron by incubating overnight with iron saturated transferrin (FeTf). Copper uptake was measured from CuHis2 complexes as previously described (McArdle et al. 1988). Cell surface NADH oxidase activity was measured by adding 0.15 mmol/L NADH in BSS containing 1 mmol/L KCN to the washed cells. The rate of oxidation was calculated using an absorption co-efficient of 6.21×10^{-3} cm^{-1} mM^{-1}.

Results

Loading the cells with iron decreased the rate of Cu uptake by the hepatocytes, while levels of Cu associated with the cell surface increased. Increasing iron resulted in an increase in K_m, the apparent affinity for uptake (Figure 1a) ($p < 0.05$), and a minor decrease in V_{max} (Figure 1b). In contrast, iron caused an increase in B_{max} (Figure 1c), the maximum surface binding value for Cu without any change in the $K_{0.5}$ (data not shown). The data suggested that copper was binding to its transporter, but, in the presence of iron, a condition necessary for uptake was not being fulfilled. We hypothesised that this condition was the reduction of Cu(II) to Cu(I) by the plasma membrane NADH oxidase. Figure 2 shows that loading the cells with iron dramatically decreases NADH oxidase activity. The effect cannot be reversed by adding NADH, which indicates that there is no competition for the active site of the enzyme.

Discussion

The data presented in this paper give further support for the importance of the NADH oxidase enzyme in the transport of copper across the hepatocyte membrane. This hypothesis (van den Berg and McArdle 1994) proposed that oxidation of NADH provided the electron necessary for reduction of Cu(II) on CuHis2 to Cu(I), which resulted in the release of Cu from its complex and uptake by the hepatocyte. The fact that changes in NADH oxidase activity are mirrored by changes in copper uptake demonstrates strong support

*This work was supported by the Wellcome Trust and the European Social Fund.

Correct citation: Whitaker, P., and McArdle, H.J. 1997. Iron inhibits copper uptake by rat hepatocytes by down-regulating the plasma membrane NADH oxidase. In *Trace Elements in Man and Animals – 9: Proceedings of the Ninth International Symposium on Trace Elements in Man and Animals. Edited by* P.W.F. Fischer, M.R. L'Abbé, K.A. Cockell, and R.S. Gibson. NRC Research Press, Ottawa, Canada. pp. 237–239.

for the theory. The second, and more important, conclusion, suggests that copper uptake by the hepatocyte is regulated by iron levels within the cell. The mechanism whereby this regulation is achieved bears striking similarity to that demonstrated in yeast, where expression of oxidase enzymes has also been shown to be involved in regulation of metal uptake (Yuan et al. 1995).

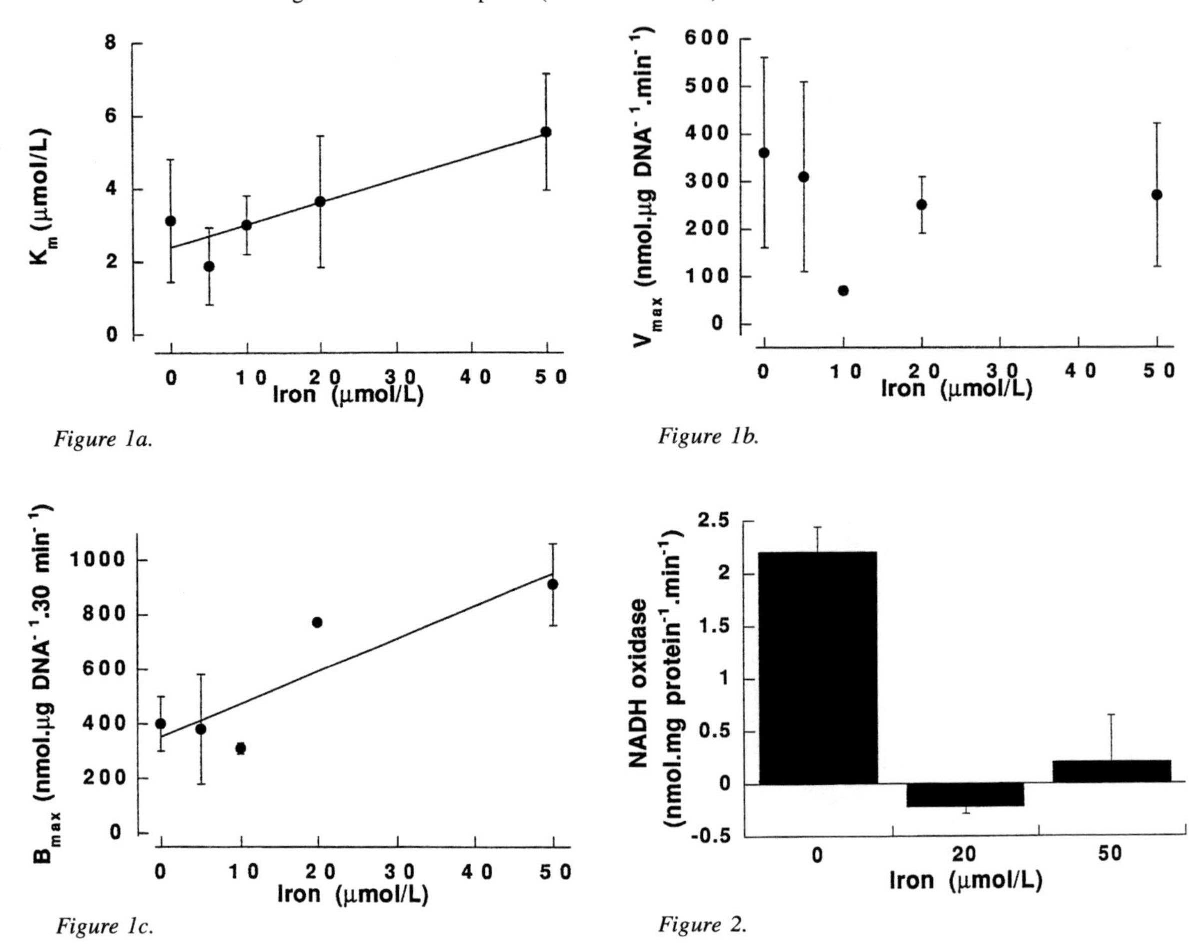

Figure 1a. *Figure 1b.*

Figure 1c. *Figure 2.*

References

Bremner, I. & Young, B.J. (1981) Effect of variation in dietary iron concentration on copper metabolism in rats. Proc. Nutr. Soc. 40: 69A.

McArdle, H.J., Gross, S.M. & Danks, D.M. (1988) Uptake of copper by mouse hepatocytes. J. Cell. Physiol. 136: 373–378.

van den Berg, G.J. & McArdle, H.J. (1994) A plasma membrane NADH oxidase is involved in copper uptake by plasma membrane vesicles isolated from rat liver. Biochim. Biophys. Acta 1195: 276–280.

Yu, S., Beems, R.B., Joles, J.A., Kaysen, G.A. & Beynen, A.C. (1994a) Copper and iron metabolism in analbuminemic rats fed a high iron diet. Comp. Biochem. Physiol. 110A: 131–138.

Yu, S., West, C.E. & Beynen, A.C. (1994b) Increasing intakes of iron reduce status, absorption and biliary excretion of copper in rats. Brit. J. Nutr. 71: 887–895.

Yuan, D.S., Stearman, R., Dancis, A., Dunn, T., Breeler, T. & Klausner, R.D. (1995) The Menkes/Wilson disease gene homologue in yeast provides copper to a Cp-like oxidase required for iron uptake. Proc. Natl. Acad. Sci. USA 92: 2632–2636.

Discussion

Q1. Ed Harris, Texas A&M University, College Station, TX, USA: You have some very interesting data there, and I think it's going to be very thought provoking, as we look a little more closely at uptake mechanisms, which we currently know very little about. You're making the assumption that you do have a certain level of Fe in the cell that is presumably at the site preventing the expression of the oxidase.

A. No, I don't. I think that what's happening is that the Fe is either up-regulating or down-regulating the enzyme itself. I don't know anybody who has cloned this enzyme unfortunately, so there's no way that I can check if there are changes in the message levels, or if the changes are in the actual protein. I think it's actually regulating it at the protein level. We have yet to do actinomycin experiments and so on to begin to address this. What I'm saying is that it may be regulating the amount of protein in the cell, but I have no information as to the mechanism by which it's doing that.

Q2. Ed Harris, Texas A&M University, College Station, TX, USA: There are many techniques for measuring Fe uptake into the cytosol. You have suggested that an oxidase is involved. Given that Cu^{2+} is stuck at the membrane you can saturate these sites with Cu^{2+}, how do you explain the stoichiometry of the reaction? NADH source has 2 electrons, are you reducing 2 Cu?

A. I beg to suggest that the stoichiometry is 1 Cu is reduced per NADH oxidized, so there's one electron missing. I would like to think that what's happening is that the transfer of Cu across membranes is electroneutral, and it's tempting to suggest that the other electron goes across as part of an electroneutral transfer process. When Cu+1 crosses the membrane, there is no evidence of any charge transfer that we know of, so what we can suggest is that the second electron is being provided. Now there's a second option that's been suggested by some people that I've talked to about this and that is that a lot of these enzymes can actually hold electrons. There are quite a lot of the NADH oxidases that will actually act as electron sinks. We are currently trying to demonstrate this directly using ESR. We are not looking at all at the forms of Fe in the cell. I have no idea which pool of iron is doing the regulation, except that it's very interesting that desferrioxamine can upregulate copper uptake. All I know is that if you increase the amount of iron inside the cell, you down-regulate copper uptake. This may be very relevant in pregnancy if Fe intake is increased and if this produces a Cu deficiency.

Q3. Maria Linder, California State University, Fullerton, CA, USA: Did you ever try to block the iron effect by having copper present at the same time? Do I read you correctly in saying that you only blocked about one-quarter of the copper uptake mechanism by this overnight incubation, and you don't get any more by adding more iron? Your data suggests that multiple Cu transport mechanisms are involved.

A. No, it doesn't. It suggests there are multiple ways of reducing Cu. So for example, you may have some ascorbate, and you may have other reducing systems.

Q4. John Sorenson, University of Arkansas, Little Rock, AR, USA: What form of iron was used?

A. Iron-transferrin was used.

GSH Levels Decrease During Fe Uptake by U937 Cells*

XAVIER ALVAREZ-HERNANDEZ, MARIA EUGENIA GOMEZ,† PAULA HUNT, KRISHNASWAMY KANNAN, AND JONATHAN GLASS

†IPN Mexico. LSUMC Center for Excellence in Cancer Research, Treatment and Education, Shreveport, Louisiana 71106, USA

Keywords iron, GSH, transport, calcein-AM

Iron (Fe) uptake and reduced glutathione (GSH) levels were measured in U937 cells using fluorescent probes. Calcein (Cal) was used as a fluorescence sensor for Fe (Breuer et al. 1995). Cal-fluorescence is stable, has a high quantum yield, is rapidly quenched when Fe is chelated and is minimally effected by variations in pH, ionic strength, Ca(II) and Mg(II). The affinity of calcein for Fe^{+2} in distilled water is 10^{14}. The use of a fluorescence sensor for metals has the advantages that cellular events can be studied in real time either by analysis in a spectrofluorimeter (millions of cells) or by fluorescence microscopy (individual cells subcellular compartmentalization). GSH is detected by the increase fluorescence of mono bromo bimane (mBrB) that is dependent of the activity of GSH-transferase. Untreated U937 cells, a human promonocyte cell line, and cells treated with MADE (a GSH-depleting agent) presented a rate of Fe incorporation of 43 vs. 11 pmol/min/10^6 cells and GSH concentration of 1.45 vs. 0.55 (mBrB fu/min/10^6 cells respectively. All experiments were terminated with the addition of the ionophore A23187 to equilibrate [Fe] across the membrane without cell lysis, and showed a rate of Fe uptake with A23187 as control of 73 vs. 177 (pmol/min/10^6 cells) for MADE treated cells. U937, HL60 and CaCo-2 cells showed a 60% diminished mBrB-GSH fluorescent product after 10 min of uptake of Fe. Both Cal and mBrB-GSH localized mainly in the cytoplasm and in vesicles resistant to digitonin lysis. The data suggest that thiols (mainly GSH) are involved in the handling of non-transferrin Fe uptake and GSH is partially depleted during this process. Non-transferrin Fe uptake may either require GSH for uptake or secondarily stresses the cell by generation of either oxygen species and/or GSH depletion.

References

Breuer, W., Epsztejn, S., Milgram, P., & Cabantchik, Z.I. (1995) Am. J. Physiol. 268: C1354-C1361.

Discussion

Q1. Ed Harris, Texas A&M University, College Station, TX, USA: I might have missed what you said regarding how the iron was presented. Did you use Fe-transferrin, free Fe or Fe-citrate?

A. We used Fe-transferrin for the confocal work.

Q2. Ed Harris, Texas A&M University, College Station, TX, USA: Would you have seen the same labelling of compartments if you had presented it as free Fe?

A. Yes, Fe-ascorbate was used in some experiments. Both forms decrease glutathione.

Q3. Bernhard Hennig, University of Kentucky, Lexington, KY, USA: What kinetics are involved in the decrease in glutathione during iron uptake? Is oxidative stress involved in the mechanism?

A. The kinetics are very fast, within minutes for a million cells. Not all cells can take up Fe^{+2}. The sequence lasts 11 seconds each.

*Supported by: Center for Excellence in Cancer Research, Treatment and Education, LSUMC.

Correct citation: Alvarez-Hernandez, X., Gomez, M.E., Hunt, P., Kannan, K., and Glass, J. 1997. GSH levels decrease during Fe uptake by U937 cells. In *Trace Elements in Man and Animals – 9: Proceedings of the Ninth International Symposium on Trace Elements in Man and Animals. Edited by* P.W.F. Fischer, M.R. L'Abbé, K.A. Cockell, and R.S. Gibson. NRC Research Press, Ottawa, Canada. pp. 240–241.

Q4. Harry McArdle, Rowett Research Institute, Aberdeen, Scotland: This is a nice method for measuring uptake. Have you tested whether the dye is altering the kinetics? Are those fluxes similar to those seen with radioactive Fe-transferrin?

A. Others have confirmed that the dye does not alter the kinetics. I like this method because it detects only what is internalized into the low molecular weight pool that is soluble in the cytosol, not what is membrane bound. I am concerned about Fe mobilized from LIP (low-molecular-weight iron pool) and label only with 20% of LIP. Eighty percent remains unlabelled.

Q5. Roger Sunde, University of Missouri, Columbia, MO, USA: Glutathione in those cells is going to be between 1 and 10 mM. What is the stoichiometry between the iron that has moved in and the quenching of glutathione?

A. We have not done the stoichiometry yet, but I expect that it will not be 1:1. It is very dynamic.

Q6. John Sorenson, University of Arkansas, Little Rock, AR, USA: Does quenching require coordination with the Fe?

A. Yes, every fluorochrome used for this kind of imaging is a chelator.

Q7. John Sorenson, University of Arkansas, Little Rock, AR, USA: Is an Fe-glutathione complex formed?

A. I don't know. There is a rate in the literature for the formation of such a complex.

Q8. John Sorenson, University of Arkansas, Little Rock, AR, USA: In the exchange process with the fluorochrome, would Fe-glutathione lose fluorescence? Is it an equilibrium between this complex and the casein?

A. We would have to look at kinetic and binding constants.

Copper Deficiency Increases Microvascular Dilation to Endotoxin*

DALE A. SCHUSCHKE, JACK T. SAARI,[†] AND FREDERICK N. MILLER

Center for Applied Microcirculatory Research, University of Louisville, Louisville, KY, and [†]USDA Human Nutrition Research Center, Grand Forks, ND, USA

Keywords copper deficiency, inflammation, prostacyclin, microcirculation

Dietary copper deficiency has long been associated with altered inflammatory responses in both humans and experimental animals. We have previously reported that there is an increased response to mast-cell mediated protein extravasation from post-capillary venules in copper-deficient rats (Schuschke et al. 1994) but a decrease in nitric oxide (NO)-mediated dilation (Schuschke et al. 1995). DiSilvestro et al. (1992) have reported that copper-deficient rats have a high mortality rate after a normally sublethal dose of endotoxin. In the current study, we have examined the microvascular reactivity to inflammatory stimuli with lipopolysaccharide (LPS) during dietary copper restriction.

Materials and Methods

Four weeks prior to experimentation, the animals were fed a purified diet which contained 9.3 μmol Cu/kg diet (copper-deficient diet; CuD) or was made copper-adequate (CuA) by the addition of 94 μmol Cu/kg diet. *In vivo* television microscopy was then used to observe and quantitate the reactivity of the rat cremaster muscle microcirculation.

The first protocol determined whether copper deficiency causes a difference in arteriolar reactivity in response to LPS stimulation. Arteriolar diameters were measured before and for 2 h after 2.5 mg/kg LPS was injected i.p. Maximum dilator capacity was then determined with 10^{-4} M papaverine.

The second protocol determined whether the increased dilation of the CuD group in protocol 1 was due to NO or prostaglandin mediation. Animals were pretreated with either L-NAME (2×10^{-4} M) or ibuprofen (9.6×10^{-5} M) 20 min prior to exposure to LPS.

The third protocol determined the sensitivity of third-order arterioles to the prostacyclin analog carbacyclin. Successive concentrations of carbacyclin were administered topically to the cremaster muscle. Arteriole diameters were measured for 10 min/concentration. After the *in vivo* experimentation, the median lobe of the liver was collected and analyzed for copper concentration.

Results

After 4 weeks on the diets hepatic copper concentrations were 175 ± 4.2 μmol/kg dry wt in the CuA group and 19 ± 2.0 in the CuD group. LPS caused a significantly greater dilation in the CuD group compared to the CuA controls (Figure 1) while maximal dilation to papaverine was not different. Initial mean arterial pressures were not different between groups (113 ± 7 in the CuD and 111 ± 3 in the CuA) but were significantly lower in the CuD group (82 ± 3 vs. 95 ± 4) after LPS.

L-NAME did not decrease the dilation to LPS in either diet group (Figure 1) while ibuprofen significantly depressed the dilator response in both groups (Figure 1).

The concentration-response curve to carbacyclin demonstrated a shift to the left in the CuD group (Figure 2). The EC_{50} is significantly lower in the CuD group (2.8 ± 0.18 nmol/L) compared to the CuA group (10.7 ± 1.8 nmol/L).

---

*Supported by USDA agreement 95-37200-1625.

Correct citation: Schuschke, D.A., Saari, J.T., and Miller, F.N. 1997. Copper deficiency increases microvascular dilation to endotoxin. In *Trace Elements in Man and Animals – 9: Proceedings of the Ninth International Symposium on Trace Elements in Man and Animals. Edited by* P.W.F. Fischer, M.R. L'Abbé, K.A. Cockell, and R.S. Gibson. NRC Research Press, Ottawa, Canada. pp. 242–243.

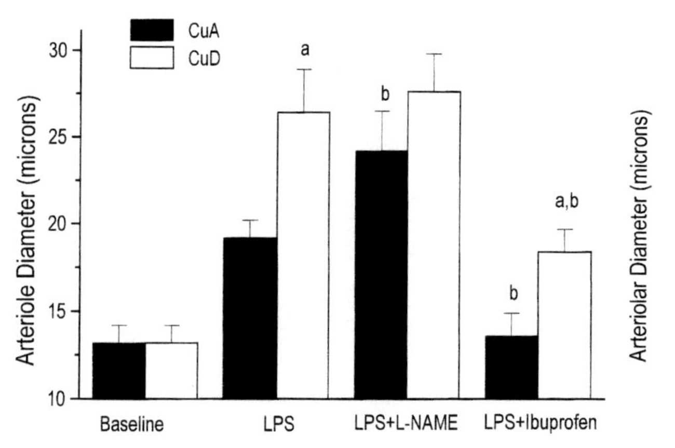

Figure 1. Comparison of arteriole dilation response to LPS in CuA and CuD rats. (a) $p < 0.05$ compared to CuA value. (b) $p < 0.05$ compared to LPS alone.

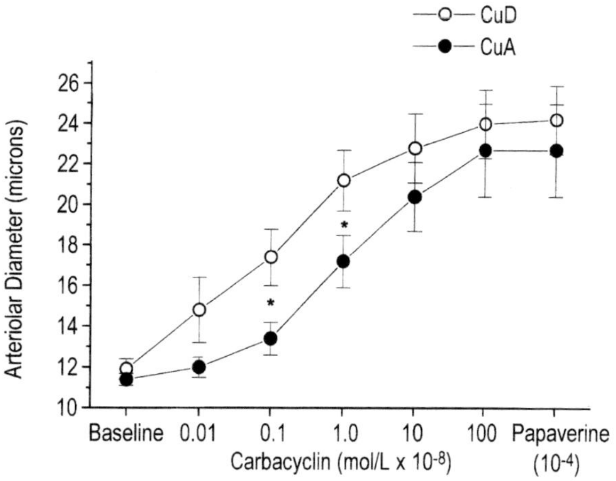

Figure 2. Comparison of concentration-response to carbacyclin in CuA and CuD rats. $p < 0.05$ compared to CuA value.*

Discussion

We have demonstrated an exaggerated arteriole dilator response to LPS in copper-deficient rats (Figure 1). Coincident with the dilation is a decreased mean arterial pressure suggesting a decrease in peripheral vascular resistance.

Blockade of the NO pathway with L-NAME did not inhibit this increased dilation in the CuD group which suggests that NO-cGMP signal transduction is not involved in the response. Significant inhibition of the dilation to LPS was achieved with the cyclooxygenase inhibitor ibuprofen in both diet groups (Figure 1). These results suggest a role for arachidonic acid metabolites such as PGI_2 which mediate vascular smooth muscle relaxation.

Dietary copper deficiency has been shown to depress production of PGI_2 in rat aorta (Nelson et al. 1992), however, our results show that copper-deficient rats have a lower threshold (Figure 2) and lower EC_{50} for the PGI_2 analog than CuA controls. These results suggest a greater sensitivity to PGI_2 and may indicate an upregulation of receptors on the vascular smooth muscle.

In conclusion, these results suggest that dietary copper deficiency increases the arteriolar dilation to LPS. The mechanism appears to involve a greater response to arachidonic acid metabolites and may result in part from a greater sensitivity to PGI_2.

Literature Cited

DiSilvestro, R.A., Joseph, E., & Yang, F.L. (1992) Copper deficiency impairs survival in endotoxin-treated rats. Nutr. Res. 12: 501–508.

Nelson, D.K., Huang, C-J., Mathias, M.M., & Allen, K.G.D. (1992) Copper-marginal and copper-deficient diets decrease aortic prostacyclin production and copper-dependent superoxide dismutase activity, and increase aortic lipid peroxidation in rats. J. Nutr. 122: 2101–2108.

Schuschke, D.A., Saari, J.T., & Miller, F.N. (1994) The role of the mast cell in acute inflammatory responses of copper deficient rats. Agents and Actions 42: 19–24.

Schuschke, D.A., Saari, J.T., & Miller, F.N. (1995) A role for dietary copper in nitric oxide-mediated vasodilation. Microcirculation 2: 371–376.

Influence of Alimentary Zinc Deficiency on the Concentration of the Second Messenger D-myo-inositol-1,4,5-trisphosphate (IP_3) in Testes, Brain and Blood Cells of Force-Fed Rats

H.-P. ROTH AND M. KIRCHGESSNER

Institute of Nutrition Physiology, Technical University of Munich, D-85350 Freising-Weihenstephan, Germany

Keywords Zn deficiency, phospholipase C, inositol-1,4,5-trisphosphate, force-feeding

Phospholipase C plays a key role in the transmission of various extracellular signals (hormones, growth factors, neurotransmitters) into the interior of the cell. The phosphatidyl-inositol-specific phospholipase C, which is presumably a Zn-metalloenzyme analogous to phospholipase C from Bacillus cereus (Ottolenghi 1965), catalyses the hydrolysis of phosphatidylinositol-4,5-bisphosphate to the second messenger inositol-1,4,5-trisphosphate (IP_3) and s,n-1,2-diacylglycerol (DAG).

The present study, therefore, set out to investigate whether alimentary zinc deficiency influences the *in vivo* activity of phosphatidylinositol-specific phospholipase, which was measured indirectly by radioimmunological determination of the IP_3 concentration because higher concentrations of DAG may also result from other metabolic processes. In order to prevent the reduced feed intake that occurs in Zn deficiency and the associated energy and protein depletion from interfering with the experimental parameters, all animals were force-fed by gastric tube.

Material and Methods

Twenty-four male Sprague-Dawley rats with an average live mass of 117 g were divided into 2 groups of 12 animals each. They received 4 mL of a semisynthetic diet pulp with casein as protein source administered four times daily (8:30, 12:30, 17:00 and 21:30) by gastric tube. The daily feed intake of the rats was 10.4 g DM/animal/d for both groups throughout the entire 12-d experimental period. The zinc content of the control diet was 115 mg Zn/kg DM and that of the zinc-deficient diet 1.6 mg Zn/kg DM. After 12 d all animals, having been fasted for 10 h, were anaesthetized with ether and decapitated. Whole blood (3 mL) was collected into EDTA-coated tubes and immediately diluted 1:3 with glucose-phosphate buffer (GPB). The blood diluted with GPB was carefully pipetted onto a Percoll density gradient and subsequently centrifuged in order to obtain platelets and monocytes. The remaining blood was transferred to Eppendorf flasks in order to extract the serum for the purpose of determining the zinc content and the alkaline phosphatase activity. Immediately after bleeding the carcases, both testes, the femur bones and the brain were removed and stored at –80°C. The Zn status of the rats was determined by measuring the alkaline phosphatase activity in the serum and the zinc content in serum and femur. The concentration of D-myo-inositol-1,4,5-trisphosphate (IP_3) in testes, brain, platelets and monocytes was determined by radioimmunological methods following extraction with perchloric acid using a test kit by Amersham Buchler, Braunschweig (IP_3 [^3H] assay system).

Correct citation: Roth, H.-P., and Kirchgessner, M. 1997. Influence of alimentary zinc deficiency on the concentration of the second messenger D-myo-inositol-1,4,5-trisphosphate (IP_3) in testes, brain and blood cells of force-fed rats. In *Trace Elements in Man and Animals – 9: Proceedings of the Ninth International Symposium on Trace Elements in Man and Animals. Edited by* P.W.F. Fischer, M.R. L'Abbé, K.A. Cockell, and R.S. Gibson. NRC Research Press, Ottawa, Canada. pp. 244–246.

Results

The Zn-deficient animals lagged behind the control animals in weight development from day 6 of the experiment, despite an identical food intake. While the control animals showed an average weight gain of 5.4 g/d, the Zn-depleted animals experienced a sudden, sharp growth depression from day 7 onwards, with an average weight gain of only 1.4 g/d.

The zinc status of the rats was measured with reference to the activity of alkaline phosphatase in the serum and the zinc content in serum and femur. The alkaline phosphatase activity in the depleted animals was reduced significantly compared to the control rats, namely by 28%. The zinc content in serum and femur was on average 74% and 43% lower than that of the control animals. These parameters of the rats' zinc status thus confirm a manifest zinc deficiency among the depleted animals.

While the testes of the zinc-depleted rats showed a significant reduction in the zinc content by 23% compared with the control animals, the zinc content in the brain did not differ between the depleted and the control group (Table 1). Likewise, the radioimmunologically determined concentration of IP_3 in the testes of the zinc-deficient rats was reduced by a significant 53% compared to the control animals, while alimentary zinc deficiency was found to have no effect on the concentration of IP_3 in the brain of the rats (Table 1). These results show that the effects of zinc depletion on the concentrations of IP_3 very depending on the sensitivity of the tissue to Zn deficiency.

The platelet fraction isolated from the whole blood of the rats exhibited a significant reduction of the IP_3 content by 55% in the Zn-deficient rats relative to the control animals (Table 2). The IP_3 content of the monocytes showed no difference between controls and depleted animals.

Table 1. Concentration of zinc and D-myo-inositol-1,4,5-trisphosphate (IP_3) in testes and brain of force-fed control and zinc-deficient rats.

Tissue	Control rats (115 ppm dietary Zn)	Zn-deficient rats (115 ppm dietary Zn)
Testes		
Zinc (μg/g DM)	141±15[a]	108±10[b]
IP_3 (nmol/g WW)	0.51±0.14[a]	0.24±0.08[b]
Brain		
Zinc (μg/g DM)	54±4	53±5
IP_3 (nmol/g WW)	1.60±0.43	1.67±0.36

Table 2. D-myo-inositol-1,4,5-trisphosphate (IP_3) concentration in platelets and monocytes (MNC) of force-fed control and Zn-deficient rats.

Zn supply	Control rats	Zn-deficient rats
Platelets (pmol/5 × 10^8)	18.8±3.1[a]	8.4±1.7[b]
MNC (pmol/10^6)	1.2±0.3	1.4±0.3

Discussion

These results show, that zinc plays a role *in vivo* in the release of IP_3 from phosphatidyl-inositol-4,5-bisphosphate following stimulation of the cells. This means that the reduction of the platelet and testicular IP_3 concentration observed in zinc deficiency has its biochemical cause in a direct zinc dependency of the phosphatidylinositol-specific phospholipase C. This would imply that phosphatidylinositol-specific phospholipase C in mammalian tissues, just like phospholipase C from Bacillus cereus, possesses catalytic or coactive zinc atoms which in zinc depletion become dissociated from the enzyme, thereby inactivating it or reducing its activity.

The results of the present study show that the effects of Zn depletion on IP_3 concentration vary depending on the sensitivity of the tissue to Zn deficiency. In the testes, an organ generally regarded as sensitive to Zn depletion, the result of Zn deficiency in growing rats is atrophy and reduced Zn levels as well as a significant reduction of the IP_3 content. A significantly lower IP_3 concentration was also observed in the platelets, which are also highly sensitive to Zn depletion. It may also be due to the fact that platelets have a relatively short livespan and hence a high turnover (O'Dell and Emery 1991). This explanation would also apply to the testes, since germ cells have a high turnover (Vallee and Falchuk 1993). The MNCs, whose IP_3 content was not affected by Zn deficiency in this study, have a lower turnover rate. The T-lymphocytes, which account for the majority of the studied monocytes, have a lifespan of a few months. The same is true for the brain, whose cells are in a more stationary, i.e., not actively dividing state, and whose IP_3 concentration was also unchanged in Zn-deficient rats here.

However, the biochemical cause of the reduced IP_3 concentration in the platelets and testes in Zn deficiency is likely to be a direct link between the zinc supply and the phosphatidylinositol-specific phospholipase C, which is presumably a Zn-metalloenzyme also in the mammalian organism.

Literature Cited

O'Dell, B.L. & Emery, M. (1991) Compromised zinc status in rats adversely affects calcium metabolism in platelets. J. Nutr. 121: 1763–1768.
Ottolenghi, A.C. (1965) Phospholipase C from Bacillus cereus, a zinc requiring metalloenzyme. Biochim. Biophys. Acta 106: 510–518.
Vallee, B.L. & Falchuk, K.H. (1993) The biochemical basis of zinc physiology. Physiol. Reviews 73: 79–118.

Analytical, Experimental and Isotopic Trace Element Techniques

Chair: R.W. Dabeka

Multi-Element and Isotope Ratio Determinations in Foods and Clinical Samples using Inductively Coupled Plasma-Mass Spectrometry*

H.M. CREWS,[†] M.J. BAXTER,[†] D.J. LEWIS,[†] W. HAVERMEISTER,[†]
S.J. FAIRWEATHER-TAIT,[‡] L.J. HARVEY,[‡] AND G. MAJSAK-NEWMAN[‡]
[†]CSL Food Science Laboratory, Norwich Research Park, Colney, Norwich NR4 7UQ, UK, and
[‡]Institute of Food Research, Norwich Research Park, Colney, Norwich NR4 7UA, UK

Keywords multi-element, copper, selenium, ICP-MS

Introduction

Inductively coupled plasma-mass spectrometry (ICP-MS) was recognised in the early 1980s as being a potentially powerful tool for the determination of both the total element concentration and the individual isotopes of elements present in samples. The technique provides rapid sample throughput with low limits of detection for multi-element analyses. In addition it can be coupled with chromatographic systems which can be used to study the chemical form or species of trace elements present in samples. This paper reports three examples of the use of ICP-MS. Thirty six elements were determined in food samples and the quality criteria used to judge the data are given. Copper (Cu) has been determined in blood serum samples using size exclusion chromatography (SEC) with ICP-MS and some of the problems encountered are described. Selenium (Se) isotope ratios have been measured for Se standards separated using ion exchange chromatography coupled to ICP-MS and a these values are discussed.

Multi-element Analysis of Food Samples

The food samples were part of the UK 1991 Total Diet Survey. Twenty food groups were collected from twenty towns in the UK. These groups are listed in Table 1 which also gives an indication of the most abundant dietary sources for some elements based on the results from this survey.

Homogenised food group samples were measured by ICP-MS following overnight digestion (6 h at 150°C) with nitric acid (5.0 mL for 0.5 g wet weight sample) in stainless steel pressure bombs with PTFE liners. Indium was used as internal standard and reagent blanks and certified reference materials were included in each batch. Additionally known amounts of the analytes were added (spiked) to replicate

*This work was supported by the Ministry of Agriculture, Fisheries and Food.

Correct citation: Crews, H.M., Baxter, M.J., Lewis, D.J., Havermeister, W., Fairweather-Tait, S.J., Harvey, L.J., and Majsak-Newman, G. 1997. Multi-element and isotope ratio determinations in foods and clinical samples using inductively coupled plasma-mass spectrometry. In *Trace Elements in Man and Animals – 9: Proceedings of the Ninth International Symposium on Trace Elements in Man and Animals. Edited by* P.W.F. Fischer, M.R. L'Abbé, K.A. Cockell, and R.S. Gibson. NRC Research Press, Ottawa, Canada. pp. 247–251.

Table 1. Element distribution among food groups showing those elements which can be present at levels 5 to 10-fold greater than the median value (elements in upper case are present at greater than 10-fold).

Food Group	Element
Bread	Manganese, Lanthanum, Neodymium
Miscellaneous Cereals	ALUMINIUM, Manganese, Mercury
Carcass meat	Zinc
Offals	Iron, Cobalt, COPPER, Zinc, SELENIUM, MOLYBDENUM, CADMIUM, MERCURY, Lead
Meat Products	MERCURY, SODIUM
Poultry	Selenium
Fish	ARSENIC, SELENIUM, MERCURY
Oils and Fats	
Eggs	Selenium
Sugars and Preserves	
Green Vegetables	
Other Vegetables	
Canned Vegetables	TIN
Fresh Fruit	
Fruit Products	TIN
Beverages	
Milk	
Dairy Products	TIN, Calcium
Nuts	BORON, MANGANESE, Cobalt, NICKEL, Copper, SELENIUM, MOLYBDENUM, SAMARIUM, EUROPIUM, MAGNESIUM
Potatoes	

samples of the reference materials and some of the unknown samples. Duplicate samples were analysed in separate batches to avoid any within batch bias. The following criteria were used to judge the quality of the data.

1. Calibration standards: linear responses should be obtained from calibration standards for all the elements with $r \geq 0.99$ for the isotopes of interest.
2. Instrument response: the response for a standard solution run as a sample during the latter half of a run should be within 10% of the original reading.
3. Certified reference materials: the data obtained for analytes with certified values should be within the certified range or within 10% of the certified value.
4. Recovery data: the recovery of added spikes should be within the range 80 to 120%.
5. Limits of detection (LOD, defined as 3 × standard deviation of the reagent blank expressed on sample weight basis by correcting for sample weight and dilution factors): the LOD should meet that required by the customer.

At the end of the survey it was found that criterion 1 was met by all analytical batches. Criterion 2 was met by 80% of batches but increasing the acceptable range to within 20% of the original reading encompassed 93% of the data. For the two reference materials used, there were a total of 28 certified and 10 reference values. The mean values ($n = 80$) for 22 of the certified values and 7 of the reference values were within tolerance, i.e., ±10%. The majority of the elements which gave less good agreement did so because the certified values were close to or below the LOD for that element. For the recovery data, values of 100% ± 25% were found for over 90% of the data, whilst 62% of the data gave recoveries of 100% ± 10%. The LODs were acceptable for all the elements measured.

In the final assessment of the data, common sense was used to judge the overall acceptability of the results. It proved impractical to expect all the data for all elements and matrices to meet all of the quality criteria in every batch. A batch would not be rejected if, for example, the recovery data did not meet the acceptance criteria for all elements. If, however, the values for reagent blanks were erratic and/or the reference data was poor the batch would be repeated. The criteria used were based on our experience of

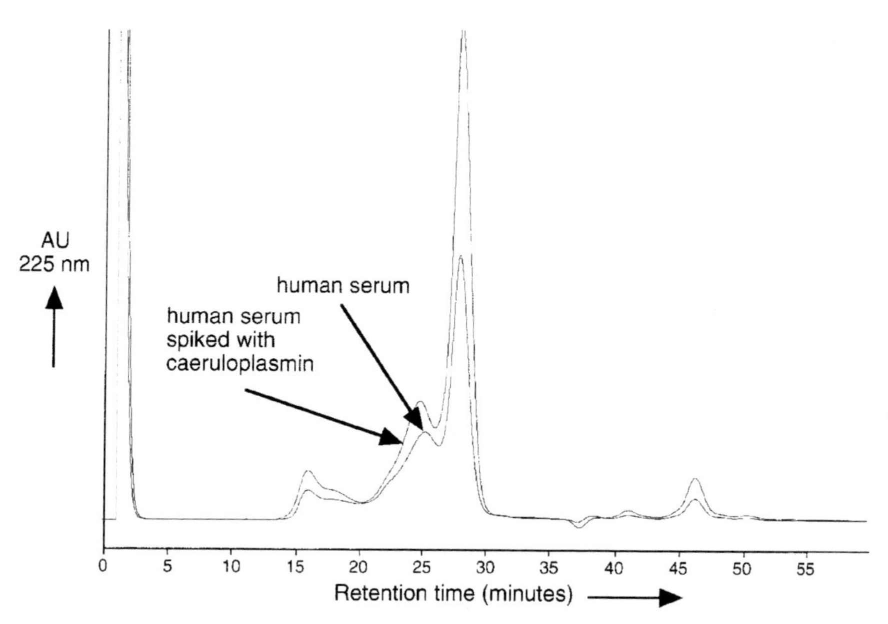

Figure 1. Chromatogram of human serum with and without added caeruloplasmin. See text for experimental conditions.

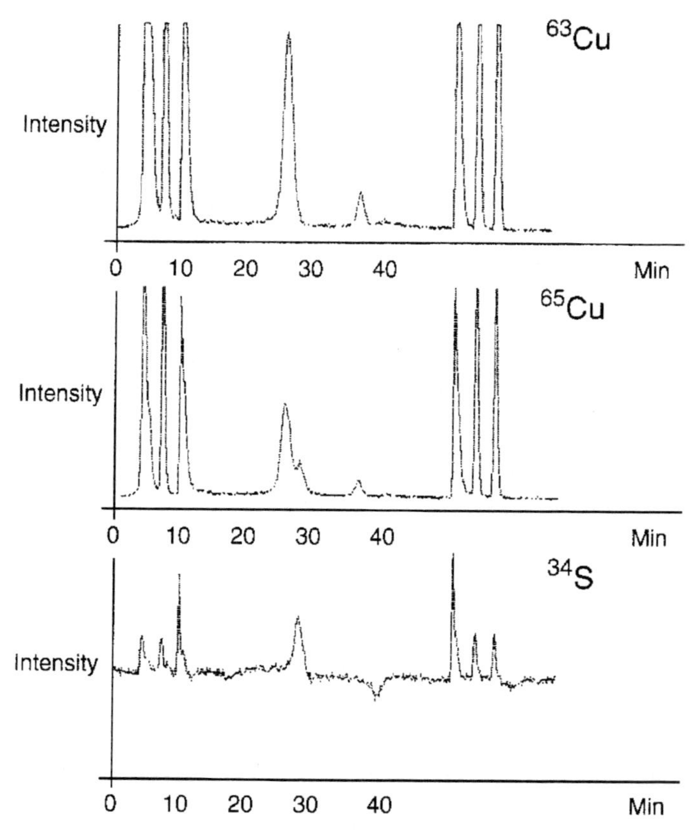

Figure 2. Copper and sulfur traces for human blood serum using SEC-ICP-MS. See text for explanation of peaks.

single element analyses. More work is required to ascertain which are the best criteria for judging multi-element data sets.

Determination of Copper in Blood Serum using SEC-ICP-MS

As part of a project investigating the metabolic and adaptive response to different dietary uptakes of copper we have established an SEC-ICP-MS method to look at the main copper-binding species in blood, i.e., caeruloplasmin. The chromatographic conditions were: Superdex 200/30 HR column with 0.1 M ammonium acetate + 1% EDTA at pH 7.0, flow rate 0.45 mL/min, with UV detection at 225 nm. Serum samples were diluted 1 + 1 with buffer prior to analysis. After injection onto the column the sample was directed firstly to the UV detector and then onto the argon plasma. Figure 1 shows the UV chromatogram of human serum and the same serum spiked with caeruloplasmin. Figure 2 shows the ICP-MS response for the human serum. The two isotopes of copper, ^{63}Cu and ^{65}Cu were monitored as well as the sulfur isotope, ^{34}S. The three peaks at the start and end of each ICP-MS trace are flow injected standards for accurate quantification of the sample peak. The large peak between 25 and 30 min is Cu associated with caeruloplasmin. However, it can be seen that at a retention time of approximately 30 min, there is a peak on both the ^{65}Cu and ^{34}S traces. This illustrates one of the problems of Cu analysis by ICP-MS; if there is S in the sample there is the possibility of an interference from this element, ($^{32}S^{33}S$), on ^{65}Cu. Therefore caution is needed when interpreting results. If more than one isotope is available for the element of interest it is worth monitoring them, at least during method development and validation stages.

Isotope Ratio Measurements Following IEX-ICP-MS Separation of Different Chemical Forms of Selenium

Four standards of selenium — selenomethionine, selenocystine, selenite and selenate — were separated using ion exchange chromatography (IEX) coupled to ICP-MS. The work is part of a project investigating the bioavailablity of Se from the diet and the role of different chemical forms of this element. A Polysphere ICAN-Z column was used, with 5 mM salicylate at pH 8.5, flow rate, 0.75 mL/min. Figure 3 shows the Se response (the same for all isotopes except ^{80}Se which cannot be monitored because of an overriding $^{40}Ar^{40}Ar$ interference from the argon plasma) against time for a 50 μL injection of either 10 or 25 ng/mL of each of the standards; i.e., for 25 ng/mL standard, 250 pg of each standard, in terms of Se, is injected onto the column. The standard species are well separated but under these conditions the percentage recovery of Se from the column was approximately 50% for each of the organo-Se species and

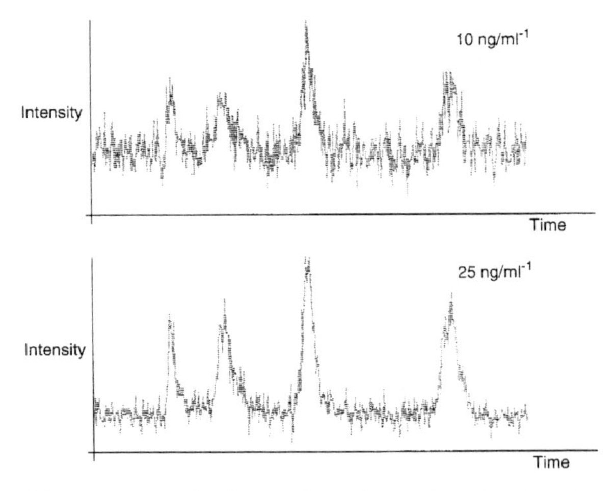

Figure 3. Chromatogram of selenium intensity against time for the separation of (from left to right in the figure) selenomethionine, selenocystine, selenite and selenate. The species were separated from two standard mixtures (10 and 25 ng/mL of each standard) using anion exchange chromatography-ICP-MS. See text for experimental conditions.

Table 2. Isotopic abundances for a selenocystine standard after IEX-ICP-MS.

Isotope	Calculated	Expected**	% Difference
74	0.85	0.96	–11
76	8.65	9.12	–5
77	7.27	7.50	–3
78	24.46	23.61	–3
82	8.84*	8.84*	–

* fixed at natural abundance, **De Bièvre and Barnes 1985.

Table 3. Isotope ratios for selenocystine after IEX-ICP-MS.

Ratio	Calculated	Expected*	% Difference
74/78	0.0348	0.041	–15
76/78	0.3536	0.386	–8
77/78	0.2972	0.318	–7
82/78	0.3610	0.374	–4

*De Bièvre et al. 1984.

100% for each of the inorganic species. It is apparent therefore that the chromatographic conditions require further optimisation to achieve good recoveries for all four species.

Tables 2 and 3 show the isotopic abundances for each isotope after IEX-ICP-MS of selenocystine and the resulting isotope ratios, and compare these data with published information (De Bièvre and Barnes 1985). The data are normalised against ^{82}Se but not corrected for mass bias.

Given the low abundance of some of the isotopes and the possibility of peak broadening during the chromatography, the results in Tables 2 and 3 are encouraging. Work is currently underway to improve the separations and to investigate the isolation of Se species from foodstuffs using both chemical extraction and *in vitro* enzyme procedures.

Conclusions

Inductively coupled plasma-mass spectrometry is a versatile technique which can be used for both total multi-element determinations and isotopic measurements. It is easily coupled to chromatography systems but care needs to be taken with the chromatographic conditions to achieve both good separations and good recoveries of the analyte from the columns. The interpretation of data, from both the determination of total analyte levels and from speciation studies, should be done with caution and with appropriate quality criteria.

References

De Bièvre, P., Gallet, M., Holden, N.E. & Barnes, I.L. (1984) Isotopic abundances and atomic weights of the elements. J. Phys. Chem. Ref. Data, 13: 809–891.

De Bièvre, P. & Barnes, I.L. (1985) Table of the isotopic composition of the elements as determined by mass spectrometry. Int. J. Mass Spec. Ion Proc. 65: 211–230.

Metal/Metalloid Speciation Analysis of Human Body Fluids by using Hydride Generation ICP-MS Methodology

J. FELDMANN,[†] E.B. WICKENHEISER,[‡] AND W.R. CULLEN[†]

[†]Department of Chemistry, University of British Columbia, Vancouver, British Columbia, Canada V6T 1Z1, and [‡]Augustana University College, Camrose, Alberta, Canada

Keywords hydride generation, GC-ICPMS, saliva, methylmercury

Introduction

The presence of different metals and metalloid species and their metabolites in the human body is mainly the results of food consumption and occupational exposure. Some excreted metabolites have been described (e.g., dimethylarsonic acid (DMA), methylarsonic acid (MAA) for As in urine, and Me_2Se in breath (Feldmann et al. 1996), PH_3 in fecal matter (Gassmann 1993). However, the contribution of implantations and dental fillings in body burdens of metal(loid) species in the human body are not well understood and there has been increased interest in metal(loid) speciation in body fluids in recent years (Das et al. 1996).

Analytical Method

Methylated or alkylated species from the following elements are able to form volatile hydrides (S, Ge, As, Se, Sn, Sb, Te, I, Hg, Tl, Pb, Bi). Consequently off-line hydride generation methodology has been coupled to gas chromatography inductively-coupled mass spectrometry (HG-GC-ICPMS) for the determination of low molecular weight metal(loid) species in body fluids. ICPMS was used as a multi-element and element-specific detection method for gas chromatography. The formed volatile species were generated by adding 0.8 mL $NaBH_4$ (6% (w/w) in deionized water) to the sample (0.1 mL to 1 mL urine or saliva diluted in 10 mL deionized water and 0.2 mL HCl (1 mol/L)) in a 50 mL glass flask. The volatile species were purged from the sample by He (80 mL/min) passed through a water trap (–78°C) and cryogenically focused (196°C) in a U-shaped tube, which was filled with SP-2100 on Supelcoport. All Teflon joints and the Teflon tubings were heated by using Nichrome wire to a temperature of 120°C to avoid condensation and adsorption. After trapping (6 min), the U-shape tube, which was wound with Nichrome as well, was heated to 200°C within 5 min by removing the liquid nitrogen Dewar and by applying an electric current. The separated species were transported into the argon plasma of the ICPMS through a quartz T-piece which mixed the gas flow with the regular nebulizer flow. The nebulizer was connected to a peristaltic pump, which transported a standard rhodium (Rh) solution to the ICPMS continuously. After treatment with $NaBH_4$ of a standard solution at pH 2 arsine, methylarsine, dimethylarsine, stibine, methylstibine, dimethylstibine, trimethylstibine, stannane, methylstannane, dimethylstannane, trimethylstannane, bismuthane, dimethylmercury, and methylmercury hydride would be generated from known precursors. A linear correlation ($R^2 = 0.9889$) between the boiling points (bp) of these volatile standards and the retention times (t) characterizes the separation. The equation ($bp = 1.598t - 125$) can be used in combination with the element-specific detection to identify the volatile derivatizated species. The detection limits were determined for Hg^0, Hg^{2+} (92 pg/mL), $MeHg^+$ (25 pg/mL), As(III/V) (34 pg/mL), MMA (39 pg/mL), DMA (33 pg/mL), in urine based on three times the standard deviation of the standard (500 pg/mL

Correct citation: Feldmann, J., Wickenheiser, E.B., and Cullen, W.R. 1997. Metal/metalloid speciation analysis of human body fluids by using hydride generation ICP-MS methodology. In *Trace Elements in Man and Animals – 9: Proceedings of the Ninth International Symposium on Trace Elements in Man and Animals. Edited by* P.W.F. Fischer, M.R. L'Abbé, K.A. Cockell, and R.S. Gibson. NRC Research Press, Ottawa, Canada. pp. 252–254.

expressed as As, and Hg, respectively). The reproducibility was for instance for Hg^0 within ±4.6%, but increased with decreasing concentration (to ± 30% for around 100 pg/mL). The major problem with the method was the high blank level. However, metal(loid) metabolites could be determined directly in urine, saliva, and skimmed milk without any clean-up or extraction procedures.

Is Dental Amalgam Methylated in the Mouth Environment?

The toxicity of dental amalgam is dependent on the release of mercury and of its metabolites. On the one hand oxidation of Hg^0 to Hg^{2+} decreases the toxicity, and on the other hand methylation of Hg^{2+} to $MeHg^+$ increases the toxicity. Methylation was observed during *in vitro* experiments with common oral bacteria (Heintze et al. 1983). Also methylmercury (MeHg) has been detected in saliva recently by Liang and Brooks (1995). However, MeHg could be transported from food (seafood is a good source) to the parotid fluid and occurs in the saliva; therefore it is necessary to monitor other metabolites to look for correlation. Thus, a study of the occurrence of Hg^{2+}, $MeHg^+$, As(III/V), MMA, DMA, TMA, $SeMe_2$ in urine and saliva has been conducted with respect to diet. Two volunteers (one amalgam bearer and one control) collected their first-morning urine and saliva before cleaning the teeth. The samples were stored immediately after sampling at –20°C. Quantification of the results of hydride generation was done by standard additions with 0.5 ng of MMA, DMA, $MeHg^+$, Hg^{2+}. In the saliva of the amalgam bearer a high amount of inorganic mercury (4.2 ng/mL) was detected. Only 10% was detected as elemental Hg^0 by purging the solution. No elemental mercury was detected in the saliva of the control and in the urine of both volunteers. $MeHg^+$ was detected in the urine and saliva of both test persons in amounts close to the detection limits. However in the urine, the amounts of MeHg and a non-identified Hg species were correlated to the amounts of DMA. DMA in urine is used as an indicator for seafood ingestion, which is also a source of MeHg in food. Therefore, the possibility that Hg^{2+} was methylated in the mouth or MeHg was introduced to the parotid fluid through food, cannot be distiguished with these limited data.

Summary

Boiling point separation combined with element-specific detection allows the identification of low molecular weight metabolites in human body fluids. The detection limits (pg/mL level) of HG-GC-ICPMS

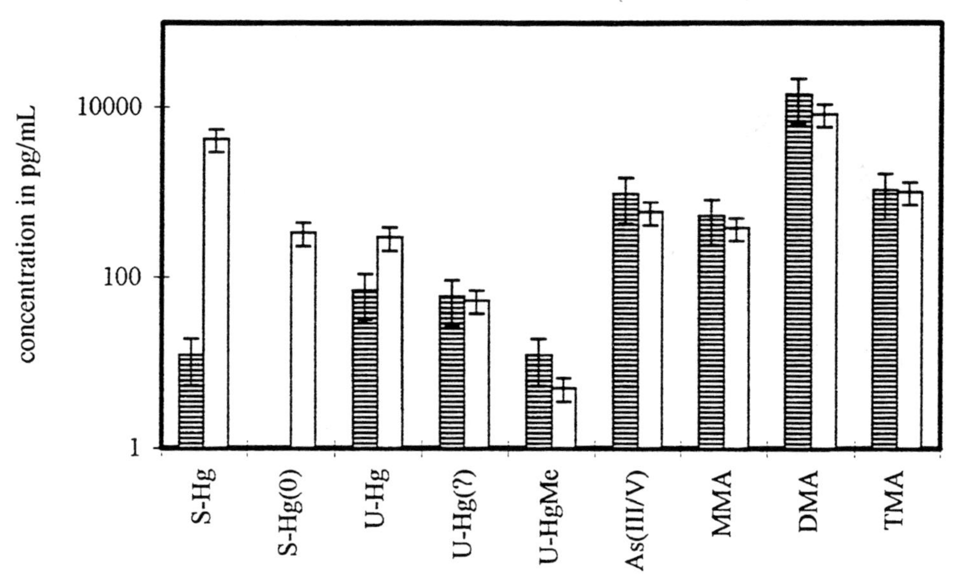

Figure 1. Concentrations of Hg^0, Hg^2, $HgMe^+$, unidentified Hg species in saliva (S-) and in urine (U-), and MMA, DMA, As(III/V) only in urine from amalgam bearer and control volunteer (dashed bars). The error bars represent the standard deviation of about 20 samples (amalgam bearer) and 10 control samples.

are low enough to determine MeHg, Hg^{2+}, Hg^0, MMA, DMA, Me^2Se, directly in saliva and in urine. Preliminary results for saliva and urine samples showed that Hgo is oxidized in saliva to Hg^{2+}. Furthermore, MeHg was detected in urine and saliva of the amalgam bearer and the control volunteer. However, regarding questions by using the data obtained to date the possibility of Hg methylation in the mouth environment cannot be answered.

References

Feldmann, J., Riechmann, T. & Hirner, A.V. (1996) Determination of organometallics in intra-oral air by LT-GC/ICPMS. Fresenius J. Anal. Chem. 354: 620–623.

Gassmann, G. (1993) Phosphane (PH_3) in the biosphere, Angew. Chem. 32: 761–763.

Das, A.K., Chakraborty, R., Cervera, M.L. & de la Guardia, M. (1996) Metal speciation in biological fluids — a review. Mikrochim. Acta 122: 209–246.

Heintze, U., Edwardsson, S., Dérand, T. & Birkhed, D. (1983) Methylation of mercury from dental amalgam and mercuric chloride by oral strepococci *in vitro*. Scandinavian J. Dent. Res. 91: 150–152.

Liang, I. & Brooks, R.J. (1995) Mercury reactions in the human mouth with dental amalgams. Water, Air, and Soil Pollution. 80: 103–107.

Discussion

Q1. Bob Dabeka, Health Canada, Ottawa, ON, Canada: How quickly did you heat up the cooling chamber and did you notice any losses?

A. With the high flow 80 mL/min, there is no loss in the water trap. If the flow is low, we get a loss, especially of elemental Hg as the BP is 320 degrees.

High Performance ICP-MS For Elemental Compositions, Isotope Ratios, and Speciation

C.B. DOUTHITT
Finnigan MAT, San Jose, CA 95134, USA

Keywords inductively coupled plasma mass spectrometry, ion trap, speciation, isotope ratio

The principal analytical challenges faced by the trace element chemist include (1) quantitation of trace and ultra trace elemental concentrations, (2) precise and accurate elemental ratios for normalization (e.g., U/Ca), (3) precise and accurate isotope ratios, at natural and enriched abundance, and (4) speciation by selective chromatographic detection. These analytical challenges have been addressed by successive generations of instruments, most recently by inductively coupled plasma mass spectrometry, ICP-MS. There are a wide variety of ICP-MS instruments, both commercialized and under development, which can be classified by the mass spectrometer that is used to separate the ions:

ICP-quadrupole MS	(nine manufacturers)
ICP-magnetic sector MS	(three manufacturers)
ICP-ion trap MS	(under development)
ICP-TOF MS	(under development)
ICP-FT MS	(being investigated)

Recent developments allow us to posit a new category, "High Performance ICP-MS" which cuts across analyzer boundaries to include both the ICP-magnetic sector MS with source at ground potential (Finnigan MAT model ELEMENT) and the ICP-ion trap MS (Finnigan MAT model PSQ). High performance ICP-MS is defined first and foremost by removal of interferences for accurate analysis, reduction or elimination of matrix effects, ultimate detection limits in the low ppt to ppq range, and accurate and precise isotopic and elemental ratios. High performance ICP-MS allows accurate analysis of complex matrices (seawater, blood) by dilution, permits methods development on novel matrices, and can be used for analysis of transient signals which result from chromatography, laser ablation, and flow injection techniques. The ELEMENT high performance ICP-MS is a superb general approach to virtually all problems of quantitative elemental analysis, high precision measurement of isotope ratios, and elemental chromatographic detection and quantification, and it offers significant performance advantages over other commercial magnetic sector ICP-MS and quadrupole ICP-MS which justify the increased cost of acquisition. The flat topped peaks and the rapid high voltage scan enable highly precise isotope ratio measurements, allowing full implementation of multi-element isotope dilution. An example of a problem that can realistically only be addressed with the ELEMENT would be the analysis of microliter samples of metallothioneins for ppb to ppt levels of transition metals and for S content. The PSQ (under development) allows the elimination of most interferences and space charge due to argon species, and is a novel, inexpensive and compact instrument for selective chromatographic detection, for analysis of natural waters for the RECRA suite of elements and in process control as an alternative to GFAA or cold plasma ICP-MS.

Discussion

Q1. Leslie Woodhouse, University of California at Berkeley, CA, USA: Have you measured calcium?

A. Oh sure, I just didn't show that because of the time constraints. We have data on virtually everything in the periodic table in most matrices. Ca is extremely difficult to determine with quadrupoles, but with high performance

Correct citation: Douthitt, C.B. 1997. High performance ICP-MS for elemental compositions, isotope ratios, and speciation. In *Trace Elements in Man and Animals – 9: Proceedings of the Ninth International Symposium on Trace Elements in Man and Animals. Edited by* P.W.F. Fischer, M.R. L'Abbé, K.A. Cockell, and R.S. Gibson. NRC Research Press, Ottawa, Canada. pp. 255–256.

systems we can measure it at very low PPT level in all matrices. The cost of the magnetic sector system is $400,000 US. We are developing an ion trap that will likely sell for $100,000 – $125,000 US.

Q2. Judith R. Turnlund, USDA-ARS, San Francisco, CA, USA: Could you address the relative ease of operation of the ICP magnetic sector versus quadrupole?

A. They are the same. The mass spectrometer is not what the operator has to learn. That's all computer controlled. It's learning the vagaries of ICP and sample preparation, but these are effectively identical for ICP-MS quadrupole and ICP-magnetic sector. So the learning curve is no steeper, and daily operation is no more difficult. They are both easier than thermal ionization.

Q3. Berislav Momcilovic, USDA-ARS, Grand Forks, ND, USA: Can you detect polonium?

A. Sure, we are working on it in sea water and we can detect 20 parts in 10^{16}.

Microwave Digestion of Food and Plant Reference Materials for Trace Multi-Element Analysis by ICP-MS with an Ultrasonic Nebulizer/Membrane Desolvator

S. WU, X. FENG, A. WITTMEIER, AND J. XU

Alberta Environmental Centre, P.O. Bag 4000, Vegreville, Alberta, Canada T9C 1T4

Keywords inductively coupled plasma mass spectrometry, microwave digestion, biological reference materials, ultrasonic nebulizer

Both the microwave sample preparation techniques and inductively coupled plasma mass spectrometry (ICP-MS) are now widely used in analytical laboratories, because of their well-known advantages. In this study, an ICP-MS system was used for trace multi-element analyses of food and plant reference materials after digestion by closed-vessel microwave heating. The digestion efficiencies with a low "high pressure" microwave system employing two digestion-acid systems: HNO_3 alone and HNO_3–HF–H_3BO_3, and the corresponding spectral and matrix interferences in the ICP-MS analyses were compared. The performance of a medium "high pressure" microwave system is also compared with the low "high pressure" system for the HNO_3 digestion. In addition, an ultrasonic nebulizer was tested to improve the detection limits.

Materials and Methods

A Perkin-Elmer Elan 5000 ICP-MS system was employed using similar operational conditions as described in Wu et al. 1996. The ultrasonic nebulizer/membrane desolvator used was a CETAC U-6000$^+$. The ICP-MS system with the pneumatic nebulizer was calibrated using undigested external standards prepared in the digestion reagent matrix, with In as the internal standard. When using the ultrasonic nebulizer, the system was calibrated using the method of standard additions.

The microwave system used was an upgraded QWAVE 1000 (Questron), which can be used at a low "high pressure" limit of 15.2 bar (220 psi) or a medium "high pressure" limit of 43.2 bar (625 psi). Sample portions of 0.5 g were digested. In the HNO_3 digestion, 5 or 2.5 mL of acid was employed under temperature/pressure limits of 165°C/13.8 bar or 200°C/41.5 bar, respectively. In the first stage of the HNO_3–HF–H_3BO_3 digestion, 5 mL of HNO_3 and 0.1 mL of HF were used under the temperature/pressure limits of 165°C/13.8 bar. In the second stage of the digestion, 6 mL of 5% H_3BO_3 was added and the vessel was heated under temperature/pressure limits of 100°C/13.8 bar. The dwell times for all digestions were 10–20 min.

The NIST Standard Reference Materials (SRM) 1575 (pine needles), 1515 (apple leaves), 1573 (tomato leaves), 1547 (peach leaves), 8433 (corn bran) and 8436 (wheat flour) were used. The latter two, distributed by NIST, were prepared and characterized by Agriculture Canada.

Results and Discussion

A portion of the ICP-MS results of the SRMs digested using the low "high pressure" microwave digestion system are presented in Tables 1 and 2, in which equations were used to correct the interferences of CaO/CaOH on ^{57}Fe, ^{59}Co and ^{60}Ni. A HF in the amount of 0.1 mL was used in preparing the standards for the results presented in the last three columns of Table 2.

Correct citation: Wu, S., Feng, X., Wittmeier, A., and Xu, J. 1997. Microwave digestion of food and plant reference materials for trace multi-element analysis by ICP-MS with an ultrasonic nebulizer/membrane desolvator. In *Trace Elements in Man and Animals – 9: Proceedings of the Ninth International Symposium on Trace Elements in Man and Animals. Edited by* P.W.F. Fischer, M.R. L'Abbé, K.A. Cockell, and R.S. Gibson. NRC Research Press, Ottawa, Canada. pp. 257–259.

Table 1. ICP-MS results (mg/g) for NIST grain reference materials by microwave digestion in HNO_3.

	8433 (Corn Bran)			8436 (Durum Wheat Flour)		
Element	Best Estimated	Found	n	Best Estimated	Found	n
As 75	0.002	0.002**	3	0.03*	0.015	4
B 11	2.8	3.45	3	–	0.72	4
Ba 137	2.4	2.35	3	2.11	2.29	3
Ca 43	420	485	3	278	260	4
Cd 114	0.012	0.012	3	0.11	0.119	3
Co 59	0.006	0.001**	3	0.008	0.001**	4
Cr 53	0.101	0.095	3	0.023	0.038	4
Cu 65	2.47	2.72	3	4.3	4.50	4
Fe 57	14.8	13.9	3	41.5	41.0	4
Mn 55	2.55	2.50	3	16	15.3	4
Mo 98	0.252	0.234	3	0.7	0.779	3
Ni 60	0.158	0.090	3	0.17	0.106	4
Pb 208	0.14	0.132	3	0.023	0.028	4
Se 77	0.045	0.061	3	1.23	1.39	4
Sr 86	4.62	4.88	3	1.19	1.21	4
V 51	0.005	0.006	3	0.021	0.021	4
Zn 66	18.6	19.0	3	22.2	23.7	3

* Estimated values.
** Less than detection limit.

Table 2. ICP-MS results (% Recoveries) for NIST SRM 1547 (peach leaves by microwave digestion in HNO_3 or HNO_3–HF–H_3BO_3).

		% Recoveries			
			HNO_3–HF–H_3BO_3		
Element	Certified Value (µg/g)	HNO_3	0.1 mL HF	0.3 mL HF	0.5 mL HF
Al 27	249	97.1	103	131	108
As	0.06	217	185	217	249
B 11	29	116	–	–	–
Ba 137	124	102	101	101	102
Ca 43	15600	109	94.3	101	107
Cd 114	0.026	105	95.6	87.8	86.2
Co 59	0.07*	79.2	129	389	915
Cr 53	1*	94.8	86.5	91.0	101
Cu 65	3.7	105	93.2	95.9	101
Mn 55	98	97.7	94.5	97.8	100
Mo 98	0.06	94.3	90.2	95.4	90.0
Ni 60	0.69	56.3	98.5	89.6	104
Pb 208	0.87	102	93.2	89.8	90.7
Se 77	0.12	425	262	269	322
Th 232	0.05*	36.6	118	114	113
U 238	0.015*	65.1	97.9	98.6	104
V 51	0.37	99.0	99.3	106	115
Zn 66	17.9	108	97.8	99.9	103

*Noncertified values.

The ICP-MS recoveries were within 85–115% for most elements determined, i.e., Al, Ba, Ca, Cd, Cr, Cu, Fe, Mn, Pb, Sr, V and Zn, in the six SRMs digested in HNO_3 with a low "high pressure" microwave system. The recoveries of As and Se varied greatly from 100–106% in NIST SRM 1575 to above 200% in NIST SRM 1547, depending upon the degree of interferences from ArCl, CaCl or S_2Cl which were difficult to eliminate. The recoveries were low for the Ni, Th and U and were not significantly improved by the microwave digestion in HNO_3 under the medium "high pressure" which, however, reduced the interference of ArC on ^{52}Cr. It appeared that Ni, Th and U were partially bound to silicon, and thus the two-stage HNO_3–HF–H_3BO_3 digestion recovered these elements successfully (Table 2). By forming tetra-fluoroboric acid, the H_3BO_3 used in the second-stage of digestion masked the free HF ions in the solution and dissolved the fluoride precipitate, and thus prevented the co-precipitation of analytes and the clogging of the ICP-MS sampling cone. However, this digestion system prevents the determination of B and causes additional isobaric polyatomic ion interferences from BO, ArB, ArF etc., which, although correctable through blank subtraction, resulted in poor detection limits for ^{27}Al, ^{51}V and ^{59}Co. A positive escalation of the ArF interference on low levels of Co is shown in Table 2.

Using an ultrasonic nebulizer/membrane desolvator, the instrumental detection limits of most elements were 5–50 fold lower, but the B signal was completely lost and the Cu signal became unstable. While the interferences from oxide and hydroxide were reduced significantly requiring no correction for ^{57}Fe, ^{59}Co and ^{60}Ni analyses, the matrix effects were so significant that the method of standard additions had to be used to obtain accurate results. By adding a small amount of polyvinyl alcohol to sample solutions as a physical modifier to carry the desolvated particles to the ICP, the matrix effects were reduced and the calibration by external standard was possible. Further study into the use of the ultrasonic nebulizer with the ICP-MS system for the analyses of food and plant materials is required.

Literature Cited

Wu. S., Zhao, Y., Feng, X. and Wittmeier A. (1996). Application of inductively coupled plasma mass spectrometry for total metal determination in silicon-containing solid samples using the microwave-assisted nitric acid – hydrofluoric acid – hydrogen peroxide – boric acid digestion system, J. Anal. At. Spectrom., 11: 287–296.

Discussion

Q1. Qianli Xie, University of Saskatchewan, Saskatoon, SK, Canada: a) What are the advantages of microwave digestion compared to regular digestion? b) The major problem you have with the ultrasonic nebulizer is the matrix effect. Is that because of the ultrasonic nebulizer or it is because of the matrix?

A. a) You do not need to use as much acid, it takes less time, there is less chance of contamination. You don't need to use the more explosive perchloric acid with plant materials. b) As far as the nebulizer goes, we get good recovery probably because we use polyvinyl alcohol as a kind of a physical modifier which acts as a carrier where the amount of water is reduced and the dissolvate is carried into the mass spectrophotometer. There are still some problems with it.

Q2. Geoffrey Judson, Central Veterinary Laboratory, Glenside, Southern Australia: Can you use the microwave digestion with high fat foods such as meat?

A. We haven't analyzed high fat foods yet, but we have looked at crude oils at 30% where we need to use the medium high pressure of the high pressure microwave. High fat foods can likely be handled by pre-digestion.

Intratumoral Carboplatin Kinetics in Patients with Malignant Melanoma: Application of Microdialysis Sampling

V. MEISINGER,[†] B. BLÖCHL-DAUM,[‡] M. MÜLLER,[‡] J. JÄGER,[†] A. FASSOLT,[‡] H.G. EICHLER,[‡] AND H. PERHAMBERGER[¶]

[†]Division of Occupational Medicine, [‡]Department of Clinical Pharmacology, and [¶]Division of General Dermatology, Vienna University Hospital, Vienna, Austria

Keywords carboplatin, microdialysis sampling, melanoma

Introduction

Generally speaking, the anti-tumour efficacy of anti-cancer drugs is a function of dose intensity, i.e., the concentration-time profile in tumour tissue. However, this is not the case in a clinical setting, when solid tumours are treated with cytotoxic drugs; the total dose administered per time period is not correlated with the dose intensity in tumour cell (Jane 1994, Eskey et al. 1992). This case may partly be due to interindividual variability of the pharmacokinetics of cytotoxic drugs and may yield highly different plasma concentrations in individual patients (De Conti et al. 1973). Besides, it may be speculated that dose intensity may be varied for different tumour lesions in the same patient according to local differences in perfusion (Coughlin et al. 1994).

Information on drug concentration profiles is of critical importance but appropiate methods for measurement are lacking. Recently, the microdialysis technique has been decribed for *in vivo* measurement of drug concentrations in the extracellular fluid (ECF) space in human tissues (Lönnroth et al. 1991, Stahle et al. 1991, Scheyer et al. 1994, Müller et al. 1995a).The purpose of this study was to obtain concentration-time profiles in the ECF of solid tumours of a model anti-cancer drug, carboplatin, by application of microdialysis sampling, and thereby to assess the scope of microdialysis for tumour pharmacokinetic studies in man.

Study Drug — Carboplatin

Carboplatin is the first of the "second generation" platinum compounds to gain widespread use in oncological practice. Carboplatin dissolved in water shows no change in its molar concentration over a period of time, however, when dissolved in dilute saline carboplatin will show an increase in molar conductance over a period of time. Therefore it is likely that the mechanism of action is similar to that of the cisplatin. At the highest dose of cisplatin there were toxic related deaths, but none occured at the highest dose of carboplatin which was also active against the tumour. Carboplatin is an effective anti-cancer drug that is mainly effective in the solid tumours of testes and ovary. There was no effect on CNS function, blood pressure or heart rate. The most important thing, there have been no recorded cases of acute overdosage. Therefore carboplatin was chosen as a model anti-cancer drug for this study. Carboplatin (Paraplatin, $C_6H_{12}N_2O_4Pt$, Bristol-Myers Squibb, Mayaguez, Puerto Rico) was administered intravenously as a single agent dose of 400 mg m^{-2}; infusion time was 20 min.

Correct citation: Meisinger, V., Blöchl-Daum, B., Müller, M., Jäger, J., Fassolt, A., Eichler, H.G., and Perhamberger, H. 1997. Intratumoral carboplatin kinetics in patients with malignant melanoma: application of microdialysis sampling. In *Trace Elements in Man and Animals – 9: Proceedings of the Ninth International Symposium on Trace Elements in Man and Animals. Edited by* P.W.F. Fischer, M.R. L'Abbé, K.A. Cockell, and R.S. Gibson. NRC Research Press, Ottawa, Canada. pp. 260–264.

Principles and Procedure of Microdialysis

The principles of microdialysis have been described in detail previously (Ungerstedt 1991, Lönnroth et al. 1987, Morrision et al. 1991, Müller et al. 1995a). Briefly, microdialysis is based on sampling of analytes from the extracellular space by diffusion through a semipermeable membrane. Our study process is accomplished *in vivo* by using a microdialysis probe (CMA 10, CMA, Stockholm, Sweden, with a molecular cut-off of 20 kDa, an outer diameter of 500 μm and a membrane length of 16 mm). The dialysis probes were implanted into malignant lesion and into nearby subcutaneous tissue. The probes were perfused by means of a precision infusion pump (Precidor, Ilfors-AG, Basle, Switzerland) at a constant flow-rate of 1.5 μL/min. Ringer's solution was used as the perfusion fluid; substances are filtered by diffusion from the extracellular fluid into the perfusion medium. Perfusate samples for measurement of drug levels were collected by using microfraction collection (CMA 120, AMA, Stockholm, Sweden). All samples were collected and stored at –20°C until analysed.

Patients and Methods

The study protocol was approved by ethics committee of the Vienna University Hospital. Six platinum naive patients (four female, two male, mean age 58 ± 3.6 years, WHO performance status <2) with metastatic malignant melanoma were included. All patients had cutaneous malignant melanoma metastases at the extremities or body trunk which were accessible to the microdialysis probe and were already scheduled to receive carboplatin (400 mg m^{-2}) intravenously as a single agent. The patients were admitted in the morning of the study day, they remained in a supine position throughout the study period. The study room temperature was kept at 25°C. One dialysis probe was inserted vertically, intratumorally into a suitable cutaneous melanoma metastasis and a second probe was inserted horizontally into healthy subcutaneous connective tissue within a 10–15 cm distance to the first microdialysis probe. After the insertion of both microdialysis probes there was a 30 min equilibrium period (Müller et al. 1995a); subsequently one 15 min microdialysis sample was taken (baseline level). Thereafter carboplatin (400 mg/m^2) was infused intravenously through a second canula over a time period of 20 min. Sampling of microdialysates and blood were performed in 15 min interval for up to 4 h, blood samples were taken at mid-time points of each microdialysis collection period.

The proper position of the probe in the tumour was confirmed by high frequency (7.5 MHz) ultrasound scanning. No local anaesthesia was used. Owing to diffusion and sampling of the dialysate there is a certain time delay before sudden concentration changes in the ambient medium are detected in the microdialysis probe. This time delay was taken into account for all experiments.

Platinum concentrations in serum and in perfusate were performed by electrothermal atomic absorption spectrometry. The method of analysis was modified according to McGahan. All measurements were carried out with AAS-Zeeman 5100, Perkin Elmer. Calibration of the microdialysis probes was carried out *in vitro* and *in vivo* by the use of the retrodialysis method.

Assessment of Probe Recovery

To characterise the transfer rate of the probes we assessed *in vitro* recovery of carboplatin. The dialysis probes were placed in glass beakers containing different concentrations of carboplatin. The probe was perfused at a flow rate of 1.5 μL/min. Analyte concentrations were measured in the dialysate and expressed as percentage of the concentration in the surrounding medium.

In vivo recovery of carboplatin was assessed according to the retrodialysis method (Stahle et al. 1991; Palmsmeier et al. 1994). Thus *in vivo* recovery value was calculated as

$$\text{Recovery (\%)} = 100 - (100 \times \text{carboplatin}_{\text{dialysate}} \times \text{carboplatin}_{\text{perfusate}} - 1)$$

In vivo recovery was assessed on separate study days by dialysing the tumour tissue with a perfusion medium containing 8 μg/mL carboplatin for 120 min. Intercellular tissue concentrations were calculated by following formula

$$\text{Tissue concentration} = 100 \times \text{dialysate concentration} \times \textit{in vivo} \text{ recovery value}^{-1}$$

As pharmacokinetic parameters did not follow a normal distribution, statistical comparisons between compartments (serum, tumour, subcutaneous tissue) were made by the Wilcoxon matched paired test. Correlations between parameters were calculated using Spearman rank order correlation (r_S). Furthermore, linear regression analyses were performed. $P < 0.05$ was considered the level of significance.

Results and Discussion

Microdialysis was well tolerated by all patients; there were no adverse events such as bleeding or pain at the site of probe insertion. In the *in vitro* experiment there was a correlation between carboplatin concentrations in the dialysate and in the surrounding medium over a wide concentration range. *In vitro* recovery at 20°C was 64% (r = 0.95). Mean *in vivo* recovery value at 37°C in tumour was 84% and in subcutaneous tissue was 74%. The time vs concentration curves for carboplatin obtained by *in vivo* sampling in serum, tumour and subcutaneous tissue are shown in Figure 1. Key pharmacokinetic parameters are presented in Table 1.

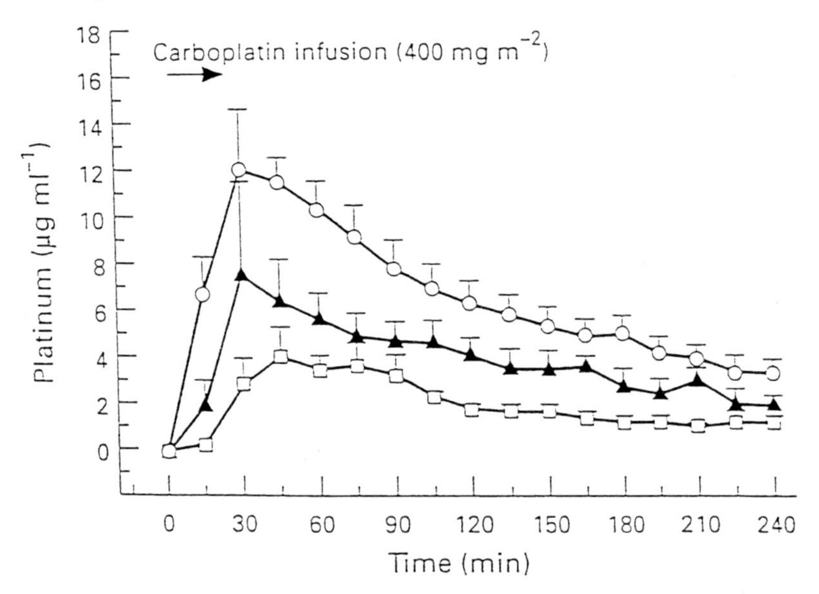

Figure 1. Concentration profile of carboplatin in serum, tumour and subcutaneous tissue. Results are presented as mean values ± s.e.m. from six patients. Time 0 is the time of the start of the carboplatin infusion (horizontal arrow). ○, serum; ▲, tumour; □, subcutaneous.

Table 1. Key pharmacokinetic parameters from serum, tumour intercellular space, and subcutaneous tissue intercellular space after carboplatin infusion (400 mg/m^2 over 20 min, i.v.). Results are expressed as means ± s.e.m. from six patients.

	Serum	Tumour	Subcutaneous
C_{max} (µg/ml)	14.6 ± 1.1	7.6 ± 2.0*	5.6 ± 1.2*
(CV%)	(17.7)	(64.5)	(52.8)
t_{max} (min)	34 ± 5	60 ± 10	54 ± 7
(CV%)	(35.3)	(41.7)	(33.3)
$t/2_{el}$ (min)	90 ± 6	99 ± 2	91 ± 23
(CV%)	(16.7)	(49.5)	(62.6)
AUC (µg/ml min)	1533 ± 189	853 ± 172*	506 ± 87*
(CV%)	(30.1)	(49.4)	(42.1)

C_{max}, maximal carboplatin concentration; t_{max}, time after start of infusion at which C_{max} occured; $t/2_{el}$, terminal elimination half-life; AUC, area under the time vs concentration curve from 0 to 4 h.
*p < 0.05 vs. serum.

Mean C_{max} and AUC levels in tumour and subcutaneous tissue were significantly lower than in serum; there was no statistically significant difference in C_{max} or AUC between tumour and subcutaneous tissue (P = 0.75 and P = 0.25 respectively). There also was no correlation between AUC_{serum} and AUC_{tumour} ($r_s = 0.66$, $P_s = 0.16$), between $C_{max\text{-}serum}$ and $C_{max\text{-}tumour}$ ($r_s = 0.26$, $P_s = 0.62$), or between AUC_{serum} and $C_{max\text{-}tumour}$ ($r_s = 0.43$, $P_s = 0.40$). Terminal elimination half-life ($t/2_{el}$) in tumour was similar to $t/2_{el}$ in serum. Mean AUC and C_{max} values of the tumour-ECF reached only 50–60% of the corresponding levels in serum (Table 1 and Figure 1). This indicates rapid but incomplete equilibration between blood and the intracellular tumour compartment. Similar results were obtained from subcutaneous tissue, although mean AUC and C_{max} values were slightly but non-significantly lower than for tumour.

By the lack of correlation between serum- and tumour- or subcutaneous-AUC and C_{max} levels in our patients indicates that serum pharmacokinetic parameters are poor predictors of carboplatin concentration in tumour and non-malignant tissue ECF. Hence, our data also suggest that the lack of correlation between serum concentrations and anti-tumour effect, which has been observed for carboplatin, may not only be a result of differing drug responsiveness of individual tumours, but also a highly variable and unpredictable tumour distribution. Distribution may not only vary between solid tumours, but also within the same tumour lesion. Clearly, the information obtained with the placement of one microdialysis probe need not be representative for the whole tumour, this may be regarded as a limitation of our study. Placement of more than one probe per tumour might be more informative, this is precluded for ethical and technical reasons.

The results of the *in vitro* calibration experiments clearly show that the process of carboplatin diffusion through the microdialysis membrane is concentration independent over a wide range of analyte concentrations, in agreement with the observations of others (Lönnroth et al. 1987, Jansson et al. 1993).

For many drugs, only the free fraction is available for equilibration with peripheral tissues. Microdialysis measures free drug concentration in the ECF compartment of target tissue. We have measured total (i.e., free and protein-bound) platinum concentration in serum. For carboplatin, protein binding is minimal during the first hours following administration (Fujiwara et al. 1988), but has been shown to increase over time (Dollery et al. 1991). Therefore we cannot estimate free carboplatin serum concentration and the "true" equilibration rate between serum and tumour may be slightly higher than is suggested by our AUC_{tumour}/AUC_{serum} ratios.

In conclusion we have demonstrated that microdialysis sampling is suitable for measuring drug concentration in the ECF space of solid tumours in human. The duration of a single microdialysis experiment is limited only by the inconvenience caused to the experimental subject, particularly the requirement for resting in a supine position, and the requirement for sensitive analytical techniques because of the small sample volumes and low concentrations obtained by microdialysis. Major limitations of the microdialysis technique are the low recovery for molecules with large molecular weights or high lipophilicity (Stahle 1991, Carnheim and Stahle 1991, Müller et al. 1995b). This technique may become a valuable addition for pharmacokinetic/pharmacodynamic study in oncology.

References

Carnheim, C. & Stahle, L. (1991) Microdialysis of lipophilic compounds: a methodological study. Pharmacol. Toxicol., 69: 378–380.

Coughlin, C.T., Richmond, R.C., & Page, R.L. (1994) Platinum drug delivery and radiation for locally advanced prostate cancer. Int. J. Radiati. Oncol. Biol. Phys., 28: 1029 –1038.

De Conti, R.C., Toftness, B.R., Lange, R.C., & Creasey, W.A. (1973) Clinical and pharmacological studies with cis-diamminedichloroplatinum II. Cancer Res., 33: 1310–1315.

Dollery, C. (1991) Therapeutic Drugs. Churchill Livingstone: Edinburgh.

Eskey, C.J., Koretsky, A.P., Domach, M.M., Jain, R.K. (1992) ^{2}H-Nuclear magnetic resonance imaging of tumour blood flow: spatial and temporal heterogeneity in a tissue-isolated mammary adenocarcinoma. Cancer Res., 52: 6010–6019.

Fujiwara, K., Miyagi, Y., Hayase, R., Yoshinouchi, M., Kobashi, Y., Kohno, I. & Sekiba, K. (1988) Pharmacokinetics of carboplatin (CBDCA) and the tissue concentration of platinum in gynecologic organs. Jpn. J. Cancer. Chemother., 15(6): 1943–1948.

Jain, R.K. (1994) Barriers to drug delivery in solid tumors. Sci. Am., 7: 42–49.

Jansson, P.A.E., Fowelin, J.P., Von Schenck, H.P., Smith, U.P., Lönnroth, P.N. (1993) Measurement by microdialysis of the insulin concentration in subcutaneous interstitial fluid. Diabetes, 42: 1469–1473.

Lönnroth, P., Jansson, P.A., & Smith U. (1987) A microdialysis method allowing characterisation of intercellular water space in humans. Am. J. Physiol., (Endocrin. Metab. 16): E228–231.

Lönnroth, P., Carlsten, J., Johnson, L., & Smith, U. (1991) Measurements by microdialysis of free tissue concentrations of propranolol. J. Chromatogr., 568: 419–425.

McGahan, Mc. & Tyczkowska, K. (1987) The determination of platinum in biological materials by electrochemical atomic absorption spectroscopy. Spectrochim. Acta. 42: 665.
Morrison, P.F., Bungay, P.M., Hsiao, J.K., Ball, B.A., Mefford, I.N., & Dedrick, R.L. (1991) Quantitative microdialysis: Analysis of transients and application to pharmacokinetics in brain. J. Neurochem., 57: 103–119.
Müller, M., Schmid. R., Georgopoulos, A,. Buxbaum, A., Wasicek, C. & Eichler, H.G. (1995a) Application of microdialysis to clinical pharmacokinetics in humans. Clin. Pharmacol. Ther., 57: 371–380.
Müller, M., Schmid, R., Wagner, O., Osten, B., Shayganfar, H., & Eichler, H.G. (1995b) in vivo characterisation of transdermal drug transport by microdialysis. J. Contr. Release, 37: 49–57.
Palsmeier, R.K. & Lunte, C.E. (1994) Microdialysis sampling in tumour and muscle: study of the disposition of 3 -amino-1,2,4-benzotriazine-1,4-DI-N-oxide (SR4233) Life Sci., 55: 815–825.
Scheyer, R.D., During, M.J., Spencer, D.D., Cramer, J.A., & Mattson, R.H. (1994) Measurement of carbamazepine and carbamazepine epoxide in the human brain using in vivo microdialysis. Neurology, 44: 1469–1472.
Stahle, L. (1991) The use of microdialysis in pharmacokinetics and pharmacodynamics. In: Microdialysis in the Neurosciences (Robinson, T.E. & Justice, Jr., J.B., eds) pp. 155–173. Elsevier Science Publishers: Amsterdam.
Stahle, L., Arner, P., & Ungerstedt, U. (1991) Drug distribution studies with microdialysis III: extracellular concentration of caffeine in adipose tissue in man. Life Sci., 49: 1853–1858.
Ungerstedt, U. (1991) Microdialysis — Principles and applications for studies in animals and man. J. Int. Med., 230: 365–373.

Discussion

Q1. Bob Dabeka, Health Canada, Ottawa, ON, Canada: Why would you apply the cis-dichloroplatinum at different areas?

A. The patients had cutaneous metastases. They looked at inserting the microprobe into the tumors. Ten to fifteen centimetres away was healthy tissue and the probe was inserted between the tumor and the healthy cells.

Distribution of Trace Elements in Bean Sprouts Determined by PIXE Analysis

MASAE YUKAWA, AYAMI KIMURA, YOSHITO WATANABE, AND SHIRO SAKURAI[†]

Division of Environmental Health, National Institute of Radiological Sciences, Chiba, 263 Japan; [†]Department of Social Information, Otsuma University, Tama, 206 Japan

Keywords bean sprout, PIXE analysis, trace elements, distribution

Introduction

Soybean is one of the major crops in the world. In Japan, soybeans have been consumed in various forms such as soy sauce, soybean paste, soybean curd and bean sprout for a long time. Since the soybean seedling is cultivated in a dark place in water without any supplement of fertilizers, growth is supported only by the nutrients stored in the seed. The elements present in the seed would be expected to migrate to the various parts during the growth process according to their physiological functions, and this transfer should easily be observed because their is no other source of these nutrients. In this study, the distribution of trace elements in bean sprouts during each growth stage was investigated by Particle Induced X-ray Emission (PIXE) analysis.

Materials and Methods

1. Cultivation of Soybean Sprout

Soybeans were germinated on a wet nylon sponge pad in a dark box at 25 ~ 26°C using ultra pure water. After 20, 40, 65 and 137 h, one of the seedlings was harvested and each spout was separated into cotyledons (sliced into the outside, interior and inside), hypocotyls (divided into the upper, middle and lower parts) and leaves. Each part of the bean spout was freeze-dried separately and applied to PIXE analysis as shown in Figure 1.

2. PIXE Analysis of Elements in Soybean Sprout

Proton irradiation and X-ray spectrometry were carried out using a 3 MV Van de Graaff accelerator as previously described (Yukawa 1991). Samples were irradiated by a proton beam of 2.3 MeV collimated to a circle of 2 mm or 0.2 mmϕ under 10^{-6} torr. Bombardment time was about 5 min or less at a rate of a few nA up to 10 nA. The concentrations of elements were determined by comparing the X-ray intensities of the elements in samples with those of Bovine Liver (Standard Reference Material supplied by NIST), which was formed into a pellet of 5 mmϕ and 0.7 mm thickness.

Results and Discussion

1. Growth of Soybean Sprout

Germination was initiated by a marked swelling of the seed. At 20 h, the primary root, the lower end of the hypocotyl, ruptured the seed coat and grew up to about 5 mm. The hypocotyl elongated rapidly and began to differentiate to a stem, plumule and root. At 65 h, the root became about 3 cm producing lateral roots and root hairs. The plumule then began active growth giving rise to the stem and the total length of the seedling reached about 15 cm at 135 h. Since the bean was not supplied with any nutrition and light

Correct citation: Yukawa, M., Kimura, A., Watanabe, Y., and Sakurai, S. 1997. Distribution of trace elements in bean sprouts determined by PIXE analysis. In *Trace Elements in Man and Animals – 9: Proceedings of the Ninth International Symposium on Trace Elements in Man and Animals. Edited by* P.W.F. Fischer, M.R. L'Abbé, K.A. Cockell, and R.S. Gibson. NRC Research Press, Ottawa, Canada. pp. 265–266.

in this experiment, the bean stopped growing after the first leaf differentiated. Dry weight of each section of the bean sprout changed during growth, indicating that the components in cotyledons moved to the root, stem and leaf through the hypocotyl.

2. Distribution of Elements in Soybean and Changes during the Growth Stages

Al, P, S, K, Ca, Ti, Cr, Mn, Fe, Ni, Cu, Zn, Br, Rb and Sr were found in the PIXE spectrum of a soybean. In all sections of bean sprout, K, Ca, Mn, Fe, Cu and Zn were detected, and these elements are essential for the plant. Concentrations of these elements were different in each part of the soybean sprout and changed during growth as shown in Figure 2. Concentrations of K and Cu in the cotyledons and hypocotyl decreased with time, while they increased in the 1st leaf. This means that K and Cu in the cotyledons moved to the hypocotyl and the 1st leaf. The decrease in the hypocotyl is due to a rapid increase in the volume as same as is the case in the 1st leaf at 137 h. Fe was present in high concentrations in the outer part of the cotyledons and the young hypocotyl at 20 h and then moved to the 1st leaf. The concentration of Fe in the hypocotyl decreased because of the volume increase and the movement to the 1st leaf. Mn showed a different pattern from K, Fe and Cu. The primary root was rich in Mn, and it moved to the 1st leaf during growth. Some of Mn also seemed to move to the cotyledons. The change in Zn concentration was intermediate between Mn and Cu. Ca hardly seemed to move to other plant parts.

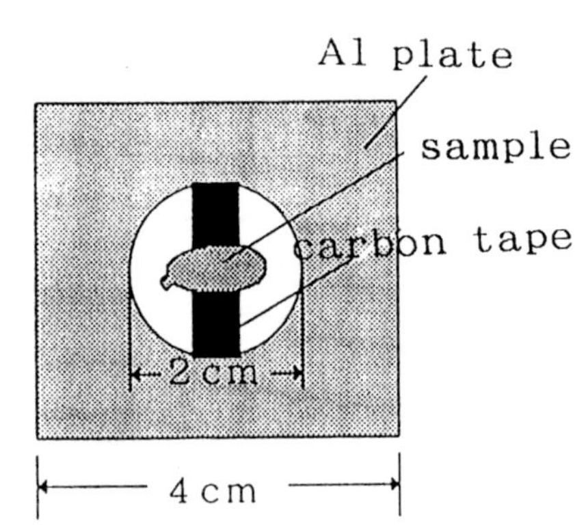

Figure 1. Sample segment for PIXE analysis.

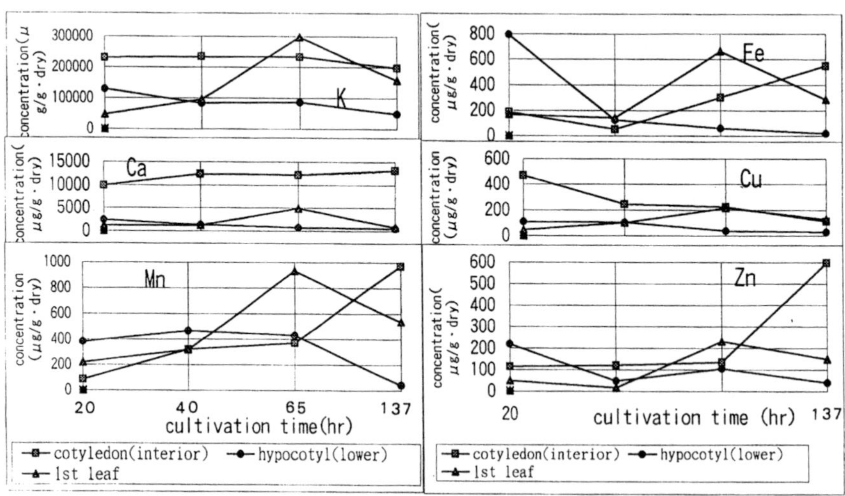

Figure 2. Changes of concentration of elements in soybean sprout along growth stage.

References

Yukawa, M. and Kitao, K. (1991) PIXE Analysis of the Distribution of Elements in the Human Kidney, Cerebrum and Breast Bone. Int. Jr. PIXE 1: 339–354.

Stable Tracers for Tracer Kinetic Investigations of Molybdenum: Intrinsic and Extrinsic Tagging

M.C. CANTONE,[†] D. DE BARTOLO,[†] A. GIUSSANI,[‡,¶] CH. HANSEN,[§] P. ROTH,[‡] P. SCHRAMEL,[||] I. WENDLER,[||] E. WERNER,[‡] AND F. NÜSSLIN[¶]

[†]Dipartimento di Fisica, Universitá degli Studi, I-20133 Milano (Italy); [§]Klinikum der J.W. Goethe-Universität, D-60596 Frankfurt am Main (Germany); [‡]Institut für Strahlenschutz and [||]Institut für Ökologische Chemie, GSF-Forschungszentrum für Umwelt und Gesundheit, D-85758 Oberschleiβheim (Germany); [¶]Abteilung für Medizinische Physik, Eberhard-Karls-Universität, D-72076 Tübingen (Germany)

Keywords molybdenum, stable isotopes, intestinal absorption, biokinetics

Molybdenum is a trace element which is considered to be essential for humans and living organisms in general, being the main constituent of the cofactor of several enzymes, such as, xanthine oxidase, sulfite oxidase and aldehyde oxidase (Rajagopalan 1987). The biokinetics and the metabolism of Mo have been widely studied in ruminants, because of the severe diseases which are caused in these animals by an imbalance in the concentrations of molybdenum, copper and sulphur in the forage (Huising and Matrone 1976). Only recently, investigations on the uptake of molybdenum into humans have been reported (Cantone et al. 1995, Cantone et al. 1996, Giussani et al. 1995, Turnlund et al. 1995a, Turnlund et al. 1995b). These investigations were carried out by administering enriched non-radioactive isotopes of molybdenum and then analysing their concentrations in blood samples and/or excreta by means of activation analysis and/or mass spectrometry. Mo was given in an aqueous solution or as an extrinsic tag of solid meals. Almost complete absorption from the GI-tract (≥90%) and rapid excretion in urine were found. When Mo was mixed with a homogeneous meal, its percentage intestinal absorption decreased to about 60% (Cantone et al. 1996).

In order to investigate the ability of Mo aqueous solutions to tag a solid meal, cress seeds (*Lepidium sativum*) were planted on blotter paper moistened with a solution labelled with ^{95}Mo.

The isotopic composition of the enriched Mo (A.Hempel, Düsseldorf, Germany) is given in Table 1. After 10 d the cress was harvested. The concentration of enriched Mo in the edible part of the cress was measured by Inductively Coupled Plasma-Mass Spectrometry (ICP-MS) and was found to be 48.5 ± 0.4 μg Mo·g^{-1} wet weight.

The experiment was conducted on 3 healthy subjects, 2 males (subject 1 and 2, age 53 and 51 years respectively) and 1 female (subject 3, age 38 years). About 10 g of cress, grown on the ^{95}Mo solution and corresponding to 510 μg of Mo, were given to each subject after mixing with 150 μL of a solution of the stable isotope ^{96}Mo (see Table 1 for its composition), corresponding to about 290 μg of Mo. ^{96}Mo was used as an extrinsic tag. Blood samples were taken at intervals up to 8 h after administration. Renal excretion was monitored over 2 d, with two 12-h-collections on the first day and one 24-h-collection on the second day. The concentrations of the Mo isotopes were measured by means of proton activation analysis in plasma and by ICP-MS in urine as described elsewhere (Cantone et al. 1995, Schramel and Wendler 1995). Figure 1 shows the results normalized to the corresponding administered amounts. The renal excretion of the administered Mo is very rapid, as has also been reported by Turnlund et al. (1995b). In each subject the kinetic behaviour of the two tracers was found to be very similar. However, both the normalized concentrations in plasma and the normalized cumulated renal excretion of the intrinsic tag ^{95}Mo are lower than the corresponding values for the extrinsic tag ^{96}Mo.

Correct citation: Cantone, M.C., de Bartolo, D., Giussani, A., Hansen, Ch., Roth, P., Schramel, P., Wendler, I., Werner, E., and Nüsslin, F. 1997. Stable tracers for tracer kinetic investigations of molybdenum: intrinsic and extrinsic tagging. In *Trace Elements in Man and Animals – 9: Proceedings of the Ninth International Symposium on Trace Elements in Man and Animals. Edited by* P.W.F. Fischer, M.R. L'Abbé, K.A. Cockell, and R.S. Gibson. NRC Research Press, Ottawa, Canada. pp. 267–269.

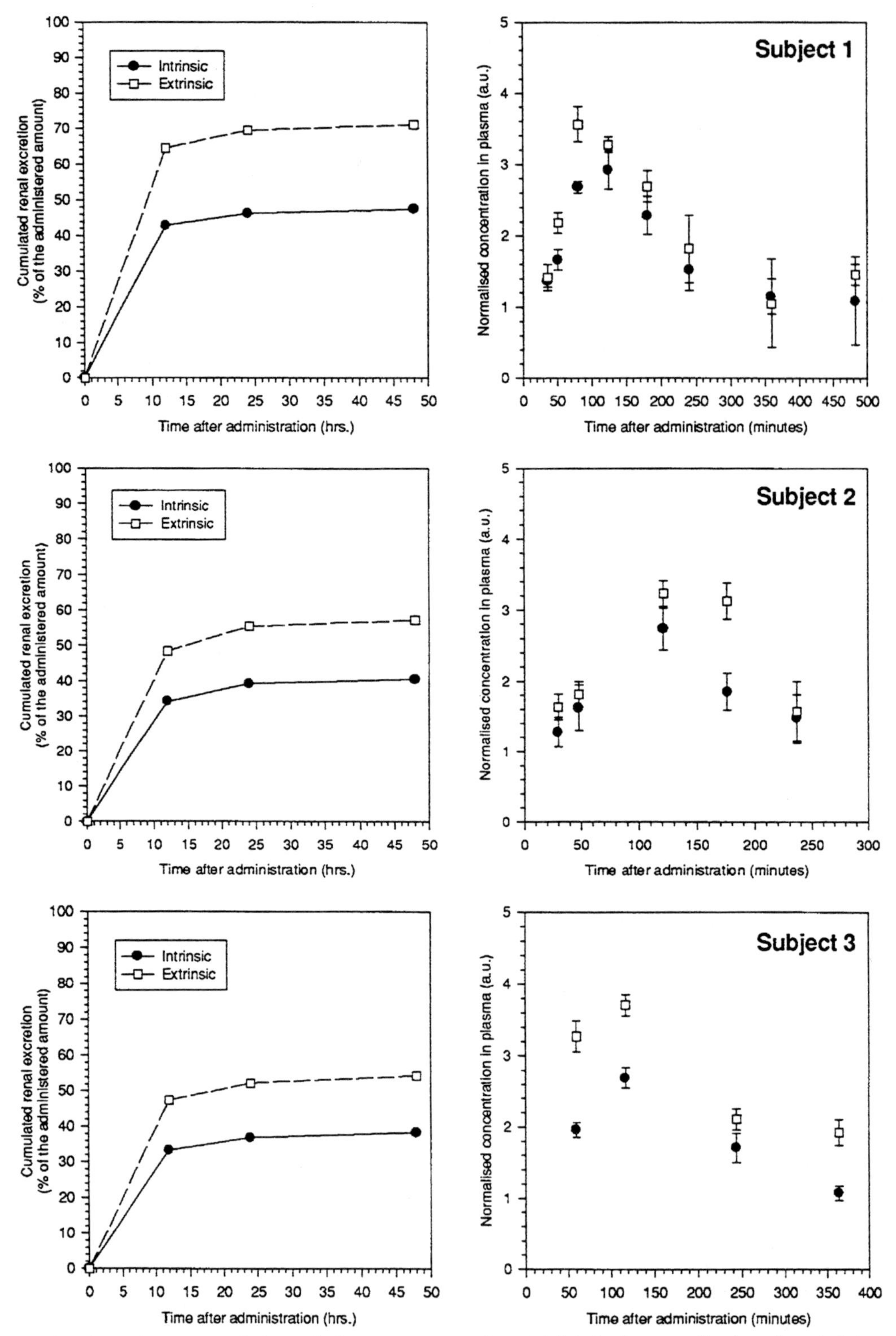

Figure 1. Cumulated renal excretion (left) and plasma concentrations (right), both normalized to the administered amount, for subjects 1 to 3.

Table 1. Composition of isotopically enriched molybdenum (as mass %).

Isotopic solution	^{92}Mo	^{94}Mo	^{95}Mo	^{96}Mo	^{97}Mo	^{98}Mo	^{100}Mo
95-enriched solution	0.0031	0.0069	**0.9540**	0.0224	0.0051	0.0065	0.0020
96-enriched solution	0.0023	0.0024	0.0133	**0.9590**	0.0097	0.0112	0.0021

Comparing the ratio ^{95}Mo/^{96}Mo in the cumulated two-day renal excretion with the corresponding ratio between the administered amounts, the absorption of the intrinsic tag turns out to be about 30% lower than that of the extrinsic tag. For subjects 1 and 3, the absorption could also be estimated from the plasma concentrations by the method of compartmental analysis. For this purpose plasma clearance curves, that had been previously obtained (Cantone et al. 1995), have been used. The fractional absorption of the extrinsic tag was calculated to be 0.90 for subject 1 and 0.73 for subject 3, and of the intrinsic tag 0.78 and 0.54 respectively. As with the data on renal excretion, a decreased absorption for the intrinsic tag can be observed.

The study has shown that extrinsic tagging of solid meals with aqueous solutions of Mo is a useful method to obtain important information concerning the biokinetics of Mo in humans. The use of an extrinsic tag may lead to an overestimation of the intestinal absorption of molybdenum from solid food.

References

Cantone, M.C., de Bartolo, D., Gambarini, G., Giussani, A., Ottolenghi, A., Pirola, L., Hansen, Ch., Roth, P. & Werner, E. (1995) Proton activation analysis of stable isotopes for a molybdenum biokinetics study in humans. Med. Phys. 22: 1293–1298.

Cantone, M.C., de Bartolo, D., Giussani, A., Ottolenghi, A., Pirola, L., Hansen, Ch., Roth, P. & Werner, E. (1996) A methodology for biokinetic studies with stable isotopes: results of repeated molybdenum investigations on a healthy volunteer, submitted for publication in Appl. Rad. Isot.

Giussani, A., Hansen, Ch., Nüsslin, F. & Werner, E. (1995) Application of thermal ionization mass spectrometry to investigations of molybdenum absorption in humans. Int. J. Mass Spect. Ion Proc. 148: 171–178.

Huisingh, J. & Matrone, G. (1976) In: Molybdenum in the environment (Chappell, W.R. & Kellogg Petersen, K., eds.), vol.1 pp. 125–148. Dekker, New York.

Rajagopalan, K.V. (1987) Molybdenum: an essential trace element. Nutr. Rev. 45: 321–328.

Schramel, P. & Wendler, I. (1995). Molybdenum determination in human serum (plasma) by ICP-MS coupled to a graphite furnace. Fresenius J. Anal. Chem. 351: 567–570.

Turnlund, J.R., Keyes, W.R., Pfeiffer, G.L. & Chiang, G. (1995a) Molybdenum absorption, excretion and retention studied with stable isotopes in young men during depletion and repletion. Am. J. Clin. Nutr. 61: 1102–1109.

Turnlund, J.R., Keyes, W.R. and Pfeiffer, G.L. (1995b) Molybdenum absorption, excretion and retention studied with stable isotopes in young men at five amounts of dietary molybdenum. Am. J. Clin. Nutr. 62: 790–796.

Can Rare Earth Elements be Used as Non-Absorbable Fecal Markers in Studies of Iron Absorption?*

S.J. FAIRWEATHER-TAIT,† A.M. MINIHANE,† J. EAGLES,† L. OWEN,‡ AND H.M. CREWS‡

†Institute of Food Research, Norwich NR4 7UA, UK, and ‡CSL Food Science Laboratory, Norwich, NR4 7UQ, UK

Keywords rare earth elements, fecal markers, iron absorption, stable isotopes

The rare earth elements (REEs) are non-essential, inert, and non-absorbable; they are present in plant and animal tissues in very low amounts (<10 ppb) and are therefore ideal biological markers. They have been successfully used as non-absorbable markers in studies of zinc and magnesium absorption with stable isotope tracers in human volunteers (Schuette et al. 1993). This study was designed to evaluate the usefulness of REEs samarium (Sm), ytterbium (Yb) and dysprosium (Dy) as fecal markers for iron in human absorption studies using stable isotopes.

Materials and Methods

The question addressed in this paper formed part of a larger study examining the effects of calcium on iron absorption (Minihane et al. 1997). Thirteen volunteers, aged 20–69 years, were recruited, all in good health and without a history of any gastrointestinal disorders. The experimental protocol is described in Figure 1, including the pattern of food, iron isotope and REE consumption. Each meal was labelled extrinsically with an isotope of iron and contained a REE. These doses were added to the drinks consumed with the meals. A capsule containing 10 radio-opaque markers was consumed with the breakfast on days 1 and 2 and with the reference dose on day 3 (to check completeness of fecal collections). All stools were collected individually commencing in the morning of day 1 until complete excretion of the carmine (given with the evening meal on day 3).

The stools were autoclaved, freeze-dried and ground to a fine powder. Fecal collection was only deemed complete when all 30 radio-opaque markers were collected. A portion of each freeze-dried sample was ashed at 480°C for 48 h in a silica crucible. The total iron and isotope ratios of the iron isotope solutions and the fecal ash were determined by flame atomic absorption spectrometry and thermal ionization quadrupole mass spectrometry (TIQMS) respectively. Prior to TIQMS the sample was passed through an AG1-X8 anion exchange column (Bio-rad Laboratories, Richmond, CA) to purify the sample. The REE concentration of each stool was determined by inductively coupled plasma mass spectrometry (ICP-MS). A 30–40 mg portion of the freeze-dried feces was placed in a 7 mL acid-washed PTFE digestion vessel and 0.5 mL of concentrated HNO_3 or 0.5 mL 50 ppb REE spike in concentrated nitric acid was added. The spike was used to determine the % recovery of the REE. The sample was digested in a microwave oven (CEM) for 15 min (30% power), allowed to cool, transferred to a test-tube, and the digestion vessel rinsed with 1.5 mL of millipore water. Internal standards were added to the sample prior to analysis (8 mL of 50 ppb aqueous solutions of indium and bismuth) and the final volume was taken to 10 mL. Standards solutions containing 0–50 ppb REE were prepared with a final concentration of HNO_3, bismuth and indium as in the samples. The REE and iron isotope concentration was determined in each stool sample and the total recovery calculated.

---

*Funding: this work was supported by the EC and BBSRC.

Correct citation: Fairweather-Tait, S.J., Minihane, A.M., Eagles, J., Owen, L., and Crews, H.M. 1997. Can rare earth elements be used as non-absorbable fecal markers in studies of iron absorption? In *Trace Elements in Man and Animals – 9: Proceedings of the Ninth International Symposium on Trace Elements in Man and Animals. Edited by* P.W.F. Fischer, M.R. L'Abbé, K.A. Cockell, and R.S. Gibson. NRC Research Press, Ottawa, Canada. pp. 270–272.

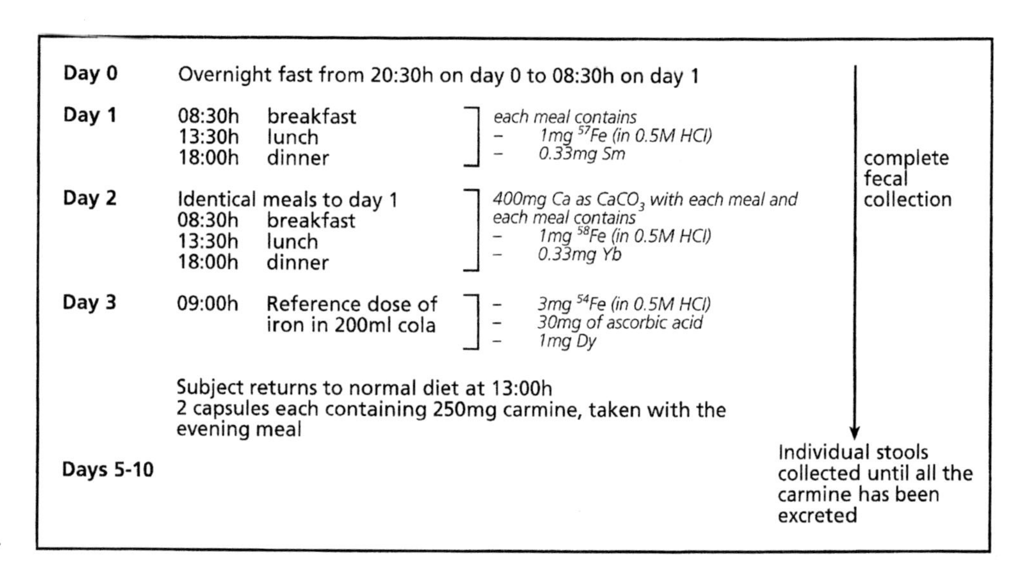

Figure 1.

Results

Complete fecal collections were carried out by 12 subjects; in one individual only 19 radio-markers were recovered, thus this volunteer was excluded from the data analysis. The % recoveries of each REE are given in Table 1.

Complete excretion of all three REEs occurred. The excretory pattern of the iron isotopes and REEs was very similar. The correlation coefficients between Sm and ^{57}Fe, Yb and ^{58}Fe, and Dy and ^{54}Fe in individual stools were 0.992, 0.989 and 0.988 respectively ($p < 0.001$).

Iron absorption was calculated as the difference between the isotope dose and fecal isotope excretion measured in stools collected for periods ranging from 5 to 10 d (depending on the transit time of the individual). The measured data were compared with predicted values estimated from the first 4 days' pooled feces, using the REE recovery data to produce corrected figures for isotope excretion. There was a highly significant correlation between measured and predicted isotope absorption at three very different levels of absorption, with mean % absorption values ranging from 4.2 to 40.6 (Table 2).

Discussion

Fecal monitoring techniques for measuring iron absorption using stable isotopes can be very prolonged in some individuals. Long collection periods increase the cost of the study, and subject compliance

Table 1. REE recoveries.

Rare earth	Mean recovery (%)	Range of individual recoveries (%)
Sm	101	96–106
Yb	101	97–106
Dy	101	92–106

Table 2. Measured and predicted figures for % iron absorption (Mean values ± SEM).

Iron isotope	REE	Measured absorption (%)	Predicted absorption (%)	Correlation (r)	p
^{57}Fe	Sm	19.1 ± 2.1	16.7 ± 2.3	0.86	<0.01
^{58}Fe	Yb	4.2 ± 1.4	4.3 ± 1.6	0.98	<0.01
^{54}Fe	Dy	40.6 ± 3.1	40.4 ± 3.1	0.99	<0.01

is more difficult, resulting in potentially greater errors in the estimate of absorption. The use of a non-absorbable fecal marker that follows the same excretory pattern as inorganic isotopes enables the calculation of luminal disappearance (apparent absorption) to be made with a smaller number of stools.

The results from this study demonstrate that it is not necessary to carry out complete fecal collections in order to measure iron absorption from foods or drinks labelled extrinsically with stable isotopes of iron. If REEs are used as non-absorbable markers a fecal collection period of 4 d is sufficient to predict iron absorption. The technique is valid over a wide absorption range, from diets of low (mean absorption 4.2%) and high bioavailability (mean absorption 19.1%), and it can also be used for measuring absorption from highly bioavailable sources of iron employed as reference doses (mean absorption 40.6%).

References

Minihane, A.M., Fairweather-Tait, S., & Eagles, J. (1997) The short and long term effects of calcium supplements on iron nutrition — is there an adaptive response? In: Trace Elements in Man and Animals – 9: Proceedings of the Ninth International Symposium on Trace Elements in Man and Animals. (Fischer, P.W.F., L'Abbé, M.R., Cockell, K.A. and Gibson, R.S., eds.). NRC Research Press, Ottawa, Canada. pp. 38–40.

Schuette, S.H., Janghorbani, M., Young, V.R., & Weaver, C.M. (1993) Dysprosium as a nonabsorbable marker for studies of mineral absorption with stable isotope tracers in human subjects. J. Am. Coll. Nutr. 12: 307–315.

Workshop: Ethics of Trace Element Research in Human and Animal Nutrition

Chair: P.R. Flanagan

Ethical Use of Animals for Experimentation and Teaching

P.R. FLANAGAN

Departments of Medicine and Biochemistry, The University of Western Ontario, London, Ontario, Canada N6A 5A5

Keywords ethics, animal, research, teaching

Introduction

What ethical issues need to be addressed when a decision is made to use animals for trace element research? Broadly, they are the same as those applying to any investigator who proposes to use animals. Does the scientific end justify the means, that is the ethical cost? In nutrition experiments, specific ethical points arise only secondarily, for example in regard to the appropriate endpoint of a harmful dietary treatment. Therefore, I will try to emphasize some general and practical ethical principles rather that specific approaches taken in various countries.

Ethics

An ethical approach directs us towards distinguishing between right and wrong, and towards determining what is right conduct. Personal moral views are not always helpful in this since they can be based on limited experience, are not always impartial and can be contaminated by self interest. Philosophers or ethicists, while not claiming to be morally wiser than the rest of society, can be a help because their knowledge of the principles of logic and ethics can help in distinguishing between sound and invalid arguments (Smith and Boyd 1991a). However, scientists may be tempted to disparage philosophy when it shows that it is virtually impossible to be consistent, coherent and disinterested in our moral thinking about animals and their use by humans (Smith and Boyd 1991b). Neither has philosophy been able to define precisely the moral standing of an animal (Beauchamp 1992). In view of these difficulties, it is prudent to proceed carefully and to examine and justify why and how we perform animal research. I believe that there are at least three reasons for a need to justify animal experimentation: a) to prevent undue harm to the animals which are sentient creatures; b) to judge the necessity of the research; and c) to obtain consent to do the research. The last point, particularly, is different from research on human animals where 'written informed consent' is an inherent part.

Correct citation: Flanagan, P.R. 1997. Ethical use of animals for experimentation and teaching. In *Trace Elements in Man and Animals – 9: Proceedings of the Ninth International Symposium on Trace Elements in Man and Animals. Edited by* P.W.F. Fischer, M.R. L'Abbé, K.A. Cockell, and R.S. Gibson. NRC Research Press, Ottawa, Canada. pp. 273–275.

Preventing Undue Harm

A useful approach to practical ethics in animal use is the '3R' concept of Russell and Burch (1959). They stressed that, wherever possible, animal use should be replaced, reduced and refined (Russell and Burch 1959). These ideas were the seed of the 'Alternatives' concept. Later Smyth (1978) defined 'Alternatives' as: 'Procedures which can completely replace the need for animal experiments, reduce the number of animals required or diminish the amount of pain or distress suffered by the animals in meeting the essential needs of man and other animals'. The term 'Alternatives' is widely accepted in terms of this broad '3R' definition. For example, Section 13 of the U.S. Animal Welfare Act (1985) requires "that the principal investigator considers alternatives to any procedure likely to produce pain to or distress in an experimental animal". Some scientists, however, respond defensively to the term because they misinterpret it too narrowly to mean only replacement alternatives. Alternatives should be broadly accepted by the scientific community as ethical and scientific methods to ensure that necessary animal use is performed properly.

Alternatives

As scientific methods have become more refined, many research animal uses have been replaced by non-animal techniques. Biological chemicals are now routinely measured directly using chromatographic, radioisotopic or immunological assays rather than the 'bioassays' of the past which measured the biological response of a compound in a living animal. Also many modern testing and screening methods now use lower organisms or culture methods. The pedagogical value of live animal use for teaching, instruction or demonstration should always be examined carefully, since some of these uses also may be replaced by computer models, video demonstrations and 'virtual' experiences. Despite all these advances, there are many areas in biological science where it is still not possible to directly replace an animal model with a non-animal one. Animal numbers should be reduced wherever possible and should be justified as a routine. For the most part, this goes hand-in-hand with good science. Optimal experimental design and the proper use of statistical techniques can considerably lower animal numbers. Nevertheless, it might be ethically correct to allow some repetition of animal use to perfect methods that otherwise would waste animals or to *increase* animal numbers in some situations in order to fully utilize data from animals already used. There are many areas where animal use can be refined to reduce anxiety, pain and suffering in the animals. Housing facilities, caging, laboratory equipment, animal care management and research personnel training all play an important role in the well-being of research animals. Where pain is unavoidable in experimental procedures, anaesthesia and analgesia must be administered to acceptable veterinary standards.

Regulations

Different methods have been implemented in different countries to ensure the ethical and humane treatment of animal use for research and teaching. In the U.K. and U.S. legislation has been updated and passed within the past decade or so to accomplish this. The Animals (Scientific Procedures) Act in the U.K. requires the licencing of premises, program and individuals performing the work. In the U.S. two laws provide the coverage: the Animal Welfare Act, (administered by the U.S. Department of Agriculture), and the Health Research Extension Act, (administered by the National Institutes of Health). Each law provides (sometimes overlapping) regulations for oversight, review of use, surveillance and inspections. No national laws exist in Canada where surveillance is accomplished by a national peer-review system overseen by the Canadian Council on Animal Care (CCAC), founded in 1968. The CCAC comprises more than 20 member organizations representing scientists, educators, and representatives of animal welfare, industry and government. The body publishes standards and guidelines for animal use and oversees a program of regular announced and unannounced inspections of facilities in Canada.

Justifying Animal Use

In order to implement the 'Alternatives' concept mentioned above, it is essential that the proposed animal use is reviewed beforehand. This provides a 'cost-benefit' analysis of the project and a judgement of the justification of thc use. Such an analysis is an inherent part of the U.K. law. Regulations in the U.S. and Canada stress the importance of a local committee to review and monitor animal use. Such committees usually comprise at least: a scientist, a non-scientist, a doctor of veterinary medicine and a community member not connected to the institution. The animal committee must weigh the benefit of the animal use against the ethical cost in order to be satisfied: a) that the goal is worthwhile; b) that it has a high moral claim to be achieved; c) that there is no less drastic method of achieving it; and d) that there actually is some reasonable possibility of achieving the goal (Smith and Boyd 1991c). Besides reviewing protocols such committees are required to carry out inspections within their institution and to make written reports of these.

Conclusion

In describing proposed use of animals for research and teaching, investigators should provide clearly-written descriptions of objectives and procedures. A summary, written in simple lay language, should provide the purpose of the project, the expected benefit, the reason for using animals (why non-animal options cannot be used), and the reason for using the requested species. The number of animals, their housing, the procedures performed on them (and by whom) and the endpoint of animal use must be described in sufficient detail to allow an assessment of ethical cost of the animal use. Russell and Burch's '3Rs' should be used as guiding moral principles toward responsible ethical treatment of animals.

References

Beauchamp, T.L. (1992) The moral standing of animals in medical research. J. Law Med. Health Care 20: 7–16.

Russell, W.M.S. & Burch, R.L. (1959) The Principles of Humane Experimental Technique. Methuen, London, UK. (Reprinted by Universities Federation for Animal Welfare, Potter's Bar, UK, 1992).

Smith, J.A. & Boyd, K.M., eds. (1991a) In: Lives in the Balance. The Ethics of Using Animals in Biomedical Research (The report of a working party of the Institute of Medical Ethics), Chapter 11, pp. 296–297, Oxford University Press, Oxford, UK.

Ibid (1991b) Chapter 11, p. 298.

Ibid (1991c) Chapter 3, p. 37.

Smyth, D.H. (1978) Alternatives to Animal Experiments. Scolar Press, London, UK.

US Animal Welfare Act (1985) 7 U.S.C. §§2131–2157. Section 13(a)(3)(B).

Ethical Considerations Involved in Trace Element Research with Humans

PHYLLIS E. JOHNSON

USDA, Agricultural Research Service, Pacific West Area Office, Albany, CA 94710, USA

Keywords ethics, human volunteers, radioisotopes

Research in nutrition and metabolism, like all areas of biomedical research, must ultimately involve experimentation with human subjects. Despite their many metabolic and physiological similarities, rats are not just little humans. The public's interest in regulating biomedical research has been strong since World War II, and continues to increase. Revelations concerning use of radioactive materials in unknowing or poorly informed subjects in the United States during the 1940's, 50's, and 60's (Mann 1994) resulted in a massive nationwide investigation that was covered not just by scientific publications, but by mass media across the country. Even popular magazines have recently carried articles advising readers of what to consider before volunteering as research subjects.

Nutrition research in humans must be carried out in accordance with international principles governing all biomedical research, such as the Nuremburg Code (Anon. 1949) and the Declaration of Helsinki (World Med. Assoc. 1964 et. seq). In addition, most nations have their own, more detailed legislation and regulations governing such work, and individual institutions further regulate the conduct of such investigations. Since a large proportion of research on trace element nutrition and metabolism is done with healthy subjects, these moral and legal codes place severe constraints on what is allowable. Research with healthy subjects cannot be legitimately carried out unless the importance of the objective is in proportion to the inherent risk to the subject, and concern for the subject must prevail over the investigator's desire for knowledge. This discussion will be limited to research with healthy humans, where the main benefit to the subject is the good feeling that comes with participation in accruing knowledge that will benefit a large number of people.

Several conditions raise general ethical concerns regarding research in healthy subjects: pregnant and lactating women; infants and children; fertile women; minors who cannot give legal consent; prisoners; adults incapable of giving informed consent. Recently, fertile men have also been added to this list. Ethical/moral issues may also arise in the context of religious objections to certain experimental procedures.

For trace element research, manipulation of dietary intakes must be within "normal" ranges of dietary intake of an element when pregnant or lactating women and infants are involved, because both deficiency and excess of many trace elements may be teratogenic, toxic, or interfere with normal growth and development. Fertile women involved in studies where dietary intakes are substantially greater or less than normal must be well informed of the potential risks of pregnancy during the study. Trace element intake continues to affect both physical and cognitive development after infancy (often irreversibly), so the constraints on dietary or other intake of trace elements that apply to pregnancy, lactation, and infancy apply to children as well.

Researchers have tended not to consider potential effects on male fertility. There is emerging appreciation that not all effects on male fertility are rapidly reversible. Hawkes (personal communication 1996) reported effects of Se intake on sperm counts and motility. Four weeks of dietary Se intake lower than average (13 μg/d) resulted in increased sperm motility and greater semen volume, while dietary Se greater than normal (356 μg/d) resulted in decreases in sperm motility. The effects were not reversed during the study, so it is not known how long they last. Thus, in some cases, it will be necessary to advise male subjects of potential negative effects on future fertility.

Correct citation: Johnson, P.E. 1997. Ethical considerations involved in trace element research with humans. In *Trace Elements in Man and Animals – 9: Proceedings of the Ninth International Symposium on Trace Elements in Man and Animals. Edited by* P.W.F. Fischer, M.R. L'Abbé, K.A. Cockell, and R.S. Gibson. NRC Research Press, Ottawa, Canada. pp. 276–278.

Two other situations that have come to the author's attention have not been much written about. The first is religious objections to experimental procedures that are not otherwise unethical. A study of Zn metabolism in young men involved collection of semen samples; members of the nursing staff objected because it involved masturbation by the subjects. The investigators obtained approval of the experimental protocol from a Roman Catholic priest. Care was taken to ensure that information for prospective volunteers was clear about the inclusion of this activity as part of the study. Staff members who objected to being involved were scheduled so as not to have to transport samples.

Another scenario that has emerged in a new guise is the possibility of completely unqualified persons carrying out research with humans, without any review of experimental procedures. The author became aware of this through a telephone call from a high school student who was planning to do a science fair project involving supplementation with Se tablets purchased at a local store. The student was unaware of any necessity for review of her experimental plans, for involvement of a medical professional, for obtaining informed consent from subjects, or any other constraints. Many of her anticipated subjects were minors. Her teacher was also unaware of these requirements.

Although there are many difficulties associated with the use of radioisotopes in healthy humans, there are some conditions under which it can be both practical and ethical. For certain trace elements, (e.g., Cu, Zn, Fe, Mn) there are radioisotopes available with suitable half-lives and gamma energies such that experiments can be done in adults. Care must be taken to ensure that female subjects are not pregnant and are using contraception. Radiation exposures similar in magnitude to those encountered in various common activities, such as jet air travel or camping in the mountains, can be justified in subjects who derive no personal benefit from the procedure.

Finally, the issue of publication of experimental results must be considered. All of the usual ethical considerations relating to authorship, etc. apply to studies with humans. I propose an additional one: It is unethical to fail to publish publishable data obtained from research with subjects who derive no personal benefit from participation in the experiment. The reasoning for this is that in order to carry out these studies, one must justify the risks to the subjects largely on the basis of the benefit that the study will be for humanity. If the results of the study are sound, but are not published because of procrastination, laziness, or other fault of the investigator, this constitutes unethical conduct because no one can benefit from unpublished information.

References

Anonymous (1949) In: Trials of War Criminals before the Nuremburg Military Tribunals under Control Council Law No. 10, vol. 2, pp. 181–182. US Govt.. Printing Office, Washington, DC.

Mann, CC. (1994) Radiation: Balancing the record. Science 263: 470–473.

World Medical Association (1964) Declaration of Helsinki: Recommendations guiding physicians in biomedical research involving human subjects. Adopted by the 18th World Medical Assembly, Helsinki, Finland, June 1964, and amended by the 29th World Medical Assembly, Tokyo, Japan, Oct. 1975; 35th World Medical Assembly, Venice, Italy, Oct. 1983; and the 41st World Medical Assembly, Hong Kong, Sept. 1989.

Workshop Discussion

Following presentations by **Peter Flanagan** (The University of Western Ontario, London, ON, Canada) on ethical use of animals (Abstract 136. Ethical use of animals for experimentation and teaching) and by **Phyllis Johnson** (USDA, Albany, CA, USA) on ethical use of humans (Abstract 137. Ethical considerations involved in trace element research with humans), a lively discussion ensued.

There was some discussion about obtaining informed consent from subjects such as terminally ill patients (**Solo Kuvibidila**, Louisiana State University, New Orleans, LA, USA) and in third world countries where literacy is a problem (**Harold Sandstead**, University of Texas Medical Branch, Galveston, TX, USA). **Kenneth Wing** (Umea University, Umea, Sweden) wondered how you can possibly get informed consent with a one-page consent from. There was also some discussion regarding the increasing length of consent forms being used.

A number of investigators discussed the ethics of treating or not treating control groups. **Rosalind Gibson** (University of Otago, Dunedin, New Zealand) had earlier commented on the difficulty of not treating control groups in a third world situation where four villages are involved in a dietary intervention study. All the children in the villages will be dewormed initially and, after the intervention period with two of the villages, the other two will be offered the same treatment as the intervention villages. **John Howell** (Murdoch University, Murdoch, Western Australia) thinks that it is appropriate to treat control if positive results are obtained.

Harold Sandstead (University of Texas Medical Branch, Galveston, TX, USA) expressed some concerns about this approach as it may be difficult to actually do the follow-up treatment with the control groups. **Kenneth Wing** (Umea University, Umea, Sweden) suggested that problems such as these could be overcome at the design stage by using, for example, a cross-over design in the aforementioned example.

Richard Black (Kellogg Canada Inc., Etobicoke, ON, Canada) commented on the manner in which control groups are treated. For example, in a colon cancer trial where fat is being lowered and fibre increased, the control group will eat as usual even though the researchers feel that a high fat diet with low fibre affects the risk of CVD. Is it ethical then not to treat the control group? **Les Klevay** (USDA, Grand Forks, ND, USA) made a comment that we don't indeed know that decreasing fat in the diet will decrease the risk of CHD.

Following the discussion on control groups, a discussion of ethics committees ensued. **Gerry Combs** (Cornell University, Ithaca, NY, USA) asked if anyone has evaluated decisions taken by investigational research boards (IRB's) and whether there are differences among various IRB'S within countries, etc. **Peter Flanagan** (The University of Western Ontario, London, ON, Canada) said that IRB's play a large role in Canada and the USA. **Phyllis Johnson** (USDA, Albany, CA, USA) pointed out that local standards are supposed to apply and that there would therefore be variability. When working with government agencies, such as the USDA, you need to go through the process more than once and one often gets different responses from each committee, thus enforcing the point of variability within a country, etc.

Judy Butler (Oregon State University, Corvallis, OR, USA) commented that a number of researchers work in a number of communities and need approval in each location. How much in the way of local standards can be imposed in another country?

Richard Black (Kellogg Canada Inc., Etobicoke, ON, Canada) commented that it seems that one can argue one's case before an ethics committee and thus get approval which may or may not be based on merit.

Wanda Chenoweth (Michigan State University, East Lansing, MI, USA) commented that if a proposal can't get past 20 people then it likely should not be done. The fine points will likely change with the composition of the committee.

Barbara Golden (University of Aberdeen, Scotland) suggested that researchers take care of their own personal moral and ethics codes and consider whether or not you or your children would undergo the intervention in a study.

John Howell (Murdoch University, Murdoch, Western Australia) pointed out that in Australia, ethics committees have been of benefit and believes that they have improved experimentation.

Gillian Lockitch (Vancouver, BC, Canada) noted that she has had trouble obtaining ethics approval for research to obtain normal values for laboratory measurements, but on the other hand, is it then ethical to do lab tests on others when there are no reference data with which to compare results?

Peter Flanagan (The University of Western Ontario, London, ON, Canada) asked about when is enough? How do you ever know. Is the quest for knowledge the driving force or are we keeping labs going by doing research?

John Sorenson (University of Arkansas, Little Rock, AR, USA) had comments to make about a few points. He feels that animal care committees are not necessary and that researchers should take it upon themselves to decide what research should be done. He doesn't feel that one needs ethics approval for research with animals raised specifically for research. There should be no interference with the investigator's plan. There was considerable disagreement with this point of view.

Joseph Prohaska (University of Minnesota, Duluth, MN, USA) says that he is in favour of academic freedom but there should be a critical review of the scientific procedure. The public needs to be served and we have to be aware of ethics and cost effectiveness.

John Sorenson (University of Arkansas, Little Rock, AR, USA) said that discoveries are made when there free intellectual pursuit and that progress will be restricted if we restrict research by stringent ethics committees.

Phyllis Johnson (USDA, Albany, CA, USA) commented that in some cases, as in government, research work is driven by the mission of the government.

Koen Wienck (University Hospital Utrecht, The Netherlands) added that responsibility should be added to the 3 R's of research which Peter Flanagan mentioned in his talk.

Gerry Combs (Cornell University, Ithaca, NY, USA) closed the discussion by saying that responsibility really is the issue and that freedom does not come without responsibility. Further, increasingly institutional liabilities are guiding ethics committees.

Defining Trace Element Requirements of Preterm Infants

Chair: S.H. Zlotkin

Adequate Zinc Status to Six Months Corrected Age in Premature Infants Fed Mother's Milk Pre- and Post-Hospital Discharge

INE P.M. WAUBEN, BOSCO PAES, JAY K. SHAH AND STEPHANIE A. ATKINSON
Department of Pediatrics, McMaster University, Hamilton, Ontario, Canada L8N 3Z5

Keywords premature infants, mother's milk, hair zinc, zinc intake

Zinc (Zn) is nutritionally essential for the maintenance of cell growth and development. Zn stores in the infant born at term are accrued in the last trimester of pregnancy. The prematurely born infant will miss this period of Zn accumulation and thus have low Zn stores (Reifen and Zlotkin 1993).

It has been demonstrated that infants born prematurely have suboptimal Zn status at six months corrected age when compared to infants born at term as measured by hair Zn (Friel et al. 1984). Low Zn status may be of functional significance to growth, since Zn intake affects post-term age predicted longitudinal growth to 12 months corrected age in premature infants (Friel et al. 1985). The Zn content of mother's milk decreases rapidly post-partum, thus it has been suggested that premature infants fed mother's milk post-term age may be in suboptimal Zn status and at risk to develop Zn deficiency (Reifen and Zlotkin 1993). Based on this supposition, Zn supplementation for breast-fed premature infants has been recommended post-hospital discharge (Nutrition Committee, Canadian Pediatric Society 1995). One limitation of previous studies describing Zn status in premature infants is that they were performed in infants fed predominantly formulas. Zn from mother's milk has superior bioavailability when compared to formulas. McDonald et al. (1982) demonstrated that term born infants fed mother's milk had greater hair Zn levels when compared to formula-fed infants at six months of age despite a significant lower Zn intake.

For these reasons, we investigated the influence of breast feeding pre- and post-hospital discharge in premature infants on Zn status to six months corrected age.

Materials and Methods

Premature infants receiving mother's milk in hospital were randomized to receive either a mother's milk fortifier (Atkinson, Wyeth-Ayerst, Canada) (protein, calcium (Ca), phosphorus (P) and providing Zn at 14 mg/L) (Pre-MM + Zn) (N = 12: birth weight (BW) = 1.4 ± 0.2 kg, gestational age (GA) = 30 ± 2 weeks) or Ca/P alone (Pre-MM) (N = 13: BW = 1.3 ± 0.2 kg, GA = 30 ± 2 weeks). A group of premature infants fed preterm formula (Preemie SMA, Wyeth-Ayerst, Canada) served as a comparison group (FF comparison) (N = 12: BW = 1.2 ± 0.2 kg, GA = 30 ± 2 weeks). The infants were followed to six months

Correct citation: Wauben, I.P.M., Paes, B., Shah, J.K., and Atkinson, S.A. 1997. Adequate zinc status to six months corrected age in premature infants fed mother's milk pre- and post-hospital discharge. In *Trace Elements in Man and Animals – 9: Proceedings of the Ninth International Symposium on Trace Elements in Man and Animals. Edited by* P.W.F. Fischer, M.R. L'Abbé, K.A. Cockell, and R.S. Gibson. NRC Research Press, Ottawa, Canada. pp. 279–281.

corrected age. Zn intake was determined at term, three and six months and hair samples were obtained from the occipital portion of the scalp at six months corrected age. Breast feeding post-hospital discharge (Post-MM) was maintained at term, three and six months in 15/25, 10/25 and 7/25 infants, respectively. Breast feeding was defined as receiving over 60% of total milk intake as breast milk. Zn intake in hospital was determined by analyzing mother's milk and formula samples for Zn combined with a three-day intake record. Zn intake post-hospital discharge was calculated from five-day dietary records for formula-fed infants (Post-FF) and test-weights of breast-fed infants using a digital electronic scale accurate to one gram (Sartorius, Germany). Hair samples were washed in detergent (Acationox), rinsed several times in distilled deionized water and dry ashed. The reconstituted ash was subsequently analyzed for Zn. Zn analysis were performed by flame atomic absorption spectrometry (Varian Spectra, Canada).

Results

Zn intake in hospital was 34.0 ± 3.2, 11.7 ± 1.8 and 23.0 ± 3.3 µmol/kg/d for Pre-MM + Zn, Pre-MM and FF comparison groups respectively ($P < 0.05$). Zn intake post-hospital discharge at three months was 3.8 ± 1.4, 11.2 ± 4.0 and 11.2 ± 1.7 and at six months was 3.5 ± 1.4, 9.3 ± 2.1 and 8.7 ± 1.8 µmol/kg/d for Post-MM, Post-FF and FF comparison, respectively. Zn intake was significantly lower in Post-MM compared to Post-FF and FF comparison and was below the premature recommended nutrient intake for all infant groups (estimate:15 µmol/kg/d, Nutrition Committee, Canadian Pediatric Society 1995). Hair Zn values at six months corrected age are shown in Figure 1.

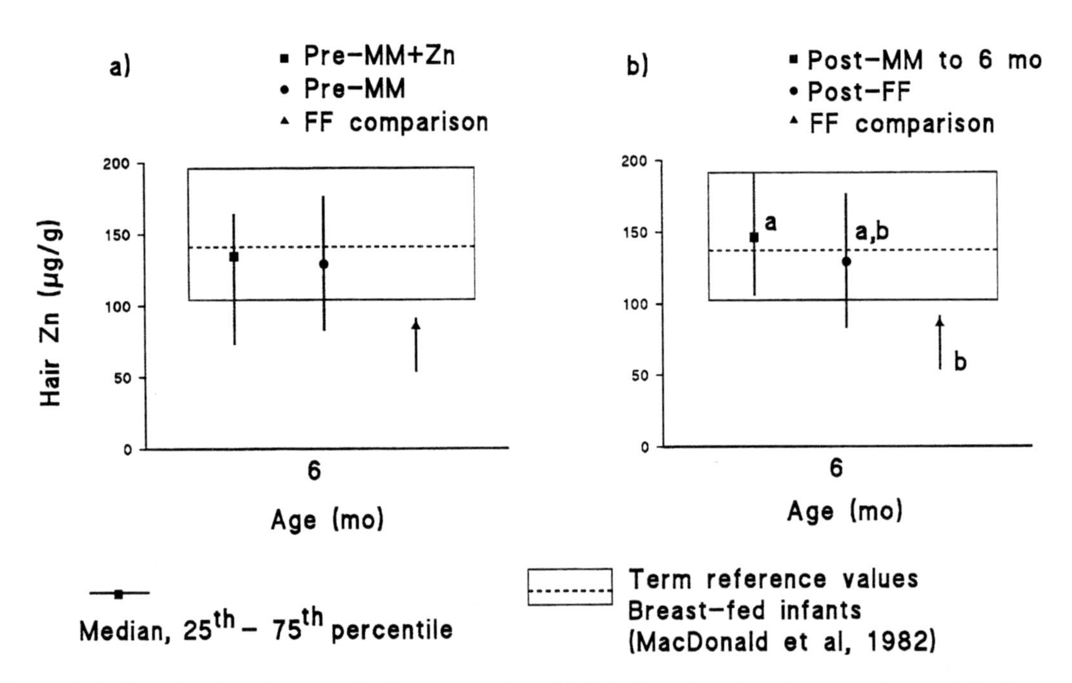

Figure 1. *a) Hair Zn values based on pre-discharge nutrition; b) Hair Zn values based on post-hospital discharge. nutrition. Values with different letters indicate $P < 0.05$.*

Summary

Zn supplementation to breast-fed infants in hospital (Pre-MM + Zn) did not improve Zn status at six months corrected age compared to breast-fed infants receiving no supplemental Zn (Pre-MM). Breast feeding in hospital resulted in improved Zn status when compared to formula feeding, although this did not reach significance (Figure 1a). Both breast feeding in hospital (Post-FF vs. FF comparison) and post-hospital discharge (Post-MM vs. FF comparison) improved Zn status which reached significance for the latter comparison (Figure 1b). Hair Zn values from preterm infants fed either mother's milk in hospital or post-hospital discharge to six months corrected age was in the normal range of previously reported values for term born breast-fed infants (McDonald et al. 1982).

In conclusion, Zn status in healthy breast-fed premature infants appears adequate and Zn supplementation is not necessary for premature infants after feeding is established at the breast. When defining optimal Zn intake, Zn bioavailability from the diet should be taken into account.

Literature Cited

Friel, J.K., Gibson, R.S., Balassa, R. & Watts, J.L. (1984) A comparison of the zinc, copper and manganese status of very low birth weight pre-term and full-term infants during the first twelve months. Acta. Ped. Scand. 73:596–601.

Friel, J.K., Gibson, R.S., Kawash G.F. & Watts, J. (1985) Dietary zinc intake and growth during infancy. J. Ped. Gastr. Nutr. 4:746–751.

MacDonald, L.D., Gibson, R.S. & Miles, J.E. (1982) Changes in hair zinc and copper concentrations of breast-fed and bottle-fed infants during the first six months. Acta Ped. Scand. 71:785–789.

Nutrition Committee, Canadian Pediatric Society. (1995) Nutrient needs and feeding of premature infants. Can. Med. Assoc. J. 152:1765–1785.

Reifen, R.M. & Zlotkin, S. (1993) Microminerals. In: Nutritional needs of the preterm infant — Scientific basis and practical guidelines. (Tsang, R.C., Lucas, A., Uauy, R., Zlotkin, S. eds.) pp:193–199. Williams & Wilkens, Baltimore, MD.

Selenium Levels in Preterm Infant Formulae and Breast Milk from the United Kingdom: A Study of Estimated Intakes

S. SUMAR,[†] B. KONDZA,[‡] AND L.H. FOSTER[†]

[†]Food Research Centre, [‡]Food Science Division, School of Applied Science, South Bank University, London SEI OAA, United Kingdom

Keywords selenium, infant formulae, breast milk

Introduction

Selenium (Se) is well recognised as an essential element in nutrition and health. It enters the food chain almost exclusively through plants, primarily in the form of selenates. Low intakes of Se have long been associated with Keshan and Kaschin-Beck disease. Recent studies have also shown Se to be involved in the possible prevention of various diseases, such as cancer (Willet and McMahon 1980), coronary heart disease (Salonen et al. 1982), multiple sclerosis (Shulka et al. 1977) and muscular dystrophy (Westermarck 1977). In addition, there is some concern that preterm and formulae fed lactose intolerant infants may also be at risk from Se deficiency due to their nutritional dependency on these products for extended time periods. Preterm infants have low hepatic stores and plasma Se concentrations, making them more vulnerable to haemolytic anemia, and possibly heart disease and cancer in later life (Reifen et al. 1993).

Limited information is available on the Se content of United Kingdom (UK) infant formulae. UK formulae are not thought to be routinely analysed for Se nor are they fortified, levels present being intrinsic to the geographic origin of the milk used in the manufacture, less possible losses due to processing. The aim of this study was to report and compare the average Se content of UK infant formulae and breast milk. The daily Se intake of infants aged 0 to 3 months consuming breast milk, infant formulae or a combination, as their sole source of nutrition, was also estimated. The Lower Reference Nutrient Intake (LRNI) and Reference Nutrient Intake (RNI) values for Se in the UK are 4 μg/d and 10 μg/d respectively. No Estimated Average Requirement (EAR) value for Se has been set in the UK.

Materials and Methods

Nine leading brands of UK powdered infant formulae samples (3 preterm, 2 term and 4 soya) were purchased from various retail outlets in London. In total, 162 infant formula samples were analysed for Se. Breast milk samples from 15 healthy London (UK) based mothers aged between 26–38 years, all of whom delivered preterm infants with a mean gestation period of 30 weeks, were also analysed for Se.

All samples including a non fat milk powder standard reference material (NIST, SRM 1549, Laboratory of Government Chemist, UK) were digested overnight with acid (nitric and perchloric). The total Se was then determined by hydride generation atomic absorption spectrometry (HGAAS). Details of the HGAAS method are described elsewhere by the authors (Foster and Sumar 1996a) as are accuracy and performance of the method.

Results and Discussion

The average Se levels found in UK preterm, term and soya formulae were 0.057 μg/g, 0.054 μg/g and 0.048 μg/g respectively compared to 0.116 μg/g for breast milk. Significant variation in Se levels were

Correct citation: Sumar, S., Kondza, B., and Foster, L.H. 1997. Selenium levels in preterm infant formulae and breast milk from the United Kingdom: A study of estimated intakes. In *Trace Elements in Man and Animals – 9: Proceedings of the Ninth International Symposium on Trace Elements in Man and Animals. Edited by* P.W.F. Fischer, M.R. L'Abbé, K.A. Cockell, and R.S. Gibson. NRC Research Press, Ottawa, Canada. pp. 282–283.

also observed between brands of infant formula powder due possibly to the geographic origin of the milk used and the varied processing methods or conditions employed in its manufacture (Foster and Sumar 1996a,b).

Figure 1 compares the daily Se intake (µg/L) of various UK preterm, term and soya infant formula with breast milk. These results show the selenium content of UK infant formula to be above the (UK) LRNI value but lower than the (UK) RNI value and those found in (UK) breast milk. A similar trend was also observed with most other countries (Figure 2). The UK average Se levels in infant formula and breast milk compare favourably with similar studies in Germany and the US and are generally higher than those in other countries.

There is little doubt that breast milk is the ideal source of nutrition, however, in the absence of further UK specific information on Se to justify fortification, these results show that it would be theoretically possible to raise the daily Se intake above the (UK) RNI value while on any formula feed (preterm, term or soya), by substituting at least one of the four formula feeds per day with breast milk. A further suggestion would be to voluntarily declare the Se levels naturally present in formula to aid selection where this may be important.

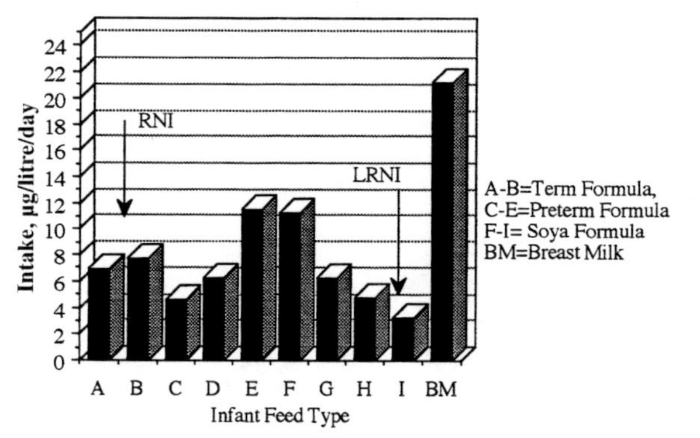

Figure 1. A comparison of Se intakes (µg/L/d) from various UK infant formulae.

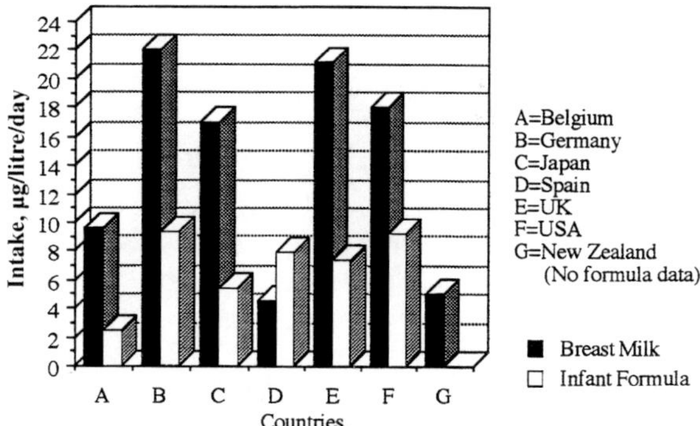

Figure 2. An estimation of daily intake of Se (µg/L/d) for infant formula and breast milk from various countries—a comparison. (Foster and Sumar 1996c).

Acknowledgements

The authors would like to express their thanks to P. Amuna, M.F. Chaplin, J.G. Davies, J.W.T. Dickerson and W. Winter for their assistance and helpful comments.

References

Foster, L.H. & Sumar, S. (1995) Methods of analysis for the determination of selenium in milk and infant formulae: a review. Food Chemistry. 53: 453–466.

Foster, L.H. & Sumar, S. (1996a) Hydride generation atomic absorption spectrometric (HGAAS) determination of selenium in term and preterm infant formulae available in the United Kingdom. Food Chemistry. 55: 193–198.

Foster, L.H. & Sumar, S. (1996b) Selenium concentration in soya based milks and infant formulae in the United Kingdom. Food Chemistry. 56: 93–98.

Foster, L.H., Kondza, B. & Sumar, S. (1996c) In: Metal Ions in Biology and Medicine, (J.L. Domingo., ed), vol. 4, in press. John Libbey Eurotext Ltd, France.

Reifen, R.M. & Zlotkin, S. (1993) In: Nutritional needs of the preterm infant. (Vavy, S., Zlotkin, S., Tsang, C., & Lucas,A., eds.), pp. 201–202. Williams and Wilkins Inc., Baltimore.

Salonen, J.K. (1982) Association between cardiovascular death and myocardial infarction in matched pair longitudinal study. Lancet. 2: 175–179.

Shulka, V.K.S., Jensen, G.E., Clausen, J. (1977) Erythrocyte glutathione peroxidase deficiency in multiple sclerosis. Acta Neurol. Scand. 52: 542–550.

Westermarck, T. (1977) Selenium content of tissue in Finnish infants and adults with various diseases and studies on the effect of selenium supplementation in neuronal ceroid lipofuscinosis patients. Acta Pharmacol. Toxicol. 41: 121–128.

Willet, W.C. & McMahon, B. (1980) Diet and cancer: an overview. N. Engl. J. Med. 310: 697–700.

Trace Element Intervention Studies: Design and Interpretation

Chair: G.F. Combs

Increases in Hemoglobin and Ferritin Resulting from Consumption of Food Containing Ferrous Amino Acid Chelate (Ferrochel) versus Ferrous Sulfate

H. DEWAYNE ASHMEAD,[†] SANDRA F.M. GUALANDRO,[‡] AND JOSE JOAO NAME[‡]
[†]Albion Labs., Inc., Clearfield, UT, USA, [‡]School of Medicine, University of Sao Paulo, Sao Paulo, Brazil

Keywords iron, anemia, hemoglobin, chelate

Iron (Fe) deficiency anemia is the most prevalent nutritional disease in the world. The most susceptible are women of childbearing age, adolescents, children, and infants (DeMaycr 1985)

Third world diets which contain little heme Fe and high vegetable fiber can impede Fe absorption. High potency Fe supplements (i.e., $FeSO_4$) can cause gastrointestinal (G.I.) problems, so supplementation compliance is generally low. Provided the Fe does not interact with dietary inhibitors (i.e., fibers, phytates, phenols, other minerals, etc.) fortification of staple foods is preferable to supplementation because no change is needed in the population's dietary habits (Cook 1983).

When Fe^{2+} is chelated to amino acids (FeAAC), bioavailability is significantly higher than $FeSO_4$ (Pineda et al. 1995) with significantly less G.I. irritability (Coplin 1991). This FeAAC is formed by reacting Fe^{2+} by coordinate covalent and ionic bonds to the alpha amino and carboxyl moieties of 2 glycine molecules so that the Fe is common to 2 heterocyclic rings. This current study compares utilization of this FeAAC to $FeSO_4$ in food fortification.

Fourteen volunteers were matched for age, sex, and initial hemoglobin (Hb) and divided into 2 groups as $FeSO_4$ ($n = 7$) and FeAAC ($n = 7$). Both groups were statistically identical. Initial ferritin iron values were also determined.

The $FeSO_4$ group received 45 mg Fe/day for 28 d. The FeAAC group received 22.5 mg Fe/day for 28 d as FeAAC (Ferrochel, Albion Labs., Inc.). This study required lower Fe intake from FeAAC. It was believed that 45 mg Fe from $FeSO_4$ was essential to reverse the Fe deficiency anemia in 28 d. Pineda et al. had previously shown that 30 mg Fe from FeAAC was as effective in elevating Hb as 120 mg Fe from $FeSO_4$. Consequently, we felt equivalent results were probable if Fe intake from FeAAC was one half that of $FeSO_4$. Furthermore, lower Fe levels would eliminate many potential problems inherent in Fe fortification of food. Except for 2 infants who each received Fe from either source in 100 ml of milk, the daily intake of Fe from either source was mixed with 20 g of a sweetened peanut butter bar (Paçoca), a common snack in Brazil. Verification of the Fe level in the Paçoca or milk was by inductive coupled plasma emission spectrometry. At the end of the 28 d fortification period, Hb and ferritin values were again determined and mean changes analyzed using paired t-test and multiple regression analysis, as summarized in Tables 1 and 2.

Correct citation: Ashmead, H.D., Gualandro, S.F.M., and Name, J.J. 1997. Increases in hemoglobin and ferritin resulting from consumption of food containing ferrous amino acid chelate (ferrochel) versus ferrous sulfate. In *Trace Elements in Man and Animals – 9: Proceedings of the Ninth International Symposium on Trace Elements in Man and Animals. Edited by* P.W.F. Fischer, M.R. L'Abbé, K.A. Cockell, and R.S. Gibson. NRC Research Press, Ottawa, Canada. pp. 284–285.

Table 1. Changes in hemoglobin values (g/dL).

	$FeSO_4$		FeAAC	
	Mean	S. D.	Mean	S.D.
Initial Hb	10.21	±1.48	10.58	±1.54
Final Hb	11.80	±1.12	12.30	±1.72
Increase	1.59	±0.71	1.72	±1.28

Table 2. Changes in ferritin values (μg/L).

	$FeSO_4$		FeAAC	
	Mean	S.D.	Mean	S.D.
Initial Ferritin	10.74	±8.96	13.90	±12.49
Final Ferritin	23.37	±13.65	27.87	±19.76
Increase	12.63	±11.54	13.97	±9.67

Both groups had significant Hb increases ($p < 0.02$). There was no significant difference between the groups. Fe bioavailability from FeAAC was at least twice that from $FeSO_4$ because only half as much Fe from FeAAC was supplied, but with equivalent results.

Both groups had significant increases in ferritin values ($p < 0.02$). The fact that the FeAAC group had a mean 1.3 μg/L greater increase in ferritin while consuming 1/2 the Fe of the $FeSO_4$ qroup, indicated that FeAAC was at least twice as bioavailable as $FeSO_4$.

Greater bioavailability suggests potentially greater toxicity. Previous studies comparing FeAAC to $FeSO_4$ have reported that FeAAC had significantly greater bowel tolerance (Coplin 1991) and 1/3 the toxicity, based on LD_{50} studies (Larsen 1982). Pathological and histopathological examinations of pig tissues from 4 filial generations to ascertain the effects of continual feeding of FeAAC, as well as the teratogenic effects of this FeAAC from multiple generations, discovered no abnormalities (Jeppsen 1993).

This current study showed significant correlation between body Fe status and the magnitude of change in Fe absorption. Multiple regression analysis determined that the magnitude of Hb change was dependent upon the form of Fe administered ($FeSO_4$ vs. FeAAC) and initial ferritin level. Hb increases were greater when the volunteers consumed food containing FeAAC, but the higher the initial ferritin value, the lower the absorption, regardless of Fe source. This suggested that in people with normal Fe metabolism, there is little danger of Fe overloading or subsequent toxicity when consuming foods fortified with either $FeSO_4$ or FeAAC. Multiple regression analysis showed that FeAAC was a more efficient source of Fe for food fortification than $FeSO_4$. The physiological response to FeAAC is such that only 1/2 the Fe dose resulted in higher mean Hb and ferritin levels. Because of a lower intake requirement and the previously determined reductions in G.I. irritability, FeAAC is a preferable source of Fe for fortification.

In conclusion, based on these data, FeAAC is a safe and more bioavailable form for fortifying food than is $FeSO_4$. Hb and ferritin values were elevated above values that could be obtained with $FeSO_4$, even though 1/2 as much Fe was supplied. It was found that absorption of Fe from FeAAC was regulated by the body, thus making FeAAC safe to include in the food. Overdosing is unlikely due both to metabolic regulation and the lower quantity required in the food.

Literature Cited

Cook, J.D. & Reusser, M.D. (1983) Iron Fortification: An update. Am. J. Clin. Nutr. 38: 648–659.

Coplin, M., Schuette, S., Leichtmann, G., & Lashner, B. (1991) Tolerability of iron: A comparison of bisglycine iron II and ferrous sulfate. Clin. Thera. 13: 606–612.

DeMayer, E. & Adelis-Teqman, M. (1985) The prevalence of anemia in the world. World Health Status Quo 315: 302–316.

Jeppsen, R.B. (1993) An assessment of long term feeding of amino acid chelates. In: The Roles of Amino Acid Chelates in Animal Nutrition (Ashmead, H.D. ed.), p. 106. Noyes, Park Ridge.

Larsen, A.E. (1982) LD_{50} studies with chelated minerals. In: Chelated Mineral Nutrition in 'Plants, Animals, and Man (Ashmead, D. Ed), pp. 163–169. Charles C Thomas, Springfield.

Pineda, O., Ashmead, H.D., and Perez, J. (1995) Effectiveness of iron amino acid chelate on the treatment of iron deficiency anemia in adolescents. J. Appl. Nutr. 46: 2–11.

Monitoring of Al, Cr, Pb, Cu and Zn in Water and in Dialysis Fluid in Haemodialysis Centres in Slovenia

M. BENEDIK[†] AND R. MILAČIČ[‡]

[†]Centre for Dialysis, Department of Nephrology, Clinical Centre, and [‡]Department of Environmental Sciences, Jožef Stefan Institute, 1000 Ljubljana, Slovenia

Keywords monitoring; Al, Cr, Pb, Cu and Zn; water and dialysis fluid; twelve haemodialysis centres in Slovenia

Introduction

Although dialysis patients could be exposed to metals from various sources, elevated concentrations in dialysis fluids due to failure in the water treatment still represents the most dangerous source of intoxication with Al and other oligoelements. Growing concern about the adequacy of water quality has resulted in many systematic studies being carried out in numerous haemodialysis (HD) centres world-wide. Water and dialysis fluids were monitored (Laurence and Lapierre 1995, Douthat et al. 1994, Lopéz Artíguez et al. 1989, Piccoli et al. 1989) and Al transfer from the dialysis fluid to dialysis patients was investigated as well (Douthat et al. 1994, Lopéz Artíguez et al. 1989, Piccoli et al. 1989). Various standards were issued which regulate upper limits for maximum concentrations of oligoelements in water used for preparation of dialysis fluids. According to the Canadian Standards Association (CSA 1986) this concentration for Al should not exceed 0.371 µmol/L. More rigorous limits for Al of 0.185 µmol/L (De Broe et al. 1993) and 0.148 µmol/L (Douthat et al. 1994) were recommended recently.

In this study systematic monitoring of Al and some other metals was carried out for more than one year in twelve HD centres in Slovenia, examining parameters which could influence the contamination of dialysis fluid. The twelve elements Na, K, Ca, Mg, Zn, Fe, Cu, Pb, Mn, Cd, Cr and Al were first measured in all twelve HD centres in tap water, water after softening, water after reverse osmosis and in dialysis fluid. On the basis of preliminary data, Al, Cr, Pb, Cu and Zn were monitored every two months in these fluids in each centre. These metals were also periodically measured in dialysis concentrates.

Apparatus

Al, Cr, Pb and Cu were determined by electrothermal atomic absorption spectroscopy (ETAAS) on a Hitachi Z-8270 Polarized Zeeman Atomic Absorption Spectrometer under clean room conditions (class 10 000). Zn was determined by flame atomic absorption spectroscopy (FAAS) on a Varian AA-5 Atomic Absorption Spectrometer in an air-acetylene flame. All measurements represented the average of three determinations.

Reagents and Sample Preparation

Merck suprapur acids and water doubly distilled in quartz were used for the preparation of samples and standard solutions. To avoid contamination by extraneous metals all laboratory ware was of polyethylene. Plastics were treated with 10% HNO_3 for 24 h and rinsed with an adequate amount of doubly distilled water before use. Samples were collected in 1 L polyethylene flasks. To 1 L of sample 1 mL of concentrated nitric acid was added to prevent adsorption of metals on the walls. With the exception of dialysis concentrate which was diluted with water (1 + 34) before measurement, all samples were measured directly without any special pre-treatment within 48 h of sampling.

Correct citation: Benedik, M., and Milačič, R. 1997. Monitoring of Al, Cr, Pb, Cu and Zn in water and in dialysis fluid in haemodialysis centres in Slovenia. In *Trace Elements in Man and Animals – 9: Proceedings of the Ninth International Symposium on Trace Elements in Man and Animals. Edited by* P.W.F. Fischer, M.R. L'Abbé, K.A. Cockell, and R.S. Gibson. NRC Research Press, Ottawa, Canada. pp. 286–287.

Results and Discussion

In Table 1 average concentrations, standard deviations and paired t-tests in tap water, water after softening, water after reverse osmosis and in dialysis fluid are presented from the data collected in 1 year's monitoring in 12 HD centres in Slovenia (number of observations = 55).

The present monitoring data indicate that reverse osmosis provided water which complied with all the requirements of the Canadian Standards Association (CSA 1986) and strategies of water treatment for dialysis fluid preparation (De Broe et al. 1993). It is evident that dialysis fluid is slightly contaminated with metals, although water after reverse osmosis and dialysis concentrates were clean. It is obvious from the high standard deviations for Al, Cu and Zn that dialysis fluid is not constantly contaminated but that contamination appears randomly. A adsorption-desorption processes, which might occur in plastic tubing inside HD monitors after treatment with contaminated water, may be responsible for this phenomenon.

*Table 1. Average concentrations ($\bar{x}$) and standard deviations (σ) (μmol/L) of oligoelements (determined by ETAAS and *FAAS) from 1 year's monitoring data in 12 HD centres in tap water, water after softening, water after reverse osmosis and in dialysis fluid (n = 55).*

	Al (μmol/L)	Cr (μmol/L)	Pb (μmol/L)	Cu (μmol/L)	*Zn (μmol/L)
Sample	$\bar{x}$ σ	$\bar{x}$ σ	$\bar{x}$ σ	$\bar{x}$ σ	$\bar{x}$ σ
T. w.	1.382±1.786	0.027±0.044	0.023±0.037	0.077±0.065	10.491±13.809
	NS	NS	+++	++	++++
S. w.	1.282±1.883	0.029±0.044	0.008±0.020	0.039±0.066	0.283±1.236
	+++	+++	+++	NS	NS
R.O. w.	0.048±0.041	0.004±0.002	0.005±0.000	0.017±0.025	0.158±0.465
	+++	+++	NS	++	NS
D. f.	0.130±0.119	0.010±0.008	0.010±0.010	0.096±0.353	0.350±0.959

paired t-test (two tail); +P < 0.05; ++ P < 0.01; +++P < 0.001; NS – nonsignificant.
T. w. – tap water; S. w. – water after softening; R.O. w. – water after reverse osmosis; D.f. – dialysis fluid.

Literature Cited

Canadian Standards Association (CSA) (1986). Exigences relatives au matériel de traitement de l'eau et à la qualité de l'eau pour l'hémodialyse, standard: # CAN3-Z364.2.-M86: 1–30.

De Broe, M.E., D'Haese, P.C., Couttenye, M.M., Van Landeghem, G.F., & Lamberts, L.V. (1993). New insights and strategies in the diagnosis treatment of aluminium overload in dialysis patients. Nephrol. Dial. Transplant. Suppl. 1, 8: 47–50.

Douthat, W., Acuña, G., Fernández Martín, J.L., Serrano M., González Carcedo, A., Canteros, A., Menéndez Fraga, P., & Cannata, J.B. (1994). Exposicíon al alumino y calidad del baño de diálisis: repercusión sorbe los niveles de alumino sérico. Nefrologia 14: 695–700.

Laurence, R.A., & Lapierre, S.T. (1995). Quality of hemodialysis water: A 7-year multicenter study. American J. Kidney Diseases 25:738–750.

López Artíguez, M., Soria, M.L., & Repetto, M. (1989). Niveles de aluminio en líquidos de diálisis y sueros de enfermos con insuficiencia renal crónica de la comunidad andaluza. Nefrologia 9: 86–90.

Piccoli, A., Andriani, M., Mattiello, G., Nordio, M., Modena, F., & Dalla Rosa, C. (1989). Serum aluminium level in the Veneto chronic haemodialysis population: Cross-sectional study on 1,026 patients. Nephron 51: 482–490.

Possible Adsorption–Desorption Processes of Al, Zn and Cu on Plastic Tubing in Haemodialysis Monitors

RADMILA MILAČIČ,[†] MIHA BENEDIK,[‡] AND SVETLANA KNEŽVIĆ[†]

[†]Department of Environmental Sciences, Jožef Stefan Institute, 1000 Ljubljana, Slovenia, and
[‡]Centre for Dialysis, Department of Nephrology, Clinical Centre, 1000 Ljubljana, Slovenia

Keywords adsorption–desorption processes; Al, Zn, Cu; plastic tubing; haemodialysis monitors

Introduction

In order to reduce the risk of exposure to Al and other metals in dialysis patients a rigorous water treatment is recommended to provide water of appropriate purity for preparation of dialysis fluid. Water monitoring quality programmes have been organised to assess the status in various haemodialysis (HD) centres (Laurence and Lapierre 1995, Douthat et al. 1994, Piccoli et al. 1989). Data on systematic monitoring in twelve haemodialysis (HD) centres in Slovenia indicated that in some centres, increased concentrations of Al, Zn and Cu appeared in dialysis fluid, although water samples after reverse osmosis and dialysis concentrates were clean and the tubing did not contain metal parts. These concentrations exceeded 0.371 µmol/L of Al, 1.57 µmol/L of Cu and 1.53 µmol/L of Zn, which were recommended as the maximum acceptable limits according to the Canadian Standards for water treatment (CSA 1986). Recently, more rigorous limits for Al of 0.185 µmol/L (De Broe et al. 1993) and 0.148 µmol/L (Douthat et al. 1994) in water used for preparation of dialysis fluid were recommended.

The increased concentrations of oligoelements which were randomly observed in dialysis fluid could be explained by adsorption-desorption processes occurring on the plastic tubing in which dialysis fluid circulates in HD monitors during the dialysis procedure. To confirm this hypothesis, contamination of the plastic tubing inside HD monitors with Al, Zn and Cu was simulated *in vitro*.

Material and Methods

Before each experiment the HD monitors (Gambro AK 10) were disinfected. Various amounts of contaminated water were then applied. An appropriate amount of Al, Zn or Cu sulphate was dissolved in doubly distilled water and fresh solutions used immediately for each particular experiment. The time of contamination of the plastic tubing of HD monitors was 0.5 or 4.5 h at a flow rate of 30 L h^{-1}, while the concentration of contaminated water ranged from 1.85 to 7.41 µmol/L for Al, 38.2 to 76.5 µmol/L for Zn (concentrations which were found in tap water) and 7.87 to 15.7 µmol/L for Cu. After contamination the tubing in the HD monitor was washed with clean water. After that dialysis fluid (in one experiment acetate and in another bicarbonate) was prepared by the conventional method for 10 min or 1 h, and samples of each dialysis fluid collected for 10 s in 1 mL fractions. Al and Cu in these fractions were measured immediately by electrothermal atomic absorption spectroscopy (ETAAS) under clean room conditions (class 10 000) and Zn by flame atomic absorption spectroscopy (FAAS).

Results and Discussion

An example of the results of our experiments is presented in Figure 1. The data from this study indicated that steady state concentrations were generally reached after 5 min. When the HD tubing was contaminated with as little as 1.85 µmol/L of Al for 4.5 h, the resulting steady state concentrations of Al in dialysis fluid were about 0.19 µmol/L for acetate and bicarbonate HD monitors. When HD tubing was

Correct citation: Milačič, R., Benedik, M., and Knežvić, S. 1997. Possible adsorption–desorption processes of Al, Zn and Cu on plastic tubing in haemodialysis monitors. In *Trace Elements in Man and Animals – 9: Proceedings of the Ninth International Symposium on Trace Elements in Man and Animals. Edited by* P.W.F. Fischer, M.R. L Abbé, K.A. Cockell, and R.S. Gibson. NRC Research Press, Ottawa, Canada. pp. 288–289.

contaminated with 7.41 µmol/L of Al for 0.5 h or 4.5 h the resulting steady state concentrations were higher (up to 0.74 µmol/L of Al) for acetate and bicarbonate HD monitors. Steady state concentrations for contamination of HD tubing with 38.2 µmol/L of Zn for 4.5 h and 76.5 µmol/L of Zn for 0.5 h were about 0.46–0.62 µmol/L of Zn for acetate and 1.53 µmol/L of Zn for bicarbonate HD monitors. These concentrations were higher when HD tubing was contaminated with 76.5 µmol/L of Zn for 4.5 h. When HD tubing was contaminated with 15.7 µmol/L of Cu for 0.5 or 4.5 h and 7.87 µmol/L of Cu for 4.5 h, the resulting Cu concentrations in dialysis fluid for both monitors ranged from 0.08 to 0.393 µmol/L.

Our *in vitro* experiment verified adsorption-desorption processes as the source of contamination of the dialysis fluid. Hence, it is of great importance to maintain adequate water quality constantly during the dialysis process. Any errors in the water treatment procedure would be later reflected in an enhanced content of oligoelements in the dialysis fluid, which could have fatal consequences for the health of patients exposed for years to long term haemodialysis.

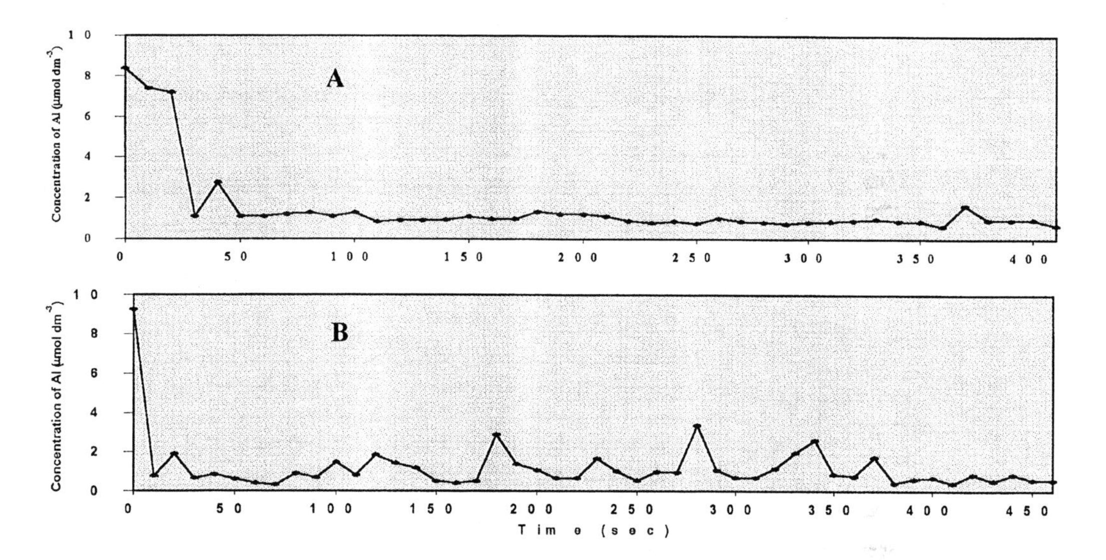

Figure 1. Desorption of Al after contamination of plastic tubing in HD monitor. Gambro AK 10 monitors: A – acetate, B – bicarbonate. Contaminated water ($Al_2(SO_4)_2 \cdot 18H_2O$, 7.41 µmol/L) passed through the monitor for 4.5 h at a flow rate of 30 L/h. After washing the tubing with clean water, clean dialysis fluid was applied (conventional HD procedure). Desorbed Al was measured in dialysis fluid by ETAAS in 1 mL fractions collected every 10 s.

Literature

Canadian Standards Association (CSA) (1986). Exigences relatives au matériel de traitement de l'eau et à la qualité de l'eau pour l'hémodialyse, standard: # CAN3-Z364.2.-M86: 1–30.

De Broe, M.E., D Haese, P.C., Couttenye, M.M., Van Landeghem, G.F., & Lamberts, L.V. (1993). New insights and strategies in the diagnosis treatment of aluminium overload in dialysis patients. Nephrol. Dial. Transplant. Suppl. 1, 8: 47–50.

Douthat, W., Acuña, G., Fernández Martín, J.L., Serrano M., González Carcedo, A., Canteros, A., Menéndez Fraga, P., & Cannata, J.B. (1994). Exposicíon al alumino y calidad del baño de diálisis: repercusión sorbe los niveles de alumino sérico. Nefrologia 14: 695–700.

Laurence, R.A., & Lapierre, S.T. (1995). Quality of Hemodialysis Water: A 7-Year Multicenter Study. American J. Kidney Diseases 25:738–750.

Piccoli, A., Andriani, M., Mattiello, G., Nordio, M., Modena, F., & Dalla Rosa, C. (1989). Serum aluminium level in the Veneto chronic haemodialysis population: Cross-sectional study on 1,026 patients. Nephron 51: 482–490.

Trace Elements and Animal Production

Chair: E.R. Chavez

Trace Elements and Animal Production

GARY L. CROMWELL

Department of Animal Sciences, University of Kentucky, Lexington, Kentucky 40546, USA

Keywords trace elements, microminerals, requirements, animals

Adequate dietary levels of certain trace elements are essential for optimal production of meat, milk, and eggs by cattle, swine, and poultry. The trace elements known to be essential for livestock and poultry are Cu, Fe, Zn, Mn, I, and Se. In addition, cattle and sheep have a dietary requirement for Co and sheep require Mo. Food-producing animals may also require certain other trace elements which have been shown to have a physiological role in laboratory species, but the requirements are so small that their dietary essentiality has not been conclusively proven.

The functions of the trace elements are quite diverse. Many are integral constituents of metalloenzymes or are essential for the activity of physiologically important enzymes. Most serve in various regulatory capacities in body tissues. The fact that most pigs and poultry are raised in confinement without access to soil or forage increases the importance of adequate trace element nutrition.

The trace element requirements are established by the National Research Council (NRC) and published in their nutrient requirement series (NRC 1984, 1985, 1988, 1989, 1994). In general, the requirements are more precisely estimated for pigs and poultry because more research dealing with trace elements has been done with these species. For example, in pigs, requirements are estimated for five weight classes (1–5, 5–10, 10–20, 20–50, 50–110 kg body weight) and for gestating and lactating sows. In poultry, estimates are established for growing chicks (broiler and egg-laying strains), layer hens, turkeys, ducks, and Japanese (Coturnix) quail. In contrast, a single requirement for each trace element is estimated for all weights and classes of cattle (except for I in dairy cattle and Fe in dairy calf milk replacers). For sheep, a range (rather than a single value) is estimated for each trace element. Table 1 shows the estimated requirements of the trace elements for the five major food-producing animals.

Table 1. Dietary trace element requirements of livestock and poultry (ppm).[a]

Element	Swine	Poultry	Beef	Dairy	Sheep
Cu	3–6	4–8	8	10	7–11
Fe	40–100	38–120	50	50–100	30–50
Zn	50–100	29–70	30	40	20–33
Mn	2–10	17–60	40	40	20–40
I	0.14	0.029–0.40	0.50	0.25–0.60	0.10–0.80
Se	0.10–0.30	0.05–0.20	0.20	0.30	0.10–0.20
Co	–	–	0.10	0.10	0.10–0.20
Mo	–	–	–	–	0.50

[a]NRC (1984, 1985, 1988, 1989, 1994).

Correct citation: Cromwell, G.L. 1997. Trace elements and animal production. In *Trace Elements in Man and Animals – 9: Proceedings of the Ninth International Symposium on Trace Elements in Man and Animals. Edited by* P.W.F. Fischer, M.R. L'Abbé, K.A. Cockell, and R.S. Gibson. NRC Research Press, Ottawa, Canada. pp. 290–293.

Table 2. Daily and dietary requirements of Zn and Fe in growing pigs and sows.[a]

		1–5	5–10	10–20	20–50	50–110	Gest.	Lact.
Zn	mg/day	25	46	76	114	155	95	265
	ppm	100	100	80	60	50	50	50
Fe	mg/day	25	46	76	114	125	152	424
	ppm	100	100	80	60	40	80	80

[a]NRC (1988). Gest. = gestation. Lact. = lactation.

The daily requirements for most of the trace elements increase as animals increase in body weight, as illustrated in Table 2. However, when expressed on a dietary concentration basis (ppm), the requirements decrease as animals increase in weight (Table 2). The requirements for reproduction and lactation for most trace elements are not greatly different than for growth when expressed on a ppm basis, but the requirements for lactation are considerably higher when expressed on a daily basis.

Trace elements are provided to animals in various ways. For pigs and poultry, they are generally included in the complete feed as a trace mineral premix. In ruminants, they are generally included in the concentrate (grain mix) as a trace mineral premix or as trace mineralized salt. In ruminants that do not receive a concentrate mix (such as breeding sheep or grazing cattle), trace elements are generally provided as trace mineralized salt, which is made available on a free-choice basis.

The remainder of this paper will give a brief overview of some of the important aspects of the trace elements required by livestock and poultry. Bioavailability in common trace mineral sources will also be addressed (Ammerman et al. 1995).

Copper. Cu plays many diverse roles in the body. It is required for hematopoiesis, collagen and elastin maturation, pigmentation of hair and feathers, keratinization of wool, integrity of the nervous system, and proper immune function. This element is required at relatively low dietary levels. Symptoms of Cu deficiency include anemia, aortic aneurysm, loss of hair and feather pigmentation, steely wool, and abnormal bone development. Cu deficiency in ruminants is common in many parts of the world (McDowell 1992).

Pharmacological levels of Cu (150–250 ppm) have been known for many years to stimulate growth and feed intake in growing pigs. Table 3 shows a recent study illustrating this effect in weanling pigs. The response to Cu is independent from the effects of antibiotics. High Cu, as Cu sulfate, is commonly included at 150–250 ppm in pig starter diets as a growth promotant. Sheep are very sensitive to high dietary Cu. Levels that are well tolerated by other animals can be toxic to sheep. An interaction between Cu and Mo has important nutritional implications for ruminants. Cattle that consume high-Mo forages are susceptible to Cu deficiency because excess Mo forms insoluble complexes with Cu and S and prevents Cu absorption. The Cu in sulfate, chloride, and carbonate forms is well utilized, but the Cu in the oxide and sulfide forms is poorly available.

Table 3. Pharmacological levels of Cu and Zn on performance of weanling pigs.[a]

	Control	High Cu	High Zn	Both	Trt effect
Daily gain, g	356	399	409	405	$P < 0.001$
Daily feed intake, g	614	651	672	665	$P < 0.001$
Feed/gain	1.74	1.65	1.66	1.65	$P < 0.001$

[a]Hill et al. (1996). Summary of 10 trials conducted at 10 universities and involving 1156 pigs from 3 to 7 weeks of age (initial wt, 6.5 kg). Antibiotics were included in all diets. Cu: 250 ppm as $CuSO_4 \cdot 5H_2O$. Zn: 3000 ppm as ZnO.

Iron. The important role of Fe in hemoglobin formation and cellular respiration is well known. The dietary requirement for Fe is relatively high compared with most of the other trace elements. Symptoms of Fe deficiency in all species is anemia. Fe is especially critical for young, nursing pigs for several reasons. Pigs are born with limited Fe reserves; the dietary Fe requirement is quite high (100 ppm); the pig's diet consists solely of milk (which is extremely low in Fe) for the first 2–3 weeks of life; newborn pigs grow very rapidly, quadrupling their weight during the first 3 weeks of life; and pigs are raised in facilities where they have no exposure to Fe-rich soil. As a result, supplemental Fe must be provided to pigs within a few days after birth to prevent anemia. Injectable Fe (100–150 mg as Fe dextran, Fe dextrin, or gleptoferron) is commonly given within 3 d after birth. Anemia is not such a problem in young calves or lambs because they do not grow as rapidly and they consume other feeds within a few days after birth.

Phosphate supplements (dicalcium phosphate, defluorinated rock phosphate) contain Fe (0.8–1.5%) which is fairly well utilized, so pigs, poultry, and ruminants receiving concentrates are able to obtain much of their dietary Fe from the phosphate supplements. The bioavailability of Fe in sulfate and chloride is high, in carbonate it is variable, and in oxide and sulfide the Fe is poorly available. Although it has no nutritional value, Fe oxide has traditionally been included in mineral mixes because of its distinctively red color.

Zinc. The requirement for Zn also is quite high for all animals as compared with the other trace elements. Zn functions as a component of many enzymes involved in energy and protein metabolism and is involved with pancreatic enzyme and insulin secretion. The classic symptom of Zn deficiency is parakeratosis in pigs. The requirement is influenced by the amount of phytic acid in the diet. For example, the Zn requirement of young pigs is 15–25 ppm when fed a phytic acid-free (starch-casein) diet, but the requirement is 80–100 ppm when fed a high phytic acid (grain-soybean meal) diet. High dietary Ca, especially when combined with phytic acid, further increases the Zn requirement. Zn sources with high bioavailability include the sulfate, chloride, and carbonate forms. The Zn in Zn oxide is less available.

Pharmacological levels of Zn (3000 ppm, as Zn oxide) have recently been shown to stimulate growth and feed intake in young pigs, similar to the effect of high Cu (Table 3). Reports from Europe suggest that high levels of Zn oxide also ameliorate diarrhea in young pigs. The mode of action of high Zn oxide is not understood.

Manganese. Most of the interest concerning Mn relates to poultry because of the effects of this trace mineral on leg soundness. The classic deficiency symptom is perosis, a condition in which the Achilles tendon slips from the condyles at the femur-tibia junction. The Mn requirement is considerably higher in poultry than in other animals. The Mn in sulfate and chloride forms is highly available, but it is intermediate in availability in oxide and carbonate forms.

Iodine. The role of I in thyroid function is well recognized. Many areas of the world are deficient in I, and both humans and animals are vulnerable to goiter if they do not receive supplemental I. In addition, some forages and concentrate feeds (rapeseed, triticale, soybean meal) consumed by livestock and poultry contain goitrogenic compounds which interfere with thyroid function. These problems are easily preventable by I supplementation. This element is commonly supplied as part of the trace mineral premix and/or as iodized salt. Most forms of I are relatively high in bioavailability.

Selenium. Se is an important part of the enzyme, glutathione peroxidase, which plays an important role in protecting membranes from peroxide damage. Se deficiency symptoms are similar to those of vitamin E deficiency in several species. The Se content of forages and grains is extremely variable. Crops grown in the northwestern, northeastern, and southeastern parts of the USA and in much of Canada are very low in Se, while certain areas in South Dakota, Wyoming, and Utah produce forage that can cause Se toxicity. Recently, corn samples from 15 midwestern states were found to vary from 0.02 to 0.20 ppm Se and soybean meal samples varied from 0.08 to 0.95 ppm Se (NCR-42 1993). Common sources of Se are Na selenate, Na selenite, and organic forms. Organic sources of Se seem to be more bioavailable than inorganic forms. In the USA, the Food and Drug Administration limits the amount of added Se in diets at 0.3 ppm.

Cobalt. This trace element is required only by ruminants for the synthesis of vitamin B_{12} by rumen microbes. Pigs and poultry have limited capacity to synthesize vitamin B_{12}, so these species are not considered to have a dietary requirement for Co.

Molybdenum. Mo is generally considered a toxic element rather than a required nutrient because of its negative effects on Cu utilization, as indicated previously. Sheep are the only species for which Mo is listed as a required nutrient. Low dietary Mo can result in Cu toxicity in sheep.

Chromium. Research studies with laboratory animals and humans have shown that Cr is involved in the regulation of glucose metabolism. This appears to be true for large animals as well. Recent studies with swine have shown that 200 ppb of Cr, as Cr picolinate, increases lean tissue accretion and reduces fat deposition (Page et al. 1993). Sufficient data are not available to establish a requirement for this element.

Literature Cited

Ammerman, C.B., Baker, D.H, & Lewis, A.J. (1995) Bioavailability of Nutrients for Animals. Academic Press, New York.

Hill, G.M., Cromwell, G.L., Crenshaw, T.D., Ewan, R.C., Knabe, K.A., Lewis, A.J., Mahan, D.C., Shurson, G.C., Southern, L.L., & Veum, T.L. NCR-42 and S-145 Regional Swine Nutrition Committees (1996) Impact of pharmacological intakes of zinc and(or) copper on performance of weanling pigs. J. Anim. Sci. 74 (Suppl. 1) (in press).

McDowell, L.R. (1992) Minerals in Animal and Human Nutrition. Academic Press, New York.

NCR-42 Committee on Swine Nutrition (1993) Variability among sources and laboratories in selenium analysis of corn and soybean meal. J. Anim. Sci. 71 (Suppl. 1): 67.

NRC (1984) Nutrient Requirements of Beef Cattle (6th Ed.). National Academy Press, Washington, DC.

NRC (1985) Nutrient Requirements of Sheep (6th Ed.). National Academy Press, Washington, DC.

NRC (1988) Nutrient Requirements of Swine (9th Ed.). National Academy Press, Washington, DC.

NRC (1989) Nutrient Requirements of Dairy Cattle (6th Ed.). National Academy Press, Washington, DC.

NRC (1994) Nutrient Requirements of Poultry (9th Ed.). National Academy Press, Washington, DC.

Page, T.G., Southern, L.L., Ward, T.L., & Thompson, Jr, D.L. (1993) Effect of chromium picolinate on growth and serum and carcass traits of growing-finishing pigs. J. Anim. Sci. 71: 656–662.

Discussion

Q1. John Howell, Murdoch University, Murdoch, Western Australia: You've certainly covered a lot of ground in a short space of time. The zinc results were interesting. The pancreas, for example in sheep, seems to be most susceptible to high levels of Zn. Have you looked at pancreas with high levels of Zn?

A. I don't know if that has been looked at, there doesn't seem to be any long term negative effects.

Q2. Nick Costa, Murdoch University, Murdoch, Western Australia: Coming back to your Cu and antibiotic data, you said there was a big response with the combined effect in the weight gain, but wasn't much in the feed:gain ratio. So it must have stimulated appetite. How do you think that works, through a direct effect or an indirect effect? How would you determine what that mechanism is?

A. I don't know that I have an answer for that, but this is generally the effect that is seen, particularly with Cu is that we see a greater response in growth rate and feed intake than we do with feed:gain. Generally there's some response, but it's not nearly to the same degree. Even with straight Cu we see the same thing, we get a bigger response in feed intake. There's been some work, at Virginia Tech. for instance they've done some studies where they've attempted to pair-feed animals, and when they do, they still get some response with Cu, though not nearly to the same degree. I think Cu does play some role in stimulating feed intake.

Q3. Manfred Anke, Friedrich Schiller University, Jena, Germany: You spoke of supplementation of sheep with molybdenum Can you give us some more information? I ask because in Europe we could not find in practice any molybdenum deficiency, since the requirement is lower than 0.1 mg dry matter in feedstuffs.

A. This is a little bit out of my area, but I would guess that the requirement set by NRC is not based on a lot of data, and is probably an estimate. There are some ruminant nutritionists in the audience here that could probably help you out.

Q4. Wilhelm Windisch, Technical University of Munich, Germany: Just a short comment to these high additions of Cu and Zn; in some countries in Europe (including Germany) this is not allowed because of environmental problems. These high levels are absolutely over the maximum allowances for Cu in the diet.

A. Good point. I didn't comment on that mainly because our next speaker is going to address the subject of mineral feeding and effects on the environment. Certainly that question is being raised in this country as well. Keep in mind that the high level of zinc, when it is fed, is usually fed for a very short period of time. Most people are using it in Phase I starter diet, where they're feeding it only for about one week and then copper is usually followed for about 3 weeks, and then most producers are not feeding high copper beyond that point.

Impact of Trace Element Nutrition of Animals on the Food Chain

J.T. YEN
U.S. Meat Animal Research Center, USDA-ARS, Clay Center, NE 68933, USA

Keywords trace elements, animal nutrition, animal production, food chain

Trace amounts of seven mineral elements including iron, zinc, copper, manganese, selenium, iodine and cobalt are usually added to the diets of food-producing animals. For monogastric animals, Co is supplied in its active form, vitamin B_{12}. Dietary supplementation of these trace elements and their accretion in body tissues assures optimal animal production, and enhances animal products as an important source of trace elements in the human food supply.

Animal products provide 36% of the Fe, over 70% of Zn and nearly 100% of vitamin B_{12} in the diets of all individuals in the United States (NRC 1988). Organ meats are the richest sources of Cu in the diet (NRC 1989). Seafoods, egg yolk, and meat are also consistently good sources of Se (NRC 1989).

Manure Trace Mineral Production by Livestock and Poultry

Not all supplemental trace elements are absorbed by animals and incorporated into edible animal products. Significant quantities of ingested trace minerals are excreted by food-producing animals and appear in the manure. Using 1994 U.S. livestock and poultry statistics (Feedstuffs 1995), along with the values of manure production and characteristics (ASAE 1993) and assumed average animal liveweights, the U.S. livestock and poultry industry was estimated to produce 1.451 billion metric tons of fresh manure with 247 million metric tons total solid of manure in 1994. This manure contained 329, 84, 12, and 37 thousand metric tons of Fe, Zn, Cu and Mn, respectively.

Environmental and Food Chain Concerns Relating to Manure Trace Minerals

Land application of animal manures is a convenient and long-used disposal method that supplies nutrients needed by crop plants and improves the physical condition of soils (CAST 1995). The use of animal manures on crop land generates concern regarding contamination of the environment and the food web on which the human food supply depends.

Compared with the concentrations of trace elements in dry biosolids (sewage sludges) the levels of trace elements in dry animal manures are lower. Concentrations of Zn and Cu in dry biosolids were 1700 and 800 ppm, respectively (Chaney 1980), and that in dry animal manures, depending on species, were 130–1190 and 20–109 ppm, respectively (ASAE 1993). Therefore, the likelihood of undesirable impacts on the food chain from land application of animal manures would appear to be less than that could possibly be caused by land application of biosolids. A great deal of research has been conducted on land application of biosolids and its influence on public health (Chaney 1990a,b). The intensive research on environmental and food chain effects of agricultural use of biosolids has demonstrated clearly that biosolids can be used safely and effectively as soil amendments.

The following discussion on the soil-plant-animal-human interrelationships relating to manure trace minerals is based on the premise that concepts developed from land application of biosolids are applicable to the assessment of potential impact of manure trace minerals from land application on the food chain. Among those trace elements that are excreted in animal manures, only Cu and Zn will be reviewed because of the relative low tolerance of Cu in sheep, the use of high dietary level of Cu as a growth stimulant in young swine and the competitive biological interactions between Cu and Zn.

Correct citation: Yen, J.T. 1997. Impact of trace element nutrition of animals on the food chain. In *Trace Elements in Man and Animals – 9: Proceedings of the Ninth International Symposium on Trace Elements in Man and Animals. Edited by* P.W.F. Fischer, M.R. L'Abbé, K.A. Cockell, and R.S. Gibson. NRC Research Press, Ottawa, Canada. pp. 294–297.

Soil-Plant Barrier for the Transfer of Manure Cu and Zn to Animals and Humans

One pathway for entry of manure and biosolids trace elements into terrestrial food chains is through the biosolids/manure-soil-plant route. A soil-plant barrier may protect the food chain from toxicity of a biosolids/manure trace mineral by limiting maximum levels of that element in edible plant tissues to levels safe for animals and humans (Chaney 1980). A soil-plant barrier can be formed by 1) insolubility of the element in soil that prevents its uptake by plants, 2) immobility of the element in plant roots that prevents its translocation to edible plant tissues, or 3) phytotoxicity of the element occurring at concentrations of the element in edible plant tissues below levels injurious to animals. Phytotoxicity can reduce the yield or even kill crops.

Unlike Fe and some other elements that are strongly bound to soil or retained in the plant roots, Cu and Zn are easily absorbed and translocated to food-chain plant tissues. However, phytotoxicity may limit plant levels of Cu and Zn to levels safe for animals and humans. Phytotoxicity of Cu occurs in most plants at about 25–40 ppm dry foliage (Chaney 1980), whereas Cu tolerance level for sensitive sheep is 25 ppm and that for cattle and monogastric animals is above 100 ppm (NRC 1980). No overt Zn toxicity would appear in most animal species when the dietary Zn level is less than 600 ppm (NRC 1980), yet phytotoxicity occurs in most plants between 100 and 400 ppm Zn (Kabata-Pendias and Pendias 1992).

The use of high dietary levels of Cu (125–250 ppm) as a growth promotent for pigs has been questioned because of the concern about potential environmental risks associated with disposal of the resulting Cu-rich manure on agricultural land. However, field experiments show that 14 annual applications of Cu-rich swine manure, at high rates (27 kg Cu · ha^{-1} · y^{-1}), were not detrimental to corn yield and Cu concentration in corn grain on three diverse Virginia soils (Reed et al. 1992). These results clearly demonstrate that Cu-rich swine manure is not phytotoxic to corn plants.

Plant–Animal Barrier for the Transfer of Biosolids/Manure Cu and Zn to Humans

The soil-plant barrier is an effective mechanism for the agricultural food chain to be protected from excessive plant uptake of biosolids/manure trace minerals. The transfer of Cu and Zn from the biosolids/manure to humans is further safeguarded by a plant-animal barrier. The plant-animal barrier functions in limiting the bioavailability of minerals from ingested plant tissues. Interactions with many other dietary factors, such as phytate and high levels of Zn, Fe, Ca, Mo and sulfate, reduce the intestinal absorption of plant Cu by the animal (McDowell 1992, Miller et al. 1991) and thus, minimize the possibility of transferring excessive quantities of biosolids/manure-derived plant Cu to humans via the biosolids/manure-soil-plant-animal pathway. The plant-animal barrier can also restrict the transfer of an unsafe amount of biosolids/manure-derived plant Zn through animals to humans, because the intestinal absorption of Zn from ingested plant tissues is decreased by phytate and high dietary levels of Cu, Ca, P, and Cr (McDowell 1992, Miller et al. 1991).

Trace minerals in biosolids and animal manures can bypass the soil-plant barrier and enter the gastrointestinal tract of animals, if there is a direct ingestion of the biosolids/manure lying on the surface soil or adhering to foliage. Application of a high rate of Cu-rich pig slurry (115 m^3/ha) for 9 times over a 3-year period added 49.2 kg Cu/ha to pasture and increased Cu content of the soil, however, it did not affect the Cu level in herbage (Poole et al. 1983). No accumulation of liver Cu occurred in sheep grazed on those plots for 15 weeks, even though fecal analysis indicted that the animals had ingested greatly increased Cu. The plant-animal barrier is definitely providing an extra safety factor to restrict the absorption and liver accumulation of manure-borne Cu in the animal and, thus lessen the potential of transferring biosolids/manure Cu to humans.

Animal–Human Barrier in the Transfer of Biosolids/Manure Cu and Zn to the Human

Upon absorption, the biosolids/manure-borne trace minerals are distributed in the body and may be accumulated differentially in various tissues. Feeding corn silage produced on biosolids -amended soil to dairy goats resulted in lower levels of Cu in liver and kidney tissues, but it had no effect on the tissue Cu content of heart and muscle (Bray et al. 1985). When cows and young steers were fed a diet containing 12% of biosolids for 9 months followed by a 4-month withdrawal period, their Cu level in liver tissue was increased during the biosolids-ingestion period and was decreased after biosolids-withdrawal but that in kidney and muscle was not affected (Baxter et al. 1982). Whereas ingestion of the biosolids increased Zn content of kidney, it had no effect on the Zn content of liver and muscle. Therefore, by consuming appropriate animal tissues, the potential of human poisoning from biosolids/manure trace minerals can be mitigated. The poor efficiency of human absorption of dietary Zn and Cu (Fairweather-Tait 1992) would further reduce chances of human toxicity from ingesting animal tissues containing manure-borne Zn or Cu.

Impact of Biosolids/Manure Trace Minerals on Marine Food Chain

When biosolids and animal manures are applied to land, contamination of ground water with biosolids/manure trace minerals is very minimum because the soil effectively filters out the trace minerals and allows very little of the trace elements to leach into ground water. However, runoff from fields and feedlots or farmyards, if not captured in retaining facilities, will pollute surface water. U.S. Environmental Protection Agency has established federal water quality policy and is actively administering several regulatory and non-regulatory federal programs for animal waste (Weitman 1995). A number of state agencies and local offices are also increasingly active in overseeing and aiding animal producers and farmers on animal waste management methods (CAST 1995). Although most regulations are concerned with nitrogen and phosphorous pollution, policy regarding land application limits for biosolids trace elements including Cu and Zn has been published by the U.S. EPA (1993). Strict adherence to the federal and state regulations on the management and disposal of animal manures can avoid animal waste contamination of surface water and any hazardous impact on marine food chain.

Nutrient Management of Food Animals

Enhancing the bioavailability of supplemental trace minerals will reduce the quantity of supplemental elements needed by food animals and decrease the amount of trace elements excreted in animal manures. The bioavailability of many minerals is adversely affected by their complexing with phytic acid. The use of supplemental phytase to hydrolyze phytate should improve the bioavailabilty of many trace elements. Indeed, recent studies showed that supplemental phytase increased the bioavailability of not only P but also Ca and several trace elements, including Cu and Zn (Adeola et al. 1995). Gastrointestinal absorption of trace minerals is further influenced by their chemical forms and interactions with other dietary components, as well as the age, sex and mineral status of animals (McDowell 1992, Miller et al. 1991). The bioavailability of many trace minerals can be improved if they are provided as suitable chemical forms with proper amounts and proportions to other antagonistic elements. Fine-tuning dietary formulation to supplement trace minerals to match animal needs for specific elements can minimize excretion of extra minerals. In conclusion, animal manures are extremely safe when applied to agricultural lands. Appropriate supplementation of trace elements in food animals does not put the food chain at risk.

Literature Cited

Adeola, O., Lawrence, B.V., Sutton, A.L. & Cline, T.R. (1995) J. Anim. Sci. 73:3384–3391.

ASAE. (1993) Manure Production and Characteristics. pp. 530–532. Standards, D384-1. Amer. Soc. Agric. Engn., St. Joseph, MI.

Baxter, J.C., Barry, B., Johnson, D.E. & Kienholz, E.W. (1982) J. Environ. Qual. 11:616–620.

Bray, B.J., Dowdy, R.H., Goodrich, R.D. & Pamp, D.E. (1985) J. Environ. Qual. 14:114–118.

CAST. (1995) Waste Management and Utilization in Food Production and Processing Rpt. 124. Council. Agric. Sci. Technol., Ames, IA.

Chaney, R.L. (1980) In: Sludge — Health Risks of Land Application (Bitton, G., Damron, B.L., Edds, G.T. and Davidson, J.M., eds.), pp 59–83 Ann Arbor Sci. Publ., Inc., Ann Arbor, MI.

Chaney, R.L. (1990a) BioCycle 31(10):69–73.

Chaney, R.L. (1990b) BioCycle 31(9):54–59.

Fairweather-Tait, S. (1992) Food Chem. 43:213–217.

Feedstuffs (1995) Reference issue. 67(30).

Kabata-Pendias, A. & Pendias, H. (1992) Trace Elements in Soils and Plants. 2nd Ed. CRC Press, Ann arbor, MI.

McDowell, L.R. (1992) Minerals in Animal and Human Nutrition. Academic Press, Inc., San Diego, CA.

Miller, E.R., Lei, X. & Ullrey, D.E. (1991) Trace elements in animal nutrition. In: Micronutrients in Agriculture. 2nd Ed. (Mortvedt, J.J. ed), pp. 593–662. Soil Sci. Soc. of Amer. Inc., Madison, WI.

NRC. (1980) Mineral Tolerance of Domestic Animals. National Academy of Sciences, Washington, DC.

NRC. (1988) Designing Foods. National Academy of Science, Washington, DC.

NRC. (1989) Recommended Dietary Allowances. 10th Ed. National Academy of Sciences, Washington, DC.
Poole, B.R., McGrath, D., Fleming, G.A. & Sinnott, J. (1983) Ir. J. Agric. Res. 22:1–10.
Reed, S.T., Martens, D.C., McKenna, J.R., Kornegay, E.T. & Lindemann, M.D. (1992) 1990–92 Virginia Tech Livestock Res. Report. 47–49.
U.S. EPA (1993) Standards for the Use and Disposal of Sewage Sludge (40 CFR Part 503). Fed. Register, Feb. 19, 1993.
Weitman, D. (1995) Proc. 7th Int. Symp. on Agricultural and Food Processing Wastes. pp. 1–5, Amer. Soc. Agric. Engn.

Discussion

Q1. Eduardo Chavez, McGill University, Ste-Anne de Bellevue, QC, Canada: What about supplementation with organic salts rather than with inorganic salts.

A. Many times bioavailability is greater with organic salts, eg chromium picolinate. The trace element and amino acid complex will cross over and enter the target tissue more readily than inorganics, and the requirement for supplementation of the trace element is then lower.

Q2. Geoffrey Judson, Central Veterinary Laboratory, Glenside, Southern Australia: In your consideration of this supplementation affecting the food chain, did you consider the isolated incidences where inappropriate use of direct supplements, for instance of injectable products into animals might affect things? There are a couple of products on the market, for instance copper supplements and long-acting selenium supplements that if inappropriately used can have quite significant effects in isolated instances. Has any of this been considered in the States, in your review of things? I'm thinking for instance of something like barium selenate, which is a very effective slow-acting product. If inappropriately administered into muscle tissue, then there's a risk in isolated cases of someone consuming that, and so consuming a high dose of selenium.

A. Not to my knowledge, but it would depend on what element we are talking about. With Se, the animal will excrete excess. The point should be considered.

Q3. Wilhelm Windisch, Technical University of Munich, Germany: You stated that the animal manures are safe with respect to the environment, because of the very effective plant barrier. We should not forget that this is due to a continuous accumulation especially of copper, but also of zinc, in the soil, so it's also a matter of time. So it may be safe for the moment, but we do not know for the future. For example, in Germany, fields used in the past for hops or vineyards where Cu was used as an antifungal agent, there still exist environmental and plant problems due to the accumulation that occurred in the soil. We have to consider the concentration in the manure, the time or duration of use, and also the intensity of the use of manure in a particularly sized space should be considered.

A. The data presented are based on continuous use for 4 to 10 years. It depends which plants you are referring to. With these plants, and the plant-soil barriers, it seems to be acceptable. The absorption and utilization of soil Cu will depend on pH, soil condition.

Q4. Nick Costa, Murdoch University, Murdoch, Western Australia: The surface water was shown to break the soil-plant barrier. Do you have data on the effects on fish populations or plant life in catchments as a result of run-off?

A. Yes, but could not cover the data in this session, due to time constraints. We should note that copper and zinc are soluble. Soluble elements will do more harm to fish than to humans. That is why the proper land application was stressed—to prevent run-off. There were several incidences of mineral spills into a creek and lots of fish were killed.

Effect of High Sulphur Intake as Elemental and Thiosulphate Sulphur on the Copper Status of Grazing Lambs

N.D. GRACE, J.R. ROUNCE, AND J. LEE

AgResearch, Grasslands, Private Bag 11008, Palmerston North, New Zealand

Keywords sheep. sulphur. copper. liver

Some New Zealand soils are deficient in both phosphorus and sulphur. The increase in the usage of S fertilisers containing a high proportion of elemental S applied at the rates of 30 to 90 kg S/ha, has been thought to be associated with an increase in the incidence of copper deficiency in ruminants, particularly cattle. Increasing S intakes as sulphate or S-amino acids have been reported to decrease the absorption and storage of Cu in sheep (Suttle 1974), while the S fertilisation (138 kg S/ha as ammonium sulphate) of sorghum resulted in a 50% decrease in the apparent absorption of Cu when the ensiled sorghum was fed to lambs (Ahmad 1995). In the pastoral situation elemental S is washed into the soil where it is only slowly oxidised to sulphate and therefore will only be consumed by grazing animals if their soil intake is high. This study was designed to determine the influence of increasing S intakes as elemental-S (E-S) and thiosulphate-S (T-S), in the presence of low dietary Mo, on the plasma and liver Cu concentrations of grazing sheep, and to relate the observed differences in total sulphur, sulphide and Cu concentrations in the gastrointestinal tract to changes in Cu status.

Materials and Methods

Animals. Forty two castrated Romney lambs, weighing 33–36 kg, were randomised into 6 groups of 7 animals. All animals were eartagged and grazed as a single mob on a ryegrass/white clover pasture and weighed at 4 weekly intervals.

Treatments. The 6 treatments were (1) control, given no supplementary S; (2) 2 g E-S/d; (3) 4 g E-S/d; (4) 4 g E-S and 15 mg Cu/d as $CuSO_4.5H_2O$; (5) 15 mg Cu/d; and (6) 4 g T-S/d. All S and Cu supplements were weighed into gelatin capsules which were then administered 3 times a week using balling gun over the 105-d study.

Collection of Samples. Pasture and blood samples were collected at monthly intervals. At the completion of the study the lambs were killed using a captive bolt gun, a ventral incision was made to expose the reticulorumen and small intestine, which were then quickly tied off and dissected out. The contents were removed, strained through cheese-cloth, subsampled, and centrifuged at 30 kg and the supernatant stored at –20°C. Cadmium acetate was added to one sub-sample as quickly as possible to trap volatile H_2S. The liver was removed, weighed, and stored at –20°C.

Analytical. Pasture, digesta and liver samples (0.5–1.0 g) were wet ashed, the residue dissolved in 2 M HCl, and S, Cu and other elements determined by inductively coupled plasma emission spectrometry (ICP). Total S, sulphide and acid volatile sulphide (AVS) fractions in the reticulorumen strained fluid were determined by the methods of Johnson and Nishita (1952) with the H_2S being liberated from CdS using 2 M HCl. Plasma Cu was determined by ICP after digesting a 1 mL sample in 1 mL concentrated AR HNO_3 and 1 mL 30% H_2O_2 in a water bath at 80°C for 4 h.

Correct citation: Grace, N.D., Rounce, J.R., and Lee, J. 1997. Effect of high sulphur intake as elemental and thiosulphate sulphur on the copper status of grazing lambs. In *Trace Elements in Man and Animals – 9: Proceedings of the Ninth International Symposium on Trace Elements in Man and Animals. Edited by* P.W.F. Fischer, M.R. L'Abbé, K.A. Cockell, and R.S. Gibson. NRC Research Press, Ottawa, Canada. pp. 298–300.

Results

The grazed pasture contained 2.6 g S/kg DM, 6.2 Cu, 141 Fe, 37 Zn and <0.5 mg Mo/kg DM. Increasing S intakes from 3.9 to 7.9 g/d had no significant effect on growth rates (90 g/d), plasma Cu concentrations (0.85 mg/L) or liver Cu concentrations (255 mg Cu/kg DM), but significantly ($P < 0.05$) increased total S, total sulphide and acid volatile sulphide concentrations in the reticulorumen and decreased soluble Cu in the small intestine (Table 1). The thiosulphate caused a much larger, but unexplained increase in total S. Increasing Cu intakes from 9.3 to 24.3 mg/d caused no significant changes in plasma Cu but increased several fold the Cu concentrations in the liver and reticulorumen contents (Table 1).

Table 1. Effect of S intake on Cu in liver, S and sulphide fractions in the reticulorumen and soluble Cu in the upper small intestine.

Treatment	Liver Cu µg/g DM	Total sulphur mg/g DM	Total sulphide µg/g DM	Acid volatile sulphide µg/g DM	Volatile sulphide µg/g DM	Soluble Cu small intestine µg Cu/mL
Control (3.9 g S/d)	173±58^{A}	4.1±0.65^{a}	138±13.1^{a}	39±8.6^{c}	99±8.9^{a}	0.11±0.02^{b}
+ 2 g S^0/d	263±60^{A}	5.9±0.43^{b}	140±9.5^{a}	34±6.2^{c}	106±14.2^{a}	0.14±0.02^{b}
+ 4 g S^0/d	274±60^{A}	7.7±0.89^{c}	189±14.9^{bc}	26±5.7^{bc}	163±15.1^{b}	0.06±0.01^{a}
+ 4 g S^0 and 15 mg Cu/d	701±75^{B}	8.4±0.93^{c}	198±19.2^{c}	27±4.5^{bc}	171±20.8^{b}	0.28±0.10^{bc}
+ 15 mg Cu/d	672±149^{B}	3.4±0.14^{a}	151±18.2^{ab}	19±4.0^{ab}	132±21.1^{ab}	0.48±0.24^{c}
+ 4 g $S_2O_3^{2-}$/d	311±60^{A}	20.4±6.7^{d}	177±29.8^{ab}	12±2.5^{a}	165±30.7^{ab}	0.05±0.006^{a}

Within a column means with different superscripts are significantly different a v b $P < 0.05$; A v B $P < 0.01$.

Discussion

Ruminants can metabolise methionine, elemental S and sulphate as the ingested S is reduced to sulphides in the reticulorumen before being used for the synthesis of microbial protein or absorbed as S^{2-} and H_2S (Bray and Till 1974). As the liver is a major storage organ for Cu in ruminants changes in liver Cu concentrations will reflect the Cu status of the sheep.

In this study increased elemental-S intakes (3.9 to 7.9 g S/d) in the presence of low and high Cu intakes decreased the concentration of soluble Cu in the small intestine but had no effect on the concentration of Cu in liver tissue. Therefore there was still adequate absorption of Cu, even though apparent CuS formation reduced levels of soluble Cu. It has been reported that increasing S intakes as sulphate or methionine from 0.04 to 4 g S/kg DM decreased the absorption of Cu (0.062 to 0.041) (Suttle 1974) and decreased liver Cu concentrations (26–30%) as well as the omasal flow of soluble Cu (5 to 2 mg/d) (Bird 1970). The relationships between dietary S and Cu absorption and rumen fluid sulphide concentrations and the daily flow of omasal soluble Cu are curvilinear with the greatest effects occurring at the lowest S intakes. The elemental-S did not have to have any significant influence on Cu absorption and storage at the high S intakes (3.9 to 7.9 g/d) fed in this study but at much lower S intakes an effect with elemental S may have been observed. Therefore the level of S intake as well as the form of dietary S in the presence of low dietary Mo (<2 mg/kg DM) appears to influence the magnitude of the Cu × S interaction.

Acknowledgments

The New Zealand Meat Research and Development Council and the Foundation for Science and Technology of New Zealand for funding.

Literature Cited

Ahmad, M.R., Allen, V.G., Fontenot, J.P., Hawkins, G.W. (1995) Effect of sulphur fertilization on chemical composition, ensiling characteristics, and utilisation by lambs of sorghum silage. J. Anim. Sci. 73: 1803–1810.

Bird, P.R. (1970) Sulphur metabolism and excretion studies in ruminants III. The effect of sulphur intake on the availability of copper in sheep. Proc. Aust. Soc. Anim. Prod. 8: 212–218.

Bray, A.C. & Till, A.R. (1975) Metabolism of sulphur in the gastro-intestinal tract. In: Digestion and Metabolism in the Ruminants (McDonald; I.W. Warner, A.C.I. eds). pp. 243–261. University of New England, Armidale.
Johnson, C.M., & Nishita, H. (1952) Microestimation of sulfur. In: Plant materials, soils and irrigation waters. Anal. Chem. *24*: 736–742.
Suttle, N.F. (1974) Effects of organic and inorganic sulphur on the availability of dietary copper to sheep. Br. J. Nutr. 32: 559–568.

Discussion

Q1. John Howell, Murdoch University, Murdoch, Western Australia: You said you were having problems with Cu deficiency in cattle, in the joint cattle-sheep enterprises. I would have thought then that there would have been a problem with the sheep, and the incidence of enzootic ataxia would have gone up. The animals this study started with were within 34 kg. It might be worthwhile repeating this experiment in younger animals that have more growing to do and therefore have a greater requirement for copper. This might give very different results.

A. There are farms out there that show cattle with disease, while the sheep are okay, based on liver copper results. In a different study but under the same conditions in the same region, we have found both sheep and cattle to be deficient.

Q2. Neville Suttle, Moredun Research Institute, Edinburgh, Scotland: You mentioned the work that we've done on predicting availability. We've got equations now for herbages, that have replaced those semi-purified diets. Pertaining to the bolus method of administration; what was the timing of the last dose before slaughter?

A. Six hours. This time point was chosen to represent a steady state of absorption after administration.

Q3. Neville Suttle, Moredun Research Institute, Edinburgh, Scotland: Inconsistent results may be due to not having taken into account diurnal patterns of Cu concentration in the digesta. Our work showed that the Cu absorptive mechanisms are saturated at quite low concentrations, and that although your sulfur supplements are bringing Cu concentrations in the digestive system down, it might not be bringing them down far enough to reduce Cu absorption.

A. I agree with that, thank you.

A Comparison of Methods of Assessing Copper Status in Cattle

ALEXANDER M. MACKENZIE, DOREEN V. ILLINGWORTH,
DAVID W. JACKSON, AND STEWART B. TELFER
Department of Animal Physiology and Nutrition, University of Leeds, Leeds, UK, LS2 9JT

Keywords cattle, copper, status, assessment

Introduction

The effects of trace element deficiencies on cattle production are of great importance to health and welfare. However, our understanding of their role and interactions are far from complete. One of the most important trace element deficiencies occurring in the United Kingdom affecting cattle production is clinical copper deficiency as it is now regarded as the Worlds second most common mineral deficiency in cattle (Wikse et al. 1992). Clinical signs of copper deficiency in cattle include alterations of hair pigmentation, bone formation, reduced growth rate and infertility. The aetiology of copper deficiency in cattle has now been shown to be due to complex interactions of copper with molybdenum and sulphur and iron and sulphur in the rumen where there is a limited supply of copper. The copper–iron–sulphur complex formed in the rumen is not absorbed by the animal and is excreted in the faeces as is the copper–molybdenum–sulphur complex. However, if there are insufficient levels of copper present in the rumen, the molybdenum–sulphur complexes (thiomolybdates) are then absorbed as ammonium thiomolybdates. Thiomolybdate then reacts with copper in the blood and tissues thus decreasing the activity of the important copper requiring enzymes. Several methods for the assessment of copper status have been advocated over the years. However, it is clear that their ability to predict the amount of biologically active copper is limited and there is evidence of decreased productivity and infertility in cattle similar to that described as clinical copper deficiency even when the copper status appears normal by measurements currently used (Phillippo et al. 1985). This study was designed to compare methods commonly used for the assessment of copper status and to describe the distribution of them in the cattle population.

Materials and Methods

Heparinised blood and serum samples were collected from 1571 cattle by Veterinary Surgeons in the period from September 1994 through to September 1995 from around the United Kingdom. These cattle comprised dairy cows at various stages of lactation, dairy heifers and suckler cattle. The blood samples were analysed for a variety of copper parameters and for selenium status at the University of Leeds. The copper parameters measured were: (1) plasma copper (Pl-Cu) by atomic absorption spectrophotometry, (2) TCA-copper (TCA-Cu) was determined as the soluble copper in 10% trichloroacetic acid as measured by atomic absorption, (3) erythrocyte superoxide dismutase (SOD) was determined by the method based on Misra and Fridovich (1977) adapted for the COBAS MIRA (Roche), (4) serum caeruloplasmin (CP) activities was determined by the method described by Henry et al. (1974) adapted for the COBAS MIRA (Roche), and (5) the ratio of caeruloplasmin activity to plasma copper concentration (CP/Pl-Cu) was calculated as described by Mackenzie et al. (1997).

Results

The results of these measurements are expressed in standard units and are presented as the distribution of the population according to whether the copper status could be classified as deficient, marginal, normal or “toxic”. For Pl-Cu and TCA-Cu concentrations, deficient is defined as being below 8 mmol/L, marginal was 8–12, normal was 12–23 and toxic levels were regarded as anything over 23 mmol/L. As shown in

Correct citation: Mackenzie, A.M., Illingworth, D.V., Jackson, D.W., and Telfer, S.B. 1997. A comparison of methods of assessing copper status in cattle. In *Trace Elements in Man and Animals –9: Proceedings of the Ninth International Symposium on Trace Elements in Man and Animals. Edited by* P.W.F. Fischer, M.R. L’Abbé, K.A. Cockell, and R.S. Gibson. NRC Research Press, Ottawa, Canada. pp. 301–302.

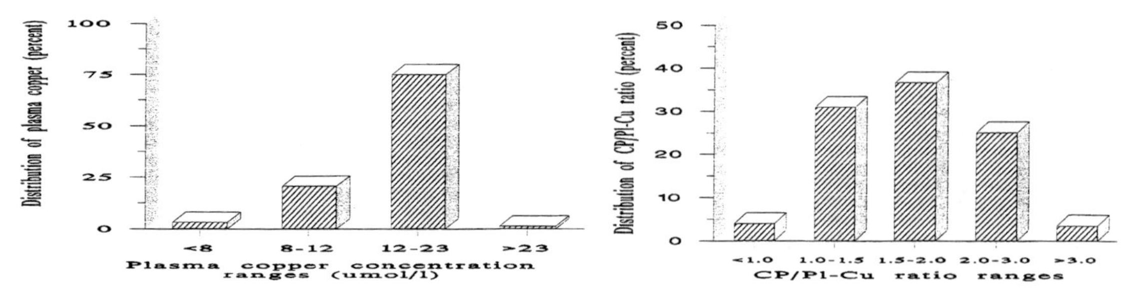

Figure 1. Distribution of plasma copper in cattle.

Figure 2. Distribution of CP/P1-Cu ratio in cattle.

Figure 1, the distribution of Pl-Cu concentrations show that 3.2% of cattle were classified as deficient, 20.7% marginal, 74.7% normal and 1.4% as toxic. The TCA-Cu concentrations showed a similar pattern with 3.5% deficient, 22.7% marginal, 72.5% normal and 1.3% toxic. Based on the erythrocyte copper enzyme, SOD, 2.3% of cattle were classified as deficient and the rest were in the normal range. The plasma copper enzyme, caeruloplasmin indicated that 11.0% of cattle were deficient whilst 89.0% were normal. However, when the copper status was expressed as the CP/Pl-Cu ratio (Figure 2), only 28.6% of cattle were regarded as being adequate and not requiring supplemental copper.

Discussion

Many of the cattle in this study were sampled by veterinary surgeons from herds that had fertility problems but were previously diagnosed to have adequate total plasma copper concentrations. The results from these cows confirm that when the copper status was measured by current conventional methods about 75% of cattle sampled were classified as being in the adequate range. There was no significant difference between the copper status when assessed by the plasma copper concentrations, TCA-copper, caeruloplasmin or erythrocyte superoxide dismutase activities. When copper status was assessed using the CP/Pl-Cu ratio the distribution of the population was altered, with just 25% of the cows being adequate. There is increasing clinical evidence from veterinary surgeons indicating that copper supplementation of these cows diagnosed as inadequate by the CP/Pl-Cu ratio have shown clinical responses to copper supplementation. It is felt that the present methods (especially total blood copper) used to evaluate copper status in cattle are inadequate and they fail to diagnose cases of thiomolybdate toxicity that will respond to copper supplementation.

References

Henry, R.J., Cannon, D.C. and Winkelman, J.W. (1974) Clinical Chemistry, Principles and Technics. Pp. 860. Harper & Row, Publishers, London.

Misra, H.P. and Fridovich, I. (1977) Arch. Biochem. Biophys., 181: 308–312.

Phillippo, M., Humphries, W.R., Bremner, I., Atkinson, T. and Henderson, G. (1985) In: TEMA-5. (Mills, C.F., Bremner, I. and Chesters, J.K., eds.). pp. 176–180. CAB, Slough.

Mackenzie, A.M., Illingworth, D.V., Jackson, D.W. and Telfer, S.B. (1997) The use of caeruloplasmin activities and plasma copper concentrations as an indicator of copper status in ruminants. In: Trace Elements in Man and Animals – 9: Proceedings of the Ninth International Symposium on Trace Elements in Man and Animals. (Fischer, P.W.F., L Abbé, M.R., Cockell, K.A. and Gibson, R.S., eds.). NRC Research Press, Ottawa, Canada. pp. 137–138.

Wikse, S.E., Herd D., Fiald, R. and Holland, P. (1992) JAVMA, 200: 1625–1629.

Discussion

Q1. Geoffrey Judson, Central Veterinary Laboratory, Glenside, Southern Australia: An interesting talk. On that last slide that you showed, if you look at the caeruloplasmin activity, the cattle in one farm were at risk of Cu deficiency, and on the other farm were Cu adequate, while caeruloplasmin activity was the same in both. How would you explain that?

A. The mechanism causing the Cu deficiency is probably via the thiomolybdate that is being absorbed through the rumen. It has previously been shown that thiomolybdate can inhibit caeruloplasmin activity and so the thiomolybdate may be causing the clinical Cu deficiency in the animals.

Q2. John Howell, Murdoch University, Murdoch, Western Australia: You said that your ratios would identify the animals that would respond to Cu treatment. Do you have figures to show the numbers of animals that did and did not respond to Cu treatment that were within those ratios?

A. At present there is only anecdotal evidence of the ratio working. We will be collecting evidence from questionnaires sent to the farmers. Just now it hasn't been fully documented.

The Use of a Soluble Glass Bolus to Prevent Zinc Deficiency in Sheep

N.R. KENDALL, A.M. MACKENZIE, AND S.B. TELFER
Department of Animal Physiology and Nutrition, University of Leeds, Leeds, LS2 9JT, UK

Keywords zinc, deficiency, bolus, sheep

Introduction

Zinc deficiency although not common in the UK in its clinical form, may be prevalent sub-clinically with supplemental zinc also possibly having a role in prophylaxis against conditions such as mastitis and footrot. However, although there is little risk of primary deficiency in the UK, there are large areas around the Mediterranean, Northern Africa and North America that are reported to be deficient in zinc. The object of this trial was to induce a zinc deficiency by feeding a low zinc diet and correct this deficiency by administration of a zinc, cobalt and selenium soluble glass bolus.

Materials and Methods

Four Suffolk lambs were fed a semi-purified zinc deficient diet (30 kg Cornflour, 16.79 kg dextrose, 30 kg ground barley straw, 11 kg Meri-white (egg albumin source), 3 kg urea, 5 L soya oil, 0.6 g vitamin A (3000 IU/kg), 0.0725 g vitamin D (360 IU/kg), 2.5 g vitamin E (α-tocopherol), 1.829 kg $Ca_3(PO_4)_2$, 0.922 kg $MgSO_4.7H_2O$, 1.024 kg $KHCO_3$, 0.391 kg NaCl, 19.17 g $FeSO_4.7H_2O$, 3.675 g $CuSO_4.5H_2O$, 16.24 g $MnSO_4.4H_2O$, 0.404 g $CoCl_2.6H_2O$, 0.169 g KIO_3, 0.0394 g $NaSeO_3$). This diet had an average zinc content of less than 2 mg/kg DM and a crude protein of approximately 80 g/kg DM (daily dietary allowance for lambs at maintenance is 66 g/d, McDonald et al. 1990). The diet was fed at a rate of 1 kg/lamb/day fed at 8:00 with refusals taken away at the next feed. The sheep had double distilled water available ad libitum throughout the trial. Once signs of zinc deficiency had been observed and these signs correlated with low plasma zinc levels, the lambs were bolused (day 0) with a cobalt, selenium and zinc soluble glass bolus (bolus weight ~33 g, 13.1% Zn, 0.5% Co, 0.15% Se). The sheep were kept in metabolism crates when on balance or housed in pairs in loose boxes when 'at rest'. Total collections of urine (into 20 ml of 25% HCl, Analar, BDH), faeces and refusals were carried out for calculation of zinc balance. Faeces, feed and refusals were dried in a 100°C oven, then microwave wet digested in 70% Nitric acid (Analar, BDH) and made up to a known volume for zinc analysis (Pye SP9 AA spectrophotometer at 213.9 nm with background correction). Cobalt status was assessed by measuring vitamin B_{12} concentrations by radioassay using a kit (Becton Dickinson, Oxford, England). Selenium status was assessed by colorimetric assay of erythrocyte glutathione peroxidase activity (Telfer et al. 1984). Plasma zincs were assessed by AA spectophotometry (Pye SP9 AA spectrophotometer at 213.9 nm with background correction) after diluting plasma 1:5 with 0.1 M HCl (Analar, BDH). Zinc binding capacity (a measure of saturation of zinc carriers in the plasma) was measured according to the method of Kincaid and Cronrath (1979) with half quantities used. Plasma copper was assayed by 1:5 dilution in 0.1 M HCl and then read on a Pye SP9 AA spectrophotometer at 324.8 nm.

Results and Discussion

In Figures 1 and 2 an initial depression in plasma zinc is accompanied by a negative zinc balance as the alimentation onto the deficient diet occurs and a peak in plasma zinc follows as available stores held

Correct citation: Kendall, N.R., Mackenzie, A.M., and Telfer, S.B. 1997. The use of a soluble glass bolus to prevent zinc deficiency in sheep. In *Trace Elements in Man and Animals – 9: Proceedings of the Ninth International Symposium on Trace Elements in Man and Animals. Edited by* P.W.F. Fischer, M.R. L'Abbé, K.A. Cockell, and R.S. Gibson. NRC Research Press, Ottawa, Canada. pp. 303–305.

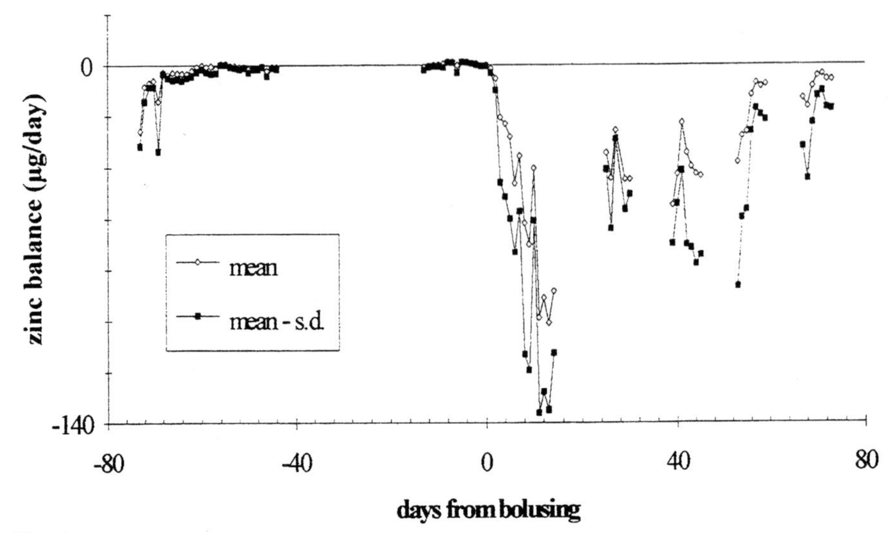

Figure 1. Zinc balance.

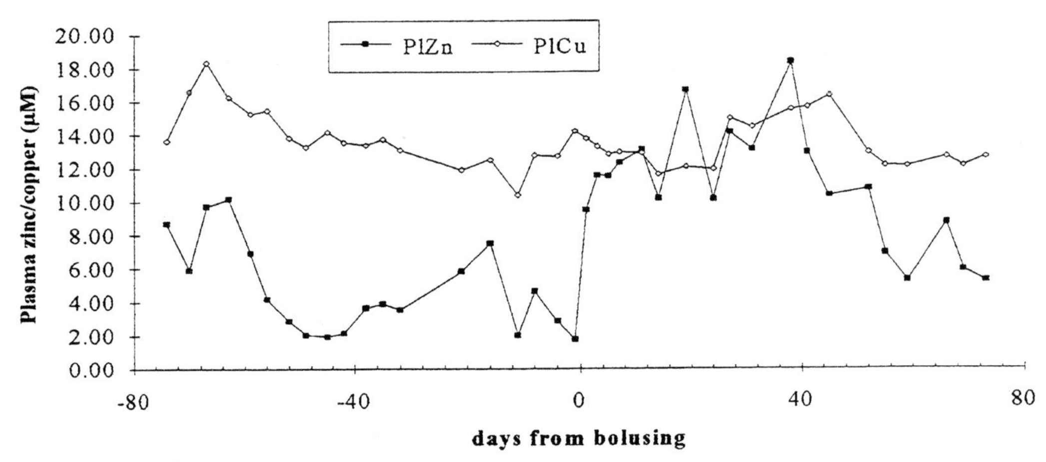

Figure 2. Plasma zinc and copper concentrations.

within the body are released. A rapid fall off in plasma zinc then occurs to a mean level of <2 µM with the zinc balance being virtually zero. At the time when low plasma zinc levels were found, some clinical signs of zinc deficiency were observed, including excessive salivation, bowing of the hind legs, a general listlessness, dulling of the coat and eyes, depression in food intake and loss of hair particularly around the eyes. A pica was also observed and this accounts for the peak in plasma zinc at around day −16 when the sheep ate a large hole in a holding pen wall and consumed high zinc (1.4 mg Zn/g wall) wall plaster and galvanised reinforcement. The sheep were removed from this source of zinc, became depleted again and were bolused (day 0). Plasma zinc responded to bolusing with a rapid rise which was maintained for 6 weeks in all four sheep and a further two weeks in two of them. The zinc balance gave large negative values due to the release of zinc from the boluses. As the boluses dissolved away the balances returned to the previous depleted values. Figure 2 also shows that although there were large alterations in plasma zinc concentrations, these were not mirrored by the plasma copper concentrations with the plasma copper remaining relatively constant within normal limits and the relatively large amounts of zinc from the bolus not depressing plasma coppers as reviewed by O'Dell (1989). The zinc binding capacity approaches complete unsaturation (100%) at the time of maximum depletion but is fully saturated after bolusing, being unable to bind significantly more zinc.

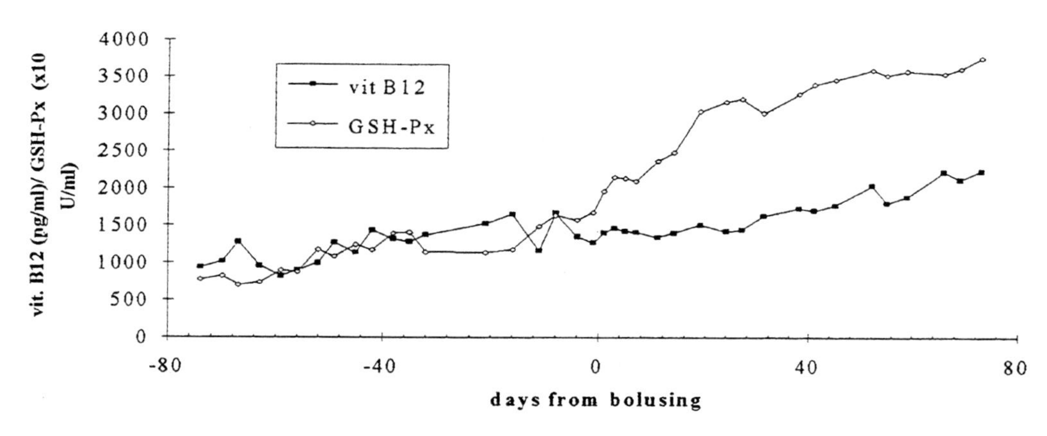

Figure 3. Vitamin B12 concentrations and glutathione peroxidase activities.

Figure 3 illustrates the available release of selenium and cobalt as shown by erythrocyte glutathione peroxidase and serum vitamin B_{12} respectively. Although the diet was formulated to give adequate levels of selenium and cobalt it is seen that the blood status of both these essential trace elements is increased by the bolus.

In conclusion the induced zinc deficiency was corrected using the soluble glass bolus for a period of at least 6 weeks with an associated elevation in cobalt and selenium status. This proves the released zinc, selenium and cobalt was biologically available to the sheep. In the case of zinc, the bolus was able to prevent the deficient state expected when the sheep were fed the low zinc diet.

Acknowledgements

We would like to acknowledge Mrs. Sue Eaton, Mark Leadbeater, Mark Townsend, Nik Ellwood and Miss Nicky Townshend for their invaluable assistance in this project.

References

Kincaid, R.L & Cronrath, J.D. (1979) J. Dairy Sci. 62(4): 572.
O'Dell, B.L. (1989) Mineral interactions relevent to nutrient requirements. J. Nutr. 119: 1832–1838.
McDonalds, P., Edwards, R.A. and Greenhalgh, J.F.D. (1990) Animal Nutrition, 4th edition, pp515–517. Longman Scientific and Technical, Essex, England.
Telfer, S.B., Zervas, G. and Carlos, G. (1984). Can. J. Anim. Sci., 64 (Suppl.1): 234–235.

Discussion

Q1. Nick Costa, Murdoch University, Murdoch, Western Australia: What was the live weight of the sheep? Weight loss may increase protein turnover in the body and therefore liberate more Zn.

A. During Zn deficiency, body weight of the sheep was maintained.

Q2. Manfred Anke, Friedrich Schiller University, Jena, Germany: What was the food intake during Zn deficiency and after giving the bolus. Did you look at the tongue? It is the first tissue to react to Zn deficiency.

A. No, did not look at the tongue. Food intake was normal during Zn deficiency until blood Zn was reduced very low (less than 4 μM) where the refusal of food was increased. The feed refusal was not total; they were still able to eat enough to maintain their weight. Intake then increased after given the bolus.

In Sacco Disappearance in the Rumen of Selenium in Feedstuffs

J.B.J. VAN RYSSEN,[†] B.J. VAN DER MERWE,[‡] AND G.E. SCHROEDER[¶]

[†]Department of Animal & Wildlife Sciences, University of Pretoria, 0002 Pretoria, [‡]Cedara Agric. Development Institute, 3200 Pietermaritzburg, and [¶]Animal Nutrition & Animal Product Institute, Agric. Research Council, 1675 Irene, South Africa

Keywords organic selenium, protein, rumen degradability

A large proportion of selenium (Se) in feedstuffs is present in the organic form, as part of protein molecules (Beilstein et al. 1991). In ruminant nutrition protein sources vary in the degree of rumen degradation. This determines the proportion of nitrogen from the feed made available for rumen microbial utilization, being lost as ammonia or escaping rumen degradation; the so-called rumen degradable (RDP) and nondegradable (UDP or by-pass) protein concept in ruminant nutrition (ARC 1980). Rumen microbes incorporate some of the dietary Se into microbial protein. Some are, however, reduced to inorganic compounds, a large proportion of which is unavailable to the animal (Aspila 1991).

It is proposed that, in ruminants, a high correlation should exist between the proportion of protein degraded in the rumen and that of the Se exposed to rumen microbial action. The proportion of Se in feedstuffs made available to the animal can vary widely depending on the degree of degradation in the rumen of protein containing the Se.

Procedure

Two studies were conducted on basic feedstuffs, mainly protein sources which contain relatively high concentrations of Se. The *in sacco* technique (AFRC 1992) was employed in which dacron bags (pore size 53 μm) containing 5 g of feedstuff were suspended in the rumens of three fistulated dairy cows. In the first investigation the bags, in triplicate, were suspended for 16 h in the rumen. After removal, the bags were washed in running water until the water from the bags was clear. A zero hour sample was washed without incubation. Disappearance of crude protein (N × 6.25) and Se from the bags was compared between 0 and 16 h incubation. In another *in sacco* study, where the effect of heat and manufacturing processes on protein degradation was investigated, Se assays were done on the feedstuff residues. The feedstuffs were incubated for 0 (washed), 2, 8, 16 and 24 h.

Results

The disappearance of Se from the bags between 0 and 16 h incubation correlated well ($r = 0.899$, $Se = 9.8 + 0.8\ CP$; $n = 67$) with the degradability of the protein in the feedstuff. Mean disappearances of protein and Se for some of the individual feedstuffs were: fish meal, 25 ± 13.3 and 33 ± 11.6%; blood meal, 19 ± 6.7 and 16 ± 6.2%; wheat and rice bran, 63 ± 8.7 and 60 ± 6.7% and oil cakes, 77 ± 16.1 and 72 ± 12.8% respectively.

In the second investigation a similarly high correlation between crude protein and Se disappearance was found at the different hours of incubation. In Figure 1a the results obtained for the two most extreme fish meal samples are depicted. Heat treatment reduced the disappearance of both protein and Se from the bags. However, the high association between the disappearance of protein and Se was not observed in some of the heat treated oil cakes, viz. very little Se disappeared from the bags at all stages of incubation (Figure 1b).

Correct citation: van Ryssen, J.B.J., van der Merwe, B.J., and Schroeder, G.E. 1997. *In Sacco* disappearance in the rumen of selenium in feedstuffs. In *Trace Elements in Man and Animals – 9: Proceedings of the Ninth International Symposium on Trace Elements in Man and Animals. Edited by* P.W.F. Fischer, M.R. L'Abbé, K.A. Cockell, and R.S. Gibson. NRC Research Press, Ottawa, Canada. pp. 306–307.

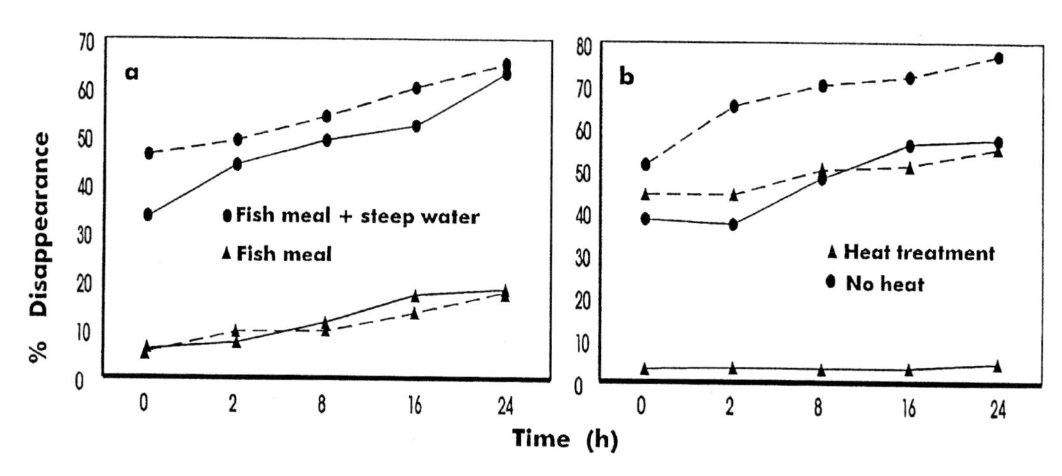

Figure 1. In sacco disappearance of protein (– – –) and selenium (———) from (a) two fish meal sources and (b) untreated and heat treated cotton seed oil cake.

Discussion

In general, it can be concluded that the extent of exposure of organic Se in feedstuffs to rumen microbial action depends on the rumen degradability of the protein source. Since protein degradability within the same feed source can vary widely (ARC 1980), e.g., as a result of heat treatment, the proportion of the Se in a feed which will be subjected to microbial action will vary likewise. Consequently, the proportion of dietary Se reaching the lower digestive tract would differ in feeds with the same concentration of Se but different RDP levels. The extent that Se is converted to reduced forms in the rumen or incorporated in microbial protein have not been measured and would influence any predictions of bioavailability of dietary Se based on RDP levels. Despite the latter, this investigation demonstrates that rumen degradation of organic Se should contribute to variations in the bio-availability of Se in ruminant feedstuffs.

References

Agricultural and Food Research Council (1992). Nutritive Requirements of Ruminant Animals: Protein. AFRC Technical Committee on Response to Nutrients. Report No 9. Nutr. Abstr. & Rev. (Series B), 12: 797–835.

Agricultural Research Council (1980). The Nutrient Requirements of Ruminant Livestock. C.A.B. International, Oxon, UK.

Aspila, P. (1991). Metabolism of selenite, selenomethionine and feed-incorporated selenium in lactating goats and dairy cows. J. Agric. Sci. Finland. 63: 1–74.

Beilstein, M.A., Whanger, P.D. & Yang, G.Q. (1991). Chemical form of selenium in corn and rice grown in a high selenium area of China. Biomed. Environ. Sci. 4: 392–398.

Discussion

Q1. Neville Suttle, Moredun Research Institute, Edinburgh, Scotland: This area of degradability and behaviour of trace elements in the rumen is very interesting, and understudied. In the cottonseed meal that you got your unusual results for, did the heat treatment change the selenium content? Is it a question of heat removing some degradable selenium component by volatilization?

A. I can't recollect the assay of the original sample just now. As far as I know there wasn't, but I could not say definitely that volatilization didn't occur.

Q2. Julian Lee, New Zealand Agricultural Research Institute, Palmerston North, New Zealand: Have you looked at the release of Se from bypass proteins? Have you looked at differences in release of Se from microbial protein versus that which may have bypassed the rumen? Is there different availability of those two sources of protein?

A. Well, we didn't do any more than what I've presented here. Se will be present as selenomethionine in the feed and this Se can be absorbed as such in the lower intestinal tract.

Bioavailability of Supplemental Sources of Manganese and Copper for Chicks and Lambs Estimated from High Dietary Additions

CLARENCE B. AMMERMAN,[†] PAMELA R. HENRY,[†] AND RICHARD D. MILES[‡]

Departments of [†]Animal Science and [‡]Dairy/Poultry Sciences, University of Florida, Gainesville, FL 32611-0900, USA

Keywords bioavailability, manganese, copper, domestic animals

It is important to know the bioavailability of the desired elements in the mineral sources which can be obtained for use as dietary supplements (Ammerman et al. 1995).

Materials and Methods

A series of studies have been conducted at the University of Florida in which accumulation of the mineral element in a target organ or organs has been used as the response criterion for estimating bioavailability. This approach was based on early observations, including those by Nesbit and Elmslie (1960) with rats and Bunch et al. (1961) with swine, which suggested that biological availability of iron and copper from various compounds was related to tissue concentrations of the element. Also, Watson et al. (1971) proposed that a biological assay for manganese utilization in the chick could be developed using accumulation of the element in bone. In the ensuing studies, one-day-old chicks have been used as the model for nonruminants and lambs weighing approximately 40 kg have been the model for ruminants. Natural diets containing adequate concentrations of manganese and copper to meet the animal requirement were used as a basal diet. Dietary additions of manganese for the chick and lamb were about 700 to as much as 4000 ppm and additions of copper ranged from 150 to 600 ppm for the chick and 20 to about 200 ppm for the lamb. Feeding periods were approximately 3 weeks in length.

Results and Discussion

Estimates of bioavailability for supplemental sources of manganese and copper obtained with chicks and lambs are presented in Table 1. The rank order for utilization of manganese sources was similar for poultry and sheep. Differences among sources were evident in these studies with high dietary additions which could not be demonstrated with purified diets and minimum dietary additions in earlier research by Watson et al. (1971). Most of the research conducted in recent years with supplemental manganese sources has been conducted at dietary levels well above the requirement (Henry et al. 1995). As illustrated in Table 1, fewer studies have been conducted with copper than with manganese. Copper as cupric oxide was uniformly less well utilized by both species and there was a suggestion that cupric carbonate (Cu_2CO_3) was less well utilized by chicks than by lambs. Other research reviewed by Baker and Ammerman (1995) indicated low utilization of copper as cupric oxide (CuO) by poultry, swine, and cattle. Copper as cuprous oxide (Cu_2O) was indicated to be well utilized.

Correct citation: Ammerman, C.B., Henry, P.R., and Miles, R.D. 1997. Bioavailability of supplemental sources of manganese and copper for chicks and lambs estimated from high dietary additions. In *Trace Elements in Man and Animals – 9: Proceedings of the Ninth International Symposium on Trace Elements in Man and Animals. Edited by* P.W.F. Fischer, M.R. L'Abbé, K.A. Cockell, and R.S. Gibson. NRC Research Press, Ottawa, Canada. pp. 308–309.

Table 1. Relative bioavailability of supplemental manganese and copper sources for poultry and sheep estimated with high dietary additions.[a]

	Poultry	Sheep
Manganese sources[b]		
Sulfate, RG[c]	100	100
Carbonate, RG	32 (1)	28 (1)
Dioxide, RG	32 (1)	33 (1)
Mn-methionine, FG	120 (1)	130 (1)
Oxide, RG	77 (4)	56 (1)
Oxide, FG	70 (4)	64 (2)
Copper sources[d]		
Cupric sulfate, RG[c]	100	–
Cupric chloride, RG[c]	–	100
Cupric acetate, RG	99 (1)	93 (1)
Cupric carbonate, FG	67 (2)	121(1)
Cu-lysine, FG	99 (1)	93 (1)
Cupric oxide, FG	<1 (1)	35 (1)

[a]Comparative values presented in this table represent data obtained by the University of Florida and taken primarily from information reviewed by Henry (1995) and Baker and Ammerman (1995). Adjustments were made if a standard other than that indicated was used. RG = Reagent grade and FG = Feed grade. Numbers of studies or samples involved indicated within parentheses.
[b]Values based on bone and kidney deposition of element.
[c]Used as standard.
[d]Values based on liver deposition of element.

Literature Cited

Ammerman, C.B., Baker, D.H. & Lewis, A.J. (Eds). (1995) Bioavailability of Nutrients for Animals: Amino Acids, Minerals, and Vitamins. Academic Press, New York, NY.

Baker, D.H. &Ammerman, C.B. (1995) Copper Bioavailability. In: Bioavailability of Nutrients for Animals: Amino Acids, Minerals, and Vitamins. (C.B. Ammerman, D.H. Baker & A.J. Lewis, eds.), Academic Press, San Diego, CA.

Bunch, R.J., Speer, V.C., Hays, V.W., Hawbaker, J.H. & Catron, D.V. (1961) Effects of copper sulfate, copper oxide and chlortetracycline on baby pig performance. J. Anim. Sci. 20: 723.

Henry, P.R. (1995) Manganese Bioavailability. In: Bioavailability of Nutrients for Animals: Amino Acids, Minerals and Vitamins. (C.B. Ammerman, D.H. Baker & A.J. Lewis, eds.), Academic Press, San Diego, CA.

Nesbit, A.H. & Elmslie, W.P. (1960) Biological availability to the rat of iron and copper from various compounds. Trans. Illinois Acad. Sci. 53: 101.

Watson, L.T., Ammerman, C.B., Miller, S.M & Harms, R.H. (1971) Biological availability to chicks of manganese from different inorganic sources. Poult. Sci. 50: 1693.

Indicators of Vitamin B_{12} Status of Cattle

G.J. JUDSON,[†] J.D. MCFARLANE,[‡] K.L. BAUMGURTEL,[†] A. MITSIOULIS,[†] R.E. NICOLSON,[†] AND P. ZVIEDRANS[†]

[†]Central Veterinary Laboratories, and [‡]South East Headquarters, Primary Industries SA, Adelaide, S.A. 5000, Australia

Keywords vitamin B_{12}, cobalt, cattle, milk

Plasma vitamin B_{12} (B_{12}) concentrations are of diagnostic value in detecting sheep at risk of cobalt (Co) deficiency (Clark et al. 1985). However, there is doubt about the usefulness of plasma B_{12} concentrations as an indicator of vitamin B_{12} status of cattle (SCA 1990). We report on the relationship of liver B_{12} concentrations with plasma and milk B_{12} concentrations and with faecal Co concentrations of cows given different oral doses of Co pellets.

Materials and Methods

Murray Grey cows were allocated to 4 treatment groups, each of 12 animals. The treatments were a single oral dose of 0, 1, 2 or 4 Co pellets: the pellet contained 9 g Co_3O_4 in a 21 g iron matrix ('Permaco', Mallinckrodt Veterinary, Australia). At the time of treatment (11 November, 1993; week 0 of the experiment) all cows received orally 1 selenium pellet ('Permasel') and a 30-g steel grubscrew.

The cattle were run as a single group on pasture of low Co content (usually <0.05 mg/kg dry matter) on a farm near Robe, South Australia. Liver biopsies and blood samples for B_{12} assay and faecal samples from the rectum for Co assay were usually obtained from 4–6 cows in each group prior to treatment and then at 13, 28, 41, 57, 75 and 96 weeks post-treatment. Milk samples for B_{12} assay were obtained from cows at weeks 28, 41, 75 and 96 of the experiment: cows calved during weeks 16–24 and 64–73 of the experiment. Analytical methods have been described by Judson et al. (1995).

Results and Discussion

Cows not given a Co pellet as well as 3 cows given 1 Co pellet and 1 cow given 2 Co pellets had on occasions liver B_{12} concentrations of less than 200 nmol/kg wet weight (Figure 1) and hence were at risk of Co deficiency (SCA 1990).

Vitamin B_{12} concentrations greater than 25 pmol/L plasma (Figure 1A) and 300 pmol/L skim milk (Figure 1B) and Co concentrations greater than 0.20 mg/kg dry faeces (Figure 1C) were usually associated with adequate liver B_{12} reserves (>200 nmol/kg wet weight). Low plasma B_{12} concentrations (<25 pmol/L) and low faecal Co concentrations (<0.20 mg/kg dry matter) were however, of limited value in detecting cows at risk of Co deficiency since these values were unrelated to liver B_{12} concentrations (Figure 1A,C).

The present study indicates that milk B_{12} concentrations may be of diagnostic value in detecting lactating cows at risk of Co deficiency since concentrations of less than 300 pmol/L were usually associated with liver B_{12} values of less than 200 nmol/kg wet weight (Figure 1B).

Acknowledgments

We thank Mr. G. Legoe for his cooperation in providing cattle and facilities to enable this work to be undertaken. Financial support for this work was provided by the South Australian Cattle Compensation Trust Fund and Mallinckrodt Veterinary Limited, Australia.

Correct citation: Judson, G.J., McFarlane, J.D., Baumgurtel, K.L., Mitsioulis, A., Nicolson, R.E., and Zviedrans, P. 1997. Indicators of vitamin B_{12} status of cattle. In *Trace Elements in Man and Animals – 9: Proceedings of the Ninth International Symposium on Trace Elements in Man and Animals. Edited by* P.W.F. Fischer, M.R. L'Abbé, K.A. Cockell, and R.S. Gibson. NRC Research Press, Ottawa, Canada. pp. 310–311.

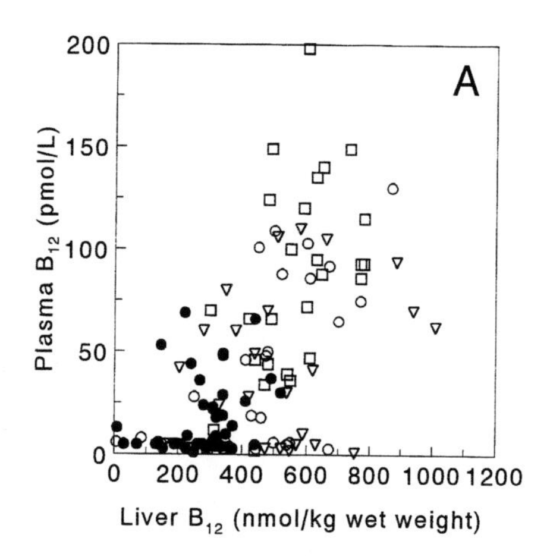

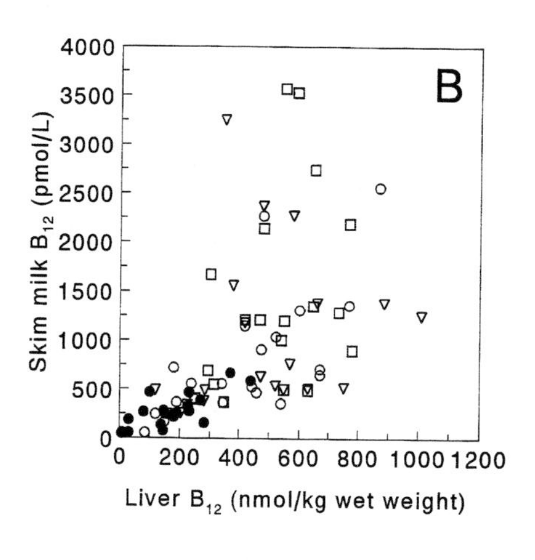

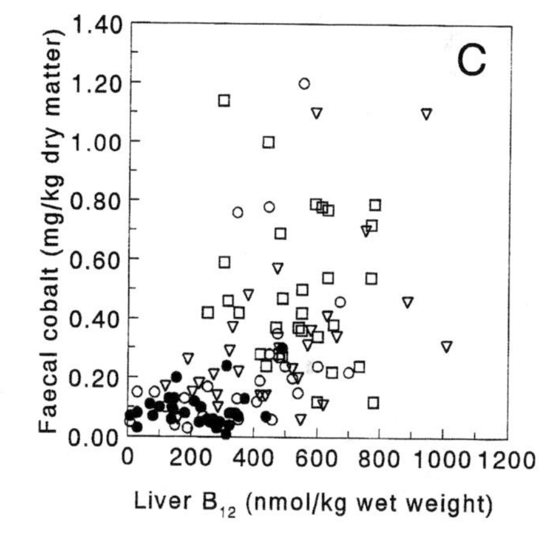

Figure 1. Relationships between liver vitamin B_{12} concentration and plasma vitamin B_{12} concentration (A), skim milk vitamin B_{12} concentration (B) and faecal cobalt concentration (C) in cows given 0 (●), 1 (○), 2 (▽) or 4 (□) Co pellets.

Literature Cited

Clark, R.G., Wright, D.F. & Millar, K.R. (1985) A proposed new approach and protocol to defining mineral deficiencies using reference curves. Cobalt deficiency in young sheep is used as a model. N.Z. Vet. J. 33: 1–5.

Judson, G.J., Woonton, T.R., McFarlane, J.D. & Mitsioulis, A. (1995) Evaluation of cobalt pellets for sheep. Aust. J. Exper. Agric. 35: 41–49.

Standing Committee on Agriculture (1990) Feeding Standards for Australian Livestock. Ruminants. CSIRO, East Melbourne, Victoria.

A New Intraruminal Controlled-Release-Capsule Providing Monensin and Selenium for Cattle

N.D. COSTA,[†] A. PURWANTORO,[†] E.G. TAYLOR,[†] AND L.B. LOWE[‡]

[†]School of Veterinary Studies, Murdoch University, Murdoch, WA 6150, Australia, and [‡] Elanco Animal Health, 112 Wharf Road, West Ryde, NSW 2114, Australia

Keywords selenium, monensin, controlled-release, cattle

A variable-geometry, slow-release intraruminal device (CRC) has been developed which measures 16 cm in length and 22 mm in diameter, and contains 32 g of monensin in a hexaglycerol distearate (45%) matrix. A constant dose of monensin is delivered by means of a spring, and regurgitation is prevented by collapsible wings on the non-releasing end. The efficacy of these CRCs in the prevention of bloat has been confirmed in dairy cattle under grazing conditions (Cameron and Malmo 1993). Monensin also increases the absorption of selenium from the gastrointestinal tract (Costa et al. 1985).

The main objective of this work was to develop a new intraruminal-CRC capable of controlled delivery of monensin and selenium. For selenium to be effectively incorporated into the hexaglycerol distearate matrix of the CRC, selenium must be present in an insoluble form. The most commonly used insoluble form in selenium supplementation in Australia is elemental selenium incorporated into intraruminal selenium pellets, consisting of 5% elemental selenium dispersed in an iron grit. Interaction between the currently available CRC containing monensin and selenium bullets was investigated in the first part of the study. In addition, we checked for interactive effects of monensin and selenium when combined in this manner.

Forty-eight yearling steers (predominantly Angus) were allocated to one of 4 treatments groups according to ranked liveweight; controls receiving a blank CRC containing only the matrix, blank CRC plus 2 selenium bullets, CRC with 32 g monensin, and CRC with 32 g monensin and 2 selenium bullets. All steers were fed a diet of chaffed hay (60%) and rolled lupins (40%) containing <0.3 μmol/kg DM Se. Liveweights of the steers were measured every 10 d, and blood samples taken at each weighing. All steers were slaughtered after 100 d. Liver and diaphragm muscle were collected at slaughter, and the CRCs were recovered and shown to have released 196 ± 4 mg of monensin per day. Selenium was assayed by the automated fluorometric method. Results from this study are shown in Table 1.

Elemental selenium was incorporated into the formulation of the CRCs as a first attempt to use a commonly available, insoluble form of selenium in the matrix. However, these CRCs did not result in any increase the concentration of selenium in plasma and tissue of recipient steers (results not shown).

Table 1. Liveweight gains and concentration of selenium in blood and tissue from steers given selenium bullets and CRCs containing monensin.

Treatment	LW gain (kg)	Plasma Se (μmol/L)	Liver Se (μmol/kg wet weight)	Muscle Se (μmol/kg wet weight)
Control	72.9	0.32[a]	2.8[a]	0.9[a]
Monensin CRC	77.1	0.45[a]	3.2[a]	1.0[a]
Se bullets	80.0	1.08[b]	4.0[ab]	1.2[b]
Monensin CRC + Se bullets	94.7	1.19[b]	4.2[ab]	1.3[b]

Values are means of 12 steers, and those with different superscripts are significantly different at $P < 0.01$.

Correct citation: Costa, N.D., Purwantoro, A., Taylor, E.G., and Lowe, L.B. 1997. A new intraruminal controlled-release-capsule providing monensin and selenium for cattle. In *Trace Elements in Man and Animals – 9: Proceedings of the Ninth International Symposium on Trace Elements in Man and Animals. Edited by* P.W.F. Fischer, M.R. L'Abbé, K.A. Cockell, and R.S. Gibson. NRC Research Press, Ottawa, Canada. pp. 312–314.

Barium selenate ($BaSeO_4$) was then used in the formulation of the monensin CRC at a rate of inclusion calculated to provide 8 mg Se per day. These CRCs showed encouraging results in an unreplicated study (results shown in abstract), so a fully replicated trial was undertaken using 47 yearling Angus × Shorthorn steers allocated to a control group of 11 and 3 groups of 12; CRC with matrix containing 1 g of selenium as $BaSeO_4$, CRC containing 32 g of monensin, and CRC containing 32 g of monensin plus 1 g of selenium as $BaSeO_4$. The steers were fed *ad libitum* on chaffed hay (40%) and 60% cubes based on lupins and barley. The overall selenium concentration of the feed was <0.6 μmol/kg DM Se. Blood samples were collected every 14 d. Liver and muscle samples were collected at slaughter, and all of the CRCs were recovered and found to have expended their matrix in most cases. The concentration of selenium and activity of glutathione peroxidase was assayed in blood and tissues.

The mean starting concentration of selenium in plasma was 1.05 ± 0.04 μmol/L which was indicative of adequate selenium status in the steers. Consequently, no Se-responsive weight gain was observed. Nevertheless, plasma selenium concentrations increased significantly ($P < 0.001$) after 30 d as shown in Figure 1. The activity of glutathione peroxidase in blood followed this response pattern.

The concentration of selenium and activity of glutathione peroxidase in liver and muscle are presented in Table 2.

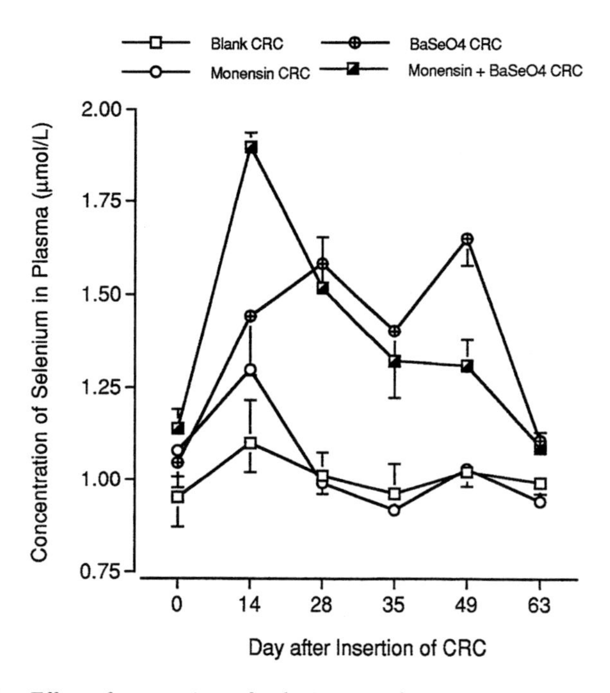

Figure 1. Effect of monensin and selenium on plasma selenium concentrations in steers. Each value is the mean ± SEM of 11 animals for the Blank CRC, and 12 animals for the other treatments.

Table 2. Concentration of selenium (Se) and activity of glutathione peroxidase (GSH.Px) in liver and muscle from steers given monensin and/or selenium as $BaSeO_4$.

Treatment	Liver Se (μmol/kg DW)	Muscle Se (μmol/kg DW)	Liver GSH.Px (EU/g protein)	Muscle GSH.Px (EU/g protein)
Control	19.0 ± 1.0[a]	6.9 ± 0.5[a]	53 ± 6	75 ± 4
Monensin	17.6 ± 1.1[a]	7.7 ± 0.3[a]	47 ± 4	68 ± 4
$BaSeO_4$	25.1 ± 1.6[b]	9.8 ± 0.6[b]	61 ± 5	79 ± 4
Monensin + $BaSeO_4$	28.1 ± 2.0[b]	10.3 ± 0.6[b]	58 ± 5	79 ± 5

Values shown are the means ± SEMs of 12 steers for all groups except controls (11).
Values with different superscripts are significantly different $P < 0.01$.

The concentration of selenium in liver and muscle increased in steers given CRCs containing $BaSeO_4$ but this increase was not reflected in the glutathione peroxidase activity. Similar results showing elevated tissue selenium not leading to increased GSH.Px activity were reported by Mallinson et al. (1985) in heifers given $BaSeO_4$ as a slow-release injection. Our results extend the findings of Archer and Judson (1994) in sheep that selenium from $BaSeO_4$ can be absorbed from the gastrointestinal tract. Thus a sparingly soluble form of selenium, $BaSeO_4$ has been successfully incorporated into the formulation for CRCs also containing monensin. The concentration of $BaSeO_4$ needed to supply supplementary doses of selenium is much greater (10×) than for soluble forms of selenium such as Na_2SeO_4. These new CRCs deliver selenium in a form that can be absorbed from the gut via the plasma to tissues, and monensin enhances the increase in tissue selenium concentrations. Monensin increased the selenium absorbed and retained in tissues when used in combination with barium selenate, but this increase was not of the same magnitude predicted from previous studies using radioisotopic forms of selenium as sodium selenite (Costa et al. 1985).

Literature Cited

Archer, J.A. and Judson, G.J. (1994) Availability of barium selenate administered orally to sheep. Proc. Aust. Soc. Anim. Prod. 20: 441.

Cameron, A.R. and Malmo, J. (1993) A survey of the efficacy of sustained-release monensin capsules in the control of bloat in dairy cattle. Aust. Vet. J. 70: 1–4.

Costa, N.D., Gleed, P.T., Sansom, B.F., Symonds, H.W. and Allen, W.M. (1985) Monensin and narasin increase selenium and zinc absorption in steers. In: Trace Elements in Man and Animals TEMA-5, (eds. Mills, C.F., Bremner, I. and Chesters, J.K.) pp 472–474. Commonwealth Agricultural Bureau, London, UK.

Mallinson, C.B., Allen, W.M. and Sansom, B.F. (1985) Barium selenate injections in cattle: Effects on selenium concentrations in plasma and liver and residues at site of injection. Vet. Record October 19: 405–407.

Study on the Herbicidal Effect of Selenium on the Dock, *Rumex obtusifolius*

S.K. WOMAK
Department of Biology, University of Swansea, South Wales, UK

Keywords herbicidal control, selenium, *Rumex obtusifolius*

Introduction

The dock *Rumex obtusifolius* is a problem in many agricultural regions. Hopkins and Peel (1986) showed that in England and Wales approximately 75,000 hectares (15%) are seriously infested with *Rumex* spp. Docks have evolved several strategies that make them extremely successful competitors. These include high fecundity, somatic seed polymorphism and high viability. The seed's ability to survive in the rumen, silage and slurry as well as its preference for acidic soils make it a particular problem for enterprises using high levels of nitrogen, such as dairy farms. Considerable financial losses are accrued as a result of lost grass production and subsequent reduction of the land's carrying capacity (Oswald and Hagger 1983). The feed value is also reduced, as *Rumex obtusifolius* has only 60% the nutritive value of grass herbage (Courtney and Johnston 1978). Both of these factors are reflected in lost milk production. Eradication of mature dock is notoriously difficult, especially on permanent lays. Traditional herbicide control can lead to the death of valuable clover and can set back grass growth. The following study was undertaken subsequent to trace element pasture amendment for animal nutritional purposes. Observations indicated that the use of selenium affected the viability of *Rumex obtusifolius*. Could selenium be used as an alternative method of dock control?

Aims

To identify the rate at which selenium is most affective in the control of *Rumex obtusifolius* and to see if selenium has a detrimental affect on other species in the sward. To assess the effectiveness of each treatment depending on whether the pasture was used for grazing or conservation.

Material and Methods

The research was undertaken in West Wales where the warm, wet climate is ideal for grass growth. The experimental trial was conducted over a four week period during August and September, the second of the dock's two flowering periods. There were two experimental sites one was cut and the other uncut pasture. The former had recently been cut for conserved grass, while the latter was continually grazed by young cattle throughout the trial. A vegetation list was compiled for each site showing percentage cover for each species.

Selenium selenite was selected in preference to the selenate species due to its stability, even though it is not as readily taken up by plants (Grant 1965, Davies and Watkins 1966). Twenty-five milligrams of selenium was applied to each of the randomly sampled one metre quadrats. There were four replicates at each site for each level of selenium concentration, including the control. The concentration levels used were 0.25, 0.50, 1.00 and 2.00 g/L. A surfactant was not added but droplet size was set to fine to ensure maximum wetting. The dock was judged to be affected by the degree of leaf necrosis. The data was ranked according to interpreted percentage of leaf necrosis: 0 = 0% leaf affected; 1 = 1–25% leaf affected; 2 = 26–50% leaf affected; 3 = 51–75% leaf affected; 4 = 76–100% leaf affected A traditional herbicide was

Correct citation: Womak, S.K. 1997. Study on the herbicidal effect of selenium on the dock, *Rumex obtusifolius*. In *Trace Elements in Man and Animals – 9: Proceedings of the Ninth International Symposium on Trace Elements in Man and Animals. Edited by* P.W.F. Fischer, M.R. L'Abbé, K.A. Cockell, and R.S. Gibson. NRC Research Press, Ottawa, Canada. pp. 315–316.

applied at the recommended dose to the cut pasture only, as it has a 7 d exclusion period, which was not possible on the cattle grazed pasture.

Results and Discussion

Parametric and non-parametric statistical analysis of the data shows that selenium is an effective control on *Rumex obtusifolius*. Damage was found to be significantly greater on cut compared to uncut pasture and increased significantly with higher concentration of selenium and with time. At all levels of selenium concentration there was no apparent effect on other species in the sward, including clover which plays an important nutritional role in dairy animal production. Interestingly, during the experiment the presence of severe dock beetle infestation was observed in four of the quadrats on the uncut pastures. To ensure that this did not distort the final results, these quadrat were separated from the trial. Later analysis shows that dock beetle could be an effective biological control.

Table 1 shows that by the second analysis selenium concentrations in the dock leaf were four times higher on the uncut pasture compared to the cut pasture. However, the levels of leaf necrosis were higher on the cut pasture. This apparent contradiction suggests a need for further investigation even though differences had levelled by the end of the trial.

Whilst it was not possible to compare seed selenium concentrations for cut pasture (as there were no seed stalks) the uncut pasture levels had risen from 0.05 ppm in the first analysis to 310 ppm by the second analysis. This may explain why the selenium treatment was less effective on the uncut pasture. It is possible the plant is able to translocate the selenium to the seed as a survival mechanism.

Table 1. Selenium levels (p.p.m.) for the seed, leaf and root of Rumex obtusifolius including soil selenium and pH values for the cut and uncut pasture.

	Analysis 1		Analysis 2		Analysis 3	
	Uncut	Cut	Uncut	Cut	Uncut	Cut
Soil pH	6.0	5.5	6.2	5.9	5.5	5.7
Soil Se	0.1	0.1	0.1	0.3	0.3	0.1
Root Se	0.05	0.05	0.3	0.1	0.1	1.3
Leaf Se	0.1	0.1	32.0	6.8	58.0	49.0
Seed Se	0.05	0.05	310.0	–	17.0	–

Conclusion

Initial indications from this study are that selenium selenate is an effective agent in controlling dock at concentrations of 1 and 2 g/L. It is thus feasible to control dock at the same time as rectifying selenium deficiencies in livestock by the use of pasture dressing.

Fruitful areas for research include

- an investigation into the viability of dock seed post selenium treatment
- the mapping of biochemical and pathological pathways for *Rumex obtusifolius* following selenium application

Literature Cited

Courtney, A. D. & Johnston, R.T. (1978) A consideration of the contribution of production of Rumex obtusifolius in a grazing system. British Crop Protection Conference-Weeds 325–332.

Davies, E.B., & Watkinson, J. H. (1966) Uptake of native and supplied selenium by pasture species. I. Uptake of selenium by browntop, ryegrass, cocksfoot and white clover from Atiamuri sand. N.Z. J. Agric. Res. 9: 317.

Grant, A.B. (1965) Pasture top dressing with selenium. N.Z. J. Agric. Res. 8: 681.

Hopkins, A. & Peel, S. (1986) The incidence of weeds on grassland. British Crop Protection Conference—Weeds 877–890.

Diurnal Pattern of Exocrine Pancreatic Secretion of Zinc and Carboxypeptidase B in Growing Pigs

M.S. JENSEN,[†] T. LARSEN,[‡] AND V.M. GABERT[†]

Danish Institute of Animal Science, Research Centre Foulum, [†]Department of Nutrition and [‡]Department of Health and Welfare, DK-8830 Tjele, Denmark

Keywords pancreatic secretion, zinc, carboxypeptidase B, pigs

Pancreatic juice has been shown to be a source of endogenous secretion of zinc into the gastrointestinal tract. However, the importance of pancreatic secretion to zinc homeostasis remains a matter of controversy. In pancreatic juice, zinc is primarily associated with the zinc dependent metalloenzymes carboxypeptidase A and B. In the present experiment the diurnal pattern of secretion of zinc and carboxypeptidase B (CPB) in pancreatic juice in fed pigs was measured.

Materials and Methods

An experiment with a total of six crossbred barrows (obtained from the Danish Institute of Animal Science swine herd) surgically prepared for continuous collection of pancreatic juice was performed. Three pigs were fitted with a pancreatic pouch re-entrant cannula according to Hee et al. (1985) (PM pigs) and three pigs had a catheter surgically inserted into the pancreatic duct and a simple T-cannula into the duodenum as described by Pierzynowski et al. (1988) (CM pigs). The pigs were fed a wheat starch and fish meal based diet with a moderate fat content. Minerals were supplemented to meet Danish standards for growing pigs. The daily intake of zinc was 182 mg. The pigs were fed 1.65 kg/d in three meals of equal amounts at 08:00, 16:00 and 24:00. Pancreatic juice was collected for a total of 24 h in 2-h intervals. The concentration of protein in pancreatic juice was measured according to the method of Lowry et al. (1951). Zinc content was analyzed by atomic absorption spectrometry (PU 9400 X, Philips Scientific, Cambridge, UK). Samples were wet ashed by addition of concentrated nitric acid (14 mol/L, 1:1 vol/vol) and exposed to increasing temperatures. AAS measurements were calibrated on commercial standards (Tritisol, Merck, Darmstadt, Germany). Standard curves were controlled by independently produced Zn-chloride solutions (pro analysi). The inactive procarboxypeptidase B in pancreatic juice was activated to carboxypeptidase B (EC 3.4.17.2) with trypsin. The method of Appel (1974) was used to determine CPB activity.

Results

The diurnal pattern of zinc secretion in pancreatic juice is shown in Figure 1. Zinc was secreted continuously in pancreatic juice. The output of zinc was maximal 2 h after feeding and declined to basal levels during the subsequent 2-h period. A distinct difference was seen between PM and CM pigs. The output of zinc in pancreatic juice was higher in PM pigs at all times. The daily secretion of zinc differed ($P < 0.001$) between PM and CM pigs and was 2.46 ± 0.41 mg and 1.46 ± 0.30 mg, respectively. The output of protein and CPB activity followed a pattern similar to zinc secretion (results not shown). Data analysis showed that protein concentration was correlated to zinc concentration, $r = 0.68$, $P < 0.001$ and $r = 0.62$, $P < 0.001$ for PM and CM pigs respectively. There was a higher correlation between protein concentration and CPB activity in CM pigs ($r = 0.62$, $P < 0.001$) than in PM pigs ($r = 0.43$, $P < 0.001$). The correlation coefficient between zinc concentration and CPB activity was 0.45 ($P < 0.001$) and 0.60 ($P < 0.001$) in PM pigs and CM pigs, respectively.

Correct citation: Jensen, M.S., Larsen, T., and Gabert, V.M. 1997. Diurnal pattern of exocrine pancreatic secretion of zinc and carboxypeptidase B in growing pigs. In *Trace Elements in Man and Animals – 9: Proceedings of the Ninth International Symposium on Trace Elements in Man and Animals. Edited by* P.W.F. Fischer, M.R. L'Abbé, K.A. Cockell, and R.S. Gibson. NRC Research Press, Ottawa, Canada. pp. 317–318.

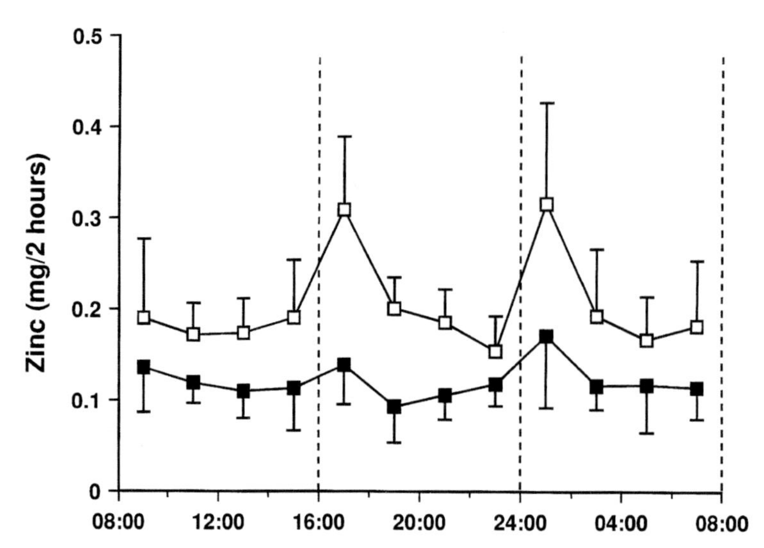

Figure 1. The pattern of secretion of zinc in pancreatic juice from pigs prepared for continuous collection of pancreatic juice using either the pouch (□) or the catheter (■) method. Feeding is indicated by broken lines. Each point is the mean of nine observations ± SE.

Discussion

The present experiment demonstrated that the secretion of zinc in pancreatic juice is stimulated by feeding. The increased zinc secretion follows the increased secretion of protein and CPB (results not shown). The correlations between zinc and protein and between zinc and CPB activity observed in the present study are in agreement with the results of previous experiments (Berger and Schneeman 1986). However, the correlation coefficients were lower in the present experiment. The differences observed between PM and CM pigs may have been due to sloughed off mucosal cells found in the pancreatic juice of PM pigs. Due to the high daily intake of zinc these cells likely contained a considerable amount of zinc bound to metallothionein and therefore contributed to the concentration of protein and zinc in the pancreatic juice of PM pigs. The percent of zinc ingested that was secreted in pancreatic juice was estimated to be 1.4 and 0.8% for PM and CM pigs, respectively, and in accordance with Reinstein et al. (1987), it is suggested that exocrine pancreatic secretion does not have a major role in zinc homeostasis in pigs.

Literature Cited

Appel, W. (1974) Carboxypeptidases. In: Methods of enzymatic analysis (Bergmeyer, U., ed.) vol. 2, pp. 986–999, Academic Press, New York.

Berger, J. & Schneeman, B.O. (1986) Stimulation of bile-pancreatic zinc, protein and carboxypeptidase secretion in response to various proteins in the rat. J. Nutr. 116: 265–272.

Hee, J.H., Sauer, W.C., Berzins, R. & Ozimek, L. (1985) Permanent re-entrant diversion of porcine pancreatic secretions Can. J. Anim. Sci. 65: 451–457.

Lowry, O.H., Rosenbrough, N.J., Farrand, A.L. & Randall, R.J. (1951) Protein measurementwith folin phenol reagent. J. Biol. Chem. 193: 265–275.

Pierzynowski, S.G., Weström, B.R., Karlsson, B.W., Svendsen, J. & Nilsson, B. (1988) Pancreatic cannulation of young pigs for long-term study of exocrine pancreatic function. Can. J. Anim. Sci. 68: 953–959.

Reinstein, N.H., Lönnerdal, B., Keen, C.L., Schneeman, B.O. & Hurley, L.S. (1987) The effect of varying dietary zinc levels on the concentration and localization of zinc in rat bile-pancreatic fluid. J. Nutr. 117: 1060–1066.

Mineral Contribution of Pancreatic Secretion in Growing Pigs Fed Two Dietary Protein Levels

EDUARDO R. CHAVEZ,† SHAOYAN LI,‡ AND WILLEM C. SAUER‡

Department of Animal Science, †McGill University, Montreal, Québec, and ‡University of Alberta, Edmonton, Alberta, Canada

Keywords pigs, pancreatic juice, protein, trace elements, macro minerals

Introduction

Metabolic studies of the availability of dietary mineral might be affected by interactions among those from exogenous sources and the endogenous digestive juices containing them. Data on mineral composition and total daily output in the gastric secretions is very limited (Thaela et al. 1995, Pohland et al. 1993). Pancreatic juice contains considerable amounts of macro and trace minerals, and considering the large volume secreted daily it may be an important mechanism for regulating the metabolic pool size of trace elements like Cu, Zn, Fe and Mn which are excreted mainly through feces. The volume and the composition of pancreatic secretion (PS) may be responsive to the dietary protein level. Thus, the main objective of this study was to evaluate the effect of a low vs a high protein diet on PS, mineral composition and total daily endogenous secretion into the intestinal tract of the pig.

Experimental Method

Six PIC barrows, average 20.2 ± 0.9 kg, were fitted with permanent pancreatic re-entrant cannulas to collect PS and determine the effect of dietary protein intake on daily volume of secretion and its mineral content. The pigs were allowed a 10 d post-operative recuperation period before the experimental diets were fed and pancreatic collection was carried out. Three pigs were fed either a high 24% (HP) or low 12% (LP) crude protein diet in a crossover design for two experimental periods of 12 d each. The diets were corn starch soybean meal (47% CP) based with a vitamin and mineral supplement to meet NRC (1988) standards. Each pig was fed twice daily, at 8:00 and 20:00 h, equal amounts each meal in the form of mash-feed at a rate of 5% of the body weight recorded at the beginning of each period. The average body weights of the pigs were 24.0 ± 1.1 and 31.3 ± 2.6 kg, respectively, and they were fed 1.2 and 1.565 kg of feed/day for periods 1 and 2, respectively. The PS was collected continuously for 12 h from 8:00 to 20:00 h on day 3, 5, 7, 10 and 12 of both periods. The collection, sampling of 10% of the h volume and return of pancreatic juice to the pigs was done each h and immediately frozen at –20°C. Freeze dried samples were used for mineral analysis by atomic absorption spectrophotometry including the macro minerals Ca, Na, K, and Mg and the trace minerals Zn, Fe, Cu and Mn. Phosphorus was also analyzed by spectrophometric methods.

Results and Discussion

Total volume per 12 h collection of PS was on average 64 and 105 mL/h for periods 1 and 2, respectively (Table 1). No significant ($p > 0.05$) changes in total volume in response to the protein level were observed but a continuous increase in volume of PS was observed with live weight gain of the pigs. Dry matter (DM) content in the PS was higher for pigs fed the HP than those fed the LP diet but the total DM secreted daily into the gastro-intestinal tract of the pigs was similar for the two groups. There were no significant ($p > 0.05$) differences between the pigs fed either of the two diets during the two phases

Correct citation: Chavez, E.R., Li, S., and Sauer, W.C. 1997. Mineral contribution of pancreatic secretion in growing pigs fed two dietary protein levels. In *Trace Elements in Man and Animals – 9: Proceedings of the Ninth International Symposium on Trace Elements in Man and Animals. Edited by* P.W.F. Fischer, M.R. L'Abbé, K.A. Cockell, and R.S. Gibson. NRC Research Press, Ottawa, Canada. pp. 319–320.

Table 1. Volume, D.M.%, and D.M., macro and trace elements secretion rate per hour in the pancreatic juice of pigs fed a HP and LP diet during two 12 h collection periods.

	Period 1		Period 2		Period 1 & 2		Diet HP+LP	
Parameter	HP	LP	HP	LP	HP	LP	P 1	P 2
Volume, mL/h	63.7	63.8	99.4	112.0	82.3	87.7	63.8*	105.5
D.M. %	1.84	1.50	1.74	1.61	1.79	1.56	1.67	1.68
DM, g/h	1.156	0.982	1.706	1.773	1.431	1.377	1.069*	1.739
Ca, mg/h	1.213	1.222	1.765	1.601	1.489	1.411	1.218*	1.683
P, µg/h	498	503	662*	399	580	451	500	531
Na, mg/h	149.7	145.8	288.0	306.2	218.8	226.0	147.8*	297.1
K, mg/h	10.08	9.33	15.9	16.6	13.0	13.0	9.75*	16.25
Zn, µg/h	191	137	162	184	176	161	164	173
Cu, µg/h	3.67	3.5	4.92	4.58	4.33	4.08	3.58*	4.75
Fe, µg/h	17.30*	26.75	43.67*	27.58	30.50	27.17	22.08*	35.67
Mn, µg/h	1.83	1.92	2.50	2.92	2.17	2.42	1.92*	2.75*

Significantly ($p < 0.05$) different from the value in the next right column.

in the total daily secretion in the PS of the macro minerals Ca, Na and K. However, it is important to point out the secretion of 5.4 g/d of Na in the PJ since it represents about 16% of its DM content and more than 3.2 times the daily Na intake (1.7 g). In contrast, total daily secretion of P and Mg in the PS of pigs although very small was significantly ($p < 0.05$) higher in the HP fed pigs. The dietary protein intake had almost no effect on the trace mineral flow per h of the PJ secreted by pigs. Although the Mn content in the PS of pigs fed the LP was higher ($p < 0.05$) than in the HP fed pigs the total secretion/d of Mn was not significantly different ($p > 0.05$). Thus, the total daily secretion of Zn, Cu, Fe and Mn in the PS of pigs was not significantly affected by the protein level. Zinc content in the PS appears to be the trace mineral with the greatest relative contribution to the nutritional balance of the pig. Total daily Zn secretion in the PS represented about 2% (4.4 mg) of the total daily dietary intake (208 mg). Other trace minerals in the PS were relatively insignificant compared to the amount of intake/day; Cu was less than 0.05% (100 µg), Mn less than 0.2% (56 µg) and Fe less than 0.3% (690 µg). Thus for the metabolic balance studies of these trace minerals the endogenous contribution of the PS to the excretion can be ignored.

Conclusions

Overall, the major effect of the dietary protein level on the mineral content in the PJ secreted by the pig appears to be related to P secretion with some minor effect on Mg secretion. The daily secretion of both macro minerals were reduced in the LP diet compared to the HP diet. Pigs receiving the HP diet secreted on average about 28% more P and 23% more Mg in the PS than pigs fed the LP diet. Other macro and trace elements showed non significant differences. There was also a significant increase in Ca, Na, K, Mg, Fe and Mn in the PS from period 1 to period 2.

References

Poland, U., Souffrant, W.B., Sauer, W.C., Mosenthin, R. & de Lange, C. F.M. (1993) J.Sci.Food Agric. 62: 229–234.
Thaela, M.J., Pierzynowski, S.G., Jensen, M.S., Jakobsen, K., Westrom, B.R. & Karlsson, B.W. (1995) J. Anim. Sci. 73: 3402–3408.

Macro and Trace Mineral Utilization in a Low-Phosphorus Diet Fed at Three Levels of Calcium to Broiler Chickens with or without Phytase

EDUARDO R. CHAVEZ, SYLVESTER SEBASTIAN, AND
SHERMAN P. TOUCHBURN
Department of Animal Science, McGill University, Montreal, Quebec, Canada

Keywords broilers, phytase, dietary calcium, trace elements, macro minerals

Introduction

Phytate is a naturally occurring complex compound which is a main natural organic source of P in animal feedstuffs of plant origin. It is well documented that supplementation of microbial phytase increases the bioavailability of phytate-P. High levels of dietary calcium can progressively precipitate all the phytate by forming the extremely insoluble calcium-phytate complex in the intestine (Nelson and Kirby 1987); consequently phytate-P, as well as calcium itself, is largely unavailable for absorption. It has also been shown that high levels of dietary Ca and Mg reduce intestinal phytase activity in chicks. A recent study with pigs suggests that supplemental microbial phytase in a corn-soybean meal diet improves phytate-P utilization more effectively at moderately low levels of dietary Ca than at normally recommended levels (Lei et al. 1994). The objective of this experiment was to study the efficacy of supplemental microbial phytase at three levels (low, recommended and high) of dietary Ca, on the performance and utilization of macro and trace minerals in broiler chickens fed a low-P corn-soybean meal diet.

Materials and Methods

A total of 240 day-old broiler chickens was housed in Petersime battery brooders with continuous light. Ten birds were assigned to each of 24 pens. All birds had *ad libitum* access to water and experimental diet throughout the 21-d trial. Each of six treatments was replicated four times. The experimental design was a completely randomized one with 3 × 2 factorial arrangements of treatments. The variables included dietary levels of Ca and phytase: Ca at 0.66% (low), 1.0% (normal), 1.25% (high); phytase at 0 and 600 phytase units (PU) per kg diet. Individual body weights of chickens and group feed consumption data were recorded on days 7, 14 and 21. On the first 3 days of the second and third week of the experiment, the daily feed consumption and total fecal output were recorded. A representative sample of excreta and feed from each pen was freeze-dried, ground, and analyzed for minerals. The difference in the mineral content of the feed consumed and of the feces was used to calculate the relative retention of minerals. On the last day of the experiment, heparinized blood was obtained by cardiac puncture, then the birds were killed and the left tibia was removed for mineral analysis.

Results and Discussion

Phytase supplementation to both the 1.0 and 1.25% Ca diets increased ($P \leq 0.05$) the body weights by 11 and 13%, respectively at 14 d; the corresponding values for 21 d were 15 and 9%. The maximum feed efficiency at 21-d was obtained on the diet containing 1.0% Ca and phytase. The improvement observed in the growth performance of chickens fed a phytase-supplemented diet is an indication of effectiveness of phytase in this experiment. The relative retentions of P, N and Mn were increased

Correct citation: Chavez, E.R., Sebastian, S., and Touchburn, S.P. 1997. Macro and trace mineral utilization in a low-phosphorus diet fed at three levels of calcium to broiler chickens with or without phytase. In *Trace Elements in Man and Animals – 9: Proceedings of the Ninth International Symposium on Trace Elements in Man and Animals. Edited by* P.W.F. Fischer, M.R. L'Abbé, K.A. Cockell, and R.S. Gibson. NRC Research Press, Ottawa, Canada. pp. 321–322.

Table 1. The effect of phytase supplementation at different levels of calcium on relative retention of minerals in 17-d-old broiler chickens fed a low-P corn-soybean diet.

Treatment		Relative retention							
Calcium (%)	Phytase (PU/kg)	P (%)	Ca (%)	Mg (%)	Zn (%)	Cu (%)	Fe (%)	Mn (%)	N (%)
0.60	0	54.7^{bc}	64.1^{a}	19.2	−11.3	-0.2^{cd}	20.0^{bc}	6.2^{b}	66.9^{b}
0.60	600	57.6^{abc}	67.6^{a}	21.2	11.2	9.1^{bc}	27.1^{b}	9.5^{b}	69.2^{ab}
1.0	0	57.5^{abc}	51.1^{b}	23.6	−1.0	25.8^{a}	22.6^{bc}	5.5^{b}	70.8^{a}
1.0	600	61.2^{a}	46.9^{bc}	22.2	−15.8	6.9^{d}	15.7^{c}	-2.3^{c}	68.6^{ab}
1.25	0	53.2^{c}	43.2^{c}	24.5	−10.5	23.4^{a}	37.2^{a}	3.7^{bc}	67.4^{b}
1.25	600	59.0^{ab}	43.6^{c}	28.1	−11.8	17.8^{ab}	29.1^{ab}	17.9^{a}	70.8^{a}

[a–d] Means within each column with no common superscripts differ significantly ($P < 0.05$).

($P \leq 0.05$) by 5.8, 3.4 and 14.2 percentage units, respectively by phytase supplementation to the 1.25% Ca diet; but at the 1.0% Ca level, phytase supplementation reduced ($P \leq 0.05$) the relative retention of Cu and Mn by 18.9 and 7.8 percentage units, respectively (Table 1). Blood analysis showed that phytase supplementation to the 1.25% Ca diet increased ($P \leq 0.05$) plasma P by 21% and reduced ($P \leq 0.05$) plasma Ca by 21% whereas other plasma minerals were not affected by phytase. Bone analysis indicated that phytase supplementation to the 1.25% Ca diet increased ($P \leq 0.05$) P and Ca contents of tibia shaft DM by 1.0 and 1.5 percentage units, respectively and to the 1.0 and 1.25% Ca diets, increased ($P \leq 0.05$) ash content of the tibia head by 3.3 and 4.8 percentage units, respectively. Phytase supplementation to the 1.25% Ca diet reduced ($P \leq 0.05$) Cu and Mn content in tibia head ash by 3.5 and 2.9 ppm, respectively but did not affect the concentration of any minerals measured in tibia shaft ash. It seems that the shaft portion of the tibia is in a more stable state and hence is less affected by the availability of minerals. Regardless of Ca level, phytase supplementation increased ($P \leq 0.036$) only total ash content (48.8 vs. 44.7% of DM) in the tibia shaft. The improvement in ash percentage in the tibia indicates an increase in bone mineralization consequent to an increase in the availability of minerals liberated by phytase from phytate-mineral complex.

Conclusions

These results clearly indicate that microbial phytase supplementation to a low-P diet improved the growth performance and mineral utilization in broiler chickens. The maximum efficacy of phytase for growth performance was obtained at the 1.0% Ca level whereas that for the retention of minerals, plasma mineral levels and bone mineral content occurred at the 1.25% Ca level. However, the optimum utilization of resources for overall performance was obtained in the 0.6% Ca diet supplemented with phytase.

References

Lei, X.G., Ku, P.K., Millar, E.R., Yokoyama, M.T. & Ullrey, D.E. (1994) Calcium level affects the efficacy of supplemental microbial phytase in corn-soybean diets of weanling pigs. J. Anim. Sci. 72: 139–143.

Nelson, T.S. & Kirby, L.K. (1987) The calcium binding properties of natural phytate inchick diets. Nutr. Rep. Int. 35: 949–956.

Trace Element Homeostasis in the Avian Embryo

MARK P. RICHARDS

U.S. Department of Agriculture, Agricultural Research Service, Livestock and Poultry Sciences Institute, Growth Biology Laboratory, Beltsville, MD 20705-2350, USA

Keywords embryo, trace elements, homeostasis, avian

A variety of trace elements including zinc, copper and iron are required in differing amounts to sustain proper growth and development of the avian embryo (Savage 1968). Deficiencies or excesses of individual trace elements can cause impaired growth, abnormal development affecting all of the major organ systems and, in extreme cases, death of the embryo (Richards and Steele 1987). Thus, a continuous and precisely regulated supply of trace elements derived from stores within the egg is essential to ensure avian embryonic survival. Appropriate amounts of each trace element required to support embryonic growth and development are transferred from hen tissue stores to specific sites in the egg at the time of its formation (Richards 1989).

Egg Stores

Table 1 lists the quantities and the percentage distribution of zinc, copper and iron in five fractions of fertile turkey eggs. Zinc and iron have similar distributions with the majority of each metal deposited in the yolk granule fraction. Copper shows a different distribution pattern with higher percentages deposited in the albumen, the eggshell, the shell membrane, and the soluble fraction of yolk. In general, the yolk constitutes the largest single available store of most trace elements deposited in the avian egg. Vitellogenin, an estrogen-induced yolk precursor protein synthesized by the liver of the hen, mediates the transfer of trace elements from hen tissue stores (i.e., liver) to the yolk of the egg (Richards and Steele 1987). The vitellogenin precursor and its two product molecules, phosvitin and lipovitellin, are trace element binding proteins (Richards 1989). Lipovitellin and phosvitin are sequestered together in the granule subfraction of yolk and these two proteins constitute the molecular basis for a major portion of trace element storage within the yolk of the avian egg (Richards 1991a).

*Table 1. Distribution of zinc, copper and iron in the turkey egg**,+

	Zinc		Copper		Iron	
Egg Fraction	(μg)	(%)	(μg)	(%)	(μg)	(%)
Yolk (Granules)	684.5	88.8	55.1	27.4	1802.8	84.6
Yolk (Soluble)	23.0	3.0	34.5	17.2	55.8	2.6
Albumen	5.4	0.6	15.6	7.8	70.4	3.3
Eggshell	53.3	6.9	67.7	33.7	198.3	9.3
Shell Membrane	5.0	0.7	28.1	14.0	4.6	0.2

*Data are from Richards (1991a) for an average egg weight of 79.66 ± 2.21 g.
+Each value represents the mean of 20 determinations.

Yolk Sac

The yolk sac, which is the extra-embryonic membrane in direct contact with yolk, plays a key role in avian embryonic trace element metabolism. Specifically, the endodermal cells which line the inner surface of the yolk sac are responsible for the mobilization and transfer to the embryo of yolk trace element

Correct citation: Richards, M.P. 1997. Trace element homeostasis in the avian embryo. In *Trace Elements in Man and Animals – 9: Proceedings of the Ninth International Symposium on Trace Elements in Man and Animals. Edited by* P.W.F. Fischer, M.R. L'Abbé, K.A. Cockell, and R.S. Gibson. NRC Research Press, Ottawa, Canada. pp. 323–325.

stores. The yolk sac has the ability to accumulate and temporarily store trace elements derived from yolk as indicated by its ability to synthesize such intracellular metal storage proteins as metallothionein and ferritin during development (Richards 1991c). Also, yolk sac endodermal cells actively synthesize plasma proteins and it has been proposed that the yolk sac might regulate the release to the embryo of trace elements mobilized from yolk by controlling the synthesis and export of specific trace element transporting plasma proteins to the vitelline circulatory system. The vitelline system, which is the embryonic equivalent of the hepatic portal system, is an important route for trace element transfer from yolk sac to the liver of the developing avian embryo (Richards and Steele 1987).

Embryonic Liver

The liver is the intra-embryonic organ responsible for the initial uptake and processing of trace elements derived from yolk stores and exported by the yolk sac. Because changes in the concentrations of zinc and iron in turkey embryo liver are highly correlated during the latter half of incubation, it has been proposed that this might reflect similar distributions of these two metals in the egg (see Table 1) as well as similar rates of mobilization by the yolk sac (Richards 1991b,c). Hepatic copper concentration, which reaches a maximum late in development of the turkey embryo, may reflect its different pattern of distribution within the egg as compared to zinc or iron (Richards and Weinland 1985, Richards 1991b). Hepatic metallothionein synthesis, which is developmentally regulated, may function in zinc and copper homeostasis by providing a short-term storage site for excess quantities of these two metals derived from yolk stores. Similarly, ferritin synthesis is linked with iron storage in the embryonic liver. The hepatocyte, like the yolk sac endodermal cell, is an active site of plasma protein synthesis and this may provide a mechanism to control the release of trace elements to the embryonic circulation and from there to all of the extra-hepatic tissues.

Conclusions

Figure 1 presents a model to describe the uptake and utilization of egg trace elements by the developing avian embryo. Yolk granules are taken up by yolk sac endodermal cells in which their constituent trace elements are released following the intralysosomal digestion of phosvitin and lipovitellin. The yolk sac acts as a short-term storage site by sequestering trace elements derived from yolk on intracellular metal-binding proteins such as metallothionein and ferritin. To regulate the release of trace elements to

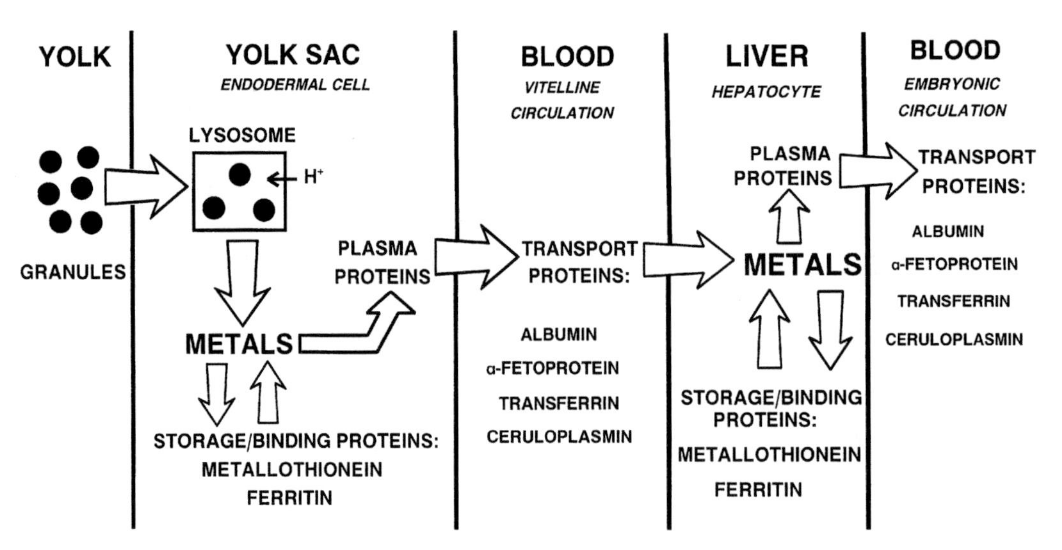

Figure 1. A proposed model to describe mobilization of egg yolk trace element stores and their subsequent transfer to the liver of the developing avian embryo.

the embryo via the vitelline circulation, the yolk sac synthesizes specific plasma proteins that function in interorgan metal transport. The liver of the embryo receives the exported trace elements and, like the yolk sac, transiently stores them on specific metal-binding proteins prior to their export to other tissues of the embryo via the embryonic circulatory system. Thus, through the coordinated actions of the yolk sac and the liver, trace element homeostasis is established in the developing avian embryo.

Literature Cited

Richards, M.P. & Weinland, B.T. (1985) Hepatic zinc, copper and iron in the developing turkey embryo and newly hatched poult. Biol. Trace Elem. Res. 7: 269–283.

Richards, M.P. & Steele, N.C. (1987) Trace element metabolism in the developing avian embryo: A review. J. Exptl. Zool. Suppl. 1: 39–51.

Richards, M.P. (1989) Influence of egg production on the zinc, copper and iron metabolism in the turkey hen (*Meleagris gallopavo*). Comp. Biochem. Physiol. 93A: 811–817.

Richards, M.P. (1991a) Distribution and storage of trace elements in the avian egg. In: Trace Elements in Man and Animals 7 (Momcilovic, B., ed.), Chapter 25, pp. 21–23. IMI Press, Zagreb, Croatia.

Richards, M.P. (1991b) Mineral metabolism in the developing turkey embryo. I. The effects of developmental age and shell-less culture on trace element contents of selected tissues. Comp. Biochem. Physiol. 100A: 1009–1016.

Richards, M.P. (1991c) Mineral metabolism in the developing turkey embryo. II. The role of the yolk sac. Comp. Biochem. Physiol. 100A: 1017–1023.

Savage, J.E. (1968) Trace minerals and avian reproduction. Fed. Proc. 27: 927–931.

Influence of Corrected Teeth on Digestibility of Macro-, and Microminerals in Pregnant Mares

D. GATTA, L. KRUSIC,[†] P. SCHRAMEL,[†] S. PINOLA, AND B. COLOMBANI
Dipartimento di Produzioni Animali, Universita di Pisa, Italy; [†]GSF, 85758 Neuherberg, Ingolstaedter Landestr. 1, Germany

Keywords retention, digestibility, teeth, horses

Introduction

Dietary supplementation of horses with minerals is mainly done according to guidelines by NRC (1989) and DLG (1994), which are based on little conclusive evidence about how well these animals digest and absorb nutrients (Pagan 1994). Little attention is given to the effect of mastication and dentition on equine digestibility. A recent digestion trial showed that the correction of occlusion improved mechanical digestion and utilization of organic nutrients (Gatta et al. 1995). In order to evaluate the effect of corrected dentition on the digestibility of minerals in pregnant mares, a digestive trial with two types of diets before and after standard correction of the teeth was conducted.

Material and Methods

Four pregnant Franches Montagnes half-bred mares, were used to evaluate mineral digestibility in two types of diets, one with hay (diet 1) and one with a hay and oats mixture (diet 2) before and after dental correction. Each digestion trial lasted four weeks. The age of the mares varied from 7 to 13 years, their average weight was 563.3 ± 46.9 kg and all were pregnant from 6 to 9 months. Correction of teeth was performed according to the technique described previously (Gatta et al. 1995). The mineral composition of the feeds is given in Table 1. Both diets were fed twice daily at 7 a.m. and 7 p.m. Feed and water consumption was measured daily. During the six day collection period the mares were housed indoors in individual stalls which allowed for the complete and separate collection of urine and faeces. Faecal samples were stored frozen for mineral analyses. Urine excretions (24 h) were measured and an aliquot was acidified and frozen for mineral analyses. Minerals were analysed using ICP-AES and ICP-MS (Schramel 1994). Digestion coefficients were calculated for each mineral measured. Statistical analyses were performed using the JUMP statistical program. Given the small number of animals in each digestive trial averages were compared with the Mann and Whitney U-Test.

Results

The daily intake of dry matter (DM kg/W 0.75) ranged from 119.9 to 93.4 (Diet 1) and from 119.5 to 90.76 (Diet 2) before and after dental correction. Mineral retention and apparent digestibility coefficients are summarized in Table 2. The apparent digestibility of calcium increased after dental treatment. Significantly increased retention of calcium ($P < 0.05$) was detected in diet 1 after dental correction. Phosphorus retention was increased ($P < 0.05$) in diet 1 after dental correction. In contrast, the difference in retention of phosphorus in diet 2 after dental correction, although increased, was not significant. The apparent digestibility of phosphorus was higher ($P < 0.05$) in diet 1, whereas in diet 2, the digestibility was lower than that found by Pagan (1994). The increased apparent absorption of phosphorus in diet 1 after dental correction may be attributed to improved mechanical grinding in the mouth as shown by smaller particles in fecal samples and better feed utilization (Gatta et al. 1995). The increased phosphorus apparent

Correct citation: Gatta, D., Krusic, L., Schramel, P., Pinola, S., and Colombani, B. 1997. Influence of corrected teeth on digestibility of macro-, and microminerals in pregnant mares. In *Trace Elements in Man and Animals–9: Proceedings of the Ninth International Symposium on Trace Elements in Man and Animals. Edited by* P.W.F. Fischer, M.R. L'Abbé, K.A. Cockell, and R.S. Gibson. NRC Research Press, Ottawa, Canada. pp. 326–328.

Table 1. Chemical mineral composition feedstuffs.

Feedstuff	Unit	Hay	Oats	Water	Unit
Al	mg/kg	417.00	2098.00	0.0168	mg/L
B	mg/kg	18.80	–	0.003	mg/L
Ba	mg/kg	19.70	9.46	0.0032	mg/L
Ca	g/kg	10.54	1.01	11.2	mg/L
Cr	mg/kg	0.91	3.40	0.001	mg/L
Cu	μg/kg	6.84	4.60	0.0067	mg/L
Fe	mg/kg	369.46	1120.00	0.0051	mg/L
K	g/kg	18.80	3.89	0.001	mg/L
Mg	g/kg	1.57	1.39	1.7	mg/L
Mn	mg/kg	39.60	49.00	0.001	mg/L
Na	g/kg	0.35	0.13	4.9	mg/L
Ni	mg/kg	1.91	5.03	0.001	mg/L
P	g/kg	1.33	3.41	0.217	mg/L
S	g/kg	1.83	1.73	2.9	mg/L
Sr	mg/kg	45.40	4.31	0.0445	mg/L
Ti	μg/kg	10.73	24.73	0.001	mg/L
Zn	mg/kg	32.93	34.50	0.0873	mg/L
Se	mg/kg	0.10	0.11	0.0001	mg/L

Table 2. Mineral retention and apparent digestibility coefficients.

	Diet	Ca-g	K-g	Mg-g	Na-g	P-g	S-g
Retained	1 – 0	58.89±8.92[a]	59.04±34.82	0.69±2.71a	1.92±0.99[a]	–0.37±3.23[a]	–11.98±2.53
	2 – 0	35.38±6.01[b]	21.71±8.85	–0.56±1.38[a]	–1.89±1.76[b]	5.14±2.86[ab]	–11.16±2.22
	1 +	87.91±13.10[c]	65.41±39.23	5.34±2.06[b]	1.44±1.20[a]	6.73±2.09[b]	–9.90±4.31
	2 +	43.99±7[b]	24.08±20.04	1.52±2.84[a]	–0.73±1.09[b]	6.04±0.71[b]	–11.31±3.65
Apparent	1 – 0	0.48±0.05	0.61±0.11	0.22±0.1[ab]	0.47±0.13[a]	–0.02±0.17[a]	0.42±0.07
Digestibility	2 – 0	0.51±0.05	0.62±0.02	0.22±0.04[a]	–0.32±0.50[bc]	0.26±0.15[ab]	0.45±0.03
	1 +	0.61±0.09	0.66±0.17	0.37±0.08[b]	0.40±0.27[ac]	0.37±0.12[b]	0.47±0.14
	2 +	0.60±0.08	0.62±0.12	0.36±0.09[b]	–0.12±0.23[b]	0.31±0.04[b]	0.54±0.08

	Diet	Cu/mg	Fe/mg	Mn/mg	Zn/mg	Se/mg
Retained	1 – 0	–39.93±17.48	–1511.03±787.1	–83.06±77.72	52.09±72.66	0.45±0.02
	2 – 0	–25.29±11.17	–2828.03±2631.88	237.31±109.52[ab]	27.31±75.54	0.37±0.11[a]
	1 +	–23.87±16.35	–756.91±1068.20	421.20±167.91[b]	13.34±90.19	0.63±0.04[b]
	2 +	–44.72±19.78	–194.70±672.86	–233.24±80.00[a]	11.67±34.36	0.34±0.16[a]
Apparent	1 – 0	–0.42±0.18	–0.29±0.15	–0.15±0.11	–0.10±0.16	0.69±0.03
Digestibility	2 – 0	–0.37±0.17	–0.48±0.44	–0.53±0.25[ab]	0.11±0.23	0.75±0.05[ab]
	1 +	–0.25±0.17	–0.15±0.21	–0.77±0.31[b]	0.04±0.20	0.80±0.04[b]
	2 +	–0.67±0.29	–0.03±0.11	–0.52±0.18[b]	0.05±0.11	0.80±0.01[b]

*0, before dental correction; +, after dental correction; a, b, $P < 0.05$.

absorption in diet 2 relative to diet 1 could be due to a higher phosphorus content and/or availability in diet 2. Magnesium balance was positive throughout the trial except in diet 2 before the correction of teeth. Apparent digestibility of magnesium was higher in both diet 1 and 2 after dental correction ($P < 0.05$). The level after dental correction was within the range reported by Pagan (1994), but was lower than that reported by Krusic et al. (1993). High concentrations of calcium and potassium in diet 1 and 2 may have depressed the absorption of magnesium (Newton et al. 1972). The amounts of retained sodium after dental treatment was not significantly different after dental correction. A positive sodium balance was observed only in diet 1. A negative sodium balance in diet 2 was due to increased fecal excretion which may have been attributed to a secretion of sodium into the small intestine (Illenser 1994).

The retention and apparent digestibility of potassium was not significantly different after dental correction in both diets. The negative balance of sulphur in this trial may be due to an elevated urine excretion. The apparent digestibility of iron before and after dental treatment was negative, although the amount retained increased after treatment. The retention and apparent digestibility remained negative after dental correction. There was a negative manganese balance throughout the trial, due to higher fecal excretion. Average apparent digestibility and retention of manganese after dental treatment appeared to be significantly decreased ($P < 0.05$) in diet 1. There were no significant changes in apparent digestibility and retention of manganese after dental correction in diet 2. Average apparent digestibility of zinc appeared increased after dental treatment. The values agree with the results of Pagan (1994), but are lower than reported by Krusic et al. (1993). There was a positive zinc balance throughout the trial except in diet 1 before dental correction. Average apparent digestibility and retention of selenium was higher ($P < 0.05$) after dental treatment only in diet 1.

Conclusion

As shown in this study, the requirement for several minerals may not be met by traditional diets such as forage or forage and oats. The most significant deficiencies were detected for phosphorus, sodium and copper. Digestibility as well as retention of some macro and microminerals were improved significantly after dental correction.

Literature

DLG (1994) Energie und Naehrstoffbedarf landwirtschaftlicher Nutztiere. DLG Verlag, Frankfurt.

Gatta, D., Krusic, L., Cassini, L. & Colombani B. (1995) Influence of corrected teeth on digestibility of two types of diets in pregnant mares. pp.326–331. Proceed. of Equine Nutrit. Physiol. Symp., Ontario, Canada.

Illenser, M. (1994) Praeileale Verdaulichkeit von Hafer-, Kartoffel- und Maniokrationen beim Pferd. Diss. TIHO, Hannover.

Krusic, L., Schramel, P., Dermelj, M., Usaj, A., Stibilj, V. & Pagan J.D. (1993) Influence of high carbohydrate and high fibre diet on mineral metabolism in horses. In: Trace Elements in Man and Animals. (Anke, M., Meissner, D., & Mills, CF., eds.), pp. 310–313. Verlag Media Touristik, Gersdorf.

Newton, G.L., Fonteno,t J.P., Tucker, R.E. & Polan C.E. (1972) Effects of high dietary potassium intake on the metabolism of magnesium by sheep. J. Anim. Sci., 35: 440–445.

NRC (1989) Nutrient requirements of horses. p 100. National Academy Press, Washington D.C.

Pagan, J. (1994) Digestibility trials provide evaluation of feedstuffs. Feedstuffs, 23: 14–15.

Schramel P. & Hasse S. (1994) Iodine determination in biological materials by ICP-MS. Research Center for Environment and Health, Institute of Ecological Chemistry, Oberschleibheim, Germany.

Trace Elements and Free Radical Mediated Disease

Chairs: T.M. Bray and M.R. L'Abbé

The Role of Iron in Free Radical Reactions

S.D. AUST AND J.-H. GUO
Biotechnology Center, Utah State University, Logan, UT 84322-4705, USA

Keywords iron, ferritin, ceruloplasmin, oxygen radicals

Iron is extensively used for hemoproteins and for non-heme iron oxidation-reduction enzymes because of its utility as a transition element. Unfortunately, transition metals can also catalyze the unintentional or deleterious oxidation of many biochemicals, both directly and indirectly, by the generation of oxygen radicals (Weinber 1966). Consequently many repair and protective enzymes have evolved to detoxify partially reduced species of oxygen, such as superoxide dismutase, catalase and glutathione peroxidase, and to metabolize lipid peroxides, repair DNA and degrade oxidized proteins. A very complex biochemical system also evolved to help control iron such that it does not catalyze these deleterious oxidations. Iron is transported bound to transferrin and its cellular uptake is controlled by a receptor for iron-loaded transferrin. Iron is stored in ferritin, in an apparently fairly stable, yet potentially large (about 2300 atoms/mole) deposit of iron. A cytoplasmic iron and a redox sensitive protein, called the iron responsive element-binding protein (IRE-BP), is a translational repressor of ferritin synthesis when bound to a regulatory region, the iron responsive element (IRE), on ferritin mRNA. When cellular levels of iron are low, the IRE-BP is bound to IRE of ferritin mRNA and ferritin is not synthesized. When intracellular iron levels rise, the IRE-BP contains iron and dissociates from the ferritin mRNA resulting in ferritin synthesis for safe storage of the iron. Conversely, association of iron with IRE-BP can affect the further uptake of iron because the 3′-end of the mRNA for the transferrin receptor also has an IRE which alters the mRNA stability and receptor translation. The net result of increased cellular iron is more storage capacity and decreased cellular iron uptake.

Ceruloplasmin seems intricately involved with iron metabolism. First, there is the proposal that ceruloplasmin has the ability to load iron into transferrin (Osaki et al. 1966). The anemia resulting from copper-deficiency cannot be corrected by the administration of iron but can be reversed by the administration of ceruloplasmin (Lee et al. 1968). Mutations in the ceruloplasmin gene appear as typical iron-deficiency anemia (Harris et al. 1995). We propose that ceruloplasmin may also be responsible for the loading of iron into ferritin (Aust 1995).

Ceruloplasmin has been considered to have broad substrate oxidase activity (Frieden 1980), however, we believe that many of the substrate oxidase activities are a result of its ferroxidase activity (Ryan et al. 1993). These chemicals are frequently contaminated with iron as they are excellent iron chelators. There are also easily oxidized by iron, i.e., they are good reductants of iron. If the ferrous state of iron is stable when bound by the biogenic amine, which is actually promoted by the nitrogen atoms of the biogenic amines, then further oxidation of the amine will be stimulated by the ferroxidase activity of ceruloplasmin (Frieden 1985, McDermott et al. 1968).

The ferroxidase activity of ceruloplasmin is thought to be responsible for loading iron into transferrin (Osaki et al. 1966). It can actually be assayed by taking advantage of this ability (Osaki 1966).

Correct citation: Aust, S.D., and Guo, J.-H. 1997. The role of iron in free radical reactions. In *Trace Elements in Man and Animals – 9: Proceedings of the Ninth International Symposium on Trace Elements in Man and Animals. Edited by* P.W.F. Fischer, M.R. L'Abbé, K.A. Cockell, and R.S. Gibson. NRC Research Press, Ottawa, Canada. pp. 329–332.

Ceruloplasmin has a very low K_m for ferrous iron and a very high V_{max} (Frieden 1980). The reaction involves the complete reduction of molecular oxygen to water without the release of partially reduced species of oxygen (Ryden 1984). Thus, by very efficient iron oxidation, ceruloplasmin may promote iron incorporation into both the iron transport and iron storage proteins.

Many investigators have studied the ferroxidase activity of (Treffry et al. 1978, Bakker and Boyer 1986), and the ferroxidase center (Lawson et al. 1986, Levi et al. 1994) in ferritin. We do not subscribe to this proposed mechanism because it results in the oxidation of amino acids in ferritin, probably due to the fact that this ferroxidase activity does not involve the complete reduction of molecular oxygen to water (DeSilva and Aust 1992). Besides, the activity cannot be observed when we use solutions in which ferrous iron does not autoxidize. For example, if Hepes buffer is replaced by histidine in physiological saline, no ferroxidase activity is observed and ferritin is not loaded (DeSilva et al. 1992). In addition, the iron loaded into the ferritin with its own "ferroxidase" activity in Hepes buffer differs dramatically from iron in ferritin loaded with ceruloplasmin (DeSilva and Aust 1992). The iron may in fact be high molecular weight polymerized ferric hydroxide only associated with ferritin. This possibility must be tested by analysis by staining for both iron and protein in native PAGE of loaded ferritin rather than the spectrophotometric assay usually employed for transferrin and ferritin loading (Yablonski and Thiel 1992).

Given the fact that ceruloplasmin is an excellent ferroxidase, for the various reasons given above, we investigated the ability of these proteins to be an effective "antioxidant". Starting with ferrous iron, which we believe would be the case *in vivo* because of the reducing nature of cells, lipid peroxidation only occurs when sufficient ferrous iron is oxidized, apparently to produce ferric iron which we (Minotti and Aust 1987) and others (Braughler et al. 1986) have proposed to be required for lipid peroxidation. The addition of ceruloplasmin simply results in a shorter lag time because ferric iron is provided faster (Figure 1) (Samokyszyn et al. 1989). However, if ferritin is also provided, such that the ferric iron produced by the ferroxidase activity of ceruloplasmin is safely sequestered, then lipid peroxidation is prevented (Figure 1) (Samokyszyn et al. 1989).

This antioxidant activity would require an "efficient" association between ceruloplasmin and ferritin. If iron were to be oxidized by ceruloplasmin and not incorporated into ferritin, then it could combine with the ferrous iron not yet oxidized to promote lipid peroxidation. The ferric iron could also be reduced by cellular reductants, which hopefully would not be detrimental, but its reduction by biochemicals not intended to be a cellular reductant would be detrimental. Thus, ceruloplasmin should not oxidize iron very

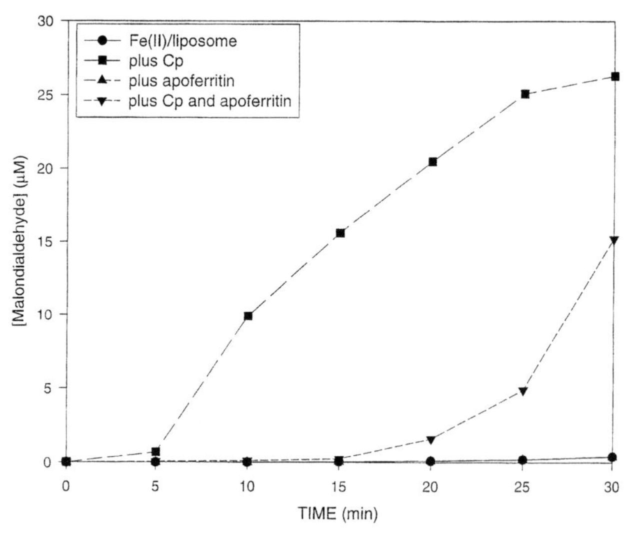

Figure 1. The effect of apoferritin on ceruloplasmin-dependent lipid peroxidation. Phospholipid liposomes (1 μmol phosphate/mL) were preincubated for 2 min at 37°C in Chelex-treated 50 mM NaCl (pH 7.0) in the presence or absence of ceruloplasmin (0.5 μM), apoferritin (50 μg/mL), or ceruloplasmin (0.5 μM) plus apoferritin (50 μg/ml) followed by the addition of ferrous chloride (200 μM).

effectively if ferritin was not available and even if it were available, the iron should be effectively loaded into the ferritin. We believe that this is the case. A complex seems to exist between ferritin and ceruloplasmin. Ferritin stimulates the ferroxidase activity of ceruloplasmin (Figure 2) and there seems to be an optimum ratio of ceruloplasmin to ferritin for the loading of iron into ferritin (Figure 3).

The association between ceruloplasmin and ferritin seems to involve the H-chain of ferritin. Ceruloplasmin does not interact with homopolymers of L-chain ferritin, prepared by recombinant DNA techniques, and only loads iron into homopolymers of H-chain.

In summary, iron can oxidize many biomolecules. It can also generate oxygen radicals or other oxidants that can damage biomolecules. Many of these biomolecules are thought to autoxidize but this is not absolutely correct. Many of these chemicals are iron reductants and the iron autoxidizes. A test was developed by Gary Buettner for chemicals that are generally thought to "autoxidize". When, for example, ascorbic acid does not promote oxygen uptake it may be considered free of transition metals (Buettner and Moseley 1992). Unfortunately many of the chemicals which can reduce iron are also iron chelators. This means that commercial preparations are often contaminated with iron or they become contaminated with iron as we work with them in the laboratory. Thus, they can appear to autoxidize. If they are ferrous chelators, that is, if the ferrous iron is stabilized by the biomolecules, the iron may be oxidized by ceruloplasmin giving the enzyme an apparent ability to oxidize an organic biochemical. We test this possibility by determining the effects of prolonged incubation of the ferrous chelator (biochemical) with desferrioxamine (Ryan et al. 1993) on the "oxidase" activity of ceruloplasmin.

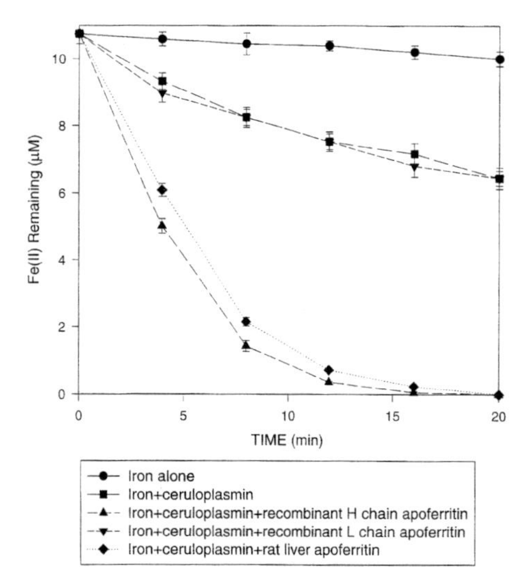

Figure 2. The effect of apoferritin on the ferroxidase activity of ceruloplasmin during iron loading. Ceruloplasmin (0.22 nmol) were preincubated at 37°C in Chelex-treated 50 mM NaCl (pH 7.0) in the absence or presence of various apoferritin (0.22 nmol) for 1 min (1 mL final volume). The reaction was initiated by addition of histidine:Fe(II) (5:1) to a final iron concentration of 11 μM. Aliquots of the reaction were removed over time and added to ferrozine.

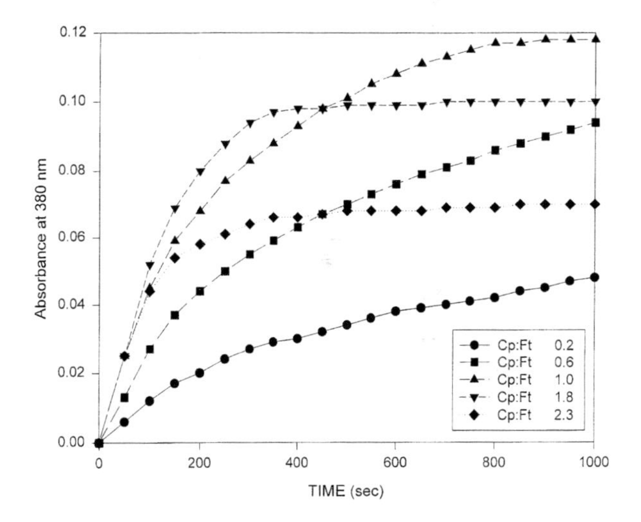

Figure 3. The effect of the molar ratio of ceruloplasmin:ferritin on iron loading into ferritin. Apoferritin (2.2 nmol/mL) was preincubated for 1 min at 37°C in Chelex-treated 50 mM NaCl (pH 7.0) with various amount of ceruloplasmin (1 mL final volume).

Literature Cited

Aust, S.D. (1995) Ferritin as a source of iron and protection from iron-induced toxicities. Toxicol. Lett. 82/93: 941–944.

Bakker, G.R. & Boyer, R.F. (1986) Iron incorporation into apoferritin. The role of apoferritin as a ferroxidase. J. Biol. Chem. 261: 13182–13185.

Braughler, J.B., Duncan, L.A., & Chase, R.L. (1986) The involvement of iron in lipid peroxidation. Importance of ferric to ferrous ratios in initiation. J. Biol. Chem. 261: 10282–10289.
Buettner, G.R. & Moseley, P.L. (1992) Ascorbate both activates and inactivates bleomucin by free radical generation. Biochemistry 31: 9784–9788.
DeSilva, D. & Aust, S.D. (1992) Stoichiometry of Fe(II) oxidation during ceruloplasmin-catalyzed loading of ferritin. Arch. Biochem. Biophys. 298: 259–264.
DeSilva, D., Miller, D.M., Reif, D.W., & Aust, S.D. (1992) In vitro loading of apoferritin. Arch. Bichem. Biophys. 293: 409–415.
Frieden, E. (1980) Caeruloplasmin: A multi-functional metalloprotein of vertebrate plasma. In: Biological Roles of Copper, p. 93–124. Elsevier/North-Holland, New York.
Frieden, E. (1985) Perspectives on copper biochemistry. Clin. Physiol. Biochem. 4: 11–19.
Harris, Z.L., Takahash, Y., & Miyajima, H. (1995) Aceruloplasminemia: Molecular characterization of this disorder of iron metabolism. Proc. Nat. Acad. Sci. USA 92: 2539–2543.
Lawson, D.M., Treffry, A., Arymiuk, P.J., Harrison, P.M., Yewdall, S.J., Luzzago, A., Cesareni, G., Levi, S., & Arosio, P. (1986) Identification of the ferroxidase center in ferritin. FEBS Lett. 254: 207–210.
Lee, C.R., Racht, S., Lukens, J.N., & Cartwright, G.E. (1968) Iron metabolism in copper-deficient swine. J. Clin. Invest. 47: 2058–2069.
Levi, S., Corsi, B., Rovida, E., Cozzi, A., Santambrogio, P., Albertini, A., & Arosio, P. (1994) Construction of a ferroxidase center in human ferritin L-chain. J. Biol. Chem. 269: 30334–30339.
McDermott, J.A., Huber, C.T., Osaki, S., & Frieden, E. (1968) Role of iron in the oxidase activity of ceruloplasmin. Biochim. Biophys. Acta 151: 541–557.
Minotti, G. & Aust, S.D. (1987) The requirement for iron(III) in the inhibition of lipid peroxidation by iron(II) and hydrogen peroxide. J. Biol. Chem. 262: 1098–1104.
Osaki, S. (1966) Kinetic studies of ferrous ion oxidation with crystalline human ferroxidase (ceruloplasmin). J. Biol. Chem. 241: 5053–5059.
Osaki, S., Johnson, D.A., & Frieden, E. (1966) The possible significance of the ferrous oxidase activity of ceruloplasmin in normal human serum. J. Biol. Chem. 241: 2746–2751.
Ryan, T.P., Miller, D.M., & Aust, S.D. (1993) The role of metals in the enzymatic and nonenzymatic oxidation of epinephrine. J. Biochem. Toxicol. 8: 33–39.
Ryden, L. (1984) Ceruloplasmin. In: Copper proteins and copper enzymes (Conti, R., ed.), vol. 3, pp. 37–100. CRC Press, Boca Raton, FL.
Samokyszyn, V.M., Miller, D.M., Reif, D.W., & Aust, S.D. (1989) Inhibition of superoxide and ferritin-dependent lipid peroxidation by ceruloplasmin. J. Biol. Chem. 264: 21–26.
Treffry, A., Sowerby, J.M., & Harrison, P.M. (1978) Variable stoichiometry of Fe(II)-oxidation in ferritin. FEBS Lett. 95: 221–224.
Weinber, E.D. (1996) The role of iron in cancer. Eur. J. Cancer Prev. 5: 19–36.
Yablonski, M.J. & Thiel, E.C. (1992) A possible role for the conserved trimer interface of ferritin in iron incorporation. Biochemistry 31: 9680–9684.

Discussion

Q1. Allan Davison, University of Northern British Columbia, Prince George, BC, USA: Steve, you have dispelled a lot of popular beliefs here, so I thought I'd give you a chance to take a tilt at another one. Can we still believe that the unloading of iron from ferritin is an important part of the defense mechanism in response to the oxidative burst?

A. That's a good question, Allan. I think that might be a possible mechanism. Now, release by superoxide is incredibly slow, so if that's the mechanism whereby the oxidative burst gets iron to catalyze for example hydroxyl radical formation, that is a possibility. I don't know how the iron is released. I'm not sure it's really proven that ferritin is a reversible source of iron. There are many people who believe that there is a form of ferritin that is strictly a storage form of iron for later use. Superoxide may be involved there, Allan, but I don't know.

Q2. Ed Harris, Texas A&M University, College Station, TX, USA: Steve, I like your ideas very much. I wanted you to expand on something a little bit. Ceruloplasmin is technically extracellular, and ferritin synthesis is intracellular. We normally don't find ceruloplasmin intracellularly. Do you feel that what you have proposed occurs entirely in the serum?

A. No, I don't know if everyone is familiar with this. This is a real problem, and is a fairly active area of research right now. Does ceruloplasmin exist in tissues? Some people believe that, at least in some tissues, it is present. We have started on that right now. Messenger RNA for ceruloplasmin exists in many different tissues. Then the question is, does the protein exist in the tissues?

Q3. Ed Harris, Texas A&M University, College Station, TX, USA: I'll support you on that last statement. We too have done a survey, and found that there are quite a few tissues that express it.

The Role of Zinc (Zn) in Free Radical Mediated Diseases

T.M. BRAY,[†] M.A. LEVY,[†] M.D. NOSEWORTHY,[‡] AND K. ILES[‡]

[†]Department of Human Nutrition, The Ohio State University, Columbus, OH 43210, USA, and
[‡]Department of Nutritional Sciences, University of Guelph, Guelph, Ontario, Canada N1G 2W1

Keywords zinc, antioxidant, free radical, oxidative stress

Zinc (Zn) is an essential nutrient for many plants (Mengel 1987), animals (Underwood 1977) and microorganisms (Failla 1977). The essentiality of Zn for humans was first demonstrated when arrested sexual development and stunted growth in adolescent children was reversed with Zn supplements (Prasad 1993). Since then dietary Zn deficiency has been well documented in humans and animals, and low Zn status has been linked to many diseases including diabetes (Car et al. 1992, Faure et al. 1992), abnormal immune function (Dardenne 1982), teratogenesis (Hurley and Swenerton 1966) and neurological disorders (Dreosti 1983).

The biochemical basis for the essentiality of Zn is not completely understood. Although it is an essential cofactor for more than 200 metalloenzymes (Vallee 1990), Zn has also been demonstrated to have roles in gene expression, neurotransmission, membrane structure and function, and as an antioxidant (Walsh et al. 1994), functions which may be independent of metalloenzyme activity and more closely related to its chemical nature. Zn is a IIb transition metal, distinct from other essential trace metals in that it has only one valence state. Since Zn does not change its valence state, it prevents the transfer of electrons from (oxidizing compounds) free radicals to organic molecules in which it is found. Thus, the most outstanding feature of the function of Zn is that it has a general stabilization effect on macromolecules (Hesketh 1985, Inouye and Kirschner 1984). However, the relationship between Zn stabilization effects and antioxidant activity is unclear; in many cases Zn appears to prevent oxidation/peroxidation of biomolecules.

Antioxidant properties of Zn have been demonstrated primarily *in vitro*. Zn may exert its antioxidant effect by decreasing the susceptibility of essential sulfhydryl groups of proteins to oxidation and by competing with prooxidant metals such as Fe and Cu for biological binding sites (Gibbs et al. 1985). Zn prevents production of the hydroxyl (OH·) and superoxide (O_2-·) free radicals through the Fenton reaction by these transition metals. Zn is also an integral part of CuZn superoxide dismutase (CuZnSOD), a cytosolic or extracellular enzyme involved in the first line of defense against O_2-·. It is difficult, however, to demonstrate the antioxidant function of Zn *in vivo*. In biological systems Zn is bound to proteins with varying degrees of affinity, and free Zn ion concentrations are very low (Bray and Bettger 1990). Nevertheless, there is *in vivo* evidence that indicates Zn deficiency pathology may be due, in part, to tissue oxidation/peroxidation. This has been demonstrated by experiments which show: i) a protective effect of antioxidants on the expression of Zn deficiency pathology (Burke and Fenton 1985) and ii) altered free radical metabolism in Zn-deficient animals (Hammermueller et al. 1987).

To investigate the antioxidant effect of Zn *in vivo*, the classical techniques in nutritional science is used. In our laboratory, we have studied the effects of dietary Zn deficiency on free radical metabolism and tissue pathology. We hypothesized that Zn deficiency impairs free radical defenses and therefore causes an animal to be more susceptible to tissue oxidation following exposure to an oxidative stress. In the absence of an oxidative stress, an animal may be able to adapt to Zn deficiency without significant signs of pathology. However, in conditions of increased oxidative stress, the defense mechanisms of a Zn deficient animal may not be able to cope with the increased flux of free radicals resulting in tissue- and organ-specific pathology. For example, one of the physiological responses or adaptations of rats to a Zn

Correct citation: Bray, T.M., Levy, M.A., Noseworthy, M.D., and Iles, K. 1997. The role of zinc (Zn) in free radical mediated diseases. In *Trace Elements in Man and Animals – 9: Proceedings of the Ninth International Symposium on Trace Elements in Man and Animals. Edited by* P.W.F. Fischer, M.R. L'Abbé, K.A. Cockell, and R.S. Gibson. NRC Research Press, Ottawa, Canada. pp. 333–336.

deficient diet is a severe voluntary reduction (>50%) of feed intake. This may be a protective mechanism which slows the animal's growth rate and oxygen consumption which may in turn reduce Zn requirements, lessen the dilution of tissue Zn concentrations and thus reduce the susceptibility of the animal to Zn deficiency pathologies. Previously it has been demonstrated that enzymatic and non-enzymatic components of the free radical defense system were not severely impaired in Zn deficient rats (Taylor et al. 1988). It is our hypothesis that Zn deficient animals are unable to cope with exposure to high concentrations of oxygen or infectious agents resulting in tissue pathologies mediated by free radicals. We have tested this deficiency hypothesis by studying animals maintained on Zn deficient and Zn adequate diets before and after exposure to an oxidative stress. Our research has focused on the lung, brain and pancreas, three organs which are particularly susceptible to oxidative stress.

The Effect of Zn Deficiency and Oxidative Stress on Tissue Pathology of Lung

We have applied a magnetic resonance imaging (MRI) technique to non-invasively measure tissue damage in Zn deficient animals before and after exposure to 5 d of hyperoxia (85% O_2) (Taylor et al. 1990). The effects of hyperoxia were investigated in rats fed either a Zn deficient (ZnDF), Zn pair-fed (ZnPF) or Zn ad-libitum (ZnAL) diet. Without hyperoxia exposure, no lung damage (i.e., lung edema) was observed, regardless of dietary treatment. Exposure to 5 d of hyperoxia, however, demonstrated that lung damage was most dramatic in ZnDF animals. A minor amount of damage was observed in the ZnPF group, and no damage was detected in the control group (ZnAL). Further, when Zn was replenished in the diet of Zn deficient rats at the beginning of hyperoxia exposure (Zn repletion group, ZnRP), lung damage was not detectable. These results, therefore, indicated that the lack of dietary Zn resulted in extensive tissue damage in animals exposed to hyperoxia.

In order to determine if the observed tissue damage resulted from tissue oxidation/peroxidation, components of the free radical defense mechanism were examined. The results demonstrated that ZnAL and ZnRP groups exposed to hyperoxia had significantly higher activities of CuZnSOD, glutathione peroxidase and catalase than their normoxia exposed counterparts. However, the activity of these enzymes was not increased in ZnDF animals when exposed to hyperoxia. This data suggested that Zn deficiency impaired the ability of the free radical defense system to cope with the increased oxidative stress of hyperoxia exposure. Therefore the ability to increase the activity of free radical defense enzymes appeared to be an important mechanism for protection against hyperoxia induced lung damage.

The Effect of Zn Deficiency and Oxidative Stress on Brain

MRI has also been applied to study the effects of Zn deficiency and hyperoxia on blood-brain barrier (BBB) permeability in rats. Again, animals were given either a ZnAL, ZnPF or ZnDF diet and exposed to either normoxia or hyperoxia. The BBB of ZnAL and ZnPF controls was not permeable without the hyperoxic stress, while the Zn deficient animal had slightly elevated BBB permeability. However, after exposure to hyperoxia, the BBB of ZnDF animals was extensively more leaky than the ZnPF and ZnAL controls (19.6% of BBB was permeable vs. 10.7% and 12.4%). It was therefore concluded that hyperoxia lead to a loss of integrity of the BBB, but this damage was exacerbated in Zn deficient animals.

Components of the free radical defense system were also examined in the brain in order to determine the contribution of tissue oxidation/peroxidation to the observed damage of the BBB. Results of this study indicated that total SOD activity was increased in the ZnAL and ZnPF controls, but remained unchanged in the ZnDF group, after exposure to hyperoxia. Notably, the increased SOD activity observed in the hyperoxia exposed ZnAL and ZnPF groups was due primarily to an approximate two-fold increase in MnSOD activity, while CuZnSOD activity was unchanged. A similar increase in MnSOD activity was observed in the hyperoxic stressed ZnDF group, however total SOD activity remained unchanged because CuZnSOD activity was dramatically decreased. Thus, as was previously noted in the lung, the ability to increase activity of free radical defense enzymes appeared to be an important protective mechanism against hyperoxia induced tissue damage in the brain.

The Effect of Zn Deficiency and Oxidative Stress on Type 1 Diabetes

Insulin dependent diabetes mellitus (IDDM) is a multifactorial disease characterized by β-cell destruction of the pancreas. Recent evidence suggests that the release of cytotoxic oxygen free radicals are a common factor in β-cell death. Two oxygen radical species, $O_2^{-\cdot}$ and nitric oxide (NO·), may be the final key mediators in this process. There has also been evidence that diabetes may be related to Zn status. Studies have demonstrated that: i) diabetes in humans and animals is characterized by reduced plasma and tissue Zn concentrations, ii) insulin therapy increases plasma Zn and suppresses hyperzincuria, and iii) Zn deficient animals show glucose intolerance. In addition, insulin is stored in the β-cells of the pancreas as preinsulin in/as Zn crystals. Zn is also known to play a key role in insulin-receptor binding at target cells. However, the potential antioxidant role of Zn in protection against diabetes, particularly free radical mediated β-cell damage, has not been investigated.

To determine the role of Zn in the pathogenesis of Type 1 diabetes, we fed a Zn deficient diet to two transgenic strains of mice, a RIPSOD strain which overexpresses CuZnSOD in pancreatic β-cells and TGHS mice which overexpress CuZnSOD in all somatic cells, and a control strain, CD1. We hypothesized that mice with overexpressed CuZnSOD would be more resistant to β-cell destruction in alloxan-induced diabetes. Alloxan is a commonly used diabetogenic compound which generates $O_2^{-\cdot}$ and causes characteristic hyperglycemia and hypoinsulinemia. We also hypothesized that Zn deficiency would impair the ability of transgenic mice to increase the transgene CuZnSOD activity and thus decrease their ability to counteract alloxan induced diabetes. Fasting blood glucose concentration was used to assess the severity of β-cell destruction and diabetes.

The results demonstrated that fasting blood glucose was elevated in all alloxan injected animals. However, the highest blood glucose levels were observed in the CD1 mice. Blood glucose levels of the TGHS and the RIPSOD mice were intermediate between the alloxan injected and the non-injected CD1 mice. Further studies demonstrated that fasting blood glucose levels were higher in ZnDF animals for each strain of mouse (RIPSOD, TGHS and CD1) compared to the ZnAL controls. Thus, these results demonstrated that: i) in ZnAL animals, enhanced CuZnSOD activity of the transgenic mice reduced fasting blood glucose levels when compared to non-transgenic mice, and ii) in alloxan induced diabetes, blood glucose levels were higher in Zn deficient animals compared to ZnAL control animals for each of the three strains of mice examined.

Summary

The mechanisms by which Zn deficiency causes free radical mediated diseases are complicated. We have demonstrated that dietary Zn deficiency causes increased susceptibility to oxidative damage in three specific tissues; lung, brain and pancreas. This may be a significant factor in free radical mediated diseases characterized by abnormal Zn status including bronchopulmonary dysplasia, cerebral palsy and Type 1 diabetes. It is believed that Zn deficiency predisposes alveolar membrane in the lung, the BBB of the brain and the β-cell in the pancreas to oxidative stress.

We have also demonstrated in the lung and brain that the ability to protect against tissue damage was associated with elevated antioxidant defenses, for example CuZnSOD activity. In the absence of oxidative stress, the Zn deficient animal may adapt and cope with the low nutritional status without observable tissue damage. It is believed that Zn deficiency predisposes alveolar membrane in the lung, the BBB of the brain and the β-cell in the pancreas to oxidative stress. Exposure to excess oxidative stress may be the main cause which lead to tissue- and organ-specific pathologies in Zn deficient animals.

Literature Cited

Bray, T. M. & Bettger, W. J. (1990) The physiological role of zinc as an antioxidant. Free Rad. Biol. Med. 8: 281–291.

Burke, J.P. & Fenton, M.R. (1985) Effect of a zinc deficient diet on lipid peroxidation in liver and tumor subcellular membranes. Proc. Soc. Exp. Biol. Med. 179:187–191.

Car, N., Car, A., Granic, M., Skrabalo, Z. & Momcilovic, B. (1992) Zinc and copper in the serum of diabetic patients. Biol. Trace Elem. Res. 32: 325–329.

Dardenne, M., Pleau, J.M., Nabarra, B., Lefrancier, P., Derrien, M., Choay, J. & Bach, J. (1982) Contribution of zinc and other metals to the biological activity of the serum thymic factor. Proc. Natl. Acad. Sci. USA. 79:5370–5373.

Dreosti, I.E. (1983) Zinc and the central nervous system. In: Neurobiology of the trace elements (Dreosti, I.E. & Smith, R.M., eds.), pp. 135–162. Humana Press, Clifton, New Jersey.
Failla, M.L. (1977) Zinc: Functions and transport in microorganisms. In: Microorganisms and minerals (Weinberg, E.D., ed), pp. 159–214. Marcel Dekker, New York.
Faure, P., Roussel, A., Coudray, C., Richard, M.J., Halimi, S., & Favier, A. (1992) Zinc and insulin sensitivity. Biol. Trace Elem. Res. 32: 305–310.
Gibbs, P.N., Gore, M.G. & Jordan, P.M. (1985) Investigation of the effect of metal ions on the reactivity of thiol groups in human δ-aminolevulinate dehydratase. Biochem. J. 225:573–580.
Hesketh, J.E. (1984) Microtubule assembly in rat brain extracts. Int. J. Biochem. 16:1331–1339.
Hammermueller, J.D., Bray, T.M., & Bettger, W.J. (1987) Effects of Zn and Cu deficiency on microsomal NADPH-dependent tissue oxygen generation in lung and liver. J. Nutr. 117:894–898.
Hurley, L.S. & Swenerton, H. (1966) Congenital malformations resulting from zinc deficiency in rats. Proc. Soc. Exp. Biol. Med. 123: 692–697.
Inouye, H. & Kirschner, D.A. (1984) Effect of $ZnCL_2$ on membrane interactions in myelin of normal and shiverer mice. Biochim. Biophys. Acta. 776: 197–208.
Mengel, K. & Kirkby, E.A. (1987) Zinc in crop nutrition. In: Principles of plant nutrition (Mengel, K. & Kirkby, E.A., eds.), pp. 533–535. International Potash Institute, Worblaufen, Switzerland.
Prasad, A.S. (1993) Biochemistry of zinc. Plenum Press, New York.
Taylor, C.G, Bettger, W.J. & Bray, T.M. (1988) Effect of dietary zinc or copper deficiency on the primary free radical defense system in rats. J. Nutr. 118:613–621.
Taylor, C.G., Towner, R.A., Janzen, E.G. & Bray, T.M. (1990) MRI detection of hyperoxia induced lung edema in zinc deficient rats. Free Rad. Biol. Med. 9:229–233.
Underwood, E.J. (1977) Zinc. In: Trace elements in human and animal nutrition (Underwood, E., ed.), pp. 196–247. Academic Press, New York.
Vallee, B.L. & Auld, D.S. (1990) Zinc coordination, function and structure of zinc enzymes and other proteins. Biochemistry 29:5647–5659.
Walsh, C.T., Sandstead, H.H., Prasad, A.S., Newberne, P.M. & Fraker, P.J. (1994) Zinc: health effects and research priorities for the 1990s. Environ. Health Perspect. 102(2): 5–46.

Discussion

Q1. Paul Saltman, University of California at San Diego, CA, USA: You had on one of your earlier slides, the ability of zinc to displace iron from a dithiol. I would like to point out that there are several lines of evidence now that say that one of the most effective ways that zinc acts as an antioxidant is by that mechanism, but the metal is not iron, it's copper. We've found this in microbial systems and in isolated hearts. You can prevent, for example in isolated hearts, reperfusion ischemia damage by the infusion of zinc-histidyl complexes. If you add more copper to those hearts, by feeding the animals copper, you can increase the severity of the ischemia induced, and it takes more zinc to restore that activity. I think that the key issue with zinc, as far as I'm concerned is not in the superoxides. You could explain your data on the basis of just having a pool of zinc available to be there in equilibrium with these other sites. The issue is the site-specificity of the free radical damage, such as Stadtman has shown, such as we have shown in many red cell systems as well.

A. I agree with you 100%. We know that free zinc, in biological systems is very low, as it's usually bound to protein. I think that there is something there as an oxygen sensor. I think that zinc is probably quite selective in terms of where it is most effective as an antioxidant.

Q2. Xavier Alvarez-Hernandez, Louisiana State University Medical Center, Shreveport, LA, USA: You mentioned effects on the leakiness of the blood-brain barrier. Is zinc involved in the synthesis of tight junctions?

A. I don't know. The experiments that we have done haven't allowed us to assess the mechanisms involved.

Q3. John Sorenson, University of Arkansas, Little Rock, AR, USA: At one point you indicated that the definition of an antioxidant is to prevent the transfer of electrons to and from oxygen? How important is copper/zinc superoxide dismutase in the pathogenesis of diabetes?

A. It appears to play a very important role in the animal models of Type I diabetes we have used. With streptozotocin I don't know, but with alloxan, we can detect the oxygen radical by spin-trapping. With streptozotocin we can't. There is indirect evidence as well. Fasting blood glucose levels are much lower in the transgenic mice compared to the control CD1 mice after alloxan injection.

Q4. Yves Rayssiguier, INRA, France: Zinc deficient rats are more susceptible to diabetes. Supplementation with tocopherol was ineffective. Why does Vitamin E not protect against lung damage in zinc deficiency?

A. It all depends on the kind of damage. Vitamin E and zinc have different roles in terms of both their antioxidant function and their cellular localization.

Transgenic Mice Overexpressing Catalase Specifically in the Heart for Studying Cardiac Oxidative Injury Induced by Copper Deficiency*

Y.J. KANG,[†] Y. CHEN,[†] AND J.T. SAARI[†]

[†]Department of Pharmacology and Toxicology, University of North Dakota School of Medicine, and [†]USDA Agricultural Research Service, Human Nutrition Research Center, Grand Forks, ND 58202, USA

Keywords antioxidant, catalase, copper deficiency, transgenic mice

Copper deficiency causes more salient pathological changes in the heart than in the liver of rats. Although oxidative stress has been implicated in copper deficiency-induced pathogenesis, little is known about the selective toxicity to the heart. Therefore, in an effort to determine the selective cardiotoxicity of copper deficiency, we examined the relationship between the capacity of antioxidant defense system and the copper deficiency induced oxidative damage in the heart compared to that in the liver (Chen et al. 1994). Weanling rats were fed a purified diet deficient in copper (0.4 μg/g diet) or one containing adequate copper (6.0 μg/g diet) for 4 weeks. Copper deficiency induced a two-fold increase in lipid peroxidation in the heart (thiobarbituric assay), but did not alter peroxidation in the liver. The antioxidant enzymatic activities of superoxide dismutase, catalase and glutathione peroxidase were, respectively, 3-, 50- and 1.5-fold lower in the heart than in the liver, although these enzymatic activities were depressed in both organs by copper deficiency. In addition, the activity of glutathione reductase was 4 times lower in the heart than in the liver. The results suggested that a weak antioxidant defense system was responsible for the relatively high degree of oxidative damage in the copper-deficient heart.

Because catalase is a major enzyme involved in detoxification of hydrogen peroxide (H_2O_2) in mammalian cells, the present study was undertaken to determine whether elevation of catalase activity specifically in the heart of transgenic mice could provide protection against oxidative cardiotoxicity. A transgene for overexpression of catalase in the heart was constructed to contain fragments from the rat catalase cDNA ligated behind the alpha cardiac myosin heavy chain promoter. The transgenic mice were identified by using Southern and dot blot, and PCR procedures. Catalase activities and mRNA concentrations in the heart and other organs were measured.

Cardiac catalase activity was analyzed in 6 animals (3 males and 3 females) of each transgenic line and non-transgenic controls. As shown in Figure 1, catalase activity was markedly elevated in the transgenic hearts. This elevation ranged from 2-fold in line 786 to 630-fold in line 784. There was no significant difference ($p > 0.10$) in the catalase activity between males and females within the same transgenic line (data not shown). The level of mRNA for catalase was also increased in the transgenic heart (data not shown). This overexpression was rather stable as evidenced by consistent results obtained in several assays performed over one year. It is clear that the increased catalase activity results from expression of the transgene. We tested whether the catalase activity is elevated to the same extent in both atria and ventricles. The results showed that the elevated catalase activity was the same in atria and ventricles from 5 representative transgenic mouse lines. We also determined whether the elevated catalase expression was specific to the heart. Catalase activities in liver, kidneys, lungs, and skeletal muscles of the transgenic mice were measured. As shown in Figure 2, catalase activities in all these tissues were the same as controls.

*Supported in part by NIH Grant CA68125 and USDA Grant 9500668.

Correct citation: Kang, Y.J., Chen, Y., and Saari, J.T. 1997. Transgenic mice overexpressing catalase specifically in the heart for studying cardiac oxidative injury induced by copper deficiency. In *Trace Elements in Man and Animals –9: Proceedings of the Ninth International Symposium on Trace Elements in Man and Animals. Edited by* P.W.F. Fischer, M.R. L'Abbé, K.A. Cockell, and R.S. Gibson. NRC Research Press, Ottawa, Canada. pp. 337–339.

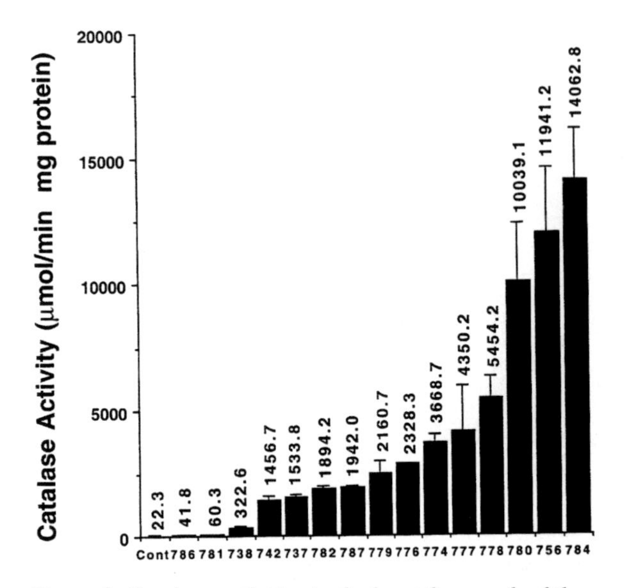

Figure 1. Catalase activities in the heart from each of the 15 different transgenic mouse lines in comparison with that of controls (Cont). Each value is the average of 6 determinations (mean ± SD).

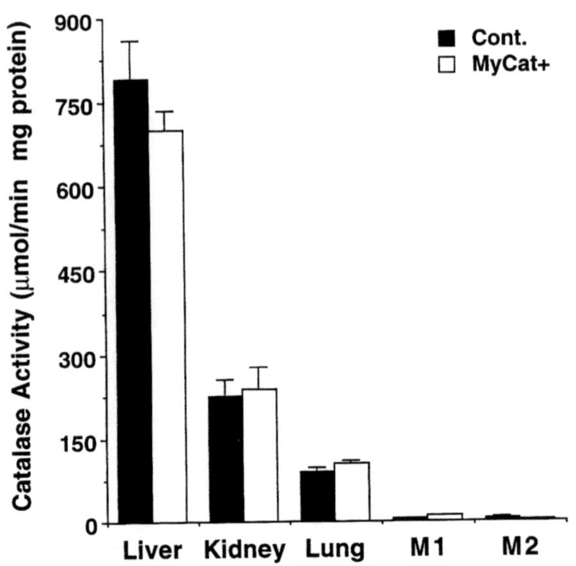

Figure 2. Comparison of catalase activities in the liver, kidney, lung, and skeletal muscles from the left rear leg (M1) and the abdominal external oblique muscles (M2) of the five transgenic mouse lines as shown in Figure 1 (lines 738, 742, 782, 776, 777). The value represents the pooled data from the 5 lines with 3 determinations of each line (mean ± SD, n = 15) for transgenic (MyCat) and from 6 control animals (Cont).

We then determined whether other antioxidant enzyme activities in the transgenic heart were altered because of changes in catalase activity. The activities of superoxide dismutase (SOD), glutathione peroxidase (GSHpx), and glutathione reductase (GR) were measured in the catalase-enriched transgenic hearts. As shown in Table 1, none of these enzyme activities were changed. We also measured the concentrations of two important nonenzymatic antioxidant components, glutathione (GSH) and metallothionein (MT). There was no alteration in either of the two components in the catalase-enriched transgenic hearts (Table 1).

These transgenic mice have been shown to be highly protective against oxidative injury in the heart induced by adriamycin, an important anticancer drug (Kang et al. 1996). This model is thus ideal for the study of the role of catalase in protection against oxidative injury to the heart induced by copper deficiency.

Table 1. Enzyme activities of SOD, GSHpx and GR, and concentrations of GSH and MT in the hearts of control and MyCat transgenic mice.

	Control	MyCat
SOD (U/g · wet wt)	8.6 ± 3.7	108.3 ± 13.4
GSHpx (nmol NADPH/min · mg protein)	29.8 ± 4.4	26.2 ± 3.0
GR (nmol NADPH/min · mg protein)	14.2 ± 1.7	16.0 ± 1.4
GSH (µg/g · wet wt)	362.2 ± 11.4	356.0 ± 10.1
MT (µg/g · wet wt)	5.1 ± 0.6	4.3 ± 0.5

The data were obtained from non-transgenic mice (Control, n = 6) and 5 representative transgenic mouse lines as shown in Figure 4 (line 738, 742, 782, 776, 777). Three mice were used from each of the 5 lines and the data were pooled together to give the average value (MyCat, n = 15).

References

Chen, Y., Saari, J.T., & Kang, Y.J. (1994) Weak antioxidant defenses make the heart a target for damage in copper-deficient rats. Free Radical Biol. Med. 17: 529–536.

Kang, Y.J., Chen, Y., & Epstein, P.N. (1996) Suppression of doxorubicin cardiotoxicity by overexpression of catalase in the heart of transgenic mice. J. Biol. Chem. 271: in press.

Discussion

Q1. Paul Saltman, University of California at San Diego, CA, USA: I think that this is absolutely spectacular work, but I want to come back to the original problem. Why does copper deficiency increase the oxidative stress and the amount of the peroxide that is in the heart? You haven't told us that yet. It seems almost counter-intuitive.

A. No, I haven't told you that yet. I think this model will give you a valuable tool to test that.

Q2. Jim Kirkland, University of Guelph, ON, Canada: Is catalase restricted usually to the peroxisome in the heart, and if you over-express it to that extent, have you forced it out of the peroxisome?

A. That's a good question. My post-doc right now is working on that localization. At this point we don't know.

Plasma Bleomycin Detectable Iron (PBDFe) and Malondialdehyde (MDA) Concentrations in Children with Kwashiorkor, Marasmus and Acute Infective Illness

A.A. SIVE, W.S. DEMPSTER, H. MALAN, AND H. DEV. HEESE
Department of Paediatrics and Child Health, Institute of Child Health, University of Cape Town, Rondebosch 7700, South Africa

Keywords kwashiorkor, iron, infection, lipid peroxidation

Decreased levels of transferrin in children with kwashiorkor may result in non-protein bound iron promoting overwhelming bacterial infection (McFarlane et al. 1970). This non-protein bound iron is available to catalyse the Haber-Weiss and Fenton-type reactions providing a mechanism for hydroxy radical formation and tissue damage (Gutteridge and Halliwell 1989). Golden has postulated that an excess of free radicals distinguishes kwashiorkor from marasmus in nutritionally deprived children (Golden et al. 1985). Previous studies in Cape Town (Dempster et al. 1995, Sive et al. 1993) have shown that children with kwashiorkor have higher plasma levels of non-protein bound iron, lower concentrations of glutathione peroxidase and elevated malondialdehyde (measured as thiobarbituric acid reactive substances, TBARS) than marasmic children suggesting that free radical injury plays a role in the pathogenesis of kwashiorkor. The majority of children with kwashiorkor have serious infections on admission to hospital. We were concerned that infection rather than nutritional state accounts for the elevated malondialdehyde and non-protein bound iron concentrations.

This study compares non-protein bound iron measured as plasma bleomycin detectable iron (PBDFe) and malondialdehyde (MDA) in children with oedematous malnutrition (Kwashiorkor), non-oedematous malnutrition (Marasmus) and normally nourished children who required admission to hospital for serious infections (Infection).

Methods

Venous blood sampes were collected on admission to a paediatric hospital from 40 consecutive kwashiorkor, 20 marasmus and 20 normally nourished patients with severe infections. The diagnosis of nutritional state conformed to accepted criteria (Gurney 1979). All samples were taken using steel needles into heparinized polystyrene containers. Haemolysed samples were discarded. Plasma was stored at –75°C until assayed. PBDFe was quantified using the basic principles developed by Gutteridge (Gutteridge et al. 1981) in an assay system which has previously been validated in our laboratory (Dempster et al. 1995). MDA was measured using the method of Lepage et al. (1991) and Hoving et al. (1992). The study was approved by the Ethics and Research Committee of the University of Cape Town and informed consent was obtained from the custodial parent of each child. Standard statistical packages were used for summary statistics, ANOVA, Chi Square and regression analysis.

Results

The results of the studies are shown are in the tables.

Correct citation: Sive, A.A., Dempster, W.S., Malan, H., and Heese, H. deV. 1997. Plasma bleomycin detectable iron (PBDFe) and malondialdehyde (MDA) concentrations in children with kwashiorkor, marasmus and acute infective illness. In *Trace Elements in Man and Animals – 9: Proceedings of the Ninth International Symposium on Trace Elements in Man and Animals. Edited by* P.W.F. Fischer, M.R. L'Abbé, K.A. Cockell, and R.S. Gibson. NRC Research Press, Ottawa, Canada. pp. 340–342.

Table 1. Mean (SD, range) age, weight, % exp wt, albumin, white cell count, PBDFe and MDA.

Mean (SD, range)	Kwashiorkor (n = 40)	Marasmus (n = 20)	Infection (n = 20)
Age (months)	17.3 (7.0, 7–37)	18.5 (14, 4–62)	25.8 (24, 3–115)
Weight (kg)	8.7 (1.7, 5.2–11.8)	5.9 (1.9, 3.0–10.7)	11.7 (4.6, 6.5–27)
Expected Wt (%)	78 (12, 52–90)	54 (6, 42–60)	100 (10, 83–120)
Albumin (g/L)	17 (3.4,11–25)	30 (7.1, 18–39)	35 (4.9, 25–45)
White cell count (10^9/L)	16.4 (6.3, 7.2–36.8)	16.7 (8.9, 9.5–50.0)	19 (9.1, 5.1– 39.6)
PBDFe(µmol/L)*	5.6 (6.5, 0–17.8)*	2.57 (5.88, 0–17)	0.29 (0.35, 0–1.27)
MDA (µmol/L)	1.36 (1.0, 0.25– 4.88)	0.98 (0.54, 0.16–2.38)	0.77 (0.52, 0.04–1.90)

*means are significantly different, p = 0.002 (ANOVA). There was no correlation between the WCC or plasma albumin and PBDFe or MDA concentration. PBDFe was >1 µmol/L in 22 kwashiorkor, 4 marasmus and 2 well-nourished children (p < 0.0001, Chi Square).

Table 2. Distribution of infective illnesses.

	Kwashiorkor (n = 40)	Marasmus (n = 20)	Infection (n=20)
Pneumonia	7	4	13
Gastroenteritis	24	14	3
Meningitis	1	0	3
Tuberculosis	1	4	0
Septicaemia	5	0	4

Discussion

The presence of PBDFe in 55% of the kwashiorkor patients is remarkably similar to the 58% published previously (Dempster et al. 1995). Three marasmics had PBDFe above 10 µmol/L. One subsequently developed oedema and clinical kwashiorkor. One of the other marasmics had life threatening gastroenteritis and severe localised herpes simplex infection. In 2 well-nourished children PBDFe, was marginally elevated at 1.10 and 1.27 µmol/L. Over 90% of the kwashiorkor and all the marasmic children had infections on admission to hospital confirming that it is vital to control for infection when comparing differences in nutritional state. Transferrin was not measured in this study but low transferrin in kwashiorkor (McFarlane et al. 1970) could lead to reduced iron binding and detectable "free" iron. Iron is securely compartmentalised to prevent tissue damage and peroxidation. We had previously reported that TBARS was elevated in kwashiorkor (Sive et al. 1993). Using a more specific method for measuring MDA in this study, we were unable to show significant differences between kwashiorkor and the other patients. Healthy controls were not included in this study, but the normal value in our laboratory for children is <0.9 µmol/L. Levels were above this cut off in 65%, 45%, and 40% respectively in the 3 groups. There was no relationship between MDA and PBDFe concentrations in any of the groups.

These findings suggest that lipid peroxidation is not a major feature of the damage in kwashiorkor, that the methods employed were not sensitive or specific enough to detect differences in lipid peroxidation, or that infection rather than the patient's nutritional state or "free" iron concentration determines MDA production. In contrast, the presence of PBDFe is associated with oedematous malnutrition rather than the presence of infection.

References

Dempster W.S., Sive A.A., Malan H. & Heese H.de.V. (1995) Misplaced iron in kwashiorkor. Eur. J. Clin. Nutr. 49: 208–210.

Golden M.H.N., Golden B.E. & Bennett F.I. (1985) Relationship of trace element deficiencies to malnutrition. In: Trace Elements in Nutrition of Children (Chandra R.K., ed.), pp. 185–207. Raven Press, NewYork.

Gurney J.M. (1979) The young child: protein energy malnutrition. In: Nutrition and Growth (Jelliffee D.B. & Jelliffee E.F.P., eds.), pp. 185–216. Plenum Press, New York.

Gutteridge J.M.C., Rowley D.A. & Halliwell B. (1981) Superoxide-dependent formation of hydroxyl radicals in the presence of iron salts. Detection of free iron in biological systems using bleomycin-dependent degradation of DNA. Biochem. J. 199: 263–265.

Gutteridge J.M.C. & Halliwell B. (1989) Iron toxicity and oxygen radicals. In: Clinical Haematology. Iron chelating therapy (Hershko C., ed.), pp 195–256. Bailliere Tindall, London.

Hoving E.B., Laing C., Rutgers H.M., Teggeler M., van Doormal J.J. & Muskiet F.A.J. (1992) Optimized determination of malondialdehyde n plasma lipid extracts using 1.3-diethyl-2-thiobarbituric acid: Influence of detection method and relations with lipids and fatty acids in plasma from healthy adults. Clin. Chem. Acta. 208: 63–76.
Lepage G., Munoz G., Champagne J. & Claude C.R. (1991) Preparative steps necessary for the accurate measurement of malondialdehyde by high-performance liquid chromatography. Anal. Biochem. 197: 277–283.
Mc Farlane H., Reddy S., Adcock K.J., Adeshina H., Cooke A.R., Akene J. (1970) Immunity transferrin and survival in kwashiorkor. Br. Med. J. 4: 268–270.
Sive A.A., Subotzky E.F., Malan H., Dempster W.S., & Heese H.deV. (1993) . Red cell antioxidant enzyme concentrations in kwashiorkor and marasmus. Ann. Trop. Paediatr. 13: 33–38.

Discussion

Q1. Steve Aust, Utah State University, Logan, UT, USA: Might the results you have found be complicated by the fact that microorganisms are removing iron and it is therefore not detectable?

A. This is perhaps a possibility, but it is difficult to prove.

Q2. (unable to identify questioner): Might copper deficiency contribute to the pathology of these diseases?

A. It is very much possible. All trace metals are decreased in acute phase diseases.

Q3. Noel Solomons, Guatemala: I'd just like to make a comment, and a word of caution at this point. We must be careful about using iron supplements in these situations, as it may in fact be deleterious to the health of the patients.

Q4. John Sorenson, University of Arkansas, Little Rock, AR, USA: Did you measure red blood cell superoxide dismutase which may have a role in free radical mediated disease? The septicemia and renal disease are consistent with a massive synthesis of nitric oxide.

A. Yes. Both catalase and SOD are normal. However, glutathione and glutathione peroxidase levels are significantly decreased.

Copper Proteins and Coronary Heart Disease

J.C. FREEBURN,[†] J.M.W. WALLACE,[†] D. SINNAMON,[‡] B. CRAIG,[‡]
W.S. GILMORE,[†] AND J.J. STRAIN[†]

[†]Human Nutrition Research Group, University of Ulster, and [‡]Causeway Health & Social Services Trust, Coleraine, Northern Ireland, BT52 1SA, UK

Keywords copper-containing proteins, antioxidant enzymes, coronary heart disease, thrombosis

Thrombosis has been identified as an important process in the etiology of coronary heart disease (CHD) and it has been suggested that copper (Cu) deficiency may result in similar pathological changes as those observed in cardiovascular disease. Cu is a component of the blood clotting factors V and VIII and some of the antioxidant enzymes (Strain 1994). In the present study the concentrations of various Cu containing proteins were compared in healthy volunteers and individuals with vascular disease.

Materials and Methods

Subjects. All subjects were Caucasian males aged 40–70 years. There were three groups: (1) those who suffered from stable angina with no recent event; (2) those who had suffered an acute myocardial infarction (MI) six weeks previously; (3) apparently healthy volunteers with no obvious signs of vascular disease (control group).

Blood Measures. Blood was collected into a sodium citrate tube and plasma separated and aliquoted (1 mL). Clotting factors V, VII, VIII and fibrinogen were measured using the ACL1000 (Instrumentation Laboratory U.K. Limited, Warrington, U.K.). Polymorphonuclear neutrophils (PMNs) were isolated from a heparinised blood sample (10 mL) using a ficoll-hypaque gradient (Histopaque-1077, Sigma, U.K.) and superoxide dismutase (SOD) and glutathione peroxidase (GSH-PX) activities were assayed on the Cobas Fara autoanalyser by the methods of Jones and Suttle (1981) and RANSEL Kit (Randox Laboratories, Antrim, U.K.) respectively.

Red cell catalase activity was measured by the method of Aebi (1984). Red cell SOD and GSH-PX were also measured and plasma caeruloplasmin (Cp) activity was measured by the method of Schosinsky et al. (1974). Tissue plasminogen activator (t-PA) and plasminogen activator inhibitor-1 (PAI-1) were determined using enzyme-linked immunosorbent assays (Diagnostica Stago, Asieres-Sur-Marne, France).

Results and Discussion

Coagulation factor V levels were significantly increased in both angina and post MI subjects ($P < 0.01$, $P < 0.001$ respectively) compared with controls. Fibrinogen levels were also higher in both the angina and post MI groups ($P < 0.01$) compared with controls. Factors VII and VIII, Cp, SOD, GSH-PX activities, however, were not significantly changed in patient groups. Red cell catalase activity was significantly ($P < 0.05$) lower in angina patients compared with controls. Both t-PA and PAI-1 were significantly ($P < 0.05$) higher in angina patients compared with controls. These results demonstrate that the levels of many Cu containing proteins are abnormal in vascular disease. The interpretation of these data is difficult because many Cu containing proteins are elevated in inflammation.

Correct citation: Freeburn, J.C., Wallace, J.M.W., Sinnamon, D., Craig, B., Gilmore, W.S., and Strain, J.J. 1997. Copper proteins and coronary heart disease. In *Trace Elements in Man and Animals – 9: Proceedings of the Ninth International Symposium on Trace Elements in Man and Animals. Edited by* P.W.F. Fischer, M.R. L'Abbé, K.A. Cockell, and R.S. Gibson. NRC Research Press, Ottawa, Canada. pp. 343–344.

Table 1. Concentration of various coagulation factors and fibrinolytic parameters and antioxidant enzyme activities in control, angina and post MI groups.

	Control mean ± se (n = 27)	Angina mean ± se (n = 26)	Post MI mean ± se (n = 28)
Fibrinogen (g/L)	2.58±0.11[a]	3.69±0.26[b**]	3.49±0.20[b**]
Factor V (%)	87.67±3.5[a]	112.32±5.5[b**]	119.04±7.5[b***]
Factor VII (%)	104.56±3.83	100.72±4.37	100.50±5.82
Factor VIII (%)	96.25±4.66	104.68±9.62	118.83±10.37
tPA (ng/mL)	6.72±0.49[a]	9.13±1.12[b*]	7.39±0.78[ab]
PAI-1 (ng/mL)	16.14±2.4[a]	33.21±7.05[b*]	25.30±4.56[ab]
Red cell catalase (U/g hb)	53.56±3.74[a]	42.67±2.93[b*]	47.10±2.59[ab]
Caeruloplasmin (U/L)	538.56±3.7	576.87±30.0	600.7±24.5
WB GSH-PX (U/g hb)	19.02±1.82	15.93±2.22	15.21±41.41
Red cell SOD (U/g hb)	852.8±46.6	864±25.6	931.2±41.4
PMN GSH-PX (U/mg protein)	0.09±0.01	0.10±0.01	0.11±0.01
PMN SOD (U/mg protein)	0.90±0.06	0.99±0.07	0.98±0.05

Values are means ± se. Different superscripts indicate significant differences (*P < 0.05; **P < 0.01; ***P < 0.001) using ANOVA and LSD.

Literature Cited

Aebi, H. (1984) Catalase. Meth. Enzymol. 105: 121–126.
Jones, D.G. and Suttle, N.F. (1981) Some effects of copper deficiency on leukocyte function in sheep and cattle. Res. Vet. Sci. 31: 151–156.
Schosinsky, K.H. (1974) Measurement of caeruloplasmin activity in serum by use of o-dianisidine dihydrochloride. Clin.Chem. 9: 1556.
Strain, J.J. (1994) Newer aspects of micronutrients in chronic disease: copper. Proc. Nutr. Soc. 53: 583–598.

Plasma Thiol and Carbonyl Content Are Not Affected by Dietary Copper Deficiency in Rat

K.A. COCKELL AND B. BELONJE

Nutrition Research Division, Health Canada, Ottawa, Ontario, Canada K1A 0L2

Keywords copper, thiol, carbonyl, rat

Introduction

Mild copper deficiency is reported to be a common condition in humans, while more severe deficiency may be seen in Menkes disease, or through total parenteral nutrition (Danks 1988). Copper deficiency is associated with reduced activities of cuproenzymes, including antioxidant enzymes such as ceruloplasmin diamine oxidase and Cu,Zn-superoxide dismutase (Cu,Zn-SOD). For example, copper-deficient rats show decreased aortic Cu,Zn-SOD activity, and increased levels of aortic peroxides (Nelson el al. 1992). Lipoproteins of copper-deficient rats are more susceptible to *in vitro* oxidation (Rayssiguier et al. 1993). *In vitro* modification of low density lipoprotein (LDL) by cells in culture has been shown to be at least partially inhibitable by SOD (Steinbrecher 1988).

We hypothesized that in conditions of copper deficiency, where Cu,Zn-SOD activity is decreased, oxidative modification of circulating lipoproteins could occur *in vivo*. Oxidized LDL *in vivo* is involved in foam cell formation in the process of atherosclerosis (Witztum and Steinberg 1991). The potential for *in vivo* lipoprotein oxidation in copper deficiency would represent an additional risk factor for atherosclerosis, over and above the known association of copper deficiency with hypercholesterolemia and hypertriglyceridemia. Lipoproteins contain antioxidants such as tocopherols, so oxidative changes may (at least initially) be subtle. Oxidative modifications of apolipoproteins may be more persistent than changes in the lipid portions of the lipoproteins. Such changes in proteins may include loss of free thiol groups, and/or increases in carbonyl (aldehyde and ketone) adducts. Preliminary measurements of thiols and carbonyls reported here were conducted on whole EDTA-plasma.

Materials and Methods

Groups of 12 weanling male Long-Evans rats were fed modified AIN-93G diets with 50% sucrose substituted for starch (13.2% corn starch, 0% dextrinized cornstarch, 49.75% sucrose, other ingredients conforming to AIN-93G formulation) with (+Cu, analyzed Cu content 5.4 μg Cu/g diet dry weight) or without copper (–Cu, 0.3 μg Cu/g diet dry weight). Half of the control (+Cu) animals were pair-fed (PF) to the copper-deficient group, while the other half were ad libitum-fed controls (AL). Final bodyweight, heart weight and cardiosomatic index (heart:bodyweight ratio) were determined after 8 weeks of feeding. Plasma thiol concentration (Hu 1994), plasma carbonyl concentration (Reznick and Packer 1994), and serum ceruloplasmin diamine oxidase activity (Schosinsky et al. 1974) were determined. Diet and tissue Cu concentrations were measured by flame atomic absorption spectrophotometry on a Perkin-Elmer 5100PC AAS. Statistical analyses were done using ANOVA and Tukey's test for unequal sample sizes. Results are reported as mean ± standard deviation.

Results and Discussion

Copper deficient rats had significantly elevated absolute and relative (to bodyweight) heart weights (Table 1). Two of 12 –Cu rats died of heart failure prior to scheduled necropsy. Plasma ceruloplasmin

Correct Citation: Cockell, K.A., and Belonje, B. 1997. Plasma thiol and carbonyl content are not affected by dietary copper deficiency in rat. In *Trace Elements in Man and Animals – 9: Proceedings of the Ninth International Symposium on Trace Elements in Man and Animals. Edited by* P.W.F. Fischer, M.R. L'Abbé, K.A. Cockell, and R.S. Gibson. NRC Research Press, Ottawa, Canada. pp. 345–346.

diamine oxidase activity was essentially abolished by the –Cu treatment. Liver and heart copper concentrations were dramatically reduced in rats fed the –Cu diet (in µg/g dry weight: liver 2.65 ± 0.73 versus 13.03 ± 1.14 and 12.55 ± 1.19; heart 3.23 ± 0.66 versus 21.49 ± 0.86 and 21.12 ± 0.76 for –Cu, pairfed control and ad libitum controls, respectively). These parameters indicate that the –Cu rats were severely copper deficient.

No significant decreases in plasma thiol content nor increases in plasma protein carbonyl content were observed in the copper-deficient group. The slight though statistically significant decrease in plasma carbonyl content in –Cu rats was inconsistent with the anticipated increase in oxidative stress in these rats, and the small magnitude of the change made it of limited biological consequence. These measures of plasma protein oxidation appeared not to be sensitive indicators of *in vivo* oxidative stress in copper-deficient rats. Follow-up experiments are underway to assess the levels of thiols and carbonyls in specific lipoprotein fractions from copper-deficient and control rats.

Table 1. Absolute and relative heart weights, serum ceruloplasmin diamine oxidase activity, and plasma thiol and carbonyl content of rats fed sucrose-based diets deficient (–Cu) or adequate (PF = pairfed, AL = ad libitum fed) in copper (see text for details).

	–Cu	PF	AL	P <
Heart weight (g)	2.05±0.23[b]	1.27±0.08[a]	1.35±0.13[a]	0.0001
Heart:Bodyweight Ratio (%)	0.49±0.06[b]	0.30±0.01[a]	0.30±0.02[a]	0.0001
Ceruloplasmin (U/mL)	0.00±0.00[a]	0.12±0.01[b]	0.12±0.02[b]	0.0001
Plasma thiols (mmol/L)	0.39±0.08	0.31±0.05	0.36±0.04	NS
Plasma carbonyls (nmol/mg protein)	0.31±0.04[a]	0.35±0.03[b]	0.35±0.01[b]	0.05

[a,b] Values within a row bearing different superscripts were statistically significantly different at the P values shown at right (NS = not statistically significant).

Literature Cited

Danks, D.M. (1988) Copper deficiency in humans. Ann. Rev. Nutr. 8: 235–257.

Hu, M.-L. (1994) Measurement of protein thiol groups and glutathione in plasma. Meth. Enzymol. 233: 380–385.

Nelson, S.K., C.-J. Huang, M.M. Mathias and K.G.D. Allen (1992) Copper-marginal and copper-deficient diets decrease aortic prostacyclin production and copper-dependent superoxide dismutase activity, and increase aortic lipid peroxidation in rats. J. Nutr. 122: 2101–2108.

Rayssiguier, Y., E. Gueux, L. Bussiere and A. Mazur (1993) Copper deficiency increases the susceptibility of lipoproteins and tissues to peroxidation in rats. J. Nutr. 123: 1343–1348.

Reznick, A. Z. and L. Packer (1994) Oxidative damage to proteins: Spectrophotometric method for carbonyl assay. Meth. Enzymol. 233: 357–363.

Schosinsky, K.H., H.P. Lehmann and M.F. Beeler (1974) Measurement of ceruloplasmin from its oxidase activity in serum by use of o-dianisidine dihydrochloride. Clin. Chem. 20(12): 1556–1563.

Steinbrecher, U.P. (1988) Role of superoxide in endothelial-cell modification of low-densitylipoproteins. Biochim. Biophys. Acta 959:20–30.

Witztum, J.L. and D. Steinberg (1991) Role of oxidized low density lipoprotein in atherogenesis. J. Clin. Invest. 88:1785–1792.

Distribution and Pathological Studies on Rare-Earth Elements in Non-Insulin Dependent Diabetes Mellitus Mice

SHUICHI ENOMOTO, BIN LIU,[†] RAJIV G. WEGINWAR,[‡] RYOHEI AMANO,[¶] SHIZUKO AMBE, AND FUMITOSHI AMBE

The Institute of Physical and Chemical Research (RIKEN), Wako, Saitama 351-01, Japan, [†]Department of Technical Physics, Peking University, Beijing 100871, China, [‡]Chandrapur Engineering College, Chandrapur-442 403, India, and [¶]School of Health Sciences, Kanazawa University, Kanazawa, Ishikawa 920, Japan

Keywords diabetes mellitus, rare-earth elements, bio-distribution, multitracer, diabetic hepato-steatosis, diabetic nephrosclerosis

Introduction

Radioactive-tracer method is a useful technique in science, technology, medicine and various other fields. A 'multitracer technique' has been developed at RIKEN. A multitracer solution contains a number of radioactive isotopes employed for tracing. The multitracer enables us to determine the characteristic behavior of various elements under an identical condition. Diabetes mellitus is one of the most general and intractable diseases of adult people, which has a lot of complications. The two most common forms of diabetes are currently referred to as insulin dependent diabetes mellitus (IDDM) and non-insulin dependent diabetes mellitus (NIDDM). The IDDM customarily occurs in child and adolescents, but it also can develop in adults. By contrast, most cases of NIDDM appear during the later decades of life, its prevalence being increased in the obese and in those of advancing age. The IDDM is characterized by few if any functional beta cells in the islets of Langerhans and substantially reduced or nonexistent insulin secretion. As a result, body fat is metabolized as a source of energy. This oxidation produces ketone bodies, which are released into the blood cosuria produce fluid and electrolyte imbalances, which can ultimately lead to coma and death. In fact, before insulin became commercially available, IDDM was usually fatal. On the other hand, the NIDDM has a share of 90% of the diabetes mellitus and is multifactorial disease with both a genetic and enviromental backgrounds. We have to solve the fundamental causes of this disease, because the mechanisms of NIDDM is not still clear. In the present study, the multitracer technique was applied to an investigation of the *in vivo* behavior of trace elements in NIDDM (KKAy/Jcl) model mice in comparison with normal one (ddY).

Experimental

A gold foil was irradiated with a C-12, N-14 or O-16 beam of 135 MeV/nucleon from RIKEN Ring Cyclotron. The foil was dissolved in aqua regia. After Au ions were removed by extraction with ethyl acetate, the rare earth elements (REE) were extracted from an HCl-multitracer solution by di(2-ethylhexyl) phosphate solution in heptane and were back-extracted with HCl. The REE multitracer solution was evaporated to dryness and the residue was dissolved in a saline solution before experiments. The solution was injected intraperitoneally to NIDDM model and normal mice (male). The mice were sacrificed several hours after injection. The tissues, organs and body fluids were weighed and their radioactivities were determined by γ-ray spectrometry. Identification and determination of tracers were done on the basis of

Correct citation: Enomoto, S., Liu, B., Weginwar, R.G., Amano, R., Ambe, S., and Ambe, F. 1997. Distribution and pathological studies on rare-earth elements in non-insulin dependent diabetes mellitus mice. In *Trace Elements in Man and Animals – 9: Proceedings of the Ninth International Symposium on Trace Elements in Man and Animals. Edited by* P.W.F. Fischer, M.R. L'Abbé, K.A. Cockell, and R.S. Gibson. NRC Research Press, Ottawa, Canada. pp. 347–348.

their energies, half-lives and peak areas. The results are given in percentage of intraperitoneally injected dose of tissues, organs and body fluids (uptake %).

Results and Discussion

From the analysis of γ-ray spectra, distributions of Y, Ce, Eu, Gd, Yb, and Lu were determined. When intraperitoneally injected, various REE exhibited quite different distributions in both normal and NIDDM model mice. The uptake of REE in the livers of both kinds of mice shown in Figure 1 was correlated with the ionic radius of REE. The uptake of "light" REE increased with increasing ionic radius; however, there was no marked tendency in that of "heavy" REE. The quantities of accumulation of REE in NIDDM model mice liver were larger than that of normal mice by a factor of about 2–5. Triglyceride and blood sugar concentrations in blood serum of NIDDM model mice were found to be remarkably higher than those of normal mice as shown in Table 1. The results of pathologic diagnosis of a NIDDM model mouse liver by a microscope revealed hepatic steatosis. The uptake of the REE in normal mice kidney, is correlated with the ionic radius of the REE. However, the uptakes in NIDDM model mice do not depend on the ionic radii. The results of pathologic diagnosis of a NIDDM model mouse kidney revealed diabetic nephrosclerosis. Mesangial cells developed vitreous structures and showed tuberous lesions. These results indicated that REE are accumulated in connective tissue of liver around fat particles in the case of NIDDM model mice and that REE are accumulated in renal corpuscles.

Table 1. Biochemical data of NIDDM model and ddY mice serum.

	NIDDM model	ddY
Blood sugar value mg/dL (GLK/G6PDH)	250.3 ± 1.6	121.4 ± 16.9
Triglyceride mg/dL	242.8 ± 14.3	95.1 ± 22.7

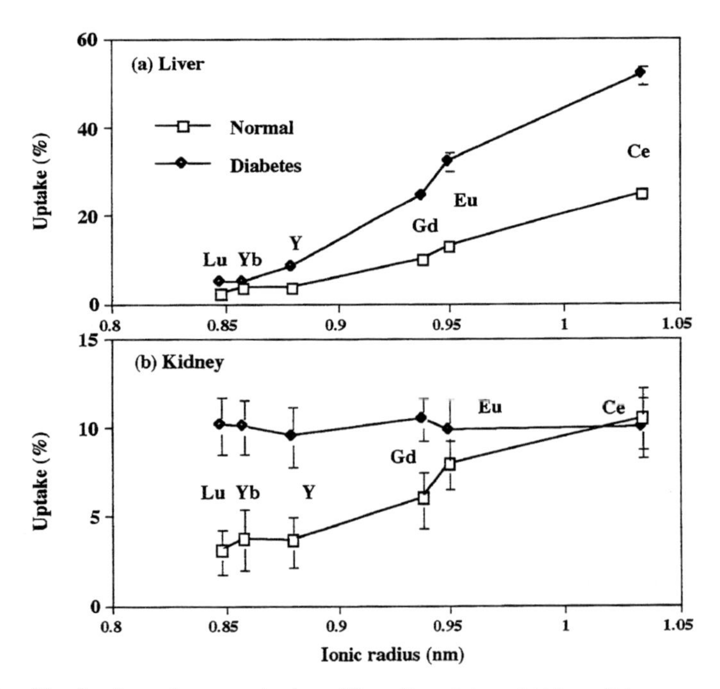

Figure 1. Uptake dependence on ionic radii on liver (a) and kidney (b).

Vanadium and Susceptibility to Peroxidative Change in STZ-diabetic Rats

K.H. THOMPSON AND J.H. MCNEILL

Faculty of Pharmaceutical Sciences, The University of British Columbia, Vancouver, British Columbia, Canada V6T 1Z3

Introduction

Orally administered vanadium salts can correct for insulin deficiency or insulin resistance in chemically-induced or spontaneously diabetic animals (for review, see Orvig et al. 1995). Vanadium is also a highly redox active transition metal. As oxidative stress is elevated in diabetes and is known to contribute to the secondary complications of diabetes, the pro- or anti-oxidant effects of chronic vanadium administration are an important consideration.

Methods

Three experiments were conducted in male, Wistar rats (Charles River, Quebec) to study the effects of varying doses of vanadyl sulfate *x* hydrate (Fisher, Fair Lawn, NJ), administered orally in the drinking water. In all experiments, rats were divided into 4 groups: control (C), non-diabetic treated with vanadium (CT), diabetic (D), and diabetic treated with vanadium (DT). Diabetes was induced in the diabetic groups using streptozotocin (STZ, Sigma Chemical Co., St. Louis, MO) 55–60 mg/kg body weight, by tail vein injection. Diabetes was confirmed by glucometer testing of tail vein blood 96 h following injection. Experiments 1 and 2 were 12 weeks in length; experiment 3, 52 weeks. Vanadyl sulfate concentration in the drinking water of diabetic animals was adjusted according to the glucose-lowering effect observed. For the CT group (n = 24) in experiment 3, rats were further subdivided into 3 groups according to level of supplementation (0.5, 0.75, or 1.25 g/L, with n = 8 in each subgroup).

Vanadium concentrations in bone, kidney, liver, testes, brain and pancreas were determined by graphite furnace atomic absorption spectrophotometer with deuterium lamp background correction (Mongold et al. 1990). Cataract formation was evaluated by visual inspection. Reduced glutathione (GSH), glutamine synthetase activity (GS), and thiobarbituric acid reactive substances (TBARS) in liver, and TBARS in erythrocytes (RBC) were assayed according to published methods (Thompson and McNeill 1993). Glutathione reductase activity in kidney homogenates was measured as rate of disappearance of NADPH (Thompson et al. 1992).

Results

The overall susceptibility of vanadium-treated rats to oxidative change indicated both pro- and anti-oxidant effects as compared to untreated (Table 1). All untreated diabetic rats developed cataracts before 20 weeks following STZ injection, while none of the vanadyl-treated rats did so. GSH levels were partially restored to normal in DT rats; while GR activity was unaffected. GS activity in liver was elevated in D compared to C and CT; but was not different in DT compared to C. TBARS were elevated in liver homogenates of DT and CT compared to D and C; however, RBC TBARS were more strongly affected by diabetes than by vanadium treatment.

Tissue vanadium levels in the 3 subgroups of CT in experiment 3 were strongly correlated with vanadium intake (Figure 1) and indicated an order of accumulation: bone > kidney > testes > liver > pancreas > brain. Tissue levels in DT animals were not different overall from CT, but were less strongly correlated with approximate dose due to greater variation in fluid intake.

Correct citation: Thompson, K.H., and McNeill, J.H. 1997. Vanadium and susceptibility to peroxidative change in STZ-diabetic rats. In *Trace Elements in Man and Animals – 9: Proceedings of the Ninth International Symposium on Trace Elements in Man and Animals. Edited by* P.W.F. Fischer, M.R. L'Abbé, K.A. Cockell, and R.S. Gibson. NRC Research Press, Ottawa, Canada. pp. 349–350.

Table 1. Summary of measures of peroxidative change.

		Cataracts <20 wks (%)	GSH Liver (µmol/g)	GS Protein (units/mg)	GR Kidney (µmol NADPH/min/g)	TBARS Liver (nmol MDA/g)	TBARS RBC (nmol MDA/g)
Non-diabetic	V (–)	0	5.0±0.2[a]	4.3±0.1[a]	4.4±0.2[a]	3.6±1.7[a]	9±3[a]
	V (+)	0	5.2±0.1[a]	5.2±0.3[b]	4.3±0.3[a]	11.4±3.1[b]	11±3[ab]
Diabetic	V (–)	100	3.8±0.3[b]	6.4±0.8[c]	5.2±0.4[a]	4.3±0.3[a]	13±2[a]
	V (+)	0	4.7±0.2[a]	4.0±0.4[a]	4.6±0.3[a]	10.0±3.3[b]	11±3[ab]

Values are means ± SEM.

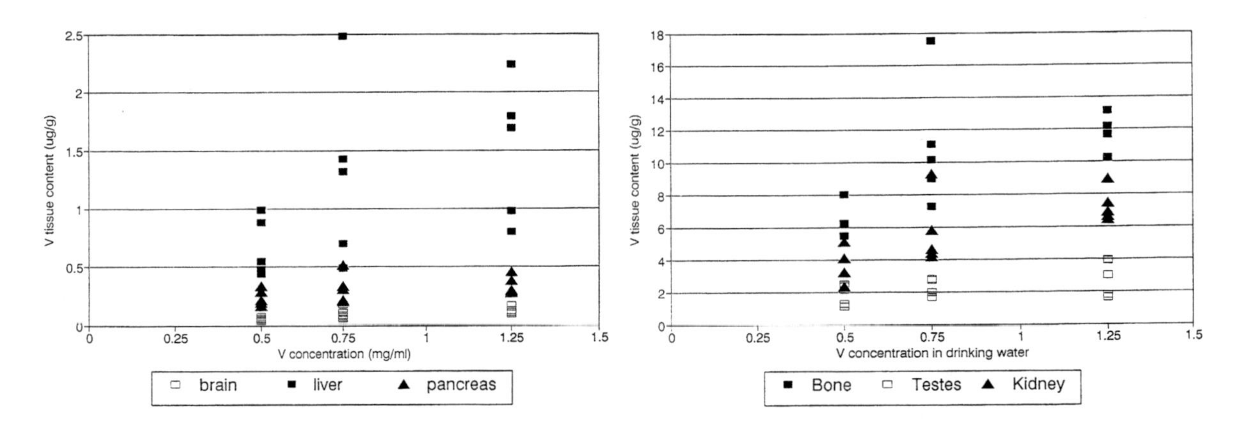

Figure 1. Vanadium content of vanadyl-treated, non-diabetic rats.

Acknowledgments

This research was supported by grants from Medical Research Council of Canada and the Canadian Diabetes Association. The technical assistance of Erika Vera, Zbysek Masin, and Violet Yuen is gratefully acknowledged.

References

Mongold, J.J., Cros, G.H., Vian, L. et al. (1990) Toxicological aspects of vanadyl sulphate on diabetic rats: effects on vanadium levels and pancreatic B-cell morphology. Pharmacol. Toxicol. 67: 192–98.

Orvig, C., Thompson, K.H., Battell, M. & McNeill, J.H. (1995) in: Metal Ions in Biological Systems (eds. H. Sigel and A. Sigel, eds.), 31: 575–94.

Thompson, K.H., Godin, D.V. & Lee, M. (1992) Tissue antioxidant status in streptozotocin-induced diabetes in rats. Effects of dietary manganese deficiency. Biol. Trace Elem. Res. 35: 213–24.

Thompson, K.H. & McNeill, J.H. (1993) Effect of vanadyl sulfate feeding on susceptibility to peroxidative change in diabetic rats. Res. Comm. Chem. Pathol. Pharmacol. 80: 187–200.

Dai, S., Thompson, K.H. & McNeill, J.H. (1994) Toxicological aspects of vanadyl sulphate on diabetic rats: effects on vanadium levels and pancreatic B-cell morphology. Pharmacol. Toxicol. 74: 101–9.

Reduced Plasma Total Antioxidant Status in Major Trauma Patients Following Trace Element Supplementation

M. BAINES,[†] C.A. WARDLE,[†] M.M. BERGER,[‡] A. PANNATIER,[‡] R. CHIOLERO,[‡] AND A. SHENKIN[†]

[†]Department of Clinical Chemistry, Royal Liverpool University Hospital, Liverpool L7 8XP, UK, [‡]Surgical ITU, CHUV CH-1011, Lausanne, Switzerland

Keywords antioxidants, trace elements, trauma

During the first week post injury, major trauma patients are often metabolically compromised, having depressed thyroid function and increased oxidative stress (Chiolero and Berger 1994). These findings may relate to negative Selenium (Se) balance (Berger et. al. 1992), since Se is an essential cofactor for both 5′ deiodinase, which catalyses T4 to T3 conversion, and glutathione peroxidase (GPx), a primary antioxidant enzyme. This study looked at the effect of Se supplementation, with or without additional copper (Cu) and Zinc (Zn), necessary cofactors to another antioxidant enzyme, superoxide dismutase (SOD), and vitamin E upon some markers of antioxidant function, including total antioxidant status (TAS). Plasma TAS is an overall indicator of the antioxidant capacity of the plasma compartment.

Twenty-four trauma patients admitted to ITU with an Injury Severity Score > 16 were randomised to be infused with either Se (500 μg/d) (Group G1), Se plus Cu (2.6 mg/d) and Zn (13 mg/d) and vitamin E (150 mg/d) (G2) or placebo (G3) for the first 5 d, starting 24 h post injury. Blood was taken at day(D) 0 (pre) and at D1, D2, and D5 post treatment and was analysed for trace elements, red cell antioxidant enzymes GPx and SOD and plasma TAS. TAS was measured by the degree of suppression by plasma of the stable radical cation, ABTS· (Randox Laboratories, Belfast, UK) (Miller et al. 1993).

With the exception of vitamin E, there were no significant differences between the supplemented groups, hence these were combined (G1 + G2) and compared with placebo (G3). Plasma Se, low in both groups at D0, was significantly higher in (G1 + G2) than G3 at D2 and D5 ($p < 0.001$). Cu and Zn did not differ significantly between the groups over the 5 days. Vitamin E was significantly higher in (G1 + G2) only on D5 ($p < 0.05$). Of the antioxidant markers, there was no significant difference between the groups for the red cell enzymes GPx and SOD. The mean (SD) plasma TAS was similar in both groups: 1.16 (0.19) vs. 1.13 (0.12) mmol/L at pre-treatment (D0), but by D1 the TAS of (G1 + G2) had fallen to 1.04 (0.17) vs. 1.17 (0.14), ($p = 0.08$) and by D2 to 0.99 (0.14) vs. 1.12 (0.11), ($p < 0.05$). By D5 both groups were similar again (1.09) (see Figure 1).

Thus supplementation with large doses of Se, with or without Cu and Zn, is able to increase significantly the plasma Se level, though this is not accompanied by increases in the activity of the red cell antioxidant enzymes GPx and SOD. There is, however, a significant fall in plasma TAS in the supplemented group, despite vitamin E supplementation. This was not due to changes in albumin or urate levels, both major contributors to the plasma antioxidant capacity. This fall may indicate an intra-cellular shift of small molecule antioxidants as cellular antioxidant defences are mobilised in a response to the trauma, possibly initiated by the trace element supplementation.

Correct citation: Baines, M., Wardle, C.A., Berger, M.M., Pannatier, A., Chiolero, R., and Shenkin, A. 1997. Reduced plasma total antioxidant status in major trauma patients following trace element supplementation. In *Trace Elements in Man and Animals – 9: Proceedings of the Ninth International Symposium on Trace Elements in Man and Animals. Edited by* P.W.F. Fischer, M.R. L'Abbé, K.A. Cockell, and R.S. Gibson. NRC Research Press, Ottawa, Canada. pp. 351–352.

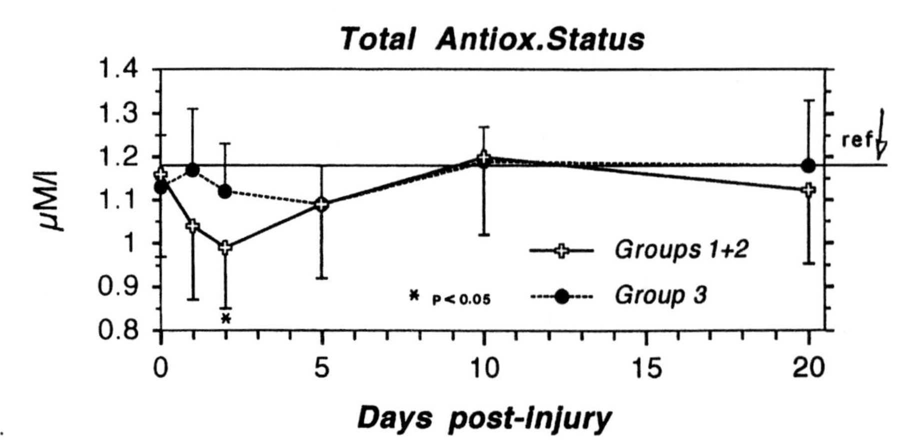

Figure 1.

References

Berger, M.M., Cavadini, C., Bart, A., Blondel, A., Bartholdi, I., Vandervale, A., Krupp, S., Chiolero, R., Freeman, J. & Dirren, H. (1992) Selenium losses in 10 burned patients. Clin. Nutr. 11: 75–82.

Chiolero, R. & Berger, M.M. (1994) Endocrine response to brain injury. New Horizons 2: 432–442.

Miller, N.J., Rice-Evans, C., Davies, M.J., Gopinathan, V. & Milner, A. (1993) A novel method for measuring antioxidant capacity and its application to monitoring the antioxidant status in premature neonates. Clin. Sci. 84: 407–412.

The Feasibility of Using Exfoliated Human Colon Cells to Monitor the Potential Pro-Oxidant Effects of Iron*

S.G. WHARF, E.K. LUND, M. PARKER, I.T. JOHNSON, AND
S.J. FAIRWEATHER-TAIT
Institute of Food Research, Norwich NR4 7UA, UK

Keywords iron, colonocyte, feces, DNA

The hypothesis that excess unabsorbed dietary iron present in the lumen of the large intestine may increase the risk of colon cancer as a result of free radical production by bacteria has been developed by a number of groups (Blakeborough et al. 1989, Babbs 1990). Iron in solution and bound to small organic molecules is able to catalyze the production of hydrogen peroxide as a result of the interaction between the free radicals formed by bacterial metabolic activity and water in the well recognised Fenton reaction. It is known that oxidative damage can be induced in cultured colon cell lines exposed to H_2O_2 in the presence of iron (Watson et al. 1994). Sufficient DNA can be extracted from human colonocytes to analyze for DNA mutations using the polymerase chain reaction (Smith-Ravin et al. 1995). The aim of this work was to assess whether DNA could be extracted from these cells in order to detect the presence of oxidised bases by GC-MS or HPLC so that this hypothesis could be tested in volunteers given an iron supplement.

Methods

Cells were isolated from feces using an adaptation of the method of Albaugh et al. (1992). Samples were kept at 4°C throughout the preparation. Fecal samples (5–10 g) were collected into 250 mL Pucks Saline G containing an antibiotic and antimycotic mixture prior to homogenising for 15 s in a stomacher bag containing an inner net bag. The media from outside the net bag was collected and centrifuged at 900 g for 10 min and the supernatant carefully layered onto Histopaque 1119 before centrifuging at 210 g for 30 min. An initial crude cell fraction was collected at the interphase between the two layers and diluted 1:1 with buffer and then layered onto a previously prepared Percoll gradient, density range 1– >1.139 and centrifuged at 800 g for a further 30 min. Cells were collected from between $\rho = 1.051–1.076$. DNA was isolated from two cell fractions, the first 'crude' preparation taken following the initial clean up with histopaque and the second after the final isolation on the percoll gradient. Cell number was initially assessed using a haemocytometer. In later cell preparations cell number and DNA content of both pellets was assessed by flow cytometry, following staining with propidium iodide.

DNA was extracted from both the crude pellet and the percoll pellet using 'Qiagen' columns following two alternative protocols described in the 'Qiagen Genomic DNA Handbook'; one designed for tissues and one for bacteria, both containing RNA'se in the lysis buffer. The quality of the DNA was then measured spectrophotometrically and the amount calculated from the absorbance at $\lambda = 260$. Further assessment of the quality was made using gel electrophoresis (1% agarose, 85 V, 2 h). The level of DNA damage in individual cells can be assessed by single cell gel electrophoresis. In this assay cells were suspended in warm 1.5% low melting point agarose which was then allowed to set on a microscope slide before being covered with 1% normal agarose. Prior to electrophoresis (25 V, 300 mA, 4°C, 20 min) the slide was exposed to an alkaline lysis buffer. DNA was stained with ethidium bromide and the slide examined by

*This work was supported by MAFF and the BBSRC.

Correct citation: Wharf, S.G., Lund, E.K., Parker, M., Johnson, I.T., and Fairweather-Tait, S.J. 1997. The feasibility of using exfoliated human colon cells to monitor the potential pro-oxidant effects of iron. In *Trace Elements in Man and Animals – 9: Proceedings of the Ninth International Symposium on Trace Elements in Man and Animals. Edited by* P.W.F. Fischer, M.R. L'Abbé, K.A. Cockell, and R.S. Gibson. NRC Research Press, Ottawa, Canada. pp. 353–355.

fluorescent microscopy. Transmission electron microscopy (TEM) of cells, embedded in Spurr resin following fixation in 3% gluteraldehyde and post fixation in 15 osmium tetroxide for 2 h and stained with uranyl acetate and lead citrate, was used to characterize the composition of the pellet. Magnetic beads coated with monoclonal antibodies (mAb) to epithelial cell surface proteins were also tried as an alternative isolation procedure for cells from the initial crude pellet in place of the percoll gradient. "Dynal beads" purchased ready coated with sheep-antimouse IgG were treated with the either carcinogenic embryonic mAb(CEA) or colon specific mAb antigen (CSA) prior to incubating with the crude cell pellet. The bead/'cell' complex was washed before examination by light and scanning electron microscopy. DNA extraction from the adherent and non-adherent fractions was performed as described above.

Results

The number of cells in the final pellet was approximately 2×10^6 cells/g feces, with similar values being found using both the flow cytometer and haemocytometer. The final cell pellet contained approximately 4 μg DNA/g feces, less than one third of the expected levels of DNA from this number of cells. Analysis of the extracted DNA by gel electrophoresis following different isolation procedures revealed that most of this DNA was probably of bacterial origin, molecular weight c. 23 kbp, and the DNA from the colonocytes was badly degraded, with a molecular weight of about 1.5 kbp. Flow cytometric analysis showed the average cell size to be small, and the DNA content very low. Under TEM the cells showed typical features of apoptosis, cell shrinkage and nuclear fragmentation. However loss of DNA from the cell implies breakdown of the cell membrane as a result of necrosis or due to membrane damage within the lumen of the large intestine.

When samples were examined using single cell gel electrophoresis, intact cells were not detectable but small dense spots of bacterial DNA were seen. Magnetic beads coated with CEA and CSA mAbs adhered to a heterogenous component of the isolated pellet including cell membrane fragments and bacteria, however insufficient DNA was extracted from the adherent material for further analysis.

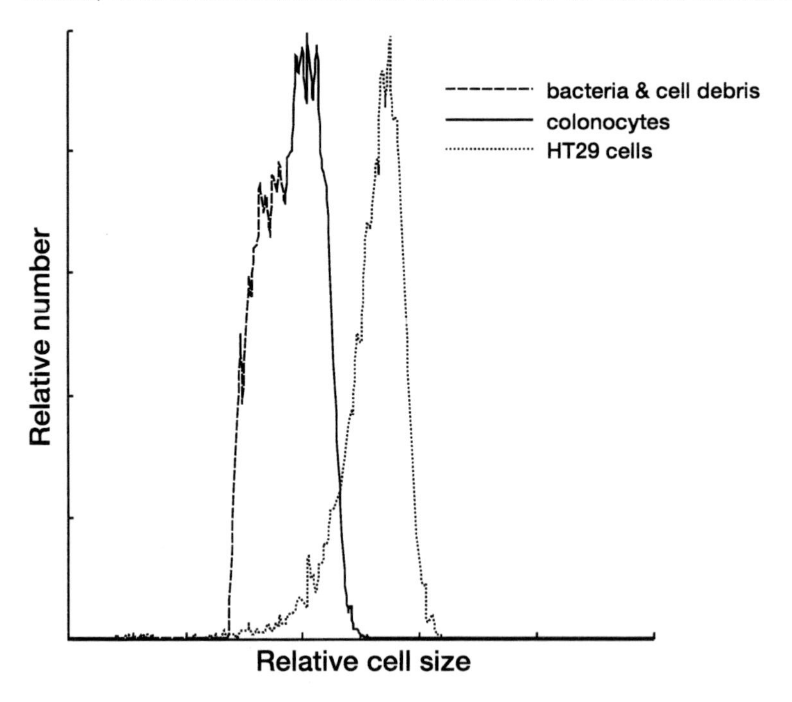

Figure 1. Flow cytometric analysis of the crude cell pellet.

Conclusion

The final colonocyte pellet obtainable from human feces contained very little, highly degraded, DNA. Those cells recognizable as epithelial cells using TEM were probably in a late stage of apoptosis or programmed cell death. Although the group who originally described the isolation of fecal colonocytes were able to characterize these cells in relation to several cell surface markers (Albaugh et al. 1992), it appears that these are associated with membrane fragments, not whole cells. The colonocytes isolated from human feces are too badly damaged to be useful in the assessment of potential oxidative DNA damage caused by excess iron in the large intestine.

References

Albaugh, G.P., Iyengar, V., Lohani, A., Malayeri, M., Bala, S. & Nair, P.P. (1992) Isolation of exfoliated colonic epithelial cells, a novel, non-invasive approach to the study of cellular markers. Int. J. Cancer 52: 347–350.

Babbs, C.F. (1990). Free radicals and the etiology of colon cancer. Free Rad. Biol. Med. 8: 191–200.

Blakeborough, M.H., Owen, R.W. & Bilton, R.F. (1989) Free radical generating mechanisms in the colon: Their role in the induction and promotion of colorectal cancer. Free Rad. Res. Com. 6: (6) 359–367.

Smith-Ravin, J., England, J., Talbot, I.C. & Bodmer, W. (1995) Detection of c-Ki-ras mutations in faecal samples from sporadic colorectal cancer patients. Gut 36: 81–86.

Watson, A.J.M., Askew, J.N., Sandle, G.I. (1994) Characterisation of oxidative injury to an intestinal cell line (HT-29) by hydrogen peroxide. Gut 35: 1575–1581.

Oral Intake of Orange Peel Prevented Cadmium-Induced Renal Dysfunction

K. NOMIYAMA,[†] H. NOMIYAMA,[†] N. KAMEDA,[†] H. SAKURAI,[‡] AND H. TAKAHASHI[†]

[†]Department of Environmental Health, Jichi Medical School, Minamikawachi-Machi, Tochigi-Ken 329-04, Japan, Department of Metabolic analysis, [‡]Kyoto Pharmaceutical University, Yamashina-Ku, Kyoto 607, Japan

Introduction

The mechanism of cadmium-induced renal dysfunction has been believed to be as follows (WHO/IPCS 1992): 1) Cadmium taken into the body accumulates in the liver, and then 2) is transferred to the kidneys via the blood stream in the chemical form of cadmium-thionein. 3) Cadmium accumulates in the kidney slowly but steadily with a constant biological half-time of 18–36 years, and 4) renal dysfunction occurs when the cadmium in the renal cortex exceeds the critical concentration, 200 μg/g. 5) Once the cadmium in the renal cortex exceeds the critical concentration, the cadmium-induced renal dysfunction is incurable. 6) The allowable daily intake of cadmium can be calculated theoretically based on fixed values for both the biological half-time and the intestinal absorption rate of cadmium.

However, the epidemiological data on residents in cadmium-polluted areas in Japan did not indicate a good dose-response relationship (Nomiyama 1988), and animal experiments indicated quite different biological half-times and critical concentrations from exposure levels of cadmium (Nomiyama et al. 1979, Nomiyama and Nomiyama 1979, Nomiyama and Nomiyama 1988).

It was necessary therefore, to find and correct the erroneous logic in the WHO criteria document on cadmium (1992) at the earliest opportunity.

New Hypothesis for Cadmium-induced Renal Dysfunction

Cadmium exposure induces excess cadmium deposition in the liver in the chemical forms of cadmium-thionein and cadmium not bound to metallothionein. When the production of metallothionein, a radical scavenger, becomes insufficient for detoxifying cadmium in the liver, the cadmium not bound to metallothionein can induce free radicals, which may in turn depress hepatic functions. Hepatic cadmium-thionein is then released into the blood stream. Apoptosis of hepatic cells may also be one of the most important sources of increased plasma cadmium-thionein when cadmium accumulates heavily within the liver. Plasma cadmium-thionein passes through the glomeruli with ease, because of its small molecular size of 6000 daltons, and reaches the proximal tubular lumen. Cadmium-thionein may be split into cadmium ion and holo-thionein on or in the brush border membrane of the proximal tubular cells by some unknown mechanism. Free radicals, which are initiated by cadmium ion, may injure the brush border membrane. Excess free radicals above a critical level may then induce dysfunctions of the brush border membrane such as low-molecular weight proteinuria, glucosuria and aminoaciduria, independent of the cadmium concentration in the renal cortex.

Working Hypothesis

1) Free cadmium not bound to metallothionein, a radical scavenger, in the liver of animals exposed to cadmium for a long period, may induce excess production of free radicals, which may directly cause hepatic dysfunction, followed by increased cadmium-thionein in the renal proximal tubular lumen; then

Correct citation: Nomiyama, K., Nomiyama, H., Kameda, N., Sakurai, H., and Takahashi, H. 1997. Oral intake of orange peel prevented cadmium-induced renal dysfunction. In *Trace Elements in Man and Animals – 9: Proceedings of the Ninth International Symposium on Trace Elements in Man and Animals. Edited by* P.W.F. Fischer, M.R. L'Abbé, K.A. Cockell, and R.S. Gibson. NRC Research Press, Ottawa, Canada. pp. 356–358.

excess cadmium-thionein may cause the dysfunction of brush border membrane of renal proximal renal tubules, independent of cadmium levels in the renal cortex.

2) Therefore, it may be difficult to induce renal dysfunction in animals chronically exposed to cadmium, by giving Valencia orange peel, which contained large amounts of vitamin C and carotene, radical scavengers.

Methods

Thirty-two male rabbits were divided into 2 groups. Rabbits of the first group were given cadmium chloride at a dose level of 0.5 mg/kg 6 times a week over a period of 13 weeks, and rabbits of another group were left untreated and served as a control group. Each group was then divided into 2: rabbits of the first subgroup were given oral 15 g/d Valencia orange peel, which contained 98 mg Vitamin C/100 g and 130 mg carotene/100 g (both radical scavengers); rabbits of another subgroup were not given orange peel and served as controls.

Results And Discussion

As shown in Table 1, cadmium intake elevated plasma lipid peroxides and induced hepatic dysfunction at 8–12th week, and then plasma cadmium increased markedly. Renal dysfunction was observed at the 12th week, after plasma cadmium exceeded 100 μg Cd/L: a critical concentration which accorded well with the value we reported before (Nomiyama and Nomiyama 1994).

Oral supplementation of orange peel, full of radical scavengers such as Vitamin C and carotene, improved cadmium-induced hepatic and renal dysfunction remarkably, as indicated in Table 1.

Cadmium levels in the liver and the renal cortex of the cadmium only group (305 and 692 μg Cd/g respectively) were the same as those of the cadmium-orange peel group (292 and 697 μg Cd/g).

Table 1. Effects of orange peel on cadmium-induced health effects.

	Biomarkers	Cadmium health effects	Orange peel effects on cadmium health effects
Plasma	Lipid peroxides	8th Week ↑↑	↓↓
Plasma	Aspartate aminotransferase	13th Week ↑	↓↓
	Alanine aminotransferase	12th Week ↑	↓↓
	Albumin	–	–
	Cholinesterase	–	–
Plasma	Cadmium	10th Week 70 μg/L	30 μg/L
		12th Week 104 μg/L	37 μg/L
		13th Week 107 μg/L	47 μg/L
Urine	Protein	13th Week ↑	↓↓
	Glucose	13th Week ↑	↓
	Amino acids	12th Week ↑	–
	Urea nitrogen	13.5th Week ↑	↓
Liver	Cadmium	305 μgCd/g	292 μgCd/g
Renal cortex	Cadmium	692 μgCd/g	697 μgCd/g

Conclusion

The above data may support our hypotheses that

1) free cadmium in the liver may induce free radicals,

2) which leads to the development of hepatic dysfunction.

3) Then hepatic cadmium-thionein is released into the blood stream upon hepatic dysfunction.

4) Cadmium-thionein reaches renal tubular lumen through glomeruli and induces a dysfunction of the brush border membrane of the proximal convoluted tubules.

5) Critical concentration of plasma cadmium to induce renal dysfunction is 100 μg Cd/L.

References

WHO/IPCS: (1992) Cadmium, WHO, Geneva.
Nomiyama, K. (1988) Food Sanit. Res. 38: 7–16.
Nomiyama, K. et al. (1979) Environ. Health Perspect. 28: 223–243.
Nomiyama, K. & Nomiyama, H. (1979) Arh. Hig. Rada Toksikol. 30(Suppl): 191–200.
Nomiyama, K. & Nomiyama, H. (1988) Kankyo Hoken Report 54: 34–38.
Nomiyama, K. & Nomiyama, H. (1994) Deut. Med. Wschr. (Jpn) 16: 189–201.

Oedema, Plasma Bleomycin Detected Iron, Transferrin, Transferrin Saturation, and Malondialdehyde in Patients with Kwashiorkor

HANS DEV. HEESE, WILLIAM S. DEMPSTER, HESTER MALAN, AND ALAN A. SIVE

Department of Paediatrics and Child Health, Institute of Child Health, University of Cape Town, South Africa

Keywords kwashiorkor, oedema, iron, lipid peroxidation

Oedema is a cardinal finding in the diagnosis of kwashiorkor. The mechanisms responsible for oedema formation remain obscure (Richardson and Iputo 1992). The role of radicals and their scavengers in its formation is still to be investigated (Kaschula 1995). Iron-catalysed chain reactions lead to, among other effects, peroxidation of lipids. Lipid hydroperoxides and other products of oxidation are toxic to cell membranes which lose their integrity and may contribute to interstitial oedema. The end product of lipid peroxidation is malondialdehyde (MDA).

The objectives of the current studies were to relate the severity of oedema in patients with kwashiorkor to plasma concentrations of bleomycin detectable iron (BDI), transferrin (TF), transferrin saturation (TS) and MDA.

Materials and Methods

Venous blood samples were collected in kwashiorkor patients on admission, at 24 h and on days 5, 10 and 30 of nutritional recovery. In 15 patients BDI (μmol/L), TF (g/L) and TS (%) concentrations were measured (Gutteridge et al. 1981, Gutteridge and Halliwell 1987). TF was measured by nephelotometry (Behring Diagnostics, Marburg, Germany) and TS calculated from plasma iron concentration (Boehringer Mannheim). Nine and 6 of the patients were grouped as having severe and moderate oedema respectively on admission. MDA (μmol/L) were measured in 27 of these patients (Lepage et al. 1991, Hoving et al. 1992). The study was approved by the Ethics and Research Committee of the Medical School of the University of Cape Town. Informed consent was obtained from the custodial parent of each child.

Results

BDI, TF and TS concentrations are given in Table 1 and those for MDA in Table 2.

Table 1. Median and interquartile ranges of plasma bleomycin detected iron (BDI), transferrin (TF) and saturation of transferrin (TS) in 6 children with kwashiorkor with moderate oedema and 9 with severe oedema.

	Oedema	Admission	24 h	10 d	30 d
BDI μmol/L	Moderate	6.75(0.5–15.5)	0(0–1.25)	0(0–0.5)	0
Ref = 0	Severe	20.5(18.5–21.0)	17.0(13.5–18.5)	0.013(0–0.88)	0
TF g/L	Moderate	0.62(0.44–1.05)	0.86(0.45–1.21)	1.97(1.88–1.98)	3.24(2.63–3.59)
Ref = 2.95	Severe	0.46(0.39–0.73)	0.42(0.38–0.5)	1.12(1.03–1.77)	2.95(2.87–3.33)
TS%	Moderate	52.2(45–67.3)	34.2(26–49.4)	9.6(7.9–14.5)	14.4(14.0–14.9)
Ref R > 16	Severe	72(60.9–80.9)	64.8(35.2–87.5)	20.2(13.8–38.6)	14.0(8.0–17.5)

Correct citation: Heese, H. deV., Dempster, W.S., Malan, H., and Sive, A.A. 1997. Oedema, plasma bleomycin detected iron, transferrin, transferrin saturation, and malondialdehyde in patients with kwashiorkor. In *Trace Elements in Man and Animals – 9: Proceedings of the Ninth International Symposium on Trace Elements in Man and Animals. Edited by* P.W.F. Fischer, M.R. L'Abbé, K.A. Cockell, and R.S. Gibson. NRC Research Press, Ottawa, Canada. pp. 359–360.

Table 2. Median and interquartile ranges of malondialdehyde (MDA) in 10 children with kwashiorkor with mild oedema, 10 with moderate oedema and 7 with severe oedema.

	Oedema	Admission	5 d	10 d	30 d
MDA	Mild	1.01(0.62–2.73)	0.82(0.59–1.12)	0.68(0.46–0.98)	0.54(0.37–0.9)
μmol/L	Moderate	0.58(0.37–1.63)	0.80(0.59–1.74)	0.69(0.29–0.97)	0.62(0.38–1.06)
Ref < 0.9	Severe	1.14(0.61–1.52)	0.87(0.69–1.82)	0.59(0.58–0.8)	0.59(0.54–0.66)

Statistical Analysis

Data was analysed using Statistica 5.1. Kruskal-Wallis one way analysis of variance was used to compare groups when there were more than 2. If this showed significant difference ($p < 0.05$) this U-test was used to determine which groups differed.

Discussion

In children who are undernourished, kwashiorkor results when there is increased radical formation and a deficiency of micronutrients essential for protective antioxidant functions (Golden and Ramadath 1987, Golden 1996). Iron is a major catalyst of radical reactions and in patients with kwashiorkor, plasma ferritin is increased (Golden et al. 1985), hepatic iron is elevated (Waterlow 1992), loosely bound iron is elevated in the plasma (Dempster et al. 1995), demonstrable iron is present in the bone marrow and elevated levels of iron are excreted in the urine following chelation with desferrioxamine (Sive et al. in press).

High concentrations of BDI were associated with low concentrations of TF which was highly saturated with iron. BDI were measured in 83% and 100% of children with moderate and severe oedema respectively. BDI was higher on admission ($p = 0.01$) and at 24 h ($p = 0.02$) in children with severe oedema compared to those with moderate oedema. There were no other differences.

MDA concentration in excess of 0.9 μmol/L were measured in 60% with mild, 40% with moderate and 72% with severe oedema. Trends in the BDI, TF, TS% and MDA concentrations and the degree of oedema suggest that the oedema parallels the laboratory findings during recovery from the acute stages of the disease. The potential for iron-catalysed radical reactions exist and the raised MDA concentrations may be regarded as probable evidence of their presence. However, whether a causal relationship between radicals and severity of oedema exists has not been answered.

References

Dempster, W.S., Sive, A.A., Malan, H. & Heese, H.de.V. (1995) Misplaced iron in kwashiorkor. Eur. J. Clin. Nutr. 49: 208–210.

Golden, M.H.N., Golden, B.E. & Bennett F.I. (1985) Relationship of trace element deficiencies to malnutrition. In: Trace Elements in Nutrition of Children. (Chandra, R.K., ed.), pp. 185–207 Raven Press, New York.

Golden, M.H.N. & Ramadath, D. (1987) Free radicals in the pathogenesis of kwashiorkor. Proc. Nutr. Soc. 46: 53–68.

Golden, M.H.N. (1996) Severe malnutrition. In: Oxford Textbook of Medicine. (Weathrall, D.J., Ledingham, J.G.G., Warrell, D.A., eds.), vol. 1, pp. 1278–1296. Oxford University Press, Oxford.

Gutteridge, J.M.C., Rowley, D.A. & Halliwell, B. (1981) Superoxide-dependent formation of hydroxyl radicals in the presence of iron salts. Detection of free iron in biological systems using bleomycin-dependent degradation of DNA. Biochem. J. 199: 263–265.

Gutteridge, J.M.C. & Halliwell, B. (1987) Radical promoting loosely-bound iron in biological fluids and bleomycin assay. Life Chem. Rep. 4: 113–114.

Hoving, E.B., Laing, C., Rutgers, H.M., Teggeler, M., van Doormal, J.J. & Muskiet, F.A.J. (1992) Optimized determination of malondialdehyde in plasma lipid extracts using 1.3-diethyl-2-thiobarbituric acid: Influence of detection method and relations with lipids and fatty acids in plasma from healthy adults. Clin. Chem. Acta. 208: 63–76.

Kaschula, R.O.C. (1995) Malnutrition and intestinal malabsorption. In: Tropical Pathology, (Doerr, W., & Seifert, G., eds.) pp. 985–1030. Springer, Berlin.

Lepage, G., Munoz, G., Champagne, J. & Claude, C.R. (1991) Preparative steps necessary for the accurate measurement of malondialdehyde by high-performance liquid chromatography. Anal. Biochem. 197: 277–283.

Richardson, D. & Iputo, J. (1992) Effects of kwashiorkor malnutrition on measured capillary filtration rate in the forearm. Am. J. Phys. 262: H496-H502.

Sive, A.A., Dempster, W.S., Rosseau, S., Kellny, M., Malan, H. & Heese, H.de.V. Bone marrow and chelatable iron in patients with protein energy malnutrition. S. Afr. Med. J. (in press).

Statsoft, Inc. (1996) Statistica for Windows.. Tulsa OK: Statsoft Inc., 2300 East 14th Street, Tulsa OK 74104.

Waterlow, J.C. (1948) Fatty liver disease in infants in the British West Indies. Med. Res. Council Spec. Rep. Series, No 263. HMSO, London.

Effect of a Maximal Effort on Blood Trace Elements and Magnesium and on some Oxidative Stress Parameters

C.P. MONTEIRO,[†] D. PEREIRA,[‡] H. RIBEIRO,[‡] C. RABAÇAL,[‡] G.M. FELISBERTO,[¶] C. MENDONÇA,[‡] L. NUNO,[‡] M. DIAS,[¶] E. CARVALHO,[‡] C. VAZ,[¶] J.S. AFONSO,[‡] J.S. FERNANDES,[‡] Z. SILVA,[§] M. BICHO,[§] A. MAZUR,[||] Y. RAYSSIGUIER,[||] AND M.J. LAIRES[†]

[†]Laboratório de Bioquímica, Faculdade de Motricidade Humana, UTL 1499 Lisboa codex, Portugal; [‡]Serviço de Cardiologia do Hospital Reynaldo dos Santos, Vila Franca de Xira, Portugal; [¶]Laboratório de Análises, IST, UTL, Lisboa, Portugal; [§]Laboratório de Genética, FML, UCL, Lisboa, Portugal; and [||]Unité Maladies Métaboliques et Micronutriments, INRA, Clermont-Theix, France

Keywords trace elements, magnesium, oxidative stress, exercise

Strenuous exercise, as it induces biochemical and endocrine changes within the body, provides an excellent model to study the dynamic balance between oxidative challenge and antioxidant defences in the biological system (Davis et al. 1982). The aim of the present study was to assess whether a maximal cycloergometer exercise influences the concentrations of trace elements in plasma and in red blood cell (RBC) and the levels of some oxidative parameters.

Methods and Materials

Fifteen well trained male athletes (24.3 ± 3.4 year, 72.3 ± 6.1 kg, 176.6 ± 5.7 cm) agreed to participate in the test. Subjects performed a maximal cycloergometer exercise, all of them having reached their respective maximal heart rate. Heart rate was continuously monitored and arterial blood pressure was measured at regular intervals. Blood was withdrawn by butterfly catheter from the anticubital vein. Samples were collected at rest, just before the exercise, and 1 min after the exercise test. Plasma and RBC Cu, Zn, Se, Fe and Mg were measured by atomic absorption spectrophotometry. Concentrations of trace elements were corrected for plasma volume variations. Oxidative stress parameters included: plasma thiobarbituric acid reactive substances (TBARS) (Uchiyama et al. 1978), adrenaline oxidase activity (Mathews et al. 1984), RBC superoxide dismutase activity and transmembrane reductase activity (Winterbourn et al. 1975, Orringer et al. 1979). LDL was isolated by sequential ultracentrifugation and submitted to *in vitro* oxidation by Cu^{2+} according Esterbauer (1989). LDL fluidity was also determined by fluorescence polarization at 37°C using DPH as a probe (Motta et al. 1996). RBC susceptibility to peroxidation *in vitro* was evaluated by the % hemolysis of RBCs induced by H_2O_2. To assess the significance of the effect of exercise on the variables, a paired t-test (one tailed) was used.

Results and Discussion

Mean effort time and heart rate were 17.9 ± 1.5 min and 161.9 ± 8.0 beats/min respectively. The mean values for the biochemical variables are shown in Table 1. Before the exercise test, the mean values of the variables studied were in the normal range, except for RBC-Mg and RBC-Cu, which were low.

Correct citation: Monteiro, C.P., Pereira, D., Ribeiro, H., Rabaçal, C., Felisberto, G.M., Mendonça, C., Nuno, L., Dias, M., Carvalho, E., Vaz, C., Afonso, J.S., Fernandes, J.S., Silva, Z., Bicho, M., Mazur, A., Rayssiguier, Y., and Laires, M.J. 1997. Effect of a maximal effort on blood trace elements and magnesium and on some oxidative stress parameters. In *Trace Elements in Man and Animals – 9: Proceedings of the Ninth International Symposium on Trace Elements in Man and Animals. Edited by* P.W.F. Fischer, M.R. L'Abbé, K.A. Cockell, and R.S. Gibson. NRC Research Press, Ottawa, Canada. pp. 361–362.

Table 1. Mean ± standard deviation and significance level of the studied parameters.

Parameter	Before effort mean ± sd	After effort mean ± sd	Variation p
RBC-Fe (mM)	15.2±1.41	16.1±1.56	p < 0.005
RBC-Cu (mM)	11.3±1.45	11.9±1.39	p < 0.05
RBC-Zn (mM)	170±24.0	170±14.5	p > 0.05 (ns)
RBC-Se (mM)	1.20±0.35	1.29±0.33	p < 0.05
RBC-Mg (mM)	1.51±0.21	1.62±0.25	p < 0.05
RBC-Trans. Red. Act. (mmol/L cel./h)	3.93±1.14	3.51±0.819	p > 0.05 (ns)
RBC hemolysis (%)	0.77±0.33	0.71±0.30	p > 0.05 (ns)
RBC-SOD (IU)	2125±345	2161±344	p > 0.05 (ns)
P-TBARS (mM)	5.4±2.1	6.5±3.0	p < 0.005
Lag time-LDL (min)	89.7±11	92.8±7.5	p > 0.05 (ns)
LDL-anisotropy (r)	0.212±0.0096	0.207±0.0054	p < 0.05
P-Adren. Oxi. Act. (mM/2 h)	33.8±12.0	23.2±11	p < 0.05
P-Mg (MM)	0.82±0.04	0.81±0.05	p < 0.05
P-Fe (mM)	25.1±10.5	25.2±12.2	p > 0.05 (ns)
P-Cu (MM)	16.4±3.09	17.0±2.50	p < 0.05
P-Zn (mM)	14.7±1.68	14.1±1.87	p > 0.05 (ns)
P-Se (mM)	1.06±0.28	1.02±0.48	p > 0.05 (ns)

These results are in agreement with those obtained for some other groups of athletes (Laires et al. 1994). The increase of TBARS in plasma after exercise may suggest an imbalance between the generation and the elimination of ROS in the plasma. However, this modification seems too weak to significantly modify the peroxidability of LDL. Regarding the variations in micronutrients with the exercise (Laires et al. 1993), it was observed that there was a significant increase in RBC-Fe. Similarly, others have observed an increase in blood copper after exhausting exercise. This increase might be due to the synthesis of ceruloplasmin by the liver. The observed increase in RBC Se might result from the mobilization of Se into the RBC. Se is an important component of GSH peroxidase. P-Mg decreased and RBC-Mg increased. The P-Mg decrease might be due to a shift of Mg into the cells, namely RBC and muscle cells (Laires et al. 1993). In conclusion, several factors can contribute to the changes observed for these elements after the exercise. These can depend on the status of the individual, on the degree of training, on the type, intensity and duration of the exercise, on environmental conditions and on the timing of the blood sampling.

Literature Cited

Davies, K.J.A., Quintanilha, A.T., Brooks, G.A. & Packer, L. (1982) Free radicals and tissue damage produced by exercise. Biochem. Biophys. Res. Commun., 107: 1198–1205.

Esterbauer, H., Striegl, G., Puhl, H. & Rotheneder, M. (1989) Continuous monitoring of *in vitro* oxidation of human low density lipoproteins. Free Rad. Res. Comms. 6: 67–75.

Laires, M.J., Monteiro, C.P., Sérgio, J., Madeira, F., Sainhas, J., Palmeira, A., Felisberto, G.M. & Vaz, C. (1994) Content of trace elements in blood before and after three standard efforts. In: Metal Ions in Biology and Medicine (Collery, P., Poirier, L.A., Littlefield, N.A., & Etienne, J.C., eds.), pp. 557–563. John Libbey Eurotext, Paris.

Laires, M.J. & Rayssiguier, Y. (1993) Magnesium, trace elements and exercise. Motricidade Humana 9: 67–75.

Mathews, S.B. & Campbell, A.K. (1984) Neutrophil activation after myocardial infarction. Lancet, I: 756–757.

Motta, C., Gueux, E., Mazur, A., & Rayssiguier, Y. (1996) Lipid fluidity of triacylglycerol-rich lipoproteins isolated from copper-deficient rats. Brit. J. Nutr. (in press).

Uchiyama, M. & Mihara, M. (1978) Determination of malonaldehyde precursor in tissues by thiobarbituric acid test. Anal. Biochem. 86: 271–278.

Winterbourn, C.C., Hawkins, R.E., Brian, M. & Carrell, R.W. (1975) The estimation of redcell superoxide dismutase activity. J. Lab. Clin. Med. 85: 337–341.

Multitracer Study on the Uptake of Inorganic Ions by Tumor

SHUICHI ENOMOTO, BIN LIU,[†] RAJIV G. WEGINWAR,[‡] SHIZUKO AMBE, AND FUMITOSHI AMBE

The Institute of Physical and Chemical Research (RIKEN), Wako, Saitama 351-01, Japan, [†]Department of Technical Physics, Peking University, Beijing 100871, China, and [‡]Chandrapur Engineering College, Chandrapur-442 403, India

Keywords multitracer, bio-distribution, tumor-bearing mouse, metabolism

Introduction

A variety of radioisotopes injected intravenously show a higher concentration in tumor than in normal tissues (Ando et al. 1985). Some of these isotopes have been used for detecting the location of tumors, although the uptake mechanism of the isotopes by tumors is still not clear. Taking advantage of the multitracer technique, we determined the bio-distribution of radioactive elements in tumor bearing mice and compared the uptake of the different elements by the tumor.

In the multitracer technique developed at RIKEN, a number of radioisotopes useful in biological research are produced by irradiation of different targets with high-energy heavy ions. The irradiated target is dissolved in an appropriate medium and the target material is chemically removed leaving the radioisotopes to be used as radioactive tracers in solution. The merit of the multitracer technique lies in its high efficiency and reproducibility among the elements used as tracers.

Experimental

For the production of multitracer, a plate of gold was irradiated with the 135 MeV/nucleon ^{12}C beam from the RIKEN Ring Cyclotron. The Au target containing various kinds of radioisotopes was dissolved in aqua regia. The solution was evaporated to near dryness under vacuum in a rotary evaporator. The residue in the evaporation flask was dissolved in 3 mol/L HCl. Gold ions were completely removed by extraction with ethyl acetate, leaving the radioisotopes as a multitracer in carrier- and salt-free states. In this study, the radioisotopes of the following 16 elements found in the multitracer were used as tracers: Be, Sc, Zn, Rb, Y, Zr, Te, Ba, Hf, Ir, Pt, Ce, Eu, Gd, Yb, and Lu.

Results and Discussion

The experimental mice underwent implantation of sarcoma in the thigh. One week later, a multitracer solution in saline was administered intravenously to the tumor-bearing mice. At 3, 24 and 48 h after injection, the mice were sacrificed and the activity of radioactive elements in the organs was measured using HpGe detectors coupled with an MCA. As a comparison, a group of tumor-bearing mice, which were administered a multitracer solution in 0.5 mol/L sodium citrate, pH 5.0, were sacrificed at 24 h after intravenous injection. The distribution data were given as injected dose %/g tissue of tumor / injected dose %/g tissue of muscle (Tumor/Muscle ratio).

Table 1 shows the Tumor/Muscle ratio of 16 elements at 3, 24, and 48 h after injection of the tumor bearing mice.

Correct citation: Enomoto, S., Liu, B., Weginwar, R.G., Ambe, S., and Ambe, F. 1997. Multitracer study on the uptake of inorganic ions by tumor. In *Trace Elements in Man and Animals – 9: Proceedings of the Ninth International Symposium on Trace Elements in Man and Animals. Edited by* P.W.F. Fischer, M.R. L'Abbé, K.A. Cockell, and R.S. Gibson. NRC Research Press, Ottawa, Canada. pp. 363–364.

Table 1. Tumor/muscle ratio of multitracer elements at 3, 24, and 48 h following injection.

	3 h Saline	24 h Saline	Citrate	48 h Saline
Be	1.48	3.07	3.94	3.21
SC	0.65	3.42	6.98	3.71
Zn	0.24	1.51	4.16	1.73
Rb	0.78	0.61	1.28	0.59
Y	2.60	17.27	30.86	10.89
Zr	0.58	0.24	6.74	0.34
Te	0.69	0.86	4.12	0.95
Ba	1.04	1.98	3.82	2.22
Hf	0.32	0.43	5.67	0.73
Ir	0.85	2.72	3.19	3.00
Pt	0.66	2.29	2.82	2.33
Ce	0.86	2.54	11.00	3.69
Eu	1.06	4.55	14.88	6.61
Gd	2.32	4.40	8.70	7.50
Yb	1.57	6.19	24.09	10.38
Lu	1.67	6.81	32.00	10.44

Generally speaking, alkaline metals such as Rb and divalent cations such as the alkaline earth metals and Zn showed a low Tumor/Muscle ratio, while the rare earth elements revealed a high tumor-uptake. On the other hand, Zr, Hf, Pt and Ir showed low Tumor/Muscle ratios. Thc mechanism of accumulation of inorganic elements in tumor is complicated. The reason for the high uptake of the trivalent cations, was presumably due to hard acids of trivalent ions replacing calcium in the calcium salts of hard bases (calcium salts of acid mucopolysaccharides, etc.) present in tumor tissue. Although Zr and Hf are hard tetravalent cations, these two elements may bind to chloride and therefore gave a lower Tumor/Muscle ratio. Also as shown in Table 1, the citrate solution gave a little higher Tumor/Muscle ratio for most of the elements studied. It is still not clear whether the citrate helps to form certain complexes, such as a ligand in serum, which then results in a higher tumor-uptake.

References

Ando, A., Ando, I., Hirak, T., & Hisada, K. (1985) Int. J. Nucl. Med. Biol. 12: 115.

Copper and Iron Thiobarbiturate Interfere with the Determination of Malondialdehyde

M. ZHOU AND J.R.J. SORENSON

Division of Medicinal Chemistry, College of Pharmacy, University of Arkansas for Medical Sciences Campus, Little Rock, Arkansas 72205, USA

Keywords copper, iron, malondialdehyde, TBA

Introduction

Unsaturated lipid oxidation is suggested to have a major pathological role in many disease states. This oxidation is often modeled in biological systems by addition of hydrogen peroxide (H_2O_2), and Fe(III) or Cu(II) to form an hypothesized aggressive oxidant most frequently suggested to be hydroxyl radical (OH) formed via Fenton homolytic cleavage of hydrogen peroxide following reduction of Fe(III) or Cu(II): Fe(II) or Cu(I) + H_2O_2 $\rightarrow$ Fe(III) or Cu(II) + HO^- + OH. Hydroxyl radical mediated unsaturated lipid oxidation is suggested to yield malondialdehyde (MDA) (1).

Very large and non-physiological concentrations of hydrogen peroxide are required in these experiments since it is known that both Cu(II) and Fe(III) have catalase-mimetic activity and can convert hydrogen peroxide to oxygen and water, and Cu(I) and Cu(II), as well as Fe(II) and Fe(III), are excellent HO scavengers, contrary to the common misunderstanding.

The Thiobarbituric Acid (TBA) assay has been widely used as a test for MDA produced by lipid oxidation in these modeled biological systems. This assay is based on the reaction of TBA with the bifunctional MDA The resulting conjugated aromatic chromophore has a characteristic absorbance maximum at 532 nm and maximum floresence at 553 nm (Janero 1990).

Additions of inorganic forms of Cu or Fe to biological systems are inappropriate since inorganic Cu and Fe salts can form many complexes that may cause system disruption, which may be interpreted as lipid oxidation. These inorganic salts are not the less reactive complexed forms of Cu and Fe normally found in biological systems. Additions of Cu and Fe may also produce an absorbance at 532 nm and an emission at 553 nm following the addition of TBA due to the formation of copper or iron complexes of TBA, which may also be falsely interpreted as lipid oxidation by Cu or Fe based upon the TBA assay. Unfortunately these TBA assays are not controlled for the addition of Cu or Fe as they may effect the TBA assay by increasing the observed absorbance due to formation of Cu or Fe complexes of TBA: Cu(II)-$(\text{thiobarbiturate})_2$ or Fe(III)-$(\text{thiobarbiturate})_3$, and/or the oxidation of TBA to its disulfide by Cu(II).

To examine the possibility that Cu and Fe form these complexes and/or TBA disulfide and interfere with the determination of the MDA-TBA adduct, a product of lipid oxidation produced by HO, both absorbance and florescence spectra were obtained for mixtures of Cu or Fe and TBA.

Materials and Methods

Cupric Chloride, [Cu(II)Cl_2] (Spectrum Chemical Corp, A. C. S. Reagent), Ferric Chloride, [Fe(III)Cl_3] (Mallinckrodt Chemical Works, Analytical Reagent), Ferric Sulfate, [$Fe_2(SO_4)_3$] (Alfa Products, Reagent), Sodium Hydroxide, (NaOH) (Aldrich Chemical Company Inc, A.C.S. Reagent), Thiobarbituric Acid, (TBA) (Sigma Chemical Co, 98%) were used without further purification. Butanol and deionized water were used to prepare solutions. Four mL quartz cuvettes were used for spectrophotometric

Correct citation: Zhou, M., and Sorenson, J.R.J. 1997. Copper and iron thiobarbiturate interfere with the determination of malondialdehyde. In *Trace Elements in Man and Animals–9: Proceedings of the Ninth International Symposium on Trace Elements in Man and Animals. Edited by* P.W.F. Fischer, M.R. L'Abbé, K.A. Cockell, and R.S. Gibson. NRC Research Press, Ottawa, Canada. pp. 365–366.

and florescence measurements. A Hewlett-Packard 8452a Diode Array Spectrophotometer with Vectna E5 Computer and Printer was used to measure absorbance. A Perkin Elmer Luminescence Spectrometer was used to measure florescence. All glassware (beakers, Erlenmeyer flasks, volumetric flasks, graduated cylinders, medium porosity fritted glass filter funnels, and suction flasks) were thoroughly cleaned with Citronox (Alconox Inc.), Acetone or Hydrochloric Acid, (HCl) (10%). Other equipment including: spatulas, test tubes, Pasteur pipettes, hand held pipettes with disposable tips, and glassine paper were metal free.

Results and Discussion

The observed linear increase in absorbance at 532 nm, when increasing concentrations of $Cu(II)Cl_2$ ranging from 0.00 mM to 3.50 mM were added to a 7 mM solution of NaTBA, would be falsely interpreted as increased MDA formation. $Cu(II)Cl_2$ above 3.5 mM resulted in the formation of a greenish-yellow precipitate, a Cu chelate of TBA or TBA disulfide.

An increase in absorbance at 532 nm and intensity of red-wine color, consistent with the false interpretation that MDA had been formed, was also found when increasing concentrations of $Fe(III)Cl_3$ ranging from 0.00 mM to 3.83 mM were added to a 35 mM solution of NaTBA. No precipitate formed when these concentrations of $Fe(III)Cl_3$ were added to 35 mM NaTBA, which was 5 times as concentrated as the NaTBA solution used for additions of $Cu(II)Cl_2$.

Ultraviolet-Visible spectrophotometric measurements were also made for a butanol extract of the mixture of 7.78 mM $Cu(II)Cl_2$ and 15.56 mM NaTBA, and the mixture of 5.11 mM $Fe(III)Cl_3$ and 15.56 mM NaTBA. Following the extraction of solutions of these mixtures it was noticed that absorbances of these extracts changed over a period of 6 h. The absorbance at 532 nm increased for the butanol extract of the $Cu(II)Cl_2$-NaTBA solution through the period of this experiment while absorbance for the $Fe(III)Cl_3$-NaTBA decreased. The observed absorbance would also be falsely interpreted as due to the formation of MDA. A linear increase in florescence at 553 nm was also observed when increasing concentrations of $Cu(II)Cl_2$ ranging from 0.00 mM to 3.50 mM were added to a 7 mM solution of NaTBA, consistent with the false interpretation that MDA had been formed. When increasing concentrations of $Fe(III)_2(SO_4)_3$ ranging from 0.00 mM to 3.83 mM were added to 35 mM solution of NaTBA the small floresence observed for the 35 mM solution of NaTBA, 0.07 ergs s^{-1}, was quenched. This quenching of florescence would provide a clue that the increase in absorbance at 532 nm following the addition of Fe is not due to the formation of MDA.

Conclusion

The addition of Cu or Fe to NaTBA or the addition of NaTBA to Cu or Fe cause an increase in absorbance at 532 nm. These results suggest that Cu and Fe TBA complexes and /or an oxidation product of TBA, TBA disulfide, have an absorbance identical to the MDA-TBA adduct. Addition of Cu to NaTBA also causes an increase in florescence at 553 nm identical to the MDA-TBA adduct. However, florescence at 553 nm decreased following the addition of Fe to NaTBA. Butanol extracts of mixtures of Cu and NaTBA and Fe and NaTBA also absorb at 532 nm with a time dependent increase probably due to complex formation with TBA and/or its oxidation product TBA disulfide. Finally, Cu or Fe complexes of TBA are products formed in systems to which either Cu or Fe and TBA are added in attempting to demonstrate lipid oxidation.

Both Cu and Fe are extremely efficient scavengers of HO with rates of removal of 10^8 to 10^9 mol^{-1} s^{-1} for Cu(II) salts and Cu(II) complexes and 10^9 to 10^{10} mol^{-1} s^{-1} for Fe(III) salts and Fe(III) complexes. Interestingly, the disappearance of floresence at 553 nm following the addition of Fe to NaTBA solutions would reveal the absence of the MDA-TBA adduct when using the TBA assay in the presence of Fe.

References

Janero, D. R. (1990) Malondialdehyde and thiobarbituric acid-reactivity as diagnostic indices of lipid peroxidation and peroxidative tissue injury. Free Rad. Biol. Med. 9: 515–540.

Trace Elements and Genetic Regulation (Part II)

Chair: W.J. Bettger

Metallothioneins 1–4

JOHN H. BEATTIE AND IAN BREMNER
Division of Biochemical Sciences, Rowett Research Institute, Aberdeen AB21 9SB, UK

Keywords metallothionein, growth inhibitory factor, transgenic mice, review

Introduction

Metallothioneins (MT) are low M_r cysteine-rich proteins which bind 7 divalent metals such as Zn in a 2-domain structure. Mammalian MT was first characterised by Kägi and Vallee (1961) and hepatic and renal MT were subsequently found to contain two 61 amino acid isoforms, MT-1 and MT-2. The genes coding for these isoforms are highly inducible by metals and also by factors such as cyclic AMP, hormones such as glucocorticoids and some cytokines. They are expressed in most tissues, but particularly in liver, kidney and pancreas. A third isoform, MT-3, has recently been isolated from brain (Uchida et al. 1991). It contains 20 cysteine residues in positions characteristic of MT-1 & 2 but has a single insertion at residue 5, and 6 insertions at residue 53. This isoform is also known as growth inhibitory factor (GIF), due to its capacity to inhibit the formation of neurofibrillary tangles characteristic of Alzheimer's disease. MT-3 was reported to be deficient in Alzheimer disease brains but this has not been confirmed. The gene sequence for MT-4 was published by Quaife et al. (1994) and the mRNA was found to be localised in squamous epithelial tissues, particularly in the tongue and upper stomach. Although MT-4 protein has yet to be isolated, translation of the gene sequence would produce a 62 amino acid protein with 20 highly conserved cysteine residues. It has a single amino acid insertion at residue 5 and should bind metals in an identical way to the other 3 isoforms.

Structure and Expression

Of the mamalian MTs, the murine family is one of the least complex and best researched. All 4 genes have 3 exons linked within a 50 kb stretch of DNA on chromosome 8. MT-1 & 2 genes are co-ordinately expressed by metal activation of the 70 kDa transcription factor MTF-1 (Radtke et al. 1993) and its subsequent binding to several metal response elements (MREs) upstream of the genes. In addition to MREs, there are response elements for other inducers such as steroids and cytokines e.g., Il-1 and Il-6; there is also a candidate antioxidant response element. Although there are concensus MREs upstream of the MT-3 gene, transcription is not greatly promoted by metals or by many of the classical MT-1 & 2 inducers (Palmiter et al. 1992). However, cerebral MT-3 is upregulated in response to brain injury (Hozumi et al. 1995) and so transcription of this gene may be promoted by quite distinct mechanisms. Unlike MT-1, MT-4 is not highly responsive to oral Zn administration (Quaife et al. 1994). Unusually, transcription of the MT-1 gene in brown adipose tissue is induced by cold stress but not by subcutaneous injection of Zn (Beattie et al. 1996a). This may reflect poor metal uptake by the tissue rather than some tissue-specific regulatory mechanism at the transcriptional level.

Correct citation: Beattie, J.H., and Bremner, I. 1997. Metallothioneins 1–4. In *Trace Elements in Man and Animals – 9: Proceedings of the Ninth International Symposium on Trace Elements in Man and Animals. Edited by* P.W.F. Fischer, M.R. L'Abbé, K.A. Cockell, and R.S. Gibson. NRC Research Press, Ottawa, Canada. pp. 367–371.

The human MT gene family is much more complex than those of rodents, with multiple forms of MT-1 giving rise to a total of at least 12 functional genes. New gene candidates are being discovered at a steady rate, but the discovery of genes and even detection of mRNAs is unlikely to be of functional relevance unless the corresponding protein products are synthesised. Post-translational modification of MT may give rise to distinct isoforms which separate by chromatography or electrophoresis and may be biologically significant (Beattie et al. 1996b). In view of the wealth of genetic data demonstrating the heterogeneity of MT gene families, surprisingly little effort has been devoted to the separation and identification of corresponding MT protein isoforms. However, rabbit liver MT has been found to contain 7 distinct isoforms, 6 of which were of the MT-2 type, as determined by the presence or absence of an acidic amino acid at residue 10 or 11 (Hunziker et al. 1995). Nevertheless, 2 MT-2 isoforms have a similar net charge to MT-1a and, for example, elute in the first fraction from a traditional low pressure anion-exchange separation. Hence any attempt to correlate gene transcription with protein synthesis requires careful selection of protein separation methodology. Due to very high resolution, fast sample turnover and the capacity to exploit differences in both protein charge and polarity, capillary electrophoresis techniques are particularly useful for resolving multiple MT-1 isoforms (Beattie and Richards 1995) as well as separating the major isoforms MT-1, MT-2 and MT-3 (Richards et al. 1996). Recent major advances in electrospray ionisation techniques and high resolution mass spectrometry of macromolecules have been exploited to confirm the existence of MT isoforms predicted from their published gene sequences (Beattie et al. 1996b). Indeed, putative sheep, chicken and rat MT isoforms, for which there are as yet no corresponding gene sequences, have been revealed using this technique (J.H. Beattie, unpublished observations).

The expression of MT genes is often pronounced in specific cell types, with MT-1 and MT-2 being most highly induced in parenchymatous tissue, certain epithelia and glial cells. MT-3 is expressed predominantly in neurons, particularly those with Zn-rich synaptic vesicles (Masters et al. 1994a), while MT-4 is expressed in the stratum spinosum of stratified squamous epithelium and is spatially distinct from MT-1, which is found in the underlying proliferating basal epithelial cells (Quaife et al. 1994).

Although MT has traditionally been regarded as a cytoplasmic protein, immunohistochemical studies have shown that it can also concentrate in the nucleus, particularly in foetal hepatocytes and proliferating cells. For example, 6–15 h after partial hepatectomy in rats, MT had completely relocalised from the cytoplasm into the nucleus (Tohyama et al. 1993). Similar relocalisation occurred with EGF-treated primary rat hepatocyte cultures and was coincident with early S-phase and rapid DNA synthesis (Tsujikawa et al. 1991). These observations imply that a specific targeting mechanism is involved; this could take the form of a localisation signal on the protein and/or relate to the localisation of MT mRNA translation. Since proteins with specific cellular localisation are thought to be translated close to their site of function, it is significant that MT, like the nuclear protein c-myc, is translated on cytoskeletal-bound polysomes in a peri-nuclear localisation (Mahon et al. 1995). This localisation of translation on the cytoskeleton may facilitate transfer of MT into the nucleus or into the cytoplasm, as required.

Function

The biological roles of MT continues to be the subject of considerable debate and conjecture. Suggested functions include the sequestering of heavy metals and reduction of their toxicity, maintaining homeostatic control of essential metals like Zn and Cu, scavenging free radicals through the thiol groups of the cysteine residues, and acting as an acute phase protein in response to stress. Until recently, it has been difficult to prove or disprove these ideas, mainly because MT-1 & 2 are expressed in many tissues in response to a wide variety of physical, chemical and emotional stresses. However, genetically modified cells and animals which over- or under-express MT are proving useful models for functional studies. When MT-1 & 2 genes were inactivated in mice by homologous recombination (Michalska and Choo 1993; Masters et al. 1994b), the resulting "MT-null" animals were phenotypically normal with no apparent impairment of reproduction, growth and development. They were, however, more sensitive to stressors and in particular, to injected heavy metals such as Cd (Liu et al. 1996a). Similarly, primary hepatocytes from these mice are more sensitive to Cd, Cu and Zn than are cells from normal mice (J.H. Beattie, unpublished observations). Conversely, mice which overexpress MT-1 are much more tolerant of Cd than normal controls (Liu et al. 1995) and are also more resistant to Zn depletion, thus reducing the incidence of teratogenic abnormalities and foetal resorption normally associated with Zn deficiency (Dalton et al. 1996). MT-null mice also show increased sensitivity to injection of pro-oxidant compounds such as paraquat (Sato et al. 1996), and primary MT-null mouse cells in culture show reduced viability when exposed to *tert*-butylhydroperoxide, paraquat (Lazo et al. 1995) and menadione (J.H. Beattie, unpublished observations). Induction of MT-1 & 2 protects against the harmful side-effects of anticancer drugs and cultured embryonic fibroblasts from MT-null mice are more sensitive to a range of these drugs including cisplatin and bleomycin (Kondo et al. 1995). Chinese Hamster Ovary cells overexpressing MT show reduced DNA damage when subjected to oxidant stress by treatment with Cu and menadione (S. Wallace, unpublished observations).

Until recently, MT-3 and MT-4 were thought to be expressed exclusively in brain and squamous epithelial tissue, respectively, and this specificity has been the basis for speculation about their function. However, all 4 mouse MT genes are expressed in maternal deciduum (Liang et al. 1996) suggesting that MT-3 and MT-4 have a generic function which is restricted to certain tissues rather than a tissue-specific role. Nevertheless, unlike MT-1 & 2, MT-3 inhibits survival of neurons cultured in medium containing Alzheimer's disease brain extract, the bio-active region of the protein being a cys-pro-cys-pro sequence at residues 6–9 in the N-terminal β-domain (Sewell et al. 1995). Substitution of the proline residues with threonine abolishes the inhibitory effect. MT-3 is more resistant to degradation than MT-1, and overexpression of MT-3 in CHO cells enhances the effects of Zn deficiency, probably due to sequestration of this metal (Palmiter 1995). MT-3 could therefore have some role in regulating cell Zn levels and availability. In the absence of evidence for MT-4 protein expression in tissues, speculation concerning its function is premature.

Conclusions

It is now clear that expression of MT-1 & 2 is not essential for growth, development or reproduction and that Zn homeostasis, for example, is remarkably unaffected by inactivation of these genes (Coyle et al. 1995). However, compensation for the lack of so-called essential protein and enzyme expression in transgenic animals underlines the potency of homeostatic mechanisms rather than the redundancy of these genes. These compensatory mechanisms may be less effective under stress, as indicated by the increased susceptibility of MT-deficient animals to Cd toxicity. However, whether MTs evolved as a heavy metal detoxification system is very doubtful, particularly as they increase the biological half-life of Cd (Liu et al. 1996b). More likely, MTs bind and detoxify Cd as a fortuitous consequence of the chemical similarity between Cd and Zn. A role of MT as an antioxidant is becoming increasingly plausible although no studies on the consequences of long-term exposure of MT-deficient animals to physiological levels of oxidant stress have been reported. The idea that MT-1 & 2 could be acute phase proteins is very attractive in view of the ubiquitous nature of their expression in response to a very wide range of stress factors. A protective mechanism in this role has not been proposed but may involve regulation of the intracullular supply of Zn, which has known antioxidant properties (Bray and Bettger 1990), or indeed a direct biological action of the protein as has been found for MT-3. Further understanding of MT function will come from the generation of additional transgenic animals in which the expression or localisation of specific isoforms has been modified.

References

Beattie J.H. & Richards M.P. (1995) The analysis of metallothionein isoforms by capillary electrophoresis: optimisation of protein separation conditions using micellar electrokinetic capillary chromatography. J. Chromatog. A 700: 95–103.

Beattie, J.H., Black, D.J., Wood, A.M. & Trayhurn, P. (1996a) Cold-induced expression of the metallothionein-1 gene in brown adipose tissue of rats. Am. J. Physiol. 270: R971–R977.

Beattie J.H., Lomax J., Richards M.P., Self R., Pesch R. & Münster H. (1996b) Metallothionein isoform analysis by liquid chromatography-electrospray ionisation mass spectrometry. Biochem. Soc. Trans. 24: 220S.

Bray, T.M. & Bettger, W.J. (1990) The physiological role of zinc as an antioxidant. Free Radical Biol. Med. 8: 281–291.

Coyle, P., Philcox, J.C. & Rofe, A.M. (1995) Hepatic zinc in metallothionein-null mice following zinc challenge: in vivo and in vitro studies. Biochem. J. 309: 25–31.

Dalton, T., Fu, K., Palmiter, R.D. & Andrews, G.K. (1996) Transgenic mice that overexpress metallothionein-1 resist dietary zinc deficiency. J. Nutr. 126: 825–833.

Hozumi, I., Inuzuka, T., Hiraiwa, M., Uchida, Y., Anezaki, T., Ishiguro, H., Kobayashi, H., Uda, Y., Miyatake, T. & Tsuji, S. (1995) Changes of growth inhibitory factor after stab wounds in rat brain. Brain Res 688: 143–148.

Hunziker, P.E., Kaur, P., Wan, M. & Kanzig, A. (1995) Primary structures of seven metallothioneins from rabbit tissue. Biochem. J. 306:265–270.

Kägi, J.H.R. & Vallee, B.L. (1961) Metallothionein: a cadmium and zinc-containing protein from equine renal cortex. II. Physicochemical properties. J. Biol. Chem. 236: 2435–2442.

Kondo, Y., Woo, E.S., Michalska, A.E., Choo K.H.A. & Lazo, J.S. (1995) Metallothionein null cells have increased sensitivity to anticancer drugs. Cancer Res. 55: 2021–2023.

Lazo, J.S., Kondo, Y., Dellapiazza, D., Michalska, A.E., Choo, K.H.A. & Pitt, B.R. (1995) Enhanced sensitivity to oxidative stress in cultured embryonic cells from transgenic mice deficient in metallothionein I and II genes. J. Biol. Chem. 270: 5506–5510.

Liang, L.C., Fu, K., Lee, D.K., Sobieski, R.J., Dalton, T. & Andrews, G.K. (1996) Activation of the complete mouse metallothionein gene locus in the maternal deciduum. Mol. Reprod. Dev. 43: 25–37.

Liu, J., Liu, Y.P., Michalska, A.E., Choo, K.H.A. & Klaassen, C.D. (1996a) Metallothionein plays less of a protective role in cadmium-metallothionein-induced nephrotoxicity than in cadmium chloride-induced hepatotoxicity. J. Pharmacol. Exp. Ther. 276: 1216–1223.

Liu, J., Liu, Y.P., Michalska, A.E., Choo, K.H.A. & Klaassen, C.D. (1996b) Distribution and retention of cadmium in metallothionein I and II null mice. Toxicol. Appl. Pharmacol. 136: 260–268.

Liu, Y.P., Liu, J., Iszard, M.B., Andrews, G.K., Palmiter, R.D. & Klaassen, C.D. (1995) Transgenic mice that overexpress metallothionein-I are protected from cadmium lethality and hepatotoxicity. Toxicol. Appl. Pharmacol. 135: 222–228.

Mahon, P., Beattie, J., Glover, L.A. & Hesketh, J. (1995) Localisation of metallothionein isoform mRNAs in rat hepatoma (H4) cells. FEBS Lett. 373: 76–80.

Masters, B.A., Quaife, C.J., Erickson, J.C., Kelly, E.J., Froelick, G.J., Zambrowicz, B.P., Brinster, R.L. & Palmiter, R.D. (1994a) Metallothionein III is expressed in neurons that sequester zinc in synaptic vesicles. J. Neurosci. 14: 5844–5857.

Masters, B.A., Kelly, E.J., Quaife, C.J., Brinster, R.L. & Palmiter, R.D. (1994b) Targeted disruption of metallothionein I and II genes increases sensitivity to cadmium. Proc. Natl. Acad. Sci. USA. 91: 584–588.

Michalska, A.E. & Choo, K.H.A. (1993) Targeting and germ-line transmission of a null mutation at the metallothionein I and II loci in mouse. Proc. Natl. Acad. Sci. USA 90: 8088–8092.

Palmiter, R.D., Findley, S.D., Whitmore, T.E. & Durnam, D.M. (1992) MT-III, a Brain-Specific Member of the Metallothionein Gene Family. Proc. Natl. Acad. Sci. USA 89: 6333–6337.

Palmiter, R.D. (1995) Constitutive expression of metallothionein-III (MT-III), but not MT-I, inhibits growth when cells become zinc deficient. Toxicol. Appl. Pharmacol. 135: 139–146.

Quaife, C.J., Findley, S.D., Erickson, J.C., Froelick, G.J., Kelly, E.J., Zambrowicz, B.P. & Palmiter, R.D. (1994) Induction of a new metallothionein isoform (MT-IV) occurs during differentiation of stratified squamous epithelia. Biochemistry 33: 7250–7259.

Radtke, F., Heuchel, R., Georgiev, O., Hergersberg, M., Gariglio, M., Dembic, Z. & Schaffner, W. (1993) Cloned transcription factor MTF-1 activates the mouse metallothionein-I promoter. EMBO J. 12: 1355–1362.

Richards, M.P., Andrews, G.K., Winge, D.R. & Beattie, J.H. (1996) Separation of three mouse metallothionein isoforms by free-solution capillary electrophoresis. J. Chromatogr. B-Bio. Med. Appl. 675: 327–331.

Sato, M., Apostolova M.D., Hamaya M., Yamaki J., Michalska A.E., Choo K.H.A., Kodama N. & Tohyama C. (1996) Susceptibility of metallothionein-null mice to paraquat. Environ. Toxicol. Pharmacol. (In Press).

Sewell, A.K., Jensen, L.T., Erickson, J.C., Palmiter, R.D. & Winge D.R. (1995) Bioactivity of metallothionein-3 correlates with its novel β domain sequence rather than metal binding properties. Biochemistry 34: 4740–4747.

Tohyama, C., Suzuki, J.S., Hemelraad, J., Nishimura, N. & Nishimura, H. (1993) Induction of metallothionein and its localization in the nucleus of rat hepatocytes after partial hepatectomy. Hepatology 18: 1193–1201.

Tsujikawa, K., Imai, T., Kakutani, M., Kayamori, Y., Mimura, T., Otaki, N., Kimura, M., Fukuyama, R. & Shimizu, N. (1991) Localization of metallothionein in nuclei of growing primary cultured adult rat hepatocytes. FEBS Lett. 283: 239–242.

Uchida, Y., Takio, K., Titani, K., Ihara, Y. & Tomonaga, M. (1991) The growth inhibitory factor that is deficient in the Alzheimer's disease brain is a 68 amino acid metallothionein-like protein. Neuron 7: 337–347.

Discussion

Q1. James Kang, University of North Dakota, Grand Forks, ND, USA: Is metallothionein found in the mitochondria and what is your hypothesis for how it gets into mitochondria?

A. Metallothionein is found in the mitochondria but we don't know the mechanism of how it gets there.

Q2. Harry McArdle, Rowett Research Institute, Aberdeen, Scotland: During pregnancy an alternative function of metallothionein may be as an reservoir for cysteine. Is there a suitable supply of cysteine in the knock-out mice?

A. Yes, a credible suggestion. There are no studies in knock-out mice. I wonder if a suitable supply of cysteine is available, given the path of degradation in lysosomes, and this casts some doubt on the suggestion.

Q3. James Kirkland, University of Guelph, ON, Canada: Could you shift the location of protein synthesis by altering cell cycle or causing DNA damage?

A. We haven't done that. We have changed localization of translation of message when the 3′ UTR is switched from other genes.

Q4. Diane Cox, University of Alberta, Edmonton, AB, Canada: If there is some significance to the store of metallothionein in the fetus, then in the knock-out mouse, is there any problem with the fetus not having that store of liver metallothionein? Does it do anything to the very early copper transport, for example?

A. The knock-out mice appear normal and if there are perturbations in Cu transport, they must be subtle.

Q5. Margaret Elmes, University Hospital Wales, Cardiff, Wales: We've found very marked differences in fetal human and rat liver metallothionein, therefore I urge you to consider species differences when addressing fetal liver metallothionein. Is the donor function of metallothionein out of favour now? Our theory as to why we get the induction of metallothionein synthesis in tumours is that it's acting as a zinc donor.

A. Yes, metallothionein probably acts as a Zn donor.

Q6. John Sorenson, University of Arkansas, Little Rock, AR, USA: Do glutathione and Cu-metallothionein deliver Cu for activation of apo-superoxide dismutase?

A. This has been proposed. Somehow, metallothionein and glutathione dock together for transport of copper. This could also happen in the opposite direction.

Metals as Regulators of the Biological and Physiological Roles of Heme Oxygenase

N.G. ABRAHAM AND A. KAPPAS

The Rockefeller University, New York, NY, USA

Keywords heme oxygenase, tin, molecular regulation, corneal epithelium

Heme oxygenase (HO) controls the initial and rate-limiting step in heme catabolism. The enzyme cleaves heme to biliverdin which is subsequently converted to bilirubin by biliverdin reductase. Iron is released when the heme ring is opened while carbon monoxide is liberated. The heme molecule plays a central role as the prosthetic moiety of hemeproteins involved in cell respiration, energy generation, oxidative biotransformation, growth differentiation processes and the generation of inflammatory mediators such as eicosanoids and nitric oxide. HO-1 activity is increased in whole animal tissues and in cultured cells following treatment with heme, metals, inflammatory cytokines as well as in hypoxic and oxidative conditions (Maines and Kappas 1977, Abraham et al. 1988, Applegate et al. 1991, Rizzardini et al. 1994). HO-1 is also induced by heat shock and the enzyme belongs to a class of macromolecules known as stress proteins, which are responsive to various types of acute cellular injuries (Shibahara et al. 1987, Mitani et al. 1989).

Induction of HO-1 is considered to be of considerable importance in the initiation of cellular protective mechanisms following exposure to various forms of cell stressful stimuli (Applegate et al. 1991, Stocker et al. 1990, Vogt et al. 1995). This idea derives in part from the fact that increased HO activity enables the removal of heme, a lipid soluble, transmissible form of the potent prooxidant iron, and results in the generation of bilirubin and biliverdin, heme metabolites with significant antioxidant and anticomplement properties (Stocker et al. 1987, Nakagami et al. 1993, Llesuy and Tomaro 1994). Indeed, a study by Nath et al. (1992) provides strong evidence that induction of HO-1 coupled to ferritin synthesis is a rapid, protective antioxidant response *in vivo* in rhabdomyolysis-induced kidney injury in the rat. Other studies have also shown that induction of HO-1 in skin fibroblasts is of value in protection against ultraviolet (UV) light-induced oxidative stress (Keyse and Tyrrell 1989, Vile and Tyrrell 1993, Vile et al. 1994) and recently we (Abraham et al. 1995a, b) have demonstrated that HO-1 gene overexpression via adenovirus-mediated HO-1 cDNA transfer protected endothelial cells from oxidative injury produced by exposure to free heme/hemoglobin. The present study was undertaken to examine the regulatory effect of metals on HO in corneal epithelial cells and to assess the relation of this enzyme activity to the development of experimentally-induced inflammation in the anterior surface of the rabbit eye.

Results and Discussion

Detection of HO-1 mRNA in a Corneal Epithelial Cell Line: Due to the limited availability of primary cell cultures from the rabbit corneal epithelia, we used the RCE cell line (Araki et al. 1993) as a model to characterize HO-1 gene expression. Upon reaching confluence, RCE cells were treated with agents known to cause induction of HO mRNA in various cell lines such as $SnCl_2$, $CoCl_2$, $ZnCl_2$ and CoPP at concentrations of 10–150 μM. Control experiments consisted of RCE cells treated with the appropriate vehicles. The effects of these agents on HO-1 mRNA levels were assessed and the results are depicted in Figure 1. A basal level of HO-1 mRNA was evident albeit to a much lesser extent than that of the rabbit liver. Treatment with all agents used resulted in an accumulation of HO-1 mRNA. A quantitative evaluation of the mRNA changes by scanning densitometry relative to G3PDH mRNA levels

Correct citation: Abraham, N.G., and Kappas, A. 1997. Metals as regulators of the biological and physiological roles of heme oxygenase. In *Trace Elements in Man and Animals – 9: Proceedings of the Ninth International Symposium on Trace Elements in Man and Animals. Edited by* P.W.F. Fischer, M.R. L'Abbé, K.A. Cockell, and R.S. Gibson. NRC Research Press, Ottawa, Canada. pp. 372–376.

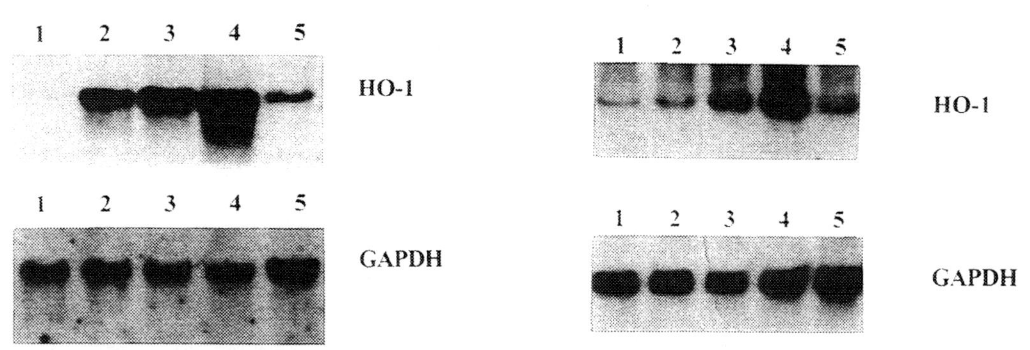

Figure 1. Effects of metals and metalloporphyrins on HO-1 mRNA expression. RCE cells were cultured and treated for 1 h with the following: lane 1, control vehicle-treated cells; lane 2, $SnCl_2$ (10 μM); lane 3, $CoCl_2$ (150 μM); lane 4, CoPP (10 μM); lane 5, $ZnCl_2$ (100 μM). Total RNA was extracted and hybridized with ^{32}P labelled HO-1 cDNA.

Figure 2. Time-dependent effect of $SnCl_2$ on the induction of HO-1 mRNA in RCE cells. Cells were treated with $SnCl_2$ (10 μM) for various lengths of time. Lane 1, control, untreated cells; lane 2, vehicle-treated cells; lane 3, $SnCl_2$, 8 h; lane 4, $SnCl_2$, 16 h; lane 5, $SnCl_2$, 24 h. Total RNA was extracted and Northern blot analysis was performed with the rat HO-1 cDNA.

indicated a 50-fold increase in HO-1 mRNA levels in RCE cells treated with $SnCl_2$ (10 μM). CoPP (10 μM) showed similar potency increasing HO-1 mRNA levels by 30-fold. On the other hand 150 μM $CoCl_2$ was needed to achieve the same induction; $ZnCl_2$ at 100 μM produce a 10-fold increase of HO-1 mRNA over control. $CoCl_2$ and $ZnCl_2$ at 10 μM did not produce an increase in HO-1 mRNA (data not shown). These results indicate that the RCE cell line responds to HO-1 inducers as do many other tissues. Since the potency of $SnCl_2$ in inducing RCE HO-1 mRNA surpassed that of the other agents studied we used $SnCl_2$ in the subsequent experiments.

Time- and Concentration-dependent Induction of HO-1 mRNA by $SnCl_2$: RCE cells were grown and maintained in 75 cm^2 flasks and upon reaching confluence were treated with $SnCl_2$ and RNA extracted. The time-dependent response to $SnCl_2$ is shown in Figure 2. The vehicle used for solubilization of $SnCl_2$ did not change HO-1 mRNA levels (lane 2) as compared to control (lane 1). Optimal induction of HO-1 in RCE was at 16 h ($SnCl_2$ 100 μM) (lane 4), with a subsequent return to control levels at 24 h (lane 5). Additional experiments were conducted to demonstrate the long term effect of $SnCl_2$ in cells grown in a medium containing $SnCl_2$ (100 μM) for up to 6 d. Results showed that the continuous presence of $SnCl_2$ in cells maintained high levels of HO-1 mRNA while no toxicity (loss of cell viability) was detected (data not shown). The dose-response to $SnCl_2$, at 16 h, is shown in Figure 3. Cells were incubated with $SnCl_2$ at a concentration of 0.1–100 μM for 16 h, after which the RNA was extracted and analyzed. Compared to untreated cells, HO-1 mRNA levels were not increased over controls (lane 1 and lane 2) after addition of $SnCl_2$ at 0.1 μM (lane 3). Induction of HO-1 mRNA by $SnCl_2$ (5–100 μM) was dose-related (lanes 4–6). Hybridization of the filters with radiolabeled G3PDH and ethidium bromide staining of RNA confirmed that similar amounts of total RNA were transferred to the filters in each lane of the paired experiments (data not shown).

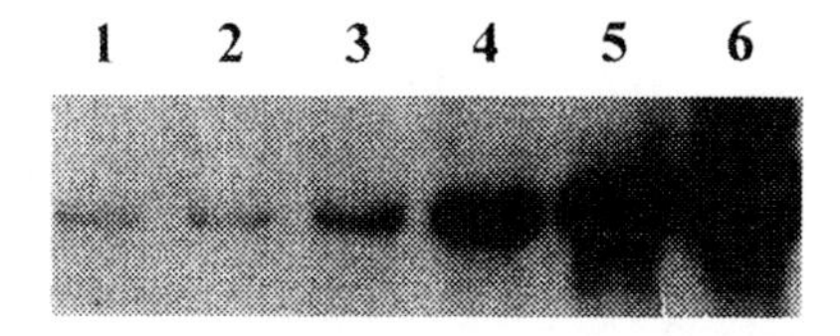

Figure 3. Dose-dependent effect of $SnCl_2$ on the induction of HO-1 mRNA in RCE cells. Cells were treated with various doses of $SnCl_2$ for 16 h and compared to untreated cells, lane 1; cells treated with the vehicle, lane 2; $SnCl_2$ (0.1 μM), lane 3; $SnCl_2$ (5 μM), lane 4; $SnCl_2$ (10 μM), lane 5; $SnCl_2$ (100 μM), lane 6. Total RNA was extracted and Northern blot analysis was performed with the rat HO-1 cDNA.

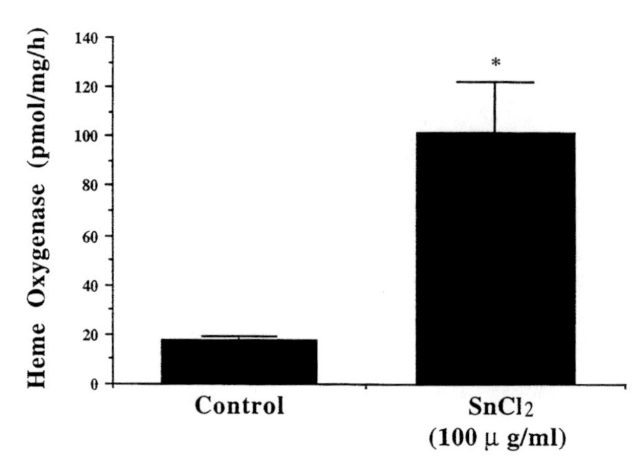

*Figure 4. Effect of $SnCl_2$ on HO activity in RCE cells. Confluent cultures were treated with $SnCl_2$ (100 μg/mL; 500 μM) or the vehicle control (0.1 M phosphate buffer, pH = 7.4) for 24 h. Microsomes were prepared and HO activity was measured. Data are expressed as specific activity in pmol/mg/h of two separate experiments measured in triplicate. The variation between triplicate of each treatment were within 5% (mean ± SEM, n = 2). * = $p < 0.05$ versus the vehicle treatment.*

Effect of mRNA-inducing Agents on HO-1 Activity: Translation of HO-1 mRNA into active HO enzyme was verified in RCE cells. RCE cells grown in 175 cm^2 flasks were treated with either vehicle control or $SnCl_2$ for 24 h. After combining 6 flasks, microsomes were prepared and HO activity was assessed. As seen in Figure 4, HO activity was detectable in untreated (control) cells (16 ± 3 pmol/mg/h, mean ± SEM, n = 3); with $SnCl_2$ treatment, HO activity significantly increased (6-fold) over the controls (103 ± 21 pmol/mg/h), mean ± SEM, n = 3). These results indicate that transcriptional activation of the HO-1 gene by $SnCl_2$ is followed by translation into functional HO enzyme in cultures.

Effect of $SnCl_2$ on Corneal Epithelial HO-1 mRNA in vivo: To determine whether induction of HO-1 would be associated with moderation or suppression of the inflammatory response to contact lens wear, total RNA was extracted from control and closed eye-hydrogel contact lens-treated corneal epithelium ± $SnCl_2$ (500 μM) after 6 d of wear. Northern blot analysis of HO-1 mRNA showed detectable levels of HO-1 mRNA for the untreated control eyes (data not shown). In corneal epithelium from vehicle-treated lenses, HO-1 mRNA was slightly elevated over the control. Corneal epithelium from $SnCl_2$-treated contact lenses in contrast demonstrated a marked increase in HO mRNA levels (data not shown). This *in vivo* induction of HO-1 mRNA at 6 d was associated with a substantial increase in HO enzyme activity and was further correlated with the severity of the *in situ* inflammatory response. Figure 5 depicts representative slit lamp photos of the ocular surfaces (n = 3–5) after 6 d of closed eye-hydrogel contact lens wear. In the vehicle treated eyes, there was progressive *in situ* inflammatory response becoming prominent at day 3 (Conners et al. 1995a) and more severe at day 6 (Figure 5A). The inflammatory response consisted of limbal vasodilation, conjunctival swelling, dilation of the iridial vessels, decreased corneal transparency (increased thickness, cloudiness and opacity) and neovascularization. One time treatment of the hydrogel lenses with $SnCl_2$ (100 μg/mL) resulted in marked attenuation of the inflammatory response (Figure 5B). Such was effective in suppressing both limbal and iridial vasodilation as well as the extent of epithelial defects and neovascularization of the cornea. Quantitative analysis by subjective inflammatory scoring (blinded to the treatment) demonstrated a significant correlation between the treatment with $SnCl_2$ and the decreased inflammatory response. Figure 6 summarizes the effect of $SnCl_2$ treatment on the *in situ* inflammatory response; corneal thickness, indicative of corneal edema, was decreased by 60% as compared to eyes with untreated contact lenses (Figure 6A), while the subjective inflammatory score was reduced by 75% (Figure 6B).

Study of the effect of increased HO-1 activity on the severity of experimentally-induced inflammation in the cornea was an important aspect of these experiments. As noted above, we have shown in earlier

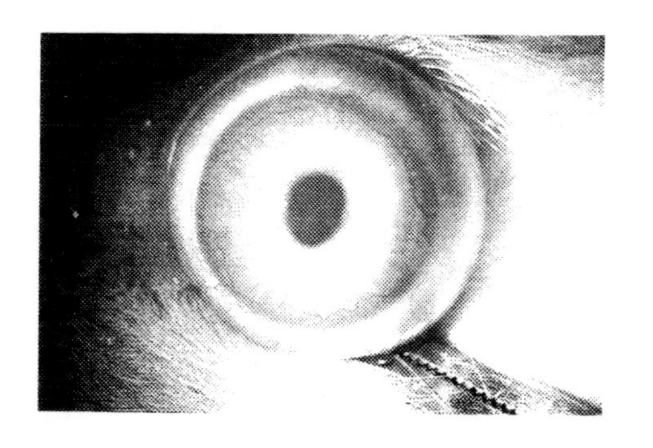
A

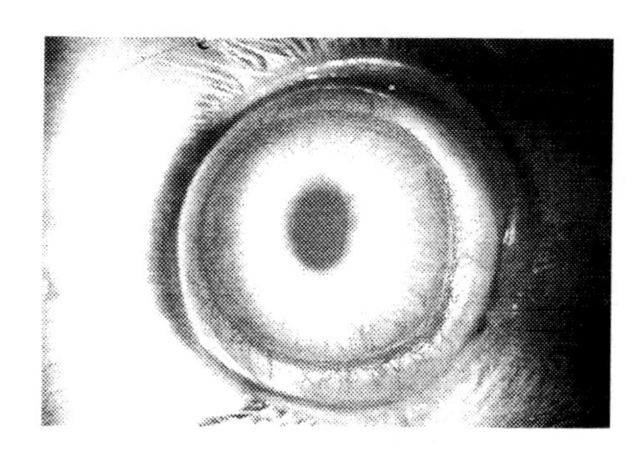
B

Figure 5. *Effect of one-time treatment with $SnCl_2$ on the in situ ocular inflammatory response at 6 d of closed eye-hydrogel contact lens wear. Lenses were treated with either vehicle (0.1 M phosphate buffer, pH = 7.4) or $SnCl_2$ (100 μg/mL) for 30 min prior to lens placement and tarsorrhaphy. At 6 d, sutures were removed and eyes examined. Biomicroscopic photos of ocular surface were taken with photographic attachment to the slit lamp. Photographs are representative of n = 8 from each group. A, untreated eye (Day 6, vehicle); and B, eye treated with $SnCl_2$-treated lenses (Day 6, $SnCl_2$).*

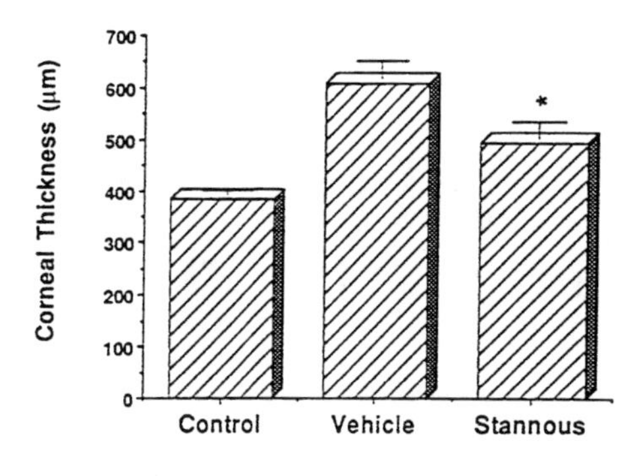

A

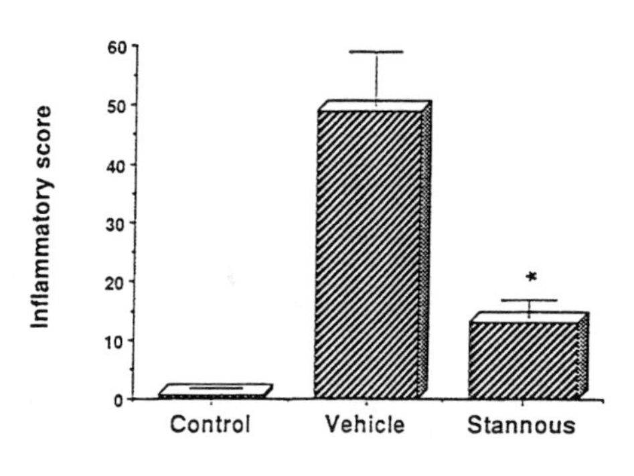

B

Figure 6. *Effect of one-time treatment with $SnCl_2$ on corneal thickness and inflammatory score at 6 d of closed eye-hydrogel contact lens wear. Lenses were treated with either vehicle (0.1 M phosphate buffer, pH = 7.4) or $SnCl_2$ (100 μg/mL) for 30 min prior to lens placement and tarsorrhaphy. At 6 d, sutures were removed and eyes examined. Corneal thickness was assessed by ultrasonic pachymetry. Biomicroscopic photos of ocular surface were taken with photographic attachment to the slit lamp. Photos were subjectively scored. Data are presented as the mean ± SEM (n = 3–5). * = $p < 0.05$.*

work that transfection of the HO-1 gene and its selective overexpression into rabbit coronary endothelial cells substantially moderated cell damage resulting from exposure to free heme/hemoglobin (Abraham et al. 1995a, b). Nath et al. (1992) also showed that HO-1 induction coupled to ferritin synthesis provided significant protection against rhabdomyolysis-induced injury to the kidney; and Willis et al. (1996) in a recent report also implicated the direct involvement of HO-1 in the resolution of complement-dependent acute inflammation.

The results of this study demonstrate that $SnCl_2$ induction of HO-1 expression in the corneal epithelium is associated with a substantial attenuation of the corneal inflammatory response elicited by extended contact lens wear; this ameliorative effect is graphically depicted in Figure 6. Its proximate mechanisms likely involve a number of factors including enhanced catabolism of the prooxidative heme to the antioxidant metabolites biliverdin and bilirubin; diminution in the cellular content and activity of hemoprotein species, such as cytochrome P450 isozymes involved in the production of pro-inflammatory mediators as shown earlier (Neil et al. 1995, Sacerdoti et al. 1989, da Silva et al. 1994, Conners et al. 1995b); and possibly the local generation of carbon monoxide which could inactivate other heme-containing proteins which contribute to the inflammatory response.

References

Abraham, N.G., Lavrovsky, Y., Schwartzman, M.L., Stoltz, R.A., Gerritsen, M.E., Shibahara, S. & Kappas, A. (1995a) Transfection of human heme oxygenase gene into rabbit coronary microvessel endothelial cells: Protective effect against heme and hemoglobin toxicity. Proc. Natl. Acad. Sci. USA 92: 6798–6802.

Abraham, N.G., da Silva, J.-L., Lavrovsky, Y., Stoltz, R.A., Kappas, A., Dunn, M.W. & Laniado-Schwartzman, M. (1995b) Adenovirus mediated heme oxygenase-1 gene transfer into rabbit ocular tissues. Invest. Ophthalmol. Vis. Sci. 36: 2202–2210.

Abraham, N.G., Lin, J.H.-C., Schwartzman, M.L., Levere, R.D. & Shibahara, S. (1988) The physiological significance of heme oxygenase. Int. J. Biochem. 20: 543–558.

Applegate, L.A., Luscher, P. & Tyrrell, R.M. (1991) Induction of heme oxygenase: A general response to oxidant stress in cultured mammalian cells. Cancer. Res. 51: 974–978.

Araki, K., Ohashi, Y., Kinoshita, S., Hayashi, K., Yang, X.Z., Hosaka, Y. & Handa, H. (1993) Immortalization of rabbit corneal epithelial cells by a recombinant SV40-adenovirus vector. Invest. Ophthalmol. Vis. Sci. 34: 2665–2671.

Conners, M.S., Stoltz, R.A., Davis, K.L., Dunn, M.W., Abraham, N.G., Levere, R.D. & Laniado-Schwartzman, M. (1995a) A closed eye contact lens model of corneal inflammation: Inhibition of cytochrome P450 arachidonic acid metabolism alleviates inflammatory sequelae. Invest. Ophthalmol. Vis. Sci. 36: 841–850.

Conners, M.S., Stoltz, R.A., Webb, S.C., Rosenberg, J., Yang, T., Dunn, M.W., Abraham, N.G. & Laniado-Schwartzman, M. (1995b) A closed eye soft contact lens model of corneal inflammation: I. Induction of cytochrome P450 arachidonic acid metabolism. Invest. Ophthalmol. Vis. Sci. 36: 828–840.

da Silva, J.-L., Tiefenthaler, M., Park, E., Escalante, B., Schwartzman, M.L., Levere, R.D. & Abraham, N.G. (1994) Tin-mediated heme oxygenase gene activation and cytochrome P450 arachidonate hydroxylase inhibition in spontaneously hypertensive rats. Am. J. Med. Sci. 307: 173–181.

Keyse, S.M. & Tyrrell, R.M. (1989) Heme oxygenase is the major 32-kDa stress protein induced in human skin fibroblasts by UVA radiation, hydrogen peroxide, and sodium arsenite. Proc. Natl. Acad. Sci. USA 86: 99–103.

Llesuy, S.F. & Tomaro, M.L. (1994) Heme oxygenase and oxidative stress. Evidence of involvement of bilirubin as physiological protector against oxidative damage. Biochim. Biophys. Acta 1223: 9–14.

Maines, M.D. & Kappas, A. (1977) Metals as regulators of heme metabolism. Science 198: 1215–1221.

Mitani, K., Fujita, H., Sassa, S. & Kappas, A. (1989) Heat shock induction of heme oxygenase mRNA in human Hep3B hepatoma cells. Biochem. Biophys. Res. Commun. 165: 437–441.

Nakagami, T., Toyomura, K., Kinoshita, T. & Morisawa, S. (1993) A beneficial role of the bile pigments as an endogenous tissue protectors: Anti-complement effects of biliverdin and conjugated bilirubin. Biochim. Biophys. Acta 1158: 189–193.

Nath, K.A., Balla, G., Vercellotti, G.M., Balla, J., Jacobs, H.S., Levitt, M.D. & Rosenberg, M.E. (1992) Induction of heme oxygenase is a rapid, protective response in rhabdomyolysis in the rat. J. Clin. Invest. 90: 267–270.

Neil, T.K., Stoltz, R.A., Jiang, S., Laniado-Schwartzman, M., Dunn, M.W., Levere, R.D., Kappas, A. & Abraham, N.G. (1995) Modulation of corneal heme oxygenase expression by oxidative stress agents. J. Ocular Pharmacol. 11: 455–468.

Rizzardini, M., Carelli, M., Cabell Porras, M.R. & Cantoni, L. (1994) Mechanisms of endotoxin-induced heme oxygenase mRNA accumulation in mouse liver: synergism by glutathione depletion and protection by N-acetylcysteine. Biochem. J. 304: 477–483.

Sacerdoti, D., Escalante, B., Abraham, N.G., McGiff, J.C., Levere, R.D. & Schwartzman, M.L. (1989) Treatment with tin prevents the development of hypertension in spontaneously hypertensive rats. Science 243: 388–390.

Shibahara, S., Muller, R.M. & Taguchi, H. (1987) Transcriptional control of rat heme oxygenase by heat shock. J. Biol. Chem. 262: 12889–12892.

Stocker, P., Yamamoto, Y., McDonagh, A.F., Glazer, A.N. & Ames, B.N. (1987) Bilirubin is an antioxidant of possible physiological importance. Science 235: 1043–1047.

Stocker, R. (1990) Induction of haem oxygenase as a defence against oxidative stress. Free Rad. Res. Commun. 9(2): 101–112.

Vile, G.F. & Tyrrell, R.M. (1993) Oxidative stress resulting from ultraviolet A irradiation of human skin fibroblasts leads to a heme oxygenase-dependent increase in ferritin. J. Biol. Chem. 268: 14678–14681.

Vile, G.F., Basu-Modak, S., Waltner, C. & Tyrrell, R.M. (1994) Heme oxygenase-1 mediates an adaptive response to oxidative stress in human skin fibroblast. Proc. Natl. Acad. Sci. USA 91: 2607–2610.

Vogt, B.A., Alam, J., Croatt, A.J., Vercellotti, G.M. & Nath, K.A. (1995) Acquired resistance to acute oxidative stress: Possible role of heme oxygenase and ferritin. Lab. Invest. 72: 474–483.

Willis, D., Moore, A.R., Frederick, R. & Willoughby, D.A. (1996) Heme oxygenase: A novel target for the modulation of the inflammatory response. Nature Medicine 2: 87–90.

Expression of Metallothionein in the Liver of Rats Fed Copper-Deficient AIN-93G Diet

Y.J. KANG,[†] Y. CHEN,[†] AND J.T. SAARI[‡]

[†]Department of Pharmacology and Toxicology, University of North Dakota School of Medicine, and [‡]USDA Agricultural Research Service, Human Nutrition Research Center, Grand Forks, ND 58202, USA

Keywords copper deficiency, iron, metallothionein, liver

Copper-loading induces metallothionein (MT) synthesis in the liver of rats. Copper-MT from the liver of copper-injected rats has been isolated and analyzed. The copper-induced MT synthesis is characterized by increased incorporation of [^{35}S]cysteine into the protein and enhanced production of the MT mRNA, suggesting that copper induction of MT occurs at the level of gene transcription. The mechanism by which copper induces MT synthesis, however, remains unsolved. It is also unknown whether copper is essential for MT production under a diversity of physiological and pathological conditions. The purpose of this study was thus to examine the effect of dietary copper deficiency on MT induction in rats.

A copper-deficient diet was formulated by replacing copper with the corresponding weight of corn starch in the AIN-93G diet. Diet analysis for copper yielded values of 5.7 mg Cu/kg diet for the copper-adequate diet and 0.4 mg Cu/kg diet for the copper-deficient diet. Male, weanling Sprague-Dawley rats (46–57 g; Sasco, Lincoln, NE, USA) were divided into two weight-matched groups having average weights of 52 g each; one had free access to the copper-adequate diet and the other to the copper-deficient diet. They also had free access to deionized water.

After 4 weeks on their respective diets and an overnight fast, each rat was anesthetized with an intraperitoneal injection of sodium pentobarbital (65 mg/kg body wt, Vet Labs, Lenexa, KS). Blood was withdrawn from the inferior vena cava for erythrocyte counting and plasma assays. The liver and heart were removed, flushed with cold 0.9% NaCl via their major vessels and divided for subsequent assays. Tissue samples including the liver, heart, and kidney were stored at –20°C for mineral assays and those for MT and mRNA assays were placed in liquid nitrogen, then stored at –80°C.

Hematocrit and hemoglobin content were determined on a cell counter. An automated analyzer was used to determine serum ceruloplasmin. Trace element contents of tissues were determined by inductively coupled argon plasma emission spectroscopy after lyophilization and digestion of the tissues with nitric acid and hydrogen peroxide.

A routine Northern blot assay was used to analyze MT-I mRNA. A probe corresponding to a 1185-base pair Hind III and Bgl II fragment of mouse MT-I cDNA was used to identify the MT transcript on the membrane. Autoradiographic images were scanned and analyzed by an imaging analyzing system. Densitometric values were then determined from digitized images of autoradiograms. MT concentrations were measured by the Cd/hemoglobin radiometric assay.

Characteristics of rats fed the copper-deficient diet were compared to those of rats fed the copper-adequate diet. Copper concentrations were significantly ($p < 0.01$) depressed in the liver, heart, and kidney of copper-deficient rats. Zinc concentrations were significantly ($p < 0.01$) decreased in the heart and kidney, but not significantly ($p > 0.05$) reduced in the liver. Iron concentrations were also significantly ($p < 0.01$) depressed in the heart and kidney, but significantly ($p < 0.01$) elevated in the liver. Other changes including reduced activity of ceruloplasmin in the plasma, depressed Cu,Zn-SOD in the tissues, and decreased hematocrit and hemoglobin concentrations in the blood were found in the rats fed the copper-deficient diet, typically indicative of severe copper deficiency.

Correct citation: Kang, Y.J., Chen, Y., and Saari, J.T. 1997. Expression of metallothionein in the liver of rats fed copper-deficient AIN-93G diet. In *Trace Elements in Man and Animals – 9: Proceedings of the Ninth International Symposium on Trace Elements in Man and Animals. Edited by* P.W.F. Fischer, M.R. L'Abbé, K.A. Cockell, and R.S. Gibson. NRC Research Press, Ottawa, Canada. pp. 377–379.

The MT-I mRNA concentrations were elevated in the copper-deficient livers (n = 15) by 75.4 ± 3.7 fold ($p < 0.01$), but were not changed in the copper-deficient hearts (n = 12) or kidneys (n = 12). MT protein concentrations in the copper-deficient liver and heart did not change, but significantly ($p < 0.01$) decreased in the copper-deficient kidney.

It has been suggested that the enhanced synthesis of MT-I mRNA in the liver of brindled mutant mice results from stress (Mercer er al. 1991). Copper deficiency induces many biochemical changes and pathophysiological consequences in many organs including liver, heart, and kidney. Several studies suggest that oxidative stress is involved in the copper deficiency-induced pathological processes. Enhanced lipid peroxidation in copper-deficient tissues and inhibition of copper deficiency-induced defects by antioxidants were observed (Lynch and Strain 1989). Because MT is an important antioxidant participating in cellular protection against oxygen free radical-induced damage, up-regulation of MT synthesis may reflect a general adaption to the copper deficiency-induced oxidative stress. However, our previous studies (Chen et al. 1994) have shown that under the same experimental conditions, a higher degree of oxidative damage occurs in the heart than in the liver of copper-deficient rats. Therefore, the extent of oxidative stress alone cannot account for the elevated MT-I mRNA concentration in the liver. Mechanisms responsible for the selective response of liver MT gene transcription to copper deficiency need to be further investigated.

A corresponding increase in MT protein in the copper-deficient liver was not detected. MT is transported from the liver into the bile and blood (Bremner 1987). Accumulation of MT in the kidney would thus be expected if such a transport occurs. However, MT transport does not account for the undetected elevation of MT content corresponding to the increased MT mRNA, because the MT concentration in the kidney did not increase, but significantly decreased. Early studies on brindled mutant mice have shown that hepatic MT synthesis was reduced in the brindled neonate (Hunt and Port 1979), which has been attributed to the low hepatic copper concentration in the mutant (Piletz and Herschman 1983). This led to a conclusion that copper is the most likely regulator of hepatic MT synthesis in the neonatal mouse liver (Hunt and Port 1979, Piletz and Herschman 1983). These early studies, however, did not examine the concentration of MT mRNA. In contrast to the early conclusion, studies by Mercer et al. (1991) suggested that hepatic copper is not regulating MT mRNA production. In the present study, we measured both MT mRNA and protein concentrations. The results, together with previous studies (Mercer et al. 1991, Hunt and Port 1979, Piletz and Herschman 1983), suggest that copper is not essential for MT mRNA production, but it may be required for MT translation.

References

Bremner, I. (1987) Involvement of metallothionein in the hepatic metabolism of copper. J. Nutr. 117: 19–29.

Chen, Y., Saari, J.T., & Kang, Y.J. (1994) Weak antioxidant defenses make the heart a target for damage in copper-deficient rats. Free Radical Biol. Med. 17: 529–536.

Hunt, D.M., & Port, A.E. (1979) Trace element binding in the copper deficient mottled mutants in the mouse. Life Sci. 24: 1453–1466.

Lynch, S.M., & Strain, J.J. (1989) Effects of copper deficiency on hepatic and cardiac antioxidant enzyme activities in lactose- and sucrose-fed rats. Br. J. Nutr. 61: 345–354.

Mercer, J.F.B., Stevenson, T., Wake, S.A, Mitropoulos, G, Camakaris, J., & Danks, D.M. (1991) Developmental variation in copper, zinc and metallothionein mRNA in brindled mutant and nutritionally copper deficient mice. Biochim. Biophys. Acta 1097: 205–211.

Piletz, J.E., & Herschman, H.R. (1983) Hepatic metallothionein synthesis in neonatal Mottled-Brindled mutant mice. Biochem. Genet. 21: 465–475.

Discussion

Q1. John Beattie, Rowett Research Institute, Aberdeen, Scotland: Can your observations be explained by increased protein degradation or turnover?

A. I don't know. I don't think so, because I don't know of any mechanism for Cu deficiency to increase degradation.

Q2. George Cherian, The University of Western Ontario, London, ON, Canada: Have you measured any cytokines in this model and could cytokines explain the induction in metallothionein?

A. That's a very good question, but I don't have an answer for you at this time. We will look at NF-kB involvement.

Q3. Harry McArdle, Rowett Research Institute, Aberdeen, Scotland: Some time ago we did some experiments where we were removing copper from hepatocytes in culture and looked at metallothionein message levels. We found an effect due to copper ions in the cells. In *in vitro* experiments, if Cu is chelated, then metallothionein decreases. Is it possible that the changes in mRNA for liver metallothionein are some kind of secondary response? Why the difference in results between *in vitro* and *in vivo* systems?

A. I agree with the possibility of a secondary response but what is it? We should note that this response is observed only in the liver and not in the heart or kidney.

Q4. Ian Bremner, Rowett Research Institute, Aberdeen, Scotland: When measured by immunoassay, increased metallothionein in Cu deficient rats is associated by decreased feed intake as the stress. How was metallothionein protein measured in your study?

A. The Cd-affinity assay was used.

The Dynamics of Metallothionein Gene Expression and Regulation in Liver and Kidney of Rats*

M. HELENA VASCONCELOS,† SHUK-CHING TAM,‡ JOHN E. HESKETH,† AND JOHN H. BEATTIE†

†Division of Biochemical Sciences, Rowett Research Institute, Aberdeen AB2 9SB, UK,
‡Department of Chemistry, University of Aberdeen, Aberdeen AB9 2UE, UK

Keywords metallothionein, cadmium, gene expression, rat

Introduction

Cadmium (Cd) is a potent inducer of metallothionein (MT)-1 and MT-2 genes, but studies on the tissue specificity of isoform gene expression following Cd treatment have been limited by the lack of specific mRNA probes. We have recently described the design and specificity of oligonucleotide probes for rat MT-1 and MT-2 mRNA (Vasconcelos 1996a). In this study we used these probes to investigate liver and kidney MT isoform gene expression, after a single Cd injection.

Materials and Methods

A group of 25 Hooded Lister rats of approximately 280 g were injected s.c. with 8.9 μmol/kg $CdCl_2.2.5H_2O$ in 0.9% NaCl. Ten control rats were injected with 0.9% NaCl, and 5 control rats received no injection. Groups of 5 animals injected with Cd and groups of two saline-injected controls were killed at 4, 6, 12, 24 and 48 h after injection. Rats with no injection were all killed at 0 h. The animals were killed by exsanguination following anaesthesia with sodium pentobarbitone and an injection of 20 mg/mL heparin (Sigma, Poole, UK) in saline, administered into the posterior vena cava. The livers and kidneys were removed from the animals and the livers were perfused with 20 mL of sterile, ice-cold 0.154 M KCl, pH 7.0. Livers and kidneys were frozen immediately in liquid nitrogen and then stored at –80°C.

The RNA extractions were carried out by the acid/guanidinium/phenol/chloroform method of Chomczynski and Sacchi (1987) and slot blots were prepared from aliquots of 2 μg of RNA as described in Vasconcelos et al. (1996a, 1996b). MT-1 and MT-2 specific oligonucleotide probes (Vasconcelos et al. 1996a) were radioactively labelled with $[^{32}P]$-ATP and hybridisation and detection was performed as described in Vasconcelos et al. (1996b). The amount of RNA loaded per slot was assessed by hybridisation of mRNA poly-A with a poly-T oligonucleotide probe.

Total MT protein from (20% w/v) liver and kidney homogenates was quantified using the silver saturation assay (Scheuhammer and Cherian 1991).

Results and Discussion

In liver, Cd treatment increased both MT isoform mRNAs in a biphasic manner. The first phase was denoted by a very rapid increase, reaching a maximum at 4 h for MT-1 and 6 h for MT-2 and the second phase was evident by 24 h (Figure 1a). In kidney, the same treatment also increased each isoforrn mRNA, both reaching a maximum at 4 h (Figure 1b). This increase, when the blots were corrected for loadings,

*This work was supported by the Scottish Office Agriculture, Environment and Fisheries Department (SOAEFD). M. Helena Vasconcelos was funded by JNICT (Junta Nacional de Investigação Científica e Tecnológica) from Portugal and by the British Council.

Correct citation: Vasconcelos, M.H., Tam, S.-C., Hesketh, J.E., and Beattie, J.H. 1997. The dynamics of metallothionein gene expression and regulation in liver and kidney of rats. In *Trace Elements in Man and Animals – 9: Proceedings of the Ninth International Symposium on Trace Elements in Man and Animals. Edited by* P.W.F. Fischer, M.R. L'Abbé, K.A. Cockell, and R.S. Gibson. NRC Research Press, Ottawa, Canada. pp. 380–383.

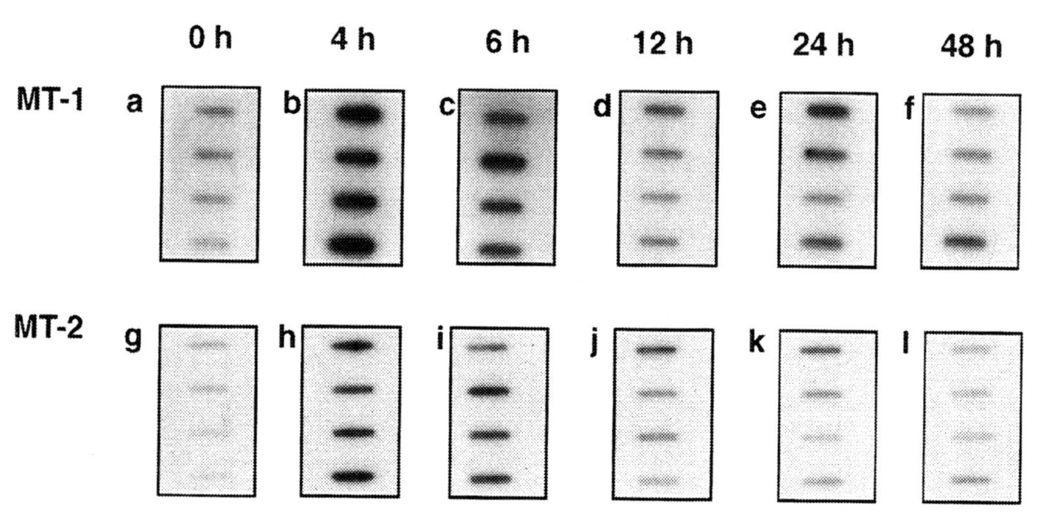

Figure 1a. Slot blots showing MT-1 and MT-2 mRNAs in liver (Figure 1a) and kidneys (Figure 1b) at different times following Cd treatment.

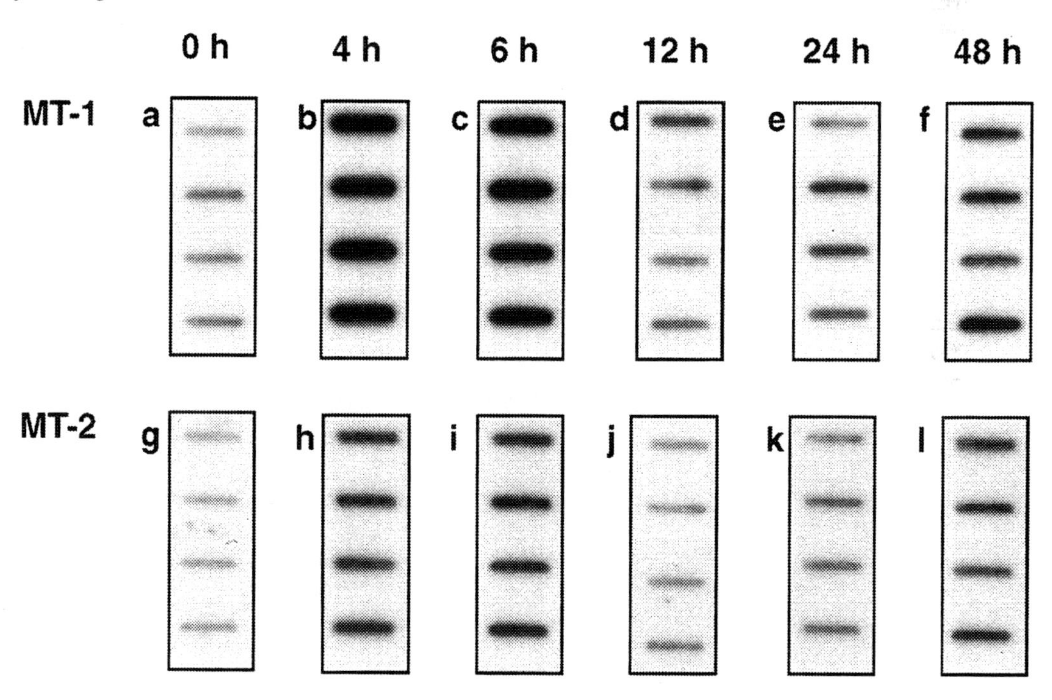

Figure 1b. Two replicate slot blots of rat liver (Figure 1a) and rat kidney (Figure 1b) RNA following injection with 8.9 μmol/kg of Cd. One blot was probed for MT-1 and the other for MT-2. Each panel shows results from 4 animals.

was monophasic for MT-2 mRNA but there was some evidence of a second induction for the MT-1 mRNA, peaking at the 24 h time point.

Results from the Ag saturation assay showed that MT-protein levels in liver increased continuously although the level of both isoform MT mRNAs decreased approximately to control levels at 12 h. This could be due to a longer half-life of the newly synthesised Cd-MT relative to Cu/Zn-MT in the control rats since it is known that the half-life of Cd-MT in rat liver is much longer (variably reported to be 67 h and 100 h) than the half-life of Cu-MT (10–12 h) or Zn-MT (19–20 h) in untreated animals (Bremner 1978).

In the kidneys, however, MT-protein levels remained constant up to 12 h and only increased slightly at 24 and 48 h. These results, together with the increase in the MT mRNA, suggest either that

i) Cd treatment causes more mRNA to be translated, but the newly synthesised Cd-MT protein is more rapidly excreted (in the urine) than the Cu/Zn-MT of the untreated animals. However, this hypothesis is not supported by the literature which states that the half-life of kidney Cd-MT is longer than our period of study (3–7 d) (Feldman et al. 1978) and that increases in MT in the urine are only detected 4 d after Cd exposure (Tohyama and Shaikh 1981)

or that

ii) Cd treatment did not result in increased translation of the kidney MT mRNA whereas transcription of the mRNA increased, during the first phase of mRNA induction. The late increase in MT-protein at 24 and 48 h could be due to an uptake of Cd-MT from other organs (possibly the liver) or a specific translation of the MT-1 mRNA that is synthesised during the second peak of mRNA induction.

Further work on liver and kidney metal levels and MT metal contents is in progress to investigate the validity of these hypotheses.

References

Bremner, I. (1978) Cadmium toxicity: Nutritional influences and the role of metallothioneins. Wld. Rev. Nutr. Diet. 32: 165–97.
Chomczynski, P. & Sacchi, N. (1987) Anal. Biochem. 162: 156–9.
Feldman, S.L., Squibb, K.S. & Cousins, R.J. (1978) Journ. Toxic. Environ. Health, 4: 805–13.
Mehra, R.K. & Bremner, I. (1983) Biochem. J. 213: 459–65.
Scheuhammer, A.M. & Cherian, M.G. (1 991) Methods Enzymol. 205: 78–83.
Tohyama, C. & Shaikh, Z.A. (1981) Fundam. Appl. Toxic. 1: 1–7.
Vasconcelos, M.H., Tam, S.C., Beattie, J.H., & Hesketh, J.E. (1996a) Biochem. Soc. Trans. 24: 225S.
Vasconcelos, M.H., Tam, S.C., Beattie, J.H., & Hesketh, J.E. (1996b) Biochem. J. 315: 665–71.

Discussion

Q1. (unable to identify questioner): Have you done any experiments with hepatocytes or other *in vitro* cell lines? Such experiments could be very helpful.

A. A good idea. This work was started in whole animals.

Q2. James Kirkland, University of Guelph, ON, Canada: In your first talk you discussed movement of metallothionein, or disappearance from the cytosol and reappearance in the nucleus over a period of about 12 h. In this talk, you said that the half life of MT was 3.5 d. In the first talk you were describing the movement in terms of the localization of synthesis of new protein. The half-life suggests that, unless there's a rapid degradation, existing metallothioneins are being moved into the nucleus. Is there a difference in half life between the two models?

A. The half life of metallothionein depends on the metals bound to the proteins. The cadmium protein has a half-life that is definitely much longer than that of the protein containing copper. In the first talk, I'm fairly certain that newly synthesized, preformed metallothionein was transported to the nucleus. There is definite evidence for translocation of protein into the nucleus.

Q3. Julian Lee, New Zealand Agricultural Research Institute, Palmerston North, New Zealand: John, I'm very interested in your results on the protein in the kidney. We found very similar results in that pharmacological amounts of Zn induced very high expression of the mRNA in liver and kidney, but MT protein was increased in liver only, and not in the kidney.

A. (no comment)

Q4. Ed Harris, Texas A&M University, College Station, TX, USA: In the kidney, is the inducing metal still in the cell after the initial induction of MT message?

A. After copper treatment and the initial response, copper levels and other metals don't change much, however, mRNA does. Results from copper studies are similar to cadmium.

Q5. Ed Harris, Texas A&M University, College Station, TX, USA: Just a comment, then. We need to get together how the protein is synthesized inside the cell when the metal is outside the cell.

Q6. Susan Haywood, University of Liverpool, England: Just a comment in regard to the kidney. Before you dismiss the excretion of MT, I think you ought to attempt to measure it in the urine. What we've done with copper in a similar situation, we found enhanced turnover with increased Cu-MT excretion in urine.

Q7. Roger Sunde, University of Missouri, Columbia, MO, USA: Was the quantity of metal in the kidney enough to sustain MT synthesis after initial upregulation of MT transcription? Others have talked about a zinc regulatory factor which may affect the MTF-type proteins in response to just zinc and not to cadmium. So you may get this cascade of metal release, such that you've got enough metal to sustain up-regulation of transcription, but not enough metal to sustain actual synthesis of metallothionein. If you don't have the metal there, then you may not detect the inducible protein.

A. I assume that induction by Zn was via displacement of MT.

Q8. Nader Abraham, The Rockefeller University, New York, NY, USA: Are there differences in nuclear binding proteins for copper in heart and kidney?

A. This has not been done.

Structure and Cloning of Serum Ferritin

M.C. LINDER, M. HAZEGH-AZAM, C.Y.J. ZHOU, K.J. SCHAFFER, C.A. GOODE, AND G.M. NAGEL

Department of Chemistry and Biochemistry, and Institute for Molecular Biology and Nutrition, California State University, Fullerton CA 92634, USA

Keywords serum ferritin, structure, regulation, cloning

Intracellular ferritins are well-studied proteins and considered the main sites for deposition of excess iron. Most intracellular ferritins are composed of 24 subunits of at least two types (known as L and H). They have molecular weights in the range of 20 k and are arranged with 4:3:2 symmetry in the form of a hollow sphere. The total molecular weight of the protein portion is about 480 k, and apoferritins have sedimentation coefficients of 16–18 S (Linder et al. 1981, 1989). Iron is deposited within the core of the molecule as ferric oxyhydroxide. Iron to protein ratios are typically 0.1–0.3 μg Fe/μg protein, and there is no evidence for the presence of carbohydrate. Tissue concentrations of the protein are in the range of tens to hundreds of μg/g, and it is found in almost all cells, across tissues and phyla.

Iron is the most important regulator of ferritin biosynthesis. Ferritin mRNAs appear to be fairly stable, and iron acts mainly by stimulating translation. It does this by binding to, and removing, repressor proteins (IRP1 or IRP2) that are otherwise bound to a stem-loop structure in the 5′UTR of the ferritin mRNAs (Leibold and Guo 1992, Klausner et al. 1993).

Serum ferritin presents a different picture and, in contrast to intracellular ferritins, has hardly been studied. It is present in trace quantities (tens of ng/mL), has a low iron content, partially binds to concanavalin A (indicative of carbohydrate), and was reported to have a 23 k glycosylated subunit (Cragg et al. 1981). It was discovered upon the development of sensitive immunoassays that used polyclonal antibodies against intracellular (spleen and liver) ferritins, and it is still measured by these procedures. Assays of serum ferritin are widely used to help diagnose iron deficiency and overload, because in the absence of disease, there is a positive correlation with the size of intracellular iron stores (which are mainly in ferritin). However, the immunoassays indicate that levels of serum ferritin rise acutely in many conditions, including liver disease, cancer, inflammation and treatment of iron deficiency (Linder et al. 1991).

We have purified ferritin from the serum of the horse, in order to better define its structure and compare it with that of intracellular ferritins. Serum ferritin was purified by a combination of heat and pH 4.8 treatments, as used for purification of other ferritins, followed by immunoaffinity chromatography with antibody against horse spleen or horse heart ferritins (Linder et al. 1996). The resulting material had a single major 7 S component in sedimentation velocity, with a molecular weight of 139 k by sedimentation equilibrium. Under the electron microscope it resembled much smaller donuts than those seen in preparations of ferritins from spleen or other tissues. A small amount of intracellular ferritin also appeared in the serum ferritin material, as judged by microcopy as well as sedimentation velocity.

SDS-polyacrylamide gel electrophoresis revealed the presence of larger subunits than seen in intracellular ferritin: one of about 26 k, and a triplet of subunits about twice as large (50, 57, and 65 k). Carbohydrate was present, as evidenced by binding to concanavalin A and that enzymatic treatment to remove asparagine-linked carbohydrate lowered the apparent molecular weights of the three larger subunits. N-terminal arnino acid sequences were obtained for three of the subunits, and these did not correspond to anything already in sequence databanks.

The sequence information was used to construct positive degenerate oligonucleotide primers that were paired with negative primers for known ferritin sequence in RT-PCR. For this we used liver RNA from rats pretreated with turpentine to induce inflammation (which increases serum ferritin). Two small,

Correct citation: Linder, M.C., Hazegh-Azam, M., Zhou, C.Y.J., Schaffer, K.J., Goode, C.A., and Nagel, G.M. 1997. Structure and cloning of serum ferritin. In *Trace Elements in Man and Animals – 9: Proceedings of the Ninth International Symposium on Trace Elements in Man and Animals. Edited by* P.W.F. Fischer, M.R. L'Abbé, K.A. Cockell, and R.S. Gibson. NRC Research Press, Ottawa, Canada. pp. 384–385.

consistent and amplifiable products were obtained and cloned. These had 40–55% homology with rat H and L ferritins at the nucleotide and amino acid levels. The homology was scattered throughout and not confined to specific regions, suggesting that we are dealing with a product/products of a different gene/ genes.

The clones we obtained were used to examine the potential tissue expression and iron responsiveness of serum ferritin mRNAs (see Schaffer et al. 1997), and to screen a rat liver cDNA library. From this it appears that several tissues, including liver, spleen and brain but not skeletal muscle, may be expressing subunits for serum ferritin, and that in all cases, iron greatly stimulates mRNA expression (which is not the case with intracellular ferritins). Stimulation of ferritin secretion by iron in cultured hepatocytes is accompanied by an increase in mRNA expression which is of the same order of magnitude (based on results with the putative 26 k serum ferritin subunit clone); and enhanced transcription appears to be involved (Schaffer et al. 1997).

Screening of the rat liver library has initially proceeded with probes made from the clone of the 26 k serum ferritin subunit. Six hundred thousand plaques were screened at high stringency. Putative positive plaques were purified and plated three times. Positive plaques surviving tertiary screening were cloned into a Bluescribe phagmid vector. Eleven positive plasmids with inserts greater than 1 kb were partially sequenced. Two of these showed potential and are being evaluated. We are also obtaining additional (internal) amino acid sequences for the subunits, to validate the identity of the clones.

Our findings indicate that serum ferritin differs markedly in structure and sequence from intracellular ferritins while retaining some homology; that it is probably the product of separate genes; that it is likely to be produced by several tissues; that expression is very sensitive to iron; but that expression is regulated differently by iron than is the expression of intracellular ferritins.

Literature Cited

Cragg, S.J., Wagstaff, M. & Worwood, M. (1981) Detection of a glycosylated subunit in human serum ferritin. Biochem J. 199: 565–571.

Klausner, R.D., Roualt, T.A., & Harford, J.B. (1993) Regulating the fate of mRNA: The control of cellular iron metabolism. Cell 72: 19–28.

Leibold, E.A. & Munro, H.N. (1988) Cytoplasmic proteins bind *in vitro* to a highly conserved sequence in the 5′ untranslated region of ferritin heavy- and light-subunit mRNAs. Proc. Natl. Acad. Sci. USA 85: 2171–2175.

Linder, M.C. (1991) Nutrition and metabolism of the trace elements. In: Nutritional Biochemistry and Metabolism, 2nd ed. (Linder, M.C., ed.), pp. 151–198, Elsevier, New York, NY.

Linder, M.C., Nagel, G.M., Roboz, M. & Hungerford, D.Jr. (1981) Heart cells contain a ferritin larger and more asymmetric than ferritins of other mammalian tissues. J. Biol. Chem. 256: 9104– 9111.

Linder, M.C., Schaffer, K.J., Hazegh-Azam, M., Zhou, C.Y.J., Tran, T.N. & Nagel, G.M. (1996) Serum ferritin: does it differ from tissue ferritin? J. Gastroenterol. Hepatol. In press.

Linder, M.C., Goode, C.A., Gonzalez, R., Gottschling, C., Gray, J. & Nagel, G.M. (1989) Heart tissue contains small and large aggregates of ferritin subunits. Arch. Biochem. Biophys. 273:34–41.

Schaffer, K.J., Tran, T.N., Hazegh-Azam, M. & Linder, M.C. (1997) Expression and regulation of the secretion of serum ferritin. In: Trace Elements in Man and Animals – 9: Proceedings of the Ninth International Symposium on Trace Elements in Man and Animals. (Fischer, P.W.F., L'Abbé, M.R., Cockell, K.A. and Gibson, R.S., eds.) NRC Research Press, Ottawa, Canada. pp. 386–387

Discussion

Q1. Ed Harris, Texas A&M University, College Station, TX, USA: Did you sequence enough protein to compare the homology with other proteins?

A. We have found no matches with anything else with high homology. The N-terminal region shows some homology with ferritin.

Q2. Roy Baynes, University of Kansas Medical School, Kansas City, KS, USA: There are a number of pseudogenes for ferritin. Your present strategy seems to be to look at conserved sequence. To what extent can you be sure that you are not looking at pseudogenes?

A. We can't be sure. However, we are isolating the protein from the serum, so if it's a pseudogene, it's an expressed one, that might even have a function. I don't know, so I'm open to all kinds of suggestion.

Expression and Regulation of the Secretion of Serum Ferritin

K.J. SCHAFFER, T.N. TRAN, M. HAZEGH-AZAM, AND M.C. LINDER

Department of Chemistry and Biochemistry, California State University, Fullerton CA 92634, USA

Keywords ferritin, serum, iron, interleukins

Intracellular ferritins are well studied proteins that sequester and detoxify excess iron. Most are comprised of 24 subunits, arranged with 4:3:2 symmetry around a hollow core within which up to 4500 atoms of iron may be deposited as ferrihydrite. Subunits of these mammalian ferritins are of at least two types, known as L and H, each about 20 kDa. mRNAs for these subunits have been cloned from many species. Iron induces increased synthesis of these subunits, and this occurs primarily by a translational mechanism, involving removal of repressor proteins from a structure in the 5′untranslated region of the mRNA (Leibold and Guo 1992, Klausner et al. 1993).

Using antibody against liver or spleen ferritins, sensitive immunoassays have been developed which show that traces of ferritin also normally appear in the serum. Moreover, concentrations of this "serum ferritin" correlate positively with body iron stores (which are in intracellular ferritin and its derivative, hemosiderin). Therefore, serum ferritin assays have been widely used to assess nutritional iron status, being very low (<10–12 ng/mL) in iron deficiency and very high (400–10,000 ng/mL) in iron overload. However, serum ferritin levels also rise in a number of other conditions that include liver disease, cancer, inflammation and iron treatment of iron deficiency (Linder 1991).

We have developed a tissue culture system that may model how serum ferritin concentrations are regulated in two of these conditions: inflammation and iron treatment. Rat hepatoma cells (H4-II-E-C3) are grown close to confluence in serum-containing Modified Eagle's Medium, then switched to protein-free hybridoma medium and incubated for an additional 24 or 48 h with various factors (Tran et al. 1996). Like normal liver cells, these cells release albumin and ceruloplasmin into the medium, and they also release ferritin. We quantitate ferritin release by rocket immunoelectrophoresis, using antibody raised against horse spleen ferritin that contains only L and H subunits, and intracellular spleen ferritin as a standard. With this procedure, we have determined that optimal doses of both IL-1-β and TNF-α (10 and 5 ng/mL, respectively) each increase release of ferritin into the medium 2–3-fold (Table 1; Tran et al. 1996), and that together, the two cytokines have at least an additive effect. Iron treatment, either with a 1:1 molar complex with nitrilotriaceate [Fe(III)-NTA] or as iron-dextran, also greatly stimulates release of serum ferritin.

Lactate dehydrogenase activity of the medium was monitored for evidence of cell damage and was only elevated with higher doses of Fe-NTA or TNF. Albumin release was not affected by iron treatment but was somewhat reduced by IL-1, mimicking what occurs *in vivo* during inflammation. Iron treatment also increased concentrations of ferritin within the hepatoma cells, but this did not occur in response to IL-1 or TNF. Our results with these cultured hepatic cells would thus explain the increases in serum ferritin observed *in vivo* in inflammation and with iron treatments.

A major question we sought to answer was whether or not the ferritin released from the hepatoma cells was being deliberately secreted, as we supposed it might be. We therefore examined the effects of the inhibitor of Golgi function, brefeldin A, on ferritiin release stimulated by IL-1, TNF, or iron (Tran et al. 1996). In all cases, treatment with 4 μg/mL brefeldin A greatly reduced ferritin release, and the same was true for albumin (used as a positive control). Hence we have concluded that, at least in response to

Correct citation: Schaffer, K.J., Tran, T.N., Hazegh-Azam, M., and Linder, M.C. 1997. Expression and regulation of the secretion of serum ferritin. In *Trace Elements in Man and Animals – 9: Proceedings of the Ninth International Symposium on Trace Elements in Man and Animals. Edited by* P.W.F. Fischer, M.R. L'Abbé, K.A. Cockell, and R.S. Gibson. NRC Research Press, Ottawa, Canada. pp. 386–387.

Table 1. Secretion of ferritin and albumin from rat hepatoma cells in culture in response to various stimulae.[1]

	Dose (per mL)	Ferritin secreted (% of control)	Albumin secreted (% of control)	Intracellular ferritin (% of control)
IL-1	10 ng	280*	88*	92
TNF	5 ng	330*	97	112
IL-1 + TNF	10 + 5 ng	530*	89*	106
Fe-NTA	10 μg	720*	93	810*

*$p < 0.001$ for difference from control.
[1]Mean values from Tran et al. (1996).

these stimulae acting upon the liver, serum ferritin is the result of regulated secretion and not random leakage.

An additional question we addressed was whether the stimulatory effects of the cytokines and iron on ferritin secretion were mediated by transcription. In the case of iron this would differentiate them from the type of regulation exerted by iron on synthesis of intracellular ferritins (which is primarily via enhancement of translation). For this we used the inhibitor of RNA polymerase II, DRB (dichlororibofuranosylbenzimidazole), which we incubated with the cells at the same time as the cytokines or iron. Each agonist had the usual effect on ferritin secretion, but the effect of each was almost completely blocked by DRB. This inhibition occurred without negative effects on cell number or integrity, and implies that the hormones and iron were acting on serum ferritin production by enhancement of production of its mRNAs.

Using what we believe are clones of portions of rat serum ferritin subunit cDNAs as probes (Linder et al. 1996a), we also slot-blot hybridized total RNA from the hepatoma cells and from several tissues of rats treated and not treated with iron. Hybridization with cDNA for 18S rRNA was used to control for RNA loading. Iron treatment increased mRNA for what we believe is serum ferritin about 10-fold (data not shown), which would fit with the >7-fold increase observed in ferritin secretion (Table 1). Serum ferritin appeared to be expressed by liver and to a lesser extent also by some other tissues, including the kidney, spleen, and brain but not skeletal muscle. In all of these tissues, iron treatment of the rats markedly enhanced expression (Linder et al. 1996b).

We conclude that serum ferritin is normally a secreted protein, produced by the liver and also some other tissues, and that its level in the serum is regulated by iron as well as by cytokines released as part of the acute phase response of inflammation.

Literature Cited

Klausner, R.D., Roualt, T.A. & Harford, J.B. (1993) Regulating the fate of mRNA: The control of cellular iron metabolim. Cell 72: 19–28.

Leibold, E.A. & Munro, H.N. (1988) Cytoplasmic proteins bind *in vitro* to a highly conserved sequence in the 5′ untranslated region of ferritin heavy- and light-subunit mRNAs. Proc. Natl. Acad. Sci. USA 85: 2171–2175.

Linder, M.C. (1991) Nutrition and metabolism of the trace elements. In: Nutritional Biochemistry and Metabolism, 2nd ed. (Linder, M.C., ed.), pp. 151–198, Elsevier, New York, NY.

Tran, T.N., Eubanks, S.K., Schaffer, K.J., Zhou, C.Y.J. & Linder, M.C. (1996) Secretion of ferritin by rat hepatoma cells and its regulation by inflammatory cytokines and iron. Submitted.

Trace Elements, Food and Nutrition Policy

Chair: M.R. L'Abbé

Considerations Regarding Future Trace Element Dietary Reference Intakes

J.C. KING
Western Human Nutrition Research Center, USDA, San Francisco, CA 94129, USA

Keywords dietary recommendations, trace elements, diet intake, nutrient function

Introduction

Since 1941, the Food and Nutrition Board of the National Academy of Sciences has issued reports periodically providing "standards to serve as a goal for good nutrition" (IOM 1994). The last edition of these standards, called the Recommended Dietary Allowances (RDAs), was published in 1989 (IOM 1989). That edition included recommendations for four trace elements — zinc, iron, iodine and selenium, and estimated safe and adequate daily dietary intakes for five other trace elements — copper, manganese, fluoride, chromium, and molybdenum. Since the standards were first proposed in 1941, their uses have expanded from guidelines for the planning and procurement of food for groups of people to standards serving as the basis of food labeling, federal food assistance programs, food fortification policies, assessment of dietary survey data, and the design of nutrient supplements and special dietary foods. The RDAs are frequently used to assess the diet intakes of individuals or to provide dietary guidance to a person. Nutrition science has also undergone a number of changes since the last edition was prepared. The role of diet in reducing the risk of chronic disease now is a major emphasis within the field. This development, along with the expanding uses of the RDAs, led the Food and Nutrition Board to conclude that a new paradigm is needed for the next set of dietary recommendations that incorporates data on the reduction of risk for chronic disease and provides several reference intakes for each nutrient in order to meet the expanding list of uses. Since the next set of standards will be based on a new paradigm, the Food and Nutrition Board proposes to change the name from "Recommended Dietary Allowances" to "Dietary Reference Intakes."

Use of Risk Reduction as a Criterion for Dietary Reference Intakes

In past editions of the RDAs, recommendations were only established for essential nutrients, which were defined as chemical substances found in food that cannot be synthesized in the body and are necessary for life, growth and tissue repair. More recent research has shown, however, that there are a number of components in food that are not considered to be essential nutrients but which are important for maintenance of health and, possibly, the risk reduction for chronic disease. Examples of those substances include dietary fiber, carotenoids, lycopenes, choline, and some of the ultra trace elements, such as boron and vanadium. Thus, new criteria need to be established for identifying those components in foods which have a Dietary Reference Intake (DRI). Criteria for making that decision may include the following questions: 1) Is the substance a normal component of the food supply? 2) Has a biological function for this substance

Correct citation: King, J.C. 1997. Considerations regarding future trace element dietary reference intakes. In *Trace Elements in Man and Animals – 9: Proceedings of the Ninth International Symposium on Trace Elements in Man and Animals. Edited by* P.W.F. Fischer, M.R. L'Abbé, K.A. Cockell, and R.S. Gibson. NRC Research Press, Ottawa, Canada. pp. 388–391.

been identified in humans? 3) Are data available on levels of intakes that maintain tissue homeostasis and biological function? 4) Is there evidence of inadequate or excessive intakes in some population groups?

Application of those criteria to boron would probably lead one to conclude that more information is needed before a DRI can be established for this element. It is known that boron is a normal component of the food supply, but an essential biological function has not been identified, intakes that maintain homeostasis and normal function are not well defined, and it is unknown if there are population groups consuming too little or too much boron because an indicator of nutritional status has not been defined. Alternatively, one might decide that sufficient data are available to establish a DRI for manganese. It is commonly found in the food supply. It is required to activate several types of enzymes and it is a constituent of several other enzymes. Some information regarding tissue homeostasis and amounts needed in the diet to maintain normal function are available, and there are reports of manganese toxicity due to the inhalation of manganese fumes and dust. The degree of risk for inadequate or excessive intakes of manganese are unknown, however, due to the lack of a biomarker of manganese nutritional status.

To establish a DRI, therefore, three types of information regarding a food component should be integrated when making a dietary recommendation. Those three types of information include: 1) Data on the biological function and the functional response to inadequate or excessive intakes, 2) Data on the mechanisms for regulating tissue nutrient homeostasis and adaptations in those mechanisms with changes in intake, and 3) Specific, quantitative data on changes in nutritional status with shifts in nutrient intake.

Functional endpoints of nutritional adequacy are needed to establish an intake that reduces the risk for chronic disease. Traditionally, functional endpoints have not been used to establish nutrient recommendations and very few quantitative, specific functional endpoints are available for assessing nutritional status. Research is needed to establish functional endpoints for nutrient requirements that reflect a dose response relationship, are consistent across a variety of studies, and are temporally correct (i.e., a change in nutrient intake occurs before the functional change).

Studies of zinc depletion in humans illustrate the need to use functional endpoints along with studies of homeostasis when estimating nutrient requirements. For example, when dietary zinc was reduced from 15 to 5.5 mg/d in a group of healthy men, zinc balance was achieved within nine days (Wada et al. 1985). Although adaptations in zinc homeostasis permitted zinc balance when low zinc diets were fed, this was not without some functional consequences. Circulating concentrations of serum albumin, pre-albumin, retinol-binding protein, thyroid stimulating hormone, and free thyroxine declined significantly (Wada et al. 1986). All of these functional changes occurred without a fall in plasma zinc concentration. Thus, without these functional endpoint data, one might conclude that the dietary zinc requirement is only 5.5 mg/d.

Establishment of DRIs For Populations And Individuals

As stated above, the use of the RDAs has expanded considerably during the last 50 years. Today, they are used to prescribe diets for populations (e.g., food planning and procurement, federal food assistance) and to assess the intakes of populations (e.g., evaluation of dietary survey data). RDAs are also used to prescribe diets for individuals and to assess the intakes of individuals. Two different types of DRIs are needed, therefore. One standard that can be used to prescribe and assess diets for populations and another set for prescribing and assessing intakes of individuals. Standards for both the population and individual require an estimation of the mean requirement for the nutrient. Establishment of an individual DRI also requires knowledge of the variance in that requirement among a group of individuals of the same gender, age, and health status. The population DRI, on the other hand, requires knowledge of the variance in intake of the nutrient in that population.

If the mean requirement for a nutrient and the variance of that requirement are known, an individual DRI can be set at the upper end of the distribution of the requirements so that the needs of practically all individuals in that population are met. This is the approach that has been used to establish RDAs in the past. This standard can be used to evaluate the intake of an individual. For example, if their intake is equal to or above the standard, one can conclude that the intake is adequate. An intake below the standard, however, does not denote an insufficient intake. An individual DRI at the upper end of the distribution of requirements provides a goal for good nutrient intakes by an individual and would be an appropriate basis for dietary guidance.

In contrast to the individual DRI, a DRI for a population, requires knowledge of the distribution of intakes in that population. Knowledge of the distribution of intakes is necessary so that an adequate intake of all members of the population can be assured, even by those who eat small amounts of food. Using probability assessment, one can determine that the probability of an inadequate intake is low (about 2–3% of the population) if the population DRI is set so that subtraction of two standard deviations of the variance in intake from that standard equals the mean requirement (NRC 1989).

Using these definitions of an individual DRI and population DRI, hypothetical estimates of zinc DRIs for individuals and populations can be estimated. Let's assume that the mean requirement for zinc is 12.5 mg/d with a standard deviation of 1.25 mg/d. The individual DRI would be the mean requirement plus two standard deviations of that requirement or 15 mg/d. Let's also assume that the variance in zinc intake in a certain population is 2.5 mg/d, or a coefficient of variation in intake of 0.20. Using the definition that the mean requirement equals the population DRI minus two standard deviations of the variance intake, one can solve for the population DRI. A hypothetical estimate of the zinc population DRI, therefore, is 20.8 mg/d.

To establish both individual and population DRIs one must answer the question, requirement for what? Requirements for nutrients will vary with the level of nutritional status desired. For example, the requirement for prevention of the clinical symptoms of nutrient deficiency is lower than the amount needed for maintenance of biological function, for provision of body stores, or for support of a pharmacological role. The need for iron among adult men to prevent anemia is thought to be only 4 mg/d whereas the amount required to provide for some body stores is at least two-fold higher. One of the biggest challenges facing DRI committees will be to determine the criterion used to establish the mean requirement.

Summary

The Food and Nutrition Board of the Institute of Medicine recommends that a new paradigm be used to establish the next set of dietary recommendations and that those new standards be called Dietary Reference Intakes, or DRIs. The Board also proposes that data regarding risk reduction for chronic disease be used to establish the reference intakes; that a model integrating information on nutrient homeostasis, function, and static tissue or circulating concentrations be used to establish the requirements; that multiple reference intakes be established for individuals; and that guidance be provided on how to estimate reference intakes for populations based on knowledge of the mean requirement and distribution of intake.

Literature Cited

Institute of Medicine. Food and Nutrition Board. (1994) How Should the Recommended Dietary Allowances be Revised? National Academy Press, Washington D.C.

Institute of Medicine. Food and Nutrition Board. (1989) Recommended Dietary Allowances. 10th edition. National Academy Press, Washington D.C.

National Research Council. Food and Nutrition Board. Subcommittee on Criteria for Dietary Evaluation. (1986) Nutrient Adequacy. National Academy Press,Washington D.C.

Wada, L. & King, J.C. (1986) Effect of low zinc intakes on basal metabolic rate, thyroid hormones and protein utilization in adult men. J. Nutr. 116: 1045–1053.

Wada, L., Turnlund, J.R. & King, J.C. (1985) Zinc utilization in young men fed adequate and low zinc intakes. J. Nutr. 115: 1345–1354.

Discussion

Q1. Leslie Klevay, USDA-ARS, Grand Forks, ND, USA: Would you infer then that if we don't have variance estimates for a nutrient will there be no DRI for it? And if this inference is correct will there be a shorter list of nutrients that will have DRIs?

A. No, we recognize that there is probably very limited information on the variance in requirement and that we will probably have to make some educated judgements of what that variance is in order to come up with DRIs.

Q2. Janet Hunt, USDA-ARS, Grand Forks, ND, USA: I may need to understand more about the population DRI calculations, but I am a little concerned about your example of 20 mg/day for zinc as a population standard. I think you can't ignore the political realities in the United States, where we have had bills proposed for dietary supplementation that criticize the scientific establishment of RDAs and claim that there is a need for additional kinds of recommendations that more optimize health, and that this movement is, I believe peddled mostly by the health food supplement industry. I think that to introduce this new concept will likely only add fuel to the fire for lobby groups critical of the process. In addition, from the scientific derivation of this intake variance, it will be very important to have good data for intake variances that do not rely on day-to-day variance in intake. For example, we should probably not use 24 hour recalls, but rather use 7- or 14-day averages. We should be ignoring the day-to-day variance.

A. First, let me clarify that we are not going to establish population DRIs, but we are establishing DRIs for individuals. We will provide guidance on how to determine what you should feed a population group, but you have the responsibility for doing that based on your knowledge of the variance of intake of nutrients within that population group. So there will be no document coming out that says the zinc population DRI is 20.8 mg/day. There will be an individual zinc DRI, and then guidance on how to convert that for the various population groups. Yes, I agree with you, we do need to have good information on variance of intake in the population group, in order to come up with appropriate standards, but that will be up to the individuals who will have to carry out that conversion. Requirements for dietary intake do vary between populations, so of course we will need good intake variance data.

Q3. Noel Solomons, Guatemala: I commend you for your clarity of explanation. However, I tend to dispute your description of Homo sapiens as having a homogeneous biology. There are differences in size and body compositions, in respect to environmental factors and not just diet (such as parasites, climate, etc.). Certainly there is genetic polymorphism, although it may be tough to identify all the details at present.

A. In the past there has not always been a clear basis of the recommendations, being sometimes based on biology, and other times on intake. My point was that variations in intake within a population will certainly be greater than variances in biology of the group of individuals. I'm sure that there will be things that will affect the requirement, as you pointed out but if we can clean it up a little bit by just focussing on the biology we've removed another major source of variability, which is intake. That's our point.

Q4. John Bogden, New Jersey Medical School, Newark, NJ, USA: What about setting numbers that represent the threshold for frank deficiency? Did the committee discuss that?

A. We did consider this. At one point, we thought of having four reference intakes, with the fourth being at a level which would be deficient, but weren't so sure it would be useful. Besides, given the mean and standard deviation of the requirement level, one can easily estimate the threshold. If we could be convinced that there is a need for it, I guess we would reconsider.

Q5. Gordon Klein, University of Texas, Galveston, TX, USA: Have you considered interactions between the nutrients? Has any consideration been given to possible effects of supplements on the bioavailability of nutrients? And finally, what effort will be made to disseminate this information to the public?

A. Yes, we have been considering nutrient interactions. That will be part of the analysis that we will have to consider in determining the mean requirements. Supplements will undoubtedly influence the variance in the intakes, but it's not clear to me right now how that might influence the nutrient requirement. And finally, the dissemination of the information, will be coming out in increments. Probably the first report will be the calcium, vitamin D, magnesium and phosphorus report, and will be publicized in the same way that all the other reports from the National Academy of Sciences are publicized.

Q6. Forrest Nielsen, USDA-ARS, Grand Forks, ND, USA: If there is not enough information on a given element to allow the setting of a DRI (e.g. for boron, vanadium or fluorine), what will be done?

A. We may use an ESADDI approach, or something similar. This hasn't really been decided yet, but we will not be ignoring those nutrients where there is some information, but insufficient to allow full determination of a DRI.

Iron Fortification in the UK: How Necessary and Effective Is It?*

S.J. FAIRWEATHER-TAIT

Institute of Food Research, Norwich NR4 7UA, UK

Keywords iron, fortification, bioavailability, United Kingdom

Fortification is one of the approaches used to modify nutrient intake. It is appropriate when a small elevation of intake is required, the target population is large, and where it is difficult to identify 'at risk' groups. Fortification policies are generally of a long-term nature. Criteria to be considered include the baseline intake of the nutrient in the population, food patterns, choice of vehicle for the fortificant, stability in the food, bioavailability, and risk of excessive intake. Fortification is generally more difficult in products with a long shelf-life, small serving size, reactive base ingredients, and a high labelling Recommended Daily Allowance (RDA).

Several foods in the UK are fortified with iron, as illustrated in Table 1. By law, since 1953 all flour except wholemeal must contain not less than 1.65 mg iron/100 g, and this must be in the form of ferric ammonium citrate, ferrous sulphate or iron powder, according to the specifications given in Schedules 1 and 2 (Statutory Instruments 1995). This is the level naturally found in flour of 80% extraction, and therefore the addition of iron to flour is strictly restoration rather than fortification. When this first became mandatory, iron deficiency was a common disorder in the UK. White flour was chosen as the vehicle because white bread provided up to one third of food energy in poorer families. However, consumption of white bread has fallen steadily and, according to data from the National Food Survey (Ministry of Agriculture Fisheries and Food 1995) it now provides less than 10% of total energy. These data indicate that the average daily consumption of iron is 9.9 mg/person, with more than 40% originating from cereal products. Information from the dietary and nutritional survey of British adults (Gregory et al. 1990) suggests that white bread provides 1.1 mg iron/day (9% of the total daily intake of 12.1 mg in adults), of which only about 0.3 mg is attributable to fortification iron.

Table 1. Iron fortified foods in the UK.

Food	Mean/typical iron concentration
(a) Statutory	
White flour	2.0 mg/100 g (Statutory minimum 1.65 mg/100 g)
Foods that simulate meat	20 mg/100 g protein
†Infant formula	0.12 mg/100 kJ (minimum)
	0.36 mg/100 kJ (maximum)
†Follow-on formula	0.25 mg/100 kJ (minimum)
	0.50 mg/100 kJ (maximum)
(b) Voluntary	
Breakfast cereals	2.3–4.6 mg/serving (17–33% of RDA)
Infant weaning foods	Variable
Meal replacements	4.6 mg/serving (33% of RDA)
Very low calorie diets	14 mg/d (100% of RDA)

†Label declaration required if iron is not added (Statutory Instrument, 1995a).

*This work was supported by MAFF, the EC and the BBSRC.

Correct citation: Fairweather-Tait, S.J. 1997. Iron fortification in the UK: How necessary and effective is it? In *Trace Elements in Man and Animals – 9: Proceedings of the Ninth International Symposium on Trace Elements in Man and Animals. Edited by* P.W.F. Fischer, M.R. L'Abbé, K.A. Cockell, and R.S. Gibson. NRC Research Press, Ottawa, Canada. pp. 392–396.

Infant formulas and follow-on formulas must meet compositional requirements (Statutory Instrument 1995a). If iron is added, the amount and type of iron is controlled by law, and if it is not added the label must carry a statement that infants must obtain iron from other sources. At present, all formulas in the UK have added iron. Currently there is a debate about the appropriate level of iron, fuelled on the one hand by concern over the high prevalence of iron deficiency in infants and toddlers, and on the other hand by worries about the pro-oxidant activity of iron and potential adverse interactions with other micronutrients, such as copper and zinc.

Breakfast cereals contribute 1.5 mg iron/day (Ministry of Agriculture Fisheries and Food 1995), but the addition of iron to these products is voluntary. The quantity of iron added is linked to marketing interests, and must be 15% or more of the RDA (14 mg) for a labelling claim to be made (Statutory Instruments 1994). Thus fortification iron generally ranges from 2.1–4.6 mg/serving, that is 15–33% of the RDA. The type of iron added is not subject to legislation, but the bioavailability of various iron compounds for humans is very different, as shown in Table 2.

Well-absorbed forms of iron such as ferrous sulphate often cause problems, including enhanced vitamin degradation (e.g., vitamin C, thiamin, retinol), oxidative rancidity, colour changes, off-flavours and precipitates. Thus food technologists often insist on the use of iron salts of lower bioavailability. Iron(III)EDTA is not a particularly well absorbed form of iron, but it has the advantage of being unaffected by dietary inhibitors often present in high amounts in high cereal and legume diets. Thus it has been used with some success in iron fortification programmes in developing countries. The efficacy of other forms of chelated iron, such as iron-glycine, are still under evaluation. An assessment of the bioavailability of a potential iron fortificant can be made by one of several methods, as summarised in Table 3.

Table 2. Relative bioavailability (RB) of different forms of iron to humans.

Form of iron	RB (%)	Reference
Ferrous sulphate	100	
Ferrous lactate	106	Hurrell (1985)
Ferrous fumarate	101	Hurrell (1985)
Ferrous succinate	92–123	Hurrell (1985)
Ferrous gluconate	89	Hurrell (1985)
Iron-glycine	85	Fox et al. (1997)
Ferrous citrate	74	Hurrell (1985)
Ferric saccharate	74	Hurrell et al. (1989)
Ferrous tartrate	62	Hurrell (1985)
Ferric pyrophosphate	39	Hurrell et al. (1989)
Haemoglobin iron	37–118	Calvo et al. (1989), MacPhail et al. (1985), Martinez et al. (1997)
Ferric suphate	34	Hurrell (1985)
Ferric citrate	31	Hurrell (1985)
Ferric orthophosphate	31	Hurrell (1985)
Iron(III)EDTA	30–290	International Nutritional Anemia Consultative Group (1993)
Sodium iron pyrophosphate	15	Hurrell (1985)
Reduced elemental iron	13–90	Hurrell (1985)
Carbonyl iron	5–20	Hallberg et al. (1986)

Table 3. Methods used to assess the bioavailability of iron fortificants.

(a) Absorption of radio- or stable isotopically-labelled compound, measured by fecal monitoring
(b) Hemoglobin incorporation of radio- or stable-isotopically labelled compound
(c) Whole-body retention of ^{59}Fe-labelled compound
(d) Rate of hemoglobin regeneration in iron-depleted individuals
(e) Plasma appearance of oral dose of iron compound (labelled or unlabelled) and disappearance of simultaneously administered i.v. dose of radio- or stable isotope iron

In every case it is essential to compare the unknown compound with a reference salt of high bioavailability, usually 3 mg ferrous ascorbate. In this way, the problems of inter-individual variation in efficiency of iron absorption, related to the level of body iron stores and erythropoietic activity, are removed. If possible, the best approach for determining iron bioavailability is to prepare isotopically-labelled material. This was first performed using radio-isotopes (^{55}Fe and ^{59}Fe), but these are gradually being replaced by stable isotopes (^{54}Fe, ^{57}Fe and ^{58}Fe) because of concern over the hazards of ionising radiation. However, it may be a very difficult task to prepare some forms of elemental iron, labelled with stable isotopes. Therefore novel ways of measuring absorption from plasma appearance/disappearance data (Costa et al. 1991) are being developed to assess the bioavailability of fortification iron. These acute measurements should be supported by longer-term studies which take into account individual adaptive responses to different levels of absorbable iron in the diet (Cook et al. 1991).

The statutory addition of iron to white flour in the UK was introduced because it was noted that during World War II a high standard of health had been maintained in the UK. This was attributed, in part, to an increase in the extraction rate of flour, and the concomitant increase in the intake of nutrients associated with the outer part of the grain, including iron. Thus when legislative control of the extraction rate of flour ceased, the addition of powdered iron became mandatory. Several studies were performed examining the absorption of this form of iron, and it was concluded that it was virtually unabsorbable (Ministry of Health 1968). Thus amendments were brought in to introduce forms of iron with a higher bioavailability (Statutory Instrument 1972), including ferrous sulphate, although the latter is not used because of technical disadvantages. Reduced iron must be of a specified particle size and solubility, but the absorption of this form of iron has not been measured. Lack of confidence in its efficacy led to a recommendation that iron should no longer be added to flour (Department of Health and Social Security 1981), but this recommendation has not yct bccn put into practice.

Iron deficiency is a common problem in infants, children, and women of child-bearing age (Table 4). According to recent nationally representative surveys, one in three women aged 18–64 years had low iron stores (serum ferritin < 25 μg/L), and 4% had hemoglobin values indicative of iron deficiency anemia (<110 g/L). The situation was even worse in pre-school children, where up to 12% had anemia. In smaller individual studies in adolescents the reported prevalence of anemia ranges from 4–20%. Thus there is a very real problem of inadequate iron supply in the UK diet for children with high physiological requirements for iron, and for women with high menstrual losses of iron.

Table 4. Iron deficiency in the UK and Ireland.

Group	Prevalence of iron deficiency	Reference
Infants < 18 months	2–20%: hemoglobin < 110 g/L	Fairweather-Tait (1996)
Pre-school children		Gregory et al. (1995)
1.5–2.5 years	28%: plasma ferritin < 10 μg/L	
	12%: hemoglobin < 110 g/L	
2.5–3.5 years	18%: plasma ferritin < 10 μg/L	
	6%: hemoglobin < 110 g/L	
3.5–4.5 years	14–16%: plasma ferritin < 10 μg/L	
	4–8%: hemoglobin < 110 g/L	
Adolescents	8–43%: serum ferritin < 10 μg/L	Fairweather-Tait (1996a)
11–18 years	4–20%: hemoglobin < 120 g/L	
Women		Gregory et al. (1990)
18–24 years	13%: serum ferritin < 13 μg/L	
	6%: hemoglobin < 110 g/L	
25–34 years	13%: serum ferritin < 13 μg/L	
	4%: hemoglobin < 110 g/L	
35–49 years	21%: serum ferritin < μg/L	
	4%: hemoglobin < 110 g/L	
50–64 years	5%: serum ferritin < μg/L	
	4%: hemoglobin < 110 g/L	

A critical evaluation of the nutritional significance of current UK iron fortification policy is needed, in order to determine to what extent at risk groups are benefiting, and to examine the relative risks of iron deficiency and iron overload (British Nutrition Foundation 1995). Although added iron makes a very small contribution to the average diet, one serving of breakfast cereal (2.4 mg) and six slices of white or brown bread (1.8 mg) could provide 4.2 mg iron, or approximately one third of the daily intake. However, little is known about the bioavailability of this added iron in the context of the total diet. Future research should be directed at assessing the efficacy of fortification iron, taking into account adaptive responses and homeostatic mechanisms.

References

British Nutrition Foundation (1995) Iron. Nutritional and physiological significance. London: Chapman & Hall.

Calvo, E., Hertrampf, E., de Pablo, S., Amar, M., & Stekel, A. (1989) Haemoglobin-fortified cereal: an alternative weaning food with high iron bioavailability. Eur. J. Clin. Nutr. 43: 237–243.

Cook, J.D., Dassenko, S.A., & Lynch, S.R. (1991) Assessment of the role of nonheme-iron availability in iron balance. Am. J. Clin. Nutr. 54: 717–722.

Costa, A., Liberato, L.N., Palestra, P., & Barosi, G. (1991) Small-dose iron tolerance test and body iron content in normal subjects. Eur. J. Haematol. 46: 152–157.

Department of Health and Social Security (1981) Nutritional aspects of bread and flour. Report on Health and Social Subjects No 23, London: HMSO.

Fairweather-Tait, S.J. (1996) Iron deficiency anaemia. In Major Controversies in Infant Nutrition (Ed. David TJ), International Congress and Symposium Series No 215, London: Royal Society of Medicine, pp. 79–89.

Fairweather-Tait, S.J. (1996a) Iron requirements and prevalence of iron deficiency. Adolescents — an overview. In Iron Nutrition in Health and Disease (Eds. Hallberg L, Asp N-G), London: John Libbey.

Fox, T.E., Eagles, J., and Fairweather-Tait, S.J. (1997) Bioavailability of an iron glycine chelate for use as a food fortificant compared with ferrous sulphate. In: Trace Elements in Man and Animals – 9: Proceedings of the Ninth International Symposium on Trace Elements in Man and Animals. (Fischer, P.W.F., L'Abbé, M.R., Cockell, K.A. and Gibson, R.S., eds.). NRC Research Press, Ottawa, Canada. pp. 460–462.

Gregory, J.R., Collins, D.L., Davies, P.S.W., Hughes, J.M., & Clarke, P.C. (1995) National Diet and Nutrition Survey: children aged $1^1/_2$ to $4^1/_2$ years. Volume 1: Report of the diet and nutrition survey. London: HMSO.

Hallberg, L., Brune, M., & Rossander, L. (1986) Low bioavailability of carbonyl iron in man: studies on iron fortification of wheat flour. Am. J. Clin. Nutr. 43: 59–67.

Hurrell, R.F. (1985) Nonelemental sources. In Iron Fortification of Foods (Eds. Clydesdale FM, Wiemer KL) London: Academic Press, pp. 39–53.

Hurrell, R.F., Furniss, D.E., Burri, J., Whittaker, P., Lynch, S.R., & Cook, J.D. (1989) Iron fortification of infant cereals: a proposal for the use of ferrous fumarate or ferrous succinate. Am. J. Clin. Nutr. 49: 1274–1282.

International Nutritional Anemia Consultative Group (1993) Iron EDTA for Food Fortification, Washington DC: The Nutrition Foundation, p. 28.

MacPhail, P., Charlton, R., Bothwell, T.H., & Bezwoda, W. (1985) Experimental fortificants. In Iron Fortification of Foods (Eds. Clydesdale FM, Wiemer KL) London: Academic Press, pp. 55–71.

Martinez, C., Fox, T., Eagles, J., & Fairweather-Tait, S.J. (1997) Evaluation of iron bioavailability in infant weaning foods fortified with heme concentrate. In: Trace Elements in Man and Animals – 9: Proceedings of the Ninth International Symposium on Trace Elements in Man and Animals. (Fischer, P.W.F., L'Abbé, M.R., Cockell, K.A. and Gibson, R.S., eds.). NRC Research Press, Ottawa, Canada. pp. 35–37.

Ministry of Agriculture, Fisheries and Food (1995) National Food Survey 1994. Annual Report on Household Food Consumption and Expenditure. London: HMSO.

Ministry of Health (1968) Iron in flour. Reports on Public Health and Medical Subjects No 117, London: HMSO.

Statutory Instrument (1972) No 1391. The Bread and Flour (Amendment) Regulations, London: HMSO.

Statutory Instrument (1994) No 804. The Food Labelling (Amendment) Regulations, London: HMSO.

Statutory Instrument (1995) No 3202. Bread and Flour Regulations, London: HMSO.

Statutory Instrument (1995a) No 77. The Infant Formula and Follow-on Formula Regulations, London: HMSO.

Discussion

Q1. Richard Black, Kellogg Canada Inc., Etobicoke, ON, Canada: I'd like to clarify one point with regard to labelling. In Canada we are not permitted to state the form of iron on the label. Is this the situation in the UK? Are you allowed to say "iron present as ferrous sulfate". I'll go on from that and say that I can speak for The Kellogg Company and say that we are very much interested in and concerned about bioavailability, but there are a couple of difficult issues: firstly, it is difficult to fortify minerals in a stable, acceptable and absorbable manner; and secondly, if for example we use the level of fortification dictated by the WIC (women, infants and children) program, this may have a significant effect on the amount of the fortificant that will be consumed by an adolescent at one sitting. It's tough to balance out the social program needs for one population against the needs of others. I wonder if you could comment about the conflict between what's good for a particular sub-population and what's good for the population as a whole, and how we address those kinds of issues with regard to food fortification.

A. It is difficult. Most people have physiological limits on uptake of iron, except of course those with hemochromatosis (and yes, possibly including heterozygotes for hemochromatosis). Some people are looking into efforts to postpone the excessive stores of iron in heterozygotes, but this is a very big issue.

Q2. Curtiss Hunt, USDA-ARS, Grand Forks, ND, USA: With regard to your comment that iron status in Asians in the UK seems to be problematic — are they really in poor iron status, or is this one of Noel Solomons' biological variants, such that they are at a level that is right for them, but appears to be poor in comparison to the average for the UK?

A. We have found that the best way to address this is to give iron, and if the status improves, then they were in poor iron status. It is true that we may need multiple cut-off points.

Universal Salt Iodization: Myth or Reality?

P.P. BOURDOUX, H.V. VAN THI, S. VAN STICHELEN, AND A.M. ERMANS

Université Libre de Bruxelles, Laboratoire de Pédiatrie, CEMUBAC, Hôpital Universitaire des Enfants, B-1020 Brussels, Belgium

Keywords iodine deficiency, iodine prophylaxis, salt iodization, thyroid

In 1990, the forty-third General Assembly of WHO has passed a resolution (WHA43.2) for virtually eliminating IDD (Iodine Deficiency Disorders) by the year 2000. A universal policy based on the iodization of salt has been chosen by international agencies (WHO, UNICEF, ICCIDD). With support and strong recommendations of these agencies, many countries have implemented programmes of iodine supplementation by promoting or enforcing the use of iodized salt. The physiological requirements of iodine are estimated to about 150 µg per day (Food and Nutrition Board 1989) corresponding to a daily consumption of 10 grams of salt containing 15 ppm iodine (DeMaeyer et al. 1979). Such a level has now been successfully enforced in Switzerland since 1980. ICCIDD (International Council for the Control of Iodine Deficiency Disorders) has prepared guidelines for obtaining appropriate levels of iodine in salt in developing countries (ICCIDD Board 1992). The recommended levels ranged between 20 and 100 ppm iodine. A footnote also indicated that an increase of up to 50% was allowable.

Over the past 4 years, 374 salt samples were randomly collected from 68 different countries throughout the world at consumer end point. This was not a systematic sampling but this mode of obtaining samples was chosen in order to avoid biased collection of salt samples.

The iodine content of these 374 salt samples ranged from less than 1 ppm to 810 ppm. Forty-five percent (170 samples) had no iodine at all and most of them (82%) were from developing countries including some presently claiming that they use the universal salt iodization. Based on daily requirements of 150 µg of iodine and a daily consumption of 10 g of salt, only 10% of the samples were appropriately iodinated, whereas another 25% would have provided a supraphysiological amount of iodine including 8% corresponding to an exaggerated intake (3 to 10 times higher than the physiological needs).

The iodine contents of the salt samples were also analyzed by country. With a few exceptions within each country, a wide range of iodine concentrations was also observed. As an example, the median, the range, and the number of samples for some Asian countries were as follows: Indonesia (5; <1–26; n = 9), Laos (29; <1–58; n = 6), Malaysia (<1; <1–77; n = 37), Myanmar (<1; <1–10; n = 10), Nepal (13; <1–28; n = 5), Pakistan (9; <1–76; n = 11), Saudi Arabia (50; <1–56; n = 5), Thailand (7; <1–64; n = 16). Homogeneous results were only observed in Bhutan (35; 34–37; n = 4). In all continents, tremendous variations were observed both in developing and developed countries.

For the last two years, another set of 294 salt samples was also collected in the Kivu in Eastern Zaire. Again, the iodine content of these samples ranged from <1 to 840 ppm, 11% had no iodine, 17% had a clearly excessive iodine content and only 18% had an iodine content in agreement with the daily physiological requirements. For a daily consumption of 10 grams of salt this corresponded to a daily iodine intake of <10 µg to 8400 µg of iodine. Among these samples, some with the same origin (same manufacturer) were identified. There were 11 samples from Malindi Industries in Mombasa (Kenya). The iodine content of these samples ranged from <1 to 67 ppm iodine with a median of 26 ppm. Four samples had no iodine at all.

The recent awareness of thyrotoxicosis in Zaire (Bourdoux et al. 1996) and Zimbabwe (Todd et al. 1995) is most likely related to the high levels of iodine which were observed in a significant number of salt samples.

Correct citation: Bourdoux, P.P., Van Thi, H.V., Van Stichelen, S., and Ermans, A.M. 1997. Universal salt iodization: Myth or reality? In *Trace Elements in Man and Animals – 9: Proceedings of the Ninth International Symposium on Trace Elements in Man and Animals. Edited by* P.W.F. Fischer, M.R. L'Abbé, K.A. Cockell, and R.S. Gibson. NRC Research Press, Ottawa, Canada. pp. 397–398.

The erratic results obtained on the 668 salt samples that we have analyzed clearly indicate that a problem still exists with iodination techniques and/or preservation of the iodine content in salt samples. These two problems certainly represent obstacles towards successful universal salt iodization.

In this context, appropriate measures should have a direct and positive impact on universal salt iodization. They include

1) a real improvement in iodization techniques for producing salt with adequate and constant levels of iodine
2) the implementation of efficient monitoring procedures at the factory level and at the consumer level.

Literature Cited

Bourdoux, P.P., Ermans, A.M., Mukalay wa Mukalay, A., Filetti, S. & Vigneri, R. (1996) Iodine-induced thyrotoxicosis in Kivu, Zaire. Lancet 347: 552–553.

DeMaeyer, E.M., Lowenstein, F.W. & Thilly, C.H. (1979) The control of endemic goitre. World Health Organization, Geneva.

Food and Nutrition Board. (1989) Committee on Dietary Allowances. US National Research Council. Iodine. In: Recommended dietary allowances, 10th ed. Washington, DC: National Academy Press Publishers, 1989: 213.

ICCIDD Board. (1992) Recommendation on iodine supplements in food aid programs. IDD Newsletter 8: 23.

Todd, C.H., Allain, T., Gomo, Z.A.R., Hasler, J.A., Ndiweni, M. & Oken, E. (1995) Increase in thyrotoxicosis associated with iodine supplements in Zimbabwe. Lancet 346: 1563–1564.

Discussion

Q1. Solo Kuvibidila, Louisiana State University, New Orleans, LA, USA: Given the prevalence and severity of IDD in developing countries (including Zaire), and the fact that many samples have high excess iodine (which will cause thyrotoxicosis), how do we deal with this problem?

A. It is difficult to deal with this. It is difficult to understand the reasoning that a whole country should be treated, in order to reach one population. It would seem more sensible to identify which local populations are at risk, and to deal with these at the local level, for example with iodized salts, or iodized oils. Why force people to buy the more expensive iodized salt, if they don't need it?

Essentiality vs. Toxicity of Essential Trace Elements: A Nutritional Toxicologist Looks at the Upper Safe Level

JAMES R. COUGHLIN

Coughlin & Associates, Laguna Niguel, California, USA

Keywords risk assessment, uncertainty factors, bioavailability, chronic disease prevention

As nutritionists face the challenge of whether to incorporate concepts beyond the alleviation of nutrient deficiencies to include the reduction of risks of chronic disease into the development of higher nutrient RDAs and Estimated Safe and Adequate Daily Dietary Intakes (ESADDIs), they quickly encounter the toxicology community struggling with the lower reaches of its dose-response curves. The nutritionists' twin goals of specifying amounts of nutrients needed to maintain adequate health by preventing deficiency diseases and providing advice to the public about nutritional avenues to chronic disease prevention often meet near the essentiality/ toxicity interface for specific nutrients, including some essential trace elements (ETEs).

In like fashion, toxicologists striving to set regulatory standards for drinking water, food and food additives, air or hazardous waste sites are often just as likely to confront extrapolated acceptable daily intakes (ADIs) which may fall near or below even current RDA/ESADDI levels for some nutrients, especially the ETEs. This is often true if these agencies apply traditional, 100-fold safety or uncertainty factors (UFs) to the toxic no-observed-adverse-effect-level (NOAEL) of a chemical. Although default UFs are generally utilized by regulators to provide maximum public health protection for most chemical toxicants, the U.S. Environmental Protection Agency (EPA) and other national and international regulatory agencies have recently begun to use reduced UFs in risk assessment when such factors are justified by pharmacokinetic, pharmacodynamic, bioavailability and speciation data on nutrients such as the ETEs. Some of these agencies have also begun to take human nutritional requirements into account when setting floors for regulatory standards for the ETEs.

Such innovative risk assessment techniques must be strongly considered by the nutrition community when it begins to undertake its revisions of current RDAs/ESADDIs for the ETEs as well as the vitamins (Food and Nutrition Board 1994). Some important risk assessment principles from the field of toxicology will be reviewed to assist nutritionists in their determination of upper safe levels for ETEs in anticipation of expanding the RDA concept to chronic disease prevention.

The FNB (1994) defined the "upper safe level" as that "level of intake of a nutrient or food component that appears to be safe for *most healthy people* and beyond which there is concern that some people will experience symptoms of toxicity over time." Such a reference level, if it can be scientifically well-determined, is at the high end of the defined "safe range of intake," the trough in the U-shaped dose-response curve above the RDA and an intake range associated with a very low probability of either nutrient inadequacy or excess. In the past, the FNB has certainly taken toxicity considerations into account when setting RDAs and ESADDIs for the vitamins and minerals, and the Board has expressed its concerns in the "Excessive Intakes and Toxicity" sections for each nutrient. However, recent RDA committees have addressed the question of the pharmacological or therapeutic effects of nutrients at levels many times the RDA to achieve health effects which are unrelated to the nutrient's essential function and achievable levels in the food supply.

U.S. and other international regulatory agencies and health advisory bodies have set many standards and guideline values for ETEs in recent years, and it is apparent from Table 1 that the oral Reference Doses (RfDs, which is EPA's term for ADIs) set by the U.S. EPA for zinc, selenium, chromium III, manganese and copper are generally much less than one order of magnitude larger than the corresponding RDAs and ESADDIs for these trace minerals. During the past five years the EPA, the European Union

Correct citation: Coughlin, J.R. 1997. Essentiality vs. Toxicity of essential trace elements: A nutritional toxicologist looks at the upper safe level. In *Trace Elements in Man and Animals – 9: Proceedings of the Ninth International Symposium on Trace Elements in Man and Animals. Edited by* P.W.F. Fischer, M.R. L'Abbé, K.A. Cockell, and R.S. Gibson. NRC Research Press, Ottawa, Canada. pp. 399–400.

Table 1. RDAs, ESADDIs and oral RfDs for adults.

Element	RDA or ESADDI	RfD
Zinc (mg/d)	12–15	21
Selenium (μg/d)	55–70	350
Chromium III (μg/d)	50–200	70 000
Manganese (mg/d)	2–5	10
Copper (mg/d)	1.5–3	5.3

and the World Health Organization have established (or are in the process of establishing) drinking water standards and guideline values for these and other ETEs based upon human or animal NOAELs for various toxic endpoints. When the animal or human toxicity data on a chemical are insufficient or when there is a high level of uncertainty associated with the data, the toxicologist generally employs default UFs, such as the classical 100-fold safety or uncertainty factors (10-fold for animal to human interspecies extrapolation and 10-fold for possible susceptibility differences among humans), to these NOAELs to arrive at RfDs and ADIs, with such levels appearing to be without appreciable risk over a lifetime.

If these agencies had used default 100-fold UFs applied against the NOAELs for the ETEs shown in Table 1, the resulting acceptable levels would have been far below the essential requirement levels for these elements. Fortunately, the agencies will or have taken advantage of increased knowledge and reduced uncertainty about the toxicity of these elements to scientifically justify the application of reduced UFs in setting more reasonable regulatory standards. These reduced UFs can be based on many factors: the use of pharmacokinetics (factors determining delivery of the chemical to the site of toxicity) and pharmacodynamics (factors determining activity or potency at the site of toxicity); the bioavailability of the element; knowledge of chemical speciation which can mitigate the severity of toxic effects; improved use of biomarkers to assess exposure and toxicities; better knowledge of customary dietary, environmental and occupational exposures; and development of less conservative risk extrapolation models to more accurately predict the effects of real-life exposures from high-dose animal testing results.

Increased communication between the nutrition and toxicology communities should be fostered as a means of assisting the nutrition community in eventually setting upper safe levels for the ETEs. Significant progress has already been made in this arena beginning with the publication of a landmark workshop monograph (Mertz et al. 1994) on assessing risk of ETEs, and this work is the starting point for a major initiative (IPCS 1996) to establish an international blueprint for future risk assessments of ETEs. However, this developing dialogue between nutritionists and toxicologists must be strengthened even further as the former grapple with the determination of upper safe levels for nutrients and the latter confront the determination of acceptable regulatory standards for ETEs.

Literature Cited

Food and Nutrition Board. (1994) How Should the Recommended Dietary Allowances Be Revised? Institute of Medicine, National Academy Press, Washington, DC.

International Programme on Chemical Safety (IPCS). (1996) Principles and Methods for Assessment of Risk From Essential Trace Elements. Concepts Meeting, March 25–27, Washington, DC.

Mertz, W., Abernathy, C.O., & Olin, S.S. (1994) Risk Assessment of Essential Elements. ILSI Press, Washington, DC.

Discussion

Q1. Janet Greger, University of Wisconsin, Madison, WI, USA: I'd just like to point out that there will be a symposium on this subject at Experimental Biology next spring.

Q2. Geoffrey Judson, Central Veterinary Laboratory, Glenside, Southern Australia: We often see more than just a single nutrient in excess. What attention is given to these interactions at the higher levels?

A. This needs to be done, but in toxicology research, a full investigation on even a single compound is extremely expensive. There are no really good ways to address interactions at present, but of course this is an area that needs attention.

Q3. Mary L'Abbe, Health Canada, Ottawa, ON, Canada: With regard to the process, is there any consideration of determining the most susceptible population, or sub-group of the population?

A. We do try to focus on this in certain cases, for example the report on Pesticides in Diets of Children, but this isn't a universal approach. This is an important consideration, however.

The AIN-93G Diet: Nephrocalcinosis, Kidney Calcium and Tissue Trace Element Levels

MARY R. L'ABBÉ,* KEITH D. TRICK, AND BARTHOLOMEUS BELONJE
Nutrition Research Division, Health Canada, Ottawa, Canada K1A 0L2

Keywords AIN-76A, AIN-93G, nephrocalcinosis, calcium/phosphorus ratio

In the 1970's, the American Institute of Nutrition established a standardized rodent diet (AIN-76A) that would meet the nutritional requirements of rats and mice and provide a consistent control diet for nutritional, toxicological and regulatory purposes (American Institute of Nutrition 1977). However, female rats fed this diet developed kidney calcium deposits (nephrocalcinosis). The purpose of this study was to evaluate the effect of the new standardized rodent diet (AIN-93G) on the incidence of nephrocalcinosis and compare it to the effects seen when rats were fed the AIN-76A diet.

Many studies have suggested that a low molar ratio of calcium to phosphorus is one of the prime contributing factors to the development of nephrocalcinosis (Shah et al. 1991). As a result, one of the major changes in the AIN-93G mineral mix was to change the chemical form of calcium and phosphorus, as well as lower the phosphorus content, thereby increasing the Ca/P molar ratio. The mineral mix in the AIN-93G diet was also formulated to take into account the contribution of casein to the P content of the final diet, another factor requiring the lowering of the P content of the mineral mix (Reeves et al. 1993a).

Materials and Methods

Weanling Sprague Dawley rats (Charles River Canada, St Constant, Que, initial body weights: male 46.6 g, female 42.5 g) were randomly divided into 3 groups and fed either AIN-76A, AIN-93G or a Laboratory Rodent Diet (LRD, powdered 5001, PMI Feeds Inc, Richmond, IN) in both short term (6 weeks) and long term (16 week) studies (8 male (M) + 8 female (F)/group per time period). At the end of each test period, rats were anesthetized with isoflurane (5% in O_2) and killed by exsanguination from the abdominal aorta. Kidneys were removed, weighed and stored at –80°C until analyzed. Kidney tissue and diet samples were dry ashed, dissolved in HNO_3 containing La and calcium was determined by atomic absorption spectroscopy. Phosphorus in the digestate was determined by a colorimetric method (Murphy and Riley 1962).

Results and Discussion

By analysis, the phosphorus content of the AIN-93G diet was significantly reduced compared the previous AIN-76A diet, thus providing a Ca/P ratio that was 65% higher than the previous ratio (Table 1).

After 6 weeks, nephrocalcinosis was observed in 1 M and 8 F fed the AIN-76A diet; no rats fed the AIN-93G diet or LRD developed nephrocalcinosis, although 2 F rats fed AIN-93G had elevated kidney Ca levels (data not shown). In the long term study (16 weeks), evidence of nephrocalcinosis was seen in 2 M and 8 F fed the AIN-76A diet and 1 F in each of the AIN-93G and LRD diet groups. In addition, elevated kidney calcium levels were observed in 1 M fed AIN-76A; 2 F fed AIN-93G, and 1 M and 2 F fed LRD (Table 2).

Thus, the AIN-93G diet significantly reduced the incidence of nephrocalcinosis in female rats compared to the AIN-76A diet, however, the diet did not totally eliminate it in the long term study (16 weeks). In contrast to our results, Reeves et al. (1993b), found no evidence of elevated kidney Ca in rats fed a

*Author to whom reprint requests should be addressed. Publication No. 488 of the Bureau of Nutritional Sciences.

Correct citation: L'Abbé, M.R., Trick, K.D., and Belonje, B. 1997. The AIN-93G diet: Nephrocalcinosis, kidney calcium and tissue trace element levels. In *Trace Elements in Man and Animals – 9: Proceedings of the Ninth International Symposium on Trace Elements in Man and Animals. Edited by* P.W.F. Fischer, M.R. L'Abbé, K.A. Cockell, and R.S. Gibson. NRC Research Press, Ottawa, Canada. pp. 401–402.

Table 1. Diet summary: (by analysis, n = 16).

	Ca (mg/kg)	P (mg/kg)	molar Ratio
AIN-76A	4,564	4,467	0.79
AIN-93G	4,481	2,658	1.31
LRD	12,098	5,859	1.60

Table 2. Kidney calcium — Week 16 (µg/g dry wt).

	Male			Female		
	Mean ± SD	Range	N/E/NC[1]	Mean ± SD	Range	N/E/NC[1]
AIN-76A	1072*	233–3709	5/1/2	32,734*	9,100–61,099	0/0/8
AIN-93G	244 ± 21	217–279	8/0/0	1,656*	235–10,289	5/2/1
LRD	240*	196–01	7/1/0	1,517*	213–7,478	4/2/1[2]

[1]Classification of kidney Ca levels. Number of rats (of 8) with Normal (N), Elevated (E) or with Nephrocalcinosis (NC), where E was kidney Ca > 350 µg/g (i.e. >Normal + 2SD), NC was kidney Ca > 10 × N.
*At least 1 rat had elevated kidney Ca or nephrocalcinosis.
[2]n = 7, 1 sample lost during analysis.

slightly modified AIN-93 diet for 16 weeks, although rats were from a different supplier and these researchers used a diet with slight differences in the composition of ultra trace elements. Additional magnesium and fluoride have been shown to have a beneficial effect in reducing nephrocalcinosis in female rats (Shah and Belonje 1983). Possibly these other elements would have to be increased in the AIN-93G diet to totally eliminate the nephrocalcinosis which we saw in the female rats in the long term study.

Literature Cited

American Institute of Nutrition (1977) Report of the American Institute of Nutrition Ad Hoc Committee on Standard for Nutritional Studies. J. Nutr. 107: 1340–1348.

Murphy, J. & Riley, J.P. (1962) A modified single solution method for the determination of phosphorus in natural waters. Anal. Chim. Acta. 27: 31–36.

Reeves, P.G., Nielsen, F.H. & Fahey, G.C. Jr. (1993a) AIN-93 Purified diets for laboratory rodents: final report of the American Institute of Nutrition ad hoc writing committee on the reformulation of the AIN-76A rodent diet. J. Nutr. 123: 1939–1951.

Reeves, P.G., Rossow, K.L. Lindlauf J. (1993b) Development and testing of the AIN-93 Purified diets for Rodents: Results on Growth, Kidney Calcification and Bone Mineralization in Rats and Mice. J. Nutr. 123: 1923–1931.

Shah, B.G. & Belonje. B. (1983) Prevention of nephrocalcinosis in male and female rats by providing fluoride and additional magnesium in the diet. Nutr. Res. 3: 749–760.

Shah, B.G., Trick, K.D. & Belonje. B. (1991) Dietary factors in rat nephrocalcinosis. Trace Elem. Med. 8: 154–160.

Discussion

Q1. Leslie Klevay, USDA-ARS, Grand Forks, ND, USA: How much calcium does it take in a kidney, for it to be visible by microscopy? Could the nephrocalcinosis be a dietary pH problem?

A. As you know, with microscopy, it all depends upon luck in sectioning, so in some ways it can be tough to score something as negative.

Q2. Solo Kuvibidila, Louisiana State University, New Orleans, LA, USA: I have just a general question. Because of the differences you report between males and females, should we be considering gender-specific diets?

A. It would add tremendous complexity to experimental designs to do that, so I think we should try to find one diet which would give acceptable results with both genders.

Q3. Janet Greger, University of Wisconsin, Madison, WI, USA: Phil, Carl and I all worked on the National Academy report on the AIN diets. The literature indicated a multifactorial nature for calcinosis. The modification to the Ca/P ration came out of our determinations, and I'm thrilled to see that it seemed to work. I would suspect that you're right, too, that it isn't just the Ca/P ratio, that there are other factors involved. It's an interesting problem. Have you done any renal function tests? We tried in our lab, and found that we couldn't really show functional changes, even when there was frank calcinosis.

A. We did a battery of clinical chemistry tests and so on, but these were not different in our studies either.

Q4. Saumya Sivaram, University of Texas Medical Branch, Galveston, TX, USA: Did the levels of any of the toxic elements change?

A. We didn't look at any of those.

Dose-Rate Idiorrhythm is a Powerful Tool for the Detection of Subtle Mineral Interactions. A Case for the Expression of Recommended Dietary Allowances (RDAs) and Safety Limits (RFDs) in the Range Format

B. MOMČILOVIĆ

USDA, ARS, Grand Forks Human Nutrition Research Center Grand Forks, ND 58202, USA, and Institute for Medical Research & Occupational Health, POB 291, Zagreb, Croatia

Keywords zinc dose-rate idiorrhythm, mineral interactions, recommended dietary allowances, safety limits

The physiological and toxicological significance of trace element (TE) interactions is sometimes debatable because they often have been demonstrated under extreme experimental conditions and have used animals whose TE status was already severely compromised (Bremner and Beattie 1995). Detection of TE interactions is customarily associated with the availability of sensitive biological indicators but no attention has been paid to the limitations of the classical dose-response model, which treat the dose as independent of time (Momčilović and Reeves 1996, Momčilović 1988). Nutrition can be described as a series of time-discrete, dose-rate events in which specific nutritional information evolves in time and space and, therefore, a new, time-sensitive idiorrhythmic experimental feeding model was proposed for the study of nutrient dose-rate impact upon growth and metabolism (Momčilović 1995, Momčilović 1993). By using zinc as a model nutrient, the aim of this experiment was to investigate the potential of idiorrhythmic dose-rate feeding for the detection of subtle TE interactions within the physiologic range of dietary intake.

The effect of idiorrhythmic dose-rate variability in dietary zinc intake on Ca, P, Mg, Na, K, Fe, Zn, Cu, and Mn deposition in the femur of weanling rats was studied. To feed by a zinc dose-rate idiorrhythm (Ix) means to administer a distinctly proportional and regularly recurring pattern of zinc dose coupled to frequency so that the preselected dose-time equivalent (Modulo, Mx) is kept constant over the duration of the experiment (Epoch, E): $Ix = [d_{nth}(Mx)]/d_{nth}$, where d_{nth} is the sequential number of zinc dosing day separated by the period of feeding the diet without zinc if $d_{nth} > 1$ (Momčilović 1995, Momčilović 1993). Low zinc M3, adequate zinc M12, and high zinc M48 provided 3, 12, and 48 mg Zn $\cdot$ kg^{-1} diet $\cdot$ d^{-1} over a 48-day idiorrhythmic epoch, respectively. Each Mx had eight analogous idiorrhythms: I = Mx/1, 2Mx/2, 3Mx/3, 4Mx/4, 5Mx/5, 6Mx/6, 7Mx/7, and 8Mx/8. The latter provided 24, 96, and 384 mg Zn $\cdot$ kg^{-1} diet $\cdot$ d_8^{-1} separated by 7 d of feeding a diet without zinc (Momčilović et al. 1976) for M3, M12, and M48, respectively. Citrus Leaves (#1572) and Bovine Liver (#1577a) (National Institute of Standards & Technology, Gaithersburg, MD) were used as quality control materials for the inductively coupled argon plasma atomic emission spectrometry analysis (Nielsen et al. 1988). The effects of Mx, Ix, and Mx·Ix interaction upon mineral deposition in the femur were analyzed by two-way ANOVA (Table 1). Only the P values of the results are shown because presentation of any single element would require two 3×8 matrices (24 cells with 8 replicates each) for the amount and concentration respectively.

Zinc dose-time equivalent modulo Mx significantly affected the amount and concentration of all the minerals studied in the femur. Zinc dose-rate idiorrhythm Ix affected all the elements in the femur except the total amount of Mn and concentrations of Mg, Fe, and Mn. An Mx·Ix interaction significantly affected the amount but not concentration of the majority of the minerals in the femur. The amount and

Correct citation: Momčilović, B. 1997. Dose-rate idiorrhythm is a powerful tool for the detection of subtle mineral interactions. A case for the expression of recommended dietary allowances (RDAs) and safety limits (RFDs) in the range format. In *Trace Elements in Man and Animals – 9: Proceedings of the Ninth International Symposium on Trace Elements in Man and Animals. Edited by* P.W.F. Fischer, M.R. L'Abbé, K.A. Cockell, and R.S. Gibson. NRC Research Press, Ottawa, Canada. pp. 403–405.

Table 1. P values for the effect of zinc dose-rate idiorrhythm (Ix), zinc dose-time equivalent modulo (Mx), and Mx · Ix interaction on Ca, P, Na, K, Mg, Fe, Zn, Cu, and Mn in the femur.[1,2]

	Amount (mg)			Concentration (mg/kg)		
	Mx	Ix	Mx · Ix	Mx	Ix	Mx · Ix
	Analysis of Variance P Values					
Ca	**0.0001**	**0.0001**	0.0011	**0.0001**	0.0157	NS
P	**0.0001**	**0.0001**	0.0009	**0.0001**	0.0200	NS
Mg	**0.0001**	**0.0001**	0.0006	**0.0001**	NS	NS
Na	**0.0001**	**0.0001**	NS	**0.0001**	0.0312	0.0178
K	**0.0001**	**0.0001**	**0.0001**	**0.0001**	0.0452	NS
Fe	**0.0001**	0.0091	NS	**0.0001**	NS	0.0006
Zn	**0.0001**	**0.0001**	**0.0001**	**0.0001**	**0.0001**	**0.0001**
Cu	**0.0001**	0.0052	0.0004	**0.0001**	0.0105	**0.0001**
Mn	**0.0001**	NS	NS	**0.0001**	NS	NS

[1]Two-way ANOVA (DF: Mx = 2, Ix = 7, Mx · Ix interaction = 14; 24 experimental groups of 8 animals each). [2]NS not significant ($p > 0.05$). Boldface indicates that the P values are < 0.0001.

concentration of zinc in the femur were particularly affected by Mx, Ix, and Mx · Ix interaction. Mx affected the amount and concentration of all the minerals in incisors whereas Ix and Mx · Ix interaction affected only total Ca, Na, K, Zn, and Cu ($p < 0.05$). This indicates that the metabolic handling of zinc by the two calcified tissues was different (data not shown).

The results of this study show the great power of the time-sensitive idiorrhythmic experimental feeding model to elucidate subtle and previously undetectable TE interactions induced by physiological amounts of dietary zinc. Femur dry weight (mg) was affected by Mx, Ix, and Mx · Ix interaction ($p < 0.05$ for each variable). Femur has already been described as the most sensitive indicator tissue for the assessment of the bioavailability of dietary zinc (Momčilović et al. 1975).

Recently, it was shown that the femur acts as an ultra-sensitive threshold switch for zinc status, i.e., more sensitive than expected from classical kinetics (Momčilović et al. 1996). In the present study, both amount and concentration of each element in femur were affected with peak dietary zinc intakes occurring below the doses known to induce a measurable biological effect. The pharmacological effects of continuous feeding of 400 mg Zn/d for over three months showed a mild adverse effect on human immunity (Fosmire 1990) but apparently is non-toxic in the rat (Vallee and Falchuk 1993). The findings suggest that the metabolic incorporation of trace elements into calcified tissues may be viewed as a set of dynamic time-dependent multi-element fluctuations that tend to equilibrate around some provisional homeostatic set point. They also support the view that homeostatic regulation is of critical importance in setting RDAs and especially RfDs, both of which state their recommendations as one number (Mertz 1995). If interactions between TE do occur within the physiological range of dietary intake, as idiorrhythmic dose-rate feeding shows, then homeostasis and threshold cannot be viewed as fixed points but should be viewed as time-dependent ranges. Both threshold and homeostasis may be viewed as a multidimensional phase space (Brosa et al. 1991); a common vector of ceaseless transactional equilibria within the network of metabolic pathways and in response to nutritional signals from the environment. Hence, recognition of the importance of homeostasis should result in RDAs and RfDs that describe a range within which homeostasis is reliably effective in maintaining adequacy and indicate the level at which it predictably breaks down.

Literature Cited

Bremner, I. & Beattie, J.H. (1995) Copper and zinc metabolism in health and disease: Speciation and interactions. Proc. Nutr. Soc. 54: 489–499.

Brosa, U., Harms, H.M., Prank, K. & Hesch, R.D. (1991) Inspection of multidimentional phase spaces with an application to the dynamics of hormone systems. J.Physiol. 1: 273–278.

Fosmire, G. (1990) Zinc toxicity. Am. J. Clin. Nutr. 51: 225–227.

Mertz, W.E. (1995) Risk assessment of essential trace elements: New approaches to setting recommended dietary allowances and safety limits. Nutr.Rev. 53: 179–185.

Momčilović, B., Belonje, B., Giroux, A. & Shah, B.G. (1975) Suitability of young rat tissue for a zinc bioassay. Nutr. Rep. Int. 12: 197–203.

Momčilović, B., Belonje, B., Giroux, A. & Shah, B.G. (1976) Bioavailability of zinc in milk and soy protein-based infant formulas. J. Nutr. 106: 913–917.

Momčilović, B. (1988) The epistemology of trace element balance and interaction. In: Trace Elements in Man and Animals – 6 (Hurley, L.S., Keen, C.L., Lonnerdal, B. & Rucker, R.B. eds.), vol. 6, pp.173–177. Plenum Press, New York, NY.

Momčilović, B. (1993) Idiorrhythmic vs. continuous zinc dietary intake — a model approach to the study of trace element dose/rate impact. In: Trace Elements in Man and Animals — TEMA 8 (Anke, M., Meissner, D. & Mills, C.F. eds.), vol. 8, pp.194–197. Verlag Media Touristik, Gersdorf, Germany.

Momčilović, B. (1995) Coupling of zinc dose to frequency in a regularly recurrent pattern shows a limited capacity of excessive dietary zinc to compensate for a previously deficient intake. J. Nutr. 125: 2687–2699.

Momčilović, B., Reeves, P.G. & Blake, M.J. (1996) Coupling of zinc dose to frequency in a regularly recurring pattern changes zinc deposition in the femur and incisor of young growing rats. (in preparation).

Momčilović, B. & Reeves, P.G. (1996) Dietary variability and metabolic availability: Effect of the idiorrhythmic zinc dose-rate variability upon zinc deposition in the femur and incisor of weanling rats by a slope-ratio assay. J. Nutr. Biochem. 7: (in press).

Nielsen, F.H., Shuler, T.R., Zimmerman, T.J. & Uthus, E.O. (1988) Magnesium and methionine deprivation affect the response of rats to boron deprivation. Biol. Trace Elem. Res. 17: 91–107.

Vallee, B.L. & Falchuk, K.H. (1993) The biochemical basis of zinc physiology. Physiol. Rev. 73: 79–118.

Analytical Quality Control

Chair: A. Chatt

Knowledge Gaps in Analytical Quality Control

ROBERT DABEKA

Food Research Division 2203D, Bureau of Chemical Safety, Health Protection Branch, Health Canada, Ottawa, Ontario, Canada K1A 0L2

Keywords lead, quality control, ICPMS

The only way to validate an analytical result obtained for an unknown sample is to use a completely independent method of analysis. As this is impractical for most situations, alternative quality control measures, such as standard reference materials (SRMs) and recovery studies are used. Despite the fact that most SRMs are sold at less than cost, their price makes them inaccessible to many laboratories. In addition, it is often impossible to purchase an SRM material which precisely simulates the samples requiring analysis. Thus, the technique of recovery studies is often the only validation given to an analytical result.

Unfortunately, recovery studies do not evaluate the accuracy of the blank signal or unspiked sample signal; they only evaluate the signal of the added analyte, thus allowing orders of magnitude errors even when giving 100% recovery. For example, recovery studies will not correct for errors due to an invalid blank, an invalid baseline, and uncorrected background absorption in atomic spectrometry. Similarly, the method of standard additions will not identify or correct for errors of these types.

Several quality control measures which are unknown to most investigators had been introduced previously (1). These are the inclusion of 2–8 reagent blanks in each analytical batch, and the sample weight test. The sample weight test is an alternative quality control measure which can identify the errors described above. It consists of taking 2 different sample weights for analysis, one weight being 2–3 times that of the other weight. If the same analytical result is obtained at each sample weight, then there is a much greater chance that analytical errors are absent. If a different result is obtained for each sample weight, then there is a clear indication that an analytical error of the type described above has occurred. Application of these techniques to the routine analysis of food samples by inductively coupled plasma mass spectrometry (ICPMS) is presented in this paper along with a comparison of the usefulness of the techniques in detecting actual analytical errors.

Materials and Methods

Instrumentation consisted of a FISONS VG PLASMAQUAD-IIe ICPMS operated under conditions recommended by the manufacturer.

Deionized water was used throughout. Reagents were generally of high-purity type and were either purchased from different manufacturers or, in the case of nitric and perchloric acids, sub-boiling distilled in quartz in the laboratory.

Most of the described analyses were on a market basket study in which 135 different food composites from 5 cities were analyzed. Lead was one of 15 elements determined. Samples (0.5–5 g) were digested in quartz flasks with nitric and perchloric acids, the digest brought to 3–5 mL perchloric acid by boiling

Correct citation: Dabeka, R. 1997. Knowledge gaps in analytical quality control. In *Trace Elements in Man and Animals – 9: Proceedings of the Ninth International Symposium on Trace Elements in Man and Animals. Edited by* P.W.F. Fischer, M.R. L'Abbé, K.A. Cockell, and R.S. Gibson. NRC Research Press, Ottawa, Canada. pp. 406–409.

and the solution was diluted to 100 mL with deionized water. Quality control parameters included in each analytical batch were

1. 7 reagent blanks run both before and after the samples
2. 2 standard blanks (standards were not digested)
3. 3 standard reference materials (SRMs), purchased from the National Institute of Standards and Technology (NIST), each run in duplicate at 2 different sample weights
4. reagent blank spike recovery (duplicates)
5. sample spike recovery (duplicates)
6. samples run in duplicate at 2 different sample weights
7. cross-check standard (standard of same elements from a completely independent source)

Results and Discussion

SRMs as a Test for other Quality Control Measures. Three SRMs with certified lead levels were included in each analytical batch. The average concentrations obtained for the SRMs over all batches agreed well with the certified levels (Table 1), although there were individual batches for which poor agreement was found. Normally, SRMs are used as a quality control measure to test for the accuracy of the unknown samples. This study uses the SRM results as a test for the efficacy of other quality control measures; i.e., can inaccurate SRM results be revealed using quality control parameters such as recovery studies.

Blank Spike Recoveries. Blank spike recovery studies serve as a check on the validity of the standards when the standards are not digested. The blank spike recoveries for lead averaged 102% and for individual batches ranged from 97 to 112%. Figure 1 shows that the blank spike recoveries reflected inaccuracy of the SRMs very poorly.

Sample Spike Recoveries. Lead spikes, equivalent to 10 ng/mL final solution concentration, were added to different samples in each batch to test for recovery. The sample spike recoveries averaged 97% and over all batches ranged from 92 to 103%. Figure 1 shows that the sample spike recoveries reflected changes in the SRM concentrations negligibly. Recovery studies are the main technique used for validating analytical accuracy, and this study shows that they are a poor quality control measure.

Reagent Blanks. Over all batches, the mean level for lead in the 7 reagent blanks run at the beginning of the batch was 0.095 ng/mL, and ranged from 0.031 to 0.238 ng/mL (Table 1). The standard deviation of these blanks averaged 0.018 ng/mL and ranged from 0.0026 to 0.062 ng/mL. Both blank level and

Table 1. Range over 16 analytical batches of quality control parameters for lead.

QC Parameter	Mean	Median	Minimum	Maximum
Reagent blank level, ng/mL[a]	0.095	0.073	0.031	0.24
Reagent blank std dev, ng/mL[a]	0.018	0.013	0.0026	0.062
Reagent blank level, ng/mL[b]	0.108	0.075	0.019	0.45
Reagent blank std dev, ng/mL[b]	0.012	0.0075	0.0025	0.041
Reagent blank spike recovery, %	102	102	97	112
Sample spike recovery, %	97	97	92	103
Agreement with standard from independent source, %	101	102	97	104
Standard sensitivity change from start to end of run, %	97	98	88	106
Bovine liver SRM, ng/g[c]	140	135	118	188
Skim milk SRM, ng/g[c]	23	21	0.7	56
Wheat flour SRM, ng/g[c]	21	15	0.1	64

[a]7 sample blanks run before samples.
[b]Same 7 sample blank solutions run in batch after samples.
[c]NIST Bovine liver SRM 1577a certified at 129 ± 4 ng/g, NIST Skim milk powder SRM 1549 certified at 19 ± 3 ng/g, Wheat flour SRM certified at 20 ± 10 ng/g.

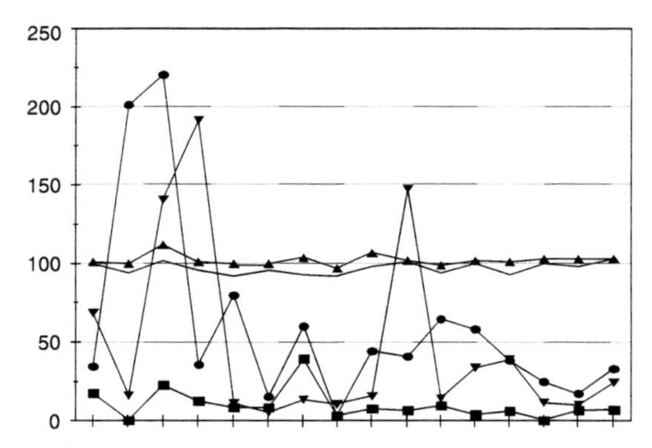

Figure 1. Batch-to-batch variation of blank (▲) and sample (+) spike recoveries, %, and lead concentrations,% error, in bovine liver (■), milk powder (▼), and wheat flour (●) SRMs.

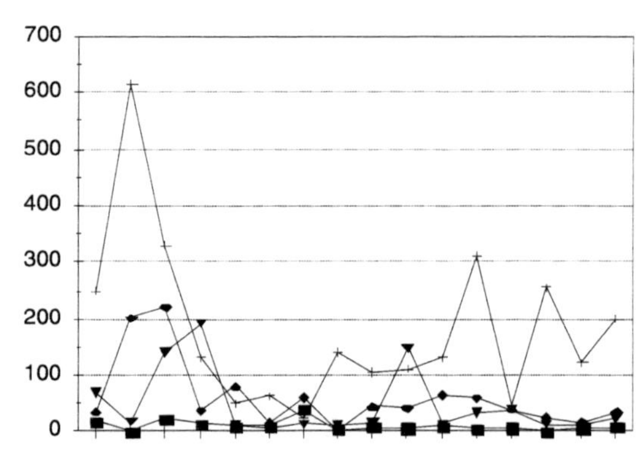

Figure 2. Batch-to-batch variation of reagent blank standard deviation times 10000 (+), and lead concentration error, %, in bovine liver (■), milk powder (▼) and wheat flour (●) SRMs.

standard deviation varied significantly from batch to batch. Thus, even when digestions were performed in one laboratory, day-to-day contamination and changes in instrument precision affected the solution detection limit (expressed as 3 times the reagent blank standard deviation) by more than an order of magnitude.

The blank standard deviation appeared to be a poor indicator of SRM inaccuracy, and only in batches 3 and 4 was a correlation found for 2 of the wheat flour results and one of the milk powder results. The high inaccuracy for milk powder in batch 4 was accompanied by a good blank standard deviation.

Sample Weight Test. The sample weight test is based on the difference in concentration found for the replicates of the same food run at the two sample weights, and is quantified according to the formula

$$\% \text{ Weight Error} = \text{ABS}\ \{100(\text{Conc. low weight} - \text{Conc. high weight})/\text{Conc. high weight}\}$$

where Conc. refers to the concentration of lead found in the reference material. The absolute value of the weight error is used only for clarity of graphical illustration. A batch-by-batch comparison of the sample weight test results with the analytical errors, based on the certified concentrations of lead in the SRMs, gave a fairly good correlation (Figure 3). Thus, it is possible to use the sample weight test to identify many of the major errors. Furthermore, this is possible, as seen from Figure 3, for individual samples within the same batch. This fundamentally differs from both recovery studies and reference materials which would simply evaluate the batch as a whole.

Conclusions

Of the tested quality control measures, the sample weight test was found most sensitive to analytical errors. Recovery studies were the least reliable indicator of accuracy. The weight test has the advantage of not requiring any additional work or reagents if more than one sample replicate is normally run. Furthermore, it evaluates the accuracy of each sample run in the batch as opposed to generalizing about the batch as a whole. It is most sensitive at low concentrations where errors due to contamination or poor sensitivity are most likely.

Acknowledgements

Thanks are due to Karen Dalglish and Art McKenzie for their excellent technical assistance.

References

Dabeka, R.W. & Hayward, S. (1993) Missing aspects in quality Control. In: Quality Assurance for Analytical Laboratories (Parkany, M., ed.), pp 67–79, Royal Society of Chemistry.

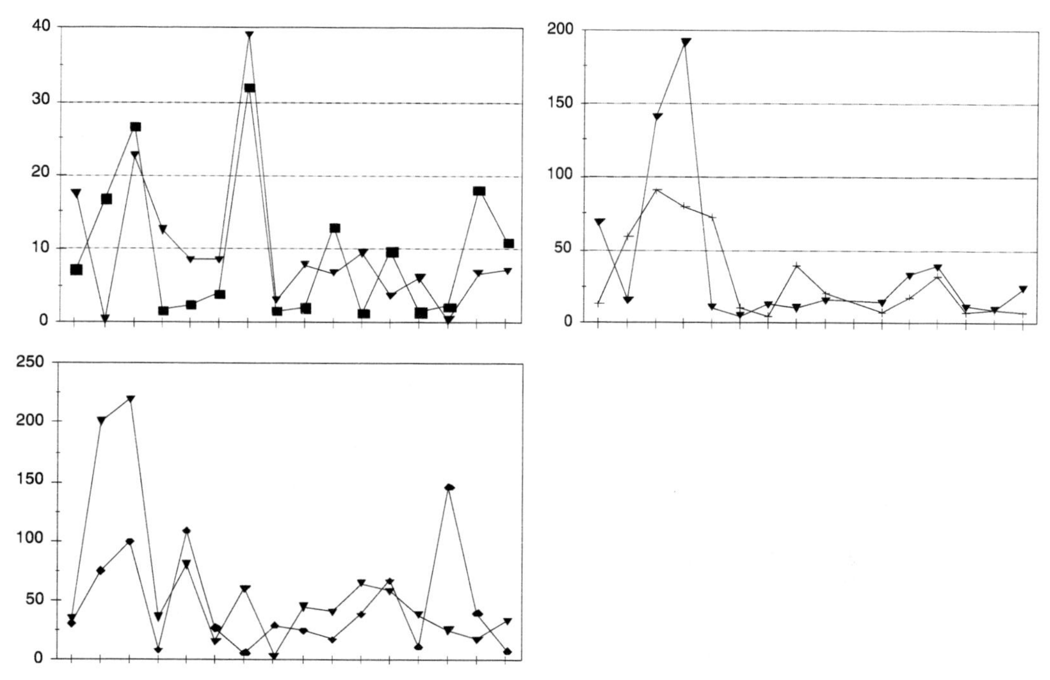

Figure 3. Batch-to-batch variation of the % weight error (▼), and the % analytical error, based on the certified levels, of lead in bovine liver (■), milk powder (+) and wheat flour (●) SRMs.

Discussion

Q1. Shaole Wu, Alberta Environmental Centre, Vegreville, AB, Canada: Why was there so much difference between results from the two labs for lead? And secondly, with the weight test method; you are assuming that the digestion acid used was suitable or adequate for both of the weights used? You are not increasing the amount of the acid used.

A. I don't know why the difference. When we told them the results, they didn't want to discuss it. We were refereeing, and we agreed with the lower concentration. With the weight test method, it is assumed that you choose your sample weights such that you do not have to vary the digestion material. If possible, you should reduce the weight of sample rather than increase it. Of course, if you're too near your detection limits, then the accuracy of the assay is not going to be very good.

Q2. Micheal Inskip, Health Canada, Ottawa, ON, Canada: We want to look at lead in other biological materials as the reference bovine liver is such a difficult matrix to work with. Are there other Standard Reference Materials that may be more suitable for biological materials?

A. Yes there are other standard materials you can work with such as skim milk powder, oyster and bovine serum.

Q3. Qianli Xie, University of Saskatchewan, Saskatoon, SK, Canada: If recovery is not a good technique for assessing accuracy, is the use of standard additions a good alternative?

A. No. You would get the same problems if you use standard additions. If the original signals are wrong it doesn't matter what you add to it. Recovery studies are a necessary part of analytical QC, but they won't reveal analytical baseline errors.

Q4. Eugene Gawalko, Canadian Grain Commission, Winnipeg, MB, Canada: Bob, are you aware of any standard reference material for vegetable oils?

A. No. (comment by Amares Chatt, Dalhousie University, Halifax, NS, Canada: What we do is to buy vegetable oils and make our own references).

Q5. Amares Chatt, Dalhousie University, Halifax, NS, Canada: Sample weight analysis is the only method that can be used when carrying out neutron activation analysis.

Q6. Shaole Wu, Alberta Environmental Centre, Vegreville, AB, Canada: I'd like to make one more comment. Spike recovery studies are still necessary for additive interference testing.

Unique Quality Assurance Aspects of Neutron Activation Analysis

ROBERT R. GREENBERG

Analytical Chemistry Division, Chemical Science and Technology Laboratory, National Institute of Standards and Technology, Gaithersburg, MD 20899, USA

Keywords neutron activation analysis, quality assurance, standard reference materials, certification

Neutron activation analysis (NAA) has become widely used, and is extremely valuable for the certification of Standard Reference Materials (SRMs) at NIST for a number of reasons. First, the method itself has characteristics which inherently provide few sources of error in common with other analytical techniques used at NIST. This is important since most certified elemental concentrations are derived from the data determined by two (and occasionally more) independent analytical techniques. Multiple techniques are typically used for SRM certification since, although each technique has previously been evaluated and shown to be accurate, unexpected problems can arise especially when analyzing new matrices. Another reason for NAA utilization is the unique quality assurance (QA) characteristics of the method which often allow the analytical values to be internally evaluated and cross checked. Characteristics of NAA leading to enhanced QA capabilities can be separated into five distinct areas: NAA is based on the nuclear properties of the elements; the method uses 3-dimensional spectroscopy; it possesses a capability for analytical redundancy; NAA is subject to very limited blank effects, and finally, this method posses a number of sample-dissolution/chemical processing advantages compared to many other analytical techniques. Fundamentals of this technique are described in Greenberg et al. (1984); specific applications of NAA to SRM certification are described in Becker et al. (1992 and 1994), Greenberg (1986) and Greenberg et al. (1988, 1990, and 1995).

Nuclear Properties: The fact that NAA is based on the nuclear properties of the elements provides a number of advantages. First, the physical basis of the technique is simple and well understood. This provides the capability of understanding and evaluating all potential sources of error. A complete list of all sources of error (uncertainty) for instrumental NAA measurements (without chemical manipulation of the samples) is given in Table 1. In addition, the analytical signal for NAA is independent of chemical form. All reactions employed in the activation and quantitation steps of this technique result from the interaction of neutrons with nuclei of the various atoms which are present in a sample. Different electronic configurations have no effect on these interactions. Thus irradiating a known quantity of an element in one chemical form will produce the identical amount of nuclear activity as will irradiating the same amount of that element in any other chemical form, as long as irradiation conditions are the same. In addition, there are only a limited number of nuclides that can be produced from a given target element via neutron irradiation. No new nuclear species will be observed during an analysis, in contrast to the potential for new or previously unrecognized interfering chemical species which may occur in many chemically-based analytical techniques. Finally, nuclear decay and counting (i.e., spectrometry) follow Poisson statistics. Therefore, an *a priori* estimate of analytical precision based upon counting statistics can be obtained. This estimate should be consistent with the observed sample-to-sample precision if all other sources of variance are under control. Thus the *a priori* estimate of precision can be used to assess the level of heterogeneity (inhomogeneity) of a material (see Becker (1993)), or alternatively if the material is known to be homogenous, to verify the absence of other, significant, sources of error for the specific analysis in progress.

Three-Dimensional Spectroscopy: Most spectroscopic techniques are two dimensional in that they separate the intensities of signals from the different excited species (analytes and interferences) by a single

Correct citation: Greenberg, R.R. 1997. Unique quality assurance aspects of neutron activation analysis. In *Trace Elements in Man and Animals – 9: Proceedings of the Ninth International Symposium on Trace Elements in Man and Animals. Edited by* P.W.F. Fischer, M.R. L'Abbé, K.A. Cockell, and R.S. Gibson. NRC Research Press, Ottawa, Canada. pp. 410–413.

Table 1. *Sources of error for INAA measurements.*

Sample Preparation (Pre-Irradiation) Errors
Mass Determination Basis Weight — Samples
Mass Determination — Comparators (Standards)
Concentrations of Comparators
Isotopic Variability
Blank Correction
Irradiation Errors
Irradiation Geometry Differences Between Samples and Comparators
Neutron Self-Shielding/Scattering Differences
Timing
Irradiation Interferences
Gamma Spectrometry (Post-Irradiation) Errors
Counting Statistics
Counting Geometry Differences
Pulse-Pileup Losses
Dead-Time Effects
Decay Timing Effects
Gamma-Ray Self Shielding
Gamma-Ray Interferences
Peak Integration Method

parameter, i.e., energy, mass, wavelength, time, etc. NAA is three dimensional since it separates the intensities of signals from excited species (radionuclides) via two different parameters, energy and time (half-life). The ability to perform gamma-ray spectrometry on the same sample after different decay times often provides the capability of verifying the absence of many sources of error such as interferences, incorrect pileup corrections, incorrect dead-time corrections, etc. For example, ^{51}Ti, the activated isotope used to quantify Ti, and ^{51}Cr, which is used to quantify Cr, have identical gamma ray energies. This is not surprising since they both decay to the same 320.08 keV excited state of ^{51}V. However, the 27.704 day half life of ^{51}Cr permits a complete separation of its signal from that of ^{51}Ti with a 5.76 minute half life. In fact, Ti can be quantified in the presence of Cr by first counting the sample shortly after irradiation to obtained the Ti signal (along with a small interference from Cr). The same sample is then recounted several hours later, after the ^{51}Ti has completely decayed away, to obtain the exact Cr interference to Ti, which can then be subtracted from the combined signal observed earlier.

Analytical Redundancy: NAA provides the capability of making many, seemingly redundant, measurements. However, these multiple measurements can provide many important QA checks of an analysis. For example, multiple gamma-ray peaks can often be determined simultaneously; quantification from two or more gamma-rays requires only slightly more effort than from one. Agreement between concentrations of an element determined from multiple gamma rays virtually eliminates the possibility of significant, unrecognized, gamma-ray interferences. Multiple counts performed at different decay times can be used to verify that corrections for dead-time and pulse pileup have been correctly applied. Such counts can also be used to verify the absence of significant interferences, since it is extremely rare that an element of interest, and one with an interfering gamma-ray, have nearly identical half lives. Multiple counts at different sample-to-detector distances can be used to quantify errors based on counting geometry. Multiple irradiations can be performed with different neutron spectra to verify the absence of significant fast neutron interferences. Use of multiple standards (of the same element) placed in different locations within the rabbit can identify irradiation geometry problems. Multiple standard mixes can be used to provide information on potential interferences from the isotopes of interest within the standards, as well as identify some types of peak integration problems. Many nuclides can be quantified using multiple activation products with different irradiation and decay characteristics. Finally, use of multiple control materials can help to identify a variety of analytical problems not observed with the use of a single control.

Limited Blanks: Chemical blanks are greatly reduced and are often insignificant for NAA compared to most other analytical techniques. Prior to, and during the irradiation step, the blank for NAA measurements is usually limited to environmental contamination (i.e., from dust or from implements used to manipulate the sample), and from direct, physical transfer of material from the irradiation container to the sample since samples can be transferred to clean containers after irradiation. After the irradiation step, the blank is only affected by contamination with radioactive species; there is no chemical-reagent blank. Since blank levels are typically small (or even negligible), their quantification and correction usually provide only an insignificant addition to the overall uncertainty of an NAA measurement. In fact when blank levels are significant, it is often possible to vary and compare different sample preparation and irradiation conditions to fully evaluate the total (not just reagent) blank values. One such example is presented in Wise et al. (1993).

Dissolution and Chemical Processing Advantages: Dissolution and chemical processing are not usually required for NAA; most analyses can be performed instrumentally (without any chemical treatment). When chemical processing is necessary after irradiation (radiochemistry), the ability to add carriers (nonradioactive (i.e., unirradiated) quantities of elements of interest) in nearly any amount (i.e., from nanograms to grams) provides a number of advantages such as enhancing separation from interfering elements, minimizing losses during dissolution and separation, and allowing the measurement of chemical yield by determining and comparing the amount of carrier present after the dissolution/separation process with the amount originally added. In addition, loss of material during the dissolution process can often be evaluated by gamma-ray spectrometric techniques. For example, potential particulate losses (undissolved material) can be observed by dissolving and filtering an irradiated sample. The resulting filter is then washed and submitted to gamma-ray spectrometry to identify and quantify any residual radioactive species left behind. Adsorption/absorption losses into/onto the walls of dissolution vessels, containers, glassware, etc. which are used for sample dissolution and/or chemical separation can be observed in a similar manner by performing gamma ray-spectrometry on each individual piece of apparatus used. More information on such use of gamma-ray spectrometry to uncover dissolution losses is provided in Greenberg et al. (1990).

In summary, the unique quality assurance capabilities of NAA often provide the means for analytical values which are determined from this technique to be internally evaluated and cross checked. This, combined with the lack of significant sources of error in common with other analytical techniques at NIST, contributes to NAA's value in certifying elemental concentrations in SRMs.

Literature Cited

Becker, D. A., Greenberg, R. R. & Stone, S. F. (1992) The use of high accuracy NAA for the certification of NIST standard reference materials. J. Radioanal. Nucl. Chem. 160: 41–53.

Becker, D. A. (1993) Unique quality assurance aspects of INAA for reference material homogeneity and certification. Fresenius J. Anal. Chem. 345: 298–301

Becker, D. A., Anderson, D. L., Lindstrom, R. M ., Greenberg, R. R., Garrity, K. M., & Mackey, E. A. (1994) Use of INAA, PGAA and RNAA to determine 30 elements for certification in an SRM: Tomato Leaves, 1573a, J. Radioanal. Nucl. Chem. 179: 149–154.

Greenberg, R. R., Fleming, R. F. & Zeisler, R. (1984) High sensitivity neutron activation analysis of biological/environmental standard reference materials. Environ. International 10: 129–136.

Greenberg, R. R. (1986) Elemental characterization of the National Bureau of Standards milk powder standard reference material by instrumental and radiochemical neutron activation analysis. Anal. Chem. 58: 2511–2516.

Greenberg, R. R., Zeisler, R., Kingston, H. M. & Sullivan, T. M., (1988) Neutron activation analysis of the NIST bovine serum standard reference material using chemical separations. Fresenius Z. Anal. Chem. 332: 652–656 .

Greenberg, R. R., Kingston, H. M., Watters, R. L. Jr. & Pratt, K. W. (1990) Dissolution problems with botanical reference materials. Fresenius. J. Anal. Chem. 338: 394–398.

Greenberg, R. R., Becker, D. A. & Mackey, E. A. (1995) The application of instrumental neutron activation analysis for the certification of the new NIST fly ash SRM. J. Radioanal. Nucl. Chem. 193: 7–14.

Wise, S. A., Schantz, M. M., Koster, B., J., Demiralp, R., Mackey, E. A., Greenberg, R. R., Burow, M., Ostapczuk, P., & Lillistolen, T. I. (1993) Development of frozen whale blubber and liver reference materials for the measurement of organic and inorganic contaminants. Fresenius Z. Anal. Chem. 345: 270–277.

Discussion

Q1. Lars Jorhem, National Food Administration, Uppsala, Sweden: You say that all sources of error are well known. How can you say that for sure?

A. The physical processes behind the technique are well known. When you irradiate a sample with neutrons you know what is going to happen. There are no molecular species which can be obtained. There are a limited number of things that can happen. The products are known, and all the sources of error are known and can be quantified.

Q2. Namik Aras, Middle East Technical University, Ankara, Turkey: One thing about chromium. You mentioned the transfer of chromium from the LPE plastic bags to the sample. Can it not be cleaned off beforehand?

A. The chromium is on the surface of the bag but it is chemically bound. Because it is chemically bound, it cannot be washed off. The recoil after neutron activation causes the transfer of chromium from the surface of the bag into the sample.

Q3. Namik Aras, Middle East Technical University, Ankara, Turkey: One other thing that I want to mention, pertaining to making corrections, for example with mercury 203 and selenium 75. They have the same gamma rays but different half lives and one can make the distinction using this. However, if you can use an alternative gamma ray, you don't have to wait to count the samples 4 months later to identify the element.

A. Let's go back to that case. Selenium is 120 days, mercury is 46 days. You can count a set of samples one month after irradiation and four months after irradiation, and you can statistically evaluate your data to determine the maximal amount of potential interference.

Q4. Bob Dabeka, Health Canada, Ottawa, ON, Canada: Is the kind of teflon used in digestion vessels the good teflon or the bad teflon?

A. It is the good (PFA) teflon and they are usually acceptable.

Q5. Bob Dabeka, Health Canada, Ottawa, ON, Canada: In the insoluble fraction determination; was that open-vessel digestion or high pressure digestion?

A. Actually, those digestions were done in two stages. First, low pressure, to break down most of the organic material, cooled, opened to relieve the pressure, closed and microwaved again. Two steps allowed us to reach a fairly high temperature. All the microwaving was done the same way. And then this one here was just diluted and transferred, these two were finished on a hot-plate with perchloric acid, this one had perchloric acid, but refluxed overnight.

Q6. Bob Dabeka, Health Canada, Ottawa, ON, Canada: Isn't there a possibility with overnight refluxing, of reducing the chromium?

A. Yes, 60% of the chromium was lost by volatility. What we did was we counted the solutions after separation, before the processing.

Comparison of Two Mineralization Procedures for Determination of Total Mercury in Biological and Clinical Materials by Cold Vapor Atomic Absorption Spectrometry*

M.C. RODRIGUEZ, J.M. SANCHEZ, H.S. CUBILLAN, M. HERNANDEZ, B.I. SEMPRUN, L.CH. BARRIOS, R.A. ROMERO AND V.A. GRANADILLO

Analytical Instrumentation Laboratory, University of Zulia, Maracaibo 4011, Venezuela

Keywords biological and clinical materials, cold vapor atomic absorption spectrometry, mercury, mineralization

Introduction

Mercury (Hg) is a potentially toxic element. As a consequence of its toxicity, there is a great interest in the determination of Hg in biological and clinical materials (Bermejo et al. 1994). All of these circumstances demand reliable analytical procedures to quantify Hg. Cold vapor atomic absorption spectrometry (CVAAS) is one of the most sensitive and simple techniques for the determination of Hg in biological and environmental materials (Adeloju et al. 1994). This technique requires a prior sample digestion to decompose organic matter. Recently, there has been a great interest in using microwaves to accelerate the mineralization of solid samples (Navarro et al. 1991). The microwave digestion decreases acid consumption, alleviates some safety risks and greatly shortens the sample preparation time (Sheppard et al. 1994). We report the successful mineralization of certified materials and real samples by two decomposition procedures.

Experimental

Mineralizations were performed in pressurized vessels (Parr 4782 microwave acid digestion bombs); heated by microwave irradiation using either a 600 W microwave oven CEM Model MDS-81D or a high intensity 950 W microwave oven CEM Model MDS-2100 (Matthews, NC). The hydride generation atomic absorption spectrometric data were measured by an atomic absorption spectrometer (Perkin-Elmer Model 460). For accuracy evaluation, standard reference materials were analyzed. Approximately 70 mg of solid sample or 0.5 mL of whole blood (1.0 mL for urine), and 2.5 mL of concentrated nitric acid were used for each mineralization. Two different microwave ovens were used: i) The system was placed into the microwave oven and irradiated for 70 s at 100% power (equivalent to 600 W and 2450 MHz), ii) The system was placed into the microwave oven and irradiated for 60 s at 100% power (equivalent to 950 W and 2450 MHz). After cooling to ambient temperature, the digestion solution was transferred into a 10-mL volumetric flask and diluted to volume with 1 M nitric acid 1 M perchloric acid solution. Sodium tetrahydroborate solution was added and the mercury vapors generated were directed to the optical cell. The absorbance reading was taken at the maximum value reached.

*Partially support by CONDES-LUZ, CONICIT and CONSEJO CENTRAL DE POST-GRADO.

Correct citation: Rodriguez, M.C., Sanchez, J.M., Cubillan, H.S., Hernandez, M., Semprun, B.I., Barrios, L.Ch., Romero, R.A., and Granadillo, V.A. 1997. Comparison of two mineralization procedures for determination of total mercury in biological and clinical materials by cold vapor atomic absorption spectrometry. In *Trace Elements in Man and Animals – 9: Proceedings of the Ninth International Symposium on Trace Elements in Man and Animals. Edited by* P.W.F. Fischer, M.R. L'Abbé, K.A. Cockell, and R.S. Gibson. NRC Research Press, Ottawa, Canada. pp. 414–415.

Results and Discussion

The mineralization procedure developed allowed us to remove concomitant substances and to produce digestion solutions that guaranteed the subsequent straightforward determination of mercury by CVAAS. Total metal concentrations found spectrometrically for the standard reference materials were statistically indistinguishable ($p > 0.01$) from the certified values (Table 1), verifying the accuracy of the analytical method. The mineralization procedure developed required a total time of 60 s. The advantage of this method is that it consumed less time than the mineralization procedure previously reported (70 s) (Tahán et al. 1993). Real samples and certified materials were mineralized by the two proposed methods and total mercury was subsequently determined by CVAAS. No significant differences ($p > 0.001$) were observed between the two decomposition procedures with a linear correlation ($Y = 0.7390 + 1.002X$; $r = 0.9944$; $n = 40$; 45°).

The reliability of the two types of mineralization was further assessed through a recovery study (range of recovery: 95–105%). The average precision obtained was better than 3% RSD in real samples of biological materials for both the within- and between run precision, for mercury concentration ranging between 0.93–38.00 µg/L. The linear range of the calibration curve was found to be 10–200 ng of Hg. Peak height absorbance reading (Ap) increased linearly in relation to the mass (in ng) of mercury (C) present according to the equation $Ap = 0.0010\ C$ (correlation coefficient 0.9999). The amount of mercury required to give a 1% absorption was 4.4 ng. The detection limit, defined as twice the standard deviation of the blank, for all samples analyzed was 53 ng/L, which correspond to 159 pg of Hg for 3 mL of solution undergoing analysis. In conclusion, the use of either mineralization procedures permitted the accurate, precise and reliable determination of mercury by CVAAS.

Table 1. Accuracy study for CVAAS determination of mercury.

		Measured value	
Reference materials	Certified value	Microwave (600 W)	Microwave (950 W)
RM 50[a]	0.95±0.10	0.94±0.05	0.96±0.01
SRM 1566a[b]	0.064±0.007	0.067±0.008	0.064±0.009
NIES No. 2[c]	1.3	1.2±0.1	1.2±0.2
OSSD 20/21[d]	10.4±0.9	11.6±0.4[e]	9.5±2
CONTOX No. 0140[f]	50±10	49±5[e]	47±5
Human whole blood[g]	–	19±5[e]	17±4
Human urine[g]	–	37±2[e]	35±2

[a]Albacore tuna from NIST, USA; [b]Oyster tissue from NIST; [c]Pond sediment from the National Institute for Environmental Studies, Japan; [d]Whole blood from Behring Institute, Germany; [e]In ng/ml; [f]Level II urine from Kaulson Laboratories, USA; [g]Real sample.

Literature Cited

Adeloju, S.B., Dhindsa, H.S., & Tandon, R.K. (1994) Evaluation of some wet decomposition methods for mercury determination in biological and environmental materials by cold vapour atomic absorption spectroscopy. Anal. Chim. Acta. 285: 359–364.

Bermejo, P., Moreda, J., Moreda, A., & Bermejo, A. (1994) Palladium as a chemical modifier for the determination of mercury in marine sediment slurries by electrothermal atomization atomic absorption spectrometry. Anal. Chim. Acta. 296: 181–193.

Navarro, M., López, M.C., Sánchez, M., & López, H. (1991) Determination of mercury in crops by cold vapor atomic absorption spectrometry after microwave dissolution. J. Agric. Food. Chem. 39: 2223–2225.

Sheppard, B.S., Heitkemper, D.T., & Gastón, C.M. (1994) Microwave digestion for the determination of arsenic, cadmium and lead in seafood products by inductively coupled emission and mass spectrometry. Analyst. 119: 1683–1686.

Determination of Ca, Fe, Mg, K and Na in Cereals by Flame Atomic Absorption Spectrometry *

M. HERNANDEZ, B.I. SEMPRUN, J.M. SANCHEZ, H.S. CUBILLAN,
L.CH. BARRIOS, M.C. RODRIGUEZ, R.A. ROMERO, AND V.A. GRANADILLO
Analytical Instrumentation Laboratory, University of Zulia, Maracaibo 4011, Venezuela

Keywords cereals, flame atomic spectrometry, metals

Introduction

Calcium (Ca), iron (Fe), magnesium (Mg), potassium (K) and sodium (Na) are essential elements for human beings which are supplied by the diet. The presence of these metals in cereal products varies, depending mainly on the type and origin of the raw material (Granadillo et al. 1995). Mineralization procedures are aimed towards the destruction of the organic matter which interferes with quantification. Flame atomic absorption spectrometry (FAAS: Ca, Fe and Mg) and flame atomic emission spectrometry (FAES: K and Na) are widely used for the determination of metals in foodstuffs (Tahán et al. 1994). They are preferred because the measurements are precise and accurate, and can be done rapidly; additionally, FAAS and FAES techniques have relatively low operating costs. This work presents the determination of Ca, Fe, Mg, Na, and K in commercial Venezuelan cereal products.

Experimental

A Perkin-Elmer Model 460 spectrometer was used for spectrometric determinations of Ca, Fe, and Mg (FAAS), and of Na and K (FAES). 1–2 kg of each type of commercial cereal product were taken at random from markets and retail outlets and kept at ambient temperature until homogenization. Samples were pooled, ground, sieved (ca. 20 mesh), mixed, and kept in polyethylene bags at 4°C until analysis. Samples were digested using high-pressure mineralization procedures previously reported (Granadillo et al. 1995). All data are reported on a dry-weight basis. Statistical analysis was carried out by conventional methods and differences were considered significant when $p < 0.05$.

Results and Discussion

The evaluation of metal contamination and losses during the grinding and sieving of the cereal products and sample pre-treatment was studied by doing carefully performed blank tests and by analyzing standard reference materials. Digestion procedures employed permitted the removal of concomitants and produced solutions that allowed straightforward determinations of the metals studied by the spectrometric techniques cited above. Accuracy was verified by analyzing two standard reference materials: Sargasso (NIES # 9) and Rice flour (NIES#10) (Table 1). Mean relative error was <4%, attesting to the excellence of the analytical methods. The reliability of the analytical methods was further assessed through recovery studies, done in triplicate on one sample of corn flour; the recovery range was 97–103%.

Precision was evaluated in the real samples under consideration; three aliquots of each material were analyzed (five runs each analysis). For all metals, precision (RSD) was better than 3%, for both the within-and between-run (day-to-day) analyses, which can be considered adequate for these types of analytical evaluations. Table 2 shows the mean concentration of Ca, Fe, Mg, Na and K found in several

*Partially supported by CONDES-LUZ, CONICIT and Consejo Central de Postgrado-LUZ.

Correct citation: Hernandez, M., Semprun, B.I., Sanchez, J.M., Cubillan, H.S., Barrios, L.Ch., Rodriguez, M.C., Romero, R.A., and Granadillo, V.A. 1997. Determination of Ca, Fe, Mg, K and Na in cereals by flame atomic absorption spectrometry. In *Trace Elements in Man and Animals – 9: Proceedings of the Ninth International Symposium on Trace Elements in Man and Animals. Edited by* P.W.F. Fischer, M.R. L'Abbé, K.A. Cockell, and R.S. Gibson. NRC Research Press, Ottawa, Canada. pp. 416–417.

Table 1. Accuracy of the spectrometric determination of metals in standard reference materials.

		Mean concentration ± 1 S.D. (µg/g of dry wt.)	
Material	Metal	Certified value	Measured value
Sargasso	Ca	13.4 ± 0.5	14.3 ± 0.1
	Fe	187 ± 6	182 ± 6
	Mg	6.5 ± 0.3	6.2 ± 0.1
Rice flour	Ca	95 ± 2	96 ± 3
	Fe	11.4 ± 0.8	11.2 ± 0.5
	Mg	1.2 ± 0.1	1.2 ± 0.02
	K	2.7 ± 0.1	2.6 ± 0.08

*Table 2. Metal concentrations (mean ± 1 SD; µg/g; *mg/g) found in several commercial cereal products.*

	Metal				
Sample	Ca	Fe	K	Mg	Na
Corn flour	66 ± 15	210 ± 41	1.3 ± 0.3	127 ± 4	8.9 ± 0.2*
Corn flour	60 ± 6	29 ± 2	1.0 ± 0.1	78 ± 3	7.5 ± 0.1*
Crop flour	3.5 ± 0.1*	137 ± 7	5.8 ± 0.1	531 ± 19	1.4 ± 0.3*
Oat flour	291 ± 20	31 ± 3	3.9 ± 0.1	1.5 ± 0.1*	51 ± 5
Corn fecula	2.8 ± 0.1*	119 ± 17	1.4 ± 0.2	528 ± 90	1.2 ± 0.1*
Rice and crop flour	4.5 ± 0.2*	192 ± 22	1.5 ± 0.1	430 ± 20	1.8 ± 0.1*

popular commercial cereal products in Venezuela. In general, the methodologies used in the laboratory to establish these values were precise, accurate and interference-free, providing full analytical confidence in the metal data for the cereal products analyzed.

It is concluded that the high-pressure mineralization methods employed allowed the adequate dissolution of the solid samples, avoided sample contamination and losses, and yielded reliable spectrometric determinations of total Ca, Fe, Mg, Na and K in the cereal products analyzed. These metal data can be useful for national nutritional recommendations and to evaluate the convenience of including these foods in therapeutic diets.

Literature Cited

Granadillo, V.A., Cubillán, H.S., Sánchez, J.M., Tahán, J.E., Márquez, E. & Romero, R.A. (1995) Three pressurized mineralization procedures that permit subsequent flame atomic spectrometric determination of Ca, Fe, K, Mg and Zn in bovine blood plasma-containing cookies and in standard reference materials. Anal Chim Acta. 306 (1): 139–147.

Tahán, J.E., Sánchez, J.M., Cubillán, H.S., & Romero, R.A. (1994) Metal content of drinking water supplied to city of Maracaibo, Venezuela. Sci. Total Environ. 144: 59–71.

Tahán, J.E., Sánchez, J.M., Granadillo, V.A., Cubillán, H.S. & Romero, R.A. (1995) Concentration of total Al, Cr Cu, Fe, Hg, Na, Pb and Zn in commercial canned seafood determined by atomic spectrometric means after mineralization by microwave heating. J. Agr. Food Chem. 43: 910–915.

Does the Cultivation System Affect the Cadmium Levels in Crops?

L. JORHEM
National Food Administration, Box 622, S-751 26 Uppsala, Sweden

Keywords ecological cultivation, conventional cultivation, cadmium

Sampling

The National Food Administration and the Agricultural University of Uppsala (UAU) have made a comparative study of products from conventional and ecological cultivation. Samples of wheat and rye were collected from UAU (2 sites) and Bollerup agricultural school, Kristianstad (KB) in the south of Sweden. Potatoes and carrots were collected from farms in the middle and south parts of Sweden.

Cultivation Systems

Wheat and rye were cultivated in frames bordering to each other as shown below, and having similar soil characteristics, pH and levels of Cd, P and K.

UAU

A: Conventional, no livestock, chemical weed control, straw ploughed in, NPK-fertilizer.	***C***: Ecological, no livestock, normal soil tillage, straw ploughed in, no nutrients.
B: Ecological, with livestock, manure, straw is removed.	***D***: Ecological, no livestock, minimal soil tillage, straw remain on field, no nutrients.

KB

A: Conventional, intensive, no livestock, harvest residues ploughed in, chemical weed control, fertilizers.
B: Conventional, with livestock, chemical weed control, fertilizers.
C: Ecological, biodynamic with livestock.
D: Ecological, other forms, with livestock.
E: Ecological, other forms, without livestock, harvest residues ploughed in.

Carrots and potatoes were collected from pairwise selected farms with similar soil properties. Each pair was juxtapositioned and consisted of one farm with conventional and one with ecological cultivation.

Analytical Methods

Wheat and rye were analysed as received. Potatoes and carrots were peeled and rinsed with deionized water, then finely chopped with a stainless steel knife. The samples were dry ashed at 450°C and Cd was determined by GFAAS (Jorhem 1993). Cd in soil was determined by flame-AAS after extraction with 2 M HNO_3 on a water bath for 2 h (Andersson 1975).

Correct citation: Jorhem, L. 1997. Does the cultivation system affect the cadmium levels in crops? In *Trace Elements in Man and Animals – 9: Proceedings of the Ninth International Symposium on Trace Elements in Man and Animals. Edited by* P.W.F. Fischer, M.R. L'Abbé, K.A. Cockell, and R.S. Gibson. NRC Research Press, Ottawa, Canada. pp. 418–419.

Analytical Quality Assurance

In parallel to the samples, three certified reference materials, wheat flour (NIST), potato powder and carrot flakes (ARC-LC, Finland) were analysed. The results of these analyses were satisfactory.

Results

The conventionally cultivated wheat from Uppsala had a significantly higher Cd-level than those produced ecologically (one way analysis of variance, $p = 0.05$). The conventionally cultivated wheat from Kristianstad, on the other hand, had significantly lower Cd-levels than those from ecological cultivation. There was no correlation between the Cd-level in wheat and soil (Table 1).

The Cd-level in 5 rye samples did not correlate with the cultivation system. The Cd-levels in potatoes and carrots (T-test of the mean, $p = 0.05$) showed no significant differences between the cultivation systems (Table 2).

Table 1. Cadmium levels in wheat and soil samples. Results in mg/kg.

Cultivation system	Wheat UAU-1	Soil UAU-1	Wheat UAU-2	Soil UAU-2	Wheat KB	Soil KB
A	0.081	0.27	0.081	0.13	0.045	0.25
B	0.032	0.19	0.035	0.13	0.042	0.22
C	0.038	0.23	0.054	0.14	0.055	0.22
D	0.035	0.19	0.061	0.12	0.054	0.22
E					0.062	0.20

Table 2. Cadmium levels in potatoes and carrots from different cultivation systems. Results in mg/kg.

Potatoes			Carrots		
	Cultivation system			Cultivation system	
Sample pair	Ecol.	Conv.	Sample pair	Ecol.	Conv.
1	0.005	0.004	11	0.070	0.036
2	0.014	0.011	12	0.12	0.033
3	0.009	0.007	13	0.018	0.039
4	0.020	0.017	14	0.028	0.007
5	0.018	0.010	15	0.022	0.010
6	0.006	0.008	16	0.010	0.008
7	0.021	0.011			
8	0.004	0.006			
9	0.018	0.033			
10	0.012	0.021			

Conclusions

This study does not indicate that the type of cultivation system has any direct effect on the Cd-level in agricultural products.

It must, however, be emphasized that this was a limited study, and too far-reaching conclusions should be avoided.

References

Jorhem, L. (1993) Determination of metals in foodstuffs by atomic absorption spectrophotometry after dry ashing: NMKL interlaboratory study of lead, cadmium, zinc, copper, iron, chromium and nickel. J. AOAC Int. 76: 798–813.

Andersson, A. (1975) Relative efficiency of nine different soil extractants. Swedish J. Agr. Res. 5: 125–135.

Trace Elements, Infection and Immune Function: Research Developments at Diverse Levels of Biological Organization

Chair: W. Woodward

What Role Does Iron Play in the Immune Response?

A.R. SHERMAN AND D. HRABINSKI

Department of Nutritional Sciences, Rutgers, The State University of New Jersey, New Brunswick, NJ 08903, USA

Keywords iron, immunity, anemia

Iron and Infection

Since the 1920's it has been appreciated that iron deficient populations have increased susceptibility to infectious illnesses (Mackay 1928). Since that time considerable evidence has been gathered from studies conducted in the field and clinic as well as controlled laboratory experiments using animal models and *in vitro* conditions to support the notion that iron plays important roles in immunity against disease. In contrast, there is also a body of evidence which indicates that supplemental iron treatment of anemia exacerbates or promotes illness even when infection was not previously diagnosed (Murray et al. 1978). The term "nutritional immunity" has been used to refer to the host's ability to redistribute iron into storage sites so that the invading organisms die off due to a lack of iron required for growth. Thus, there appears to be a delicate balance between iron's effect on microbial growth and iron's effect on the host's immune system.

Immune Response

The immune response is a complex interactive system which enables the host to distinguish self from non-self or foreignness. With no central organ of control, the immune response is an integrated system of lymphoid organs, 1 trillion white blood cells, and the secretary products of these cells and organs. The system is very responsive to stimuli including bacteria, virus, cancer cells, chemicals, and particulates. Once activated, both nonspecific components (mucosal, phagocytic, and inflammatory) and specific components (humoral and cell-mediated) come into action locally and systemically to remove the stimulus. The immune system operates under strict feedback regulation and has memory for stimuli previously encountered. Considering the nature of immunity there are numerous modes by which trace elements could function in immunity including: anatomical development of lymphoid tissue; mucous production; skin health; synthesis and function of immunologically active proteins; cell proliferation; cellular functioning and movement; intracellular killing; modulation and regulation of immune processes. A number of nutritional states have been shown to alter immunity. As will be discussed here, many aspects of immunity have been shown to be altered by iron deficiency. Detailed reviews of the subject have been published (Sherman 1984, Sherman and Helyar 1989, Sherman and Spear 1993).

Correct citation: Sherman, A.R., and Hrabinski, D. 1997. What role does iron play in the immune response? In *Trace Elements in Man and Animals – 9: Proceedings of the Ninth International Symposium on Trace Elements in Man and Animals. Edited by* P.W.F. Fischer, M.R. L'Abbé, K.A. Cockell, and R.S. Gibson. NRC Research Press, Ottawa, Canada. pp. 420–424.

Cell Mediated Immunity

Exposure to "foreignness" brings into action both cell mediated (largely involving T cells) and humoral (primarily B cell) components of the immune response. T lymphocytes originate from stem cells in the bone marrow and travel to the thymus for maturation and proliferation. In rodent iron deficiency, there is thymic atrophy (Rothenbacher and Sherman 1980) characterized by reduced cellularity, cellular hypertrophy and thymic involution suggested by increased lipid content (Kochanowski and Sherman 1982, 1985b). The percentage of T lymphocytes is markedly reduced in iron deficient mice (Helyar and Sherman 1992) and children (Chandra 1975, Macdougall et al. 1975). Several subsets of T cells have been found to be affected by iron deficiency. Splenic T helper cells, which display surface receptors for specific antigens, and T suppressor cells, which "turn off" immune activation once an infection has been squelched, are reduced in both moderately and severely iron deficient mice (Helyar and Sherman 1992, Kuvibidila et al. 1990). While the percentage of splenic B cells is also lowered by iron deficiency, the null cell population appears to be increased.

Cell-mediated immunity is often measured as the ability of B and T cells to proliferate when stimulated *in vitro*. Iron deficiency is associated with depressed proliferative responses in humans (Krantman et al. 1982) and rodents (Kuvibidila et al. 1983). Delayed type hypersensitivity to skin test antigens, a functional *in vivo* measure of cell mediated immunity, is also depressed in iron deficient subjects (Krantman et al. 1982).

Cell mediated immune functions include several types of cells which are cytotoxic to cancer cells. Cytolytic T lymphocytes have been reported to be less active in severely iron deficient mice (Kuvibidila et al. 1983a); but not in moderate iron deficiency (Helyar and Sherman 1993). Natural killer cells are large granular lymphocytes of unknown ontogeny which are cytotoxic to cancer cells and virally infected cells. In iron deficient neonatal and adult rats, natural killer cell cytotoxicity is impaired (Sherman and Lockwood 1987, Hallquist and Sherman 1989, Spear and Sherman 1992, Hallquist et al. 1992). Even when natural killer cells are stimulated *in vitro* with rat interferon γ or allogeneic macrophage derived interferon, activity is not completely restored (Lockwood and Sherman 1988, Hallquist and Sherman 1989). Macrophages, also cytotoxic against tumors, are impaired in iron deficient rat pups (Hallquist et al. 1992).

Besides their direct roles in cell mediated immune functions, many types of cells produce and secrete immunologically active substances such as cytokines. Activated macrophages from iron deficient rodents produce low levels of interleukin 1 (Helyar and Sherman 1987). Production of interleukin 2, a cytokine produced by T helper cells, is also impaired by iron deficiency (Kuvibidila et al. 1992). The interferons are also cytokines produced by activated macrophages and T cells. Interferons aid in B cell maturation, stimulate T cells, stimulate natural killer cells and inhibit tumor growth. During iron deficiency, rats produce less interferon than normal (Spear and Sherman 1991).

Humoral Immunity

B lymphocytes are stimulated to mature and proliferate in response to soluble factors secreted by T cells, by macrophages and by direct contact with T cells. Under the influence of the spleen and bone marrow, B cells proliferate and differentiate into plasma cells which secrete one of five classes of immunoglobulins which have antibody function and constitute a major component of the humoral immunity. Measures of humoral responses include circulating B cell count, serum immunoglobulin levels, and measurement of antibody production in response to an antigenic challenge. The effects of iron deficiency on humoral immunity has been assessed in clinical, field, and experimental research and has yielded conflicting results. In earliest studies of serum immunoglobulin levels and antibody response to immunization in iron deficient children (Chandra 1975, Macdougall et al. 1975, Bagchi et al. 1980, Krantman et al. 1982), normal responses were observed. In human subjects these measures assess previous antigenic exposure.

Use of rodents permits measurement of responsiveness to current antigen exposure and reveals effects of iron deficiency in the absence of past exposures, intercurrent infections, multiple nutritional deficiencies and other health problems. Use of the Jerne plaque assay permits measurement of antibody secretion by individual splenocytes in rats immunized with sheep red blood cells (SRBC). Both neonates born to iron deficient dams (Kochanowski and Sherman 1985a) and young growing rats (Sherman 1990) show impaired IgG and IgM responses using this very sensitive and specific assay. The importance of adequate iron status during the pre- and post-natal periods on development of immunity is underscored by the observation that iron repletion of deficient pups at weaning does not correct impaired humoral responses (Kochanowski and Sherman 1985a) or impaired cellular growth of immunocompetent organs (Kochanowski and Sherman 1985b). Defective splenic production of antibody against SRBC has been demonstrated in moderate iron depletion even before hematocrit is affected (Sherman 1990).

What Mechanisms are Responsible for the Widespread Effects of Iron on Immunity?

From the available evidence it appears that many components of the immune response system are altered by iron deficiency. Is there one underlying cellular or molecular mechanism which explains all of these diverse effects? Analysis of the knowledge reveals that there are 4 major roles of iron which contribute to the complex interrelated immune responses: protein synthesis, cell cycle, signal transduction and gene expression. As discussed above, immune reactions include many secretary proteins released from lymphocytes and macrophages which stimulate other branches of immunity. In addition, protein synthesis is involved in cellular replication and antibody production. Protein synthesis is impaired in spleens of both severely and moderately iron deficient rats pups and in thymus of severely deficient rats (Rosch et al. 1987). While splenic protein synthesis nearly doubled in moderately deficient and normal pups after immunization with SRBC, protein synthesis in severe deficiency did not increase significantly. While the specific site for this impairment in protein synthesis during iron deficiency has not been identified, a role for iron in protein synthesis provides an explanation for many of the defects in immunity observed.

Cellular proliferation is an early and sustained characteristic of all phases of the immune response. Roles for iron in G1, S and G2 phases of the cell cycle are suggested by recent studies using *in vitro* depletion of cell lines and iron deficient animals. Helyar and Sherman (1991) found that the percentage of splenic lymphocytes in G2-M phase was lower in iron deficient mice that in controls or mice restricted in food. This suggests a delay in the S phase, possibly related to iron's role in DNA synthesis via ribonucleotide reductase. Further support for the importance of iron in S phase is provided by *in vitro* treatment of cells with transferrin receptor antibody resulting in accumulation of cells in S phase (Trowbridge and Lopez 1982).

In vitro iron depletion of cultured T lymphocytes (Terata et al. 1993) and neuroblastoma cells (Brodie et al. 1993) leads to blockage at or near the G1/S border. This suggests a role for iron before DNA replication in S phase. Cell cycle transitions are regulated by a family of serine protein kinases which are cyclin dependent. Of these proteins, p34cdc2, active in late G1 before S phase, is inhibited by iron depletion of cells in culture (Terada et al. 1993). Iron chelation by desferrioxamine inhibits the transcription of the cdc2 gene responsible for the p34cdc2 protein (Lucas et al. 1995). The p34cdc2 protein is also active in M phase and may represent an additional site for iron in the cell cycle. Reddel et al. (1985) showed that iron and/or transferrin depletion of T47D breast cancer cells blocks at the G2 phase.

For proliferation, lymphocytes require activation by an antigen or mitogen which sets into motion secretion of cytokines and activation of protein kinase C. Protein kinase C (PKC) phosphorylates proteins involved in signal transduction which lead to biochemical changes and ultimately cellular proliferation. Kuvibidila et al. (1991) have shown that PKC activity is low in splenocytes from iron deficient mice. While the food restriction of iron deficiency contributed somewhat to the reduction in PKC, the correlation of iron status with activity of this important component of signal transduction was positive.

Activation of protein kinase C in many lymphoid and hematopoietic cells requires iron-transferrin. Alcantara et al. (1991, 1994) have shown that delivery of transferrin iron to cultured lymphoblastoid T-cells stimulates transcription of the PKC-β gene while treatment with desferrioxamine was inhibitory. Transcription of other gene subspecies, PKC-α and -γ, were not affected by iron treatment. Transcriptional upregulation of PKC-β by iron-transferrin appears to be mediated by DNA sequences located between –2200 bp and –587 bp in the 5′-flanking region of the human PKC-b gene. These *in vitro* studies provide evidence that iron is involved in the expression of specific genes which are central to many of the biochemical control points in cellular function including immune responses.

Taken together, current knowledge at the cellular, molecular, and genetic levels suggests that iron has multiple roles in immunity. The challenge for future research is to integrate this knowledge at the whole animal and clinical level and to pinpoint specific uses of iron in correcting immunodeficiencies.

Literature Cited

Alcantara, O., Javors, M. & Boldt, D.H. (1991) Induction of protein kinase C mRNA in cultured lymphoblastoid T cells by iron-transferrin but not soluble iron. Blood 77: 1290–1297.

Alcantara, O., Obeid, L., Hannun, Y., Ponka, P. & Boldt, D.H. (1995) Regulation of protein kinase C (PKC) expression by iron: effect of different iron compounds on PKC-β and PKC-α gene expression and the role of the 5′-flanking region of the PKC-β gene in the response to ferric transferrin. Blood 84: 3510–3517.

Bagchi, K., Mohanram, M. & Reddy, V. (1980) Humoral immune response in children with iron deficiency anemia. Br. Med. J. 1: 1249–1251.

Brodie, C., Siriwardana, G., Lucas, J., Schleicher, R., Terada, N., Szepesi, A., Glefand, E. & Seligman, P. (1993) Neuroblastoma sensitivity to growth inhibition by desferrioxamine: evidence for a block in G1 phase of the cell cycle. Cancer Res. 53: 3968–3975.

Chandra, R. K. (1975) Impaired immunocompetence associated with iron deficiency. J. Pediatrics 86: 899–902.

Hallquist, N. A., McNeil, L. K., Lockwood, J. F., Sherman, A. R. (1992) Effect of maternal iron deficiency on peritoneal macrophage and peritoneal natural killer cell cytotoxicity in rat pups. Am. J. Clin. Nutr. 55: 741–746.

Hallquist, N. A., Sherman, A. R. (1989) Effect of iron deficiency on the stimulation of natural killer cells by macrophage-produced interferon. Nutr. Res. 9: 282–292.

Helyar, L., Sherman, A. R. (1987) The effect of iron deficiency on interleukin 1 production by rat leukocytes. Am. J. Clin. Nutr. 46: 346–452.

Helyar, L., Sherman, A. R. (1991) Iron deficiency impairs lymphocyte activation marker expression and cell cycle shift in response to concanavalin A stimulation. FASEB J. 5: Al292.

Helyar, L., Sherman, A.R. (1992) Moderate and severe iron deficiency lowers numbers of spleen T-lymphocyte and B-lymphocyte subsets in the C57/1316 mouse. Nutr. Res. 12: 1113–1122.

Helyar, L., Sherman, A.R. (1993) Cell-mediated cytotoxicity is protected in moderately iron-deficient C5BL/6 mice. Nutr. Res. 13: 1313–1323.

Kochanowski, B. A., Sherman, A. R. (1982) Cellular growth in iron-deficient rat pups. Growth 46: 126–134.

Kochanowski, B. A., Sherman, A. R. (1985a) Decreased antibody formation in iron-deficient rat pups-effect of iron repletion. Am. J. Clin. Nutr. 41: 278–284.

Kochanowski, B. A., Sherman, A. R. (1985b) Cellular growth in iron-deficient rats: effect of pre- and postweaning iron repletion. J. Nutr. 115: 279–287.

Krantman, H.J., Young, S.R., Ank, B.J., O'Donnell, C.M., Rachelefsky & Stiehm, E.R. (1982) Immune function in pure iron deficiency. Am. J. Dis. Child. 136: 840–844.

Kuvibidila, S., Baliga, B.S. & Suskind, R.M. (1983a) The effect of iron-deficiency anemia on cytolytic activity of mice spleen and peritoneal cells against tumor cells. Am. J. Clin. Nutr. 38: 239–244.

Kuvibidila, S., Nauss, K.M., Baliga, B.S. & Suskind, R.M. (1983b) Impairment of blastogenic response of splenic lymphocytes from iron-deficient mice: in vivo repletion. Am. J. Clin. Nutr. 37: 15–25.

Kuvibidila, S., Dardenne, M., Savino, W. & Lepault, F. (1990) Influence of iron-deficiency anemia on selected thymus functions in mice: thymulin activity, T-cell subsets, and thymocyte proliferation. Am. J. Clin. Nutr. 51: 228–232.

Kuvibidila, S., Baliga, B.S. & Murthy, K.K. (1991) Impaired protein kinase C activation as one of the possible mechanisms of reduced lymphocyte proliferation in iron deficiency in mice. Am. J. Clin. Nutr. 54: 944–950.

Kuvibidila, S., Murthy, K.K. & Suskind, R.M. (1992) Alteration of interleukin-2 production in iron deficiency anemia. J. Nutr. Immunol. 1: 81–98.

Lockwood, J., Sherman, A. R. (1988) Spleen natural killer cells from iron-deficient deficiency. FASEB J. 5: Al 293.

Spear, A. T. & Sherman, A. R. (1992) Iron deficiency alters DMBA-induced tumor burden and natural killer cell cytotoxicity in rats. J. Nutr. 122: 46–55.

Sherman, A. R. & Spear, A. T. (1993) Iron and immunity. In: Nutrition and Immunology, (Klurfeld, D. M., ed.), pp. 285–308. Plenum Publishing Corp., New York, NY.

Terada, N., Or, R., Szepesi, A., Lucas, J.J. & Gelfand, E.W. (1993) Definition of the roles of iron and essential fatty acids in cell cycle progression of normal human T lymphocytes. Exper. Cell Res. 204: 260–267.

Trowbridge, I.S. & Lopez, F. (1982) Monoclonal antibody to transferrin receptor blocks transferrin binding and inhibits human tumor cell growth *in vitro*. Proc. Natl. Acad. Sci. USA 79: 1175–1179.

Discussion

Q1. John Sorenson, University of Arkansas, Little Rock, AR, USA: With regard to the rationale that in infection the decrease in plasma iron is a physiological mechanism to withhold iron from microorganisms, is iron undergoing tissue redistribution to facilitate activation of the immune system, or is it being withdrawn from microorganisms?

A. That's an interesting point that you raise, and something of a philosophical issue that would be difficult to put one way or the other. Certainly it would be efficacious to the organism to do both, i.e., to be able to withhold iron, and to be able to shunt iron to the lymphoid tissues so that the immune system can be activated.

Q2. John Sorenson, University of Arkansas, Little Rock, AR, USA: Can we offer that in non-infectious diseases the same decrease in plasma iron occurs, and that this may be evidence of a redistribution of iron in activation of the immune system.

A. Thank you, that's an interesting possibility. I would also like to point out that the data from animal models on the function of iron in immunity was gathered in the absence of any infectious diseases.

Q3. Maryam Hazegh-Azam, California State University at Fullerton, CA, USA: Have you examined the effects of iron supplementation?

A. We have done experiments in neonates and in older animals and there is a partial restoration of immunity, but not always a complete restoration.

Q4. John Bogden, New Jersey Medical School, Newark, NJ, USA: Adria, you showed some very interesting data on the influence of iron deficiency on thymus. Of course the thymus involutes with age and has virtually disappeared in humans by age 20. Are there differences in the way people respond to iron deficiency when they're young versus older, and therefore have a totally involuted thymus?

A. That's a difficult question to answer, because the clinical work and the field studies represent one point in time, so it's difficult to know what the previous history was. I think that our animal data show that the neonatal period is a critical period in establishing the thymic structure, so I would speculate based on that, that if an individual had been iron deficient during this critical period that there would be an underlying lack of immunocompetence later on in life. If the iron deficiency is first manifest later in life, then I would speculate that the effects would be less, but still evident.

Q5. John Bogden, New Jersey Medical School, Newark, NJ, USA: Are you saying that we don't know whether children respond differently than adults?

A. I think that immunity seems to be more impaired as a result of iron deficiency in children than in adults, but the ultimate answer to that will have to await the results of some longitudinal study, which is unlikely to happen.

Q6. Harold Sandstead, University of Texas Medical School, Galveston, TX, USA: I would like to challenge Dr. Bogden. I don't really think that the term involution means necessarily that certain critical cells in the thymus have disappeared in adults, because most of us have a fairly competent immune system, and if those cells did not exist, I venture to say that we would not have as competent an immune system. The thymus does get very small, and this is what people are talking about when they refer to involution of the thymus, but if you examine the tissue, you can still find these very essential cells, so I challenge you in terms of the idea that this critical organ ceases to function.

A. Well also, during childhood, the functional T-cells have been educated, and have been seeded into various tissues so that a large thymus is no longer necessary.

Copper and Immunocompetence

M.L. FAILLA AND R.G. HOPKINS

Department of Food, Nutrition and Food Service Management, The University of North Carolina at Greensboro, Greensboro, NC 27412-5001, USA

Keywords copper, immunity, interleukin-2, T-lymphocytes

One of the important challenges for investigators in the area of nutritional immunology is to define the specific biochemical and molecular roles of the essential nutrients in the development, activation, differentiation and effector functions of the diverse cell types that constitute the host defense system. The essentiality of copper (Cu) for the innate and acquired branches of the immune system has been demonstrated by numerous *in vitro* and more limited *in vivo* observations that have considered the impact of experimentally induced, acquired, and genetic Cu deficiency in animals and humans. Examples of the changes that occur in response to Cu deficiency are listed in Table 1. Some data suggest that the degree of suppression of the immune system that is associated with an inadequate supply of Cu is sufficient to increase susceptibility to infection and, possibly, the development and growth of tumors. A comprehensive review of available literature related to Cu and the immune system was published several years ago (Prohaska and Failla 1993) and should be consulted for details. In addition to the need to elucidate the specific functions of Cu in immune cell activities, the relevance of findings using severely Cu deficient rodents for humans who rarely develop frank Cu deficiency continues to be questioned. This article summarizes our attempts to address such problems by examining the role of Cu in the activation and proliferation of quiescent T-lymphocytes of rat and human origins.

Cu Deficiency Impairs Mitogen-induced Proliferation of T-cells by Decreasing Production of Interleukin-2. Because of its reliability and technical simplicity, lymphocyte proliferation is one of the most commonly used *in vitro* indicators for assessing the impact of dietary and environmental factors on immunocompetence. The plant lectins phytohemagglutinin (PHA) and concanavalin A (Con A) are frequently employed as mitogens that nonspecifically activate unprimed T-lymphocytes by a process that requires antigen-presenting cells (APC). Numerous investigators have reported that severe Cu deficiency suppresses PHA- and Con A-induced DNA synthesis in cultures of mononuclear cells (MNC) prepared from the whole blood or spleens of laboratory and domestic animals (Prohaska and Failla 1993). Because the outcome of this procedure is dependent upon both the numbers and the activities of T-lymphocytes and APC, the impact of Cu status on the phenotypic profile of immune cells was examined. Analysis of cell specific surface antigens by flow cytometry revealed that Cu deficiency decreased the relative number of T-lymphocytes and especially the CD4 (T helper) population in murine (Lukasewycz et al. 1985; Mulhern and Koller 1987) and rat (Bala et al. 1991) spleen, and in rat peripheral blood (Bala et al. 1990). However, the extent of the decline in the *in vitro* mitogenic reactivity of T-cells greatly exceeded the reduction in the relative number of T-cells. Moreover, several T-cell dependent activities were reported to

Table 1. Abnormalities in immune system associated with copper deficiency.

Innate immunity	Acquired immunity
↓ neutrophils	↓ T-lymphocytes
↓ respiratory burst activity	↓ T-cell dependent antibody production
↓ killing of microorganisms by phagocytic cells	↓ mitogen-induced proliferation of T-cells
↓ natural killer cell activity	↓ delayed type hypersensitivity response
	↓ mixed lymphocyte reaction

Correct citation: Failla, M.L., and Hopkins, R.G., 1997. Copper and immunocompetence. In *Trace Elements in Man and Animals – 9: Proceedings of the Ninth International Symposium on Trace Elements in Man and Animals. Edited by* P.W.F. Fischer, M.R. L'Abbé, K.A. Cockell, and R.S. Gibson. NRC Research Press, Ottawa, Canada. pp. 425–428.

be inhibited by Cu deficiency (see Table 1). Together, these observations suggested that Cu is required for both the activation and effector functions of T-lymphocytes.

To attempt to define the role(s) of Cu in the activation and proliferation of T-lymphocytes, we decided to examine the mitogenic reactivity of splenic SMC from Cu deficient rats more closely. Initiation of the activation of quiescent T-cells requires contact between a small antigen fragment in the pocket of a class II major histocompatibility complex on the surface of the APC with its specific antigen receptor on the T-cell surface, as well as additional interactions between other plasma membrane proteins of the respective cells (Weiss 1993; Figure 1). Such contact initiates a series of cytoplasmic and nuclear events that result in the synthesis of various surface receptors and cytokines. The expression of molecules such as the transferrin receptor and the interleukin-2 (IL-2) receptor on the cell surface reflects the cell's "competent" state. Binding of the secreted cytokine IL-2 to its receptor provides the key signal for the activated cell to progress to the proliferation stage. Other cytokines affect the subsequent differentiation and clonal expansion of the TH_1 and TH_2 subclasses that play central roles in modulating the specific type of immune response.

Initial studies revealed that Cu deficiency did not influence the up-regulation of transferrin and IL-2 receptors on the surface of mitogen-treated T-lymphocytes isolated from the spleens of control and Cu deficient rats (Bala et al. 1991). That is, Cu deficiency did not prevent T-lymphocytes from becoming "competent". Attention was next directed to the levels of IL-2 in mitogen-treated cultures. PHA-treated cultures of splenic MNC from Cu-deficient rats contained 40–50% as much IL-2 activity as identically treated cultures with cells from copper adequate rats (Bala and Failla 1992). Similarly Lukasewycz and Prohaska (1990) had reported that cultures of activated splenocytes from Cu deficient mice contained lower levels of IL-2 than control cultures. We next found that the addition of rat IL-2 to PHA-treated cultures of splenic MNC from Cu deficient rats increased the incorporation of ^{3}H-thymidine into DNA to control levels (Bala and Failla 1992). Moreover, the addition of physiological levels of Cu, but not Zn or Mn, to cultures containing PHA-treated splenic MNC from Cu deficient rats increased IL-2 activity and restored DNA synthesis to control levels. Thus, the impaired *in vitro* mitogenic reactivity of T-lymphocytes from Cu deficient rats is due to decreased production of IL-2 which likely results from an inadequate availability of Cu within the lymphocyte.

In vitro Cu Deficiency Reduces the Level of IL-2 mRNA in Human T-cells. IL-2 has a role in a variety of cellular immune processes, including the proliferation and clonal expansion of T and B

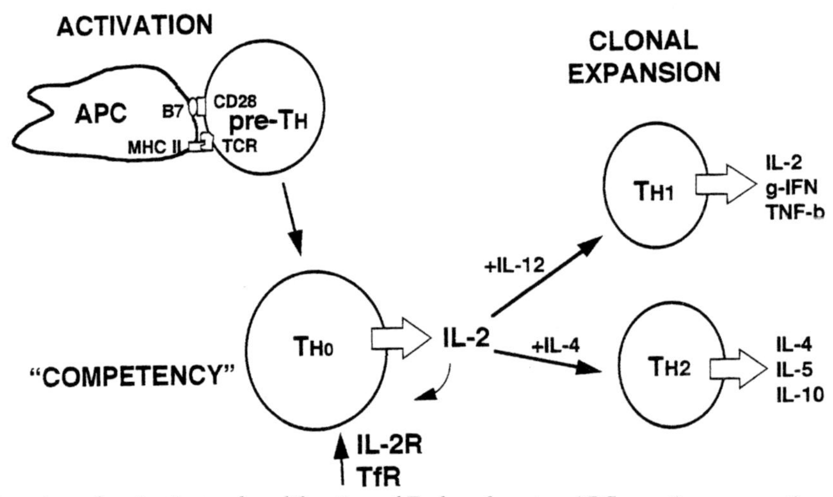

Figure 1. Overview of activation and proliferation of TH-lymphocytes. APC = antigen presenting cell, TH = T helper cell, MHC II = major histocompatibility complex II, TCR = T-cell receptor, IL-2R = interleukin-2 receptor, TfR = transferrin receptor, g-IFN = gamma-interferon, TNF-β = tumor necrosis factor-beta, and B7 and CD28 = surface antigens.

lymphocytes, the activities of natural killer and cytotoxic T-lymphocytes, and activation-induced death of effector T-cells once foreign materials have been neutralized (Willerford et al. 1995). Therefore, determination of the role of Cu in IL-2 production may provide insights about the immuno-suppressed status of Cu deficient animals and, perhaps, humans. An *in vitro* model of human T-cell Cu deficiency has been established (Hopkins and Failla 1996). Jurkat, a human leukemic T-cell line that is widely used to study the regulation of IL-2 gene expression (e.g., Schwartz et al. 1993) was selected as the cell model. IL-2 synthesis is induced by treating Jurkat cells with PHA and phorbol myristate acetate. Pretreatment of Jurkat cells with as low as 5 μmol/L N,N′-bis(2-aminoethyl)-1,3-propanediamine, a high affinity chelator of Cu (Fawcett et al. 1980), for ≥ 24 h decreases the activity of Cu,Zn-superoxide dismutase by approximately 30% without affecting cell viability, replication, respiration or protein synthesis. When activated, chelator-treated cells produce 70–80% less IL-2 and contain about 50% as much IL-2 mRNA as control cells. This impact of the chelator on cellular Cu status and IL-2 production is blocked by the addition of a slight molar excess of Cu, but not Zn or Fe, to medium containing the chelator. Moreover, the chelator does not induce secondary Fe deficiency since the binding of Fe_2-transferrin and the uptake of transferrin-Fe are similar in control and chelator-treated cells. Studies are ongoing to determine if the adverse impact of Cu deficiency on IL-2 expression occurs at the transcriptional or post-transcriptional levels. Direct effects of Cu on the transcription of mammalian genes have not been reported previously.

T-cell Reactivity is Suppressed by Marginal Cu Deficiency. The vast majority of our current understanding of the roles of Cu in the immune system has resulted from studies in which Cu has been severely restricted in the diet fed to young animals. However, frank Cu deficiency is extremely rare in humans. Because the typical western diet provides less Cu than the ESADDI for all age groups (Pennington et al. 1989), it is possible that some populations may be at risk for developing marginal Cu deficiency. This situation has provided the impetus for examining the impact of marginal Cu deficiency on the activities of rat neutrophils and T-cells. In our first study, weaned male rats were fed diets containing either adequate (7 mg/kg) or low Cu (2.7 mg/kg, i.e., 55% of the NRC requirement) for 5 weeks. Tissue levels of Cu, erythrocyte Cu,Zn-superoxide dismutase activity, relative heart size and hematological indices were similar for the treatment groups. However, the levels of Cu and Cu,Zn-superoxide dismutase activity and the candidacidal activity of elicited neutrophils isolated from rats fed the marginally low Cu diet were all significantly below that of cells isolated from the adequate Cu group (Babu and Failla 1990). More recently, we examined the impact of chronically feeding a diet marginally low in Cu on the *in vitro* mitogenic reactivity of splenic MNC of adult rats (Hopkins and Failla 1995). Dams were fed diets with either adequate (6.7 mg/kg) or marginally low Cu (2.7 mg/kg) from the latter half of pregnancy through lactation. Offspring were weaned to the diet fed to their mothers. At 5–6 months of age, traditional indicators of Cu status were similar for the two groups. However, ^3H-thymidine incorporation and IL-2 production were markedly decreased in mitogen-treated cultures of splenic MNC isolated from male, but not female, rats fed the diet marginally low in Cu. Also, PMA-stimulated respiratory burst activity of neutrophils elicited from male rats fed the diet with marginally low Cu was 60% that of cells isolated from control animals. These data demonstrate the sensitivity of cells in both branches of the immune system of the rat to sub-optimal Cu nutriture and support the need to further consider the relationship between Cu intake and immunocompetence in humans.

Cu Deficiency in Humans Suppresses Mitogen-induced Activation of T-lymphocytes. We recently had the opportunity to evaluate the impact of acquired Cu deficiency on the *in vitro* mitogenic reactivity of T-lymphocytes obtained from a Cu deficient adult female (Smith et al. 1994). The patient originally presented with anemia and neutropenia. The anemia did not respond to Fe or vitamin B12 supplementation. Evaluation of indicators of Cu status revealed low levels of plasma Cu, plasma ceruloplasmin activity and protein, erythrocyte superoxide dismutase activity, and platelet cytochrome c oxidase; lymphocytes and CD4 cells were within the normal range. ^3H-thymidine incorporation by whole blood cultures treated with various doses of PHA and Con A was markedly suppressed compared to that in cultures with cells from normal subjects. Treatment to elevate Cu status to the normal range was associated with a restoration of *in vitro* mitogenic reactivity. On several occasions, the subject relapsed to the Cu deficient state. This was associated with a decline in T-cell mitogenic reactivity that was enhanced following Cu supplementation.

Acknowledgement

We are grateful for financial support provided by USDA NRICGP (92–57200–7544), the North Carolina Institute of Nutrition and North Carolina Agriculture Experiment Station.

Literature Cited

Babu, U. & Failla, M.L. (1990) Copper status and function of neutrophils are reversibly depressed in marginal and severely copper-deficient rats. J. Nutr. 120: 1700–1709.

Bala, S., Failla, M.L. & Lunney, J.K. (1990) T-cell numbers and mitogenic responsiveness of peripheral blood mononuclear cells are decreased in copper deficient rats. Nutr. Res. 10: 749–760.

Bala, S., Failla, M.L. & Lunney, J.K. (1991) Alterations in splenic lymphoid cell subsets and activation antigens in copper-deficient rats. J. Nutr. 121: 745–753.

Bala, S. & Failla, M.L. (1992) Copper deficiency reversibly impairs DNA synthesis in activated T-lymphocytes. Proc. Natl. Acad. Sci. USA 89: 6794–6797.

Fawcett, T.G., Rudich, S.M., Toby, B.H., Lalancette, R., Potenza, J.A. & Schugar, H.J. (1980) Studies of chelation therapy. Crystal and molecular structure of $Cu[H_2N(CH_2)_2NH(CH_2)_3NH(CH_2)_2NH_2](ClO_4)_2$, $Cu(2,3,2\text{-tet})(ClO_4)$. Inorg. Chem. 19: 940–945.

Hopkins, R.G. & Failla, M.L. (1995) Chronic intake of a marginally low copper diet impairs *in vitro* activities of lymphocytes and neutrophils from male rats despite minimal impact on conventional indicators of copper status. J. Nutr. 125: 2658–2668.

Hopkins, R.G. & Failla, M.L. (1996) Copper deficiency decreases interleukin-2 production and IL-2 mRNA in human T-lymphocytes. FASEB J. 10: A293.

Lukasewycz, O.A. & Prohaska, J.R. (1990) The immune response in copper deficiency. Ann. N.Y. Acad. Sci. 587: 147–159.

Lukasewycz, O.A., Prohaska, J.R., Meyer, S.G., Schmidtke, J.R., Hatfield, S.M. & Marder, P. (1985) Alterations in lymphocyte subpopulations in copper-deficient mice. Infect. Immun. 48: 644–647.

Mulhern, S.A. & Koller, L.D. (1988) Severe or marginal copper deficiency results in a graded reduction in immune status in mice. J. Nutr. 118: 1041–1047.

Pennington, J.A., Young, B.E. & Wilson, D.B. (1989) Nutritional elements in the U.S. diets: Results from the Total Diet Study 1982–86. J. Amer. Diet. Assn. 89: 659–664.

Schwartz, E.M., Salgame, P. & Bloom, B.R. (1993) Molecular regulation of human interleukin-2 and T-cell function by interleukin-4. Proc. Natl. Acad. Sci. USA 90: 7734–7738.

Smith, D., Hopkins, R.G., Kutlar, A. & Failla, M.L. (1994) Diagnosis and treatment of copper deficiency in adult humans. FASEB J. 8: A4754.

Weiss, A. (1993) T-lymphocyte activation. In: Fundamental Immunology (Paul, W.E., ed.) pp. 467–504. Raven Press, Ltd., New York, NY.

Willerford, D.M., Chen, J., Ferrg, J.A., Davidson, L., Ma, A., & Alt, F.W. (1995) Interleukin-2 receptor α-chain regulates the size and content of the peripheral lymphoid compartment. Immunity 3: 521–530.

Discussion

Q1. Harold Sandstead, University of Texas Medical School, Galveston, TX, USA: Mark, I don't disagree with you. We certainly know the mechanisms from an intellectual point of view, if for no other reason. I think we can take advantage of your unexplained biological observations (because we don't know the mechanisms yet) for looking at humans by properly designed, double-blind randomized control trial. We don't have to know the details of the chemistry. That's true of many of the things that physiologists and biologists have been using long before they ever understood the chemistry. Biochemists spend their lifetimes trying to figure out how these things work, and these two go hand in hand. In the meantime, we can take advantage of your observations, probably in the clinical setting.

A. Well, I'm glad to hear that. And I would agree. I'm not saying that immune assessment should not be a part of the assessment of the individuals. I think it's an important parameter to look at, but what I'm saying is that if someone asks me to find out if a group of people has copper deficiency, that just because I see a change does not mean that there's copper deficiency, it means that there's some suboptimal response in the cells, so if I overstated it, I apologize.

Q2. Harold Sandstead, University of Texas Medical School, Galveston, TX, USA: I didn't make my point correctly. What I'm saying is that in a randomized, controlled, double-blind trial, with appropriate controls, where there are no other deficiencies, just as you do in animal studies, and the only thing you treat with in that setting is copper, for instance, and the function changes, then one can presume that copper was the cause. But you have to design the experiment correctly. That is very difficult, and requires appropriate controls. In the clinical setting, unfortunately many people do not understand, or they chose to ignore proper experimental design. You can take advantage of your observations, mechanism unknown, and begin to find out if copper deficiency is occurring in people. Your outcome may be more sensitive as an indicator than the so-called traditional things. And the male:female thing may be very critical as well.

Reduced Hydrolysis of Cell Membrane Phospholipids in Murine Splenic Lymphocytes in Iron Deficiency: Implication on Impaired Blastogenesis*

S. KUVIBIDILA,[†] S.B. BALIGA,[‡] R.P. WARRIER,[†] AND R.M. SUSKIND[†]
[†]Pediatrics, LSUMC, New Orleans, LA 70112, and [‡]Pediatrics, University of South Alabama, Collegee of Medicine, Mobile, AL 36617, USA

Keywords iron deficiency, blastogenesis, phosphatidylinositol, lymphocytes

We as well as other investigators have previously observed that iron deficiency, the most common single nutrient deficiency, impairs lymphocyte proliferation in humans and laboratory animals (reviewed by Kuvibidila et al. 1989). We have also observed that protein kinase C (PKC) activation, an early event in lymphocyte proliferation, was impaired in murine spleen cells by iron deficiency but not undernutrition (Kuvibidila et al. 1991). We concluded that inefficient PKC translocation was in part responsible for reduced blastogenesis associated with iron deficiency. In addition to calcium, PKC is activated by diacylglycerol, one of the two end products of cell membrane phosphatidylinositol 4,5 bisphosphate (PIP_2) hydrolysis (Abbas et al. 1991). The second end product, inositol 1,4,5 triphosphate or IP_3 increases free cytoplasmic calcium concentration. The question is, "could inefficient PIP_2 hydrolysis explain, at least in part, poor PKC activation and reduced blastogenesis in lymphocytes from iron deficient mice?" This study was therefore designed to determine the effects of iron deficiency on PIP_2 hydrolysis.

Materials and Methods

The study involved three groups of weanling C57BL/6 female mice: the control (C, n = 10), the pairfed (PF, n = 8) and the iron deficient (ID, n = 21). ID mice were fed a diet that contained only 5 ppm of iron. C and PF mice received the same diet that has been supplemented with ferrous sulfate to given 50 ppm iron. Iron deficiency was induced as we previously reported (Kuvibidila et al. 1991). When the hematocrit of iron deficient (ID) mice decreased from ≥40% to ≤25%, 8 ID mice were given the control diet for 14 d before being sacrificed (Repleted or R group). At the time of killing, hemoglobin (Hb) and liver iron stores were measured by standard techniques. Mitogenic response and PIP_2 hydrolysis were studied in concanavalin A (5 µg/mL)-treated and untreated splenic lymphocytes as described in the literature (Kuvibidila et al. 1991, van Tits et al. 1991). The data were analyzed by analysis of variance.

Results and Discussion

At the beginning of the feeding period, the mean Hb levels were between 15.0 and 16.6 g/dL in all groups. At the time of killing, mean Hb levels were significantly lower ($p < 0.005$) in ID (6.08 g/dL) than in control groups (17.34 g/dL, 17.6 g/dL, 15.3 g/dL, in C, PF, and R, respectively). Liver iron stores were also lower ($p < 0.05$) in ID than in control or repleted mice (3.5; 7.5; 9.4; 5.2 µg/g liver, in ID, C, PF, and R respectively). No significant difference was observed between the three iron sufficient groups.

Five important observations were made on PIP_2 hydrolysis (Table 1). First, PIP_2 hydrolysis increased with incubation time. The mean ratios of IP_3 in Con A-treated over untreated cells were <200% at 1 min of activation, but were approximately 300 to 600% after 60 min. Second, at each time point, iron deficiency

*This work was supported by NIH Grant # K14HL03144.

Correct citation: Kuvibidila, S., Baliga, S.B., Warrier, R.P., and Suskind, R.M. 1997. Reduced hydrolysis of cell membrane phospholipids in murine splenic lymphocytes in iron deficiency: Implication on impaired blastogenesis. In *Trace Elements in Man and Animals – 9: Proceedings of the Ninth International Symposium on Trace Elements in Man and Animals. Edited by* P.W.F. Fischer, M.R. L'Abbé, K.A. Cockell, and R.S. Gibson. NRC Research Press, Ottawa, Canada. pp. 429–430.

*Table 1. Mean ± SEM of IP_3 Ratios = (CPM + Con A/CPM – Con A) × 100. (*p < 0.05 compared to ID).*

Incubation time in min	ID (13)	C (10)	PF (8)	R (8)
1	113 ± 5	147 ± 26	107 ± 6	188 ± 48
2	123 ± 5	226 ± 89	125 ± 7	192 ± 34*
5	147 ± 8	277 ± 75*	153 ± 10	295 ± 108*
10	190 ± 17	277 ± 44*	192 ± 20	266 ± 119*
30	208 ± 16	300 ± 45*	210 ± 7	326 ± 77*
60	304 ± 40	447 ± 58*	583 ± 85*	490 ± 17*

reduced PIP_2 hydrolysis by 23–54% compared to control mice. The difference between both groups was statistically significant at some, but not all time points. Third, refeeding ID mice the iron supplemented diet restored PIP_2 hydrolysis to normal; and the difference between ID and R mice was statistically significant at almost all time points studied. Fourth, the ratios obtained in lymphocytes from R mice were in general, slightly though not significantly higher than those obtained in cells from C mice. Next, during the initial activation period, PIP_2 hydrolysis in cells from PF mice was slightly less efficient than those from control or R mice, and it was not significantly different from that of ID mice. However, after 60 min of incubation, PIP_2 hydrolysis was more efficient in cells from PF mice than those from the other three groups. In fact the difference between PF and ID mice was statistically significant ($p < 0.05$). In parallel to reduced PIP_2 hydrolysis, blastogenic response to Con A was significantly ($p < 0.05$) reduced in cells from ID mice (mean 34,849 cpm) compared to C (45,078 cpm) and PF mice (61,107 cpm). After 14 d of repletion, the response returned to normal (mean: 55,427 cpm). Both PIP_2 hydrolysis and blastogenic response positively ($r = 0.185$ to 0.525) and in many instances significantly ($p < 0.05$) correlated with indices of iron status.

The present study confirms our previous observation as well as that of other investigators that iron deficiency impairs lymphocyte proliferation (Kuvibidila et al. 1991). However, we are unaware of any study showing reduced PIP_2 hydrolysis in lymphocytes from ID subjects. Despite delayed rise in PIP_2 hydrolysis in cells from PF mice, after 60 min of activation, the hydrolysis was above that of control mice. Our data certainly suggest that the major factor in reduced PIP_2 hydrolysis is the lack of iron but not undernutrition per se. Further experiments are being conducted for incubation periods longer than 60 min. In summary, although other mechanisms including decreased activity of iron dependent enzymes are not ruled out, our data suggest that reduced PIP_2 hydrolysis contributes to impaired blastogenesis associated with iron deficiency.

Literature Cited

Abbas, K., Lichtman, A.H. & Jordan J.S. (1991) Molecular basis of T cell antigen recognition and activation. In Cellular and Molecular Immunology; Saunders, Philadelphia; page 139–167.

Kuvibidila, S., Baliga, B.S. & Suskind, R.M. (1989) Consequences of iron deficiency on infection and immunity. In, Textbook of Gastroenterology and Nutrition in Infancy, Second Edition, E. Lebenthal, Ed. Raven Press, Ltd., New York, page 423–431.

Kuvibidila, S., Baliga, B.S. & Murthy, K.K. (1991) Impaired protein kinase C activation as one of the possible mechanisms of reduced lymphocyte proliferation in iron deficiency. Am. J. Clin. Nutr. 54: 944–950.

van Tits, L.J.H., Michel, M.C., Motulsky, H.J., et al. (1991) Cyclic AMP counteracts mitogen-induced inositol phosphate generation and increases in intracellular Ca^{2+} concentrations in human lymphocytes. Br. J. Pharmacol. 103: 1288–1294.

Discussion

Q1. Bill Woodward, University of Guelph, ON, Canada: Do you have any information on the hydrolysis of phosphatidyl choline in these cells?

A. I have no data and there is none in the literature.

Candidastatic Activity of Human Plasma: Its Relation to Zinc and Iron Availability*

B.E. GOLDEN AND P.A. CLOHESSY

Department of Child Health, University of Aberdeen, Aberdeen, Scotland AB9 2ZD

Keywords zinc, iron, infection, human

During infection in man, plasma zinc and iron concentrations fall. This is associated with growth inhibition of invading microorganisms. There is little evidence that this inhibition is due to hypozincemia (Tocco-Bradley and Kluger 1984) but considerable evidence that it is due to hypoferremia and, particularly, reduced availability of iron to the invaders. The latter results mainly from transferrin's ability to bind iron more successfully than microorganisms (Weinberg 1978). However, there are no obvious, comparable zinc-binding agents either circulating in the host or in microorganisms. This apparent lack of control of zinc availability to microorganisms suggests they could be either more or less sensitive to hypozincemia than to hypoferremia.

Calprotectin (MW 36.5 kDa) is a calcium and zinc binding protein of the S-100 family. It comprises 60% of cytosolic protein in neutrophils (Fagerhol et al. 1990). Its plasma concentration rises early from <27 nmol/L to up to 1300 nmol/L during bacterial infection (Sander et al. 1984). In culture medium, calprotectin inhibits the growth of microorganisms, including *Candida* species, at concentrations found in plasma during bacterial infection (Steinbakk et al. 1990). We demonstrated that inhibition of *Candida albicans* growth in Sabouraud's medium follows binding of zinc by calprotectin thereby reducing its availability to the yeast (Clohessy and Golden 1995). Thus, we hypothesised that, during infection, calprotectin in plasma binds zinc, in the same way as transferrin binds iron, and hence, helps inhibit the growth of invading microorganisms. To test this, we compared the growth of *Candida* in plasma samples of varying calprotectin, zinc and iron concentrations. We also tested the effect of supplementing the plasma samples with physiological quantities of zinc and iron.

Materials and Methods

On admission to hospital, 3 mL blood was obtained from each of 49 children with bacterial infections and 20 non-infected controls of similar age (1–11 years) and male/female ratio. Informed written consent was obtained from parents and older children. Ethical approval was obtained from the Joint Ethics Committee of the University of Aberdeen and Grampian Health Board.

Plasma calprotectin was determined by a non-competitive ELISA (Golden et al. 1996) using an IgG fraction of rabbit polyclonal anticalprotectin antibody donated by M.K. Fagerhol, Oslo. Plasma transferrin was measured by nephelometry (Unimate 3 TRSF, Roche) and plasma zinc and iron were determined by flame atomic absorption spectrophotometry. For the microbiological studies, 85 μL plasma (from heparinised blood) mixed with 10 μL deionised water or 10 μL of a 100 μmol/L solution of $ZnSO_4$ or $FeCl_3$ was inoculated with 5 μL of a diluted, washed culture of *Candida kefyr* (strain 3898) (Public Health Laboratory, Bristol) containing about 800 colony forming units. *Candida kefyr* growth was assessed as the 24 h change in total viable count (TVC) determined by the Miles and Misra method.

Data were analysed using SPSS for Windows on an IBM computer. Non-normal distributions were compared using Mann Whitney U and Wilcoxon tests, and normal distributions were compared using Student's t-tests, for unpaired and paired data respectively. Differences were considered significant when $P < 0.05$.

*Supported by a Wellcome Trust project grant.

Correct citation: Golden, B.E., and Clohessy, P.A. 1997. Candidastatic activity of human plasma: Its relation to zinc and iron availability. In *Trace Elements in Man and Animals – 9: Proceedings of the Ninth International Symposium on Trace Elements in Man and Animals. Edited by* P.W.F. Fischer, M.R. L'Abbé, K.A. Cockell, and R.S. Gibson. NRC Research Press, Ottawa, Canada. pp. 431–432.

Results and Discussion

As expected, in samples from infected children, plasma zinc was lower (10.9 vs. 13.2 μmol/L, $P < 0.01$) (median values, infected vs. controls), plasma iron was lower (8.1 vs. 23.6 μmol/L, $P < 0.0001$) and transferrin saturation was lower (12 vs. 22%, $P < 0.001$) than in controls. Plasma calprotectin was higher (130 vs. 19 nmol/L, $P < 0.001$).

Candida kefyr grew successfully in the plasma samples but less in plasma from infected children than from controls (4.52 vs. 4.79, $P < 0.001$) (mean values of $\log_{10}$ 24 h increase in TVC, infected vs. controls). Its growth was not related to plasma zinc concentration and was not affected by adding zinc to the medium. However, the higher the plasma calprotectin, the lower the growth ($r = -0.33$, $P < 0.005$). In contrast, *Candida kefyr* growth correlated positively with plasma iron concentration ($r = +0.33$, $P < 0.01$) and transferrin saturation ($r = +0.40$, $P < 0.001$). When iron was added to the plasma from controls, *Candida* growth was not affected. However, when added to plasma from infected children, *Candida* growth increased (4.78 vs. 4.52, $P < 0.001$) (Fe supplemented vs. unsupplemented) to values not different from its growth in plasma from controls.

Thus, in this study, growth inhibition of *Candida kefyr* in plasma from infected children was closely related to the availability of iron but not related to that of zinc. The hypothesis that calprotectin has a zinc-binding role in plasma analogous to the iron-binding role of transferrin has not been supported. This is in contrast to the findings in abscess fluid in which calprotectin concentrations are much higher (Sohnle et al. 1991).

Acknowledgements

The authors thank Professor M.K. Fagerhol for the gift of calprotectin and relevant antibodies.

Literature Cited

Clohessy, P.A. & Golden, B.E. (1995) Calprotectin-mediated zinc chelation as a biostatic mechanism in host defence. Scand. J. Immunol. 42: 551–556.

Fagerhol, M.K., Andersson, K.B., Naess-Andresen, C-F., Brandtzaeg, P & Dale, I. (1990) Calprotectin (The leucocyte protein). In: Stimulus Response Coupling: The role of Intracellular Calcium-Binding Proteins. (Smith,V.L. & Dedman, J.R., eds.) pp. 187–210. CRC Press, Boca Raton, FL.

Golden, B.E., Clohessy, P.A., Russell, G. & Fagerhol, M.K. (1996) Calprotectin as a marker of inflammation in cystic fibrosis. Arch. Dis. Child. 74: 136–139.

Sander, J., Fagerhol, M.K., Bakken, J.S. & Dale, I. (1984) Plasma levels of the leucocyte L1 protein in febrile conditions: relation to aetiology, number of leucocytes inblood, blood sedimentation reaction and C-reactive protein. Scand. J. Clin. Lab. Invest. 44: 357–362.

Sohnle, P.G., Collins-Lech, C. & Wiessner, J.H. (1991) The zinc-reversible antimicrobial activity of neutrophil lysates and abscess fluid supernatants. J. Infect. Dis. 164: 137–142.

Steinbakk, M., Naess-Andresen, C-F., Lingaas, E., Dale, I., Brandtzaeg, P. & Fagerhol, M.K. (1990) Antimicrobial actions of calcium binding leucocyte L1 protein, calprotectin. Lancet 336: 763–765.

Tocco-Bradley, R. & Kluger, M.J. (1984) Zinc concentration and survival in rats infected with Salmonella typhimurium. Infect. Immunol. 45: 332–338.

Weinberg, E.D. (1978) Iron and infection. Microbiol. Rev. 42: 45–66.

Discussion

Q1. Bill Woodward, University of Guelph, ON, Canada: I wonder if you could comment: In relation to your iron study, as I look at that graph rather quickly just now, it seems to me that perhaps the iron supplementation study using the plasma from affected children indicates that zinc levels might not be low enough to begin with and therefore this might not provide the kind of system you would need to test the efficacy of calprotectin?

A. We wanted to try at physiological levels first. More studies are needed, for example to add calprotectin to the plasma, to see what it can do there.

Q2. Bill Woodward, University of Guelph, ON, Canada: So perhaps this is a protein which is effective at low levels of zinc, but you have to have something to lower zinc levels before calprotectin can really be useful. Is that possible?

Q3. Tammy Bray, The Ohio State University, Columbus, OH, USA: Could you use another endpoint, other than growth of bacteria? Because I can see the effect of zinc and of iron were different, and another endpoint may have allowed you to better differentiate between these.

A. Could you suggest a better alternative? (Reply by **Tammy Bray** "I don't know.")

Q4. John Sorenson, University of Arkansas, Little Rock, AR, USA: Are either the zinc- or iron-calprotectin complexes useful forms of the mineral for host cells?

A. Well, nobody knows that yet. There aren't iron-calprotectin complexes as far as we know.

Large Trace Element Supplements Reduce the Number of Infectious Episodes after Burns

M.M. BERGER,[†] A. SHENKIN,[‡] F. SPERTINI,[¶] C.A. WARDLE,[‡] A. CLOUGH,[‡] C. SCHINDLER,[§] L. WIESNER,[||] C. CAVADINI,[+] C. CAYEUX,[†] AND R. CHIOLERO[†]

[†]Surgical ICU & Burns Centre, [¶]Immunology & Allergy, [§]Pharmacy, and [||]Plastic & Reconstructive Surgery, CHUV, CH-1011-Lausanne, Switzerland, [‡]Department of Clinical Chemistry, Royal Liverpool University Hospital, PO Box 147, UK – Liverpool L69 3BX, United Kingdom, [+]Nestlé Research Center, CH-1000-Lausanne 26, Switzerland

Keywords burns, trace element, supplementation, infection

Burn patients differ from other trauma patients by the intensity and the duration of their responses to injury, with the most extensive hypermetabolic response and the largest protein losses, due in part to exudative leakage from their wounds. The magnitude of their acute phase response leads to increased nutritional requirements (Cunningham et al. 1989). Burn patients are thus liable to develop acute nutritional deficiencies. Under such circumstances, increased energy and macronutrient supplies (carbohydrates, lipids, proteins) must be provided. The exact requirements of trace elements (TE), which are essential cofactors, remain largely unknown. Moreover infectious complications remain a leading cause of mortality after major burns, and their origin is multifactorial. The TEs copper (Cu), selenium (Se) and zinc (Zn) are involved in both humoral and cellular immunity (Percival 1995, Sandstead 1994). TE supplementation has been associated with positive immune effects and reduction of infections in elderly subjects (Chandra 1992), but there are no such studies available for acute pathologies like burns. The plasma concentrations of various trace elements (TE) are severely depressed in this type of injury, due mainly to exudative losses (Berger et al. 1992a, Berger et al. 1992b). In a preliminary open study, TE supplementation after burn injuries has been associated with increased leukocyte counts and shortened hospital stay (Berger et al. 1994). The present trial investigated TE metabolism and immune responses in severely burnt patients receiving either the usually recommended parenteral TE supplements or early large intravenous supplements.

Patients and Methods

Twenty patients, burnt 30 to 87% of body surface were studied from day (D) 0 (= admission) to D30 post-injury.

Intervention and Nutrition: The study was designed as a double blind placebo controlled supplementation trial. The patients were randomised to receive either the recommended parenteral TE supplements (1 amp Addamel®-N/day, Pharmacia, Stockholm, Sweden) providing 1.3 mg Cu, 32 µg Se and 6.5 mg Zn together with a placebo saline infusion (control group), or an additional supplement of 1.3 mg Cu, 200 µg Se and 20 mg Zn (+TE group) as a 12 h infusion from D1 to D8. Enteral nutrition (jejunal) was started within 12 h of injury in all patients: energy intake was adapted to the resting energy expenditure (1.5 times REE) measured by indirect calorimetry on D2-D3 after injury (Deltatrac II Metabolic monitor, Datex, Finland). All the patients received intravenous recommended vitamin intakes provided as 1 amp Cernevit® (Clintec, Plessis, France) plus 500 mg vitamin C/day (Redoxon®, Roche, Basel, Switzerland).

Analysis: Plasma Se concentrations (furnace AAS), plasma Cu and Zn concentrations (flame AAS), C-reactive protein (CRP) and albumin (nephelometry) and leukocytes (Coulter Counter) were determined

Correct citation: Berger, M.M., Shenkin, A., Spertini, F., Wardle, C.A., Clough, A., Schindler, C., Wiesner, L., Cavadini, C., Cayeux, C., and Chiolero, R. 1997. Large trace element supplements reduce the number of infectious episodes after burns. In *Trace Elements in Man and Animals – 9: Proceedings of the Ninth International Symposium on Trace Elements in Man and Animals. Edited by* P.W.F. Fischer, M.R. L'Abbé, K.A. Cockell, and R.S. Gibson. NRC Research Press, Ottawa, Canada. pp. 433–435.

on D0, D1, D5, D10, D15, D20 and D30. Reference ranges for TE were: Se 0.8–1.9 µmol/L, Cu 14–22 µmol/L, and Zn 12–19 µmol/L. Immunological parameters were measured at D10 and D20.

Infections and micro-organisms were monitored until D30 (urinary infection, pneumonia, wound infection). Diagnosis was based on blood leukocyte counts, CRP levels, bacterial cultures of urine, sputum, skin or blood, and chest X-ray. Pneumonia was defined as the combination of purulent sputum, fever, leukocytosis, new infiltrate on chest X-ray, and bronchoalveolar lavage. A new infection was defined as one fulfilling the above mentioned criteria and requiring the introduction of an antibiotic, or a major change in antibiotherapy.

Statistics: Results are expressed as mean ± SD, and Mann-Withney U tests were used for comparisons.

Results

The patients were aged 41 ± 15 years, with burns 49 ± 17% of body surface. Demographic data were similar in both groups.

Mean plasma Cu concentrations were below the reference range in both groups and reverted to within reference between D15 and D20: they did not differ between groups. Mean Se levels decreased below the reference range in controls until D15, but stayed within normal ranges in the +TE group. On D5, the Se levels were 0.49 ± 0.15 µmol/L in control versus 0.91 ± 0.27 µmol/L in the +TEs ($p < 0.03$), and on D10, 0.58 ± 0.20 µmol/L in control versus 0.88 ± 0.33 pmol/L in +TE ($p < 0.05$). Mean Zn levels were below the reference range until D15 in both groups, with a trend towards higher levels in the +TE group ($p = 0.10$: NS).

C-reactive protein was strongly increased in both groups of patients until D30, peaking between D2 and D10. Albumin, initially normal, decreased and remained below reference ranges in both groups.

Total leukocyte counts were higher in group +TE between D11 and D19. Total lymphocyte counts did not differ between groups, but neutrophil counts were significantly higher in the +TE group. The CD14+ (monocytes), CD3+ (lympho T) and CD19+ (lympho B) cell counts did not differ significantly between the groups. Proliferation to mitogens was significantly depressed in all patients compared to healthy controls (no significant difference between groups). Chemotactism was not altered.

The number of infectious episodes was significantly lower ($p < 0.02$) in the +TE group (1.9 ± 0.8 infections in +TE versus 3.0 ± 1.0 in control group during the first 30 d). The types of infections did not differ between the groups.

Discussion

The early large TE supplements were associated with an earlier normalisation of plasma TE levels levels, but especially of Se. The Cu levels were the least modified by the supplementation. This can in part be explained by the relatively small difference between the 2 groups in the total Cu amounts received by the patients: 1) the differences in prescription levels were not as different between groups as for Se and Zn; and 2) the large TE intakes resulting from the delivery of enteral nutrients further reduced the difference between the groups. As for Zn, the plasma levels were higher, although not significantly in the supplemented group: the acute phase response induces a marked redistribution and probably explains the persistence of low levels until D15 in both groups.

The TE supplementation was also associated with a significant decrease in the number of severe infections. The total leukocytes, mainly the neutrophils were increased in the supplemented group, possibly reflecting an improved non specific immune defence. Further studies in larger groups of patients are required to determine the precise mechanism of this improvement.

References

Berger, M., Cavadini, C., Bart, A., Blondel, A., Bartholdi, I., Vandervale, A., Krupp, S., Chioléro, R., Freeman, J., & Dirren, H. (1992a) Selenium losses in 10 burned patient. Clin. Nutr., 11: 75–82.

Berger, M., Cavadini, C., Bart, A., Mansourian, R., Guinchard, S., Bartholdi, I., Vandervale, A., Dirren, H., Krupp, S., Chioléro, R., & Freeman, J. (1992b) Cutaneous zinc and copper losses in burns. Burns, 18: 373–380.

Berger, M., Cavadini, C., Chioléro, R., Guinchard, S., Krupp, S., & Dirren, H. (1994) Influence of large intakes of trace elements on recovery after major burns. Nutrition, 10: 327–334.
Chandra, R. (1992) Effect of vitamin and trace-element supplementation on immune responses and infection in elderly subjects. Lancet, 340: 1124–1127.
Cunningham, J., Hegarty, M., Meara, P., & Burke, J. (1989) Measured and predicted calorie requirements of adults during recovery from severe burn trauma. Am. J. Clin. Nutr., 49: 404–408.
Percival, S. (1995) Neutropenia caused by copper deficiency: Possible mechanisms of action. Nutr. Rev., 53: 59–66.
Sandstead, H. (1994) Understanding zinc: Recent observations and interpretations. J. Lab. Clin. Med., 124: 322–327.

Discussion

Q1. John Sorenson, University of Arkansas, Little Rock, AR, USA: What were the compounds used for supplementation?

A. The copper was in the form of cupric sulfate, I think. The zinc was certainly zinc sulfate and the selenium was sodium selenite.

Q2. John Sorenson, University of Arkansas, Little Rock, AR, USA: I would suggest that you use a more bioavailable form if you continue these studies, so that you get better absorption and tissue distribution.

A. Right.

Q3. Margaret Elmes, University Hospital Wales, Cardiff, Wales: A delightful study. Most interesting. How are you going to convince the surgical manager that this is cost effective?

A. It's very cost effective. Of all the things we do to intensive care patients, this is the cheapest.

Q3. Harry McArdle, Rowett Research Institute, Aberdeen, Scotland: I'd just like to follow up on Sorenson's comments earlier. You're giving these elements intravenously, aren't you, so there's no problem with absorption as far as getting across the gut is concerned. As soon as the metals get into the plasma, then they're going to be complexed by amino acids, so the original form in which they are given is probably largely irrelevant, as long as there are complexing agents available in the plasma and in the i.v. drip to aid them. What I want to ask you is that this is a very constant effect, the depletion of serum zinc and so on in response to a remarkable number of stresses. I wonder whether or not that's actually an evolutionarily advantageous effect, and that by providing supplements, you may actually be causing problems because you're bypassing a protective effect.

A. You're right. This is what worried me in particular, when we were planning these studies. It's well recognized, as we've heard this morning, the balance of benefits in a situation. The convincing thing as far as I was concerned was the ground studies which Mette Berger performed, demonstrating a marked negative balance due to loss of these elements through the skin particularly. All we're trying to do here is to replace, and maintain a zero balance. We're not giving pharmacological doses, we're just replacing apparent nutrient losses. I think we've demonstrated enough benefit to justify continuing these studies.

Q5. Noel Solomons, Guatemala: I wonder what was the total number of patients that you were able to enroll, and whether or not you have the statistical power to find the survival time as a significant outcome? Everything is very nice, with the directions that things are going, but if death continues to be predominant in the treatment groups, reaching a statistical level, then presumably all nutritional bets are off.

A. The Swiss group had no deaths. It's impossible to run a study like this with mortality as an endpoint. It's just as important an endpoint to reduce the number of days spent in hospital, and especially the days in the intensive care unit, which will have major financial as well as clinical benefits.

Q6. Noel Solomons, Guatemala: You've missed my point. Over a long period of time, despite the excellence of the treatment, and the low mortality rates of today, if at some point, death is significantly greater, although albeit low, in the treated, then we have a serious problem. It will take a long time to achieve those numbers, but later on, people may say "But Dr. Shenkin...". So, we have to keep it as an endpoint, even if it's not a frequent event.

Neutropenia in a Perinatal Copper Deficient Mouse Model

S.S. PERCIVAL AND B. LANGKAMP-HENKEN
Food Science and Human Nutrition Department, University of Florida, Gainesville, Florida 32611, USA

Keywords copper, neutropenia, mouse, immunity

Introduction

Neutropenia is known to occur during Cu deficiency (Graham and Cordano 1976, Williams 1983). In humans, a deficient level of dietary copper, usually through unsupplemented enteral or parenteral solutions, leads to a reduction in the number of circulating neutrophils (see review: Percival 1995). The exact biochemical mechanism for this has not been determined.

Bone marrow aspirates from copper deficient humans show an increased number of promyelocytes (young cells) and a decreased number of cells beyond the myelocyte state (mature cells) (Dunlap et al. 1974, Zidar et al. 1977). This has been interpreted as an arrest of maturation, however, other potential causes of neutropenia have not been explored. Our objective was to establish neutropenia in a copper deficient animal model in order to further examine mechanisms behind the loss of peripheral neutrophils.

Methods

On the day of parturition, 3 dams were placed on a copper deficient diet and three dams served as controls. The diets were formulated as the AIN-93G diet and contained 0.7 and 6 ppm copper as determined by graphite furnace atomic absorption spectrophotometry. Pups (n = 6 males per each group) were weaned at three weeks of age to their respective diets and then sacrificed four weeks later. Whole blood (100 μL) was analyzed by flow cytometry to count neutrophils. The remaining blood was separated into plasma, leukocytes and erythrocytes. Bone marrow cells (BMC) were obtained from the femur. Plasma was analyzed for copper and ceruloplasmin (Cp) activity; erythrocytes, leukocytes and BMC were analyzed for CuZn-SOD activity. The lung was homogenized and extracted to analyze for neutrophil myeloperoxidase activity.

Results

Body weight was significantly smaller in copper deficient animals (Table 1). Heart weight and heart:body weight ratios were significantly greater. Relative to body weight, the spleen and the intestine were significantly greater in copper deficiency and the thymus was significantly smaller. The lung to body weight ratio was not significantly different.

Plasma copper and Cp activity was, as expected, significantly reduced by copper deficiency (Table 2). Cp activity was not detectable in 3 of the 6 copper deficient pups. CuZn-SOD activity in erythrocytes, peripheral leukocytes and bone marrow cells was significantly reduced due to copper deficiency. Erythrocytes contained the most CuZn-SOD activity when comparing cell types. In copper adequate animals, bone marrow cell CuZn-SOD was 10-fold greater than in the peripheral leukocytes, which is expected of more immature cells (Auwerx et al. 1989) . In the copper deficient animal, bone marrow CuZn-SOD activity was 20-fold greater than the leukocyte.

Control animals had 17% neutrophils of the total leukocytes while the copper deficient animals had 6.8% neutrophils. When the lung was analyzed for myeloperoxidase activity, the copper deficient mice ($5.9 \pm 0.9 \times 10^{-6}$ Units/min/mg protein) had almost 2-fold more activity per mg tissue protein than the copper adequate animals ($3.1 \pm 0.5 \times 10^{-6}$ Units/min/mg protein) (Figure 1). This suggests that some

Correct citation: Percival, S.S., and Langkamp-Henken, B. 1997. Neutropenia in a perinatal copper deficient mouse model. In *Trace Elements in Man and Animals – 9: Proceedings of the Ninth International Symposium on Trace Elements in Man and Animals. Edited by* P.W.F. Fischer, M.R. L'Abbé, K.A. Cockell, and R.S. Gibson. NRC Research Press, Ottawa, Canada. pp. 436–437.

Table 1. Organ weights of copper adequate and deficient mice.

	Copper adequate	Copper deficient	p value
Body Weight (g)	30.9 ± 2.9	20.8 ± 2.6	<0.001
Heart (mg/g BW)	5.2 ± 0.5	10.6 ± 2.0	<0.001
Thymus (mg/g BW)	2.65 ± 0.50	1.99 ± 0.52	0.045
Spleen (mg/g BW)	3.48 ± 0.39	6.34 ± 1.18	0.0022(est)
Intestine (mg/g BW)	30.3 ± 2.64	37.9 ± 4.23	0.0039
Lung (mg/g BW)	7.89 ± 0.981	7.32 ± 1.26	0.406

Table 2. Indicators of copper status in copper adequate and copper deficient mice.

	Copper adequate	Copper deficient	p value
Plasma Cu (μmol/L)	12.0 ± 4.2	1.22 ± 0.71	<0.0001
Cp Activity (Units/L)	17.08 ± 4.52	0.16 ± 0.31	<0.0001
Marrow CuZn-SOD	1533.4 ± 384.1	934.7 ± 271.5	0.022
Leukocyte CuZn-SOD	122.7 ± 29.8	44.5 ± 17.7	0.005
Erythrocyte CuZn-SOD	399,900 ± 19,000	43,655 ± 14,300	0.0002

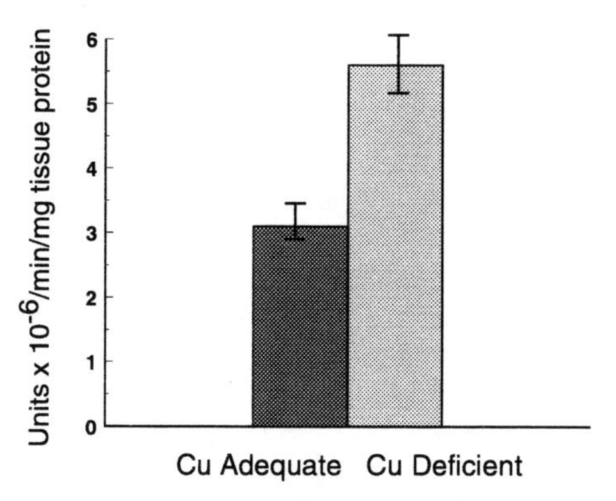

Figure 1. Myeloperoxidase activity in the lungs of mice.

neutrophils have marginated and/or infiltrated into the lung. We do not know whether the increase in lung neutrophils can account for the loss in the periphery.

Summary

This is the first time that neutropenia has been demonstrated in a copper deficient rodent. In this severe model of copper deficiency, common indicators of copper status are low or non-detectable. Neutrophils appeared to infiltrate and/or marginate into the lung of the copper deficient animals. Future research will focus on the mechanism by which neutrophils marginate into tissues in copper deficiency as well as ability of copper deficient bone marrow cells to differentiate into mature neutrophils.

Literature Cited

Auwerx, J. H., Chait, A., Wolfbauer, G. & Deeb, S. S. (1989) Loss of copper-zinc superoxide dismutase gene expression in differentiated cells of myelo-monocytic origin. Blood 74: 1807–1810.

Dunlap, W. M., James, G. W. & Hume, D. M. (1974) Anemia and neutropenia caused by copper deficiency. Ann.Intern.Med. 80: 470–476.

Graham, G. G. & Cordano, A. (1976) Copper deficiency in human subjects. In: Trace Elements in Human Health and Disease. Volume 1 Zinc and Copper. (Prasad, A. S. & Oberleas, D. eds.), pp. 363–372. Academic Press, New York, New York.

Percival, S. S. (1995) Neutropenia caused by copper deficiency: Possible mechanism of action. Nutr.Rev. 53: 59–66.

Williams, D. (1983) Copper deficiency in humans. Semin.Hematol. 20: 118–128.

Zidar, B. L., Shadduck, R. K., Zeigler, Z. & Winkelstein, A. (1977) Observations on the anemia and neutropenia of human copper deficiency. Am.J.Hematol. 37: 177–185.

Effects of Marginal Copper Deficiency on the Outcome of a *Nippostrongylus brasiliensis* Infection in Rats

N. SANGWAN,[†] N.F. SUTTLE,[‡] D.P. KNOX,[‡] AND K. MCLEAN[‡]

[†]Veterinary Parasitology Research Station, Uchani, Karnal 132001, India; [‡]Parasitology Division, Moredun Research Institute, Edinburgh EH17 7JH

Keywords copper deficiency, parasitism, infection, *Nippostrongylus brasiliensis*

There is growing evidence that marginal copper deficiency can impair some properties of cellular components of the immune system when they are isolated and challenged *in vitro*. Effects on susceptibility to infection have been more difficult to demonstrate (Stabel et al. 1993) and where they have occurred, the context has been one of severe Cu deficiency and microbial challenge. With parasitic nematode infections of the gut in sheep, addition of a Cu antagonist (molybdenum) to the diet has improved resistance to infection and enhanced the inflammatory responses which lead to nematode rejection (Suttle et al. 1992). This paper describes the effects of a simple, marginal copper deficiency on the outcome of a parasitic infection of the gut by *Nippostrongylus brasiliensis* in rats.

Materials and Methods

Two groups of 12 male, Wistar rats, 8 weeks old and weighing 216 g initially, were given a partially-purified low Cu diet consisting of (g/kg): casein, 200; sucrose, 310; corn starch, 310; mixed vegetable oil, 50; wheat bran, 80; complete mineral mix, 40; complete vitamin mix, 10: it contained 17.04 µmol Cu/kg DM. One group (+Cu) was given supplementary Cu via the drinking water (94.4 µmol Cu/L). After 50 d, 8 rats from each group were infected by injecting 2000 L3 larvae of *N. brasiliensis* subcutaneously. Daily faecal collections were made from moistened trays beneath individual cages with raised, mesh floors from the 5th to 9th days post-infection (d.p.i.) for faecal egg counts, performed by NaCl-flotation. Sub-groups of four were slaughtered 7 and 9 d.p.i., the normal period of worm expulsion by the host, and adult worms retrieved from the small intestine and counted (Nawa and Miller 1978). Two members of the non-infected sub-group were also slaughtered 7 and 9 d.p.i. Final Cu status was assessed by plasma and liver Cu analyses of samples procured at slaughter. All parasite data were transformed to $\log_e$ prior to the comparison of group means because variances were heterogenous.

Results

There were no significant differences in mean (± s.e.) liveweight gain (166 and 160 ± 10 g) or haematocrit (48.3 and 48.0 ± 0.59%) between Cu-depleted and Cu-supplemented rats at or during the time of infection but liver and plasma Cu concentrations at slaughter were 31% lower in Cu-depleted rats ($P < 0.001$) (Table 1). There were marked differences in the patterns shown by FEC (Table 1): eggs appeared on 6 d.p.i. when there were more than twice as many in the +Cu as in the –Cu group ($P < 0.05$) but by 7 d.p.i. FEC were similar in the two groups. A marked fall in FEC from an earlier peak (7 v 8 d.p.i.) in the –Cu group gave lower FEC again by 9 d.p.i. than those in the +Cu group ($P < 0.05$). There were no marked differences in worm counts at 7 d.p.i. (Table 2) but by 9 d.p.i. there were more female worms in the –Cu than in the +Cu group, indicating a major reduction in fecundity (eggs produced/female worm) (Table 2) in parasites from Cu-depleted hosts.

Correct citation: Sangwan, N., Suttle, N.F., Knox, D.P., and McLean, K. 1997. Effects of marginal copper deficiency on the outcome of a *Nippostrongylus brasiliensis* infection in rats. In *Trace Elements in Man and Animals – 9: Proceedings of the Ninth International Symposium on Trace Elements in Man and Animals. Edited by* P.W.F. Fischer, M.R. L'Abbé, K.A. Cockell, and R.S. Gibson. NRC Research Press, Ottawa, Canada. pp. 438–439.

Table 1. Effects of copper depletion and infection with N. brasiliensis on copper concentrations in plasma and liver of male Wistar rats (pooled values for 7 and 9 d.p.i.).

		Liver Cu (μmol/kg DM)			Plasma Cu (μmol/L)		
Cu group	Infection	0	+	s.e.d.m.	0	+	s.e.d.m.
–		158	140		7.5	8.4	
(n)		(4)	(8)	12.3	(4)	(8)	1.22
+		221	204		11.2	11.7	
(n)		(4)	(8)		(4)	(8)	
	s.e.d.m.	14.2	10.0		1.41	1.00	

Table 2. Effects of copper depletion on faecal egg count (FEC) female worm count (FWC) and fecundity (F, FEC ÷ FWC) (all data are transformed to $\log_e$).

	FEC				FWC		F	
D.p.i.	6	7	8	9	7	9	7	9
Cu group								
–	8.61^b	10.15^a	9.77^e	7.78^b	6.13^a	4.49^a	3.33^a	3.28^a
+	9.73^a	9.87^a	10.13^a	8.44^a	6.54^a	1.83^b	4.02^a	6.61^b
s.e.d.m.	0.408	0.428	0.574	0.217	0.218	0.913	0.436	1.095

Superscript differences within columns denote group differences ($P < 0.05$).

Discussion

The contrasting patterns of FEC could reflect three important influences: firstly, effects of Cu depletion on host resistance; secondly, effects of Cu depletion on the egg-laying female parasite; thirdly, interactions within the parasite population.

Direct effects of copper depletion on the parasite seem unlikely because there were no effects of dietary treatment on the activity of superoxide dismutase in parasite homogenates at 7 d.p.i. (unpublished data). The initial reduction in FEC in the –Cu group may reflect a host-induced delay in the arrival of mature worms in the intestinal tract. The final reduction in FEC may reflect continued impairment of parasite development which had yet to be reflected by a reduction in worm numbers in the –Cu group. Inhibition of egg-laying amongst the larger population of –Cu females seems unlikely at that late stage. The study needs to be repeated with more frequent slaughter groups to confirm these suggestions but appear to be consistent with previous observations in parasitised sheep.

Literature Cited

Nawa, Y. and Miller, H.R.P. (1978) Protection against *Nippostrongylus brasiliensis* by adoptive communisation with immune thoracic duct lymphocytes. Immunol. 37: 57–60.

Stabel, J.R., Spears, J.W. and Brown, T.T. (1993) Effect of copper deficiency on tissue, blood characteristics and immune function of calves challenged with Infectious Rhinotracheitis Virus and *Pasteurella haemolytica*. J. Anim. Sci. 74: 1247–1255.

Suttle, N.F., Knox, D.P., Angus, K.W., Jackson, F. and Coop, R.L. (1992) The effects of dietary molybdenum on nematode and host during *Haemonchus contortus* infection in lambs. Res. Vet. Sci. 52: 230–235.

Role of Zinc in Systemic and Local Immunity

E. MOCCHEGIANI, L. SANTARELLI, AND N. FABRIS
Immunol. Centre, Research Department, I.N.R.C.A., 60100 Ancona, Italy

Keywords zinc, infections, thymulin, immunity

Introduction

Zinc is one of the most relevant trace elements in the body (Fabris and Mocchegiani 1995). Zinc is required as a catalytic component of more than 200 enzymes, as a structural constituent of many proteins (zinc fingers) and, likely, is able to prevent free radical formation (see review Fabris and Mocchegiani 1995). The zinc requirement for cell division and differentiation, as well as for programmed cell death, for gene transcription, for biomembrane functioning and for many enzymatic activities, has made this element a leading one in assuring the correct functioning of various tissues, organs and systems, including the immune system (Chandra 1983). Zinc is required in the defence against infectious diseases such as those caused by viruses, bacteria, fungi and protozoa (Chandra 1983). The present paper aims to summarize the role of zinc and the possible beneficial effect of zinc supplementation in systemic and local immunity, both in congenital and acquired human pathologies characterised by zinc deficiency and by the presence of an immunodeficiency state associated with an increased susceptibility to infections due to opportunistic agents.

Role of Zinc in Systemic and Local Immunity

Zinc is essential for immunocompetence in man and animals both at the systemic and local level (Chandra 1983). Zinc deficiency both in humans and in animals causes thymic and lymphonode atrophy and impaired T-cell-mediated immunity (Chandra 1983). With regard to local immunity, the role of zinc on wound-healing has been recognized (Sandstead et al. 1982). Zinc deficiency in laboratory animals impairs neutrofil chemiotaxis and macrophage functions and causes increased susceptibility to gut infections by Salmonella (Chandra 1983). Zinc deficiency in these human conditions is also associated with increased susceptibility to infectious episodes frequently due to opportunistic microorganisms (see review Fabris and Mocchegiani 1995).

Mechanism of Action of Zinc on the Immune System

The mechanism by which zinc may affect the immune system is certainly multifaceted, due to the widespread action of zinc on different enzymes and cytokines required for proliferation and function of immune cells. The role of zinc in cell division is primarily based on its requirement by DNA polymerase and thymidine kinase, both enzymes being zinc-dependent (Pardee et al. 1986). Zinc is also involved in promoting programmed cell death (apoptosis). A peculiar characteristic of apoptosis is the early activation of an endogenous endonuclease which affects internucleosomal DNA cleavage. These endonucleases are activated by calcium and magnesium and are strongly inhibited by zinc (Martin et al. 1991). Zinc may modulate the proliferation rate of lymphoid cells through its effect on the production of cytokines, such as interleukin-1 (IL-1), interleukin-6 (IL-6), and interferon (IFN). The first one is an obligatory step in T cell proliferation (Scuderi 1990). Zinc is required for the activation of one of the best known thymic peptides called thymulin in its zinc-bound form (ZnFTS), which is responsible for cell mediated immunity (Dardenne et al. 1982), whereas the peptide without bound zinc (FTS) is inactive and likely prevents the active form exerting its action (Fabris et al. 1984). The "*in vitro*" addition of zinc ions to plasma samples

Correct citation: Mocchegiani, E., Santarelli, L., and Fabris, N. 1997. Role of zinc in systemic and local immunity. In *Trace Elements in Man and Animals – 9: Proceedings of the Ninth International Symposium on Trace Elements in Man and Animals. Edited by* P.W.F. Fischer, M.R. L'Abbé, K.A. Cockell, and R.S. Gibson. NRC Research Press, Ottawa, Canada. pp. 440–441.

unmasks the presence of inactive hormone showing the total amount of the thymic hormone produced (zinc bound ZnFTS + zinc unbound FTS) (Fabris et al. 1984). This occurs in conditions characterized by zinc deficiencies, such as Down syndrome, acrodermatitis enteropathica, cystic fybrosis, Duchenne syndromes and infectious diseases, such as AIDS (see review Fabris and Mocchegiani 1995). Thus one of the mechanisms of action of zinc on the immune system may be the direct activation of thymic peptides. The variety of mechanisms by which zinc modulates the efficiency of the immune system may explain its involvement either in local and systemic immune reactions, as well as, it role in central thymic functions.

Effect of Zinc Supplementation on Infections in Down's Syndrome and AIDS

We have investigated the effect of a zinc supplementation on immune dysfunctions in Down's syndrome, and in an acquired zinc-deficiency, such as AIDS, both being pathologies frequently associated with infectious episodes by opportunistic agents. Zinc supplementation in Down's syndrome subjects (1 mg Zn^{++}/kg/body weight/day, for 2–4 months per year) is able to: a) activate thymulin, b) increase T cell subsets in peripheral blood, c) increment neutrophil chemotaxis, and d) decrease the incidence of infectious diseases (Mocchegiani et al. 1991). In AIDS pathology, a reduced zinc bioavailability has been demonstrated in stage IV Cl (full-blown AIDS) compared to stage III of the disease. This is accompanied with a reduction in thymic endocrine function (Fabris et al. 1988). Zinc supplementation for 1 month (45 mg of Zn^{++}/day) both in stage III and stage IV C1 subjects causes: a) a full restoration of thymic endocrine activity, b) an increment in the number of $CD4^{+}$ cells, c) a greater than 30% reduction in the occurrence of opportunistic infectious episodes in the 24 months following entry into the study in stage IV C1 subjects, and a consistent delay in the appearance of the first opportunistic infection in stage III subjects (Mocchegiani et al. 1995). The effect of zinc is restricted to infections due to Pneumocystis Carinii (pneumonitis) and Candida Albicans (aesophagitis) (Mocchegiani et al. 1995). The effect of zinc supplementation on local infections has been also observed in Down's syndrome (cutaneous and mucosal infections) (Mocchegiani et al. 1991). The cause for such an effect may be due to the known role of zinc to help lung macrophage function (Chandra 1990). From the data presented the following points may be stressed: a) Zinc is a crucial trace element for the efficiency of the whole immune system; its action may be quite wide-spread since it may modulate both central and peripheral immune functions. b) Zinc deficiencies are associated with increased susceptibility to infectious episodes at the local level; zinc supplementation may reduce such a susceptibility both in Down's syndrome and in AIDS.

References

Chandra, R.K. (1983) Trace elements and immune response. Immunol. Today 4: 322–325.

Dardenne, M., Pleau, J.M., Nabama, B., Lefancier, P., Denien, M., Choay, J. & Bach, J.F. (1982) Contribution of zinc and other metals to the biological activity of the serum thymic factor. Proc. Nat. Acad. Sci. USA. 79: 5370–5373.

Fabris, N., Mocchegiani, E., Aniadio, L., Zannotti, M., Licastro, F. & Franceschi, C. (1984) Thymic hormone deficiency in normal, aging and Down's syndrome. Is there a primary failure of the thymus? Lancet 1: 983–986.

Fabris, N., Mocchegiani, E., Galli, M., Irato, L., Lazzarin, A. & Moroni, M. (1988) AIDS, zinc deficiency, and thymic hormone failure. JAMA 259: 839–840.

Fabris, N & Mocchegiani, E. (1995) Zinc, human diseases and aging. A review. Aging. Clin. and Exp. Res. 7: 77–93.

Martin, S.J., Mazdai, G., Strain, J.J., Cotter, T.G. & Hannigam, B.M. (1991) Programmed cell death (apoptosis) in lymphoid and myeloid cell lines during zinc deficiency. Clin. Exp. Immunol. 8: 338–343.

Mocchegiani, E., Licastro, F., Franceschi, C. & Fabris, N. (1991) Immunological function in Down's syndrome before and after zinc supplementation. J. Chemiotherapy 3: 170–175.

Mocchegiani, E., Veccia, S., Ancarani, F., Scalise, G. & Fabris, N. (1995) Benefit of oral zinc supplementation as an adjunct to AZT therapy against opportunistic infections in AIDS. Int. J. Immunopharmacol. 17: 719–727.

Pardee, A.B., Coopock. D.L. & Yang, H.C. (1986) Regulation of cell proliferation at the onset of DNA synthesis. J. Cell. Sci. 4: 171–180.

Sandstead, H.H., Henriksen, L.K., Greger J.L., Prasad, A.S. & Good, R.A. (1982) Zinc nutriture in the elderly in relation to taste acuity, immune response and wound healing. Am. J. Clin. Nutr. 36: 1056–1059.

Scuderi, P. (1990) Differential effect of copper and zinc on human peripheral blood macrocyte cytokine secretion. Cell. Immunol. 126: 391–405.

Effect of a Soluble Cobalt, Selenium and Zinc Glass Bolus on Humoral Immune Response and Trace Element Status in Lambs

N.R. KENDALL, A.M. MACKENZIE, AND S.B. TELFER
Department of Animal Physiology and Nutrition, University of Leeds, Leeds, LS2 9JT, UK

Keywords trace element status, humoral immune response, bolus, sheep

Introduction

Trace element deficiencies are common in many countries and affect animal health, productivity and welfare. Trace elements that can be limiting in animal nutrition include copper, cobalt, selenium and zinc. Deficiency diseases may be manifested as a consequence of a single or a multiple element deficiency. Deficiency in any of the above trace elements can result in an increased disease susceptibility and a decreased immune function. A soluble glass bolus has been developed for the sustained release of cobalt, selenium and zinc to sheep. This bolus is similar to the commercial Co/Se/Cu soluble glass (Cosecure©) but has zinc replacing copper.

Materials and Methods

Thirty-four Suffolk cross store lambs were allocated to two groups by restricted randomisation of body weight (mean 27.8 kg, s.d. ± 3.6). and were kept at pasture on the University of Leeds Farms throughout the trial. On day 0, one group was bolused with the cobalt, selenium and zinc soluble glass boluses (bolused), whilst the other group was left unbolused (controls). The boluses weighed approximately 33g with a composition of 13.1% zinc, 0.5% cobalt and 0.15% selenium. Immune function was measured in the lambs by measuring their humoral immune response to a novel antigen, keyhole limpet haemocyanin (KLH). On day 34, lambs were immunised with 1 mg KLH (Calbiochem, San Diego, California, USA) precipitated in alum and given subcutaneously at a site over the ribs. Blood samples were taken for the antibody response on days 20, 42, 49 and 63 and serum was stored at –20°C. Lamb anti-KLH IgG responses were measured by a direct ELISA method (Mackenzie et al. 1996). Blood samples were taken for assessment of trace element status on days 0, 20, 42 and 63. Cobalt status was assessed by measuring vitamin B_{12} concentrations by radioassay kit (Becton Dickinson, Oxford England). Selenium status was assessed by colorimetric assay of erythrocyte glutathione peroxidase activity (Telfer et al. 1984). Plasma zincs and coppers were diluted 1:5 with 0.1 M HCl (Analar, BDH) and read by Atomic Absorption spectrophotometry (Pye SP9 AA spectrophotometer, zinc at 213.9 nm with background correction, copper at 324.8 nm). The sheep were slaughtered for recovery of boluses and collection of liver samples. Livers were analysed after freeze drying and microwave wet digestion (70% HNO_3, Analar) using the same instrument parameters as for plasma. Recovered boluses were oven dried and weighed for calculation of dissolution rates.

Table 1. Effect of bolusing on lamb liveweight (kg).

Day	Bolused	Controls	SE
0	27.77	27.79	0.896
20	32.32	33.29	1.006
42	35.27	35.71	0.993
63	37.32	36.24	1.046

Correct citation: Kendall, N.R., Mackenzie, A.M., and Telfer, S.B. 1997. Effect of a soluble cobalt, selenium and zinc glass bolus on humoral immune response and trace element status in lambs. In *Trace Elements in Man and Animals – 9: Proceedings of the Ninth International Symposium on Trace Elements in Man and Animals. Edited by* P.W.F. Fischer, M.R. L'Abbé, K.A. Cockell, and R.S. Gibson. NRC Research Press, Ottawa, Canada. pp. 442–444.

Results

Figure 1 shows the bolused lambs having significantly greater anti-KLH IgG levels on day 42 ($p < 0.05$) and day 63 ($p < 0.01$). Serum vitamin B_{12} concentrations are illustrated in figure 2 and were significantly greater ($p < 0.001$) in the bolused group on all post-bolusing samplings with many of the controls being deficient (mean ~260 pg/mL) at day 42. Erythrocyte glutathione peroxidase activities are illustrated in figure 3 and were significantly increased ($p < 0.001$) in the bolused group on all post-bolusing samplings. Figure 4 shows the bolused lambs having higher plasma zinc concentrations on day 42 ($p < 0.05$) and day 63 ($p < 0.01$).

There was no significant effect of bolusing on liver copper and zinc concentrations, plasma copper concentrations or liveweight of the lambs. The average bolus dissolution rate was 326 mg glass/ day (s.d 30 mg/day) giving a daily release of 45.6 mg zinc, 1.6 mg cobalt and 0.5 mg selenium.

Discussion

The bolus release rate (quoted in parentheses) was adequate for all three trace elements with dietary requirements of 20–30 (45.6) mg Zn/d, 0.1–0.2 (0.5) mg Se/d and 0.1–0.2 (1.6) mg Co/d given a dietary intake of ~1 kg dry matter. These release rates are below the maximum tolerable levels of 750, 2 and 10 mg/kg DM for zinc, selenium and cobalt respectively (N.R.C. 1985). The bolused lambs had significantly higher serum vitamin B_{12} concentrations and erythrocyte glutathione peroxidase activities on all

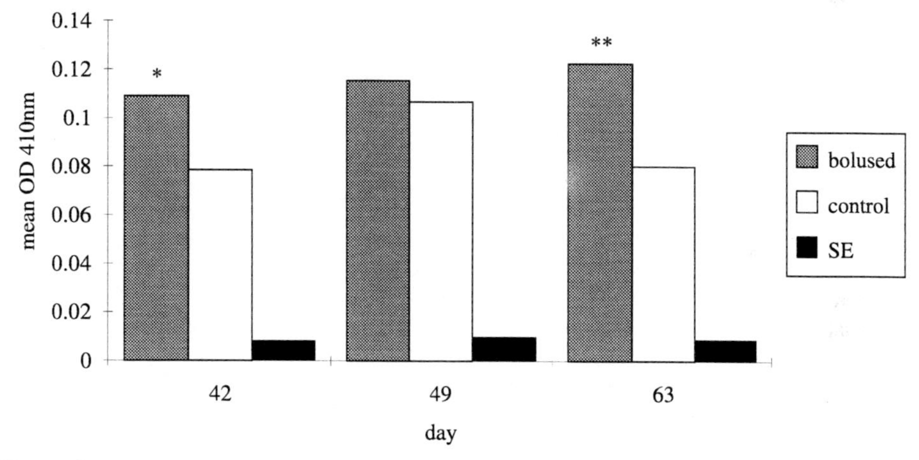

Figure 1. Humoral immune response after challenge on day 34 after bolusing.

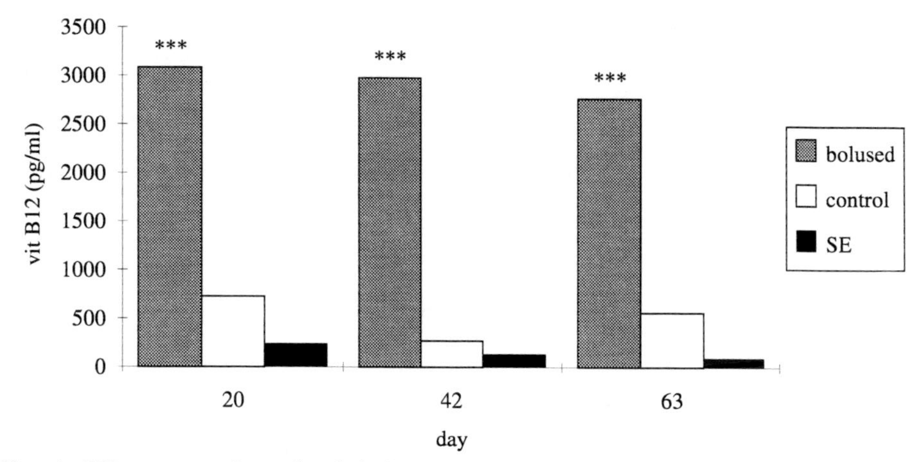

Figure 2. Vitamin B12 concentrations after bolusing.

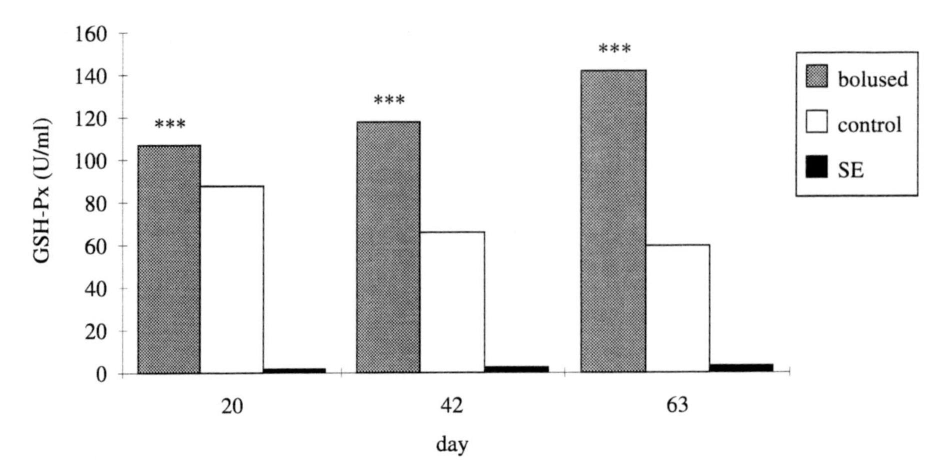

Figure 3. Glutathione peroxidase activities after bolusing.

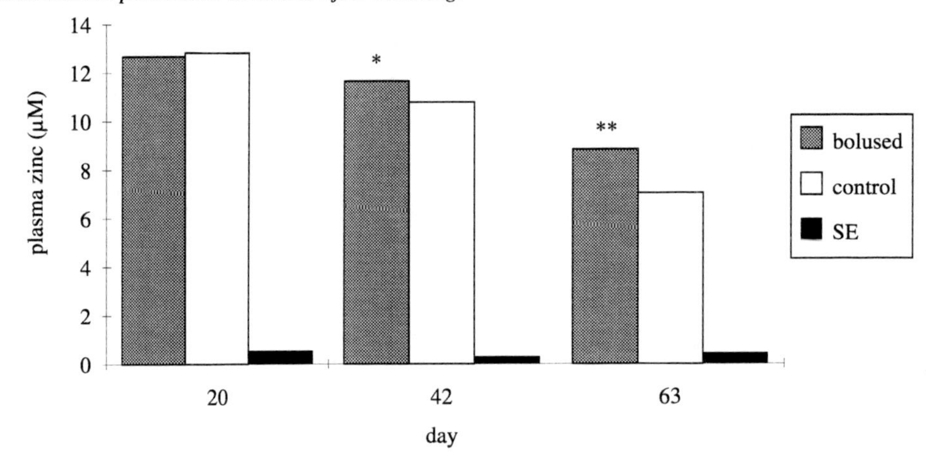

Figure 4 Plasma zinc concentrations after bolusing.

samplings after bolusing, with significantly greater plasma zinc concentrations on days 42 and 63 with no depression in plasma copper. The significant increase in the levels of anti-KLH IgG in the bolused lambs may be due to an increase in the secretion rate of the antibody or an increase in the number of plasma cells. Previous reports have shown that cobalt, selenium and zinc affect the immune response in ruminants (MacPherson et al. 1987, Pollock et al. 1994). In this study it is not clear if the effect of the bolus on lymphocyte function is due to a single element or a combination of two or three elements. To conclude, bolusing with the cobalt, selenium and zinc bolus resulted in an increased antibody response and an elevated cobalt, selenium and zinc status of the lambs.

Acknowledgements

We are gratified for the assistance of D.V. Illingworth and D.W. Jackson.

References

Mackenzie, A.M., Drennan, M., Rowen, T., Carter, S. & Dixon, J. (1996) Effect of transportation and weaning on humoral immune response of calves. Res. Vet. Sci (in press).

MacPherson, A., Gray, D., Mitchel, G.B.B. & Taylor, C.N. (1987) Br. Vet. J. 132: 348–353

N.R.C. (1985) Nutrient requirements of sheep. 6th edition. National Academy Press, Washington, D.C. USA.

Pollock, J.M., McNair, J., Kennedy, S., Walsh, D.M., Goodall, E.A., Mackie, D.P. & Crockard, A.D. (1994) Res. Vet. Sci. 56: 100–107.

Telfer, S.B., Zervas, G. & Carlos, G. (1984). Can. J. Anim. Sci., 64 (Suppl.1): 234–235.

Lead Poisoning and Renal Function: Relationship with Trace Elements and Immune Response

N. RESTEK-SAMARZIJA, B. MOMČILOVIĆ, AND M. BLANUŠA

Department of Plant Science, University of Adelaide, Glen Osmond 5064, Australia, and US Plant, Soil and Nutrition Laboratory, Tower Road, Ithaca, NY 14953, USA

Keywords lead, renal function, trace elements, immunity

Introduction

Although lead toxicity has been known for centuries, the chronic lead effects on kidneys and blood pressure remain a controversy (Castellino et al. 1995). The aim of this study was to investigate the relation between chronic, recurrent lead poisoning, kidney function, blood pressure, anaemia, immunological parameters and interaction with trace elements.

Patients and Methods

The investigation was carried out in 66 lead exposed workers. In 53 of them excessive lead exposure had caused one or even several episodes of lead poisoning. The remaining 13 subjects were currently exposed to lead but never poisoned. In each subject the glomerular filtration rate (GFR) was measured by creatinine and DTPA clearances and tubular function by renal hipurane flow, urine osmolality and proteinuria. Blood count and haemoglobin concentrations were measured. Immunological status was evaluated by determination of the percentage of T and B lymphocytes in blood. Concentration of zinc (Zn), copper (Cu), calcium (Ca), iron (Fe) and magnesium (Mg) in serum and blood lead (PbB), erythrocyte protoporphyrin (EP) and aminolevulinic acid dehydratase activity (ALAD) was also measured. Data were statisticaly evaluated by factor analysis.

Results and Discussion

Four significant factors obtained are shown in Table 1. In the first factor a significant correlation between renal function and blood pressure was found. The second factor was defined by haematological parameters, levels of Zn and Ca in serum and age. The third factor showed a significant association of birth, number of past lead poisonings, DTPA clearance, hipurane renal flow, T and B lymphocytes and serum magnesum level. In the fourth factor PbB, EP, ALAD and serum Zn level were correlated. The results of our study showed that chronic lead poisoning, due to an increase in a lead body burden, can cause an impairment of renal function with a concomitant suppression of cellular immunity and stimulation of humoral immunity. These findings are in agreement with some previous studies (Wedeen et al. 1979, Coscia et al. 1987) suggesting the possible role of the immune system in the nephrotoxic effects of lead. In our study hypertension was not directly associated with chronic lead poisoning, duration of lead exposure or biological exposure indices. However, our results showed an important role of trace elements in haematological and immunological changes in chronic lead poisoning.

Correct citation: Restek-Samarzija, N., Momčilović, B., and Blanuša, M. 1997. Lead poisoning and renal function: Relationship with trace elements and immune response. In *Trace Elements in Man and Animals – 9: Proceedings of the Ninth International Symposium on Trace Elements in Man and Animals. Edited by* P.W.F. Fischer, M.R. L'Abbé, K.A. Cockell, and R.S. Gibson. NRC Research Press, Ottawa, Canada. pp. 445–446.

Table 1. Four significant factors (F) showing the correlations between variables.

Variable	F 1	F 2	F 3	F 4
Year of Birth	–0.23	**0.50**	**–0.51**	–0.03
Exposure	0.02	–0.20	–0.23	0.20
No. Of Poisonings	0.18	–0.04	**0.47**	0.32
PB Blood	–0.13	0.12	0.02	**0.89**
EP	0.05	–0.10	–0.13	**0.84**
Alad	–0.11	0.06	–0.03	**–0.84**
Uric Acid	**0.75**	0.17	–0.19	0.09
Creatinine	**0.59**	–0.39	0.06	0.18
Creatinine Clearance	**–0.69**	0.21	0.00	–0.08
DTPA Clearance	**–0.49**	0.17	**–0.58**	–0.05
Hipurane Flow	–0.24	–0.01	**–0.55**	–0.19
Urine Osmolality	**–0.51**	0.21	–0.20	0.04
Proteinuria	**0.51**	0.07	0.08	–0.01
Blood Pressure	**0.57**	–0.12	–0.30	–0.21
Erythrocyte	–0.28	**0.55**	–0.07	–0.08
Hemoglobin	–0.07	**0.61**	0.16	–0.14
B Lymphocyte	0.01	0.16	**0.72**	0.05
T Lymphocyte	0.17	0.08	**–0.62**	0.09
Zinc	–0.19	**0.43**	0.04	–0.41
Copper	0.16	–0.26	–0.14	–0.13
Calcium	–0.01	**0.58**	–0.02	0.28
Iron	–0.31	0.20	0.21	–0.27
Magnesium	0.28	0.12	**–0.54**	0.13

Loadings >0.40 are significant.

Literature

Castellino, N., Castellino P., & Sannolo, N. (1995) Inorganic lead exposure. Lewis Publishers, Boca Raton.
Coscia, C.G., Discalzi, G., & Ponzetti, C. (1987) Immunological aspects of occupational lead exposure. Med. Lav. 78: 360–364.
Wedeen, R.P., Mallik, D.K., & Batuman, V. (1979) Detection and treatment of occupational lead exposure. Arch. Intern. Med. 139: 53–57.

Trace Element Intervention Strategies in Animal and Human Nutrition

Chairs: R.S. Gibson and E. Chavez

A Strategy for Breeding Staple-Food Crops with High Micronutrient Density

ROBIN D. GRAHAM AND ROSS M. WELCH

Department of Plant Science, University of Adelaide, Glen Osmond 5064, Australia, and US Plant, Soil and Nutrition Laboratory, Tower Road, Ithaca, NY 14953, USA

Keywords iron, zinc, β-carotene, wheat

Food-based approaches to addressing micronutrient malnutrition in much of the human population have hitherto been based mainly on balancing cereal-based diets with vegetables, and to a lesser extent, animal products. Although high in micronutrients, these commodities are more expensive foods than staples, and also more seasonal, subject to spoilage, and difficult to store and transport. Moreover, their availability in some countries is only 10–15% of what is required to meet the needs of the people. Because of the wide availability of staples, their high proportion in the diet of the most malnourished, because field fortification has several advantages over fortification during processing, and because staples are known for their low content of these essentials, we have studied the possibility of breeding to improve plant staples as sources of micronutrients for humans. This is in response to the WHO-FAO call for a food-based solution to micronutrient malnutrition which is considered now to affect more than 2 billion people world-wide (FAO-WHO 1992).

Over the last two years a set of criteria has been established that needs to be satisfied if such a breeding program is to be successful. The criteria are

1. Adequate genetic variability within crops for genotypes with high micronutrient density in seeds, roots or tubers
2. Compelling reasons for incurring the extra breeding cost associated with the objectives
3. Compatibility of high density with high yield and quality to ensure market acceptability
4. High bioavailability of the micronutrient in high-density types
5. Sufficiently high heritability of traits that the rate of progress in breeding is reasonable
6. Suitable selection criteria for the traits in question
7. Capacity of soils to supply the extra micronutrient required by these high density types

Criterion 1. Important genetic variability in the nutritional value of plant foods exists among varieties of the one crop, as well as between crops, and in levels of micronutrients just as it does for the major components of food. Cereals, from a 1% sample of the germplasm collections, show a range of concentrations in grain of the essential transition metals of up to four-fold from the lowest to the highest; the best are about twice that of standard high-yielding varieties (Table 1).

We therefore expect to find higher density types. Rice lines which had about twice the mean iron density had a frequency of 1% in a sample of 1100 lines, but the frequency of high zinc types was 15%,

Correct citation: Graham, R.D., and Welch, R.M. 1997. A strategy for breeding staple-food crops with high micronutrient density. In *Trace Elements in Man and Animals – 9: Proceedings of the Ninth International Symposium on Trace Elements in Man and Animals. Edited by* P.W.F. Fischer, M.R. L'Abbé, K.A. Cockell, and R.S. Gibson. NRC Research Press, Ottawa, Canada. pp. 447–450.

Table 1. The iron and zinc density in seeds of selected wheat genotypes grown together in the same field experiment at El Batan, Mexico in 1994 (data of I. Monasterio and R.D. Graham, unpublished). Some high yielding lines such as 439 CMH 84 are also high in micronutrient density, but most high yielding released varieties tended to be around the grand mean.

Line identification	Iron (mg/kg)	Zinc (mg/kg)
149 Turkish	59	44
439 CMH 84	58	67
486 Spot-2	20	17
94 Pitic 62	32	64
360 Lira/Tan	41	25
Grand mean (n = 505)	35	34

reflecting the benefits of past selection of zinc-efficient varieties by breeders. β-carotene is not essential in storage tissues, so the range of β-carotene in maize and cassava can be much greater than four-fold. The prospects of breeding for greater β-carotene content and for alleviating vitamin A deficiency has been established already, as this deficiency was eliminated in grain-fed hogs 30 years ago by a simple breeding effort in maize. In cassava, β-carotene in roots varied 0–25 mg/kg, the highest enough to provide the RDA of a child from 100 g fresh root.

Criterion 2. The most compelling reason for the breeding program outlined in this talk is the extent of micronutrient malnutrition both in the developing and the developed world.

Criterion 3. Unlike protein, micronutrient density appears to be positively correlated with yield. The domestication of beans does not seem to have decreased the iron density in that crop, and like wheat, high yielding lines with high density are found. Most farmers, whether subsistence or commercial, are more likely to grow a nutritionally superior variety if it also is high yielding and otherwise acceptable in the market-place. Another important reason is the progress we can make in yield and disease resistance by breeding better adapted varieties that fortify themselves through their ability to absorb more of the limiting micronutrients from deficient soils (termed micronutrient-efficient varieties). A trait to enhance yield, disease resistance and nutritional value all at once seems entirely feasible where the soils are low in soluble micronutrient. This possibility seems most likely on soils low in zinc, globally about 49% of all soils.

Criterion 4. While our initial strategy is to enhance the absolute content of micronutrients in staples, our real objective is bioavailable micronutrient content. This may be achieved by increased concentrations at the same bioavailability (%), by higher bioavailability, or both; the constraints and opportunities are addressed elsewhere in this symposium. The urgent need for the success of this project, as in human nutrition generally, is faster yet reliable methods of assessing bioavailability, which are relevant to the deficient individuals in our target group. Current research suggests that certain amino acids are important in zinc and iron bioavailability, and screening for amino acid quality of the grain protein is also an objective of the project.

Criterion 5. The genetics of micronutrient traits appear to be relatively simple, single major genes in some cases, making the breeders' task easier. For β-carotene, many additive genes are involved but selection can still be rapid as trait expression can be detected by eye in segregating populations. For iron, efficiency is a matter of stacking additive minor genes, since most viable cultivars already have the major gene present. Many commercial varieties, however, are low in zinc efficiency, and in this case there appear to be three or four genes necessary to attain full agronomic and nutritional advantage. On top of these genes for agronomic efficiency, we expect to find major genes for retranslocational efficiency — that is, for the ability to remobilise iron and zinc (and maybe also β-carotene and iodine) from vegetative parts to grain. In paddy rice, this is a particularly good prospect as the vegetative iron pool is high, and only the grain content is low.

Criterion 6. Faster and cheaper means of selecting for efficiency and loading traits for these micronutrients are needed. Clearly, selection for yield in deficient soil in order to identify efficient segregants is too slow for effective breeding in the future. The best prospects are molecular markers (cDNAs, RFLPs, AFLPs and RAPDs) for the mRNAs expressed by efficiency and seed loading traits, these will materially increase the prospects of efficient breeding in future food systems approaches to nutritional balance.

***Criterion** 7*. Scientists and farmers alike are concerned about depletion of poor soils by nutrient-efficient crops able to extract more micronutrients from these soils, probably because agricultural pursuits can and do induce zinc deficiency. Almost invariably however, this is due to an induced imbalance rather than to depletion. Calculations show that absolute depletion of even the poorest soils would take more than a thousand years under current cropping practices (Graham and Welch 1996). As pro-vitamin A carotenoids are synthesised de novo in the plant, depletion is again not an issue. Only for iodine is depletion an issue, owing to leaching by rain, since iodine does not bind strongly to soil constituents. In dry areas, however, soils may contain appreciable iodine which could be exploited by breeding for uptake efficiency. It is possible to apply zinc and iodine fertilisers to improve the content in grain but this basic agronomic approach does not work for iron or β-carotene so breeding is the only approach available to crop scientists to improve staples for these nutrients.

The economics of a food-based approach to overcoming micronutrient malnutrition are also quite favourable, the cost being perhaps 1% of alternative non-food strategies (Bouis 1995). This paper advocates the exploration of the potential of staples to provide more than energy and protein as a matter of urgency, before the extent of malnutrition gets worse, as it is apparent that the human population must depend even more on staples as the population grows towards the 8 billion expected in 2020.

Literature Cited

Bouis, H. E. (1995) Plant Breeding and Food Policy to Reduce Micronutrient Malnutrition: CGIAR Strategies for Improving Dietary Quality. A Proposal. International Food Policy Research Institute, Washington, DC. August 1995.

Food and Agriculture Organisation and World Health Organisation (1992) Preventing Specific Micronutrient Deficiencies. Theme paper No. 6. International Conference on Nutrition. Rome.

Graham, R.D. & Welch, R.M. (1996) Breeding for staple-food crops with high micronutrient density. Agricultural Strategies for Micronutrients. Working Paper No 3. International Food Policy Research Institute, Washington, DC.

Discussion

Q1. Richard Black, Kellogg Canada Inc., Etobicoke, ON, Canada: Let me play devil's advocate for just a moment. If I was working for a large chemical company, I would say why should you bother doing this when I could probably put together a fertilizer which is tailored specifically to the needs of the soil. I can try a number of micronutrients, as opposed to trying to breed a specific ability to effectively absorb some individual nutrient. And I can do that today, and give it to you today, so why should we continue with your program.

A. Yes, of course that is a point. My work does not really address that question. In our environment in particular, as in many environments in Third World countries, fertilizers are not as effective as we think they are. They appear effective but have never taken it to the full potential yield. There's always an advantage to having an efficient genotype there. For practical reasons, we can only put nutrients in the topsoil. It is very expensive and takes an enormous amount of energy to put them in the subsoil, which is where most of the water is found. The subsoils also tend to have higher pH than the surface soils. so the availability of micronutrients is less in the subsoils. Wherever you've got topsoils drying with just a week or two of dry weather, plants are depending on the subsoil for nutrients. Therefore there is an advantage for use of efficient genotypes.

Q2. Richard Black, Kellogg Canada Inc., Etobicoke, ON, Canada: The second question I would have would be if I was a food processor. Have you looked at the food science behind these different genotypes? Are they essentially the same? We use specific wheats in our cereal products, and not other wheats. Do the kind of manipulations you're conducting alter the food science involved in processing these plants?

A. We haven't addressed this a great deal yet, but all the evidence we do have so far says no. Trace elements change but macronutrients and quality do not change.

Q3. Nick Costa, Murdoch University, Western Australia: What was the distribution of trace elements within the plant? We have high concentrations in the seed, does that mean there are equally high concentrations in the leaf and stem material, or is it mobilized from there to the seed? The reason that I ask that is you could be creating a bit of a problem for the animal production people.

A. All our evidence is that there is no problem there. Quite the opposite in fact. Those nutrients are remobilized from the straw to the grain, but the total plant contents are higher in these efficient users, so the residues of the plant are high in micronutrients as well.

Q4. Manfred Anke, Friedrich Schiller University, Jena, Germany: Do you have information on the uptake of more toxic elements, for example cadmium, in these plants? I ask you because in Germany we have some problems with the cadmium content of flax seeds, or in poppy, and also in wheat, and it is affected by the breeding of new sorts of wheat, and also in vegetables, lettuce for example.

A. We are very interested in cadmium in our Institute. We just seem to be getting evidence that zinc-efficient varieties actually transport less cadmium to the grain. I think there's a paper which suggests that there's an effect of zinc on phloem transport. I think we're very lucky in this respect.

Dietary Strategies to Help Combat Trace Mineral Deficiencies in Weanlings from Sub-Saharan Africa

R.S. GIBSON,[†] J.KÖPF,[‡] J. LEHRFELD,[¶] W.J. BETTGER,[‡] AND E.L. FERGUSON[†]
[†]Department of Human Nutrition, University of Otago, PO Box 56, Dunedin, New Zealand,
[‡]University of Guelph, Ontario, Canada N1G 2W1, and [¶]USDA, Peoria, Illinois, USA

Keywords trace minerals, deficiencies, phytate, weaning foods

Deficiencies of trace minerals (TM) have far-reaching consequences on growth, development, and health of infants and children, contributing to impairments in immune competence and cognitive function, growth failure, and increased mortality and morbidity.

Infants from less industrialized countries are especially vulnerable to TM deficiencies. Their supply of TMs from breast milk may be compromised, possibly by an inherently low TM content coupled with a reduction in breast milk volume induced by the early introduction of weaning foods that replace rather than complement breast milk. In Sub-Saharan Africa, these weaning foods are generally thin porridges prepared fom unrefined cereal flours, sometimes with added legumes, starchy roots or tubers. Hence, they generally have a low energy density, a low TM content, but a high content of dietary fibre and phytic acid, two components known to inhibit TM absorption. As a result, the bioavailability of TM in these weaning porridges is often poor, especially when they are unfermented. Table 1 summarizes the TM and antinutrient content of four unfermented Malawian weaning porridges (10% dry flour) prepared from unrefined (U) or refined (R) maize with or without full-fat soya (Soya) flour, groundnut (GN), and sorghum (Sorg) flours and assuming an intake of 750 mL/d. Data are based on TM analyses conducted in our laboratory. Note the energy densities are very low, ranging from 1.52–1.59 kJ/g, whereas the phytate: zinc (Phy:Zn) molar ratios are high (23–30.4); ratios above 12–20 have been associated with suboptimal zinc deficiency in humans (Sandström and Lönnerdal 1989).

Even if these same flour blends are made into porridges with a higher content of dry flours (approx 25%), intakes of energy, calcium, trace minerals, with the exception of copper, and protein, fall below the estimated requirements, using basal requirements for zinc (WHO 1992) and iron (FAO/WHO 1988)

Table 1. Energy, protein, and selected nutrient and antinutrient content of four porridges (10% dry flour) used for infant feeding in rural Malawi.

750 mL/d	100% U-Maize	80% R-Maize + 20% Soya	80% U-Maize + 20% Soya	70% U-Maize, 20% GN + 10% Sorg
Energy (kJ)	1136	1145	1190	1149
Energy (kJ/g)	1.52	1.53	1.59	1.53
Protein (g)	7	10	11	12
Ca (mg)	6	33	32	13
Fe (mg)	1.9	2	3	2
Zn (mg)	1.7	1	2	2
Cu (mg)	0.2	0.6	1.0	0
Mn (mg)	0.2	0.4	0.5	0.4
NSP (g)	6	3	6	6
Phytate (mg)	449	301	584	386
Phy/Zn	26.8	28.0	30.4	23.0

Correct citation: Gibson, R.S., Köpf, J., Lehrfeld, J., Bettger, W.J., and Ferguson, E.L. 1997. Dietary strategies to help combat trace mineral deficiencies in weanlings from Sub-Saharan Africa. In *Trace Elements in Man and Animals – 9: Proceedings of the Ninth International Symposium on Trace Elements in Man and Animals. Edited by* P.W.F. Fischer, M.R. L'Abbé, K.A. Cockell, and R.S. Gibson. NRC Research Press, Ottawa, Canada. pp. 451–454.

and assuming either low or moderate bioavailability factors for iron and zinc absorption. Clearly, dietary strategies must be devised to enhance the content and bioavailability of TM in porridges for infant and child feeding in Sub-Saharan Africa. Some examples are outlined below.

1. Incorporate germinated cereals (at level 5–10% total flour) to enhance TM content. During germination of cereals but not legumes, alpha-amylase activity is increased resulting in hydrolysis of amylose and amylopectin to dextrins and maltose. Hence, if germinated cereal flours are added as an additive (at level 5–10% flour), the viscosity of thick cereal porridges (i.e., 20–25% flour) with a viscosity of 50,000 cP can be reduced to a semi-liquid consistency (viscosity 3000 cP) suitable for infant feeding without diluting with water (Alnwick et al. 1988). As a result, energy density and trace mineral content are increased. As well, the taste becomes sweeter and more appealing to infants.

2. Incorporate germinated cereals to improve TM bioavailability. Germination also enhances phytase activity as a result of de novo synthesis and/or activation of endogenous phytase; levels may increase by 23–588% after 2–3 d, depending on the cereal. Phytase enzymes hydrolyze phytic acid to yield inorganic orthophosphate and myo-inositol via intermediate myoinositol phosphates (penta to monophosphates (IP-5 to IP-1). Only the higher inositol phosphates (IP-5 and IP-6) inhibit zinc and non-heme iron absorption (Alnwick et al. 1988, Lönnerdal et al. 1989). Reduction of Phy:Zn molar ratios to below 10 will enhance zinc bioavailability (Sandstrom and Lonnerdal 1989).

3. Use lactic fermentation to enhance TM bioavailability in cereal-based porridges via phytate hydrolysis induced by microbial phytases and by formation of lactic acid which forms soluble ligands with iron and zinc. Antimicrobial substances may also be produced. Microbial fermentation can occur spontaneously from the microflora on the surface of the seeds, or by inoculating the flour slurries with microbial starter cultures before cooking. Fermentation should be carried out for at least 16–24 h at 25–30°C, to ensure that the pH of the cooked porridge falls to a level (i.e., <pH 3.8) that reduces the growth of diarrheal pathogens (Alnwick et al. 1988). Phytate degradation by microbial phytase is less in high-tannin cereals possibly because of the inhibitory effect of polyphenols on phytase activity (Svanberg et al. 1993).

The trace mineral content of weaning porridges is not increased by lactic acid fermentation. Lactic acid bacteria are apparently not amylolytic and thus do not hydrolyze amylose and amylopectins to dextrins and maltose and reduce the viscosity of cereal porridges. Indeed, viscosity is increased; porridges prepared in our laboratory prepared from lactic acid fermented flour slurries with a standard viscosity, had a lower dry matter content (i.e., 17–19%) compared to those made from unfermented slurries (i.e., 17–19% vs. 22–28%) and the same viscosity.

4. Use a combination of soaking flours containing germinated flour as an additive, followed by fermentation using a microbial starter culture. Prior soaking flour blends containing germinated flour (at level of 5–10%) in water at room temperature for 24–48 h activates endogenous phytases and allows a longer time at the optimal pH for cereal phytases (i.e., pH 5.0–4.5) to hydrolyze phytate before the starter culture is added. Addition of a starter culture provides a source of exogenous microbial phytases which act over a wider pH range (i.e., 2.5–5.5) than cereal phytases. Thus hydrolysis of phytic acid can proceed during the lactic fermentation until the pH falls to <3.8. This takes between 16–24 h, after which the fermented flours are mixed with water to form 20% slurries, and then cooked to form porridges. This combination of strategies can reduce the phytic acid content by as much as 90% (Svanberg et al. 1993) while simultaneously enhancing trace mineral bioavailability, protein quality and digestibility, microbiological safety, and keeping quality, yet decreasing toxins such as haemagluttinins and cyanide. Commercial phytase enzymes prepared from *Aspergillus oxyzae* or *A. niger* which are stable over a wider pH (3.5–7.8) and temperature range could also be used, although their high cost probably precludes their use in LICs.

Prior soaking of the cereal flours in water at room temperature for 24–48 h *before* the addition of the starter culture activates endogenous cereal phytases and allows a longer time to hydrolyze phytic acid at the optimal pH for cereal phytases (i.e., 5.0–4.5). Addition of the starter culture by the traditional back-slopping technique without soaking the flours first is not recommended because the time period when the pH is at the optimal for phytate hydrolysis by cereal phytases will be too short.

5. Multi-micronutrient fortification of cereal/legume based weaning foods enriched with germinated sorghum. This strategy could be used for weaning foods used in supplementary feeding programs. The fortificants selected must be safe, stable, acceptable, bioavailable, and at levels which do not induce any adverse nutrient-nutrient interactions, or influence the organoleptic qualities and shelf-life of the weaning food. Ideally, protected fortificants such as NaFe EDTA which are resistant to common inhibitors of TM absorption should be used. In some cases, organic chelators such as cysteine could be added to improve the solubility of the fortificants (e.g., zinc).

To date, a mineral pre-mix suitable for refugee populations in which levels are designed for the needs of mild/moderately rather than severely malnourished children has been formulated (Golden et al. 1995). Levels are expressed as desirable nutrient densities and are based on nutrient intakes expressed as a function of age/sex specific energy requirements, and adjusted upwards by 10% to allow for reduced availability with gastrointestinal disease, an initial deficit, and increased rate of weight gain. An additional increase of 20% was then applied to the minerals to compensate for the inhibitory effect of phytic acid on mineral bioavailability. Beaton (1995) has provided cost estimates for vitamin/mineral premixes ranging from US $1.86 to $2.25/kg depending on the levels used.

6. Mix foods known to promote TM absorption such as ascorbic-acid-rich foods and powdered fish into cereal porridges. Ascorbic acid enhances non-heme iron absorption; the effect is dose-related. Likewise, certain amino acids and cysteine-containing peptides and organic acids produced during fermentation form soluble ligands with iron and zinc, thus enhancing absorption.

In conclusion, practical dietary strategies which are economically feasible and sustainable can be used to prevent TM deficiencies in LICs. They involve agricultural and horticultural activities as well as some simple modifications to food preparation and processing techniques used for local staple foods. To enhance their likelihood of being adopted and implemented by local communities, the dietary strategies should be combined with well designed nutrition education, communication, and social marketing strategies aimed to change attitudes and food-related behaviors and practices. For supplementary feeding programs for weanlings and refugees, dietary strategies can be combined with fortification programs, preferably mandated by government regulations that eliminate competition with unfortified products.

Literature Cited

Alnwick, D., Moses, S., & Schmidt, O.G., eds.(1988) Improving young child feeding in Eastern and Southern Africa. Household-level food technologies. Nairobi: IDRC 265e.

Ashworth, A. & Draper, A. (1992) The potential of traditional technologies for increasing the energy density of weaning foods. Geneva; Diarrhoeal Disease Control Programme WHO/CDD/EDP/92.4.

Beaton, G.H. (1995) Fortification of foods for refugee feeding. Final Report to the Canadian International Development Agency. Toronto, Canada.

FAO/WHO (Food and Agricultural Organization/World Health Organization).(1988) Requirements for vitamin A, iron, folate and vitamin B-12. Report of a joint FAO/WHO expert consultation. Food and Nutrition Series, FAO, Rome.

FAO/WHO/ILEA (Food and Agricultural Organization/World Health Organization/International Atomic Energy Agency).(In press) Trace elements in human nutrition. WHO, Geneva.

Gibson, R.S. & Ferguson, E.L. (1994) Dietary strategies for preventing iron and zinc deficiency in African children. Procedings of the XVth International Congress of Nutrition: IUNS Adelaide. Walhlqvist, M.L., Truswell, A.S., Smith, R, Nestel, P.J. (eds). pp. 301–303.

Golden, M.H.N., Briend, A., & Grellety, Y. (1995) Report of meeting on supplementary feeding programmes with particular reference to refugee populations. Eur. J. Clin. Nutr. 49: 137–145.

Lönnerdal, B., Sandberg, A-S., Sandstrom, B, Kunz, C. (1989) Inhibitory effects of phytic acid and other inositol phosphates on zinc and calcium absorption in sucking rats. J. Nutr. 119: 211–214.

Sandström, B., & Lönnerdal, B. (1989) Promoters and antagonists of zinc absorption. In: Mills, C.F. ed. Zinc in human biology. International Life Sciences Institute. Human Nutrition Reviews, London: Springe-Verlag, pp. 57–78.

Svanberg, U., Lorri, W., & Sandberg, A-S. (1993) Lactic fermentation of non-tannin and high-tannin cereals: effects on *in vitro* estimation of iron availability and phytate hydrolysis. J. Food Sci. 58: 408–412.

Discussion

Q1. Julian Lee, New Zealand Agriculture Research Institute, Palmerston, New Zealand: Early on in your talk you mentioned the sulfur amino acids. Is the content of sulfur amino acids in these grain products high or low? If it is low, would improving sulfur amino acid content be an effective strategy?

A. Earlier in the week, Dr. Ross Welch addressed this issue, and showed some data that convincingly showed that increasing the methionine content of certain cereals enhances the absorption of non-heme iron and zinc. Now, that work was done in rats, and there was some discussion about how translatable that is to humans, but it does appear that this is probably one strategy you could use. In relation to your question about the sulfur amino acid content in these particular weaning foods, I'm afraid I don't have those data, but I do know that if you add the legumes to the cereals, that will certainly give enhancement.

Q2. John Sorenson, University of Arkansas, Little Rock, AR, USA: Do you have government support in these countries for this research?

A. Yes, it is absolutely essential to have government support for a sustainable program. We collaborated with the Department of Community Health and the medical college, the University of Malawi, the Ministries of Health and Agriculture and Community Development. We have also implemented a rural extension program and they are involved to the extent that their agents will be able to continue to administer the program after the research program is done.

Q3. Barbara Golden, University of Aberdeen, Scotland: What do you gain from adding cooked green leaves to the cereals and are you not worried about possible toxicity?

A. I feel that it is advisable to add green leaves to the weaning porridges, in concert with oil, because in Malawi, the vitamin A problem is not as bad as the lack of fat in the diet. We have looked at the possibility of toxicity and have no problem with adding the leaves. We are using white sorghum rather than the red which should be less of a problem.

Q4. Noel Solomons, Guatemala: In reference to the changes in viscosity of the porridges: By increasing the solids, one appears to eat less but has consumed a lot more. You have shown that the least of our nutrient concerns is that of adequate protein intake. There might have been more concentration on amino acid rations vis a vis bioavailability of trace elements (protein quality) as opposed to protein concentration.

Trace Element and Dietary Fiber Intake in Hungary*

ANNA GERGELY AND MARIANN KONTRASZTI

National Institute of Food Hygiene and Nutrition, Budapest 100 P.O. Box 52, 1476 Hungary

Keywords human diet, diet analysis, trace element intake, dietary fiber intake

The trace element and dietary fiber content of several foodstuffs were analysed for years in our Institute. The results summarized in the Hungarian Food CompositionTable (Biró et al. 1995) indicate that the nutritionally well formed diets prepared from the foodstuffs and raw materials available in Hungary may contain the trace elements and dietary fiber at the recommended levels. However several efforts are made by the importers and manufacturers in Hungary to introduce new products containing trace elements and fiber as enrichment or supplementation. The present study deals with the determination of actual trace element and dietary fiber intake in Hungary with the aim of achieving a realistic answer as to whether or not enrichment or supplementation of certain trace elements and fiber in human nutrition is needed.

Materials and Methods

Total diets prepared for adults in state institutions were collected three times on 15 consecutive days (1, 1995 May; 2, 1995 September; 3, 1996 January) from three parts of Hungary (A, Budapest, the capital; B, Pécs, city in an industrial surrounding; C, Szolnok, town in an agricultural province). The subsamples of each homogenized sample were analysed as follows: Cu, Zn, Fe, Mn, Cr, Pb, Cd – dry ashing, flame AAS; Se – wet digestion, hydride generation FIAS-AAS; Al – dry ashing, HGA-AAS (Perkin Elmer 1990); dietary fiber – enzymatic treatment, gravimetry (Prosky et al. 1985).

Some calculations were made based on the amounts of the food components of the menus and the trace element and dietary fiber content of these components.

Results and Discussion

The calculated and measured trace element or dietary fiber values were generally in good concordance (e.g., Cu measured 1.14 ± 0.50 mg/d, calculated 1.46 ± 0.31 mg/d; Fe measured 9.06 ± 3.99 mg/d; calculated 10.86 ± 1.96 mg/d; Mn measured 2.98 ± 0.89 mg/d, calculated 2.02 ± 0.53 mg/d; dietary fiber measured 27.3 ± 7.1 g/d, calculated 26.4 ± 4.5 g/d).

The Pb, Cd, Al, content of the diets were determined and compared to the tolerable values calculated for adults of 65 kg body weight. The Pb, Cd, Al intakes were under the tolerable values (Pb tolerable 232 μg/d, measured max. 107 μg/d; Cd tolerable 56 μg/d, measured max. 10 μg/d; Al tolerable 65 mg/d, measured max. 8.1 mg/d).

The results of total diet trace element and dietary fiber analysis (x ± SD) are compared to Hungarian recommended safe daily intake values (Biró et al. 1995) and are presented in Table 1 and Table 2.

Among the studied places or time periods tendencious differences could not be observed. The Cu intakes were generally around the recommendation except place C in period 1. The Zn daily intake proved to be adequate. Great variability could be found from day to day in iron content of the diets in place B in period 1. The values covered 4–8 times daily fluctuation in connection of meat and liver content of the menus. Mn and Cr intakes by several diets are only half of the recommendations. Se supply with the diets are generally sufficient. The dietary fiber intake is adequate compared to the recommended value.

Special attention must be taken in enrichment of foodstuffs and supplementation in Hungarian diets. According to our data Cr, Mn and in some cases Fe supplementation must be taken into consideration.

---

*Supported by Hungarian Ministry of Welfare ETT 02 176/1993.

Correct citation: Gergely, A., and Kontraszti, M. 1997. Trace element and dietary fiber intake in Hungary. In *Trace Elements in Man and Animals – 9: Proceedings of the Ninth International Symposium on Trace Elements in Man and Animals. Edited by* P.W.F. Fischer, M.R. L'Abbé, K.A. Cockell, and R.S. Gibson. NRC Research Press, Ottawa, Canada. pp. 455–457.

Table 1. Trace element intake in Hungary.

Place	Period 1		Period 2		Period 3	
	X	SD	X	SD	X	SD
COPPER, mg/day recommendation 1.4						
A	1.14	0.50	1.15	0.41	1.30	0.41
B	1.34	0.99	1.49	1.71	1.53	0.70
C	0.89	0.37	1.28	1.15	1.05	0.23
ZINC, mg/day recommendation man 10 women 9						
A	9.78	2.84	10.71	4.16	9.44	2.84
B	13.05	4.82	12.44	4.37	11.05	2.15
C	11.97	2.74	10.27	2.48	10.38	2.95
IRON, mg/day recommendation man 12 women 15						
A	9.33	2.80	7.18	2.23	9.43	2.53
B	17.64	18.29	10.37	3.71	12.06	4.68
C	9.06	3.99	12.12	3.27	9.75	2.83
MANGANESE, mg/day recommendation 4						
A	2.50	0.80	2.09	0.90	3.14	0.83
B	2.98	0.89	1.66	0.34	2.78	0.58
C	2.39	0.45	2.55	0.46	2.12	0.31
CHROMIUM, µg/day recommendation 120						
A	44.0	13.0	102.0	32.0	59.0	16.0
B	81.0	26.0	35.0	8.0	68.0	20.0
C	42.0	9.0	35.0	6.0	101.0	22.0
SELENIUM, µg/day recommendation man 75 women 60						
A	66.0	25.0	89.0	32.0	65.0	30.0
B	92.0	44.0	53.0	30.0	68.0	27.0
C	76.0	35.0	49.0	16.0	62.0	17.0

Table 2. Dietary fiber intake in Hungary g/d, recommendation 25.

Place	Period 1		Period 2		Period 3	
	X	SD	X	SD	X	SD
A	34.8	9.3	27.8	7.8	28.0	7.8
B	27.3	7.1	33.3	11.1	28.1	4.9
C	25.6	7.9	23.3	6.9	27.4	7.6

Literature Cited

Biró, G., Lindner, K. (1995) Tápanyagtáblázat (Food Composition Tables, 12 th edition) Medicina Publ. Co, Budapest.

Perkin Elmer (1990) Analytical Methods for Atomic Absorption Spectrophotometry, Bodensewerk, Perkin Elmer GMBH, Germany.

Prosky, L., Asp, N. G., Furda, J., De Kries, J., Schweizer, T.F., Harland, B. (1985) The determination of total dietary fiber in foods and food products: collaborative study. J. Assoc. Off. Agric. Chem. 68: 677–681.

Discussion

Q1. Richard Black, Kellogg Canada Inc., Etobicoke, ON, Canada: Very interesting data. I haven't seen the Hungary data before. Can you clarify for me what populations you were examining at each of those different sites? You said faculty and students at one institution in one example. Certainly in North America, while diets in institutional settings like that are based on the traditional foods of the country, they generally have a better profile of nutrient intake than you would in other situations. So I wonder in fact, if you would be able to model, for instance, rural intakes. I also thought the dietary fibre was extraordinarily high, and I'm glad to see that. I'm wondering if you know what the dietary fat content for this population was, in comparison to the very high dietary fibre?

A. These diets were collected from these institutions, where the students were having breakfast, lunch and dinner. These may be representative of the diets of the communities in Hungary, because the menus are based on some descriptions of the needs of these adults for protein and energy content in the diets. The diets were very fatty. The energy intake was large which may account for the high levels of nutrient and dietary fibre intakes.

Selenium Supplementation in Sub-Fertile Human Males

R. SCOTT,[†] A. MACPHERSON,[‡] AND R. YATES[¶]

[†]Department of Urology, Glasgow Royal Infirmary, [‡]SAC, Auchincruive, AYR, [¶]Department of Gynaecology, Glasgow Royal Infirmary, 16 Alexandra Parade, Glasgow, G31 2ER

Keywords selenium, fertility, male, human

The present communication reports an extension of the original work presented at TEMA-8 in which male subfertile patients are given selenium in an attempt to improve sperm motility (MacPherson et al. 1993).

Materials and Methods

Male patients attending the combined subfertility clinic in Glasgow Royal Infirmary are routinely examined and measurements made of the blood luteinizing hormone (LH), follicle stimulating hormone (FSH), prolactin and testosterone. Each individual is requested to produce two semen samples for analysis. The subjects where the motility of the sperm were shown to be reduced were invited after full explanation and informed consent to participate in a trial involving selenium supplements (L-Selenomethionine 100 µg/d) versus a placebo. The selenium was taken as a tablet at night and the supplementation was continued for three months. At the end of this time the subject was asked to produce another semen sample for analysis. Blood selenium was measured pre- and post-treatment. Samples were retained in heparinised tubes stored in a refrigerator at 4°C before analysis in the laboratory (A. MacP.). The analysis was undertaken using a PS Analytical hydride generator and a PS Analytical Excalibur atomic fluorescence spectrophotometer.

Parallel with the main subject study, seminal plasma was retained from specimens after a basic analysis with respect to fertility. Seventy-five samples were obtained.

Ninety-four subjects have now entered the selenium supplementation study. Of these 94 subjects, two who received the selenium and one who received the placebo, had to be withdrawn because of upper gastrointestinal irritation. Of the 94 subjects, at the time of analysis, 58 had completed the three month trial period. The mean age was 33.5 years. Those receiving selenium (n = 38) were classified as group one subjects and those on placebo (n = 20) group two. There was no statistically significant demonstrable difference initially between the two groups with respect to sperm density (group 1, average 48.0 ± 10.5 million per mL, group 2, 31.0 ± 5.88 million per mL). Plasma selenium levels likewise were comparable between the two groups; (group 1, 82.9 ± 2.77 versus group 2, 79. 5 ± 3. 37 µg/L). Motility in the two groups was again not found to be statistically significantly different; (18.4% ± 2.219 group 1, 12.6% ± 3.49 group 2). Sperm density did not change after three months therapy (38.8 ± 6.94 versus 41.0 ± 11.1 million per mL). Comparison of the groups however revealed that the plasma selenium 124.8 ± 6.7 (group 1) versus 70.9 ± 8.13, (group 2) and sperm motility 35.1% ± 4.43 versus 17.5% ± 4.04 were statistically highly significantly ($P < 0.001$ and $P < 0.01$ respectively) different for both of these measurements.

Analysis of Selenium Content of Seminal Plasma

The selenium content of the seminal plasma indicated that 40 of 75 subjects (64%) had seminal plasma selenium values of less than 40 µg/L, the mean value being 38.14.

Correct citation: Scott, R., MacPherson, A., and Yates, R. 1997. Selenium supplementation in sub-fertile human males. In *Trace Elements in Man and Animals – 9: Proceedings of the Ninth International Symposium on Trace Elements in Man and Animals. Edited by* P.W.F. Fischer, M.R. L'Abbé, K.A. Cockell, and R.S. Gibson. NRC Research Press, Ottawa, Canada. pp. 458–459.

Discussion

There is evidence in the literature that sperm counts and the quality of semen is falling in the Western world, although some authorities question this statement (Farrow 1994). It is clear however that infertility in married couples may be as high as 15% (Templeton 1995). Recent studies indicate that there are possible reasons for a reduction in sperm quality and included in the factors which might be specific is the possibility that chemicals with an oestrogenic effect may enter the food chain. These chemicals include polychlorinated biphenyls, chlorinated hydrocarbons and phyto-oestrogens (Jones 1989, Sharpe 1993). It is recognised that illness such as mumps and chemotherapeutic agents, such as cyclophosphamide can cause reduction of sperm counts and even azoospermia (Farley et al. 1992).

There has been very little attention paid to the potential effects of dietary deficiency in human subjects. It has been demonstrated in experimental animals that reduction in selenium causes a reduction in the selenium protein in the flagellum of the sperm with the consequent reduction in motility.

The present authors have already observed a reduction in the blood selenium levels in the West central belt of Scotland and the current study of the selenium content of seminal plasma indicates that over 60% of subjects are below the normal limit of 40.70 with respect to that trace metal (Bleau et al. 1984). Of interest to those who have to cope with the problem of reduced fertility, the present study confirms that supplementation of the subfertile individual with 100 µg of selenium causes a significant rise in the blood selenium levels which is mirrored in a statistically significant improvement in sperm motility.

Literature Cited

Bleau G., Lemarbe J., Faucher G., Roberts K.D. & Chapdelaine A. (1984) Semen selenium and human fertility. Fertil. Steril. 42:890–894.

Farley K.F., Barrie J.U. & Johnson W. (1992) Sterility and testicular atrophy related to cyclophosphamide therapy. Lancet 1:568–569.

Farrow S. (1994) Falling sperm count and quality: fact or fiction. Brit. Med. J. 309:1–2.

Jones G.R.M. (1989) Polychlorinated biphenyls: where do we stand now? Lancet 2:791–794.

MacPherson A., Scott R. & Yates R. (1993) The effects of selenium supplementation in subfertile males. Proceedings of the eight international symposium on trace elements in man and animals. (Anke M., Meissner D. & Mills C.F., eds.) Verlag Media Touristik, Jena, Germany, pp. 566–569.

Sharpe R.M. (1993) Declining sperm counts in man - is there an endocrine cause? J. Endocrinol. 136:357–360.

Tempelton A. (1995) Infertility—epidemiology, aetiology and effective management. Health Bulletin 153:294–298.

Discussion

Q1. John Sorenson, University of Arkansas, Little Rock, AR, USA: How long has this study been going on? Were there any pregnancies in the placebo group? Is there any evidence that giving selenite is beneficial?

A. The study has been going on for at least 6 months. At the time we prepared this talk we only wished to speak of the ones who had completed the therapy. There were no pregnancies in the placebo group. I am not aware of any evidence that selenite is beneficial. We have stuck to the selenomethionine in this study.

Bioavailability of an Iron Glycine Chelate for use as a Food Fortificant Compared with Ferrous Sulphate*

T.E. FOX, J. EAGLES, AND S.J. FAIRWEATHER-TAIT

Institute of Food Research, Norwich NR4 7UA, UK

Keywords iron glycinate, stable isotopes, hemoglobin incorporation

Iron deficiency is a worldwide problem, and is usually the result of low dietary bioavailability of iron. Public health measures to improve the iron status of vulnerable groups often include food fortification and the iron compounds are primarily selected according to stability and low reactivity with other constituents of the food matrix. Unfortunately, these properties are associated with low solubility and hence limited absorption from the gastrointestinal tract. Attempts have been made to overcome these problems, such as using chelating agents to protect minerals. Iron EDTA has recently been approved by the joint FAO/WHO Expert Committee on Food Additives for use as a food fortificant in programmes to improve iron status (International Nutritional Anemia Consultative Group 1993), but EDTA can complex other minerals, such as calcium, copper and zinc. Thus alternative iron chelates, where the iron molecule is held between amino acids by ionic covalent bonds or ligands, such as iron glycine or iron glycine sulphate, have been developed (Ashmead 1991, Galdi et al. 1988), but there is little peer-reviewed published information on the absorption of chelated iron by humans. This study was designed to measure hemoglobin incorporation of isotopically-labelled iron from iron glycinate and to compare it with ferrous sulphate in weaning infants.

Materials and Methods

Isotopically-enriched iron sulphate was prepared from elemental ^{57}Fe and ^{58}Fe (Isotec Division, Nippon Sanso Europe GMBH, Germany) by dissolving the metal in concentrated nitric acid, adding concentrated sulphuric acid, and heating the mixture slowly to dryness to evaporate the acids. The resulting powder was placed in a muffle furnace at 500°C for 30 min. Sterile, sealed ampoules containing 1.4 mg of iron either as ^{57}Fe or ^{58}Fe sulphate solution were prepared from the powder by dissolving it in 0.5 M sulphuric acid. The iron was maintained in its reduced form by the addition of 0.83 mg of ascorbic acid per mg of iron present. The iron glycine chelate solution was prepared by Albion Laboratories (Clearfield, Utah, USA) at Norwich, from the iron isotopes and was dispensed into sealed vials containing 1.4 mg of iron as either ^{57}Fe or ^{58}Fe glycinate.

Twelve infants aged 9 months were recruited to the study and allocated to one of two groups. One group was given ^{57}Fe-labelled sulphate and ^{58}Fe-labelled glycinate and the other group ^{58}Fe-labelled sulphate and ^{57}Fe-labelled glycinate. The labelled iron solution was weighed and mixed into an unfortified commercial weaning food made from pureed parsnips, potatoes, cauliflower and milk (Boots PLC, Nottingham, UK) immediately before consumption on eight separate days. The two isotopically labelled compounds of iron were given on alternate days. Any waste food was recorded and analysed for total iron and iron isotopes by thermal ionization quadrupole mass spectrometry (TIQMS, Finnegan Mat, Bremen Germany). Fourteen days after the last test meal, a heel-prick blood sample was taken and iron isotope ratios measured by TIQMS. The infants were weighed at the time of the heel-prick and blood volume estimated.

---

*This work was supported by the MRC (ROPA) and the BBSRC.

Correct citation: Fox, T.E., Eagles, J., and Fairweather-Tait, S.J. 1997. Bioavailability of an iron glycine chelate for use as a food fortificant compared with ferrous sulphate. In *Trace Elements in Man and Animals – 9: Proceedings of the Ninth International Symposium on Trace Elements in Man and Animals. Edited by* P.W.F. Fischer, M.R. L'Abbé, K.A. Cockell, and R.S. Gibson. NRC Research Press, Ottawa, Canada. pp. 460–462.

The blood sample was divided into two weighed portions of 100 mg each and 200 μg of $^{54}FeCl_3$ was added to one portion (for total iron determination by TIQMS). The solutions were digested in a microwave, purified by anion exchange, and isotope ratios and total iron measured by TIQMS (Fairweather-Tait et al. 1995).

The abundances (atom%) of all isotopes from each iron compound were calculated from the measured isotopic ratios, and average molecular weights were calculated. The quantity of labelled iron in the blood from both of the labelled iron compounds (Y and Z) was calculated using the following equations:

$$^{57/56}R = \frac{(\%nat57/aMWnat)X + (\%Gly57/aMWGly)Y + (\%SO_457/aMWSO_4)Z}{(\%nat56/aMWnat)X + (\%Gly56/aMWGly)Y + (\%SO_456/aMWSO_4)Z}$$

$$^{58/56}R = \frac{(\%nat58/aMWnat)X + (\%Gly58/aMWGly)Y + (\%SO_458/aMWSO_4)Z}{(\%nat56/aMWnat)X + (\%Gly56/aMWGly)Y + (\%SO_456/aMWSO_4)Z}$$

$$T = X + Y + Z$$

where R = ratio of isotopes measured, aMW = average molecular weight, X = wt of natural Fe, Y = wt of iron glycinate, Z = wt of iron sulphate, Gly = Fe label from glycinate, SO_4 = Fe label from sulphate, and T = wt of total iron measured (μg). Similar calculations were made for the ^{57}Fe-enriched glycinate and ^{58}Fe-enriched sulphate. Analysis of variance was performed on the hemoglobin incorporation data.

Results and Discussion

Mean hemoglobin incorporation (± SEM) was 13.2% (0.87) for ferrous sulphate and 11.3% (0.59) for iron glycine. There was no significant difference between the two iron compounds.

Earlier studies in rats suggested that iron glycine is more bioavailable than ferrous sulphate. Fairweather-Tait et al. (1992) reported higher hemoglobin values in young rats fed iron glycine than those given ferrous sulphate for 4 weeks, but no difference in liver iron concentrations. Langini et al. (1988) also reported higher iron absorption in weanling rats given infant formula labelled with ^{59}Fe glycinate compared with ^{59}Fe sulphate. Amino acid chelated iron was reported to be as effective as ferrous sulphate in the treatment of iron deficiency anemia in adolescents, but was associated with fewer complaints of gastric distress (Pineda et al. 1994).

It has been suggested that the higher bioavailability of iron glycinate is due to the fact that it is absorbed intact by the epithelial cells of the gut (Ashmead and Jeppsen 1993), probably via an active transport mechanism. The results obtained from the present study do not lend support to this hypothesis, since hemoglobin incorporation of the iron from both the sulphate and glycinate compounds was similar. Further research is underway in our laboratory in which the iron chelate is being compared with ferrous sulphate under adverse dietary conditions to test further its usefulness as a fortificant; any differences in transport mechanism, if correct, should favour the glycine chelate. However, even if iron dissociates from the complex under acid conditions of the stomach, the fact that it is as well absorbed as ferrous sulphate and is less reactive than other iron compounds makes it a very useful and effective iron fortificant.

References

Ashmead, H.D. (1991) Comparative intestinal absorption and consequent metabolism of metal amino acid chelates and inorganic metal salts. In: Biological Trace Element Research (Subramanian, K.S., Iyengar, G.V. and Okamoto, K., eds.) American Chemical Society Symposium Series 445, Washington DC.

Ashmead, H.D., & Jeppsen, R.B. (1993) Enhanced tissue metabolism of minerals chelated to amino acids. In: Bioavailability '93. (Schlemmer U., ed.) Part 2. Bundesforschungsanstalt fur Ernahrung, Ettlingen, Germany, pp. 63–67.

Fairweather-Tait, S.J., Fox, T.E., Wharf, S.G., & Ghani, N.A. (1992) A preliminary study of the bioavailability of iron- and zinc-glycine chelates. Food Add. Contam. 9: 97–101.

Fairweather-Tait, S.J., Fox, T.E., Wharf, S.J., & Eagles, J. (1995) The bioavailability of iron in different weaning foods and the enhancing effect of a fruit drink containing ascorbic acid. Ped. Res. 37: 389–394.

International Nutritional Anemia Consultative Group (1994) Iron EDTA for Food Fortification, Washington DC: The Nutrition Foundation.

Langini, S., Galdi, M., Carbone, N., Barrio Rendo, M.E., Portela, M.L., Caro, R., & Valencia, M.E. (1988) Ferric glycinate iron bioavailability to rats as determined by isotopic labelling in an infant formula. Nutr. Rep. Int. 38:729.

Pineda, O., Ashmead, H.D., Perez, J., & Lemus, C.P. (1994) Effectiveness of iron amino acid chelate on the treatment of iron deficiency anemia in adolescents. J. Appl. Nutr. 46:1–13.

Discussion

Q1. Neville Suttle, Moredun Research Institute, Edinburgh, Scotland: A very interesting result, Tom, but it's completely at odds with the message from a poster presented on Tuesday by Ashmead et al. with what seemed to be a similar chelate. The chelate was reported to have 100% greater availability than ferrous sulfate. Can you explain?

A. The chelate was made in the same laboratory. I must point out that in that poster, the actual amount of iron that was given as a dose, half of it was actually the iron chelate, and therefore double the dose was given for the non-chelated form. Because the two doses were not comparable, I can't say that the actual results were quite reliable.

Q2. John Sorenson, University of Arkansas, Little Rock, AR, USA: Were the iron glycinate and sulphate mixed with food? How were they given? and for how long? Would the iron in ferrous sulfate form complexes with food components when it was mixed?

A. The iron sulfate was mixed just prior to consumption by the infant. I must say that in order to make sure that all of the isotope was consumed, it was added to half of what the mother estimated the infant would consume, and then once that was fed, the other half was mixed in the same bowl and fed. The iron glycinate was given in the same way. I would think that the iron chelate would actually be relatively stable and would not form complexes in the food. The iron sulfate may of course be open to that, but the time that the iron sulfate was added to that food matrix and consumed was within the space of 5 minutes, so I don't think that there was any food-label interaction. (Sorenson disagreed.)

Q3. Janet Hunt, USDA-ARS, Grand Forks, ND, USA: What is the shelf life of the iron glycinate and does it confer a particular colour or taste?

A. The iron did impart some colour to the food, but as far as the actual taste of the food, I don't think there was any difference between the iron sulfate or the iron chelate. And the amounts added were maybe not enough to cause any kind of a problem this way.

Ammonium Tetrathiomolybdate (TTM) and Metal Distribution in Copper Supplemented Sheep

S. HAYWOOD AND Z. DINCER

Department of Veterinary Pathology, University of Liverpool, England L69 3BX

Keywords ammonium tetrathiomolybdate, copper, molybdenum

Ammonium tetrathiomolybdate (TTM) is the recommended treatment for copper poisoning in sheep (Humphries et al. 1988). TTM has also been used in penicillamine intolerant Wilsons disease patients (Brewer et al. 1991).

However a flock of TTM treated, Cu-poisoned Bleu du Maine sheep subsequently developed an endocrinopathy with elevated brain Cu and Mo (Haywood et al. 1993, Dincer 1994) suggesting possible disruption of the neuroendocrine axis and raising questions as to the long term safety of the drug.

The purpose of this study was to evaluate the systemic distribution and retention of metals (Cu, Mo, Zn) after TTM administration to sheep with high liver Cu.

Materials and Methods

Animals and Protocol. Cambridge-cross sheep (12), aged 9–10 months, equal numbers of either sex, were fed in addition to hay, controlled quantities of a high Cu (150–200 ppm) pelleted diet for 2.5 months, when serum liver enzymes indicated sustained liver damage. Four sheep were killed and necropsied and the other 8 animals were administered TTM subcutaneously (3.4 mg/kg on 3 alternate days). This dose was repeated every month for 5 months after which 4 more TTM sheep were killed and the remaining sheep 4 months after discontinuation of treatment.

A parallel group of 6 North Ronaldsay sheep with naturally-acquired high liver Cu and fed hay alone received the 5-course TTM treatment. Sheep were then killed immediately and 7 months afterwards. Untreated Ronaldsays were used as controls.

Necropsy and Sampling Procedures. Samples of liver, kidney, heart, skeletal muscle, ovaries/testes, brain and endocrine organs were retained for routine histological examination and metal analysis.

Metal Analysis. Triplicate samples of dried tissue were digested with concentrated nitric and perchloric acids (2:1). Cu and Zn concentrations were assessed using an IL 157 Atomic Absorption Spectrophotometer (AAS). Mo was measured in a graphite furnace AAS. Metal contents are expressed in µg/g dry weight. Statistical analysis used Students T test and the Pearson Product-Moment Correlation.

Results

Metal Content (Table 1). Mo concentrations: Mo rose in all organs studied and especially in the pituitaries after TTM administration ($p < 0.05$) in both breeds of sheep, and was retained after discontinuation of treatment.

Cu Concentrations. Liver Cu fell in the TTM treated Cu-supplemented groups ($p < 0.05$). Kidney Cu rose temporarily and brain Cu rose more permanently ($p < 0.05$). Cu and Mo showed a positive correlation ($r = + 0.7$) in the brains of the TTM-treated Ronaldsays.

Zn concentrations. Zn content was not changed overall.

Correct citation: Haywood, S., and Dincer, Z. 1997. Ammonium tetrathiomolybdate (TTM) and metal distribution in copper supplemented sheep. In *Trace Elements in Man and Animals – 9: Proceedings of the Ninth International Symposium on Trace Elements in Man and Animals. Edited by* P.W.F. Fischer, M.R. L'Abbé, K.A. Cockell, and R.S. Gibson. NRC Research Press, Ottawa, Canada. pp. 463–465.

*Table 1. Cu and Mo (µg/g M ± SD) of organs of (1) Cu supplemented Cambridge-cross sheep, (2) after TTM administration and (3) 4 months later. * P < 0.05.*

	Liver	Kidney	Heart	Pituitary	Brain (cerebellum)
Mo					
(1)	1.7 ± 0.3	1.7 ± 0.3	0.05 ± 0.01	0.2 ± 0.1	0.03 ± 0.01
(2)	*27.8 ± 3.5	*28.0 ± 4.0	0.43 ± 0.11	*3.3 ± 1.3	*0.23 ± 0.11
(3)	*6.5 ± 2.3	6.5 ± 2.3	*0.19 ± 0.01	*3.3 ± 0.8	*9.12 ± 0.01
Cu					
(1)	932 ± 327	18 ± 5	18 ± 0.4	73 ± 22	17 ± 1
(2)	*412 ± 165	*66 ± 21	19 ± 0.5	41 ± 15	*24 ± 3
(3)	*388 ± 146	34 ± 12	18 ± 1.3	110 ± 24	*27 ± 6

Histopathology. Focal hepatic necrosis/chronic hepatitis was present in the high Cu livers of untreated groups but was not present/less evident in the TTM treated animals. No other significant changes were identified in other organs.

Discussion and Conclusions

The widespread systemic retention of Mo in unbound or complexed form indicates that TTM is not all excrcted after its administration. Moreover the seemingly preferential retention of Mo by the pituitary makes it a potential target organ for Mo toxicity. This is in accordance with earlier observations that Mo excess causes a disturbance of reproductive maturation mediated possibly by failure of luteinizing hormone (Phillipo et al. 1987). Although signs of endocrine disturbance were not identified in the limited time scale of this study, the pituitary failure, so marked a finding of the more protracted clinical investigation (Dincer 1994, Haywood et al. 1993) supports this hypothesis.

The redistribution of Cu from the liver to the brain is also in line with earlier observations (Dincer 1994, Haywood et al. 1993) and supports the view that Cu-TTM is selectively redistributed to this site as has been reported with diethyldithiocarbamate, another Cu chelating agent (Allain and Krari 1991).

These findings question the long-term safety of TTM as a chelating agent for therapeutic purposes.

Acknowledgements

We wish to thank Mr. W.R. Humphries of the Rowett Research Institute for providing TTM, Mr. J.D. Holding of the Department of Clinical Chemistry, Liverpool and Miss N. Parry, the recipient of a Wellcome Trust Vacation Scholarship for the Mo analyses. We also wish to acknowledge the University of Selcuk, Turkey for the award of a graduate studentship (ZD) and SmithKline Beecham for a research fellowship (SH).

Literature Cited

Allain P. & Krari N. (1991) Diethyldithiocarbamate, copper and neurological disorders. Life Sci. 48: 291–299.

Brewer G.J., Dick R.D., Yuzbasiyan-Gurkan V., Tankanow R., Young A.B. & Kluin K.J. (1991) Initial therapy of patients with Wilsons disease with tetrathiomolybdate.

Dincer Z. (1994) Copper Toxicity in Sheep: Studies on Copper Chelation by Ammonium tetrathiomolybdate (TTM) and Metallothionein. PhD Thesis, University of Liverpool.

Haywood S., Dincer Z. & Humphries W.R. (1993) Endocrinopathy and brain copper elevation in tetrathiomolybdate treated sheep. In: Trace Elements in Man and Animals – TEMA 8. (M. Anke, D. Meissner & C.F. Mills, eds). Verlag Media Touristik.

Humphries W.R., Morrice P.C. & Bremner I. (1988) A convenient method for the treatment of chronic copper poisoning in sheep using subcutaneous ammonium tetrathiomolybdate. Vet. Rec. 9: 51–53.

Phillippo M., Humphries W.R., Atkinson T., Henderson G.D. & Garthwaite P.H. (1987) The effect of dietary molybdenum and iron on copper status, puberty, fertility and oestrus cycles in cattle. J. Agric. Sci. 109: 321–336.

Discussion

Q1. Neville Suttle, Moredun Research Institute, Edinburgh, Scotland: These findings don't bring into question the long term safety of TTM as an antidote for copper poisoning. In our experience with the use of this compound (about 50 cases of copper poisoning over the past three years), no complications have been reported. The difference between the general use of TTM and the use in these experimental simulations is that the course of the injections was repeated 5 times and this was gross folly. Now, we had an incident with copper poisoning in our own flock, and we took the opportunity to compare the two dose rates for thiomolybdate, at 1.7 and 3.4 mg/kg. We did that because I shared Susan's fears about possible safety, and I drew down from the dose recommended by the Rowett Institute. And I have to say that they were absolutely right, you need 3.4 mg/kg in 3 doses to provide complete recovery. The use of this dosing regime, without repetition, should not be prejudiced by a rather extreme use of the antidote as we have seen reported here.

A. Thank you very much, Neville. I do accept these comments, and I do accept that this was a particularly high dose. But I do want to repeat that this was monitored by the Rowett Research Institute, and it was a dose that I have on verbal confirmation from the supplier that would be equivalent to a dose that we had tried out on copper deficient sheep for a year, pregnant sheep in normal copper status for a year, but never on copper supplemented sheep. Yes, it was a high dose, but even so, I'm putting it forward that more work needs to be done.

Q2. Danny Goodwin-Jones, Trace Element Services Limited, Dyfed, Wales: I'm a sheep farmer, and have seen many sheep in my lifetime. May I just comment on your choice of animal for your experiments. The Cambridge is notoriously a genetic muddle. It was designed by certain people to ostensibly produce a lot of animals, for fecundity. You would have been better off using a Suffolk animal, which needs buckets of copper to survive and do very well. In 20 years, we have never seen copper toxicity because we use selenium with it.

A. Thank you for your comment. We were not looking at a specific breed sensitivity or response to this, we thought it was possibly a general principle, and we did, in all, use 3 breeds. Secondly, copper toxicity is not a widespread problem. We have managed to prevent it very well.

Q3. Barbara Golden, University of Aberdeen, Scotland: Where do you think the molybdenum ended up in the pituitary or hypothalamus? Do you have any information that it might bind to a particular fraction?

A. This is a very interesting issue. We do not know at this point, but we are hoping to do some immunochemistry to find out.

Q4. John Sorenson, University of Arkansas, Little Rock, AR, USA: Do you have any ideas as to the structure of the Cu-Mo complex formed *in vivo* that would allow transport across the blood-brain barrier?

A. This again is an interesting point. The blood-brain barrier should be impervious to copper and its complexes. Nevertheless, we have information from the Chemistry Department that this is a lipophilic compound that can cross the blood-brain barrier, and release Cu ions.

Effects of Selenium on Chemically-Induced Gallbladder Carcinogenesis in Hamsters

H. NAKADAIRA, T. ISHIZU, AND M. YAMAMOTO

Department of Hygiene & Preventive Medicine, Niigata University School of Medicine, Niigata, Japan 951

Keywords selenium, gallbladder cancer, hamsters, 3-methylcholanthrene

Introduction

Se has been shown to be an effective chemopreventive agent in many animal experiments. On the other hand, there have been studies indicating the lack of inhibitory effects of Se in chemical carcinogenesis. The standardized mortality ratios for gallbladder cancer of Japanese were high in the world. However, little is known about the role of Se in gallbladder carcinogenesis. We investigated the effects of Se on chemically induced gallbladder cancer in hamsters.

Materials and Methods

A total of 100 weaned female Syrian golden hamsters at 4 weeks of age were randomly assigned to four experimental groups. They were caged under controlled conditions (temperature, 24 ± 1°C; humidity, 76 ± 7%; 14 h-light – 10 h-dark cycle) and fed a purified casein-based diet containing the AIN-76 mineral mix. Sodium selenite was added to distilled water to the final concentrations of 0.0, 0.55 2.0, and 4.0 ppm. Each of the groups was given one of the Se-added distilled water throughout the experiment. Gallbladder tumours were induced by inserting a Beeswax pellet containing 3-methylcholanthrene (3-MC) into the gallbladder at week 4 as described by Suzuki et al. (1981). An 8.0 mg pellet contained 3.0 mg of 3-MC. At week 24, the abdominal cavity was opened. The liver, gallbladder, pancreas, extrahepatic bile duct and duodenum were removed and then examined microscopically. Serum Se was determined by a graphite-furnace atomic absorption spectrometric method.

Results

Ninety-four animals survived for the entire experimental period. Body weights at each of the monthly periods were not influenced by Se addition. Final mean body weights and mean concentrations of serum Se of the four groups and the estimated amount of Se intake from both water and diet per animal are shown in Table 1. Organs other than the liver and gallbladder showed no macroscopic change. The histological types of gallbladder tumours were benign, precancerous and malignant. A normal mucosa was found in the 4.0 ppm Se-group. The benign lesion was a granuloma in the 0.0 ppm Se-group. The malignant tumours were carcinoma *in situ* (CIS) and adenocarcinoma. All the cases of carcinoma but two were considered to develop through the sequence from dysplasia to CIS and from CIS to invasive adenocarcinoma (Albores-Saavedra et al. 1984). The exceptional two cases had adenocarcinoma in adenoma (Kozuka et al. 1982). There was no significant difference in the incidence of malignant tumours between the groups (Table 2). However, the incidence of CIS was lower ($p < 0.05$) and that of invasive adenocarcinoma was higher ($p < 0.05$) in the groups treated without Se than in those treated with Se (Table 2).

Correct citation: Nakadaira, H., Ishizu, T., and Yamamoto, M. 1997. Effects of selenium on chemically-induced gallbladder carcinogenesis in hamsters. In *Trace Elements in Man and Animals – 9: Proceedings of the Ninth International Symposium on Trace Elements in Man and Animals. Edited by* P.W.F. Fischer, M.R. L'Abbé, K.A. Cockell, and R.S. Gibson. NRC Research Press, Ottawa, Canada. pp. 466–467.

Table 1. Estimated selenium (Se) intake, mean body weight and mean serum Se of four experimental groups of female Syrian golden hamsters.

Se (ppm)	Effective no. of hamsters	Estimated Se intake (µg/day/animal)	Mean body weight at week 24[1] (g)	Mean serum Se at week 24[2] (ng/ml)
0.0	25	1	195 ± 16	152.8 ± 58.62[a] (n = 24)
0.5	23	5	187 ± 15	226.5 ± 88.16[b] (n = 20)
2.0	24	15	185 ± 21	283.3 ± 62.42[b,c] (n = 21)
4.0	22	29	195 ± 19	317.1 ± 89.53[c] (n = 22)

[1]Values are mean ± SD. [2]Values are mean± SD. Values with different superscripts are significantly different by the chi-squared test ($p < 0.05$). Numbers of hamsters are in parentheses.

Table 2. Incidence[1] of gallbladder tumours induced by a Beeswax pellet containing 3-methyl-cholanthrene in female Syrian golden hamsters receiving selenium (Se) in drinking water.

				Malignant			
Se (ppm)	Effective no. of hamsters	Benign	Dysplasia	CIS[2]	Invasive adenoca.[3]	Adenoca. in adenoma	Total
0.0	24 (100%)	1(4)	2(8)	0(0)[a]	21(88)[a]	0(0)	21(88)
0.5	20 (100)	0(0)	5(25)	3(15)[a,b]	11(55)[b]	1(5)	15(75)
2.0	21 (100)	0(0)	4(19)	7(33)[b]	10(48)[b]	0(0)	17(81)
4.0	22 (100)	0(0)	3(14)	5(23)[b]	12(52)[b]	1(5)	18(82)

[1]Values with different superscripts are significantly different by the chi-squared test ($p < 0.05$). [2]Carcinoma in Situ. [3] Invasive adenocarcinoma into the liver tissue.

Conclusions

The present study provided no evidence for the inhibition by Se of total gallbladder cancer induced by a 3-MC Beeswax pellet. There were two main sequences of the development of adenocarcinoma in hamsters. In the present study, all the cases of adenocarcinoma but two corresponded to the sequence from dysplasia to CIS and from CIS to invasive adenocarcinoma. The retardation of the progression of adenocarcinoma by Se was indicated by the higher incidence of invasive adenocarcinoma and the lower incidence of CIS in hamsters without Se than in those treated with Se.

Literature Cited

Albores-Saavedra, J., Angeles-Angeles, A., Manrique, J.J. & Hemson, D.E. (1984) Carcinoma in situ of the gallbladder. A clinicopathologic study of 18 cases. Am. J. Surg. Pathol. 8: 323–333.

Kozuka, S., Tsubone, M., Yasui, A. & Hachisuka, K. (1982) Relation of adenoma to carcinoma in the gallbladder. Cancer 50: 2226–2234.

Suzuki, A. & Takahashi, T. (1981) Induction of carcinoma of the gall in hamsters by intracholecystic methylcholanthrene Beeswax pellets. Jpn. J. Gastroenterol. 78: 66–71.

The Effect of Level and Form of Supplementary Selenium on Tissue Concentrations in Lambs and their Importance in Raising Human Dietary Selenium Intake

J. MOLNÁR, S. DRUSCH, AND A. MACPHERSON

Biochemical Sciences Department, Scottish Agricultural College, Auchincruive, Ayr KA6 5HW, UK

Keywords selenium, lamb, dietary intake, tissue concentration

Several diseases resulting from selenium deficiency have been described both in humans and animals, e.g. white muscle disease in lambs and calves, hepatosis dietetica in pigs, Keshan disease, Kashin-Beck disease, and an increased risk of cardiovascular disease and cancer in humans. Since over the past decade the dietary intake of selenium in Scotland has fallen dramatically and is considerably below the recommended intakes, an approach to the problem would be to increase the selenium concentration of carcass meat by supplementation of meat animals for a limited period prior to slaughter in order to raise the concentration of the element in the human diet.

Materials and Methods

In a first trial 16 Scottish Blackface lambs were stratified according to liveweight and then randomly allocated to one of 4 treatments as follows: nil, 3.5, 7.0 or 10.5 mg Se/head/week. The selenium was administered orally as sodium selenite in aqueous solution. In a subsequent study with 44 Texel and Suffolk lambs the efficiency of sodium selenite and L-selenomethionine at a supplementary level of 3.5 mg/week was compared. The lambs were fed a diet of hay, barley, maize and milk powder. After 14 (1st trial) and 7 (2nd trial) weeks respectively, the lambs were sacrificed and samples of shoulder and thigh muscle, liver and kidney were obtained for analysis. The tissue samples were dried at 60°C and the results expressed on a dry matter basis. Samples for selenium determination were digested with nitric, perchloric and sulphuric acids and the concentration measured by atomic absorption spectrometry following hydride generation (Hershey and Oostdyk 1988). The analytical method was verified by analysis of several reference samples. Glutathione peroxidase was measured in whole blood by the method of Paglia and Valentine (1967) as modified in the Ransel kit.

Results

Mean plasma selenium concentrations were measured in weeks 2, 8 and 14 (Figure 1). All three treatments effected a significant increase in plasma selenium over the control group which was evident at 8 weeks and maintained until the end of the experiment. Both muscle and liver selenium concentrations responded linearly to the level of supplementation with sodium selenite, whereas kidney, possibly due to its much higher initial concentration, showed only small responses to supplementation with no significant difference among the three levels (Figures 2 and 3). Glutathione peroxidase activity increased in the supplemented groups over the whole period of the experiment with no significant differences among the groups (Figure 4). The efficiency of L-selenomethionine and sodium selenite differed only in muscle tissue, possibly due to non-specific incorporation into tissue proteins (Figure 5).

Correct citation: Molnár, J., Drusch, S., and MacPherson, A. 1997. The effect of level and form of supplementary selenium on tissue concentrations in lambs and their importance in raising human dietary selenium intake. In *Trace Elements in Man and Animals – 9: Proceedings of the Ninth International Symposium on Trace Elements in Man and Animals. Edited by* P.W.F. Fischer, M.R. L'Abbé, K.A. Cockell, and R.S. Gibson. NRC Research Press, Ottawa, Canada. pp. 468–470.

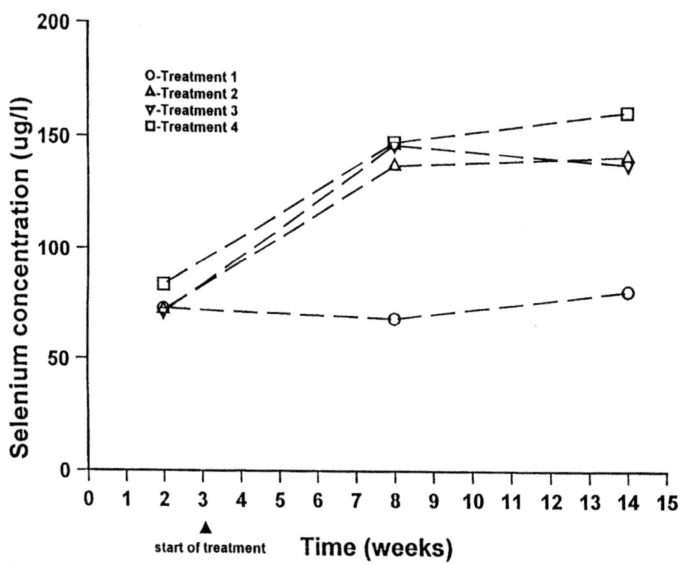

Figure 1. Changes in lamb plasma selenium concentration with treatment.

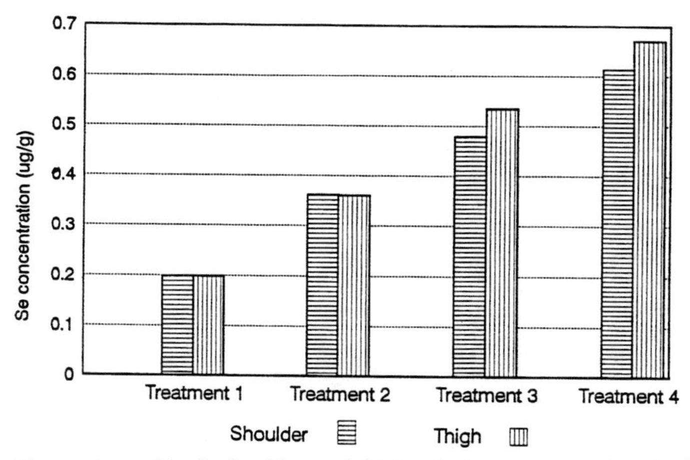

Figure 2. Comparison of lamb shoulder, and thigh selenium concentrations in the four treatment groups.

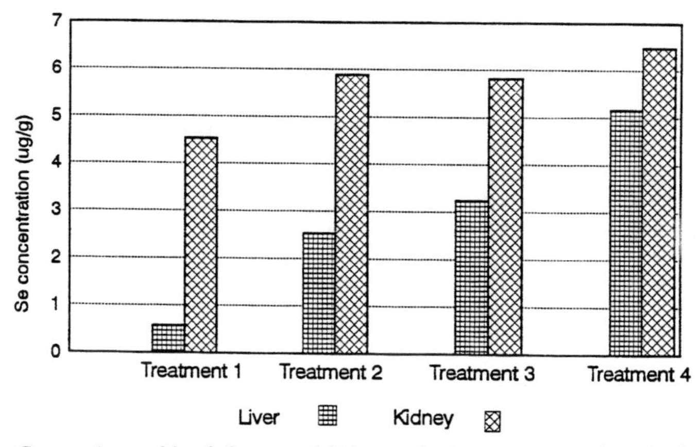

Figure 3. Comparison of lamb liver, and kidney selenium concentrations in the four treatment groups.

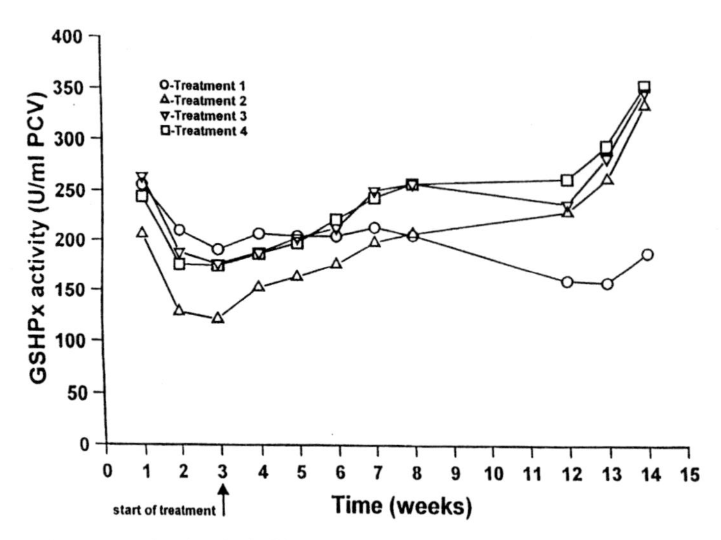

Figure 4. Changes in lamb whole blood glutathione peroxidase activity with treatment.

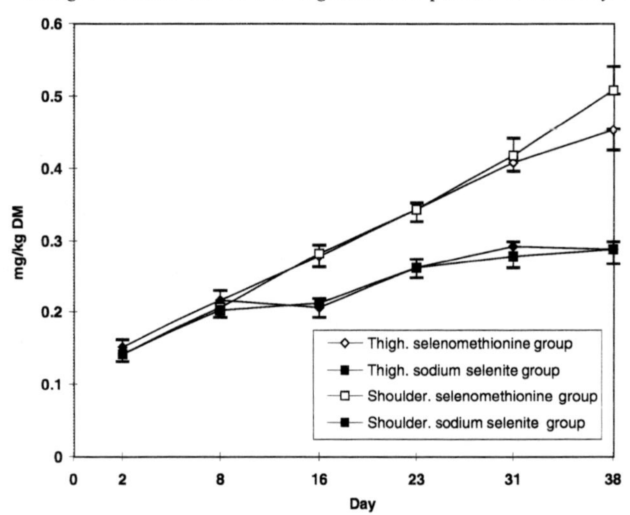

Figure 5. Changes in muscle selenium concentrations over time (mg/kg DM).

Discussion

Summarizing the results all three levels of treatment were effective in raising plasma and tissue selenium concentrations above those of the unsupplemented animals. Both sodium selenite and L-selenomethionine were suitable to raise tissue selenium concentrations and consequently human dietary selenium intake. Cost effectiveness of the different supplements and toxicological aspects have also to be taken into account. This approach therefore offers an alternative strategy to the one adopted in Finland of the fertilization of all crop growing grounds with selenium supplemented fertilisers. It has the added advantage that it can be reversed at any time if necessary or desirable.

Literature Cited

Hershey, J. W. & Oostdyk, T. (1988) Determination of arsenic and selenium in environmental and agricultural samples by hydride generation atomic absorption spectrometry. J. Assoc. Off. Anal. Chem. 71: 1090–1093.

Paglia, D. E. & Valentine, W. N. (1967) Studies on the quantitative and qualitative characterization of erythrocyte glutathione peroxidase. J. Lab. Clin. Med. 70: 158–159.

Trace Element Binding Proteins: Their Regulation and What They Regulate

Chair: S.A. Atkinson

The Soluble Form of the Transferrin Receptor and Iron Status

ROY D. BAYNES
Division of Hematology, Department of Medicine, University of Kansas Medical Center, Kansas City, Kansas 66160, USA

Keywords transferrin, receptor, soluble, iron

Introduction

The biological requirement for iron and its reactivity and toxicity dictate cellular iron delivery and storage be highly regulated. At the cellular level, iron delivery is mediated by transferrin receptor (TfR) and intracellular storage by ferritin (Ferr). These proteins are precisely and reciprocally regulated, at post-transcriptional level, by interaction of an iron response element binding protein (IREBP) with iron responsive elements (IREs) of mRNA (Klausner et al. 1993). Non iron saturated IREBP has IREBP activity. Saturated, it has aconitase activity but lacks IREBP function. IREBP interaction with the IRE on the 5′ end of Ferr mRNA limits Ferr synthesis by preventing polysome formation. IREBP binding to IREs at the 3′ end of TfR mRNA stabilizes mRNA, increasing translation. This mechanism can precisely and reciprocally modulate Ferr and TfR production such that at any iron status a particular ratio of cell Ferr to cell TfR is expected. Ferr is released from cells proportionately to cellular Ferr. Consequently serum Ferr is an excellent indicator of body iron stores. A soluble form of TfR is also released from cells in direct proportion to the number of intact receptors expressed. Since cellular TfR and Ferr are reciprocally regulated it is predictable that a close inverse relationship will exist between serum TfR and serum Ferr.

Serum TfR: Molecular Identity

Cellular TfR is a transmembrane glycoprotein comprising identical 95kDa monomers linked by a pair of disulfide bridges. Each comprises 786 amino acids organized into an amino terminal cytoplasmic domain of 61 amino acids, 28 amino acid membrane spanning segment and large extracellular domain of 671 amino acids. The cytoplasmic domain is required for intracellular trafficking. The transmembrane segment, consisting of hydrophobic residues, functions as a signal peptide and membrane anchor. The transferrin binding site is in the extracellular domain. Cysteines at positions 89 and 98 are sites of disulfide linking. N-linked glycosylation sites have been identified at residues 251, 317, and 727 while threonine 104 has been identified as the site of o-linked glycosylation. Each receptor dimer can bind two transferrin molecules.

Soluble TfR isolated from human sera by immunoaffinity consists of a single 85 kDa protein. By amino acid sequencing it is an abbreviated monomeric extracellular domain with the truncation event

Correct citation: Baynes, R.D. 1997. The soluble form of the transferrin receptor and iron status. In *Trace Elements in Man and Animals – 9: Proceedings of the Ninth International Symposium on Trace Elements in Man and Animals. Edited by* P.W.F. Fischer, M.R. L'Abbé, K.A. Cockell, and R.S. Gibson. NRC Research Press, Ottawa, Canada. pp. 471–475.

occurring between arginine 100 and leucine 101. An identical soluble moiety is in the supernatant of human cell cultures providing an *in vitro* model to evaluate mechanism of production (Baynes et al. 1994).

Serum TfR: Mechanism of Production

Cellular transferrin receptor binds diferric transferrin and internalizes it by receptor mediated endocytosis. Affinity of receptor for ligand varies at physiological pH proportionately to transferrin iron content. After binding, the receptor-ligand complex is internalized in an endocytic vesicle. With protonation, iron dissociates from transferrin and undergoes transmembrane transport into the cytoplasm. The remaining apo moiety has at this reduced pH high affinity for the receptor and bound to it escapes degradation. Most returns to the cell surface via recycling endosomes and at physiological pH is released to participate in further cycles of iron transport. A minor amount of endocytosed transferrin receptor follows another pathway. In this, the endosomal vesicle membrane undergoes internal bleb formation yielding multiple 50 nm vesicles (exosomes) within a larger limiting membrane (multi vesicular endosome). Exosome surfaces carry TfR and are extruded by a process of exocytosis. The majority of TfR in man are expressed by erythroid progenitor cells.

Molecular identity of serum TfR is consistent with it arising by proteolytic cleavage. In cell culture models this phenomenon appears mediated by a surface membrane associated serine protease (Baynes et al. 1993). Remnant tracking experiments indicate maximal proteolysis at the surface of the exosome within the multi vesicular endosome (Baynes et al. 1994). Important questions relate to the identity of the serine protease and whether activity is regulated.

Serum TfR: Factors Modulating Production

Factors targeting receptors for truncation are incompletely defined but, clearly at the core of whether soluble receptor production is a random event or related to specific properties of the particular receptor, eg. aging. That diferric transferrin reduces soluble receptor production suggests occupancy protects against truncation or directs the receptor away from the multi vesicular endosome (Baynes 1995). Site directed mutagenesis to substitute threonine 104 abrogates o-linked glycosylation while increasing soluble receptor production suggesting alterations to the o-linked glycan may be key to soluble receptor generation (Rutledge et al. 1994). This implies that o-linked glycan protects the biologically relevant cleavage site or directs the receptor away from the subcellular location of proteolytic activity. Changes in the o-linked glycan may occur as part of receptor aging.

Serum TfR: Determinants of Serum Concentration

In vitro data clearly establish that soluble TfR production is proportional to total TfR expressed (Baynes et al. 1991). Ferrokinetic observations in man establish this *in vivo* (Huebers et al. 1990). Therefore two factors influence the number of TfR expressed. First, total numbers of growing cells contribute to the total receptor pool. Therefore cellular hyperplasia and particularly expansion of erythroid progenitor mass will increase serum concentration of TfR. Second, total cellular TfR may be increased by reduction in cellular iron content by regulatory mechanisms already outlined. Cellular iron deficiency increases serum concentration of TfR.

Serum TfR: Measurement

Data presented subsequently are based on monoclonal enzyme linked immunoassay (EIA) reported by Flowers and co-workers (Flowers et al. 1989). Prior immunoradiometric assay (IRMA) using commercially available monoclonal antibodies, OKT9 and B3-25 reported values some 20-fold lower (Kohgo et al. 1987). Polyclonal EIA reported values in the same concentration range as the monoclonal EIA outlined above (Huebers et al. 1990). Systematic difference between EIAs and IRMA relate to the source of antigen used to raise antibodies. Commercially available monoclonals were prepared against membrane bound transferrin receptor largely free of ligand. Reagents for EIAs were produced against TfR complexed with transferrin. Our isolation studies show this is the form in which serum TfR circulates. Modest differences between monoclonal and polyclonal EIAs relate to differences in purified transferrin receptor standards rather than to a systematic difference in immunological reagents.

Serum TfR: Assessment of Iron Status

Body iron is found in 3 compartments. Metabolically active iron associated with iron-requiring proteins and enzymes is termed the functional compartment. Deficiency in this compartment is associated with liabilities. Iron absorbed in excess of that required for functional compartment is incorporated in storage proteins and constitutes the storage compartment. These are linked by iron in a small transport compartment (transferrin). Storage compartment is accurately reflected by serum ferritin concentration. Once stores are depleted serum ferritin values remain about the threshold value of 12 ng/mL. Serum ferritin provides little information about size of the functional compartment. The transport compartment is best reflected in percentage saturation of transferrin. This is too labile an index for accurate assessment of iron status.

All serum TfR assays report iron deficiency associated with a 3–5 fold increase in concentration. In careful phlebotomy study to reduce body iron, serum TfR concentrations remain stable until storage compartment is exhausted. This corresponds to a serum ferritin concentration of 12 ng/mL. As functional compartment becomes progressively depleted serum TfR shows a predictable increase proportional to the magnitude of deficit (Skikne et al. 1990). As functional compartment depletion progresses anemia supervenes. Between storage depletion and development of anemia serum TfR provides the only accurate assessment of functional compartment depletion. Other indicators including mean cell volume, red cell distribution width and free erythrocyte protoporphyrin are less sensitive, less predictable and late indicators compared with serum TfR.

Serum TfR proves exceptionally valuable in situations where evaluation of functional compartment is problematic. These include growing children, pregnant women (Carriaga et al. 1991) and anemia of chronic disease (Ferguson et al. 1992). In the first, two stores are modest or absent resulting in loss of any discriminant value for serum Ferr, while in the latter serum Ferr is spuriously elevated secondary to behaving as an acute phase reactant. Serum TfR is raised in pregnancy only in functional compartment deficiency.

Before development of serum TfR measurement the only reliable way of distinguishing iron deficiency anemia from anemia of chronic disease was by assessing iron content on a bone marrow aspirate and/or biopsy. Serum TfR remains normal in inflammation such that it is the only non-invasive test to reliably distinguish these two conditions (Colton et al. 1995, Ferguson et al. 1992, Pettersson et al. 1994). Combining data available for anemia of chronic disease, serum TfR gives positive predictive value for iron deficiency of 91% and negative predictive value of 95%. When anemia of chronic disease and iron deficiency anemia coexist serum TfR tracks functional iron deficit.

Raised serum TfR reflects functional iron depletion if conditions associated with expanded erythropoiesis are excluded. Such conditions might include hemolytic anemias or ineffective erythropoiesis associated with megaloblastic anemia, myelodysplasia and hemoglobinopathies such as thalassemia. In most of these, serum Ferr is normal or increased. Expanded erythropoiesis in these conditions may be associated with functional compartment depletion despite expanded iron stores.

A novel application was recently described in which measurements of serum TfR and Ferr could reliably detect erythropoietin abuse by elite athletes and could distinguish this from high altitude conditioning (Gareau et al. 1996).

The Serum TfR: Serum Ferr Ratio

Body iron status can be defined by serum TfR and Ferr. Since serum TfR remains unaffected by storage compartment depletion while serum Ferr progressively declines and since serum TfR increases progressively with functional compartment depletion while serum Ferr remains constant at its reduced level it is logical to combine the measurements in a ratio. This ratio is logical on theoretical grounds based on the post-transcriptional reciprocal regulation previously outlined. The previously mentioned phlebotomy study showed a linear inverse relationship between the ratio and body iron content. The ratio increases from <100 in the presence of adequate stores to >2000 once significant functional depletion has occurred. A median ratio of 500 corresponds to the point of depleted stores. The ratio was recently evaluated in an iron supplementation study in pregnant women in Jamaica (Simmons et al. 1993). The mean ratio observed in control women receiving no supplement was 1200 compared with 470 in the supplemented group. The cumulative frequency distribution of log TfR:Ferr was linear in both indicating that in any given population iron status shows normal distribution without discreet deficient or overloaded sub-populations.

Literature Cited

Baynes, R.D. (1995) Transferrin reduces the production of soluble transferrin receptor. Proc. Soc. Exp. Biol. Med. 209: 286–294.

Baynes, R.D., Shih, Y.J., & Cook, J.D. (1991) Production of soluble transferrin receptor by K562 erythroleukemia cells. Br. J. Haematol. 78: 450–455.

Baynes, R.D., Shih, Y.J., Hudson, B.G., & Cook, J.D. (1993) Production of the serum form of the transferrin receptor by a cell membrane-associated serine protease. Proc. Soc. Exp. Biol. Med. 204, 65–69.

Baynes, R.D., Shih, Y.J., Hudson, B.G., & Cook, J.D. (1994) Identification of the membrane remnants of transferrin receptor with domain-specific antibodies. J. Lab. Clin. Med. 123: 407–414.

Baynes, R.D., Skikne, B.S., & Cook, J.D. (1994) Circulating transferrin receptors and assessment of iron status. J. Nutr. Biochem. 5: 322–330.

Carriaga, M.T., Skikne, B.S., Finley, B., Cutler, B., & Cook, J.D. (1991) Serum transferrin receptor for the detection of iron deficiency in pregnancy. Am. J. Clin. Nutr. 54: 1077–1081.

Colton, C., Geraci, K., Tahsildar, H., Flowers, C., Cook, J., Cooper, G., Hornbuckle, K. (1995) Serum transferrin receptor/ferritin ratio is a reliable predictor of iron deficiency anemia in inflammatory bowel disease patients. Am. J. Gastroenterol. 108A: 800.

Ferguson, B.J., Skikne, B.S., Simpson, K.M., Baynes, R.D., Cook, J.D. (1992) Serum transferrin receptor distinguishes the anemia of chronic disease from iron deficiency anemia. J. Lab. Clin. Med. 119: 385–390.

Flowers, C.H., Skikne, B.S., Covell, A.M., & Cook, J.D. (1989) The clinical measurement of serum transferrin receptor. J. Lab. Clin. Med. 114: 368–377.

Gareau, R., Audran, M., Baynes, R.D., Flowers, C.H., DuVallet, A., Senecal, L., & Bruisson, G.R. (1996) Erythropoietin abuse in athletes. Nature 380: 118.

Huebers, H.A., Beguin, Y., Pootrakul, P., Einspahr, D., & Finch, C.A. (1990) Intact transferrin receptors in human plasma and their relation to erythropoiesis. Blood 75: 102–107.

Klausner, R.D., Rouault, T.A., & Harford, J.B. (1993) Regulating the fate of mRNA: The control of cellular iron metabolism. Cell 72: 19–28.

Kohgo, Y., Niitsu, Y., Kondo, H., Kato, J., Tsushima, N., Sasaki, K., Hirayama, M., Numata, T., Nishisato, T., & Urushizaki, I. (1987) Serum transferrin receptor as a new index of erythropoiesis. Blood 70: 1955–1958.

Pettersson, T., Kivivuori, S.M., & Siimes, M.A. (1994) Is serum transferrin receptor useful for detecting iron-deficiency in anaemic patients with chronic inflammatory diseases? Br. J. Rheumatol. 33: 740–744.

Rutledge, E.A., Root, B.J., Lucas, J.J., Enns, C.A. (1994) Elimination of the O-linked glycosylation site at Thr 104 results in the generation of a soluble human transferrin receptor. Blood 83: 580–586.

Simmons, W.K., Cook, J.D., Bingham, K.C., Thomas, M., Jackson, J., Jackson, M., Ahluwalia, N., Kahn, S.G., Patterson, A.W. (1993) Evaluation of a gastric delivery system for iron supplementation in pregnancy. Am. J. Clin. Nutr. 58: 622–626.

Skikne, B.S., Flowers, C.H., Cook, J.D. (1990) Serum transferrin receptor: A quantitative measure of tissue iron deficiency. Blood 75: 1870–1876.

Discussion

Q1. F.W. Sunderman, University of Connecticut, Farmington, CT, USA: This is a beautiful demonstration of the involvement of serine proteases. Have you done any work on the characterization of the specific protease(s) involved in this system?

A. I have no information. Does iron deficiency up- or down-regulate serine proteases? I don't have an answer at this time.

Q2. Ed Harris, Texas A&M University, College Station, TX, USA: In the turnover of the receptors, the state of glycosylation would seem to be quite important. I'm very interested in the fact that you've identified the serine protease that may be involved in clipping. Have you identified a glycosylase or a deglycosylase that would prepare this protein for the protease?

A. We have not yet identified a deglycosylation enzyme. That is obviously an important question. People who are working on aging of proteins are very interested in generation of advanced glycation products and those kind of issues. This needs a lot more work.

Q3. Ed Harris, Texas A&M University, College Station, TX, USA: Let me just raise another point with you. Of course, during iron repletion, we're dealing with an excessive synthesis of that receptor, through the transgolgi and so forth. Is it possible that the cell is starting to make some receptor that doesn't contain the carbohydrate, and thus becomes more labile?

A. This is an important question which has not been addressed, to my knowledge, at this point in time.

Q4. Lindsay Allen, University of California at Davis, CA, USA: How specific is your assay? You have published a major effect of vitamin B_{12} deficiency on these receptors. What about folate, vitamin A and so on?

A. We haven't looked at folate. As we looked at inefficient erythropoietic states we're expecting a standard erythroid mass, and so you would predict that you wouldn't have an increased receptor. The problem is that this receptor is not specific for iron because of the fact that it is also determined by the number of erythroid progenitors. That kind of begs the question of how to get around that problem, and that's the second area we're looking at, which is the soluble form of the erythropoietin receptor. The soluble form correlates very well with erythropoietic stimulation, but there's no effect of iron deficiency on it.

Q5. Kenneth Wing, Umea University, Umea, Sweden: Do you think that the ratio that you form between the transferrin receptor and the ferritin concentration will get around the plasma volume increase in pregnancy, and similar dilutional effects that complicate other measures of iron status in pregnancy?

A. I didn't mention it before, but we have published a study where we have looked at pregnant women. We found exactly that. We could get rid of the plasma dilution influences by looking at the ratio. The ratio seems to be a very good indicator of iron deficiency.

Q6. Stephanie Atkinson, McMaster University, Hamilton, ON, Canada: From a pediatric perspective, has anyone studied the ontogeny of the transferrin receptor? Do you have any comments on the use or applicability of the assay in pediatric situations?

A. We've not done any of that work at all, but it appears from work of others that rapid periods of growth increase levels of the serum receptor. This requires further study.

Involvement of Lactoferrin and Lactoferrin Receptors in Iron Absorption in Infants

B. LÖNNERDAL

Department of Nutrition, University of California, Davis, CA 95616, USA

Keywords lactoferrin, lactoferrin receptors, iron, breast milk

Scenario Supporting Concept of Lactoferrin Receptors

It is well-known that term, exclusively breast-fed infants rarely develop iron deficiency anemia. Several studies have shown that iron status of exclusively breast-fed infants is satisfactory up to 6–9 months (Siimes et al. 1984, Duncan et al. 1985, Hernell & Lönnerdal 1994) and perhaps even 12 months (Pisacane et al. 1995). This is surprising considering that breast milk has a low concentration of iron (0.2–0.4 mg/L) as compared to iron-fortified formula (7–12 mg/L) and that the total amount of iron provided by breast milk is lower than the estimated requirement for absorbed iron (Lönnerdal 1984). The fact that infants fed unfortified formula, which contains 1–2 mg iron/L, more frequently develop anemia than breast-fed infants (Saarinen and Siimes 1977), strongly suggests that the bioavailability of iron from breast milk is high as compared with formula. Radioisotope studies in human infants (Saarinen et al. 1977) show that iron is better absorbed from breast milk (~50%) than from formula (10–20%), provided that the method of extrinsic labeling is valid.

Human milk has a very high concentration of lactoferrin, an iron-binding protein synthesized by the mammary gland (Lönnerdal and Iyer 1995). Lactoferrin is a major constituent of the protein fraction of human milk, constituting ~10–20% of the total protein content. This protein has a molecular weight of ~80 kD, has two glycans per molecule and has the capacity to bind two atoms of ferric iron per molecule, although only a small fraction (4–10%) of this iron-binding capacity is utilized in breast milk (Fransson and Lönnerdal 1980). Although there are other ligands in human milk binding iron, we have found that the major part of iron is bound to lactoferrin (Fransson and Lönnerdal 1980). Lactoferrin has a very high affinity for iron (K_{ass} ~10^{30}) and does not release its iron until the pH is substantially less than 4. Since gastric acid secretion of infants is lower than later in life, the pH of the stomach is often 4–5, and lactoferrin is capable of retaining its iron during passage of the gastrointestinal tract, perhaps binding even more iron than initially in the milk. Lactoferrin is also resistant against proteolytic degradation *in vitro* (Brines and Brock 1983). This is also likely to be the case *in vivo* as exclusively breast-fed infants have significant amounts of undigested lactoferrin in feces (Davidson and Lönnerdal 1987). Thus, it is entirely possible that intact human lactoferrin from breast milk is present in the small intestine of the infant and that it carries the major part of milk iron.

Evidence for Presence of Intestinal Lactoferrin Receptors

There are now several lines of evidence for the existence of lactoferrin receptors in the small intestine. The first suggestion of mucosal lactoferrin receptors came from work by Cox et al. (1979), which showed binding of lactoferrin but not transferrin to human intestinal biopsies. We explored this notion in an experiment, in which suckling piglets were given radiolabeled milk formula with ferrous sulfate or bovine lactoferrin (Fransson et al. 1983). Iron uptake into blood was higher and more rapid when iron was provided as lactoferrin as compared to ferrous sulfate. We then postulated that lactoferrin binds to membrane-bound lactoferrin receptors and that the lactoferrin-lactoferrin receptor complex is internalized (Lönnerdal 1984).

Correct citation: Lönnerdal, B. 1997. Involvement of lactoferrin and lactoferrin receptors in iron absorption in infants. In *Trace Elements in Man and Animals – 9: Proceedings of the Ninth International Symposium on Trace Elements in Man and Animals. Edited by* P.W.F. Fischer, M.R. L'Abbé, K.A. Cockell, and R.S. Gibson. NRC Research Press, Ottawa, Canada. pp. 476–480.

In order to develop an animal model which is as close to human infants as possible and in which homologous lactoferrin can be used for radioisotope studies, we isolated lactoferrin from rhesus monkey milk (Davidson and Lönnerdal 1985). The protein was remarkably similar to human lactoferrin and even cross-reacted with the antibody to human lactoferrin. We then used brush border membrane vesicles from infant rhesus monkeys to perform kinetic studies of the receptor. Binding of lactoferrin to the membrane vesicles was saturable and we found that the receptor is specific for human and monkey lactoferrin, has a binding constant (K_d) of $\sim 10^{-6}$ M, and that bovine lactoferrin and human transferrin do not bind to the receptor (Davidson and Lönnerdal 1988). The receptor was found at all ages, but the largest number of receptors per tissue weight was found in intestine from infant monkeys. A major fragment of lactoferrin, produced by *in vitro* proteolysis, also bound to the receptor, showing that only part of the molecule is needed for binding and that partially digested lactoferrin may also aid in iron absorption in infants. Deglycosylation of lactoferrin did not alter its binding characteristics, demonstrating that the glycan chain(s) is not needed for binding to the receptor. We also showed that lactoferrin can deliver both iron and manganese to the membrane vesicles, but that zinc does not bind to lactoferrin under physiological conditions (Davidson and Lönnerdal 1989). Lactoferrin receptors were also shown to be present in the small intestine of the mouse (Hu et al. 1988, 1990).

The lactoferrin receptor was subsequently purified from human infant small intestine by immunoaffinity chromatography using immobilized human lactoferrin (Kawakami and Lönnerdal 1991). The receptor was found to have a molecular weight of ~115 kD, with a subunit size of 38 kD. Enzymatic deglycosylation reduced the subunit size to ~34 kD and demonstrated that the receptor is glycosylated. Deglycosylation of lactoferrin did not affect its binding to the receptor. Subsequent studies in suckling piglets (Gislason et al. 1993) showed that the lactoferrin receptor is present in similar numbers in all parts of the small intestine and that the number of receptors or its affinity to lactoferrin do not vary during the suckling period. Thus, there are now both kinetic and biochemical data supporting the existence of intestinal lactoferrin receptors.

Fate of Lactoferrin in the Enterocyte

The kinetic studies provided no information about the further fate of lactoferrin within the enterocyte. In order to obtain such information, studies in human enterocytes were needed. We used the human colon carcinoma cell line, Caco-2, which in culture spontaneously differentiates to enterocytes similar to those of the small intestine, and found specific and saturable binding of lactoferrin (Iyer et al. 1994). We found that lactoferrin is degraded intracellularly and that iron is delivered to ferritin, possibly for further export out of the cell. Sánchez et al. (1996) found that lactoferrin enhanced iron transport across Caco-2 cells, but that it was degraded. Mikagami et al. (1994) used a similar human carcinoma cell line, HT29, in monolayers and found that most lactoferrin is degraded, but that a very small fraction is transported across the cell with iron. This small amount of lactoferrin may explain the finding of small amounts of lactoferrin in the urine of preterm infants (Hutchens et al. 1991). In any case, it is evident that most of the iron is released intracellularly and that the further fate of this iron will be regulated by factors other than lactoferrin. Intracellular iron concentration, the individual's iron status and other intracellular factors will ultimately determine if iron will be transported further or lost during cell sloughing. Recent studies on iron-depleted cells in culture demonstrate increased uptake of lactoferrin and iron (Mikagami et al. 1995), emphasizing the importance of iron status on regulation of cellular uptake of iron via lactoferrin.

Effect of Lactoferrin on Iron Absorption in Infants

Bovine lactoferrin is commercially available in quantities allowing inclusion into infant formula, provided that positive effects can be demonstrated. Several studies have evaluated the effect of lactoferrin-fortified formula on iron absorption, balance and status in infants. Fairweather-Tait et al. (1987) used a stable isotope of iron and studied iron absorption in 7-d-old infants. The two formulas tested contained iron as ferrous sulfate or in lactoferrin. There was no significant difference in iron absorption between the two groups. In a study by Schulz-Lell et al. (1991), two different levels of lactoferrin were added to infant formula. There was no significant difference in iron balance between term infants fed the unfortified formula and those fed the lower level of lactoferrin, but infants fed formula with the higher level had a marginally higher iron balance. However, the formula with the higher lactoferrin concentration also had a higher concentration of iron and it is possible that the higher iron content was the cause of the higher values. A study by Chierici et al. (1992) also failed to demonstrate any effect of lactoferrin fortification of formula on iron status. We have recently conducted a study on term infants in which part (Lönnerdal and Hernell 1994) or all (Hernell and Lönnerdal 1996) of the iron was provided in the form of lactoferrin. We found no difference in iron status at 6 months of age between the groups given lactoferrin and those given the same formulas with iron as ferrous sulfate.

We have also evaluated the effect of human lactoferrin on iron absorption in human infants using stable isotopes (Davidsson et al. 1994). Rather than adding human lactoferrin to infant formula, we removed lactoferrin selectively and examined the effect of human milk and lactoferrin-free human milk on iron incorporation into erythrocytes. The expressed breast milk was separated into fat, casein and whey, lactoferrin was removed by affinity adsorption and the fractions combined before being fed to the infants. Somewhat surprisingly, mean iron absorption was slightly higher from lactoferrin-free milk than from breast milk.

Reasons for Lack of *in vivo* Evidence for Lactoferrin Receptors

It is apparent that studies on formulas fortified with bovine lactoferrin so far have failed to demonstrate any advantages on iron status and iron absorption in infants as compared to ferrous sulfate, the iron compound presently used to fortify infant formula. It is likely that the bovine lactoferrin is not recognized by the mucosal receptor for lactoferrin and that there is no receptor-mediated uptake of iron. On the other hand, it should be noted that iron in lactoferrin is provided in the ferric form, which is much less prone to oxidative reactions and formation of free radicals. Thus, this may be a way to present iron in a biologically relatively "safe" form. Eventually, though, it is likely that iron is released from bovine lactoferrin by proteolysis and that this iron is reduced to ferrous iron by ascorbic acid. In contrast, iron may be taken up into the mucosal cell directly via human lactoferrin, thereby further protecting the gut of breast-fed infants.

There are several possible reasons why we found no reduction in iron absorption when lactoferrin was removed from breast milk. One possibility is that the milk was processed and that the milk as fed was not the same as "native" breast milk. Another, and perhaps more likely explanation, is that the infants we studied were too old to observe an effect of lactoferrin. For methodological reasons, all infants in the study except one were 4 months old or older. We found earlier that up to 4 months of age there is a significant proportion of lactoferrin that survives digestion, but the amount becomes very small after that (Davidson and Lönnerdal 1987). Thus, it is possible that by far the major part of lactoferrin had been digested and that no receptor-mediated uptake of iron via lactoferrin would take place. Interestingly, one infant was only two months old and this infant showed higher iron absorption from human milk than from lactoferrin-free milk. It is evident that further studies are needed to evaluate the effect of human lactoferrin in young infants. In addition, long-term studies are needed to evaluate the effect of human lactoferrin on iron status. It may soon be possible to conduct such studies with recombinant human lactoferrin, which now can be produced. However, the efficacy of recombinant human lactoferrin needs to be evaluated very carefully as it may not be completely identical to native lactoferrin, particularly with regard to glycosylation.

References

Brines, R.D. and Brock, J.H. (1983) The effect of trypsin and chymotrypsin on the in vitro antimicrobial and iron-binding properties of lactoferrin in human milk and bovine colostrum: unusual resistance of human lactoferrin to proteolytic digestion. Biochim. Biophys. Acta 759: 229–235.

Chierici, R., Sawatzki, G., Tamisari, L., Volpato, S. and Vigi, V. (1992) Supplementation of an adapted formula with bovine lactoferrin. 2. Effects on serum iron, ferritin and zinc levels. Acta Paediatr. 81: 475–479.

Cox, T.M., Mazurier, J., Spik, G., Montreuil, J. and Peters, T.J. (1979) Iron binding proteins and influx of iron across the duodenal brush-border. Evidence for specific lactotransferrin receptors in the human intestine. Biochim. Biophys. Acta 558: 129–141.

Davidson, L.A. and Lönnerdal, B. (1986) Isolation and characterization of monkey milk lactoferrin. Pediatr. Res. 20: 197–201.

Davidson, L.A. and Lönnerdal, B. (1987) Persistence of human milk proteins in the breast-fed infant. Acta Pediatr. Scand. 76: 733–740.

Davidson, L.A. and Lönnerdal, B. (1988) Specific binding of lactoferrin to brush-border membrane: ontogeny and effect of glycan chain. Am. J. Physiol. 254: G580-G585.

Davidson, L.A. and Lönnerdal, B. (1989) Fe-saturation and proteolysis of human lactoferrin: effect on brush-border receptor-mediated uptake of Fe and Mn. Am. J. Physiol. 257: G930-G934.

Davidsson, L., Kastenmayer, P., Yuen, M., Lönnerdal, B. and Hurrell, R.F. (1994) Influence of lactoferrin on iron absorption from human milk in infants. Pediatr. Res. 35: 117–124.

Duncan, B., Schifman, R.B., Corrigan, J.J. and Schaefer, C. (1985) Iron and the exclusively breast-fed infant from birth to 6 months. J. Pediatr. Gastroenterol. Nutr. 4: 421–425.

Fairweather-Tait, S.J., Balmer, S.E., Scott, P.H. and Ninski, M.J. (1987) Lactoferrin and iron absorption in newborn infants. Pediatr. Res. 22: 651–654.

Fransson, G.-B. and Lönnerdal, B. (1980) Iron in human milk. J. Pediatr.2: 693–701.

Fransson, G.-B., Thoren-Tolling, K., Jones, B., Hambraeus, L. and Lönnerdal, B. (1983) Absorption of lactoferrin-iron in suckling pigs. Nutr. Res. 3: 373–384.

Gislason, J., Iyer, S., Hutchens, T.W. and Lönnerdal, B. (1993) Lactoferrin receptors in piglet small intestine: binding kinetics, specificity, ontogeny and regional distribution. J. Nutr. Biochem. 4:528–33.

Hernell, O. and Lönnerdal, B. (1996) Iron requirements and prevalence of iron deficiency in term infants during the first 6 months of life. In: Iron Nutrition in Health and Disease (L. Hallberg and N.G. Asp, eds.), Libby Publishing Co. (in press).

Hu, W.L., Mazurier, J., Sawatzki, G., Montreuil, J. and Spik, G. (1988) Lactotransferrin receptor of mouse small intestinal brush border. Biochem. J. 248: 435–441.

Hu, W.L., Mazurier, J., Montreuil, J. and Spik, G. (1990) Isolation and partial characterization of a lactotransferrin receptor from mouse intestinal brush border. Biochemistry 29: 535–541.

Hutchens, T.W., Henry, J.F., Yip, T.T., Hachey, D.I., Schanler, R.J., Motil, K.J. and Garza, C. (1991) Origin of intact lactoferrin and its DNA-binding fragments found in the urine of human-milk fed preterm infants. Evaluation of stable isotope enrichment. Pediatr. Res. 29: 243–250.

Iyer, S., Yuen, M. and Lönnerdal, B. (1993) Binding and uptake of human lactoferrin by its intestinal receptor studied in Caco-2 cells. FASEB J. 7: A64.

Kawakami, H. and Lönnerdal, B. (1991) Isolation and function of a receptor for human lactoferrin in human fetal intestinal brush-border membranes. Am. J. Physiol. 261: G841-G846.

Lönnerdal, B. (1984) Iron in breast milk. In: Iron Nutrition in Infancy and Childhood (Stekel, A., ed.), vol. 4, pp. 95–118. Raven Press, New York.

Lönnerdal, B. and Hernell, O. (1994) Iron, zinc, copper and selenium status of breast-fed infants and infants fed trace element fortified milk-based infant formula. Acta Paediatr. 83: 367–373.

Lönnerdal, B. and Iyer, S. (1995) Lactoferrin: molecular structure and biological function. Annu. Rev. Nutr. 15: 93–110.

Mikogami, T., Heyman, M., Spik, G. and Desjeux, J.-F. (1994) Apical-to-basolateral transepithelial transport of human lactoferrin in the intestinal cell line HT-29cl.19A. Am. J. Physiol. 267: G308-G315.

Mikogami, T., Marianne, T. and Spik, G. (1995) Effect of intracellular iron depletion by picolinic acid on expression of the lactoferrin receptor in the human colon carcinoma cell subclone HT29–18-C_1. Biochem. J. 308: 391–397.

Pisacane, A., De Vizia, B., Valiante, A., Vaccaro, A., Russo, M., Grillo, G. and Giustardi, A. (1995) Iron status in breast-fed infants. J. Pediatr. 127: 429–431.

Saarinen, U.M. and Siimes, M.A. (1977) Iron absorption from infant milk formula and the optimal level of iron supplementation. Acta Paediatr. Scand. 66: 719–722.

Saarinen, U., Siimes, M. and Dallman, P. (1977) Iron absorption in infants: High bioavailability of breast-milk iron as indicated by the extrinsic tag method of iron absorption and by the concentration of serum ferritin. J. Pediatr. 91: 36–39.

Sánchez, L., Ismail, M., Liew, F.Y. and Brock, J.H. (1996) Iron transport across Caco-2 cell monolayers. Effect of transferrin, lactoferrin and nitric oxide. Biochim. Biophys. Acta 1289: 291–297.

Schulz-Lell, G., Dörner, K., Oldigs, H.D., Sievers, E. and Schaub, J. (1991) Iron availability from an infant formula supplemented with bovine lactoferrin. Acta Paediatr. Scand. 80: 155–158.

Siimes, M.A., Salmenperä, L. and Perheentupa, J. (1984) Exclusive breast-feeding for 9 months: risk of iron deficiency. J. Pediatr. 104: 196–199.

Discussion

Q1. Roy Baynes, University of Kansas Medical Center, Kansas City, KS, USA: I seem to recall something about a small peptide sequence in the lactoferrin molecule which was active against gram negative bacteria. Could it be that the receptor, and internalization and breakdown is more a way of delivering this very active peptide to infants at a time when they may need protection against gram negative infections, rather than having anything to do with iron?

A. It is a possibility. There is no doubt that the iron is latching on and coming in at the same point. The quantitative significance of that I'm not sure of. There are many other aspects of lactoferrin that I could not talk to you about here. For example, lactoferrin has been shown to get into the nucleus and bind to very specific segments on DNA, and thereby affect gene expression.

Q2. Xavier Alvarez-Hernandez, LSUMC, Shreveport, LA, USA: Why are you not quoting one relevant paper from the Glasgow group (Sanchez was the first author), which showed what is happening to lactoferrin and iron in Caco-2 cells. The main point of that paper was that lactoferrin was taken up, degraded, and the degraded lactoferrin and iron both come out the apical side of the cells.

A. There are similar papers, derived from work in Boston, which were cited in the text of this presentation, but not in the lecture, due to the constraints of time.

Q3. Ed Harris, Texas A&M University, College Station, TX, USA: Can you draw a parallel for us between the lactoferrin receptor and the transferrin receptor? You've got the sequence. Do they look alike? Do you see similar kinds of structure or is it a totally different protein? With regard to Dr. Baynes' question, do you need iron on the protein in order for it to be endocytosed?

A. We don't see any similarity whatsoever. As for the second question, it appears that there may be some differences between the systems, as apo-lactoferrin is capable of binding and internalizing.

Metallothionein (MT) Enhancement in Brains of Ammonium Tetrathiomolybdate (TTM) Treated Copper Supplemented Sheep

Z. DINCER AND S. HAYWOOD
Department of Veterinary Pathology, University of Liverpool, England L69 3BX

Keywords metallothionein, ammonium tetrathiomolybdate, copper, brain

Introduction

Ammonium tetrathiomolybdate (TTM) treated sheep with high liver copper (Cu) show some redistribution of this excess Cu (Cu-TTM) to the brain (Dincer 1994, Haywood and Dincer 1997). Notwithstanding brain Cu concentrations similar to Wilsons disease in man, comparable neuronal damage has not been identified in the sheep brains suggesting that the excess metal is inactivated or rendered innocuous.

Metallothionein (MT), a metal binding protein, is known to modify the toxicity of Cu in liver and kidney (Evering et al. 1991) and it was decided to investigate whether MT functions likewise in the brain, using immunocytochemistry for MT localisation.

Materials and Methods

Sheep Brain Samples. High Cu brains from TTM treated Bleu du Maine sheep were obtained from a field case of Cu poisoning (Dincer 1994, Haywood et al. 1993); TTM treated Cu-loaded Cambridge sheep and TTM, high liver Cu, Ronaldsay sheep (Dincer 1994, Haywood and Dincer 1997).

Non-treated low Cu sheep brains were used for comparison.

Immunocytochemistry. Metallothionein was identified in paraffin-embedded sections using the DNP localisation system applied in conjunction with a mouse IgG monoclonal antibody (E9) to MT-1 and MT-2 (Evering et al. 1990).

MT immunoreactivity was graded weak (+) to marked (+++).

Results

Increased MT immunoreactivity was found in all the high-Cu brains from TTM treated sheep, irrespective of breed, in direct proportion to the brain Cu concentrations.

Thus MT immunostaining was greater in the copper poisoned Bleu du Maine sheep and the Ronaldsay sheep brains (Cu concentrations 34–60 μg/g) than the Cu supplemented Cambridge group (Cu concentrations 19–22 μg/g). Untreated sheep brains reacted weakly (Cu concentrations 13–15 μg/g). Zn brain concentrations were unchanged.

The enhanced MT immunoreactivity occurred mainly in the astrocytes of the white matter of the cerebellum, but also in hypothalamus, cerebrum and medulla oblongata. Enhanced MT immunostaining was also found in vascular endothelium and pia mater, less strikingly in ependymal cells.

Discussion and Conclusions

This study has shown that excess brain copper (Cu^{2+}) in sheep is apparently inactivated and sequestered by astrocytes as a metal chelate (Cu-MT) thus protecting vulnerable neurones. A similar occurrence

Correct citation: Dincer, Z., and Haywood, S. 1997. Metallothionein (MT) enhancement in brains of ammonium tetrathiomolybdate (TTM) treated copper supplemented sheep. In *Trace Elements in Man and Animals – 9: Proceedings of the Ninth International Symposium on Trace Elements in Man and Animals. Edited by* P.W.F. Fischer, M.R. L'Abbé, K.A. Cockell, and R.S. Gibson. NRC Research Press, Ottawa, Canada. pp. 481–483.

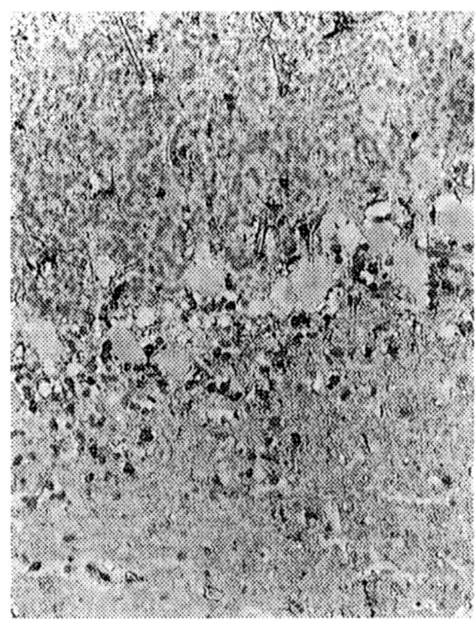

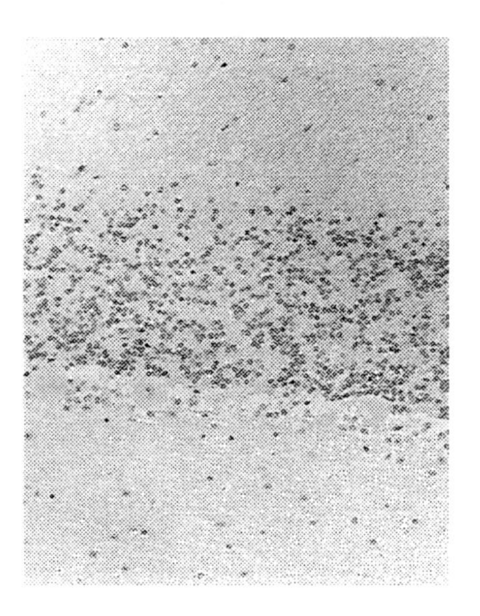

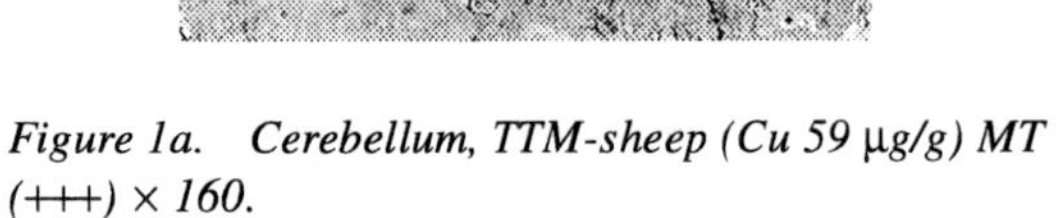

Figure 1a. Cerebellum, TTM-sheep (Cu 59 μg/g) MT (+++) × 160.

Figure 1b. Cerebellum, control sheep (Cu 13 μg/g) MT (+) × 160.

probably takes place in rat brain in which Hg and Cd is associated with increased MT immunoreactivity (Nishimura et al. 1992).

Metal induced MT in astrocytes supports the hypothesis that these cells are modulators of intracerebral metal homeostasis, acting to detoxify, sequester and even possibly excrete heavy metals.

Furthermore induction of MT at the blood-brain barrier suggests that these also are sites of stabilisation and containment for Cu ions in transport.

Acknowledgements

We wish to thank Dr. B. Jasani, Department of Pathology, University of Wales, Cardiff, for his gift of E9 monoclonal antibody. We also wish to acknowledge the University of Selcuk, Turkey, for the award of a graduate studentship (ZD) and SmithKline Beecham for a research fellowship (SH).

Literature Cited

Dincer Z. (1994) Copper Toxicity in Sheep: Studies on Copper Chelation by Ammonium tetrathiomolybdate (TTM) and Metallothionein. PhD Thesis, University of Liverpool.

Haywood S., Dincer Z. and Humphries W.R. (1993) Endocrinopathy and brain copper elevation in tetrathiomolybdate treated sheep. pp 601–602 in Trace Elements in Man and Animals TEMA 8. (M. Anke, D. Meissner and C.F. Mills, eds). Verlag Media Touristik.

Haywood S. and Dincer Z. (1997) Ammonium tetrathiomolybdate (TTM) and metal distribution in copper supplemented sheep. In: Trace Elements in Man and Animals – 9: Proceedings of the Ninth International Symposium on Trace Elements in Man and Animals. (Fischer, P.W.F., L'Abbé, M.R., Cockell, K.A. and Gibson, R.S., eds.). NRC Research Press, Ottawa, Canada. pp. 463–465.

Evering W.E.N.D., Haywood S., Elmes M.E., Jasani B. and Trafford J. (1990) Histochemical and immunocytochemical evaluation of copper and metallothionein in the liver and kidney of copper-loaded rats. J. Pathol. 160: 305–312.

Evering W.E.N.D., Haywood S., Bremner I. and Trafford J. (1993) The protective role of metallothionein in copper overload: 1. Differential distribution of immunoreactive metallothionein in copper-loaded rat liver and kidney. Chem. Biol. Interact. 78: 283–295.

Nishimura N., Nishimura H., Ghaffer A. and Tohyama C. (1992) Localisation of metallothionein in the brain of rat and mouse. J. Histochem. Cytochem. 40: 309–315.

Discussion

Q1. Arturo Leone, University of Salerno, Italy: Are you using a monoclonal antibody against the conserved amino-terminal region of metallothionein? Can you distinguish metallothionein isoforms in the astrocytes (i.e., MT-1 versus MT-2 versus the astrocyte-specific neuronal MT-3)?

A. The monoclonal antibody is specific for MT-1 and MT-2 and doesn't differentiate between those two isoforms. I can't answer with regard to is interaction with MT-3.

Q2. George Cherian, University of Western Ontario, London, ON, Canada: This is an interesting observation, but it is the astrocyte processes that are staining. This suggests that MT-1 and MT-2 are being recognized. But you showed one neuron staining, which would suggest MT-3 also. No one has really shown detection of MT-3 in normal brains so far.

A. Thank you for your comments.

Q3. Ed Harris, Texas A&M University, College Station, TX, USA: I'm particularly interested in your last comment, where you said that you feel that the astrocytes are involved in metal ion excretion. I just want to support that concept, because we have been looking at the C6 glial cells, and we definitely do see the Menkes factor being expressed, and also a functional efflux system for copper in these cells. So what you're saying is supported by biochemical evidence from a pure cell line. My other comment was, I was curious why we're not seeing copper going into those neurons, which apparently have an extremely effective blocking system, but yet we know that neurons require copper? Of course, you were looking at metallothionein, but do you still think we're seeing passage of copper into those cells?

A. It must be getting in. How it differentiates between excess copper and normally-acquired copper must be interesting too.

Q4. Ian Bremner, Rowett Research Institute, Aberdeen, Scotland: Can you clarify please the control animals you used? Were these normal sheep, or were they copper-loaded sheep? Has the metallothionein distribution in the brains of Wilson's disease patients ever been reported?

A. They were copper-loaded, non-TTM-treated sheep of all three breeds. I'm sorry if I didn't make that quite clear initially. I'm not aware that anyone has looked at metallothionein distribution in brains of Wilson's patients. We have looked quite extensively in the literature for this, as it is obviously something that we would like to see, but as far as I know, it hasn't been done yet. Ourselves, we have not had access to brains from any patients with Wilson's disease. We have done a couple of brains from people with sclerosing panencephalitis and Alzheimers and shown the astrocytes normal.

Investigations in a Child with Hyperzincaemia: Partial Characterisation of an Abnormal Zinc Binding Protein, Kinetics and Studies of Liver Pathology

B. SAMPSON,[†] I.Z. KOVAR,[‡] J.H. BEATTIE,[¶] H.J. MCARDLE,[§] A. RAUSCHER,[||] S.J. FAIRWEATHER-TAIT,[||] AND B. JASANI[+]

[†]Department of Chemical Pathology, Charing Cross Hospital, London, W6 8RF, UK, [‡]Department of Pediatrics, Chelsea & Westminster Hospital, London, SW10 9NH, UK, [¶]Division of Biochemical Sciences, Rowett Research Institute, Bucksburn, Aberdeen, AB2 9SB, UK, [§]Department of Child Health, Ninewells Hospital & Medical School, Dundee, DD1 9SY, UK, [||]Nutrition, Diet & Health Department, Institute of Food Research, Norwich, NR4 7UA, UK, and [+]Department of Pathology, University of Wales College of Medicine, Cardiff, CF4 4XN, UK

Keywords zinc, zinc binding proteins, liver, growth disorders

We have previously reported the case of a child with a plasma Zn concentration of 200 mmol/L (Sampson et al. 1995). Many of his symptoms are reminiscent of the classic description of Zn deficiency given by Prasad et al. (1963). The child is now 12 years old, but has the stature of a 4 year old. He has hepatosplenomegaly, vasculitis, anaemia, recurrent infections and osteoporosis. Liver function tests suggest a hepatocellular disorder. Urinary Zn excretion is negligible, but erythrocyte Zn concentration is normal (230 μmol/L). Plasma Cu concentration is 25–30 μmol/L and plasma Fe 1–2 μmol/L. Measurements of serum metal binding proteins show that caeruloplasmin concentration is appropriately increased (0.4–0.45 g/L, ref range 0.2–0.4) and α_2-macroglobulin concentration is normal (1.9 g/L, ref range 2.8–6.7).

Stable isotope studies with intravenously administered ^{70}Zn show that there are demonstrable abnormalities of Zn metabolism (Kovar et al. 1995). The results are clearly abnormal, with increased plasma (10.2 mg) and exchangeable (230 mg) Zn pools, rapid movement from the plasma to the exchangeable pool and a relatively slow movement out of the system. The exchangeable pool has been tentatively identified as liver.

Size exclusion chromatography shows the presence of a Zn binding protein with mass 180–300 kda. There is no Zn bound to albumin, and exogenous Zn binds *in vivo* to the unidentified protein. We have now partially purified the protein by a combination of size exclusion and ion exchange chromatography and partially characterised it by electrophoretic techniques and immunological methods. The protein consistently shows the presence of several components on agarose and polyacrylamide gel electrophoresis and capillary electrophoresis. These components are present in similar proportions in several preparations and in fractions across eluates from chromatographic columns. The major component reacts with antibody to α_2-macroglobulin. Immunodiffusion also shows reactivity to α_1-acid glycoprotein. Antibodies raised in rabbits to the protein show cross reactivity to α_2-macroglobulin, but not to α_1-acid glycoprotein. There is no reactivity to any other common plasma proteins or to histidine-proline rich glycoprotein (Morgan 1978). Mass spectra obtained by MALDI/MS showed 4 prominent components of m/z 10,421–13,195. In addition, a component of m/z 34,081 was found. Although no prominent components of m/z greater than 34,081 were observed, the sensitivity of MALDI/MS decreases with increasing analyte mass and it is possible that larger molecules were present but not detected.

Correct citation: Sampson, B., Kovar, I.Z., Beattie, J.H., McArdle, H.J., Rauscher, A., Fairweather-Tait, S.J., and Jasani, B. 1997. Investigations in a child with hyperzincaemia: Partial characterisation of an abnormal zinc binding protein, kinetics and studies of liver pathology. In *Trace Elements in Man and Animals – 9: Proceedings of the Ninth International Symposium on Trace Elements in Man and Animals. Edited by* P.W.F. Fischer, M.R. L'Abbé, K.A. Cockell, and R.S. Gibson. NRC Research Press, Ottawa, Canada. pp. 484–486.

We suggest that the protein may be an aggregate of several loosely associated proteins or sub-units which are sufficiently stable in column separations but dissociate during electrophoresis. One problem is that there may be a loss of Zn during the protein isolation, and that the Zn binding protein isolated as α_2-macroglobulin reactive material may simply represent a more stable Zn-binding protein and not be the abnormal protein.

The results of a liver biopsy have not confirmed the supposition that the liver is a major repository of Zn. The liver Zn concentration is 87 µg/g (ref range 50–70) and muscle Zn 95 µg/g (ref range 50–70). Liver Cu concentration is 31 µg/g (ref range 5–10) and muscle Cu 18 µg/g (ref range 1.3–2.7). Liver histology shows no evidence of liver damage. Staining for metallothionein (Elmes et al. 1987) shows a generalised increase in basal MT expression, with accentuation in the periportal region.

We have considered several therapeutic options. Any possible therapy must take into account the full range of symptoms, especially his deteriorating lung function and progressing osteoporosis. Chelation therapy, to reduce the plasma Zn concentration and remove any stored Zn, may be difficult to justify in the absence of any positive evidence for systemic Zn accumulation. Administration of chelated Zn to correct systemic Zn deficiency may be a means of overcoming a possible transport defect, but in the absence of any evidence for such a defect there can be no basis for such a treatment. Plasmapheresis could reduce the plasma concentration of the Zn-binding protein, but may only be an effective course of action if there were a supposition that the presence of this protein in the plasma were itself causing symptoms. A more drastic option to consider is a liver transplant, as this may allow normal Zn metabolism and permit a 'cure' of the condition. However, the patient's other problems dictate that if this were to be contemplated a lung transplant would also be required, and this is unlikely to occur.

We suggest that the patient suffers from a previously unreported disorder of Zn metabolism. It was initially suggested that this may be similar to Wilson's Disease or haemochromatosis, with the Zn pool identified by the stable isotope studies located in the liver. The liver Zn concentration and liver histology would appear to preclude this suggestion, and it is now apparent that, although the liver is enlarged, splenomegaly may contribute more to the abdominal enlargement. Further progress in identifying the nature of the defect depends on identification of the protein. Cultures of fibroblasts and hepatocytes have been established, and we hope to be able to use these for future metabolic studies.

Literature Cited

Clarkson, J.P., Elmes, M.E. & Jasani, B. (1985) Histological demonstration of immunoreactive zinc metallothionein liver and ileum of rat and man. Histochem. J. 17: 343–352.

Kovar, I.Z., Rauscher, A., Sampson, B., Fairweather-Tait, S. & Aggett, P. (1995) Zinc metabolism in a child with hyperzincaemia. Proceedings of the European Society for Paediatric Research, Ped. Res. 38: 440.

Morgan, W.T. (1978) Human serum histidine-rich glycoprotein. I. Interactions with heme, metal ions and organic ligands. Biochim. Biophys. Acta; 535: 319–333.

Prasad, A.S., Miale, A., Farid, Z., Schulert, A. & Sandstead, H.H. (1963) Zinc metabolism in patients with the syndrome of Fe deficiency anemia, hepatosplenomegaly, dwarfism & hypogonadism. J. Lab. Clin. Med. 61: 537–547.

Sampson, B., Kovar, I.Z., Beattie, J., Diment, J., McArdle, H.J., Ahmed, R. & Green, C. (1995) Hyperzincaemia with functional zinc depletion: a new syndrome. Proceedings of Vth COMTOX Symposium, Ann. Clin. Lab. Sci. 25: 423.

Discussion

Q1. John Chesters, Rowett Research Institute, Aberdeen, Scotland: This is absolutely fascinating. I've seen the same kinds of symptoms, but in a pig. Your Sephadex chromatography investigation was on serum or plasma?

A. We did the chromatography on heparinized plasma and with serum, and got the same results with both.

Q2. John Chesters, Rowett Research Institute, Aberdeen, Scotland: What I found with the pig, when you get this high molecular weight exchangeable zinc protein, is that if you use fluoride, it shifts completely down almost the same as albumin, but not quite. That doesn't help you at all, but it might be a partial explanation for your electrophoretic events, because I have a feeling that this high complex was an association between the heparin and another protein.

A. Yes, we thought of that one, but when we did the tests with serum, we did get the same result, so this seems unlikely.

Q3. Stephanie Atkinson, McMaster University, Hamilton, ON, Canada: Speaking of pigs, studies we'll report on this afternoon with a piglet model, trying to reverse growth and bone abnormalities that occur with dexamethasone by using growth hormone and IGF-1, we found that we could reverse the effects and attenuate the abnormalities in bone and growth with 100 μg/kg/day of growth hormone.

A. We worry about using growth hormone, and causing too much growth.

Q4. Harry McArdle, Rowett Research Institute, Aberdeen, Scotland: I can't recall if you assayed zinc enzymes, like alkaline phosphatase? Have you tried assaying Cu,Zn-SOD for example, in liver samples or something like that, or in white cell samples?

A. We found a slight elevation in plasma alkaline phosphatase. I'd be very pessimistic about getting enough blood sample to assay white cells.

Metallothionein mRNA Induction Associated with Nitric Oxide Related Pathway

K. ARIZONO, Y. MIMAMI, H. HAMADA, AND T. ARIYOSHI

Faculty of Pharmaceutical Sciences, Nagasaki University, Nagasaki 852, Japan

Keywords calcium, lipopolysaccharide, metallothionein, nitric oxide

Introduction

Metallothionein (MT) is induced by a wide variety of internal and external stimuli including metals, glucocorticoids, catecholamines, cytokines and acute phase agents. There is increasing evidence that MT can act as a free radical scavenger (Sato and Bremner 1993). Recently, we have investigated the role of calcium on the induction mechanisms of MT mRNA in several cell lines. We have revealed that calcium signaling strongly involved the mechanisms of MT mRNA induction via crosstalk among transduction pathways such as protein kinase A, protein kinase C and calmodulin dependent protein kinase (Arizono et al. 1993). Lipopolysaccharide (LPS) is known to be produced by acute stress agents and is an effective inducer of hepatic MT. Induction of MT by LPS is mediated by cytokines such as IL-1, IL-6 and TNF (Liu et al. 1991, Sato et al. 1995).

It is reported that nitric oxide (NO) is a potent biological mediator produced by hepatocytes following exposure to cytokines and LPS and that the production of NO plays a protective role against the hepatic acute phase response (Nathan 1992).

The objective of this study was to clarify that NO acts as the mediator of MT expression. The induction of MT mRNA mediated by NO associated with calcium and/or other cellular transduction mechanisms has been studied in rat primary cell cultures.

Materials and Methods

Preparation of Primary Cultures of Rat Hepatocytes

Primary cultures of rat hepatocytes were obtained essentially according to the method of Seglen (1976). Cell viability was >90%, using Trypan Blue exclusion and hemocytometer. Cells were plated on collagen-coated tissue culture plates (100 mm dia, 8×10^6 cells/plate), using serum Eagle's minimum essential medium (MEM). The cells were cultured for 48 h at 5% CO_2 – 95% Air atmosphere 37°C. Serum MEM contained 90% Eagle's minimum essential medium, 10% newborn calf serum, 10^{-7}M dexamethazone. Experimental MEM lacked serum. Medium was renewed every day, and after renewal with serum free medium, treatment with chemicals was started. Chemicals were dissolved in DMSO or distilled water.

Concentration of Chemicals Used

LPS: 0.1, 1.0, 3.0, 5.0 and 10.0 mg/L; W-7: 5.0 and 50.0 μM; TMB-8: 5.0 and 50.0 μM; H-7: 5.0 and 50.0 μM; L-NAME: 0.1 and 1.0 mM; PTIO: 5.0, 25.0 and 50.0 μM.

Metallothionein-1 mRNA Assays

Total RNA was prepared from the hepatocytes by ISOGEN™ according to the acid guanidinium thiocyanate/phenol/chloroform extraction method (Chomczynski and Sacchi 1987) after 1, 3, 6 and 12 h incubation with each chemical. Cells from triplicate plates were pooled for RNA isolation. MT-1 mRNA

Correct citation: Arizono, K., Mimami, Y., Hamada, H., and Ariyoshi, T. 1997. Metallothionein mRNA induction associated with nitric oxide related pathway. In *Trace Elements in Man and Animals – 9: Proceedings of the Ninth International Symposium on Trace Elements in Man and Animals. Edited by* P.W.F. Fischer, M.R. L'Abbé, K.A. Cockell, and R.S. Gibson. NRC Research Press, Ottawa, Canada. pp. 487–488.

was analyzed by dot-blot and/or RT-PCR analysis. To characterize induction of MT-1 mRNA, the mRNA signals on X-ray film were quantitated by NIH image after scanning.

Results

A23187 and LPS exposure led to a rapid increase of MT mRNA and peak level revealed 2.0- (A23187) and 2.5- (LPS) fold induction as compared to control at 6 h incubation. Dose dependent effects of A23187 and LPS on MT mRNA in primary cultures of rat hepatocytes were also observed. The highest induction level (2.0–2.5-fold) was seen at 3.0 μM (A23187) and 3.0 mg/L (LPS).

Protein kinase C inhibitor (H-7) did not affect the induction of MT mRNA by A23187. On the other hand, protein kinase C inhibitor (H-7), calmodulin-dependent protein kinase inhibitor (W-7) and cytosolic calcium chelator (TMB-8) did not affect the induction of MT mRNA by LPS.

MT mRNA induction by A23187 and LPS were reduced by L-NAME (a competitive inhibitor of NOS activity) and PTIO (a scavenger of NO), whereas these inhibitors alone had no effect.

Discussion

It is generally believed that hepatic MT induction is not directly caused by free radicals and that MT induction by cytokines in the liver exerts an antioxidative role during acute phase response, therefore protecting tissues from injury by oxidative stress (Sato et al. 1995). Thus prevention of tissue injury and inflammation by oxidative stress is due to the radical scavenging capability of MT in tissues and cells. It is speculated that induction of MT can be caused through NO and is related to the active oxygens produced by iNOS as one of the acute phase proteins.

In the present study, we demonstrate that MT mRNA expression by LPS is not a part of a calcium dependent pathway, and MT mRNA expression by A23187 also has a ability on the induction pathway related to NO. Finally, NO may play an important role in the complicated second messenger pathways in the MT induction. It is possible that MT induction is one of the physiological effects of NO in cytoplasm.

Literature Cited

Arizono, K., Peterson, K.L. & Brady, F.O. (1993) Inhibitors of Ca channels, calmodulin and protein kinases prevent A23187 and other inducers of metallothionein mRNA in EC3 rat hepatoma cells. Life Sciences 53: 1031–1037.

Chomczynski, P. & Sacchi, N. (1987) Single-step method of RNA isolation by acid guanidinium thiocyanate-phenol-chloroform extraction. Anal. Biochem. 162: 156–159.

Liu, L., Liu, L.P., Sendelbach, L.E. & Klaassen, C.D. (1991) Endotoxin induction of hepatic metallothionein is mediated through cytokines. Toxicol. Appl. Pharmacol. 109: 235–240.

Nathan, C. (1992) Nitric oxide as a secretary product of mammalian cells. FASEB J. 3051–3064.

Sato, M. & Bremner, I. (1993) Oxygen free radicals and metallothionein. Free Radical Biol. Med.14: 325–337.

Sato, M., Sasaki, M. & Hojo, H. (1995) Antioxidative roles of metallothionein and manganese superoxide dismutase induced by tumor necrosis factor-alpha and interleukin-1. Arch. Biochem. Biophys. 316: 738–744.

Seglen, P.O. (1976) Preparation of isolated rat liver cells. Meth. Cell Biol. 13: 29–83.

Accumulation of Mercury and Selenium in Human Thyroid

I. FALNOGA,[†] I. KREGAR,[‡] P. STEGNAR,[†] AND M. TUŠEK-ZNIDARIČ[†]

[†]Department of Environmental Sciences, and [‡]Department of Biochemistry and Molecular Biology, J. Stefan Institute, 1111 Ljubljana, Slovenia

Keywords mercury vapour, selenium, metallothionein, human thyroid

It is well known that thyroid can avidly accumulate mercury together with selenium and that molar ratios of the elements can be close to one (Byrne et al. 1995, Kosta et al. 1975). In our study accumulation of mercury was investigated in thyroid autopsy samples from one active and three retired Idrija mercury mine workers. We were interested in the cellular distribution of accumulated (retained) Hg and Se, particularly between supernatant and pellets, and the presence of metallothionein-like proteins (MT-LP).

Materials and Methods

Autopsy Cases: Thyroids were obtained at autopsy from 4 Idrija mercury mine employees, 1 with current (C) and 3 with previous (P) occupational exposure to Hg^0 vapour. The active worker (1C) aged 33 was exposed for about 3 years and died in an accident. Three retired workers (1P, 2P, 3P) aged 63–68 were exposed for 28–24 years and retired for 13–22 years. Samples were obtained less than 48 h after death and kept at –20°C until use.

Sample Preparation: At 4°C, 25% or 18% homogenates of thyroid tissues (1C, 1P, 2P, 3P) were prepared in N_2-saturated Tris HCl buffer (10 mM, pH 7.6) containing 1 mM dithiothreitol (DTT), strained through a 0.25 μm nylon net, and centrifuged at 12,000 g, 28,000 g or 100,000 g. Supernatants 1C, 1P (cytosol with microsomes) and 2P (pure cytosol) were applied to a calibrated Sephadex G-75 column (2.5 × 65 cm or 1.6 × 60 cm), and eluted with N_2-saturated Tris HCl buffer (10 mM, pH 7.6) containing 1 mM DTT at a flow rate of 13.8 or 9.8 mL/h.

Metal Analysis: Hg and Se were determined in wet tissue and cell fractions by radiochemical neutron activation analysis (RNAA) (Byrne and Kosta 1974). In gel fractions determinations of Hg by RNAA, Cu and Cd by ETAAS and Zn by flame AAS were performed.

Results

In all thyroid samples mercury was accumulated and in all analyzed pellets (with or without microsomes) the Hg/Se molar ratio was near one (Table 1). Recovery of tissue Hg and Se from thyroid with low tissue Hg concentration (current exposure 1C; active worker) and from thyroids with higher tissue Hg concentrations (previous exposure 1P, 2P, 3P; retired persons) differed significantly (Table 2). In cases with previous exposure tissue Hg and Se were retained above all in the pellet and connective tissue. In three thyroid supernatants (1C, 1P, 2P) cytosol MT was isolated and partially characterised (Table 2).

Discussion

Comparing the situation in human thyroid tissues between current and previous exposure to elemetal mercury vapour, it was found that

Correct citation: Falnoga, I., Kregar, I., Stegnar, P., and Tušek-Znidarič, M. 1997. Accumulation of mercury and selenium in human thyroid. In *Trace Elements in Man and Animals – 9: Proceedings of the Ninth International Symposium on Trace Elements in Man and Animals. Edited by* P.W.F. Fischer, M.R. L'Abbé, K.A. Cockell, and R.S. Gibson. NRC Research Press, Ottawa, Canada. pp. 489–490.

Table 1. Hg and Se in human thyroids: tissue concentrations, distribution[a] between supernatant and pellet, Hg/Se molar ratio in tissue and pellet.

Case No.	Tissue Hg μg/g	Se μg/g	Hg/Se mol.r.	Sup.[a] Hg %	Se %	Pellet[a] Hg %	Se %	Hg/Se mol.r.
1C[b]	0.51	0.28	0.72	50	47	50	53	1.16
1P[c]	3.37	1.47	0.90	26	23	74	77	1.06
2P[d]	8.48	1.95	1.71	11	31	89	69	1.19
3P[d]	25.70	5.43	1.86	–	–	–		–

C – current exp. to Hg^0 vapour, P – previous exp. to Hg^0 vapour; a – percentage defined as element in sup. or pellet over element in (sup. + pellet); b – 12 000 g, c – 23 000 g, d – 100 000 g centrifugation of homogenate.

Table 2. % of thyroid tissue Hg and Se recovered in cytosol with microsomes or in pure cytosol; the presence of cytosol MT-LP.

Case No.	Supernatant	Hg (%)	Se (%)	MT-LP
1C	cytosol + microsomes	40	23	Zn,Cu,Cd,Hg MT
1P	cytosol + microsomes	12	10	Cd,Cu,Zn MT
2P	pure cytosol	1.3	6.8	Cd,Cu,Zn MT
3P	pure cytosol	3.1	5.3	–

C – current exposure to Hg^0 vapour, P – previous exposure to Hg^0 vapour.

1. In the case of current exposure Hg and Se were almost evenly distributed between supernatant and sediment (disregarding sieved connective tissue). About 40% of tissue Hg was recovered in the supernatant and a part of it was bound to cytosol MT-LP (Table 2).
2. In the case of previous exposure Hg and Se were mostly retained in sediment (disregarding sieved connective tissue). Just a minor part of tissue Hg was recovered in supernatant (1% to about 12%) and it was not bound to cytosol MT-LP (Table 2).
3. In all cases, regardless of the kind of exposure, the Hg/Se molar ratio was close to 1 in pellet, and part of the supernatant cadmium, copper and zinc was bound to cytosol MT-LP. From these results it could be concluded that cytosol MT-LP are important for Hg accumulation in current exposure only, and that after cessation of exposure (on retirement) Hg together with Se is found mainly in cellular organelles (pellet), probably in lysosomes and/or nuclei (Eley et al. 1990). Regarding the high Hg concentrations in sieved connective tissues (results not shown), this compartment is obviously also involved in Hg and Se accumulation (retention).

References

Byrne, A.R. & Kosta, L. (1974) Simultaneous neutron-activation determination of selenium and mercury in biological samples by volatilisation. Talanta 21: 1083–1090.

Byrne, A.R., Škreblin, M., Falnoga, Al-Sabti, K., Stegnar, P. & Horvat, M. (1995) Mercury and selenium: perspectives from Idrija. Acta Chim. Slov. 42: 175–198.

Kosta, L., Byrne, A.R. & Zelenko, V. (1975) Correlation between selenium and mercury in man following exposure to elemental mercury vapour. Nature 254: 238–239.

Eley, B.M. (1990) A study of mercury redistribution, excretion and renal pathology in guinea-pigs implanted with powdered dental amalgam for between 2 and 4 years. J. Exp. Path. 71: 375–393.

Ferritins of Different Sizes and Structures in Horse Heart and Serum

L. BUTCHER, J.R. WANG, C. AMBROSI, AND M.C. LINDER
Department of Chemistry and Biochemistry, California State University, Fullerton CA 92634, USA

Keywords ferritin, heart, serum, oligomers

Ferritins are well known as the site of deposition of excess iron in cells and organisms. Most are large proteins, composed of 24 subunits arranged as a shell with 4:3:2 symmetry (Ford et al. 1984, Theil 1987). Subunits are of at least two types (L and H), each about 20 kD in size, and there is high homology between the H and L subunits that have been characterized. Amino acid sequences appear to be highly conserved not only among related organisms but across many species and phyla. Thus, polyclonal antibodies raised against one kind of ferritin usually cross-react with ferritins in other tissues of the same organism as well as with ferritins from other organisms. Southern blots of human DNA have indicated that ferritins comprise a large multigene family. Some are pseudogenes, but the products of only a few of the genes have been described and functionally assigned. This leaves room for potentially wide variations in structure and expression of ferritins by different cells in different tissues.

We have previously explored the ferritins of heart tissue and have found that it is particularly rich in terms of the variety of ferritins it expresses. As summarized in Table 1, when we analysed heart ferritins from the horse, rat and human, we discovered a larger, iron-rich ferritin, with a total molecular weight 50% greater than that of spleen and liver ferritins, as determined by sedimentation equilibrium (Linder et al. 1981). Composed mainly of H subunits of about 20 kD, it must contain 36 rather than 24 subunits. By electron microscopy it looks the same as spleen ferritin but has a diameter that is 25–30% greater (J.R. Wang and M.C. Linder, unpublished).

During these initial studies of heart ferritins, we also identified two slower-sedimenting forms, with little or no iron, that had molecular weights corresponding to that of two and eight or nine L or H-type subunits, respectively (Linder et al. 1989; Table 1). The mixture of these low molecular weight heart ferritins, however, contained not only the usual H and L subunits of 20 kD but also some larger ones that corresponded roughly in size to subunit dimers. These larger subunits are also seen in serum ferritin (Linder et al. 1997). They all react with antibody against spleen ferritin, and both sets of larger subunits appear to be glycosylated (Linder et al. 1997, Campbell et al. 1993).

In addition to these forms of ferritin, it has been known for a long time that there are dimers, trimers and oligomers of the 24-subunit ferritins in spleen and liver (Munro and Linder 1978). These appear as clean bands in non-denaturing polyacrylamide gel electrophoresis (PAGE), migrating more slowly than the bulk of the ferritin "monomer" (which has 24 subunits) (Figure 1). By electron microscopy, the ferritin extracted from these bands appears to be composed of pairs, triplets, and higher aggregates of the donut-like 24-subunit ferritin. The functions and origins of these ferritins has never been clear, nor have

Table 1. Forms of ferritin in horse heart and serum.

	Fe:Protein (μg/μg)	Sedimentation coefficient	Molecular weight	
			Total	Subunits
Heart apoferritins	0.25–0.3	18 S	480 & 700 k	20 & 22 k
	0.01	3 S & 7 S	40 & 160 k	20, 22, 43, 55, 65 k
Serum ferritin	0.0006	7 S	140 k	26, 50, 57, 65 k

Correct citation: Butcher, L., Wang, J.R., Ambrosi, C., and Linder, M.C. 1997. Ferritins of different sizes and structures in horse heart and serum. In *Trace Elements in Man and Animals – 9: Proceedings of the Ninth International Symposium on Trace Elements in Man and Animals. Edited by* P.W.F. Fischer, M.R. L'Abbé, K.A. Cockell, and R.S. Gibson. NRC Research Press, Ottawa, Canada. pp. 491–492.

Figure 1. Non-denaturing PAGE of purified horse spleen ferritin, showing the migration of monomers, dimers and trimers. From left to right, bands seen are the trimer, dimer and monomer.

disparate findings about the nature of the bonds between them ever fully been resolved (see Munro and Linder 1978). In the studies here reported, we attempted (a) to establish whether heart ferritins also have these larger ferritin dimers, trimers, and oligomers; (b) to further characterize the nature of the bonds holding them together; and (c) to determine whether dimers were comprised of the same kinds of ferritin units as the monomers.

Horse heart and spleen ferritins were purified by classical procedures, involving 70° heat and pH 4.8 treatments, followed by ammonium sulfate precipitation and several rounds of size exclusion chromatography on large-pore gels. Figure 1 shows a non-denaturing PAGE gel of the leading edge of the spleen ferritin peak eluting from size exclusion FPLC. The most prominent band is the "monomer". That migrating half as fast is the dimer, and that migrating half again as fast is the trimer (as determined by electron microscopy). Purified heart ferritin had the same kinds of bands (data not shown), indicating that it also formed these larger aggregates. Attempts to fully separate trimers and dimers from each other and from monomers, on various size exclusion gels, were unsuccessful. We therefore electroeluted them from unstained, non-denaturing tube gels. This preparative PAGE procedure was carried out at both pH 8.8 and pH 7.0, with the same results. Re-electrophoresis (in the same non-denaturing system) determined that the trimers and dimers could be partly disaggregated, even with these mild treatments. Most of the trimers disappeared, with dimers appearing to be more stable. This indicates that some of the ferritin molecules are linked via non-covalent interactions. On the other hand, dilution of the ferritin samples up to 16-fold in pH 7 phosphate prior to electrophoresis, failed to change the ratios of trimers, dimers and monomers, and there appeared to be a residue of dimers that did not dissociate.

Dimer/trimer-enriched preparations were then compared with monomers in SDS-PAGE. Some additional larger subunits were observed, and we hypothesized that they might represent the the covalently cross-linked units that were holding the undenatured dimers and trimers together. However, peptide mapping of TPCK-trypsin digests gave identical profiles, showing the 21 peaks expected for horse spleen ferritin. Although the work is continuing, the results so far suggest that most of the binding between the dimers and trimers is due to non-covalent interactions.

Literature Cited

Campbell, C.H. Crocker, D., Gruntmeir, J.J., Head, M., Kelly, T., Langfur, M. & Leimer, A.H. (1993) Purification of a novel glycosylated ferritin from horse heart. J. Cell. Biochem. 53: 420–432.

Ford G.C., Harrison, P.M., Rice, D.W., Smith, J.M.A., Treffry, A., White, J.L. & Yariv, J. (1984) Ferritin: design and formation of an iron-storage molecule. Phil. Trans. R. Soc. London Ser. B 304: 551–565.

Linder, M.C., Hazegh-Azam, M., Zhou, C.Y.J. & Nagel, G.M. (1997) Structure and cloning of serum ferritin. In: Trace Elements in Man and Animals – 9: Proceedings of the Ninth International Symposium on Trace Elements in Man and Animals. (Fischer, P.W.F., L'Abbé, M.R., Cockell, K.A., and Gibson, R.S., eds.). NRC Research Press, Ottawa, Canada. pp. 384–385.

Munro, H.N. & Linder, M.C. (1978) Ferritin: Structure, biosynthesis, and role in iron metabolism. Physiol. Rev. 58: 317–396.

Theil, E.C. (1987) Ferritin: structure, gene regulation and cellular function in animals, plants and microorganisms. Annu. Rev. Biochem. 56: 289–315.

Role of Trace Elements in Growth and Development

Chairs: S.A. Atkinson and J.K. Friel

Zinc Homeostasis in Infants*

N.F. KREBS

Section of Nutrition, Department of Pediatrics, University of Colorado School of Medicine, Denver, CO 80262, USA

Keywords zinc, homeostasis, breastfeeding, isotopes

Consideration of zinc homeostasis in the infant must include consideration of the amount retained for growth, which differs at different ages; dietary intake, both amount and form; and the metabolic adaptations in absorption, secretion and excretion to changes in either demands or supply. This paper will review each of these areas, with emphasis on results from stable isotope studies performed in normal infants.

Because of the role of zinc in the synthesis of lean tissue, the rate of growth, specifically accretion of lean body mass, has critical implications for net absorption of zinc and ultimately dietary zinc requirements. Thus the normally growing young infant, gaining initially an average 30 g/d, has exceptionally high zinc requirements and is dependent on a generous intake of bioavailable zinc to achieve normal growth rates. Growth velocity steadily declines, however, and the estimated zinc required for growth thus also declines. The composition of the weight gain also affects the zinc requirements, with less need for fat accumulation.

The concentration of Zn in human milk briskly declines over the early weeks of lactation and then more slowly declines after 5 months post partum. In well nourished populations, milk zinc concentrations are not very sensitive to maternal intake. Supplementation of maternal diet with zinc is unlikely to impact milk zinc concentrations unless there is an underlying maternal zinc deficiency (Krebs et al. 1995a, Krebs et al. 1985). The decline in concentrations is not entirely compensated for by increase in volume of milk consumed. Exclusively breastfed infants in Denver were found, on the basis of 72 h test weighing, to have intakes of zinc from human milk decline from 2.3 ± 0.68 mg/d at 2 week to 1.0 ± 0.43 and 0.81 ± 0.42 mg/d at 3 and 5 months, respectively. On a body weight basis, the zinc intakes at these time points were approximately 0.6, 0.16 and 0.12 mg/kg/d, respectively (Krebs et al. 1994). In contrast, zinc intakes from formulas, with typical fortification in North American products at levels of ≥5 mg/L, are several fold higher than that of breastfed infants beyond the very early weeks of life. Extrapolation from reported volume intakes from a large group of formula fed infants yields daily intakes of approximately 0.75 and 0.60 mg/kg/d at 3 and 6 months, respectively (Heinig et al. 1993). The highest intakes relative to body weight are in premature infants because of the specially designed formulations with very high zinc fortification, resulting in intakes at about 2 mg/kg/d (Friel et al. 1996 and Krebs, unpublished data).

What are the metabolic adaptations to such ranges in intake? The net or overall response will depend on the needs of the infant and its rate and composition of growth, which is obviously related to many other factors, including dietary intake of energy, protein and other nutrients. The major site of homeostatic regulation for zinc is in the gastrointestinal tract, specifically in the small intestine, through modulation

*Supported in part by National Institutes of Health grants DK12432, DK07658, DK02240, DK48520, RR00069 and by the Pew Charitable Trusts Nutrition Fellowship Program.

Correct citation: Krebs, N.F. 1997. Zinc homeostasis in infants. In *Trace Elements in Man and Animals – 9: Proceedings of the Ninth International Symposium on Trace Elements in Man and Animals. Edited by* P.W.F. Fischer, M.R. L'Abbé, K.A. Cockell, and R.S. Gibson. NRC Research Press, Ottawa, Canada. pp. 493–496.

of fractional absorption of dietary zinc and through secretion and reabsorption of endogenous zinc. To a lesser extent, the kidney can conserve or excrete zinc. The latter may be influenced by medications, disease states and, in the young (<2 months old) premature infant, by renal immaturity (Hambidge, unpublished data). In the normal term infant, however, urine excretion of zinc is likely <10 μg/kg/d (Krebs and Hambidge 1986).

A number of studies in our lab have been undertaken to examine and quantify the gastrointestinal homeostatic responses under different feeding conditions. These investigations have used stable isotope techniques first validated in adult subjects and then applied to studies in infants. The specifics of the methodology have been published (Krebs et al. 1995b, Krebs et al. 1996). Briefly, accurately measured quantities of zinc stable isotope were administered with feeds over 8–24 h as an extrinsic label in human milk or formula. Complete fecal collections were made for 8 d after administration of isotope; urine was collected for the last 3–4 d of the collection period; intake of milk or formula was determined by 72 h test weighing for breast fed infants or by weighing bottles before and after feeds for the formula fed infants; and samples of milk or formula for zinc analysis were obtained with each feed. After careful sample preparation including ashing and isolation of zinc, isotopic enrichment was determined by measurement of isotope ratios by fast atom bombardment-induced secondary ion mass spectrometry on a double-focussing mass spectrometer equipped with an atom gun (Krebs et al. 1996). Calculations from these studies yielded dietary zinc intake, fractional absorption, total absorbed zinc, endogenous fecal zinc, and net absorption (Krebs et al. 1995b, Krebs et al. 1996).

Results for normal 3–5 month breastfed and formula fed infants indicate that although the dietary intake of the formula fed infants, was ~5.5 fold that of the breastfed infants (1.2 ± 0.45 mg/d), the fractional absorption of 0.52 ± 0.06 vs. 0.25 ± 0.07, for the breastfed and formula fed groups, respectively, substantially reduces the difference in total absorbed zinc, but the formula fed group still averages about 2.5 times greater absorbed zinc. The breastfed infant with the lowest dietary zinc intake exhibited a fractional absorption of 0.88, but this still resulted in the lowest total absorbed zinc. The differences in fractional absorption between the types of feeds are likely due to a number of factors, including the difference in quantity of natural abundance zinc; the difference in the dose of isotope administered (although that was approximately an equal percentage of the total zinc intake for both groups); the differences in the composition of the formula and human milk, including type and quantity of protein, micronutrient composition, and other factors. Since all of the infants in these groups were approximately 4 months of age, maturational differences of the gastrointestinal tract are unlikely to have played a role.

Further adjustment in zinc homeostasis is seen for these infants with the figures for endogenous fecal excretion of zinc, which at 0.31 ± 0.15 mg/d for the breastfed infants, was only about 25% that of the formula fed infants. For both groups, the total absorbed zinc is significantly positively correlated with the endogenous fecal zinc. The mean net absorption for the formula fed infants was only ~1.5 times greater than that of the breastfed infants. Thus, the ability of the breastfed infants to achieve positive net absorption is dependent on the combination of both high fractional absorption and very low excretion of endogenous zinc. In contrast, the formula fed infants predictably respond to the relatively generous intake of dietary zinc by both reducing fractional absorption and increasing excretion of endogenous zinc.

Evaluation of the net absorption figures for the breastfed and formula fed infants indicates that the latter, despite lower fractional absorption and higher excretion of endogenous zinc, still retain significantly more zinc. The adequacy of the retention must be judged in relation to needs, ie in relation to the zinc needed to support growth and to replace urine and integumental losses. The average net absorbed zinc of the breastfed infants appears just adequate to maintain normal growth. For example, the mean rate of growth for these infants was 16 g/d. Applying a figure of 20 μg zinc/g of total weight gain, the mean net absorption of 290 seems quite plausible. The mean daily increments in weight gain for a larger group of exclusively breastfed male infants at this age mean was 14 ± 5.7 g/d (Krebs et al. 1994). A projected need for retained zinc for that rate of growth would be ~0.28 mg/d. Thus, for the exclusively breastfed infant at 4–5 months of age, the amount of zinc for retention and replacement of urine and integumental losses, i.e., net absorption, is approximately 0.30 mg/d (Figure 1). Although the absorption studies in normal infants have all been performed in males, the figure for females is likely to be similar since growth rates were quite similar at this interval in two large studies (Dewey et al. 1992, Krebs et al. 1994). Applying

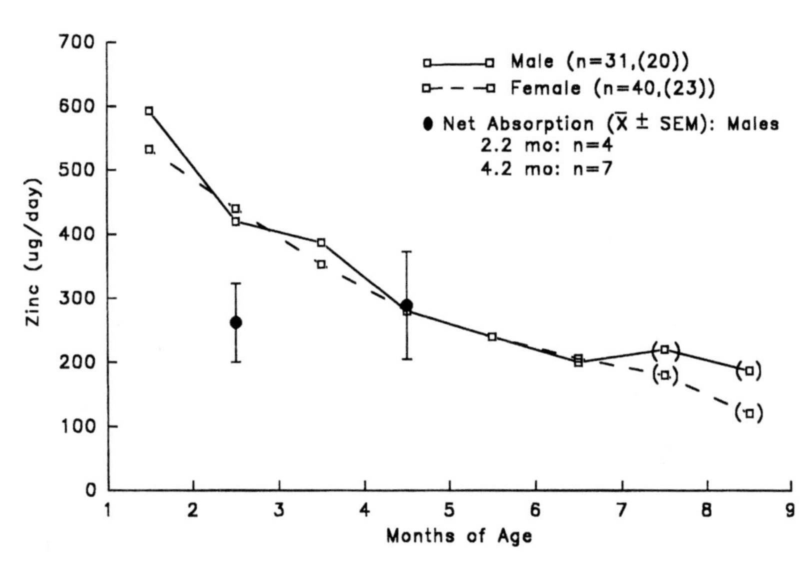

Figure 1. Estimated requirements of zinc for retention for growth (reference Krebs et al. 1994) plus mean net absorption figures for 2 and 4 mo old breastfed male infants.

similar calculations to the formula fed infants, reported mean daily weight increments at 4–5 months of age were approximately 18 g/d (Dewey et al. 1992), which would result in a projected need for zinc retention for growth of ~0.37 mg/d. The mean net absorption of the formula fed infants would comfortably meet this with additional available to replace losses and perhaps result in additional tissue "stores."

Absorption studies in younger (2.2 ± 0.2 months) exclusively breastfed infants, yielded a net absorption of 0.26 ± 0.12 μg/d. As shown in Figure 1, that amount alone seems unlikely to be sufficient to support typical growth rates of 20–25 g/d (Dewey et al. 1992, Krebs et al. 1994). Several considerations may be relevant. Liver metallothionein may provide an additional source of zinc to meet the infant's needs. The rate of growth of these infants (during the study period) averaged 25 g/d, making unusually slow growth an unlikely explanation, but the composition of the weight gain may have favored fat deposition rather than gain in lean body mass. Technical difficulties or perturbation with normal intake must always be considered and studies on a larger number of subjects may have yielded different results.

In summary, the application of stable isotope techniques to metabolic studies in infants can provide new insight into zinc homeostasis and how it is maintained by infants under different conditions. Consideration of the adequacy of net absorption of zinc must account for not only endogenous losses but also the amount of zinc needed for retention to support growth, which will differ according to age and the composition of weight gain.

Literature Cited

Dewey K.G., Heinig M.J., Nommsen L.A., Peerson J.M. & Lonnerdal B. (1992) Growth of breast-fed and formula-fed infants from 0 to 18 months: The DARLING study. Pediatrics 89:1035–1041.

Friel J.K., Andrews W.L., Simmons B.S., Miller L.V. & Longerich H.P. (1996) Zinc absorption in premature infants: comparison of two isotopic methods. Am. J. Clin. Nutr. 63: 342–347.

Heinig M.J., Nommsen L.A., Peerson J.M., Lonnerdal B. & Dewey K.G. (1993) Energy and protein intakes of breast-fed and formula-fed infants during the first year of life and their association with growth velocity: the DARLING study. Am. J. Clin. Nutr. 58: 152–161.

Krebs N.F., Reidinger C.J., Miller L.V. & Hambidge K.M. (1996) Zinc homeostasis in breast-fed infants. Pediatr. Res. 39: 661–665.

Krebs N.F., Reidinger C.J., Hartley S., Robertson A.D. & Hambidge K.M. (1995a) Zinc supplementation during lactation: effects on maternal status and milk zinc concentrations. Amer. J. Clin. Nutr. 61: 1030–1036.

Krebs N.F., Miller L.V., Naake V.L., Sian L., Westcott J.E., Fennessey P.V. & Hambidge K.M. (1995b) The use of stable isotope techniques to assess zinc metabolism. J. Nutr. Biochem. 6:292–301.

Krebs N.F., Reidinger C.J., Robertson A.D. & Hambidge K.M. (1994) Growth and intakes of energy and zinc in infants fed human milk. J. Pediatr. 124:32–39.

Krebs N.F. & Hambidge K.M. (1986) Zinc requirements and zinc intakes of breast-fed infants. Am. J. Clin. Nutr. 43:288–292.
Zlotkin S.H. & Cherian G.M. (1988) Hepatic metallothionein as a source of zinc and cysteine during the first year of life. Pediatr. Res. 24: 326–329.

Discussion

Q1. Ine Wauben, McMaster University, Hamilton, ON, Canada: I have a little bit of a problem with your rationale for supplementing zinc to healthy term babies. I assume the rationale is that the babies were growing slower than the NCHS references, but as I understand it those references were developed on formula-fed infants. Do you think that the nutritional quality of breast milk is declining? Is that why you want to supplement with zinc? Breast milk is the best thing, so why do you want to change Mother Nature?

A. I have no intention of changing Mother Nature. I think, Ine, that you've put your finger right on the controversy, that growth charts are developed from formula-fed infants. We have to struggle with the idea of maximal growth to full genetic potential, or some notion of optimal growth pattern. Traditionally, all the zinc deficiency studies in infants and children have been based on growth response. If you're going to accept that we should grow to meet potential, then you may have to accept that some of these infants are not getting optimal amounts. I really think that this is a very controversial concept. Clearly breast milk is the best way to feed a human infant, but I think that there are probably many people in this room, and many public health people, and many pediatricians who are happy to believe that iron intake is not adequate after the middle of the first year of life. I think that it's much harder to study zinc. All I'm trying to say is that probably the zinc intake by that stage is also in need of help.

Q2. Stephanie Atkinson, McMaster University, Hamilton, ON, Canada: Have you plotted your data on the new WHO growth charts for breast-fed infants?

A. No, I haven't. Our data are in that pool, but I haven't actually plotted these results onto that.

Q3. Barbara Golden, University of Aberdeen, Scotland: You showed that females had a response, but males didn't. You've shown the opposite before, and others too have shown that males seemed to be more sensitive. But you didn't tell us what the somatomedin concentrations were in males versus females.

A. The numbers are quite small. We haven't done enough yet to really break it down by gender. In terms of why we didn't see a difference in the males, I don't know. It's not without precedent, although it is unprecedented in Denver. I think James has seen some effects only in females, and other investigators have as well. The growth rates don't actually back up our thinking that the males have higher requirements, because they were about the same.

Zinc–Iron Interaction in the Premature Infant*

J.K. FRIEL,[†] P.V. FENNESSEY,[‡] L.V. MILLER,[‡] W.L. ANDREWS,[†] B.S. SIMMONS,[†] AND R.E. SERFASS[¶]

[†]Departments of Biochemistry, Pediatrics, Memorial University, St. John's, Newfoundland, Canada A1B 3X9; [‡]University of Colorado, Denver, Colorado 80262, USA; [¶]Iowa State University, Ames, Iowa 50010, USA

Keywords zinc, iron, interaction, infant

Introduction

Zinc and iron are both essential for growth and development in the premature infant yet elevated levels of either element in the diet may interfere with absorption of the other (Davis 1980). Since 1976, the zinc content of some formulas has increased ten-fold while the concentration of iron in many cases has remained the same so that the zinc:iron ratios are approximately 4:1 (Table 1).

Most investigators to date have been concerned with the effects of excess iron on zinc status particularly those who may be given iron supplements routinely (Solomons 1986). However, the interaction of these two metals can be bi-directional. Animal studies have shown distinct effects of excess zinc on iron metabolism within the intestinal mucosa such that incorporation of iron into or release from ferritin is impaired (Bafundo et al. 1984). The little work that has been done in humans supports excess zinc inhibiting iron absorption (Aggett et al. 1983). Therefore the purpose of this project was to determine if zinc to iron ratios of 4:1 would inhibit iron absorption when given both with and without regular foods for infants of this age.

Table 1. The zinc and iron content of premature formulas (24 kcal).[1]

	SCC	ENF	SMA
ZINC	12	8	8
IRON	3	2	3

[1]Similac Special Care, Ross Labs, Columbus, OH; Enfamil Premature, Mead Johnson, Evansville, IL; SMA Preemie, Wyeth Ltd, ON.

Subjects and Methods

Subjects

Full ethical approval was obtained from Hospital and University Ethics committees. Informed consent was obtained from guardians.

Experiment 1. Five premature infants (<2500 g birth weight) were enrolled once stable on full oral feeds (100–120 kcal kg^{-1} d^{-1}) and receiving uniform management in the neonatal intensive care unit. These infants received either high (1200 µg/kg) or low (300 µg/kg) doses of oral zinc sulphate, concurrently with 300 µg/kg oral ^{58}Fe as chloride with 10 mg/kg vitamin C. The vitamin C keeps iron in the reduced state and 300 µg/kg oral ^{58}Fe should provide enough enrichment for analytical precision (Ehrenkranz et al. 1992). Because of the known interindividual variation in iron absorption, in this crossover study, each infant served as its own control. Subjects were randomized to receive either the high or low zinc

*Supported by the Medical Research Council of Canada.

Correct citation: Friel, J.K., Fennessey, P.V., Miller, L.V., Andrews, W.L., Simmons, B.S., and Serfass, R.E. 1997. Zinc–iron interaction in the premature infant. In *Trace Elements in Man and Animals – 9: Proceedings of the Ninth International Symposium on Trace Elements in Man and Animals. Edited by* P.W.F. Fischer, M.R. L'Abbé, K.A. Cockell, and R.S. Gibson. NRC Research Press, Ottawa, Canada. pp. 497–500.

doses, 2 weeks apart, with the same amount of iron per kilogram bodyweight. Zinc and iron were given at baseline and two weeks later, by nasogastric tube in between feeds flushed with saline.

Experiment 2. Eight further premature infants were enrolled and followed the same protocol except the zinc and iron were given with and equilibrated in normal oral feeds (formula or human milk) prior to feeding.

Methods

Enriched isotope as the oxide was purchased from two sources (Cambridge Isotope Laboratories, MA, Atomergic Chemetals Corp, NY.). The enriched oxide was dissolved in a small amount of Aqua Regia, pH adjusted to 5, filtered through a 0.2 micron filter, tested for pyrogens and stored in glass vials until use. Iron concentration was determined by AAS and enrichment calculated from manufacturers data. For Experiment 2, the amounts of zinc intrinsic to the formula were accounted for in preparing the hypothesized ratios.

Blood samples (1 mL) were collected at baseline, 2 and 4 weeks later. Blood samples were prepared in the following format; Experiment 1) Iron was extracted from blood samples by ether extraction and ion exchange chromatography, then analysed by FAB-SIMS (Denver) for $^{58}Fe/^{54}Fe$ ratios (Flory et al. 1993). Experiment 2) Iron was extracted using thenoyltrifluoroacetone xylene and analysed by ICP-MS (Ames) for $^{58}Fe/^{57}Fe$ ratios (Fomon et al. 1995). For both preparations the amount of iron was determined by AAS and enrichment determined from manufacturers specifications.

The following assumptions were made: (1) Of the iron that is absorbed after 14 d 90% is incorporated into red blood cells (RBC), (2) RBC enrichment 14 d after the first dose remains constant for the next 14 d (Fairweather-Tait et al. 1992).

The quantity of administered ^{58}Fe label incorporated into erythrocytes ($^{58}Fe^*_{inc}$) 14 d after each dose was calculated as follows (Ehrenkranz et al. 1992);

$$^{58}Fe^*_{inc} = IR_t - IR_0 / IR_0 \times Fe_{circ} \times 0.0029244$$

IR_t is the ratio 14 d after isotope dosing, IR_0 is the baseline ratio, Fe_{circ} is total circulating iron at time "t", 0.0029244 is derived from the atomic weight of iron using atom% (Fomon et al. 1995).

Total circulating iron was estimated as

$$Fe_{circ} = BV \times Hb \times 3.47$$

BV is blood volume, assumed to be 0.85 L/kg, Hb is hemoglobin concentration in g/L, 3.47 is the concentration of iron in Hb (mg/g).

Differences between high and low zinc intakes were assessed with crossover ANOVA (SAS Version 6, SAS Institute, Cary NC). Probability was assigned to $P < 0.05$.

Table 2. Mass and abundance (%) of natural iron and enriched ^{58}Fe.[1]

Isotope mass	Natural abundance[2]	Enriched abundance
54	5.61	0.001
56	91.9	0.045
57	2.18	6.5
58	0.322	93.05

[1]Cambridge Isotope Laboratories, MA, Atomergic Chemetals Corp, NY.
[2]Taylor et al. (1992).

Results

Infants in the first group weighed 1068 ± 356 g at birth and were 28 ± 3 weeks gestational age. Infants in the second group weighed 1262 ± 397 g at birth and were 31 ± 2 weeks gestational age. Infants in the second group were heavier at study entry (1947 ± 253 g) compared to infants fed between feeds (1563 ± 209 g). The average feed volume at baseline was 39 ± 14 mL and at week 2 was 55 ± 21 mL. All infants gained about 30 g/d during the study period.

Table 3. *$^{58}Fe/^{54}Fe$ ratios and erythrocyte incorporation on high (HI) and low (LO) zinc:iron intakes given in between feeds.*

Subj #	Zinc Dose	Base $^{58}Fe/^{54}Fe$	Week 2 $^{58}Fe/^{54}Fe$	Week 4 $^{58}Fe/^{54}Fe$	^{58}Fe Inc (Low)	^{58}Fe Inc (High)
1	Low/High	0.0552	0.0656	0.0700	7.5%	2.3%
2	Low/High	0.0552	0.0657	0.0748	6.8%	5.1%
3	High/Low	0.0549	0.0600	0.0756	11.6%	5%
4	High/Low	0.0543	0.0593	0.0712	7.6%	3.2%
5	Low/High	0.0550	0.0609	0.0666	5.3%	3.4%

Table 4. *$^{58}Fe/^{57}Fe$ ratios and erythrocyte incorporation on high (HI) and low (LO) zinc:iron intakes given with feeds.*

Subj #	Zinc Dose	Base $^{58}Fe/^{57}Fe$	Week 2 $^{58}Fe/^{57}Fe$	Week 4 $^{58}Fe/^{57}Fe$	^{58}Fe Inc (Low)	^{58}Fe Inc (High)
1	High/Low	0.1341	0.1980	0.2460	6.7%	14.7%
2	Low/High	0.1341	0.1956	0.3167	18%	17%
3	Low/High	0.1334	0.1395	0.1460	1.5%	1.2%
4	High/Low	0.1354	0.1441	0.1602	3.2%	2.4%
5	Low/High	0.1238	0.1538	0.2008	4.9%	8.3%
6	High/Low	0.1309	0.2765	0.2925	1.8%	32%[1]
7	Low/High	0.1334	0.2760	0.3153	33%[2]	4%
8	Low/High	0.1295	0.1896	0.2316	14.1%	6.7%

[1]This infant was given human milk.
[2]Taylor et al. (1992).

From Table 3, it can be seen that high zinc intakes significantly inhibited iron absorption when given in between feeds. The amount of ^{58}Fe incorporation with the low zinc:iron intake was 8 ± 2.3%. The amount of ^{58}Fe incorporation on the high zinc:iron intake was 4 ± 1.2%.

From Table 4, it can be seen that high zinc intakes did not inhibit iron absorption when given with feeds. The amount of ^{58}Fe incorporation on the low zinc:iron intake was 10 ± 11%. The amount of ^{58}Fe incorporation on the high zinc:iron intake was 11 ± 10%.

Discussion

From the above results it can be seen that zinc and iron given together, between feeds will interact with each other. Similar elements may share common pathways for absorption so that the co-ingestion of two or more elements will result in competition for uptake into the mucosal cells (Aggett et al. 1983, Solomons 1986). Zinc has been shown to interfere with the intestinal uptake of iron in adults (Crofton et al. 1989).

Levels of RBC incorporation are similar to those reported by Fomon et al. (1989) in term infants of 7.9% (3.2–16%).

Once feeds are introduced this interaction is negated, probably due to the inclusion of other nutrients and dietary factors present in milk feeds. Levels of RBC incorporation from enriched formula are reported to be 7% (Fomon et al. 1993), 4.4 ± 3.9% (Fomon et al. 1995) and 12 ± 10% (Ehrenkranz et al. 1992). It is noteworthy that the absorption of the extrinsic iron increased when mixed with human milk. There are lesser quantities of inhibitors of iron absorption (casein, calcium, whey proteins) in human milk than formula (Fomon et al. 1993). Thus the whole diet must be taken into consideration when evaluating the effect of one nutrient upon another. In conclusion, iron should be given to infants in between feeds with vitamin C alone.

Literature Cited

Aggett, P., Crofton, R.J., Khin, C., Gvozdanovic, S., & Gvozdanovic, D. (1983) The mutual inhibitory effects on their bioavailability of inorganic zinc and iron. In: Zinc deficiency in human subjects. (Prasad, A., Cavdar, A., Brewer, G., & Aggett, P., eds), pp 117–124, Alan R. Liss, NY.

Bafundo, K.W., Baker, D.H., & Fitzgerald. (1984) The iron-zinc relationship in the chick as influenced by Eimeria ascervulina infection J. Nutr. 114: 1306–1312.

Crofton, R.W., Gvozdanovic, D., Gvozdanovic, S., Khin, C.C., & Aggett, P.J. (1989) Inorganic zinc and the intestinal absorption of ferrous iron. Am. J. Clin. Nutr. 50: 141–144.

Davis, N. (1980) Trace element interactions in man. N.Y. Acad. Sci 56: 342–345.

Ehrenkranz, R.A., Gettner, P.A., Nelli, C., Sherwonit, E.A., Williams, J.E., Pearson, H.A., Ting, B.T.G., & Janghorbani, M. (1992) Iron absorption and incorporation into red blood cells by very low birthweight infants: Studies with the stable isotope ^{58}Fe. J. Pediatr. Gastrol. Nutr. 15: 270–78.

Fairweather-Tait, S.J., Powers, H.J., Minski, M.J., Whitehead, J., & Downes, R. (1992) Riboflavin deficiency and iron absorption in Gambian men. Ann. Nutr. Met. 36: 34–40.

Fairweather-Tait, S.J. (1995) Iron-zinc and calcium interactions in relation to Zn and Fe absorption. Proc. Nutr. Soc. 54: 456–473.

Flory, D.R., Miller, L.V., & Fennessey, P.V. (1993) Development of techniques for the isolation of iron from biological material for measurement of isotope ratios by fast atom bombardment mass spectrometry. Anal. Chem. 65: 3501–3504.

Fomon, S.J., Janghorbani, M., Ting, B.T.G., Ziegler, E.E., Rogers, R.R., Nelson, S.E., Ostedgaard, L.S., & Edwards, B.B. (1988) Erythrocyte incorporation of ingested 58-iron by infants. Ped. Res. 24(1): 20–24.

Fomon, S.J., Ziegler, E.E., Nelson, S.E., Serfass, R.E., & Frantz, J.A. (1995) Erythrocyte incorporation of iron by 56-day-old infants fed a ^{58}Fe-labeled supplement. Ped. Res. 38: 373–378.

Fomon, S.J., Ziegler, E.E., & Nelson, S.E. (1993) Erythrocyte incorporation of ingested ^{58}Fe by 56-day-old breast-fed and formula-fed infants Ped. Res. 33: 573–576.

Solomons, N.W. (1986) Competitive interactions of iron and zinc in the diet: Consequences for human nutrition. J. Nutr 116: 927–935.

Taylor, P.D.P., Maeck, R., & De Bievre, P. (1992) Determination of the absolute isotopic composition and atomic weight of natural iron. Int. J. Mass Spec. Ion Proc. 121: 111–125.

Discussion

Q1. Lena Davidsson, Nestle Research Centre, Lausanne, Switzerland: James, don't you think it's difficult to present data like you did just now, when you say that there's such a big difference between breast-fed infants and formula-fed infants, but you grouped them together for the mean values? Don't you need to add more infants to your study, and to look at breast-fed infants and formula-fed infants to give more strength to your data?

A. Yes, it would be a good idea and we are considering doing more infants. We did not get consistent results with respect to high iron absorption. The results are very puzzling indeed. That is what makes science so exciting.

Q2. John Sorenson, University of Arkansas, Little Rock, AR, USA: Is the role of vitamin C in absorption to reduce Fe(III) to Fe (II) and to keep it in that oxidation state, or is there something known about Fe (II) forming a complex with vitamin C, and that the complex is more readily absorbed?

A. I think that you're more of an expert on that than me. My assumption was from the literature that it keeps it in a reduced state and that is why we give it. I've never thought of an iron-vitamin C ligand that would stimulate absorption.

(comment by Paul Saltman, University of California at San Diego, CA, USA: It does. It's been well characterized. Diketogulonate as well as ascorbate itself are both very good chelators for the +++, and nobody has measured the ++. As long as there's enough vitamin C around to keep it reduced, you've got that constant issue of the solubility of Fe (II) versus Fe(III), the reactivity in the presence of oxygen, you could have a whole bunch of ligands... . If you feed only ascorbate, then it's the only ligand there, plus the endogenous ligands secreted by the gut, not well characterized. And so it can act in both ways.

Q3. John Sorenson, University of Arkansas, Little Rock, AR, USA: The absorption of iron then should depend upon the amount of vitamin C if vitamin C is limiting?

(comment by **Paul Saltman**, University of California at San Diego, CA, USA: That's been shown too. The problem is, everybody always wants the bumper-sticker answer, but all these things are happening at once.

Q4. John Sorenson, University of Arkansas, Little Rock, AR, USA: So why doesn't somebody feed the iron complex of vitamin C?

(comment by **Paul Saltman**, University of California at San Diego, CA, USA: Which one?

Q5. John Sorenson, University of Arkansas, Little Rock, AR, USA: Well, you'd imagine that in the gastrointestinal tract, that's aerobic, so the iron would continually be re-oxidized once it's been reduced, so it would consume vitamin C. If you could determine how much vitamin C it took for maximal absorption, so you'd have some stoichiometry with regard to what's happening, with regard to absorption.

(comment by **Paul Saltman**, University of California at San Diego, CA, USA: We tried to do the in vitro experiment with C and iron in various rations, and tried to factor in all of the problems when you get diketogulonate being synthesized as part of that process. And the answer is it's changing all of the time, and what ultimately happens when you run out of ascorbate is it's all ferric iron, and it's all bound to diketogulonate or dehydroascorbate. But again, we can't tell you the difference. There's just no simple answer to this issue.

Zinc Needs of Premature Infants to One Year of Age

STEPHANIE A. ATKINSON, INE P.M. WAUBEN, AND BOSCO PAES
Department of Pediatrics, McMaster University, Hamilton, Ontario, Canada L8N 3Z5

Keywords zinc, premature infants, mother's milk, hair zinc

The basis of establishing optimal zinc needs of prematurely born infants should be the amount of dietary zinc required to meet physiological needs for growth and fetal development. Other factors to be considered in estimating zinc needs of premature infants are: variations in bioavailability of zinc due to the chemical form of zinc in specific dietary sources; interactions of zinc with other nutrients, especially other dication metals; and the impact of drugs used in the neonatal period, such as steroids, on zinc absorption and utilization.

Nutritional zinc deficiency has been observed in premature infants (summarized in Atkinson et al. 1989). These infants presented with delayed growth, perioral and perineal acrodermatitis, irritability, delayed wound healing, hypozincemia and low plasma alkaline phosphatase. The etiology was clearly a dietary deficiency since infants displayed a dramatic and complete response both biochemically and clinically to oral supplements of zinc sulphate (even when only 15 µmol (1 mg)/kg/d of elemental zinc was prescribed (Atkinson et al. 1989)). The uniqueness of the reported cases is that all infants were receiving their mothers' milk; however, the concentration of milk zinc in these cases was abnormally low (Figure 1). Although relatively rare, the reported cases of zinc deficiency in premature infants emphasizes the essentiality of zinc to the rapidly growing infant and the need to accurately define zinc needs of such infants during the early critical stages of growth and development.

Intrauterine Accretion of Zinc to Estimate Preterm Infant Zinc Needs

The greatest proportion of body zinc is accumulated during the third trimester of pregnancy; thus, prematurely born infants have lower hepatic stores of zinc on a body weight basis than term born infants (Zlotkin and Cherian 1988). Such a storage pool of zinc could easily be depleted during early neonatal life of the premature infant. If Widdowson's (1988) revised predictions of zinc accretion between 24 and 34 weeks of gestation of 13 µmol (850 µg)/d are used as a basis to derive recommended intakes, zinc accretion would be 22 µmol/kg/d at 24 weeks of gestation declining to 8 to 5 µmol/kg/d at 32 to 34 weeks. When these values are adjusted for efficiency of dietary zinc absorption (14% from preterm formula and 36% from fortified preterm human milk (Ehrenkranz et al. 1989) and urine losses, recommended intakes at 24 weeks of gestation would be of the order of 160 and 60 µmol/kg/d from preterm formula and fortified human milk, respectively. At 34 weeks of gestational age, the predicted zinc intakes would be 36 and 14 µmol/kg/d from preterm formula and human milk, respectively. This wide range of predicted zinc needs would be difficult to adopt as guidelines for recommended intakes for premature infants during early neonatal life. The current Premature-Recommended Nutrient Intake (P-RNI, Nutrition Committee, Canadian Pediatric Society 1995) used the zinc content of human milk as a basis of their recommended zinc intake for premature infants of 8 to 12 µmol (500–800 µg)/kg/d and with an upper limit of 17 µmol(1.1 mg)kg/d for infants fed fortified mother's milk or formula.

Clinical Evidence of Zinc Needs of Premature Infants

Zinc Needs from Birth to Term Age

Premature infants fed their mother's milk would benefit from the higher zinc content of milk during the first four to six weeks of life (Figure 1). Previously published zinc balance studies in premature infants

Correct citation: Atkinson, S.A., Wauben, I.P.M., and Paes, B. 1997. Zinc needs of premature infants to one year of age. In *Trace Elements in Man and Animals – 9: Proceedings of the Ninth International Symposium on Trace Elements in Man and Animals. Edited by* P.W.F. Fischer, M.R. L'Abbé, K.A. Cockell, and R.S. Gibson. NRC Research Press, Ottawa, Canada. pp. 501–505.

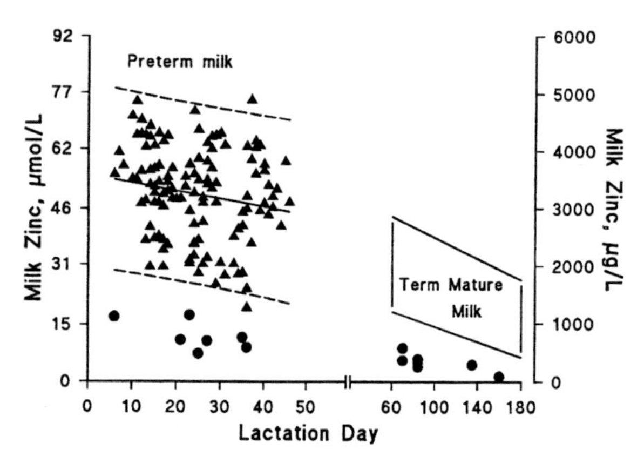

Figure 1. Zinc concentration in milk from mothers of premature infants (▲) which declines over the first six weeks of lactation (y = 55.3 – 14.0x, r2 = –0.2, p < 0.05) compared to the range for mature milk from mothers of term born infants in the boxed area (Krebs et al. 1995). The solid ● represent values for milk zinc from reported cases of premature infants who developed nutritional zinc deficiency (summarized in Atkinson et al. 1988).

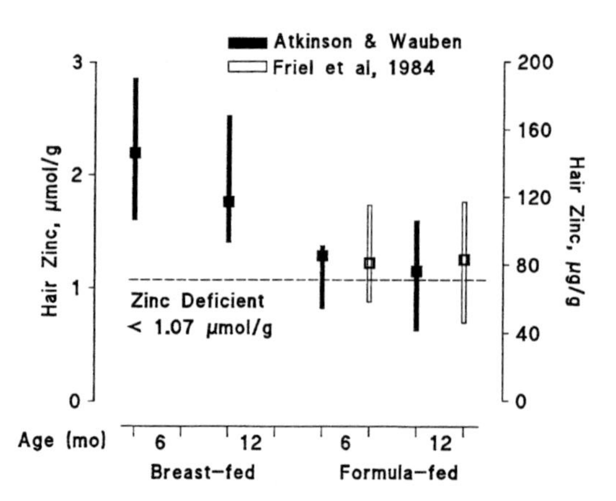

Figure 2. Hair zinc (median with 25–75th%ile values indicated by bars) of breast-fed compared to formula-fed infants at 6 and 12 months corrected age as observed by Wauben et al. (1996) and by Friel et al. (1984) for infants from the same community. The hatched line indicates the hair zinc (1.07 µmol/g) below which is considered to reflect zinc deficiency in older children (Strain et al. 1966).

had conflicting results with respect to absorption and retention of zinc from mother's milk and formula with both positive and negative zinc retention reported (summarized in Zlotkin et al. 1995). Such variable findings may relate to methods used (mass balance versus stable isotope tracers), gestational age of the infants at study, or differences in bioavailability of human milk compared to cow's milk-based formula (Sandstrom et al. 1983).

In infants of 1.2 to 1.4 kg birth weight, studied at about 35 weeks post-menstrual age, zinc intakes from mother's milk appeared adequate (Wauben et al. 1997) which supports the current estimates for P-RNI for zinc (Nutrition Committee, Canadian Pediatric Society 1995). At zinc intakes (µmol/kg/d) of 11 from mother's milk, 33 from zinc-fortified mother's milk and 23 from preterm formula (Preemie SMA, Wyeth-Ayerst Ltd, Montreal, Canada), mean retention of zinc met or exceeded intrauterine accretion values of 5 to 8 µmol/kg/d (Wauben et al. 1997) as estimated based on Widdowson et al's data (1988).

Zinc Needs from Term Corrected Age to One Year

Our recent studies in premature infants show that prolonged breast feeding in premature infants to six months of age adjusted for prematurity resulted in a significantly better zinc status as measured by hair zinc than in premature infants fed standard term formulas (Figure 2, Wauben et al. 1997). Hair zinc concentrations in the zinc deficient range (<1.07 µmol/g, Strain et al. 1966) were observed in 3/7 formula-fed and 1/5 breast-fed infants at six months corrected age. Similar values for hair zinc in formula-fed premature infants from the Hamilton-Wentworth region (Figure 2) were previously observed (Friel et al. 1984).

Despite the well known higher bioavailability of zinc from human milk as compared to formula (Sandstrom et al. 1983), the Canadian Pediatric Society currently recommends a zinc intake of 15 µmol/kg/d; and they suggest a supplement of 7.5 µmol/kg/d of zinc for breast-fed premature infants after discharge from hospital until they are weaned from breast feeding. Since growth was adequate in the breast-fed infants over the first six months and hair zinc was maintained above the zinc deficient range for all but one infant (Wauben et al. 1994, 1997), we suggest that the supplemental zinc (7.5 µmol/kg/d) currently recommended for premature breast-fed infants (Nutrition Committee, Canadian Pediatric Society 1995) is not necessary to maintain adequate zinc status in healthy infants of birth weight >1.2 kg.

Conversely, our observations of lower hair zinc in preterm infants fed standard infant formula (average intake of 11–12 μmol/kg/d), some of which were in the zinc deficient range (Wauben et al. 1997), may suggest that standard term formulas do not provide optimal zinc intakes for this group of infants.

Influence of Dietary Minerals and Steroid Drugs on Zinc Needs

Calcium/Phosphorus:Zinc Interaction

Calcium and phosphorus are known to interact antagonistically with zinc at the intestinal level in animals (Zemel and Bedari 1983) and adult humans (Greger and Snedecker 1980). In piglets, a good animal model for rapidly growing infants, a five-fold addition of calcium/phosphorus as calcium glycerophosphate to the control diet resulted in a reduction in the true fractional absorption of zinc from 98 ± 1% to 85 ± 4% (Atkinson et al. 1990). Using isolated brush border membrane vesicles from piglets, we demonstrated a calcium:zinc ratio of 20:1 (4 mmol calcium/L) reduced mucosal zinc uptake of zinc to less than 50% of control values (Bertolo, Bettger and Atkinson, unpublished). While it is difficult to measure metal interactions at the intestinal level in human infants, previous observations suggested that addition of calcium supplements of 4 mmol/kg/d to premature infants fed mother's milk reduced zinc absorption (Atkinson and Shah 1990).

Iron:Zinc Interaction

Iron is known to act antagonistically with zinc in adults (Solomons 1986). There is the potential for concern of an iron:zinc interaction in premature infant feeding since high iron intakes (2–4 mg/kg/d) are recommended for this infant population from shortly after birth to one year of age (Nutrition Committee, Canadian Pediatric Society 1995). In vitro, in brush border membranes isolated from young piglets, we observed that at iron concentrations comparable to the upper concentrations found in infant formulas (0.25 mmol iron/L), zinc uptake was reduced by 20 to 30% (Bertolo, Bettger and Atkinson, unpublished). Iron:zinc interactions in premature infant have not been well studied.

Steroid (Dexamethasone) Interactions with Zinc

The therapeutic use of dexamethasone in premature infants to induce lung maturation and to improve dynamic lung compliance may impair normal zinc absorption and storage (Wang et al. 1992). Glucocorticoids have been implicated in the regulation of zinc homeostasis by inducing transcription of the metallothionein gene via interaction with specific promoters located upstream from the metallothionein gene (Hager and Palmiter 1981). Using the dexamethasone-treated piglet model, we have shown steroid-induced alterations in zinc uptake by brush border membrane vesicles (Wang et al. 1992). Since dexamethasone treatment up-regulated intestinal metallothionein synthesis but did not alter hepatic metallothionein, it may be that absorbed zinc (and likely copper) are trapped in the intestinal mucosa, so less of the trace elements are absorbed into the blood (Wang et al. 1993).

Summary

The zinc needs of premature infants should allow for accumulation of zinc to meet needs for growth and to meet zinc storage similar to that achieved by infants remaining in utero until term gestation. The ability for prematurely born infants to attain this goal depends on the bioavailability of the source of dietary zinc provided (breast milk likely being the preferable source); the influence of other dietary metals in interfering with zinc absorption (calcium and iron are of particular concern); and the potential detrimental effect on zinc absorption of exogenous dexamethasone, a drug commonly used to treat severe lung disease in premature infants. These factors must be considered in future evaluations of the zinc needs of premature infant during early growth and development.

Literature Cited

Atkinson, S.A., Shah, J., Gibson, R.S., Gibson, I.L. & Webber, C.E. (1993) A multi-element isotopic tracer assessment of true fractional absorption of minerals from formula with additives of calcium, phosphorus, zinc, copper and iron in young piglets. J. Nutr. 123: 1586–1593.

Atkinson, S.A. & Shah J.K. (1990) Calcium and phosphorus fortification of preterm formulas: Drug-mineral and mineral-mineral interactions. In: Mineral Requirements for the Premature Infant. (Hillman J., ed.), Wyeth-Ayerst Seminar Series, pp. 58–75. Elsevier Press, NY.

Atkinson S.A., Whelan D., Whyte R.K., et al. (1989) Abnormal zinc content in human milk. Am. J. Dis. Child. 43: 608–611.

Ehrenkranz, R.A., Gettner, P.A., Nelli, C.M. et al. (1989) Zinc and copper nutritional studies in very low birthweight infants: Comparison of stable isotopic extrinsic tag and chemical balance methods. Pediatr. Res. 26: 298–344.

Friel, J.K., Gibson, R.S., Balassa, R. & Watts, J.L. (1984) A comparison of the zinc, copper and manganese status of very low birth weight pre-term and full-term infants during the first twelve months. Acta. Pediatr. Scand.73: 596–601.

Greger, J.L. & Snedeker, S.M. (1980) Effect of dietary protein and phosphorus levels on the utilization of zinc, copper and manganese by adult males. J. Nutr. 110: 2242–2253.

Hager, L.J. & Palmiter, R.D. (1981) Transcriptional regulation of mouse liver metallothionein-1 gene by glucocorticoids. Nature 291: 340–342.

Krebs N.F., Reidinger C.J., Hartley S., Robertson A.D. & Hambidge K.M. (1995) Zinc supplementation during lactation: effects on maternal status and milk zinc concentrations. Am. J. Clin. Nutr. 61: 1030–1036.

Nutrition Committee. Canadian Pediatric Society. (1995) Nutrient needs and feeding of premature infants. Can. Med. Assoc. J. 152: 1765–1785.

Sandstrom, B., Davidson, L., Lederblad, A. & Lonnerdal, B. (1983) Zinc absorption from human milk, cow's milk and infant formulas. Am. J. Dis. Child. 137: 726–729.

Solomons, N.W. (1986) Competitive interaction of iron and zinc in the diet: consequences for human nutrition. J. Nutr. 116: 927–935.

Shaw, J.C. (1979) Trace elements in the fetus and young infant. Am. J. Dis. Child. 133: 1260–1268.

Strain, W.H., Steadman, L.T., Lankau, C.A. Jr. et al. (1966) Analysis of zinc levels in hair for the diagnosis of zinc deficiency in man. J. Lab. Clin. Med. 68: 244–255.

Wang, Z., Atkinson, S.A., Bertolo, R., Polberger, S. & Lonnerdal, B. (1992) Alterations in intestinal uptake and compartmentalization of zinc in response to short-term dexamethasone therapy or excess dietary zinc in piglets. Pediatr. Res. 33: 118–124.

Wang, Z., Weiler, H.A. & Atkinson, S.A. (1993) Long term dexamethasone therapy + high dietary zinc alter intestinal zinc uptake in the piglet model. Proc. Can. Fed. Biol. Sci. 36: 144.

Widdowson, E.M., Southgate, D.A.T. & Hey, E. (1988) Fetal growth and body composition, In: Perinatal Nutrition (Lindblad, B.S., ed.), pp. 3–14, Academic Press, New York.

Wauben, I.P.M., Atkinson, S.A., Grad, T.L., Shah, J.K. & Paes, B. (1994) Calcium glycerophosphate (CaGP) as Ca/P source in a new fortifier for preterm infants: measurement of mineral balances, bone mineralization and growth. Pediatr. Res. 35(4);323A:A1920.

Wauben, I.P.M., Paes, B., Shah, J.K. & Atkinson, S.A. (1997) Adequate zinc status to six months corrected age in premature infants fed mother's milk pre- and post-hospital discharge. In: Trace Elements in Man and Animals – 9: Proceedings of the Ninth International Symposium on Trace Elements in Man and Animals. (Fischer, P.W.F., L'Abbé, M.R., Cockell, K.A. and Gibson, R.S., eds.), National Research Council, Ottawa, Canada. pp. 279–281.

Zemel, M.B. & Bedari, M.T. (1983) Zinc, iron and copper availability as affected by orthophosphates, polyphosphates and calcium. J. Food. Sci. 48: 567–573.

Zlotkin, S.H., Atkinson, S.A. &, Lockitch, G. (1995) Trace elements in nutrition for premature infants. Clin. Perinatol. 22(1): 223–240.

Zlotkin, S.H. & Cherian, G. (1988) Hepatic metallothionein as a source of zinc and cysteine during the first year of life. Pediatr. Res. 24: 326–329.

Discussion

Q1. Harold Sandstead, University of Texas Medical Branch, Galveston, TX, USA: This really is very interesting, and certainly extremely difficult work to do. It caused me to think about some experiments we've done in animals long ago. It seems to me the organ system that's at greatest risk here in these babies, if they survive, is the nervous system. There is a substantial body of knowledge now, that even very mild zinc deficiency during early development, either intra-uterine or post-uterine, in animals will result in very long-term irreversible damage. With such mild deficiency that the pregnancy continues normally, and the pups look normal when they're born, but when they're 300 d of age, they can't do a standard maze. They just can't learn. They have problems in cognition that are irreversible. It seems to me that if the hair zinc less than 100, or 75, or whatever the cutoff is, is a reflector of decreased body stores, whatever those stores are, that those babies are probably at significant risk of later mental retardation. At least, that would be my hypothesis. Some people would say that this means that they need supplementation, or they may not be able to function in our modern world, in 2025. I have a question now, you referred to Zn stores. I would like to know, what stores do you mean, and what tissues? Soft tissues conserve zinc very closely, even in zinc deficiency, and you can have severe disruption of function, and no change in the concentration in soft tissue. Are you talking about bone, because that is a store? How is it available for use?

A. I don't know how labile zinc bone stores are. I certainly accept your point there. When we are talking about zinc stores, we're talking about that little bit of hepatic zinc that's available. Your point about cognition is also well taken. We have thought of that, and are doing cognitive function studies in our ongoing studies. We have a little bit of data from this study, but it's not a sufficient sample size to do a lot with. It is very important, and we are following up on that.

Q2. Harold Sandstead, University of Texas Medical Branch, Galveston, TX, USA: I think your work is extremely important, because there probably are not a lot of people taking care of pre-term infants, who are worrying about this problem. It is well known that babies that are born pre-term have a much higher rate of all sorts of neuropsychological problems later on.

A. Thank you.

Q3. Bob Smith, Great Smokies Diagnostics Lab., Asheville, NC, USA: Hair zinc goes down very rapidly, and upon birth drops very rapidly. How long a segment of hair are you using? I would be very pleased to do hair, not just looking for zinc, but also other elements. Specifically, I'd like to do the mothers, and look at lead and cadmium in the mothers to see if you've got something there. We'd like to get the mothers as well as the children, and to segment this hair very, very closely. With the mass spec, we can get down to some pretty low numbers.

A. These are premature babies. They hardly have any hair at six months, and we take the maximum. We've even been told by some parents that we do a shaving job. We are already at the limits of our the amount of hair that it's possible to analyze with any confidence.

Q4. Tammy Bray, Ohio State University, Columbus, OH, USA: I was thinking about the animal model we use with hyperoxia. Using 85% oxygen for three days, we can clearly show the permeability of the blood:brain barrier, and also see lung damage. I wonder, with the premature infant, if we don't have to just consider the requirement for growth, but also for prevention of tissue damage, particularly in these target organs affected by high oxygen.

A. We are very aware of your work, Tammy, and would love to pursue this line of research in the piglet model.

Q5. Harry McArdle, Rowett Research Institute, Aberdeen, Scotland: This related to the group of mothers who had very low zinc milk. I was wondering, you know the lethal milk mouse mutation is related to zinc? I was wondering if you've had a look at these mothers, to see whether or not there was any alteration in their zinc binding in any other tissues, and whether you could identify any of those at risk.

A. I have not done further studies in these women, and I'm not aware of anyone else having done so in the literature.

Effects of Zinc (Zn) and Micronutrient (M) Repletion on Chinese Children*

H.H. SANDSTEAD,[†] N.W. ALCOCK,[†] H.H. DAYAL,[†] J.G. PENLAND,[‡] X.C. CHEN,[¶] J.S. LI,[§] F.J. ZHAO,[||] AND J.J. YANG[+]

[†]Division of Human Nutrition, Department of Preventative Medicine, University of Texas Medical Branch, Galveston, TX 77555-1109 and [‡]USDA ARS Human Nutrition Research Center, Grand Forks, ND 58202, USA; [¶]Chinese Academy of Preventative Medicine, Beijing, [§]Qingdao Medical College, Qingdao, [||]2nd Military Medical University, Shanghai, and [+]3rd Military Medical University, Chongqing, PRC

Keywords zinc, lead, growth, stunting

Zn deficiency was suspected 50 years ago among malnourished adult Chinese (Eggleton 1940). Chen et al. (1985) suggested Zn deficiency is common among Chinese children. Therefore we conducted a 10 week double-blind treatment (Rx) trial of Zn, during the spring of 1994 in urban first grade children, from Chongqing (CQ), and Qingdao (QD) and Shanghai (SH), to ascertain effects on growth, cognition and lead (Pb) status.

Methods

The Rxs were: (1) 20 mg Zn, (2) 20 mg Zn with selected M, and (3) selected M alone. The M mixture provided 50% of the 1989 US NRC RDA or mean ESADDI, except that folate was included at 25% of the RDA, and iron, calcium, magnesium and phosphate were excluded. Classes of about 40 children each were assigned an Rx. Here we report height (Ht), knee height (KH) (Cronk et al. 1989), hemoglobin (HB), serum and hair Zn (s Zn, h Zn), and whole blood and hair lead (b Pb, h Pb). Data were assessed by ANOVA and two tailed t-test.

Results and Discussion

Data are summarized in Table 1. Ages (years) were CQ 6–7 ; QD, 7–8; SH, 8–9. Mean Hts (cm) were: CQ, 116.5; QD, 121; and SH, 119.0. Rx 2 and 3 significantly increased Ht in CQ and SH compared to Rx1, but changes after Rx 2 and Rx3 were similar. Changes in Ht were similar across Rxs in QD. Changes in KH reflected Rx in subjects from CQ and QD. Growth after Rx2 > Rx3 > Rx1 (all Rx comparisons: $p < 0.001$). KH findings from SH appeared technically compromised.

Laboratory analysis found mean base HB was normal in CQ and SH. After Rx changes occurred at all locations. The significance of the changes in HB is unclear. Base s Zn was similar in CQ and QD, as were responses to Rx. Increases occurred after Rx 2 and 3, with Rx2 > Rx3. The base s Zn levels suggested many subjects had low Zn status, an impression confirmed by the increases in KH after Rx. Base s Zn was higher in SH. Small and similar (to CQ and QD) increases occurred after Rx 2 and 3. The lack of a substantial increase in s Zn in SH after Rx may indicate subjects did not receive Rx consistently. Base h Zn was also consistent with many subjects low in Zn. Decreases in h Zn after Rx were without a clear pattern; they may reflect the limited Zn available to the growing hair follicle. A similar decrease in h Zn after Zn Rx was reported in a subject with low cellular immunity from Zn deficiency (Pekarek et al. 1979).

---

*Support: International Lead Zinc Research Organization & General Nutrition Products Co.

Correct citation: Sandstead, H.H., Alcock, N.W., Dayal, H.H., Penland, J.G., Chen, X.C., Li, J.S., Zhao, F.J., and Yang, J.J. 1997. Effects of zinc (Zn) and micronutrient (M) repletion on Chinese children. In *Trace Elements in Man and Animals – 9: Proceedings of the Ninth International Symposium on Trace Elements in Man and Animals. Edited by* P.W.F. Fischer, M.R. L'Abbé, K.A. Cockell, and R.S. Gibson. NRC Research Press, Ottawa, Canada. pp. 506–508.

Table 1. KH (mm), HB (g/L), s Zn (μmol/L), h Zn (μg/g), b Pb (μmol/L) and h Pb (μg/g).

City	Inde	n	base	CV	delta	CV	p <	Rx1	Rx2	Rx3
CQ	KH	123	340.4	5	8.4	19	0.001	5.4[a]	10.9[b]	9.1[c]
QD	KH	120	375.2	5	6.6	35	0.002	5.5[a]	7.5[b]	6.7[c]
CQ	HB	118	145	7	–14	–75	0.04	–13[abc]	–17[ab]	–10[c]
QD	HB	121	113	6	6	98	0.001	2[a]	9[b]	6[c]
SH	HB	117	124	11	–16	–122	0.003	–20[ab]	–20[ab]	–7[c]
CQ	s Zn	110	12.15	32	6.92	46	0.001	2.00[a]	11.69[b]	7.84[c]
QD	s Zn	114	12.15	32	7.07	49	0.001	2.31[a]	11.07[b]	8.46[c]
SH	s Zn	111	15.75	16	1.38	235	0.001	–0.45[a]	2.32[bc]	2.25[bc]
CQ	h Zn	113	128	25	–29	–109	0.3	–33.1	–32.1	–21.6
QD	h Zn	119	109	30	–29	–93	0.2	–26	–37	–26
SH	h Zn	119	107	26	–44	–71	0.02	–50[ab]	–51[ab]	–32[c]
CQ	b Pb	100	0.487	27	0.024	509	0.06	0.042[abc]	–0.017[b]	0.063[c]
QD	b Pb	116	0.462	36	–0.023	–737	0.06	–0.024	–0.030	–0.021
SH	b Pb	117	0.433	31	0.084	132	0.001	–0.057[a]	0.055[b]	0.295[c]
CQ	h Pb	112	0.77	99	4.32	116	0.2	3.07[a]	5.45[bc]	4.68[bc]
QD	h Pb	118	0.43	133	1.14	160	0.06	0.68[a]	1.56[bc]	1.54[bc]
SH	h Pb	112	0.31	187	2.55	123	0.05	3.26[ac]	1.59[b]	2.95[ac]

Different superscripts in the same row indicate significant differences ($p < 0.05$).

Base b Pb concentrations were > USPHS CDC safe guidelines in nearly half of the subjects. Changes in b Pb after Rx were inconsistent across locations. Mean base h Pb concentrations were 2x as high in CQ as SH. After Rx h Pb increased, perhaps by Pb mobilization from bone.

The efficacy of KH for measuring growth over short intervals of Rx was confirmed. The findings were consistent with the Chen et al. (1985) hypothesis. KH changes after M Rx indicated the presence of multiple deficiencies. The findings also suggest many children were exposed to increased environmental lead.

Literature Cited

Chen, X.C., Yin, T.A. and He, J.S. (1985) Low levels of zinc in hair and blood, pica, anorexia, and poor growth in Chinese preschool children. Am. J. Clin. Nutr. 42: 694–700.

Cronk, C.E., Stallings, V.A., Spender, Q.W., Ross, J.L. and Widdoes, H.D. (1989) Measurement of short-term growth with a new knee height measuring device. Am. J. Human Biol. 1: 421–428.

Eggleton, W.G.E. (1940) The zinc and copper contents of the organs and tissues of Chinese subjects. Biochem. J. 34: 991–997.

Pekarek, R., Sandstead, H.H., Jacob, R. and Barcome, D. (1979) Abnormal cellular immune responses during acquired zinc deficiency. Am. J. Clin. Nutr. 32: 1466–1471.

Discussion

Q1. Bob Smith, Great Smokies Diagnostics Lab., Asheville, NC, USA: I've done segmented hair analyses on people who use chelators, EDTA, DMPS and so on. Without question, the level of lead goes up when you give a chelating agent. I've also seen many times with supplementation, if you take a hair sample, the level of lead goes up in hair. Also, the levels of zinc do go down when you have someone who has had long-term zinc deprivation. Zinc goes down, and then goes back up.

A. Thank you.

Q2. Les Klevay, USDA-ARS, Grand Forks, ND, USA: There is known sex difference between males and females in hair lead, and I can't remember for sure, but I think females have higher leads. Did you find this as well?

A. We haven't yet looked at this as the results have only just been completed.

Q3. John Sorenson, University of Arkansas, Little Rock, AR, USA: Did you say where you sampled the hair?

A. We asked for the proximal 2 cm from the occipital area of the scalp. Sometimes the hair sample was a little longer than 2 cm, however. The duration of the intervention was 10 weeks and the sample was taken right at the end of the treatment.

Q4. Janet King, USDA, San Francisco, CA, USA: Did you make any observations on the morbidity of the children?

A. We have collected the data, and we have finally received the forms, but it is not yet analyzed.

Q5. Noel Solomons, Guatemala: There is a message here for those of us who do field supplementation and try to draw conclusions. I'd like to try to draw some conclusions from the results you've presented us with. It seems to me that you had two important deltas to look at: nothing versus zinc alone, and zinc on top of micronutrients. The hypothesis then should be that there would be a greater response to zinc than to its control. That is, that zinc should be better than nothing, and micronutrients plus zinc should be better than micronutrients. The conclusion that I could draw statistically is that you should accept both of null hypotheses that there is no independent effect of zinc, either on top of micronutrients, or on top of nothing. I think that is consistent with many other studies. Changes in knee height are very small.

Zinc is better than nothing and zinc plus micronutrients is better than micronutrients. Micronutrients are better than anything.

A. Well, our statistics didn't agree with your analysis. In fact, if you looked at the figures, you'd conclude that these effects were very highly significant. We used an ANOVA, which was very highly significant, and then everything else was compared with a two-tailed t-test which is very conservative. And those were also extremely significant. That was presented to you. And in fact, our statistician at our place told me I shouldn't put any more information on these charts, because anybody who knows anything could observe the changes. (Please, everyone understand that Noel and I are very good friends). We conclude that Zn is better than nothing and that zinc plus micronutrients is better than micronutrients alone. You are absolutely right that micronutrients are best of all. The conclusion is that if you're going to treat people, you've got to give them micronutrients and zinc.

Q6. Susan Kaup, Wyeth-Ayerst Research, Philadelphia, PA, USA: To follow up on that, if I remember right from Penland's abstract on cognitive development, zinc seemed to stand out markedly, whereas micronutrients did not. Yet you've demonstrated that growth responded great to micronutrients, and zinc didn't do to well, so which do we rely on, growth, or cognitive performance?

A. Well, that's an interesting thing for us to think about. Our current hypothesis is that neuropsychological indices may be better than growth to evaluate but growth is easier to measure. You do get different sorts of answers, that's exactly right.

Q7. Barbara Golden, University of Aberdeen, Scotland: I am impressed with the micronutrients but not necessarily with zinc. I would like you to speculate on what micronutrients might have affected growth?

A. Okay, it will be a wild speculation. There are data from Central America. Jim Smith has done a randomized trial down there, and one of his groups was vitamin A, one of his groups was vitamin A plus zinc and so forth. He found a substantial growth response to vitamin A in children. So one wonders, there's probably a whole bunch of things in there, but probably the vitamin A was important here, because they were all given vitamin A.

Endogenous Iron Losses in 6–12 Month Old Infants*

J.L. BELSTEN, C.E. PHILLIPS, S.J. FAIRWEATHER-TAIT, J. EAGLES, AND P.J. AGGETT

Institute of Food Research, Norwich NR4 7UA, UK

Keywords iron, infant, blood loss, stable isotopes

Iron deficiency in infants is a widespread problem and can lead to impairment of intellectual development, which may be irreversible. However, physiological requirements are unclear due to the lack of information on iron loss in this age group. The routes of iron loss via the gastrointestinal (GI) tract are digestive secretions, mucosal cell exfoliation and bleeding. The latter has been suggested as contributing to infantile iron-deficiency anemia. The possibility that cows' milk increases bleeding has led to the endorsement in the UK of the recommendation that it should not be the main source of milk before one year of age (Department of Health 1994). This is the first study in which the measurement of endogenous iron losses in healthy infants using stable isotopes has been attempted. Infants were given an intravenous dose of ^{58}Fe, and loss of labelled iron from the GI tract was measured.

Materials and Methods

At approximately 6, 9 and 12 months of age, a venous blood sample was removed and ^{58}Fe citrate administered intravenously to healthy, non-anemic, full-term infants. The dose was half that required to saturate plasma transferrin. A parent-selected diet, free of red meat (ie. low in heme iron), was provided for two weeks following injection. Infants were given 100 mg carmine as a fecal marker 7 and 14 d post-injection to delineate the fecal collection period. Parents completed a weighed intake record for this period (data to be presented in a separate publication).

Blood was analyzed for hemoglobin (Hb), plasma ferritin and plasma transferrin receptor levels (R&D Systems, Abingdon, UK). A sub-sample of each bulked 7-day fecal collection was retained for fecal blood analysis. The remainder was analyzed for total iron and isotopic enrichment by thermal ionization quadrupole mass spectrometry.

Results

There were no significant changes in any of the measured hematological indices and no effect of initial milk source on iron status (Figure 1).

Mean daily fecal iron excretion fell by 1 mg/d from 6 to 12 months of age, and varied with type of milk (Figure 2).

There was a high inter-individual variation in GI blood loss (18–114 μL/d). The data suggest that cows' milk only increased GI bleeding when first introduced (Figure 3). Mean blood losses equate to 16, 23 and 20 μg iron/d at 6, 9 and 12 months respectively. The amount of ^{58}Fe in the feces was variable but very small (Figure 4). The mean fecal losses of ^{58}Fe were 0.23, 0.20 and 0.19% of administered dose at 6, 9 and 12 months respectively.

Endogenous losses of iron were calculated using the method of Garby et al. (1964), which involves estimating exchangeable body iron. Two approaches were taken

(1) storage iron and Hb iron (0.347 × Hb concentration × blood volume) were deducted from total body iron (38 mg/kg, Smith and Rios 1974); storage iron was calculated as suggested by Finch

*This work was funded by MAFF.

Correct citation: Belsten, J.L., Phillips, C.E., Fairweather-Tait, S.J., Eagles, J., and Aggett, P.J. 1997. Endogenous iron losses in 6–12 month old infants. In *Trace Elements in Man and Animals – 9: Proceedings of the Ninth International Symposium on Trace Elements in Man and Animals. Edited by* P.W.F. Fischer, M.R. L'Abbé, K.A. Cockell, and R.S. Gibson. NRC Research Press, Ottawa, Canada. pp. 509–511.

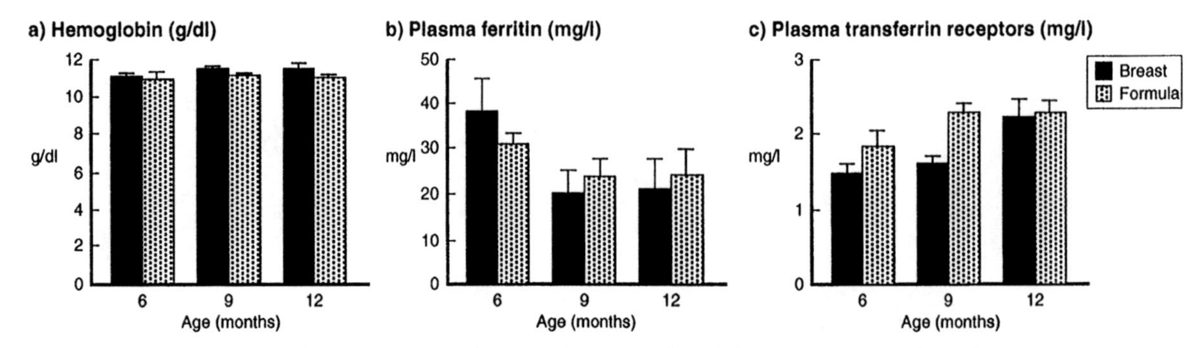

Figure 1. Hematological indices, grouped by initial source of milk (breast n=6, formula n=3).

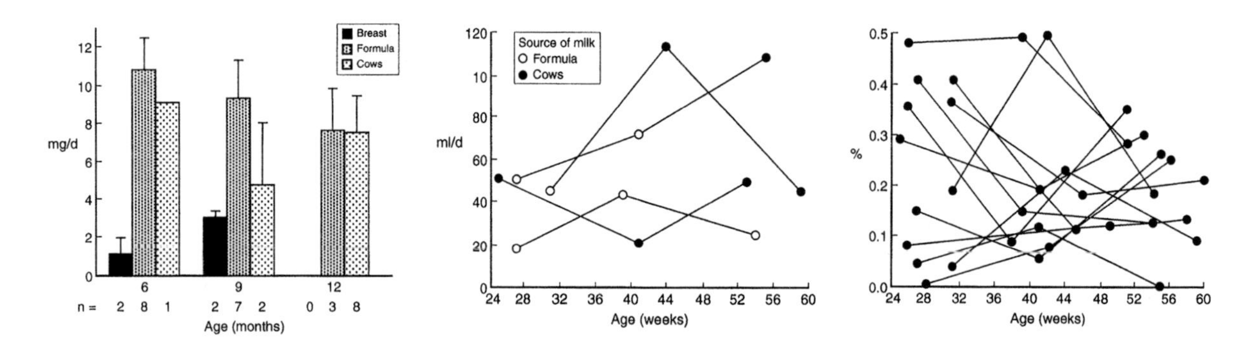

Figure 2. Fecal total iron excretion, grouped by source of milk at each age.

Figure 3. Gastro-intestinal blood loss for 4 infants.

Figure 4. Appearance of ^{58}Fe in feces for 12 infants.

(Siimes 1984) whereby 1 µg/L plasma ferritin is equivalent to 140 µg storage iron/kg body weight; the blood volume estimate was from Lentner (1984).

(2) exchangeable iron was estimated to be 4 mg/kg according to the calculations for iron distribution of Smith and Rios (1974). Exchangeable pools were assumed to be present in bone marrow (2 mg/kg) and muscles and enzymes (2 mg/kg).

Endogenous losses from the GI tract (Table 1) amounted to 10 µg/kg/d at 6m falling to 5–7 µg/kg/d at 9 months and 3–8 µg/kg/d at 12 months, depending on the calculation of exchangeable body iron.

Discussion

Initial milk type has no effect on later iron status. GI bleeding does not make a significant contribution to iron loss in healthy infants, irrespective of the main source of milk, averaging only 16, 23 and 20 µg of iron/d at 6, 9 and 12 months respectively. Earlier published estimates of endogenous iron losses were about 15 µg/kg/d between 2 months and 4 years (Josephs 1958), but higher values have also been proposed, for example Garby et al. (1964) reported GI losses to be 30 µg/kg/d in a radio-isotopic study of

Table 1. Mean estimated endogenous losses in infants aged 6 to 12 months.

	Exchangeable iron (mg)		Endogenous loss (µg/kg/d)	
Age (m)	(1)	(2)	(1)	(2)
6	32.6	32.0	11	9
9	28.0	36.0	5	7
12	20.1	39.3	3	8

three infants of indeterminate age fed "cow-milk mixtures". Our research, however, suggests that iron is conserved more efficiently. The mean total GI iron loss (endogenous iron plus occult blood) for all infants was 12 μg/kg/d at 6 months and 10 μg/kg/d at 9–12 months. These figures are approximately one-third of previous estimates, and have important consequences for the calculation of physiological requirements needed to derive dietary recommendations.

References

Department of Health (1994). Weaning and the weaning diet. London: HMSO.

Garby, L., Sjölin, S. & Vuille, J.C. (1964). Studies on erythrokinetics in infancy. IV. The long term behaviour of radio-iron in circulating foetal and adult haemoglobin, and its faecal excretion. Acta Paediatr. 53: 33–41.

Josephs, J.W. (1958). Absorption of iron as a problem in human physiology. Blood 13: 1.

Lentner, C. (1984). Ed. Geigy Scientific Tables 3: 66.

Siimes, M.A. (1984). Iron nutrition in low-birth-weight infants. In: Iron Nutrition in Infancy and Childhood (Stekel, A., ed). New York: Raven. 75–94.

Smith, N.J. & Rios, E. (1974). Iron metabolism and iron deficiency in infancy and childhood. Adv. Pediatr. 21: 239–280.

Influence of Alimentary Zinc Deficiency on the Somatotropic Axis of Rats

H.-P. ROTH AND M. KIRCHGESSNER

Institute of Nutrition Physiology, Technical University of Munich, D-85350 Freising-Weihenstephan

Keywords Zn-deficiency, growth hormone, IGF-1, insulin

Alimentary Zn deficiency reduced the concentration of growth hormone (GH) in the serum of growing rats in comparison with ad libitum fed control rats, but not compared with pair-fed control rats (Roth and Kirchgessner 1982). This means that the reduced GH concentration in the serum of Zn-deficient and pair-fed control rats is the result of the reduced feed intake and the associated protein and energy depletion rather than being caused by Zn deficiency per se. But in situations of marginal Zn depletion, when rat's feed intake was unaffected, the GH concentration in the serum of Zn-deficient rats was significantly reduced compared with pair-fed control rats (Kirchgessner and Roth 1985).

In the present study, rats were fed by gastric tube. This made it possible to supply the rats with adequate nutrients even in Zn deficiency. The experimental design enabled us to demonstrate the influence of a specific Zn deficiency on the somatotropic axis (GH and IGF-1 concentration) and insulin in the serum of rats.

Material and Methods

Twenty-four male Sprague-Dawley rats with an average body weight of 108 g were divided into 2 groups of 12 animals each. The Zn-deficient group and the control group received a semisynthetic diet with casein as protein source with a Zn content of 1.3 and 25 ppm respectively. All animals received 4 mL of a thoroughly homogenised food pulp with a dry matter content of 58.5% administered 4 times daily (8:30, 12:30, 17:00 and 21:30) by stomach tube. The specific weight of the food pulp was 1.24 g/mL, resulting in a daily feed intake of 11.6 g DM for each animal. After 12 d all animals having been fasted for 10 h, were decapitated under ether anaesthesia and the blood collected into polyethylene tubes. The serum extracted from the blood was aliquoted and, together with the dissected femur bones, frozen at –20°C for the subsequent analysis. The Zn status of the rats was determined by measuring the alkaline phosphatase activity [EC 3.1.3.1] in the serum and the Zn concentration in serum and femurs, following dry ashing in platinum dishes at 480°C in a muffle furnace, by means of AAS. For the determination of the serum concentrations of growth hormone and insulin special species-specific radioimmunoassay [^{125}I] systems for rat growth hormone and rat insulin by Amersham were used. IGF-1 (insulin-like growth factor 1) was also determined with a [^{125}I] RIA kit by Amersham.

Results

The rats on the control diet with a zinc content of 25 ppm increased their body weight from 108 g at the start of the experiment to 160 g after 12 d feeding by gastric tube. In the Zn-deficient rats, where the dietary Zn concentration was 1.3 ppm, reduced growth was recorded from day 8 onwards, despite an identical quantitative feed intake, followed by a reduction in bodyweight during the subsequent course of the experiment. The mean bodyweight of the Zn-deficient rats at that time was only 124 g.

The reduction in the zinc concentration in the serum and the femur by 62% and 44% respectively and the significantly reduced alkaline phosphatase activity (by 70%) in the serum of the Zn-deficient rats (Table 1) proved that after 12 d the experimental animals were in a state of severe Zn deficiency.

Correct citation: Roth, H.-P., and Kirchgessner, M. 1997. Influence of alimentary zinc deficiency on the somatotropic axis of rats. In *Trace Elements in Man and Animals – 9: Proceedings of the Ninth International Symposium on Trace Elements in Man and Animals. Edited by* P.W.F. Fischer, M.R. L'Abbé, K.A. Cockell, and R.S. Gibson. NRC Research Press, Ottawa, Canada. pp. 512–514.

Table 1. Zinc concentration and activity of alkaline phosphatase [EC 3.1.3.1] in serum and femur of force-fed zinc-deficient and control rats.

Tissue	Control rats (25 ppm dietary Zn)	Zn-deficient rats (1.3 ppm dietary Zn)
Serum		
Zn concentration (μg/mL)	1.25 ± 0.09[a]	0.47 ± 0.08[b]
Alkaline phosphatase activity (U/L)	357 ± 69[a]	109 ± 17[b]
Femur		
Zn concentration (μg/g DM)	202 ± 16[a]	113 ± 12[b]

Table 2. Concentration of growth hormone (GH), insulin-like growth factor I (IGF-1) and insulin in serum of force-fed zinc-deficient and control rats.

Serum	Control rats (25 ppm dietary Zn)	Zn-deficient rats (1.3 ppm dietary Zn)
GH (ng/mL)	22.6 ± 14.8[b]	40.2 ± 15.6[a]
IGF-1 (ng/mL)	214 ± 17[a]	155 ± 45[b]
Insulin (ng/mL)	1.08 ± 0.13[a]	0.81 ± 0.07[b]

After 12 d into the experiment the Zn-deficient rats showed to our surprise a significant 78% increase in the serum GH concentration compared with the control animals (Table 2). By contrast, the concentration of IGF-1 in the serum of the Zn-deficient rats was significantly lower, namely by 28%, compared to the control animals, despite elevated GH levels. The serum insulin concentration in the Zn-deficient rats, similar to the IGF-1, was also significantly reduced (by 25%).

Discussion

Histochemical analysis of the anterior pituitary have shown that Zn^{2+} is present in high concentrations in the GH secretory granules. Cunningham et al. (1991) showed that Zn^{2+} ions induce a dimerisation of human GH. The formation of this dimeric Zn-GH complex might have a bearing on the storage of growth hormone in the secretory granules as the dimeric form is more stable than monomeric GH. The release of growth hormone can presumably be regulated via the Zn^{2+} concentration because high levels of Zn^{2+} ions inhibit the release of growth hormone (Lorenson et al. 1983). One might hypothesize that the circulating Zn^{2+} concentration, which is already considerably reduced in incipient Zn deficiency, first leads to an enhanced release of GH from the secretory granules while at the same time reducing the storage capacity for newly synthesised GH. As the experiment progresses and the severity of the Zn deficiency increases, this can ultimately also lead to a reduction in the concentration of growth hormone in the serum as shown by us in several conventional Zn deficiency experiments with a duration of 4–6 weeks.

The major target organ for GH is the liver, which is considered to be the principal producer of IGF-1. Synthesis and secretion of IGF-1 occur after binding of the growth hormone to specific liver receptors. The growth-promoting effect of GH is thus mediated to a large extent indirectly via insulin-like growth factors, notably IGF-1, which stimulate anabolic processes such as cell division, skeletal growth and protein synthesis. However, in the study described here the concentration of IGF-1 in the serum of the Zn-deficient rats was reduced by 28% despite elevated GH levels. The cause of the reduced IGF-1 levels in Zn deficiency while GH concentrations in the serum are elevated is believed to be an impairment of the stimulating effect of GH on IGF-1 synthesis. Several reasons may be advanced for this phenomena, such as a reduced biological activity of the growth hormone, a receptor defect, or a reduced sensitivity of the tissue to GH.

As the IGF-1 synthesis correlates positively with the insulin content, the formation of IGF-1 is also dependent on the insulin concentration, which in the present study was reduced significantly by 25% in the Zn-deficient rats. The diminished serum IGF-1 concentration in the Zn-depleted rats, despite increased GH levels, might therefore also be the result of the reduced insulin concentration in the Zn depleted rats because insulin deficiency can also lead to reduced IGF-1 production in the liver. It is therefore conceivable that, as well as the lowering IGF-1 concentration, the reduced serum insulin concentration might also be responsible for the growth depression observed in young Zn-deficient rats.

Literature Cited

Cunningham, B.C., Mulkerrin, M.G. & Wells, J.A. (1991) Dimerization of human growth hormone by zinc. Science, 253: 545–548.

Lorenson, M.Y., Robson, D.L. & Jacobs, L.S. (1983) Divalent cation inhibition of hormone release from isolated adenohypophysial secretory granules. J. Biol. Chem. 258: 8618–8622.

Roth, H.-P. & Kirchgessner, M. (1982) Gehalte von Wachstumshormon (GH) im Rattenserum bei Zn-Mangel. Z. Tierphysiol., Tierernährg., Futtermittelkde. 47: 197–200.

Kirchgessner, M. & Roth, H.-P. (1985) Influence of zinc depletion and zinc status on growth hormone levels in rats. Biol. Trace Elem. Res. 7: 263–268.

Effects of Copper Deficiency on the Tibia of the Neonatal Rat. Morphologic and Histologic Survey

I. CHAVARRI,[†] I. NAVARRO,[†] AND I. VILLA[‡]

[†]Pediatric Research Unit, University of Navarra, 31.080 Pamplona, and [‡]Department of Pediatrics, Gregorio Marañón Hospital, 28.009 Madrid, Spain

Keywords copper, deficiency, neonate, ossification, cartilage epifiso-metafisario

Introduction

Copper deficiency results, either directly or indirectly, in growth delay, as well as number of other characteristic signs (structural abnormalities, irregular calcification, etc.) (Conlan et al. 1990, Shaw 1988, Hurley 1981). The delay in growth, as well as the other symptoms of copper deficiency, are reversible with the subsequent supplementation of the metal. Maternal nutrition is fundamentally important in the development of the embryo and of the fetus. Deficiencies or excesses of given trace elements can produce anomalies in prenatal development, such as functional defects and growth delay, which can in some cases lead to the death of the fetus (Linder 1979). The purpose of this study was to observe the alterations in the epifiso-metafisario cartilage of the tibias of newborn rats following maternal copper deficiency.

Material and Methods

The tibia from neonates of three groups of female Sprague-Dawley rat, consisting of ten rats each, were analyzed. The copper deficient group was fed with a copper free diet ad libitum (AIN-76) (n = 7). The control group, received a diet supplemented with copper at 6 ppm ad libitum (n = 9). The pair-fed group (n = 7), received a similar diet but with calories equivalent to the copper-deficient group. The neonatal tibias were dissected from five newborn rats which were previously chosen from the litter. The length of the tibias were measured with a precision of 0.05 mm with the help of a stereomicroscope. The microscopic study of the epifiso-metafisario cartilage was done using an Olympus light microscope at 40x magnification and a visual grid.

Results and Discussion

A significant difference in the length of the tibias was observed between the control group and the copper deficient group; the latter being shorter. The tibias of the pair-fed group were intermediate in length (Table 1). In all the bones from the control group, the proliferated and hypertrophic zones are present in an ordered column arrangement. The morphology of the cartilage and other structures of the tibias of the pair-fed group is identical to that observed in the control group. The cells in the proliferated and hypertrophic zones were present in a disorganized arrangement in three of the seven tibias from the copper deficient group. Table 2 presents the relationship between both zone cells and the measurement of the medullar cavity in the tibia of all the groups studied. The values are means ± standard deviations. Also shown are the probabilities obtained with the ANOVA test and the results of the Scheffe test. A smaller number of proliferated and hypertrophic condrocytes were found in the copper deficient group, which would result in a delay in cellular production in the tibia of the deficient animals. However, there were also similar differences between the pair-fed group and the control group, which does not suggest a specific action for copper at the level of proliferation of the condrocytes. The morphologic data indicate an abnormal structure of the epifiso-metafisario cartilage, resulting in impaired growth and ossification.

Correct citation: Chavarri, I., Navarro, I., and Villa, I. 1997. Effects of copper deficiency on the tibia of the neonatal rat. Morphologic and histologic survey. In *Trace Elements in Man and Animals – 9: Proceedings of the Ninth International Symposium on Trace Elements in Man and Animals. Edited by* P.W.F. Fischer, M.R. L'Abbé, K.A. Cockell, and R.S. Gibson. NRC Research Press, Ottawa, Canada. pp. 515–516.

Table 1. Neonatal tibia length of the control (C), deficient (D) and pair-fed (P) groups.

Parameter	Deficient (n = 10)	Control (n = 10)	Pair-fed (n = 10)	ANOVA (P)	SCHEFFE D/C	SCHEFFE D/P	SCHEFFE C/P
Length (mm)	5.47±0.63	6.00±0.57	5.95±0.49	0.001	**	**	ns

Table 2. The relationship between proliferated and hypertrophic zone cells and the medullar cavity of the control (C), deficient (D) and pair-fed (P) groups.

Zone cells	Deficient (n = 15)	Control (n = 18)	Pair-fed (n = 20)	ANOVA (p)	SCHEFFE D/C	SCHEFFE D/P	SCHEFFE C/P
Proliferated (P)	41.71±0.80	47.05±4.10	50.28±1.89	0.0001	**	**	*
Hypertrophic (H)	17.64±1.08	18.35±2.85	21.08±1.77	0.0157	ns	**	*
P + H	59.35	65.40	71.36	0.0001	**	**	**
P/H	2.36	2.56	2.38	0.0163	**	ns	*
P (%)	70.3	71.9	70.5	0.0001	**	ns	*
H (%)	29.7	28.1	29.5	0.0157	ns	**	*
Med. cav. leng.†	129.38±9.16	140.89±13.0	126.86±8.47	0.0195	*	ns	*

†Medullar cavity length (μm).

References

Conlan, D., Korula, R. & Tallentira D. (1990) Serum copper levels in elderly patients with femoral-neck fractures. Age Ageing, 19: 212–214.
Hurley, L.S. (1981) Teratogenic aspects of manganese, zinc and copper nutrition. Physiol. Rev. 61: 249–295.
Linder, M.C. (1979) Copper regulation of ceruloplasmin in copper deficient rat. Enzyme 24: 23–25.
Shaw, J.L.C. (1988) Copper deficiency and non-accidental injury. Arch. Dis. Child. 63: 448–455.

Trace Elements in Bone Metabolism

Chair: P. Paterson

More Precious than Gold: Trace Elements in Osteoporosis

P. SALTMAN
Department of Biology, University of California San Diego, La Jolla, California 92093-0322, USA

Keywords osteoporosis, bone density, trace elements, calcium

In the classical Japanese film *Rashomon* an encounter involving three individuals is recreated through the mind's eye of each — the Prince, the Princess, and the Bandit. Three divergent accounts are presented. A dispassionate fourth observer, the Woodcutter, tells the tale from yet another perspective. The 'truth' is difficult to discern.

Osteoporosis is viewed from differing vantage points. To clinicians, osteoporosis is one of the most debilitating public health problems concerning both developed and developing nations. The disease profoundly affects not only post-menopausal women, but elderly men as well. Increasing numbers of young women and men develop fragile bones as a direct consequence of dietary inadequacies and physical stress. Genetics and ethnicity are at play. Endocrinologists provide powerful strategies of hormone replacement therapy and calcitonin. Pharmacologists develop bis-phosphonate treatments. Physiologists extol the skeletal virtues of weight-bearing work. Nutritionists and dietitians focus on food habits with respect to Ca, F, Mg, P, Vitamin D, and now trace elements. In fact, each of these perspectives is vital to optimal skeletal development and maintenance. No single approach is sufficient. All must be integrated and applied in a most efficient and timely fashion.

The trace element view of bone health begins with observations of bone malformation in Cu-, Zn- and Mn-deficient animals and humans (Saltman and Strause 1993). A dynamic equilibrium exists between the mineral and organic matrices of this tissue, as well as between the osteoblastic synthesis and osteoclastic breakdown of bone. Trace elements participate as enzymatic cofactors in both formation and degradation of the organic matrix. These effects have been documented *in vitro*, in cell cultures and tissues, and in normal and genetically impaired intact animals including humans. Several trace elements, particularly Zn and F, function in the formation and structure of the mineral matrix.

The recent comprehensive and articulate review by Beattie and Avenell (1992) describes and evaluates the published observations and conclusions for the participation of Zn, Cd, Cu, Fe, Mn V, B, Al, Ga, Si, Pb and F in bone metabolism. All trace elements are not equal. The "some more equal than others" are Zn, Mn, Cu and F . Certainly Si influences bone at the level of collagen and matrix aminoglycan synthesis. Under normal dietary and environmental circumstances Si is not of major concern.

Some trace elements are quite toxic and interfere with bone development and health. Cd and Ca are antagonistic. Cd toxicity is manifest as a debilitating condition resulting in fragile and brittle bones. It has been proposed that Cd compromises renal regulation of Ca, as well as competes with Cu and Zn for many metalloenzyme sites. Pb is concentrated in bone, and may play a significant role in long term bone toxicity. Manifestation of Al inhibition of bone growth and development has been observed in dialysis patients and

Correct citation: Saltman, P. 1997. More precious than gold: Trace elements in osteoporosis. In *Trace Elements in Man and Animals – 9: Proceedings of the Ninth International Symposium on Trace Elements in Man and Animals. Edited by* P.W.F. Fischer, M.R. L'Abbé, K.A. Cockell, and R.S. Gibson. NRC Research Press, Ottawa, Canada. pp. 517–520.

attributed to the blocking of osteoblast activity. Several elements including Ga, B, V and Fe manifest relatively minor direct effects in bone.

Our interest in the relation between trace elements and osteoporosis began with the speculation that slow healing fractures in a star athlete might be correlated with his neglect of adequate Ca and trace elements in his macrobiotic diet (Gold 1980). This was confirmed by our measurement of low serum levels of Cu, Zn and Mn, as well as high levels of Ca. Trace element and Ca supplementation was effective in healing the bones and permitting resumption of his career.

To pursue these observations we developed a trace element deficient rat model low in Cu and Mn which manifested porotic bones over a two year period (Strause et al. 1986). Cu is a well known cofactor for lysine oxidation and pyridinium cross-linking of collagen and elastin. Mn is required for glycosyl-transferases that are essential for the synthesis of glycosaminoglycan chondroitin sulfate, which is required for bone growth and development. In collaboration with Dr. Mark Grynpas we isolated total and bone-specific proteoglycan extracts from bones of deficient and control rats. The proteoglycans were significantly lower in the trace element deficient animals. The cellular basis of porosis was found both in osteoblasts and osteoclasts. By comparing demineralized bone powder implants with whole bone powders using Cu/Mn deficient and normal rats we found that osteoblasts were more severely inhibited than osteoclasts, leading to net bone loss (Strause et al. 1987).

To extend our observations from the animal model to the human we began a fruitful collaboration with Dr. Jean-Yves Reginster (Reginster et al. 1987). We found a clear implication that trace elements, particularly Mn, were involved in osteoporosis. Serum and bone biopsy analysis from normal and osteoporotic women demonstrated that osteoporotics had low trabecular bone volume, low bone mineral content, and low bone mineral density. These women also had significantly lower serum Mn levels than the controls.

We embarked upon a prospective, double-blind, placebo-controlled study to evaluate the efficacy of supplementary Ca with and without a combination of Cu, Mn, and Zn. Our subjects were 137 post-menopausal Caucasian volunteers, at least 50 years old and in good health who completed the study with 90% compliance. Ca-citrate-malate was used to provide 1000 mg/d of bioavailable Ca in four 250 mg doses. The trace element supplement (TE) was given as a single daily pill containing Cu (5.0 mg), Mn (2.5 mg) and Zn (15 mg). Food frequency questionnaires were also administered. Quantitative digital radiographic absorptiometry (Hologic) was used to compute bone mineral density in the lumbar (L2-L4) vertebral area. Women were divided into four groups for the study: Ca-placebo/TE-placebo; Ca/TE-placebo; Ca-placebo/TE; Ca/TE. Women receiving estrogen were not affected by any of the treatments. Their bone density remained constant. Women without hormone therapy showed significant changes in bone density which is shown clearly in Figure 1. As observed by others, post menopausal women lose

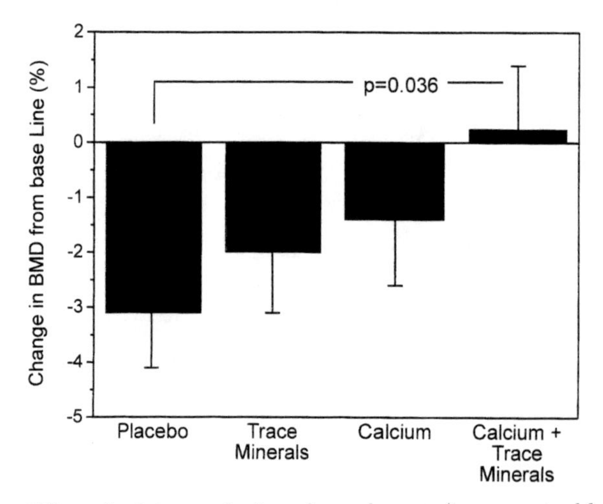

Figure 1. Effect of calcium and mineral supplementation on spinal bone mineral density in postmenopausal women. Figure values are mean ± SE. Sample sizes are: placebo n = 22, trace minerals n = 19, calcium n = 15, calcium plus trace minerals n = 20.

about 1.0–1.5% of their bone mineral density per year. Ca alone appears to somewhat ameliorate this loss. However, Ca and TE together prevented bone loss (Strause et al. 1994).

Medical science is vigorously involved in the search for 'magic bullets' to cure the ills besetting our society. Frequently the fundamental issues are of prevention rather than cure. We have extensive knowledge concerning the nutritional importance of 1000 mg/d of Ca in our diets for bone health. Yet in our studies the women were 'content' to ingest about 600 mg/d by their own food and supplement choices. If one includes the benign neglect of the average person for Ca and TE in their diets, not to mention the conventional wisdom of gurus and pyramids limiting consumption of dairy products and meats, it is no wonder that the incidence of osteoporosis is so high.

Perhaps the classical fairy tale recounting how for the want of a nail the kingdom was lost should be paraphrased — for the want of trace elements the skeleton was lost. Application of present nutritional strategies for adequate dietary and supplementary intake of Ca, and TE, as well as Vitamin D and F, combined with hormone replacement therapy and exercise would significantly improve our bone health in general, and our individual well being in particular.

Literature Cited

Beattie, J.H. & Avenell, A. (1992) Trace element nutrition and bone metabolism. Nutr. Rev. 5: 167–188.

Gold, M. (1980) Basketball bones. Science 80: 101–102.

Reginster, J.Y., Strause, L., Saltman, P. & Franchimont, P. (1988) Trace elements and postmenopausal osteoporosis: A preliminary study of decreased serum manganese. Med. Sci. Res. 16: 337–338.

Saltman, P. & Strause, L. (1993) The role of trace minerals in osteoporosis J. Am. Coll. Nutr. 12: 384–389.

Strause, L., Hegenauer, J., Saltman, P., Cone, R. & Resnick, D. (1986) Effects of long-term dietary manganese and copper deficiency on rat skeleton. J. Nutr. 116: 135–141.

Strause, L., Saltman, P. & Glowacki, J. (1987) The effect of deficiencies of manganese and copper on resorption of bone particles in rats. Calcif. Tissue. Int. 41: 145–150.

Strause, L., Saltman, P., Smith, K., Bracker, M. & Andon, M.B. (1994) Spinal bone loss in postmenopausal women supplemented with calcium and trace elements. J. Nutr. 124: 1060–1064.

Discussion

Q1. Alan Shenkin, Royal Liverpool University Hospital, Liverpool, England: The numbers in your trial were large enough for you to stratify them into those who had adequate calcium intake in the placebo group, and in the supplemented group, into those who had relatively crude intake, and relatively good intake. Did you separately analyze the totally supplemented group to those who had a crude intake versus those who had a good intake without the supplement. Did those who had a good intake have a better response? You should have been able to analyze each of the groups separately.

A. We did the best we could to balance them all the way across the four groups. I don't have those numbers. We had about 25 in each group at the end of the game, and so it wasn't that big. The fact of the matter was that we felt that to slice that salami any thinner would not get us anywhere. The nutrient value of the salami would be diminished. We were just pleased to get the numbers that we did. There it is, clear and unambiguous. But I hear what you're saying. Let me go back and see what I can squeeze out of the numbers.

Q2. Curtiss Hunt, USDA-ARS, Grand Forks, ND, USA: I appreciate your appreciation of the relationship between the organic and the inorganic components of bone. The assay you used will show you an increase in density related to the mineral. My question is on all these experiments, is it good bone, just because there's more density as revealed by this assay. In other words, did you see any change in the fracture rate?

A. You've got to talk to my orthopedic mavens. They are very happy to see this, because that's what they go with. I don't know who among the osteoporotic gurus are doing "is it good bone or not". It's certainly, from everything that we are told by them, good bone. We didn't have enough women, and we didn't include fracture rates as an endpoint. We have not done any follow-up, and there have been no resources to do so.

Q3. Mary L'Abbé, Health Canada, Ottawa, ON, Canada: Were any of the women on estrogen therapy?

A. We had a group of them who were, and we separated them. None of those that you saw in today's result were. What I said before, I want to re-emphasize, if they've been on hormone replacement therapy, we see no effect. HRT is good stuff.

Q4. John Bogden, New Jersey Medical School, Newark, NJ, USA: Paul, we have a large enough sample size to stratify on the basis of age. Did age matter? For example, did the younger women do better than the older women?

A. No, they did not. The minimum age here was 50, John. The mean age was 65 in each group. We looked for that a little bit, and there was not enough that we could see any pattern evolving. I was impressed at the fact that these were older women, and you could manipulate them. You know, everybody has kind of written them off at this point, and they're going to the bis-phosphonates and the calcitonins and God knows what else, and the fluoride treatments, and I don't want to start preaching, but the fact is, we've got a hell of a lot of knowledge in our society about nutrition, and it's not being used. And everybody around here is jerking around on whether it's zinc or manganese. Get a U.S. RDA out to everybody as the U.S. RDA stands now, and I don't want to put Janet out of business, but in point of fact we'd have a healthier America. It's self abuse that's killing us. It's ignorance that's killing us, not the lack of knowledge.

The Role of Aluminum in the Bone Disease of Parenterally Fed Patients

G.L. KLEIN

Department of Pediatrics, University of Texas Medical Branch, Galveston, Texas 77555-0352, USA

Keywords aluminum, total parenteral nutrition, osteomalacia, bone formation

Adults treated with total parenteral nutrition (TPN) for 4–6 weeks experience the insidious onset of progressive periarticular pain in weight-bearing joints that is unresponsive to narcotic analgesics. Moreover, these patients had low serum concentrations of calcitriol (1,25-dihydroxyvitamin D) with normal levels of its precursor, 25-hydroxyvitamin D (Klein et al. 1981, Shike et al. 1981). They were also shown to have low-normal serum concentrations of parathyroid hormone (PTH) and hypercalciuria (Klein et al. 1980, Shike et al. 1980). Iliac crest bone biopsies demonstrated low bone formation, patchy osteomalacia, and aluminum (Al) accumulation at the mineralization front, the region of new bone formation. Furthermore, Ott et al. (1983) demonstrated an inverse relationship between quantitative Al at the mineralization front and the rate of new bone formation. Al was also shown to be elevated in blood and urine of these patients (Klein et al. 1982) and subsequently in the bone, blood, and urine of infants receiving TPN treatment (Sedman et al. 1985). Similar observations were made in uremic patients receiving dialysis therapy with Al-treated water (Platts et al. 1977, Parkinson et al. 1979, Ott et al. 1982) or Al-containing phosphate binding gels (Bournerais et al. 1983, Andreoli et al. 1984). Identification of the main source of Al contamination of TPN solutions at that time, casein hydrolysate (Klein et al. 1982), its removal from the market, and introduction of chelation therapy using deferoxamine in uremic patients resulted in clinical, histomorphometric, and biochemical resolution (Klein and Coburn 1994, Ott et al. 1986). Experimental Al loading of rats (Ellis et al. 1979, Goodman et al. 1984a), piglets (Sedman et al. 1987), and dogs (Goodman et al. 1984b) provided supportive evidence for Al-associated reduced bone formation and osteomalacia.

Independent of bone formation, Al may adversely affect mineralization. Thus, the report of an infant with high serum Al levels while receiving long-term therapy with TPN, who developed sustained hypocalcemia following deferoxamine chelation, suggested that chelation of Al from bone allowed osteopenic bone to take up calcium at the expense of serum (Klein et al. 1989). Hypocalcemia was also reported in uremic patients receiving deferoxamine for Al chelation (Ott et al. 1986) and experimental studies *in vitro* (Meyer and Thomas 1986) and *in vivo* in rats (Rodriguez et al. 1990) provide further support for an Al-associated mineralization defect.

Al may also indirectly affect bone by accumulating in the parathyroid glands (Cann et al. 1979) and by inhibiting PTH secretion (Morrissey et al. 1983). However, a hypocalcemic stimulus in an Al-loaded TPN patient can elicit a PTH response (Klein et al. 1981). In addition, Al may inhibit conversion of 25-hydroxyvitamin D to calcitriol by the renal enzyme 25-hydroxyvitamin D-1-alpha hydroxylase by indirect and as yet undefined mechanisms (Henry and Norman 1985).

Another way in which Al may contribute to the bone loss in TPN patients is by aggravating hypercalciuria. A corollary to the Al impairment of bone calcium uptake is that calcium infused in the TPN solution will not be taken up by the skeleton and will be excreted in the urine. There is a direct relationship between urinary calcium excretion and both absorbed calcium and intravenously infused calcium (Klein and Coburn 1994). Linear regression of urinary calcium excretion plotted against both absorbed and infused calcium showed virtual overlap (Klein and Coburn 1994). Al loading appears to exacerbate the calciuria by increasing the amount of calcium excreted per amount of calcium infused (Klein and Coburn 1994).

Correct citation: Klein, G.L. 1997. The role of aluminum in the bone disease of parenterally fed patients. In *Trace Elements in Man and Animals – 9: Proceedings of the Ninth International Symposium on Trace Elements in Man and Animals. Edited by* P.W.F. Fischer, M.R. L'Abbé, K.A. Cockell, and R.S. Gibson. NRC Research Press, Ottawa, Canada. pp. 521–523.

However, despite the abundance of clinical observations and experimental studies demonstrating adverse effects of Al on bone, several experimental studies both *in vivo* (Quarles et al. 1989) and *in vitro* alone (Lau et al. 1991) or in combination with fluoride (Caversazio et al. 1996) argue for a stimulatory effect of Al on osteoblast recruitment and new bone formation. The explanation for these discrepancies is not at hand and future studies of Al toxicity to bone must address these apparently antithetical findings.

Literature Cited

Andreoli, S.P., Bergstein, J.M. & Sherrard, D.J. (1984) Aluminum intoxication from aluminum containing phosphate binders in children with azotemia not undergoing dialysis. N. Engl. J. Med. 310: 1079–1084.

Bournerais, F., Monnier, N. & Reveillaud, R.J. (1983) Risks of orally administered aluminum hydroxide and results of withdrawal. Proc. Europ. Dial. Transp. Assoc. 20: 207–212.

Cann, C.E., Prussin, S.G. & Gordan, G.S. (1979) Aluminum uptake by the parathyroid glands. J. Clin. Endocrinol. Metab. 49: 543–545.

Caversazio, J., Imai, T., Ammann, P., Burgener, D. & Bonjour, J.-P. (1996) Aluminum potentiates the effect of fluoride on tyrosine phosphorylation and osteoblast replication *in vitro* and bone mass *in vivo*. J. Bone Miner. Res. 11: 46–55.

Ellis, H.A., Mc Carthy, J.H. & Herrington, J. (1979) Bone aluminum in hemodialyzed patients and in rats injected with aluminum chloride: relationship to impaired bone mineralization. J. Clin. Pathol. 32: 832–844.

Goodman, W.G., Gilligan, J. & Horst, R.L. (1984a) Short term aluminum administration in the rat: effects on bone formation and relationship to osteomalacia. J. Clin. Invest. 73: 171–181.

Goodman, W.G., Henry, D.A., Horst, R., Nudelman, R.K., Alfrey, A.C. & Coburn, J.W. (1984b) Parenteral aluminum administration in the dog. II. Induction of osteomalacia and effect on vitamin D metabolism. Kidney Int. 25: 370–375.

Henry, H.L. & Norman, A.W. (1985) Interactions between aluminum and the actions and metabolism of vitamin D3 in the chick. Calcif. Tissue Int. 37: 484–487.

Klein, G.L., Alfrey, A.C., Miller, N.L., Sherrard, D.J., Hazlet, T.K, Ament, M.E. & Coburn, J.W. (1982) Aluminum loading during total parenteral nutrition. Am. J. Clin. Nutr. 35: 1425–1429.

Klein, G.L. & Coburn, J.W. (1994) Total parenteral nutrition and its effects on bone metabolism. Crit. Rev. Clin. Lab. Sci. 31: 135–167.

Klein, G.L., Horst, R.L., Norman, A.W., Ament, M.E., Slatopolsky, E. & Coburn, J.W. (1981) Reduced serum levels of 1-alpha, 25-dihydroxy-vitamin D during long-term parenteral nutrition. Ann. Intern. Med. 94: 638–643.

Klein, G.L., Snodgrass, W.R., Griffin, M.P., Miller, N.L. & Alfrey, A.C. (1989) Hypocalcemia complicating deferoxamine therapy in an infant with parenteral nutrition-associated aluminum overload: evidence for a role of aluminum in the bone disease of infants. J. Pediatr. Gastroenterol. Nutr. 9: 400–403.

Klein, G.L., Targoff, C.M., Ament, M.E., Sherrard, D.J., Bluestone, R., Young, J.H., Norman, A.W. & Coburn, J.W. (1980) Bone disease associated with parenteral nutrition. Lancet 2: 1041–1044.

Lau, K.H.W., Yoo, A. & Wang, S.P. (1991) Aluminum stimulates the proliferation and differentiation of osteoblasts *in vitro* by a mechanism that is different from fluoride. Mol. Cell. Biochem. 105: 93–105.

Morrissey, J., Rothstein, M., Mayor, G. & Slatopolsky, E. (1983) Suppression of parathyroid hormone secretion by aluminum. Kidney Int. 23: 699–704.

Ott, S.M., Andress, D.L., Nebeker, H.G., Milliner, D.S., Maloney, N.A., Coburn, J.W. & Sherrard, D.J. (1986) Changes in bone histology after treatment with desferrioxamine. Kidney Int. 29(sup 18): S108-S113.

Ott, S.M., Maloney, N.A., Coburn, J.W., Alfrey, A.C. & Sherrard, D.J. (1982) The prevalence of bone aluminum deposition in renal osteodystrophy and its relation to the response to calcitriol therapy. N. Engl. J. Med. 307: 709–713.

Ott, S.M., Maloney, N.A., Klein, G.L., Alfrey, A.C., Ament, M.E., Coburn, J.W. & Sherrard, D.J. (1983) Aluminum is associated with low bone formation in patients receiving chronic parenteral nutrition. Ann. Intern. Med. 98: 910–914.

Parkinson, I.S., Ward, M.K., Feest, T.G., Fawcett, R.W.P. & Kerr, D.N.S. (1979) Fracturing dialysis osteodystrophy and dialysis encephalopathy. An epidemiological survey. Lancet 1: 406–409.

Platts, M.M., Goode, G.C. & Hislop, J.S. (1977) Composition of domestic water supply and the incidence of fractures and encephalopathy in patients on home dialysis. Br. Med. J. 2: 657–660.

Quarles, L.D., Gitelman, H.J. & Drezner, M.K. (1989) Aluminum-induced de novo bone formation in the beagle: a parathyroid hormone-dependent event. J. Clin. Invest. 83: 1644–1650.

Rodriguez, M., Felsenfeld, A.J. & Llach, F. (1990) Aluminum administration in the rat separately affects the osteoblast and bone mineralization. J. Bone Miner. Res. 5: 59–67.

Sedman, A.B., Alfrey, A.C., Miller,N.L. & Goodman, W.G. (1987) Tissue and cellular basis for impaired bone formation in aluminum-related osteomalacia in the pig. J. Clin. Invest. 79: 86–92.

Sedman, A.B., Klein, G.L., Merritt, R.J., Miller, N.L., Weber, K.O., Gill, W.L., Anand, H. & Alfrey, A.C. (1985) Evidence of aluminum loading in infants receiving intravenous therapy. N. Engl. J. Med. 312: 1337–1343.

Discussion

Q1. Curtiss Hunt, USDA-ARS, Grand Forks, ND, USA: If we say that the dose makes the poison, and knowing that TPN is circumventing the normal way in which aluminum would be taken into the body, is there a chance here that what you're seeing is simply a toxic effect, or is it that in a normal situation, at much lower levels, that aluminum really could have a toxic effect?

A. Well, there's always that possibility. I guess what I would say is that a lot of the experiments have been done with parenteral aluminum. It's difficult to speculate on that issue, because nobody has been clear about it.

Q2. (unable to identify questioner): Given what we know about the aluminum content of these fluids, is there any regulation now to minimize aluminum content?

A. In a nutshell, no. It's not for lack of trying. Things have moved, but very slowly. This is not just a problem in the United States, but internationally. People are looking to try to get regulations, but I can't be very optimistic about that happening anytime soon.

Serum Fluoride and Bone Metabolism in Patients with Long Term Hemodialysis

K. KONO, Y. ORITA, T. DOTE, Y. TAKAHASHI, K. USUDA, Y. SUMI, M. SAITO, M. SHIMAHARA,[†] AND N. HASHIGUCHI[†]

Department of Hygiene and Public Health, and [†]Department of Oral Surgery, Osaka Medical College, 2-7 Daigakumachi, Takatsuki City, Osaka 569, Japan

Keywords serum fluoride, bone metabolism, patients with hemodialysis

With growing experience of the long-term treatment of patients with end stage renal disease by hemodialysis, the safety of fluoridated water supply for dialysate and the effect on bone metabolism has been discussed. In Europe and the USA, administration of sodium fluoride (NaF) has been attempted both independently and concurrently with calcium and vitamin D (Franke et al. 1974, Hasling et al. 1987). The authors conducted the current study to examine bone metabolism and the behavior of fluoride in cases of chronic renal failure.

Materials and Methods

In the present study, concentrations of fluoride (F), calcium (Ca), aluminium (Al), and biochemical indices of bone metabolism, such as bone gla protein (BGP), parathyroid hormone (PTH), alkaline phosphatase (ALP) in serum, and the bone mineral density of both radius (radial-BD) and lumbar spine (spinal-BD) were analyzed in 95 patients (45 males and 50 females) with hemodialysis to clarify the combined effects of F concentration and treatment of hemodialysis on the bone metabolism in those patients. The average length of hemodialysis treatment was 5 years and 8 months for males, and 8 years 5 months for females. Patients received hemodialysis treatment in 4-h sessions 3 times a week. As the control group, 240 healthy subjects (100 males and 140 females) with normal kidney functions were chosen.

Results

Table 1 shows the mean values of observation points in the subjects. The average values of BGP, F, PTH, ALP and inorganic phosphate (i-P) were much higher than those in healthy subjects. Although Radial-BD values for the patients were lower than those for healthy subjects, Spinal-BD values were nearly the same as those for the controls.

Table 2 shows correlation matrices of the observation points in the patients. F and BGP in males, F and ALP in females, and PTH and BGP in both subjects showed direct correlations.

Discussion

It has been discussed that F causes hardening of the bone and promotes formation of the osteoid, especially in the backbone of patients undergoing hemodialysis (Riggs et al. 1990).

In this study, the duration of hemodialysis treatment and radial-BD displayed a significant negative correlation, though spinal-BD did not decrease with duration of hemodialysis. The administration of NaF has been practiced as the therapy for osteoporosis in some countries (Pak et al. 1990). The dosage is 4 to 5 mg/kg body weight, and serum F concentration of patients in the current study is near to the peak in serum concentration seen with NaF therapy. Although bone morbidity is not caused by any single factor, this study suggests that F may suppress the decrease in spinal-BD among the patients.

Correct citation: Kono, K., Orita, Y., Dote, T., Takahashi, Y., Usuda, K., Sumi, Y., Saito, M., Shimahara, M., and Hashiguchi, N. 1997. Serum fluoride and bone metabolism in patients with long term hemodialysis. In *Trace Elements in Man and Animals – 9: Proceedings of the Ninth International Symposium on Trace Elements in Man and Animals. Edited by* P.W.F. Fischer, M.R. L'Abbé, K.A. Cockell, and R.S. Gibson. NRC Research Press, Ottawa, Canada. pp. 524–526.

Table 1. Mean values of the observation items in patients with HD (Mean ± SD).

Items	male (n = 45)	female (n = 50)
Age (year)	58.2 ± 12.4	56.6 ± 19.5
Duration (year)	5.7 ± 3.8	8.4 ± 5.3
BGP (ng/mL)	33.2± 24.8	35.6 ± 16.8
Spinal-BD (g/cm^2)	0.96 ± 0.16	0.91 ± 0.17
Radial-BD (g/cm^2)	0.71 ± 0.13	0.58 ± 0.26
PTH-HS (pg/mL)	5733.8 ± 4822.5	6510.2 ± 5622.7
PTH-intact (pg/mL)	90.6 ± 74.5	96.9 ± 69.2
ALP (IU)	228 ± 110.9	222 ± 85.8
F (μmol/L)	12.0 ± 10.3	11.8 ± 8.62
Ca (mmol/L)	2.15 ± 0.28	2.23 ± 0.60
P (mmol/L)	1.84 ± 0.84	2.03 ± 0.61
Al (μmol/L)	66.7 ± 48.1	74.1 ± 33.3

Table 2. The correlation matrices of observation items in patients with HD.

Male (n = 45)

	BGP	F	Spinal BD	Radial BD	PTH HS	PTH intact	P	Ca	ALP	Al	Age
BGP	...	0.3	–0.15	–0.25	0.54	0.53	–0.12	–0.27	0.13	0.14	–0.12
F	0.12	...	0.2	0.05	–0.14	–0.08	–0.12	–0.15	0.1	–0.09	0.05
Spinal BD	–0.4	0.23	...	0.23	–0.31	–0.48	0.02	0.25	–0.33	0.17	0.07
Radial BD	–0.14	0.02	0.37	...	–0.18	–0.38	–0.08	0.03	–0.34	–0.1	–0.41
PTH HS	0.61	–0.06	–0.28	–0.09	...	0.73	0.1	–0.17	0.05	–0.01	–0.23
PTH intact	0.48	–0.02	–0.31	–0.1	0.52	...	–0.03	–0.36	0.24	0.03	–0.04
P	0.29	0.02	0.06	–0.12	0.18	0.3	...	0.05	–0.04	–0.13	0.01
Ca	–0.18	–0.2	0.12	–0.19	–0.11	–0.58	–0.16	...	–0.15	–0.07	–0.34
ALP	0.19	0.69	–0.24	–0.03	0.03	0.2	–0.19	–0.38	...	0.31	0.05
Al	–0.23	–0.04	0.01	0.08	–0.21	–0.23	–0.18	–0.03	–0.04	...	0.01
Age	–0.03	0.02	–0.14	–0.33	0.1	0.12	–0.06	0.15	0.16	–0.17	...

Female (n = 50)

Literature Cited

Franke, J., Rempel, H. & Franke, M. (1974) Three years experience with sodium fluoride therapy of osteoporosis. Acta Orthop. Scand. 45: 1–2.

Hasling, G., Neilsen, H.E., Meleson, F. & Mosekilde, L. (1987) Safety of osteoporosis treatment with sodium fluoride, calcium phosphate and vitamin D. Mineral Electrolyte Metab. 13: 96–103.

Pak, C.Y.C., Sakhaee, K. & Zerwekh, J.E. (1990) Effect on intermittent therapy with slow release fluoride preparation. J. Bone Mineral Res. 5: 5149–5155.

Riggs, B.L., Hodgoon, S.F. & Ofallon, W.M. (1990) Effect of fluoride treatment on the fracture rate in postmenopausal women with osteoporosis. N. Engl. J. Med. 322: 802–809.

Discussion

Q1. Phyllis Paterson, University of Saskatchewan, Saskatoon, SK, Canada: You are proposing that fluoride has a key role, however fluoride correlated with spinal bone density and was not associated with hemodialysis. Can you explain this further?

A. The serum fluoride concentration is a little bit high, compared to other countries, because we have so much fluoride from seafood. Fluoride maybe strongly correlated to bone density. In animal experiments that we have done, the fluoride concentration in the spinal bone is very high. So maybe fluoride does affect bone metabolism.

Q2. Phyllis Paterson, University of Saskatchewan, Saskatoon, SK, Canada: Has the aluminum problem been sorted out?

A. Yes, aluminum has also been correlated to bone metabolism. But in Japan, the concentrations of aluminum are not so high.

Q3. Gordon Klein, University of Texas Medical Branch, Galveston, TX, USA: Just a comment, and that is that there are some recent data from Switzerland that suggest that aluminum and fluoride together actually may stimulate bone formation.

A. Yes, that may be so.

Effect of Cadmium Exposure and Diet with Lower Calcium on Bone in Female Rats

M. PIASEK, N. SCHÖNWALD, M. BLANUŠA, K. KOSTIAL, AND B. MOMČILOVIĆ
Laboratory of Mineral Metabolism, Institute for Medical Research and Occupational Health, 10001 Zagreb, Republic of Croatia

Keywords bone, cadmium exposure, calcium, female rats

Due to increasing environmental pollution with cadmium (Cd), the potential Cd effects with regard to so called Itai-Itai disease, which is characterized by extensive bone demineralization in multiparous women, has attracted considerable attention. Recent work of Bhattacharyya (1991) in mice and beagles supports the view that Cd exposure contributes to the pathogenesis of the Itai-Itai disease providing evidence that Cd causes increased Ca bone loss in midlactating animals.

The aim of this study was to evaluate Cd effects on the bone of rats in relation to lower Ca in the diet in nonpregnant vs. lactating dams, and also in suckling and adolescent animals.

Materials and Methods

Female rats (Wistar) were exposed to Cd in drinking water (0.27 mmol/L, Cd as chloride). The animals were fed either high or low calcium rat feed (Domžale, Slovenia) with 1% and 0.3% calcium (Ca), respectively, during three weeks of gestation and 17 d of lactation (dams), or in total 5.5 weeks (nonpregnant) and then killed. Some of the female offspring, weaned at the age of 17 d, continued to receive the same Cd exposure and feeding regimen till the age of 10 weeks, and then killed. General health (body weight, appearance and mortality) and reproductive parameters (fertility index, litter size, and pup survival) were recorded throughout the experiment. The bone parameters evaluated in the femur at the end of experiment were: dry and ash femur weights, and Ca content. Femur Ca concentrations were measured by flame atomic absorption spectrometry (Varian, AA 375, Australia). All numeric data were statistically analyzed by ANOVA/MANOVA and Duncan's multiple range test at level of significance $P < 0.05$ (CSS Biostatistica, rel. 3.1, Statsoft 1991). Cadmium effect was evaluated within the group of animals on the same Ca feeding regimen. Also overall main Cd effect irrespective of Ca intake, main Ca effect irrespective to Cd exposure, and Cd × Ca interactions were evaluated.

Results and Discussion

Subchronic oral exposure to Cd and/or to lower Ca in feed had no influence on either general health or reproductive outcome in adult females. However, Cd significantly reduced pup weight, and both Cd and lower Ca reduced body weights in 10-week-old adolescent female offspring (Schönwald 1995). Table 1 shows that Cd exposure reduced dry weight, ash weight, and Ca content of the femur in lactating dams and not in nonpregnat rats. No effects on bone parameters were found in 17-day-old sucklings. In 10-week-old adolescent females, both Cd exposure and lower Ca intake decreased dry and ash weights and Ca content of the femur.

In conclusion, our results show that lactating rat dams present a group with an increased sensitivity to Cd-induced bone mass loss, as was observed earlier in nursing mice (Bhattacharyya 1991). In addition, adolescent females present a group with increased sensitivity to Cd-induced reduction of bone mass formation, especially under the condition of a low Ca intake.

Correct citation: Piasek, M., Schönwald, N., Blanuša, M., Kostial, K., and Momčilović, B. 1997. Effect of cadmium exposure and diet with lower calcium on bone in female rats. In *Trace Elements in Man and Animals – 9: Proceedings of the Ninth International Symposium on Trace Elements in Man and Animals. Edited by* P.W.F. Fischer, M.R. L'Abbé, K.A. Cockell, and R.S. Gibson. NRC Research Press, Ottawa, Canada. pp. 527–528.

Table 1. Dry and ash weights and Ca content of the femur in nonpregnant rats, lactating dams, and their 17-d-old suckling pups after 5.5-week exposure to Cd (0.27 mmol/L in drinking water) and feeding either a high (1%) or low (0.3%) Ca diet, and in adolescent 10-wk-old female offspring who continued with the same Cd exposure and Ca feeding regimen after weaning at age of 17 d.[1]

	High Ca Diet (mg)		Low Ca Diet (mg)	
Group	Control	Cd Exposed	Control	Cd Exposed
Dry Femur Weight				
Nonpregnant	0.469±0.041	0.478±0.036	0.484±0.044	0.476±0.042
Dams [b]	0.453±0.029	0.346±0.047*	0.417±0.024	0.371±0.045*
Sucklings	0.070±0.007	0.067±0.008	0.064±0.018	0.068±0.007
Adolescent [a,b]	0.323±0.026	0.278±0.031*	0.301±0.015	0.241±0.015*
Femur Ash Weight				
Nonpregnant	0.332±0.032	0.328±0.024	0.336±0.035	0.327±0.027
Dams [b]	0.307±0.022	0.235±0.034*	0.273±0.021	0.240±0.032*
Sucklings	0.024±0.003	0.023±0.002	0.021±0.007	0.023±0.003
Adolescent [a,b]	0.206±0.020	0.175±0.020*	0.184±0.009	0.149±0.011*
Femur Ca Content				
Nonpregnant	157 ±15.7	157 ± 10.0	159 ±16.6	159 ±12.7
Dams [b]	143 ±7.57	115 ± 15.9*	129 ±11.7	114 ±16.6
Sucklings	10.5±1.16	9.83± 0.903	9.30± 4.03	10.4±1.17
Adolescent [a,b]	98.6±10.8	80.1±9.24*	92.4±5.67	65.5±4.91*

[1]Values are means ± SD, n = 8–10. *Significantly different from respective control ($P < 0.05$).
[a]Main Ca effect; [b]Main Cd effect. No Ca × Cd interactions were found.

Literature Cited

Bhattacharyya, M.H. (1991) Cadmium-induced bone loss: increased susceptibility in females. Water Air Soil Poll. 57–58, pp. 665–673.

Schönwald, N. (1995) Health effects of cadmium on rats fed low iron or calcium diet (in Croatian). Ph.D. Thesis. University of Zagreb, Zagreb, Republic of Croatia.

Relationships Between Indicators of Mineral Status and Bone Mineral Hormones in Adults Supplemented with Zinc, Copper or Chromium*

A.B. ARQUITT, J.R. HERMANN, AND B.O. ADELEYE†

Nutritional Sciences Department, Oklahoma State University, Stillwater, OK 74078-6141, USA; †Present address: College of Applied Sciences, University of Southwest Louisiana, Lafayette, LA 70504-0399, USA

Keywords bone, trace elements, copper, iron

Osteoporosis is a significant health problem in the United States particularly as the mean age of the population increases. Preserving bone density or preventing further deterioration has been the focus of drug intervention therapies and dietary advice particularly for women. Calcium has long been identified as important in bone health with trace elements receiving less attention. This study investigated the effects of supplementation with zinc, copper, chromium or placebo in adults over the age of 50 on bone mineral hormones and enzymes.

Materials and Methods

Thirty-four apparently healthy, independently living adult volunteers over the age of 50 were randomly assigned to one of four treatment groups which were balanced for sex and mean serum cholesterol. Capsules containing either lactose only (placebo), 15.2 mg zinc as zinc sulfate, 1.7 mg copper as copper carbonate, or 120.5 μg chromium as chromium chloride were taken twice a day for eight weeks with meals. Males in the chromium supplemented group were dropped from all analyses due to initial high serum ferritin concentrations. The subjects recorded three day dietary intakes immediately prior to each data collection period.

Blood samples were collected in plastic syringes (Sarstedt, Newton, NC). Serum samples were held on ice for two hours prior to separation to allow for clot retraction without affecting alkaline phosphatase activity. Hemoglobin in whole blood and erythrocytes was analyzed using the cyanmethemoglobin method (Sigma). Plasma samples were separated within 30 min of collection and stored frozen until analyses. Radioimmunoassays were used for analysis of osteocalcin (Nichols Institute) and parathyroid hormone, ferritin and estradiol (Diagnostic Products Corp.). Alkaline phosphatase was determined by a quantitative kinetic method (Sigma). Plasma and erythrocyte minerals were wet and dry ashed and analyzed by atomic absorption spectroscopy using a Perkin-Elmer 5100PC (Norwalk, CT) instrument. Plasma and erythrocyte iron and zinc were analyzed using an air–acetylene flame with deuterium background correction; copper samples were analyzed by graphite furnace AAS with Zeeman background correction. All mineral analyses were completed using dilutions of the same ashed sample. Erythrocyte mineral concentrations are expressed per gram of hemoglobin.

Results

Correlations between changes in mineral status indicators and bone mineral metabolism occurred most commonly in those subjects supplemented with copper (Table 1). In the zinc supplemented group change

*Supported by Hatch OKLO 2132 and 2142.

Correct citation: Arquitt, A.B., Hermann, J.R., and Adeleye, B.O. 1997. Relationships between indicators of mineral status and bone mineral hormones in adults supplemented with zinc, copper or chromium. In *Trace Elements in Man and Animals – 9: Proceedings of the Ninth International Symposium on Trace Elements in Man and Animals. Edited by* P.W.F. Fischer, M.R. L'Abbé, K.A. Cockell, and R.S. Gibson. NRC Research Press, Ottawa, Canada. pp. 529–531.

Table 1. Correlations between differences in pre and post supplementation measures of bone metabolism and mineral status indicators in copper supplemented adults.[1]

	Osteocalcin		PTH	ALP	
	Male	Female	Female	Male	Female
Erc Cu		0.966		–0.983	
P Zn			–0.959		–0.966
P Fe	0.999				
Hemoglobin	0.997		–0.996		

[1] Correlations are significant at $p < 0.05$

Table 2. Regression coefficients for prediction of change in osteocalcin in subjects supplemented with copper or chromium.

	Cu-Male	Cu-Female	Cr-Female
Intercept	3.440	–0.868	–0.209
Δ			
P Fe, μmol/L	0.926		0.051
P Zn, μmol/L			0.237
Erc Cu, μg/g Hb		–1.163	
Hemoglobin, g/L	–0.016		
Estradiol, pmol/L			–0.014
ALP, μkat/L			0.002
R^2	0.999	0.934	0.999
p<	0.01	0.04	0.04

Table 3. Regression coefficients for prediction of change in PTH in subjects supplemented with copper or chromium.

	Cu-Female	Cr-Female
Intercept	–1.945	3.315
Δ		
P Zn, μmol/L	0.516	
Ferritin, μg/L		0.121
Hemoglobin, g/L	–0.277	
ALP, μkat/L		0.041
R^2	0.999	0.929
p<	0.01	0.02

Table 4. Regression coefficients for prediction of change in ALP in subjects supplemented with copper or chromium.

	Cu-Male	Cu-Female	Cr-Female
Intercept	–50.180	–51.352	57.341
Δ			
P Fe, μmol/L		–0.569	–5.113
P Zn, μmol/L		2.761	
Erc Cu, μmol/L	–13.765		
Hemoglobin, g/L			–0.807
PTH, μg/L			16.892
Estradiol, pmol/L			–0.248
R^2	0.966	0.999	0.999
p<	0.02	0.01	0.01

in serum ferritin was negatively correlated with change in PTH in both women and men but was positively correlated with parathyroid hormone (PTH) in the chromium supplemented females. In zinc supplemented females change in plasma copper was positively correlated with change in osteocalcin. In the placebo group no significant correlations between mineral status indicators and bone metabolism were found.

Stepwise regression equations for osteocalcin, PTH and alkaline phosphatase (ALP) are shown in Tables 2–4 for subjects in the copper and chromium supplementation groups. In the zinc group only for PTH in males did the data support a significant regression equation. In this case change in serum ferritin was the only significant predictor for change in PTH ($R^2 = 0.997$, $p < 0.05$).

Discussion

Bone mineral hormones and enzymes are affected by supplementation with Cr, Cu and Zn. These supplements have different effects in males and in females. As shown by the significant correlations between osteocalcin, parathyroid hormone and alkaline phosphatase and mineral status indicators as well as the regression equations, these minerals appear to interact through their effects on indicators of iron status.

Osteopenia Induced by Long-Term, Low and High-Level Exposure of the Adult Rat to Lead

H.E. GRUBER,[†] F. KHALIL-MANESH,[‡] T.V. SANCHEZ,[¶] M.H. MEYER,[†] AND H.C. GONICK[‡]

[†]Orthopaedic Research, Carolinas Medical Center, Charlotte, NC 28232, [‡]Department of Medicine, Cedars-Sinai Medical Center, Los Angeles, CA 90048, and [¶]Norland, Fort Atkinson, WI 53538, USA

The skeleton, the major site for lead (Pb) accumulation, is responsible for the largest fraction of the total body burden, but long-term effects of low level exposure in adults remain unclear. Most previous studies of Pb and the skeleton have focused upon deleterious effects of Pb on bone growth and development in the young. In addition, skeletal changes with Pb exposure have rarely been correlated with Pb nephropathy. We present here findings in an adult rat model in which: 1) a low Pb exposure (100 ppm) produced blood Pb values comparable to those seen in human populations with modest environmental exposure; this treatment resulted in no overt renal functional abnormalities over 12 months (Khalil-Manesh et al. 1993); 2) a higher Pb exposure (5000 ppm) produced blood Pb levels comparable to those seen with industrial exposure (prior to 1978 U.S. standards). This higher Pb exposure resulted in Pb nephropathy which advanced in severity from the sixth to the twelfth month (Khalil-Manesh et al. 1992). Skeletal changes were assessed by bone mineral densitometry, chemical bone ash evaluation and quantitative bone histomorphometry.

Materials and Methods

Animal studies were approved by the Institutional Animal Care and Use Committee of Cedars-Sinai Medical Center. Male Sprague-Dawley rats weighed approximately 200 g at study initiation. Both high Pb (HiPb) and low Pb (LoPb) rats were fed a semi-synthetic purified rat diet mixture (normal protein test diet, I.C.N. Pharmaceuticals, Cleveland, OH) containing 1.1% calcium and 2.2% vitamin fortification mixture (which contained 0.125 g vitamin D_2 in a vitamin mix, added as 1 kg/45.45 kg of diet). This diet was chosen because of observations by Mylorie et al. (1978) that it produced enhanced susceptibility to lead toxicity. Animals were given either distilled water (controls) or water supplemented with Pb as lead acetate. High lead animals received 5000 ppm lead acetate in drinking water, reduced to 1000 ppm after 6 months exposure because of the early development of central nervous system toxicity and blood lead levels in excess of 100 µg/dL. Low lead animals received 100 ppm lead acetate in drinking water throughout this study. Control animals were pair-fed with their respective experimental group. Rats were euthanized at 1, 3, 6, 9 and 12 months. The right femorae were dissected free of soft tissue and stored cold in 100% ethanol for bone densitometry (using the Norland XR-26 Dual Energy X-ray Densitometer) or for ashing and atomic absorption mineral analyses. Contralateral femora were fixed and embedded undecalcified in methacrylate, and assessed with the OsteoMetrics, Inc. software and hardware system.

Results

Both LoPb and HiPb animals showed significant 12 month blood Pb levels (LoPb 21 ± 3 µg/dL; HiPb 59 ± 18; controls, 3 ± 1 (mean ± sem) $p = 0.001$). Dual energy X-ray densitometry of the femur detected a significant decrease in bone density in HiPb animals by 3 months which remained significantly lowered through 12 months (3 months: HiPb: 0.498 ± 0.011 grams (6) vs. control: 0.546 ± 0.012 (6), $p < 0.003$). By 12 months density was significantly lowered in LoPb animals ($p = 0.001$). Mineral analyses of ashed

Correct citation: Gruber, H.E., Khalil-Manesh, F., Sanchez, T.V., Meyer, M.H., and Gonick, H.C. 1997. Osteopenia induced by long-term, low and high-level exposure of the adult rat to lead. In *Trace Elements in Man and Animals – 9: Proceedings of the Ninth International Symposium on Trace Elements in Man and Animals. Edited by* P.W.F. Fischer, M.R. L'Abbé, K.A. Cockell, and R.S. Gibson. NRC Research Press, Ottawa, Canada. pp. 532–533.

femurs showed a significant lead content after 1, 3, 9 and 12 months of exposure (1 month: LoPb, 0.020 ± 0.002 (4) (% ash weight) vs. control, 0.008 ± 0.0004 (4); HiPb, 0.016 ± 0.001 (8); control, 0.007 ± 0.0004 (6) ($p < 0.002$; 12 months: LoPb, 0.062 ± 0.002 (9) vs. control 0.006 ± 0.0001 (5); HiPb, 0.090 ± 0.006 (8) vs. control 0.006 ± 0.0001 (8), $p < 0.0001$). Ca levels (% ash weight) were significantly lowered at 9 months in HiPb and 12 months in both groups; P was significantly lowered at 12 months in both groups ($p \leq 0.003$). At 12 months in the LoPb group, trabecular bone volume in the distal femur was decreased to only 20% of control ($p = 0.002$), osteoid indices were significantly elevated for both osteoid surface, volume and seam thickness, and resorbing surface was significantly elevated. In the HiPb group at 12 months, endosteal bone surface findings were similar to those seen in the 12 month LoPb group; osteoclast numbers and surface were significantly elevated.

Discussion

These findings show that 1) Pb is incorporated into bone mineral after only 1 month exposure to LoPb; 2) Significant osteopenia (osteoporosis) occurred following 12 months exposure to LoPb; HiPb caused osteopenia after 3 months. Osteopenia was supported by both bone densitometry and quantitation histomorphometry. 3) No normal compensatory mechanism was elicited to maintain bone mass in the presence of Pb exposure. 4) Bone histology points to the presence of a mineralization defect (osteomalacia) indicated by widened osteoid seams and increased osteoid surface and seam widths. 4) The elevations in resorption indices point to the presence of secondary hyperparathyroidism.

Goyer et al. (1994) and Silbergeld et al. (1988) have alerted us to the important fact that we should consider lead as a serious potential risk factor for osteoporosis in the elderly population and in other individuals at risk because of Pb exposure. The data presented here reinforce that need and also raise the possibility of co-existing osteomalacia and secondary hyperparathyroidism.

Acknowledgement

International Lead Zinc Research Organization.

References

Goyer, R. A., Epstein, S., Bhattacharyya, M., Korach, K. S. & Pounds, J. (1994) Environmental risk factors for osteoporosis. Environmental Health Perspectives 102: 390–394.

Khalil-Manesh, F., Gonick, H. C. & Cohen, A. H. (1993) Experimental model of lead nephropathy. III. Continuous low lead administration. Arch. Envirn. Health 48: 271–278.

Khalil-Manesh, F., Gonick, H. C., Cohen, A. H., Alinovi, R., Bergamaschi, E., Mutti, A. & Rosen, V. J. (1992) Experimental model of lead nephropathy. I. Continuous high-dose lead administration. Kid. Int. 41: 1192–1203.

Mylorie, A. A., Moore, L., Olyai, B. & Anderson, M. (1978) Increased susceptibility of lead toxicity in rats fed semipurified diets. Environ. Res. 15: 57–64.

Silbergeld, E. K., Schwartz, J. & Mahaffey, K. (1988) Lead and osteoporosis: Mobilization of lead from bone in postmenopausal women. Environ. Res. 47: 79–94.

Mechanisms of Trace Element Toxicity

Chair: M.G. Cherian

Epigenetic Mechanisms of Nickel Carcinogenesis

MAX COSTA

The Nelson Institute of Environmental Medicine and The Kaplan Comprehensive Cancer Center, New York University Medical Center, New York, NY 10016, USA

Keywords nickel, carcinogenesis, cancer

Epidemiological studies have clearly implicated certain nickel compounds as human carcinogens based upon a higher incidence of lung and nasal cancer among nickel refinery workers (IARC 1990). Particulate nickel compounds were the agents thought to be most carcinogenic to humans (IARC 1990). Certain nickel compounds (crystalline nickel subsulfide and nickel sulfide) are highly carcinogenic in animal models, inducing cancer at virtually any site of administration, whereas similar administration of amorphous nickel sulfide or water-soluble nickel salts generally did not yield tumors (IARC 1990, Costa 1991). A recent National Toxicology Program (NTP) bioassay has been conducted on water-soluble nickel sulfate, crystalline nickel subsulfide, and green nickel oxide (calcined at high temperatures) (Dunnick and Elwell 1995). Following a two year, 5 d per week inhalation study, rats and mice did not develop tumors with water-soluble nickel sulfate. However, crystalline nickel subsulfide and nickel oxide induced lung tumors (Dunnick and Elwell 1995).

In 1980 I proposed that the carcinogenic activity of finely powdered nickel compounds was proportional to target cell uptake by phagocytosis since carcinogenic crystalline nickel sulfide and subsulfide particles were phagocytized by fibroblasts undergoing cellular transformation while non-carcinogenic amorphous nickel sulfide particles were not (Costa and Mollenhauer 1980). Water-soluble nickel salts were less transforming compared to the phagocytized crystalline nickel sulfide particles because the uptake of nickel ions in solution by cells was poor in comparison to phagocytized crystalline nickel sulfides. Results of the recent NTP bioassay supports this data in demonstrating that crystalline nickel subsulfide, a compound that has a dissolution half-life of approximately 30 d was very carcinogenic. Retention of nickel in the lungs was almost the same as water-soluble $NiSO_4$, but not as high as nickel oxide which has an 11 year dissolution half-life, however, NiO was not phagocytized as readily by fibroblasts as the highly carcinogenic nickel subsulfide (Costa 1991, Dunnick and Elwell 1995, Costa and Mollenhauer 1980). From these and other studies, the hypothesis was put forth that the carcinogenic potency of a nickel compound depended upon the ability of that compound to elevate levels of Ni(II) in target cells. Animal studies which employed Mg^{2+} as an antagonist of Ni_3S_2 carcinogenesis showed that these ions could only inhibit carcinogenesis if given within the first day after Ni_3S_2 administration. When given after this critical time interval, tumor formation in animals was not attenuated (Kasprzak et al. 1985). Such results suggest that early phagocytosis was a critical event in Ni_3S_2 carcinogenesis using animal models and support the model that I proposed based upon *in vitro* studies (Costa and Mollenhauer 1980).

During the course of nickel-induced transformation of Chinese hamster embryo cells, we found that Ni(II) selectively interacted with the heterochromatic long arm of the X-chromosome in Chinese hamster embryo cells, probably because Ni(II) favored binding to protein, and in particular, it coordinated well

Correct citation: Costa, M. 1997. Epigenetic mechanisms of nickel carcinogenesis. In *Trace Elements in Man and Animals – 9: Proceedings of the Ninth International Symposium on Trace Elements in Man and Animals. Edited by* P.W.F. Fischer, M.R. L'Abbé, K.A. Cockell, and R.S. Gibson. NRC Research Press, Ottawa, Canada. pp. 534–537.

with histones (Costa 1991, Margerum and Anliker 1988). Binding of Ni(II) to histone HI (which is abundant in heterochromatin) has been shown to render it quite redox active (Costa 1991). Ni(II) forms octahedral or square planar complexes with proteins which are more efficient in coordinating nickel complexes than high molecular weight, double-stranded DNA (Margerum and Anliker 1988). Mg^{2+} is capable of antagonizing Ni(II)-induced damage to heterochromatin and also inhibits nickel-induced cell transformation and carcinogenesis in animal models (Costa 1991, Kasprzak et al. 1985). Ni(II) and Mg^{2+} have similar atomic radii and Ni(II) may substitute for Mg^{2+} in its function of maintaining a higher order of chromatin structure (Margerum and Anliker 1988, Borochov et al. 1984). However, based on circular dichroism studies, Ni(II) is known to be more efficient at condensing chromatin compared with Mg^{2+} (Borochov et al. 1984).

Carcinogenic nickel compounds transformed male Chinese hamster embryo cells much more readily than female cells and about half of the male nickel-transformed cells exhibited deletions of the heterochromatic long arm of the X-chromosome (Conway and Costa 1989). The long arm of the X-chromosome (q) is the longest contiguous region of heterochromatin in the Chinese hamster genome and provides a favorable site for nickel binding. Experiments transferring a normal X-chromosome into nickel-transformed cells suggested that nickel had inactivated a senescence gene on the X-chromosome (Klein et al. 1991). Further work revealed similar results with the human X-chromosome (Wang et al. 1992).

The senescing activity of either the Chinese hamster or human X-chromosome was found to be regulated by DNA methylation (Klein et al. 1991, Wang et al. 1992). Microcell fusion transfer of chromosomes from one cell to another showed that the Chinese hamster's X-chromosome lost senescing activity with time as its host mouse A-9 cells were passaged in tissue culture (Klein et al. 1991, Wang et al. 1992). The drug 5-azacytidine incorporated into DNA in place of cytosine, but unlike the normal cytosine base analog could not be methylated. 5-azacytidine treatment was found to reactivate the X-linked senescence gene (Klein et al. 1991, Wang et al. 1992). During the course of cellular transformation, nickel was found to inactivate an X-linked Chinese hamster senescence gene by DNA methylation (Klein et al. 1991). Although the senescence gene is thought to reside on the short arm of the X-chromosome, Ni(II) was directed towards inactivating the senescence gene by its favorable binding to the entirely heterochromatic-long arm of the Chinese hamster X-chromosome (the largest region of heterochromatin in the hamster genome) (Costa 1991).

Thus, our hypothesis was formulated to provide an alternative to a possible mutagenic mechanism by which carcinogenic nickel compounds might induce cellular transformation by inactivating the transcription of senescence and other critical tumor suppressor genes that maintain normal cellular homeostasis. Since de novo DNA methylation is inherited, the methylation-induced loss in transcription of these genes would remain inactive for multiple cell generations and is consistent with the monoclonal nature of cancer cells. These studies, reported in 1991 were the first to point to DNA hypermethylation as important in regulating expression of tumor suppressor genes (a senescence gene) (Klein et al. 1991). Recent studies by other investigators have confirmed our studies in hamster cells, and in fact, nickel compounds are discussed as being the most active in immortalizing Syrian hamster cells compared to a number of other agents tested (i.e., X-ray, fast neutron, DMN, BPDE) (Trott et al. 1995).

Transgenic Chinese hamster V79 cell lines have been utilized as a model to study the mechanism of gene inactivation by carcinogenic nickel compounds but because these results are discussed quite extensively in the Progress Report section, they will only be summarized here (Kargacin et al. 1993, Lee et al. 1995). The G12 as well as other transgenic lines were derived by inactivating the endogenous hgprt gene by an irreversible mutation and transfecting a bacterial gpt gene under SV40 promoter control (Klein & Rossman 1990). Cells with a single copy of the gpt gene integrated at different chromosomal positions were selected, and the inactivation of the target gene as a function of location was studied following exposure to carcinogenic nickel compounds (Lee et al. 1995). The G12 clone was extraordinarily responsive (frequency of up to 1×10^{-3}) to carcinogenic nickel-induced inactivation of the gpt gene (Lee et al. 1995).

The variability in response of the G12 depended upon the intracellular bioavailability of Ni(II) since finely powdered crystalline nickel subsulfide or sulfide preparations exhibit different phagocytic activity depending upon surface properties (Costa 1991). The G12 cell line had the gpt target gene positioned very close to a dense heterochromatic region of chromosome 1 (Lee et al. 1995). In contrast, in the G10 line the same gpt target gene was placed away from any dense heterochromatin region (chromosome 6) and was not responsive to carcinogenic nickel-induced inactivation of gpt activity (Kargacin et al. 1993). It was hypothesized that nickel inactivated the gpt gene by inducing an initial increased chromatin condensation and DNA methylation of the flanking and coding regions of the G12 gpt target gene (Lee et al. 1995). Since the G12 cell line has been used to study mutagenesis by numerous other agents including X-rays, bleomycin, alkylating agents etc., and yielded typical point mutations and deletions, the inactivation of the gpt by nickel-induced DNA methylation is unique and not an expected response of this cell line to most carcinogenic and toxic agents (Klein and Rossman 1990, Klein et al. 1994). These studies solidify the model that Ni^{2+} induces heterochromatinization of genes that are located near heterochromatin. The inactivation of tumor suppressor genes and senescence genes is an advantage for cell proliferation and thus has a selective advantage.

Literature Cited

Borochov, N., Ausio, J. & Eisenberg, H. (1984) Interaction and conformational changes of chromatin with divalent ions. Nucleic Acid Res. 12: 3089–3096.

Conway, K. & Costa, M. (1989) Nonrandom chromosomal alterations in nickel-transformed Chinese hamster embryo cells. Cancer Res. 49: 6032–6038.

Costa, M. (1991) Molecular mechanisms of nickel carcinogenesis. Annu. Rev. Pharmacol. Toxicol. 31: 321–337.

Costa, M. & Mollenhauer, H.H. (1980) Carcinogenic activity of particulate metal compounds is proportional to their cellular uptake. Science 209: 515–517.

Dunnick, J.K. & Elwell, M.R. (1995) Experimental studies of the toxicity and carcinogenicity of nickel sulfate, nickel subsulfide, and nickel oxide. J. Cell. Biochem. Supp. 19A: 196.

IARC (1990). IARC Monographs on the Evaluation of Carcinogenic Risk to Humans: Chromium, Nickel and Welding, vol. 49, Lyon, France.

Kargacin, B., Klein, C.B. & Costa, M. (1993) Mutagenic responses of nickel oxides and nickel sulfides in Chinese hamster V79 cell lines at the xanthine-guanine phosphoribosyl transferase locus. Mutation Res. 300: 63–72.

Kasprzak, K.S., Quander, R.V. & Poirier, L.A. (1985) Effects of calcium and magnesium salts on nickel subsulfide carcinogenicity in Fischer rats. Carcinogenesis 6: 1161–1164.

Klein, C.B. & Rossman, T. (1990) Transgenic Chinese hamster V79 cell lines which exhibit variable levels of gpt mutagenesis. Environ. Mol. Mutagen. 16: 1–12.

Klein, C.B., Conway, K., Wang, X-W., Bhamra, R.K., Lin, X.H., Cohen, M.D., Annab, L., Barrett, J.C. & Costa, M. (1991) Senescence of nickel-transformed cells by an X-chromosome: Possible epigenetic control. Science 251: 796–799.

Klein, C.B., Su, L., Rossman, T.G. & Snow, E.T. (1994) Transgenic gpt+ V79 cell lines differ in their mutagenic response to clastogens. Mutation Res. 304: 217–228.

Lee, Y-W., Klein, C.B., Kargacin, B., Salnikow, K., Kitahara, J., Dowjat, K., Zhitkovich, A., Christie, N.T. & Costa, M. (1995) Carcinogenic nickel silences gene expression by chromatin condensation and DNA methylation: A new model for epigenetic carcinogens. Mol. Cell. Biol. 15: 2547–2557.

Margerum, D.W. & Anliker, S.L. (1988) Nickel chemistry and properties of the peptide complexes of Ni(II) and Ni(III). In: The Bioinorganic Chemistry of Nickel (Lancaster, Jr., J.R. (ed.)), pp. 29–551. VCH Publishers, New York.

Trott, D.A., Cuthbert, A.P., Overell, R.W., Russo, I. & Newbold, R.F. (1995) Mechanisms involved in the immortalization of mammalian cells by ionizing radiation and chemical carcinogens. Carcinogenesis 16: 193–204.

Wang, X-W., Lin, X., Klein, C.B., Bhamra, R.K., Lee, Y-W., & Costa, M. (1992). A conserved region in human and Chinese hamster X-chromosomes can induce cellular senescence of nickel-transformed Chinese hamster cell lines. Carcinogenesis 13: 555–561.

Discussion

Q1. Allan Davison, University of Northern British Columbia, Prince George, BC, Canada: I am curious, as a neophyte in this area, why nickel is unique in this regard. Do other carcinogenic metals act in this way?

A. No they don't. We've looked at the G12 model. The ability of nickel to inactivate that is very striking. We've looked with cadmium, arsenic, mercury, zinc, copper or chromate, none of them work by inactivating the gpt gene of G12. I believe nickel is unique because it is so similar to magnesium, which is critical in maintaining DNA heterochromatin structure. Nickel also has the ability to bind well to histone-H1, unlike other metals, its ability to coordinate with lysines is probably also unique, and its ability to become redox active when it binds to these lysines may also be important.

Q2. James Kirkland, University of Guelph, ON, Canada: Can all cells phagocytize these particles, or are there differences between cell types?

A. The nickel sulfide particles have been looked at with many cell types in culture, and also *in vivo* in epithelial cells. Respiratory cells, epithelial cells and all cultured cells examined so far can readily phagocytize these particles.

Q3. James Kirkland, University of Guelph, ON, Canada: Another thing I was curious about is why a basic amino acid like lysine, that would carry a positive charge, would interact with a positively charged metal?

A. The charges may differ within the protein environment, but I don't know enough about the chemistry at this point. I was just at Barcelona, and there were a couple of papers presented there on the importance of lysine in nickel enzymes, such as urease and the hydrogenases. It seems to be more and more recognized that lysine is an important coordination site for nickel. We're hoping to do some studies of nickel binding to H1 to find out just what the sites are. Right now I'm just proposing that it's probably lysine, but I don't know for sure. The other thing that's coming out with a lot of these enzymes that utilize nickel, is that the nickel doesn't get oxidized itself, but it's a secondary site that undergoes oxidation/reduction, and so it's probably very complicated.

Studies of the Embryotoxicity and Teratogenicity of Metals in Xenopus Laevis

F.W. SUNDERMAN, JR.

Departments of Laboratory Medicine, Nutritional Sciences, and Pharmacology, University of Connecticut, Farmington, CT 06030-2225, USA

Keywords FETAX assay, metallothionein, lipovitellins, pNiXa

The studies summarized in this report were performed in collaboration with post-doctoral fellows, visiting scientists, research associates, and members of the faculty of the University of Connecticut, as well as several colleagues at other institutions. Their individual contributions are identified by the authorship of the cited papers, and are all gratefully acknowledged.

The basic experimental system for our studies has been the FETAX protocol, which stands for "Frog Embryo Teratogenesis Assay: Xenopus," a four-day teratogenesis assay that has been thoroughly standardized by interlaboratory trials and widely adopted for teratogenic research using the South African clawed frog, Xenopus laevis (Davies and Freeman 1995). Our studies have shown that the divalent chloride salts of Ni, Co, Cd, Zn, and Cu are all potent teratogens for Xenopus, with median teratogenic concentrations ranging from 2.5 μM for Ni and Cu to 40 μM for Zn (Hauptman et al. 1993, Hopfer et al. 1991, Luo et al. 1993a, Plowman et al. 1991, Sunderman et al. 1991). To determine if the malformations of Xenopus embryos observed in FETAX assays persist after the tadpoles undergo metamorphosis to juvenile frogs, tadpoles that had been exposed for 4 d to median teratogenic concentrations of Ni, Co, or Cd were reared under normal laboratory conditions for three months and then killed and examined for developmental anomalies. The incidences of anomalies in the Ni-, Co-, or Cd-exposed frogs at 13 weeks of age were 55, 40, and 51%, respectively, versus 3% in controls (Plowman et al. 1994). In FETAX assays, the teratogenic effects of metals are strongly influenced by either decreasing or increasing the concentration of Mg(II) in the medium, which normally has a magnesium concentration of 620 μM. Mg-deprivation greatly enhanced and Mg-supplementation reduced the incidence and severity of the teratogenic and embryotoxic effects of divalent Ni, Co, Zn, and Cd in Xenopus embryos (Luo et al. 1993b). These observations can probably be attributed to competition of Mg(II) with the respective test metals ions for a carrier mechanism that is involved in metal absorption or cellular uptake, or for binding to critical molecular targets.

The FETAX protocol was used to study the influence of exposures to metals on metallothionein (MT) levels in Xenopus embryos. At specified intervals after fertilization, groups of control and metal-exposed embryos were analyzed for MT content by the silver-saturation method and for MT-mRNA content by a quantitative technique involving the reverse transcriptase and polymerase chain reactions; the localization of MT was established by immunoperoxidase staining of whole-mount embryos (Sunderman et al. 1995b, 1996a). Zn and Cd exposures caused progressive increases of MT contents of Xenopus embryos, while the MT contents of Ni-, Co-, and Cu-exposed embryos did not differ significantly from the controls, when the respective metals were tested at concentrations that cause >95% incidence of malformations, but <5% mortality. Concentration-response studies in Zn- and Cd-exposed embryos showed that Cd is three to five times more potent than Zn for enhancing the MT-mRNA and MT contents of embryos. However, since the threshold for mortality from exposure to Cd is much lower than from Zn, the peak MT-mRNA and MT levels that could be induced by sublethal exposures were higher for Zn than Cd (Sunderman et al. 1995b). Immunoperoxidase staining with MT-antibody showed that MT is distinctly localized during early embryogenesis in the nuclei of myotomal cells in the developing somites, at a stage well before MT becomes evident in the gills, mesonephros, and hepatic anlage (Sunderman et al. 1996b).

Correct citation: Sunderman, F.W., Jr. 1997. Studies of the embryotoxicity and teratogenicity of metals in Xenopus laevis. In *Trace Elements in Man and Animals – 9: Proceedings of the Ninth International Symposium on Trace Elements in Man and Animals. Edited by* P.W.F. Fischer, M.R. L'Abbé, K.A. Cockell, and R.S. Gibson. NRC Research Press, Ottawa, Canada. pp. 538–541.

A brief discussion of vitellogenin and related yolk proteins may be helpful, prior to considering their roles in trace metal nutrition and toxicology (Grbac-Ivankovic et al. 1994, Sunderman et al. 1995a). Vitellogenin is the major precursor of the yolk proteins in Xenopus, as in most oviparous animals. Four vitellogenin genes are transcribed in hepatocytes in response to estrogens, forming a family of vitellogenin precursors that undergo post-translational glycosylation, lipidation, and phosphorylation. After removal of a signal peptide, vitellogenins are secreted into the plasma, where they circulate as a family of homo- and hetero-dimers. A surface receptor on oocytes mediates the selective uptake of vitellogenin by endocytosis. The receptor is released and vitellogenin is promptly cleaved to yield lipovitellin 1, lipovitellin 2, and phosvitin, which are the principal proteins of Xenopus yolk platelets. Montorzi et al. (1994) showed that vitellogenin is a Zn-protein and Sunderman et al. (1995a) showed that one of its products, lipovitellin 1, avidly binds Zn(II) and Cd(II) in vitro. Similarly, lipovitellin 2b avidly binds Ni(II) (Grbac-Ivankovic et al. 1994), and the phosphoserine-rich protein, phosvitin, is an Fe(II)-binding protein (Taborsky 1980). Unpublished studies (Kotyzova and Sunderman, in preparation) show that the total Zn content of control Xenopus embryos diminishes slightly during the FETAX assay, and that a substantial fraction of the Zn burden is progressively transferred from the yolk mass to the somatic tissues, consistent with gradual consumption of yolk platelets, which constitute the embryo's primary source of protein, carbohydrate, lipid, vitamins, and trace elements until the ingestion of food begins on the fifth day post-fertilization. The consumption of yolk proteins evidently involves chymotrypsin-like and cathepsin L-like proteinases, but the identity of the enzymes, or the inhibitors that modulate their activities, has not been established.

Our studies point to pNiXa, an abundant Ni(II)-binding serpin of Xenopus oocytes and embryos, as a factor that may influence the proteolysis of yolk proteins during embryogenesis (Beck et al. 1992, Sunderman et al. 1996b, Kotyza et al. 1996). The amino acid sequences of peptide fragments of pNiXa, obtained by digestion of the purified protein with cyanogen bromide or formic acid, show identity to the deduced sequence of Ep45, a serpin that is synthesized in Xenopus liver in response to estrogens (Holland et al. 1992). The Ni-binding domain of pNiXa is a histidine-rich segment near the N-terminus. This is a unique feature of pNiXa, since no such histidine-rich segment is present in the numerous serpins that are listed in the protein-sequence data banks. Metal-blot competition assays showed that pNiXa also binds divalent Co, Zn, and Cu, although with less avidity than Ni(II). pNiXa was tested for an inhibitory effect on serine proteinases, and found to be a potent inhibitor of chymotrypsin, a weak inhibitor of elastase, and a non-inhibitor of trypsin. The kinetics of the pNiXa effects on chymotrypsin and elastase activities are typical of reversible, slow-binding inhibition, since the steady-state for enzyme inhibition is not attained for several minutes after mixing pNiXa with the enzymes (Sunderman et al. 1996b). pNiXa was also shown to be a slow-binding inhibitor of one of the cysteine proteinases, cathepsin L, but it had no inhibitory effect on cathepsins B or H (Kotyza et al. 1996). The inhibition constants (Ki) for pNiXa inhibition of chymotrypsin and cathepsin L are very low, ranging from 1 to 3 nM, depending on the pH. pNiXa is one of the few serpins that are known to cause cross-class inhibition of both cysteine and serine proteinases.

Although pNiXa has high affinity for Ni(II), the pNiXa inhibition of chymotrypsin and cathepsin L activities is not influenced by Ni(II). Ni(II) concentrations as high as 50 μM have no effect on pNiXa inhibition of either enzyme (Sunderman et al. 1996b, Kotyza et al. 1996). The failure of Ni(II) to alter the proteinase inhibitory activities of pNiXa is not surprising, since the histidine-rich Ni-binding domain is located near the N-terminus, while the putative active site for proteinase inhibition is near the C-terminus. Binding of Ni(II) to pNiXa may, however, stimulate strong oxidative activity, based on experiments with a 17-residue, N-acetylated, synthetic peptide with amino acid sequence identical to the histidine-rich domain of pNiXa (Sunderman et al. 1996b). UV-visible and circular dichroism spectra indicated that Ni(II) was bound to the peptide by six-coordinate octahedral geometry, with Ni-coordination to imidazole nitrogens of histidyl residues. In the presence of Ni(II) and H_2O_2, the pNiXa peptide caused a Fenton-type reaction that oxidized 2′-deoxyguanosine to mutagenic 8-hydroxy-2′-deoxyguanosine. This observation was confirmed with various reactant concentrations, incubation temperatures, pH's, and reaction times (Sunderman et al. 1996b). Thus, Ni-binding to pNiXa may hypothetically induce teratogenesis by causing oxidative damage of nucleic acids and other macromolecules that are needed for normal embryonic development.

An antibody to pNiXa was prepared by conjugating the synthetic histidine-rich peptide with keyhole limpet hemocyanin and using the peptide complex to immunize rabbits. The antibody, after purification by antigen-affinity chromatography, did not cross-react with other serpins, or with other Xenopus proteins, based on immunoblotting of purified proteins and embryo homogenates. Immunoperoxidase staining of whole-mount embryos with the pNiXa-antibody showed pNiXa to be localized in the mesencephalon, spinal cord, tail-bud, hepatic anlage, and anal pore. Thus, the distribution of pNiXa during organogenesis matches the sites of the common malformations in Ni-exposed embryos. This finding is consistent with the premise that pNiXa is a target for Ni-binding and may be involved in teratogenesis by provoking oxidative damage at specific loci (Sunderman et al. 1996b).

Acknowledgement

This research was supported by a NIH grant from the National Institute of Environmental Health Sciences (ES-05331–06) and by an endowed professorship from Northeast Utilities, Inc.

References

Beck, B.L., Henjum, D.C., Antonijczuk, K., Zaharia, O., Korza, G., Ozols, J., Hopfer, S.M., Barber, A.M., & Sunderman, F.W., Jr. (1992). pNiXa, a Ni^{2+}-binding protein in Xenopus oocytes and embryos, shows identity to Ep45, an estrogen-regulated hepatic serpin. Res. Commun. Chem. Pathol. Pharmacol. 77: 3–16.

Davies, W.J. & Freeman, S.J. (1995). Frog embryo teratogenesis assay Xenopus (FETAX). Meth. Mol. Biol. 43: 311–316.

Grbac-Ivankovic, S., Antonijczuk, K., Varghese, A.H., Plowman, M.C., Antonijczuk, A., Korza, G., Ozols, J., & Sunderman, F.W., Jr. (1994). Lipovitellin 2b is the 31 kD Ni^{2+}-binding protein (pNiXb) in Xenopus oocytes and embryos. Mol. Reprod. Develop. 38: 256–263.

Hauptman, O., Albert, D.M., Plowman, M.C., Hopfer, S.M., & Sunderman, F.W., Jr. (1993). Ocular malformations in frogs (Xenopus laevis) exposed to nickel during embryogenesis. Ann. Clin. Lab. Sci. 23: 397–406.

Holland, L.J., Suksang, C., Wall, A.A., Roberts, L.R., Moser, D.R., & Bhattacharya, A. (1992). A major estrogen-regulated protein secreted from the liver of Xenopus laevis is a member of the serpin superfamily. J. Biol. Chem. 267: 7053–7059.

Hopfer, S.M., Plowman, M.C., Sweeney, K.R., Bantle, J.A., & Sunderman, F.W., Jr. (1991). Teratogenicity of Ni^{2+} in Xenopus laevis, assayed by the FETAX procedure. Biol. Trace Elem. Res. 29: 203–216.

Kotyza, F., Hiwasa, T., Makino, T., Varghese, A.H., & Sunderman, F.W., Jr. (1996). pNiXa, a serpin of Xenopus oocytes and embryos, inhibits cathepsin L: an example of cross-class inhibition. Biochim. Biophys. Acta (submitted).

Luo, S.Q., Plowman, M.C., Hopfer, S.M., & Sunderman, F.W., Jr. (1993a). Embryotoxicity and teratogenicity of Cu^{2+} and Zn^{2+} for Xenopus laevis, assayed by the FETAX procedure. Ann. Clin. Lab. Sci. 23: 111–120.

Luo, S.Q., Plowman, M.C., Hopfer, S.M., & Sunderman, F.W., Jr. (1993b). Mg^{2+}-deprivation enhances and Mg^{2+}-supplementation diminishes the embryotoxic and teratogenic effects of Ni^{2+}, Co^{2+}, Zn^{2+}, and Cd^{2+} for frog embryos in the FETAX assay. Ann. Clin. Lab. Sci. 23: 121–129.

Montorzi, M., Falchuk, K.H., & Vallee, B.L. (1994). Xenopus laevis vitellogenin is a zinc protein. Biochem. Biophys. Res. Commun. 200: 1407–1413.

Plowman, M.C., Peracha, H., Hopfer, S.M., & Sunderman, F.W., Jr. (1991). Teratogenicity of cobalt chloride in Xenopus laevis, assayed by the FETAX procedure. Teratogen. Carcinogen. Mutagen. 11: 83–92.

Plowman, M.C., Grbac-Ivankovic, S., Martin, J., Hopfer, S.M., & Sunderman, F.W., Jr. (1994). Malformations persist after metamorphosis of Xenopus laevis tadpoles exposed to Ni^{2+}, Co^{2+}, or Cd^{2+} in FETAX assays. Teratogen. Carcinogen. Mutagen. 14: 135–144.

Sunderman, F.W., Jr., Plowman, M.C., & Hopfer, S.M. (1991). Embryotoxicity and teratogenicity of cadmium chloride in Xenopus laevis, assayed by the FETAX procedure. Ann. Clin. Lab. Sci. 21: 381–391.

Sunderman, F.W., Jr., Antonijczuk, K, Antonijczuk, A., Grbac-Ivankovic, S., Varghese, A.H., Korza, G., & Ozols, J. (1995a). Xenopus lipovitellin 1 is a Zn^{2+}-and Cd^{2+}-binding protein. Mol. Reprod. Develop. 42: 180–187.

Sunderman, F.W., Jr., Plowman, M.C., Kroftova, O.S., Grbac-Ivankovic, S., Foglia, L., & Crivello, J.F. (1995b). Effects of teratogenic exposures to Zn^{2+}, Cd^{2+}, Ni^{2+}, Co^{2+}, and Cu^{2+} on metallothionein and metallothionein-mRNA contents of Xenopus embryos. Pharmcol. Toxicol. 76: 178–184.

Sunderman, F.W., Jr., Grbac-Ivankovic, S., Plowman, M.R., & Davis, M. (1996a). Zn^{2+}-induction of metallothionein in myotomal cell nuclei during somitogenesis of Xenopus laevis. Mol. Reprod. Develop. 43: 444–451.

Sunderman, F.W., Jr., Varghese, A.H., Kroftova, O.S., Grbac-Ivankovic, S., Kotyza, J., Datta, A.K., Davis, M., Bal, W., & Kasprzak, K.S. (1996b). Characterization of pNiXa, a serpin of Xenopus laevis oocytes and embryos, and its histidine-rich, Ni(II)-binding domain. Molec. Reprod. Develop. (in press).

Taborsky, G. (1980). Iron binding by phosvitin and its conformational consequences. J. Biol. Chem. 255: 2976–2985.

Discussion

Q1. Susan Haywood, University of Liverpool, England: As a copper person, I was fascinated to hear you say that copper induction of metallothionein was very poor. Do you have any explanation for this? And are you aware of a comparable system in mammalian embryos which would test this?

A. We were astounded by the lack of metallothionein response to copper in this system. I'm not aware, though others in the audience, or our chairman may be aware, but I'm not aware of other examples of copper non-responsive systems. Dr. Wiggins in Paris has found some response to copper in tissue cultures of Xenopus cells, with very high copper concentrations, concentrations that would be lethal in the embryos. I suspect that the lack of any response due to copper is the fact that copper toxicity for the embryo, i.e., embryolethality is so great that it prevents our using sufficient concentrations, that it may be a dose-dependent relationship. But it may be that there is simply a difference in the regulation of metallothionein biosynthesis in Xenopus embryo.

Investigation of the Relationship Between Hypermanganesaemia and Liver Disease

C.A. WARDLE,[†] A. CLOUGH,[†] N.B. ROBERTS,[†] A. JAWAHARI,[‡] A. FORBES,[‡] AND A. SHENKIN[†]

[†]Department of Clinical Chemistry, Royal Liverpool University Hospital, Liverpool, L7 8XP, UK, and [‡]St Mark's Hospital, Middlesex, HA1 3UJ, UK

Keywords manganese, toxicity, liver disease, parenteral nutrition

Introduction

Whole blood manganese (Mn) concentrations have been shown to be increased in some neonates receiving total parenteral nutrition (TPN). The same patients had evidence of cholestatic liver disease (Reynolds et al. 1994). Withdrawal of, or reduction of Mn supplements given to children receiving TPN resulted in a decline in serum bilirubin concentrations (Fell et al. 1996), suggesting that increased circulating Mn concentrations may play a role in the development of cholestasis. Manganese is known to be excreted primarily by the liver (Papavasiliou et al. 1966), and it is as yet unknown whether TPN-induced cholestasis leads to Mn accumulation, or whether Mn toxicity causes cholestasis. Mn toxicity may play a role in chronic hepatic encephalopathy (Krieger et al. 1995) and patients with primary biliary cirrhosis have been shown to have elevated serum Mn-superoxide dismutase activities (Ono et al. 1991). The aims of the present study were firstly to investigate the incidence of increased blood Mn concentrations both in adults receiving long-term parenteral nutrition and in patients with chronic liver disease, and secondly to investigate the relationship between whole blood Mn concentrations and the severity of cholestasis in both groups of patients.

Methods

The following groups of patients were included in the study

a) Patients with chronic liver disease (CLD) and biochemical evidence of cholestasis (Serum gamma-glutamyl transferase (GGT) activity >50 IU/L and alkaline phosphatase (ALP) activity >125 IU/L; n = 10).

b) Patients receiving long-term (range 3–168 months) home parenteral nutrition (HPN; n = 30). Up to 50% of nutrition was given orally and intravenous feeds were given 3–7 times per week. A multitrace element preparation was used with a Mn content of 5 µmol per feed.

c) Control subjects without biochemical or clinical evidence of liver disease (n = 10).

Whole blood Mn concentrations were measured using a Varian SpectrAA-400 graphite furnace spectrophotometer. Serum GGT and ALP activities and bilirubin concentration were measured in all subjects in order to assess the severity of cholestasis.

Results

All patients with CLD exhibited some degree of cholestasis, as demonstrated by elevated serum GGT and ALP activities. Despite this, the whole blood Mn concentrations of all patients with CLD were within the reference range (73–210 nmol/L), and were not significantly different than those of the control group (CLD 151 ± 44 nmol/L vs. control 155 ± 35 nmol/L).

Correct citation: Wardle, C.A., Clough, A., Roberts, N.B., Jawahari, A., Forbes, A., and Shenkin· A. 1997. Investigation of the relationship between hypermanganesaemia and liver disease. In *Trace Elements in Man and Animals – 9: Proceedings of the Ninth International Symposium on Trace Elements in Man and Animals. Edited by* P.W.F. Fischer, M.R. L'Abbé, K.A. Cockell, and R.S. Gibson. NRC Research Press, Ottawa, Canada. pp. 542–543.

Serum ALP and GGT activities were increased in 11 HPN patients, and were greater than 2× the upper limit of normal in a further 7 HPN patients. These values were stable, and were not associated with clinical evidence of cholestatic liver disease. The remaining 12 HPN patients showed neither clinical nor biochemical evidence of cholestatic liver disease. Serum bilirubin concentrations of all HPN patients were <20 µmol/L. Of the patients receiving HPN, 26 had increased whole blood Mn concentrations (>210 nmol/L), whilst 7 of these had levels in the toxic range (>360 nmol/L). There was no correlation between whole blood Mn concentration and either serum ALP or GGT activity (ALP vs. Mn r = –0.21; GGT vs. Mn r = –0.19). None of the HPN patients exhibited neurological signs of Mn toxicity.

Discussion

The results of this study show that some adult patients receiving long-term TPN have increased whole blood Mn concentrations. This is not related to biochemical measurements of cholestasis. Furthermore, patients with CLD and biochemical evidence of cholestasis do not appear to have increased blood Mn concentrations. These findings suggest that hypermanganesaemia in patients receiving long-term TPN is more likely to be a result of excessive Mn supply, rather than decreased excretion. This is in contrast to the finding in children where all neonates on long-term TPN with evidence of cholestatic liver disease in one study were found to have increased whole blood Mn concentrations (Reynolds et al. 1994). Decreased excretion may therefore be a more important factor in children and neonates than in adults.

References

Fell, J.M.E., Reynolds, A.P., Meadows, N., Khan, K., Long, S.G., Quaghebeur, G., Taylor, W.J. & Milla, P.J. (1996) Manganese toxicity in children receiving long-term parenteral nutrition. Lancet. 347: 1218–1221.

Krieger, D., Krieger, S., Jansen, O., Gass, P., Theilmann, L. & Lichtnecker, H. (1995) Manganese and chronic hepatic encephalopathy. Lancet 346: 370–274.

Ono, M., Sekiya, C., Ohhira, M., Ohhira, M., Namiki, M., Endo, Y., Suzuki, K., Matsuda, Y. & Taniguchi, N. (1991) Elevated serum Mn-superoxide dismutase in patients with primary biliary cirrhosis: Possible involvement of free radicals in the pathogenesis in primary biliary cirrhosis. J. Lab. Clin. Med. 118: 476–483.

Papavasilou, P.S., Miller, S.T. & Cotzias, G.C. (1966) Role of liver in regulating distribution and excretion of manganese. Am. J. Physiol. 211: 211–216.

Reynolds, A.P., Kiely, E. & Meadows, N. (1994) Manganese in long term paediatric parenteral nutrition. Arch. Dis. Childhood. 71: 527–528.

Discussion

Q1. Ed Harris, Texas A&M University, College Station, TX, USA: I'm curious about that wide spread that you saw in manganese levels. Were the patients all put on TPN for the same reasons? These were not normal patients, they were all suffering some other problem that caused them to go on TPN.

A. Most of them had Crohn's disease. There were some others, but most were suffering from malabsorption.

Q2. Margaret Elmes, University Hospital Wales, Cardiff, Wales: Are you concerned about abnormal liver function tests and do you think they were linked to the high manganese levels? Liver function tests do tend to be abnormal with TPN.

A. Obviously this is a concern, though none of our patients showed clinical symptoms of liver disease. I don't think it's causal, as it was not correlated with manganese concentration. However, as I said before, manganese is excreted by the liver, so it might be one factor out of a spectrum of factors.

Q3. Roy Scott, Royal Infirmary, Glasgow, Scotland: You had mostly Crohn's patients — had they undergone surgery? That is the most common reason for TPN treatment.

A. To be honest, I don't have that information, as the TPN patients were looked after by the St. Mark's group of collaborators. I was dealing more with the analytical side of things. (comment by **Alan Shenkin**, Liverpool, England: Yes, most of these patients were post-operative, but were stable, without ongoing fistulae or infection.)

Bile is an Important Route of Elimination of Ingested Aluminum by Unanesthetized Rats

J.L. GREGER[†] AND J.E. SUTHERLAND[‡]

[†]Department of Nutritional Sciences and [‡]Environmental Toxicology Center, UW-Madison, Madison, WI 53706, USA

Keywords aluminum, bile, liver, rats

Biliary clearance of Al is believed to be less important than renal clearance when Al is administered intravenously (Xu et al. 1991). However, the relative importance of biliary secretion when Al is administered orally is less clear. Ishishara and Matsushiro (1986) reported that biliary Al secretion sometimes exceeded urinary Al excretion in patients whose main source of Al exposure was diet. The purpose of this work was to characterize biliary Al secretion by unanesthetized rats given graded oral doses of Al. In previous studies, including those of Klein et al. (1989), rats were anesthetized, which could affect biliary flow.

Methods

Bile ducts of male, Sprague Dawley rats (weighing 201 ± 5 g in Study 1 and 239 ± 1 g in Study 2) were cannulated to allow both bile collection and reinfusion of bile acids (Sutherland et al. 1996). Prior to and after surgery, rats were fed a semipurified diet containing 8 μg Al/g diet. On the fifth day after surgery in Study 1, thirty unanesthetized rats were gavaged with 1 mL of 16% citrate solution with 0, 1, 2, or 4 mmol Al/kg rat as Al lactate. In Study 2, twenty-six rats were gavaged with 0, 0.25, 0.5, or 1 mmol Al/kg rat. In Study 3, serum and liver Al concentrations were monitored at 15 min, 30 min, 1 h, and 2 h after dosing 84 rats (weighing 231 ± 1 g) with 0, 0.25, 0.5, or 1 mmol Al/kg rat.

Results

In Study 1, rats given all three doses of Al secreted more Al in bile during 1–7 h after dosing than did control rats which received the citrate vehicle only (Figure 1). Biliary Al secretion did not vary among rats given 1, 2, or 4 mmol Al/kg body weight. This suggests that the process was saturated at these doses.

Although biliary secretion of Al appeared saturated at the 1, 2, and 4 mmol Al/kg doses, serum Al generally increased in proportion to the dose. Liver Al concentrations (7-h after dosing) were increased above control levels only in rats dosed with 4 mmol Al/kg.

In Study 2, biliary Al secretion was already elevated 15 min after oral dosing. Biliary Al secretion by rats dosed with 0.5 and 1 mmol Al/kg (140 and 157 nmol Al/4-h, respectively) was greater than biliary Al secretion by rats dosed with 0.25 mmol Al/kg (97 nmol Al/4-h) which was greater than secretion by control rats (33 nmol Al/4-h). This suggests that biliary secretion of Al would be saturated when a 200-g rat consumed as little as 0.1 mmol Al in a single meal.

Bile secretion by control rats was constant throughout the collection period. Thus, the significant decrease in biliary Al during the 4-h collection period reflected clearance of the dosed Al not limitations of the model system.

Correct citation: Greger, J.L., and Sutherland, J.E. 1997. Bile is an important route of elimination of ingested aluminum by unanesthetized rats. In *Trace Elements in Man and Animals – 9: Proceedings of the Ninth International Symposium on Trace Elements in Man and Animals. Edited by* P.W.F. Fischer, M.R. L'Abbé, K.A. Cockell, and R.S. Gibson. NRC Research Press, Ottawa, Canada. pp. 544–545.

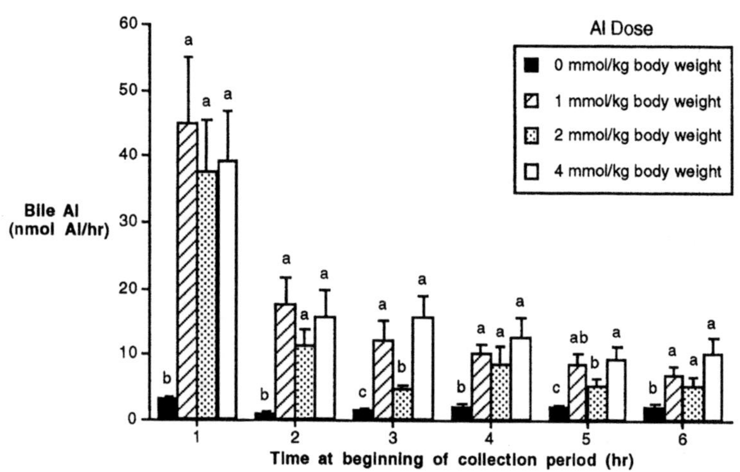

Figure 1. Bile Al secretion in Study 1. At each time point, means without common superscripts are significantly ($p < 0.05$) different.

Summary

Biliary secretion of ingested Al occurs rapidly and is saturated by moderate oral doses of Al. Greger and Powers (1992) found that rats fed control diets secreted 0.6 nmol Al/h in urine; our control rats secreted 8.1 nmol Al/h in bile.

Literature Cited

Greger, J.L. & Powers, C.F. (1992) Assessment of exposure to parenteral and oral aluminum with and without citrate using a desferrioxamine test in rats. Toxicology 76: 119–132.

Ishishara, N. & Matsushiro, T. (1986) Biliary and urinary excretion of metals in humans. Arch. Environ. Health 41: 324–330.

Klein, G. L., Lee, T. C., Mann, P. A., Miller, N. L. & Alfrey, A. C. (1989) Effects of aluminum on the liver following high-dose enteral administration to rats. J. Pediatr. Gastroenterol. Nutr. 9: 105–107.

Sutherland, J.E., Radzanowski, G.M. & Greger, J.L. (1996) Bile is an important route of elimination of ingested aluminum by conscious male Sprague Dawley rats. Toxicology (In Press).

Xu, Z.-X., Pai, S. M. & Melethil, S. (1991) Kinetics of aluminum in rats. II: Dose-dependent urinary and biliary excretion. J. Pharm. Sci. 80: 946–951.

Discussion

Q1. George Cherian, The University of Western Ontario, London, ON, Canada: Is the form of aluminum important?

A. We gave aluminum lactate in a citrate solution for both animals and humans, to allow us to keep it in solution for dosing and because it provided the highest absorption. Yes, obviously form is important. Aluminum hydroxide without the citrate would be only very poorly absorbed. And of course, to make this work, you need to have aluminum absorption. So of course, it makes a big difference, what form you choose.

Q2. Ed Harris, Texas A&M University, College Station, TX, USA: You said that you don't believe aluminum is essential. If you had to make a statement as to what type of system you think aluminum is paralleling, it would seem to be more like copper in the way it's being handled, than iron. Is that how you see this?

A. I think there's amazing parallels with manganese, in some ways. I wouldn't say copper, but I think that characterizing in this way may be one of the problems in the aluminum field. It gets in the way of all the things we should know. There is some evidence that aluminum interacts with iron, or with calcium and vitamin D. In terms of how aluminum is transported and excreted, all the things we should know, we really don't.

Q3. Margaret Elmes, University Hospital Wales, Cardiff, Wales: Just because it's in bile doesn't mean it's been lost from the body. What happens to it after that? How much of the biliary aluminum is reabsorbed?

A. That's a good question. How much is reabsorbed? We haven't done those studies yet. Aluminum radionuclide half-lives are too short (in the order of 28 seconds), and we don't have a stable isotope, so there are technical problems in doing the experiments necessary to distinguish fecal losses due to biliary excretion.

Ultrastructural Changes in the Duodenum of Copper-Loaded Rats

I. CARMEN FUENTEALBA AND DOROTA WADOWSKA

Atlantic Veterinary College, University of Prince Edward Island, Charlottetown, PEI, Canada C1A 4P3

Keywords copper, toxicity, ultrastructure, intestine

Copper (Cu) toxicosis associated with elevated hepatic copper occurs in Wilson's disease as a familial disorder in humans (Underwood 1977) and in inherited Cu storage disease in Bedlington Terriers (Hardy et al. 1975) and West Highland White Terriers (Thornburg et al. 1986). Experimental Cu toxicosis in rats results in morphologic changes similar to those described in copper-associated diseases in humans (Fuentealba et al. 1987, Goldfischer et al. 1980).

The role of the intestine in spontaneous and experimental Cu toxicosis has not been adequately explored (Hair-Bejo et al. 1990, Fuentealba et al. 1994). Hence, the purpose of this study was to study the ultrastructural changes due to Cu accumulation in the duodenum.

Material and Methods

Thirty-two male Wistar rats approximately 10 weeks of age and 125 g were randomly divided into two groups: Group A was fed a normal rodent diet (Purina, St. Louis, Missouri) that contained 18 ppm Cu and Group B was fed a pelleted high Cu diet (Teklad Diets, Madison, Wisconsin) which contained 1500 mg/kg Cu added as $CuSO_4$. Rats were housed two/cage and given free access to food and tap water.

Duodenum samples were collected at 1, 5, 10 and 15 weeks and processed for transmission electron microscopy and copper analysis by atomic absorption spectrophotometry as described elsewhere (Fuentealba et al. 1989, Fuentealba et al. 1994).

Analysis of variance was used for statistical analysis and the level of significance was set at $p \leq 0.05$.

Results

Copper was significantly higher ($p < 0.05$) in rats exposed to the elevated copper diet compared to the controls. Duodenal copper content was 50 ± 8 µg/g Cu (wet weight) at week 1 and 134 ± 49 µg/g Cu at week 5. A significant decrease in tissue copper was detected after 10 weeks (29 ± 8 Mg/g Cu) and 15 weeks of copper-loading (24 ± 5 µg/g Cu). In control rats, (group A), copper content in the duodenum was 7 ± 1 Mg/g Cu at week 1 and 3 ± 0.2 µg/g Cu at week 15.

Ultrastructural changes at week 1 consisted of slight shortening of villi, peripheralization of chromatin and presence of vesicles at the periphery of the nuclei in most epithelial and endothelial cells (Figure 1). Loss of villi and presence of membranes blebs, in addition to misshapen nuclei, were seen at week 5 (Figure 2). At this time, there also was an increase in the number of electron-dense membrane-bound structures, identified as lysosomes. Goblet cell and Paneth cell hyperplasia were very marked after 5 weeks of copper-loading (Figure 3) and persisted throughout the sampling period. Morphology of villi and nuclei returned to normal after 10 weeks of copper-loading (Figure 4).

Discussion

The mechanisms of copper toxicity are poorly understood. It is possible that the cytotoxic effects of the metal occur as a consequence of nuclear disorganization (Haywood et al. 1985). Ultrastructural changes

Correct citation: Fuentealba, I.C., and Wadowska, D. 1997. Ultrastructural changes in the duodenum of copper-loaded rats. In *Trace Elements in Man and Animals – 9: Proceedings of the Ninth International Symposium on Trace Elements in Man and Animals. Edited by* P.W.F. Fischer, M.R. L'Abbé, K.A. Cockell, and R.S. Gibson. NRC Research Press, Ottawa, Canada. pp. 546–548.

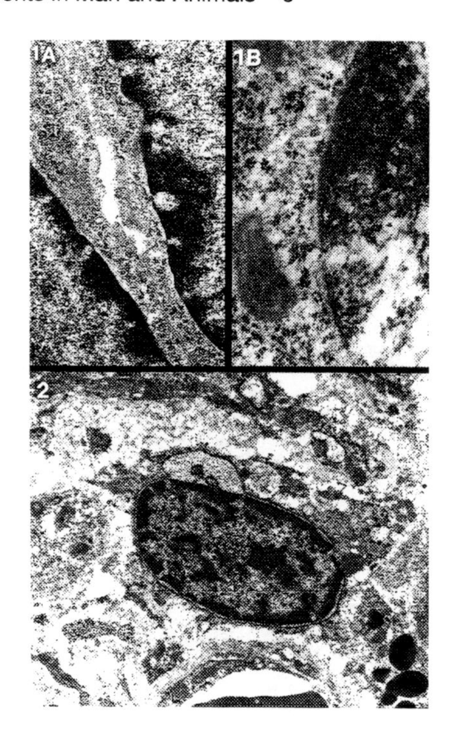

Figure 1. Week 1. (A) Duodenum from a rat receiving 1500 ppm copper. Presence of vesicles at the periphery of the nuclei. TEM × 26,000. (b) Nucleus from a control rat. TEM × 48,000.

Figure 2. Week 5. Duodenum (lamina propria) from a rat receiving 1,500 ppm copper. Membrane blebs in the nucleus. TEM × 19,200.

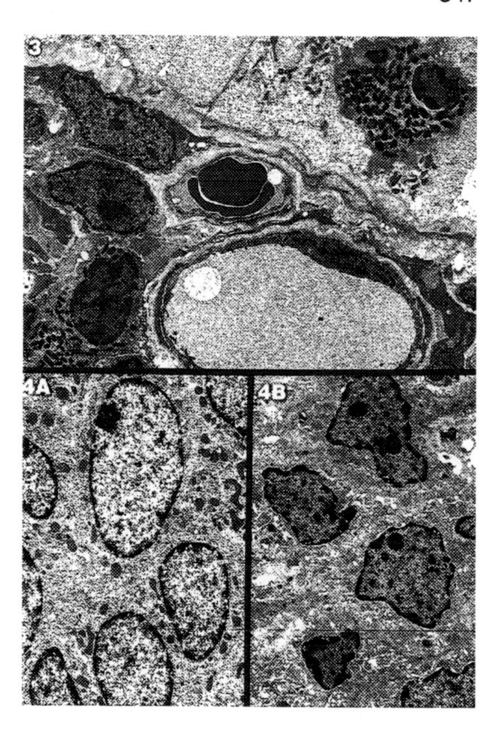

Figure 3. Week 5. Duodenum (lamina propria) from a rat receiving 1500 ppm copper. Increase in the numbers of Paneth cells (arrow). TEM × 4800.

Figure 4. (A). Duodenum (epithelial cells) from a rat receiving 1500 ppm copper during 10 weeks. Note normal nuclear morphology. TEM × 6300. (B) Duodenum (epithelial cells) from a rat receiving 1,500 ppm copper during 1 week. Note misshapen nuclei. TEM × 6300.

in the nucleus observed in this study add support to this theory, and are similar to those previously reported in the copper loaded rat liver (Fuentealba et al. 1989).

Loss of villi and shortening of villi during the initial stages of copper-loading are consistent with the hypothesis that a malabsorption enteropathy occurs as a result of Cu loading (Fuentealba and Bratton 1994) and may explain the failure of animals on copper-loaded diets to gain weight (Fuentealba and Bratton 1994). A decrease in duodenal Cu, associated with a return to normal morphology, was observed after 10 weeks of copper-loading. The rat duodenum seems to have an adaptation mechanism to Cu toxicity, allowing the release of excess Cu from the epithelium. Goblet cell hyperplasia appears to contribute to this adaptation. Hair-Bejo et al. (1990) suggested that excess mucous secretion might cause luminal copper to become less toxic and less available for absorption. Paneth cell hyperplasia is thought to represent an additional route for excretion of excess Cu (Hair-Bejo et al. 1990). It has also been suggested that intestinal metallothionein has a role in Cu homeostasis by providing binding sites for Cu within the intestinal mucosa and maintaining a reserve to protect against absorption of toxic levels of Cu and other trace elements (Evans 1973).

In contrast to the copper-unloading observed in rats, carriers of Wilson's disease (Strickland et al. 1972) and patients with other liver diseases have a reduced capacity for gastrointestinal excretion of copper (Ritland et al. 1977). Therefore, we postulate that copper-tolerance is due to an unusual capacity of the rat to efficiently transport and unload copper, and the ability to trigger other protective mechanisms, such as synthesis of intestinal metallothionein.

References

Evans, G.W. (1973) Copper homeostasis in the mammalian system. Physiol. Rev. 53: 535–570.
Fuentealba, I.C., Haywood, S. & Trafford, J. (1987) Evaluation of histochemical methods for the detection of copper overload in the liver of the rat. Liver 7: 277–282.
Fuentealba, I.C., Haywood, S. & Foster, J. (1989) Cellular mechanisms of toxicity and tolerance in the copper-loaded rat. II. Pathogenesis of copper toxicity in the liver. Exp. Mol. Pathol. 50: 26–37.
Fuentealba, I.C. & Bratton, G.R. (1994) The role of the liver, kidney and duodenum in tolerance in the copper-loaded rat. Analyt. Cell Pathol. 6: 345–358.
Goldfischer, S., Popper, H. & Sternlieb, I. (1980) The significance of variations in the distribution of copper in liver disease. Am. J. Pathol. 99: 715–724.
Hair-Bejo, M., Haywood, S. & Foster, J. (1990) Gastrointestinal response to high dietary copper in the rat. Proc. Forty-first Annual Meeting The American College of Veterinary Pathologists. pp 201.
Hardy, R.M., Stevens, J.B. & Stowe, C.M. (1975) Chronic progressive hepatitis in Bedlington Terriers associated with elevated liver copper concentration. Minn. Vet. 15: 13–24.
Haywood, S., Loughran, M. & Batt, R.M. (1985) Copper toxicosis and tolerance in the rat, III. Intracellular localization of copper in the liver and kidney. Exp. Mol. Path. 43: 209–219.
Ritland, S., Steinnes, E. & Skrede, S. (1977) Hepatic copper content, urinary copper excretion and serum ceruloplasmin in liver disease. Scand. J. Gastroenterol. 12: 81–88.
Strickland, G.T., Beckner, W.M., Leu, M.L. & O'Reilly, S. (1972) Absorption of copper in homozygotes and heterozygotes of Wilson's disease and controls: isotope tracer studies with 67 Cu and 67 Cu. Clin. Sci. 43: 610–616.
Thornburg, L.P., Shaw, D., Dolan, M., Raisbeck, M., Crawford, S., Dennis G.L. & Olwin, D.B. (1986). Hereditary copper toxicosis in West Highland White Terriers. Vet. Pathol. 23: 148–154.
Underwood, E.J. (1977) Copper II. In: Trace elements in human and animal nutrition, 4th ed. Academic Press, London, pp 58–59, 94–95.

Discussion

Q1. Ed Harris, Texas A&M University, College Station, TX, USA: Regarding the Teklad diet with 1500 ppm copper. The cells look normal on the outside but are damaged on the inside. Is it reversible? There are a number of reports in the literature, that it is possible to raise a cell line that is more tolerant to copper by inducing genes or whatever it is that is involved. Metallothionein, of course, does come into play here. But another one that has recently been reported is in Chinese hamster ovary cells, where one can induce an increased capacity to excrete copper by way of the Menkes factor. I would be intrigued to find out if these cells are increasing production of that excreting ATPase. I think that would be very interesting.

A. That is a possibility. In a separate experiment that used high copper in the diet, then normal copper, then returned to high copper again — when we measured copper levels in intestinal cells they went back to normal.

Q2. Maria Laura Scarino, Instituto Nazionale della Nutrizione, Rome, Italy: I was very interested in the appearance of the nuclei. Some of them look very much like apoptotic cell nuclei. I did some work on Caco-2 cells, an intestinal human cell line, treated with copper, and they did go apoptotic. So I think part of your cell population may actually be apoptotic. As for villar shortening, I did some cadmium treatment on these same cells, and they shortened the villi too.

A. Thank you.

Q3. Margaret Elmes, University Hospital Wales, Cardiff, Wales: There's some really old work that says that Paneth cells are active metal secretors. Does your work support this?

A. I would like to think so. I don't know exactly what the role of Paneth cells is. We've looked in the literature, and couldn't find much there either to explain the role of Paneth cells.

Q4. Margaret Elmes, University Hospital Wales, Cardiff, Wales: You said the microvilli were shortened. Did you see whether the villi were shortened as well?

A. We did some morphometry of the microvilli and villi, and showed that both were shortened but it depends at what time one looks at it.

Q5. John Howell, Murdoch University, Murdoch, Western Australia: With the involutions of the nuclear membrane, we see that both in copper toxicity in sheep, and in sheep poisoned with pyrrolizidine alkaloids and copper. It may be that the changes you saw were an earlier version of the changes that we saw. I wonder if you could tell us anything about the feed consumption during the course of the experiment. This is a dramatic change, was it related in any way to changes in feed consumption?

A. No, actually it was quite interesting. In the first week there was a drop in food consumption, related to the abrupt change in diet, then they got used to it and at week 3 food consumption reached that of normal rats, and by week 7 they consumed more than controls.

Thiomolybdate and Zinc Therapy in Copper Overload Syndromes

J. MCC. HOWELL

School of Veterinary Studies, Murdoch University, Perth, Western Australia 6150

Keywords copper overload, thiomolybdate, zinc

Sheep readily accumulate excessive amounts of copper (Cu) and after weeks of clinical normality suddenly become ill and develop haemolysis. Most of the excess Cu accumulates in the liver within the lysosomes of hepatocytes and the proliferation and increase in size of the lysosomes is directly related to the concentration of Cu in the liver of the sheep. On the first day of haemolysis the numerical density of lysosomes is reduced but the volume density is little changed. This indicates that in heavily Cu loaded hepatocytes new small lysosomes are not being formed, the system for producing lysosomes has been exhausted and that large Cu loaded lysosomes are left (Howell and Gooneratne 1987, Kumaratilake and Howell 1989a, 189b).

The efficiency of tetrathiomolybdate (TTM) as a treatment or preventative agent for Cu poisoning in sheep was tested following intravenous administration. When given to clinically normal but Cu loaded sheep TTM prevented the development of haemolysis. When given to animals in haemolysis a significant number failed to develop further episodes of haemolysis and when the liver of these animals was examined after eleven weeks of TTM treatment the histological structure had returned to normal. In Cu loaded sheep significant necrosis of hepatocytes occurs just prior to haemolysis. Copper is released from the dead hepatocytes, the concentration of Cu first increases in the plasma, then in the red blood cells and this is followed by haemolysis. After giving TTM the concentration of Cu in the plasma again increases but it is in a Cu-molybdenum-albumin complex which does not enter red blood cells and haemolysis does not occur. In sheep in which TTM administration commenced on the first day of haemolysis a reduction in the volume density and mean volume of hepatocyte lysosomes was found when they were measured eleven weeks later. The concentration of Cu in the liver and cytosol of Cu loaded sheep was much reduced after eleven weeks of TTM treatment. This occurred whether the treatment commenced at the first rise of acid phosphatase activity, indicating that significant liver damage occurred, or on the first day of haemolysis (Kumaratilake and Howell 1989c).

TTM had been successfully used to treat the LEC rat model in which excess Cu accumulates in the liver and is largely bound to metallothionein. The results of the treatment have confirmed that TTM is a very rapid decoppering agent that removes Cu from metallothionein (see Suzuki 1995). A series of studies has recently commenced on the effect of TTM on the toxic milk mouse, a mouse in which there is a defect in Cu metabolism in that Cu accumulates in tissues with age but is not transported to pups which are deficient in Cu when born and during lactation (Howell and Mercer 1994). The TTM was given in the drinking water or by intraperitoneal injection for varying periods of time and it was found that the concentration of Cu was reduced in the liver and increased in the kidney. The concentration of molybdenum was increased in both liver and kidney and in an experiment in which TTM was given in the drinking water for 28 weeks about one third of the Cu was removed from the brain (Howell and Mercer, unpublished).

When the two toxins, Cu and pyrrolizidine alkaloids, were ingested together by sheep the excessive accumulation of Cu in the liver and the degree of liver damage was much greater than it would have been if each toxin had been given on its own (Howell et al. 1991a, 1991b). The administration of zinc to these animals greatly reduced the accumulation of Cu in and damage to the liver and reduced the number of sheep with clinical signs. The concentration of zinc and metallothionein in the liver were increased

Correct citation: Howell, J. McC. 1997. Thiomolybdate and zinc therapy in copper overload syndromes. In *Trace Elements in Man and Animals – 9: Proceedings of the Ninth International Symposium on Trace Elements in Man and Animals. Edited by* P.W.F. Fischer, M.R. L'Abbé, K.A. Cockell, and R.S. Gibson. NRC Research Press, Ottawa, Canada. pp. 549–550.

(Noordin et al. 1993). It would appear that in these sheep zinc had induced metallothionein formation in a way that Cu was not capable of and that the metallothionein had a protective action in these sheep.

The studies of treatment regimes in Cu overload syndromes in sheep and the rat and mouse models have provided new evidence for the importance of lysosomes and metallothionein as well as indicating the usefulness of TTM and zinc as preventative agents.

It is most probable that our understanding of Cu metabolism and the pathogenesis of Cu overload syndromes will be further enhanced by continued studies of the effects and mechanisms of treatments.

Literature Cited

Howell, J. McC. & Gooneratne, S.R. (1987) The pathology of copper toxicity in animals. In: Copper in Animals and Man (Howell, J. McC. and Gawthorne, J.M. eds)., vol. 11, pp. 53–78. CRC Press, Boca Raton, Florida.

Howell, J. McC. & Mercer, J.F.B. (1994) The pathology and trace element status of the toxic milk mutant mouse. J. Comp. Path. 110: 37–47.

Kumaratilake, J.S. & Howell, J. McC. (1989a) Lysosomes in the pathogenesis of liver injury in chronic copper poisoned sheep: an ultrastructural and morphometric study. J. Comp. Path. 100: 381–390.

Kumaratilake, J.S. & Howell, J.McC. (1989b) Intracellular distribution of copper in the liver of copper-loaded sheep — a subcellular fractionation study. J. Comp. Path. 101: 161–176.

Kumaratilake, J.S. & Howell, J.McC. (1989c) Intravenously administered tetrathiomolybdate and the removal of copper from the liver of copper-loaded sheep. J. Comp. Path. 101: 178–199.

Noordin, M.M., Howell, J. McC. & Dorling, P.R. (1993) The effects of zinc on copper and heliotrope poisoning in sheep. In: Proceedings of the Eighth International Symposium on Trace Elements in Man and Animals (Anke, M., Meissner, D. And Mills, C.F. eds), pp. 279–282. Verlag Media Touristik, Gersdorf.

Suzuki, K.T. (1975) Disordered copper metabolism in LEC rats, an animal model of Wilson's Disease : roles of metallothionein. Res. Com. Mol. Path. Pharm. 89: 221–240.

Discussion

Q1. Ian Bremner, Rowett Research Institute, Aberdeen, Scotland: In your last slide, John, you suggest that the effect of zinc relates to metallothionein production. I calculate that at most, 10% of copper in the liver is bound to metallothionein. Is that significant?

A. I think we'll have to talk about this after the session.

The Function of Thiols in the Toxicity of Selenite and Nuclear Cataract

G.E. BUNCE, J.L. HESS, K.D. DELL, AND K. SMITH

Department of Biochemistry, Virginia Polytechnic Institute and State University, Blacksburg, VA 24061-0308, USA

Keywords cataract, selenite, free radical, thiols

Bilateral nuclear cataract forms in lenses of preweanling rats 96 h after administering a single dose of sodium selenite (Bunce and Hess 1981). Nuclear opacity results from a 3-to 5-fold increase in lens calcium and consequent activation of calpain II (David and Shearer 1984). Uncoupling of the plasma membrane calcium pump, a transient increase in membrane calcium permeability and a decrease in Ca^{+}-dependent ATPase (Wang et al. 1993) contribute to the loss of calcium homeostasis in the lens.

Glutathione effectively reduces selenite (Tsen and Tappel 1959)

$$4GSH + SeO_3^{2-} \rightarrow GSSG + GS\text{-}Se\text{-}SG + 2OH^- + H_2O \quad (1)$$

$$GS\text{-}Se\text{-}SG \rightarrow GS\text{-}SG + SeO \quad (2)$$

$$GS\text{-}Se\text{-}SG + O_2 + 2OH^- \rightarrow GSSG + SeO_3^{2-} + H_2O \quad (3)$$

In the reoxidation of selenium single electron transfer and formation of superoxide may occur in reaction (3) (Seko et al. 1989). Hence, at the time of maximum selenium content oxidative stress causes irreversible changes to membranes and proteins in the lens.

Methods

Breeding pairs of Sprague-Dawley rats were maintained at 24°C and 80% relative humidity with a 12-hour light/dark cycle. Animals were fed Agway 3000 rat chow [Agway Inc., Syracuse, N.Y.] and distilled water. Ten-day-old animals were injected subcutaneously with 20 mM sodium selenite [$Na2SeO_3$], 0.9% NaCl for a total dose of 30 nmol g^{-1} body weight. Animals from the same litter were not treated and served as age-matched controls (Bunce and Hess 1981). All experiments adhered to the ARVO Statement for the Use of Animals in Ophthalmic and Vision Research.

At designated times after treatment animals were decapitated, lenses were removed by a posterior approach into 0.9% saline at 30–32°C, and attached tissue was rapidly separated. Lenses were quickly frozen in liquid nitrogen and extracted in 10% trichloro-acetic acid for quantifying glutathione content (Brehe and Burch 1976).

Selenite-dependent oxidation of glutathione (GSH) was quantified by monitoring, with reverse phase HPLC at 210 nm (Bjornstedt et al. 1992), GSH and oxidized glutathione (GSSG) and selenodiglutathione (GSSeSG) concentrations. Reactions occurred in 50 mM phosphate, 100 μM EDTA, 0.1–2 mM GSH and 1.5–50 μM sodium selenite at pH 7.4. Selenite/glutathione-dependent reduction of 10 μM cytochrome *c* or 100 μM nitrobluetetrazolium (NBT) was quantified by absorbance changes at 550 nm or 560 nm respectively.

Correct citation: Bunce, G.E., Hess, J.L., Dell, K.D., and Smith, K. 1997. The function of thiols in the toxicity of selenite and nuclear cataract. In *Trace Elements in Man and Animals – 9: Proceedings of the Ninth International Symposium on Trace Elements in Man and Animals. Edited by* P.W.F. Fischer, M.R. L'Abbé, K.A. Cockell, and R.S. Gibson. NRC Research Press, Ottawa, Canada. pp. 551–553.

Results

The decrease in glutathione content in rat lens following a single, systemic dose of sodium selenite occurred uniformly in both the nuclear and cortical regions of the preweanling rat lens. A maximum decrease occurred 24 h after treatment with selenite.

In the presence of μM quantities of sodium selenite, GSH was oxidized at significant rates when pH > 7.0 (Figure 1). Loss of GSH and formation of GSSG were detected as was transient formation of the selenodiglutathione (GSSeSG) intermediate. Cytochrome *c* or NBT addition shifted the equilibrium of the reaction.

Oxygen significantly inhibited NBT reduction but had little effect on cytochrome *c* reduction under aerobic contitions (Table 2). Further, rate of NBT reduction decreased in the presence of added superoxide dismutase (SOD). A similar role for superoxide was not observed when cytochrome *c* served as electron acceptor, since SOD did not affect GSH/selenite-dependent cytochrome *c* reduction (Table 2). A longer lifetime of the GSSeSG intermediate routinely occurred in the presence of cytochrome *c* (Figure 1).

Table I. Response of GSH content in lenses from control and selenite treated rats. Mean values reported as nmol GSH/lens ± SE, n = 8 lenses. Lenses, maintained frozen in liquid N_2, were separated into a cortex/epithelium and a nuclear fraction.

Time	Cortex	SE	Nucleus	SE
Control 12	33.9	5.96	27.8	2.61
Treated 12	25.0	2.42	21.5	2.93
Control 24	31.6	3.94	35.6	4.99
Treated 24	7.7	2.40	9.8	2.99

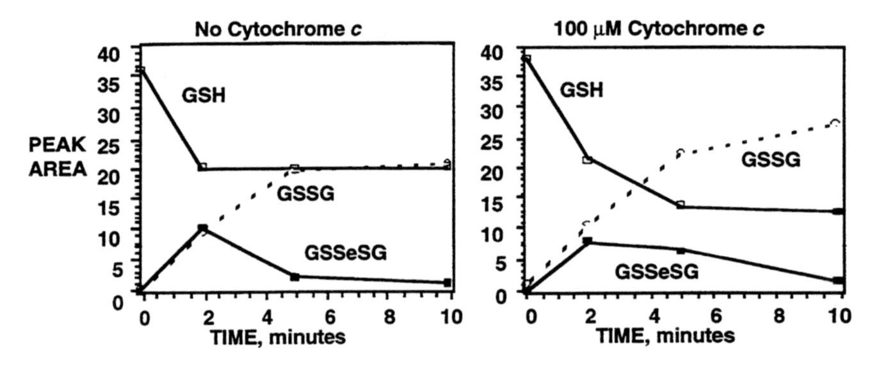

Figure 1. Reactions at 25°C in 50 mM phosphate, 10 mM EDTA at pH 7.4. contained 200 μM GSH. Reactions were initiated with addition of 100 μM sodium selenite and aliquots removed at designated times and chromatographed (Bjornstedt et al. 1992).

Table 2. Glutathione/selenite-dependent reduction of cytochrome c and NBT. Rates of increased absorbance were measured for reactions at 25°C in 50 mM phosphate and 10 mM EDTA at pH 7.4. Mean reaction rates were reported for 3 experiments; SE was <10% for all values. Rates under anaerobic conditions were measured following purging solutions with argon for at least 30 min. All reactions were initiated with addition of selenite.

Condition	NBT reduction ΔA_{560} min^{-1}		Cytochrome *c* reduction ΔA_{550} min^{-1}	
	Anaerobic	Aerobic	Anaerobic	Aerobic
Control	0.323	0.167	0.065	0.057
Plus SOD (400 units)	0.398	0.026	0.064	N.D.

Discussion

Systemic delivery of selenite may provide a potent oxidative stress to the lens of preweanling rats. The significant, consistent decrease in glutathione concentration throughout the lens (Table 1) reveals enhanced oxidative burden that affects lens cell physiology, altered fiber cell development and calcium homeostasis (Shearer et al. 1987).

Free radical intermediates form in the redox cycle of glutathione oxidation in the presence of selenite. Selenite is not reacting in a stoichiometric relationship with the electron donor. Hence, reoxidation of reactive selenium intermediates occurs in these reactions. The impact of oxygen and SOD on the selenite-dependent reduction of NBT by GSH is consistent with a function for superoxide anion. Further, the lack of any effect of oxygen on cytochrome *c* reduction reveals that other single electron donors may function as potent reducing agents. The diglutathiyl radical ($GSSG^{\circ -}$) is a potent reducing agent.

As a prooxidant, selenite generates free radicals in the presence of thiol compounds like GSH. Furthermore, in the presence of O_2, superoxide anion forms and contributes to the oxidative burden in cells with elevated selenite content and metabolism. Toxicity of selenite is manifested as nuclear cataract in the lens of the preweanling rat.

Literature Cited

Bjornstedt, M., Kumar, S., & Holmgren, A. (1992) Selenodiglutathione is a highly efficient oxidant of reduced thioredoxin and a substrate for mammalian thioredoxin reductase. J. Biol. Chem. 267: 8030–8034.

Brehe, J.E. & Burch, H.B. (1976) Enzymatic assays for glutathione. Anal. Biochem. 74: 189–197.

Bunce G.E. & Hess, J.L. (1981) Biochemical changes associated with selenite-induced cataract in the rat. Exp Eye Res. 33: 505–514.

David, L.L. & Shearer, T.R. (1984) Calcium-activated proteolysis in the lens nucleus during selenite cataractogenesis. Invest. Ophthalmol. Vis. Sci. 25: 1275–1283.

Seko, Y., Saito, Y., Kitahara, J. & Imura, I. (1989) Active oxygen generation by the reaction of selenite with reduced glutathione in vitro. In: Selenium in Biology and Medicine, ed. Wendel, A. Springer-Verlag, New York, NY. pp.70–73.

Shearer, T.R., David, L.L., & Anderson, R.S. (1987) Selenite cataract: a review. Curr. Eye Res. 6:289–300.

Tsen, C.C. & Tappel, A.L. (1958) Catalytic oxidation of glutathione and other sulfhydryl compounds by selenite. J. Biol. Chem. 233: 1230–1232.

Wang, Z., Bunce, G.E., & Hess J.L. (1993) Selenite and Ca^{2+} homeostasis in the rat lens: effect on Ca-ATPase and passive Ca^{2+} transport. Curr. Eye Res. 12: 213–218.

Toxicity of Inorganic and Organic Selenium to Chickens

M.B. MIHAILOVIĆ,[†] M. TODOROVIĆ,[‡] M. JOVANOVIĆ,[†] T. PALIĆ,[†] I.B. JOVANOVIĆ,[†] O.J. PEŠUT,[†] AND M. KOSANOVIĆ,[¶]

[†]Department of Physiology and Biochemistry, Faculty of Veterinary Medicine, [‡]Department of Physiology, Faculty of Agriculture, University of Belgrade and [¶]Military Medical Academy, Belgrade, Yugoslavia

Keywords sodium selenite, selenized yeast, toxicity, chicken

Although naturally occurring and experimentally induced selenium toxicosis has been reported in virtually all domestic animals there have been only limited reports of experimentally induced selenium toxicosis in broiler chickens. The level at which selenium becomes toxic to chickens is suggested to be approximately 4 μg/g of diet (Cantor et al. 1984). Several factors are known to reduce the toxicity of selenium, including diets containing crude casein (Moxon 1941), linseed meal (Jensen et al. 1976), methionine (Levander et al. 1970) and a number of trace elements such as: As, Cu, Cd, Ag and Hg (El-Begearmi et al. 1977).

This research was conducted to determine the effects of toxic levels of Se from sodium selenite and selenized yeast to chickens.

Materials and Methods

One hundred and fifty unsexed Hybro chickens, divided into 15 groups, were fed from hatching to six weeks of age, with dietary concentrations of 0, 2, 5, 10, 15, 20 or 30 mg Se/kg as sodium selenite or selenized yeast or 30 mg Se/kg + 0.2% clinoptiolite (minazel). Heparinized blood samples were collected weekly and used for plasma Se analyses (using atomic absorption spectrometry). Two animals from each group were killed at 21 or 42 d of experiment and samples of liver and kidney were taken and used for light microscopic examination. Chicken body weights were determined at weekly intervals.

Results and Discussion

Body weights did not differ for the 6-week period in broiler chickens fed the basal diet or the diet with 2 mg Se/kg as selenite (Figure 1A). Animal weight gains were lower as broilers were provided with higher sodium selenite-containing diets. This was evident from the end of the 3rd or 5th week for broilers provided 10 or 5 mg Se/kg diets, respectively, and within the first week for broilers fed 15, 20, 30 mg Se/kg or 30 mg Se/kg + 0.2% clinoptiolite diets. Similarly, Cantor et al. (1984) found that intake of 2 mg Se/L drinking water had no adverse effects in broilers, and intake of 4 mgSe/L produced lower gain and feed intake. Mortality was higher as broilers were provided with the higher selenite-containing diets. In groups of chickens fed diets containing 15, 20, 30 mg Se/kg or 30 mg Se/kg + 0.2% clinoptiolite, 41.7, 58.3, 91.7 and 100% of animals died respectively. Jensen et al. (1977) found increased mortality in chickens fed 40 mgSe/kg.

Selenium from selenized yeast showed significantly lower toxicity to broilers. Only chickens fed 20 or 30 mg Se/kg had lower weights (from 3rd week, Figure 1B). Not one chicken from groups supplemented even with the highest levels of this form of selenium died during the experimental period.

Correct citation: Mihailović, M.B., Todorović, M., Jovanović, M., Palić, T., Jovanović, I.B., Pešut, O.J., and Kosanović, M. 1997. Toxicity of inorganic and organic selenium to chickens. In *Trace Elements in Man and Animals – 9: Proceedings of the Ninth International Symposium on Trace Elements in Man and Animals. Edited by* P.W.F. Fischer, M.R. L'Abbé, K.A. Cockell, and R.S. Gibson. NRC Research Press, Ottawa, Canada. pp. 554–555.

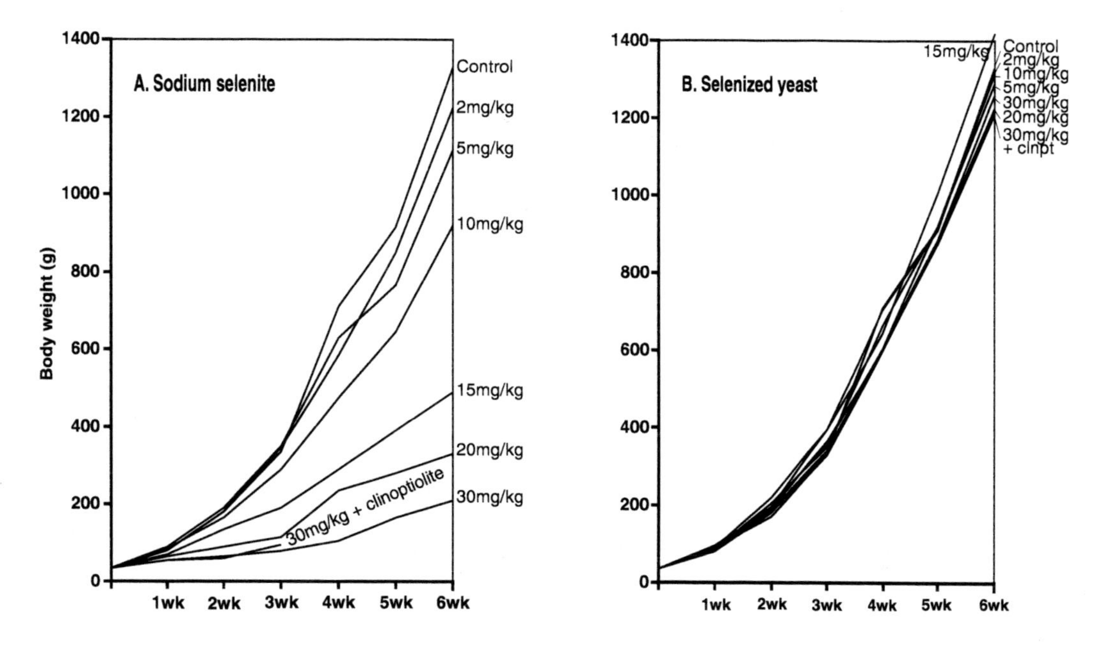

Figure1. The effects of dietary selenium (given as sodium selenite—A, or selenized—B) on chicken body weight gains.

Plasma selenium concentrations of chickens fed the unsupplemental Se basal diet were unchanged, while they increased for the groups fed Se-supplemented diets. Broilers fed diets supplemented with selenite had plasma selenium levels slightly, but not significantly higher than those fed diets with corresponding levels of selenium as selenized yeast.

Histological examinations revealed alterations in liver and kidney that were of similar quality, but differed in quantity, according to the form of selenium, the selenium dietary level and duration of exposure. Liver lesions included an intracellular edema, fat accumulation and different stages of necrosis. Kidney lesions were characterized by intracellular edema and vacuolar dystrophy of the renal tubular epithelia, while fatty degeneration was observed only at higher doses of inorganic selenium. Intensity of lesions was much higher in chickens treated with inorganic selenium than in animals treated with corresponding doses of organic selenium. Lesions in liver and kidney in chickens fed diets supplemented with clinoptiolite were less intensive than in chickens without clinoptiolite. Lower toxicity of selenium from selenized yeast in chickens may have been the result of lower availability or different tissue distribution of this form of selenium for chickens.

References

Cantor, A.H., Nash, D.M. & Johnson, T.H. (1984) Toxicity of selenium in drinking water of poultry. Nutr. Rep. Int. 29: 683–688.

El-Begearmi, M.M., Sunde, M.L. & Ganther, H.E. (1977) A mutual protective effects of mercury and selenium in Japanese quail. Poultry Sci. 56: 313–322.

Jensen, L.S. & Chang, C.H. (1976) Fractionation studies on a factor in linseed meal protecting against selenosis in chicks. Poultry Sci. 55: 594–599.

Levander, O.A. & Morris, V.C. (1970) Interactions of methionine, vitamin E and antioxidants in selenium toxicity in rat. J. Nutr. 100: 1111–1118.

Moxon, A.L. (1941) The intact influence of some proteins on toxicity of selenium. Ph.D. thesis, Univ. Wisconsin, Medison.

Influence of New Carbodithioates on the Tissue Cadmium and Trace Elements in Mice — Interaction with Selenium*

V. EYBL,[†] D. KOTYZOVÁ,[†] J. KOUTENSKÝ,[†] V. MÍČKOVÁ,[†] M.M. JONES,[‡] AND P.K. SINGH[‡]

[†]Department of Pharmacology and Toxicology, Charles University, Faculty of Medicine, 301 66 Pilsen, Czech Republic, and [‡]Department of Chemistry, Nashville, TN 37235, USA

Keywords cadmium, carbodithioates (dithiocarbamates), selenium, trace elements

Previous studies have demonstrated the interaction of selenium with metals *in vivo*. Various metals influence the kinetics and toxicity of selenium and vice versa, selenium modifies the toxicity and tissue distribution of several metals (Eybl et al. 1969, Eybl et al. 1975, Eybl et al. 1986, Eybl et al. 1992).

In the present study the new effective cadmium mobilizing agents — carbodithioate derivatives (Jones et al. 1991) were used to examine their influence on the distribution of cadmium and the level of several trace elements in the organs of animals treated simultaneously with selenium. The following compounds, as sodium salts, were administered: BGDTC (*N*-benzyl-D-glucamine-*N*-carbodithioate), MeOBGDTC (*N*-4-methoxybenzyl)-D-glucamine-*N*-carbodithioate), BLDTC (*N*-benzyl-4-O-(beta-O-galactopyranosyl)-D-glucamine-*N*-carbodithioate .

Experiments were performed in male mice (CD-1, Charles River, FRG) injected with a single iv dose of $CdCl_2.2.5\ H_2O$ (anal. gr., LACHEMA) (0.0175 mmol/kg) alone or in combination with the simultaneous sc administration of $Na_2SeO_3.5H_2O$ (anal. gr., MERCK) (dose corresponding to a Cd:Se molar ratio 1:2). The antidotes were injected in a single dose (1:50) at the 24th h. The following groups of animals (n = 5–7 in each group) were used: I, control; II, cadmium; III, selenium; and IV, cadmium + selenium treated. The level of elements in the liver, kidneys, testes and in the brain was determined at the 48th h of the experiments using flame- or GF-AAS. Student's t-test was used for statistical evaluation and the results were considered significant at $p < 0.05$.

Selenium changed the distribution of the cadmium in the body. The cadmium concentration in the liver decreased to 79.9%, in the kidneys to 60.6%. In the brain and in the testes the cadmium concentration increased to 185.7% or 869% respectively. The mobilizing effect of antidotes on Cd deposits was decreased in selenium treated mice. MeBLDTC decreased the Cd level in the liver to 70.5% of the controls. In the selenium treated mice the best result in the liver was reached by BLDTC (90.1% of the controls). Cd increased the content of zinc and calcium. The effect of cadmium on zinc was corrected partially by selenium. Selenium in the presence of cadmium decreased the concentration of magnesium in the liver by 4.1%.

The cadmium content in the kidneys was also decreased by dithioates.The most effective appeared to be MeOBGDTC (decrease to 51.5% of the controls). Selenium also decreased the cadmium content (to 60.8% of the controls). In the selenium pretreated mice the most effective was BGDTC (decrease to 84%). Cadmium increased the concentration of zinc (to 111.1%). This effect was not corrected by selenium. Calcium level was increased due to cadmium in all groups. The increase in the concentration of copper caused by cadmium was prevented by selenium. The increase in the concentration of calcium and iron in the testes caused by cadmium was prevented by selenium.

*Supported by the grant No. 07/1996 from the Charles University Grant Agency and No. 307/94/0118 from GACzR.

Correct citation: Eybl, V., Kotyzová, D., Koutenský, J., Míčková, V., Jones, M.M., and Singh, P.K. 1997. Influence of new carbodithioates on the tissue cadmium and trace elements level in mice — Interaction with selenium. In *Trace Elements in Man and Animals – 9: Proceedings of the Ninth International Symposium on Trace Elements in Man and Animals. Edited by* P.W.F. Fischer, M.R. L'Abbé, K.A. Cockell, and R.S. Gibson. NRC Research Press, Ottawa, Canada. pp. 556–557.

Antidotes administered alone caused only negligible changes in zinc and copper concentration in the organs. The level of zinc in the liver was increased in mice pretreated by selenium. A remarkable increase in the level of iron in the kidneys (to 125%) after the administration of carbodithioates with/without selenium was shown. BLDTC and BGDTC decreased the level of copper in the kidneys also in the presence of selenium.

Carbodithioates represent the most effective agents which are able to remove cadmium deposits. Selenium decreases the availability of cadmium for drugs administered. Carbodithioates alone cause only negligible changes of trace elements tissue level.

References

Eybl, V., Sýkora, J., & Mertl, F. (1969). Über den Einfluss von Natriumsulfit, Natriumtellurit und Natriumselenit auf die Retention von Zink, Kadmium und Quecksilber im Organismus. Experientia 25: 504–505.

Eybl, V., Sýkora J., Koutenský, J., Koutenská, M., & Mertl, F. (1975) *In vivo* interaction of mercury and cadmium with sodium selenite. Proc. Euro. Soc. Toxicol., 16: 262–259.

Eybl, V., Sýkora, J., & Mertl, F. (1986) *In vivo* interaction of selenium with zinc. Acta Pharmacol. et Toxicol. 59,S7: 547–548.

Eybl, V., Koutenská, M., Koutenský, J., Sýkora, J., Bláhová R., & Smolíková, V. (1992) Selenium-silver interaction in mice. Arch. of Toxicology, Suppl. 15: 160–163.

Jones, M.M., Singh, P.K., Jones, S.G., & Holscher, M.A. (1991) Dithiocarbamates of improved efficacy for the mobilization of retained cadmium from renal and hepatic deposits. Pharmacology & Toxicology, 68: 115–120.

Effect of Simultaneous Pretreatment with Selenium and Cadmium on Testicular Injury by a Subsequent Challenge Dose of Cadmium in Mice

MICHIHIRO KATAKURA, NAOKI SUGAWARA, AND HIROTSUGU MIYAKE
Department of Public Health, Sapporo Medical University, Sapporo 060, Japan

Keywords selenium, cadmium, testicular injury, mouse

Introduction

It is well-known that pretreatment with Cd decreases the testicular damage caused by a subsquent challenge dose of Cd (Yoshikawa 1973). When Se is injected concurrently with Cd, Se prevents the testicular injury caused by Cd (Katakura and Sugawara 1995). This is most effective in the case of simultaneous injection of Cd and Se. We reported previously that when Cd and Se were injected simultaneously, although the second administration of Cd followed after a long period of time, its deleterious action was not found in the testis (Katakara and Sugawara 1994). The purpose of the study was to clarify the protective combination effect of Cd and Se on the testicular injury normally caused by a subsequent challenge dose of Cd.

Experiment Design

ICR strain male mice (7 weeks old) were divided into 6 groups as follows:

1) Control group, the mice were given s.c. 0.2 mL deionized water.
2) Cd group, the mice were given s.c. 1.4 mg Cd/kg.
3) Cd + Se → Cd (24) group, Cd (1.0 mg/kg) and Se (1.4 mg/kg) were injected s.c. simultaneously and at different sites. A high dose of Cd (1.4 mg/kg) was injected 24 h after the combinated injection.
4) Cd + Se → Cd (72) group, Cd (1.0 mg/kg) and Se (1.4 mg/kg) were injected s.c simultaneously and at different sites. A high dose of Cd (1.4 mg/kg) was injected 72 h after the combinated injection.
5) Se → Cd (24) group, Se (1.4 mg/kg) was injected s.c. 24 h prior to an injection of Cd (1.4 mg/kg).
6) Se → Cd (72) group, Se (1.4 mg/kg) was injected s.c. 72 h prior to an injection of Cd (1.4 mg/kg).

All mice were sacrificed at 24 h after the final injection. Testicular damage was assessed from concentrations of testicular TBARS (Thiobarbituric acid-reactive substances), hemoglobin and metal, and from histological findings of testis (stain with hematoxylin-eosine).

Results and Discussion

The concentration of TBARS and hemoglobin in the testis was greatly increased in the Se → Cd (24 and 72) groups and Cd group compared to the Cd + Se → Cd (24 and 72) groups and control group (Table 1). The concentration of MT in the testis was lower in the Se → Cd (24 and 72) groups and Cd group than in the Cd + Se → Cd (24 and 72) groups and control group (Table 1).

The concentration of Cd in the testes and serum were significantly higher in the Cd + Se → Cd (24 and 72) groups than in the Se → Cd (24 and 72) groups (Tables 2 and 3). The histology of the testes in the Cd + Se → Cd (24 and 72) groups was very similar to that in the control group. Concentration of

---

Correct citation: Katakura, M., Sugawara, N., and Miyake, H. 1997. Effect of simultaneous pretreatment with selenium and cadmium on testicular injury by a subsequent challenge dose of cadmium in mice. In *Trace Elements in Man and Animals – 9: Proceedings of the Ninth International Symposium on Trace Elements in Man and Animals. Edited by* P.W.F. Fischer, M.R. L'Abbé, K.A. Cockell, and R.S. Gibson. NRC Research Press, Ottawa, Canada. pp. 558–559.

Table 1. Hemoglobin, TBARS and MT concentrations in testis.

	#	Hemoglobin	TBARS	MT
Control	4	0.017±0.003a	42.8± 4.9a	92.7± 8.8a
Cd	4	0.264±0.056b	125.8±13.6b	9.6± 3.5b
Cd + Se → Cd(24)	4	0.053±0.005a	67.5± 6.1a	91.8±11.0a
Cd + Se → Cd(72)	4	0.053±0.010a	62.0± 9.7a	98.9± 7.3a
Se → Cd (24)	4	0.102±0.020c	88.0±14.8c	90.2±16.3a
Se → Cd (72)	4	0.198±0.099b	104.4±12.2b	36.6±10.7c

Hemoglobin: absorbance Δ A414.
TBARS: thiobarbituric acid reactive substances (TEP:n mol/g tissue).
MT: metallothionein (μg/g tissue).
Data: mean ± SD. *p < 0.05 (Significantly different between a and b).

Table 2. Cd and Se concentrations in serum.

		Cd	Se
Control	(8)	–	112.5± 22.7a
Cd	(8)	64.1± 58.4a	105.0± 24.4a
Cd + Se → Cd(24)	(8)	1858.0±632.2b	413.5±112.8b
Cd + Se → Cd(72)	(8)	654.0±775.2c	219.8± 49.0c
Se → Cd(24)	(8)	418.3±182.3c	185.0± 26.4c
Se → Cd(72)	(8)	345.8±230.4c	158.5± 41.5c

Data: mean ± SD. Metal: ng/ml serum.
* p < 0.05 (Significantly different between a and b).

Table 3. Metal concentrations in testes.

		Cd	Se	Zn	Fe
Control	(4)	–	11.4±0.6a	30.9±1.3a	20.1± 1.2a
Cd	(4)	0.35±0.03a	12.1±0.8a	24.9±1.9b	56.2± 8.3b
Cd + Se → Cd(24)	(4)	0.88±0.07b	16.1±0.3b	28.5±2.0a	25.6± 1.6a
Cd + Se → Cd(72)	(4)	0.77±0.03b	15.8±0.4b	30.0±1.3a	24.8± 1.9a
Se →Cd (24)	(4)	0.34±0.01a	14.5±1.0c	27.2±1.1a	32.9± 7.3c
Se → Cd (72)	(4)	0.33±0.02a	13.6±0.6c	26.2±1.5c	44.8±15.2c

* p < 0.05 (Significantly different between a and b). Metal: μg/g testes.

Cd in the testes was significantly higher in the Cd + Se → Cd (24 and 72) groups than in the Cd group (Table 3). Our results showed that although concentration of Cd in the testes was significantly higher in the Cd + Se → Cd (24 and 72) groups than in the Cd group, testicular injury seen in the latter was not observed in former. The Se → Cd (24 and 72) groups, however, was less efficient than the Cd+ Se → Cd (24 and 72) groups in preventing injury.

References

Yoshikawa H. (1973) Ind. Health 11: 113.
Katakura M., and Sugawara N. (1995) J. Nor. Occ. Health 40: 14
Katakura M., and Sugawara N. (1994) Biomed. Res. Trace Elements 5: 207.

Effect of Lead, Magnesium and Selenium on the Lysosomes of the Carp (*Cyprinus carpio* L.)

N. KRALJ-KLOBUČAR AND A. STUNJA
Department of Biology, Faculty of Science, 10000 Zagreb, Croatia

Keywords elements, lysosomes, acid phosphatase

Low concentrations of metals are permanently present in the environment, and long term exposure induces their accumulation in organisms and may be lethal. Cell defence mechanisms that protect against excessive intakes of trace elements include their incorporation into lysosomes and inactivation by lysosomal enzymes. For this reason, lysosomal enzyme localization and activity can serve as a marker of changes in lysosomal population. The aim of our research was to observe the effects of chronic intoxication by low metal concentrations.

The experimental model used was the carp (*Cyprinus carpio* L.). The animals were treated idividually during the 80 d experimantal period with 0.1 mg/L Mg ($MgCl_2$), 0.1 mg/L Se (Na_2SeO_3), and 0.3 mg/L Pb ($Pb(NO_3)_2$). Samples of liver tissue were taken at 7 d intervals, fixed for 24 h in 4% neutral formalin solution, and 9 μm thick slices were incubated in the medium for histochemical determination of acid phosphatase (Barka and Anderson 1962). Morphometric analysis were performed by an image analyser using the Lucia M measurement program and a Nikon FXA microscope with a magnification of 2590.

The results show differences between the control group and treated animals, as indicated by the lysosomal number (Figure 1), and the area they cover (Figure 2). In animals treated with magnesium, the lysosomal number is slightly higher then in the control group but the area they cover is significantly reduced, suggesting that the lysosomes are smaller. Lead treatment induced increase in the number of lysosomes, but their size is not significantly different compared to the control group. The most significant change in lysosomes is seen in the animals treated with selenium. Their lysosomes are very numerous and cover a large area, indicating a large size. The results are in agreement with histochemical results on the effect of elements on the lysosomes of carp (Kralj-Klobučar and Stunja 1994).

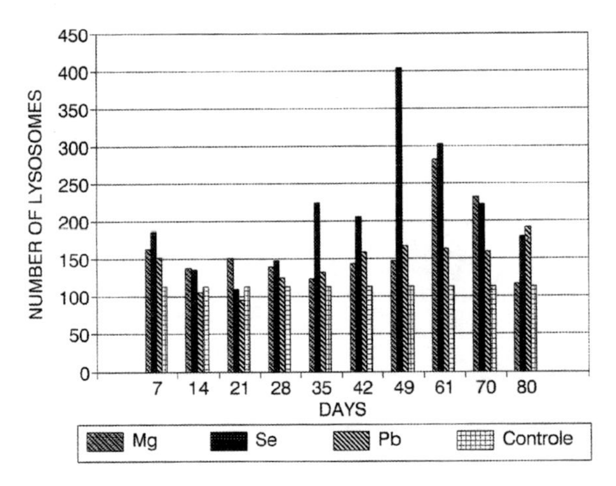

Figure 1. Comparative analysis of the lysosome numbers depending on the trace element and duration of the treatment.

Correct citation: Kralj-Klobučar, N., and Stunja, A. 1997. Effect of lead, magnesium and selenium on the lysosomes of the carp (*Cyprinus carpio* L.). In *Trace Elements in Man and Animals – 9: Proceedings of the Ninth International Symposium on Trace Elements in Man and Animals. Edited by* P.W.F. Fischer, M.R. L'Abbé, K.A. Cockell, and R.S. Gibson. NRC Research Press, Ottawa, Canada. pp. 560–561.

Statistical analysis of variance between treated groups show that changes in the number of lysosomes per field were not statistically significant. Moreover, the area covered by lysosomes (Table 1) was statistically significant at a level of 5% (F-ratio).

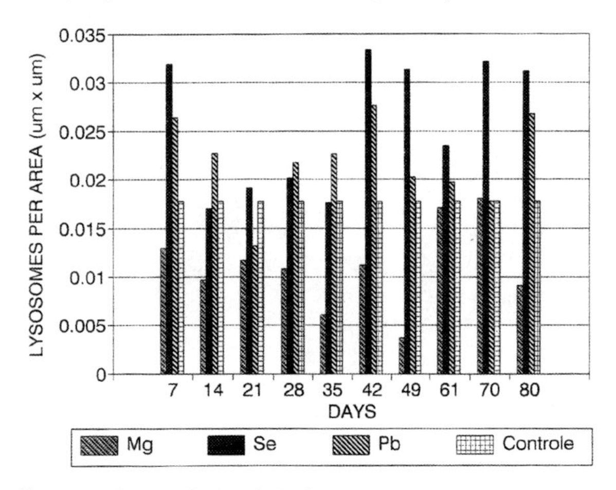

Figure 2. Comparative analysis of the lysosomal size depending on the trace element and duration of the treatment.

Table 1. Comparative statistical analysis of the lysosomal size differences between the three treatment groups.

Source of Variation	Sum of Squares	d.f.	Mean Square	F-Ratio
Between Groups	0.0011557	2	0.0005778	20.238
Within Groups	0.0007709	27	0.000029	
Total (corrected)	0.0019266	29		

References

Barka, T. & Anderson, P.J. (1962) Histochemical methods for acid phosphatase using hexazonium pararosaniline as a coupler. J. Histochem. Cytochem. 10: 741–753.

Kralj-Klobučar, N. & Stunja, A. (1994) Effect of selenium, lead and magnesium on the activity of hydrolitic enzymes in kidney of the carp (*Cyprinus carpio* L.). Period. Biol. 96: 496–498.

Plasma and Cellular Lead in Hypertension

K. KISTERS, W. ZIDEK, AND C. SPIEKER
Med. Univ.-Poliklinik, 48149 Münster, Albert-Schweitzer-Str. 33, Germany

Keywords lead, hypertension, lymphocytes, plasma

In various studies it has been shown, that commonly encountered lead exposure may be directly related to elevated blood pressure (Beevers et al. 1976, Harlan 1985). Among workers being exposed to lead a high incidence of high blood pressure has been noted (Choie and Richter 1980), although the topic is still somewhat controversial (Cramer and Dahlberg 1966). From animal studies it is known that lead exposure may play a pathogenetic role in the development of high blood pressure (Cooper and Steinberg 1977, Mulvany et al. 1980, Piccinini et al. 1977, Webb et al. 1981, Williams et al. 1977). Therefore plasma and cellular lead was of interest in normotensives and hypertensives.

Patients and Methods

Fifteen normotensive patients with normal renal function served as controls. Sixteen patients with essential hypertension and normal renal function were examined.

Heparinized blood was drawn from each patient. Plasma lead concentrations (normal range: 20–150 μg Pb/L) were determined by atomic absorption spectroscopy directly in each blood sample by a Perkin-Elmer 500 HGA apparatus. Intracellular lead concentrations were performed in lymphocytes of each patient. Intralymphocytic lead concentration is difficult to determine since lymphocytic weight or volume cannot be assessed directly. For these reasons, we referred intralymphocytic lead concentration to lymphocytic protein content. Lymphocytic protein concentration was determined by Bradford's method (Bradford 1976) using Coomassie Blue (Serva, Deisenhofen). The lymphocytes were isolated by the density gradient method using Lymphoprep. The suspension of lymphocytes was then washed twice in bi-distilled water. For lysis of the cells, the suspension was frozen to –18°C and thereafter rethawed.The measurements of lead concentrations in lymphocytes were analogously performed by means of atomic absorption spectroscopy. Then lymphocytic lead concentrations were referred to lymphocytic protein content. For statistical analysis the Mann–Whitney–Wilcoxon test was used.

Results

In 15 controls plasma lead concentrations were measured 52.1 ± 18.9 μg Pb/L versus 66.6 ± 31.9 μg Pb/L in 16 patients with high blood pressure (means ± SD). In normotensive patients intralymphocytic lead content was 10.8 ± 4.9 μg Pb/g lymphocytic protein content. In patients with hypertension intracellular lead content was significantly increased (29.7 ± 19.8 μg Pb/g lymphocytic protein, means ± SD) as compared to controls ($p < 0.05$).

Analysis of the hypertensive group showed one subgroup ($n = 6$) with intracellular lead contents between 28 and 81 μg Pb/g lymphocytic protein content, and another, resembling the normotensive group, with values between 4 and 11 μg Pb/g lymphocytic protein content.

Discussion

Our results show cellular lead concentrations in patients with hypertension being significantly increased as compared to normotensive controls with normal renal function. Cellular measurements were performed in lymphocytes, since lymphocytes can easily be obtained. From previous studies it seems likely that the results obtained in human blood cells are similar to those in smooth muscle cells, which are of

Correct citation: Kisters, K., Zidek, W., and Spieker, C. 1997. Plasma and cellular lead in hypertension. In *Trace Elements in Man and Animals – 9: Proceedings of the Ninth International Symposium on Trace Elements in Man and Animals. Edited by* P.W.F. Fischer, M.R. L'Abbé, K.A. Cockell, and R.S. Gibson. NRC Research Press, Ottawa, Canada. pp. 562–563.

special interest in the development of primary hypertension. There was no significant difference in plasma lead concentrations in both groups ($p < 0.05$). Our results are similar to those in previous studies, finding lead exposure being directly related to hypertension (Beevers et al. 1976, McAllister et al. 1971). There was no significant difference in plasma or cellular lead concentrations in men or women in our study corresponding with previous findings (Harlan 1985). The role of lead in the development of hypertension is still controversely discussed (Cramer and Dahlberg 1966). Increased blood pressure may partially be due to the increased reactivity of the vasculature to adrenergic receptor activation. The mechanisms of this increased reactivity appear to be related to a larger pool of intracellular calcium available for activation of contraction (Webb et al. 1981). Some authors (Piccinini et al. 1977) found out that *in vitro* treatment with lead increased the tissue content of radioactive calcium in rat tail artery. Furthermore catecholamine metabolism may be influenced by lead exposure (Silbergeld and Chisolm 1976).

References

Beevers, D.G., Erskine E. & Robertson M. (1976) Blood lead and hypertension. Lancet 2: 1–3.

Bradford, M.A. (1976) Rapid and sensitive method for the quantitation of microgram quantities of protein utilizing the principle of protein-dye binding. Anal. Biochem. 72: 248–254.

Choie, D.D. & Richter, G.W. (1980) Effects of lead on the kidney. In R.L. Singhal and J.A. Thomas (eds.), Lead toxicity, Urban & Schwarzenberg, Baltimore: 187–212.

Cooper, G.P. & Steinberg, D. (1977) Effects of cadmium and lead on adrenergic neuro-muscular transmission in the rabbit. Am. J. Physiol. 232: C128-C131.

Cramer, K. & Dahlberg, L. (1966) Incidence of hypertension among lead workers: a follow-up study on regular control over 20 years. Br. J. Ind. Med. 23: 101–104.

Harlan, W.R. (1985) Blood lead and blood pressure. J. Amer. Med. Assoc. 253: 530–534.

McAllister, R.G., Michelakis, A.M. & Sandstead, H.H. (1971) Plasma renin activity in chronic plumbism. Arch. Intern. Med. 127: 919–923.

Mulvany, M.C., Aalkjaer, C. & Christensen, J. (1980) Changes in noradrenaline sensitivity and morphology of arterial resistance vessels during development of high blood pressure in spontaneously hypertensive rats. Hypertension 2: 664–671.

Piccinini, F., Favalli L. & Chiari, M. (1977) Experimental investigations on the concentration induced by lead in arterial smooth muscle. Toxicology 8: 43–51.

Silbergeld, E.K. & Chisolm, J.J. (1976) Lead poisoning: altered urinary catecholamine metabolites as indicators of intoxication in mice and children. Science 192: 153–155.

Webb, B.C., Winquist, R.J., Vicerty, W. & Vander, A.J. (1981) *In vivo* and *in vitro* effects of lead on vascular reactivity in rats. Am. J. Physiol. 241: H211-H216.

Williams, B.J., Griffith, W.H.,Albrecht, C.N., Pirch, J.H. & Hajtmancik, R.H. (1977) Effects of chronic lead treatment on some cardiovascular responses to norepinephrine in the rat. Toxicol. Appl. Pharmacol. 40: 407–413.

Methylmercury Toxicity and Transport in a Human Intestinal Cell Line in Culture (Caco-2)

M.L. SCARINO, C. DI NITTO, M. DI FELICE, AND G. ROTILIO
Istituto Nazionale della Nutrizione, I - 00179 Rome, Italy

Keywords methylmercury, Caco-2, cytotoxicity, intestinal absorption

Mercury is a toxic metal present in food as contaminant and mainly in seafood, were it is accumulated as methylmercury (MeHg) (Miura et al. 1995). Still little is known on the absorption and intestinal toxicity of this compound. The Caco-2 cell line, derived from a human colon adenocarcinoma, spontaneously differentiates in culture into cells closely resembling small intestinal absorptive enterocytes (Neutra and Louvard 1989). Caco-2 cells can be grown on permeable supports reproducing *in vitro* the polarized organization of the intestinal mucosa. This cell line is a well established model for studies on the absorption and toxicity of nutrients and xenobiotics in the human intestine.

Materials and Methods

Caco-2 cells were grown and allowed to differentiate on either plastic or on permeable (polycarbonate) filter inserts (Transwells, Costar) for 14 to 16 d after confluency in complete D-MEM medium containing 10% fetal calf serum (Rossi et al. 1996). The MeHg cytotoxicity was assessed in complete medium by dose-response curves using nearly confluent, undifferentiated (UD) and 14–16 d confluent differentiated (D) cells grown on plastic. Cell survival was assessed by the neutral red uptake test and expressed as percent of control. Total protein was measured by modified Lowry method (Rossi et al. 1996). Transport experiments were done in Hank's balanced salt solution (HBSS) at pH 7.4 on D cells grown on permeable filters: MeHg was added to the upper chamber medium (apical, AP) and after 4 h the lower chamber (basolateral, BL) medium was collected. MeHg was detected in cells and media, after extraction (IAEA/UNEO/FAO/IOC 1991) by a gas-liquid chromatograph equipped with a capillary column and an electron capture detection system (Perkin Elmer, AutoSystem).

Data are the mean of three experiments done in triplicate ± SD.

Results and Discussion

Dose-response curves of UD and D Caco-2 cells treated with increasing doses of MeHg for 24 h (Figure 1) show that D cells are more resistant than UD cells at low concentrations (1–10 μmol/L). However, the inhibition dose 50 (ID 50) is the same for both UD and D cells and is close to 12.5 μmol/L. Under the same conditions and at concentrations lower than ID 50, the uptake of MeHg is higher in UD than in D cells (Figure 2). These data support and partially account for the observed differences in toxicity. Transport experiments (Figure 3) show that the uptake of MeHg in D Caco-2 increases linearly with the dose (1 to 3 μmol/L), starting to level off when transport to the opposite compartment starts (3 to 6 μmol/L). Concentrations used were not interfering with tight junction functionality, as shown by unchanged transepithelial electrical resistance (data not shown). In these experiments we demonstrate that the degree of differentiation of intestinal cells influences the uptake of MeHg at low concentrations. This finding is in agreement with the high lipophilicity of MeHg: this compound should diffuse better across less specialized membranes (van Meer 1988) resulting in a possible damage to the staminal, undifferentiated compartment (crypts) in the adult intestine. In D Caco-2 grown on filters, MeHg is accumulated into the cells before being transported out from the BL membrane (equivalent to the serosal side of the

Correct citation: Scarino, M.L., Di Nitto, C., Di Felice, M., and Rotilio, G. 1997. Methylmercury toxicity and transport in a human intestinal cell line in culture (Caco-2). In *Trace Elements in Man and Animals – 9: Proceedings of the Ninth International Symposium on Trace Elements in Man and Animals. Edited by* P.W.F. Fischer, M.R. L'Abbé, K.A. Cockell, and R.S. Gibson. NRC Research Press, Ottawa, Canada. pp. 564–565.

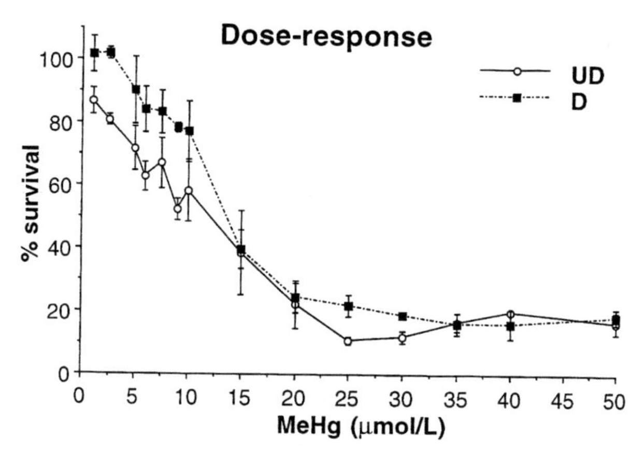

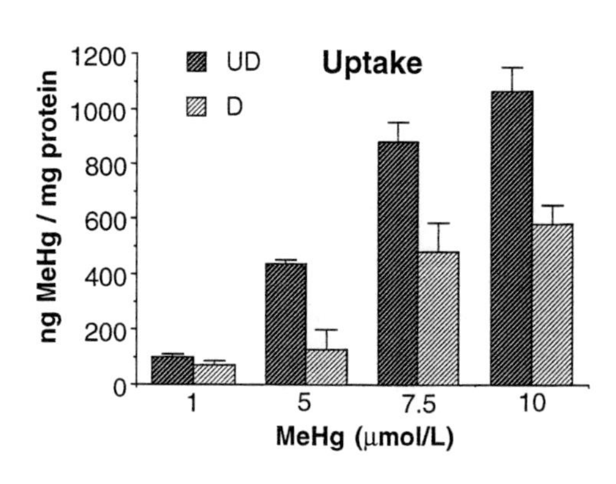

Figures 1 and 2. Undifferentiated (UD) and differentiated (D) Caco-2 cells were grown on plastic in complete medium and treated with MeHg for 24 h. Percentage of survival in dose-effect (Figure 1) was measured by neutral red uptake test. Uptake (Figure 2) was determined in cell lysates.

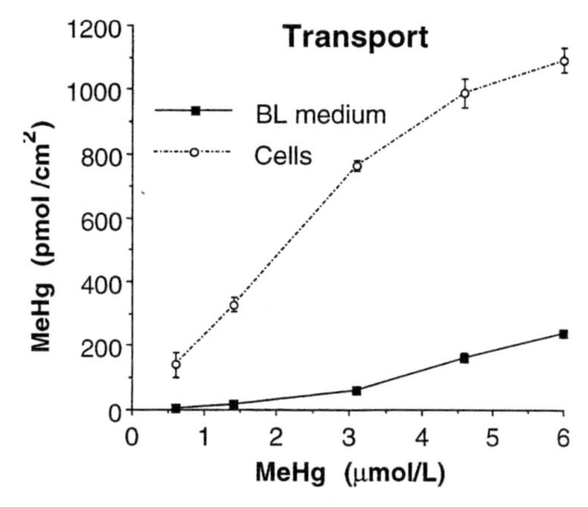

Figure 3. Transport experiments were done in Caco-2 cells differentiated on permeable filters. 1 cm^2 of cell monolayer corresponds to 0.25 mg of protein.

enterocyte *in vivo*.) This result is in agreement with our previous data on copper and cadmium transport in Caco-2 demonstrating the accumulation of these heavy metals in the D cells on filters (Rossi et al. 1996). These results are consistent with a protective role of intestinal cells to MeHg at least at or below the concentration tested. This finding is of interest if related to the average ingestion of MeHg in the Italian high fish consumers calculated as 100–200 μg/(person·day) (P. Santaroni, personal communication), with an estimated postprandial concentration of 0.1–0.3 μmol/L in the intestinal lumen.

Literature Cited

IAEA/UNEO/FAO/IOC (1991) Determination of methylmercury in marine organism. Reference Methods for Marine Pollution Studies No 13, Rev. 1 UNEP.

Miura K., Naganuma A., Himeno S. & Imura N. (1995) Mercury Toxicity. In: Handbook of Experimental Pharmacology Vol.115: Toxicology of Metals. R.A. Goyer and M.G. Cherian Eds, Springer-Verlag, Berlin Heidelberg New York.

Neutra M. & Louvard D. (1989) Differentiation of intestinal cells *in vitro*. In: Functional Epithelial Cells in Culture, K.S. Matlin and J. Valurtich Eds, pp 363–398, Alan R. Liss, New York.

Rossi A., Poverini R., Di Lullo G., Modesti A., Modica A. and Scarino M.L. (1996) Heavy metal toxicity following apical and basolateral exposure in the human intestinal cell line Caco-2 Toxicol. In Vitro 10, 27–36.

van Meer, G. (1989) How epithelia grease their microvilli, TIBS,13, 242–243.

Changes in the Blood and Dialysate Concentration of Trace Metals During the Management of Desferrioxamine B Therapy in Peritoneal Dialysis*

J.M. SANCHEZ,[†] H.S. CUBILLAN,[†] B.I. SEMPRUN,[†] M. HERNADEZ,[†] L.CH. BARRIOS,[†] M.C. RODRIGUEZ,[†] L.E. ELEJALDE,[¶] J.E. TAHAN,[†] AND V.A. GRANADILLO[†]

[†]Analytical Instrumentation Laboratory, University of Zulia, [‡]Renal Service, University Hospital and [¶]Statistic Department, Faculty of Veterinary, Maracaibo 4011, Venezuela

Keywords blood, dialysate, desferrioxamine B, peritoneal dialysis, trace metals

Introduction

Alterations in the blood and tissue concentration of aluminum (Al), iron (Fe), copper (Cu) and zinc (Zn) may induce severe clinical disturbances, particularly, in patients with chronic renal failure (CRF) undergoing periodic hemodialysis. Dialysis encephalopathy, dialysis osteomalacia, microcytic anemia, copper fever, hypogonadism and sexual dysfunction, delayed growth and wound healing, etc., can be associated with incorporation of or depletion of these elements (Afrey 1986). We studied the changes occurring in blood and dialysate Al, calcium (Ca), chromium (Cr), Cu, Fe, magnesium (Mg) and Zn of an azotemic patient undergoing continuous ambulatory peritoneal dialysis (CAPD) and desferrioxamine B (DFO) therapy for a period of 4 months.

Patient and Methods

A 30-year-old male patient, on CADP, with severe aluminum-related osteomalacia was studied in this research. At the time of physical examination, he was not mobile, and DFO therapy was initiated. The patient received a single weekly dose of 2 g intravenously on Mondays for a year. Trace metal determinations were done every 4 weeks in whole blood and in drainage peritoneal dialysate, before each weekly DFO dose was given, and then, 24, 48 and 72 after DFO. The patient's CAPD daily regimen consisted in 4 exchanges of 2.0 L dialysis solution each. The patient was ordered to collect the containers filled with drainage dialysate for trace metal determination. Trace metals in whole blood samples and CAPD dialysate were determined by electrothermal atomization (for Al, Cr, Cu and Zn) and flame atomic absorption spectrometry (for Ca, Fe and Mg) following methodologies reported previously (Romero et al. 1990, Granadillo et al. 1994). For all metals, the average RSD was <3%, for both the within- and between-run precision, and accuracy was verified by analyzing several standard reference materials; a mean relative error <4% between analyzed and certified values obtained. Recoveries were in the range of 94 to 102%. These analytical figures indicated that the methodologies were precise, accurate and interference-free.

Results and Discussion

Table 1 shows the concentrations of metals in samples of whole blood and drainage peritoneal dialysate during the DFO therapy. At the 1st and 4th weeks, DFO administration induced increments in

*Partially supported by CONDES-LUZ, CONICIT and Consejo Central de Pos grado-LUZ.

Correct citation: Sanchez, J.M., Cubillan, H.S., Semprun, B.I., Hernadez, M., Barrios, L.Ch., Rodriguez, M.C., Elejalde, L.E., Tahan, J.E., and Granadillo, V.A. 1997. Changes in the blood and dialysate concentration of trace metals during the management of desferrioxamine B therapy in peritoneal dialysis. In *Trace Elements in Man and Animals – 9: Proceedings of the Ninth International Symposium on Trace Elements in Man and Animals. Edited by* P.W.F. Fischer, M.R. L'Abbé, K.A. Cockell, and R.S. Gibson. NRC Research Press, Ottawa, Canada. pp. 566–567.

blood Al (~420%), Fe (~18%) and Ca (~15%). At the same time, significant removal ($p < 0.005$) of Al was observed. In addition, blood Cr levels increased at the end of treatment (~275%). These facts correlated statistically with Ca reabsorption into the bone structure (Semprún et al. submitted). Other metals evaluated did not change significantly after DFO sequestering therapy (Table 1). We must point out that whole blood metal data were used since they are more representative of the total metal burden (Romero et al. 1990) and important increments were able to be quantified. These results are expected in patients with toxic Al tissue levels (Milliner et al. 1984). Depletion of Ca and Cu was observed in whole blood but no losses were detected in dialysate. Hence, these results suggested a possible redistribution of the essential elements from accumulation pools (e.g., soft tissues, bone, etc.) to blood and vice versa. For Zn, very low blood levels were observed from the beginning. In contrast, bone Zn was higher from the start of DFO treatment, suggesting a possible underlying alteration of Zn homeostasis. By the 13th week, blood Zn was about in normal range. At the end of the treatment, DPD concentrations of Al and Fe confirmed the changes observed in blood levels. Total Al and Fe removal in the period of study were 22.3 mg and 19.5 mg, respectively. Significant removal of other metals was not observed. Clinically, a substantial improvement of the patient's general conditions and symptoms could be observed during and after chelation treatment. No side-effects of DFO therapy were documented. DFO chelation therapy proved to be useful and safe for toxic metal removal in peritoneal dialysis patient.

Table 1. Metal concentrations (mean ± 1 SEM, μg/L) in samples of whole blood and drainage peritoneal dialysate (in parenthesis) during desferrioxamine B (DFO) chelation therapy.

		DFO administration			
Metal	Pre-DFO	1st week	4th week	8th week	13th week
Al	24 ± 1	120 ± 4 **	129 ± 8**	71 ±14	173 ±28**
	(UD)	(375 ±12)	(302 ±46)	(43 ± 8)	(72 ±12)
Fe	237 ± 3[a]	244 ± 3[a]	314 ± 5[a],**	268 ±11[a],**	273 ± 8[a],**
	(26 ±11)	(273 ±52)**	(177 ±66)*	(174 ±85)**	(186 ±76)**
Ca[a]	86 ± 1	101 ± 1**	96 ± 2**	58 ± 3**	38 ± 1**
	(64 ± 2)	(62 ± 1)	(65 ± 0.3)	(36 ± 1)**	(18 ± 1)**
Cr	0.8±0.4	UD	UD	UD	3 ± 1**
	(UD)	(2 ± 0.4)	(1 ± 0.1)	(1 ± 0.1)	(1 ± 0.04)
Cu	803 ±46	864 ± 9	760 ±17	803 ±35	554 ± 10**
	(25 ± 5)	(17 ± 2)	(14 ± 4)	(8 ± 1)**	(11 ± 3)*
Mg[a]	38 ± 1	39 ± 1	35 ± 1	39 ± 1	38 ± 0.3
	(21 ± 1)	(19 ± 0.4)	(18 ± 1)	(20 ± 3)	(21 ± 0.3)
Zn	73 ± 1	72 ± 1	72 ± 1	62 ± 2**	66.2± 0.2[a],**
	(4 ± 1)	(2 ± 1)	(3 ± 1)	(9 ± 2)	(284 ± 42)

[a]In mg/L. UD = Undetectable. Differences with respect to Pre-DFO values as follows: * $p < 0.05$; ** $p < 0.01$.

Literature Cited

Alfrey, A.C. (1986) Aluminum Metabolism. Kidney Int. 29: S8–S11.

Granadillo, V.A., Parra de Machado, Ll., & Romero, R.A. (1994) Determination of total chromium in whole blood, blood components, bone, and urine by fast furnace program electrothermal atomization AAS and using neither analyte isoformation nor background correction. Anal. Chem. 66: 3624–3631.

Milliner, D.S., Nebeker, H.G., Ott, S.M., Andress D.L., Sherrard, D.J., Alfrey, A.C., Slatopolsky, E.A., & Coburn, J.W. (1984) Use of the desferrioxamine infusion test in the diagnosis of aluminum-related osteodystrophy. Ann. Intern. Med. 101: 775–780.

Romero, R.A., Navarro, J.A., Rodríguez-Iturbe, B., Garcias, R., Parra O.E., & Granadillo, V.A. (1990) Distribution of trace metals in blood components of patients with chronic renal failure undergoing periodical hemodialysis treatment. Trace Elem. Med. 7: 176–181.

Biodistribution of Au(III) Ion and its Complexes in Mice

SHUICHI ENOMOTO,[†] RAJIV G. WEGINWAR,[‡] BIN LIU,[¶] SHIZUKO AMBE,[†] AND FUMITOSHI AMBE[†]

[†]The Institute of Physical and Chemical Research (RIKEN), Wako, Saitama 351-01, Japan; [‡]Chandrapur Engineering College, Chandrapur-442 403, India; and [¶]Department Technical Physics, Peking University, Beijing 100871, China

Keywords gold(III) metabolism, gold(III) ion, 2-amino-2-hydroxymethyl-1,3 propandiol, sodium diethyldithiocarbamate, gold radioactive tracer

Introduction

Gold has no known natural biological function but influences some side effects like skin rashes, diarrhea and nephrosis. Many gold containing drugs are effective against rheumatoid arthritis but their mode of action is still unclear (Ni Dhubhghaill et al. 1993). Au(III) is readily reduced by cysteine or methionine in cell culture media. Some carboxylates are oxidized by $[AuCl_4]^-$ to give CO_2 and metallic gold (Parish et al. 1987). This probably means Au(III) would never reach the desired site and may instead be toxic, hence it is necessary to choose the ligands very carefully to prevent this. It is important to know the exact metabolic rate of Au(III) in the living body. There may be sites of proteins which could stabilize the square planar coordination environment which gold(III) prefers. A few Au(III) complexes have been shown to possess antiviral activity, hence there is much scope for gold(III) complexes as pharmaceuticals. In the present study, we have determined the biodistribution of radioactive gold(III) isotope in a physiological saline solution and its complexes in various tissues, organs and blood of normal ddY mice to clarify the biobehavior of gold(III).

Experimental

A thick gold foil was irradiated with 135 MeV/nucleon C-12 ion beam at RIKEN Ring Cyclotron. The target containing Au-196 tracer was extracted with ethyl acetate. This organic phase was evaporated and an equimolar solution of 2-amino-2-hydroxymethyl-1,3-propandiol was added to form a colorless complex (Au(III)-amino complex). Similarly, an equimolar solution of sodium diethyl dithiocarbamate was added to form a yellow colored complex (Au(III)-dedtc complex). Also Au(III) tracer with carrier was prepared in a physiological saline solution (Au(III)-ion). These solutions were prepared as 5% ethyl alcohol, and injected intraperitoneally (12 mg Au(III)/kg mice) to the normal ddY mice. The mice were sacrificed at 3, 24 and 48 h after injection; tissues, organs and blood were weighed and the activities were determined by gamma ray spectrometry using HPGe detectors. The results are given in percentage of injected dose (I.D.) to the weight of tissues, organs and blood (ID %/g) of an average of at least three mice.

Results and Discussion

After injection, the complexes begin to be absorbed and metabolized. Our results give the distribution of Au(III)-ion, Au(III)-amino complex and Au(III)-dedtc complex in different organs, tissues and blood of mice 3 h after injection, as shown in Figure 1. In comparison to the 3-hour data, Au(III)-ion, Au(III)-amino

Correct citation: Enomoto, S., Weginwar, R.G., Liu, B., Ambe, S., and Ambe, F. 1997. Biodistribution of Au(III) ion and its complexes in mice. In *Trace Elements in Man and Animals – 9: Proceedings of the Ninth International Symposium on Trace Elements in Man and Animals. Edited by* P.W.F. Fischer, M.R. L'Abbé, K.A. Cockell, and R.S. Gibson. NRC Research Press, Ottawa, Canada. pp. 568–569.

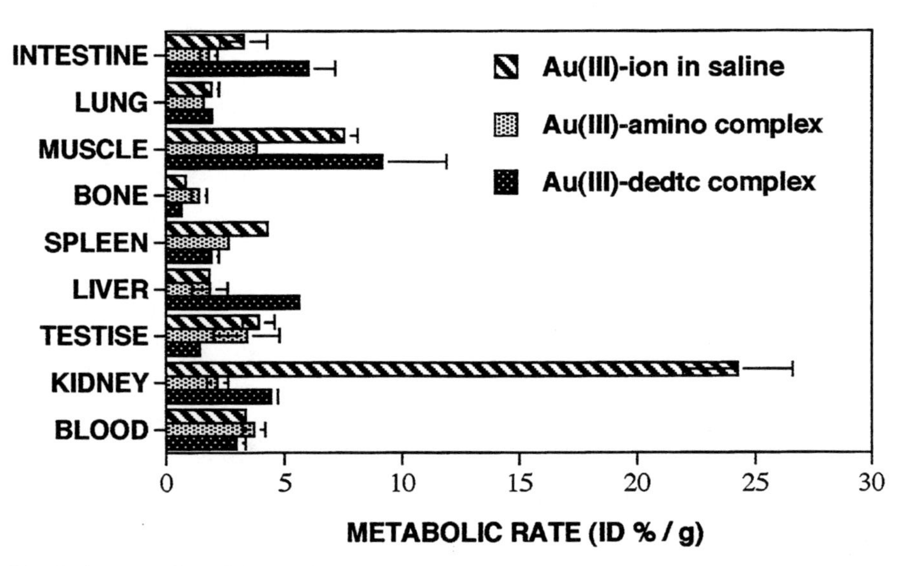

Figure 1. Metabolic rate of Au(III)-ion, Au(III)-amino complex and AU(III)-dedtc complex after 3 h of injection.

and Au(III)-dedtc complex were absorbed almost instantly into the blood, from which they are rapidly transported throughout the body. Clearance of gold from the blood to all organ begins immediately. Of particular note, 25% of Au(III)ion, is present in kidney; it may be because of unbound gold(III) which is cleared by urinary excretion and may be associated with albumins. It has been reported that Au(III) possibly forms a simple complex with cysteine or glutathione, and one of the others is probably a metallothionein complex. Their metabolic rate for intestine, lung, muscle and bone is almost similar 3 h after injection except for Au(III)-amino complex. It has been reported that nitrogen ligands such as amine and porphyrins tends to stabilize Au(III) ions which are very soft with high affinity for nitrogen. In the case of spleen and testis, metabolic rate is in a descending order of Au(III)-ion, Au(III)-amino and Au(III)-dedtc.

In comparison to the 3, 24 and 48 h data for Au(Ill)-ion in intestine, lung, muscle, spleen, testis and blood; the metabolic rate decreases gradually with time but it is almost constant for bone and liver. On the contrary, kidney metabolic rate is maximum after 24 h of injection and sharply decreased by about 80% after 48 h.

In the case of Au(Ill)-amino complex, concentration of Au in blood is very high initially, it decreases with time. A gradual decreasing trend of metabolic rate is also observed for bone and testis, while spleen show almost the same pattern of metabolism; livers, however, show marginal increases with time. Kidney metabolic rate is high after 24 h but decreases by about 35% after 48 h of injection.

The present data suggest that there must be a fundamental difference between these complexes either in the absorption of gold from the blood or its binding property in the organs. From these observations it is clear that the absorption and distribution of the Au(III) ion and its complexes in the different organs, tissues and body fluids depends markedly on the role of ligands used to stabilize Au(III) and the carrier of gold(III).

References

Ni Dhubhghaill, O.M. & Sadler, P.J., (1993) In: Metal Complexes in Cancer Chemotherapy (Keppler B.K.ed.), pp. 21–248. VCH, Verlagsgesellschaft, mbH, Weinheim, Germany.

Parish, R. V. & Cotnill, S. M. (1987) Medicinal gold compounds. Gold Bull. 20: 3–12.

Does TTM-Treatment Alleviate Hepatitis Occurring Naturally in Long-Evans Cinnamon (LEC) Rats Fed a Copper Normal Diet?

NAOKI SUGAWARA, YU-RONG LAI, AND CHIEKO SUGAWARA
Department of Public Health, Sapporo Medical University, Sapporo 060 Japan

Introduction

Recently, LEC (Long-Evans Cinnamon) rats have been established from a closed colony of Long-Evans rats. The LEC rat is a mutant strain that develops fulminant hepatitis and severe jaundice at around 4 months of age (Sasaki et al. 1985). The hepatitis is due to an abnormal accumulation of hepatic Cu led by a reduced secretion of bile and serum Cu (Sugawara et al. 1991). Sasaki et al. (1994) identified the rat homolog to the human Wilson disease (WD) gene as the gene (hts gene) responsible for hepatitis in LEC rats. The hepatic failure and neurological disorder in patients with WD have been due to the gross deposition of Cu in their organs. Accordingly, a key factor in therapy is the induction of negative Cu-balance in the body by promoting removal of excess Cu from the tissues. The purpose of this study, was to clarify the effect of TTM injected after the onset of jaundice on its improvement as well as investigate its effect on the metabolism of Cu in the liver. LEC rats which had already developed jaundice were divided into two groups, one injected with TTM and other one without TTM.

Animals

Male LEC and Fischer rats were weaned at 30 d after birth. They were maintained on a basal diet (containing 7 ppm Cu) and tap water ad lib for 30 d. At 60 d after birth, eight LEC rats were switched a diet containing 30 ppm Cu. Three other LEC rats and four Fischer rats were fed the basal diet. Forty days after the switching (at 90 d after birth), eight LEC rats were returned to the basal diet and divided into two groups in the order of body weight. Immediately, four rats, composed of two jaundice rats, were injected i.p. with TTM (10 mg/kg) for two consecutive days (LEC + TTM group). The other rats were treated with 0.9% NaC1 solution (LEC – TTM group). Thereafter, four LEC rats without TTM (LEC – TTM) who showed a rapid decrease in body weight and severe jaundice were necropsied at 110 d after birth. The other rats, Fischer, LEC and LEC + TTM were necropsied at 125 d.

Results and Discussion

Body weight (BW) on day 60 was expressed as 100 (Figure 1). From 92 d, BW was not increased in the LEC rats fed the diet containing 30 ppm Cu. At 100 d, 2 of 8 rats suffered from jaundice. This dietary concentration of Cu is not toxic for experimental rats (Sugawara et al. 1995). Accordingly, our result suggests that Cu is toxic for the LEC rat. Probably, hepatic Cu level regulates the onset of hepatitis in LEC rat. Eight LEC rats were divided into two groups. In one group, composed of 4 rats injected with TTM (10 mg/kg), body weight increased rapidly and jaundice disappeared soon. However, the other group without TTM, showed decreased body weight and severe jaundice within one week after the division. The other two groups, Fischer and LEC were growing normally (Figure 1). It seems likely that, TTM injection is effective for the improvement and prevention of jaundice. We measured enzyme activity in serum and Cu concentration in the liver (Table 1). In the LEC – TTM group (rats without TTM), activities of LDH and GOT, and BB (bilirubin) concentration were increased. The TTM injection prevented the increases.

Correct citation: Sugawara, N., Lai, Y.-R., and Sugawara, C. 1997. Does TTM-treatment alleviate hepatitis occurring naturally in Long-Evans Cinnamon (LEC) rats fed a copper normal diet? In *Trace Elements in Man and Animals – 9: Proceedings of the Ninth International Symposium on Trace Elements in Man and Animals. Edited by* P.W.F. Fischer, M.R. L'Abbé, K.A. Cockell, and R.S. Gibson. NRC Research Press, Ottawa, Canada. pp. 570–571.

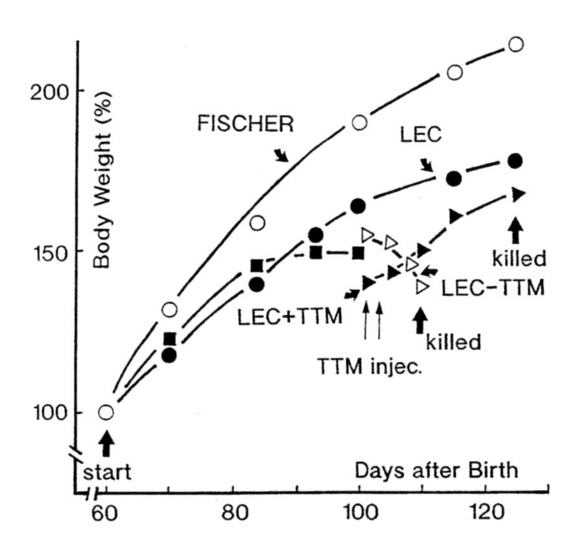

Figure 1. Growth curve for experiment.

Table 1. LDH, GOT, BB and Cu in serum, and Cu in liver.

			Serum			Liver
		LDH	GOT	BB	Cu	Cu
Fischer	(5)	29± 8	9± 1	0.1±0.0	0.95±0.10	5.7± 0.5
LEC	(3)	51± 26	22± 2	0.2±0.0	0.10±0.02	328.4±23.0
LEC – TTM	(4)	377±131	40± 9	56.7±6.1	1.21±0.11	305.0±32.4
LEC + TTM	(4)	45± 14	23±10	0.2±0.0	0.58±0.05	459.1±38.7

Mean ± SD. Units: LDH and GOT: IU; BB (bilirubin): mg/L; Cu: pg/mL serum and pg/g liver.

Serum Cu increased only in the LEC – TTM group, but the increase was prevented by the TTM injection. Hepatic Cu concentration was significantly higher in the three LEC groups than in the Fischer group. However, the LEC + TTM group did not show the decreased concentration of Cu in the liver. Our results suggest that the preventative action of TTM is due to not only the deprivation of Cu from the liver but also due to a modified mode of Cu metabolism.

References

Sasaki, M. et al. (1985) Rat News Lett. 14: 4–.

Sugawara, N., Sugawara, C., Katakura, M., Takahashi, H. & Mori, M. (1991) Copper metabolism in the LEC rat: involvement of induction of metallothionein and disposition of zinc and iron. Experientia 47: 1060–1063.

Sasaki, N., Hayashizaki, Y., Muramatsu, M., Matsuda, Y., Ando, Y., Kuramoto, T., Serikawa, T., Azuma, T., Naito, A., Agui, T. et al. (1994) The gene responsible for LEC hepatitis, located on rat chromosome 16, is the homolog to the human Wilson disease gene. Biochem. Biophys. Res. Commun. 202: 512–518.

Sugawara, N., Li, D., Sugawara, C. & Miyake, H. (1995) Response of hepatic function to hepatic copper deposition in rats fed a diet containing copper. Biol. Trace Elem. Res. 49: 161–169.

Protective Role of Metallothionein in Cisplatin-Caused Renal Toxicity

M. SATOH, Y. AOKI, AND C. TOHYAMA

Environmental Health Sciences Division, National Institute for Environmental Studies, Tsukuba, Ibaraki 305, Japan

Keywords cisplatin, metallothionein, bismuth, zinc, renal toxicity

Introduction

Cisplatin (*cis*-DDP) has potent antitumor activity with a broad anticancer spectrum against certain human neoplasms, but also has significant renal toxicity. The pre-administration of metallothionein (MT)-inducing metals such as bismuth and zinc can prevent the renal toxicity of *cis*-DDP (Naganuma et al. 1987, Satoh et al. 1993). However, it is still unclear whether MT induced by bismuth or zinc directly prevents the renal toxicity of *cis*-DDP because other factors besides MT could not be ruled out. The question of how basal levels of MT are associated with the *cis*-DDP toxicity must still be clarified. To study the possible physiological roles of MT, transgenic mice deficient in the MT-I and MT-II genes (MT-null mice) have been established (Michalska and Choo 1993, Masters et al. 1994). Using these mice, we have investigated whether and how MT is involved in the lethal and renal toxicity caused by *cis*-DDP.

Materials and Methods

Animals

MT-null mice whose MT-I and II genes had a null mutation were produced and kindly provided to use by Dr. A. Choo (Murdoch Institute for Research into Birth Defects, Royal Children's Hospital, Australia) and were routinely bled in the vivarium of NIES. Since the MT-null mice were of a mixed genetic background of 129 Ola and C57BL/6 strains, female C57BL/6J and 129/Sv mice, purchased from Japan Clea Co. (Tokyo, Japan), were used for wild-type controls.

Treatments and Analyses

Nine-week-old female MT-null, C57BL/6J and 129/Sv mice were randomized into control and experimental groups. MT-null and C57BL/6J mice were given i.p. injections of *cis*-DDP at doses between 20 and 60 μmol/kg. To examine the renal toxicity of *cis*-DDP, blood was collected from each mouse while under diethylether anesthesia 4 d after the injection. The survival rate of these mice (7 mice/dose group) was determined 20 d after the injection of *cis*-DDP (30, 40, 50 and 60 μmol/kg). To examine the effect of induced MT in *cis*-DDP toxicity, mice were given s.c. injections of $Bi(NO_3)_3$ (50 μmol/kg), $ZnSO_4$ (100 μmol/kg) or saline once a day for 2 d. These mice were given by i.p. injection *cis*-DDP (30 and 40 μmol/kg) on day 2 (24 h after the last injection of metal compounds). A portion of these treated-mice were sacrificed by cervical dislocation to determine renal MT levels on day 2 (at the time of *cis*-DDP injection). On day 6 (4 d after the *cis*-DDP injection), blood was collected from each mouse while under diethylether anesthesia to evaluate renal toxicity. Mice were handled humanely according to the NIES guidelines.

BUN and plasma creatinine values were determined using an automatic dry-chemistry analyzer system (Spotchem SP-4410, Kyoto Daiichikagaku, Kyoto, Japan). MT concentrations in the kidneys were measured by radioimmunoassay (Nishimura et al. 1991). The data were analyzed by Student's t-test.

Correct citation: Satoh, M., Aoki, Y., and Tohyama, C. 1997. Protective role of metallothionein in cisplatin-caused renal toxicity. In *Trace Elements in Man and Animals – 9: Proceedings of the Ninth International Symposium on Trace Elements in Man and Animals. Edited by* P.W.F. Fischer, M.R. L'Abbé, K.A. Cockell, and R.S. Gibson. NRC Research Press, Ottawa, Canada. pp. 572–573.

Results and Discussion

The basal MT levels in the kidneys of the C57BL/6J and 129/Sv mice were similar to each other, approximately 3.5 μg/g tissue on average, whereas the basal MT level of the MT-null mice was below the detection limit (<0.2 μg/g tissue). Since no significant difference in the *cis*-DDP nephrotoxicity, estimated by BUN and plasma creatinine values, between the C57BL/6J and 129/Sv mice was found, C57BL/6J mice were used as a wild type control in the rest of the study. BUN and creatinine values of the *cis*-DDP injected MT-null mice were markedly increased at doses between 30 and 45 μmol/kg, but the C57BL/6J mice were tolerant against *cis*-DDP up to a dose of 40 μmol/kg. The more susceptible nature of the MT-null mice compared to the C57BL/6J animals was also apparent from LD_{50} values; those for the MT-null and C57BL/6J mice were approximately 45 and 55 μmol/kg respectively.

MT was induced in the kidneys of the C57BL/6J mice by administration of $Bi(NO_3)_3$ or $ZnSO_4$, but not in those of the MT-null mice. BUN and creatinine values in the C57BL/6J mice were significantly increased by the injection of *cis*-DDP at a dose of 40 μmol/kg but not 30 μmol/kg, as in the experiment described above. Pretreatment with $Bi(NO_3)_3$ or $ZnSO_4$ clearly canceled the *cis*-DDP effects on BUN and creatinine levels. In contrast, the pretreatment of MT-null mice with the either compound did not affect the elevated levels of BUN and creatinine caused by *cis*-DDP doses of 30 and 40 μmol/kg.

Earlier *in vitro* studies suggest that cellular MT levels determine the sensitivity of mammalian cells to *cis*-DDP (Basu and Lazo 1990, Kasahara et al. 1991). Overexpressed MT prevented the cytotoxicity caused by *cis*-DDP, while the *cis*-DDP resistance was reversed by a decrease in MT content. Recently, MT-null fibroblasts were found to be more sensitive to *cis*-DDP than MT-positive wild type (Kondo et al. 1995). The present *in vivo* studies support these *in vitro* results and show that endogenous and induced MT in the kidney protects mice against the lethal and renal toxicity of *cis*-DDP. These results clearly indicate, therefore, that preinduction of MT synthesis in the target organs protects mice against the renal toxicity of *cis*-DDP.

It is still not clear how endogenous and induced MTs protect mice from renal and lethal toxicity. Two possible mechanisms have been presented so far. First, MT lessens *cis*-DDP toxicity by binding the platinum as shown in the liver and kidney of *cis*-DDP-treated rats (Zelazowski et al. 1984). Apparently, zinc or bismuth bound to MT is replaced by platinum. The second hypothesis is that MT acts as a free radical scavenger (Sato and Bremner 1993) and protects against oxidative stress due to free radicals produced by *cis*-DDP (Boogaard et al. 1991). We speculate that MT may exert its detoxifying role through these two mechanisms.

References

Basu, A. & Lazo, J.S. (1990) A hypothesis regarding the protective role of metallothioneins against the toxicity of DNA interactive anticancer drums. Toxicol. Lett. 50: 123–135.

Boogaard, P.J., Slikkerveer, A., Nagelkerke, J.F. & Mulder, G.J. (1991) The role of metallothionein in the reduction of cisplatin-induced nephrotoxicity by Bi^{3+}-pretreatment in the rat *in vivo* and *in vitro*. Biochem. Pharmacol. 41: 369–375.

Kasahara, K., Fujiwara, Y., Nishino, K., Ohmori, T., Sugimoto, Y., Komiya, K., Matsuda, T. & Saijo, N. (1991) Metallothionein content correlates with the sensitivity of human small cell lung cancer cell lines to *cis*platin. Cancer Res. 51: 3237–3242.

Kondo, Y., Woo, E.S., Michalska, A.E., Choo, K.H.A. & Lazo, J.S. (1995) Metallothionein null cells have increased sensitivity to anticancer drugs. Cancer Res. 55: 2021–2023.

Masters, B.A., Kelly, E.J., Qualfe, C.J., Brinster, R.L. & Palmiter, R.D. (1994) Targeted disruption of metallothionein I and II genes increases sensitivity to cadmium. Proc. Natl. Acad. Sci. USA 91: 584–588.

Michalska, A.E. & Choo, K.H.A. (1993) Targeting and germ-line transmission of a null mutation at the metallothionein I and II loci in mouse. Proc. Nad. Acad. Sci. USA 90: 8088–8092.

Naganuma, A., Satoh, M. & Imura, N. (1987) Prevention of lethal and renal toxicity of *cis*-diamminedichloroplatinum(II) by induction of metallothionein synthesis without compromising its antitumor activity in mice. Cancer Res. 47: 983–987.

Nishimura, H., Nishimura, N., Kobayashi, S. & Tohyama, C. (1991) Immunohistochemical localization of metallothionein in the eye of rats. Histochemistry 95: 535–539.

Sato, M. & Bremner, I. (1993) Oxygen free radicals and metallothionein. Free Rad. Biol. Med. 14: 325–337.

Satoh, M., Kloth, D.M., Kadhim, S.A., Chin, J.L., Naganuma, A., Imura, N. & Cherian, M.G. (1993) Modulation of both cisplatin nephrotoxicity and drug resistance in murine bladder tumor by controlling metallothionein synthesis. Cancer Res. 53: 1829–1832.

Zelazowski, A.J., Garvey, J.S. & Hoeschele, J.D. (1984) In vivo and in vitro binding of platinum to metallothionein. Arch. Biochem. Biophys. 229: 246–252.

Newer Trace Element Interaction in Animal and Human Nutrition

Chair: T.M. Bray

Selenium and Iodine Deficiencies and the Control of Selenoprotein Expression

J.R. ARTHUR,[†] F. NICOL,[†] J.H. MITCHELL,[†] AND G.J. BECKETT[‡]

[†]Division of Biochemical Sciences, Rowett Research Institute, Bucksburn, Aberdeen AB21 9SB and [‡]Cellular Endocrinology Unit, University Department of Clinical Biochemistry, The Royal Infirmary, Edinburgh EH3 9YW, UK

Keywords selenium, iodine, selenoproteins, thyroid

Selenoproteins

Selenium is recognised as playing an important role in cell metabolism in eucaryotes and is thus essential for normal function and health. Most of the biological functions of selenium are probably mediated by selenoproteins. The selenoproteins which have so far been identified contain selenium as the amino acid selenocysteine, which acts as an efficient redox catalyst. Specific mechanisms exist which allow the incorporation of selenocysteine into proteins. These mechanisms which were initially characterised in prokaryotes, rely on four gene products. These are, a specific tRNA which binds serine and recognises the UGA stop codon in mRNA that specifies the incorporation of selenocysteine and two proteins which convert the serine to selenocysteine using a selenophosphate intermediate. The fourth component is an elongation factor which recognises "stem loop" structures in the mRNA which allow the stop codon to code for selenocysteine (Heider and Bock 1993). In eucaryotes the mechanisms for incorporation of selenium into selenocysteine are less well defined. Mammalian selenophosphate synthetase has been cloned and has some similarity to the corresponding bacterial enzyme (Low et al. 1995). There are also specific mammalian tRNAs which specify selenocysteine incorporation at the UGA stop codon. In contrast to prokaryotes, where the stem loop structures are in the reading frame of the mRNA, in eucaryotes the stem loop structures are in the 3′ untranslated region (3′utr) of the mRNA. These stem loops vary in sequence and structure between selenoproteins. Thus, in cells transfected with chimeric constructs of deiodinase mRNA coding regions with 3′utrs from different selenoproteins, type one deiodinase (IDI) activity varies with the 3′utr providing a mechanism for differential control of selenoprotein expression (Salvatore et al. 1995).

Control of Selenoenzyne Expression

In addition to selenocysteine synthesis, there are many levels at which selenium metabolism could be regulated and controlled. Availability of selenium from the diet can have a general effect on all selenium metabolism however, within animals there are several further potential levels of control (Table 1).

Correct citation: Arthur, J.R., Nicol, F., Mitchell, J.H., and Beckett, G.J. 1997. Selenium and iodine deficiencies and the control of selenoprotein expression. In *Trace Elements in Man and Animals – 9: Proceedings of the Ninth International Symposium on Trace Elements in Man and Animals. Edited by* P.W.F. Fischer, M.R. L'Abbé, K.A. Cockell, and R.S. Gibson. NRC Research Press, Ottawa, Canada. pp. 574–578.

Table 1. Control of selenoenzyme metabolism.

Organ uptake of selenium (particularly in deficiency)
Variation of organ selenium uptake with form of the element?
Distribution of selenium within the organ
Availability of co-factors for selenocysteine synthesis?
Transcriptional or translational control
Selenoenzyme stability

Uptake and retention of selenium varies between organs, particularly during deficiency. For instance, the brain and thyroid retain selenium more efficiently than liver or kidney (Behne et al. 1988). Additionally, organ uptake of selenium may vary with the chemical form of the element. Thus a tissue which takes up methionine efficiently may also selectively absorb selenomethionine. Once selenium is absorbed into organs little is known about how it becomes distributed in selenoproteins within subcellular organelles and cell compartments. Other controls of selenoenzyme expression may arise from availability of as yet unidentified cofactors for the enzymes of selenocysteine synthesis. However, the most sensitive control of selenoenzyme expression is likely to be at transcriptional or translational levels. Transcriptional and translational modulation of selenoprotein levels could be further refined by differences in stabilities of the individual proteins.

Selenium and Thyroid Metabolism

The importance of selenium in thyroid hormone metabolism is now well established. Three deiodinase enzymes (IDI, IDII and IDIII) which control the formation and metabolism of triiodothyronine (T3) (Table 2) have been cloned and each cDNA contains a stop codon in the reading frame specifying selenocysteine (Davey et al. 1995, Larsen and Berry 1995).

In selenium deficiency, IDI activity in liver and kidney decreases more than IDII and IDIII activities and thus circulating concentrations of T4 increase and T3 decrease. However, the decrease in T3 concentrations is less than the change in T4 due to several compensatory mechanisms which may include increases in thyroidal IDI activity (Chanoine et al. 1992, Beckett et al. 1992).

In iodine deficiency, thyroid hormone metabolism is also impaired and decreased plasma thyroid hormones cause increases in circulating TSH. This stimulates the thyroid which may increase in size and become goitrous in an attempt to synthesise more thyroid hormones. As well, as its involvement in thyroid hormone deiodination, selenium in glutathione peroxidases can protect the thyroid gland from hydrogen peroxide and other peroxides produced during thyroid hormone synthesis (Mitchell et al. 1996). Additionally, extracellular glutathione peroxidase may regulate levels of hydrogen peroxide in the colloid of the thyroid follicle, thus providing a mechanism to control thyroid hormone synthesis (Howie et al. 1995).

Table 2. Thyroid hormones and deiodinases. Type 1 (IDI), type 2 (IDII) and type 3 (IDIII) iodothyronine deiodinases catalyse the interconversions of T4, T3, reverse T3 (rT3) and diiodothyronine (T2), see Beckett and Arthur (1994).

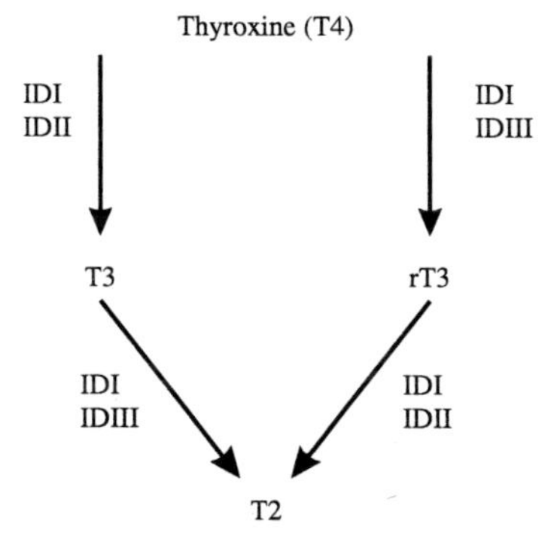

Since both selenium and iodine deficiencies can have adverse effects on thyroid hormone metabolism through impairment of intra- and extra-thyroidal hormone synthesis, it is important to understand the interaction between these micronutrient deficiencies since they can occur concurrently in some parts of the world.

Where the human population have a concurrent selenium and iodine deficiency, the selenium deficiency has been proposed to play a role in thyroid atrophy associated with myxedematous cretinism. However, selenium supplementation of subjects in the selenium and iodine deficient areas caused further decreases in already low plasma T4 concentrations. Thus it was postulated that the concurrent selenium deficiency protected against some of the adverse effects of iodine deficiency (Contempre et al. 1991). This is in contrast to studies with rats in which selenium deficiency exacerbated some of the hypothyroid effects due to the iodine deficiency (Beckett et al. 1991, 1993). In particular, hormonal changes and brown adipose tissue uncoupling protein were more severely affected in combined selenium and iodine deficiencies compared with either of the single deficiencies (Geloen et al. 1990). Thus there is conflicting evidence as to the consequences of interactions between selenium and iodine deficiencies. The work discussed in the remainder of this review was designed to identify the consequences of concurrent selenium and iodine deficiencies on selenoprotein expression in liver, brain and thyroid of rats.

Experimental Studies

In experiments, female rats were fed one of four diets from weaning, namely, control or selenium-deficient or iodine-deficient or both selenium- and iodine-deficient. These animals were mated with normal male rats, since selenium deficiency can impair sperm structure and thus fertility. The pups from each group of rats were examined at 19 d of age and selenoprotein activity and messenger RNA levels were determined in brain, liver and thyroid. The selenoproteins examined were cytosolic glutathione peroxidase (cGSHPx), phospholipid hydroperoxide glutathione peroxidase (phGSHPx), IDI and IDII. cDNAs were available for cGSHPx, PHGSHPx and IDI. The methods used are described in full in Bermano et al. (1995, 1996).

Summary of Enzyme Changes

In selenium-deficient rat pups, hepatic IDI and cGSHPx activities decreased to less than 10% of control levels and PHGSHPx activity decreased by 50%. Hepatic mRNA levels for cGSHPx decreased by 90% and for IDI decreased by 50% and PHGSHPx by 40% (Table 3). In marked contrast, thyroidal

Table 3. Selenoproteins in liver, thyroid and brain of selenium- and iodine-deficient rat pups.

	control	-Se	-I	-Se-I
Liver				
IDI mRNA	100	50	55	25
activity	100	11	60	12
cGSHPx mRNA	100	5	95	5
activity	100	<1	58	<1
phGSHPx mRNA	100	74	86	82
activity	100	57	92	60
Thyroid				
IDI mRNA	100	105	225	175
activity	100	75	330	250
cGSHPx mRNA	100	100	500	260
activity	100	55	400	100
phGSHPx mRNA	100	120	275	200
activity	100	25	50	33
Brain (cerebellum)				
IDII activity	100	100	392	557

Data are expressed as percentage of enzyme activity or mRNA abundance in control pups (selenium- and iodine-sufficient). Data recalculated from Mitchell et al., 1996.

selenoprotein messenger RNAs were unchanged or increased in selenium deficiency (Table 3). In brain, selenoenzyme mRNAs and activities were unchanged in selenium deficiency (not shown).

The most dramatic effect of iodine deficiency was to induce thyroidal IDI and cGSHPx activity and mRNAs. Additionally, IDII activity was increased four-fold in brain.

Compared with controls, in combined selenium and iodine deficiency, thyroidal IDI activity was induced whereas thyroidal cGSHPx and phGSHPx activities were unchanged or decreased. Brain IDII activity increased up to six-fold in combined selenium and iodine deficiency (Table 3) whilst cGSHPx and phGSHPx expression was unchanged (not shown).

Conclusions

Compared with liver, there is little effect of selenium deficiency on brain and thyroid selenoenzyme expression. As the brain in the rat may weight 100 times more than the thyroid, the former represents a significant pool of selenium in the selenium-deficient animal. Within the brain and thyroid there is preferential induction of IDII and IDI to maintain thyroid hormone metabolism and in brain local T3 supply for normal development in selenium and iodine deficiencies. These compensatory mechanisms act to protect the animal from some of the damaging consequences of the micronutrient deficiencies. The induction of selenoperoxidases in the thyroid gland during iodine deficiency emphasises the importance of protection of the gland from excessive hydrogen peroxide production during TSH stimulation.

Acknowledgements

We are grateful to The Scottish Office Agriculture Environment and Fisheries Department (SOAEFD) for financial support.

Literature Cited

Beckett, G.J., Peterson, F.E., Choudhury, K., Rae, P.W.H., Nicol, F., Wu, P.S.C., Toft, A.D., Smith, A.F. & Arthur, J.R. (1991) Inter-relationships between selenium and thyroid hormone metabolism in the rat and man. J. Trace Elem. and Elect. in Health and Disease. 5: 265–267.

Beckett, G.J., Russell, A., Nicol, F., Sahu, P., Wolf, C.R. & Arthur, J.R. (1992) Effect of selenium deficiency on hepatic type-I 5-iodothyronine deiodinase and hepatic thyroid hormone levels in the rat. Biochem. J. 282: 483–486.

Beckett, G.J., Nicol, F., Rae, P.W.H., Beech, S., Guo, Y. & Arthur, J.R. (1993) Effects of combined iodine and selenium deficiency on thyroid hormone metabolism in rats. Am. J. Clin. Nutr. 57: S240-S243.

Beckett, G.J. & Arthur, J.R. (1994) The iodothyronine deiodinases and 5′-deiodination. Baillieres-Clinical-Endocrinology-and-Metabolism 8: 285–304.

Behne, D., Hilmert, H., Scheid, S., Gessner, H. & Elger, W. (1988) Evidence for specific selenium target tissues and new biologically important selenoproteins. Biochim. Biophys. Acta. 966: 12–21.

Bermano, G., Nicol, F., Dyer, J.A., Sunde, R.A., Beckett, G.J., Arthur, J.R. & Hesketh, J.E. (1995) Tissue-specific regulation of selenoenzyme gene expression during selenium deficiency in rats. Biochem. J. 311: 425–430.

Bermano, G., Nicol, F., Dyer, J.A., Sunde, R.A., Beckett, G.J., Arthur, J.R. & Hesketh, J.E. (1996) Selenoprotein gene expression during selenium-repletion of selenium-deficient rats. Biol. Trace Elem. Res. 51: 211–223.

Chanoine, J.P., Safran, M., Farwell, A.P., Dubord, S., Alex, S., Stone, S., Arthur, J.R., Braverman, L.E. & Leonard, J.L. (1992) Effects of selenium deficiency on thyroid hormone economy in rats. Endocrinology 131: 1787–1792.

Contempre, B., Dumont, J.E., Ngo, B., Thilly, C.H., Diplock, A.T. & Vanderpas, J. (1991) Effect of selenium supplementation in hypothyroid subjects of an iodine and selenium deficient area — the possible danger of indiscriminate supplementation of iodine-deficient subjects with selenium. J. Clin. Endocrinol. Metab. 73: 213–215.

Davey, J.C., Becker, K.B., Schneider, M.J., St. Germain, D.L. & Galton, V.A. (1995) Cloning of a cDNA for the type II iodothyronine deiodinase. J. Biol. Chem. 270: 26786–26789.

Geloen, A., Arthur, J.R., Beckett, G.J. & Trayhurn, P. (1990) Effect of selenium and iodine deficiency on the level of uncoupling protein in brown adipose tissue of rats. Biochem. Soc. Trans. 18: 1269–1270.

Heider, J. & Bock, A. (1993) Selenium metabolism in micro-organisms. Adv. Micro. Physiol, 35: 71–109.

Howie, A.F., Walker, S.W., Akesson, B., Arthur, J.R. & Beckett, G.J. (1995) Thyroidal extracellular glutathione peroxidase: A potential regulator of thyroid-hormone synthesis. Biochem. J. 308: 713–717.

Larsen, P.R. & Berry, M.J. (1995) Nutritional and hormonal regulation of thyroid hormone deiodinases. Ann. Rev Nutr. 15: 323–352.

Low, S.C., Harney, J.W. & Berry, M.J. (1995) Cloning and functional characterization of human selenophosphate synthetase, an essential component of selenoprotein synthesis. J. Biol. Chem. 270: 21659–21664.

Mitchell J.H., Nicol, F., Beckett, G.J., Arthur, J.R.. Selenoenzyme expression in thyroid and liver of second generation selenium and iodine deficient rats. J. Mol. Endoc. 1996: 16: in press.

Salvatore, D., Low, S.C., Berry, M., Maia, A.L., Harney, J.W., Croteau, W., St. Germain, D.L. & Larsen, P.R. (1995) Type 3 iodothyronine deiodinase: Cloning, in vitro expression, and functional analysis of the placental selenoenzyme. J. Clin. Invest. 96: 2421–2430.

Discussion

Q1. Roger Sunde, University of Missouri, Columbia, MO, USA: John, this is elegant work, showing this regulation. I think this is pretty exciting, what you're unravelling. My question for you is, in something like brown adipose tissue, let's take the deiodinase II, it looks like if you've got a +Se/–I situation, then you get these elevated levels. That would suggest maybe that there might be some transcriptional upregulation and expression, and if there's enough selenium, you can accommodate it, but in the double deficiency that doesn't happen. Are you looking at a combination, then, of transcriptional regulation, is this how you're modelling this with iodine deficiency? And then you've got to judge what happens with activity depending on whether or not the selenium is present?

A. Yes, I think that describes exactly how we think it is operating. The brown adipose tissue is sensitive to adrenergic stimulation, and that could have some effect on the gene transcription. Once we get the oligos to the Type II deiodinase, at least we'll be able to look at the message, assuming of course it's the same protein. I think it is, but I think we'll soon be able to tell by looking at the message levels. Also, I think we can get enough brown adipose tissue to prepare some nuclei, and start doing some run-on assays as well. I think that will be essential in this case, because I think the mechanisms are very specific to each different tissue. I think the brown adipose tissue will be different from the tyroid, which is different from the brain, different from the liver, and I hope that eventually we'll find one or two tissues which behave in a similar fashion. But if you look in the kidney, it has a very different suite of selenoenzymes than the liver. The other thing is we have a unique stress we can put on the brown adipose tissue, which is to cold-stress the animal. We've done a little bit of work on this, and that also makes big changes in other messages within the brown adipose tissue, so we should be able to look at the messages for the selenoenzymes as well. I think that may provide us with a very powerful tool to try and unravel some of these questions.

Biological Interactions of Iron and Copper. New Revelations

EDWARD D. HARRIS

Department of Biochemistry and Biophysics and the Faculty of Nutrition, Texas A&M University, College Station, Texas 77843-2128, USA

Keywords copper-iron interaction, ceruloplasmin, ferroxidase activity, iron transport in yeast

This paper explores newer discoveries and concepts that have emerged regarding Cu–Fe interactions in living systems. Two stand out as seminal: (1) the discovery that humans with a mutation in the ceruloplasmin gene are prone to abnormal tissue Fe storage with acquired diabetes and neural degeneration, and (2) the identification of a Cu oxidases in yeast that may play a crucial role in high affinity iron transport in yeast. These developments represent a fairly strong leap forward in nutrient interactions in general and iron-copper interactions in particular.

Introduction

A list of the biological functions of Cu seldom denotes a role in iron metabolism. Cu, in fact, is essential if not obligatory for Fe utilization in biological systems. The link that established this role in animals goes back nearly 60 years (Hart et al. 1928), but only recently has the understanding caught up with the phenomenon. A early rationale rested on the realization that both metals were required for safe transport and utilization of molecular oxygen. Today, however, there is a newer appreciation for a requirement for Cu in iron transport, storage, and incorporation into hemoglobin. In essence, we have now come to realize that both metals show an interdependence that outweighs antagonism.

Chemical Features of Iron and Copper

Whether two metals will show antagonism or cooperativity will depend on their respective chemical properties. Cu and Fe belong to the first transition series metals and, as members of this family, are capable of specific electron transitions between unfilled 3d orbitals. Photons of light, for example, induce specific electron transitions that impart the familiar colorful displays to Fe and Cu complexes. Biological Fe is either ferrous [Fe(II)] or ferric [Fe(III)]; Cu is cuprous [Cu(I)] or cupric Cu [Cu(II)]. Both have higher oxidation states, but these tend to be of less prominence in biological reactions. Cu favors square planar or distorted tetrahedral complexes. Fe, in contrast, is almost always present as an octahedral complex. The two metals have sharply different ionic radii ranging from 97 pm for Cu(I) to 67 pm for Fe(III) (Figure 1). Thus, ferric iron, the most oxidized form, is only about two-thirds the size of a cuprous ion. The two metals, therefore, differ structurally and form different complexes in living systems, an observation that weakens arguments of interactions based on chemical similarity. The two metals do not have the design to be competitors.

Both Cu and Fe have powerful redox capabilities. Such reactions take the form of single electron transfers and hence open the possibility to generating products that are free radicals. The propensity for single electron transitions makes its imperative that neither metal be allowed to exist as a free ion or to interact with biological molecules. Of the two oxidation states, the reduced form is the more dangerous because of the potential transfer of a single electron to a substrate. One tends to feel that an interaction between Cu and Fe, as redox pairs, is designed to keep either from reaching the more dangerous reduced state. Hence, Cu(II) has the potential to oxidize Fe(II) to Fe(III). This type of direct interplay may have an important bearing on the biological system effects.

Correct citation: Harris, E.D. 1997. Biological interactions of iron and copper. New revelations. In *Trace Elements in Man and Animals – 9: Proceedings of the Ninth International Symposium on Trace Elements in Man and Animals. Edited by* P.W.F. Fischer, M.R. L'Abbé, K.A. Cockell, and R.S. Gibson. NRC Research Press, Ottawa, Canada. pp. 579–583.

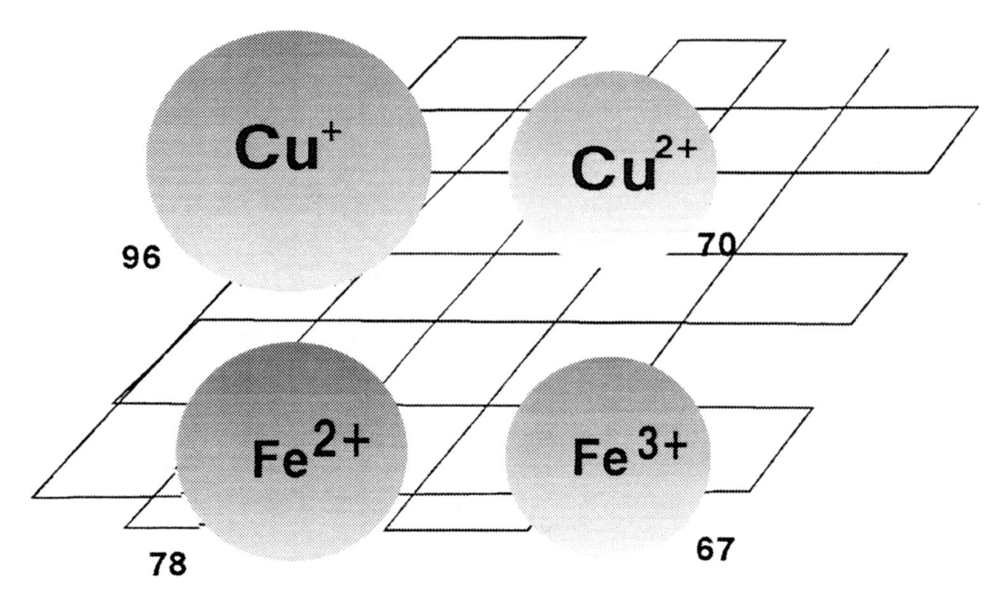

Figure 1. Relative sizes of iron and copper ions. Numbers stand for ionic radii of each of the respective ions as expressed in picometres. Note the relative sizes of Cu(I) as compared to Fe(III).

Diabetes and Neural Degeneration in Humans

Synergistic interactions between Cu and Fe have become a focal point in clinical diagnosis. Human subjects lacking the ability to express a functioning ceruloplasmin (Cp), the plasma Cu protein, accumulate large amounts of iron in the liver, pancreas, and other organs while displaying low iron in serum (Miyajima et al. 1987, Logan et al. 1994). Ferritin levels in the plasma are also elevated and TIBC which measures transferrin saturation is also increased. The propensity for Fe to accumulate in the liver of Cu-deficient animals stifles the flow of Fe to the reticuloendothelium for incorporation into hemoglobin. This could explain the anemia that develops in some patients. Unusually high Fe storage will eventually cause pathologenic changes in the storage tissues.

The biochemical basis for the defect in ceruloplasmin biosynthesis is known. In a Japanese family where hereditary aceruloplasminemia, diabetes and retinal degenerations were common symptoms, the defect in the Cp gene has been traced to a splicing error in exon 7 that results in a five base inclusion in the primary transcript (Harris et al. 1995). The five added bases give an out-of-frame reference which eventually leads to the nascent Cp molecule on the ribosome to terminate prematurely before the Cu-binding sites have been formed. Such a Cp lacks the ferroxidase activity needed to mobilize Fe.

The Ferroxidase Reaction of Ceruloplasmin

Uncommon Fe storage in humans lacking an ability to synthesize Cp is similar to the symptoms that had been seen earlier in animals suffering severe Cu deficiency (Frieden 1979). The observations tended to support an explanation that linked the ferroxidase activity of Cp to Fe mobility. The ferroxidase hypothesis, as it became known, expanded the earlier work of Curzon and O'Reilly (1960) who showed that Cp was capable of catalyzing the oxidation of Fe(II) to Fe(III). Osaki, Frieden and colleagues later used this reaction to predict how transferrin acquired Fe(III) for transport to tissues, particularly the blood forming organs (Mcdermott et al. 1968). Ceruloplasmin was believed to be the factor that catalyzed the oxidation.

A test of the ferroxidase hypothesis in animals confirmed that Fe that had accumulated in the livers of Cu-deficient swine, dogs, and rats could be released into the serum as a complex with transferrin (Ragan et al. 1969, Roser et al. 1970, Evans and Abraham 1973, Williams et al. 1974). Only ceruloplasmin as opposed to $CuSO_4$ and reducing agents was capable of causing rapid release of the stored iron.

Copper–Iron Interactions in Yeast

The yeast system has contributed substantially to our current understanding of Fe and Cu interactions. As shown in Figure 2, the products of at least five genes take an active part in the transport of Fe into the yeast cell. Three of the genes code for compounds that are concerned directly with Cu metabolism. Events that transpired that led to the discoveries are documented below.

Yeast are capable of absorbing Fe(III) from the immediate environment of the cells. In *Saccharomyces cerevisiae*, Fe^{2+} uptake is a function of a transporter that is believed to recognize only reduced iron. A ferric ion reductase, a product of the *FRE1* gene, is required to prepare the Fe for transport. *FRE1* expression is suppressed when internal Fe concentrations are high. Some mutants that fail to suppress *FRE1* expression, therefore, lack the ability to transport ferrous iron into the cell. While the cause of the defect may appear to be in the ferrous transporter itself, the defective gene when cloned and sequenced was found to code for a protein that was structurally similar to Cu-binding proteins in bacteria (Dancis et al. 1994). The role of Cu was confirmed by observing that the mutant transported Cu poorly. The defective gene in this mutant, referred to as *CTR1* (Copper TRansport 1), denoted a defect associated with input of Cu, not Fe. Mutants that lacked a functional CTR1, therefore, can transport neither Cu nor Fe. Subsequent studies revealed that the two metals had separate and distinct transport systems. This observation linked a Cu transport defect with Fe uptake.

A second mutant that failed to suppress FRE1 was found to be more sensitive to Fe transport and less sensitive to Cu. Complementation analysis in this mutant led to the discovery of a gene that coded for a protein with a partial sequence that bore a striking similarity to the Cu-binding regions of ceruloplasmin and laccase (Askwith et al. 1994). The gene, referred to as *FET3*, coded for a protein that was likely to be a copper oxidase and to have ferroxidase activity towards Fe^{2+}.

The last major component of the yeast system was discovered by tracing the gene responsible for Ca sensitivity in yeast. By pure happenstance, this gene referred to as CCC2 (calcium-sensitive crosslinking complemented 2) was found to have a close sequence similarity to the gene for Wilson disease (Yuan et al. 1995). In effect, the gene product was a P-type ATPase with a heavy metal binding domain for Cu. In all probability, the CCC2 gene product is a protein that delivers Cu to FET3 protein. Figure 2 shows how

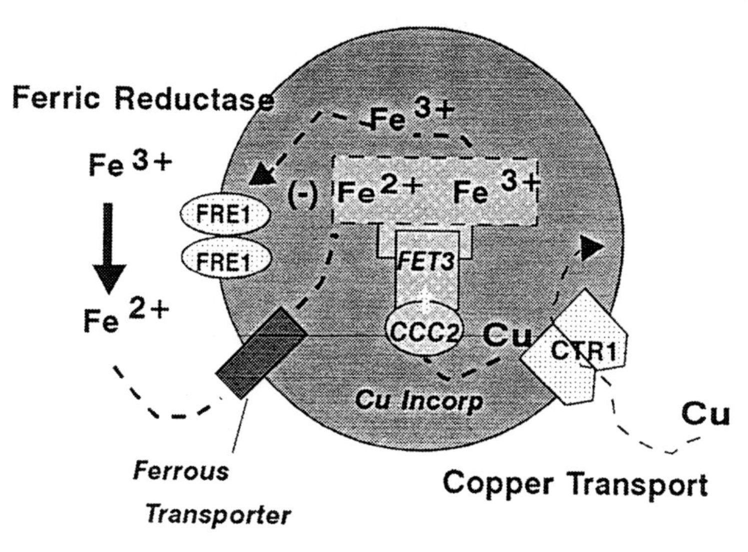

Figure 2: Iron transport system in yeast. Yeast depend on a ferric reductase to reduce Fe^{2+} to Fe^{3+} in the first stage of the transport process. As a cofactor for FET3 protein, Cu is responsible for oxidizing Fe^{2+} back to Fe^{3+} in order to recognize a second transporter either on the inner membrane or within the cytosol. CCC2 refers to a P-type ATPase that functions to transport Cu, obtained from the Cu transporter (CTR1), into the FET3 gene protein. Failure to transport Cu into the cytsol compromises the yeast cell's ability to take up iron.

the major Cu and Fe components of the yeast system interact to assure the safe delivery of Fe to the yeast cell.

Conclusions

We now have an unprecedented opportunity to learn the biochemical basis for Fe–Cu interactions in biological systems. Fe and Cu compounds in tissues and fluids appear to exist in a cooperative state and to have a specific interdependence. For too long, metals have been regarded as antagonistic or detrimental to one anothers actions. If Fe and Cu have taught us a lesson, it is that some nutritionally essential metals rely on one anothers properties in certain phases of transport, storage, and catalysis. One may suspect that oxygen was a driving force that mandated systems to acquire both metals for survival; one (either Fe or Cu) did not suffice and else would have created a dangerous scenario. Thus, one may look at the functions of Fe and Cu as being subservient, yet mutually vigilant to one anothers biological use.

Bibliography

Askwith, C., Eide, D., Van Ho, A., et al. (1994) The *FET3* gene of S. cerevisiae encodes a multicopper oxidase required for ferrous iron uptake. Cell 76: 403–410.

Curzon, G., & O'Reilly, S. (1960) A coupled iron-ceruloplasmin oxidation system. Biochem. Biophys. Res. Comm. 2: 284–286.

Dancis, A., Haile, D., Yuan, D.S., & Klausner, R.D. (1994) The *Saccharomyces cerevisiae* copper transport protein (Ctr1p). Biochemical characterization, regulation by copper, and physiologic role in copper uptake. J. Biol. Chem. 269: 25660–25667.

Evans, J.L., & Abraham, P.A. (1973) Anemia, iron storage and ceruloplasmin in copper nutrition in the growing rat. J. Nutr. 103: 196–201.

Frieden, E. (1979) Ceruloplasmin: the serum copper transport protein with oxidase activity. In: Copper in the Environment, Part II (Nriagu, J.O., ed.), pp.241–284. Wiley and Sons, Inc., New York.

Harris, Z.L., Takahashi, Y., Miyajima, H., Serizawa, M., MacGillivray, R.T.A., & Gitlin, J.D. (1995) Aceruloplasminemia: Molecular characterization of this disorder of iron metabolism. Proc. Natl. Acad. Sci. (USA) 92: 2539–2543.

Hart, E.B., Steenbock, H., Waddell, J., & Elvehjem, C.A. (1928) Iron in nutrition. VII. Copper as a supplement to iron for hemoglobin building in the rat. J. Biol. Chem. 77: 797–812.

Logan, J.I., Harveyson, K.B., Wisdom, G.B., Hughes, A.E., & Archbold, G.P.R. (1994) Hereditary caeruloplasmin deficiency, dementia and diabetes mellitus. Q. J. Med. 87: 663–670.

Mcdermott, J.A., Huber, C.T., Osaki, S., & Frieden, E. (1968) Role of iron in the oxidase activity of ceruloplasmin. Biochim. Biophys. Acta 151: 541–557.

Miyajima, H., Nishimura, Y., Mizoguchi, K., Sakamoto, M., Shimizu, T., & Honda, N. (1987) Familial apoceruloplasmin deficiency associated with blepharospasm and retinal degeneration. Neurology 37: 761–767.

Ragan, H.A., Nacht, S., Lee, G.R., Bishop, C.R., & Cartwright, G.E. (1969) Effect of ceruloplasmin on plasma iron in copper-deficient swine. Am. J. Physiol. 217: 1320–1323.

Roeser, H.P., Lee, G.R., Nacht, S., & Cartwright, G.E. (1970) The role of ceruloplasmin in iron metabolism. J. Clin. Invest. 49: 2408–2417.

Williams, D.M., Lee, G.R., & Cartwright, G.E. (1974) Ferroxidase activity of rat ceruloplasmin. Am. J. Physiol. 227: 1094–1097.

Yuan, D.S., Stearman, R., Dancis, A., Dunn, T., Beeler, T., & Klausner, R.D. (1995) The Menkes/Wilson disease gene homologue in yeast provides copper to a ceruloplasmin-like oxidase required for iron uptake. Proc. Natl. Acad. Sci. (USA) 92: 2632–2636.

Discussion

Q1. Joe Prohaska, University of Minnesota, Duluth, MN, USA: I would argue against a role for ceruloplasmin in anemia. I don't think those patients that lack that protein are anemic, for one, even though they have accumulated iron levels. It's true in copper deficiency, you develop anemia, but it's not because of the lack of ferroxidase activity, there's another role.

A. That is obviously a controversial issue right now. I'm hoping that these aceruloplasminemic patients will be a little more convincing in that regard.

Q2. Susan Haywood, University of Liverpool, England: Fascinating interactions between liver iron and copper, Ed. Can I just give you an example which has explained the interaction more clearly to me? In Bedlington terrier toxicosis in the liver, as you know there are very high accumulations of copper, and for a number of years, we've seen iron also, and this has just been dismissed as "one of those things". And now you would say that the transport mechanisms for both iron and copper are interfered with. That is an example, would you agree?

A. Yes, I would agree, but for different reasons. In other words, we might see redox interference, rather than any type of antagonism based on similar chemistry. I think that's one thing I've tried to bring out here.

Q3. John Sorenson, University of Arkansas, Little Rock, AR, USA: I don't understand how ceruloplasmin, with a molecular mass of 130,000 can serve as an oxidizing agent to oxidize ferrous iron to ferric iron, because if it's a plasma protein it has to undergo cell wall translocation. I have a problem with that. And another question—Steve Aust showed ferritin iron as iron (III), and I'd always understood it as iron (II). Is there some resolution of that?

A. First of all, ferritin iron is iron (III). The other point, regarding ceruloplasmin acting extracellularly—that has yet to be worked out. It's a good point, and I think right now we're not quite sure exactly what this means, if the iron is more available on the outside of the cell, which some people feel. It's certainly not internal, we don't think.

Q4. Maria Linder, California State University, Fullerton, CA, USA: Thanks, Ed, for reviewing a lot of important information from the past. Just one minor point, because I don't want to take much more time, but there's more than one ferroxidase in the blood plasma that are copper enzymes, and it could be that the FET3 analogue is the ferroxidase II. It hasn't been cloned yet, so we may not be able to find out right away.

A. Yes, that would be very interesting to look at.

Interactive Effects of Iodine and Vitamin A Supplementations on Measures of Thyroid and Retinol Metabolism in Rural Tanzanian Children*

NIGEL C. ROLLINS,† J.J. STRAIN,‡ JENNY CRESSWELL,‡ K. MTEBE,¶ AND N. BANGU¶

†Nuffield Department of Child Health, The Queen's University of Belfast, ‡Human Nutrition Research Group, University of Ulster at Coleraine, Northern Ireland, ¶Department of Food Science and Technology, Sokoine University of Agriculture, Morogoro, Tanzania

Keywords iodine, vitamin A, supplementation, interaction

Animal experiments suggest that thyroxine (T4) may have a regulatory role by promoting vitamin A metabolism, particularly when demand is increased under physiological conditions (Rajguru et al. 1989). In addition, gene receptor studies suggest that retinol can compete with T_4 for common sites and may have common outcomes (Umescono et al. 1988). A link between thyroid and retinol metabolism is also indicated by observational epidemiology, as endemic goitre and increased goitre prevalence is inversely correlated with dietary protein and vitamin A intakes and with serum retinol concentrations in a number of populations (Horvat and Mower 1958, Ingenbleek and De Visscher 1979, Kimiagar et al. 1990).

The aim of the current study, therefore, was to evaluate the clinical significance of retinol supplements on thyroid metabolism, and iodine supplements on retinol metabolism in the context of concurrent iodine and retinol deficiencies in rural Tanzanian children and to determine if there were any significant interactions between the respective supplements following their combined administration.

Materials and Methods

The study was conducted in Tanzania over the period January to June 1991, which corresponded with the mid-late dry season and with an unusually severe drought. A factorial design allocated at random (stratified into two age groups, 2–3 years and 4–7 years) a total of 718 children (338 boys, 380 girls) from three villages of the Wakaguru tribe in Gairo sub-district of the Morogoro Region, to receive either iodine 1 mL (400 mg) Lipiodol (Laboratoire Guerbet) only by intramuscular injection, retinol (60 mg) only by oral capsule, iodine and retinol together, or oral lactose capsule placebo. At baseline, children were assessed clinically by the paediatrician (NCR) assisted by a trained nurse. Particular attention was paid to the detection of goitre by observation and palpation of the neck and goitres were classified according to WHO criteria. Eyes were examined for evidence of vitamin A deficiency, previous scars or present inflammatory/infective processes and cytological samples were collected for a modified conjunctival impression cytology (CIC) technique-impression cytology with transfer (Luzeau et al. 1987). Children in the initial screen were excluded in the event of overt evidence of vitamin A or protein-energy malnutrition, current measles infection or having had vitamin A supplementation within the preceding 3 months.

A venous blood sample (5 mL) was also taken into a clotting tube and serum was separated on site, aliquoted, protected from light, and transported back to Gairo for storage overnight in a kerosene freezer at +2–4°C and then in Morogoro town at –30°C for a maximum of 3 months. The samples were transported under dry ice to Coleraine, UK by air for subsequent laboratory analyses. A repeat blood sample was taken

*This study was partially funded by the European Commission, Second Research and Development Programme "Science and Technology for Development" (STD 2), Contract No TS2-0307-UK.

Correct citation: Rollins, N.C., Strain, J.J., Cresswell, J., Mtebe, K., and Bangu, N. 1997. Interactive effects of iodine and vitamin A supplementations on measures of thyroid and retinol metabolism in rural Tanzanian children. In *Trace Elements in Man and Animals – 9: Proceedings of the Ninth International Symposium on Trace Elements in Man and Animals. Edited by* P.W.F. Fischer, M.R. L'Abbé, K.A. Cockell, and R.S. Gibson. NRC Research Press, Ottawa, Canada. pp. 584–586.

and clinical assessments were performed 2–3 months later on 592 children who returned for follow-up. Serum T_4, tri-iodothyronine (T_3), thyroxine binding globulin (TBG), and thyroid stimulating hormone (TSH) were measured using radioimmunoassay kits. Prealbumin and retinol binding protein (RBP) were measured by radial immunodiffusion and enzyme-linked immunoabsorbent assay respectively, while retinol was measured by reverse-phase high performance liquid chromatography.

Ethical approval for the project was obtained from the Research and Ethical Committee of the Tanzanian Food and Nutrition Centre, Dar es Salaam.

Results and Discussion

There was a significant degree of chronic malnutrition in all children, but it was most marked in children 24–48 months old and those living in Masenge village. Goitre was diagnosed in 28.8% of all children at baseline. Vitamin A deficiency was also prevalent, with 35.5% of children having serum retinol concentrations indicative of marginal deficiency (<70 µmol/L) and 4.7% indicative of overt deficiency (<35 µmol/L); 27.7% of children had abnormal CIC.

Iodine supplements produced changes in thyroid hormones typically found in iodine sparing states (see Table 1). During the study period, however, there was a significant increase in TSH in both the placebo group and in the retinol supplemented group. There was no change in TBG, prealbumin or RBP over the study period. Retinol supplements also benefited T_4 levels ($P = 0.043$, two way analysis of variance; ANOVA) when the magnitude of change (i.e., follow up minus baseline) was assessed (not shown). The augmentary effect of retinol was additive to that of iodine but no synergistic interaction was evident. Supplementations with retinol ($P = 0.006$) alone or with iodine ($P = 0.039$) alone gave higher follow-up serum retinol compared with the placebo group which, in turn, experienced a marked fall in follow-up serum retinol compared with baseline. The group given retinol in conjunction with iodine, however, had mean follow-up serum retinol significantly ($P = 0.033$) lower than would have been expected when retinol or iodine were given alone.

In conclusion, therefore, iodine supplementation benefits retinol status as well as thyroid hormone profiles and retinol supplementation improved thyroid hormone profiles in addition to the expected improvement in retinol status. The clinical significance seems to be determined by the adaptive mechanisms which come into play when iodine and retinol are both deficient from the diet. It seems probable that the

Table 1. Mean (SD) of thyroid measurements and mean (median) values for retinol.

	Group			
Parameter	Retinol n = 149	Iodine n = 149	Retinol + Iodine n = 146	Placebo n = 148
T_4 (nmol/L)				
Baseline	105 (21)	110 (20)	108 (19)	109 (20)
Follow-up	107 (20)	119 (25)	122 (23)	107 (23)
(Paired t-test)	NS[1]	P = 0.0001	P = 0.0001	NS[1]
T_3 (nmol/L)				
Baseline	2.60 (0.46)	2.68 (0.53)	2.68 (0.46)	2.60 (0.53)
Follow-up	2.68 (0.49)	2.51 (0.50)	2.52 (0.44)	2.70 (0.49)
(Paired t-test)	NS[1]	P = 0.0015	P = 0.0025	NS[1]
TSH (mU/L)				
Baseline	3.19 (1.87)	2.97 (1.56)	2.94 (1.46)	3.04 (2.04)
Follow-up	3.64 (2.03)	2.99 (1.91)	2.68 (1.50)	3.55 (2.15)
(Paired t-test)	P = 0.0076	NS[1]	NS[1]	P = 0.0019
Retinol (µmol/L)				
Baseline	92.9 (80.1)	85.2 (80.5)	95.2 (79.5)	97.5 (78.8)
Follow-up	90.0 (77.0)	84.8 (77.9)	80.4 (77.6)	76.8 (70.8)
(2 way ANOVA)	P = 0.006	P = 0.040	NS[2]	

[1]NS, not significantly different from baseline value; [2]NS, not significantly different from placebo group.

carrier protein complex common to retinol and T_4 is an important mediator of these effects, although not all of the results from the study can be explained in terms of either changes in protein levels or the respective affinities for the carrier sites.

References

Horvat, A. & Mower, H. (1958) The role of vitamin A in the occurrence of goitre on the island of Krk, Yugoslavia. J. Nutr. 66: 189–203.

Ingenbleek, Y. & De Visscher, M. (1979) Hormonal and nutritional status: critical conditions for endemic goitre epidemiology. Metabolism 28: 9–19.

Kimiagar, M., Azizi, F., Navai, L., Yassai, M. & Nafarabadi, J. (1990) Survey of iodine deficiency in a rural area near Tehran: association of food intake and endemic goitre. Eur. J. Clin. Nutr. 44 : 17–22.

Luzeau, R., Carlier, C., Elbrodt, A. & Amadee-Manesme, O. (1987) Impression cytology with transfer : an easy method of detection of vitamin A deficiency. Internat. J. Vit. Nutr. Res. 58: 166–179.

Rajguru S., Nahavandi, M. & Ahluwalia, B. (1989) Facilitatory role of thyroid hormone in vitamin A uptake in rat testes. Internat. J. Vit. Nutr. Res. 59: 107–112.

Umescono, K., Giguere, V., Glass, C.K. & Rosenfeld, M.G. (1988) Retinoic acid and thyroid hormone induce gene expression through a common responsive element. Nature 336: 262–265.

Discussion

Q1. Pierre Bourdoux, University of Brussels, Belgium: Have you any idea about the urinary iodine excretion of those children, because you didn't show those data?

A. I didn't show those data, and in fact you measured them for us. They were 36 µg/L, I believe, which is moderate deficiency. But again, I would disagree with urinary iodine as a measure of iodine status. It's more a measure of dietary iodine. It's also a little bit above what you'd get with smokers.

Q2. Pierre Bourdoux, University of Brussels, Belgium: I have a second question or comment. You have an increase in the prevalence of goitre during your study period, and I'm just wondering whether this is not due to an excess of iodine which was administered to those children. This has been reported in a large number of studies also, and you did not observe that when you decreased the amount of iodine which was administered.

A. Yes, I've heard of the thyrotoxicosis with iodine supplementations in Zaire, which was causing a lot of concern because children were actually dying. It may have been a factor in our study, but there certainly wasn't any noticeable thyrotoxicosis there.

Zinc–Selenium–Iodine Interactions in Rats*

M. RUZ,[†] J. CODOCEO,[†] J. GALGANI,[†] L. MUÑOZ,[‡] N. GRAS,[‡] S. MUZZO,[¶] AND C. BOSCO[§]

[†]Department of Nutrition, Faculty of Medicine, University of Chile, [‡]Chilean Commission for Nuclear Energy, [¶]Institute of Nutrition and Food Technology, and [§]Department of Experimental Morphology, Faculty of Medicine, University of Chile, Santiago, Chile

Keywords zinc, selenium, iodine, micronutrient interactions

Several micronutrients are involved in thyroid hormone metabolism. Among these, iodine, selenium, and zinc have the most critical roles. Iodine is crucial for the formation of the hormones at the thyroid gland (Clugston and Hetzel 1994). Selenium participates in the extrathyroidal deiodination of T4 to the active form T3 (Arthur et al. 1993). Zinc, in addition to its participation in protein synthesis, is involved in T3 binding to its nuclear receptor (Miyamoto et al. 1991).

Current evidence indicates that the simultaneous occurrence of nutritional deficiencies of more than one micronutrient can be a situation more common than previously considered. Hence, the need to explore the effects of nutritional deficiencies of iodine, selenium, and zinc, either alone or in combination. In this report, some effects of the interaction among zinc, selenium, and iodine in a rat model are presented.

Material and Methods

Wistar rats were fed amino acid based diets during 4 (pilot trial) or 6 weeks (study) starting from weaning. Diets were made in accordance with the AIN guidelines for preparing diets for experimental animals. All nutrients were kept constant except Zn, Se, and I: low Zn, <1 ppm (pilot trial) or 3.5 ppm (study); adequate Zn, 38 ppm; low Se, <0.05 ppm; adequate Se, 0.18 ppm; low I, <0.05 ppm; adequate I, 0.20 ppm. Thus, seven groups were formed with all possible combinations: Group A: Se–; Group B: I–; Group C: Zn–; Group D; Se– I–; Group E: Se– I– Zn–; Group F: I– Zn–; Group G: Se– Zn–. In addition, two groups were used as controls, ad libitum (Group H), and pair-fed (Group Z). The animals were kept in stainless steel cages and had free access to deionized drinking water.

The pilot trial, although originally proposed to last for six weeks, had to be terminated at the fourth week due to the extreme zinc deficiency (including cases of spontaneous death) developed by the Zn- groups. In consequence, a second experiment (study) was performed increasing slightly the amount of zinc of the diets from <1 ppm to 3.5 ppm. The results presented here correspond to the findings of the latter.

A series of determinations were carried out in all experimental groups. In plasma: T3, T4, TSH, Zn, and Se concentrations; alkaline phosphatase and glutathione peroxidase activities. In erythrocytes: glutathione peroxidase activity; histological examination of the thyroid gland.

Results

A summary of the main findings related to the parameters evaluated is presented in Table 1. The histological examination showed that all groups with I deficiency presented clear evidence of strong TSH stimulation, which is compatible with the high TSH values observed in plasma. In terms of their morphological characteristics the iodine-deficient groups however, did not show differences with respect to the presence of simultaneous Zn or Se deficiency. The Zn deficient alone group presented histological patterns

*This study was funded by FONDECYT research grant Nº 1950734.

Correct citation: Ruz, M., Codoceo, J., Galgani, J., Muñoz, L., Gras, N., Muzzo, S., and Bosco, C. 1997. Zinc–selenium–iodine interactions in rats. In *Trace Elements in Man and Animals – 9: Proceedings of the Ninth International Symposium on Trace Elements in Man and Animals. Edited by* P.W.F. Fischer, M.R. L'Abbé, K.A. Cockell, and R.S. Gibson. NRC Research Press, Ottawa, Canada. pp. 587–588.

Table 1. Thyroid weight, thyroid hormone concentrations, plasma alkaline phosphatase and erythrocyte glutathione peroxidase activities in the experimental groups.

Group	Thyroid weight (mg)	T 3* (nmol/L)	T 4* (nmol/L)	TSH* (mU/mL)	Plasma A P (U/L)	RBC GSH-Px (U/g Hb)
Se–	25 ± 6.6	1.1	80	35	190 ± 77	36 ± 12
I–	53 ± 18.7	1.4	15	467	148 ± 61	260 ± 20
Zn–	22 ± 5.3	1.1	55	25	116 ± 27	250 ± 38
Se– I–	49 ± 14.6	0.9	22	126	151 ± 56	36 ± 8
Se– I– Zn–	40 ± 15.4	1.2	13	245	115 ± 60	42 ± 13.6
I– Zn–	33 ± 8.6	1.7	24	122	103 ± 32	284 ± 38
Se– Zn–	19 ± 5.2	1.0	85	7	130 ± 83	37 ± 16
Control ad libitum	23 ± 2.7	1.2	66	41	158 ± 40	250 ± 31
Control pair-fed	18 ± 3.1	1.3	69	35	158 ± 36	305 ± 55

*Median.

compatible with hypothyroidism. Se deficient only group showed morphological characteristics similar to controls.

References

Arthur, J.R., Nicol, F. & Beckett, G.J. (1993) Selenium deficiency, thyroid hormone metabolism, and thyroid hormone deiodinases. Am. J. Clin. Nutr. 57 (suppl): 236S- 239S.

Clugston, G.A. & Hetzel, B.S. (1994) Iodine. In: Modern nutrition in health and disease, 8th edition (Shils, M.E., Olson, J.A. & Shike, M., eds). pp 252–263. Lea & Febiger, Philadelphia.

Miyamoto, T., Sakurai, A. & DeGroot, L.J. (1991) Effects of zinc and other divalent metals on deoxyribonucleic acid binding and hormone-binding activity of human alpha-1 thyroid hormone receptor expressed in Escherichia coli. Endocrinology 129: 3027–3033.

Discussion

Q1. Roger Sunde, University of Missouri, Columbia, MO, USA: Do you have a specific molecular mechanism for explaining how zinc would be interacting directly in either the selenium or the iodine deficiency?

A. No, the information relating to zinc is sparse. For instance, there is one study conducted a few years ago, and they also suggested that zinc might be involved in the conversion of T4 to T3 in some unknown way. There are some opposite forces at work with these interactions. I cannot give you more than that at the moment.

Q2. Roger Sunde, University of Missouri, Columbia, MO, USA: The only interaction that I can think of is that as you decrease the growth of an animal, you might in fact improve the selenium status, or other nutrient status, just because there wasn't as much dilution.

Q3. Jim Kirkland, University of Guelph, ON, Canada: Along those same lines, I may have missed it in your slides, but did you look at the actual size of the gland? And did zinc deficiency decrease the growth of the gland, when compared to iodine deficiency alone, or selenium alone? What about with all three deficiencies together? Perhaps zinc deficiency is impeding cell division in the gland as the gland tries to adapt through hyperplasia to produce more T4. I wondered if zinc deficiency were to interfere with that cell division, and prevent the formation of the goitre, which is actually a helpful adaptation to iodine deficiency?

A. With zinc deficiency alone, the size is normal, compared with the control. With iodine deficiency, of course, it is much larger. With combined deficiencies, the size is larger than control, but lower than with iodine deficiency alone.

Effects of Aspirin, Chromium and Iron Interactions on Insulin, Cholesterol and Tissue Minerals in Mice

B.J. STOECKER,[†] M.L. DAVIS-WHITENACK,[†] AND B.O. ADELEYE[‡]

[†]Department of Nutritional Sciences, Oklahoma State University, Stillwater, Oklahoma 74078-6141 and [‡]School of Human Resources, University of Southwestern Louisiana, Lafayette, Louisiana 70504-0399, USA

Keywords aspirin, chromium, iron, drug-nutrient interactions

Absorption of ^{51}Cr is markedly enhanced by aspirin (Davis et al. 1995); Fe absorption, likewise, increases with aspirin. The effects of aspirin may be due to chelation (Mertz 1969) or inhibition of gastrointestinal prostaglandins (Kamath et al. 1995). However, Fe and Cr compete for binding on transferrin (Ani and Moshtaghie 1992) and other Fe-Cr interactions have been proposed (Sargent et al. 1979) which have not been explored adequately *in vivo*.

Materials and Methods

A three-factor central composite rotatable response surface experimental design (Table 1) was used to evaluate interactions among aspirin, Cr and Fe (Myers & Montgomery 1995).

Eighty-four C57BL/6J lean mice were randomly assigned to one of fifteen experimental diets supplemented with aspirin (0 to 1500 mg/kg), Cr (20 to 1000 μg/kg) as $CrCl_3$ and Fe (20 to 200 mg/kg) as $FeCl_2$. The basal diet was the AIN-93G (Reeves et al. 1993). After 24 d, mice were anesthetized and exsanguinated.

Table 1. Levels of dietary chromium (μg/kg diet), iron (mg/kg diet) and aspirin (mg/kg diet).

Coded Levels	−1.68	−1.00	0.00	+1.00	+1.68
X1 (Cr)	20	219	510	801	1000
X2 (Fe)	20	56	110	163	200
X3 (Aspirin)	0	304	750	1196	1500

Results and Discussion

Weight gain was not significantly affected by the interactions between Cr, Fe and aspirin. The dietary variables also did not have significant effects on serum glucose, cholesterol, triglycerides, or fructosamine, a compound used to assess short-term glucose control. The whole model graph for insulin is shown (Figure 1).

Both Cr and Fe had significant effects of hemoglobin (Figure 2). Also, there was an significant interaction between Fe and aspirin that produced similar effects on hemoglobin, hematocrit and number of red blood cells. The best stepwise model for number of white blood cells showed a Cr- aspirin interaction as did the best stepwise model for number of platelets.

In this study, more dietary Fe tended to increase hepatic Fe but Fe and Cr did not have significant interactive effects on hepatic Fe concentrations. Dietary Fe had significant linear and quadratic effects on iron Fe in spleen and tended to affect hepatic Cr. Magnesium concentrations in spleen, kidney and heart were significantly affected by dietary variables. In the spleen, there was a significant interaction between

Correct citation: Stoecker, B.J., Davis-Whitenack, M.L., and Adeleye, B.O. 1997. Effects of aspirin, chromium and iron interactions on insulin, cholesterol and tissue minerals in mice. In *Trace Elements in Man and Animals – 9: Proceedings of the Ninth International Symposium on Trace Elements in Man and Animals. Edited by* P.W.F. Fischer, M.R. L'Abbé, K.A. Cockell, and R.S. Gibson. NRC Research Press, Ottawa, Canada. pp. 589–591.

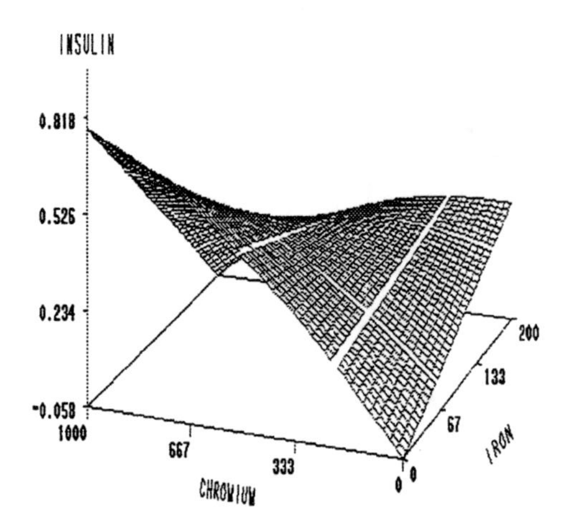

Figure 1. Full model plot of serum insulin (pmol/L) as a function of dietary chromium (µg/kg diet), iron (mg/kg) and aspirin (mg/kg).

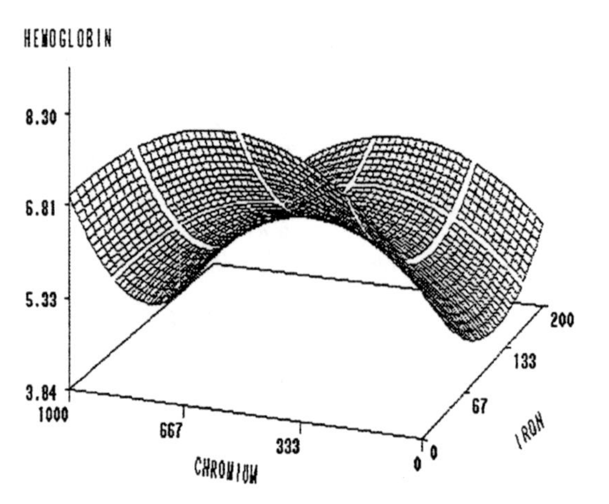

Figure 2. Full model plot of hemoglobin (mg/dL) as a function of dietary chromium (µg/kg diet), iron (mg/kg) and aspirin (mg/kg).

Fe and aspirin. In heart, both Cr and Fe had significant quadratic effects on Mg. The stationary point was a minimum. Both Cr and Fe and Cr and aspirin had significant interactive effects on kidney Mg.

Interactions among aspirin and variable intakes of chromium and iron affect a wide range of tissue mineral concentrations and clinical parameters.

Literature Cited

Ani, M. & Moshtaghie, A.A. (1992) The effect of chromium on parameters related to iron metabolism. Biol. Trace Elem. Res. 32: 57–64.

Davis, M.L., Seaborn, C.D. & Stoecker, B.J. (1995) Effects of over-the-counter drugs on ^{51}chromium retention and urinary excretion in rats. Nutr. Res. 15: 201–210.

Kamath, S.M., Stoecker, B.J., Whitenack, M.D., Smith, M., Adeleye, B.O. & Sangiah, S. (1995) Indomethacin and prostaglandin E_2 analogue effects on absorption, retention, and urinary excretion of ^{51}chromium. FASEB J. 9: A577 (abst.).

Mertz, W. (1969) Chromium occurrence and function in biological systems. Physiol. Rev. 49: 163–239.

Myers, R.H. & Montgomery, D.C. (1995) Response Surface Methodology: Process and Product Optimization Using Designed Experiments. John Wiley & Sons, Inc., New York.

Reeves, P.G., Nielsen, F.H. & Fahey, G.C., Jr. (1993) AIN-93 purified diets for laboratory rodents: Final report of the American Institute of Nutrition ad hoc writing committee on the reformulation of the AIN-76A rodent diet. J. Nutr. 123: 1939–1951.

Sargent, T. III, Lim, T.H. & Jenson, R.L. (1979) Reduced chromium retention in patients with hemochromatosis, a possible basis of hemochromatotic diabetes. Metabolism 28: 70–79.

Discussion

Q1. John Sorenson, University of Arkansas, Little Rock, AR, USA: How much in terms of grams of diet does a mouse consume per day? The aspirin was given in mg/kg diet? It seems like the amount of aspirin consumed is not going to be very large.

A. The mouse will consume something between 5 and 10 grams of food per day. Closer to 5, actually, they spill the rest. However, even if we express our aspirin dose per kilogram of bodyweight, it's still a fairly high dose. The dose that somebody with arthritis would be taking would be down between 0 and 500 mg/kg diet, if you're trying to extrapolate straight from a rodent. So our levels, from above 500 up to 1500 are really quite high.

Q2. John Sorenson, University of Arkansas, Little Rock, AR, USA: But an arthritic consumes what, 15 tablets per day, and each tablet contains 600 mg, so that's 9000 mg aspirin/day. Divide that by 60 kg for body mass... Still, your 1500 sounds like a reasonable number. It's not a huge amount.

A. There are papers in the literature that use more, like 2.5 or 3. Beyond that you really have quite an ulcerigenic effect.

Q3. Richard Anderson, Beltsville Human Nutrition Research Center, MD, USA: I enjoyed your presentation, Barbara. I was glad to see all the variables where you could see effects of chromium, but what struck me the most about it is when you do human studies, lots of times you see certain people that don't respond, and you really don't know why. It just looks to me like, for example, when you looked at the iron, you had basically, depending upon the iron and the chromium, you had no response to chromium, and then you would get a very high response to chromium. It seemed to depend on iron level, or iron status, or whatever it would be in humans. If we look at the people who respond, obviously we can't set up a study like this in humans. If we did some sort of retro-analysis, and determined the iron status and the aspirin intakes, do you think that would help us predict who responds to chromium?

A. I don't know about that. With the aspirin, there's no evidence in the literature. Others have speculated that hemochromatosis might influence chromium. We get very strong effects of aspirin in enhancing chromium absorption. We have had some studies where intake of drugs seemed to be a variable, and I think that others have reported this too. That's not quite as specific as one would like. I think there's a whole lot of work that needs to be done, but I would not do another human study on chromium status without keeping a very close eye on what the drug intake was—aspirin and other non-steroidal antiinflammatories. Prostaglandin analogues are being used; we have some data that they inhibit chromium absorption.

Analysis of Serum Boron Distribution among Inhabitants of an Urban Area in Japan

K. USUDA,[†] K. KONO,[†] M. WATANABE,[†] K. NISHIURA,[†] K. IGUCHI,[†] H. NAGAIE,[†] T. TAGAWA,[†] E. GOTO,[†] M. SHIMAHARA,[‡] AND J. SENDA[‡]

[†]Department of Hygiene and Public Health, and [‡]Department of Oral Surgery, Osaka Medical College, 2-7 Daigakumachi, Takatsuki City, Osaka, 569 Japan

Keywords boron, ICPES, human serum, log-normal distribution

Although the recent development of industry has brought about widespread applications of boron compounds (Miyazaki et al. 1986), the value of boron in human samples was determined in too small number of subjects to establish reference values (Kobori et al. 1983, Nancy et al. 1994). Therefore, it was considered important to define the reference value and determine the index of boron exposure. In this study, operating the Inductively-Coupled Plasma Emission Spectrometry (ICPES) system, we determined serum boron concentrations of healthy subjects.

Materials and Methods

The study was carried out on a healthy group of 397 male (mean age 49 years with a range from 0 to 91 years), and 583 female (mean age 45 years with a range from 0 to 85 years) from an urban area in Japan. One millilitre of serum samples was diluted with 1 mL of 200 mM nitric acid without heating. After agitation, samples were immediately analyzed. Differences between the two groups were examined on the log-transformed value, using Student's if the variances were equal and Welch's t-test if the variances were unequal (F-test), and a correlation coefficient of p value <0.05 was considered statistically significant.

Results

The frequency distribution of serum boron is shown in Figure 1. It shows an apparent log-normal distribution in both sexes. As shown in Table 1, the subjects were divided by age into 8 subgroups. In all subgroups, geometric mean value of male serum boron is higher than that of female subjects, with or without significance. In male subjects, serum boron showed a radical increase up to the age of 49 years,

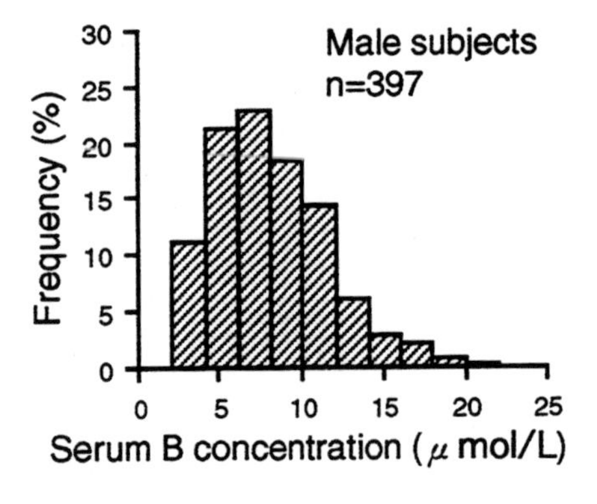

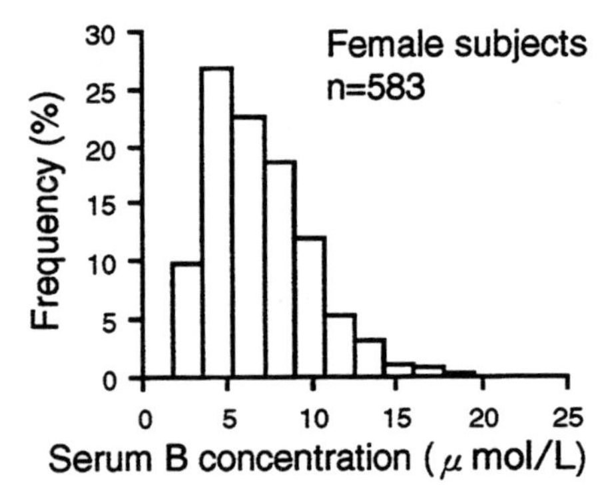

Figure 1. Frequency distribution for serum boron concentration in male and female subjects.

Correct citation: Usuda, K., Kono, K., Watanabe, M., Nishiura, K., Iguchi, K. Nagaie, H., Tagawa, T., Goto, E., Shimahara, M., and Senda, J. 1997. Analysis of serum boron distribution among inhabitants of an urban area in Japan. In *Trace Elements in Man and Animals – 9: Proceedings of the Ninth International Symposium on Trace Elements in Man and Animals. Edited by* P.W.F. Fischer, M.R. L'Abbé, K.A. Cockell, and R.S. Gibson. NRC Research Press, Ottawa, Canada. pp. 592–594.

Table 1 Serum boron concentration(µmol/L) from healthy inhabitants of an urban area in Japan.

	Male			Female		
Age	n	G.M.	C.I.	n	G.M.	C.I.
<20	45	5.90	2.21–15.7	51	5.49	2.21–13.8
20–29	45	6.15	2.82–13.4	104	5.48	2.37–12.2
30–39	32	6.96**	2.65–18.3	88	5.83	2.47–13.6
40–49	41	8.00	3.40–18.8	93	6.24	2.59–15.0
50–59	71	7.76*	3.38–17.8	89	6.67	2.87–15.5
60–69	111	7.87	3.66–16.9	87	6.94	3.10–15.5
70–79	38	8.11	3.44–19.1	49	7.59	3.89–14.8
79<	14	9.15***	3.33–23.7	22	7.62	3.55–16.4
Total	397	7.38	3.06–17.7	583	6.28	2.73–14.3

n, number of subjects; G.M., geometric mean; C.I., 5%–95% confidential interval.
Asterisks indicate significant differences: *$p < 0.05$, ** $p < 0.01$, ***$p < 0.01$ with respect to the female subjects.

reaching a plateau at age 50 years to 69 years, and increasing gradually over the age of 70 years. Female subjects showed a gradual increase up to the age of 70 years.

Discussion

The frequency distribution of serum boron was log-normal (Figure 1), so the geometric means were used. As shown in Table 1, serum boron in male subjects is higher than in female subjects, and an increase of serum boron with the age in both sexes and the plateau interval seen in male subjects was not observed in female subjects. A similar phenomenom has been observed in blood cadmium (Roggi et al. 1995) or lead (Silbergeld et al. 1988). The blood boron values by Imbus et al. (1963) have been regarded as the reference value. More recent studies have all reported lower values (Kobori et al. 1983, Nancy et al. 1994). Making allowance for these values and their quality, it is evident that serum boron exhibits wide variations. Taking into consideration previously reported values and our own results, serum boron reference values of 7.38 µmol/L and 6.28 µmol/L, and reference interval of 3.08–17.7 µmol/L and 2.73–14.3 µmol/L, for male and female subjects, respectively, is suggested. It is concluded that our present reference values may be useful in health screening for boron exposure.

Literature Cited

Imbus, H.R., Cholak, J., Miller, L.H., & Sterling, T. (1963) Boron, cadmium, chromium, and nickel in blood and urine, Arch. Environ. Health 6: 112–121.

Kobori, K., Mise, Y., Takata, N., Sakakibara, H., Maruyama, K., & Kobayashi, T. (1983) Simultaneous multielement determination of metals in biomaterials by quick digester-inductively coupled plasma emission spectroscopy system. Eiseikensa 33: 119–126.

Miyazaki, A. & Bansho, K. (1986) Determination of trace boron in natural waters by inductively coupled plasma emission spectrometry combined with solvent extraction. Anal. Sci. 2: 451–455.

Nancy, R.G., & Ferrando, A.A. (1994) Plasma boron and the effects of boron supplementation in males. Environ. Health Perspect. 102 (Suppl 7) : 73–77.

Roggi, C., Sabbioni, E., Minoia, C., Ronchi, A., Gatti, A., Hansen, B., Silva, S. & Maccarini, L. (1995) Trace element reference values in tissues from inhabitants of the European Union. IX. Harmonization of statistical treatment: blood cadmium in Italian subjects. Sci. Total Environ. 166: 235–243.

Silbergeld, E.K., Schwartz, J. & Mahaffey, K. (1988) Lead and osteoporosis: Mobilization of lead from bone in postmenopausal women, Environ. Res. 47: 79–94.

Discussion

Q1. Curtiss Hunt, USDA-ARS, Grand Forks, ND, USA: The study that we carried out in Grand Forks, feeding 3 mg/d was fairly similar to what your mean values were. Do you have a feel for what the mean boron intake was in your subjects?

A. The intakes in Japan are not so high, maybe 3 or 4 mg/d.

Q2. Julian Lee, New Zealand Agricultural Research Institute, Palmerston, New Zealand: Would you care to comment on the speciation of boron in serum. And secondly, what was the clearance rate from plasma?

A. The chemical form of boron in plasma was borate anion. We did not measure plasma clearance rate.

Q3. James Coughlin, Coughlin & Associates, Laguna Niguel, CA, USA: You think that boron is low in Japan, at 3–4 mg/d. We just completed a study using USDA total diet surveys, and we think the mean for the U.S. population is 1 mg/d. Is it the seafood, tea, soybeans, miso, those kinds of products? One other question, is osteoporosis a big problem in Japan, like it is in the West?

A. Yes, those are foods that are higher in boron. Osteoporosis is a problem in Japan, too.

Q4. John Sorenson, University of Arkansas, Little Rock, AR, USA: Is boron thought to have some useful pharmacological effect in Japan?

A. Maybe, I think so, in combination with fluoride.

Q5. Phil Strong, U.S. Borax Inc., Valencia, CA, USA: I'd just like to clear up something on speciation. At the pH of the body, serum's about 7.5, boric acid is 98%. The problem is that borate chelates very easily with all kinds of biological molecules. We don't have good techniques yet to allow speciation, especially at low levels. That's one place where a lot of work needs to be done.

Q6. John Sorenson, University of Arkansas, Little Rock, AR, USA: Does it form transition metallo-element complexes? Like with copper, zinc, or whatever?

(comment by **Phil Strong**, U.S. Borax Inc., Valencia, CA, USA: No, you don't find boron as an ion, a +1 or +2, it's electronegative, so you'll always find it complexed, with water, or with some other negative ion. So you'll either find boric acid, where it's in the +3 state, or you'll find the anion, where it's tetrahedral. It will form a variety of complexes, for example with thiols, so there are all sorts of biological molecules which are just right, like riboflavin, and a lot of different things.)

Q7. John Sorenson, University of Arkansas, Little Rock, AR, USA: Is borate BO_4?

(comment by **Phil Strong**, U.S. Borax Inc., Valencia, CA, USA: Boric acid is BOH_3, neutral. The borate anion is BOH_4^{-1}.)

(comment by **John Sorenson**, University of Arkansas, Little Rock, AR, USA: Well, that should coordinate with the metallo-elements. Thank you.)

Varying Concentrations of Dietary Iodine and Selenium Interact in Thyroid Hormone Metabolism

C.S. HOTZ,[†‡] M.R. L'ABBÉ,[†‡] D.W. FITZPATRICK,[†] AND K.D. TRICK[‡]

[†]Department of Foods and Nutrition, University of Manitoba, Winnipeg, Manitoba, Canada R3T 2N2. [‡]Nutrition Research Division, Food Directorate, Health Protection Branch, Health Canada 2203C, Ottawa, Ontario, Canada K1A 0L2

Keywords iodine, selenium, selenoenzymes, thyroid hormones

Adequate amounts of both I and Se are required for optimal thyroid hormone (TH) metabolism. As a structural component of TH, I is a primary requirement for TH synthesis. One form of TH, thyroxine (T4), is produced solely by the thyroid and is relatively biologically inactive. Triiodothyronine (T3), the biologically active TH form, is mostly produced by the deiodination of T4 in peripheral tissues. Se plays a secondary role in the control of TH metabolism. The deiodinating enzyme which produces T3, type I iodothyronine 5′-deiodinase (DI-I), is a selenoenzyme with most activity occurring in liver, kidney, and thyroid. Se may also have an indirect role in the control of TH synthesis via the cytosolic selenoenzyme, glutathione peroxidase (GSH-Px). The antioxidant activity of GSH-Px within the thyroid is thought to be the main mechanism of neutralization of the cytotoxic H_2O_2 that is produced by the thyroid as a cofactor in TH synthesis.

Observations of the complex relationship between I and Se in TH metabolism have raised questions pertaining to the interaction effects of varying levels of dietary I and Se. In some studies, Se deficiency has been observed to compound the adverse effects of an I deficiency (Arthur et al. 1992) while in other studies it has not (Golstein et al. 1988). Contempré et al. (1995) have demonstrated that a high I dose given to rats deficient in both I and Se produced greater thyroid tissue damage than when rats were only deficient in I. The present study was designed to investigate these interactions by studying the effects of various combinations of dietary Se and I on the activities of the selenoenzymes GSH-Px and DI-I, circulating T4 and T3, and thyroid stimulating hormone (TSH).

For 6 weeks, male weanling Sprague-Dawley rats were fed an AIN-93G diet with modified selenium and iodine content: 3 levels each of iodine (low, no added I; normal, 0.2 mg I/kg diet added; high, 1.0 mg I/kg diet added) and selenium (low, no added Se; normal, 0.18 mg Se/kg diet added; high, 1.0 mg Se/kg diet added) were used in a 3 × 3 factorial design. Tissue DI-I activity was determined in liver and kidney using ^{125}I-reverse T3 as substrate and separating the reaction products using ion-exchange chromatography. An automated method was used to determine GSH-Px activity in thyroid using *t*-butyl hydroperoxide as substrate and measuring the disappearance of NADPH. TH were measured using Beckman automated T4 and T3-uptake reagent kits and TSH was measured using a commercial RIA kit. Statistical analyses used repeated measures ANOVA and LSD post-hoc analysis.

Considering the effects of dietary I level alone, low I produced the typical effects of hypothyroidism: increased thyroid weight, elevated TSH, and decreased circulating T4 were observed. Additionally, low dietary I increased the activity of thyroidal GSH-Px. No parameters differed significantly from controls with high I intake. With low Se intake, DI-I activity was reduced in kidney but not liver, and GSH-Px activity in thyroid was unaffected. Plasma T4 was increased with low Se intake. High Se intake had no significant effect on the parameters measured, as compared to controls. Both plasma T4 and thyroidal GSH-Px activity responded interactively to varying levels of dietary I and Se (Figures 1 and 2). Combined I and Se deficiency produced a control level of plasma T4. Thyroidal GSH-Px activity was reduced

Correct citation: Hotz, C.S., L'Abbé, M.R., Fitzpatrick, D.W., and Trick, K.D. 1997. Varying concentrations of dietary iodine and selenium interact in thyroid hormone metabolism. In *Trace Elements in Man and Animals – 9: Proceedings of the Ninth International Symposium on Trace Elements in Man and Animals. Edited by* P.W.F. Fischer, M.R. L'Abbé, K.A. Cockell, and R.S. Gibson. NRC Research Press, Ottawa, Canada. pp. 595–596.

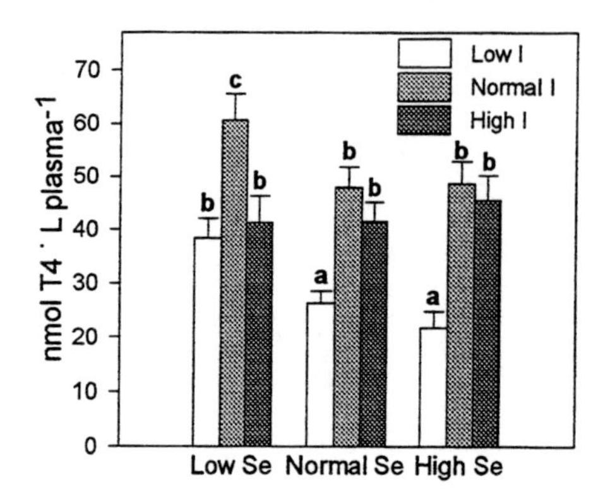

Figure 1. Interaction of I and Se on plasma T4.

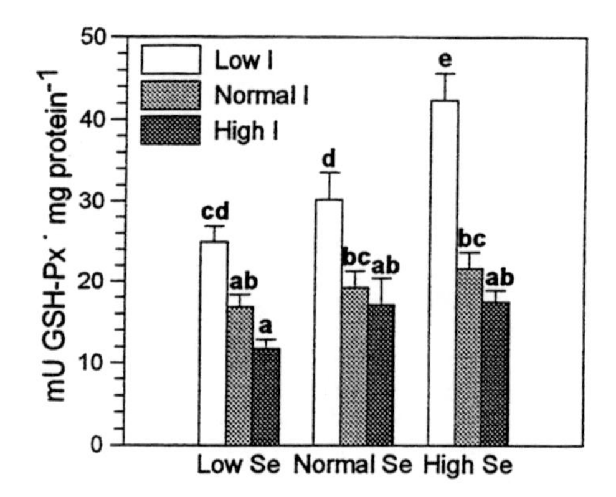

Figure 2. Thyroid GSH-Px activity.

significantly from control levels when high I intake was combined with low Se intake. The ability of low I intake to increase thyroidal GSH-Px activity appeared to be limited by the availability of dietary Se.

The limiting effect of Se availability on thyroidal GSH-Px activity during I deficiency may suggest that antioxidant activity is inadequate and tissue damage may result from the incomplete neutralization of H_2O_2. Caution may also be needed in giving high doses of I when Se is deficient; the toxic effect that excess I has on thyroid tissue (Contempré 1995) may be increased when GSH-Px activity is low, as occurred in the high I and low Se intake groups. In this study, combined I and Se deficiency did not further antagonize TH metabolism beyond what was observed with I deficiency alone. In fact, a control level of plasma T4 with the dual deficiency occurred. This normalizing effect is thought to be protective in pregnancy against neurological damage to the fetus which results when maternal T4 is low (Dumont 1994). It seems that the differences in results between this study and previous studies with combined I and Se deficiencies are related to differences in study design, such as length and severity of the mineral deficiencies.

References

Arthur, J.R., Nicol, F. & Beckett, G.J. (1992) The role of selenium in thyroid hormone metabolism and effects of selenium deficiency on thyroid hormone and iodine metabolism. Biol. Trace Elem. Res. 34: 321–325.

Contempré, B., Dumont, J.E., Denef, J.-F. & Many, M.-C. (1995) Effects of selenium deficiency on thyroid necrosis, fibrosis and proliferation: a possible role in myxoedematous cretinism. Eur. J. Endocrinol. 133: 99–109.

Dumont, J.E., Corvilain, B. & Contempré, B. (1994) The biochemistry of endemic cretinism: roles of iodine and selenium deficiency and goitrogens. Mol. Cell. Endocrinol. 100: 163–166.

Golstein, J., Corvilain, B., Lamy, F., Paquer, D. & Dumont, J.E. (1988) Effects of a selenium deficient diet on thyroid function of normal and perchlorate treated rats. Acta. Endocrinol. 118: 495–502.

Interactions Between Vitamin A Deficiency and Iron Metabolism: Clues as to Underlying Mechanisms

A.J.C. ROODENBURG,[†‡] C.E. WEST,[†] AND A.C. BEYNEN[‡]

[†]Department of Human Nutrition, Wageningen Agricultural University, PO Box 8129, 6700 EV Wageningen and [‡]Department of Laboratory Animal Science, Utrecht University, PO Box 80.166, 3508 TD Utrecht, The Netherlands

Keywords vitamin A, iron, rat, red blood cell

In third world countries vitamin A deficiency is associated with impaired iron status in women and children. To simulate the interrelationship between vitamin A and iron metabolism and to elucidate the underlying mechanisms, experiments with rats have been carried out.

Influence of Vitamin A on Iron Metabolism in Rats

Effects of vitamin A and iron deficiency on iron status were studied in rats. Mild vitamin A deficiency produced anemia and increased iron absorption (Figure 1). Moderate vitamin A deficiency caused iron accumulation in spleen and bone and reduced total iron-binding capacity (measure for circulating transferrin) (Table 1). Severe vitamin A deficiency raised hemoglobin concentrations due to hemoconcentration and reduced growth.

Mechanisms

The data indicated that vitamin A deficiency might affect red blood cell metabolism, and iron mobilization.

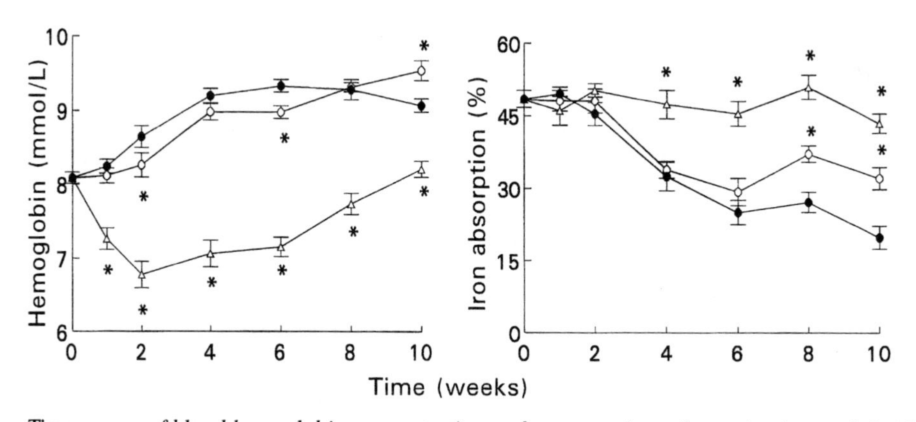

Figure 1. *Time course of blood hemoglobin concentration and apparent iron absorption in rats fed either the control diet (●, 1200 RE/kg vitamin A and 35 mg Fe/kg), vitamin A deficient diet (○, 0 RE/kg, 35 mg Fe/kg), iron deficient diet (△, 1200 RE/kg and 3.5 mg Fe/kg). Male Wistar rats (Cpb:WU) were 5 weeks of age at week 0. Means of 12 animals per group ± SE. *, $P < 0.05$ vs. control group. (Roodenburg et al. 1994).*

Correct citation: Roodenburg, A.J.C., West, C.E., and Beynen, A.C. 1997. Interactions between vitamin A deficiency and iron metabolism: Clues as to underlying mechanisms. In *Trace Elements in Man and Animals – 9: Proceedings of the Ninth International Symposium on Trace Elements in Man and Animals. Edited by* P.W.F. Fischer, M.R. L'Abbé, K.A. Cockell, and R.S. Gibson. NRC Research Press, Ottawa, Canada. pp. 597–598.

Table 1. Iron in organs (spleen, tibia) and plasma levels of retinol, transferrin receptors, total iron-binding capacity and total bilirubin in rats fed one of the three experimental diets[1] for 10 weeks.

	Control	Vitamin A deficient	Iron deficient
Iron in spleen (μmol/g dry wt)	29.8 ± 1.8	44.6 ± 6.4*	10.6 ± 0.3*
Iron in bone (μmol/g dry wt)	1.28 ± 0.04	1.55 ± 0.08*	0.80 ± 0.05*
Plasma retinol (μmol/L)	1.47 ± 0.04	0.12 ± 0.02*	1.24 ± 0.04*
Plasma transferrin receptors (ng/mL)	2356 ± 193	2216 ± 139	3733 ± 261*
Plasma total iron-binding capacity (μmol/L)	97.7 ± 1.9	90.1 ± 1.7*	121.2 ± 2.1*
Plasma total bilirubin (μmol/L)	13.9 ± 2.2	8.1 ± 0.9*	11.4 ± 1.1

[1]See Figure 1. Means of 12 animals per group ± SE. *, $P<0.05$ vs. control group. (Roodenburg et al., submitted.)

Red Blood Cell Metabolism

Vitamin A deficiency in man and rats is associated with anemia. It is hypothesized that the primary effect of vitamin A deficiency is impairment of red blood cell synthesis, leading to abnormally shaped cells and mild anemia. A potential role for vitamin A in erythropoietin gene transcription corroborates this hypothesis (Table 2). An increased destruction of the abnormal red blood cells could cause iron accumulation in the macrophages of spleen and bone marrow, in turn leading to stimulation of the erythropoietic tissue. However in vitamin A deficient rats plasma bilirubin decreased and plasma transferrin receptors were unchanged. Thus, there is no evidence for enhanced red blood cell degradation or raised erythropoietic activity in vitamin A deficiency (Table 1).

Iron Mobilization

The accumulation of iron in tissues and the reduction in total iron-binding capacity (Table 1) indicate that iron mobilization might be impaired in vitamin A deficiency. Synthesis of transferrin, the iron transport protein, might be reduced at the level of gene transcription. There is evidence that transferrin gene transcription was stimulated *in vitro* by vitamin A retinoic acid (Hsu et al. 1992). Thus vitamin A deficiency could inhibit transferrin synthesis. With the use of a gene bank, candidate retinoic acid response elements for transferrin and erythropoietin were located (Table 2). The occurrence of the response elements for the transferrin and erythropoietin genes supports the hypothesis that vitamin A affects both iron transport and red blood cell synthesis at the level of gene transcription.

In conclusion, we hypothesized that vitamin A deficiency influences red blood cell metabolism and transferrin synthesis. However indirect parameters of red blood cell synthesis and degradation in rats could not confirm this. Possibly, both synthesis of red blood cells and transferrin are affected by vitamin A on the level of gene transcription.

Table 2. Candidate retinoic acid response elements[1] in non-coding DNA sequences of transferrin and erythropoietin, obtained from the EMBL-Genbank.

Protein of interest	AGGTCA-analogue	location
Erythropoietin	AGGTCAggAGTTCA	5′ flank, ≈–200 bp
Transferrin	GGGTCAaatgaGGGTCA	intron 1, +611 bp

[1]Retinoic acid response elements regulate gene transcription by binding receptors and their ligand (retinoic acid). The sequences listed are direct repeats, analogues of sequence AGGTCA spaced by 1,2 or 5 basepairs.

Literature Cited

Hsu, S.L., Lin, Y.F. & Chou, C.K. (1992) Transcriptional regulation of transferring and albumin genes by retinoic acid in human hepatoma cell line Hep3B. Biochem. J. 283: 611–615.

Roodenburg, A.J.C., West, C.E., Yu, S. & Beynen, A.C. (1994) Comparison between time-dependent changes in iron metabolism of rats as induced by marginal deficiency of either vitamin A or iron. Br. J. Nutr. 71: 687–699.

Roodenburg, A.J.C. et al. (1996) Indicators of erythrocyte formation and degradation in rats with either vitamin A or iron deficiency. (submitted).

Dietary Boron and Vitamin D Affect Hepatic Glycolytic Metabolite Concentrations in the Chick

C.D. HUNT

United States Department of Agriculture, Agricultural Research Service, Grand Forks Human Nutrition Research Center, Grand Forks, North Dakota 58202, USA*

Keywords boron, vitamin D, glycolysis, chick

Dietary boron, fed at physiological concentrations, affects energy substrate utilization and the effects are especially pronounced during concomittant vitamin D deficiency. For example, in vitamin D-deficient chicks fed a boron-low diet, supplemental dietary boron decreases the elevated plasma concentrations of glucose, pyruvate, and β-hydroxybutyrate that are typically associated with vitamin D deficiency in chicks (Hunt et al. 1994). In adult volunteers fed a boron-low diet, supplemental boron decreased serum glucose concentrations within a range of normal values (Nielsen 1994). These findings suggest that boron has a role in energy substrate metabolism, perhaps through involvement in vitamin D metabolism.

Materials and Methods

To characterize further the influence of boron on energy substrate utilization, hepatic glycolytic metabolite concentrations were determined in three cohorts with the same fully crossed two-factor design. Day-old cockerel chicks (18 per group per cohort) were fed a ground corn, high protein casein, and corn oil based diet (~0.20 mg B/kg) supplemented with boron at 0 or 1.3 mg/kg and cholecalciferol (vitamin D_3) at 3.13 μg/kg (125 IU/kg; inadequate) or 15.63 (625 IU/kg; adequate) μg/kg. The chicks were provided 24 h of light daily by using fluorescent lighting filtered through acrylic plastic and 0.25 in plate glass. Liver samples were isolated from each chick (non-fasted) and freeze clamped for subsequent measurement of all metabolites in the glycolytic pathway from glycogen to pyruvate and glucose and lactate according to established protocols (Bergmeyer 1974). Samples of poor integrity were discarded and the data were analyzed by using 2×2 ANOVA with cohort as a blocking factor (NS = $p > 0.17$). The average concentration of selected metabolites are shown in Table 1.

Results

At age 28 d, the dietary treatments did not affect hepatic glucose concentrations (Table 1). Dietary vitamin D supplementation (DVDS) increased glycogen concentrations. Hexose phosphate pool metabolite concentrations were affected mainly by DVDS; DVDS decreased the concentrations of fructose 6-phosphate (not shown) and glucose 6-P. Concentrations of metabolites within the fructose 1,6-biphosphate and triose phosphate pool were affected mainly by dietary boron supplementation (DBS); for example, DBS decreased dihydroxyacetone phosphate concentrations. The concentration of 1,3-bisphosphoglycerate was not affected by the dietary treatments. The treatments strongly, but inversely, affected the concentrations of metabolites within the three-carbon phosphorylated acid pool. For example, DVDS increased the concentrations of 2-phosphoglycerate and phosphoenolpyruvate (PEP) and tended to increase the concentrations of 3-phosphoglycerate (not shown). However, DBS decreased the concentrations of 2-phosphoglycerate and tended to decrease the concentrations of PEP. Pyruvate concentrations were not influenced

*U.S. Department of Agriculture, Agricultural Research Service, Northern Plains Area, is an equal opportunity/affirmative action employer and all agency services are available without discrimination.

Correct citation: Hunt, C.D. 1997. Dietary boron and vitamin D affect hepatic glycolytic metabolite concentrations in the chick. In *Trace Elements in Man and Animals – 9: Proceedings of the Ninth International Symposium on Trace Elements in Man and Animals. Edited by* P.W.F. Fischer, M.R. L'Abbé, K.A. Cockell, and R.S. Gibson. NRC Research Press, Ottawa, Canada. pp. 599–601.

Table 1. Effects of dietary boron, vitamin D and their interaction on chick body weight and hepatic glycolytic substrate concentrations.

Treatments							
Inadequate vitamin D3		Adequate vitimin D3			P values		
0.2 mg B/kg	1.5 mg B/kg	0.2 mg B/kg	1.5 mg B/kg	Pooled SP	B	D_3	$B \times D_3$
Glycogen, µmol/g							
41.3	36.0	44.8	43.0	14.7	NS	0.05	NS
Glucose, µmol/g							
3 .88	3.92	3.88	3.91	0.31	NS	NS	NS
Glucose 6-phosphate, µmol/g							
0.243	0.254	0.209	0.220	0.080	NS	0.005	NS
Dihydroxyacetone phosphate, µmol/g							
0.057	0.051	0.070	0.052	0.018	0.02	NS	NS
1,3-bisphosphoglycerate, µmol/g							
0.248	0.286	0.277	0.239	0.117	NS	NS	NS
2-phosphoglycerate, µmol/g							
0.055	0.048	0.065	0.057	0.013	0.003	0.0002	NS
Phosphoenolpyruvate, µmol/g							
0.090	0.077	0.103	0.100	0.032	NS	0.002	NS
Pyruvate, µmol/g							
0.077	0.066	0.076	0.088	0.035	NS	NS	NS
Lactate, µmol/g							
0.844	0. 783	1.010	1.012	0.336	NS	0.0001	NS
Bodyweight, g							
668	736	877	921	145	0.008	0.0001	NS

by either dietary treatment and DVDS increased lactate concentrations. Both supplemental DVDS and DBS increased body weight.

Discussion

Previous findings showing that dietary boron influenced circulating energy substrate metabolites led to the hypothesis that, in the vitamin D-deficient chick, boron restores hepatic glycolytic metabolites to concentrations exhibited by vitamin D-adequate chicks. However, the present findings indicate that boron does not act directly or indirectly on vitamin D to modify intermediate energy metabolism in the hepatocyte. This is of special interest because recent findings indicate that boron supplementation markedly increases 1,25-dihydroxyvitamin D_3 concentrations in the vitamin D-deficient chick (Bakken 1995). Regardless of the mechanism by which boron influences hepatic glycolytic concentrations, the element is apparently beneficial because chick growth was improved by boron supplementation in the current study. Because boron affects circulating insulin concentrations (Hunt and Herbel 1991–1992) and peak insulin production (Bakken 1995), efforts are underway to determine possible relations among insulin, hepatic glycolysis and dietary boron.

Literature Cited

Bakken, N. (1995) Dietary boron modifies the effects of vitamin D nutriture on energy metabolism and bone morphology in the chick. Masters of Science thesis, University of North Dakota, Grand Forks, ND.

Bergmeyer, H. U. (1974) Methods of enzymatic analysis, 2nd ed., Vol. 1–7, Academic Press, New York.

Hunt, C. D. & Herbel, J. L. (1991–1992) Boron affects energy metabolism in the streptozotocin-injected, vitamin D_3-deprived rat. Magn. Trace Elem. 10: 374–386.

Hunt, C. D., Herbel, J. L. & Idso, J. P. (1994) Dietary boron modifies the effects of vitamin D_3 nutriture on indices of energy substrate utilization and mineral metabolism in the chick. J. Min. Bone Res. 9: 171–181.

Nielsen, F. H. (1994) Biochemical and physiologic consequences of boron deprivation in humans. Environ. Health Perspect. 102: 59–63.

Iodine and Selenium in Budgerigar (*Melopsittacus undulatus*) Juvenile Plumage during Moulting Accelerated by L-Thyroxine

Z. GOLOB,[†] V. STIBILJ,[‡] M. DERMELJ,[‡] AND S.V. BAVDEK[†]

[†]Institute of Anatomy, Histology and Embryology, Veterinary Faculty, and [‡]J. Stefan Institute, University of Ljubljana, Slovenia

Keywords selenium, iodine, L-thyroxine, budgerigar

The iodination rate of thyroglobulin molecules in birds is appreciably higher than in mammals; iodine comprises about 1.5% of the thyroglobulin mass in birds (0.5% in mammals), and in one avian thyroglobulin molecule there can be as many as 50–90 atoms of iodine (Wenworth and Ringer 1986). Iodine is strongly accumulated in the thyroid gland; however, when its supply is elevated in the food, iodine excretion through the kidneys is more expressed (Ritchie et al. 1994). Selenium is important for conversion of T4 into T3, particularly via iodothyronine deiodinase I, which is a selenoenzyme. A deficiency of selenium may also reduce the iodothyronine deiodinase II activity (Arthur et al.1993). Selenium appears to deposit in the feathers (Goede and Bruin 1984). In budgerigars (*Melopsittacus undulatus*) during the moulting period the elimination of feathers can be forced with L-thyroxine (Golob and Bavdek 1995).

To our knowledge no data on the concentrations of iodine and selenium in budgerigar plumage are described in the literature. The objective of this study was to determine the influence of thyroxine (L-T4) on the iodine and selenium content in excreta and in plumage during moulting.

Material and Methods

During the period of juvenile moult 66 budgerigars were divided into three groups: a group of non-treated (control) animals, and two groups of animals treated with thyroxine (L-T4, LEK, Ljubljana, Slovenia) at doses of 500 $\mu g\ kg^{-1}$ and of 1000 $\mu g\ kg^{-1}$ body weight, respectively. L-thyroxine was placed directly into the crop, every day for a period of 25 d. Moulted downy, cover, flying and steering feathers and excreta were collected every day; at the end of the experiment the animals were sacrificed and all their feathers were pulled out for study.

The plumage samples were prepared according to a standard protocol for hair samples (IAEA 1978). Excrement was dried at 40°C and then homogenised in an agate mill. Iodine and selenium were determined from one sample aliquot by the use of the LICSIR technique — Long Irradiation (for selenium), Cooling (minimum one week), Short Irradiation (for iodine), and Radiochemistry. The method was described more in details elsewhere (Dermelj et al. 1991, Stibilj et al. 1994). Both elements were determined in four replicates of all samples.

Results and Discussion

The animals were fed with food especially prepared for budgerigars, in which we determined the iodine (0.813 ± 0.028 mg kg^{-1}) and selenium (0.397 ± 0.037 mg kg^{-1}) contents by radiochemical neutron activation analysis (LICSIR technique); the contents were in the range of reference values for poultry (Puls 1988).

The accuracy and reproducibility of the results obtained in this study was checked by the analysis of certified reference materials. Results for selenium of 0.388 ± 0.028 mg kg^{-1} in NIST Rice Flour 1568a

Correct citation: Golob, Z., Stibilj, V., Dermelj, M., and Bavdek, S.V. 1997. Iodine and selenium in budgerigar (*Melopsittacus undulatus*) juvenile plumage during moulting accelerated by L-thyroxine. In *Trace Elements in Man and Animals – 9: Proceedings of the Ninth International Symposium on Trace Elements in Man and Animals. Edited by* P.W.F. Fischer, M.R. L'Abbé, K.A. Cockell, and R.S. Gibson. NRC Research Press, Ottawa, Canada. pp. 602–603.

Table 1. Results for iodine and selenium (μg g^{-1} dry weight) in excreta and plumage of control and treated groups.

Group	Excreta		Plumage	
	Iodine	Selenium	Iodine	Selenium
Control	0.314 ± 0.007	1.46 ± 0.22	0.559 ± 0.033	1.485 ± 0.275
500 mg L-T4 kg^{-1}	8.81 ± 2.29	1.56 ± 0.15	0.705 ± 0.083	0.914 ± 0.051
1000 mg L-T4 kg^{-1}	4.53 ± 0.47	2.79 ± 0.40	2.33 ± 0.27	0.531 ± 0.03

and 1.88 ± 0.08 mg kg^{-1} in BCR Human Hair 397 showed good agreement with the certified values of 0.380 ± 0.080 mg kg^{-1} and 2.00 ± 0.08 mg kg^{-1} respectively. Since certified values for iodine in these two CRM are not given, quality control for iodine was performed with NIST Oyster Tissue 1566a (4.56 ± 0.34 mg kg^{-1}) and good agreement obtained with certified value (4.46 ± 0.42 mg kg^{-1}).

In animals treated with thyroxine at a dose of 500 μg kg^{-1} the highest concentration of iodine was found in excreta (Table 1). This is in accordance with data in the literature, that during moderate elevation of iodine in food it appears to eliminate more efficiently through the kidneys (Ritchie et al. 1994). In animals treated with thyroxine at a dose of 1000 μg kg^{-1} iodine accumulation in the plumage was noticed, while its concentration in excreta was only about a half of that found in animals treated at a dose of 500 μg kg^{-1}.

The results for selenium in plumage and excreta show that treatment with the highest dose of thyroxine is followed by a drop of selenium content in the plumage and by its increase in excreta. This could be related to the increased activation of selenoenzymes.

Literature Cited

Arthur, J.R., Nicol F. & Beckett G.J. (1993) Selenium deficiency, thyroid hormone metabolism and thyroid hormone deiodinases. Am. Nutr. Suppl. 57: 236S–239S.

Dermelj, M., Stibilj, V., Stekar, J. & Byrne, A.R. (1991) Simultaneous determination of iodine and selenium in biological samples by radiochemical neutron activation analysis. Fresenius J. Anal. Chem. 340: 258–261.

Goede, A.A. & Bruin, M. (1984) The use of bird feather parts as a monitor for metal pollution. Environ. Poll. (series B). 8: 281–298.

Golob, Z. & Bavdek, S.V. (1995) Response of the skin to moult induction with thyroxine in the budgerigar (*Melopsittacus undulatus*). The 3rd Conference of the European Committee of the Association of Avian Veterinarians. Israel. 212–216.

Puls, R. (1988) In: Mineral levels in animal health. Diagnostic data. Canada: Sherpa International Clearbook. 104–105, 194–196.

Report IAEA/RL/50, 1978.

Ritchie, B.W., Harrison, G.J. & Harrison, L.R. (1994) In: Avian Medicine: Principles and Applications. pp.84–639. Wingers Publishing, Florida.

Stibilj, V., Dermelj, M., Franko, M. & Byrne, A.R. (1994) Spectrophotometric determination of the chemical yield in radiochemical neutron activation analysis of selenium. Analytical Sciences. 10: 789–793.

Wenworth, B.C. & Ringer, R.K. (1986) Thyroids. In: Sturkie PD. Avian Physiology 4th ed. pp. 452–466. Springer Verlag, New York.

Metabolic and Physiological Consequences of Marginal Trace Element Deficiencies

Chairs: S.C. Cunnane and L.M. Klevay

Disruption of the Metabolism of Polyunsaturates During Moderate Zinc Deficiency

S.C. CUNNANE[†] AND J. YANG[‡]

[†]Department of Nutritional Sciences, Faculty of Medicine, University of Toronto, Toronto, Ontario M5S 3E2 and [‡]Hospital for Sick Children, Toronto, Canada

Keywords zinc, polyunsaturates, linoleate, food intake

Attempts to understand and define the physiological role of zinc in mammals frequently involve assessing the impact of dietary zinc deficiency, most often in rodent models, particularly the rat. The commonest approach to maximizing the intended metabolic or structural effect of zinc deficiency involves providing a diet with a very low level of residual zinc to immature, rapidly growing rats. One undesirable outcome of maximal zinc depletion in the immature rat is impaired food intake. This is generally controlled for by food-restricting (pair-feeding) control animals. If the effect being sought is still observed in comparison with the pair-fed controls, then it is considered to be a consequence of zinc deficiency *per se*; if not, then it is viewed as a non-specific effect of food restriction. This approach has been used over several decades and has helped define many biological parameters influenced by zinc.

Zinc and Metabolism of Polyunsaturates

In principle, the system of pair-feeding control animals to their zinc deficient counterparts should reveal most significant metabolic influences of zinc. In practice, differences in energy and fatty acid metabolism between pair-fed (energy-restricted) and zinc deficient young rats may be subtle and pair-feeding may mask real effects of zinc deficiency on fatty acid metabolism. The gross symptoms of a dietary deficiency of polyunsaturates (mainly linoleate, 18:2n – 6) or of zinc are at least superficially similar and include growth retardation, dry, scaly skin lesions, reproductive impairment in both sexes and decreased tissue levels of arachidonate (20:4n – 6) which is the main n – 6 polyunsaturate derived from linoleate. The possible connection between zinc and metabolism of polyunsaturates was first made by Ralph Holman who was studying the effects of a deficiency of polyunsaturates and noted that zinc deficiency and calcium excess both independently exacerbated this condition. This was later confirmed (Bettger et al. 1980) and the ameliorating influence of linoleate and possibly gamma-linolenate (18:3n – 6) on gross symptoms of zinc deficiency was demonstrated (reviewed by Cunnane 1988a).

The changes reported in tissue and blood levels of polyunsaturates induced by zinc deficiency in rodent models have been inconsistent over the 15 y that this has been studied. On the one hand, reduced *in vitro* activity of the desaturases responsible for converting linoleate to arachidonate seems clear-cut (reviewed by Cunnane 1988a); on the other hand, the expected decrease in tissue levels of arachidonate that would occur with impaired linoleate desaturation has been observed by some but does not occur

Correct citation: Cunnane, S.C., and Yang, J. 1997. Disruption of the metabolism of polyunsaturates during moderate zinc deficiency. In *Trace Elements in Man and Animals – 9: Proceedings of the Ninth International Symposium on Trace Elements in Man and Animals.* *Edited by* P.W.F. Fischer, M.R. L'Abbé, K.A. Cockell, and R.S. Gibson. NRC Research Press, Ottawa, Canada. pp. 604–608.

consistently. The inconsistent changes in tissue fatty acid profiles in zinc deficiency may be due to a variety of methodological differences including dietary fat and protein sources, age, and duration of study. Perhaps the two biggest confounding factors are the severity of zinc depletion, and degree of food intake restriction in the pair-fed controls.

Adjusting the age of study towards a more mature, slower growing animal and adjusting the amount of residual zinc in the diet to induce more moderate zinc deficiency have two important advantages in determining the applicability of animal models to human cases of zinc deficiency: (1) reduced food intake in the zinc deficient group does not need to be corrected for by food restriction in the controls and (2) more moderate zinc depletion is more clinically-relevant. We review here two studies in which different models of moderate zinc deficiency in rats are shown to have effects in metabolism of polyunsaturates that are consistent with impaired metabolism of linoleate to arachidonate. Neither involves significantly reduced food intake in the experimental group or pair-feeding in the controls (Cunnane 1988b; Cunnane and Yang 1995).

Model 1: Young Male Rats

Weanling male Sprague-Dawley rats were provided with a powdered semi-purified diet from weaning. The diet contained an adequate zinc level (36 mg/kg diet) and was based on egg white, corn starch, cellulose, and 10% corn oil to supply fatty acids (Cunnane 1988b). At 10 d post-weaning, two groups of 10 rats each either continued with the same diet or were switched to one that was identical except for lower zinc (3.4 mg/kg diet). They remained on these diets for 10 weeks. Final body weight and body weight gain was lower in the zinc deficient group but food intake did not differ statistically between the two groups. Plasma zinc was 60% lower in the zinc deficient group. Total lipids were extracted from plasma and liver and analysed for fatty acid profiles. In plasma phospholipids, linoleate was higher and arachidonate was lower in the zinc deficient group, leading to an increase of the linoleate/arachidonate ratio of 20–50%. This ratio is an estimate of desaturation and chain elongation of linoleate and, in this case, shows that zinc deficiency impaired this metabolic process. In liver phospholipids, linoleate was also raised but arachidonate was not significantly decreased in the zinc deficient group. The linoleate/arachidonate ratio was nevertheless raised in liver phospholipids by 30%.

The key outcome of this study was that the changes in fatty acid profiles or linoleate/arachidonate ratio were consistent with the concept of impaired desaturation of linoleate but were observed in the absence of significantly impaired food intake. Raising the age of introduction of the zinc deficient diet by 10 d and increasing the residual zinc in the zinc deficient diet by 10 fold were probably important in sustaining normal food intake and yet this more physiological model of moderate zinc deficiency produced the intended effect on fatty acid metabolism without pair-feeding the controls.

Model 2: Pregnancy

A more difficult but perhaps more relevant animal model of zinc deficiency involves studying fetal development and fatty acid accumulation during pregnancy. In this model, it has proven more difficult to titrate dietary zinc levels to a point at which food intake is normal yet zinc deficiency symptoms are induced. After attempting two approaches in which food intake was still impaired even though the rats were receiving 3–5 mg/kg zinc in the diet and for periods of less than 10 d during pregnancy, we resorted to the tube-feeding approach (Flanagan 1984). From our other fatty acid studies, it had also become apparent that more complete information about fatty acid metabolism than tissue fatty acid profiles was needed to establish how moderate zinc deficiency altered fatty acid utilization. In particular, we had developed the whole body fatty acid balance method which allows one to measure whole body fatty acid accumulation relative to intake and excretion and to estimate oxidation of linoleate or α-linolenate (18:3n – 3) indirectly and we wanted to apply it in this model. This requires that the rat be adapted to the dietary fatty acid profile under study for about 2 weeks before the first group of animals is killed for baseline samples.

First parity Sprague-Dawley rats were given a semi-purified, powdered diet based on egg white, corn starch and 5% soybean oil that contained 26 mg/kg zinc. They consumed this diet for 2 weeks prior to mating and for the first 8 d of gestation. At day 8 of gestation, one group of rats was killed for baseline data (n = 3) and the remainder were switched to being tube-fed the same dietary zinc level (n = 5) or 3.2 mg/kg zinc (n = 5). The powdered diet was made into a water-based slurry and gavaged in equal amounts to both groups 8 times/d (Yang and Cunnane 1994). At day 20 of gestation, the rats were each given an i.p. injection of 5 μCi of [1-^{14}C]-linoleate and killed 12 h later. Food consumption was monitored throughout and fatty acid accumulation in maternal and fetal organs was determined relative to the baseline values obtained at day 8 of gestation. Distribution of the ^{14}C in maternal and fetal organs, lipid classes and fatty acids separated by degree of unsaturation was also determined (Cunnane and Yang 1995).

Body weights were the same in both groups but plasma zinc was reduced by 34% in the zinc deficient group. In general, fatty acid profiles (% composition) did not differ significantly in the two groups although n – 6 and n – 3 long chain polyunsaturates were lower in the zinc deficient fetuses (mg/g and mg/fetus; Yang and Cunnane 1994). Whole body fatty acid balance was measured over the 11 d study period. Since linoleate and α-linolenate cannot be synthesized by animals consuming semi-purified diets (devoid of 16:2n – 6 or 16:3n – 3), comparison of intake to excretion and accumulation of these fatty acids or their longer chain products provides an estimate of whole body oxidation by difference. Despite identical food intake, whole body (including the conceptus) accumulation of all polyunsaturates decreased by 20–70% in the zinc deficient group and, by difference, oxidation of linoleate increased by 32% while that of α-linolenate increased by 77% (both $p < 0.01$; Cunnane and Yang 1995). The ratio of accumulated linoleate to n – 6 long chain polyunsaturates was not different in the two groups but was 23% lower for α-linolenate compared to n – 3 long chain polyunsaturates.

In the zinc deficient group, ^{14}C recovery in total lipids of maternal uterus, placenta, liver, and periuterine fat was reduced by 33–59%, but was unchanged relative to controls in the fetuses and was 61% higher in the carcass lipids. ^{14}C recovered in water-soluble metabolites of these organs was 70–98% higher in the maternal organs of the zinc deficient group (Yang and Cunnane 1994). Distribution of the ^{14}C in lipid classes was largely unaffected by zinc deficiency with the exception of higher labelling of free cholesterol in uterus, fat and carcass lipids. In the zinc deficient group, there was lower recovery of ^{14}C in liver polyunsaturates (all fatty acids with 2 or more double bonds) and higher ^{14}C recovery in carcass saturates, monounsaturates and polyunsaturates (Cunnane and Yang 1995). Overall, the radiotracer data indicate that, in the zinc deficient group, less ^{14}C was retained in the original [1-^{14}C]-linoleate or its longer chain products, more ^{14}C was recycled into lipids synthesized *de novo* (free cholesterol, saturates, monounsaturates) and more was present in water-soluble metabolites (primarily ketones.

Conclusion

Progress in establishing the role of zinc in fatty acid metabolism has developed slowly over the past two decades. The potential role of zinc in metabolism of polyunsaturates has received the most attention but desaturation of stearate (18:0) has also been studied. Zinc deficiency effects on *de novo* lipid synthesis are receiving greater attention (Cunnane and Yang 1995; Eder and Kirchgessner 1995). In fact, the interaction with metabolism of polyunsaturates and in lipogenesis may be inter-related because impaired desaturation of linoleate seems to be associated with recycling of linoleate carbon into lipids synthesized *de novo*.

Attention has been focussed on zinc deficiency as the predominant model for evaluating the potential role of zinc in fatty acid metabolism. However, zinc supplementation at 411 mg/kg of diet has effects on the linoleate/arachidonate ratio consistent with *increased* desaturation and chain elongation in this pathway, suggesting that the effects of zinc deficiency are specific and that inconsistent results are more a function of the model than of the role of zinc in this pathway (Cunnane 1988b). Since the desaturases and the related cytochromes are iron-containing proteins, the role of zinc has to be ancillary to the desaturases *per se*. Cytochrome B5 reductase activity is impaired in isolated liver microsomes from zinc deficient rats (Hammermueller et al. 1987), and this is presently the most likely explanation for impaired desaturase activity and linoleate metabolism in zinc deficiency.

References

Bettger, W.J., Reeves, P.G., Moscatelli, E.A., Savage, J.E. & O'Dell, B.L. (1980) Interaction of zinc and polyunsaturated fatty acids in the chick. J. Nutr. 110: 50–58.

Cunnane, S.C. (1988a) The role of zinc in lipid and fatty acid metabolism and in membranes. Prog. in Food and Nutr. Sci. 12: 151–188.

Cunnane, S.C. (1988b) Evidence that the adverse effects of zinc deficiency on essential fatty acid composition in rats are independent of food intake. Br. J. Nutr. 59: 273–278.

Cunnane, S.C. & Yang, J. (1995) Zinc deficiency impairs whole body accumulation of polyunsaturates and increases the utilization of [1-^{14}C] linoleate for *de novo* lipid synthesis in pregnant rats. Can. J. Physiol. Pharmacol. 73: 1246–1252.

Eder, K. & Kirchgessner, M. (1995) Zinc deficiency and activities of lipogenic and glycolytic enzymes in liver of rats fed coconut oil or linseed oil. Lipids 30: 63–69.

Flanagan, P. (1984) A model to produce pure zinc deficiency in rats and its use to demonstrate that dietary phytate increases the excretion of endogenous zinc. J. Nutr 114: 493–502.

Hammermueller, J.D., Bray, T.M. & Bettger, W.J. (1987) Effect of zinc and copper deficiency on microsomal NADPH-dependent active oxygen generation in rat lung and liver. J. Nutr. 117: 894–899.

Yang, J. & Cunnane, S.C. (1994) Quantitative measurements of dietary and [1-^{14}C]-linoleate metabolism in pregnant rats: Specific influence of moderate zinc depletion independent of food intake. Can. J. Physiol. Pharmacol. 72: 1080–1085.

Discussion

Q1. John Chesters, Rowett Research Institute, Aberdeen, Scotland: Steve, I'm very intrigued by the data, but I'm really at sea as to what the critical parameter that you've been looking at is. You've stressed that percent differences are not all that great, whereas some of these quantitative differences are. If one is thinking of membrane composition, for example, it's not likely to be the percentage composition that likely to be the functionally critical factor? I don't know which is the more crucial from a functional point of view.

A. Well that's a fair question, John, and it's something that people in the lipid field go around in circles with all the time. I can't say that I know what the functional significance of a lower DHA is, except that the clinical outcomes are associated with impaired neural function during early development. Is that because, even though the percentage composition is not different, the total is lower? Nobody can really account for that. I don't have some of the tools that other people in this area have, and they haven't figured it out. To some extent I think that we need to establish that these effects are real, and then let the people who can do that sort of work get involved. I think that we've started to show that this effect really does exist.

Q2. Les Klevay, USDA-ARS, Grand Forks, ND, USA: Well I have a good feel for the zinc intakes. Surely there are some people who are pregnant who are eating not very much zinc. Are there any fatty acid data in difficult pregnancies, that would be related to this?

A. I don't have any for pregnancies, Les, but we have had a chance with a pediatrician in Detroit to study two cases of neonatal zinc deficiency. One was what you would call the transient form, where there were no genetic problems per se, while the other was acrodermatitis, and the arachidonic and DHA levels were as low as we can get in any other condition experimentally, and they respond to zinc supplementation. That's not directly addressing your point, but it certainly suggests that this effect is a zinc effect. Of course, in acrodermatitis, there are probably quite a few other problems besides zinc deficiency. I recognize that that's not a clean example.

Q3. Les Klevay, USDA-ARS, Grand Forks, ND, USA: You mentioned acrodermatitis, but what about other infants that don't thrive very well? Has anybody looked at those for zinc intakes?

A. Well, failure to thrive is a multi-factorial thing, so we might see changes in the fatty acids that would be just due to low energy intake. It's just not a clear situation. What we're trying to do is, in fact, remove all the other factors, and look at a clean zinc deficiency.

Q4. Klaus Eder, Institute of Nutrition Physiology, Munich, Germany: Steve, you know there are a lot of similarities between zinc deficiency and essential fatty acid deficiency. Do you think the fatty acid changes you have shown us are responsible for this, or would you speculate that metabolism of eicosanoids is the main reason for this similarity?

A. Well, eicosanoids are important compounds, there is no question. But they occur in amounts that are at least two orders of magnitude smaller than the amounts of fatty acid we're looking at. So you've got a problem of being able to manipulate these eicosanoids and their importance. It just doesn't make a measurable dent in the level of fatty acids that are precursors to them, in membranes. It's somewhat analogous to looking at lipid peroxides, and the fatty acids that produce them. You can raise the peroxides by a large amount without actually changing the fatty acid content very much. So I struggle with the idea that eicosanoids are responsible for what we see. But the bottom line is I don't have much experience with eicosanoid metabolism, and it's not been studied much in this context over the past few years. So I don't really have a feel for how it would be involved.

Q5. Janet King, USDA-ARS, San Francisco, CA, USA: I might just comment, Steve, we don't have data nearly as sophisticated as what you have, but several years ago we fed a group of young men 5 mg of zinc for about 55 d, and we saw a decrease in serum levels of arachidonic acid. We also saw a drop in their RQ, meaning that they were metabolising fat as their fuel, more than carbohydrate. So with that model we saw the same thing that you saw here.

A. That's nice confirmation. Have you published that, Janet?

Q6. Les Klevay, USDA-ARS, Grand Forks, ND, USA: You did mention iron. Can we ask about other metals? Have you done any work with other metals?

A. Yes, we have. In fact, iron has a similar effect. Desaturase is an iron-containing enzyme, and arachidonic acid levels are lower in the iron-deficient rats. With copper, it's dramatically the opposite effect. Copper deficiency will raise the longer-chain polyunsaturates, and lower the amount of linoleic acid, so all metals aren't doing the same thing.

The Effect of Zinc Depletion on the Concentration of Circulating Hormones and Transport Proteins

B. SUTHERLAND,[†] N.M. LOWE,[†] B.J. BURRI,[‡] AND J.C. KING[‡]

[†]Department of Nutritional Sciences, U.C. Berkeley, Berkeley, CA 94720, and the [‡]USDA Western Human Nutrition Research Center, San Francisco, CA 94129, USA

Keywords zinc deficiency, hormones, protein, men

The pathologies that develop with prolonged zinc (Zn) deficiency may be explained in part by changes in protein metabolism, for instance growth retardation, dermatitis and abnormal immune function. Several hormones perform key roles in the regulation of protein metabolism: insulin, cortisol, testosterone and the thyroid hormones. Reduced plasma levels of testosterone, thyroid hormones and plasma transport proteins, as well as altered cortisol and insulin response to stimuli, have been reported to occur with Zn deficiency (Wada and King 1986, Baer et al. 1985, Guigliano and Millward 1987). The purpose of this study was to determine if acute Zn depletion alters plasma concentrations of the thyroid hormones, insulin, cortisol, testosterone, and the transport proteins (albumin, transthyretin, transferrin and retinol binding protein).

Method

Eight men participated in the 85 d Zn depletion/repletion study which was divided into three metabolic periods: a 16 d baseline (MP1), a 40 d depletion (MP2) and a 29 d repletion (MP3). Zinc intake was 12 mg/d during baseline and repletion, and 1 mg/d during depletion. The subjects were fed a synthetic formula; energy and protein intakes were held constant after adjustments during the baseline period. The subjects were confined to a metabolic unit. Their daily exercise consisted of two 3-mile, moderately-paced walks.

Fasting plasma Zn was measured weekly to monitor the change in Zn status. The plasma proteins measured were albumin, transthyretin, transferrin and retinol binding protein (RBP); the hormones were insulin, cortisol, testosterone and the thyroid hormones. Fasting plasma proteins and hormones were measured on day 6 of MP1, days 6 and 27 of MP2 for two subjects who depleted rapidly, days 6 and 35 of MP2 for the other six subjects, and on day 21 of MP3 for all subjects. Plasma Zn was measured by atomic absorption spectrophotometry (AAS). Plasma albumin, transthyretin and transferrin were measured on an automated centrifugal analyzer; albumin by a bromocresol green binding assay, transthyretin and transferrin by immunoprecipitation. RBP was measured by high pressure liquid chromatography (Burri and Kutnink 1989). All hormones were measured with commercially available radioimmunoassay kits. Statistical analysis of data was performed using a one factor repeated measures ANOVA and Tukey's follow up test.

Results

The ages of the 8 subjects ranged from 20–35 years; they had a height of 1.78 ± 0.06 m (mean ± S.D.), weight of 77 ± 10 kg and BMI of 24.3 ± 2.5 kg/m^2. They were all healthy with no recent change in weight. Their usual diet contained meat at least twice a week and provided more than 8 mg Zn/d. They did not take mineral supplements for at least six weeks prior to the study and they all reported no use of cigarettes or drugs.

By the end of depletion mean plasma Zn concentration was 5 µmol/L; with Zn repletion the level returned to baseline value (12 µmol/L). RBP concentration was significantly lower ($p < 0.05$) by the end

Correct citation: Sutherland, B., Lowe, N.M., Burri, B.J., and King, J.C. 1997. The effect of zinc depletion on the concentration of circulating hormones and transport proteins. In *Trace Elements in Man and Animals – 9: Proceedings of the Ninth International Symposium on Trace Elements in Man and Animals. Edited by* P.W.F. Fischer, M.R. L'Abbé, K.A. Cockell, and R.S. Gibson. NRC Research Press, Ottawa, Canada. pp. 609–611.

Table 1. Response of plasma hormones and proteins to Zn depletion/repletion.[1]

Proteins and hormones	Metabolic Period			
	Baseline	Early Depletion	Late Depletion	Repletion
Albumin (g/L)	42±1	41±1	44±1	41±1
Transthyretin (g/L)	0.21±0.13	0.19±0.24	0.21±0.19	0.23±0.13
Transferrin (g/L)	2.1±0.09[ab]	2.0±0.07[a]	2.1±0.08[b]	2.2±0.08[b]
Retinol BindingProtein (g/L)	0.06±0.01[b]	0.05±0.002[ab]	0.05±0.002[a]	0.05±0.002[a]
Total T4 (nmol/L)	80±4[b]	73±3[a]	81±4[b]	79±4[ab]
Free T4 (nmol/L)	15±1[b]	14±1[a]	14±1[a]	14±1[a]
Total T3 (nmol/L)	2.2±0.2	2.0±0.1	2.1±0.2	2.1±0.1
Free T3 (pmol/L)	4.1±0.3	4.3±0.3	4.6±0.3	4.8±0.2
rT3 (nmol/L)	0.37±0.05[b]	0.29±0.03[a]	0.29±0.05[a]	0.29±0.02[a]
TSH (mU/L)	1.3±0.2	1.3±0.2	1.5±0.3	1.5±0.2
Insulin (pmol/L)	66±11	67±7	77±9	65±6
C-peptide (μg/L)	1.7±0.2	1.9±0.2	2.1±0.2	2.0±0.2
Cortisol (nmol/L)	408±44[b]	309±38[a]	342±50[a]	315±36[a]
Total Testosterone (ng/dL)	571.8±48.6	568.8±50.3	594.2±55.5	577.1±58.3
Free Testosterone (pg/mL)	25.4±1.7[b]	20.4±1.0[a]	20.7±1.3[a]	21.0±1.2

[1]Values are mean ± SEM. Values in a row not sharing similar superscripts are significantly different at $p < 0.05$.

of depletion and remained lower at day 21 of repletion (Table 1). An additional RBP measurement was done on repletion day 30 at which point the level had returned to baseline. Plasma albumin and transthyretin concentrations were unchanged with Zn depletion, transferrin increased significantly ($p < 0.05$) from early to late depletion but did not decline in repletion. The following significant changes ($p < 0.05$) in plasma hormone concentrations were observed (Table 1): total and free thyroxine (T4) and reverse triiodothyronine (rT3) decreased by day 6 of depletion; total T4 returned to baseline by the end of depletion while free T4 and rT3 remained low. Free testosterone and cortisol levels also declined by day 6 of depletion and remained low throughout the rest of the study. Toward the end of depletion the following clinical manifestations of Zn deficiency were observed: skin rashes on the perioral and groin regions, lethargy and moodiness. These signs were reversed within a few days of Zn repletion.

Conclusions

Acute Zn depletion was associated with a decrease in plasma levels of cortisol, total and free T4, reverse T3 and free testosterone. This decrease was seen within 6 d of depletion suggesting an early adaptive response to the acute decline in Zn intake. Several of the changes observed suggest that acute Zn depletion altered protein synthesis: the rapid decline in total and free T4, reverse T3 and free testosterone, and the fall in plasma RBP by the end of depletion. The decrease in thyroid hormones was not associated with a change in resting metabolic rate.

Literature Cited

Baer, M.T., King, J.C., Tamura, T., Margen S., Bradfield, R.B., Weston, W.L., & Daugherty, N.A. (1985) Nitrogen utilization, enzyme activity, glucose intolerance and leukocyte chemotaxis in human experimental zinc depletion. Amer. J. Clin. Nutr. 41:1200–1235.

Burri, B.J. & Kutnink, M.A. (1989) Liquid-chromatography assay for free and transthyretin-bound retinol binding protein in serum from normal humans. Clin. Chem. 35:582–586.

Guigliano, R. & Millward, D.J. (1987) The effects of severe zinc deficiency on protein turnover in muscle and thymus. Br. J. Nutr. 57:139–155.

Wada, L. & King, J.C. (1986) Effects of low zinc intakes on basal metabolic rate, thyroid hormones and protein utilization in adult men. J. Nutr. 116:1045–1053.

Discussion

Q1. Steve Cunnane, University of Toronto, ON, Canada: Are the data consistent with Janet's report of the 5 mg study, where RQs declined?

A. We didn't see any changes in basal metabolic rate in this study. However, the 5 mg study didn't show the same changes in T3 levels that we saw. No changes were noted in that study, but they didn't separate total versus free T3. The differences between these two studies may represent differences in response to acute depletion and chronic.

Q2. Hans-Peter Roth, Technical University of Munich, Germany: How do you explain the observation that insulin levels decreased, then increased late in zinc deficiency?

A. This was seen in our previous study also, though it was actually reflected in that study as impaired glucose tolerance, as we didn't measure insulin levels directly in that study.

Q3. Les Klevay, USDA-ARS, Grand Forks, ND, USA: Your data were expressed in terms of percent change? Were any of your measurements outside of the clinically normal ranges for these parameters?

A. No, all of our results were within normal ranges for humans, at all times.

Q4. Klaus Eder, Technical University of Munich, Germany: What were the sources of fats in the diet? New studies have indicated that dietary fats can influence serum thyroid hormones.

A. We used safflower oil, at 30% of the diet. I'm not familiar with that literature, but we'll certainly look into it. Thank you.

Copper and Zinc Levels in Patients with Coronary Heart Disease (CHD)

G.W. MIELCARZ,[†] N.R. WILLIAMS,[‡] J. RAJPUT-WILLIAMS,[‡] S.V. NIGDIKAR,[‡] A.N. HOWARD,[‡] P.G. MULLINS,[¶] AND D.L. STONE[¶]

[†]K. Marcinkowski University of Medical Sciences, Department of General Chemistry, 60-780 Poznan, Poland; [‡]COAG Trace Elements Laboratory, Pathology Department and [¶]Cardiac Unit, Papworth Hospital, Cambs. CB38RE, UK

Keywords copper, zinc, humans, coronary heart disease

Copper and zinc are essential nutrients and an integral part of the antioxidant enzymes copper,zinc-superoxide dismutase (SOD) and ceruloplasmin. The heart is particularly sensitive to oxidant stress, and cardiac and vascular abnormalities have been observed in humans exposed to low copper diets. Also, the mortality ratio of ischemic heart disease in populations with low copper intake is significantly increased (Mielcarz et al. 1994, Mielcarz et al. 1995). The detection of marginally inadequate trace element nutrition is difficult, and not all the trace element present in cell or tissue is metabolically active. Thus, the leucocyte copper or zinc is promising indicator of nucleated tissue status in adults. Leucocytes are readily accessible nucleated haemopoetic tissue, and have a high rate of turnover (Howard 1995). The purpose of this study was to evaluate the hypothesis that low copper body status promotes progression of CHD.

Methods

The subjects were 23 females and 57 males, aged between 30–70, due to have a cardiogram. Individual angiograms were scored by combining the degree of CHD in all the major arteries into one score of a scale of 1.00 for patency to 0.00 for severe CHD according to the method of Stone et al. (1980). Leucocyte samples were isolated using an amended version of Hinks et al. (1983), and analysed for copper and zinc by GFAAS. Leucocyte protein was determined by Baumgarten's (1985) method, and DNA by the method of Switzer and Summer (1971). Leucocyte results are expressed as ratios (e.g. μg Cu per μg DNA).

Table 1. Comparison of patients with high and low angiogram scores.

Parameters	Advanced CHD (angiogram score <0.40) n = 22	Relatively normal arteries (angiogram score >0.60) n = 44
Plasma copper (μmol/L)	13.3 ± 4.0	13.9 ± 4.1
Plasma zinc (μmol/L)	10.9 ± 1.1	11.3 ± 1.4
Leucocyte Cu/DNA ($\times 10^{-2}$)	0.91 ± 0.40	1.23 ± 0.78*
Leucocute Cu/protein($\times 10^{-4}$)	6.32 ± 2.58	8.80 ± 4.89*
Leucocyte Zn/DNA	0.34 ± 0.22	0.35 ± 0.24
Leucocyte Zn/protein ($\times 10^{-2}$)	2.14 ± 1.29	2.69 ± 1.77
Total cholesterol (mmol/L)	6.5 ± 0.9	6.3 ± 1.0
LDL-cholesterol (mmol/L)	4.8 ± 0.6	4.3 ± 1.1
HDL-cholesterol (mmol/L)	0.87 ± 0.28	0.89 ± 0.35
Triglycerides (mmol/L)	1.8 ± 0.9	2.1 ± 1.1

x ± SD; *$p < 0.05$, significance of difference estimated by t-test.

Correct citation: Mielcarz, G.W., Williams, N.R., Rajput-Williams, J., Nigdikar, S.V., Howard, A.N., Mullins, P.G., and Stone, D.L. 1997. Copper and zinc levels in patients with coronary heart disease (CHD). In *Trace Elements in Man and Animals – 9: Proceedings of the Ninth International Symposium on Trace Elements in Man and Animals. Edited by* P.W.F. Fischer, M.R. L'Abbé, K.A. Cockell, and R.S. Gibson. NRC Research Press, Ottawa, Canada. pp. 612–613.

Table 2. Comparison of sexes using the pooled data from all patients.

Parameters	Females n = 23	Males n = 57	p value*
Plasma copper (μmol/L)	15.6 ± 3.9	13.1 ± 3.6	<0.001
Plasma zinc (μmol/L)	10.9 ± 1.1	11.4 ± 1.4	N.S.
Leucocyte Cu/DNA (×10^{-2})	1.07 ± 0.44	1.19 ± 0.75	N.S.
Leucocyte Cu/protein (×10^{-4})	8.24 ± 3.80	7.57 ± 7.57	N.S.
Leucocyte Zn/DNA (×10^{-2})	0.34 ± 0.18	0.40 ± 0.37	N.S.
Leucocyte Zn/protein (×10^{-4})	2.63 ± 1.45	2.44 ± 1.75	N.S.
Total cholesterol (mmol/L)	6.7 ± 0.8	6.1 ± 0.9	<0.05
LDL-cholesterol (mmol/L)	4.8 ± 0.9	4.3 ± 0.9	N.S.
HDL-cholesterol (mmol/L)	1.0 ± 0.3	0.8 ± 0.2	<0.005
Triglycerides (mmol/L)	1.9 ± 0.9	2.1 ± 1.1	N.S.

x ± SD; *Significance of difference estimated by t-test.

Results and Discussion

The results of the lipid profiles, copper and zinc concentration in plasma and leucocyte depending on severity of angiogram score are presented in Table 1 and the pooled data according to sex in Table 2.

There was a linear relationship between a number of major risk factors and the degree of decreasing leucocyte concentrations and angiogram score. Leucocyte copper had a significant link with the severity of atherosclerosis which was independent of sex. Prolonged sub-optimal levels of copper as indicated by leucocyte levels could impair copper-enzyme dependent function and may be involved in mechanisms associated with CHD.

Literature Cited

Baumgarten, H. (1985) A simple microplate assay for the determination of cellular protein. J. Immunol. Methods 82(1): 25–37.

Howard, A.N. (1995) Copper and zinc status in cardiovascular disease. In: Trace elements in medicine, health and atherosclersis. (Reis, F.M., Miguel, J.M.P., Machado, A.A.S.C., Abdulla, M., eds.) pp. 51–54. Smith-Gordon, Nishimura Company Ltd., London, UK.

Mielcarz, G., & Howard, A.N. (1994) The concentration of copper and zinc in foodstuffs in two populations with different prevalence of ischemic heart disease. In: Defizite und Uberschusse an Mengen- und Spurenelementen in der Ernahrung. Verlag Harald Schubert, Leipzig, Germany

Mielcarz, G.W., Howard, A.N., Williams, N.R., Kinsman, G.D., Mielcarz, B., Moriguchi, E., Moriguchi, Y., Sizuschima, S., & Yamori Y. (1995) Copper and zinc status in two Japanese populations from Brazil and Okinawa, with different cardiovascular mortality. In: Trace elements in medicine, health and atherosclersis. (Reis, F.M., Miguel, J.M.P., Machado A.A.S.C., Abdulla M., eds.) Smith-Gordon, Nishimura Company Ltd., London, UK.

Stone, D., Dymond, D., Elliott, A.T., Britton, K.E., & Banim, S.O. (1980) Exercise first-pass radionuclide ventriculography detection of coronary artery disease. Brit. Heart J. 44: 208–214.

Switzer, B.R., & Summer, G.K. (1971) A modified fluorometric micromethod for DNA. Clin. Chim. Acta. 32: 203–206.

Discussion

Q1. Richard Black, Kellogg Canada Inc., Etobicoke, ON, Canada: Could you put this in perspective for me please? I think what I hear with the cholesterol story, I'm always told that for every one percent drop in cholesterol level, there's a two percent reduction in coronary heart disease risk. Are you able to look at your data, and say "for every x amount of change in copper, risk increases by y"?

A. We have tried to find such a correlation, but couldn't find any relation between plasma copper and leukocyte copper. I think that it is not possible to determine what you want from these data. Individual angiograms are the only way we could assess the degree of disease. We are suggesting that sub-optimal intake of copper may provoke changes in arteries in coronary heart disease, because we found this relationship between these two populations. Afterward, to be more sure, we did individual angiograms to score the degree of disease, and this correlated with leukocyte copper.

Q2. Judy Turnlund, USDA-ARS, San Francisco, CA, USA: I just wanted to comment on that leukocyte copper. We have hypothesized that leukocyte copper might be a good index in the presence of disease. We studied human subjects, and saw plasma copper go down, ceruloplasmin go down, and urinary copper go down, as well as the leukocyte copper. However, in this situation, you didn't see all those changes because of the presence of disease. This is good evidence that leukocyte copper has real value in those situations.

A. We didn't find this relationship in plasma copper. In other investigations we did in Poland with femoral atherosclerosis, we found that plasma copper was higher in patients with disease than in controls. So it seems that leukocyte copper may be a better index.

Iron Can be Cholesterotropic and Cuprotropic

LESLIE M. KLEVAY

USDA-ARS, Grand Forks Human Nutrition Research Center, Grand Forks, ND 58202 USA*

Keywords cholesterotropic, copper, cuprotropic, deficiency, iron

Cholesterotropic and cuprotropic chemicals (Klevay 1985, 1987, 1989) are those that reciprocally modify the metabolism of both cholesterol and copper simultaneously. About a dozen of these chemicals, which vary considerably in structure, have been identified. Half of them increase cholesterol in plasma and inhibit copper utilization; the others decrease cholesterol and enhance copper utilization. Some data from an experiment on iron overload are presented to illustrate these principles.

Methods

Thirty male, weanling rats of the Sprague-Dawley (Sasco, Omaha (Mention of a trademark or proprietary product does not constitute a guarantee or warranty of the product by the U.S. Department of Agriculture and does not imply its approval to the exclusion of other products that may also be suitable.)) strain were matched by mean weight (51 g) into two equal groups. They were given a purified diet based on sucrose (62%), egg white protein (20%) and corn oil (10%) containing all nutrients known to be essential for rats (Klevay 1973). This diet has been in continuous use since 1968, with minimal modification (Klevay 1990). Iron in the diet was either 217 or 35 mg/kg, the latter being the U.S. National Research Council suggestion. Dietary iron was controlled by two batches of copper- and zinc-free salt mix obtained commercially (ICN, Irvine, CA) made by using identical ingredients and differing only in iron content. Copper and zinc were added by thorough mixing of finely ground zinc acetate and cupric sulfate. Dietary zinc was slightly greater than 13 mg/kg and copper was 2.0 mg/kg. The amount of copper added was based on analysis of diet by atomic absorption spectrometry. Copper in organs and diet was measured after destruction of organic matter (Analytical Methods Committee 1960). Cholesterol was measured by fluorometry (Carpenter et al. 1957).

Results

Hypercholesterolemia was detected and rats were killed with pentobarbital and pneumothorax at 42 d (Table 1). Higher dietary iron decreased hepatic copper by 36%; cardiac weight increased 13%, but cardiac copper decreased 43%.

A similar experiment was done later with dietary copper at 2.5 mg/kg. Cholesterol was measured periodically, but the difference between groups never exceeded 12 mg/dL. After 96 d, cholesterol in the lower and higher iron groups was 155 and 165 mg/dL, respectively, a non-significant difference. Cardiac

Table 1. Copper indicators, Mean ± SE, at 2.0 mg copper/kg diet.

	Lower Fe	Higher Fe	p
Cardiac copper, µg/g	12.9 ± 1.1	7.4 ± 0.5	<0.0002
Cardiac weight, g	1.22 ± 0.02	1.38 ± 0.04	<0.005
Hepatic copper, µg/g	8.7 ± 0.55	5.6 ± 0.71	<0.002
Plasma cholesterol, mg/dL	127 ± 4.2	141 ± 3.9	<0.02

*The U.S. Department of Agriculture, Agricultural Research Service, Northern Plains Area, is an equal opportunity/affirmative action employer and all agency services are available without discrimination.

Correct citation: Klevay, L.M. 1997. Iron can be cholesterotropic and cuprotropic. In *Trace Elements in Man and Animals – 9: Proceedings of the Ninth International Symposium on Trace Elements in Man and Animals. Edited by* P.W.F. Fischer, M.R. L'Abbé, K.A. Cockell, and R.S. Gibson. NRC Research Press, Ottawa, Canada. pp. 614–615.

copper was lower with higher iron (17.3 vs. 22.0, $p < 0.0025$), but dietary iron had no effect on either cardiac weight or hepatic copper.

Discussion

Increased cholesterolemia and decreased copper in heart and liver induced by higher iron show that iron can be cholesterotropic and cuprotropic. Groups of rats fed this diet usually have mean cholesterol values less than 100 mg/dL when copper is adequate. The increased heart weight induced by higher iron is another sign that iron can induce copper deficiency.

Hypercholesterolemia occurs at both 2.0 and 2.5 mg copper/kg confirming a shorter experiment where 3 mg/kg produced lower cholesterol than 2 mg/kg (Klevay and Saari 1993). Although the experiments were done at different times and had different durations, it seems likely that the 2.5 mg copper/kg was able to prevent some of the ill effects of higher iron. Some of the harmful effects of iron overload may be caused by the induction of mild copper deficiency.

References

Analytical Methods Committee (1960) Methods for the destruction of organic matter. Analyst 85: 643–656.

Carpenter, K.J., Gotsis, A., and Hegsted, D.M. (1957) Estimation of total cholesterol in serum by a micro method. Clin. Chem. 3: 233–238.

Klevay, L.M. (1973) Hypercholesterolemia in rats produced by an increase in the ratio of zinc to copper ingested. Am. J. Clin. Nutr. 26: 1060–1068.

Klevay, L.M. (1985) Cholesterotropic and cuprotropic chemicals. In: Trace elements in man and animals-TEMA 5 (Mills, C. F., Bremner, I. & Chesters, J. K. eds.), pp. 180–183. Commonwealth Agricultural Bureaux, Farnham Royal, U.K.

Klevay, L.M. (1987) Dietary copper: a powerful determinant of cholesterolemia. Med.Hypotheses. 24: 111–119.

Klevay, L.M. (1989) Ischemic heart disease as copper deficiency. In: Copper bioavailability and metabolism (Adv. Exp. Med. Biol. Vol. 258) (Kies, C. ed.), pp. 197–208. Plenum Press, New York.

Klevay, L.M. (1990) Ischemic heart disease: toward a unified theory. In: Role of Copper in Lipid Metabolism (Lei, K. Y. & Carr, T. P. eds.), pp. 233–267. CRC Press, Boca Raton.

Klevay, L.M. and Saari, J.T. (1993) Comparative responses of rats to different copper intakes and modes of supplementation. Proc. Soc. Exp. Biol. Med. 203: 214–220.

Discussion

Q1. Manfred Anke, Friedrich Schiller University, Jena, Germany: Do you have any idea how the Germans can live with 0.9 and 0.2 mg/d for women and men on an average. I ask because we have a large difference in intake, and we couldn't find deficiency symptoms of copper in these populations.

A. Well, I'm glad you asked that. I think that your data on dietary intakes of copper are among the lower, perhaps the lowest I've seen in the industrialized world. I don't know how much beer they drink in your part of Germany, but I'm told that in some parts of Germany, they drink one-third of their energy as beer. And if you feed rats a diet deficient in copper, and let half of them drink beer, they live six times as long, with less cardiac damage, and lower cholesterols. Somehow they accomplish this by utilizing copper better. If you do the same experiment (and I've published this in the American Journal of Clinical Nutrition a few years ago) with a similar amount of alcohol in water, none of those benefits occur. So there's some compound in beer, I think, that assists in the utilization of copper. In fact, in Germany, if you look at their heart disease rate, it is lower than one would expect based on the amount of fat that you all eat.

(comment by **Steve Cunnane**, University of Toronto, ON, Canada: I think we've been doing an uncontrolled trial of copper supplementation here this week.)

Well, it's not the copper in the beer, even though the beer is brewed and fermented and so forth in big copper tanks. The amount of copper in beer is vanishingly small. In my rat experiment, in fact, the rats that had the damage and drank the water actually ate more copper than those that had less damage, and drank the beer.

Contrasting Responses of Hearts from Rats Fed Diets Deficient or Marginal in Copper

DENIS M. MEDEIROS, THUNDER JALILI, AND LAURA SHIRY

Department of Human Nutrition and Food Management, The Ohio State University, Columbus, Ohio 43210-1295, USA

Keywords copper, cardiac, cytochrome C oxidase, pathology

Marginal intake of dietary copper is more likely to occur among humans and information on the impact upon cardiac integrity is limited. Two studies were conducted over longer periods of time than traditional deficiency trials (up to 10 months of age) in rats. We also studied the effect of marginal copper consumption in a genetic rat strain that develops cardiac hypertrophy and failure, to determine if the diet influences the expression of the pathology, either in terms of early onset or severity of the disease. The expression and activity of the nuclear encoded subunits of cytochrome C oxidase (CCO; EC 1.9.3.1) are significantly decreased in copper deficiency. Therefore, a second study was conducted to determine if a similar effect results from marginal copper intake.

Methods

Semi-purified diets formulated after recommendations of the American Institute of Nutrition (AIN-76a) were used in both studies. Two strains of rats were used in the first study: 1) male Long-Evans hooded rats, and 2) male Spontaneously Hypertensive Heart Failure (SHHF) rats (Medeiros et al. 1991). From weaning, rats were fed diets containing either marginal copper (2 mg Cu/kg diet) or adequate copper levels (6 mg/kg diet), for 3, 6 or 10 months of age. Fifteen rats of each strain were placed on the respective diets. At each age, 5 animals had electrocardiograms assessed and heart tissue was processed for electron microscopy analysis at each age. In the second study, 7 male Long-Evans rats were assigned to each diet at weaning and maintained to 6 months of age. At this time, the rats were sacrificed and hearts were removed to determine CCO activity. Western blots were used to determine the amount of CCO subunits. In both studies, liver superoxide dismutase (SOD) (EC 1.15.1.1) activity, body weights and hearts weights were recorded to assess relative degree of copper status.

Results

For both studies, rats fed marginal copper had significantly lower liver SOD activity compared to the adequate copper group. For study one, this occurred in both rat strains. Neither heart weight nor body weight were influenced by diet. In both rat strains, mitochondria and myofibril volume densities were not influenced by dietary copper treatment. However, SHHF rats did develop cardiac hypertrophy independent of dietary copper treatment. Neither the onset of the hypertrophy nor the ultrastructure patterns of this strain appeared to be influenced by marginal copper intake. However, in both models studied and at all three ages, there was a significant increase in lipid droplet volume density ($P < 0.05$) compared to the adequate copper group (Figure 1). Furthermore, at 6 months of age, ECG analysis revealed decreased T wave amplitude, but increased duration in rats fed a marginal Cu diet when compared with those fed adequate diets. At 10 months of age, rats fed the marginal diets had significantly increased QRS amplitude. These changes were evident in both rat strains. The second study revealed that at various substrate levels of cytochrome C, CCO activity was significantly ($P < 0.05$) decreased (Figure 2). However, Western blots

Correct citation: Medeiros, D.M., Jalili, T., and Shiry, L. 1997. Contrasting responses of hearts from rats fed diets deficient or marginal in copper. In *Trace Elements in Man and Animals – 9: Proceedings of the Ninth International Symposium on Trace Elements in Man and Animals. Edited by* P.W.F. Fischer, M.R. L'Abbé, K.A. Cockell, and R.S. Gibson. NRC Research Press, Ottawa, Canada. pp. 616–618.

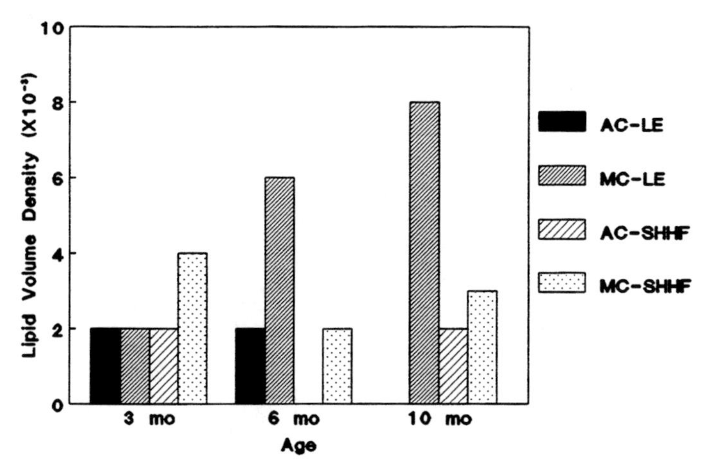

Figure 1. Lipid volume density of rats fed adequate or marginal copper until 3, 6 or 10 months of age. LE: Long-Evans strain; SHHF; Spontaneously Hypertensive Heart Failure Strain.

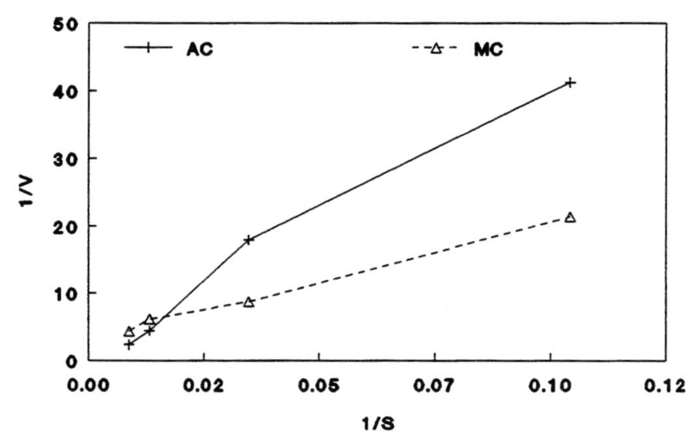

Figure 2. Cytochrome C Oxidase activity in hearts from rats fed adequate or marginal copper until 6 months of age at various substrate concentrations. V = U/mg nonmyofibrillar protein; S = µmol.

showed no apparent difference ($P > 0.05$) in the level of nuclear encoded subunits of CCO in the marginal group as compared to the adequate copper group.

Discussion

Marginal copper intake did not appear to exacerbate cardiac pathology or ECG in a genetic strain of rats which develop heart disease. Consistent with a previous study (Wildman et al. 1995), marginal dietary copper intake resulted in enhanced lipid droplet deposition in cardiac tissue in the present study. Therefore a common finding between marginal copper intake and outright copper-deficiency appears to be lipid droplet accumulation in the hearts (Medeiros et al. 1991). This could be due to the decrease activity of CCO, resulting in decreased electron transport to the extent that normal lipid oxidation in heart tissue is impaired. The decrease in CCO activity was not accompanied by a decrease in nuclear encoded subunits of CCO. The catalytic activity of CCO is due to the mitochondrial encoded subunits. However, in both studies, cardiac hypertrophy was absent. In studies with outright copper deficiency, cardiac hypertrophy in tandem with a decrease in the nuclear encoded subunits has been shown to occur in several studies from our lab (Medeiros et al. 1993, Liao et al. 1995, Chao et al. 1994). The fact that the nuclear encoded subunits of CCO did not decrease with marginal copper may explain the absence of cardiac hypertrophy.

The CCO peptides form part of the inner mitochondrial membrane. The hearts of copper deficient rats have increased mitochondrial volume density and fragmented cristae. This enhanced volume density in copper deficient hearts may be an attempt to replace the compromised mitochondria to maintain energy supply.

References

Chao, C.J.J., Medeiros, D.M., Davidson, J. & Shiry, L. (1994) Decreased levels of ATP synthase and cytochrome c oxidase subunit peptide from hearts of copper-deficient rats are not altered by the administration of dimethyl sulfoxide. J. Nutr. 124: 789–803.

Liao, Z., Medeiros, D.M., McCune, S.A., & Prohaska, L.J. (1995) Cardiac levels of fibronectin, laminin, isomyosins, and cytochrome c oxidase of weanling rats are more vulnerable to copper-deficiency than those of postweanling rats. J. Nutr. Biochem. 6: 385–391.

Medeiros, D.M., Bagby, D., Ovecka, G. & McCormick, R. (1991) Myofibrillar, mitochondrial and valvular morphological alterations in cardiac hypertrophy among copper deficient rats. J. Nutr. 121: 815–824.

Medeiros, D.M., Shiry L., Lincoln, A.J. & Prochaska, L. (1993) Cardiac non-myofibrillar proteins in copper-deficient rats:amino acid sequencing and Western blotting of altered proteins. Biol. Trace Elem. Res. 36: 271–282.

Wildman, R.E.C., Hopkins, R., Failla, M.L. & Medeiros, D.M. (1995) Marginal copper-restricted diets produce altered cardiac ultrastructure in the rat. Proc. Soc. Exp. Biol. 210: 43–49.

Discussion

Q1. Joe Prohaska, University of Minnesota, Duluth, MN, USA: In your second rat study, at the six month time point, you showed us a decrease in cytochrome oxidase activity, but no change in the subunits. Can you explain how the activity was decreased?

A. Well, my explanation would be that the activity is decreasing because of the decreased copper binding to the catalytic subunits, which are the mitochondrial-encoded subunits 1 and 2. I believe what is happening is that as the dietary copper stores are declining, you have a decrease in catalytic copper in those subunits. When it declines further, only in overt deficiency do you get an effect upon the nuclear-encoded subunits.

Q2. Joe Prohaska, University of Minnesota, Duluth, MN, USA: So the mitochondrial copper concentrations were affected by your marginal copper deficiency?

A. Yes, that's correct.

Adult Responses to Various Short-Term Dietary Copper Intakes: Insights on Human Requirements and Indicators of Status

D.B. MILNE
USDA-ARS, Grand Forks Human Nutrition Research Center, Grand Forks, ND 58202-9034, USA*

Keywords human requirements, nutritional assessment, cytochrome c oxidase, superoxide dismutase

Over the past decade there have been 12 experiments reported on experimental copper (Cu) deprivation and Cu interactions with other nutrients in adult humans (Table 1). Cu intakes have ranged between 0.4 and 1.25 mg Cu/d for periods of 36–165 d. These were alternated with or followed by intakes of 2.6 to 6.4 mg Cu/d for 29 to 48 d. Biochemical and physiological changes caused by Cu deprivation were noted in many, but not all, subjects consuming less than 1.2 mg Cu/d. These include changes in electrocardiograms (ECG) in 4 of 24 men receiving 1.02 mg Cu/d and a high fructose diet after 77 d, 1 of 7 men fed 0.8 mg Cu/d for 105 d, 3 of 6 men fed 0.6 mg Cu/d for 49 d, and 3 of 13 postmenopausal women fed 0.57 mg Cu/d for 105 d. Plasma Cu decreased with low Cu intake in only 4 of the 12 studies; it seemed to be related more to weight loss or fructose intake in two of these studies and was not as sensitive to Cu intake as Cu enzymes in blood cells. Erythrocyte superoxide dismutase (ESOD) and cytochrome c oxidase (cyt. oxid.) in platelets (PLT) or mononucleated white cells (MNC) seemed to be the most sensitive indicators to changes in Cu status. The appearance of signs of Cu deprivation were affected by other dietary factors including fructose, zinc, and sulfur amino acids. The time on the low Cu diet, amount of Cu in the diet, age and gender also seemed to affect the response to Cu deprivation.

Recovery from Cu deprivation, as indicated by recovery of ESOD activity was documented when 3 to 6.4 mg Cu/d were fed for 30 d or more, but not when 2.6 mg or less/d were fed for up to 42 d. These studies indicate that between 1.0–1.25 mg Cu/d is needed for maintenance for periods of up to 6 months and that 2.6 mg Cu/d, or less, for periods of up to 42 d is not sufficient for recovery from Cu deprivation.

Table 1. Adult responses to various Cu intakes.

Cu intake mg/d	Time d	n	Sex	Response[1]	Reference
1.25	165 d	7	F	nil	Johnson et al. 1988
0.78–0.99	108–120	7	M	4/7 ↓SOD, 2/7 ↓pl Cu, 3/7 ↓RID Cp, 1/7 ↓ENZ Cp	Milne et al. 1990
0.83	105	1	M	↓pl Cu, ↓SOD, ↑cholesterol, abnormal ECG	Klevay et al. 1984
0.78	120	2	M	↓glucose tolerance	Klevay et al. 1986
0.67	42	8	F	↓ENZ Cp, ENZ/RID Cp, ↓PLT, MNC cyt oxid, hypertension (exercise)	Milne et al. 1988; Lukaski et al. 1988
0.6–0.7	36	10	M	↓ENZ Cp, 6/10 ↓MNC cyt oxid	Milne et al. 1991
0.6	49	6	M	3/6 abnormal ECG,[2] ↓pl Cu, SOD, RID Cp, ↑glutathione	Milne, Nielsen 1993
0.57	105	13	F[3]	3/13 abnormal ECG, ↓PLT cyt oxid, ↓SOD ↓PLT Cu	Milne, Nielsen 1996

[1]ENZ Cp is ceruloplasmin measured enzymatically; RID Cp is ceruloplasmin measured by radial immunodiffusion; pl Cu is plasma copper.
[2]High fructose.
[3]Postmenopausal women.

*The U.S. Department of Agriculture, Agricultural Research Service, Northern Plains Area, is an equal opportunity/affirmative action employer and all agency services are available without discrimination.

Correct citation: Milne, D.B. 1997. Adult responses to various short-term dietary copper intakes: Insights on human requirements and indicators of status. In *Trace Elements in Man and Animals – 9: Proceedings of the Ninth International Symposium on Trace Elements in Man and Animals. Edited by* P.W.F. Fischer, M.R. L'Abbé, K.A. Cockell, and R.S. Gibson. NRC Research Press, Ottawa, Canada. pp. 619–620.

These observations are consistent with previous estimates of copper requirements of between 1.2 and 2.0 mg/d as estimated by Cu balance (Cartwright 1950, Klevay et al. 1980, Mills 1991).

Literature Cited

Cartwright, G.E. (1950) Copper metabolism in human subjects. In: Copper metabolism (McElroy, W.D., Glass, B., eds.), pp 274–314. John Hopkins Press, Baltimore.

Johnson, P.E., Stuart, M.A., Hunt, J.R., Mullen, L. & Starks, T.L. (1988) ^{64}Copper absorption by women fed intrinsically and extrinsically labeled goosemeat, goose liver, peanut butter, and sunflower butter. J. Nutr. 118: 1522–1528.

Klevay, L.M., Canfield, W.K., Gallagher, S.K., et al. (1986) Decreased glucose tolerance in two men during experimental copper depletion. Nutr. Rep. Int. 33: 371–382.

Klevay, L.M., Inman, L., Johnson, L.K., et al. (1984) Increased cholesterol in plasma in a young man during experimental copper depletion. Metabolism 33: 1112–1118.

Klevay, L.M., Reck, S.J., Jacob, R.A., et al. (1980) The human requirement for copper. I. Healthy men fed conventional American diets. Am. J. Clin. Nutr. 33: 45–50.

Lukaski, H.C., Klevay, L.M. & Milne, D.B. (1988) Effects of copper on human autonomic cardiovascular function. Eur. J. Appl. Physiol. 58: 74–80.

Mills, C.F. (1991) The significance of copper deficiency in human nutrition and health. In: Trace Elements in Man and Animals. TEMA-7, (Momcilovic, B., ed.), pp 5.1–5.4. IMI, Zagreb, Yugoslavia.

Milne, D.B., Johnson, P.E., Klevay, L.M. & Sandstead, H.H. (1990) Effect of copper intake on balance, absorption, and status indices of copper in men. Nutr. Res. 10: 975–86.

Milne, D.B., Klevay, L.M. & Hunt, J.R. (1988) Effects of ascorbic acid supplements and a diet marginal in copper on indices of copper nutriture in women. Nutr. Res. 8: 865–873.

Milne, D.B. & Nielsen, F.H. (1993) Effect of high dietary fructose on Cu homeostasis and status indicators in men during Cu deprivation. In: Trace elements in man and animals TEMA-8 (Anke, M., Meisner, D., Mills, C.F., eds.), pp 370–3. Verlag Media Touristik, Bersdorf, Germany.

Milne, D.B. & Nielsen, F.H. (1996) Effects of a diet low in copper on copper status indicators in postmenopausal women. Am. J. Clin. Nutr. 63: 358–364.

Milne, D.B., Nielsen, F.H. & Lykken, G.I. (1991) Effects of dietary copper and sulfur amino acids on copper homeostasis and selected indices of copper status in men. In: Trace elements in man and animals TEMA-7 (Momcilovic, B., ed.), pp 5.12–5.13. IMI, Zagreb, Yugoslavia.

Discussion

Q1. Joe Prohaska, University of Minnesota, Duluth, MN, USA: What if those high levels of copper were somehow inducing some superoxide dismutase in a pool that really was way beyond what we need, so that your statement that we need more than 2.6 to replete may be misleading. I would suggest that you find several ways to look at that. Since you're just using that one marker, I'd be cautious about over-interpreting your results. People with Down's Syndrome are worried about having too much superoxide dismutase.

A. Well actually, my thought was that we're looking at two mechanisms, like you suggest. I think that probably the more realistic would be the 2.6, but it would have to be taken over a much longer time. At the higher level, there is apparently a different mechanism inducing the activity.

Effects of Selenium Repletion and Depletion on Selenium Status and Thyroid Hormone Metabolism in Rats with Mild Iodine Deficiency

MUQING YI AND YIMING XIA

Department of Trace Element Nutrition, Institute of Nutrition and Food Hygiene, CAPM, Beijing 100050, China

Keywords selenium status, iodine deficiency, thyroid hormone, Keshan disease

In the 1970s it was demonstrated that Keshan disease (KD) is a cardiomyopathy located in selenium (Se) deficient areas. Se deficiency is one of the main causes for occurrence of KD. KD can be prevented by Se supplementation (Xia et al. 1994).

Since type I deiodinase (IDI) was identified as selenoenzyme (Arthur et al. 1990), after reviewing the records of KD, it was found that 80% of KD patients by autopsy suffered from thyroid gland enlargement (Zhang et al. 1983). In order to know the effects of Se deficiency on thyroid metabolism in KD patients, one of the studies was conducted as follows.

Wistar weaning rats were divided into two groups and fed diet from KD area with and without Se supplementation (0.25 and 0.01 mg Se/kg diet). The iodine contents were 0.059 mg/kg in diet and 0.010 mg/L in drinking water. The two group diets were exchanged at the end of the fourth week as the experiment start point (zero week), then continued to further 8 weeks to see the effects of Se repletion and depletion on Se status and thyroid metabolism. We supposed that the depletion group represents an impaired model by Se deficiency in KD areas; the repletion group represents an improving model by Se supplementation.

The Se contents and glutathione peroxidase (GPX) activities in RBC, plasma, liver and kidney increased or decreased in different rates with Se repletion and depletion. However, Se and GPX in brain did not decrease with Se depletion; GPX in thyroid even increased with depletion. These may indicate that Se is maintained in relatively important tissues by homeostasis and GPX keeps at high level by compensation.

The IDI in liver, kidney and thyroid increased with Se repletion. The peak points were reached in liver at the 4th week, in kidney and thyroid at the 2nd week. The IDI in thyroid increased much faster and kept at stable status after 2 weeks (Figure 1).

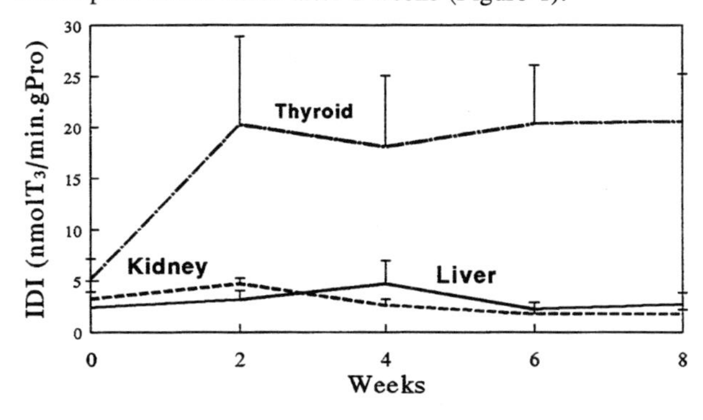

Figure 1. Type I deiodinase (IDI) in liver, kidney and thyroid during 8 weeks of selenium repletion in rats.

Correct citation: Yi, M., and Xia, Y. 1997. Effects of selenium repletion and depletion on selenium status and thyroid hormone metabolism in rats with mild iodine deficiency. In *Trace Elements in Man and Animals – 9: Proceedings of the Ninth International Symposium on Trace Elements in Man and Animals. Edited by* P.W.F. Fischer, M.R. L'Abbé, K.A. Cockell, and R.S. Gibson. NRC Research Press, Ottawa, Canada. pp. 621–623.

The IDI in liver and kidney decreased with Se depletion and reached the lowest point at the 6th week. It did not decrease in thyroid, but rather it increased. That also indicates that IDI in thyroid is more important than that in other tissues (Figure 2).

The thyroid weight (TW) and thyroidal peroxidase (TPO) activity increased gradually either in Se repletion or depletion (Figure 3). This indicates that the function of thyroid gland was impaired in both groups due to iodine deficiency.

The thyroxin (T4) decreased with Se repletion, perhaps due to the increased ID in tissues (Figure 4). But T4 also decreased with Se depletion, which may indicate that the effect of iodine deficiency was

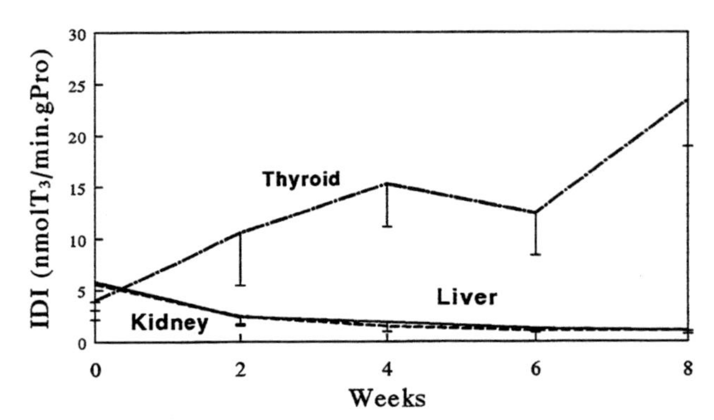

Figure 2. Type I deiodinase (IDI) in liver, kidney and thyroid during 8 weeks of selenium depletion in rats.

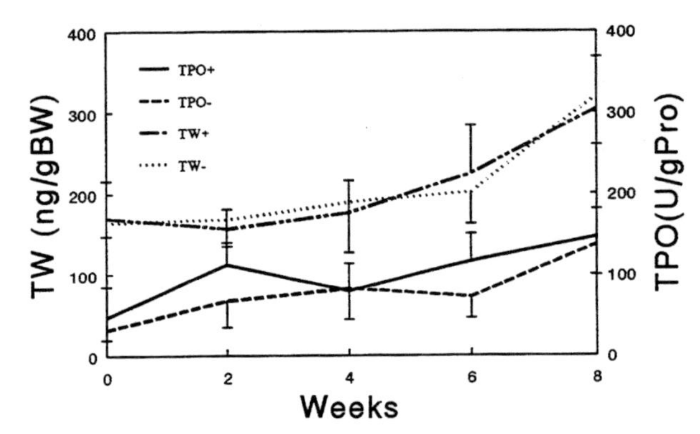

Figure 3. Thyroid relative weight (TW) and thyroidal peroxidase (TPO) during 8 weeks of selenium repletion and depletion in rats.

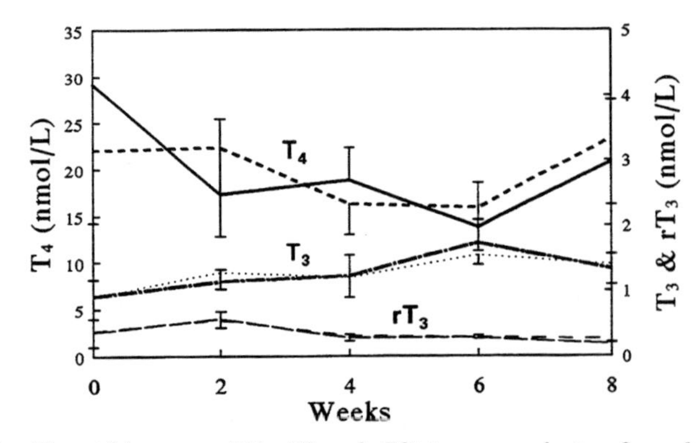

Figure 4. Thyroid hormone (T4 , T3 and rT3) in serum during 8 weeks of selenium repletion and depletion in rats.

stronger than the effect of decreased ID in tissues. However, T4 increased in both groups after the 6th week, which may indicate a compensation in response to the increased thyroid weight and thyroidal peroxidase.

The changes of T3 and rT3 in both groups were almost the same (Figure 4). The increase of T3 may have been related to decrease of T4 and was in response to the increase of IDI in thyroid. There was a significant correlation between T3 in serum and IDI in thyroid ($P < 0.001$).

Comparing the data in these two groups, it seems that TW, TPO were affected by lack of iodine. The TH was affected by both Se and iodine deficiency (Beckett et al. 1993).

In summary, Se supplementation in KD areas is necessary not only for enhancement of antioxidant defenses, but also for improvement of thyroid hormone metabolism by increasing IDI activities. However iodine supplementation in KD areas is important for treatment of thyroid function disorder and is of benefit for preventing KD. This study implicates that mild iodine deficiency may be a co-factor in the etiology of KD.

Literature Cited

Arthur, J. R., Nicol, F. & Beckett, G. J. (1990) Hepatic iodothyronine deiodinase : the role of selenium. Biochem. J. 272: 537–540.

Beckett, G. J., Nicol, F., Rae, P. W. H., Beech, S. & Arthur, J. R. (1993) Effects of combined iodine and selenium deficiency on thyroid hormone metabolism in rat. Am. J. Clin. Nutr. 57 : 240s–243s.

Xia, Y., Piao, J., Hill, K. E. & Burk, R. F. (1994) Keshan disease and selenium status of populations in China. In : Selenium and Human Health (Burk, R.F., ed.), pp. 181–196. Springer-Verlag, New York, NY.

Zhang, H., Bu, X. & Wei, Y. (1983) RIA of serum T4 among Keshan disease patients. Bull. Univ. Med. Sci. Mem. Dr. Norman Bethune 9: 50–51.

Discussion

Q1. Roger Sunde, University of Missouri, Columbia, MO, USA: Could you expand on the analysis of the tissue-bank samples? How compelling was that evidence?

A. We originally didn't think about iodine in Keshan Disease. This idea came from the discovery of the Type I deiodinase as a selenoenzyme, by Dr. Arthur. This prompted us to take a retrospective look, and we found that 80% of Keshan Disease cases reported thyroid enlargement. The problem is, we don't know the exact iodine requirement for humans, so it's a little more difficult to say if people are deficient.

Q2. John Arthur, Rowett Research Institute, Aberdeen, Scotland: Could you comment, please, on the hearts of the animals in your experiment?

A. We are now working on that part, and will report those results at the next meeting. All I can tell you right now is that we just saw the seleno-glutathione peroxidase decreased, and increased as expected from depletion and repletion.

Copper Deficiency Causes *in vivo* Oxidative Modification of Erythrocyte Membrane Proteins in Rats

W.T. JOHNSON,[†] T.P. LABERGE,[‡] AND K.A. SUKALSKI[‡]

[†]USDA, ARS, Grand Forks Human Nutrition Research Center, and [‡]University of North Dakota School of Medicine and Health Sciences, Grand Forks, North Dakota, 58202, USA[*]

Keywords copper deficiency, erythrocyte, protein carbonyl, oxidant injury

Erythrocytes are exposed not only to high concentrations of oxygen but also to superoxide anion that is constantly generated from the autoxidation of hemoglobin (Misra et al. 1972). Although the intracellular concentration of superoxide anion in erythrocytes is normally controlled by Cu,Zn superoxide dismutase (SOD), the protection provided by this enzyme against harmful effects of superoxide may be compromised when its activity is reduced by copper deficiency. Thus, cellular components of erythrocytes may be targets for oxidative modification during copper deficiency. Although previous studies have indicated that lipid peroxidation and the susceptibility of lipids to oxidation are increased during copper deficiency (Johnson et al. 1992), the possibility that copper deficiency may promote the oxidation of membrane proteins has not been investigated. The objective of this study was to determine if proteins associated with erythrocyte membranes could be damaged by the oxidative stress produced by copper deficiency.

Materials and Methods

Two groups of male, Sprague-Dawley rats (mean weights 52 g) were fed casein-based diets containing either 0.5 μg Cu/g (CuD) or 6.0 μg Cu/g (CuA) for 35 d. Erythrocytes obtained from these animals were separated into low and high density cells by centrifugation through a density gradient formed with Percoll and erythrocyte membranes were prepared from the two cell populations by hypotonic lysis (Dodge et al. 1965). Erythrocyte membrane proteins were treated with 2,4-dinitrophenylhydrazine and dinitrophenyl (DNP) derivatives of the protein carbonyls were detected on Western blots by screening with antibodies against the DNP groups (Shacter et al. 1994). Chemiluminescence was used to visualize the DNP-proteins on the blots. Proteins were also visualized by staining the blots with Amido Black. Areas of the peaks representing DNP-proteins and proteins stained with Amido Black were obtained from densitometry scans. The integrated areas of bands for both the chemiluminescent exposures and the corresponding amido black-stained proteins were normalized to a standard oxidized lactate dehydrogenase sample detected by the corresponding procedure. The relative carbonyl content of a specific protein was calculated by subtracting the ratio of normalized chemiluminescence area to normalized amido black area for the non-derivatized protein from the ratio for the DNP-protein.

Results and Discussion

DNP derivatized carbonyls were detected primarily in two erythrocyte membrane proteins whose molecular weights, determined from their electrophoretic migration, were 240 kDa and 210 kDa. Also, these two proteins constituted 30% of the total erythrocyte membrane protein. Based on their molecular

*U.S. Department of Agriculture, Agricultural Research Service, Northern Plains Area is an equal opportunity/affirmative action employer and all agency services are available without discrimination.

Correct citation: Johnson, W.T., LaBerge, T.P., and Sukalski, K.A. 1997. Copper deficiency causes *in vivo* oxidative modification of erythrocyte membrane proteins in rats. In *Trace Elements in Man and Animals – 9: Proceedings of the Ninth International Symposium on Trace Elements in Man and Animals. Edited by* P.W.F. Fischer, M.R. L'Abbé, K.A. Cockell, and R.S. Gibson. NRC Research Press, Ottawa, Canada. pp. 624–625.

weights and abundance, the 240 kDa and 210 kDa proteins are most likely the α- and β-chains, respectively, of spectrin (Cohen 1983). Table 1 shows the effects of copper deficiency and cell density on α- and β-spectrin. Regardless of cell density or dietary treatment, the carbonyl content of α-spectrin was higher ($P < 0.05$, Friedman's ANOVA) than the carbonyl content of β-spectrin which suggests that α-spectrin is more susceptible to oxidative damage than β-spectrin. The carbonyl content of α-spectrin tended to be higher in high density cells than in low density cells, but this difference was not statistically significant. Because high density erythrocytes represent an older cell population than low density erythrocytes, the similarity in the carbonyl content of high and low density cells suggests that protein oxidation is well controlled during erythrocyte aging. However, regardless of cell density, copper deficiency increased ($P < 0.05$, Friedman's ANOVA) the carbonyl content of both α- and β-spectrin.

These data indicate that erythrocyte spectrin is particularly susceptible to oxidative modification during copper deficiency. Because copper deficiency reduces SOD activity in erythrocytes, the present findings suggest that spectrin may be a specific target for oxidative damage caused by superoxide anion.

Table 1. Relative carbonyl content of α- and β-spectrin in high and low density erythrocytes from rats fed either copper-deficient (CuD) or copper-adequate (CuA) diets.

	α-Spectrin	β-Spectrin
Low Density Erythrocytes		
CuD	1.53 ± 0.78[a,b]	0.45 ± 0.22[b]
CuA	1.04 ± 0.47[a]	0.31 ± 0.32
High Density Erythrocytes		
CuD	1.88 ± 1.45[a,b]	0.48 ± 0.24[b]
CuA	1.58 ± 1.26[a]	0.31 ± 0.19

Values are means ± SD for six membrane preparations in each group. [a]Significantly greater than β-spectrin. [b]Significantly greater than CuA.

References

Cohen, C.M. (1983) The molecular organization of the red cell membrane skeleton. Semin. Hematol. 20: 141–158.

Dodge, J.T., Mitchell, C. & Hanahan, D.J. (1963) The preparation and chemical characteristics of hemoglobin-free ghosts of human erythrocytes. Arch. Biochem. Biophys. 100: 119–130.

Johnson, M.A., Fischer, J.G. & Kays, S.E. (1992) Is copper an antioxidant nutrient? Critical Reviews in Food Science and Nutrition 32: 1–31.

Misra, H.P. & Fridovich, I. (1972) The generation of superoxide radical during the autoxidation of hemoglobin. J. Biol. Chem. 247: 6960–6962.

Shacter, E., Williams, J.A., Lim, M. & Levine, R.L. (1994) Differential susceptibility of plasma proteins to oxidative modification: examination by Western blot immunoassay. Free Rad. Biol. Med. 17: 429–437.

Absence of Hypertension in Adult Copper-Deficient Rats Despite Vascular Alteration: Potential Renal Compensation*

JACK T. SAARI

U.S. Department of Agriculture, Agricultural Research Service, Grand Forks Human Nutrition Research Center, Grand Forks, North Dakota 58202, USA

Keywords copper deficiency, blood vessel, kidney, blood pressure

Prior studies have indicated that dietary copper (Cu) restriction causes high blood pressure in adult rats (Klevay 1987, Medeiros 1987). The aim of this study was to examine the cause of this Cu deficiency-induced hypertension. Two ways that Cu deficiency may affect blood pressure are by effects on blood vessel vasoactivity and on renal control of blood volume. These possibilities are suggested by studies indicating that nitric oxide mediated vasodilation is impaired (Saari 1992) and that kidney function is altered in male, weanling Cu-deficient rats (Reeves et al. 1990). The possibility that these defects of Cu deficiency contribute to development of hypertension in older rats was examined by testing aortic response to a nitric oxide donor and by measurement of kidney variables related to blood pressure regulation.

Materials and Methods

Male Sprague-Dawley rats (180–200 g) were fed a Cu-deficient (0.4–0.6 µg/g diet; n = 5) or a Cu-adequate diet (5.5–6.0 µg/g diet; n = 5) for 29 weeks. Blood pressure was measured by tail cuff at 2–3 week intervals between weeks 15 and 29. Urine was collected from rats placed in metabolic cages the night before blood vessel testing. Rats were anesthetized, blood was drawn from the vena cava, and hearts, livers and segments of descending thoracic aorta were excised. Each aorta was placed in a tissue bath and attached to a force transducer. Vasodilatory response was tested by first contracting the vessel with phenylephrine (3×10^{-7} M) and then adding increasing concentrations of the nitric oxide donor sodium nitroprusside (10^{-9}–10^{-6} M). Blood cells were counted with a cell counter. Urine and serum sodium (Na) were determined by inductively coupled argon plasma emission spectroscopy (ICAP) and, with urine volume, were used to calculate Na clearance (C_{Na}) and excretion (Ex_{Na}). Urine and serum creatinine, used to calculate glomerular filtration rate (GFR), were determined by reaction with picrate. Livers were digested with nitric acid and hydrogen peroxide and assayed for copper by ICAP.

Results

Cu deficiency was confirmed by observation of reduced liver Cu concentrations and anemia (Table 1). Mean systolic blood pressures measured between weeks 15 and 29 were in the range of 115–137 mm Hg and failed to show a relative hypertension in Cu-deficient rats (Table 2; repeated-measures ANOVA, $p > 0.05$). Isolated aortas of Cu-deficient rats showed a depressed relaxation to nitroprusside (Figure 1). Although GFR was unchanged in Cu-deficient rats, both C_{Na} and Ex_{Na} were higher than in Cu-adequate rats (Table 1).

*The U.S. Department of Agriculture, Agricultural Research Service, Northern Plains Area, is an equal opportunity/affirmative action employer and all agency services are available without discrimination.

Correct citation: Saari, J.T. 1997. Absence of hypertension in adult copper-deficient rats despite vascular alteration: Potential renal compensation. In *Trace Elements in Man and Animals – 9: Proceedings of the Ninth International Symposium on Trace Elements in Man and Animals. Edited by* P.W.F. Fischer, M.R. L'Abbé, K.A. Cockell, and R.S. Gibson. NRC Research Press, Ottawa, Canada. pp. 626–627.

Table 1. Characteristics (mean ± SE) of rats fed Cu-adequate and Cu-deficient diets.

Diet	Bodywt g	Liver Cu nmol/g	Hct %	GFR mL/h	C_{Na} µL/h	Ex_{Na} µmol/h
Cu-ad	469 ± 15	256 ± 18	46 ± 1	94 ± 20	26 ± 11	3.5 ± 1.5
Cu-def	462 ± 13	94 ± 7*	34 ± 2*	76 ± 15	119 ± 30*	15.5 ± 3.9*

* Significantly different from value for Cu-adequate rats, $p<0.05$.

Table 2. Systolic blood pressures (mm Hg, mean ± SE) in Cu-adequate and Cu-deficient rats.

	Week of experiment						
Diet	15	17	19	22	24	27	28–29
Cu-ad	125 ± 5	127 ± 3	115 ± 6	128 ± 4	131 ± 7	131 ± 4	137 ± 2
Cu-def	129 ± 4	129 ± 4	128 ± 2	136 ± 4	138 ± 4	130 ± 3	126 ± 4

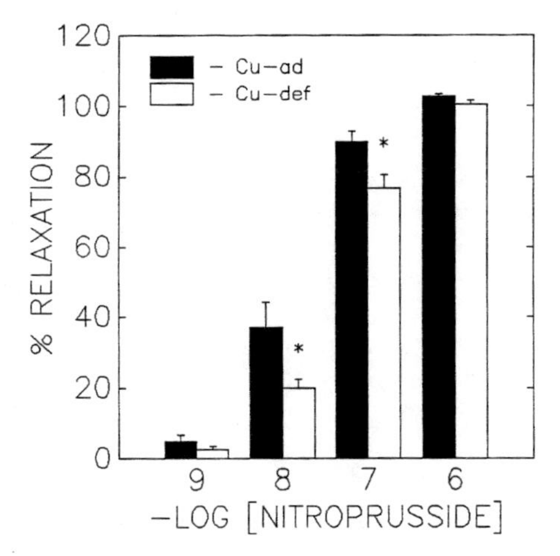

*Figure 1. Effect of Cu-adequate and Cu-deficient diets on the percentage of relaxation (± SE) to sodium nitroprusside in phenylephrine-contracted aortas. * Significantly different from Cu-adequate value, $p < 0.05$.*

Discussion

Depressed blood vessel reactivity to nitroprusside in Cu-deficient rats suggested that higher blood vessel resistance should have produced higher blood pressures than in Cu-adequate rats. However, the finding of higher sodium clearance and sodium excretion in Cu-deficient rats suggests that absence of high blood pressure may be explainable in part by blood volume reduction caused by loss of extracellular sodium.

Literature Cited

Klevay, L.M. (1987) Hypertension in rats due to copper deficiency. Nutr. Rep. Int. 35: 999–1005.

Medeiros, D.M. (1987) Hypertension in the Wistar-Kyoto rat as a result of post-weaning copper restriction. Nutr. Res. 7: 231–235.

Reeves, P.G., Noordewier, B. & Saari, J.T. (1990) Effect of copper deficiency and cis-diaminedichloroplatinum (II) treatment on the activities of renal microvillar enzymes in rats. J. Trace Elem. Electrolytes Health Dis. 4: 11–19.

Saari, J.T. (1992) Dietary copper deficiency and endothelium-dependent relaxation of rat aorta. Proc. Soc. Exp. Biol. Med. 200: 19–24.

Copper and Zinc Levels in Patients with Chronic Pancreatitis

K. LINKE[†] AND G.W. MIELCARZ[‡]

[†]Department of Gastroenterology and [‡]Department of General Chemistry, K. Marcinkowski University of Medical Sciences, 60-780 Poznan, Poland

Keywords copper, zinc, humans, pancreatic juice

Numerous essential biochemical reactions generate highly reactive, oxygen-containing molecules called free radicals, which have been implicated in beta cell destruction. Copper and zinc are essential nutrients and an integral part of the antioxidant enzymes copper,zinc-superoxide dismutase (SOD) and ceruloplasmin. Pancreatic antioxidant enzymatic defences provide significant endogenous protection from these free radicals (Taylor et al. 1988). Inflammatory diseases such as chronic pancreatitis may increase copper and zinc requirements. Some studies have attributed an important role to the pancreas in the homeostasis of zinc (McClain 1990). It has been shown that patients with chronic pancreatitis have decreased levels of zinc in pancreatic juice (Sullivan et al. 1965, Gjorup et al. 1991 and Tillman et al. 1981). However, many results are contradictory. Ishihara (1983) did not find significant changes in these metals between males and females or with pathological changes in pancreas. The purpose of this study was to evaluate the hypothesis that low copper and zinc status promote progression of chronic pancreatitis (CP).

Methods

Human pancreatic juice was sampled by endoscopic cannulation after administration of secretin and pancreozymin (intravenous [i.v.], 1 mu/kg). There were 21 subjects (9 control and 12 with chronic pancreatitis) and the means of their ages were: in the control group 42.3 ± 9.0 and in chronic pancreatitis group 52.6 ± 10.2. None had symptoms suggestive of disturbances in endocrine or exocrine functions of the pancreas. The concentrations of copper and zinc in plasma and pancreatic juice were determined by atomic absorption spectroscopy (Varian GTA-96). Vitamin C was determined by the method of Homolka (1971), and immunoglobulin G (IgG), and albumin by the turbidometry method using conventional kits (Boehringer Mannheim GmbH).

Results

The concentrations of copper and zinc, immunoglobulin G (IgG), albumin and vitamin C in plasma and pancreatic juice (PJ), and sodium, potassium and bicarbonate in pancreatic juice are presented in Table 1.

Discussion

The results indicated that under fasting conditions, zinc in pancreatic juice was significantly lower ($p < 0.05$) in patients with chronic pancreatitis than in the control group. The low level of zinc in pancreatic juice has also been observed by other authors. Plasma zinc was higher in the control group ($p < 0.05$). Low serum levels of zinc have also been observed in patients with chronic alcoholism. However patients with chronic alcoholic pancreatitis had not significant or even higher levels of zinc in serum (Kondo et al. 1989). Copper in pancreatic juice was higher in normal subjects but the differences were not significant. Alvarez (1989) reported that copper deficiency caused a slight reduction in the flow of pancreatic juice and significantly decreased chloride output in rabbit. Kimura (1987) found that copper deficiency caused

Correct citation: Linke, K., and Mielcarz, G.W. 1997. Copper and zinc levels in patients with chronic pancreatitis. In *Trace Elements in Man and Animals – 9: Proceedings of the Ninth International Symposium on Trace Elements in Man and Animals. Edited by* P.W.F. Fischer, M.R. L'Abbé, K.A. Cockell, and R.S. Gibson. NRC Research Press, Ottawa, Canada. pp. 628–629.

Table 1.

Variables	Control n = 9	Chronic Pancreatitis n = 12	p-value*
Plasma zinc (μmol/L)	13.2 ± 1.4	11.4 ± 1.7	0.0194
Plasma copper (μmol/L)	14.8 ± 2.5	16.1 ± 2.5	NS
PJ zinc (μmol/L)	11.7 ± 8.1	4.8 ± 4.6	0.0241
PJ copper (μmol/L)	1.5 ± 1.9	0.77 ± 0.6	NS
IgG plasma (g/L)	11.7 ± 1.9	14.4 ± 4.8	0.0358
Plasma albumin (g/L)	47 ± 4	48 ± 5	NS
PJ albumin (g/L)	0.02 ± 0.01	0.10 ± 0.08	0.0196
Plasma vitamin C (μmol/L)	38.2 ± 7.9	41.2 ± 11.1	NS
PJ pH	8.23 ± 0.14	8.31 ± 0.16	NS
PJ sodium (mmol/L)	138 ± 4	134 ± 8	NS
PJ potassium (mmol/L)	3.5 ± 0.3	3.7 ± 0.8	NS
PJ bicarbonate (mmol/L)	113.9 ± 11.2	92.8 ± 16.8	0.0070

x ± SD; *significance of difference estimated by t-test.

disturbances in enzyme synthesis, and selective destruction of acinar cells in rat. We found significantly increased level of albumin and immunoglobulin G in pancreatic juice of patients with CP ($p < 0.05$). The concentration of zinc in pancreatic juice was well correlated with plasma albumin ($r = 0.68$ $p < 0.05$) in the control group, but in patients with CP this correlation was not significant. The results suggest that the decreased level of zinc may be involved in mechanisms associated with chronic pancreatitis.

Literature Cited

Alvarez C., Garcia J.F. & Lopez M.A. (1989) Exocrine pancreas in rabbits fed a copper-deficient diet: structural and functional studies. Pancreas 4(5): 543–549.

Gjorup I., Petronijevic L., Rubinstein E., Andersen B., Worning H. & Burchart F. (1991) Pancreatic secretion of zinc and copper in normal subjects and in patients with chronic pancreatitis. Digestion 49: 161–166.

Homolka J. (1971) Biochemia kliniczna interpretacje i metodyka. Warsaw, PZWL 542–543.

Ishihara N., Yoshida A. & Koizumi M. (1987) Metal concentrations in human pancreatic juice. Arch. Environ. Health 42(6): 356–360.

Kimura T., Sumii T., Oogami Y., Yamaguchi H., Ibayashi H. & Kinjo M. (1987) Morphological and biochemical changes in the pancreas of copper deficient rats. Gastroenterol-Japn 22(4): 480–486.

McClain C.J. (1990). The pancreas and zinc homeostasis. J. Lab. Clin. Med. 116: 275–276.

Sullivan J.F., O'Grady J. & Lankford H. G. (1965) The zinc content of pancreatic secretion. Gastroenterology 48: 438–443.

Taylor C.G., Bettger W.J. & Bray T.M. (1988) Effect of dietary zinc or copper deficiency on the primary free radical defence system in rats. J.Nutr. 118: 613–621.

Tillman R., Steinberg W., Ahmed E., Duncan D. & Toskes P. (1981) Zinc content of duodenal secretions following stimulation with secretin-pancreozymin: a new sensitive test of pancreatic function. Gastroenterology, 80: 1304.

Effect of Low Copper Diet on Opioid Peptides in Young Men

S.J. BHATHENA,[†] M.J. WERMAN,[‡] AND J.R. TURNLUND[¶]

[†]USDA, ARS, Beltsville Human Nutrition Research Center, Beltsville, MD 20705, USA; [‡]Department of Food Engineering and Biotechnology, Technion-Israel Institute of Technology, Haifa 32000, Israel; and [¶]Western Human Nutrition Research Center, San Francisco, CA 94129, USA

Keywords low copper diet, opioid peptides, cardiac abnormality, humans

Both excess and very low intakes of copper affect the central nervous system and the functions of several other organs via alterations in the endocrine and neuroendocrine systems. In animals copper deficiency syndrome is characterized by biochemical, morphological and endocrine alterations accompanied by neurological and cardiac disorders. Several symptoms of relative copper deficiency including cardiac abnormalities have been observed (Klevay 1983, 1984; Reiser et al. 1985) in humans fed low copper diets. The activity of several copper-dependent enzymes involved in catecholamine, opioid peptide and neuropeptide synthesis and processing is decreased in copper deficiency. Since opioid peptides and neuropeptides are involved in cardiac function (Bhathena 1989), it is possible that some of the cardiac abnormalities in copper deficiency could be due in part to alterations in one or more of the opioid peptides or neuropeptides.

In animals, the severity of copper deficiency syndrome is aggravated by fructose compared to starch. Several metal ions, especially zinc and iron, antagonize the action of copper, and diets high in zinc and iron aggravate the signs of copper deficiency. In an earlier study, feeding low copper diets with high zinc and high fructose to humans altered plasma opioid peptide levels and produced cardiac abnormalities in 4 out of 24 subjects (Bhathena et al. 1986). Whether these changes were due to copper deficiency alone or the result of interaction between low copper and high zinc and high fructose is not clear. To address this question, in the present study 11 healthy young men were fed low copper diets with adequate zinc and no added fructose.

The subjects were fed diets with 0.66 mg/d copper for 24 d (marginal copper period) (MP1) followed by 0.38 mg/d copper for 42 d (low copper period) (MP2) and then replenished with 2.49 mg/d copper for the last 24 d (MP3). Blood was collected at the beginning of the study (baseline) and at the end of each feeding period. Copper status was assessed by measuring plasma copper and ceruloplasmin and urinary copper excretion. Plasma opioid peptides, B-endorphin (BEN), leu-enkephalin (LE) and met-enkephalin (ME) were measured by radioimmunoassay.

Plasma copper and ceruloplasmin and urinary copper excretion were lower at the end of MP2 compared to baseline and MP1. All parameters increased with copper repletion (MP3) indicating that the subjects responded to the level of copper in the diet (Turnlund et al. 1994). No significant differences were observed in any of the opioid peptides during any of the periods. Thus the changes in plasma opioid peptides did not respond to dietary copper as did other parameters of copper status such as plasma copper and ceruloplasmin and urinary copper excretion. Further, none of the subjects showed any sign of cardiac abnormalities.

In a previous study we observed significant decreases in plasma LE and ME and a small increase in BEN in middle aged male subjects fed marginally low copper diets (Bhathena et al. 1986). The dietary copper was lowered from 1.3 mg/d to 1.03 mg/d. There was however no significant alteration in copper status as measured by plasma copper and ceruloplasmin. Only erythrocyte superoxide dismutase activity

---

Correct citation: Bhathena, S.J., Werman, M.J., and Turnlund, J.R. 1997. Effect of low copper diet on opioid peptides in young men. In *Trace Elements in Man and Animals – 9: Proceedings of the Ninth International Symposium on Trace Elements in Man and Animals. Edited by* P.W.F. Fischer, M.R. L'Abbé, K.A. Cockell, and R.S. Gibson. NRC Research Press, Ottawa, Canada. pp. 630–631.

showed small but significant decrease when subjects consumed fructose and 1.03 mg/d copper (Reiser et al. 1985). Four out of 24 subjects also showed cardiac abnormalities including abnormal electrocardiogram, acute myocardial infarction and tachycardia. It is important to note that in the present study the subjects were younger and the level of dietary copper was lowered to 0.38 mg/d but the zinc content was adequate and no fructose was added. There was significant decline in usual parameters of copper status such as plasma copper and ceruloplasmin and urinary copper excretion (Turnlund et al. 1994) but no significant changes in opioid peptides were observed.

Thus the results from these two studies suggest that cardiac abnormalities occur only when low copper diets are fed with high zinc and high fructose levels even though the decrease in copper is only marginal. Secondly the changes in plasma opioid peptides occur only when the low copper diet is fed along with high zinc and high fructose. It is not clear whether changes in opioid peptides cause cardiac abnormalities or vice versa. However since opioid peptides are involved in cardiac function, it is possible that the changes in opioid peptides may be primary which then leads to cardiac abnormalities. This is further supported by the fact that the severity of copper deficiency in animals including cardiac abnormalities is worsened by high fructose (Fields et al. 1984, Redman et al. 1988) and that the high ratio of zinc to copper is possibly more responsible for cardiac abnormality (Klevay 1975, 1983).

References

Bhathena, S.J. (1989) Recent advances on the role of copper in neuroendocrine and central nervous system. Med. Sci. Res. 17: 537–542.

Bhathena, S.J., Recant, L., Voyles, N.R., Timmers, K.I., Reiser, S., Smith, Jr., J.C. & Powell, A.S. (1986) Decreased plasma enkephalins in copper deficiency in man. Am. J. Clin. Nutr. 43: 42–46.

Fields, M., Ferretti, R.J., Reiser, S. & Smith, Jr., J.C. (1984) The severity of copper deficiency in rats is determined by the type of dietary carbohydrate. Proc. Soc. Exp. Biol. Med. 175: 530–537.

Klevay, L.M. (1975) Coronary heart diseases: the zinc/copper hypothesis. Am. J. Clin. Nutr. 28: 764–774.

Klevay, L.M. (1983) Copper and ischemic heart disease. Biol. Trace Elem. Res. 5: 254–255.

Redman, R.S., Fields, M., Reiser, S. & Smith, Jr., J.C. (1988) Dietary fructose exacerbates the cardiac abnormalities of copper deficiency in rats. Atherosclerosis 74: 203–214.

Reiser, S., Smith, Jr., J.C., Mertz, W., Holbrook, J.T., Scholfield, D.J., Powell, A.S., Canfield, W.K. & Canary, J.J. (1985) Indices of copper status in humans consuming a typical American diet containing either fructose or starch. Am. J. Clin. Nutr. 42: 242–251.

Turnlund, J.R., Keen, C.L., Sakanashi, T.M., Jang, A.M., Keyes, W.R. & Peiffer, G.L. (1994) Low dietary copper and the copper status of young men. FASEB J. 8: A820 (Abstract).

The Use of Stable Isotopes to Determine Adaptive Responses to Different Dietary Intakes of Copper*

L.J. HARVEY,† G. MAJSAK-NEWMAN,† S.J. FAIRWEATHER-TAIT,† J. EAGLES,† D.J. LEWIS,‡ AND H. CREWS‡

†Nutrition, Diet & Health Department, Institute of Food Research, Norwich Research Park, Colney, Norwich NR4 7UA, UK, and ‡CSL Food Science Laboratory, Norwich Research Park, Colney, Norwich NR4 7UQ, UK

Keywords copper, stable isotope, cardiovascular health, ICP-MS

Extremes of copper intake may be associated with metabolic perturbations which increase the risk of cardiovascular disease, and a combination of low intake of copper and high intake of fructose may exacerbate these effects. The levels of copper which impair health are uncertain, as highlighted in the latest reports of committees making dietary recommendations. This study was designed to characterise the effects of different intakes of copper in order to assess dietary requirements.

Methods

In this study, 12 healthy male volunteers (aged 18–60) are participating in a longitudinal dietary intervention investigating the metabolic effects of medium (1.6 mg/d), low (0.7 mg/d) and high (6.0 mg/d) intakes of copper over eight week periods. The volunteers live in a residential Human Nutrition Unit and are fed low copper diets supplemented with copper as appropriate. All other nutrients meet the Dietary Reference Values and other dietary variables are within customary limits. During the course of each dietary period, risk factors for cardiovascular disease and putative measures of copper homoeostasis are monitored. After 6 weeks of each diet the volunteers receive an oral dose of ^{65}Cu (3 mg), either with or without fructose in order to measure the efficiency of copper absorption, endogenous excretion, and the interaction between copper and fructose. Blood samples are taken just before the administration of the isotope, and 90 min and 3 d afterwards, and complete collections of urine and faeces are made for a period of 2 weeks. Previous studies (Lyon et al. 1995) have shown that ^{65}Cu enrichment in the serum peaks 1–2 h after oral administration, with a secondary peak approximately 3 d later.

The total quantity in urine and faeces and the form and concentration of copper in the blood are determined by inductively coupled plasma-mass spectrometry (ICP-MS) (Lyon and Fell 1990). Prior to administration of the ^{65}Cu stable isotope the volunteers undergo an oral fructose tolerance test, receiving a 25 g dose of fructose labelled with 100 mg of U-^{13}C-fructose. Breath samples are taken for 6 h and $^{13}CO_2/^{12}CO_2$ ratios are determined by gas isotope ratio mass spectrometry.

Results and Discussion

Preliminary data have been obtained for two volunteers. Total caeruloplasmin was found to be unaffected by copper intake in either volunteer. The maximum 5 μM ADP-stimulated platelet aggregation was greater during the low copper intake than during the medium or high intakes (Table 1) (Milne and Gallagher 1991).

The values for plasma lipoproteins and triglycerides are shown in Figure 1.

The LDL/HDL ratios are shown in Figure 2.

*Funding: Ministry of Agriculture Fisheries & Food and European Commission.

Correct citation: Harvey, L.J., Majsak-Newman, G., Fairweather-Tait, S.J., Eagles, J., Lewis, D.J., and Crews, H. 1997. The use of stable isotopes to determine adaptive responses to different dietary intakes of copper. In *Trace Elements in Man and Animals – 9: Proceedings of the Ninth International Symposium on Trace Elements in Man and Animals. Edited by* P.W.F. Fischer, M.R. L'Abbé, K.A. Cockell, and R.S. Gibson. NRC Research Press, Ottawa, Canada. pp. 632–635.

Table 1. Biochemical data after 6 weeks equilibration at each dietary intake level of copper.

	Subject No.*	Medium Cu	Low Cu	High Cu
Maximum 5 μM ADP-stimulated platelet aggregation (%)	1	82	93	81
	2	74	77	72
Plasma copper conc. (μmol/L)	1	16.0	17.2	17.9
	2	13.5	12.0	14.5
Total caeruloplasmin conc. (g/L)	1	0.23	0.25	0.24
	2	0.15	0.16	0.17

*1: 28 years, 98 kg; 2: 21 years, 68 kg.

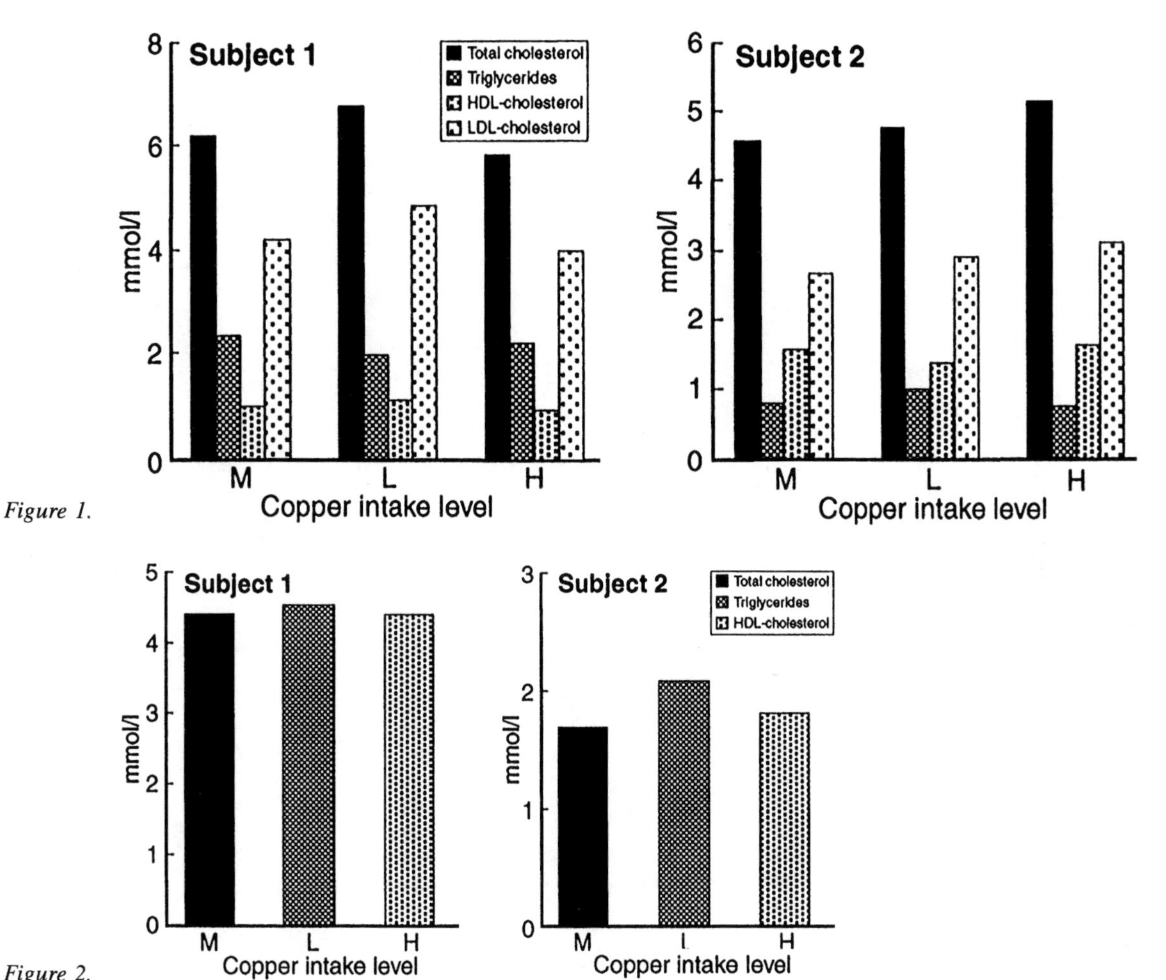

Figure 1.

Figure 2.

Both volunteers had a higher ratio during the low copper dietary period. After a period of 6 weeks on the medium copper diet, subject 1 was given an oral dose of ^{65}Cu (3.0 mg) without fructose. The appearance of the ^{65}Cu oral dose in the faeces is shown in Figure 3.

Approximately 35% of the unabsorbed ^{65}Cu was excreted the day after ingestion and the remainder within 5 d. However, the $^{65}Cu/^{63}Cu$ ratio remained above the baseline level for at least 43 d, presumably reflecting endogenous excretion of absorbed ^{65}Cu. Endogenous losses from absorbed ^{65}Cu will be followed in future volunteers by giving 1 mg of a non-absorbable rare earth element (holmium) with the ^{65}Cu oral dose and dysprosium 5 d later. Other studies with isotopes of zinc (Schuette et al. 1993) and iron (Fairweather-Tait et al. 1997) have shown that the rare earth element excretory pattern is similar to inorganic elements.The appearance of the second dose will be used to mark the period for calculation of endogenous losses of copper. Prior to receiving the ^{65}Cu oral dose, subject 1 underwent an oral fructose tolerance test. Maximum enrichment of $^{13}CO_2$ in breath samples occurred 120 min after administration of fructose (Figure 4).

Conclusions

This study will provide information on the effect of copper intake on risk factors for cardiovascular health. The efficiency of copper absorption can be measured using ^{65}Cu, and in future it will be possible to calculate endogenous copper losses by collecting faecal samples after administration of an oral dose of ^{65}Cu stable isotope with rare earth elements. Interactions between copper and fructose can be determined using oral doses of U-^{13}C-fructose.

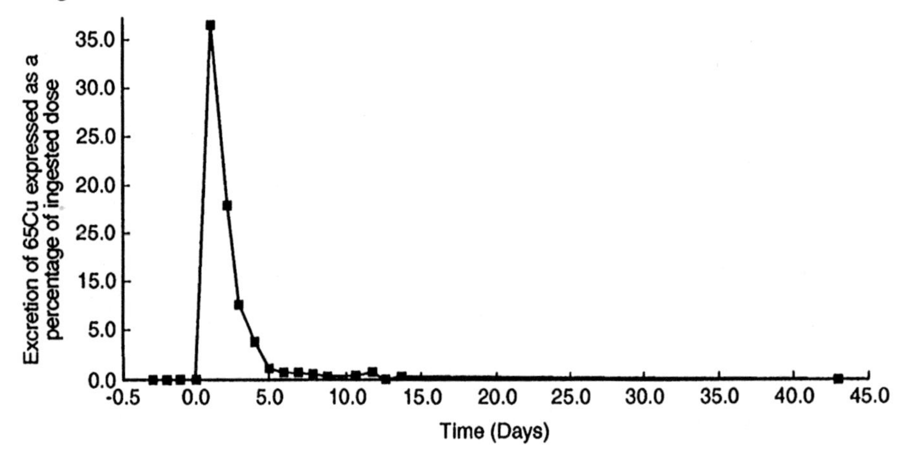

Figure 3.

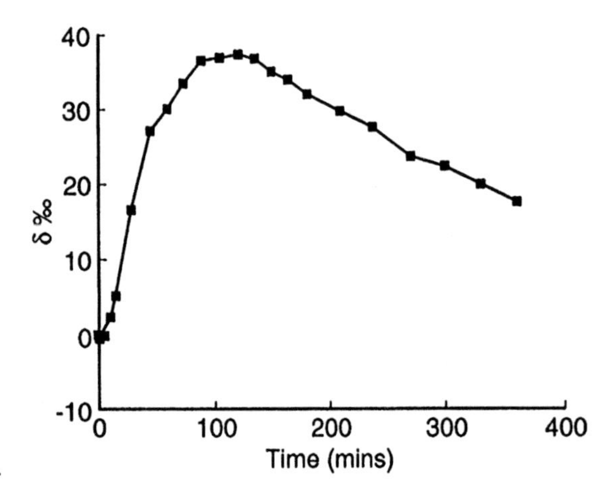

Figure 4.

References

Fairweather-Tait, S.J., Minihane, A.M., Eagles, J., Owen, L. & Crews, H. (1997) Can rare earth elements be used as non-absorbable faecal markers in studies of iron absorption? In: Trace Elements in Man and Animals – 9: Proceedings of the Ninth International Symposium on Trace Elements in Man and Animals. (Fischer, P.W.F., L'Abbé, M.R., Cockell, K.A., and Gibson, R.S., eds.). NRC Research Press, Ottawa, Canada. pp. 270–272.

Lyon, T.D.B. & Fell, G.S. (1990) Isotopic composition of copper in serum by inductively coupled plasma mass spectrometry. J. Anal. Atom. Spec. 5, 135–137.

Lyon, T.D.B., Fell, G.S., Gaffney, D., McGaw, B.A., Russell, R.I., Park, R.H.R., Beattie, A.D., Curry, G., Crofton, R.J., Gunn, I., Sturniolo, G.S., D'Inca, R. & Patriarca, M. (1995) Use of copper stable isotope (^{65}Cu) in the differential diagnosis of Wilson's disease. Clin. Sci. 88, 727–732.

Milne, D.B. & Gallagher, S.K. (1991) Effect of dietary copper on platelet volume and aggregation in men. FASEB J. 5:A1076

Schuette, S.A., Janghorbani, M., Young, V.R. & Weaver, C.M. (1993) Dysprosium as a nonabsorbable marker for studies of mineral absorption with stable isotope tracers in human subjects. J. Am. Coll. Nutr. 12 (3) 307–315.

The Effects of Zinc Status on the Osmotic Fragility of Human Erythrocytes

L.R. WOODHOUSE,[†] L.J. LEDERER,[†] N.M. LOWE,[†] AND J.C. KING[‡]

[†]Department of Nutritional Sciences, U.C. Berkeley, Berkeley, CA 94720, USA; and [‡]USDA Western Human Nutrition Research Center, San Francisco, CA 94129, USA

Keywords zinc depletion, erythrocytes, osmotic fragility

There is increasing evidence that zinc plays an important role in the structure and function of biomembranes. The erythrocyte membrane has been widely used to study membrane biochemistry, and the response of erythrocytes to hypotonic solutions, measured by hemoglobin release, is often used to assess membrane fragility. Zinc deficiency in rats leads to increased erythrocyte osmotic fragility, which is reversed after two days of zinc repletion (O'Dell et al. 1987). The effect of zinc status on erythrocyte osmotic fragility has not been studied in man.

An 85-day study was conducted to measure the metabolic and physiological effects of zinc depletion and repletion in men. Erythrocyte fragility was affected by zinc depletion, and the method of repletion, either by dietary or intravenous zinc, affected recovery of erythrocyte fragility.

Methods

Nine healthy men (27 ± 6 y) participated in a metabolic study consisting of a baseline period of 16 d, a depletion period of 33 d for 2 subjects followed by a 36 d repletion, or a depletion period of 40 d followed by a 30 d repletion for the remaining subjects. Using a semi-synthetic formula, 12 mg zinc per day was fed during baseline and repletion, and about 1.0 mg zinc per day during depletion. Subjects were divided into 2 groups for repletion: Group A subjects (n = 5) received two intravenous infusions of 66 mg zinc on day 1 and day 12 of repletion, while dietary zinc remained at 1.0 mg until after the day 12 zinc infusion, at which time dietary zinc was increased to 12 mg/d; Group B subjects (n = 4) were fed 12 mg zinc per day starting on day 1 of repletion.

Plasma zinc was monitored weekly throughout the study. Two of the subjects were repleted one week early due to the appearance of clinical signs of zinc depletion (perioral and groin skin rashes, lethargy, moodiness). Total body water was measured at 4 timepoints using deuterium dilution. Fasting blood for erythrocyte osmotic fragility testing was taken 3 times: late depletion, and days 8 and 21 of repletion. The erythrocytes were tested for fragility in hypotonic saline solutions based on a modification of the method of O'Dell et al. (1987). Isolated erythrocytes were diluted 1:1 with saline and added to various hypotonic solutions of NaCl in 0.05 mol/L Na_2HPO_4, pH 7.4. Isotonic NaCl, and 0% NaCl were used for 0% and 100% hemolysis values, respectively. After incubation, solutions were centrifuged and the hemoglobin released from the erythrocytes was measured at 540 nm. The absorbance data were expressed as percent hemolysis based on the absorbance value for 100% hemolysis. Normal human erythrocytes yield about 50% hemolysis in 0.38% NaCl, therefore the percent hemolysis at 0.38% NaCl was used as the comparison value for the erythrocyte fragility measurements. Analysis of variance repeated measures, followed by the Scheffe F-test was used to determine differences at the 3 timepoints. Unpaired t-tests were used to compare differences between Groups A and B.

Correct citation: Woodhouse, L.R., Lederer, L.J., Lowe, N.M., and King, J.C. 1997. The effects of zinc status on the osmotic fragility of human erythrocytes. In *Trace Elements in Man and Animals – 9: Proceedings of the Ninth International Symposium on Trace Elements in Man and Animals. Edited by* P.W.F. Fischer, M.R. L'Abbé, K.A. Cockell, and R.S. Gibson. NRC Research Press, Ottawa, Canada. pp. 636–638.

Results

Plasma zinc values declined significantly during depletion from 10.7 ± 1.4 μmol/L to 3.7 ± 2.2 μmol/L, and increased to 11.7 ± 1.2 μmol/L by the end of repletion. Total body water significantly increased 7% above baseline values in late depletion, returning to baseline values during repletion. Table 1 shows the erythrocyte fragility results for all subjects. Erythrocyte hemolysis was highest at the early repletion measurement, and then significantly decreased after three weeks of repletion. Table 1 also shows a measure of "improvement" in erythrocyte stability, expressed as the ratio of erythrocyte hemolysis at early repletion to erythrocyte hemolysis at late repletion. There was a significant difference observed in the improvement of erythrocyte fragility between the intravenous zinc repletion and the oral zinc repletion. Erythrocyte fragility improved 100% compared to the early repletion measurement with oral zinc repletion, whereas erythrocyte fragility improved only 50% with the intravenous zinc repletion. Preliminary zinc balance data showed that zinc retention was higher in those subjects repleted with oral zinc early in repletion.

Discussion

The turnover of human erythrocytes (120 d) may explain the lag time required before increased erythrocyte fragility was observed, even though plasma zinc values had returned to baseline levels. Also, a slower rate of erythropoiesis induced by zinc depletion may slow erythrocyte production and contribute to the lag time (Morgan et al. 1995). The turnover rate of erythrocytes in the rat is about 30 d, thus accounting for the more rapid reversal of erythrocyte fragility with zinc repletion in the rat model. The change observed in total body water may be related to an increase in intracellular water due to the effects of zinc depletion on cellular membranes (Sutherland et al. 1995).

Altered structure and function of cell plasma membranes has been proposed as a biochemical basis for the pathologies of zinc deficiency (Bettger et al. 1993). The loss of zinc from specific proteins in cell plasma membranes appears to alter membrane structure and function and reduce the integrity of the plasma membrane. The specific biochemical changes responsible for the altered physical and functional properties of the erythrocyte membrane are unknown, although recent research (Xia et al. 1995) suggests that the sulfhydryl content of plasma membrane proteins may be the basis for zinc deficiency pathology. Zinc

Table 1. Percent erythrocyte hemolysis at 0.38% NaCl.

Subject	Late Depletion (LD)	Early Repletion (ER)**	Late Repletion (LR)†	ER:LR
Group A:				
1	43.0	45.7	36.7	1.25
2	72.1	81.3	74.3	1.09
3	82.4	85.5	59.0	1.45
4	68.2	90.2	52.2	1.73
5	75.1	83.3	49.5	1.68
Mean ± SEM	68.1 ± 6.7[a]	77.2 ± 8.0[b]	54.3 ± 6.2[c]	1.44 ± 0.3[d]
Group B:				
7	40.9	73.5	33.0	2.23
10	57.7	78.1	38.4	2.03
11	74.4	89.3	58.1	1.54
12	52.4	72.0	35.9	2.01
Mean ± SEM	56.3 ± 7.0[a]	78.3 ± 3.9[b]	41.3 ± 5.7[c]	1.95 ± 0.3[e]
Overall Mean ± SEM	62.9 ± 5.0[a]	77.7 ± 4.5[b]	48.6 ± 4.6[c]	

a, b, c: Values differ at p = 0.01; d, e: Values differ at p = 0.03.
*Subjects 1 & 2: day 28; remainder: day 36.
**Subjects 1 & 2: day 3; remainder: day 7.
†All subjects: day 21.

deficiency also alters the phospholipid fatty acid composition of erythrocyte membranes (Driscoll et al. 1992). Several other mechanisms have been proposed to explain the altered properties of erythrocytes in zinc deficiency (Bettger et al. 1993).

In summary, acute zinc depletion increased the osmotic fragility of erythrocyte membranes, and this *in vivo* effect was reversed slowly with zinc repletion. Further studies are needed to determine if erythrocyte osmotic fragility is affected by marginal zinc intakes typically found in the general population.

Literature Cited

Bettger, W.J. & O'Dell, B.L. (1993) Physiological roles of zinc in the plasma membrane of mammalian cells. J. Nutr. Biochem. 4: 194–207.

Driscoll, E.R. & Bettger, W.J. (1992) Zinc deficiency in the rat alters the lipid composition of the erythrocyte membrane Triton shell. Lipids 27: 972–977.

Morgan, P.N., Wehr, C.M., MacGregor, J.T., Woodhouse, L.R. & King, J.C. (1995) Zinc deficiency, erythrocyte production, and chromosomal damage in pregnant rats and their fetuses. J. Nutr. Biochem. 6: 263–268.

O'Dell, B.L., Browning, J.D. & Reeves, P.G. (1987) Zinc deficiency increases the osmotic fragility of rat erythrocytes. J. Nutr. 117: 1883–1889.

Sutherland, B., Lowe, N.M., Van Loan, M., King, J.C. & Turnlund, J.R. (1995) Effect of experimental zinc depletion on body composition and basal metabolism in men. FASEB J. 9: A736, (abs).

Xia, J., Browning, J.D. & O'Dell, B.L. (1995) Relationship of plasma membrane sulfhydryl concentration to zinc deficiency pathology. FASEB J. 9:A867, (abs).

The Effect of Zinc Deficiency on Desaturation of Fatty Acids

KLAUS EDER AND MANFRED KIRCHGESSNER

Institute of Nutrition Physiology, Technical University of Munich, 85350 Freising, Germany

Keywords zinc deficiency, desaturation, fatty acids, rat

The clinical features of zinc deficiency in rats are similar to those of essential fatty acid deficiency, and therefore, it has been suggested that zinc deficiency impairs desaturation of essential fatty acids. However, results of zinc deficiency on fatty acid desaturation are contradictory, and it seems that the reduced food intake in zinc-deficient rats confounds the effect of zinc deficiency on fatty acid metabolism. In order to investigate the effects of zinc deficiency on fatty acid desaturation without the distortion by low food intake, we have carried out two experiments in which rats were fed sufficient quantities of a zinc-deficient diet by gastric tube.

Material and Methods

In the first experiment, the effect of zinc deficiency on fatty acid desaturation was investigated in rats fed two different types of fat, either a mixture of coconut oil and safflower oil (7:1, w/w, "CO diet") or linseed oil ("LO diet"). Thirty-six male Sprague-Dawley rats (121 ± 4 g) were divided into four groups: (I) Zinc-adequate (ZA, 30 mg Zn/kg), CO diet; (II) ZA, LO diet; (III) zinc-deficient (ZD, 0.5 mg Zn/kg), CO diet; (IV) ZD, LO diet. The rats were fed by gastric tube 11.6 g of food per day for 10 d. In that experiment, activities of fatty acid desaturases and the fatty acid composition of liver phospholipids were determined. Details about animals, diets, tube-feeding, and analytical methods are given by Eder and Kirchgessner (1995).

In the second experiment, 44 male Sprague-Dawley rats were divided into two groups and force-fed initially fat-free diets with (40 mg Zn/kg) or without zinc supplementation (0.5 mg Zn/kg) for 6 d in order to stimulate the activity of fatty acid desaturases. Thereafter, the diets were supplemented with 5% safflower oil with a high level of linoleic acid. This feeding period lasted 3.5 d. The conversion of linoleic acid into arachidonic acid was assessed by determining the fatty acid composition of liver phospholipids. Details about animals, diets and analytical methods are given by Eder and Kirchgessner (1996).

Results and Discussion

In both experiments, feeding the ZD diets reduced body weight gain as well as zinc concentrations and activities of alkaline phosphatase in serum, proving the severe zinc-deficient state of those rats.

In the first experiment, zinc deficiency did not influence the activities of $\Delta 5$ and $\Delta 6$ desaturases in liver microsomes, but reduced the activity of $\Delta 9$ desaturase in the rats fed either the CO diet or the LO diet. In agreement with this result, the levels of monounsaturated fatty acids in liver phosphatidylcholine, phosphatidylethanolamine and total lipids of hepatic microsomes were reduced by zinc deficiency in the rats fed either type of dietary fat, whereas those of saturated fatty acids were elevated by zinc deficiency. In contrast, levels of fatty acids derived from desaturation of linoleic acid (20:4 n – 6) and α-linolenic acid (20:5 n – 3, 22:5 n – 3 and 22:6 n – 3) were not reduced by zinc deficiency, regardless of the dietary fat. The results of that experiment suggest that zinc deficieny impairs $\Delta 9$ desaturation of saturated fatty acids, but does not affect desaturation of essential fatty acids (18:2 n – 6, 18:3 n – 3) if the diet contains sufficient quantities of polyunsaturated fatty acids. In the rats fed the LO diet, zinc deficiency even elevated the levels of 20:5 n – 3 in liver phosphatidylcholine and total lipids of hepatic microsomes. This suggests

Correct citation: Eder, K., and Kirchgessner, M. 1997. The effect of zinc deficiency on desaturation of fatty acids. In *Trace Elements in Man and Animals – 9: Proceedings of the Ninth International Symposium on Trace Elements in Man and Animals. Edited by* P.W.F. Fischer, M.R. L'Abbé, K.A. Cockell, and R.S. Gibson. NRC Research Press, Ottawa, Canada. pp. 639–640.

that zinc deficiency causes preferential incorporation of that fatty acids into liver phosphatidylcholine (results in detail are shown by Eder and Kirchgessner 1995).

In the second experiment, zinc deficiency impaired the conversion of linoleic acid into higher unsaturated fatty acids after the activities of desaturases were stimulated by feeding a fat-free diet. This has been shown by determining the fatty acid composition of liver phospholipids (Table 1). In zinc-deficient rats, the levels of precursors of Δ5 and Δ6 desaturation (18:2 n – 6, 20:3 n – 6) were higher and those of products of Δ5 and Δ6 desaturation (20:4 n – 6, 22:4 n – 6 and 22:5 n – 6) were lower than in zinc-adequate rats. This suggests that zinc deficiency affects desaturation of linoleic acid. However, in our studies this was evident only if desaturation was initially stimulated by feeding a fat-free diet. An impaired desaturation of linoleic acid may account for the similarities between zinc deficiency and essential fatty acid deficiency observed in former studies.

Table 1. Fatty acid composition of liver phosholipids of zinc-adequate and zinc-deficient rats fed initially a fat-free diet and thereafter the same diet supplemented with 5% safflower oil (moles/100 mol fatty acids).

	18:2 n – 6	20:3 n – 6	20:4 n – 6	22:4 n – 6	22:5 n – 6
PC					
Zinc-adequate	10.9	1.2	23.5	0.6	2.7
Zinc-deficient	12.3*	2.3*	17.4*	0.5*	1.5*
PE					
Zinc-adequate	4.0	0.5	28.1	1.0	5.6
Zinc-deficient	4.7*	0.8*	25.1*	0.8*	3.3*
PS					
Zinc-adequate	2.7	1.2	28.7	1.3	3.4
Zinc-deficient	2.7	1.7*	23.7*	1.1*	3.5

Abbreviations: PC, phosphatidylcholine; PE, phosphatidylethanolamine; PS, phosphatidylserine.
*indicates significant ($P < 0.05$) difference between zinc-adequate and zinc-deficient group.

Literature Cited

Eder, K. & Kirchgessner, M. (1995) Activities of liver microsomal fatty acid desaturases in zinc-deficient rats force-fed diets with a coconut oil/safflower oil mixture or linseed oil. Biol. Trace Elem. Res. 48: 215–230.

Eder, K. & Kirchgessner, M. (1996) Zinc deficiency and the desaturation of linoleic acid in rats force-fed fat-free diets. Biol. Trace Elem. Res. (in press).

The Effect of Zinc Deficiency on Regulation of Hepatic Lipogenic Enzymes by Dietary Polyunsaturated Fatty Acids in Rats

KLAUS EDER AND MANFRED KIRCHGESSNER
Institute of Nutrition Physiology, Technical University of Munich, 85350 Freising, Germany

Keywords zinc deficiency, lipogenic enzymes, liver, rat

Recent studies revealed that zinc deficiency causes markedly elevated activities of hepatic lipogenic enzymes and a fatty liver in rats force-fed sufficient quantities of a diet containing predominately coconut oil as dietary fat. In contrast, when linseed oil or fish oil was used as dietary fat, zinc deficiency did not produce a fatty liver, and activities of hepatic lipogenic enzymes were only slightly elevated (Eder and Kirchgessner 1994, 1995). Those studies suggested that the dietary fat is involved in the effects of zinc deficiency on lipogenic enzymes. The dietary fat in general plays an important role for the regulation of hepatic lipogenic enzymes. Feeding fat-free diets elevates the activities of hepatic lipogenic enzymes whereas supplementation with polyunsaturated fatty acids suppresses the activities of hepatic lipogenic enzymes. We hypothesized that zinc deficiency impairs the suppression of hepatic lipogenic enzymes by dietary polyunsaturated fatty acids. Therefore, zinc-adequate (ZA) and zinc-deficient (ZD) rats were fed initially fat-free diets in order to elevate the activities of hepatic lipogenic enzymes. Then the rats were switched to diets supplemented with 5% safflower oil (SO) and the reaction of lipogenic enzymes on fat supplementation was investigated.

Material and Methods

Forty-four male Sprague-Dawley rats (122 ± 5 g) were divided into two groups and force-fed fat-free diets with (40 mg Zn/kg) or without zinc supplementation (0.5 mg Zn/kg) for 6 d. Thereafter, the groups were divided and one half remained on the ZA or ZD fat-free diets whereas the other half was switched to the same diet supplemented with 5% safflower oil. This feeding period lasted 3.5 d, and then the rats were killed after a light anesthesia. Details about animals, diets, tube-feeding and analytical methods are given by Eder and Kirchgessner (1996).

Results and Discussion

Zinc deficiency caused a reduced body weight gain as well as largely reduced zinc concentrations and activities of alkaline phosphatase in both the rats fed the fat-free and the 5% SO diet, proving the severe zinc-deficient state of those rats (for results see Eder and Kirchgessner 1996).

Additionally, like in former studies, zinc deficiency caused a fatty liver in both, the rats fed the fat-free diet and the diet supplemented with 5% SO, which was characterized by elevated concentrations of triacylglycerols with predominately saturated and monounsaturated fatty acids. The concentration of cholesterol in the liver was not influenced by zinc deficiency whereas the concentrations of cholesterol, triacylglycerols and phospholipids in serum were markedly elevated by zinc deficiency in the rats fed the fat-free diet and in the rats fed the 5% SO diet.

The activities of lipogenic enzymes in liver and adipose tissue are shown in Table 1. ZD rats fed the fat-free diet had slightly higher activities of hepatic lipogenic enzymes than ZA rats fed the fat-free diet. Addition of 5% SO reduced the activities of hepatic lipogenic enzymes in both ZD and ZA rats. However,

Correct citation: Eder, K., and Kirchgessner, M. 1997. The effect of zinc deficiency on regulation of hepatic lipogenic enzymes by dietary polyunsaturated fatty acids in rats. In *Trace Elements in Man and Animals – 9: Proceedings of the Ninth International Symposium on Trace Elements in Man and Animals. Edited by* P.W.F. Fischer, M.R. L'Abbé, K.A. Cockell, and R.S. Gibson. NRC Research Press, Ottawa, Canada. pp. 641–642.

Table 1. Activities of lipogenic enzymes in liver and adipose tissue.[§]

Enzyme	ZA,* fat-free	ZD, fat-free	ZA, 5% SO	ZD, 5% SO
Liver				
FAS	21[a]	23[a]	17[b] (–21%)[#]	20[ab](–14%)
G6PDH	125[b]	155[a]	78[c] (–37%)	132[a] (–15%)
6PGDH	112[b]	143[a]	100[b] (–11%)	135[a] (–5%)
ME	100[a]	100[a]	70[b] (–30%)	77[b] (–23%)
CCE	55[ab]	63[a]	45[b] (–21%)	60[a] (–5%)
Adipose tissue				
FAS	0.61[b]	0.53[b]	0.85[a] (+39%)	0.44[b] (–17%)
G6PDH	4.00[ab]	3.57[b]	4.80[a] (+20%)	2.74[c] (–23%)
6PGDH	1.66[ab]	1.58[b]	2.18[a] (+31%)	1.33[b] (–16%)
ME	9.42[ab]	8.70[bc]	11.3[a] (+20%)	6.34[c] (–27%)
CCE	1.49[ab]	1.14[bc]	1.81[a] (+22%)	1.03[c] (–10%)

[§]Activities are given as nmoles NAD(P, H) oxidized or reduced per min and mg protein (liver) or mg fresh matter (adipose tissue) at 37°C. *Abbreviations: ZA, zinc-adequate; ZD, zinc-deficient; SO, safflower oil; FAS, fatty acid synthase; G6PDH, Glucose-6-phosphate dehydrogenase; 6PGDH, 6-Phosphogluconate dehydrogenase; ME, malic enzyme; CCE, citrate cleavage enzyme.
[#]relative to the equivalent group fed the fat-free diet.

the reduction was more pronounced in ZA than in ZD rats. Therefore, the activities were clearly higher in ZD rats fed the 5% SO diet than in the ZA rats fed 5% SO diet. In adipose tissue, the addition of 5% SO elevated the activities of lipogenic enzymes in ZA rats whereas it lowered the activities of lipogenic enzymes in ZD rats. Therefore, ZA rats fed the 5% SO diet had markedly higher activities of lipogenic enzymes in adipose tissue than ZD rats fed the 5% SO diet. The results suggest that in ZA rats the supplementation of 5% SO shifts lipogenesis from liver into the adipose tissue. In contrast, in ZD rats supplementation with 5% SO only partially suppresses hepatic lipogenesis. This may be responsible for elevated activities of lipogenic enzymes and elevated concentrations of triacylglycerols in liver of ZD rats. Elevated concentrations of serum lipids observed in this and former studies also may be due to increased hepatic lipogenesis in ZD rats.

Literature Cited

Eder, K. & Kirchgessner, M. (1994) Dietary fat influences the effect of zinc deficiency on liver lipids and fatty acids in rats force-fed equal quantities of diet. J. Nutr. 124: 1917–1926.

Eder, K. & Kirchgessner, M. (1995) Zinc deficiency and activities of lipogenic and glycolytic enzymes in liver of rats fed coconut oil or linseed oil. Lipids 30: 63–69.

Eder, K. & Kirchgessner, M. (1996) The effect of dietary fat on lipogenic enzymes in liver and adipose tissue of zinc-adequate and zinc-deficient rats. J. Nutr. Biochem. (in press, April issue).

Effect of Moderate Zinc Deficiency on Vitamin B_6 Metabolism in Rats

FLORIAN L. CERKLEWSKI AND JAMES E. LEKLEM

Department of Nutrition and Food Management, Oregon State University, Corvallis, OR 97331-5103, USA

Keywords zinc, vitamin B_6, pyridoxal phosphate, 4-pyridoxic acid

Present knowledge suggests that pyridoxal-5′-phosphate (PLP) released into the general circulation by the liver is hydrolyzed by the membrane-bound zinc metalloenzyme alkaline phosphatase (Leklem 1994). Thus any significant reduction in alkaline phosphatase activity should be reflected in increased plasma PLP and by a decrease in the major catabolic end-product of vitamin B_6 metabolism, urinary 4-pyridoxic acid (4-PA) as seen in individuals genetically deficient in alkaline phosphatase (Whyte et al. 1985).

Materials and Methods

In experiment 1, male Sprague-Dawley rats initially weighing about 80 g were fed a purified diet similar to AIN 93G providing either 5 or 30 mg Zn/kg and pyridoxine at a moderately-low level of 3 mg/kg. This 6-week study was repeated in a factorial design with two levels of pyridoxine (0.8 and 6.0 mg/kg) versus two levels of zinc (6 and 30 mg/kg).

Plasma was analyzed for zinc by atomic absorption spectrophotometry (AAS) (Smith et al. 1979), alkaline phosphatase activity, and PLP by a modified radiometric-enzyme method (Chabner and Livingston 1970). Urine was analyzed for 4-pyridoxic acid by HPLC (Gregory and Kirk 1979). Femur was defatted, ashed and analyzed for Zn by AAS.

Results

In experiment 1, the zinc deficiency produced, evidenced by zinc content of plasma and femur (Table 1), significantly disrupted vitamin B_6 metabolism.

In experiment 2 (Table 2), where plasma zinc was only reduced by 10% in comparison to the 25% reduction in experiment 1, neither indicator of vitamin B_6 metabolism was affected by zinc. Plasma alkaline phosphatase activity was unaffected by zinc status in either experiment 1 or 2.

Table 1. Effect of a moderate dietary zinc deficiency on vitamin B_6 metabolism at a single dietary level of vitamin B_6 (3 mg/kg).

	mg Zn/kg diet		
Measure	30	5	p-value
Plasma Zn, µmol/L	18.8 ± 2.4	14.3 ± 2.2	<0.01
Femur Zn, µmol/g ash	6.6 ± 0.3	2.4 ± 0.4	<0.001
Plasma PLP, nmol/L	390 ± 40	500 ± 91	<0.02
Urinary 4-PA, nmol/mg creatinine	20.8 ± 4.0	12.6 ± 5.7	<0.01

*mean ± SD (n = 7).

Correct citation: Cerklewski, F.L., and Leklem, J.E. 1997. Effect of moderate zinc deficiency on vitamin B_6 metabolism in rats. In *Trace Elements in Man and Animals – 9: Proceedings of the Ninth International Symposium on Trace Elements in Man and Animals. Edited by* P.W.F. Fischer, M.R. L'Abbé, K.A. Cockell, and R.S. Gibson. NRC Research Press, Ottawa, Canada. pp. 643–644.

Table 2. Effect of dietary zinc (6 or 30 mg/kg) on vitamin B_6 metabolism at two dietary levels of vitamin B_6 (0.8 or 6.0 mg/kg).

	Dietary treatments, mg/kg			
	6.0 B_6		0.8 B_6	
Measure	30 Zn	6 Zn	30 Zn	6 Zn
Plasma PLP, nmol/L	618 ±114[a]	634 ± 94[a]	175 ± 38[b]	183 ± 47[b]
Urinary 4-PA, nmol/mg creatinine	32.3 ± 3.0[a]	30.5 ± 5.3[a]	3.4 ± 0.8[b]	2.6 ± 0.6[b]
Plasma Zn, µmol/L	17.5 ± 1.1[a]	15.9 ± 2.3[b]	17.3 ± 1.0[a]	15.2 ± 1.2[b]
Femur Zn, µmol/g ash	6.2 ± 0.2[a]	2.7 ± 0.6[b]	6.2 ± 0.5[a]	2.4 ± 0.3[b]

*mean ± SD (n = 8); means with non-matching superscripts are $P < 0.05$.

Discussion

(1) A dietary zinc deficiency can disrupt vitamin B_6 metabolism when plasma zinc concentration is reduced by 25%, but not by 10%. Therefore, vitamin B_6 metabolism is unlikely to be severely compromised by the degree of zinc deficiency commonly encountered in a normal healthy population living in the United States. (2) The mechanism by which zinc deficiency impairs vitamin B_6 metabolism is not a simple inhibition of alkaline phosphatase activity. (3) A pre-existing zinc deficiency could mask a vitamin B_6 deficiency by elevating plasma PLP concentration.

Literature Cited

Chabner, B. & Livingston, D. (1970) A simple enzymic assay for pyridoxal phosphate. Anal. Biochem. 34: 413–425.

Gregory, J.F. & Kirk, J.R. (1979) Determination of urinary 4-pyridoxic acid using high performance liquid chromatography. Am. J. Clin. Nutr. 32: 879–873.

Leklem, J.E. (1994) Vitamin B-6. In: Modern Nutrition in Health and Disease (Shils, M.E., Olson J.A. and Shike, M., eds.), 8th edition, vol 1, pp. 383–394. Lea & Febiger, Philadelphia.

Smith, J.C., Jr., Butrimovitz, G.P., Purdy, W.C., Boeckx, R.L., Chu, R., McIntosh, M.E., Lee, K-D., Lynn J.K., Dinovo, E.C., Prasad, A.S. & Spencer, H. (1979) Direct measurement of zinc in plasma by atomic absorption spectroscopy. Clin. Chem. 25: 1487–1491.

Whyte, M.P., Mahuren J.D., Vrabel, L.A. & Coburn S.P. (1985) Markedly increased circulating pyridoxal-5′-phosphate levels in hypophosphatasia. J. Clin. Invest. 76: 752–756.

Effect of Nickel Deficiency on the Activities of Enzymes from Energy Metabolism

GABRIELE I. STANGL AND MANFRED KIRCHGESSNER

Institute of Nutrition Physiology, Technical University of Munich, 85350 Freising, Germany

Keywords nickel deficiency, citric cycle, cytochrome-c-oxidase, serum variables

Nickel deficiency has been shown to influence the metabolism of rats in various ways. Some of these studies were carried out with rats, whose growth was found to be suboptimal after nickel deprivation (Schnegg and Kirchgessner 1975). Nickel deficiency also appeared to depress the ATP level in serum (Schnegg and Kirchgessner 1977). Those results elucidate that nickel deficiency impairs the energy metabolism in animals. Both the citric cycle and the oxidative phosphorylation pathway in mitochondria play a considerable role in the synthesis of energy. Therefore, this investigation was designed to examine the effect of nickel deficiency on the activities of some enzymes from the citric cycle and the cytochrome-c-oxidase as well as various clinical-chemistry variables. As proof of nickel deficiency symptoms weight gain as well as nickel and iron status were determined.

Materials and Methods

A study over two generations was conducted feeding a nickel-deficient semisynthetic diet based on casein containing 13 μg/kg nickel and a nickel-adequate diet supplemented with 1 mg/kg nickel as $NiSO_4 \cdot 6H_2O$. In order to intensify the nickel deficiency in the second generation, nickel-depleted dams were mated twice and therefore passed two successive gestations and lactation cycles to depress body nickel. All measurements were taken from the 7-week-old pups of the second mating. The detailed description of the experiment, the composition of the diet, and the care and treatment of the rats were as previously described by Stangl and Kirchgessner (1996a). For determining the nickel deficiency criteria, nickel in liver and kidney, iron in serum, and hematological variables including erythrocyte count, hematocrit and hemoglobin were measured. The activities of hepatic malate dehydrogenase, isocitric dehydrogenase, aconitase and fumarase, and cytochrome-c-oxidase in liver and kidney, as well as serum substrate concentrations and enzymes were determined as previously described by Stangl and Kirchgessner (1996a,b).

Results

Pups fed a diet poor in nickel tended to have a depressed weight gain (–6.6%), nickel concentration in liver (–58%) and iron level in serum (–18%), and were nickel-deficient on the basis of a significantly reduced nickel concentration in kidney (–39%) and hematological parameters. Furthermore nickel deficiency altered the activities of enzymes from the hepatic citric cycle (Table 1). Nickel deficiency significantly reduced the activity of malate dehydrogenase (–21%) and tended to lower the activity of isocitric dehydrogenase (–25%). Additionally, nickel-depleted pups had a significantly higher activity of hepatic aconitase (+11%) than nickel-adequate rats. Nickel deficiency tended to lower the activity of cytochrome-c-oxidase from oxidative phosphorylation pathway in the liver (–14%), but not in kidney. Moreover, nickel-deficient pups had significantly increased activities of aspartate aminotransferase (+12%), alanine aminotransferase (+18%), and a somewhat higher activity of glutamate dehydrogenase (+11%) in serum. Moreover, nickel-depleted pups showed a somewhat reduced concentration of glucose (–6.3%), and

Correct citation: Stangl, G.I., and Kirchgessner, M. 1997. Effect of nickel deficiency on the activities of enzymes from energy metabolism. In *Trace Elements in Man and Animals – 9: Proceedings of the Ninth International Symposium on Trace Elements in Man and Animals. Edited by* P.W.F. Fischer, M.R. L'Abbé, K.A. Cockell, and R.S. Gibson. NRC Research Press, Ottawa, Canada. pp. 645–646.

Table 1. The activities§ of hepatic enzymes from citric cycle and cytochrome-c-oxidase from 7-week-old rats from the second mating fed a nickel-deficient or a nickel-adequate diet.

Enzyme	Nickel-deficient	Nickel-adequate	p<
Malate dehydrogenase in liver	10.6 ± 0.7	13.4 ± 0.5	0.01
Isocitric dehydrogenase in liver	2.03 ± 2.0	2.72 ± 0.34	0.15
Fumarase in liver	26.0 ± 2.0	29.1 ± 1.2	n.s.
Aconitase in liver	507 ± 13	457 ± 14	0.05
Cytochrome-c-oxidase in liver	0.79 ± 0.05	0.92 ± 0.05	0.10
Cytochrome-c-oxidase in kidney	1.23 ± 0.13	1.22 ± 0.14	n.s.

§Activities are expressed as changes in optical density·mg protein^{-1}·min^{-1} at 25°C. +Data are means ± SEM, n = 24 for Ni-depletion group and n = 23 for Ni-adequate group

slightly elevated concentrations of total protein and albumin in serum. The levels of creatinine and urea in serum did not differ between the groups.

Discussion

It is very difficult to establish criteria for nickel deficiency, because no nickel metalloenzyme has yet been described in animals, whose activity can be used as marker for nickel deficiency. Therefore, it was necessary to determine other parameters which have been shown to be altered in other successful nickel depletion studies. These criteria are a small reduction in growth (Schnegg and Kirchgessner 1975), in the concentration of body nickel (Sunderman et al. 1972), and a reduction in hematological variables (Nielsen et al. 1975, Schnegg and Kirchgessner 1975). Therefore, the present experiment has been shown to be successful in producing nickel deficiency in rat offspring. Moreover, this study demonstrates that nickel deficiency slightly alters the activities of hepatic enzymes, which are involved in energy metabolism. Since the cytochrome-c-oxidase activity in kidney was not impaired by nickel deficiency, it was suggested that kidney does not react to nickel deficiency in such a sensitive way as liver. Apart from an increase in serum alanine and aspartate aminotransferase and glutamate dehydrogenase activities, which are indicative of liver damage, there are other changes in liver metabolism, including an elevated triacylglycerol concentration, reduced activities of lipogenic enzymes and slight changes in fatty acid composition of phospholipids (Stangl and Kirchgessner 1996c). Additionally, Nielsen et al. (1975) reported that the signs of nickel deprivation in chicks also included ultrastructural abnormalities with swollen mitochondria in liver. These impairments in the liver may be associated with an alteration in the activities of membrane-bound enzymes such as those from the citric cycle and the terminal cytochrome-c-oxidase.

Literature Cited

Nielsen, F.H., Myron, D.R., Givand, S.H. & Ollerich, D.A. (1975) Nickel deficiency and nickel-rhodium interactions in chicks. J. Nutr. 105: 1607–1619.

Schnegg, A. & Kirchgessner, M. (1975) Veränderungen des Hämoglobingehaltes, der Erythrozytenzahl und des Hämatokrits bei Nickelmangel. Nutr. Metabol. 19: 268–278.

Schnegg, A. & Kirchgessner, M. (1977) Konzentrationsänderungen einiger Substrate im Serum und Leber bei Ni- bzw. Fe-Mangel. Z. Tierphysiol. Tierernährg. Futtermittelkde. 39: 247–251.

Stangl, G.I. & Kirchgessner, M. (1996a) Effect of nickel deficiency on various metabolic parameters of rats. J. Anim. Physiol. Anim. Nutr. 75 (in press).

Stangl, G.I. & Kirchgessner, M. (1996b) Effect of nickel deficiency on the activity of cytochrome-c-oxidase in rats. J. Anim. Physiol. Anim. Nutr. 75 (in press).

Stangl, G.I. & Kirchgessner, M. (1996c) Nickel deficiency alters liver lipid metabolism in rats. J. Nutr. (accepted).

Sunderman, F.W., Nomoto, S., Mornag, R., Nechay, M.W., Burke, C.N. & Nielsen, S.W. (1972) Nickel deprivation in chicks. J. Nutr. 102: 259–268.

Nickel Deficiency Alters Liver Lipid Metabolism in Rats

GABRIELE I. STANGL AND MANFRED KIRCHGESSNER
Institute of Nutrition Physiology, Technical University of Munich, 85350 Freising, Germany

Keywords nickel deficiency, liver lipids, lipogenic enzymes, lipoproteins

Nickel deficiency causes general disturbances in lipid metabolism of animals (Nielsen et al. 1975b, Schnegg and Kirchgessner 1977). In 1971 Nielsen described a dependence of the plasma cholesterol concentration and the liver fat content upon nickel supply in chicks. In the studies by Schnegg and Kirchgessner (1977), hepatic triacylglycerol concentration of nickel-deprived rats was reduced by 27% from the control level and the activity of glucose-6-phosphate dehydrogenase in liver, a key enzyme in lipogenesis, was lowered (Kirchgessner and Schnegg 1976). Those results obtained from several experiments elucidate that nickel plays a considerable role in lipid metabolism, particularly regulation of synthesis and concentration of lipids in organs. This investigation was designed to examine the effect of nickel deficiency on hepatic lipid metabolism and serum lipids of rats.

Materials and Methods

A study over two generations was conducted feeding a nickel-deficient semisynthetic diet based on casein containing 13 μg/kg nickel and a nickel-adequate diet supplemented with 1 mg/kg nickel as $NiSO_4 \cdot 6H_2O$. In order to intensify the nickel deficiency in the second generation, nickel-depleted dams were mated twice and therefore passed two successive gestations and lactation cycles to depress body nickel. All measurements were taken from the pups of the second mating, raised for a total of 7 weeks with the nickel-deficient or the nickel-adequate diet. The detailed description of the experiment, the composition of the diet, care and treatment of the rats as well as measurement of liver lipids, hepatic lipogenic enzymes, fatty acid composition of liver lipids and serum lipoproteins were previously described by Stangl and Kirchgessner (1996). For determining nickel deficiency, nickel in liver and kidney, and iron in serum, and hematological variables including erythrocyte count, hematocrit and hemoglobin were measured. The effect of nickel concentration in the diet was compared for statistical significance ($p < 0.05$) by the Student's t test. All data in the text are expressed as means ± SEM.

Results and Discussion

It has been shown that pups fed a diet poor in nickel tended to have a lower body weight gain (184 ± 4 g vs. 197 ± 7 g, $p < 0.15$), nickel concentration in liver (0.33 ± 0.09 μmol/kg vs. 0.77 ± 0.15 μmol/kg, $p = 0.10$), and iron level in serum (16.1 ± 1.10 μmol/L vs. 19.7 ± 1.12 μmol/L, $p < 0.10$) and were nickel-deficient on the basis of a significantly reduced nickel concentration in kidney (1.70 ± 0.23 μmol/kg vs. 2.78 ± 0.27 μmol/kg) and significantly reduced hematological measurements ($6.8 \pm 0.1 \cdot 10^{12}$/L vs. $7.3 \pm 0.1 \cdot 10^{12}$/L for erythrocyte count, 85.0 ± 1.33 g/L vs. 94.6 ± 1.29 g/L for hemoglobin and 275 ± 4.29 mL/L vs. 307 ± 3.54 mL/L for hematocrit) relative to nickel-adequate animals.

Nickel deficiency also caused a significant lipid accumulation in liver, which occurred exclusively in the triacylglycerol fraction (45.3 ± 5.0 μmol/g liver vs. 31.4 ± 2.0 μmol/g liver) with increased concentrations of saturated, monounsaturated, and polyunsaturated fatty acids derived from the diet. However, the concentrations of total cholesterol, phosphatidylcholine, phosphatidylethanolamine (PE) and phosphatidylserine remained unchanged in nickel-depleted pups from this study compared to nickel-adequate animals. To test whether the triacylglycerol accumulation in nickel deficiency was caused by an enhanced

Correct citation: Stangl, G.I., and Kirchgessner, M. 1997. Nickel deficiency alters liver lipid metabolism in rats. In *Trace Elements in Man and Animals – 9: Proceedings of the Ninth International Symposium on Trace Elements in Man and Animals. Edited by* P.W.F. Fischer, M.R. L'Abbé, K.A. Cockell, and R.S. Gibson. NRC Research Press, Ottawa, Canada. pp. 647–648.

hepatic fatty acid synthesis the activities of the lipogenic enzymes in the liver were determined. From the results obtained the liver lipid accumulation might not be based on an enhanced lipid synthesis, because the activities of the lipogenic enzymes glucose-6-phosphate dehydrogenase (–32%), 6-phosphogluconate dehydrogenase (–18%), malic enzyme (–33%) and fatty acid synthase (–26%) in liver were significantly and that of the citrate cleavage enzyme (–13%) nominally reduced by nickel deficiency. Only the activity of acetyl-CoA-carboxylase did not differ between the two groups.

Another possible cause of the triacylglycerol accumulation in liver might be an impaired secretion of triacylglycerols from liver into blood. If this would be the case, reduced lipid levels in serum would be expected. However, nickel-deficient pups had higher triacylglycerol (+19%) and phospholipid concentrations (+31%) in serum VLDL as well as cholesterol concentration (+39%) in serum LDL, which were both secreted from liver, than nickel-adequate rats. Thus, it is unlikely that hepatic lipid accumulation in nickel deficiency is caused by an impaired lipid secretion from liver into blood.

Moreover, in contrast to its concentration, the fatty acid composition of the individual hepatic phospholipids was slightly altered in nickel-depleted rats compared to nickel-adequate animals. Among these alterations in fatty acid composition, slightly elevated levels of 18:2 (n – 6) and 18:3 (n – 3) and reduced levels of 20:4 (n – 6) and 22:6 (n – 3) were most important, and indicative of a depressed desaturation. The effect on other long-chain desaturation products was inconsistent, in which the most marked alterations occurred in liver PE.

Former studies with rats clearly demonstrated that nickel deficiency causes a reduced iron status by impairing the iron absorption from the intestine (Nielsen et al. 1975a, Schnegg and Kirchgessner 1976). In this study, nickel-deficient rats had also a lower iron status than nickel-adequate animals. Apart from that, nickel-deficient rats had significantly reduced hematological parameters, including erythrocyte count, hemoglobin and hematocrit. The same variables also have been shown to be affected by iron deficiency (Lee et al. 1981). Since most of the alterations in lipid metabolism obtained from this study were similar to those obtained in several iron deficiency studies, it is therefore possible that at least some of the observed changes are due to the moderate iron deficiency observed in nickel-depleted pups relative to nickel-adequate pups.

Literature Cited

Kirchgessner, M. & Schnegg, A. (1976) Malate dehydrogenase and glucose-6-phosphate dehydrogenase activity in livers of Ni-deficient rats. Bioinorg. Chem. 6: 155–161.

Lee, Y.H., Layman, D.K. & Bell, R.R. (1981) Glutathione peroxidase activity in iron-deficient rats. J. Nutr. 111: 194–200.

Nielsen, F.H. (1971) "Studies on the essentiality of nickel". In: Mertz, W. & Cornatzer, W.E. (eds.) Newer trace elements in nutrition. Marcel Dekker, New York, pp. 215–253.

Nielsen, F.H., Myron, D.R., Givand, S.H. & Ollerich, D.A. (1975a) Nickel deficiency and nickel-rhodium interactions in chicks. J. Nutr. 105: 1607–1619.

Nielsen, F.H., Myron, D.R., Givand, S.H., Zimmerman, T.J. & Ollerich, D.A. (1975b) Nickel deficiency in rats. J. Nutr. 105: 1620–1630.

Schnegg, A. & Kirchgessner, M. (1976) Zur Absorption und Verfügbarkeit von Eisen bei Nickel-Mangel. Intern. Z. Vit. Ern. Forschung 46: 96–99.

Schnegg, A. & Kirchgessner, M. (1977) Konzentrationsänderungen einiger Substrate in Serum und Leber bei Ni- bzw. Fe-Mangel. Z. Tierphysiol. Tierernährg. Futtermittelkde. 39: 247–251.

Stangl, G.I. & Kirchgessner, M. (1996) Nickel deficiency alters liver lipid metabolism in rats. J. Nutr. (accepted).

Trace Element Research in the Year 2000

Chair: W.J. Bettger

Differential mRNA Display, Competitive Polymerase Chain Reaction and Transgenic Approaches to Investigate Zinc Responsive Genes in Animals and Man

ROBERT J. COUSINS

Center for Nutritional Sciences and Food Science and Human Nutrition Department, University of Florida, Gainesville, FL 32611, USA

Keywords cysteine-rich intestinal protein, differential display, metallothionein, technology

In all fields of science, it is difficult to project with certainty what technologies will be used in the future. Nevertheless, there are trends emerging that indicate certain technologies and methodologies will be of value to investigators studying trace elements in man and animals in the year 2000 (the topic of this part of our meeting). Table 1 provides a list of some technologies and methods this investigator believes will be important for studies on metabolism, nutrient functions, or status assessment in the years to come. This list is certainly not all inclusive.

It is envisioned that trace elements influence expression of a given cassette of genes which ultimately describe the physiologic functions for that trace element (Figure 1). The effect can be one that is a metal response element-regulated change in transcription which influences the cellular abundance of a specific mRNA population and the amount of the specific protein produced. The other general mechanism by which trace elements could influence gene expression and/or the production of specific protein products is for the trace element to perturb a given physiologic system which, in turn, influences the production of one or more secondary mediators or modulators which then influence the transcription/translation of specific genes and mRNAs.

Our recent research which focuses on zinc has used some of the methods listed in Table 1. The first approach to be described centers on two of the fundamental roles for zinc, i.e., structural and regulatory. Despite years of research, there seems to be no general consensus as to the basic biochemical mechanism(s) which result in failure of various processes which collectively is described as zinc deficiency (Cousins 1996). It was reasoned that, through either role, a dietary zinc restriction could sufficiently perturb regulation of various genes that could be identified through the technique of differential display of mRNA (Liang and Pardee 1992).

The differential RNA display technology combines reverse transcriptase polymerase chain reaction (RT-PCR) and random amplified polymorphic DNA (RAPD) genomic mapping to sequentially select a population of mRNAs which are then amplified and displayed by polyacrylamide gel electrophoresis (PAGE). The technique uses a 3′ PCR primer as one of four anchored oligo-dT primers which selects 1/4 of the population of mRNAs while a total of 26 individual 5′ PCR primers of arbitrary sequence (decamers)

Correct citation: Cousins, R.J. 1997. Differential mRNA display, competitive polymerase chain reaction and transgenic approaches to investigate zinc responsive genes in animals and man. In *Trace Elements in Man and Animals – 9: Proceedings of the Ninth International Symposium on Trace Elements in Man and Animals. Edited by* P.W.F. Fischer, M.R. L'Abbé, K.A. Cockell, and R.S. Gibson. NRC Research Press, Ottawa, Canada. pp. 649–652.

Table 1. Trace element research in the year 2000 — some technologies and potential uses.

Enzyme-Linked Immunosorbent Assay (ELISA)
Status Assessment
Competitive Reverse Transcriptase-Polymerase Chain Reaction
Status Assessment
Differential Display of mRNA
Nutrient Functions
Fluorescent Microscopy
Specific metal chelator
Fluorescent antibodies
Metabolism
Status Assessment
Fluorescence-Activated Cell Sorting (FACS)
Nutrient Functions
Status Assessment
Transgenic Animals
Metabolism
Nutrient Functions
Null (Knock Out) Animals
Metabolism
Nutrient Functions

provide analysis of the various mRNAs within a subpopulation. This is analogous to a library which has four floors as described by the 3′ PCR primer, with each floor having 26 stacks which contain all of the books on that particular floor. Once all floors have been searched, virtually all of the library's collection has been identified. In actual practice, each arbitrary decamer provides enough specificity that only about 50–100 product bands are produced. Collectively, the 4 anchored primers and 26 arbitrary decamers can screen an estimated population of 20,000 expressed RNAs.

For these experiments, male rats were maintained as previously described by our laboratory (Cousins and Lee-Ambrose 1992) and were fed a purified diet adjusted to be either zinc-deficient (<1 mg Zn/kg) or -adequate (30 mg Zn/kg). A pair-fed group was included to allow for the effects of anorexia in the –Zn rats. At the end of an 18 d comparison period, the animals were zinc depleted based upon depressed serum zinc concentrations, and no detectable kidney metallothionein mRNA levels as determined by Northern blot analysis. The validity of the technique was first verified using kidney RNA as the source of RNA for the RT-PCR. Subsequent display of the cDNAs produced would be expected to show down-regulation of metallothionein in zinc depleted rats. Indeed, a 208 base cDNA sequence derived by RT-PCR of kidney RNA was found to be rat metallothionein-1, thus confirming the ability of the differential display technique

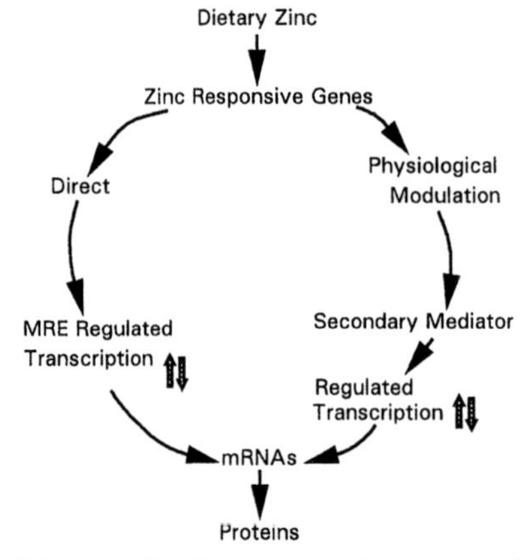

Figure 1. Direct and indirect routes for zinc to alter gene expression.

to identify a zinc-regulated gene (Blanchard and Cousins 1996).

The first use of this technology was applied to intestinal RNA since zinc restriction might produce changes in intestinal epithelium and/or alter expression of a zinc transporter protein (Blanchard and Cousins 1996). Anchored primers with $T_{12}MC$ and $T_{12}MG$ sequences and the battery of arbitrary decamers yielded 47 cDNA bands with visible differences in intensity. Thirty of the differentially expressed cDNAs increased in zinc deficiency while 17 decreased. The sequences were compared using DNA data banks and were found to represent previously cloned genes, expressed sequence tags (sequences that provide coding elements for previously reported sequences for unisolated proteins), or novel genes. In all cases, the cDNAs produced were used in the Northern blot analyses of RNA from other rats subjected to the same dietary treatment to confirm that response of these RNAs is regulated during zinc deficiency.

The differential display technique identified two cDNAs that are of particular importance in zinc deficiency. The first gene identified as being increased in expression in intestine during zinc deficiency was cholecystokinin (CCK). This intestinal peptide has endocrine and neuroendocrine functions as a satiety factor. The increase in CCK expression could explain in part the anorexia observed in zinc deficiency (O'Dell and Reeves 1989). Of equal interest was the up-regulation of uroguanylin RNA expression in zinc deficient rats. Uroguanylate cyclase-activating peptide II is a recently discovered peptide hormone (Forte and Currie 1995). When uroguanylin binds to the GC-C receptor of intestinal cells, fluid balance is altered, producing secretory diarrhea. Over-production of this hormone could be directly related to another clinical sign of zinc deficiency in humans, secretory diarrhea (Aggett 1989), and the reversal of that clinical problem by supplemental zinc (Sazawal et al. 1995). Experiments to further explore the potential mechanisms in anorexia and secretory diarrhea associated with zinc deficiency are in progress. Nevertheless, the power of the differential RNA display technique has been clearly identified through these experiments (Blanchard and Cousins 1996).

Another technology that has potential applications, particularly in nutrient status assessment, is the technique of competitive RT-PCR. In this approach, RNA is converted to cDNA by reverse transcription. The cDNA produced (target DNA) is then combined with a competitor cDNA of known concentration and dilutions and, along with appropriate 3′ and 5′ primers, the PCR produces two cDNA products (target cDNA and the competitor cDNA) which are separable by PAGE. Knowing the competitor concentration, the concentration of the target cDNA can be determined. Target cDNA is directly proportional to the mRNA for the given sequence in the cells from which the RNA was originally obtained. With this approach, one can take RNA from cells (e.g., leukocytes), carry out the competitive RT-PCR procedure, and determine the level of a given mRNA. The method could be used as an approach to assessment of zinc status in humans.

The competitive RT-PCR method was used to study the influence of zinc supplementation on metallothionein mRNA levels (Sullivan et al. 1996). Total RNA from purified monocytes was converted to cDNA by reverse transcription. The primers for competitive RT-PCR produced a 201 base pair cDNA which is shown by sequencing to be human metallothionein cDNA. When human subjects are supplemented for a period of 10 d with 50 mg of zinc, the competitive RT-PCR procedure showed there was a five-fold difference in the amount of metallothionein cDNA produced ($P < 0.05$). The sensitivity of the RT-PCR method is such that it is particularly attractive for status assessment work.

A third technology that has already exquisitely been utilized in trace element research is the production of transgenic animals. We have applied this technology and the corollary technique of cell transfection as well as ELISA technology to study expression of the zinc finger protein, cysteine-rich intestinal protein (CRIP).

During the course of experiments to study the transcellular movement of zinc in the rat small intestine, we noticed with high resolution gel filtration chromatography that a large portion of the ^{65}Zn in the cytosol associated with a species of smaller molecular size than metallothionein (Hempe and Cousins 1991). Subsequently, sufficient amounts of this material was obtained and, upon sequence analysis, was identified as CRIP. CRIP has a double zinc-finger motif called a LIM domain, that probably does not function in DNA binding. We sequenced 2.7 kb of the rCRIP promoter and identified a number of glucocorticoid and cytokine response elements (Levenson et al. 1994). The promoter and a reporter gene were transfected into IEC-6 cells. We established that, in both the transfected cells as well as in intact animals, CRIP is not regulated by zinc (Levenson et al. 1994). We have cloned 1.8 kb of genomic cDNA that contains the rCRIP coding region (Davis et al. 1996). The rat CRIP gene has five exons of varying sizes. Neither of the two zinc fingers is coded within one exon. This entire construct plus an additional DNA fragment for the purposes of transgenic identification have been introduced successfully into mice. These transgenic animals over express CRIP in a tissue-specific fashion as determined by Northern analysis and by analysis of tissue samples via ELISA. The results suggest that over-production of this zinc finger protein does not create a lethal situation and that a normal supply of dietary zinc seems adequate to maintain the higher levels of CRIP expression. We have also cloned the human CRIP cDNA and expressed the protein in E. coli. rhCRIP differs from rCRIP by two amino acids (Khoo et al. 1996).

The above briefly summarizes our recent experience with a number of technologies that hopefully will be used in the years ahead by other investigators in the field of trace element research.

Acknowledgements

The author wishes to acknowledge support from the NIH (DK 31127) and Boston Family Endowment funds. Appreciation is also extended to the author's colleagues R.K. Blanchard, B.A. Davis, C. Khoo and V.K. Sullivan who conducted some of the experiments briefly described above.

References

Aggett, P.J. (1989) Severe zinc deficiency. In: Zinc in Human Biology (Mills, C.F., ed.), pp. 259–279. Springer-Verlag, New York, NY.

Blanchard, R.K. & Cousins, R.J. (1996) Differential display of intestinal mRNAs regulated by dietary zinc. Proc. Natl. Acad. Sci. USA (in press).

Cousins, R.J. (1996) Zinc. In: Present Knowledge in Nutrition (Filer, L.J. & Ziegler, E.E., eds.), 7th Edition, in press. Internat. Life Sci. Inst.-Nutr. Foundation, Washington, DC.

Cousins, R.J. & Lee-Ambrose, L.M. (1992) Nuclear zinc uptake and interactions and metallothionein gene expression are influenced by dietary zinc in rats. J. Nutr. 122: 56–64.

Davis, B.A., Blanchard, R.K., & Cousins, R.J. (1996) Genomic organization of the rat cysteine-rich intestinal protein. FASEB J. 10: A758.

Forte, L.R. & Currie, M.G (1995) Guanylin: a peptide regulator of epithelial transport. FASEB J. 9: 643–650.

Hempe, J.M. & Cousins, R.J. (1991) Cysteine-rich intestinal protein binds zinc during transmucosal zinc transport. Proc. Natl. Acad. Sci. USA 88: 9671–9674.

Khoo, C., Blanchard, R.K., Sullivan, V.K. & Cousins, R.J. (1996) Cloning of human cysteine-rich intestinal protein: Identification in plasma and expression in human peripheral blood mononuclear cells. FASEB J. 10: A192.

Levenson, C.W., Shay, N.F., Hempe, J.M., & Cousins, R.J. (1994) Expression of cysteine-rich intestinal protein in rat intestine and transfected cells is not zinc dependent. J. Nutr. 124: 13–17.

Liang, P. & Pardee, A.B. (1992) Differential display of eukaryotic messenger RNA by means of the polymerase chain reaction. Science 257: 967–971.

O'Dell, B.L. & Reeves, P.G. (1989) Zinc status and food intake. In: Zinc in Human Biology (Mills, C.F., ed.), pp. 173–181. Springer-Verlag, New York, NY.

Sazawal, S., Black, R.E., Bhan, M.K., Bhandari, N., Sinha, A., & Jalla S. (1995) Zinc supplementation in young children with acute diarrhea in India. New Engl. J. Med. 333: 839–844.

Sullivan, V.K., Blanchard, R.K., & Cousins, R.J. (1996) Competitive RT-PCR shows that zinc supplementation up-regulates metallothionein mRNA levels in human monocytes and THP-1 cells. FASEB J. 10: A192.

Beyond Copper, Iodine, Selenium and Zinc: Other Elements That Will be Found Important in Human Nutrition by the Year 2000

F.H. NIELSEN

USDA, ARS, Grand Forks Human Nutrition Research Center, Grand Forks, ND 58202, USA*

Keywords boron, chromium, ultratrace elements, human nutrition

Over 30 years ago, excitement abounded about the trace elements which were considered the new breakthrough for better health. One trace element after another was identified as essential, or at least suggested to be essential. As these elements were identified, hopes were raised that trace element nutriture was the answer to the puzzle of the cause of some diseases, especially chronic diseases associated with aging. Unfortunately, those hopes have not been realized. Except for iodine, iron, and possibly selenium, trace elements, especially the newer, or ultratrace elements, are not generally thought of as having any widespread clinical importance.

There are probably several reasons that the vision of the past for the trace elements has not materialized; among them would be the following: 1) There is a paucity of clinically useful methods for assessing human trace element status. 2) There is a lack of appreciation for the paradigm that trace and ultratrace elements are most important, or will produce marked pathology only when a nutritional, metabolic, hormonal, or physiological stressor is present that enhances the need, or interferes with the utilization of an element. 3) There is a lack of recognition that the response to a low dietary intake of a mineral element will vary in extent and nature among individuals. Of course, the reason for the unfulfillment of the vision could be that the trace elements really are of minor clinical importance; however, I believe this is not true and that research through the year 2000 will overcome the shortcomings listed above and bring about the realization that trace elements are important for health and well-being; this includes showing that copper and zinc are more important than currently acknowledged.

Perhaps the major reason that the newer, or ultratrace elements, are thought to be nutritionally unimportant is that a specific biochemical function has not been defined for any of the at least 15 elements suggested to be essential since 1959. That is, the lack of a defined function has inhibited the acceptance of any of the ultratrace elements as being nutritionally essential and thus results in the opinion that they do not prevent deficiency diseases; therefore none are important. This viewpoint is often expressed although several of the ultratrace elements have been shown to have effects suggestive of an essential function, and if not this, to have beneficial actions in humans or animal models. In other words, this viewpoint ignores past history which has taught us that a defined essential function is not necessary for a substance to be nutritionally important (e.g., fiber has health benefits). Additionally, it ignores an emerging paradigm in which a nutrient does not have to prevent a deficiency disease to be clinically important. This paradigm is exemplified by a number of recent nutritional recommendations that are based on total health effects, not on the basis of deficiency or essentiality paradigms, including the high intake of antioxidants to protect against the deleterious effects of reactive oxygen radicals, and high intakes of nutrients to help prevent pathological conditions (e.g., calcium for osteoporosis).

Those who research the possible nutritional importance of elements other than copper, iodine, iron, zinc and selenium should be encouraged by, and those who criticize or scorn the use of resources to study

---

*U.S. Department of Agriculture, Agricultural Research Service, Northern Plains Area, is an equal opportunity/affirmative action employer and all agency services are available without discrimination.

Correct citation: Nielsen, F.H. 1997. Beyond copper, iodine, selenium and zinc: Other elements that will be found important in human nutrition by the year 2000. In *Trace Elements in Man and Animals – 9: Proceedings of the Ninth International Symposium on Trace Elements in Man and Animals. Edited by* P.W.F. Fischer, M.R. L'Abbé, K.A. Cockell, and R.S. Gibson. NRC Research Press, Ottawa, Canada. pp. 653–655.

the ultratrace elements should contemplate the selenium story. Selenium was first suggested to be essential in 1957 based on findings with vitamin E-deficient animals. Even today it is difficult to produce signs of pathology caused by a simple dietary selenium deficiency in animals or humans; a stressor such as vitamin E deficiency or a viral infection is needed to obtain marked pathology. Between 1957 and 1972 selenium was viewed mainly as being of toxicological, especially carcinogenic, concern. Fifteen years after it was suggested to be essential a biochemical function was finally identified for selenium. Since then, many resources and numerous researchers have been devoted to the study of selenium. Yet, only recently, or about 40 years after being suggested as being essential, has selenium become recognized as possibly having widespread clinical or nutritional importance, based upon its effect on viral virulence (Beck et al. 1994) and its protective effect against certain cancers (Clark et al. 1996).

Based on the findings coming forth about the ultratrace elements, I believe in the year 2000 we will become cognizant of other elements with a chronicle similar to that of selenium. These elements probably are among a disparate group of 17 for which there are reports suggesting that they are nutritionally important. The 17 are manganese and molybdenum which have known essential functions but no unequivocally identified practical nutritional importance; boron and chromium which have apparent beneficial, if not essential, actions in humans, but no specifically identified biochemical roles; nickel and vanadium which have identified biochemical roles in lower forms of life and an impressive number of reports describing deprivation signs in animals; arsenic and silicon for which there are reported deprivation signs for animals, and an arsenic-methylating enzyme has been identified in higher animals including humans while silicon has been found to be essential for some lower forms of life; fluorine and lithium which have beneficial pharmacologic actions, but rather limited evidence for essentiality; and finally, aluminum, bromine, cadmium, germanium, lead, rubidium, and tin which have some bits of credible evidence from animal models to suggest that they have at least positive effects under some situations.

Based on the findings reported to date, two elements of the 17 listed above that research in the year 2000 will most likely find to be of practical nutritional or clinical importance are boron and chromium; they will be discussed briefly here. Other good candidates that will not be discussed include arsenic, manganese, silicon and vanadium.

Chromium. At present, chromium is a very controversial element in regards to its nutritional significance. On one side are the zealots who claim that ingesting chromium, most often as a chromium supplement such as chromium picolinate, has numerous beneficial effects; these include building muscle, losing weight without dieting and exercising, and preventing diabetes, osteoporosis, heart disease and aging. On the other side are the disbelievers who feel there is no credible evidence to support most of these claims, and that chromium nutriture, if chromium is an essential nutrient, is an inconsequential concern for almost everyone. However, evidence accumulating since 1959 indicates that the true situation for chromium is somewhere between these two positions.

A recent review (Nielsen 1994a) discloses that there is a large amount of circumstantial evidence which supports the view that chromium is an essential nutrient. The evidence includes the finding that humans on long-term total parenteral nutrition containing a low amount of chromium developed impaired glucose tolerance, or hyperglycemia with glucose spilling into the urine, and a resistance to insulin action; these abnormalities were reversed by chromium supplementation. Additionally, there have been a number of reports from numerous research groups that describe beneficial effects of chromium supplementation of subjects with varying degrees of glucose intolerance ranging from hypoglycemia to insulin-dependent diabetes (Anderson 1993, Ravina et al. 1995). Beneficial effects of chromium supplementation on blood lipid profiles also have been reported. Thus, even the skeptics usually accept the fact that chromium can at least be a beneficial element in some situations.

Although a dietary recommendation of 50 µg/d has been established for chromium in the U.S. (National Research Council 1989), balance studies have demonstrated that people maintain metabolic chromium balance despite consuming diets supplying much less than this (Offenbacher 1992). Additionally, the average daily intake of chromium in the U.S. apparently is well below 50 µg and may be closer to 25 µg, yet widespread apparent cases of chromium deficiency have not been documented. That is, supplemental chromium fed to many individuals apparently consuming chromium at these low amounts did not result in any beneficial effect. Thus, a daily intake of 25 to 35 µg of chromium may be adequate.

On the other hand, some data suggest that an intake of less than 20 μg/d is inadequate. Based on dietary surveys, there are a significant number of people consuming less than 20 μg Cr/d. As a result, it is not surprising that a large number of studies have found individuals that respond to chromium supplementation. Hopefully, research by the year 2000 will identify the mechanism through which chromium affects glucose and lipid metabolism because this knowledge is likely to help confirm the apparent need for concern about inadequate chromium nutriture affecting health and well being of a significant number of people.

Boron. Since 1981 circumstantial evidence has been accumulating which suggests that boron is an element of nutritional concern. In humans and animal models a dietary deprivation of boron has consistently resulted in changed biological functions that could be construed as detrimental, and were preventable or reversible by an intake of physiological amounts of boron. Findings involving boron deprivation of humans have come mainly from two studies (Nielsen 1994b) in which men over the age of 45, postmenopausal women and postmenopausal women on estrogen therapy were fed a diet containing about 0.25 mg B/2000 kcal for 63 d, and then fed the same diet supplemented with 3 mg B/d for 49 d. The effects of boron supplementation after depletion included an effect on calcium metabolism evidenced by increased serum 25-hydroxycholecalciferol; an effect on energy metabolism evidenced by increased serum triglycerides; an effect on nitrogen metabolism indicated by decreased blood urea nitrogen and serum creatinine; and an effect on oxidative metabolism indicated by increased erythrocyte superoxide dismutase and serum ceruloplasmin. Boron repletion after depletion also enhanced the elevation in serum 17β-estradiol and plasma copper caused by estrogen ingestion (Nielsen 1994b), altered encephalograms such that they suggested improved behavioral activation (e.g., less drowsiness) and mental alertness, and improved psychomotor skills and the cognitive processes of attention and memory (Penland 1994). Changes similar to those found in humans also have been found in animal models (Hunt 1994).

The findings described above indicate that people consuming about 0.25 mg B/d respond positively to boron supplementation, which suggests that boron intakes should be higher than this. Extrapolations from animal data have resulted in the suggestion that humans may benefit or have a boron requirement between 0.5 and 1.0 mg/d (Nielsen 1992). Because many people consistently consume less than this amount, by the year 2000 boron most likely will be recognized as an element of clinical and nutritional importance.

References

Anderson, R.A. (1993) Recent advances in the clinical and biochemical effects of chromium deficiency. In: Essential and Toxic Trace Elements: An Update. (Prasad, A.S., ed.), pp. 221–234, Wiley-Liss, New York, NY.

Beck, M.A., Kolbeck, P.C., Rohr, L.H., Shi, Q., Morris, V.C. & Levander, O.A. (1994) Benign human enterovirus becomes virulent in selenium-deficient mice. J. Med. Virol. 43:166–170.

Clark, L.C., Combs, G.F., Jr. & Turnbull, B.W. (1996) The nutritional prevention of cancer with selenium 1983–1993: A randomized clinical trial. FASEB J. 10: A550.

Hunt, C.D. (1994) The biochemical effects of physiologic amounts of dietary boron in animal nutrition models. Environ. Health Perspect. 102 (Suppl): 35–43.

Nielsen, F.H. (1992) Facts and fallacies about boron. Nutr. Today 27: 6–12.

Nielsen, F.H. (1994a) Chromium. In: Modern Nutrition in Health and Disease, 8th Ed. (Shils, M.E., Olson, J.A., & Shike, M., eds.), vol. 1, pp.268–286, Lea & Febiger, Philadelphia, PA.

Nielsen, F.H. (1994b) Biochemical and physiologic consequences of boron deprivation in humans. Environ. Health Perspect. 102 (Suppl): 59–63.

NRC (National Research Council) (1989) Recommended Dietary Allowances, 10th ed., Food and Nutrition Board, National Academy of Sciences, National Academy Press, Washington, D.C.

Offenbacher, E.G. (1992) Chromium in the elderly. Biol. Trace Elem. Res. 32: 123–131.

Penland J.G. (1994) Dietary boron, brain function, and cognitive performance. Environ. Health Perspect. 102 (Suppl): 65–72.

Ravina, A., Slezak, L., Rubal, A. & Mirsky, N. (1995) Clinical use of the trace element chromium (III) in the treatment of diabetes mellitus. J. Trace Elem. Exp. Med. 8: 183–190.

List of Participants

ABRAHAM, Nader G., The Rockefeller University, 1230 York Avenue, New York, NY 10021, USA, Tel: 212-327-8762, Fax: 212-327-8510

AGADJANYAN, Nicolai A., Department of Normal Physiology, Russian Peoples' Friendship University, 8 Miclukho-Mayklay Str., Moscow 117198, Russia, Tel: 095-433-1022, Fax: 095-433-1022

AGUILAR, A.E., Dept. Biologia, UNAM, México, D.F., Facultad de Quimica. Cd. Universitaria, México, D.F., 04510, México, Tel: 622-3740, Fax: 622-3696

ALLEN, Lindsay, Department of Nutrition, University of California, Davis, CA 95616, USA, Tel: 916-752-5920, Fax: 916-752-3406, lhallen@ucdavis.edu

ALCOCK, Nancy, Preventive Medicine & Community Health, University of Texas, Medical Branch, 700 Harborside Drive, Galveston, TX 77555-1109, USA, Tel: 409-772-4661, Fax: 409-772-6287

ALVAREZ-HERNANDEZ, Xavier, Department of Medicine – Cancer Center, LSU Medical Center, 1505 Kings Highway, PO Box 33932, Shreveport, LA, USA, Tel: 318-675-5970, Fax: 318-675-5944

AMBE, Fumitoshi, Institute of Physical & Chemical Research (RIKEN), Wako-shi, Saitama 35101, Japan, Tel: 81-48-462-1111, Fax: 81-48-462-4662

AMBE, Shizuko, Institute of Physical & Chemical Research (RIKEN), Wako-Shi, Saitama 351-01, Japan, Tel: 81-48-462-1111, Fax: 81-48-462-4662

AMMERMAN, Clarence B., Department of Animal Science, University of Florida, Gainesville, FL 32611-0910, USA, Tel: 352-392-9635, Fax: 352-392-7652, cba@gnv.ifas.ufl.edu

ANDERSON, Richard A., Nutr. Requirements and Function, Beltsville Nutr. Research Center, USDA, Building 307, Room 224, Beltsville, MD 20705, USA, Tel: 301-504-8091, Fax: 301-504-9062

ANKE, Manfred, Friedrich-Schiller-University, Biological-Pharmaceutical Faculty, Institute for Nutrition and Environment, Dornburger Str. 24, D-07743 Jena, Germany, Tel: 03647-637082, Fax: 03647-637008

ARAS, Namik K., Department of Chemistry, Middle East Technical University, Ankara, 06531, Turkey, Tel: 90-312-2101000, Fax: 90-312-2101280, aras@rorqual.cc.metu.edu.tr

ARIZONO, Koji, Department Hygienic Chemistry, Faculty of Pharmaceutical Sciences, Nagasaki University, 1-14 Bunkyo-cho, Nagasaki/852, Japan, Tel: 81-958-47-1111, Fax: 81-958-44-6774, kjarzn@net.nagasaki-u.ac.jp

ARQUITT, Andrea B., Department Nutritional Sciences, Oklahoma State University, 425 HES, Stillwater, OK 74074-6141, USA, Tel: 405-744-8285, Fax: 405-744-6113, abarqut@okway.okstate.edu

ARTHUR, John R., Division of Biochemical Sciences, Rowett Research Institute, Bucksburn, Aberdeen, Scotland, AB2 9SB, UK, Tel: 44-1224-716630, Fax: 44-1224-716687, jra@rri.sari.ac.uk

ASHMEAD, H. DeWayne, President, Albion Laboratories, Inc., 101 North Main Street, Clearfield, UT 84037, USA, Tel: 801-773-4631, Fax: 801-773-4633

ATKINSON, Stephanie A., Department Pediatrics, McMaster University, 1200 Main Street West, Room #V42, Hamilton, ON, Canada L8N 3Z5, Tel: 905-521-2100, Fax: 905-521-1703, satkins@fhs.mcmaster.ca

AUDETTE, Robert J., Laboratory Medicine and Pathology, University of Alberta Hospitals, 8440-112 Street, Edmonton, AB, Canada T6G 2B7, Tel: 403-492-6648, Fax: 403-492-6267, robert.audette@ualberta.ca

AUST, Steven D., Biotechnology Center, Utah State University, Logan, UT 84322-4705, USA, Tel: 801-797-2730, Fax: 801-797-2755, sdaust@cc.usu.edu

BÄCKSTRÖM, Åsa, Department of Environmental Health, S-90187 Umeå, Sweden, Tel: 46-90-165002, Fax: 46-90-135636, asa.backstrom@hs.umu.se

BARCLAY, Denis, Department of Nutrition, Nestlé Research Center, PO Box 44, Vers-chez-les-Blanc, 1000 Lausanne 26, Switzerland, Tel: 41-21-785-8626, Fax: 41-21-785-8556, denis.barclay@chlsnr.nestrd.ch

BARNES, Ramon, Department of Chemistry, University of Massachusetts, LGRC Tower, Box 34510, Amherst, MA 01003-4510, USA, Tel: 413-545-2294, Fax: 413-545-3757, ramon.m.barnes@chemistry.umass.edu

BAYNES, Roy D., Division of Hematology, Internal Medicine, University of Kansas Medical Center, 3901 Rainbow Boulevard, Kansas City, KS 66160, USA

BEATTIE, John H., Department of Biochemical Sciences, Rowett Research Institute, Greenburn Road, Bucksburn, Aberdeen, Scotland, AB2 9SB, UK, Tel: 44-1224-716631, Fax: 44-1224-716622, j.beattie@rri.sari.ac.uk

BELSTEN, Joanne L., Nutrition, Diet and Health, Institute of Food Research, Norwich Research Park, Colney, Norwich, NR4 7UA, UK, Tel: 44-1603-255000, Fax: 44-1603-452578, jo.belsten@bbsrc.ac.uk

BENEDIK, Miha, Department of Environ. Sci., Jožef Stefan Institute, Jamova 39, Ljubljana 61111, Slovenia, Tel: 386-61-1773-900, Fax: 386-61-219-385

BERMANO, Giovanna, Biochemistry, Rowett Research Institute, Greenburn Road, Bucksburn, Aberdeen AB2 9SB, UK, Tel: 01224-712751 ext. 3255, Fax: 01224-716687, gb@rri.sari.ac.uk

BETTGER, William, Human Biology and Nutritional Sciences, University of Guelph, Guelph, ON, Canada N1G 2W1, Tel: 519-824-4120, ext. 3747, Fax: 519-763-5902, wbettger.ns@aps.uoguelph.ca

BHATHENA, Sam J., BHNRC, Metabolism & Nutrient Laboratory, USDA, ARS, B307, R-323, BARC-East, Beltsville, MD 20705, USA, Tel: 301-504-8422, Fax: 301-504-9456

BLACK, Richard M., Corporate Affairs, Kellogg Canada Inc., 6700 Finch Avenue West, Etobicoke, ON, Canada M9W 5P2, Tel: 416-675-5347, Fax: 416-675-5243, richard.black@kellogg.com

BOGDEN, John, UMDNJ – New Jersey Medical School, 185 South Orange Avenue, Newark, NJ 07103-2714, USA, Tel: 201-982-5432, Fax: 201-982-7625, bogden@umdnj.edu

BORSCHEL, Marlene, Pediatric Nutrition R&D, 625 Cleveland Avenue, Columbus, OH 43215, USA, Tel: 614-624-7578, Fax: 614-624-3453, usabthwm@ibmmail.com

BOURDOUX, Pierre, Laboratory of Pediatrics, Free University of Brussels, Hopital Universitaire des Enfants, Avenue J.J. Crocq 15, B-1020 Brussels, Belgium, Tel: 32-2-477-2581, Fax: 32-2-477-2563, pbourdou@resulb.ulb.ac.be

BRAY, Tammy M., Human Nutrition and Food Management, The Ohio State University, 347 Campbell Hall, 1787 Neil Avenue, Columbus, OH 43210-1295, USA, Tel: 614-292-5504, Fax: 614-292-8880, bray.21@osu.edu

BREMNER, Ian, Rowett Research Institute, Greenburn Road, Bucksburn, Aberdeen, AB2 9SB, Scotland, Tel: 44-1224-716602, Fax: 44-1224-716622, i.bremner@rri.sari.ac.uk

BRODNER, Wolfram, Department of Orthopedics, University of Vienna, Wahringer Gurtel 18-20, Vienna, Austria A-1090, Tel: 43140400-4079, Fax: 43140400-4077

BUNCE, George E., Department of Biochemistry and Anaerobic Microbiology, Virginia Tech, Engel Hall, Blacksburg, VA 24061-0308, USA, Tel: 540-231-9684, Fax: 540-231-4522, gebunce@vt.edu

BURGESS, Ellen, Department of Medicine, Foothills Hospital, 1403-29th Street, NW, Calgary, AB, Canada T2N 2T9, Tel: 403-670-1598, Fax: 403-283-2494

BUTCHER, Linda, Department of Chemistry & Biochemistry, California State University, Fullerton, 800 N. State College Boulevard, Fullerton, CA 92634, USA, Tel: 714-773-2475, Fax: 714-449-5316

BUTLER, Judy A., Department of Agricultural Chemistry, Oregon State University, Room 1007, Ag. & Life Sciences II, Corvallis, OR 97331-7301, USA, Tel: 541-737-1809, Fax: 541-737-0497, butlerj@bcc.orst.edu

CARLSON, Marcia, Department of Animal Science, Michigan State University, 2721-1C Trappers Cove Trail, Lansing, MI 48910, USA, Tel: 517-887-8116, Fax: 517-432-0190, carlsom@pilot.msu.edu

CERKLEWSKI, Florian, L., Department Nutrition and Food Management, Oregon State University, 108 Milam Hall, Corvallis, OR 97331-5103, USA, Tel: 541-737-0964, Fax: 541-737-6914, cerklewf@ccmail.orst.edu

CHATT, Amares, Department of Chemistry, Dalhousie University, Trace Analysis Research Centre, Halifax, NS, Canada B3H 4J3, Tel: 902-494-2474, Fax: 902-494-2474, -1310, a.chatt@dal.ca

CHAUDHARI, Panna, Technical Department, Fortitech, Inc., Riverside Technology Park, 2105 Technology Drive, Schenectady, NY 12308, USA, Tel: 518-372-5155, Fax: 518-372-5599

CHAUDHARI, Ram, Department of Quality Control, Foritech. Inc., Riverside Park, 2105 Technology Drive, Schenectady, NY 12308, USA, Tel: 518-372-5155, Fax: 518-372-5599

CHAVEZ, Eduardo, R., Department of Animal Science, MacDonald Campus of McGill University, 21111 Lakeshore Road, Ste. Anne de Bellevue, QC, Canada H9X 3V9, Tel: 514-398-7795

CHENOWETH, Wanda, Department of Food Science & Human Nutrition, Michigan State University, 208 Trout FSHN Building, East Lansing, MI, USA, Tel: 517-353-9606, Fax: 517-353-8963, chenowe1@pilot.msu.edu

CHERIAN, George, Department of Pathology, Dental Sciences Building, The University of Western Ontario, London, ON, Canada N6A 5C1, Tel: 519-661-2030, Fax: 519-661-3370, mcherian@julian.uwo.ca

CHESTERS, John K., Department of Biochemistry, Rowett Research Institute, Bucksburn, Aberdeen, AB2 9SB, UK, Tel: 44-1224-716633, Fax: 44-1224-716622, jkc@rri.sari.ac.uk

CHO, Kyung H., Division of Chemistry & Radiation, Korea Research Institute of Standards & Science, PO Box 102, Yusung, Taejon, 305-600, Republic of Korea, Tel: 82-42-868-5363, Fax: 82-42-868-5042

COCKELL, Kevin A., Nutrition Research Division, Health Canada, 2203C Banting Research Centre, Ross Avenue, Ottawa, ON, Canada K1A 0L2, Tel: 613-957-0923, Fax: 613-941-6182, kcockell@hpb.hwc.ca

COMBS, Gerald, Division of Nutritional Sciences, Cornell University, 122 Savage Hall, Ithaca, NY, USA, Tel: 607-255-2140, Fax: 607-255-1033, gfc2@cornell.edu

COSTA, Max, Department of Environmental Medicine, New York University Medical Center, 550 First Avenue, New York, NY 10016, USA, Tel: 914-351-2368, Fax: 914-351-2118, costam@charlotte.med.nyu.edu

COSTA, Nick D., Department of Veterinary Studies, Murdoch University, South Street, Murdoch, WA 6150, Australia, Tel: 61-9-3602485, Fax: 61-9-3104144, costa@numbat.murdoch.edu.au

COUGHLIN, James R., Coughlin & Associates, 2 Compadre Circle, Laguna Niguel, CA 92677, USA, Tel: 714-363-1612, Fax: 714-363-6445

COUSINS, Robert J., Centre for Nutritional Sciences, University of Florida, Gainesville, FL 32611, USA, Tel: 352-392-2133, Fax: 352-392-1008, cousins@gnv.ifas.ufl.edu

COX, Diane W., Department of Genetics, Hospital for Sick Children, 555 University Avenue, Toronto, ON, Canada, Tel: 416-813-6384, Fax: 416-813-4931, dcox@sickkids.ca

CREWS, Helen M., Chemical Safety of Food Group, MAFF, CSL, CSL Food Science Laboratory, Norwich Research Park, Colney, Norwich NR4 7UQ, UK, Tel: 00-44-1603-259350, Fax: 00-44-1603-501123, h.crews@cst.gov.uk

CROMWELL, Gary L., Animal Sciences Department, University of Kentucky, 609 W.P. Garrigus Building, Lexington, KY 40546-0215, USA, Tel: 606-257-7534, Fax: 606-323-1027, gcromwel@ca.uky.edu

CUBILLÁN, Hernán S., Laboratorio de Inst. Analitica, Universidad del Zulia, Apartado Postal 15308 Las Delicias, Maracaibo, Zulia 4003-A, Venezuela, Tel: 00-58-61-516868/596763, Fax: 00-58-61-516868/52688, vgrana@solidos.cien.luz.ve

CUNNANE, Stephen C., Department of Nutritional Sciences, University of Toronto, Toronto, ON, Canada M5S 1A8, Tel: 416-978-8356, Fax: 416-978-5882, scunnane@utoronto.ca

DABEKA, Robert, Food Research Division 2203D, Health Canada, Ottawa, ON, Canada K1A 0L2, Tel: 613-957-0951, Fax: 613-941-4775, bdabeka@hpb.hwc.ca

DAVIDSSON, Lena, Department of Food Science, Human Nutrition, Eth Zurich, PO Box 474, CH-8803 Ruschlikon, Switzerland, Tel: Int 411-724-2144, Fax: Int 411-724-0183, davidsson@ilw.agrl.ethz.ch

DAVISON, Allan J., Department of Chemistry, University of Northern British Columbia, 3333 University Way, Prince George, BC, Canada V2N 4Z9, Tel: 604-960-6674, Fax: 604-960-5545, adavison@ubc.ca

DE VRESE, Michael, Institut fur Physiologie und Biochemie, Bundesanstalt fur Milchforschung, Hermann-Weigmann-Str. 1, D-24103, Kiel, Germany, Tel: 011-49-431-609471, Fax: 011-49-431-609472

DE KIMPE, Jürgen, Laboratory for Analytical Chemistry, University of Ghent, Proeftuinstraat 86, B-9000 Ghent, Ghent/East-Flanders/B-9000, Belgium, Tel: 32-9-264-6624, Fax: 32-9-264-6699, dekimpe@inwchem.rug.ac.be

DEVLETI, Behzad, Med. and Pharm. Chemistry, University of Istanbul, Department of Medicinal & Pharmaceutical Chemistry, Faculty of Pharmacy, Istanbul/Beyasit 34452, Turkey, Tel: 212-519-0812, Fax: 212-528-0788, dowlatab@hisar.cc.boun.edu.tr

DIPLOCK, Anthony, Department of Biochemistry & Molecular Biology, UMDS, Guy's Hospital, St. Thomas Street, London SE1 9RT, England, Tel: 44-171-955-4521, Fax: 44-171-403-7195

DJUKIĆ, Mirjana, Department of Toxicology, School of Pharmacy, University of Belgrade, Vojvode Stepe 450, 11000 Belgrade, Yugoslavia, Tel: 381-11-107-258, Fax: 381-11-107-258, edjukicm.ubbg

DOUTHITT, Charles, Finnigan Corporation, 9412 Rocky Branch Drive, Dallas, TX 75243, USA, Tel: 214-348-8330, Fax: 214-348-8810

DRASCH, Gustav, Rechtsmedizin, Universitat, Frauenlob Str. 7A, 80337 München, Germany, Tel: 49-89-51605111, Fax: 49-89-51605144

DREBICKAS, V., Vilnius Pedagogical University, Studentu 39, Vilnius 2034, Lithuania

EDER, Klaus, Technical University of Munich, Institute of Nutrition Physiology, TV-Muenchen-Weihenstephan, Freising, 85350, Germany, Tel: 49-8161-713877, Fax: 49-8161-713999

EIDE, David, Department of Biochemistry and Molecular Biology, University of Minnesota-Duluth, 10 University Drive, Duluth, MN 55812, USA, Tel: 218-726-6508, Fax: 218-726-8014, deide@d.umn.edu

EL-AMRI, Fathi A., Department of Chemistry, University of Al-Fateh, PO Box 13361, Tripoli, Libya, Fax: 00-218-21-603068

ELMES, Margaret, Department of Pathology, University Hospital of Wales, Heath Park, Cardiff CF4 4XN, Wales, UK, Tel: 01222-742703, Fax: 01222-742701

ENOMOTO, Shuichi, Nuclear Chemistry Laboratory, The Institute of Physical and Chemical Research (RIKEN), Wako, Saitama 351-01, Japan, Tel: 81-48-4621111 ext. 3643, Fax: 81-48-4624462; 81-48-4624624, semo@postman.riken.go.jp

EYBL, V., Department of Pharmacology and Toxicology, Charles Univ. Med. Fac. in Pilsen, Karkovarska 48, Pilsen CZ-30166, Czech Republic, Tel: 42-19-557-224, Fax: 42-19-521-943

FAILLA, Mark L., Department of Food, Nutrition & Food Service Mgmt., University of North Carolina at Greensboro, 318 Stone Building, 1000 Spring Garden St., Greensboro, NC 27412-5001, USA, Tel: 910-334-5313, Fax: 910-334-4129, faillam@iris.uncg.edu

FAIRWEATHER-TAIT, Susan, Department of Nutrition Diet & Health, Institute of Food Research, Norwich Research Park, Colney, Norwich NR4 7UA, UK, Tel: 44-1603-255306, Fax: 44-1603-452578, sue.fairweather-tait@bbsrc.ac.uk

FALNOGA, Ingrid, Jamova 39, Ljubljana 1111, Slovenia, Tel: 386-61-1885-450

FELDMANN, Jörg, Department of Chemistry, University of British Columbia, 2036 Main Mall, Vancouver, BC, Canada V6T 1Z1, Tel: 604-822-2938, Fax: 604-822-2847, joerg@chem.ubc.ca

FENTON, Jennifer, Department of Animal Science, Michigan State University, 1023 Beech St., East Lansing, MI 48823, USA, Tel: 517-432-1448, Fax: 517-432-0190, imigjeni@pilot.msu.edu

FINLAY, D'Ann, Department of Nutrition, University of California, 704 Elmwood Drive, Davis, CA 95616, USA, Tel: 510-642-2552, Fax: 510-642-5867, dfinley@uclink.berkeley.edu

FISCHER, Peter, Health Canada, Nutrition Research Division, AL2203C Sir F. Banting Building, Tunney's Pasture, Ottawa, ON, Canada K1A 0L2, Tel: 613-957-0918, Fax: 613-941-6182, pfischer@hpb.hwc.ca

FLANAGAN, Peter R., Department of Medicine, The University of Western Ontario, Room 5-0F13, University Hospital, London, ON, Canada N6A 5A5, Tel: 519-663-3574, Fax: 519-663-3232, flan@julian.uwo.ca

FOX, Tom, Department Nutrition, Diet and Health, Institute of Food Research, Norwich Research Park, Colney, Norwich, Norfolk NR4 7UA, UK, Tel: 1603-255307, Fax: 1603-452578, tom.fox@bbsrc.ac.uk

FREEBURN, Jenny C., Department Human Nutrition Research Group, University of Ulster, Cromore Road, Coleraine, Co Londonderry BT52 1SA, Northern Ireland, Tel: 1265-324418, Fax: 1265-324965

FRIEL, James K., Department of Biochemistry, Memorial University, St. John's, NF, Canada A1B 3X9, Tel: 709-737-7954, Fax: 709-737-2422, jfriel@kean.ucs.mun.ca

FREER, Dennis E, 18A Regent Park Blvd, Asheville, NC 28801, USA

FUENTEALBA, Carmen, Department Pathology & Microbiology, Atlantic Veterinary College, UPEI, 550 University Avenue, Charlottetown, PE, Canada C1A 4P3, Tel: 902-560-0868, Fax: 902-560-0958, cfuentealba@upei.ca

FUNG, Ellen B., Department Nutritional Sciences, 316 Morgan Hall, University of California at Berkeley, Berkeley, CA 94720, USA, Tel: 510-642-7389, Fax: 510-642-0535, ignuf@nature.berkeley.edu

GABERT, Vince M., Agricultural, Food and Nutritional Science, University of Alberta, Edmonton, AB, Canada T6G 2P5, Tel: 403-492-5784; 403-435-4199, Fax: 403-492-9130; 403-492-4265, vgabert@afns.ualberta.ca

GAWALKO, Eugene, Grain Research Laboratory, Canadian Grain Commission, 1404-303 Main Street, Winnipeg, MB, Canada R3C 3G8, Tel: 204-983-8995, Fax: 204 983-0724

GERGELY, Anna, Trace Element Department, National Institute of Food Hygiene and Nutrition, Gyali ut 3/A, Budapest H-1097, Hungary, Tel: 361-215-4130, Fax: 361-215-1545

GIBSON, Rosalind S., Department of Human Nutrition, University of Otago, PO Box 56, Dunedin, New Zealand, Tel: 64-3-479-7957, Fax: 64-3-479-7958, rosalind.gibson@stonebow.otago.ac.nz

GIUSSANI, Augusto Marco, Institut fur Drahlenschutz, GSF-Forschungszentrum fur Umwelt und Gesundheit GMBH, Postfach 1129, 85758 Oberschleißheim, Germany, Tel: 49-89-31874247, Fax: 49-89-31872517, augusto@gretel.gsf.de

GOLDEN, Barbara, Department of Child Health, University of Aberdeen, Medical School, Foresterhill, Aberdeen, AB9 2ZD, Scotland, UK, Tel: 44-1224-681818 ext. 53894, Fax: 44-1224-663658, b.e.golden@abdn.ac.uk

GOLIGHTLY, Danold, Ross Division, Abbott Laboratories, 105686-RP43, 625 Cleveland Avenue, Columbus, OH 43215, USA, Tel: 614-624-3655, Fax: 614-624-3570, dgolightly@sr-server.rossnutrition.com

GONICK, Harvey, Department of Medicine/Nephrology, Cedars-Sinai Medical Center, 522 South Sepulveda Boulevard, Suite 207, Los Angeles, CA 90049, USA, Tel: 310-471-6646, Fax: 310-203-8640, hgonick@ucla.edu or, 72134.2021@compuserve.com

GOODWIN-JONES, Danny, Trace Elements Services Ltd., Abergorlech Road, Carmarthen. DYFED, SA32 7BA, UK, Tel: 01267-290229, Fax: 01267-290112

GOTTSCHALL-PASS, Katherine T., College of Pharmacy & Nutrition, University of Saskatchewan, 110 Science Place, Saskatoon, SK, Canada S7N 5C9, Tel: 306-966-5834, Fax: 306-966-6377, gottschall@sask.usask.ca

GRAHAM, Robin D., Department of Plant Science, University of Adelaide, Glen Osmond, SA, 5064, Australia, Tel: 618-303-7297, Fax: 618-303-7109, rgraham@waite.adelaide.edu.au

GRALAK, Mikolaj A., Department of Animal Physiology, Warsaw Agricultural University, Nowoursynowska 166, 02-787 Warszawa, Poland, Tel: 4822-439041 ext. 1507, Fax: 4822-472452, gralak@alpha.sggw.waw.pl

GRANADILLO, Victor A., Laboratorio de Inst. Analitica, Apartado Postal 15308 Las Delicias, Maracaibo, Zulia 4003-A, Venezuela, Tel: 00-58-61-516868/596763, Fax: 00-58-61-516868/526885, vgrana@solidos.cien.luz.ve

GREENBERG, Robert R., Dept. Analytical Chemistry Division, NIST, Building 235, Room B108, Gaithersburg, MD 20899, USA, Tel: 301-975-6285, Fax: 301-208-9279, greenber@micf.nist.gov

GREGER, Janet L., Department of Nutritional Sciences, University of Wisconsin, 1415 Linden Drive, Madison, WI 53706, USA, Tel: 608-262-9972, Fax: 608-262-5860

GUNSHIN, Hiromi, Renal Division, Brigham & Women's Hospital, and Harvard Medical School, 75 Francis Street, Boston, MA 02115, USA, Tel: 617-732-6677, Fax: 617-732-6392, hgunshin@bics.bwh.harvard.edu

HANNING, Rhona M., Department of Nutritional Sciences, University of Toronto, St. Michael's Hospital, Room 6C, 30 Bond Street, Toronto, ON, Canada M5B 1W8, Tel: 416-864-5551, Fax: 416-864-5414, hanningr@smh.toronto.on.ca

HANSEN, Marianne, Res. Dept. Human Nutr., Royal Vet. Agric. Univ., Rolighedsvej 30, 1959 Frederiksberg C, Denmark, Tel: 45-35282490, Fax: 45-35282483, mha@kvl.dk

HARLAND, Barbara, Department of Nutritional Sciences, Howard University, 6th and Bryant Streets N.W., Washington, DC, USA, Tel: 202-806-5656, Fax: 202-806-9233

HARRIS, Edward D., Department of Biochemistry/Biophysics, Texas A&M University, College Station, TX 77843-2128, USA, Tel: 409-845-3642, Fax: 409-845-9274, eharris@bioch.tamu.edu

HARTMAN, Henrick A., Department of Pathology, University of Wisconsin, 1300 University Avenue, Madison, WI 53706, USA, Tel: 608-262-0372, Fax: 608-265-3301

HARVEY, Linda, Department of Nutrition, Diet & Health, Institute of Food Research, Norwich Research Park, Colney, Norwich, Norfolk NR4 7VA, UK, Tel: 44-(0)1603-255308, Fax: 44-(0)1603-452578, linda.harvey@bbsrc.ac.uk

HAYWOOD, Susan, Department of Veterinary Pathology, University of Liverpool, PO Box 147, Liverpool, Merseyside, L69 3BX, England, Tel: 0151-794-4265, Fax: 0151-794-4268

HAZEGH-AZAM, Maryam, Department of Chemistry & Biochemistry, California State University, Fullerton, 800 N. State College Boulevard, Fullerton, CA 92634, USA, Tel: 714-773-2475, Fax: 714-449-5316, mazam@fullerton.edu

HEESE, Hans deV., Department of Paediatrics and Child Health, Institute of Child Health, University of Cape Town, Red Cross War Memorial Children's Hospital, Kipfontein Road, Rondebosch 7700, South Africa, Tel: 27-21-685-6529, Fax: 27-21-689-1287, lhever@ich.uct.ac.za

HEESE, Margaret, Department of Paediatrics and Child Health, Institute of Child Health, University of Cape Town, Red Cross War Memorial Children's Hospital, Kipfontein Road, Rondebosch 7700, South Africa, Tel: 27-21-685-6529, Fax: 27-21-689-1287, lhever@ich.uct.ac.za

HENNIG, Bernhard, Department of Nutrition and Food Science, University of Kentucky, 204 Funkhouser Building, Lexington, KY 40506-0054, USA, Tel: 606-257-6880, Fax: 606-257-3707, nfsbh@ukcc.uky.edu

HONGO, Tetsuro, Department of Human Ecology, Faculty of Medicine, University of Tokyo, 7-3-1 Hongo Bunkyo-ku, Tokyo 113, Japan, Tel: 81-3-3812-2111 ext. 3531, Fax: 81-3-5684-2739

HOTZ, Christine, Nutrition Research, Health Canada, 2203C Banting Research Center, Ross Avenue, Ottawa, ON, Canada K1A 0L2, Tel: 613-957-0925, Fax: 613-941-6182

HOWELL, John McC., School of Veterinary Studies, Murdoch University, South Street, Murdoch, Western Australia 6150, Australia, Tel: 09-360-2477, Fax: 09-3104144, howell@numbat.murdoch.edu.au

HUDDLE, Janet-Marie, Applied Human Nutrition, University of Guelph, Guelph, ON, Canada N1G 2W1, Tel: 519-823-9141, Fax: 519-766-0691, jhuddle@uoguelph.ca

HUNT, C.D., USDA-ARS, Grand Forks Human Nutrition Research Center, PO Box 9034, Grand Forks, ND 58202-9034, USA, Tel: 701-795-8423, Fax: 701-795-8395, chunt@badlands.nodak.edu

HUNT, Janet R., USDA-ARS, Grand Forks Human Nutrition Research Center, PO Box 9034, Grand Forks, ND 58202, USA, Tel: 701-795-8328, Fax: 701-795-8395, jhunt@badlands.nodak.edu

INSKIP, Michael J., Bureau of Human Prescription Drugs, Health Canada, 2(W) Banting Research Center, Ross Avenue, Ottawa, ON, Canada K1A 0L2, Tel: 613-957-1885, Fax: 613-941-5034, mike_inskip@isdtcp3.hwc.ca

JENSEN, Mette S., Department of Nutrition, Danish Institute of Animal Science, Research Centre Foulum, P.O. Box 39, DK-8830 Tjele, Denmark, Tel: 45-89-99-11-18, Fax: 45-89-99-11-66, mettes.jensen@sh1.foulum.min.dk

JOHNSON, A. Bruce, Research & Development, Zinpro Corporation, 6500 City West Pkwy, Suite 300, Eden Prairie, MN 55344, USA, Tel: 612-944-2736, Fax: 612-944-2749

JOHNSON, Phyllis E., Agricultural Research Service, Department of Agriculture, 800 Buchanan Street, Albany, CA 94710, USA, Tel: 510-559-6071, Fax: 510-559-5779, a03assoc@attmail.com

JOHNSON, W. Thomas, USDA-ARS, Grand Forks Human Nutrition Research Center, PO Box 9034, Grand Forks, ND 58202-9034, USA

JORHEM, Lars, Chemistry Division 2, National Food Administration, Box 662, S-751 26 Uppsala, Sweden, Tel: 46-18175500, Fax: 46-18105848, lajo@slv.se

JORY, Joan S., Applied Human Nutrition, University of Guelph, 268 Kathleen Street, Guelph, ON, Canada N1H 4Y5, Tel: 519-763-7834, Fax: 519-837-8575, jjory@uoguelph.ca

JUDSON, Geoffrey, Central Veterinary Laboratories, Primary Industries SA, 33 Flemington Street, Glenside, SA 5065, Australia, Tel: 618-2077979, Fax: 618-2077854, judson.geoffrey@pi.sa.gov.au

KANG, Y. James, Department of Pharmacology & Toxicology, University of North Dakota, School of Medicine, 501 North Columbia Road, Grand Forks, ND 58202-9037, USA, Tel: 701-777-2295, Fax: 701-777-6124, james.kang@medicine.und.nodak.edu

KATAKURA, Michihiro, Department of Public Health, Sapporo Medical University, S-1, W-17, Chuo-ku, Sapporo 060, Japan, Tel: 011-611-2111 ext. 2361, Fax: 011-641-8101

KAUP, Susan, Department of Nutritional Research, Wyeth-Ayerst Research, PO Box 8299, Philadelphia, PA 19101, USA, Tel: 610-341-2336, Fax: 610-989-4856, kaups@war.wyeth.com

KAWAMURA, Mieko, Department of Applied Biochemistry, Faculty of Agriculture, Tohoku University, 1-1, Tsutsumidori-Amamiyamachi, Aoba-ku, Sendai 981, Japan, Tel: 022-272-4321 ext. 326, Fax: 022-263-5358

KEEN, Carl L., Department of Nutrition, University of California, Davis, 3135 Meyer Hall, Davis, CA 95616, USA

KENDALL, Nigel R., Department of Animal Physiology & Nutrition, University of Leeds, Leeds LS2 9JT, UK, Tel: 0113-233-3068, Fax: 0113-233-3072, apnnrk@leeds.ac.uk

KESSLER, Jürg, Department of Physiology, Swiss Federal Research Station for Animal Production, CH-1725 Posieux, Switzerland, Tel: 037-877-275, Fax: 037-877-400

KING, Janet C., USDA, ARS, WHNRC, P.O. Box 29997, Presidio of San Francisco, CA 94129, USA, Tel: 415-556-9697, Fax: 415-556-1432, jking@nature.berkeley.edu

KIRKLAND, James B., Human Biology and Nutritional Sciences, University of Guelph, Guelph, ON, Canada N1G 2W1, Tel: 519-824-4120 ext. 6693, Fax: 519-763-5902, jkirkland.ns@aps.uoguelph.ca

KISTERS, Klaus, Med. Univ. Policlinic, Wilhems-Universität, Albert-Schweifzen-Str. 33, 48149 Münster, Germany, Tel/Fax: 0049/251/837528

KLEIN, Catherine, Human Nutrition and Food Systems, University of Maryland, 3304 Marie Mount Hall, College Park, MD 20742, USA, Tel: 410-964-9477, Fax: 301-854-9050

KLEIN, Gordon L., Department of Pediatrics (Gastroenterology), University of Texas Medical Branch, 301 University Boulevard, Room 3-240B, Galveston, TX 77555-0352, USA, Tel: 409-772-1689, Fax: 409-772-4599

KLEVAY, Leslie M., USDA-ARS, Grand Forks Human Nutrition Research Center, PO Box 9034, Grand Forks, ND 58202, USA, Tel: 701-795-8464, Fax: 701-795-8395

KONO, Koichi, Department of Hygiene and Public Health, Osaka Medical College, 2-7 Daigakumachi, Takatsuki City, Osaka 569, Japan, Tel: 0726-84-6419, Fax: 0726-84-6519

KRALJ-KLOBUČAR, Nada, Department of Zoology, Faculty of Science, Rooseveltoz trg 6, 10000 Zagreb, Croatia, Tel: 385-1-442604, Fax: 385-1-4552645

KREBS, Nancy, F., Department of Pediatrics/Center for Human Nutrition, University of Colorado Health Sciences Center, 4200 E 9th Avenue, C225, Denver, CO 80262, USA, Tel: 303-270-7037, Fax: 303-270-3273, nancy.krebs@uchsc.edu

KRUSIC, Luka, Department of Nutrition, Equinutri, Wolfratshauser str. 25, 82538 Geretsried, Germany, Tel: 49-171-5102221, Fax: 49-8171-72970, rfc-822:luka.krusic@guest.ornes.si

KUVIBIDILA, Solo, Department of Pediatrics, Louisiana State University, Medical School, 1542 Tulane Avenue, New Orleans, LA 70112, USA, Tel: 594-568-3990, Fax: 594-568-7532

L'ABBÉ, Mary R., Nutrition Research, Health Canada, AL2203C Banting Research Centre, Ross Avenue, Ottawa, ON, Canada K1A 0L2, Tel: 613-957-0924, Fax: 613-941-6182, mlabbe@hpb.hwc.ca

LAMBERT, Elizabeth, Department of Chemical Pathology, Guy's Hospital, Ground Floor, Medical School, London SE1 9RT, England, Tel: 0171-955-5000 ext. 3018, Fax: 0171-403-9810

LARSEN, Torben, Department of Animal Health and Welfare, Danish Institute of Anim. Sci., PO Box 39, 8830, Foulom, Tjele, Denmark, Tel: 45-89-991900, Fax: 45-89-991919

LASTRA, M.D., Dept. Biologia, UNAM, Facultad de Quimica, Cd. Universitaria, D.F. 04510, México

LEE, Julian, Dairy & Beef Division, AgResearch, Private Bag 11008, Palmerston North, New Zealand, Tel: 0061-6-356-8019, Fax: 0061-6-351-8003, leej@agresearch.cri.nz

LEE, Melvin, Department of Nutrition, Hiroshima Joqakuin University, 4-13-1 Ushitahigashi, Higashiku, Hiroshima 732, Japan, Tel: 082-228-0386, Fax: 082-228-2924

LEVENSON, Cathy, Department of Nutrition, Florida State University, Tallahassee, FL, USA, Tel: 904-644-4122, Fax: 904-644-0700, levenson@neuro.fsu.edu

LEONE, Arturo, Department of Biochemistry & Medical Biotechnology, University of Naples, Via S. Pansini 5, Naples I-80131, Italy, Tel: 39-81-746-3200, Fax: 39-81-746-3150

LEVY, Mark, 668 Stinchomb Avenue, Apt. #1, Columbus, OH 43210, USA, Tel: 614-292-4751, Fax: 614-292-8880

LINDER, Maria C., Department of Chemistry & Biochemistry, Calif. State University, Fullerton, 800 N. State College Boulevard, Fullerton, CA 92634, USA, Tel: 714-773-3621, Fax: 714-449-5316, mlinder@fulerton.edu

LITTLEFIELD, Neil A., Division of Nutritional Toxicology, National Center for Toxicological Research, 3900 NCTR Drive, Jefferson, AR 72079, USA, Tel: 501-543-7551, Fax: 501-543-7662

LOCKITCH, Gillian, Department of Pathology & Lab Medicine, B.C. Children's & Womens Hospital, 409 Staulo Crescent, Vancouver, BC, Canada V6N 3S1, Tel: 604-261-0394, Fax: 604-263-1402, kidsug@unixg.ubc.ca

LOMBECK, Ingrid, University Children's Hospital, Henrich-Heine University, Universitatsstr. 1, Geb, 2312.02, 40627 Düsseldorf, Germany, Tel: 0211/8113319, Fax: 0211/3190910

LÖNNERDAL, Bo, Department of Nutrition, University of California, Davis, CA 95616, USA

LOTFI, Mahshid, Micro Nutrient Initiative (MI), 250 Albert Street, PO Box 8500, Ottawa, ON, Canada K1G 3H9, Tel: 613-236-6163 ext. 2482, Fax: 613-236-9579, mlotfi@idrc.ca

LOWE, Nicola M., Department of Nutritional Sciences, Morgan Hall, University of California at Berkeley, Berkeley, CA 94720, USA, Tel: 510-642-7389, Fax: 510-642-0535

MACKENZIE, Alexander M., Department of Animal Physiology and Nutrition, University of Leeds, Leeds LS2 9JT, UK, Tel: 0113-233-3068, Fax: 0113-233-3072, apnamm@leeds.ac.uk

MACPHERSON, Allan, Department of Biochemical Sciences, SAC, Auchincruive, Ayr, KA6 5HW, Scotland, Tel: 004401292-525156, Fax: 004401292-525177, a.macpherson@au.sac.ac.uk

MADAN, Jagmeet, Department of Foods & Nutrition, SVT College of Home Science, 11, Merry-Niketan, Mount Mary Road, Bandra (W), Bombay 400050, India, Tel: 6425118/6102504, Fax: 9122-6107673

MAJSAK-NEWMAN, Gosia, Department of Nutrition, Diet & Health, Institute of Food Research, Norwich Research Park, Colney, Norwich NR4 7UA, UK, Tel: 44-0-1603-255-308, Fax: 44-0-1603-452-578

MATOVIC, Vesna, Department of Toxicology, Faculty of Pharmacy, Vojvode Stepe 450, 11000 Belgrade, Yugoslavia, Tel: 38111-23591-647, Fax: 38111-2372-840

MAZUR, André, Lab. Maladies Métaboliques, INRA, Theix/Clermont Fd, St. Genès Champanelle 63 122, France, Tel: 33-736-24234, Fax: 33-736-24638, mazur@clermont.inra.fr

MCARDLE, Harry, Department of Child Health, University of Dundee, Nine Hills Hospital, Dundee, DD1 9SY, UK, Tel: 1382-660111 ext. 3308, Fax: 1382-645783, h.j.mcarde@dux.dundee.ac.uk

MEDEIROS, Denis M., Department of Human Nutrition & Food Management, The Ohio State University, 345 Campbell Hall, 1787 Neil Avenue, Columbus, OH 43210-1295, USA, Tel: 614-292-5575, Fax: 614-292-7536, medeiros2@osu.edu

MEISINGER, Vanee, Department of Occupational Medicine, University Clinic of Internal Medicine IV, Wahringer Gurtel 18-20, A-1090 Vienna, Austria, Tel: 431-404004701, Fax: 431-4088011

MIELCARZ, Grzegorz W., Department of General Chemistry, K. Marcinkowski University of Medical Sciences, Grunwaldzka 6, Poznan, Poland, Tel: 48-61-658619, Fax: 48-61-520455

MIHAILOVIĆ, B. Momčilo, Department of Physiology and Biochemistry, Faculty of Veterinary Medicine, Bulevar JNA 18, 11000 Belgrade, Yugoslavia, Tel: 301-11-685-261, Fax: 381-11-685-936

MILAČIČ, Radmila, Department of Environ. Sci., Jožef Stefan Institute, Jamova 39, Ljubljana 61111, Slovenia, Tel: 386-61-1773-900, Fax: 386-61-219-385, radmila.milacic@ijs.si

MILLER, Leland V., Department of Pediatrics, University of Colorado Health Sciences Center, 4200 E. 9th Avenue, Box C232, Denver, CO 80262, USA, Tel: 303-270-8126, Fax: 303-270-7097, lealand.miller@uchsc.edu

MILNE, David B., USDA-ARS, Grand Forks Human Nutrition Research Center, PO Box 9034, Grand Forks, ND 58202, USA, Tel: 701-795-8424, Fax: 701-795-8395, milne@plains.nodak.edu

MINAMI, Takeshi, Department of Clinical Chemistry, Fac. Pharmaceut. Sci., Kinki University, 3-4-1 Kowakae, Higashi-Osaka, 577, Japan, Tel: 6-721-2332 ext. 3812, Fax: 6-730 1394, minamita@phar.kindai.ac.jp

MOLNAR, Jeannette, Semmelweiss University, II. Department of Internal Medicine, H-1088 Budapest, Szentkiralyi utca 46, Tel: 36-1-210-1225, Fax: 36-1-210-0799

MOMČILOVIĆ, Berislav, USDA-ARS, Grand Forks Human Nutrition Research Center, PO Box 9034, Grand Forks, ND 58202-9034, USA

NAKADAIRA, Hiroto, Department of Hygiene & Preventive Medicine, Niigata University School of Medicine, 1-757 Asahimachi-dori, Niigata 951, Japan, Tel: 81-25-223-6161 ext. 2337, Fax: 81-25-223-7971, hiroto@med.niigata-u.ac.jp

NASU, Tameyuki, Tokyo College of Welfare, 1-14-26 Kyodo, Setagaya-ku, Tokyo 156, Japan, Tel: 033-795-8456, Fax: 033-429-0265

NIELSEN, Forrest H., USDA-ARS, Grand Forks Human Nutrition Research Center, PO Box 9034, Grand Forks, ND 58202-9045, USA, Tel: 701-795-8456, Fax: 701-795-8395, a03dirgfhnrc@attmail.com

NOMIYAMA, Hiroko, Department of Environmental Health, Jichi Medical School, Minamikawachi-Machi, Tochigi-Kan 329-04, Japan, Tel: 81-285-44-2111 ext. 3139, Fax: 81-285-44-8465

NOMIYAMA, Kazuo, Department of Environmental Health, Jichi Medical School, Minamikawachi-Machi, Tochigi-Ken 329-04, Japan, Tel: 81-285-44-2111 ext. 3139, Fax: 81-285-44-8465

NOORDIN, Mustapha M., Department of Veterinary Pathology & Microbiology, Universiti Pertanian Malaysia, Faculty of Veterinary Medicine & Animal Science, 43400, UPM, Selangor DE, Malaysia, Tel: 603-9486101 ext. 1486, Fax: 603-9486317, noordin@upm.edu.my

NOROSE, Noboru, Department of Pediatrics, Matsumoto National Hospital, 1209 Yoshikawa, Murai, Matsumoto, Nagano, 399, Japan, Tel: 263-58-4567, Fax: 263-86-3183

NOSEWORTHY, Mike, Human Biology and Nutritional Sciences, University of Guelph, Guelph, ON, Canada N1G 2W1, Tel: 519-824-4120 ext. 3725, Fax: 519-763-5902

OBERLEAS, Donald, Department of Food & Nutrition, Texas Tech University, PO Box 41162, Lubbock, TX 79409-1162, USA, Tel: 806-742-3068, Fax: 806-742-3042

OKAZAKI, Jiro, Department of Clinical Chemistry, Fac. Pharmaceut. Sci., Kinki University, 3-4-1 Kowakae, Higashi-Osaka, 577, Japan, Tel: 6-721-2332 ext. 3812, Fax: 6-730-1394

OLDFIELD, Jim, Department of Animal Sciences, Oregon State University, Withycombe Hall 200, Corvallis, OR 97331-6702, USA, Tel: 403-762-6308, Fax: 403-762-7502

OLIVERO, Jesus, Environmental Chemistry Group, Universidad de Cartagena, AA 6541, Cartagena, Columbia

PALLAUF, Joseph, Department of Nutritional Sciences, Justus-Liebig-University Giessen, Institute of Animal Nutrition, Senckenbergstrasse 5, D-35390 Giessen, Germany, Tel: 011-49-641-702-85700, Fax: 011-49-641-702-85709

PATERSON, Phyllis G., College of Pharmacy and Nutrition, University of Saskatchewan, 110 Science Place, Saskatoon, SK, Canada S7N 5C9, Tel: 306-966-5838, Fax: 306-966-6377

PATTERSON, Carla, Centre for Public Health Research, QUT, Locked Bag No. 2, Red Hill, Queensland, 4059, Australia, Tel: 07-3864-5795, Fax: 07-3864-5830, cpatterson@qut.edu.au

PENLAND, James G., USDA-ARS, Grand Forks Human Nutrition Research Center, PO Box 9034, Grand Forks, ND 58202-9034, USA, Tel: 701-795-8471, Fax: 701-795-8395, penland@badlands.nodak.edu

PERCIVAL, Susan S., Food Science & Human Nutrition, University of Florida, Box 110370, Gainesville, FL 32611, USA, Tel: 904-392-1991, Fax: 904-392-9467, ssp@gnv.ifas.ufl.edu

PROHASKA, Joseph R., Biochemistry & Molecular Biology, University of Minnesota, Duluth, School of Medicine, 10 University Drive, Duluth, MN 55812, USA, Tel: 218-726-7502, Fax: 218-726-6181, jprohask@d.umn.edu

PURWANTORO, Aris, Department of Veterinary Studies, Murdoch University, Veterinary Biology, VB 3-26, South Street, Murdoch, W.A. 6150, Australia, Tel: 61-9-3602485, Fax: 61-9-3104144, aris@numbat.murdoch.edu.au

RAINEY, Charlie, Nutrition Research Group, 4199 Campus Drive, Suite 550, Irvine, CA 92715, USA, Tel: 714-509-6537, Fax: 714-509-6538

RANDALL-SIMPSON, Janis, Department of Pediatrics, McMaster University, 1200 Main Street West, Room 3V42, Hamilton, ON, Canada L8N 3Z5, Tel: 905-525-9140 ext. 22729, Fax: 905-525-1703, simpsonj@fhs.csu.mcmaster.ca

RAYMAN, Margaret P., Chemistry Department, University of Surrey, Guildford, GU2 5XH, UK, Tel: 44-1483-259583, Fax: 44-1483-259514

RAYSSIGUIER, Yves, Unité des Maladies Métaboliques, INRA, Centre de Recherche de Clermont-Fd-Theix, Saint Genès Champanelle 63122, France, Tel: 33-736-24230, Fax: 33-736-24638

REEVES, Philip, USDA-ARS Human Nutrition Research Center, PO Box 9034, Grand Forks, ND 58202-9034, USA, Tel: 701-795-8497, Fax: 701-795-8395

RHODES, Joseph, Department of Agriculture, Marine Natural Products, Kingston House, Gripps Common, Cotgrave, Nottingham, UK, Fax: 011-44-0115-989-9588

RICHARDS, Mark, Growth Biology Laboratory, USDA, Agricultural Research Service, Building 200, Room 201, BARC-East, Beltsville, MD 20705-2350, USA, Tel: 301-504-8892, Fax: 301-504-8623, richards@ggpl.arsusda.gov

RIMBACH, Gerald, Institute of Animal Nutrition, Justus-Liebig-University Giessen, Senckenbergstrasse 5, D-35390 Giessen, Germany, Tel: 011-49-641-702-85700, Fax: 011-49-641-702-85709

ROBINSON, Lynne J., College of Pharmacy and Nutrition, University of Saskatchewan, 110 Science Place, Saskatoon, SK, Canada S7N 5C9, Tel: 306-966-5834, Fax: 306-966-6377, robinson@duke.usask.ca

ROODENBURG, Annet J.C., Department of Human Nutrition, Wageningen Agricultural University, PO Box 812G, 6700 EV Wageningen, Netherlands, Tel: 31-317-484298, Fax: 31-317-483342, annet.roodenburg@et2.voed.wau.nl

ROTH, Hans-Peter, Institute of Nutrition Physiology, Technical University Munich, D-85350 Freising-Weihenstephan, Germany, Tel: 08161/713879, Fax: 08161/713999

ROUSSEL, Anne Marie, Department of Biochemistry, J. Fourier University, Pharmacie Domaine de la Merci, 38700, France, Tel: 33-76-663-7131, Fax: 33-76-51-86-67

RUSKAN, Bronia

RUZ, Manuel, Nutrition Fac. Medicine, University of Chile, 1027 Independencia, Santiago, Chile, Tel: 562-678-6134, Fax: 562-735-5581, mruz@machi.med.uchile.cl

SAARI, Jack T., USDA-ARS, Grand Forks Human Nutrition Research Center, PO Box 9034, Grand Forks, ND 58202-9034, USA, Tel: 701-795-8499, Fax: 701-795-8395

SALTMAN, Paul, Department of Biology, University of California San Diego, La Jolla, CA 92093-0322, USA, Tel: 619-534-3824, Fax: 619-534-0936, psaltman@ucsd.edu

SAMPSON, Barry, Department of Chemical Pathology, Charring Cross Hospital, Fulham Palace Road, London W6 8RF, UK, Tel: 0181-846-7080, Fax: 0181-567-0302, bsampson@cxwms.ac.uk

SANDBERG, Ann-Sofie, Department of Food Science, Chalmers University of Technology, c/o SIK, PO Box 5401, S-402 29 Göteborg, Sweden, Tel: 46-31-35-56-30, Fax: 46-31-83-37-82, ann@sik.se

SANDSTEAD, Harold H., Preventive Medicine & Community Health, University of Texas, Medical Branch, Route 1109 @ UTMB, Galveston, TX 77555-1109, USA, Tel: 409-772-4661, Fax: 409-772-6287, hsandste@pmchpo.pmchs1.utmb.edu

SCARINO, Maria Laura, Dept. Nutrizione Sperimentale, Instituto Nazionale della Nutrizione, Via Ardeatina 546, Rome 00179, Italy, Tel: 396-504-2589, Fax: 396-5031592

SCHAFFER, Kenneth J., Department of Chemistry & Biochemistry, California State University, Fullerton, 800 N. State College Boulevard, Fullerton, CA 92634, USA, Tel: 714-773-2475, Fax: 714-449-5316

SCHUSCHKE, Dale A., Center for Applied Microcirculatory Research, University of Louisville, 500 S. Preston, Louisville, KY 40292, USA, Tel: 502-852-7553, Fax: 502-852-6239, daschu01@ulkyum.louisville.edu

SCOTT, Roy, Department of Urology & Male Infertility, Glasgow Royal Infirmary, 16 Alexandra Parade, Glasgow G31 2ER, Scotland, Tel: 0141-211-4000, Fax: 0141-211-4461

SHENKIN, Alan, Department of Clinical Chemistry, Royal Liverpool University Hospital, 4th Floor Duncan Building, Royal Liverpool University Hospital, Liverpool, L69 3BX, UK, Tel: 44-0151-706-4232, Fax: 44-0151-706-5813, shenkin@liverpool.ac.uk

SHERMAN, Adria Rothman, Department of Nutritional Sciences, Rutgers, The State University of New Jersey, PO Box 231, New Brunswick, NJ 08903-0231, USA, Tel: 908-932-9379, Fax: 908-932-6837, asherman@aesop.rutgers.edu

SHULZE, Roman A., Department of Chemistry & Biochemistry, California State University, Fullerton, 800 N. State College Boulevard, Fullerton, CA 92634, USA, Tel: 714-773-3912, Fax: 714-449-5316

SIPS, Adrienne, Laboratory of Exposure Assessment, National Institute of Public Health and Environmental Protection, PO Box 1, 3720 BA Bilthoven, The Netherlands, Tel: 030-274-2043, Fax: 030-270-4447

SIVA SUBRAMANIAN, K.N., Department of Pediatrics/Neonatology, Georgetown University Hospital, 3800 Reservoir Rd, NW, #M3400, Washington, DC 20007, USA, Tel: 202-687-8569, Fax: 202-784-4747

SIVARAM, Saumya A., Department of Preventive Medicine & Community Health, 700 Harborside Drive, 3.102 Ewing Hall, Galveston, TX 77555-1109, USA, Tel: 409-772-4661, Fax: 409-772-6287

SMITH, Bob, Department of Elemental Analysis, Great Smokies Diagnostic Lab, 18 A. Regent Park Boulevard, Asheville, NC 28806, USA, Tel: 704-253-0621, Fax: 704-253-1127

SMITH, Karol R., Department of Food Science & Human Nutrition, University of Florida, PO Box 110370, Gainesville, FL 32611, USA, Tel: 904-392-1991, Fax: 904-392-9467, krsm@gnv.ifas.ufl.edu

SOLOMONS, Noel W., CESSIAM Hosp de Ojos-Oidos, Dr. Rodolfo Robles V, Diagonal 21 y 19 Calle-Zona 11, Guatemala City, Guatemala, Tel: 502-473-0258 ext 110, Fax: 502-473-3906

SONG, Li, Department of Geological Sciences, University of Saskatchewan, 114 Science Place, Saskatoon, SK, Canada S7N 5E2, Tel: 306-966-8587, Fax: 306-906-8593, song@pangea.usask.ca

SORENSON, John, Department of Medical Chemistry, University of Arkansas for Medical Sciences Campus, College of Pharmacy, Slot 522-3, 4301 W. Markham, Little Rock, AR 72205, USA, Tel: 501-686-6494, Fax: 501-686-6057, jrjsorenson@life.uams.edu

SPEARS, Jerry, Department of Animal Science, North Carolina State University, Box 7621, Raleigh, NC, USA, Tel: 919-515-4008, Fax: 919-515-4463, jerry_spears@ncsu.edu

STIBILJ, Vekoslava, Department of Environmental Sciences, J. Stefan Institute, Jamone 39, 61111 Ljubljana, Slovenia, Tel: 386-61-137-25-40 or 386-61-188-54-50, Fax: 386-61-374-919

STOECKER, Barbara J., Department of Nutritional Sciences, Oklahoma State University, 425 Human Environmental Sciences, Stillwater, OK 74078, USA, Tel: 405-744-5040, Fax: 405-744-7113, chrom@okway.okstate.edu

STRAIN, J.J. (Sean), Human Nutrition Research Group, University of Ulster, Coleraine, BT52 1SA Northern Ireland, UK, Tel: 01265-324795, Fax: 1265-324965

STRONG, Phil, Department of Occupational Health, US Borax Inc., 26877 Tourney Rd., Valencia, CA 91355, USA, Tel: 805-287-5634, Fax: 805-287-5542

ŠTUPAR, Janez, Department of Environment Science, J. Stefan Institute, Jamova 39 Ljubljana, Slovenia, Tel: 386-61-1773-900, Fax: 386-61-219-385

SUGAWARA, Naoki, Department of Public Health, Sapporo Medical University, S-1, W-17 Central West, Sapporo 060, Japan, Tel: 011-611-2111 ext. 2361, Fax: 011-641-8101

SUMAR, Salim, Food Research Centre, South Bank University, 103 Borough Rd., London SE1 0AA, UK, Tel: 0-171-815-7969, Fax: 0-171-815-7999, sumars@vax.sbu.ac.uk

SUNDE, Roger A., Nutritional Sciences & Biochemistry, University of Missouri-Columbia, 217 Gwynn Hall, Columbia, MO 65211, USA, Tel: 314-882-4526, Fax: 314-882-0185, roger_sunde@muccmail.missouri.edu

SUNDERMAN, F.W., Jr., Department of Laboratory Medicine, University of Connecticut Medical School, 263 Farmington Avenue, Mail Code 2225, Farmington, CT 06030-2225, USA, Tel: 203-679-2328, Fax: 203-679-2154, sunderman@nso1.uchc.edu

SUTHERLAND, Barbara, Department of Nutritional Sciences, Morgan Hall, University of California at Berkeley, Berkeley, CA 94720, USA, Tel: 510-642-7389

SUTTLE, Neville F., Department of Parasitology, Moredun Research Institute, 408, Gilmerton Rd., Edinburgh EH17 7JH, Scotland, Tel: 0131-664-3262, Fax: 0131-664-8001

SUZUKI, Kazumasa, National Institute of Nutrition, 3-1-8 Tokiwadai Itabashi-ku, Tokyo 174, Japan, Tel: 03-3968-3170

SUZUKI, Sachiko, Department of Toxicology, National Institute of Health Sciences, Kamiyoga 1-18-1 Setagaya-ku, Tokyo 158, Japan, Tel: 03-3700-1141, Fax: 03-3700-2348, ssuzuki@nihs.go.jp

TANAKA, Masaki, Department of Clinical Chemistry, Fac. Pharmaceut. Sci., Kinki University, 3-4-1 Kowakae, Higashi-Osaka, 577, Japan, Tel: 6-721-2332, Fax: 6-730-1394

TANAKA, Hidenori, Department of Clinical Chemistry, Fac. Pharmaceut. Sci., Kinki University, 3-4-1 Kowakae, Higashi-Osaka, 577, Japan, Tel: 6-721-2332 ext. 3812, Fax: 6-730-1394

TAYLOR, Carla, Department of Foods & Nutrition, University of Manitoba, Winnipeg, MB, Canada R3T 2N2, Tel: 204-474-8079, Fax: 204-261-0372, taylor@bldghumec.lanl.umanitoba.ca

TEUCHER, Sirgit, Department of Nutrition, Diet & Health, Institute of Food Research, Norwich Research Park, Colney, Norwich, Norfolk, NR4 7UA, UK, Tel: 44-0-1603-255000, Fax: 44-1603-4522578, teucher@bbsrc.ac.uk

THOMPSON, Katherine H., Faculty of Pharmaceutical Science, The University of British Columbia, 2145 East Mall, Vancouver, BC, Canada V6T 1Z3, Tel: 604-822-2343, Fax: 604-822-3035, wa.thompson@aol.com

THOMSON, Christine D., Department of Human Nutrition, University of Otago, PO Box 56, Dunedin, New Zealand, Tel: 64-3-4797943, Fax: 64-3-4797958, christine.thomson@stonebow.otago.ac.nz

TOHYAMA, Chiharu, Environmental Health Sciences, National Institute for Environmental Studies, 16-2 Onogawa, Tsukuba/Ibaraki/305, Japan, Tel: 81-298-50-2336, Fax: 81-298-56-4678, ctohyama@nies.go.jp

TROW, Liam, Department of Nutrition, Diet & Health, Institute of Food Research, Norwich Research Park, Norwich Laboratory, Colney Lane, Norwich, NR5 8NW, UK, Fax: 44-0-1603-255307

TURNLUND, Judith R., USDA/ARS Western Human Nutrition Research Center, Building 1100, PO Box 29997, Presidio of San Francisco, CA 94129, USA, Tel: 415-556-5662, Fax: 415-556-1432, a03rlmicronu@attmail.com

USUDA, Kan, Department of Hygiene and Public Health, Osaka Medical College, 2-7 Daigakumachi, Takatsuki City, Osaka 569, Japan, Tel: 0726-83-1221 ext. 2661, Fax: 0726-84-6519

VALENTINE, Jane L., Department of Environmental Health Sciences, UCLA/School of Public Health, 10833 Le Conte Avenue, Los Angeles, CA 90095, USA, Tel: 310-825-8751, Fax: 310-825-8440

VAN RYSSEN, Jannes B.J., Department of Animal and Wildlife Sciences, University of Pretoria, Pretoria 0002, South Africa

VILLA ELIZAGA, Ignacio, Unidad de Investigación de Oligoelementos, Dpto Química, Universidad de Navarra, C/Irunlarrea s/n, 31.080 Pamplona (Navarra), Spain, Tel: 5868702, Fax: 5868630

VRTOVEC, Matjaz, Department of Endocrinology & Metabolism, University Medical Centre, Zaloska 7, 61000 Ljubljana, Slovenia, Tel: +386-61-13-17-224, Fax: +386-61-13-21-178

WALTER, Tomas, Department of Hematology, Institute of Nutrition, University of Chile, 5540 Macul, Santiago 138-11, Chile, Tel: 56-2-678-1480, Fax: 221-4030, twalter@vec.inta.uchile.cl

WARDLE, Catherine A., Department of Clinical Chemistry, Royal Liverpool University Hospital, Prescot Street, Liverpool L7 8XP, UK, Tel: 0151-706-4246, Fax: 44-0151-706-5813

WASTNEY, Meryl E., Department of Pediatrics/Neonatology, Georgetown University Medical Center, 3800 Reservoir Road N.W. M3400, Washington, DC 20007, USA, Tel: 202-687-5004, Fax: 202-784-4747, mwastn01@gumedlib.dml.georgetown.edu

WAUBEN, Ine P.M., Department of Pediatrics, McMaster University, HSC Room 3V42, 1200 Main St. West, Hamilton, ON, Canada L8N 3Z5, Tel: 905-521-2100 ext. 5644, Fax: 905-521-1703, waubeni@fhs.csu.mcmaster.ca

WEDEKIND, Karen, Department of Advanced Research, Hill's Pet Nutrition, Inc., PO Box 1658, Topeka, KS 66601, USA, Tel: 913-286-8095, Fax: 913-286-8014, karen_wedekind@hillspet.com

WELCH, Ross M., US Plant, Soil & Nutrition Lab, USDA-ARS, Tower Road, Ithaca, NY 14853, USA, Tel: 607-255-5434, Fax: 607-255-2459, rmw1@cornell.edu

WHARF, S. Gabrielle, Department of Nutrition, Diet & Health, Institute of Food Research, Norwich Research Park, Colney, Norwich, Norfolk, NR4 7VA, UK, Tel: 44-0-1603-255304, Fax: 44-0-1603-452578

WIENK, Koen J.H., Laboratory Animal Science, Utrecht University, PO Box 80.166, 3508 TD Utrecht, Netherlands, Tel: 31-30-2532033, Fax: 31-30-2537997, wieter@pobox.ruu

WINDISCH, Wilhelm, Technical University of Munich, Institute of Nutrition Physiology, Tu Munchen-Weihenstephan, Freising, 85350, Germany, Tel: 49-8161-713673, Fax: 49-8161-713999

WING, Kenneth, Oral Cell Biology, Umeå University, Umeå, S-901 87, Sweden, Tel: 46-90-167886, Fax: 46-90-135636, kenneth.wing@mhs.umu.se

WOOD, Marjorie, 18 Friar's Way, London, ON, Canada N6G 2A8

WOODHOUSE, Leslie R., Department of Nutritional Sciences, University of California at Berkeley, Morgan Hall, Berkeley, CA 94720, USA, Tel: 510-642-7389, Fax: 510-642-0535, lrw@nature.berkeley.edu

WOODWARD, William (Bill), Human Biology and Nutritional Sciences, University of Guelph, Guelph, ON, Canada N1G 2W1, Tel: 519-824-4120 ext. 3741, Fax: 519-763-5902, wwoodward.ns@aps.uoguelph.ca

WOMACK, Sian, Trace Element Services Ltd., Ceffyl Mor. St. Margaret's Park, Rending, Carmarthenshire, SA33 4PQ Wales, UK, Tel: 01994-453643, Fax: 01267-290112

WU, Shaole, Department of Environmental Chemistry, Alberta Environmental Centre, PO Box 4000, Vegreville, AB, Canada T9C 2C1, Tel: 403-632-8444, Fax: 403-632-8379, shaole@aec.env.gov.ab.ca

XIA, Qianli, Geo. Sci./University of Saskatchewan, 114 Science Place, Saskatoon, SK, Canada S7N 5E2, Tel: 306-966-5737, Fax: 306-966-8593, xie@pangea.usask.ca

YEN, Jong-Tseng, Nutrition Research Unit, US Meat Animal Research Center, PO Box 166, Clay Center, NB 68933-0166, USA, Tel: 402-762-4206, Fax: 402-762-4148, yen_j@marcvm.marc.usda.gov

XIA, Yiming, Department of Trace Element Nutrition, Institute of Nutrition & Food Hygiene, Chinese Academy of Preventative Medicine, 29 Nam Wei Rd., Beijing 100050, China, Tel: 86-10-63040634, Fax: 86-10-3011875

YUKAWA, Masae, Division of Environmental Health, National Institute of Radiological Sciences, 9-1, Anagawa-4-chome, Inage-ku, Chiba-shi 263, Japan, Tel: 043-251-2111, Fax: 043-284-1769, yukawa@nirs.go.jp

ZACHARA, B.A., Department of Biochemistry, University School of Medical Sciences, 24 Karłowicza Street, Bydgoszcz, 85-092, Poland

ZIEGLER, E.E., Department of Pediatrics, University of Iowa Hospitals and Clinics, 200 Hawkins Drive, Iowa City, IA 52241, USA, Tel: 319-356-2836, Fax: 319-356-8669

ZLOTZKIN, S.H., Pediatrics & Nutritional Sciences, University of Toronto, Hospital Sick Children, 555 University Avenue, Toronto, ON, Canada M5G 1X8, Tel: 416-813-6171, Fax: 416-813-4972, zlotkin@sickkids.on.ca

Author Index

Keyword Index